CONTENTS

PART 1: THE FOUNDATION

CHAPTER 1: THE ATOM

CHAPTER 2: THE ATOM: THE ARRANGEMENT OF ELECTRONS

CHAPTER 3: EQUATIONS AND EQUILIBRIA

CHAPTER 4: THE CHEMICAL BOND

CHAPTER 5: THE SHAPES OF MOLECULES

CHAPTER 6: CHEMICAL BONDING AND THE STRUCTURE OF SOLIDS

PART 2: PHYSICAL CHEMISTRY

CHAPTER 7: GASES

CHAPTER 8: LIQUIDS

CHAPTER 9: SOLUTIONS

CHAPTER 10: THERMOCHEMISTRY

CHAPTER 11: CHEMICAL EQUILIBRIUM

CHAPTER 12: ELECTROCHEMISTRY

CHAPTER 13: OXIDATION–REDUCTION EQUILIBRIA

CHAPTER 14: REACTION KINETICS

PART 3: INORGANIC CHEMISTRY

CHAPTER 15: PATTERNS OF CHANGE IN THE PERIODIC TABLE

CHAPTER 16: THE NOBLE GASES

CHAPTER 17: HYDROGEN

CHAPTER 18: THE s BLOCK METAL: GROUPS 1 AND 2

CHAPTER 19: GROUP 3

CHAPTER 20: GROUP 7: THE HALOGENS

CHAPTER 21: GROUP 6

CHAPTER 22: GROUP 5

CHAPTER 23: GROUP 4

CHAPTER 24: THE TRANSITION METALS

PART 4: ORGANIC CHEMISTRY

CHAPTER 25: ORGANIC CHEMISTRY

CHAPTER 26: THE ALKANES

CHAPTER 27: ALKENES AND ALKYNES

CHAPTER 28: BENZENE AND OTHER ARENES

CHAPTER 29: HALOGENOALKANES AND HALOGENOARENES

CHAPTER 30: ALCOHOLS, PHENOLS AND ETHERS

CHAPTER 31: ALDEHYDES AND KETONES

CHAPTER 32: AMINES

CHAPTER 33: ORGANIC ACIDS AND THEIR DERIVATIVES

CHAPTER 34: THE IDENTIFICATION OF ORGANIC COMPOUNDS

CHAPTER 35: SOME GENERAL TOPICS

NOTE Sections and questions marked with an asterisk cover material which is not required by all the Examination Boards. A dagger denotes a question which should be tackled on re-reading, as it involves material which is covered in later chapters.

PREFACE

This text was written to prepare students for A-level, AS-level and S-level examinations in Chemistry. The third edition has been prepared in response to the change in the examination which students take before beginning A-level studies. In the third edition, it is assumed that students are approaching A-level from a study of GCSE Science: Double Award. Additional material has been included in a number of chapters to provide a more gradual transition to advanced studies.

Some topics are not common to all Examination Boards. These topics are marked with asterisks so that readers can check their own syllabuses and omit those topics if they wish. Each topic is treated from the beginning without assuming that GCSE work has been remembered.

I do not wish to prescribe an order in which teachers take classes through their chemistry syllabus. Nevertheless, every book has to have its contents arranged in a linear manner and I have had to order mine. I have presented atomic structure, chemical bonding, equations and equilibria early in the text. Since studies of systems at equilibrium pervade all chemical topics, I have found it necessary to put some qualitative work on equilibria into the introductory section. Once this foundation has been laid, I envisage students being taken through the organic, inorganic and physical sections of the subject simultaneously. Since thermodynamic considerations throw light on much inorganic and organic chemistry, students will find it an advantage to take this chapter early in their course.

At intervals in each chapter, 'checkpoints' are included so that students can pause and test their understanding and, if necessary, revise a section before they pass on to new material. The A-level syllabus covers so much ground that many teachers find it difficult to take their classes through all the material, while still leaving time for practical work. I hope that teachers will be able to allow students to cover parts of the syllabus on their own from this text, assisted by the checkpoints, thus reducing the amount of note-making which needs to be done in class and releasing time for discussion, reinforcement and practical work. Each chapter ends with a searching set of questions, including some from examination papers. At the end of each section is a collection of questions which span the chapters in that section.

The margin carries a summary of the text. On reaching the end of a chapter, a reader can glance back through the summary to see whether he or she has assimilated all the material. If the reader notes any points which need further study, he or she has only to glance at the text alongside the summary to find the relevant passage.

I hope that students will like the technique I have devised for integrating descriptive material with diagrams, so that the reader's eyes do not have to travel constantly to and fro between the diagram and the text. I have used this technique largely in the physical chemistry section of the book. The annotated diagrams have been consumer-tested and approved by sixth formers in my own school.

Much of the detailed inorganic chemistry is summarised in the form of tables and reaction schemes. Students find these helpful for revision. My preference is to take the s block metals first, follow them with the halogens, and then work through the non-metallic elements of Groups 6 and 5 to arrive at Group 4, with its interesting gradation from non-metallic to metallic behaviour, and end with the transition metals. Teachers who prefer a different order will find no difficulty in taking the inorganic chapters in a different sequence.

In my experience, even students with a fair knowledge of the various series of organic compounds find difficulty in tackling problems which require a knowledge of several series of compounds. The method of converting, say, **A** into **D** via the route

$$\mathbf{A} \rightarrow \mathbf{B} \rightarrow \mathbf{C} \rightarrow \mathbf{D}$$

may well be difficult to formulate. I have tackled this problem in stages by summarising at the end of each chapter the relationships between the series of compounds studied in that chapter and those covered in previous chapters. At the end of the section on organic chemistry, all the synthetic routes are summarised in a few reaction schemes. A number of threads which run through the separate chapters of organic chemistry are also drawn together at the end of this section.

There is more to chemistry than the content of any A-level syllabus. I assume that all readers will be following a course of practical work, but I have not found room in the text for instructions for experiments. I would also like to think that A-level students are reading outside the confines of their syllabus. At A-level, their understanding is sufficiently advanced to open the door to the study of many fascinating topics. Many chapters include examples of the importance of chemistry in medicine (e.g. anaesthetics and drugs) and in agriculture (e.g. pesticides), the balance of economic and environmental factors (e.g. the salt-based industries), the history of science (e.g. the 'new gas'), the problems faced by the chemical industry (e.g. the disasters at Seveso and Bhopal), environmental concerns (e.g. the greenhouse effect and the ozone layer). Since I wrote the account of liquid crystals, I have read that liquid crystals have now moved into pocket-sized colour television sets. To keep up to date with developments in chemistry, students will have to read newspapers and periodicals. Their advanced chemistry course equips them to understand many of the scientific issues which they will see reported. I hope they will continue to take an interest in scientific topics long after their examinations are behind them.

E. N. Ramsden,
Oxford, 1993

Acknowledgements

My task has been made possible through my being able to draw on the counsel of the staff of the Chemistry Department of the University of Hull. I am indebted to Professor R R Baldwin, Professor N B Chapman, Dr P J Francis, Professor G W Gray, FRS, Professor W C E Higginson and Dr J R Shorter for excellent advice. I have been fortunate in receiving once again the guidance of my former supervisor, Professor R P Bell, FRS. My work has benefited from the advice on content and presentation which I received from Mr G H Davies, Dr J J Guy and Dr G H Pratt.

The numerical values in the text have been taken largely from *Chemistry Data* by J G Stark and H G Wallace (John Murray, 1982).

I thank the following examination boards for permission to reprint questions from their papers.
The Associated Examining Board (AEB)
The Northern Examinations and Assessment Board (formerly Joint Matriculation Board) (NEAB)
The Northern Ireland Schools Examination and Assessment Council (NI)
The Oxford and Cambridge Schools Examination Board (O & C)
The Oxford Delegacy of Local Examinations (O)
The University of Cambridge Local Examinations Syndicate (C)
The University of London School Examinations Council (L)
The Welsh Joint Education Committee (WJEC)

The following people and organisations have kindly supplied me with photographs and given permission for their inclusion.

Bristol Uniforms Ltd	Figure 19.3
British Aerospace plc	Figure 19.4
British Petroleum plc	Figures 6.12, 26.1, 26.2 and back cover (left and right)
British Steel Corporation	Figures 24.18(b), 24.20(b) and (c), 24.26(b) and back cover (centre)
British Telecom plc	Figures 19.5
Capper Pass	Figure 24.31
Chubb Security Services Ltd	Figure 23.10
De Beers Consolidated Mines Ltd	Figure 6.13
Hanna Instruments	Figure 13.5
ICI plc Agricultural Division	Figures 22.3(b) and 23.11
ICI plc Mond Division	Figures 12.3, 18.4(b) and 18.7
IMI Refiners Ltd	Figure 12.4
Ind Coope Burton Brewery	Figures 19.1 and 19.2
J Allan Cash Photolibrary	Figure 24.22
Perkin Elmer	Figure 34.6
Pilkington Brothers plc	Figure 23.15
RTZ Services Ltd	Figures 24.21, 28 and 24.33
STEAM ICI	Figure 34.2
Dr H Sutherland	Figure 6.2
United Kingdom Atomic Energy Authority	Figures 1.21 and 1.22
Vidocq Photo Library	Figure 24.19

Illustration Acknowledgements

Figures 4.13, 4.14 and 4.28 after H Witte and E Wolfel, *Reviews of Modern Physics,* 30, 51–5, used by permission of the American Physical Society.

Figure 4.23, C A Coulson, *Proc. Cam. Phil. Soc.* 34, 210 (1938) used by permission of the Cambridge Philosophical Society.
Figures 4.27, 15.6, Linus Pauling, *The Nature of the Chemical Bond,* Second Edition (1939), used by permission of Cornell University Press.
Figure 4.39 adapted from Pauling, Corey and Branson *Proc. Natl Acad. Sci.,* US37, 205 (1951).
Figure 6.19 after G W Gray.

I thank Stanley Thornes (Publishers) for the commitment which they have shown to the production of this volume and my family for the encouragement which has sustained me during its preparation.

E N Ramsden,
1993

Part 1

THE FOUNDATION

1

THE ATOM

1.1 THE ATOMIC THEORY

The idea that matter consists of atoms dates from ancient times

Chemistry is the study of how matter behaves. One of the oldest ideas in science is that matter can be divided and further divided into smaller and smaller particles until eventually the smallest possible particles are obtained; these particles cannot be further divided. This idea was put forward by the Greek philosopher Democritus in 400 BC. He called the particles **atoms** (Greek: *atomos*, indivisible). This early idea was not based on experimental results and did not influence science.

Dalton advanced the theory. He postulated that atoms cannot be created or destroyed or split

In 1803, the atomic theory was revived by John Dalton. By this time, a large number of experimental results had been built up. The atomic theory offered explanations for the observations which chemists had made. The atomic theory was able to explain the chemical laws which had been formulated to describe the behaviour of matter. Chemists needed to account for the relationships between the masses of reactants and products which they observed in their experiments. All the calculations on reacting masses and volumes of chemicals which you see in Chapter 3 are based on the idea that each chemical element has atoms with a characteristic atomic mass.

The main points in Dalton's theory can be summarised as follows.

1. Matter is composed of tiny particles called atoms, which cannot be created or destroyed or split.

2. All the atoms of any one element are identical: they have the same mass and the same chemical properties. They differ from the atoms of all other elements.

3. A chemical reaction consists of rearranging atoms from one combination into another. The individual atoms remain intact. When elements combine to form compounds, small whole numbers of atoms combine to form compound atoms (as Dalton called them or **molecules** as we call them).

Some points in the atomic theory have been modified since Dalton's time. The atoms of some elements, e.g. uranium, can be split [see § 1.9.10]. Some elements have atoms of more than one kind, which differ slightly in mass; we call these atoms **isotopes** [see § 1.7].

Evidence for the particulate nature of matter

You will be familiar from your earlier work with evidence to support the atomic theory. Checkpoint 1A will allow you to check up on how much you remember.

CHECKPOINT 1A

1. A purple crystal is dropped into a beaker full of water. After a time, a pink solution has formed. Explain, on the basis of the particle theory of matter, how this happens.

2. Chlorine is denser than air. Yet when air and chlorine are mixed, a homogeneous mixture is obtained [see Figure 1.1]. Explain, on the basis of the particulate theory of matter, how this happens.

3. A smoke cell enables one to view the paths of particles of smoke. Figure 1.2 shows the path of a single particle. Explain, on the basis of the particle theory of matter, why the particle moves in this way, constantly changing direction.

4. When ice is heated, it melts to form water. When water is heated, it forms water vapour. Explain what happens to particles in the transitions (*a*) solid → liquid and (*b*) liquid → gas.

5. What has happened to make us disagree with Dalton on the question of whether atoms can be created or destroyed or split?

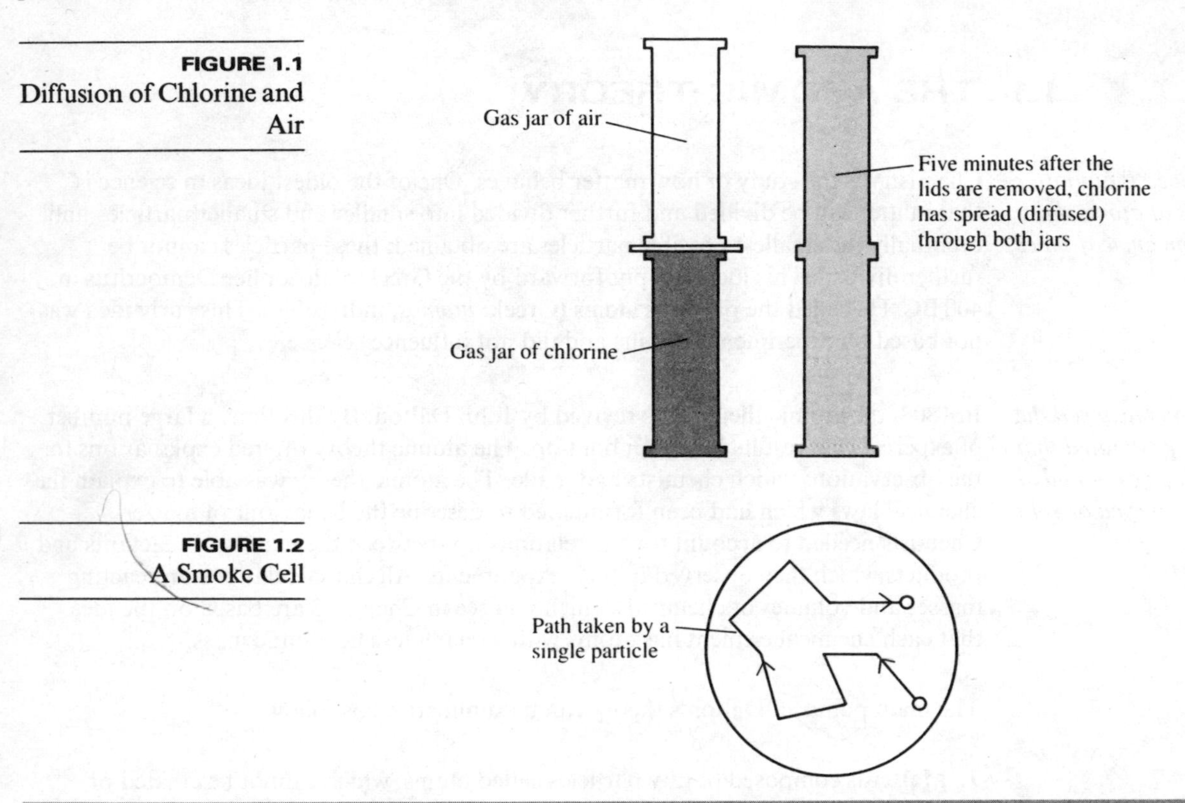

FIGURE 1.1 Diffusion of Chlorine and Air

FIGURE 1.2 A Smoke Cell

Evidence from X-ray diffraction and electron microscope studies

Dalton's atomic theory was quickly accepted by scientists. It was a working theory, but there was no direct proof that the theory was correct. Modern instruments have provided startling evidence for the theory. X ray diffraction studies [see § 6.1] have provided first-hand evidence for the existence of atoms.

You know from your previous studies that different kinds of matter consist of different kinds of particles. An element is a substance which cannot be split into simpler substances. Some elements consist of single atoms, e.g. helium consists of helium atoms, He. Other elements consist of molecules, particles which are composed of two or more atoms of the same kind, e.g. nitrogen consists of N_2 molecules, phosphorus consists of P_4 molecules and sulphur consists of S_8 molecules. A compound is a substance which is composed of two or more elements, chemically combined. Some compounds consist of molecules; these particles consist of two or more atoms bonded together. Other compounds consist of charged particles called ions. The numbers of positive and negative charges in a compound are equal. This topic is taken further in Chapters 4 and 6.

1.1.1 PARTICLES IN MOTION

The particulate theoy of matter explains the difference in behaviour between solids, liquids and gases.

Particles in solids, liquids and gases

In a solid, the particles are very close together, and their only motion is vibration about a mean position [see Figure 1.3(a)]. In a liquid the particles are further apart than in a solid. The particles move slowly until they collide with another particle or with the container [see Figure 1.3(b)]. In a gas the particles are far apart. They move in straight lines, changing direction when they collide with another particle or with the walls of the container [see Figure 1.3(c)]. Particles of a gas move at speeds of the order of 1000 km h^{-1} [see § 7.3.2].

FIGURE 1.3
Particles in a Solid, a Liquid and a Gas

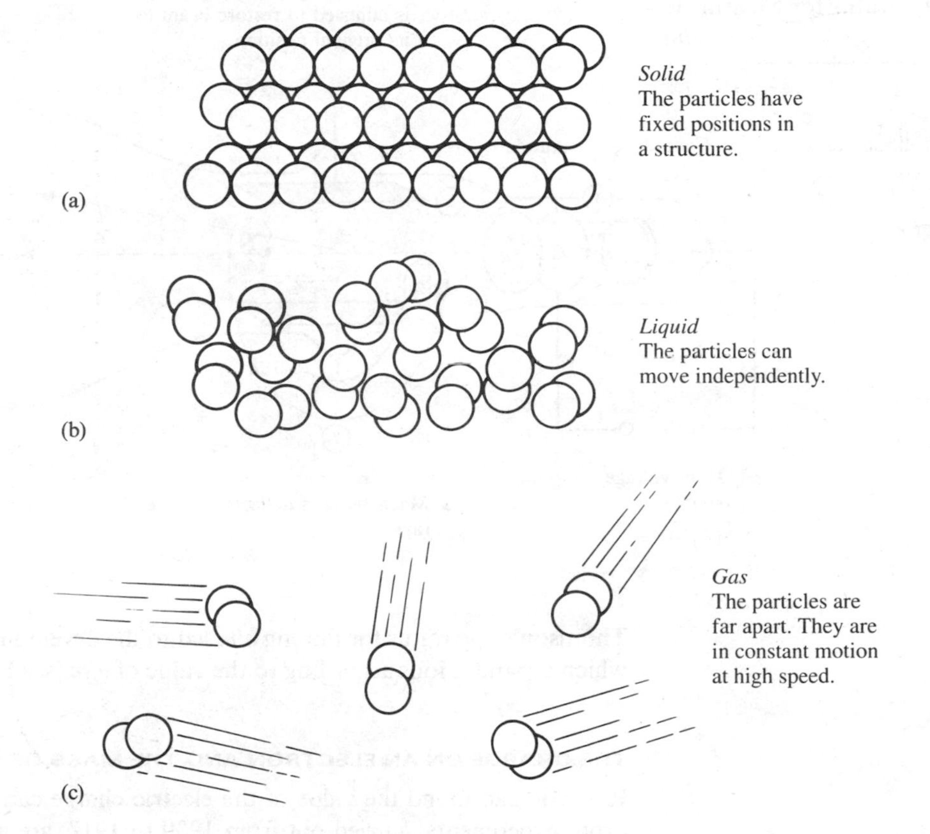

(a)
Solid
The particles have fixed positions in a structure.
(b)
Liquid
The particles can move independently.
(c)
Gas
The particles are far apart. They are in constant motion at high speed.

1.2 THE SIZE OF THE ATOM

Twentieth-century X ray work [§ 6.1] has shown that the diameters of atoms are of the order of 2×10^{-10} m, which is 0.2 nm (1 nm = 1 nanometre = 10^{-9} m).

The masses of atoms [§ 1.8] range from 10^{-27} to 10^{-25} kg. They are often stated in atomic mass units, u (where 1 u = 1.661×10^{-27} kg).

1.3 THE ELECTRON

Around the year 1900, physicists began to find evidence that atoms are made up of smaller particles. Sir William Crookes was experimenting in 1895 on the discharge of electricity through gases at low pressure. He discovered that a beam of rays was given off by the cathode (the negative electrode). Crookes called the rays **cathode rays**.

Crookes detected cathode rays

The rays also have the properties of particles

Crookes showed that cathode rays also behave like negatively charged particles.

The ratio e/m for cathode ray particles

Sir J J Thomson studied the deflection of cathode rays in electric and magnetic fields; Figure 1.4 shows the kind of apparatus he used. From his measurements he calculated that the ratio of charge/mass, *e/m*, was -1.76×10^{11} C kg^{-1} (C = coulomb, the SI unit of charge). Since he obtained the same value, regardless of what gas was used or what kind of electrodes were used, he deduced that these negatively charged particles are present in all matter. They were named **electrons,** and were recognised as the particles of which an electric current is composed.

FIGURE 1.4 Thomson's Apparatus for Measuring *e/m*

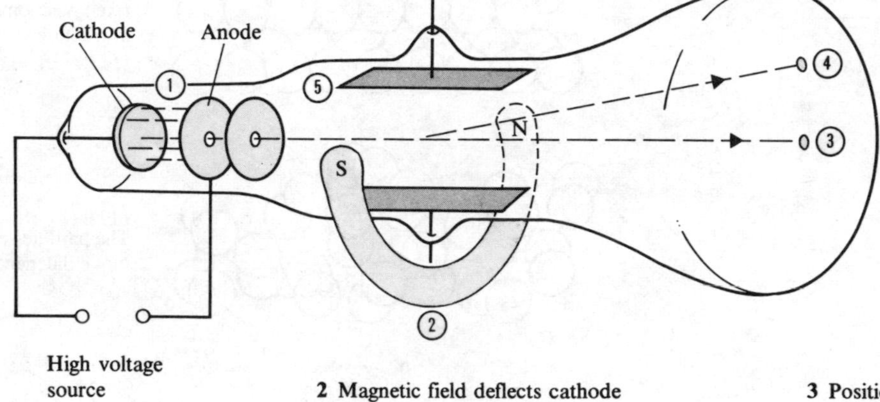
5 Electric field opposes magnetic field. Strength of electric field is adjusted to restore beam to its original position.
4 Deflection of beam produced by magnetic field alone.
1 Cathode rays.
Cathode
Anode
N
S
High voltage source
2 Magnetic field deflects cathode rays.
3 Position of beam is shown by luminescence of phosphor-treated glass.

Thomson's apparatus for finding *e/m* led to the development of the mass spectrometer, which separates ions according to the value of *e/m* [see Figures 1.8, 1.9, § 1.8].

THE CHARGE ON AN ELECTRON AND THE MASS OF AN ELECTRON

The charge on an electron and its mass

R A Millikan found the value of the electric charge carried by an electron. His 'oil drop' experiments, carried out from 1909 to 1917, are described in many physics books*. From his experiments, he obtained the value of -1.60×10^{-19} C. This amount of charge is called 1 elementary charge unit. Combining this value of charge with Thomson's value of charge/mass gave a value of 9.11×10^{-31} kg for the mass of an electron. This is 5×10^{-4} times the mass of a hydrogen atom.

1.4 THE ATOMIC NUCLEUS

The Thomson model of the atom

In 1898, Thomson surveyed all the evidence that atoms consist of charged particles. He described an atom as a sphere of positive electricity, in which negative electrons are embedded. Other people described this as the 'plum pudding' picture of the atom!

*See, e.g., R Muncaster, *A-Level Physics* (Stanley Thornes)

Geiger and Marsden tested the model...

If this model of the atom is correct, then a metal foil is a film of positive electricity containing electrons. A beam of α particles (helium nuclei, see § 1.7) fired at it should pass straight through. In 1909, Lord Rutherford's colleagues, Geiger and Marsden, tested this prediction [see Figure 1.5].

FIGURE 1.5 Illustration of Geiger–Marsden Experiment

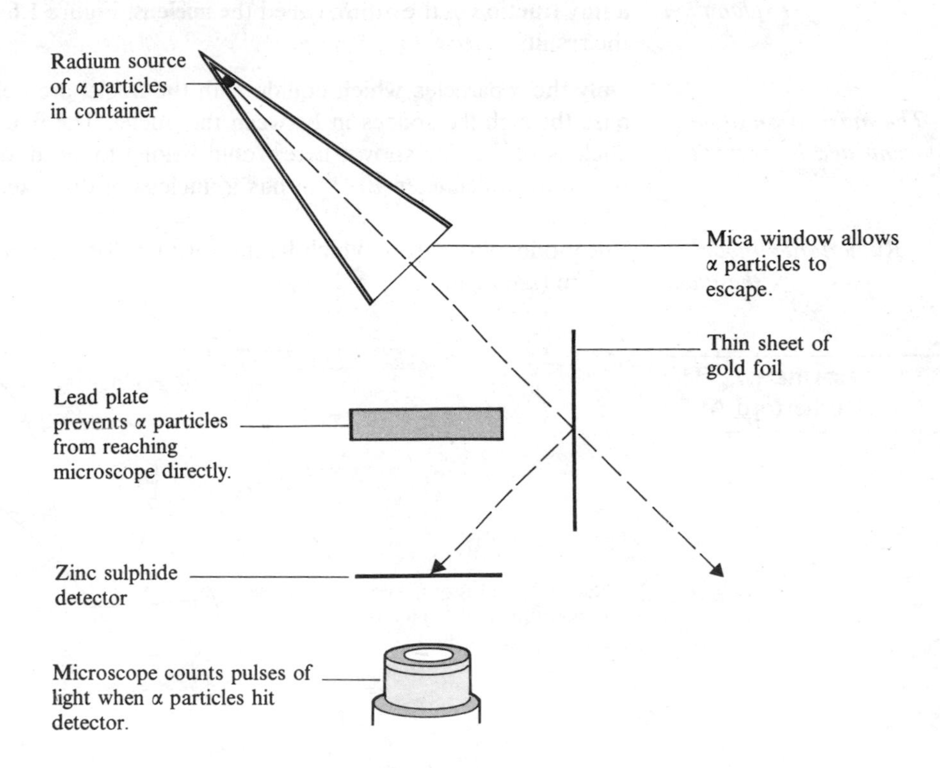

FIGURE 1.6 Scattering of α particles by the Nuclei of Metal Atoms

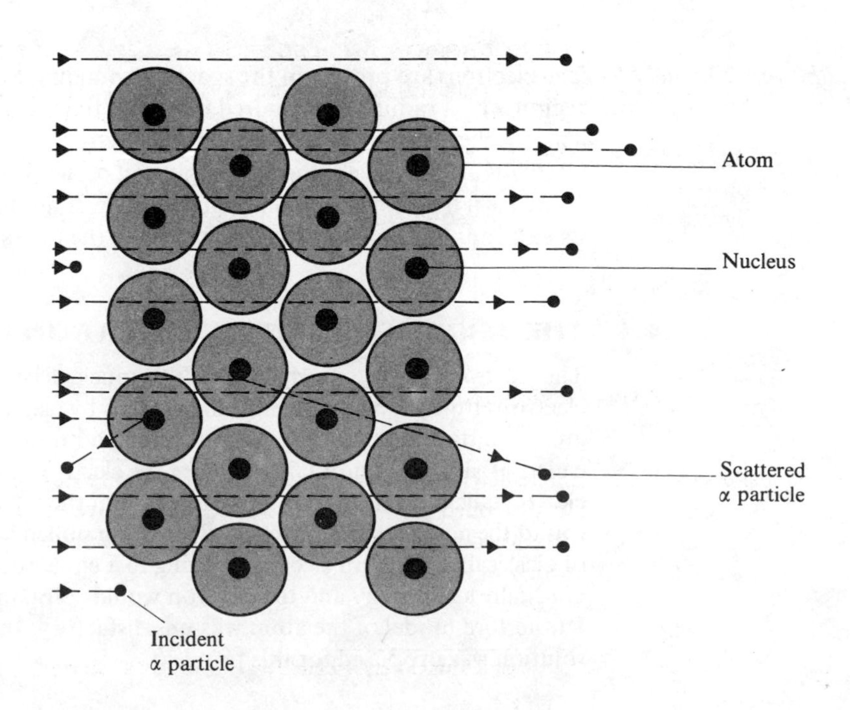

... their results ...

They found, as they expected, that α particles penetrated the gold foil. They also found, to their amazement, that a small fraction (about 1 in 8000) of the α particles were deflected through large angles and even turned back on their tracks. Rutherford described this as 'about as incredible as if you fired a 15-inch shell at a piece of tissue paper and it came back and hit you'.

... Rutherford's explanation

Rutherford deduced that the mass and the positive charge must be concentrated in a tiny fraction of the atom, called the **nucleus**. Figure 1.6 shows his interpretation of the results.

The mass of an atom is concentrated in its nucleus

Only the α particles which collide with the nuclei are deflected; the vast majority pass through the spaces in between the nuclei. The figure is not drawn to scale: a nucleus of the size shown here would belong to an atom the size of your classroom. An atom of diameter 10^{-10} m has a nucleus of diameter 10^{-15} m.

Rutherford's model of the atom

The model of the atom which Rutherford put forward in 1911 was like the solar system (see Figure 1.7).

FIGURE 1.7 The Rutherford Atom

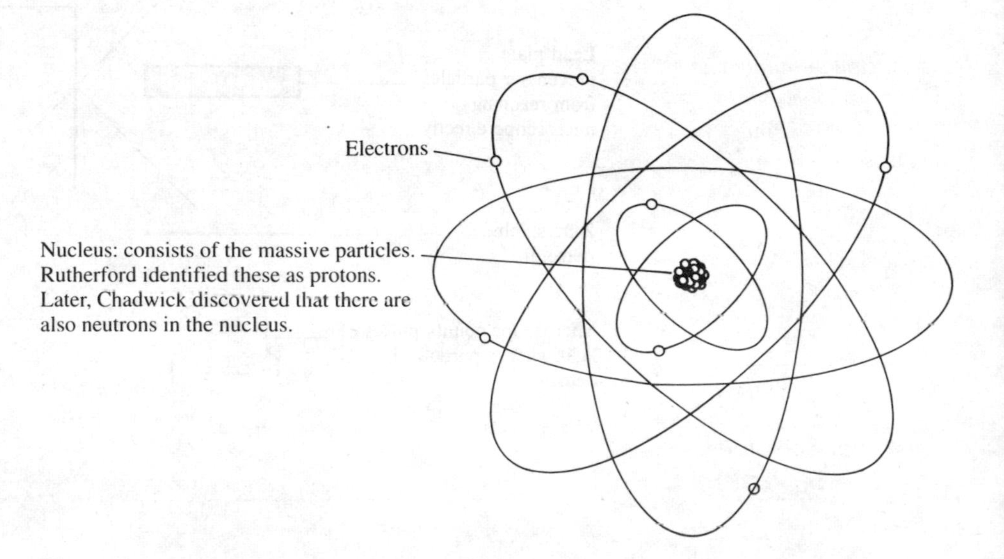
Electrons
Nucleus: consists of the massive particles.
Rutherford identified these as protons.
Later, Chadwick discovered that there are
also neutrons in the nucleus.

Rutherford's model of the atom

The electrons are present in the space surrounding the nucleus. The electrons inhabit a region with a radius one hundred thousand times greater than that of the nucleus. They occupy this space by repelling the electrons of neighbouring atoms. Since electrons are negatively charged, repulsion occurs between the like charges on two electrons if they come close enough together. If another atom approaches too closely, its electrons are repelled by the electrons of the first atom.

1.4.1 THE ARRANGEMENT OF ELECTRONS IN THE ATOM

The electrons are negatively charged; the nucleus is positively charged. What stops the electrons from being pulled into the nucleus by electrostatic attraction? The electrons are in constant motion. They move round and round the nucleus in circular paths called orbits [see Figure 1.7]. Why are the electrons not pulled into the nucleus by electrostatic attraction? It was suggested that the movement of electrons in orbits round the nucleus would prevent their being pulled in. However, according to the laws of classical physics, an electron moving in a circle round a positive nucleus would gradually lose energy and the electron would spiral into the nucleus. In this way, the Rutherford model of the atom was unsatisfactory. In § 2.2.1, you will see what solution was proposed for this problem.

1.5 THE NEUTRON

The atomic number is the number of protons in the nucleus which equals the number of electrons in the atom

H G Moseley suggested in 1913 that the multiple charge on the nucleus arose from the presence of **protons**, which contribute the charge. Since atoms are neutral, the number of electrons must be the same as the number of protons. Atomic masses are greater than the mass of the protons in the atom. To make up the extra mass, the existence of **neutrons** was postulated. These particles should have the same mass as a proton and zero charge. The search for the neutron began.

Neutrons have mass but no charge

It was a member of Rutherford's team, J Chadwick, who established the existence of the neutron in 1934. He found that beryllium emitted uncharged particles when it was bombarded with α particles (which are helium nuclei). The uncharged radiation was a stream of neutrons. The equation for the reaction between α particles and beryllium is shown in § 1.7.

1.6 THE FUNDAMENTAL PARTICLES

Proton number or atomic number. Nucleon number or mass number

The nucleus was thus shown to consist of protons and neutrons. The number of protons is called the **atomic number** or **proton number.** Protons and neutrons are both **nucleons**. The number of protons and neutrons is called the **nucleon number**, or, alternatively, the **mass number**.

TABLE 1.1 The Mass and Charge of Sub-atomic Particles

Particle	*Charge/C*	*Relative charge*	*Mass/kg*	*Mass/u*
Proton	$+1.6022 \times 10^{-19}$	+1	1.6726×10^{-27}	1.0073
Neutron	0	0	1.6750×10^{-27}	1.0087
Electron	-1.6022×10^{-19}	−1	9.1095×10^{-31}	5.4858×10^{-4}

1.7 NUCLIDES AND ISOTOPES

Notation for nuclides

The word **nuclide** is used to describe any atomic species of which the proton number and the nucleon number are specified. Nuclides are written as $^{\text{nucleon number}}_{\text{proton number}}\text{Symbol}$ (i.e. $^{\text{mass number}}_{\text{atomic number}}\text{Symbol}$). The species $^{12}_{6}C$ and $^{9}_{4}Be$ are nuclides. Protons are represented as $^{1}_{1}H$, neutrons as $^{1}_{0}n$, α particles as $^{4}_{2}He$ and electrons as $^{0}_{-1}e$. Using this notation, the equation for Chadwick's reaction is

Equation for the production of neutrons

$$^{4}_{2}He + ^{9}_{4}Be \rightarrow ^{1}_{0}n + ^{12}_{6}C$$

Isotopes contain the same number of protons and different numbers of neutrons

When an element has a relative atomic mass [§ 3.3] which is not a whole number, it is because it consists of a mixture of **isotopes**. Isotopes are nuclides of the same element. They have the same atomic number but different mass numbers, i.e. they differ in the number of neutrons in the nucleus. Since chemical properties depend upon the nuclear charge and electronic structure of an atom, with mass having little effect, isotopes show the same chemical behaviour. The isotopes of chlorine, $^{35}_{17}Cl$ and $^{37}_{17}Cl$, have the same atomic number, 17. The difference between the mass numbers shows that one isotope has 18 neutrons and the other has 20 neutrons. The chemical reactions of the two isotopes are identical. Their names can be written as chlorine-35 and chlorine-37. The isotopes of hydrogen differ more than isotopes of other elements [§ 17.8].

CHECKPOINT 1B: ATOMS

1. State the number of protons, neutrons and electrons in the following atoms:
(*a*) $^{39}_{19}K$ (*b*) $^{27}_{13}Al$ (*c*) $^{137}_{56}Ba$ (*d*) $^{226}_{88}Ra$

2. State the number of protons, neutrons and electrons in the following atoms:
(*a*) $^{12}_{6}C$ (*b*) $^{12}_{6}C$ (*c*) $^{1}_{1}H$ (*d*) $^{2}_{1}H$ (*e*) $^{3}_{1}H$ (*f*) $^{87}_{38}Sr$ (*g*) $^{90}_{38}Sr$
(*h*) $^{235}_{92}U$ (*i*) $^{238}_{92}U$
What is the relationship between the different atoms of strontium?

3. 'Dalton was incorrect in saying that all the atoms of a particular element are identical.' Discuss this statement.

4. Why did Chadwick look for the neutron? Why was it hard to find?

1.8 MASS SPECTROMETRY

In mass spectrometry...

Atomic masses are determined by **mass spectrometry**. The mass spectrometer was developed by F W Aston from J J Thomson's apparatus for measuring the ratio e/m for a particle. In the mass spectrometer, atoms and molecules are converted into

FIGURE 1.4 A Mass Spectrometer

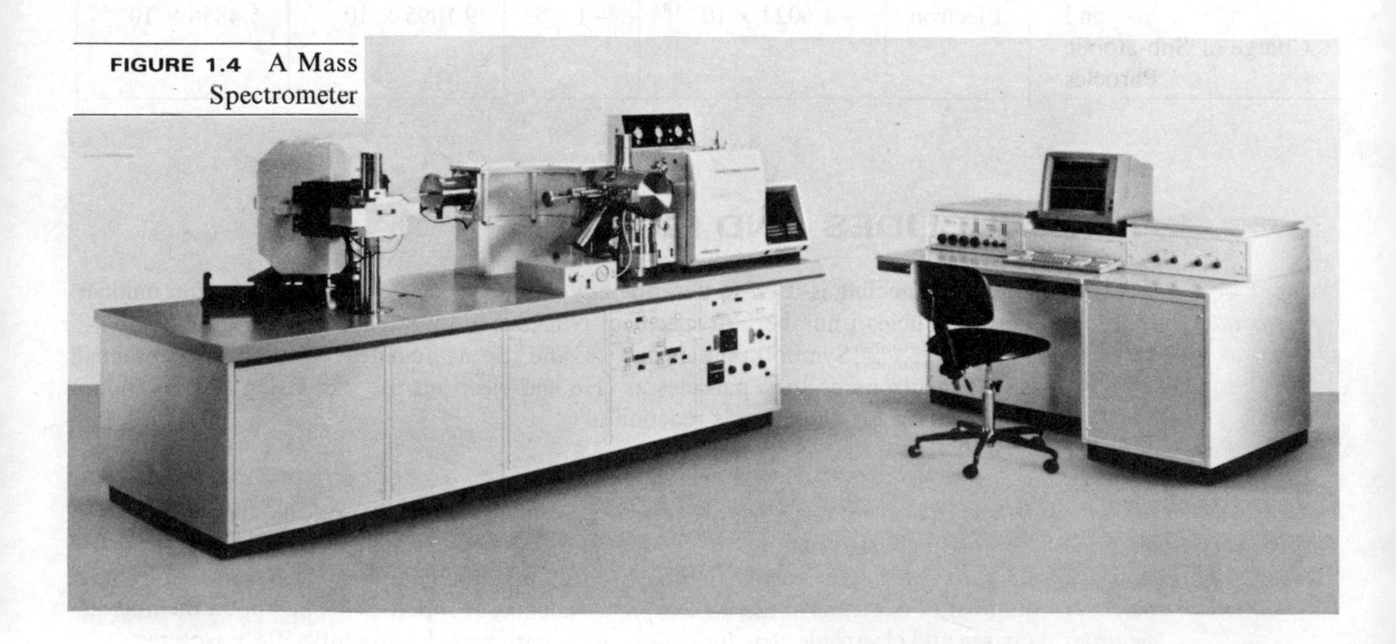

... ions are deflected in a magnetic field

ions. The ions are separated as a result of the deflection which occurs in a magnetic field. Figure 1.9 shows how a mass spectrometer operates, and Figure 1.8 is a photograph of an instrument. Figure 1.10 shows a mass spectrometer trace for copper(II) nitrate.

FIGURE 1.9 A Mass Spectometer: How it Works

1 Heated filament gives electrons. They pass into the ionisation chamber.

2 The sample is injected as a gas into the ionisation chamber. Electrons collide with molecules of the sample and remove electrons to give positive ions. Some molecules break into fragments. The largest ion is the molecular ion.

3 To this plate, a negative potential is applied (about 8000 V). The electric field accelerates the positive ions.

4 An electromagnet produces a magnetic field. The field deflects the beam of ions into circular paths. Ions with a high ratio of mass/charge are deflected less than those with a low ratio of mass/charge.

5 These ions have the correct ratio of mass/charge to pass through the slit and arrive at the collector.

6 Amplifier. Here the charge received by the collector is turned into a sizeable electric current.

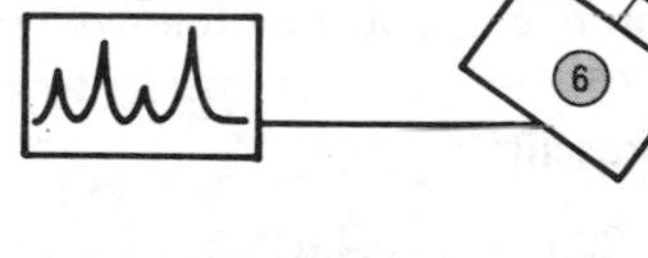

7 Recorder. The electric current operates a pen which traces a peak on a recording.

8 If the magnetic field is kept constant while the accelerating voltage is continuously varied, one species after another is deflected into the ion collector. A trace such as that in Figure 1.10 is obtained.

FIGURE 1.10 The Mass Spectrum of Copper (II) Nitrate

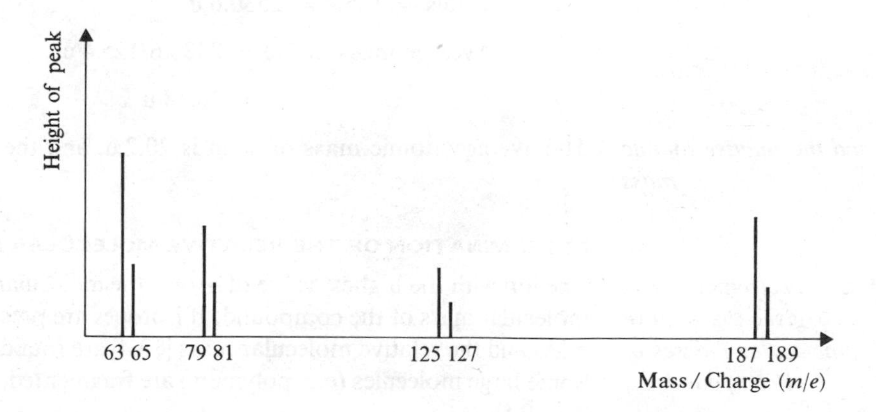

Notes

(*1*) The height of each peak measures the relative abundance of the ion which gives rise to that peak.

(*2*) The ratio of mass/charge for each species is found from the value of the accelerating voltage associated with a particular peak. Many ions have a charge of +1 **elementary charge unit**, and the ratio m/e is numerically equal to $\bar{m}$, the mass of the ion (1 elementary charge unit = 1.60×10^{-19} C).

... deflection of ion depends on ratio m/e

(*3*) The peaks on this trace correspond to the ions

$$63 = {}^{63}Cu^+,\ 65 = {}^{65}Cu^+,\ 79 = {}^{63}CuO^+,\ 81 = {}^{65}CuO^+,$$

$$125 = {}^{63}CuNO_3{}^+,\ 127 = {}^{65}CuNO_3{}^+,\ 187 = {}^{63}Cu(NO_3)_2{}^+,$$

$$189 = {}^{65}Cu(NO_3)_2{}^+$$

1.8.1 USES OF MASS SPECTROMETRY

DETERMINATION OF THE RELATIVE ATOMIC MASS OF AN ELEMENT

Mass spectrometry is used for the determination of relative atomic mass

Relative atomic mass, A_r, is defined in § 3.3. Figure 1.11 shows the mass spectrum of neon.

FIGURE 1.11 The Mass Spectrum of Neon

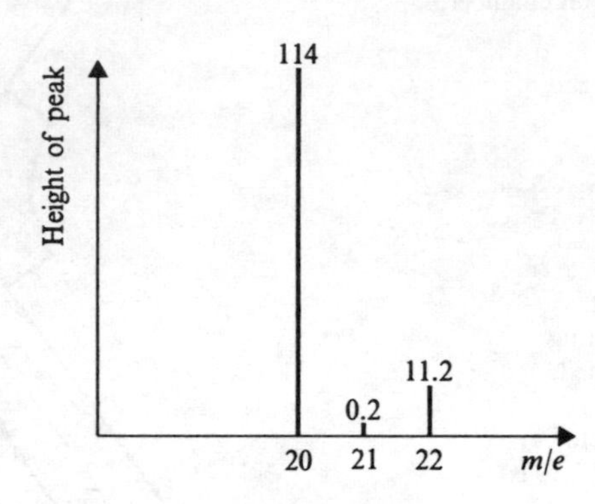

The calculation of the average atomic mass of neon...

The average atomic mass of neon is calculated as follows.
Multiply the relative abundance (the height of the peak) by the mass number to find the total mass of each isotope present:

$$\text{Mass of } ^{22}\text{Ne} = 11.2 \times 22.0 = 246.4 \text{ u}$$

$$\text{Mass of } ^{21}\text{Ne} = 0.2 \times 21.0 = 4.2 \text{ u}$$

$$\text{Mass of } ^{20}\text{Ne} = 114 \times 20.0 = 2280.0 \text{ u}$$

$$\text{Totals} = 125.4 = 2530.6 \text{ u}$$

$$\text{Average mass of Ne} = 2530.6/125.4 \text{ u}$$

$$= 20.18 \text{ u}$$

...and the relative atomic mass

The average atomic mass of neon is 20.2 u, and the relative atomic mass is 20.2.

DETERMINATION OF THE RELATIVE MOLECULAR MASS OF A COMPOUND

Mass spectrometry also gives the relative molecular masses of compounds...

The ion with the highest value of m/e is the molecular ion, and its mass gives the molecular mass of the compound. If isotopes are present, the average molecular mass and the relative molecular mass [§ 3.4] are found as in the neon example. Some large molecules (e.g. polymers) are fragmented, and do not give molecular ions [§ 34.9.5].

IDENTIFICATION OF COMPOUNDS

...and can be used for the identification of compounds...

A mass spectrum is obtained, and information about the peak heights and m/e values is fed into a computer. The computer compares the spectrum of the unknown compound with those in its data bank, and thus identifies the compound.

FORENSIC SCIENCE

...and in forensic science where it is useful as a small sample is enough to give results

The sensitivity of the mass spectrometer makes it an admirable tool for forensic scientists. The size of sample which they receive for analysis is often very small. A mass spectrum can be obtained on as little as 10^{-12} g. Small amounts of drugs can be identified by mass spectrometry. A fibre left at the scene of a crime can be compared by mass spectrometry with a fibre from a suspect's clothing.

CHECKPOINT 1C: MASS SPECTROMETRY

1. Describe how, in a mass spectrometer, ions are (*a*) formed, (*b*) accelerated, (*c*) separated and (*d*) detected.

2. Define the terms mass number, isotope, relative atomic mass.

Chlorine has two isotopes of relative atomic masses 34.97 and 36.96 and relative abundance 75.77% and 24.23% respectively.

Calculate the mean relative atomic mass of naturally occurring chlorine.

3. The mass spectrum of dichloromethane shows peaks at 84, 86 and 88. The intensities of the lines at 84, 86 and 88m_u are in the ratio 9 : 6 : 1. What species give rise to these lines? How do you account for the relative intensities of the lines?

***4.** Carbon consists of 99% of ^{12}C and 1% of ^{13}C. In the mass spectrum of a hydrocarbon, the peak with the highest mass number occurs at mass $M + 1$. It corresponds to the molecular ion containing one ^{13}C atom. The peak at mass M corresponds to the molecular ion containing only ^{12}C. The peak at mass M is 16.5 times as intense as that at mass $M + 1$. How many C atoms are there in a molecule of the hydrocarbon?

5. Figure 1.12 shows the mass spectrum of zirconium, and Figure 1.13 shows that of lead. The heights of the peaks and the mass numbers of the isotopes are shown on the figures. Calculate the average atomic masses of (*a*) zirconium and (*b*) lead.

**Note* An asterisk denotes a question on a topic which is not included in *all* the examination syllabuses.

FIGURE 1.12
The Mass Spectrum of Zirconium

Height of peak
51.5
11.2
17.1
17.4
2.8
90 91 92 94 96
Mass number

FIGURE 1.13
The Mass Spectrum of Lead

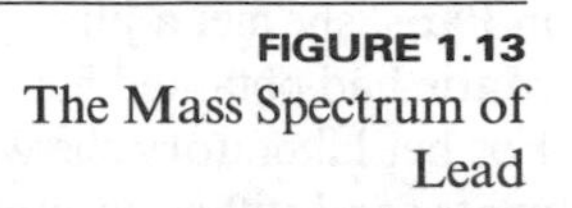

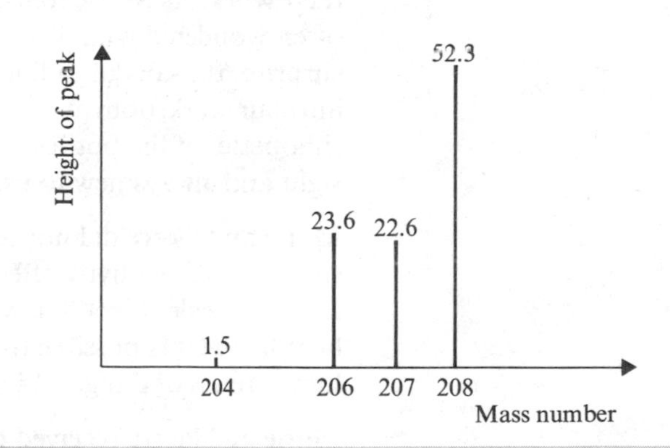

1.9 NUCLEAR REACTIONS

In nuclear reactions new elements are formed

A nuclear reaction is different from a chemical reaction. In a chemical reaction, the atoms which make up the reactants enter into different combinations to form the products, but the nuclei of the atoms are unchanged. In a nuclear reaction, a rearrangement of the protons and neutrons in the nucleus of the atom takes place, and new elements are formed.

1.9.1 RADIOACTIVITY

DISCOVERY

Becquerel discovered a new type of radiation

In 1896, a French physicist called A H Becquerel was experimenting on salts which fluoresced (glowed in the dark). One day he developed a photographic plate which had been left wrapped in a drawer and found to his surprise that the plate had been exposed. He knew that no light could penetrate the wrapping, so Becquerel concluded that the uranium salts in the drawer were to blame. Perhaps the plate had been fogged by some rays coming from the uranium salt. However, there was no known type of radiation that had this effect. Becquerel's instinct told him that it would be worth while investigating this mysterious radiation. He gave the problem to a young research worker called Marie Curie.

Marie Curie found that the radiation was a property of uranium atoms, which she named radioactivity

Marie Curie soon found that the strange effect happened with all uranium salts. It depended on the amount of uranium present but not on the type of compound. She realised that the ability to give off the radiation must be a property of the *atoms* of uranium, and was independent of any chemical bonds which they formed. She realised that this was a completely new type of property, quite different from a chemical reaction. Marie Curie called this property of the uranium atom **radioactivity**.

Marie's husband, Pierre, left his own research work to help her with her exciting new discovery of radioactivity. In 1898 they discovered two new radioactive elements. They called one **polonium** after Marie's native country, Poland. The other element they named **radium**, meaning 'giver of rays'.

Profile: Marie Curie (1867–1934)

Marie Sklodowska left Poland to study in France. At the University of the Sorbonne in Paris, she met a physics lecturer called Pierre Curie. After they had married and Marie had obtained her first degree, she became a research worker in the university. For her laboratory she was allocated a miserable shed which was freezing cold in winter and stifling in summer. In spite of the wretched conditions, the Curies enjoyed their work. As Marie toiled to extract the element which she later named radium, she often wondered what it would look like, what colour its salts would be. To her surprise, the salts of radium glowed in the dark. She wrote 'One of our joys was to go into our workroom at night; we then perceived on all sides the feebly luminous silhouettes of the bottles and capsules containing our products. It really was a lovely sight and always new to us. The glowing tubes looked like fairy lights.'

Marie and Pierre did not realise that one of the effects of radioactivity is far more sinister. Radioactivity affects the blood cells and causes leukaemia. Pierre was killed in a street accident in 1906, when he walked into the path of a bus. He was not in good health, and it is possible that radiation had already begun to take its toll and to make him careless of danger. Marie died of leukaemia at the age of 67.

Marie and Pierre received the Nobel Prize for physics in 1903 for their discovery of radioactivity. Marie received the Nobel Prize for chemistry in 1911 for the discovery of polonium and radium. Had Pierre been alive, he would have shared the prize with her. Their daughter, Irène Joliot-Curie won the Nobel Prize jointly with her husband in 1936 for creating the first artificial radioactive elements.

FIGURE 1.14 Marie Curie in Her Laboratory

Madame Curie is often praised for the arduous work she did in isolating a small quantity of radium from a tonne of the uranium ore **pitchblende**. The work involved back-breaking handling of the ore, crushing it and stirring it with reagents to remove the large mass of unwanted substances. Her real claim to fame, however, is the insight she showed in realising that she was dealing with a new phenomenon which was completely uncharted in either physics or chemistry.

1.9.2 TYPES OF RADIATION

Rutherford suggested that radioactivity is the result of atoms splitting

Why do the atoms of uranium, polonium and radium give off these rays which Marie Curie named radioactivity? The explanation came from the New Zealand-born British physicist Ernest (Lord) Rutherford in 1902. His suggestion was that the atoms split and in the process energy is released. This energy is in the form of radiation. As you can appreciate, this idea was revolutionary. For a century, scientists had accepted Dalton's view that atoms could not be created or destroyed or split. We can now list a large number of elements that have unstable atoms which split to form smaller particles.

Some nuclides are unstable and split up to form smaller atoms

When an atom splits, the nucleus divides and the protons and neutrons in it form two new nuclei. The electrons divide themselves between the two. Sometimes, protons, neutrons and electrons fly out when the original nucleus divides. The process is called **radioactive decay**, and the element is said to be **radioactive**. The particles and energy are called **radioactivity**. Radioactive isotopes have unstable nuclei.

Radioactive substances give three types of radiation... ...α, β and γ rays

Three types of radiation are given off by radioactive substances. They all cause certain substances, such as zinc sulphide, to luminesce, and they all ionise gases through which they pass. They differ in their response to an electric field, in the manner shown in Figure 1.15. The uncharged rays, γ (**gamma**) rays, are similar to X rays. They have high penetrating power, being able to pass through 0.1 m of metal. Measurements of e/m identified α (**alpha**) rays as the nuclei of helium atoms and β (**beta**) rays as electrons. β rays can pass through 0.01 m of metal, and α rays can penetrate no more than 0.01 mm of metal.

FIGURE 1.15 Effect of an Electric Field on Radiation

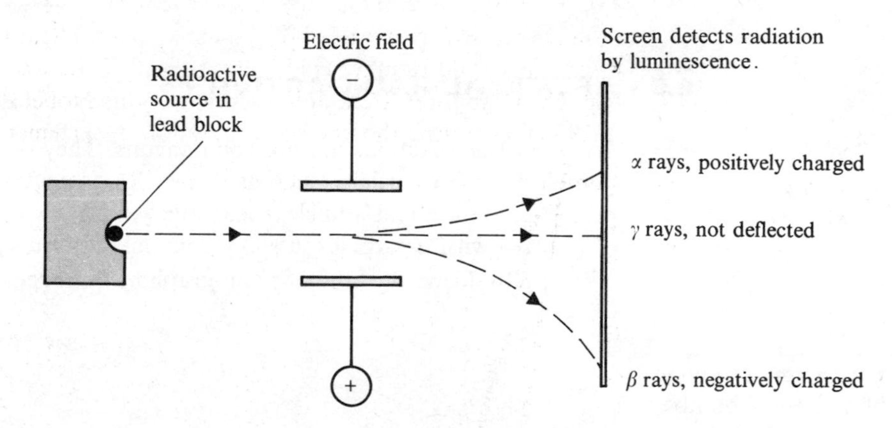

1.9.3 NATURAL RADIOACTIVITY

The uranium series

There are three naturally occurring series of radioactive elements. The **uranium series** starts with $^{238}_{92}U$ and decays through a series of unstable isotopes to $^{206}_{82}Pb$. The first two steps in the decay are:

$$^{238}_{92}U \rightarrow ^{234}_{90}Th + ^{4}_{2}He \quad (\alpha \text{ decay})$$

$$^{234}_{90}Th \rightarrow ^{234}_{91}Pa + ^{0}_{-1}e \quad (\beta \text{ decay})$$

α decay and β decay

When an isotope undergoes α decay (with the emission of an α particle), its atomic number decreases by 2, and its mass number decreases by 4. The isotope produced is two groups to the left in the Periodic Table. When an isotope undergoes β decay (with the emission of an electron), its atomic number increases by 1, and its mass number is unchanged. The isotope produced is one group to the right in the Periodic Table.

Group 4 Group 5 Group 6

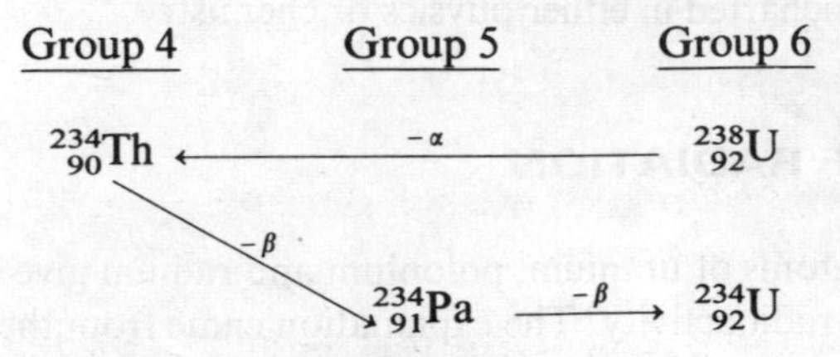

The actinium and thorium series

The other naturally occurring series is the **actinium series**, which decays from $^{235}_{92}U$ to $^{207}_{82}Pb$, and the **thorium series**, which starts with $^{232}_{90}Th$ and ends with $^{208}_{82}Pb$.

There is a fourth series of radioisotopes, which do not occur in nature but have been made by nuclear reactions.

1.9.4 BALANCING NUCLEAR EQUATIONS

Balancing nucleon (mass) numbers and proton (atomic) numbers

In the equation for a nuclear reaction, the sum of the nucleon numbers (mass numbers) is the same on both sides, and the sum of the proton numbers (atomic numbers) is the same on both sides of the equation. For example, when nitrogen-16 undergoes β decay

$$^{16}_{7}N \rightarrow \, ^{a}_{b}O + \, ^{0}_{-1}e$$

Considering mass numbers gives $16 = a + 0 \therefore a = 16$
Considering atomic numbers gives $7 = b + (-1) \therefore b = 8$
The isotope produced is $^{167}_{8}O$
The equation is $^{16}_{7}N \rightarrow \, ^{16}_{8}O + \, ^{0}_{-1}e$

1.9.5 ARTIFICIAL RADIOACTIVITY

Radioactivity can be induced

Some nuclear reactions are not spontaneous. They occur when stable isotopes are bombarded with particles such as α particles or neutrons. Rutherford was the first person to bring about a nuclear reaction. He was experimenting on the bombardment of nitrogen with α particles in a cloud chamber of the type invented by C T R Wilson. Figure 1.16 shows the kind of photograph he obtained.

FIGURE 1.16 Drawing of a Cloud Chamber Photograph of a Nuclear Reaction

1 The cloud chamber is filled with air which is supersaturated with water vapour. If any ions are produced, they cause condensation.

2 Radioactive source of α particles

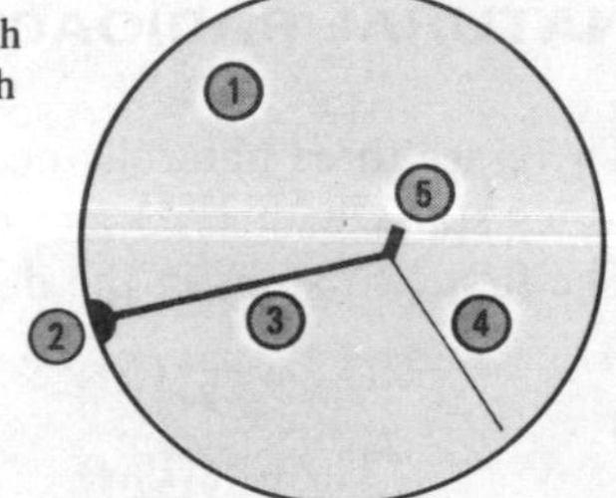

5 Short, thick track of $^{17}_{8}O$

4 Long, thin track of $^{1}_{1}H$

3 This track is a trail of condensation produced by an α particle.

Rutherford used bombardment by α particles

Figure 1.16 shows the track of an α particle coming to an end and being replaced by a short, thick track and a long, thin track. Rutherford realised that two particles had been formed in a nuclear reaction. He attributed the short, thick track to $^{17}_{8}O$ and the long, thin track to $^{1}_{1}H$. He proposed that they had been formed by the nuclear reaction

Nuclear reactions

$$^{14}_{7}N + ^{4}_{2}He \rightarrow ^{17}_{8}O + ^{1}_{1}H \qquad (\alpha, p)$$

... (α, p)

This is classified as an (α, p) reaction since the projectile is an α particle and a proton (p) is produced in the reaction.

Other bombarding particles were used, and more nuclear reactions were observed. Examples are

... (α, n) $$^{9}_{4}Be + ^{4}_{2}He \rightarrow ^{12}_{6}C + ^{1}_{0}n \qquad (\alpha, n)$$

... (p, α) $$^{7}_{3}Li + ^{1}_{1}H \rightarrow 2^{4}_{2}He \qquad (p, \alpha)$$

... (n, α) $$^{16}_{8}O + ^{1}_{0}n \rightarrow ^{13}_{6}C + ^{4}_{2}He \qquad (n, \alpha)$$

Elements with Z greater than 92 are artificially made

Neutrons (n) have the advantage over α particles and protons in that, being uncharged, they are not repelled by the positive nuclei of the bombarded atoms. Since 1940, a set of new elements with proton numbers greater than 92, the proton number of the heaviest naturally occurring element, uranium, have been made. They are called the **transuranium elements**. The element neptunium is made by neutron bombardment of $^{238}_{92}U$, followed by radioactive decay of the isotope formed:

$$^{238}_{92}U + ^{1}_{0}n \rightarrow ^{239}_{92}U$$

$$^{239}_{92}U \rightarrow ^{239}_{93}Np + ^{0}_{-1}e$$

The first artificially made radioisotope

The first artificially produced **radioisotope** was made by Irene Curie and J Joliot in 1934 by an (α, n) reaction. An isotope of boron was converted into a radioactive isotope of nitrogen. This decayed by emitting **positrons** (positive electrons):

$$^{10}_{5}B + ^{4}_{2}He \rightarrow ^{13}_{7}N + ^{1}_{0}n \quad (\alpha, n)$$

$$^{13}_{7}N \rightarrow ^{13}_{6}C + ^{0}_{+1}e \qquad \text{(positron emission)}$$

1.9.6 RATE OF RADIOACTIVE DECAY

The rate of radioactive decay is proportional to the number of radioactive atoms present

The half-life is the time taken to decay to half the number of radioactive atoms

The rate at which a radioactive isotope decays cannot be speeded up or slowed down. It depends only on the identity of the isotope and the amount of isotope present. The nature of **nuclear decay** is illustrated in Figure 1.17. The time taken for a number N_0 of radioactive atoms to decay to $N_0/2$ atoms is called the **half-life**, $t_{1/2}$, of the radioactive isotope. The times taken for $N_0/2$ atoms to decay to $N_0/4$ and for $N_0/4$ atoms to decay to $N_0/8$ atoms are the same and have the same value as $t_{1/2}$. The rate of decay is thus proportional to the number of atoms present. Such reactions are described as **first-order** reactions [§ 14.4.2].

FIGURE 1.17
Radioactive Decay

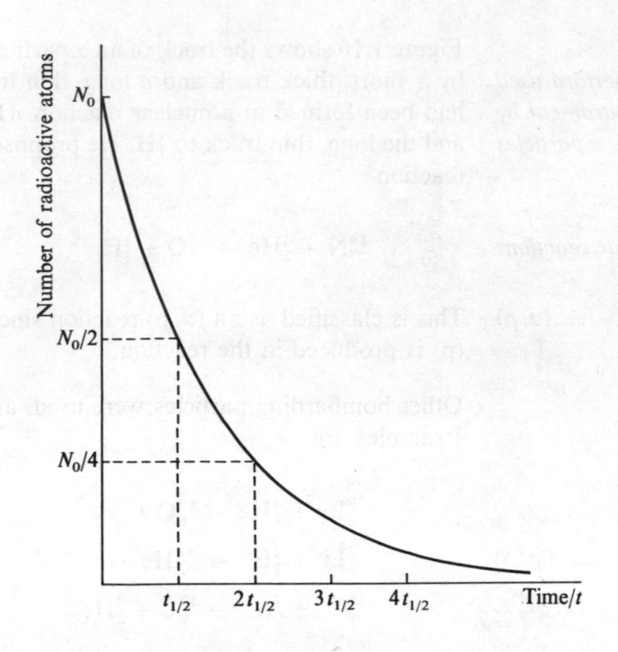

1.9.7 METHODS OF DETECTING AND MEASURING RADIOACTIVITY

WILSON CLOUD CHAMBER

Radioactivity is detected by means of a cloud chamber...

The tracks of condensation produced by ionising radiations in the Wilson cloud chamber can be photographed [see Figure 1.16, § 1.9.5].

THE GEIGER–MÜLLER TUBE

... and measured in a Geiger–Müller counter...

H Geiger and F M Müller invented the device shown in Figure 1.18. It enables the particles emitted in radioactive decay to be counted as pulses of electric current.

FIGURE 1.18
A Geiger–Müller Tube

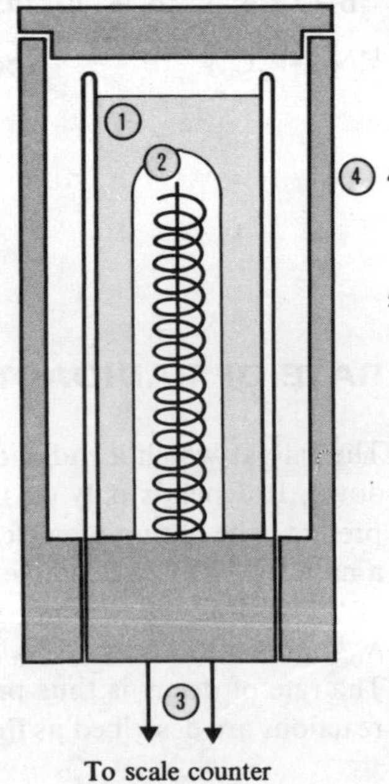

1 Liquid under investigation sends β particle or γ ray into the Geiger-Müller tube through a thin-walled glass tube.

2 Geiger-Müller tube contains a gas under reduced pressure. Two electrodes are at a voltage just less than that which will allow an electric current to pass. Each time radiation ionises the gas for a fraction of a second, a pulse of electric current flows.

3 The pulses are fed into a loudspeaker to produce a series of clicks, or are counted on an electronic counter.

4 Lead 'castle' shields Geiger-Müller tube from background radiation.

5 For α particle emitters, a different design of counter is used. A very thin window is used to allow penetration.

PHOTOGRAPHIC FILM

...or by a photographic film...

Photographic film is exposed by radioactivity. It was this which led to the discovery of radioactivity by Becquerel. It is used as a protective device in the pieces of photographic film worn by people who work with radioactivity. The 'fogging' of the photographic film is a measure of the amount of radioactivity to which they have been exposed.

SCINTILLATION COUNTER

...or in a scintillation counter

There are substances, such as zinc sulphide, which phosphoresce when affected by radioactivity. As each particle from the radioactive source hits the **phosphor**, a flash of light is emitted. In a scintillation counter, each flash of light gives rise to a pulse of current. A digital counter records the pulses of current.

1.9.8 USES OF RADIOACTIVE ISOTOPES

Radioactivity is used to destroy cancer cells...

1. Cancerous tissue is destroyed by radioactivity in preference to healthy tissue. A cobalt-60 source (a γ emitter, $t_{1/2}$ = 5 years) is used to irradiate cancer patients. The dose which the patient receives must be carefully calculated to destroy only the cancer cells without harming the patient's healthy tissues.

...is used in surgery...

2. Surgical instruments can be sterilised more effectively by radioactivity than by boiling.

...and in factories

3. A production line use for radioactivity is to check whether cans have been correctly filled [see Figure 1.19].

FIGURE 1.19
Monitoring a Production Line

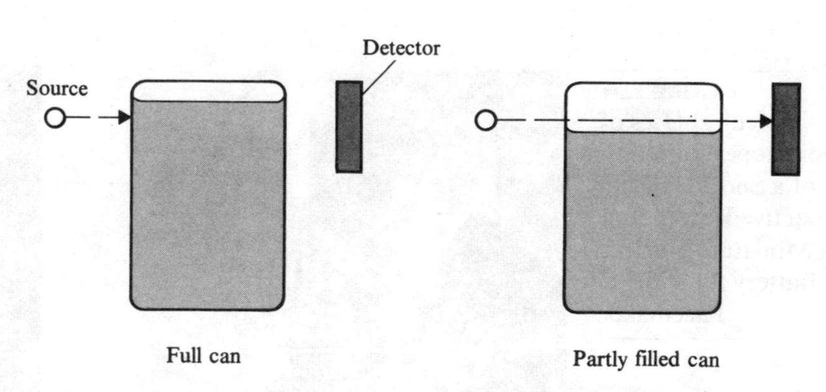

Detector
Source
Full can
Partly filled can
Radioactivity reaches the detector because the can is not full. The detector can be made to operate an arm on the production line to reject the can.

4. Figure 1.20 shows a method of using a radioactive source and detector to regulate the thickness of aluminium foil.

Radioactivity is also used for the detection of leaks...

5. Underground leaks in water or fuel pipes can be detected by introducing a short-lived radioisotope into the pipe. The level of radioactivity on the surface can be monitored. A sudden increase of surface radioactivity shows where water or fuel is escaping.

...for measuring engine wear...

6. Engine wear can be measured by using radioactive piston rings. As the piston rings wear away, the lubricating oil becomes radioactive. In this way, the efficiency of various lubricating oils can be tested.

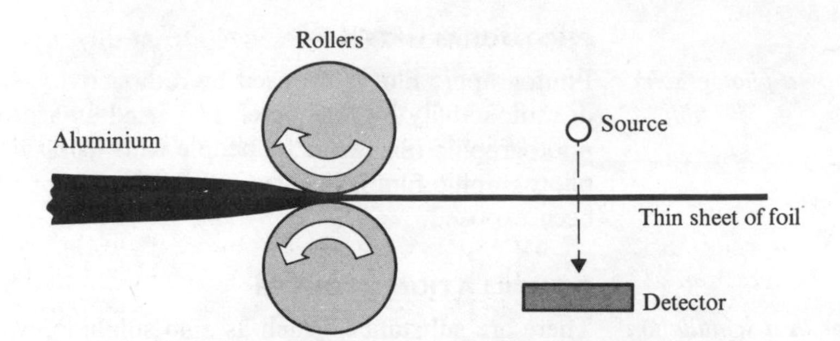
Rollers
Aluminium
Source
Thin sheet of foil
Detector
If the amount of radiation reaching the detector increases, the detector operates a mechanism for moving the rollers further apart, and vice versa.

FIGURE 1.20 Monitoring the Thickness of Metal Foil

... in carbon-14 dating ...

7. Carbon-14 dating can be used to calculate the age of plant and animal remains. Living plants and animals take in carbon, which includes a small proportion of the radioactive isotope carbon-14. When a plant or animal dies, it takes in no more carbon-14, and that which is already present decays. The rate of decay decreases over the years, and the activity that remains can be used to calculate the age of the plant or animal material [see example, § 14.5.3].

... in medicine ...

8. Tracer studies use radioactive isotopes to track the path of an element through the body. Radioactive iodine (iodine-131) is administered to patients with defective thyroids to enable doctors to follow the path of iodine through the body. As the half-life is only 8 days, the radioactivity soon falls to a low level. [See Figure 1.21.]

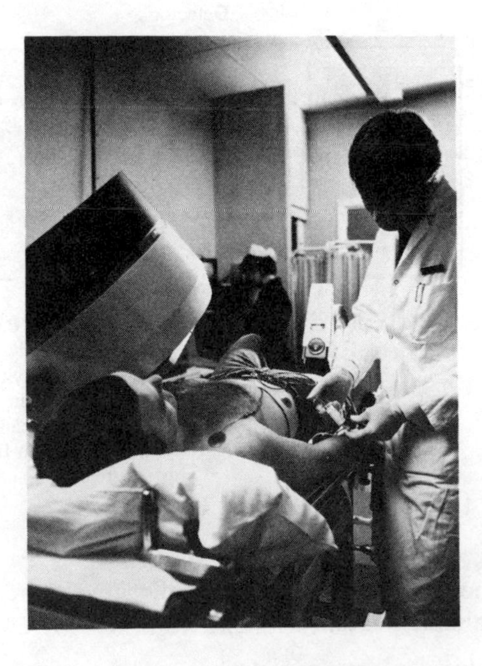

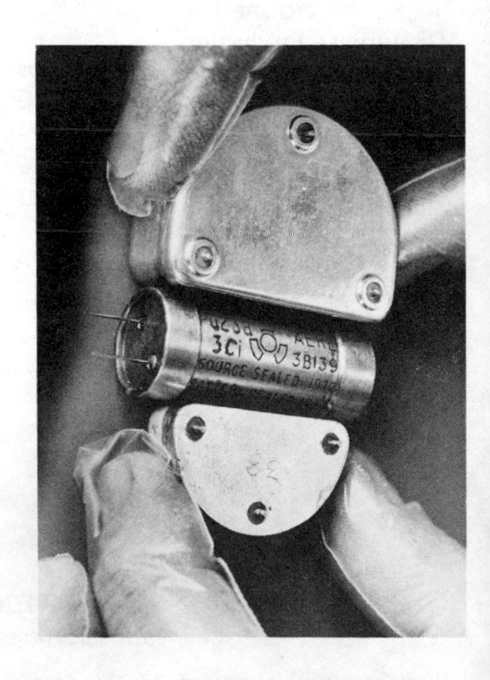
3Ci
3B139

FIGURE 1.21 Medical Uses of Radioisotopes. Injection of a Short Half-life Radioactive Isotope and a Miniature Nuclear Battery for a Heart Pacemaker

... in analysis ...

9. Dilution analysis is the name given to the use of labelled compounds in analysis.

Example The problem is to find the mass of a substance **X** present in a mixture. An isotopically labelled form of **X** has an activity of $3 \times 10^6\,\text{cpm}\,\text{g}^{-1}$ (counts per minute per gram). 1 μg of this tracer is added to the mixture. After thorough mixing,

a pure specimen of **X** is isolated from the mixture. It is assayed and found to have an activity of $30\,\text{cpm}\,\text{g}^{-1}$.
The dilution of radioactivity is $1/10^5$; therefore $1\,\mu g$ is present in 100 mg of the specimen. The specimen contains 100 mg of the substance **X**.

... in elucidating structure ...

10. Structural studies sometimes call on the use of radioisotopes. A question asked about the structure of the thiosulphate ion, $S_2O_3^{2-}$, was whether the two sulphur atoms occupy equivalent positions in the ion. A radioactive isotope of sulphur can be used in the preparation of thiosulphate:

$$^{35}S(s) + SO_3^{2-}(aq) \rightarrow {}^{35}SSO_3^{2-}(aq)$$

When dilute acid is added to the thiosulphate formed, sulphur is precipitated, and it is found that all the radioactivity is present in the precipitate of sulphur and none in the sulphur dioxide:

$$^{35}SSO_3^{2-}(aq) + 2H^+(aq) \rightarrow {}^{35}S(s) + SO_2(g) + H_2O(l)$$

The two sulphur atoms in the thiosulphate ion cannot occupy the same type of position in the ion.

... and in studies of reaction mechanisms

Further work has shown that the structure of the thiosulphate ion is

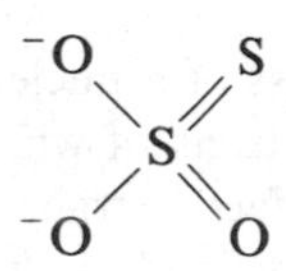

11. Mechanistic studies sometimes employ radioisotopes.
The path of a labelled atom in a molecule can be followed through a sequence of reactions [see esterification, § 33.8.2].

12. People who work with radioactive materials take precautions to ensure that they do not receive a high dose of radiation. A radioactive source is surrounded by a wall of lead bricks, except for an outlet through which a beam of radiation can emerge. To handle a powerful source of radiation, people use long-handled tongs [see Figure 1.22]. Since radioactivity fogs photographic film, workers exposed to radioactivity wear badges containing film, which is examined periodically so that the dose of radiation they are receiving can be monitored.

CHECKPOINT 1D: NUCLEAR REACTIONS I

1. Write the symbols for the isotopes of chlorine (proton number 17, nucleon numbers 35 and 37).

2. If 8 g of a radioactive isotope decay in a year to 4 g, will 6 g of the same isotope decay to 2 g in the same time? Explain your answer.

3. Supply the missing proton numbers and nucleon numbers:

(*a*) ${}^{14}_{6}C \rightarrow N + {}^{0}_{-1}e$

(*b*) ${}_{10}Ne \rightarrow {}^{19}F + {}^{0}_{+1}e$

(*c*) ${}_{88}Ra \rightarrow {}^{4}_{2}He + {}^{222}Rn$

(*d*) ${}^{73}As + {}^{0}_{-1}e \rightarrow {}_{32}Ge$

(*e*) ${}^{24}_{12}Mg + {}^{4}_{2}He \rightarrow Si + {}^{1}_{0}n$

(*f*) ${}^{19}_{9}F + \quad \rightarrow {}^{16}_{7}N + {}^{4}_{2}He$

FIGURE 1.22 A Scientist at the Atomic Energy Research Establishment Using a Remote Handling Technique with a Radioactive Isotope

1.9.9 MASS CHANGES IN NUCLEAR REACTIONS; BINDING ENERGIES

The mass of the nucleus is less than the sum of the nucleon masses...

The mass of a nucleus is slightly less than the sum of the masses of the protons and neutrons of which it is composed. The difference in mass, which is called the **mass defect**, is transformed into the **binding energy** of the nucleus. The binding energy can be defined as the energy required to separate a nucleus into individual nucleons. The connection between mass and energy is given in Einstein's equation

$$E = mc^2$$

...The mass defect is the source of the binding energy of the nucleus

where E = energy released, m = loss in mass, and c = velocity of light. Since the constant c^2 has a large numerical value, even a very small loss in mass is equivalent to the loss (or release) of a large amount of energy. This is the origin of the substantial binding energies of atomic nuclei, and it is also the reason why nuclear reactions are such an important source of energy. Figure 1.23 shows a plot of binding energy per nucleon against mass number.

FIGURE 1.23 Graph of Binding Energy per Nucleon against Mass Number

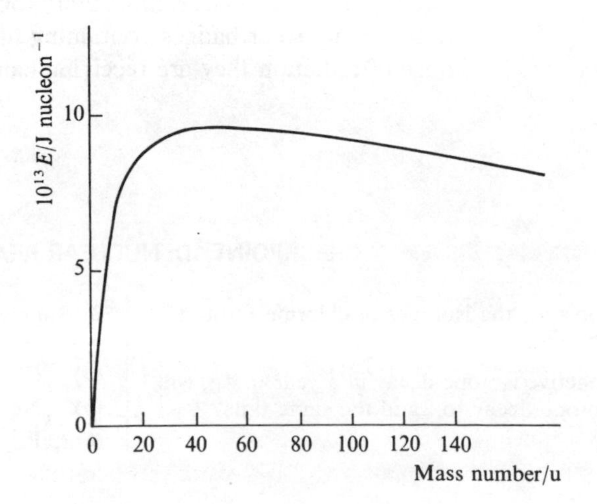

Nuclei of mass greater than 60 u should be able to split...

... lighter nuclei should be able to combine

You can see from the graph that elements with mass numbers of around 60 are the most stable. Elements with nuclei heavier than this should be able to split up to form lighter, more stable nuclei with the release of energy. Elements with nuclei lighter than 60 u should be able to combine, if the repulsion between nuclear charges can be overcome, to form heavier nuclei with the release of energy. These processes are known as **fission** and **fusion** respectively.

1.9.10 NUCLEAR FISSION

The first person to obtain energy from '**splitting the atom**' was O Hahn, in 1937. He bombarded uranium-235 with neutrons. Atoms of ^{235}U split into two smaller atoms and two neutrons, with the release of energy:

Splitting the atom of $^{235}_{92}U$

$$^{235}_{92}U + ^{1}_{0}n \rightarrow ^{144}_{56}Ba + ^{90}_{36}Kr + 2^{1}_{0}n$$

Mass is converted into energy

The sum of the masses of the fission products is less than the mass of the $^{235}_{92}U$ atom

If the sample of uranium-235 is smaller than a certain size, called the **critical mass**, neutrons will escape from the surface. In a large block of uranium-235, neutrons are more likely to meet uranium-235 atoms and produce fission than to escape. Since each nuclear fission produces two neutrons, as shown in Figure 1.24, a chain reaction is set up. Each time an atom of uranium-235 is split, the mass of the atoms produced is 0.2 u less than the mass of an atom of $^{235}_{92}U$. The lost mass is converted into energy. This is where the energy of the atomic bomb comes from. The atomic bomb consists of two blocks of uranium-235, each smaller than the critical mass. On detonating the bomb, one mass is fired into the other to make a single block larger than the critical mass. The detonation is followed by an **atomic explosion**.

FIGURE 1.24
The Chain Reaction in Fission of Uranium-235

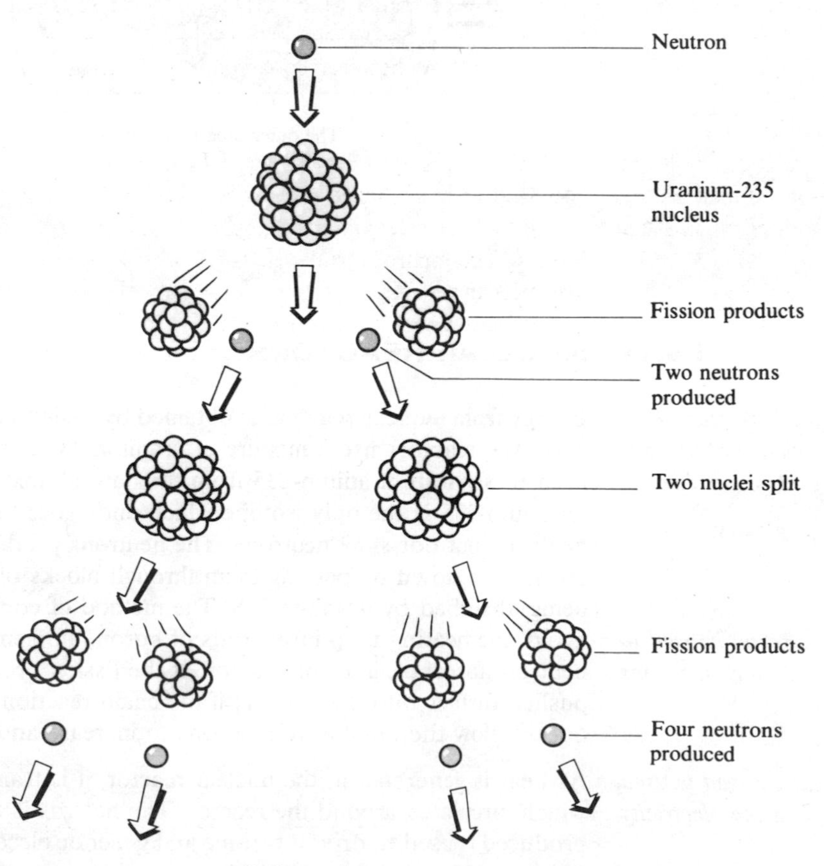

Atomic bombs

The only time atomic bombs have been used in warfare was when two cities in Japan, Hiroshima and Nagasaki, were destroyed in 1945. The death and destruction which followed were on such a terrible scale that nations fighting subsequent wars have avoided using atomic weapons.

SEPARATION OF URANIUM ISOTOPES

Separation of ^{235}U from ^{238}U by gaseous effusion of $UF_6(g)$

Natural uranium is a mixture of $^{235}_{92}U$ and $^{238}_{92}U$. Since $^{238}_{92}U$ absorbs neutrons without undergoing fission, the $^{235}_{92}U$ used in atomic bombs must be separated from the other isotopes. This is achieved by means of gaseous **effusion** [§ 7.2.4]. Uranium(VI) fluoride, UF_6 (g), is prepared and passed along a porous pipe which is surrounded by a larger concentric pipe [see Figure 1.25]. As $^{235}UF_6$ (g) effuses out of the pipe faster than $^{238}UF_6$ (g), the ratio of $^{235}UF_6$ to $^{238}UF_6$ in the outer pipe gradually increases. The rates of effusion are related by the equation [§ 7.2.4]

$$\frac{\text{Rate of effusion of } ^{235}UF_6(g)}{\text{Rate of effusion of } ^{238}UF_6(g)} = \sqrt{\frac{M_r(^{238}UF_6)}{M_r(^{235}UF_6)}} = 1.004$$

Miles of piping must be used to give a substantial separation.

FIGURE 1.25 Separation of Uranium Isotopes

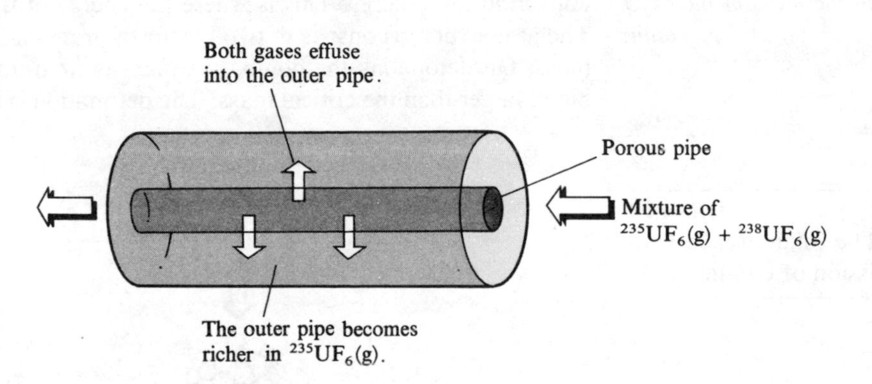
Both gases effuse into the outer pipe.
Porous pipe
Mixture of $^{235}UF_6(g)$ + $^{238}UF_6(g)$
The outer pipe becomes richer in $^{235}UF_6(g)$.

1.9.11 NUCLEAR REACTORS

In a nuclear reactor... ...the release of nuclear energy is controlled...

Energy from **nuclear reactors** is obtained by fission of ^{235}U, carried out in a controlled way. The reactors use a mixture of uranium-235 and uranium-238, avoiding the need to separate uranium-235 from the natural mixture of uranium isotopes. Uranium-235 is the only isotope which undergoes fission. Uranium-238 absorbs fast neutrons but not slow neutrons. The neutrons produced from uranium-235 fission are slowed down by passing them through blocks of graphite to prevent them from being absorbed by uranium-238. The method of controlling the rate of fission to avoid overheating is to insert rods of boron, an element which is a very good neutron-absorber, into the reactor. If the fission process speeds up, the rods are pushed further into the reactor; if the chain reaction slows down, the rods are pulled out to allow the number of neutrons to increase and speed up the reaction.

...Boron is used to absorb neutrons

The heat evolved is used to generate electricity

As heat is generated in the nuclear reactor, it is transferred to a stream of gas which circulates around the reactor. The hot gas is used to boil water; the steam produced is used to drive a turbine and generate electricity. [See Figure 1.26.] Other coolants, e.g. water and liquid sodium, are used in different designs.

FIGURE 1.26
A Nuclear Reactor

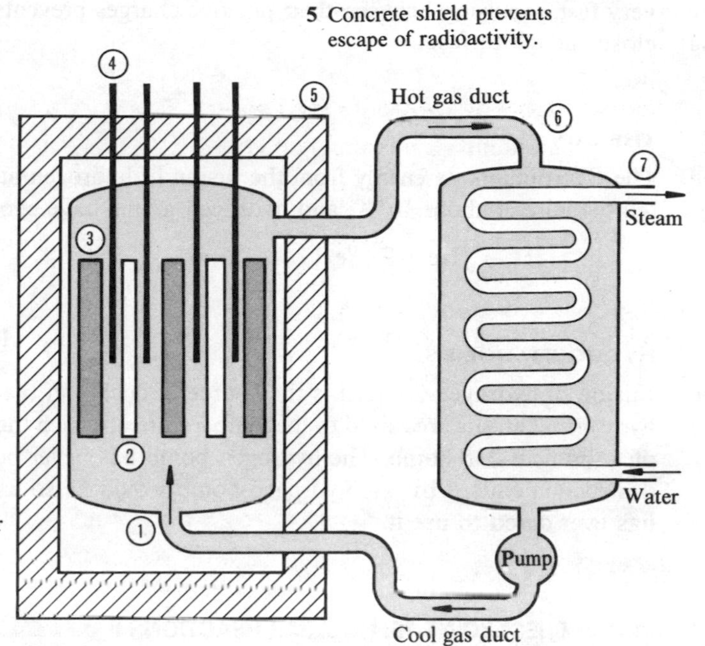
5 Concrete shield prevents escape of radioactivity.
Hot gas duct
Steam
Water
Pump
Cool gas duct
4 Rods of boron, a good neutron-absorber, regulate the supply of neutrons.
3 Graphite rods slow neutrons and prevent capture by ^{238}U.
2 Uranium rods.
1 Cool gas is passed into reactor to absorb heat produced.
6 Hot gas is used to boil water.
7 Steam is used to drive a turbine which generates electricity.

1.9.12 RADIOISOTOPES

New isotopes are made in reactors

Many radioisotopes have been made by putting stable isotopes into a nuclear reactor. They are bombarded with neutrons and absorb one or two neutrons into the nucleus. The new isotope may well be radioactive because it has an excessive number of neutrons. The number of elements in the universe has been extended from 92 to 105 by the formation of artificially produced radioisotopes [see Periodic Table, p. 792].

1.9.13 NUCLEAR FUSION

When two atoms of hydrogen-2 (**deuterium**) collide at high speed, they may interact by either of the reactions

In theory, fusion is a source of nuclear energy

$$2\,{}^{2}_{1}H \rightarrow {}^{3}_{2}He + {}^{1}_{0}n$$

$$2\,{}^{2}_{1}H \rightarrow {}^{3}_{1}H + {}^{1}_{1}H$$

For the second of these reactions, there is a release of 97 TJ (9.7×10^{7} MJ) for the fusion of 1.00 kg of deuterium. The energy produced by the combustion of 1 tonne of coal is 33 MJ.

The enormous release of energy from fusion gives more ${}^{2}_{1}H$ nuclei the energy they need to fuse, and a chain reaction is set up. Hydrogen-2 (deuterium) is present together with the normal isotope, ${}^{1}_{1}H$, in all natural compounds of hydrogen, e.g. water. It is therefore an inexhaustible and a more readily accessible starting material than

Fusion requires hydrogen-2
The ^2H atoms must be accelerated before they can fuse

uranium-235. This is why the fusion process is expected eventually to replace fission as the source of nuclear energy. The problem of accelerating the hydrogen-2 atoms sufficiently to initiate the reaction has still to be solved. Unless the atoms are moving very fast, repulsion between their positive charges prevents the nuclei from getting close enough to fuse.

THE SUN

Fusion takes place in the Sun...

The Sun obtains its energy from the fusion of hydrogen atoms. In the Sun, the temperature is about 10^7 K, and hydrogen atoms have enough energy to fuse:

$$4\,{}^{1}_{1}\text{H} \rightarrow {}^{4}_{2}\text{He} + 2\,{}^{\ 0}_{+1}\text{e}$$

HYDROGEN BOMBS

...and in a hydrogen bomb

Fusion of hydrogen-2 nuclei is the source of energy in the hydrogen bomb. The hydrogen-2 atoms are raised to the temperature at which they can fuse by the explosion of a uranium-235 bomb. The hydrogen bomb has never been used in war. The destruction caused by one hydrogen bomb would be so catastrophic that no nation has ever dared to use it.

CHECKPOINT 1E: NUCLEAR REACTIONS II

1. How does the size of the nucleus compare with that of the whole atom? What is the binding energy of the nucleus?

2. How does nuclear fission differ from nuclear decay? How does nuclear fission differ from nuclear fusion? What are the difficulties that have to be overcome before energy from nuclear fusion becomes a commercial proposition?

3. What is a nuclear chain reaction? What is done to stop a nuclear reactor from becoming dangerously hot?

4. What is the source of the energy that can be obtained from uranium-235? Why must uranium-235 be separated from uranium-238 when it is used for making atomic bombs but not when it is used in nuclear power stations?

1.10 CHERNOBYL

TOPIC

WHAT HAPPENED?

It was the world's worst nuclear accident. A reactor at Chernobyl in the former USSR blew up and caught fire on 26 April 1986. The energy released lifted the 1000 tonne lid of the reactor, exposing the highly radioactive core to the air. The explosion and fire sent streams of radioactive material into the atmosphere above the USSR and across Europe for more than a week. For nine days, it was touch and go whether the accident could be controlled.

HOW DID IT HAPPEN?

In Chernobyl, USSR, in 1986, a nuclear reactor exploded, causing a massive release of radioactive material

It was ironic that the Chernobyl accident occurred in the course of a safety test. A problem with the Russian RBMK reactors is that, should the reactor lose power, the electricity supply to the cooling pumps fails. Standby generators are ready to take over and supply electricity to the cooling pumps. These standby generators are diesel generators and take 50 seconds to reach full power — too long for the reactor to be without fully working cooling pumps. The experiment was to see whether the inertia, the latent mechanical power in the spinning turbines, would bridge the gap. Contrary to operating instructions, the team set the reactor at low power output, making it more difficult to control. They withdrew most of the control rods from the core. To prevent the automatic safety systems from interfering with the experiment, the technicians disconnected them. They switched off the devices that shut down the reactor if the turbines stop and the safety devices that shut down the reactor if the steam pressure rises. In all, they made six fatal errors.

A lack of coolant caused the reactor to overheat. Chemical explosions were caused by reactions between steam and zirconium and between steam and graphite

After the flow of cooling water had been reduced and safety devices had been disengaged, a surge of power caused rapid overheating. The operators could not shut down the reactor quickly because most of the control rods had been withdrawn from the core. Soviet experts calculate that the reactor came close to exploding like a nuclear bomb. The energy released caused a steam explosion that lifted the lid off the reactor. With the reactor core now open to the sky, there was a further explosion. Reactions between steam and zirconium and between steam and graphite produced hydrogen which started a fire. Some people died in the initial explosion. Of the firemen who fought the fire in the reactor hall and stopped it spreading to the neighbouring reactor, many died from the dose of radiation they received.

The damaged reactor released a stream of radioactive material into the air. This was detected in Sweden two days later

The damaged reactor core and the graphite surrounding it began burning at a temperature of about 1600 °C. The reactor core heated up further because of radioactive decay. As the reactor core was open to the sky, the volatile fission products, e.g. iodine-131, caesium-134, caesium-137 and strontium-90 vaporised and were carried away with the combustion gases. They were detected in Sweden on the morning of 28 April. This was the first that the rest of the world knew of the Russian accident.

HOW WAS THE ACCIDENT CONTAINED?

Soviet firefighters dropped boron carbide, lead, sand, clay and dolomite from helicopters on to the blazing reactor. They finally extinguished the fire with cold nitrogen

A graphite fire cannot be put out with water. Soviet firefighters in helicopters frantically tried to extinguish the blaze by dumping 5000 tonnes of boron carbide, lead, sand, clay and dolomite on to the reactor core. They used boron carbide to absorb neutrons and slow down the fission reaction in the core. They dropped dolomite (magnesium carbonate) to generate carbon dioxide. The clay and sand helped to filter escaping radioactive substances and to quench the fire. Lead absorbed heat, melting to form a liquid layer which in time solidified and shielded the top of the core. Nine days after the explosion, the firefighters decided to feed cold nitrogen into the reactor space to provide additional cooling of the core debris and to blanket it against oxygen. The technique worked, but they did not succeed in putting out the fire until 12 days after the accident. They tunnelled under the reactor and pumped in concrete to prevent radioactive material reaching the water table.

In some reactors, the fuel rods are made of uranium, which melts at 1130 °C. The Chernobyl fuel rods were made of uranium oxide, which melts at over 2000 °C, and fortunately the fuel rods did not melt in the accident. Had the nuclear fuel melted, its high density and high temperature would have made it burn through the floor of the reactor and into the earth beneath.

There were 31 deaths and hundreds of people hospitalised. Thousands of people received high doses of radiation which are likely to cause delayed damage. Land all over Europe in the path of the radioactive cloud was contaminated

WHAT WAS THE EXTENT OF THE DAMAGE?

This was the world's worst nuclear accident: 31 people died, 200 were hospitalised with severe radiation sickness, 135 000 people were evacuated from 179 towns and villages in the area and have not been allowed back home. It is estimated that 5000 people will die prematurely from the excessive doses of radiation they received. The region surrounding the plant will continue to be dangerous for years to come.

Sweden estimated that the cloud of radioactive material that fell on Sweden cost at least £100 million in contaminated food. The 15 000 Lapps who live in central Sweden depend on reindeer, berries and fish; all these have been seriously contaminated by radioactive substances. British agriculture lost over £10 million when farmers could not bring their sheep to market because they had eaten contaminated grass.

WAS THE ACCIDENT CAUSED BY HUMAN ERROR?

Officials at the power station were found guilty of negligence and sentenced to years in a labour camp

After a trial the following year, six officials were found guilty of criminal negligence. The plant director, the chief engineer and his deputy were sentenced to 10 years in a labour camp, and three other officials got lighter sentences. Some people saw in these sentences the desire of the USSR to attribute the cause of the disaster to human error, rather than faulty design. Russia does not like to be shown as technically inferior to the West. The question that cannot be ignored is: if human failing caused the accident at Chernobyl, how can Western plants be safe from similar human errors?

The design of the Russian reactor is more accident-prone than others. It is a water-cooled reactor which operates with graphite at a high temperature

The reactor was one of four RBMK-1000 nuclear reactors at Chernobyl power station. It was a graphite-moderated, water-cooled reactor which contained 190 tonnes of uranium as uranium oxide and 1700 tonnes of graphite. The cooling water circulated under 140 atm pressure through zirconium alloy tubes round the fuel rods [see Figure 1.27], taking in heat and producing steam for driving the turbines. In most reactor designs, there is a primary cooling circuit which runs through the reactor and transfers its heat to a secondary circuit which drives the turbines. With a direct steam circuit, as in the RBMK, a leak anywhere in the circuit can result in overheating of the reactor core. In the RBMK reactor, the temperature was 750 °C: the graphite was red-hot. With this design, any break in one of the 1600 cooling tubes would lead to superheated steam at high pressure meeting red-hot graphite. The result is the production of carbon monoxide and hydrogen. The RBMK reactor does not meet Western safety standards. The main disadvantage is that it is moderated by graphite which can catch fire if an accident happens in a water-cooled reactor.

COULD IT HAPPEN HERE?

British reactors operate with the graphite at a lower temperature and use carbon dioxide as a coolant. This is a safer design

In Britain's Magnox reactors, the graphite is at about 400 °C. In the Advanced Gas-Cooled Reactors, the graphite is kept cooler than in the RBMK because carbon dioxide used to cool the fuel is first made to percolate through the graphite. The gas also provides a chemically inert atmosphere. In the Soviet reactor, the graphite is blanketed in a mixture of helium and nitrogen [see Figure 1.27]. At the temperature of the reactor, 750 °C, the graphite is above its ignition temperature. Safety depends on the ability of the gas mixture to exclude air.

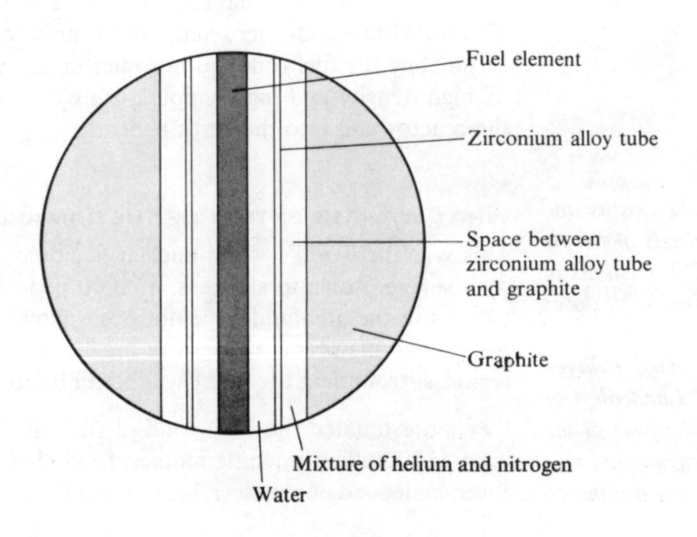

FIGURE 1.27 A Fuel Rod in an RBMK Reactor

CHECKPOINT 1F: CHERNOBYL

1. In April 1987, the Soviet press reported that the concrete and steel tomb encasing the remains of Unit 4 at the Chernobyl power plant was capped with a thick layer of snow. Why was this report so significant?

2. What is the difficulty in trying to put out a graphite fire with water?

3. Write equations for the reactions between (*a*) steam and zirconium to form hydrogen and ZrO_2 and (*b*) steam and graphite.

4. Reindeer eat grass in summer, and in winter they eat lichen, plants without roots, which grow on rocks and absorb nutrients from the air. Cows eat grass. Why did reindeer absorb more radioactive substances than cows did?

5. Explain what is wrong with each of the following statements.

(*a*) 'Iodine-131 has a short life.' (*The Guardian*, 6 May 1986)

(*b*) 'Plutonium retains its maximum radioactivity for 24 000 years.' (*The Times*, 15 May 1986)

(*c*) 'The massive radiation leak was caused by a nuclear core meltdown.' (*The Daily Mirror*, 30 April 1986)

(*d*) 'Radiation is still pouring into the air from a fire at the plant.' (*The Daily Mail*, 1 May 1986)

(*e*) 'Radiation from the Russian nuclear disaster has been found in British milk.' (*The Star*, 5 May 1986)

(*f*) 'People can take iodine tablets to counter radioactive poisoning.' (*The Sunday Times*, 18 May 1986)

(*g*) 'Liquid iodine is an antidote against radiation sickness.' (*The Daily Mirror*, 1 May 1986)

6. Could it happen here? Read as much as you can in the newspapers, *The New Scientist*, etc. about the Chernobyl explosion. Perhaps different students could prepare summaries of different articles and then present them to the class. Find out about the Three Mile Island accident in the USA in 1983 and the Sellafield (formerly called Windscale) accident in the UK in 1958. Have a discussion to try to answer these difficult questions.

(*a*) Were the accidents at Chernobyl, Three Mile Island and Sellafield due to human error or another cause?

(*b*) Could the errors at Chernobyl, Three Mile Island and Sellafield have been avoided by improved technology?

(*c*) Could an accident like Chernobyl happen again?

(*d*) Where are the UK nuclear reactors situated? Which areas would be most seriously contaminated if there were an accident at a UK nuclear power station? What would be the effects on parts of the UK distant from the accident?

(*e*) Where are the French nuclear power stations? Which areas would be most seriously affected by an accident in a French nuclear power station?

QUESTIONS ON CHAPTER 1

†**1.** Explain how a mass spectrometer is used to measure molecular mass. Are there any limits to its use for molecular mass determination?
[For help with second part, see § 1.8.1.]

2. Identify the emitted particles (1) and (2), and state in which groups of the Periodic Table the elements Pb, **X**, **Y** and **Z** occur.

$$^{212}_{82}\text{Pb} \xrightarrow{(1)} {}^{212}_{83}\mathbf{X} \xrightarrow{(2)} {}^{208}_{81}\mathbf{Y} \xrightarrow{\beta\text{ particle}} \mathbf{Z}$$

3. In the case of lighter elements, a nuclide in which the ratio neutrons/protons is greater than 1 is likely to emit radiation to bring the ratio nearer to 1. Calculate the ratio of neutrons/protons for the radioactive isotope $^{32}_{15}P$ and the stable isotope $^{31}_{15}P$. Write the equation for the decay of $^{32}_{15}P$ by α particle emission. Calculate the ratio of neutrons/protons in the nuclide formed.

4. Plot the number of neutrons against the number of protons in the nuclides ^{1}H, ^{4}He, ^{7}Li, ^{9}Be, ^{11}B, ^{12}C, ^{14}C, ^{14}N, ^{16}O, ^{19}F, ^{23}Na, ^{24}Na, ^{24}Mg, ^{27}Al.

In which two nuclides is the neutron/proton ratio greater than the stable ratio? [See Question 3.] Write equations for the decay by β emission of these radioactive isotopes.

5. Write equations for

(*a*) β particle decay of $^{27}_{13}Al$

(*b*) α particle decay of $^{27}_{13}Al$

(*c*) neutron capture by $^{27}_{13}Al$ followed by β emission

(*d*) a possible path for

$$^{211}_{83}\text{Bi} \rightarrow {}^{207}_{83}\text{Bi}$$

6. The mass spectrum of C_2H_5Cl shows peaks corresponding to ^{1}H, ^{2}H, ^{12}C, ^{13}C, ^{35}Cl and ^{37}Cl. Calculate the mass numbers of the most abundant molecular ion and the heaviest molecular ion. Write the formulae of all the possible ions that contribute to the peak at a mass number of 66.

7. Imagine you have a mixture of hydrogen-1, hydrogen-2 and hydrogen-3, (hydrogen, deuterium and tritium) present as diatomic molecules and that the numbers of atoms of the three species are the same. Sketch the mass spectrum.

8. Give values for *a*, *b*, *c* and *d*, and the symbols for **X** and **Y** in the equations

(*a*) $^{35}_{17}\text{Cl} + {}^{1}_{0}\text{n} \rightarrow {}^{a}_{b}\text{X} + {}^{1}_{1}\text{H}$

(*b*) $^{7}_{3}\text{Li} + {}^{2}_{1}\text{H} \rightarrow 2\,{}^{c}_{d}\text{Y} + {}^{1}_{0}\text{n}$

9. (*a*) How are the procedures of (i) refluxing and (ii) distillation carried out in the laboratory? Explain the purpose of carrying out the procedures by reference to two different chemical reactions.
(*b*) What do you understand by the term *relative atomic mass*?
(*c*) The mass spectrum of an organic compound which can be obtained by the oxidation of an alcohol is shown in Figure 1.28.

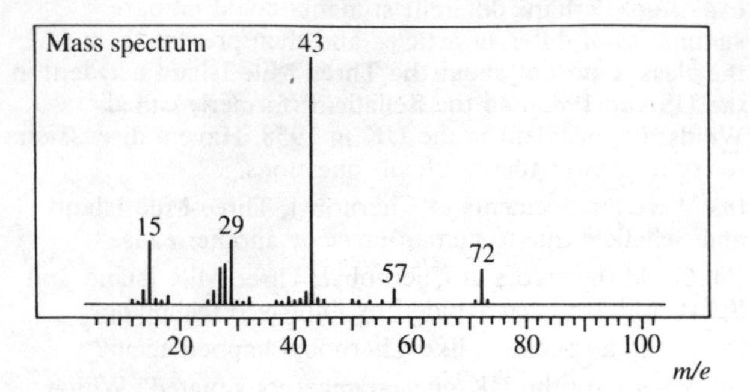

FIGURE 1.28

The compound has the following composition by mass: C 66.7%, H 11.1%, O 22.2%.

Calculate the empirical formula of the compound and by interpreting the labelled peaks on the mass spectrum determine the molecular formula of the compound.

(L 92, AS)

10. In 1909, Geiger and Marsden reported the amazing results of their experiments on α particles and thin metal foils.
(*a*) What is an α particle?
(*b*) Why did most α particles pass through the foils?
(*c*) Why were some α particles scattered backwards?
(*d*) What did they infer from their results about the structure of metal atoms?

11. A series of radioactive decays can be represented

$$^{232}_{90}\text{Th} \xrightarrow{\alpha\text{ emission}} \mathbf{X} \xrightarrow{\beta\text{ emission}} \mathbf{Y} \xrightarrow{\beta\text{ emission}} \mathbf{Z}$$

State the mass number and atomic number of the element **Z**.

12. When chlorine is bubbled through a concentrated aqueous solution of ammonium chloride, a yellow oily liquid, nitrogen trichloride, NCl_3, is formed, together with a solution of hydrochloric acid. Nitrogen trichloride is hydrolysed by aqueous sodium hydroxide, producing ammonia gas and a solution of sodium chlorate(I).

†(*a*) Write balanced equations for the formation and hydrolysis of nitrogen trichloride.

†(*b*) Draw the shape of the nitrogen trichloride molecule.

(*c*) Apart from peaks associated with solitary nitrogen atoms (at $m/e = 14$) and chlorine atoms (at $m/e = 35$ and $m/e = 37$), the mass spectrum of nitrogen trichloride contains 9 peaks arranged in 3 groups, ranging from $m/e = 49$ to $m/e = 125$. Predict the m/e values of all 9 peaks, and suggest a formula for the species responsible for each one.

[For (*a*) see § 22.5.]
[For (*b*) see § 5.1.]

13. (*a*) The relative atomic mass of antimony is 121.75. Explain carefully the meaning of this statement.
(*b*) A radioactive isotope of the element thorium $^{232}_{90}$Th decays according to the following scheme:

$$^{232}_{90}\text{Th} \xrightarrow[\text{emission}]{\alpha\text{-particle}} \mathbf{X} \xrightarrow[\text{emission}]{\beta\text{-particle}} \mathbf{Y} \xrightarrow[\text{emission}]{\beta\text{-particle}} \mathbf{Z}$$

Deduce the mass numbers and atomic numbers of **X**, **Y** and **Z**.
(*c*) A sample of carbon dioxide was prepared from carbon (assume 100% ^{12}C) and isotopically enriched oxygen containing $^{16}O_2$ and $^{18}O_2$ in the molar ratio 4 : 1. This sample gave a mass spectrum containing three peaks associated with singly charged molecular ions.

Deduce the relative molecular masses associated with these peaks and their relative intensities.

(O 91)

14. The mass spectrum of an element enables the relative abundance of each isotope of the element to be determined. Data relating to the mass spectrum of bromine, atomic number 35, appear below.

Mass number of isotope	*Relative abundance (%)*
79	50.5
81	49.5

(*a*) Define the term *isotope*.

(*b*) Write down the conventional symbols for the two isotopes of bromine.

(*c*) Calculate the relative atomic mass of bromine to three significant figures.

(*d*) (i) On a copy of the axes in Figure 1.29, sketch the expected mass spectrum of bromine vapour in the m/e region shown.

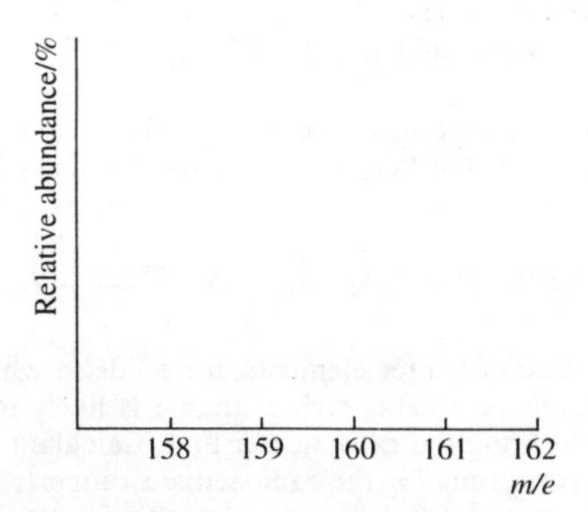

FIGURE 1.29

(ii) Explain the origin of each peak in your spectrum.

(L 92)

†*Note* A dagger denotes a question which should be tackled on re-reading, as it involves material which is covered in later chapters.

2

THE ATOM: THE ARRANGEMENT OF ELECTRONS

2.1 ATOMIC SPECTRA

In this chapter we look at atomic spectra. We see how they pose a problem which the Rutherford picture of the atom does not solve.

Isaac Newton (1642–1727) showed that sunlight is composed of many colours.

If sunlight or light from an electric light bulb is formed into a beam by a slit and passed through a prism on to a screen, a rainbow of separated colours is seen. The spectrum of colours is composed of visible light of all wavelengths and is called a **continuous spectrum** [see Figure 2.1].

FIGURE 2.1
A Continuous Spectrum

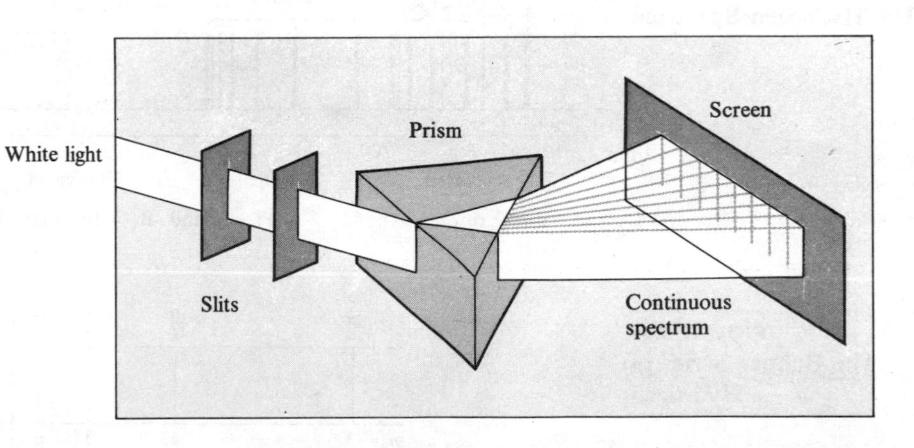

Absorption spectra are black lines on a bright background

Light of all visible wavelengths is called white light. All atoms and molecules absorb light of certain wavelengths. When white light is passed through a substance, black lines appear in the spectrum where light of some wavelengths has been absorbed by the substance. Spectrometers are instruments used for viewing absorption spectra. The pattern of frequencies absorbed by a substance is called its absorption spectrum. It can be used to identify a substance.

If atoms and molecules are heated to sufficiently high temperatures, they emit light of certain wavelengths. Figure 2.2 shows a discharge tube containing a gaseous element. The observed spectrum consists of a number of coloured lines on a black background. The spectrum is called an **atomic emission spectrum** or **line spectrum**.

Elements have emission spectra in the visible and ultraviolet region

All substances give emission spectra when they are excited in some way, by the passage of an electric discharge or by a flame. The atomic emission spectra of elements are in the visible and ultraviolet region of the spectrum. When sodium or a sodium compound is put into a flame, it emits light with a wavelength of 590 nm, and colours the flame yellow. A tube of hydrogen gas which has been excited by an electric discharge glows a reddish-pink colour.

FIGURE 2.2
An Emission Spectrum (or Line Spectrum)

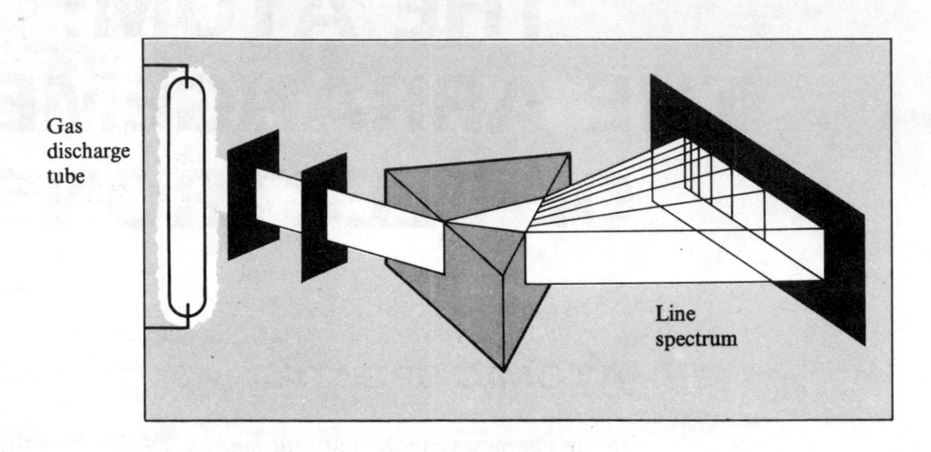

Through a spectrometer, the hydrogen emission spectrum is seen to consist of series of lines

Viewed through a spectrometer, the emission spectrum of hydrogen is seen to be a number of separate sets of lines or **series** of lines. These series of lines are named after their discoverers, as shown in Figure 2.3. The Balmer series, in the visible part of the spectrum, is shown in Figure 2.4.

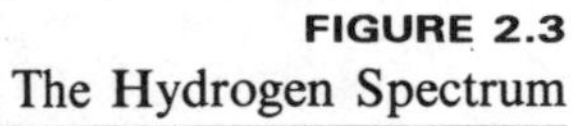

FIGURE 2.3
The Hydrogen Spectrum

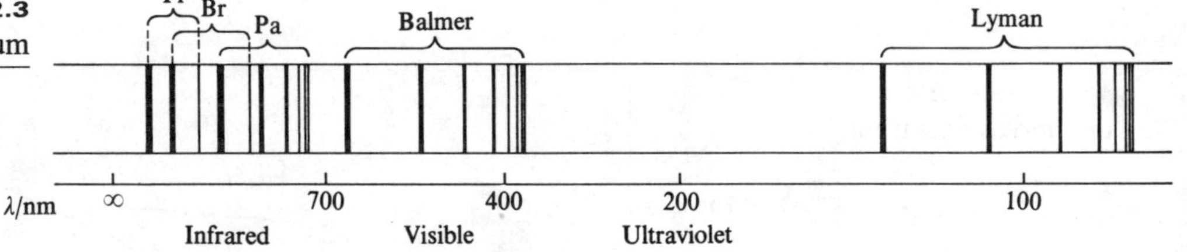

FIGURE 2.4
The Balmer Series of Hydrogen

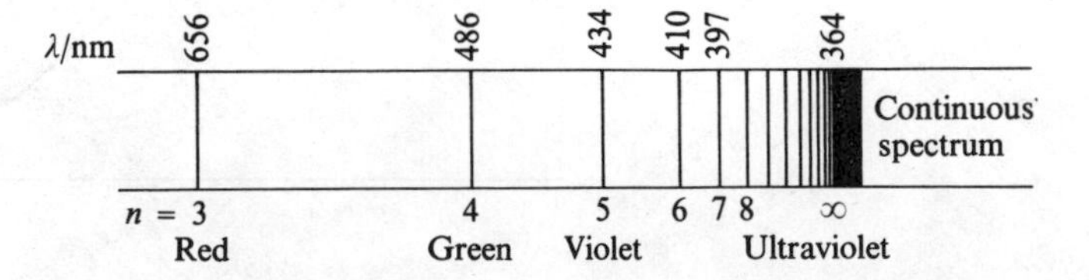

In each series, the lines become closer together as the frequency increases until at high frequency the lines coalesce

In each series, the intervals between the frequencies of the lines become smaller and smaller towards the high frequency end of the spectrum until the lines run together or **converge** to form a **continuum** of light.

The Rutherford model of the atom does not explain spectral lines

Why do atomic spectra consist of **discrete** (separate) lines? Why do atoms absorb or emit light of certain frequencies? Why do the spectral lines converge to form a continuum? The Rutherford picture of the atom offers no explanation.

2.2 THE BOHR–SOMMERFELD ATOM

2.2.1 THE BOHR MODEL

Planck theorised that energy is quantised. Bohr suggested that electrons can have only certain amounts of energy...

...and their orbits can have only certain radii

In 1913, Niels Bohr (1885–1962) put forward his picture of the atom. He solved the problem of the instability of Rutherford's atom and gave a complete explanation of the emission and absorption spectra of elements. Bohr referred to Max Planck's recently developed **quantum theory**, according to which energy can be absorbed or emitted in certain amounts, like separate packets of energy, called **quanta**. Bohr suggested that an electron moving in an orbit can have only certain amounts of energy, not an infinite number of values: its energy is **quantised**. The energy that an electron needs in order to move in a particular orbit depends on the radius of the orbit. An electron in an orbit distant from the nucleus requires higher energy than an electron in an orbit near the nucleus. If the energy of the electron is quantised, the radius of the orbit also must be quantised. There is a restricted number of orbits with certain radii, not an infinite number of orbits.

Electrons which absorb photons move to higher orbits

An electron moving in one of these orbits does not emit energy. In order to move to an orbit farther away from the nucleus, the electron must absorb energy to do work against the attraction of the nucleus. If an atom absorbs a **photon** (a quantum of light energy), it can promote an electron from an inner orbit to an outer orbit. If sufficient photons are absorbed, a black line appears in the absorption spectrum.

Electrons which fall to lower orbits emit photons of light...

According to the quantum theory, the energy contained in a photon of light of frequency ν is $h\nu$, h being Planck's constant (6.626×10^{-34} J s). For an electron to move from an orbit of energy E_1 to one of energy E_2, the light absorbed must have a frequency given by **Planck's equation**:

$$h\nu = E_2 - E_1$$

The emission spectrum arises when electrons which have been excited (raised to orbits of high energy) drop back to orbits of lower energy. They emit energy as light with a frequency given by Planck's equation. [See Figure 2.5.]

FIGURE 2.5 The Origin of Spectral Lines

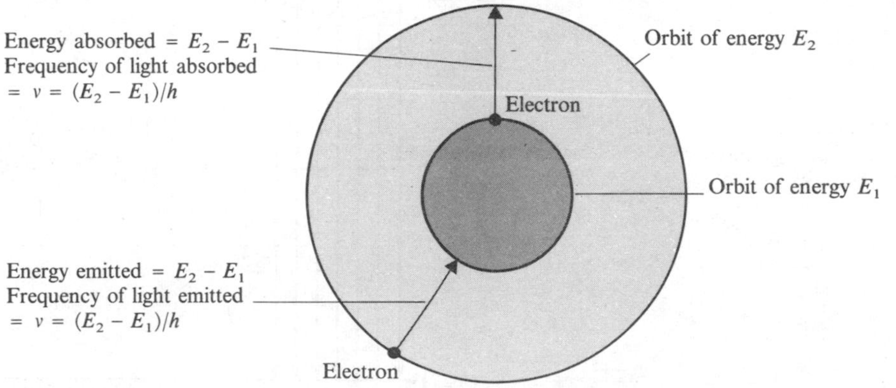

Planck's equation gives the frequencies of light emitted

Bohr gave orbits of different energy different quantum numbers

Bohr assigned **quantum numbers** to the orbits. He gave the orbit of lowest energy (nearest to the nucleus) the quantum number 1. An electron in this orbit is in its **ground state**. The next energy level has quantum number 2 and so on [see Figure 2.6]. If the electron receives enough energy to remove it from the attraction of the nucleus completely, the atom is **ionised**.

FIGURE 2.6 The Energy Levels at Various Values of the Quantum Number, n

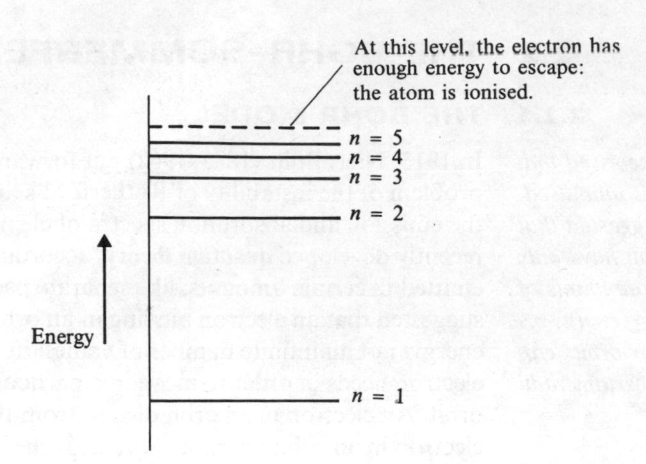

The hydrogen emission spectrum arises as electrons move from orbits of high quantum number to orbits of lower quantum number

Figure 2.7 shows how the lines in the hydrogen emission spectrum arise from transitions between orbits. The Lyman series in the emission spectrum arise when the electron moves to the $n = 1$ orbit (the ground state) from any of the other orbits. The Balmer series arise from transitions to the $n = 2$ orbit from the $n = 3$, $n = 4$ etc. orbits. The Paschen, Brackett and Pfund series arise from transitions to the $n = 3$, $n = 4$ and $n = 5$ orbits from higher orbits.

The frequency of the convergence of spectral lines can be used to give the ionisation energy

FIGURE 2.7 Energy Transitions in the Hydrogen Atom

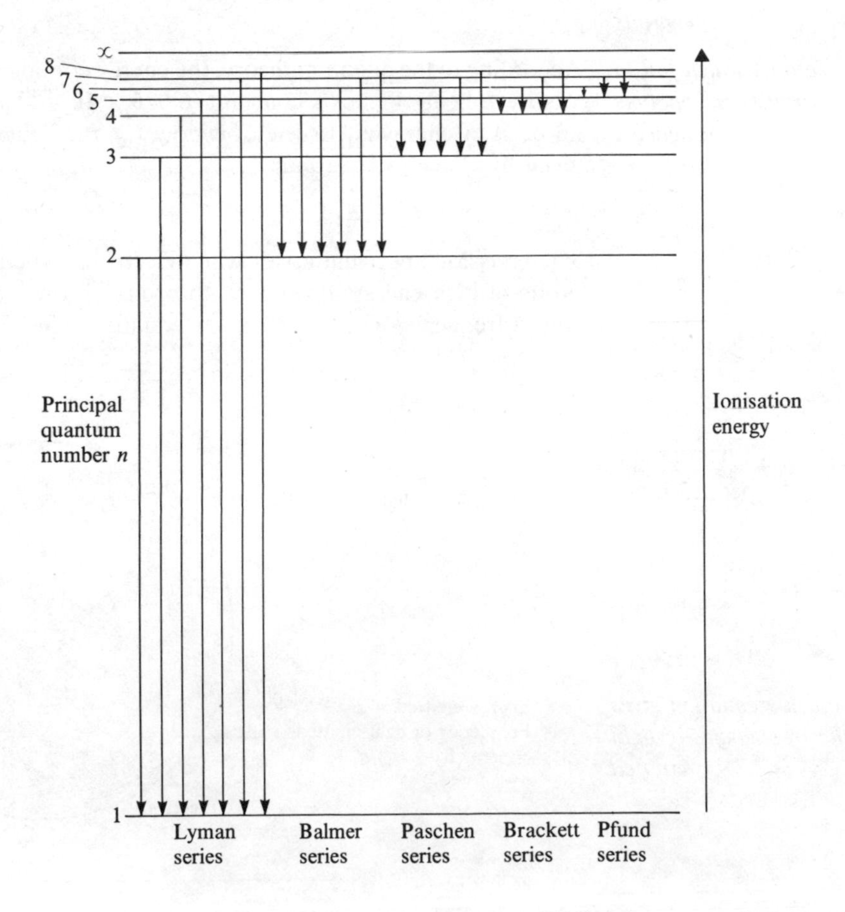

In each series of lines, as the frequency increases, each line becomes closer to the previous line until the lines converge, and the spectrum becomes continuous. The Lyman series arises from transitions to the ground state from higher energy levels. The highest frequency lines relate to the highest energy levels. The limit of the Lyman series (the convergence of the lines) corresponds to a transition from the $n = \infty$

When the lines in the spectrum converge it means that the atom has ionised

orbit (i.e. from an energy level where the electron has escaped from the atom, and the atom has ionised) to the $n = 1$ orbit (the ground state):

$$A^+ + e^- \rightarrow A$$

This transition happens when an electron collides with an ion and returns to the ground state. The convergence frequency can be used to find the **ionisation energy** of the atom.

The nature of scientific theories

There are two ways of proposing a new theory in science. One way is to collect a great quantity of experimental results so that the new theory becomes self-evident to anyone who weighs up the results. The theory is almost a summary of the results. The other way is to make a bold new assertion which is not based on experimental results and then to demonstrate that this assertion can be utilised to explain many experimental observations. Bohr's theory is of this second type. He wanted to answer the question of why the electron does not spiral into the nucleus. He simply postulated that classical physics did not apply to the electron, so the electron does not lose energy and spiral into the nucleus. This bold postulate was successful in explaining the wavelengths of the lines in the atomic spectrum of hydrogen.

2.2.2 DETERMINATION OF IONISATION ENERGY

The definition of the first ionisation energy

The first ionisation energy of an element is the energy required to remove one electron from each of a mole of atoms in the gas phase to form a mole of cations in the gas phase:

$$A(g) \rightarrow A^+(g) + e^-$$

A graphical method can be used to find the value of the ionisation energy from the emission spectrum. The interval between the frequencies of spectral lines becomes smaller and smaller as they approach the continuum.

(*a*) The frequencies of the first lines in the Lyman series are measured. If these are ν_1, ν_2, ν_3, ν_4, etc., the intervals $\Delta\nu = (\nu_2 - \nu_1)$, $(\nu_3 - \nu_2)$, $(\nu_4 - \nu_3)$, etc., can be calculated.

... by a graphical method...

(*b*) A graph of ν (the lower frequency) against $\Delta\nu$ is shown in Figure 2.8. It can be extrapolated back to $\Delta\nu = 0$. If there is no interval between lines, this is the beginning of the continuum.

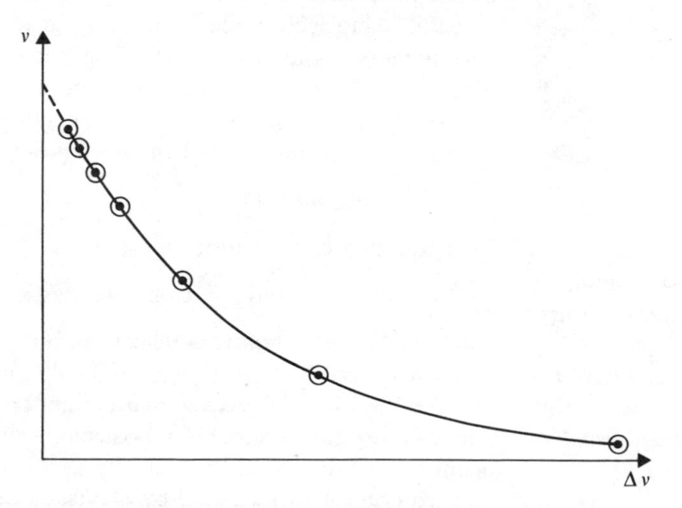

FIGURE 2.8 Finding the Convergence Frequency by a Graphical Method

(*c*) The value of ν at $\Delta\nu = 0$ is read off and inserted in Planck's equation

$$\Delta E = h\nu$$

(*d*) The value of ΔE is multiplied by Avogadro's constant to give the first ionisation energy for a mole of atoms.

Example The value of the wavelength at the start of the continuum in the sodium emission spectrum is 242 nm. Calculate the first ionisation energy of sodium.

Method

...the calculation

Since $\Delta E = h\nu$

$= hc/\lambda$

where $c = 2.998 \times 10^8\,\text{m s}^{-1}$, $h = 6.626 \times 10^{-34}\,\text{J s}$ and $L = 6.022 \times 10^{23}\,\text{mol}^{-1}$.

$$\begin{aligned}\text{Ionisation energy} &= L\Delta E \\ &= Lhc/\lambda \\ &= 6.022 \times 10^{23} \times 6.626 \times 10^{-34} \times 2.998 \times 10^8/(242 \times 10^{-9}) \\ &= 494\,300\,\text{J mol}^{-1} \\ &= 494\,\text{kJ mol}^{-1}\end{aligned}$$

Ionisation energy can also be determined by an electrical method which measures the potential difference at which ionisation takes place.

2.2.3 SOMMERFELD'S QUANTUM NUMBERS

Sommerfeld's second quantum number

Sommerfeld elaborated Bohr's theory in 1916. He proposed that each quantum number governed the energy of a circular orbit and also a set of elliptical orbits of similar energy. He called n the **principal quantum number** and introduced a second quantum number which describes the shape of the elliptical orbits (the degree of eccentricity). The **second quantum number**, l, can have values from $(n - 1)$ down to 0. If $n = 4$, $l = 3, 2, 1$ and 0.

2.3 THE WAVE THEORY OF THE ATOM

2.3.1 PARTICLES AND WAVES

TOPIC

According to the wave theory of light, refraction* and diffraction* can be explained by the properties of waves. Other properties of light, such as the origin of line spectra and the photoelectric effect*, need a particle or photon theory for their explanation. The success of the dual theory of light led Louis de Broglie to speculate in 1924 on whether particles might have wave properties. He postulated that a particle of mass m moving with velocity v has a wavelength λ associated with its motion. He predicted the value of λ from the **de Broglie equation**

$$\lambda = h/mv$$

Light is a wave motion with the properties of particles also

This equation has the form

Wave property = Constant/Particle property

Do electrons have the properties of waves as well as being particles?

Applying this equation to an electron, and inserting the values $v = 6 \times 10^6\,\text{m s}^{-1}$ for an electron accelerated through 100 volts, $m = 9 \times 10^{-31}\,\text{kg}$, and $h = 6.63 \times 10^{-34}\,\text{J s}$ gives $\lambda = 0.12\,\text{nm}$. This is the sort of distance between the ions in a crystal. It should be possible, according to de Broglie's theory, to send a beam of electrons with this velocity at a crystal and obtain a diffraction pattern. The experiment was tackled by Davisson and Germer. They succeeded in showing that a crystal could act as a diffraction grating for a beam of electrons. This experimental evidence was strong support for de Broglie's bold prediction.

Like waves, electrons can give diffraction patterns

*See R Muncaster, *A-Level Physics* (Stanley Thornes)

De Broglie theorised... ...a moving particle has a wavelength associated with its motion

The de Broglie equation can be applied to more massive particles. The wavelengths associated with macroscopic bodies cannot be detected because they are very much less than the spacing in any diffraction grating. This is why de Broglie's equation is of greatest importance when applied to the least massive of particles.

Heisenberg's Uncertainty Principle... ...states that it is not possible to measure both the position and the velocity of an electron accurately

The Bohr–Sommerfeld picture of the atom specifies the velocity of an electron and the orbit occupied by an electron. To find out the position of an electron, light of short wavelength (comparable with the size of the electron) must be used. Since $E = h\nu = hc/\lambda$, light of short wavelength has photons of high energy. When these interact with an electron, they change its velocity. To avoid changing the velocity of an electron, light of longer wavelength could be used, but the position of the electron would not be found accurately. The idea that it is impossible to measure accurately both the velocity and the position of a particle was expressed by W K Heisenberg and termed the **Uncertainty Principle**. From the viewpoint of Heisenberg's Uncertainty Principle, the Bohr picture of the atom does not appear satisfactory. The electrons are moving in orbits of specified radii at specified velocities, and these quantities cannot both be measured experimentally. A theory which involves quantities which cannot be measured does not follow the tradition of scientific work. Advances in science have never come through theories which cannot be tested by quantitative measurement. Scientists were dissatisfied with the Bohr–Sommerfeld theory of the atom on these grounds.

2.3.2 ATOMIC ORBITALS

The Schrödinger wave equation...

De Broglie attributed both the properties of particles and the properties of waves to the electron. Erwin Scrödinger used this model to work out a wave theory of the atom. The Schrödinger treatment sets up a wave equation for the atom. Solutions to the Schrödinger wave equation can be obtained only under certain conditions. If the electron is to be treated as a wave, then an integral number of wavelengths must be fitted into one circuit of the electron. [See Figure 2.9.]

FIGURE 2.9
The Wave Nature of the Electron

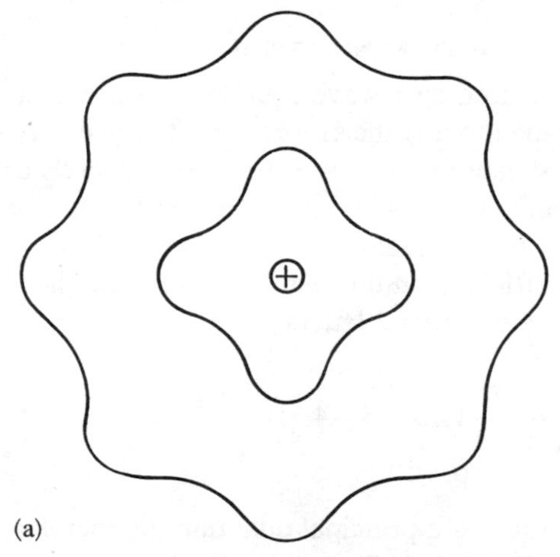

(a)

The waves shown in (a) are standing waves. The maxima and minima of the waves match exactly for consecutive circuits of the electron about the nucleus. (The figure is a two-dimensional representation of three-dimensional waves.)

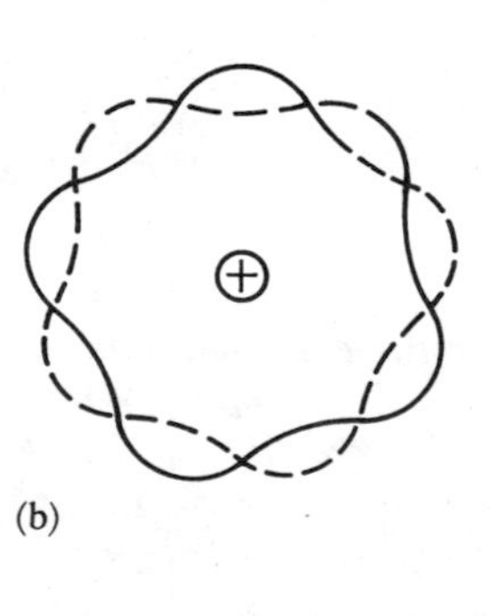

(b)

The pattern in (b) is not a solution. A maximum in one circuit is cancelled by a minimum in the next.

...Its solution gives the probability of finding the electron at any distance from the nucleus

The solution of the wave equation gives the **probability density** of the electron. This is the probability that the electron is present in a given small region of space. The probability that the electron is at a distance, r, from the nucleus is plotted against r for the hydrogen atom in its ground state in Figure 2.10. The maximum probability of finding the electron is at a distance of 0.053 nm. This is the same as the radius of the orbit occupied by the electron in its ground state according to the Bohr–Sommerfeld theory.

FIGURE 2.10
A Probability Density Diagram for the Hydrogen Atom in its Ground State

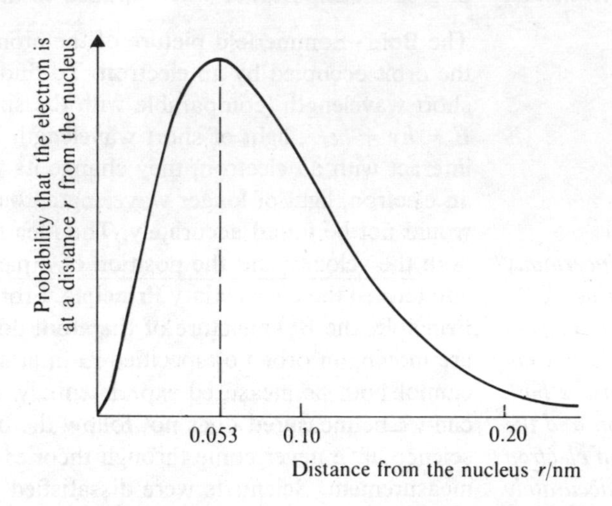

There is a possibility that the electron will be either closer to the nucleus or outside the radius of 0.053 nm. The probability of finding the electron decreases sharply, however, as the distance from the nucleus increases beyond $3r$. The volume of space in which there is a 95% chance of finding the electron is called the **atomic orbital**. There is a 5% probability that the electron will be outside this volume of space at a given instant. On this model, the electron is not described as revolving in an orbit. The electron is said to occupy a three-dimensional space around the nucleus called an atomic orbital. The nucleus is described as being surrounded by a three-dimensional 'cloud of charge' or 'electron cloud'.

The volume of space in which there is a 95% chance of finding the electron is called the atomic orbital

THE FOUR QUANTUM NUMBERS

Since the Schrödinger wave equation has a limited number of solutions, the total energy of the atom is therefore found to have only certain values. This agrees with Bohr's postulate that the electronic energy levels are quantised. Solutions of the wave equation can be obtained if the orbitals are described by four quantum numbers. The first is Bohr's quantum number, n. The second quantum number, l, corresponds to Sommerfeld's quantum number describing the shape of elliptical orbits. The values of l are assigned letters

Bohr's quantum number, n.
Sommerfeld's quantum number, l

l =	0	1	2	3	4
	s	p	d	f	g

If an electron has a principal quantum number $n = 2$ and a second quantum number $l = 0$, it is said to be a 2s electron. For various values of n, the different combinations of the two quantum numbers are

1s

2s 2p

3s 3p 3d

4s 4p 4d 4f

5s 5p 5d 5f 5g

A third quantum number, m_l...

The wave equation leads to a **third quantum number**, m_l. This gives the maximum number of orbitals for the different values of l as

one s orbital

three p orbitals

five d orbitals

seven f orbitals

...and a spin quantum number, m_s

The **fourth quantum number** is called the **spin quantum number**, m_s. It has values of $+\frac{1}{2}$ and $-\frac{1}{2}$. It represents the spin of an electron on its own axis, which can be clockwise or anticlockwise, relative to the orbital of the electron.

The Pauli Exclusion Principle

In his Exclusion Principle, W Pauli stated that no two electrons in an atom can have the same four quantum numbers. It follows that, if two electrons in an atom have the same values of n, l and m_l, they must have different values of m_s. Their spins must be opposed. Each orbital can hold two electrons with opposed spins.

2.3.3 SHAPES OF ATOMIC ORBITALS

FIGURE 2.11 The Shapes of s Orbitals

(a) The shape of a 1s orbital

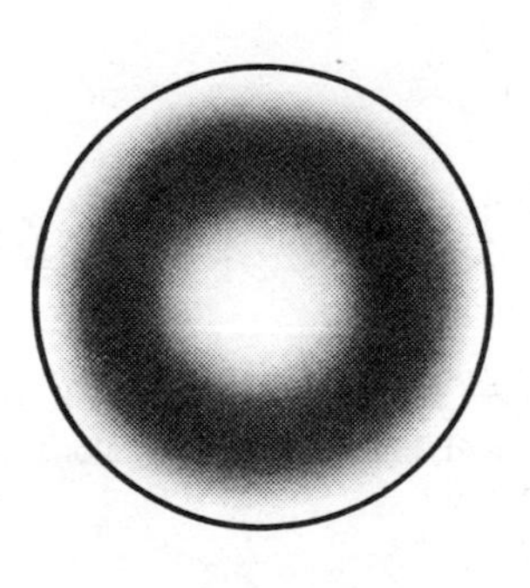

(b) The shape of a 2s orbital

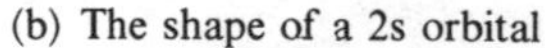

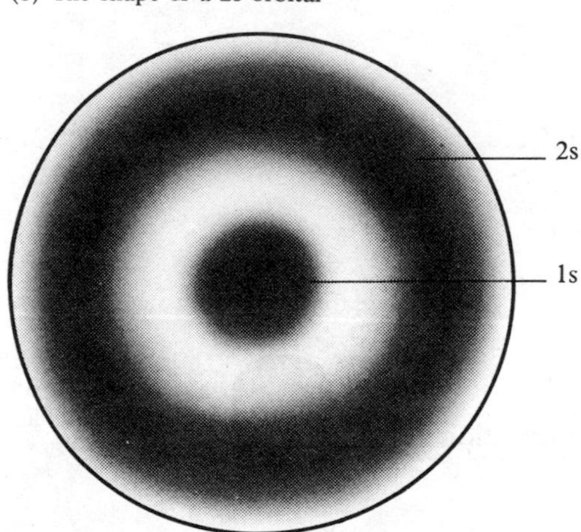

(The density of the shading is a measure of the probability of finding an electron at that distance from the nucleus.)

s orbitals are spherical

p orbitals have an hourglass shape

The shape of an s orbital is spherically symmetrical about the nucleus. The orbital has no preferred direction. The probability of finding an electron at a distance r from the nucleus is the same in all directions. [See Figure 2.11.] A p orbital is not symmetrical: it is concentrated in certain directions. The electron density is shaped like an hourglass. [See Figure 2.12.] The shapes of d orbitals are shown in Figure 2.13.

FIGURE 2.12
The Shape and Orientation of p Orbitals

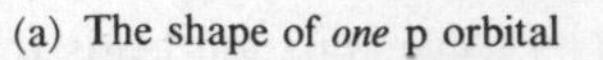

(a) The shape of *one* p orbital

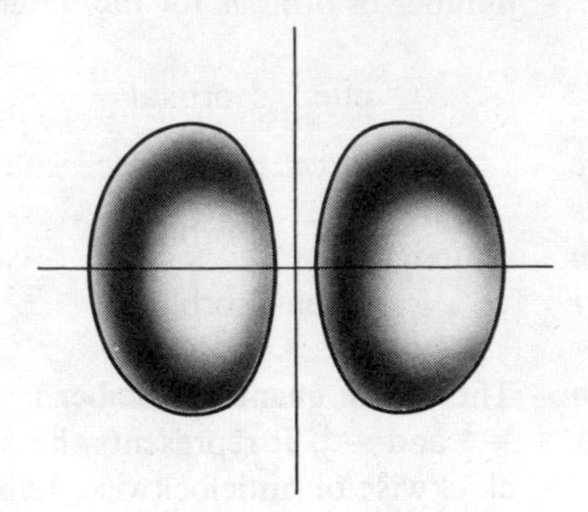

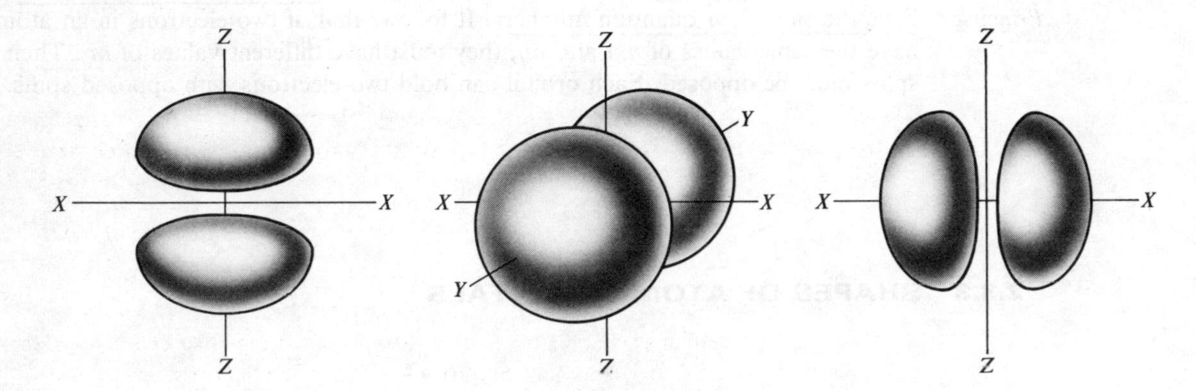

(b) The orientation in space of *three* p orbitals. (Each of the three p orbitals is perpendicular to both of the others. The shape of each orbital is as shown in (a).)

FIGURE 2.13
The Shape and Orientation of d Orbitals

There are four d orbitals of shape (a). The lobes lie between the X–Y axes as shown in (a), between the X–Z axes, between the Y–Z axes and, in the fourth case, along the axes X and Y as shown in (b). The fifth orbital has the shape shown in (c).

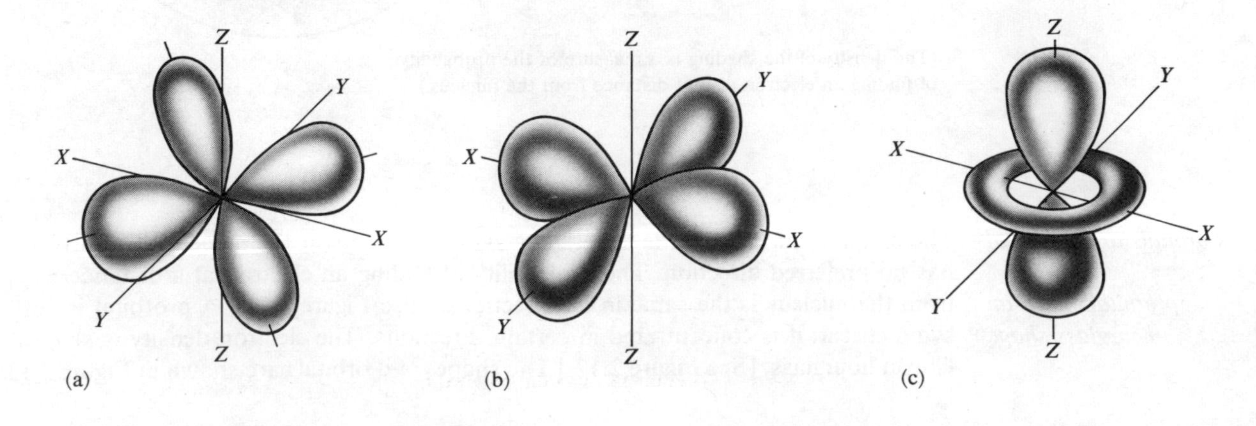

2.4 ELECTRONIC CONFIGURATION OF ATOMS

In atoms with more than one electron, there are shells of orbitals with the same principal quantum number

The energy levels of the orbitals of the hydrogen atom are illustrated in Figure 2.7 [§2.2.1]. For atoms with more than one electron, there are different values of the quantum number l for each value of the principal quantum number n. The energy levels for each value of n are split between orbitals with different values of l. The relative energy levels of the orbitals are shown in Figure 2.14. The term **shell** is used for a group of orbitals with the same principal quantum number. The $n = 1$ shell is termed the K shell, $n = 2$ the L shell, $n = 3$ the M shell and so on. A **subshell** is a group of orbitals with the same principal and second quantum numbers, e.g. the 3p subshell.

The orbitals of lowest energy are filled first

The arrangement of electrons in atomic orbitals is governed by two factors. The first is that in a normal atom the electrons are arranged so that the energy is at a minimum. Any other arrangement would make the atom an excited atom, which could emit energy and pass to its ground state. The second factor is the Pauli Exclusion Principle, i.e. no two electrons can have the same four quantum numbers.

It is convenient to draw an 'electrons-in-boxes' diagram to show the arrangement of electrons in orbitals. A box represents one orbital and can contain two electrons with opposite spins. The electrons are represented by arrows, pointing upwards for $m_s = +\frac{1}{2}$ and downwards for $m_s = -\frac{1}{2}$:

'Electrons-in-boxes'

↑↓

An s subshell consists of one box, a p subshell of *three* boxes, a d subshell of *five* boxes and an f subshell of *seven* boxes. The boxes are arranged in order of energy in Figure 2.15. An aid to help you remember the order in which the levels are filled is shown in Figure 2.16.

Electrons occupy lowest energy 'boxes' first

To work out the arrangement of electrons in an atom with 12 electrons, the electrons must be put into the lowest energy boxes first, two to a box, until all the electrons are accommodated [see Figure 2.15]. It is interesting to do this for the elements in order of atomic number, the order in which they appear in the Periodic Table [p. 792].

FIGURE 2.14 The Relative Energy Levels of Atomic Orbitals (not to scale)

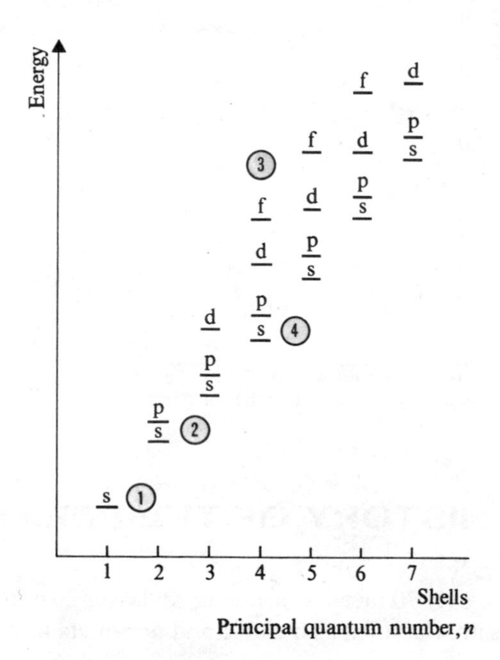

3 Note 4f > 4d > 4p > 4s in energy.

4 Note 4s < 3d in energy. The orbitals of $n = 4$ overlap those with $n = 3$.

2 Note 2p electrons have more energy than 2s.

1 Lowest energy: $n = 1$

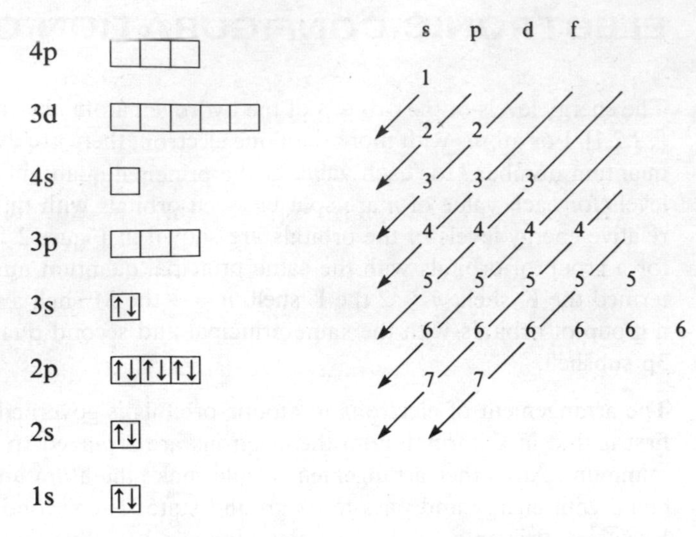

(*a*) 1 appears once, 2 appears twice and so on.

(*b*) Letters, s p d f, read across. Numbers read downwards.

(*c*) To obtain the order, read diagonally downwards from right to left: 1s, 2s, 2p, 3s, 3p, 4s, 3d, 4p, 5s, 4d, 5p, 6s, 4f, 5d, 6p, 7s, 5f, 6d, 7p etc.

FIGURE 2.15
'Electrons-in-boxes' Diagram for an Atom with 12 Electrons

FIGURE 2.16
How to Remember the Order in which Boxes are Filled

Measurements of successive ionisation energies support the idea of shells

Evidence for the arrangement of electrons in shells of different energies is provided by values of successive **ionisation energies** for elements. Figure 2.17 shows a graph of the logarithm of the ionisation energy required for the removal of one electron after another from a potassium atom. A logarithmic plot is used in order to give a condensed graph. You can see that the electrons fall into four groups. The higher the ionisation energy, the more difficult the electrons are to remove and the nearer they must be to the nucleus.

FIGURE 2.17 Graph of lg (Ionisation Energy) against Number of Electron Removed for the Potassium Atom

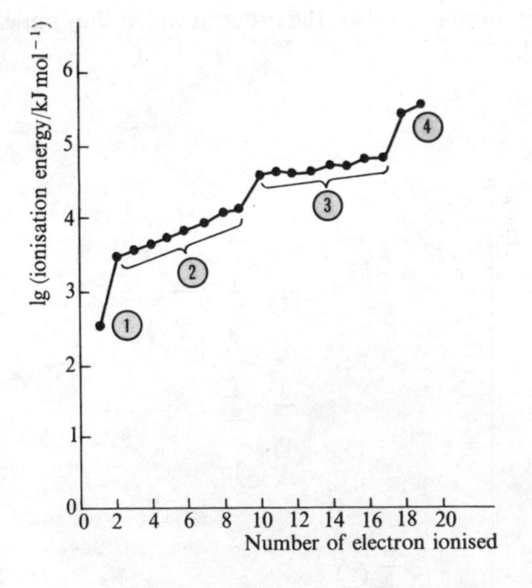

4 The two electrons with the highest ionisation energies are closest to the nucleus, and form the K shell.

3 These eight electrons are in the next shell, the L shell.

2 These eight electrons are in the M shell.

1 This electron has the lowest ionisation energy. It is the easiest to remove. It is in the N (n = 4) shell.

2.5 THE HISTORY OF THE PERIODIC TABLE

There are over 70 metallic elements, and over 20 non-metallic elements. As you know, elements are classified as metallic and non-metallic elements.

Metallic elements	*Non-metallic elements*
Solids (except mercury, a liquid)	Solids or gases (except bromine, a liquid)
A fresh surface is shiny; corrosion can occur.	Have no one characteristic appearance.
Malleable (can be hammered) and ductile (can be drawn into wire)	Shatter when attempts are made to change the shape.
Conduct heat and electricity.	Are poor thermal conductors and electrical conductors, with exceptions.
The oxides are basic.	The oxides are acidic or neutral.

TABLE 2.1
Properties of Metallic and Non-metallic Elements

Classification of elements . . .

. . . as metallic and non-metallic . . .

. . . and metalloid

For a long time chemists looked at ways of dividing up the two big groups, metallic and non metallic elements, into smaller sub-groups. They drew up groups of similar elements, such as the very reactive metals lithium, sodium and potassium. They grouped together a set of slightly less reactive metals, calcium, strontium and barium. Another such group was the very reactive non-metals, chlorine, bromine and iodine. Some elements were discovered which had properties in between metallic and non-metallic. These elements, e.g. silicon, were described as **metalloids**.

Newlands arranged the elements in order of relative atomic mass

In 1866, a British chemist called John Newlands had the idea of arranging the elements in order of their relative atomic masses:

H Li Be B C N O F Na Mg Al Si P S Cl K Ca

[For the symbols of elements see p. 792.]

Newlands noticed that similar elements appeared at regular intervals in the list. He arranged the elements in columns [Figure 2.18].

FIGURE 2.18
Newlands' Octaves of Elements

H	Li	Be	B	C	N	O
F	Na	Mg	Al	Si	P	S
Cl	K	Ca	Cr	Ti	Mn	Fe

He drew up a Law of Octaves . . .

In the first column were hydrogen and the very reactive non-metallic elements fluorine and chlorine. In the second column were the very reactive metals sodium, lithium and potassium. In the third column were the metals beryllium, magnesium and calcium. Carbon and silicon both fell into the fifth column, and oxygen and sulphur both fell into the sixth column.

. . . but his ideas were scorned

Newlands compared his chemical 'octaves' with musical octaves, and called the resemblance the **Law of Octaves**. The comparison was unfortunate: people poured scorn on his ideas.

Mendeleev carried on the work on classification . . .

It was a Russian chemist, Dmitri Mendeleev, who developed Newlands' idea and persuaded chemists to use it. Mendeleev had an encyclopaedic knowledge of chemistry, e.g. the types and formulae of the compounds which an element formed with hydrogen and oxygen, physical properties such as appearance, melting and boiling temperatures, density and heat of vaporisation. Careful experiments were done to find the masses of elements which combined with 1 g of hydrogen or 8 g of oxygen, and these led to a set of relative atomic masses. In 1869, Mendeleev summarised his **periodic law** in the statement: The properties of chemical elements are not arbitrary, but vary with their relative atomic masses in a systematic way. He arranged the elements in order of increasing relative atomic mass [see Figure 2.19]. A modern version of his classification, which we call the Periodic Table, is shown on p. 792. You will notice that Mendeleev's Periodic Table lacks the noble gases (helium, neon, argon, etc. in Group 0) because they had not yet been discovered! A vertical row of elements is called a **group** and a horizontal row is called a **period**.

. . . and drew up his Periodic Table

FIGURE 2.19 Part of Mendeleev's Periodic Table of 1871

	Gp 1	*Gp 2*	*Gp 3*	*Gp 4*	*Gp 5*	*Gp 6*	*Gp 7*	*Gp 8*
Row 1	H							
Row 2	Li	Be	B	C	N	O	F	
Row 3	Na	Mg	Al	Si	P	S	Cl	
Row 4	K	Ca	–	–	–	–	–	Ti V Cr Mn Fe Co Ni
Row 5	Cu	Zn	–	–	As	Se	Br	

For the full Periodic Table, see p. 792

Mendeleev made various improvements on Newlands' system.

Mendeleev's improvements included long periods to accommodate transition metals . . .

1. Mendeleev introduced long rows or **periods** for the elements we now call **transition metals**. This meant that the metals Ti, Mn, Fe were no longer placed under the non-metals Si, P, S [see Figure 2.18].

. . . spaces . . .

2. He left spaces. When he saw that arsenic fitted naturally into Group 5 he left two spaces between zinc and arsenic.

. . . new values of relative atomic mass . . .

3. When elements did not fit comfortably into the slots in the Periodic Table dictated by their relative atomic masses, Mendeleev instigated new determinations of relative atomic mass. In each case (Cr, In, Pt, Au) the new value justified the arrangement in Mendeleev's Periodic Table.

. . . predictions about undiscovered elements

4. Where he had left gaps in the Periodic Table, Mendeleev predicted that new elements would be discovered to fill the gaps. He had some outstanding successes in predicting the properties of elements. When elements were discovered and found to have the relative atomic mass and the physical and chemical properties Mendeleev had predicted, faith in the Periodic Table soared.

The Periodic Table helped chemists in their search for the elements which were still to be discovered. For example, Mendeleev predicted that an element would be discovered to fill the space under silicon and above tin. Table 2.2 shows the predictions for the element which he called ekasilicon (below silicon) and the properties of the element germanium, which was discovered in 1886 by Winkler. Similar agreement was found between the predicted properties of eka-aluminium and gallium, which was discovered in 1875 and between ekaboron and scandium, discovered in 1879.

TABLE 2.2 Mendeleev's predictions for 'Ekasilicon'

Property	*Ekasilicon*	*Germanium*
Relative atomic mass	72	72.3
Density (g cm^{-3})	5.5	5.36
Appearance	Grey metal	Grey metal
Melting point (°C)	High	958
Reactions	Resistant to attack by acids and alkalis	Not attacked by hydrochloric acid and sodium hydroxide; attacked by nitric acid
Oxide, density (g cm^{-3})	EO_2, 4.7	GeO_2, 4.70
Chloride, density (g cm^{-3})	ECl_4, 1.9	$GeCl_4$, 1.887
Boiling point (°C)	< 100 °C	86 °C
Ethyl compound, b.p. (°C), density (g cm^{-3})	$E(C_2H_5)_4$, 160, 0.96	$Ge(C_2H_5)_4$, 160, 1.00

The noble gases were discovered and found to fit into a new group of the Periodic Table . . .

The **noble gases** had not been discovered when the Periodic Table was drawn up. As they were discovered one by one, they were found to fit in between the halogens in Group 7 and the alkali metals in Group 1. A separate Group 0 was added to the right-hand side of the table. Argon, however, has a higher relative atomic mass than potassium ($A_r(Ar) = 40$; $A_r(K) = 39$) [§ 3.3]. It made more sense chemically to put potassium with the alkali metals, rather than keep to the order of relative atomic masses. Another example of this kind was the positions of tellurium and iodine. Relative atomic masses placed tellurium under bromine, and iodine under sulphur and selenium; chemical properties placed them in the reverse order.

. . . Discrepancies were resolved by arranging elements in order of atomic number

Moseley's work on X rays, in 1914, solved this problem. He showed that the atomic numbers (proton numbers) of elements are more significant than their relative atomic masses [see § 1.6]. This discovery was the final step in the validation of the Periodic Table. In the modern Periodic Table elements are arranged in order of proton number (atomic number).

2.6 FEATURES OF THE PERIODIC TABLE

What patterns can be seen in the arrangement of the elements in the Periodic Table? First, note the positions occupied by metallic and non-metallic elements. The reactive metals are at the left-hand side of the table, less reactive metallic elements in the middle block and non-metallic elements at the right-hand side.

Looking more closely, what is the difference between the metals in Group 1, those in Group 2 and the transition metals? The metals in the block between Group 2 and Group 3 are called **transition metals**. They include iron and copper. You will be studying the reactions of metals later in Chapters 18, 19, 23 and 24. An outline, summarising some of their reactions, is given in Table 2.3. (**M** stands for the symbol of a metallic element.)

Element	*Reaction with air*	*Reaction with water*	*Reaction with dilute acids*
Group 1			
Lithium Sodium Potassium Rubidium Caesium	Burn vigorously to form the strongly basic oxide, $\mathbf{M_2O}$	React vigorously to form hydrogen and a solution of the strong alkali, **MOH**	The reaction is dangerously violent. The vigour of these reactions increases down the group.
Group 2			
Beryllium	Burn to form the strongly basic oxide, **MO**	Reacts very slowly	React readily to give hydrogen and a salt, e.g. $\mathbf{MCl_2}$. The vigour of these reactions increases down the group.
Magnesium		Burns in steam	
Calcium Strontium Barium		React readily to form hydrogen and the alkali $\mathbf{M(OH)_2}$	
Transition metals			
Iron	When heated, form oxides, without burning.	Rusts slowly, reacts with steam to form hydrogen and iron oxide.	Reacts to give hydrogen and a salt.
Copper		Does not react.	Does not react.

TABLE 2.3
Some Reactions of Metals

2.6.1 LOOKING AT THE GROUPS

The very reactive alkali metals of Group 1 ...

From the summary in Table 2.3, you can see how the Periodic Table makes it easier to learn about all the elements. Look at the elements in Group 1: lithium, sodium, potassium, rubidium and caesium. They are all very reactive metals. Their oxides and hydroxides have the general formulae $\mathbf{M_2O}$ and **MOH** (where **M** is the symbol for the metallic element) and are strongly basic. The oxides and hydroxides dissolve in water to give strongly alkaline solutions. The metals in Group 1 are called the **alkali metals**. Their reactivity increases as you pass down the group. If you know these facts, you do not need to learn the properties of all the metals separately. If you know the properties of sodium, you can predict those of potassium and lithium. Think of having to learn the properties of 106 elements separately! The Periodic Table saves you from this.

... the alkaline earth metals of Group 2 ...

The metals in Group 2 are less reactive than those in Group 1. They form basic oxides and hydroxides with the general formulae **MO** and $\mathbf{M(OH)_2}$. Their oxides and hydroxides are either sparingly soluble or insoluble. These elements are called the **alkaline earths**. Their reactivity increases as you pass down the group. Again, if you know the chemical reactions of one element, you can predict the reactions of other elements in the group.

... and the transition metals

The transition metals are less reactive than those in Groups 1 and 2. Their oxides and hydroxides are less strongly basic and are insoluble.

The noble gases of Group 0

The elements in Group 0 are the **noble gases**, formerly called the **inert gases**. They are present in air. They are the least reactive of the elements. For many years, it seemed as though they took part in no chemical reactions. However, in 1960, two of them, krypton and xenon, were made to combine with the very reactive element, fluorine.

The halogens of Group 7

Group 7 precedes Group 0. It contains a set of very reactive non-metallic elements: fluorine, chlorine, bromine, iodine and astatine. They are called the **halogens** (derived from the Greek: halogen = salt-former) because they react with metals to form salts. Examples of their chemical reactions are given in Table 2.4.

TABLE 2.4 Some Reactions of the Halogens

Halogen	*State at room temperature*	*Reaction with sodium*	*Reaction with iron*	*Trend*
Fluorine	Gas	Explosive	Explosive	The vigour of these reactions decreases down the group.
Chlorine	Gas	Heated sodium burns in chlorine to form sodium chloride.	Reacts vigorously with hot iron to form iron(III) chloride.	
Bromine	Liquid	Reacts less vigorously to form sodium bromide.	Reacts less vigorously to form iron(III) bromide.	
Iodine	Solid	Reacts less vigorously than bromine to form sodium iodide.	Reacts less vigorously than bromine to form iron(II) iodide.	

Group 6

At the right-hand side of each period, preceding the halogens, are the non-metallic elements of Group 6: oxygen, sulphur, selenium and tellurium. They are a set of elements with the ability to form two chemical bonds. They show an increase in metallic character from oxygen (a non-metal) to tellurium (a semi-metal or metalloid).

Group 5

The elements nitrogen, phosphorus, arsenic, antimony and bismuth form a group with the ability to form three or five chemical bonds. They show a gradation in properties from nitrogen and phosphorus (non-metals) through metalloid arsenic to antimony and bismuth (metals). These elements are in Group 5.

Group 4 shows the largest gradation in properties from top to bottom of the group

The elements carbon, silicon, germanium, tin and lead have the ability to form four chemical bonds. A marked gradation in properties occurs from carbon (non-metal) through silicon and germanium (metalloids) to tin and lead (metals). In Group 4, there is the maximum gradation in properties down the group.

Group 3

The elements aluminium, gallium, indium and thallium form ions with a charge of +3. Boron is a metalloid, while the rest of the elements in Group 3 are metals.

CHECKPOINT 2A: THE PERIODIC TABLE I

Refer to the complete Periodic Table on p. 792.

1. Locate the position in the Periodic Table of francium (Fr, atomic number 87). Make as many predictions as you can about the properties of francium, e.g. whether it is a solid or a liquid or a gas, whether it is a metal or a non-metal, its chemical reactions with air, water and dilute acids.

2. Use the Periodic Table to predict the properties of astatine (At, atomic number 85), e.g. whether it is a solid or a liquid or a gas, whether it is a metal or a non-metal, its chemical reactions.

3. Find the position of beryllium, Be, in the Periodic Table. Say what you think will happen when beryllium is added to cold water.

4. Locate radium (Ra, atomic number 88) in the Periodic Table. Predict (*a*) the formula of radium chloride, (*b*) the nature and formula of radium oxide, and (*c*) the products and speed of the reaction of radium with water.

2.6.2 LOOKING AT THE PERIODS

Period 1 contains hydrogen and helium. Period 2 contains the elements from lithium to neon. Table 2.5 shows some of the patterns which are seen in passing across Period 3. The properties shown are metallic and non-metallic character, the charge on the ions which the element forms, and the structure of the element and its oxide. You will be learning more about structure in Chapters 4 and 6. This table simply tells whether the substance consists of

1. individual atoms, e.g. argon which consists of Ar atoms.

2. individual molecules, e.g. sulphur dioxide which consists of SO_2 molecules.

3. a giant ionic structure containing millions of ions bonded together, e.g. sodium chloride which consists of Na^+ ions and Cl^- ions in equal numbers.

From left to right across a period, there is a transition from metals to non-metals, from elements which form positive ions to elements which form negative ions, from giant structures to molecular structures, from basic, ionic oxides to acidic, molecular oxides.

TABLE 2.5 Trends across Period 3

Group	*1*	*2*	*3*	*4*	*5*	*6*	*7*	*0*
Element	Na	Mg	Al	Si	P	S	Cl	Ar
Character	Metallic			Metalloid	Non-metallic			Noble gas
Reactivity	—Decreases →			–	← Decreases—			–
Structure of element	Giant metallic			Giant covalent	Molecular			Atomic
Ion	Na^+	Mg^{2+}	Al^{3+}	None	P^{3-}	S^{2-}	Cl^-	None
Oxide	Na_2O	MgO	Al_2O_3	SiO_2	P_2O_5 P_2O_3	SO_2 SO_3	Cl_2O Cl_2O_7 etc.	None
Type of oxide	Strongly basic		Some acidic and some basic character		Acidic			None
Structure of oxide	Giant ionic			Giant covalent	Molecular			None

4. a giant covalent structure containing millions of atoms bonded together, e.g. silicon(IV) oxide, SiO_2 which contains silicon atoms and oxygen atoms in the ratio of 2 atoms of oxygen for every silicon atom.

5. a giant metallic structure, which consists of millions of atoms of a metallic element bonded together by the metallic bond [see § 6.2].

CHECKPOINT 2B: THE PERIODIC TABLE II

1. Look at the elements in Period 2. Say what you would expect to be the acidic or basic nature of the oxides of lithium, beryllium, boron and carbon.

2. State what types of structure you would expect for the elements lithium, beryllium, nitrogen, oxygen, fluorine and neon.

3. Look at the elements of Period 4. Say what formula or formulae you would expect for the oxides of gallium, germanium, arsenic and selenium.

2.7 THE ELECTRONIC CONFIGURATIONS OF THE ELEMENTS

In the Periodic Table, elements are arranged in order of their proton numbers (atomic numbers). The proton number, Z, is the nuclear charge and also the number of electrons in an atom of the element.

Hydrogen has one electron. It occupies the lowest energy level, 1s.	H	1s [↑] This can be written as 1s
Helium has two electrons which can occupy the same 1s orbital with opposing spins.	He	1s [↑↓] written as $1s^2$ The superscript 2 denotes the number of electrons in the subshell.
Lithium has three electrons. Two occupy the 1s box, which is then full. The third must go into a box at the next level.	Li	2s [↑] 1s [↑↓] written as $1s^2 2s$
Beryllium has four electrons.	Be	2s [↑↓] 1s [↑↓] $1s^2 2s^2$
Boron has five electrons. The fifth electron cannot enter the 1s or 2s boxes: it occupies an orbital in the 2p subshell.	B	2p [↑][ ][ ] 2s [↑↓] 1s [↑↓] $1s^2 2s^2 2p$
Carbon has six electrons. There are three ways in which the fifth and sixth electrons can be accommodated in the 2p subshell. The two 2p electrons may be (*a*) in the same box or in different boxes with the spins (*b*) parallel or (*c*) opposed. In fact arrangement (*b*) is favoured.	C	(*a*) 2p [↑↓][ ][ ] (*b*) 2p [↑][↑][ ] (*c*) 2p [↑][↓][ ] 2p [↑][↑][ ] 2s [↑↓] 1s [↑↓] $1s^2 2s^2 2p^2$

According to Hund's Rule, only when all the orbitals in a subshell contain an electron do electrons begin to occupy orbitals in pairs

Hund's Multiplicity Rule comes into play here. F Hund stated that the favoured configuration is the one in which the electrons occupy different boxes and have the same spins. This arrangement puts the electrons further apart than the others. According to Hund's rule, electrons do not pair in an orbital until all the other orbitals in the subshell have been occupied by a single electron.

Nitrogen (Z = 7) obeys Hund's rule by accommodating the three 2p electrons in different boxes with the same spins.

N
2p [↑|↑|↑]
2s [↑↓]
1s [↑↓] $1s^22s^22p^3$

Oxygen (Z = 8). The fourth 2p electron pairs with one of the other three 2p electrons.

O
2p [↑↓|↑|↑]
2s [↑↓]
1s [↑↓] $1s^22s^22p^4$

Fluorine (Z = 9) has the arrangement $1s^22s^22p^5$.

F
2p [↑↓|↑↓|↑]
2s [↑↓]
1s [↑↓] $1s^22s^22p^5$

Neon (Z = 10) $1s^22s^22p^6$, has a full 2p subshell, thus completing the L shell. The next element, **sodium** (Z = 11) has to utilise the M shell, starting with the 3s subshell. The diagrams for sodium and the twelfth element, **magnesium**, are

Na
3s [↑]
2p [↑↓|↑↓|↑↓]
2s [↑↓]
1s [↑↓] $1s^22s^22p^63s$

Mg
3s [↑↓]
2p [↑↓|↑↓|↑↓]
2s [↑↓]
1s [↑↓] $1s^22s^22p^63s^2$

The following six elements have electrons in the 3p subshell. The configurations of the next six elements are (writing (Ne) for $1s^22s^22p^6$)

Al (Z = 13) (Ne)$3s^23p$

Si (Z = 14) (Ne)$3s^23p^2$

P (Z = 15) (Ne)$3s^23p^3$

S (Z = 16) (Ne)$3s^23p^4$

Cl (Z = 17) (Ne)$3s^23p^5$

Ar (Z = 18) (Ne)$3s^23p^6$

How do the elements fit into the Periodic Table? So far we have:

Electron configuration and the Periodic Table

Filling the K shell: H and He: First period

Filling the L shell: Li to Ne: Second period (first short period)

Filling the M shell: Na to Ar: Third period (second short period)

The next elements start the fourth period. They are

K (Z = 19) (Ar)$4s$

Ca (Z = 20) (Ar)$4s^2$

Once the 4s subshell is full, the 3d orbitals are filled [see Figure 2.15, § 2.4]. Over the next 10 elements, electrons enter the 3d subshell. These elements are

The fourth period

$$\text{Sc}\,(Z = 21) \quad (\text{Ar})4\text{s}^2 3\text{d}$$

$$\text{to Zn}\,(Z = 30) \quad (\text{Ar})4\text{s}^2 3\text{d}^{10}$$

Filling the d subshell...

While the d subshell fills, the chemistry of the elements is not greatly affected. The metals scandium to zinc are a very similar set of metals, called **transition metals** [Chapter 24]. The elements gallium (Z = 31) to krypton (Z = 36) complete the M shell by filling the 4p orbitals. The 18 elements from potassium to krypton comprise the **first long period**.

Electronic configurations can be written as e.g. B (2.3), C (2.4), N (2.5), O (2.6), F (2.7), Mg (2.8.2), where the numbers give the numbers of electrons in the K, L and M shells.

TABLE 2.6 Electronic Configurations of the Atoms of the Elements

	Z	*1s*	*2s*	*2p*	*3s*	*3p*	*3d*	*4s*	*4p*	*4d*	*4f*	*5s*	*5p*	*5d*	*5f*	*6s*	*6p*	*6d*	*6f*	*7s*
H	1	1																		
He	2	2																		
Li	3	2	1																	
Be	4	2	2																	
B	5	2	2	1																
C	6	2	2	2																
N	7	2	2	3																
O	8	2	2	4																
F	9	2	2	5																
Ne	10	2	2	6																
Na	11	2	2	6	1															
Mg	12				2															
Al	13				2	1														
Si	14				2	2														
P	15		10 electrons		2	3														
S	16				2	4														
Cl	17				2	5														
Ar	18	2	2	6	2	6														
K	19	2	2	6	2	6		1												
Ca	20							2												
Sc	21						1	2												
Ti	22						2	2												
V	23						3	2												
Cr	24						5	1												
Mn	25						5	2												
Fe	26						6	2												
Co	27			18 electrons			7	2												
Ni	28						8	2												
Cu	29						10	1												
Zn	30						10	2												
Ga	31						10	2	1											
Ge	32						10	2	2											
As	33						10	2	3											
Se	34						10	2	4											
Br	35						10	2	5											
Kr	36	2	2	6	2	6	10	2	6											

The fifth period

The fifth period is a **second long period**, extending from rubidium (Z = 37) to xenon (Z = 54). This period includes a **second transition series**.

	Z	1s	2s	2p	3s	3p	3d	4s	4p	4d	4f	5s	5p	5d	5f	6s	6p	6d	6f	7s
Rb	37	2	2	6	2	6	10	2	6			1								
Sr	38											2								
Y	39									1		2								
Zr	40									2		2								
Nb	41									4		1								
Mo	42									5		1								
Tc	43									5		2								
Ru	44									7		1								
Rh	45									8		1								
Pd	46									10		0								
Ag	47									10		1								
Cd	48									10		2								
In	49									10		2	1							
Sn	50									10		2	2							
Sb	51									10		2	3							
Te	52									10		2	4							
I	53									10		2	5							
Xe	54	2	2	6	2	6	10	2	6	10		2	6							

The sixth period

The sixth period starts with caesium (Z = 55) and ends with radon (Z = 86). During this period, the 4f orbitals are filled between cerium (Z = 58) and ytterbium (Z = 70). The **lanthanons** (from lanthanum (Z = 57) to lutetium (Z = 71)) are the first set of **rare earth elements**, and are even more similar to one another than are the transition metals. The elements from lutetium to mercury, (Z = 80), comprise a second transition series in which the 5d subshell is filled.

The rare earth elements

	Z	1s	2s	2p	3s	3p	3d	4s	4p	4d	4f	5s	5p	5d	5f	6s	6p	6d	6f	7s
Cs	55	2	2	6	2	6	10	2	6	10		2	6			1				
Ba	56											2	6			2				
La	57											2	6	1		2				
Ce	58										2	2	6			2				
Pr	59										3	2	6			2				
Nd	60										4	2	6			2				
Pm	61										5	2	6			2				
Sm	62										6	2	6			2				
Eu	63										7	2	6			2				
Gd	64										7	2	6	1		2				
Tb	65										8	2	6	1		2				
Dy	66										10	2	6			2				
Ho	67										11	2	6			2				
Er	68										12	2	6			2				
Tm	69										13	2	6			2				
Yb	70					46 electrons					14	2	6			2				
Lu	71										14	2	6	1		2				
Hf	72										14	2	6	2		2				
Ta	73										14	2	6	3		2				
W	74										14	2	6	4		2				
Re	75										14	2	6	5		2				
Os	76										14	2	6	6		2				
Ir	77										14	2	6	7		2				
Pt	78										14	2	6	9		1				
Au	79										14	2	6	10		1				
Hg	80										14	2	6	10		2				
Tl	81										14	2	6	10		2	1			
Pb	82										14	2	6	10		2	2			
Bi	83										14	2	6	10		2	3			
Po	84										14	2	6	10		2	4			
At	85										14	2	6	10		2	5			
Rn	86	2	2	6	2	6	10	2	6	10	14	2	6	10		2	6			

The seventh period The seventh and final period starts at francium ($Z = 87$). The last naturally occurring element, uranium, has $Z = 92$. The electron configurations of all the elements are shown in Table 2.6. The elements which follow uranium are artificially made and are radioactive [§ 1.9].

	Z	1s	2s	2p	3s	3p	3d	4s	4p	4d	4f	5s	5p	5d	5f	6s	6p	6d	6f	7s
Fr	87	2	2	6	2	6	10	2	6	10	14	2	6	10		2	6			1
Ra	88															2	6			2
Ac	89															2	6	1		2
Th	90															2	6	2		2
Pa	91														2	2	6	1		2
U	92														3	2	6	1		2
Np	93														5	2	6			2
Pu	94														6	2	6			2
Am	95							78 electrons							7	2	6			2
Cm	96														7	2	6	1		2
Bk	97														7	2	6	2		2
Cf	98														9	2	6	1		2
Es	99																			
Fm	100																			
Md	101																			
No	102																			
Lr	103																			

CHECKPOINT 2C: ELECTRONIC CONFIGURATIONS

1. There are six calcium isotopes, of nucleon number 40, 42, 43, 44, 46 and 48. How many protons and neutrons are there in the nuclei?

2. Draw 'electrons-in-boxes' diagrams of the electronic configurations of the following atoms, given the proton number (Z): boron (5), fluorine (9), aluminium (13) and potassium (19).

3. Draw diagrams to show the electronic configurations of the ions K^+, Cl^-, Ca^{2+}, O^{2-}, Al^{3+}, H^-. (Proton numbers are K = 19, Cl = 17, Ca = 20, O = 8, Al = 13 and H = 1.)

4. Write down the electronic configurations of the atoms with the proton numbers 4, 7, 18, 27, 37. State to which Group of the Periodic Table each element belongs.

5. Write the electronic configurations of the following species (e.g., Li = $1s^22s$). Their proton numbers range from Na = 11 to Ar = 18.

Na^+, Mg^{2+}, Al, Si, P, S, S^{2-}, Cl, Cl^-, Ar

2.8 THE REPEATING PATTERN OF THE ELEMENTS

We have spent time looking at the structure of the atom [Chapter 1] and the electron configurations of the elements [§§ 2.4 and 2.7]. How does this study fit in with the Periodic Table? If you look at the electron configurations of the atoms, some interesting points strike you.

The noble gases have an octet of electrons in the outermost shell

First, notice the elements with a full outer shell of electrons. These are helium (2), neon (2.8), argon (2.8.8), krypton and xenon. These elements are the noble gases (Group 0). Their lack of chemical reactivity has been mentioned. They exist as single atoms. Their atoms do not combine in pairs to form molecules as do the atoms of most gaseous elements (e.g. O_2, H_2). It seems only logical to suppose that it is the full outer shell of electrons that makes the noble gases chemically unreactive.

For members of other groups, the number of electrons in the outermost shell equals the group number

Following each noble gas (that is, with atomic number 1 greater than each noble gas) is an alkali metal [see Table 2.7]. These elements are lithium (2.1), sodium (2.8.1), potassium (2.8.8.1), rubidium and caesium. We can infer that it is because the alkali metals all have a single electron in the outer shell that they all behave in a very similar way.

Following each alkali metal is an alkaline earth metal [see Table 2.7]. The alkaline earths, beryllium (2.2), magnesium (2.8.2), calcium (2.8.8.2), strontium and barium all have two electrons in the outer shell. It seems logical to suppose that it is the similar configuration of electrons that gives the elements similar properties.

Preceding each noble gas (with atomic number 1 less than the noble gas) are the halogens of Group 7: fluorine (2.7), chlorine (2.8.7), bromine and iodine [see Table 2.4]. We can infer that the halogens all have similar chemical reactions because they all have the same number of electrons in the outer shell.

								Group 0
Period 1	H (1)							He (2)
	Group 1	*Group 2*	*Group 3*	*Group 4*	*Group 5*	*Group 6*	*Group 7*	
Period 2	Li (2.1)	Be (2.2)	B (2.3)	C (2.4)	N (2.5)	O (2.6)	F (2.7)	Ne (2.8)
Period 3	Na (2.8.1)	Mg (2.8.2)	Al (2.8.3)	Si (2.8.4)	P (2.8.5)	S (2.8.6)	Cl (2.8.7)	Ar (2.8.8)
Period 4	K (2.8.8.1)	Ca (2.8.8.2)						

TABLE 2.7
A Section of the Periodic Table

You can see the following features:

1. The elements are listed in order of increasing atomic number.

2. Elements which have the same number of electrons in the outermost shell fall into the same group of the Periodic Table.

3. The first period contains only hydrogen and helium. The second period contains the elements lithium to neon. The third period contains the elements sodium to argon.

4. The chemical properties of elements depend on the electron configurations of their atoms.

5. The electron configuration of its atoms is related to the position of an element in the Periodic Table.

6. The noble gases (Group 0) all have a full outer shell of electrons.

7. The reactive alkali metals (Group 1) have a single electron in the outer shell.

8. The alkaline earth metals (Group 2) have two electrons in the outer shell.

9. The halogens (Group 7) have 7 electrons in the outer shell.

CHECKPOINT 2D: THE PERIODIC TABLE III

1. (*a*) What are the 'noble gases'?
(*b*) In which group of the Periodic Table are they?
(*c*) What do the noble gases have in common (i) regarding their electron configurations and (ii) regarding their chemical reactions?

2. X is a metallic element. It reacts slowly with water to give a strongly alkaline solution. In which group of the Periodic Table would you place **X**?

3. Y is a non-metallic element. It reacts vigorously with sodium to give a salt of formula Na**Y**. In which group of the Periodic Table would you place **Y**?

4. Z is a metallic element which reacts vigorously with water to give a strongly alkaline solution. In which group of the Periodic Table would you place **Z**?

QUESTIONS ON CHAPTER 2

1.

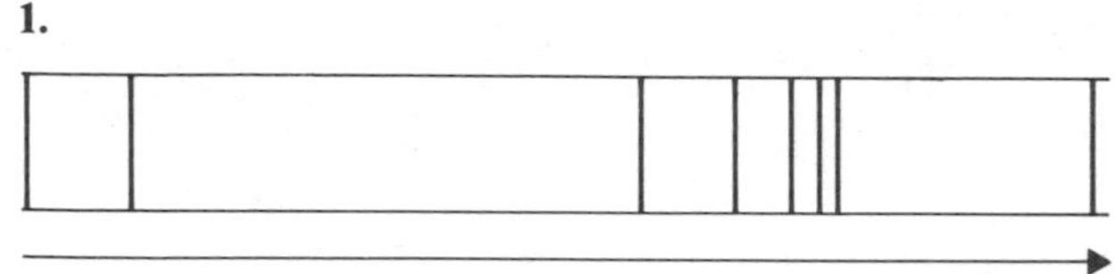

Increasing energy

FIGURE 2.20

Figure 2.20 represents the atomic emission spectrum of hydrogen. Explain why it is composed of lines, and say what each line indicates. Why do the lines become closer together as you read from left to right?

2. 'In its ground state, the electron in a hydrogen atom is in a 1s orbital.' Explain this statement.

3. (*a*) Describe how the first ionisation energy of an element can be determined experimentally.
(*b*) Results obtained for the ionisation energies of boron are

Electron number	*1st*	*2nd*	*3rd*	*4th*	*5th*
Ionisation energy/ kJ mol^{-1}	800	2400	3700	25 000	32 800

On graph paper, plot lg (ionisation energy/kJ mol^{-1}) against the number of electron removed. From the graph, deduce the most likely formula of boron chloride.

4. The following questions refer to the *emission* spectrum of the hydrogen atom. The spectrum consists of several series of lines. Each line results from an electron transition between energy levels characterised by different values of *n*.

Part of the highest energy (Lyman) series in the hydrogen atom spectrum is shown in Figure 2.21. Line **A** is the first line of this series.

A B C D E F G

FIGURE 2.21

(*a*) (i) Below a copy of the diagram, draw an arrow to show the direction of increasing energy and label this arrow 'energy'.

(ii) Below a copy of the diagram, draw a second arrow to show the direction of increasing frequency and label this arrow 'frequency'

(*b*) Why does the spectrum consist of lines?

(*c*) Name the symbol *n*.

(*d*) What do the transitions in the same series all have in common?

(*e*) Which, if any, of the lines A to G shown above corresponds to each of the following transitions? If none of the lines corresponds to the given transition, answer 'none'. In each case, explain your reasoning.

(i) the transition $n = 3$ to $n = 1$

(ii) the transition $n = 3$ to $n = 2$

(iii) the transition $n = 1$ to $n = 2$ (NEAB 92)

5. (*a*) Hydrogen forms diatomic molecules. How can the emission spectrum of the hydrogen atom be obtained?

(*b*) How does this spectrum differ from that of the light emitted from a tungsten filament light bulb?

(*c*) Explain how the appearance of the atomic spectrum of hydrogen can be accounted for in terms of quantised energy levels.

(*d*) Define the ionisation energy of hydrogen, and explain how its value can be obtained by measurements on the atomic spectrum of hydrogen.

(*e*) The emission spectrum of atomic hydrogen obtained at high temperatures is more complex and covers a larger part of the electromagnetic spectrum than the emission spectrum of atomic hydrogen recorded at lower temperatures. Suggest why this is so.

(*f*) The unusual ions He^{+} and C^{5+} have been detected spectroscopically in light from stars. Comment on the spectra which are observed, by referring to the sub-atomic particles present in each ion, and on why the two spectra are not identical. (O 92, S)

6. (*a*) Explain how the atomic spectrum of hydrogen can be used to measure the first ionisation energy of hydrogen.

(*b*) Sketch a graph of first ionisation energy against atomic number for the eleven elements hydrogen to sodium. Account for the shape of your graph including any small irregularities.

(*c*) State and explain how you would expect the second ionisation energy of sodium to compare with

(i) the first ionisation energy of neon

(ii) the second ionisation energy of magnesium

(*d*) ^{24}Na decays by β^{-} emission with a half-life of 15 hours. Calculate the composition by mass of 200 mg of ^{24}Na after 45 hours. (NEAB 91)

3

EQUATIONS AND EQUILIBRIA

3.1 FORMULAE

Every element has a symbol [see Periodic Table, p. 792]. A symbol is a letter or two letters which stand for one atom of the element. Formulae are written for compounds. The formula of a compound consists of the symbols of the elements present and the numbers which show the ratio in which the atoms are present. The compound sulphur dioxide has the formula SO_2. The compound sulphur trioxide has the formula SO_3. The formulae tell you the difference between them. Sulphur dioxide contains 2 oxygen atoms for every sulphur atom: the 2 below the line multiplies the O in front of it. Sulphur trioxide contains 3 oxygen atoms for every sulphur atom. Sulphur dioxide and sulphur trioxide consists of molecules [see Figure 3.1]. To show three molecules of sulphur dioxide, you write $3SO_2$.

(a)

(b)

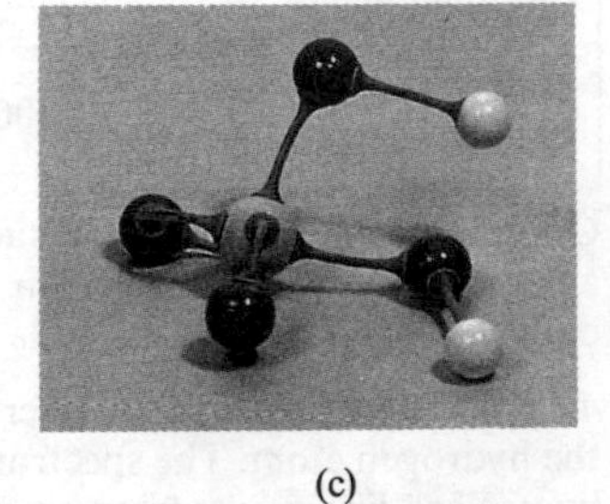

(c)

FIGURE 3.1 Models of Molecules of (a) Sulphur Dioxide, (b) Sulphur Trioxide, (c) Sulphuric Acid

The formula of a compound is a set of symbols and numbers. The symbols say what elements are present in the compound. The numbers give the ratio of the numbers of atoms of the different elements in the compound

The formula of sulphuric acid is H_2SO_4. The compound contains two hydrogen atoms and four oxygen atoms for every sulphur atom; to write three molecules, you write $3H_2SO_4$. The 3 in front of the formula multiplies everything after it. In $3H_2SO_4$, there are 6 H, 3 S and 12 O atoms, a total of 21 atoms.

Many compounds do not consist of molecules; they consist of ions. The compound calcium hydroxide is composed of calcium ions, Ca^{2+}, and hydroxide ions, OH^-. There are twice as many hydroxide ions as calcium ions, so the formula for calcium hydroxide is $Ca(OH)_2$. The 2 multiplies the symbols in the brackets. There are 2 oxygen atoms, 2 hydrogen atoms and 1 calcium atom. This is not a molecule of calcium hydroxide; it is a formula unit of calcium hydroxide: one calcium ion and 2 hydroxide ions. A piece of calcium hydroxide contains this formula unit repeated many times. To write $4Ca(OH)_2$ means that the whole of the formula is multiplied by 4. It means 4 Ca, 8 O and 8 H atoms. § 4.2.7 deals with how to work out the formula of a compound. Table 3.1 lists the formulae of some common compounds.

TABLE 3.1 The Formulae of some Compounds

Compound	Formula	Compound	Formula
Water	H_2O	Aluminium chloride	$AlCl_3$
Sodium hydroxide	$NaOH$	Aluminium oxide	Al_2O_3
Sodium chloride	$NaCl$	Carbon monoxide	CO
Sodium sulphate	Na_2SO_4	Carbon dioxide	CO_2
Sodium nitrate	$NaNO_3$	Sulphur dioxide	SO_2
Sodium carbonate	Na_2CO_3	Ammonia	NH_3
Sodium hydrogencarbonate	$NaHCO_3$	Ammonium chloride	NH_4Cl
Calcium oxide	CaO	Hydrogen chloride	HCl
Calcium hydroxide	$Ca(OH)_2$	Hydrochloric acid	$HCl(aq)$
Calcium chloride	$CaCl_2$	Sulphuric acid	$H_2SO_4(aq)$
Calcium sulphate	$CaSO_4$	Nitric acid	$HNO_3(aq)$
Calcium carbonate	$CaCO_3$	Copper(II) oxide	CuO
		Copper(II) sulphate	$CuSO_4$

CHECKPOINT 3A: FORMULAE

1. How many atoms are present in the following?
(*a*) C_6H_6 (*b*) P_4O_{10} (*c*) SO_2Cl_2 (*d*) $C_2H_4Cl_2$
(*e*) $2ZnSO_4$ (*f*) $5CuSO_4$ (*g*) $Al(NO_3)_3$ (*h*) $2Al(OH)_3$
(*i*) $Fe_2(SO_4)_3$ (*j*) $3Fe(NO_3)_3$

2. Give the formula of:
(*a*) sodium hydroxide, (*b*) hydrochloric acid, (*c*) ammonia, (*d*) sodium chloride, (*e*) calcium oxide, (*f*) calcium hydroxide, (*g*) calcium carbonate, (*h*) sulphuric acid, (*i*) nitric acid.

3.2 EQUATIONS

You have studied symbols for elements and formulae for compounds [§ 3.1]. These enable you to write equations for chemical reactions.

Example (a) Calcium carbonate decomposes to give calcium oxide and carbon dioxide.

Calcium carbonate → Calcium oxide + Carbon dioxide

Writing an equation . . .

Replacing names with formulae, you can write the chemical equation for the reaction:

$$CaCO_3 \rightarrow CaO + CO_2$$

On the left-hand side, you have 1 atom of calcium, 1 atom of carbon and 3 atoms of oxygen combined as calcium carbonate. On the right-hand side, you have 1 atom of calcium and 1 atom of oxygen combined as calcium oxide and 1 atom of carbon and 2 atoms of oxygen combined as carbon dioxide. The two sides are equal, and this is why the expression is called an equation.

. . . adding state symbols

You can give more information if you include state symbols in the equation. These are (s) = solid, l = liquid, (g) = gas, and (aq) = in aqueous (water) solution. Putting in the state symbols,

$$CaCO_3(s) \rightarrow CaO(s) + CO_2(g)$$

tells you that solid calcium carbonate decomposes to form solid calcium oxide and carbon dioxide gas.

Example (b) Magnesium reacts with sulphuric acid to give hydrogen and a solution of magnesium sulphate.

Magnesium + Sulphuric acid → Hydrogen + Magnesium sulphate

The chemical equation is

$$Mg(s) + H_2SO_4(aq) \rightarrow H_2(g) + MgSO_4(aq)$$

Hydrogen is written as H_2 because hydrogen gas consists of molecules containing two atoms.

Example (c) Hydrogen and oxygen combine to form water. The word equation is

Hydrogen + Oxygen → Water

The chemical equation could be

$$H_2(g) + O_2(g) \rightarrow H_2O(l)$$

This equation is not balanced. There are 2 oxygen atoms on the left-hand side (LHS) and only 1 oxygen atom on the right-hand side (RHS). To balance the O atoms, multiply H_2O on the RHS by 2.

$$H_2(g) + O_2(g) \rightarrow 2H_2O(l)$$

The O atoms are now balanced, but there are 4H atoms on the RHS and only 2H on the LHS. Multiplying H on the LHS by 2,

$$2H_2(g) + O_2(g) \rightarrow 2H_2O(l)$$

The equation is now balanced.

Number of atoms on LHS = 4H + 2O

Number of atoms on RHS = 4H + 2O

Number of atoms of each element on LHS = Number of atoms of each element on RHS

The total mass of the reactants = The total mass of the products [see Figure 3.2]

FIGURE 3.2
A Balanced Equation (Example (c))

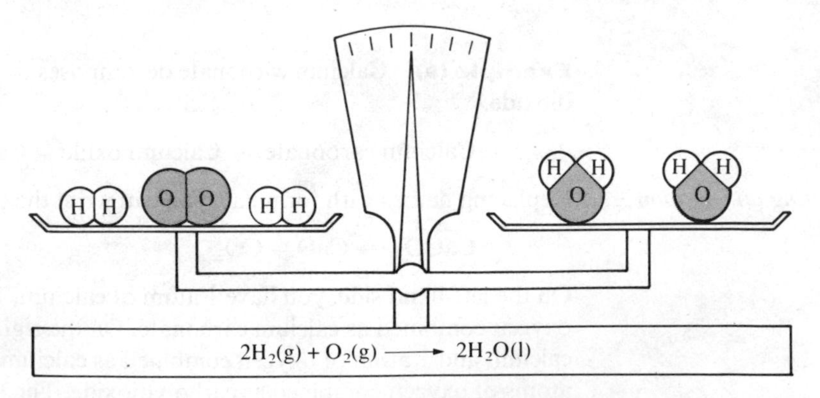

Example (d) Sulphur dioxide is oxidised by oxygen to sulphur trioxide.

Sulphur dioxide + Oxygen → Sulphur trioxide

$$SO_2(g) + O_2(g) \rightarrow SO_3(g)$$

Balancing an equation

You can see that the equation is not balanced. There are 4O on the LHS and 3O on the RHS. It is tempting to write O for oxygen on the LHS. You must not do this. Never change a formula. All you can do to balance an equation is to multiply formulae. Instead of changing O_2 to O, multiply SO_2 and SO_3 by 2.

$$2SO_2(g) + O_2(g) \rightarrow 2SO_3(g)$$

The equation is now balanced: 2S + 6O on the LHS; 2S + 6O on the RHS [see Figure 3.3].

FIGURE 3.3
A Balanced Equation (Example (d))

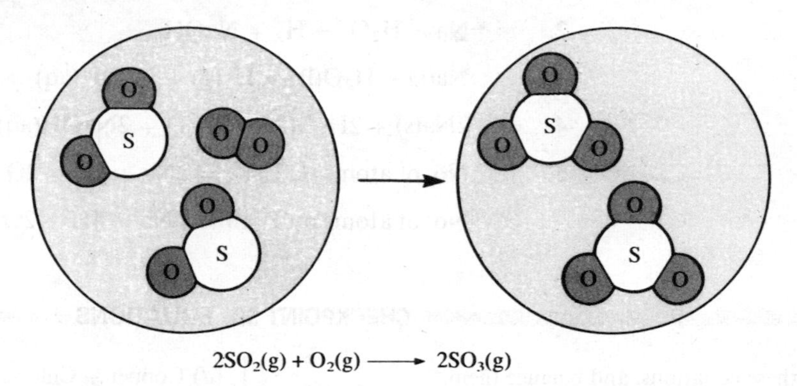

$$2SO_2(g) + O_2(g) \longrightarrow 2SO_3(g)$$

Example (e) Sodium carbonate reacts with dilute hydrochloric acid to give carbon dioxide and a solution of sodium chloride.

Sodium carbonate + Hydrochloric acid → Carbon dioxide + Sodium chloride + Water

The chemical equation could be

$$Na_2CO_3(s) + HCl(aq) \rightarrow CO_2(g) + NaCl(aq) + H_2O(l)$$

When you add up the atoms on the RHS, you find that they are not equal to the atoms on the LHS. The equation is not balanced. Start by balancing Na atoms. Multiplying NaCl by 2,

$$Na_2CO_3(s) + HCl(aq) \rightarrow CO_2(g) + 2NaCl(aq) + H_2O(l)$$

Now, there are 2 Na atoms on the RHS and 2Na atoms on the LHS. With 2Cl atoms on the RHS, the HCl on the LHS must be multiplied by 2.

$$Na_2CO_3(s) + 2HCl(aq) \rightarrow CO_2(g) + 2NaCl(aq) + H_2O(l)$$

The equation is now balanced. Check again:

Number of atoms on LHS = 2Na + C + 3O + 2H + 2Cl

Number of atoms on RHS = 2Na + C + 3O + 2H + 2Cl

When you are balancing a chemical equation, the only way to do it is to put a number in front of a formula. You never try to alter a formula. In this example, you got 2Cl atoms by multiplying HCl by 2, not by altering the formula to HCl_2, which does not exist. You can multiply a formula, but you cannot change it.

The steps in writing a chemical equation are:

1. Write a word equation for the reaction.
2. Write the symbols and formulae for the reactants and products.
3. Add the state symbols.
4. Balance the equation. Multiply the formulae if necessary. Never change a formula.
5. Check again:

No. of atoms of each element on LHS = No. of atoms of each element on RHS

Take the reaction between sodium and water to form hydrogen and sodium hydroxide solution. Work through the five steps:

1. Sodium + Water → Hydrogen + Sodium hydroxide solution
2. $Na + H_2O \rightarrow H_2 + NaOH$
3. $Na(s) + H_2O(l) \rightarrow H_2(g) + NaOH(aq)$
4. $2Na(s) + 2H_2O(l) \rightarrow H_2(g) + 2NaOH(aq)$
5. No. of atoms on LHS = 2Na + 4H + 2O

   No. of atoms on RHS = 2Na + 4H + 2O

CHECKPOINT 3B: EQUATIONS

1. Copy these equations, and balance them.

(*a*) $Fe_2O_3(s) + C(s) \rightarrow Fe(s) + CO(g)$

(*b*) $Fe_2O_3(s) + CO(g) \rightarrow Fe(s) + CO_2(g)$

(*c*) $NH_3(g) + O_2(g) \rightarrow NO(g) + H_2O(l)$

(*d*) $Cr(s) + HCl(aq) \rightarrow CrCl_3(aq) + H_2(g)$

(*e*) $Fe_3O_4(s) + H_2(g) \rightarrow Fe(s) + H_2O(l)$

(*f*) $C_3H_8(g) + O_2(g) \rightarrow CO_2(g) + H_2O(l)$

2. Try writing equations for the reactions:

(*a*) Hydrogen + Copper(II) oxide → Copper + Water

(*b*) Carbon + Carbon dioxide → Carbon monoxide

(*c*) Magnesium + Sulphuric acid → Hydrogen + Magnesium sulphate

(*d*) Copper + Chlorine → Copper(II) chloride

(*e*) Mercury + Oxygen → Mercury(II) oxide

(*f*) Iron + Sulphur → Iron(II) sulphide

3. Write balanced chemical equations for the reactions:

(*a*) Calcium + Water → Hydrogen + Calcium hydroxide solution

(*b*) Iron + Hydrochloric acid → Iron(II) chloride solution + Hydrogen

(*c*) Iron + Chlorine → Iron(III) chloride

(*d*) Aluminium + Chlorine → Aluminium chloride

(*e*) Zinc + Steam → Zinc oxide + Hydrogen

(*f*) Sodium + Oxygen → Sodium oxide

3.3 RELATIVE ATOMIC MASS

The masses of atoms are very small, from 10^{-24} to 10^{-22} grams. Instead of using the actual masses of atoms, **relative atomic masses** (A_r) are used. Originally, they were defined as

Atoms range in mass from 10^{-24} g to 10^{-22} g

$$\text{Original relative atomic mass} = \frac{\text{Mass of one atom of an element}}{\text{Mass of one atom of hydrogen}}$$

Since relative atomic masses are now determined by mass spectrometry, and since volatile carbon compounds are much used in mass spectrometry, the mass of an atom of ${}^{12}_{6}C$ is now taken as the standard of reference:

The definition of relative atomic mass...

$$\text{Modern relative atomic mass} = \frac{\text{Mass of one atom of an element}}{1/12 \text{ the mass of one atom of carbon-12}}$$

The difference between the two scales is small. On the carbon-12 scale, the relative atomic mass of ${}^{12}_{6}C$ is 12.0000, and the relative atomic mass of ${}^{1}_{1}H$ is 1.0078. The mass of a ${}^{12}_{6}C$ atom is 12.0000 u, and the mass of a ${}^{1}_{1}H$ atom is 1.0078 u [§ 1.2].

3.4 RELATIVE MOLECULAR MASS

The relative molecular mass, M_r, of a compound is the sum of the relative atomic masses of all the atoms in one molecule of a covalent compound

The mass of a molecule is the sum of the masses of all the atoms in it. The **relative molecular mass**, M_r, of a compound is the sum of the relative atomic masses of all the atoms in a molecule of the compound. For example, you find the relative molecular mass of sulphuric acid in this way:

Formula of compound is H_2SO_4

2 atoms of H ($A_r = 1$) $= 2$

1 atom of S ($A_r = 32$) $= 32$

4 atoms of O ($A_r = 16$) $= 64$

Total $= 98$

Relative molecular mass, M_r, of $H_2SO_4 = 98$

The relative formula mass of a compound, with the same symbol, M_r, is the sum of the relative atomic masses of all the atoms in one formula unit of an ionic compound

Many compounds consist of ions, not molecules. For ionic compounds, the formula represents a formula unit, rather than a molecule of the compound. A formula unit of sodium sulphate is Na_2SO_4. The term **relative formula mass**, symbol M_r, can be used for ionic compounds. Many people use the term relative molecular mass, M_r, for ionic compounds as well as molecular compounds.

CHECKPOINT 3C: RELATIVE MOLECULAR MASS

1. Work out the relative molecular masses of these compounds:

$NaOH$, KCl, MgO, $Ca(OH)_2$, HNO_3, $CuCO_3$, NH_4NO_3, $CuSO_4$, $CuSO_4 \cdot 5H_2O$, $Mg(HCO_3)_2$

3.5 THE MOLE

Very often chemists want to measure out the exact quantities of substances that will react together. What is really useful is to be able to work out these quantities on the basis of the number of atoms (or molecules) of substance **A** that will react with a certain number of atoms of substance **B**. Counting out atoms sounds a tricky business, but, thanks to the mole concept, it can be done! How can we count out numbers of atoms by measuring masses? The key to the calculation is the idea which chemists call the **mole concept**. The origin of the mole concept was the work of a nineteenth century Italian chemist called Amadeo Avogadro. This is how he argued:

We know from their relative atomic masses that one atom of magnesium is twice as heavy as one atom of carbon: $A_r(\text{Mg}) = 24$, $A_r(\text{C}) = 12$.
Therefore we can say:

If 1 atom of magnesium is twice as heavy as 1 atom of carbon,
then 1 hundred Mg atoms are twice as heavy as 1 hundred C atoms,
and 5 million Mg atoms are twice as heavy as 5 million C atoms,
and it follows that, if we have a piece of magnesium which has twice the mass of a piece of carbon, the two masses must contain equal numbers of atoms.
2 grams of magnesium and 1 gram of carbon contain the same number of atoms;
10 tonnes of magnesium and 5 tonnes of carbon contain the same number of atoms.

The same argument applies to the other elements. Take the relative atomic mass in grams of any element:

12 g Carbon	24 g Magnesium	56 g Iron	40 g Calcium	108 g Silver	238 g Uranium	207 g Lead

All these masses contain the same number of atoms. The number is 6.022×10^{23}.

The amount of an element that contains 6.022×10^{23} atoms (the same number of atoms as 12 g of carbon-12) is called **one mole** of that element.

The symbol for mole is **mol**. The ratio 6.022×10^{23} mol^{-1} is called the **Avogadro constant**. When you weigh out 12 g of carbon, you are counting out 6×10^{23} atoms of carbon. This amount of carbon is one mole (1 mol) of carbon atoms. Similarly, 48 g of magnesium is two moles (2 mol) of magnesium atoms. You can say that the **amount** of magnesium is two moles (2 mol).

You can have a mole of magnesium atoms, Mg, a mole of magnesium ions, Mg^{2+}, a mole of sulphuric acid molecules, H_2SO_4. One mole of sulphuric acid contains 6×10^{23} molecules of H_2SO_4, that is, 98 g of H_2SO_4 (the molar mass in grams). To write 'a mole of nitrogen' is imprecise: one mole of nitrogen atoms, N, has a mass of 14 grams; one mole of nitrogen molecules, N_2, has a mass of 28 grams.

CHECKPOINT 3D: THE AVOGADRO CONSTANT

(Take the Avogadro constant to be 6×10^{23} mol^{-1}.)

1. There are 4 billion people in the world. If you had one mole of £1 coins to distribute equally between them, how much would each person receive?

2. What mass of potassium contains (*a*) 6×10^{23} atoms, (*b*) 2×10^{25} atoms?

3. The price of gold is £8.20 per gram; $A_r(Au) = 198$. Calculate the price of 1 million million atoms of gold.

3.6 MOLAR MASS

Molar mass is defined

The mass of one mole of a substance is called the **molar mass**, symbol M, unit g mol^{-1}. The molar mass of carbon is 12 g mol^{-1}; that is the relative atomic mass expressed in grams per mole. The term molar mass applies to compounds as well as elements. The molar mass of a compound is the relative molecular mass expressed in grams per mole. Sulphuric acid, H_2SO_4, has a relative molecular mass of 98; its molar mass is 98 g mol^{-1}. Notice the units: relative molecular mass has no unit; molar mass has the unit g mol^{-1}.

$$\text{Amount (in moles) of substance} = \frac{\text{Mass of substance}}{\text{Molar mass of substance}}$$

Molar mass of element = Relative atomic mass in grams per mole

Molar mass of compound = Relative molecular mass in grams per mole

Sample calculations of amount of substance

Example What is the amount of calcium present in 120 g of calcium?

Method

A_r of calcium = 40

Molar mass of calcium = $40\,\text{g mol}^{-1}$

$$\text{Amount of calcium} = \frac{\text{Mass of calcium}}{\text{Molar mass of calcium}} = \frac{120\,\text{g}}{40\,\text{g mol}^{-1}}$$

$$= 3.0\,\text{mol}$$

The amount (number of moles) of calcium is 3.0 mol.

Example If you need 2.50 mol of sodium hydrogencarbonate, what mass of the substance do you have to weigh out?

Method

Relative molecular mass of $NaHCO_3 = 23 + 1 + 12 + (3 \times 16) = 84$

Molar mass of $NaHCO_3 = 84\,\text{g mol}^{-1}$

$$\text{Amount of substance} = \frac{\text{Mass of substance}}{\text{Molar mass of substance}}$$

$$2.50\,\text{mol} = \frac{\text{Mass}}{84\,\text{g mol}^{-1}}$$

$$\text{Mass} = 84\,\text{g mol}^{-1} \times 2.50\,\text{mol}$$

$$= 210\,\text{g}$$

You need to weigh out 210 g of sodium hydrogencarbonate.

CHECKPOINT 3E: THE MOLE

1. State the mass of:
(*a*) 3 mol of magnesium ions, Mg^{2+}
(*b*) 0.50 mol of oxygen atoms, O
(*c*) 0.50 mol of oxygen molecules, O_2
(*d*) 0.25 mol of sulphur atoms, S
(*e*) 0.25 mol of sulphur molecules, S_8

2. Find the amount (moles) of each element present in:
(*a*) 69 g of lead, Pb
(*b*) 14 g of iron, Fe
(*c*) 56 g of nitrogen, N_2
(*d*) 2.0 g of mercury, Hg
(*e*) 9.0 g of aluminium, Al

3. State the mass of
(*a*) 2.0 mol of carbon dioxide molecules, CO_2
(*b*) 10 mol of sulphuric acid, H_2SO_4
(*c*) 2.0 mol of sodium chloride, NaCl
(*d*) 0.50 mol of calcium hydroxide, $Ca(OH)_2$

4. Calculate the molar masses of the following:
(*a*) $NH_4Fe(SO_4)_2{\cdot}12H_2O$
(*b*) $Al_2(SO_4)_3$
(*c*) $K_4Fe(CN)_6$

5. How many moles of substance are present in the following?
(*a*) 0.250 g of calcium carbonate
(*b*) 5.30 g of anhydrous sodium carbonate
(*c*) 5.72 g of sodium carbonate-10-water crystals

6. Use the value of $6.0 \times 10^{23}\,\text{mol}^{-1}$ for the Avogadro constant to find the number of atoms in
(*a*) 2.0×10^{-3} g of calcium
(*b*) 5.0×10^{-6} g of argon
(*c*) 1.00×10^{-10} g of mercury

3.7 EMPIRICAL FORMULAE

The **empirical formula** of a compound is the simplest formula which represents its composition. It shows the elements present and the ratio of the amounts of elements present.

Finding an empirical formula...

To find an empirical formula, you need to work out the ratio of the amounts of the elements present.

Example A 0.4764 g sample of an oxide of iron was reduced by a stream of carbon monoxide. The mass of iron that remained was 0.3450 g. Find the empirical formula of the oxide.

Method

...A worked example

Elements present	*Iron*		*Oxygen*
Mass/g	0.3450		0.1314
A_r	56		16
Amount/mol	0.3450/56		0.1314/16
	$= 6.16 \times 10^{-3}$		$= 8.21 \times 10^{-3}$
Ratio of amounts	1	:	$\frac{8.21 \times 10^{-3}}{6.16 \times 10^{-3}}$
	1	:	1.33
	3	:	4

Empirical formula is Fe_3O_4.

3.8 MOLECULAR FORMULAE

Finding a molecular formula...

The **molecular formula** is a simple multiple of the empirical formula. If the empirical formula is CH_2O, the molecular formula may be CH_2O, $C_2H_4O_2$, $C_3H_6O_3$ and so on.

The way to find out which molecular formula is correct is to find out which gives the correct molar mass.

Example A polymer of empirical formula CH_2 has a molar mass of $28\,000\,g\,mol^{-1}$. What is its molecular formula?

Method

...A worked example

Empirical formula mass $= 14\,g\,mol^{-1}$

Molar mass $= 28\,000\,g\,mol^{-1}$

The molar mass is 2000 times the empirical formula mass; therefore the molecular formula is $(CH_2)_{2000}$.

3.9 CALCULATION OF PERCENTAGE COMPOSITION

The empirical formula shows percentage by mass composition...

From the formula of a compound and the relative atomic masses of the elements in it, the percentage of each element in the compound can be calculated. This is called the **percentage composition by mass**.

Example Calculate the percentage mass of water of crystallisation in copper(II) sulphate-5-water.

...A worked example

Method

Formula is $CuSO_4 \cdot 5H_2O$

Relative atomic masses are Cu = 63.5 S = 32 O = 16 H = 1

$$\text{Molar mass} = 63.5 + 32 + (4 \times 16) + (5 \times 18) = 249.5\,\text{g mol}^{-1}$$

$$\text{Percentage of water} = \frac{90}{249.5} \times 100 = 36\%$$

CHECKPOINT 3F: FORMULAE AND PERCENTAGE COMPOSITION

1. Calculate the percentage by mass of the named element in the compound listed:

(*a*) Mg in Mg_3N_2

(*b*) Na in NaCl

(*c*) Br in $CaBr_2$

2. Calculate the empirical formulae of the compounds for which the following analytical results were obtained:

(*a*) 27.3% C, 72.7% O

(*b*) 53.0% C, 47.0% O

(*c*) 29.1% Na, 40.5% S, 30.4% O

(*d*) 32.4% Na, 22.6% S, 45.0% O

3. Find the empirical formulae of the compounds formed in the reactions described below:

(*a*) 10.800 g magnesium form 18.000 g of an oxide.

(*b*) 3.400 g calcium form 9.435 g of a chloride.

(*c*) 3.528 g iron form 10.237 g of a chloride.

4. Weighed samples of the following crystals were heated to drive off the water of crystallisation. When they reached constant mass, the following masses were recorded. Deduce the empirical formulae of the hydrates:

(*a*) 0.942 g of $MgSO_4 \cdot a\,H_2O$ gave 0.461 g of residue

(*b*) 1.124 g of $CaSO_4 \cdot b\,H_2O$ gave 0.889 g of residue

(*c*) 1.203 g of $Hg(NO_3)_2 \cdot c\,H_2O$ gave 1.172 g of residue

3.10 EQUATIONS FOR THE REACTIONS OF SOLIDS

Equations give us much information

Equations tell us not only what substances react together but also what amounts of substances react together. The equation for the action of heat on sodium hydrogencarbonate

$$2NaHCO_3(s) \rightarrow Na_2CO_3(s) + CO_2(g) + H_2O(g)$$

tells us that 2 moles of sodium hydrogencarbonate give 1 mole of sodium carbonate. Since the molar masses are $NaHCO_3 = 84\,\text{g mol}^{-1}$ and $Na_2CO_3 = 106\,\text{g mol}^{-1}$, it follows that 168 g of sodium hydrogencarbonate give 106 g of sodium carbonate.

The stoichiometry of a reaction is the relationship between the amounts of reactants and products

The amounts of substances undergoing reaction, as given by the balanced chemical equation, are called the **stoichiometric** amounts. **Stoichiometry** is the relationship between the amounts of reactants and products in a chemical reaction. If one reactant is present in excess of the stoichiometric amount required to react with another of the reactants, then the excess of one reactant will be left unused at the end of the reaction.

A worked example of a calculation based on a stoichiometric equation

Example What mass of zinc can be obtained from the reduction of 10.00 tonnes of zinc oxide by 10.00 tonnes of charcoal? (1 tonne = 10^3 kg)

Method Write the equation

$$ZnO(s) + C(s) \rightarrow Zn(s) + CO(g)$$

$$\text{Amount of ZnO} = 10.00 \times 10^6/(65.4 + 16.00) = 1.23 \times 10^5\,\text{mol}$$

$$\text{Amount of C} = 10.00 \times 10^6/12.00 = 8.33 \times 10^5\,\text{mol}$$

Since zinc oxide is present in the smaller amount, the amount of zinc formed is limited by the amount of zinc oxide. From the equation, you can see that 1 mole of ZnO forms 1 mole of Zn.

$$\text{Amount of Zn} = 1.23 \times 10^5\,\text{mol}$$

$$\text{Mass of Zn} = 1.23 \times 10^5 \times 65.4 \times 10^{-6}\,\text{tonne}$$

$$\text{Mass of Zn} = 8.04\,\text{tonne}$$

CHECKPOINT 3G: MASSES OF REACTING SOLIDS

1. What mass of pure aluminium oxide must be electrolysed to give 50 tonnes of aluminium?

2. The sulphur present in 0.1000 g of an organic compound is converted into barium sulphate. A precipitate of 0.1852 g of dry $BaSO_4$ is obtained. Calculate the percentage by mass of sulphur in the compound.

3. What is the maximum mass of 2,4,6-trichlorophenol, $C_6H_2Cl_3OH$, that can be obtained from 10.00 g of phenol, C_6H_5OH? A chemist who carried out this conversion obtained 19.54 g of the product. What percentage yield did he obtain?

4. What is the maximum mass of N-benzoylphenylamine, $C_6H_5NHCOC_6H_5$, that can be obtained from 1.00 g of phenylamine, $C_6H_5NH_2$? A chemist who made this derivative obtained 2.04 g. What percentage yield did she obtain?

3.11 EQUATIONS FOR REACTIONS OF GASES

The volume of 1 mole of any ideal gas is the same

For reactions of gases, it is more usual to consider the volumes of reactants and products, rather than their masses. The volume of 1 mole of any ideal gas is the same; 22.414 dm^3 at 0 °C and 1 atm (standard temperature and pressure). The gas molar volume is 22.414 dm^3 at stp.

3.12 EQUATIONS FOR REACTIONS OF SOLIDS AND GASES

When a reaction involves both solids and gases, the solids are usually measured by mass and the gases by volume.

A worked example of a reaction of solids and gases

Example What mass of potassium chlorate(V) must be decomposed to supply 200 cm^3 of oxygen (measured at stp)? In the presence of a catalyst, decomposition proceeds according to the equation

$$2KClO_3(s) \xrightarrow{MnO_2} 2KCl(s) + 3O_2(g)$$

Method From the equation you can see that

2 mol of $KClO_3$ give 3 mol of O_2

Molar mass of $KClO_3$ = 92.5 g mol^{-1}

Therefore $2 \times 92.5\,\text{g}\,KClO_3 \rightarrow 3 \times 22.4\,\text{dm}^3\,O_2$

To supply 200 cm^3 O_2 you need $\dfrac{2 \times 92.5}{3 \times 22.4} \times 200 \times 10^{-3}\,\text{g}\,KClO_3$

Mass of $KClO_3$ decomposed = 0.551 g

CHECKPOINT 3H: REACTING VOLUMES OF GASES

1. What volume of hydrogen is formed when 3.00 g of magnesium react with an excess of dilute sulphuric acid?

2. Carbon dioxide is obtained by the fermentation of glucose:

$$C_6H_{12}O_6(aq) \rightarrow 6CO_2(g) + 6H_2O(l)$$

If 20.0 dm^3 of carbon dioxide (at stp) are collected, what mass of glucose has reacted?

3. In the preparation of hydrogen chloride by the reaction

$$NaCl(s) + H_2SO_4(l) \rightarrow HCl(g) + NaHSO_4(s)$$

what masses of sodium chloride and sulphuric acid are required for the production of 10.0 dm^3 of hydrogen chloride (at stp)?

3.13 CONCENTRATION

The concentration of a solution can be stated...

One way of stating the concentration of a solution is to state the **mass** of solute present in 1 cubic decimetre of solution, e.g. grams per cubic decimetre ($g\,dm^{-3}$). There is another method which is more convenient when it comes to chemical reactions. This is to state the **amount** in moles of a solute present in 1 dm^3 of solution.

...either in grams of solute per cubic decimetre of solution ($g\,dm^{-3}$)...

If 1 mole of solute is present in 1 dm^3 of solution, the concentration of solute is 1 mole per dm^3 (1 $mol\,dm^{-3}$). The solution is a 1 $mol\,dm^{-3}$ solution or, for short, a 1 M solution (see Figure 3.4).

...or in moles of solute per cubic decimetre of solution ($mol\,dm^{-3}$)

2 moles of solute in 1 dm^3 of solution: concentration = 2 $mol\,dm^{-3}$ or 2 M.

2 moles of solute in 250 cm^3 of solution: concentration = 8 $mol\,dm^{-3}$ or 8 M.

1.5 moles of solute in 250 cm^3: concentration = 6 $mol\,dm^{-3}$ or 6 M.

One cubic decimetre = 1000 cubic centimetres.

A cubic decimetre is also known as a litre, l.

$$1\,dm^3 = 1000\,cm^3 = 1\text{ litre}$$

FIGURE 3.4
Solutions of Known Concentration

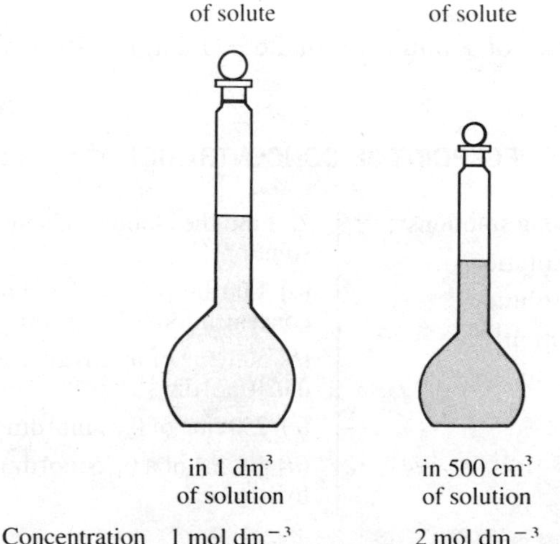
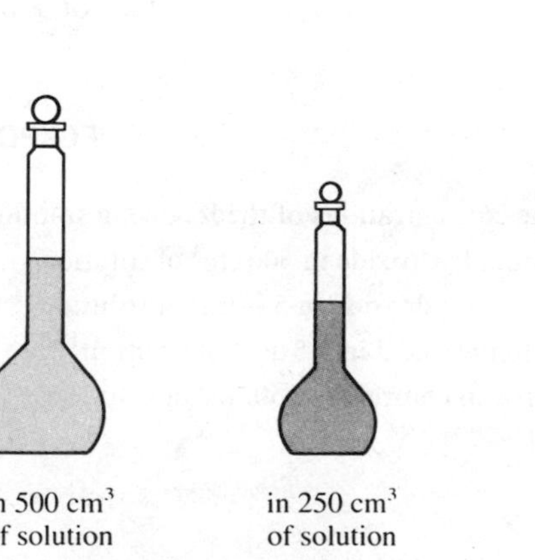

A standard solution is a solution of known concentration

A solution of known concentration is called a **standard solution**. In strict SI units, concentration is expressed in $mol\,m^{-3}$ ($1\,m^3 = 10^3\,dm^3$).

$$\text{Concentration in moles per litre} = \frac{\text{Amount of solute in moles}}{\text{Volume of solution in dm}^3}$$

Rearranging,

$$\underset{(\text{mol})}{\text{Amount of solute}} = \underset{(\text{dm}^3)}{\text{Volume of solution}} \times \underset{(\text{mol dm}^{-3})}{\text{Concentration}}$$

Example Calculate the amount of solute present in $250\,cm^3$ of a solution of hydrochloric acid which has a concentration of $2.0\,mol\,dm^{-3}$.

Method

$$\text{Amount (mol)} = \text{Volume (dm}^3) \times \text{Concentration (mol dm}^{-3})$$

$$\text{Amount of solute, HCl} = 250 \times 10^{-3}\,\text{dm}^3 \times 2.0\,\text{mol dm}^{-3}$$

$$= 0.50\,\text{mol}$$

Note that when you are given the volume in cm^3, you have to change it into dm^3.

Then the units are coherent:

$$\text{Amount (mol)} = \text{Volume (dm}^3) \times \text{Concentration (mol dm}^{-3})$$

Example What mass of sodium carbonate must be dissolved in $1\,dm^3$ of solution to give a solution of concentration 1.5 M (a 1.5 M solution)?

Method

$$\text{Amount (mol)} = \text{Volume (dm}^3) \times \text{Concentration (mol dm}^{-3})$$

$$= 1.00\,\text{dm}^3 \times 1.5\,\text{mol dm}^{-3}$$

$$= 1.5\,\text{mol}$$

$$\text{Molar mass of sodium carbonate, } Na_2CO_3 = (2 \times 23) + 12 + (3 \times 16)$$

$$= 106\,\text{g mol}^{-1}$$

$$\text{Mass of sodium carbonate} = 1.5\,\text{mol} \times 106\,\text{g mol}^{-1} = 159\,\text{g}$$

CHECKPOINT 3I: CONCENTRATION

1. Calculate the concentrations of the following solutions:
(*a*) 4.0 g of sodium hydroxide in $500\,cm^3$ of solution
(*b*) 7.4 g of calcium hydroxide in $5.0\,dm^3$ of solution
(*c*) 49.0 g of sulphuric acid in $2.5\,dm^3$ of solution
(*d*) 73 g of hydrogen chloride in $250\,cm^3$ of solution

2. Find the amount of solute present in the following solutions:
(*a*) $1.00\,dm^3$ of a solution of sodium hydroxide of concentration $0.25\,mol\,dm^{-3}$
(*b*) $500\,cm^3$ of hydrochloric acid of concentration $0.020\,mol\,dm^{-3}$
(*c*) $250\,cm^3$ of $0.20\,mol\,dm^{-3}$ sulphuric acid
(*d*) $10\,cm^3$ of a $0.25\,mol\,dm^{-3}$ solution of potassium hydroxide

3.13.1 PREPARING A STANDARD SOLUTION BY WEIGHING

A standard solution is made from a primary standard

Now you know how to calculate the mass of solid which you need to make a standard solution. A standard solution can only be made from a solid which can be obtained 100% (almost) pure. Anhydrous sodium carbonate and sodium hydrogencarbonate can be used to make standard solutions. They are called **primary standards**. Ethanedioic acid, $C_2H_2O_4$, and butanedioic acid, $C_4H_6O_4$, are primary standards which can be used to make standard acid solutions. For other substances, a solution of approximately known concentration is made and then the solution is standardised against a primary standard. You could not make a standard solution of sodium hydroxide. As you were weighing it out, it would absorb water vapour from the air and react with carbon dioxide in the air. You would have to make a solution of approximately known concentration and titrate it against, for example, a standard solution of ethanedioic acid to find its exact concentration.

METHOD OF PREPARING A STANDARD SOLUTION OF SODIUM CARBONATE

A known mass of the primary standard is dissolved in distilled water . . .

1. Calculate the mass of sodium carbonate needed, m_1.

2. Weigh a clean weighing bottle, and record its mass, m_2. [See Figure 3.5(a).] With a clean spatula, add pure anhydrous sodium carbonate until the combined mass of weighing bottle and sodium carbonate is $m_1 + m_2$.

3. Transfer the sodium carbonate carefully into a clean beaker. [See Figure 3.5(b).] Use a wash bottle of distilled water so that all the washings run into the beaker. Add about 100 cm^3 of distilled water. Stir with a glass rod until all the solid has dissolved [Figure 3.5(c)].

FIGURE 3.5 Preparing a Standard Solution

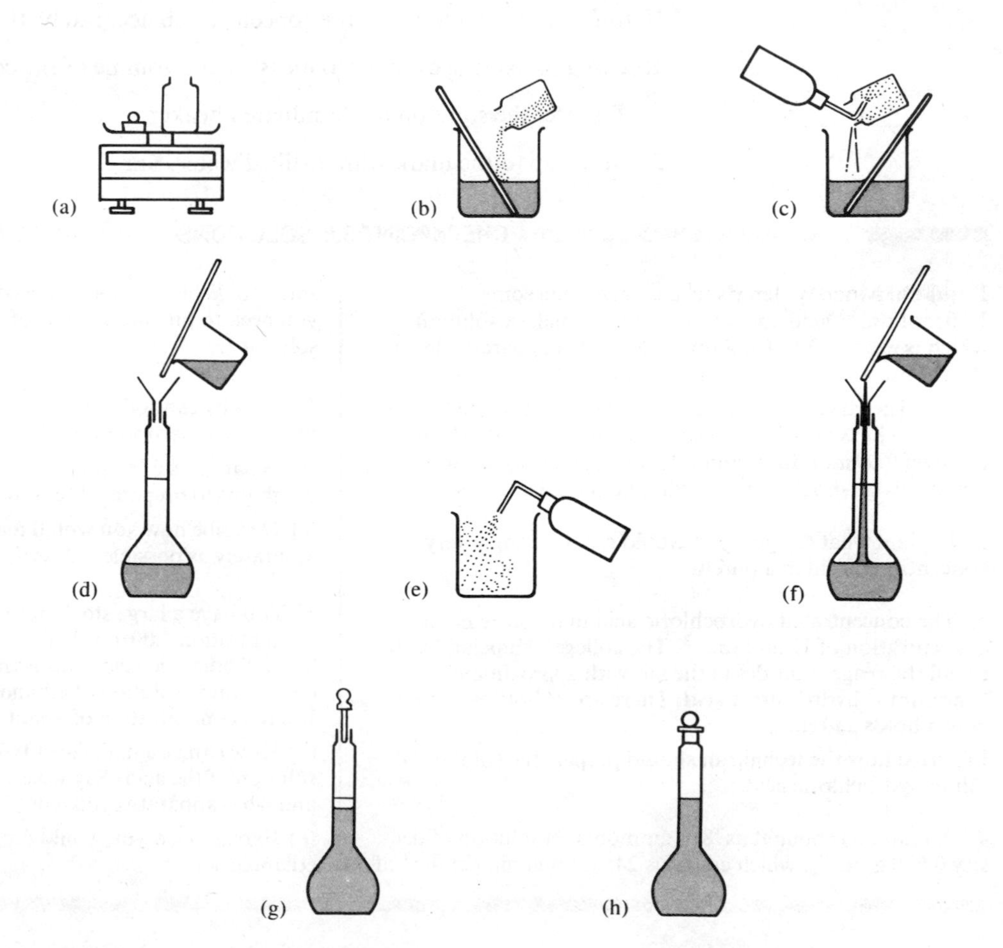

4. Pour all the solution carefully through a filter funnel into a graduated flask [Figure 3.5(d)]. Wash all the solution out of the beaker and off the glass rod [Figure 3.5(e), (f)].

... and the volume of the solution is made up to a known volume

5. Add distilled water until the level is about 2 cm below the graduation mark on the graduated flask. Add the rest of the distilled water drop by drop from a dropping pipette until the bottom of the meniscus is level with the graduation mark when viewed at eye level [Figure 3.5(g)]. Insert the stopper of the flask and invert the flask several times to mix the solution [Figure 3.5(h)].

3.13.2 PREPARING A STANDARD SOLUTION BY DILUTION

A standard solution can be made by diluting a more concentrated standard solution

You can prepare a dilute standard solution by diluting a more concentrated standard solution in a measured way.

If you want to know the concentration accurately, you use a burette and a volumetric flask. You would not be able to do this with a very concentrated solution, e.g. concentrated sulphuric acid or glacial ethanoic acid. You could use the method to prepare, for instance, a 0.1 mol dm^{-3} solution from a 2 mol dm^{-3} solution. The steps you would follow are:

1. Fill a clean, dry burette with the more concentrated standard solution.

2. Run the calculated volume of the more concentrated solution into a volumetric flask.

3. Make the solution up to the mark with distilled water. Shake.

(*Note* You cannot use very concentrated acids and alkalis in burettes.)

If you do not need to know the concentration accurately, the steps to follow are:

1. Use a measuring cylinder to measure the volume of the concentrated solution.

2. Transfer the solution to a graduated beaker.

3. Make up to the mark with distilled water. Stir.

CHECKPOINT 3J: SOLUTIONS

1. (*a*) On Monday, Jerry's teacher gives him some 1.00 mol dm^{-3} acid and instructs him to make a solution which is exactly 0.100 mol dm^{-3}. Say what apparatus Jerry should use and describe what he should do.

(*b*) On Tuesday, Jerry is given the same 1.00 mol dm^{-3} acid. This time he is asked to prepare quickly a solution which is between 0.09 mol dm^{-3} and 0.11 mol dm^{-3}. Say what apparatus he should use and what he should do.

2. Explain what dangers you would risk by using a very concentrated acid in a burette.

3. The concentrated hydrochloric acid in the store has a concentration of 12 mol dm^{-3}. The college technician has to fill all the reagent bottles in the lab with approximately 2 mol dm^{-3} hydrochloric acid. There are 24 bottles, each of which holds 250 cm^3.

Describe how the technician should prepare the solution of dilute hydrochloric acid.

4. Ammonia is bought as '880 ammonia' (a solution of density 0.880 g cm^{-3}), which contains 245 g ammonia per dm^3 of solution. What volume of the concentrated solution would you need to prepare 1.0 dm^3 of 2.0 mol dm^{-3} ammonia solution?

5. (*a*) Why can sodium hydrogencarbonate be used to prepare standard solutions?

(*b*) What mass of sodium hydrogencarbonate would you weigh out to prepare 500 cm^3 of a 0.0100 M solution?

(*c*) Describe how you would make up the solution as accurately as possible.

6. You have a large stock bottle of ethanoic acid of concentration 4.00 mol dm^{-3}. You also have a large bottle of 'glacial' ethanoic acid. This is the name given to a concentrated solution of ethanoic acid which freezes at 10 °C. It has a concentration of about 17 mol dm^{-3}.

(*a*) How can you make up 1.00 dm^3 of a 0.250 mol dm^{-3} solution of the acid? Say what quantities you would measure and what apparatus you would use.

(*b*) Explain how you would prepare 2 dm^3 of 2 mol dm^{-3} ethanoic acid.

3.14 VOLUMETRIC ANALYSIS

The concentration of a solution can be found by volumetric analysis

The method of titration is used...

...for example an acid of unknown concentration is titrated against a measured volume of a standard solution of a base

Volumetric analysis is a means of finding the concentration of a solution. The method is to add a solution of, say, an acid to a solution of, say, a base, in a measured way until there is just enough of the acid to neutralise the base. This method is called **titration** [see Figures 3.6 and 3.7]. The concentration of one of the two solutions must be known, and the volumes of both must be measured. You can use a standard solution of a base to find out the concentration of a solution of an acid. You have to find out what volume of the acid solution of unknown concentration is needed to neutralise a known volume, usually 25.0 cm^3, of the standard solution of a base. An indicator tells when exactly the right volume of solution has been added to achieve neutralisation. You will learn titration in your laboratory periods. Here is a reminder of the practical details:

1. Use a pipette to deliver 25.0 cm^3 of the alkali solution into a clean conical flask [see Figure 3.6]. Add a few drops of indicator.

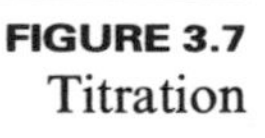

FIGURE 3..6
Using a Pipette

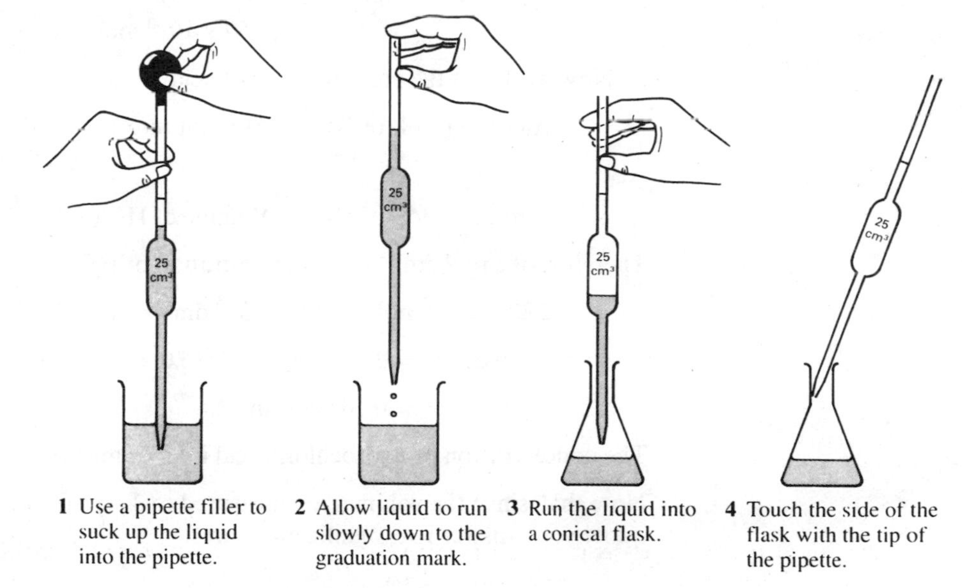

25 cm³
25 cm³
25 cm³
25 cm³
1 Use a pipette filler to suck up the liquid into the pipette.
2 Allow liquid to run slowly down to the graduation mark.
3 Run the liquid into a conical flask.
4 Touch the side of the flask with the tip of the pipette.

FIGURE 3.7
Titration

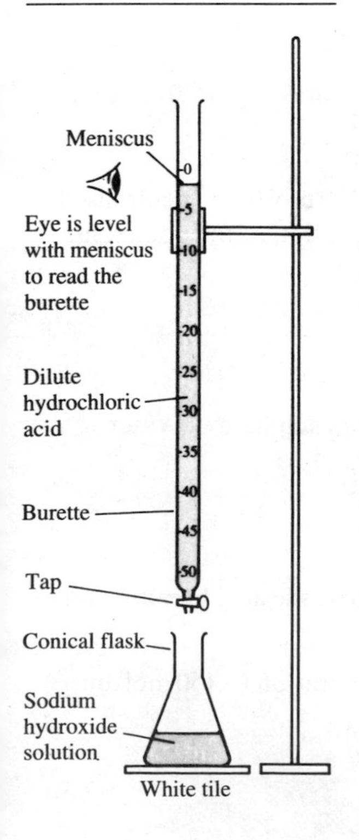

2. Wash the burette with a little of the acid solution. Allow the solution to run into the tip of the burette. Read the burette (V_1 cm^3, the bottom of the meniscus) [see Figure 3.7].

3. Arrange the apparatus as shown in Figure 3.7. Run the acid solution from the burette dropwise. Use your left hand to open the tap and your right hand to swirl the conical flask (unless you are left-handed). Stop when the indicator just changes colour. This is the 'end-point' of the titration.

4. Read the burette again (V_2 cm^3). Subtract to find the volume of acid used, $(V_2 - V_1)$ cm^3. This 'titre' is the volume of acid needed to neutralise 25.0 cm^3 of alkali.

5. Repeat the titration. Obtain an average titre. From this volume, you can calculate the unknown concentration.

The method of calculating concentration from the results of titration

Example By titration, you find that $15.0\,cm^3$ of hydrochloric acid neutralise $25.0\,cm^3$ of a $0.100\,mol\,dm^{-3}$ solution of sodium hydroxide. What is the concentration of hydrochloric acid?

Method

1. The equation for the reaction,

Hydrochloric acid + Sodium hydroxide → Sodium chloride + Water

$$HCl(aq) + NaOH(aq) \rightarrow NaCl(aq) + H_2O(l)$$

tells you that 1 mole of HCl neutralises 1 mole of NaOH.

2. Now work out the amount (mol) of base. You must start with the base because you know the concentration of base, and you do not know the concentration of acid.

$$\text{Amount (mol)} = \text{Volume (dm}^3) \times \text{Concentration (mol dm}^{-3})$$

$$\text{Amount (mol) NaOH} = \text{Volume (25.0) cm}^3 \times \text{Concentration (0.100 mol dm}^{-3})$$

$$= 25.0 \times 10^{-3}\,dm^3 \times 0.100\,mol\,dm^{-3}$$

$$= 2.50 \times 10^{-3}\,mol$$

3. Now work out the concentration of acid.

$$\text{Amount (mol) of HCl} = \text{Amount (mol) of NaOH} = 2.50 \times 10^{-3}\,mol$$

Also

$$\text{Amount (mol) of HCl} = \text{Volume of HCl(aq)} \times \text{Concentration of HCl(aq)}$$

Therefore, if $c\,mol\,dm^{-3}$ is the concentration of HCl,

$$25.0 \times 10^{-3}\,mol = 15.0 \times 10^{-3}\,dm^3 \times c\,mol\,dm^{-3}$$

$$c\,mol\,dm^{-3} = 2.50 \times 10^{-3}\,mol / 15.0 \times 10^{-3}\,dm^{-3}$$

$$= 0.167\,mol\,dm^{-3}$$

The concentration of hydrochloric acid is $0.167\,mol\,dm^{-3}$.

Note that, since the volumes are measured to 3 significant figures, e.g. $15.0\,cm^3$, you quote your answer to 3 significant figures.

A second example

Example $25.0\,cm^3$ of sulphuric acid of concentration $0.150\,mol\,dm^{-3}$ neutralised $31.2\,cm^3$ of potassium hydroxide solution. Find the concentration of the potassium hydroxide solution.

Method

1. The equation,

Sulphuric acid + Potassium hydroxide → Potassium sulphate + Water

$$H_2SO_4(aq) + 2KOH(aq) \rightarrow K_2SO_4(aq) + 2H_2O(l)$$

tells you that 1 mole of H_2SO_4 neutralises 2 moles of KOH.

2. Now work out the amount (mol) of acid. You must choose the acid because you do not know the concentration of the base.

$$\text{Amount (mol) acid} = \text{Volume (25.0 cm}^3) \times \text{Concentration (0.150 mol dm}^{-3})$$

$$= 25.0 \times 10^{-3}\,dm^3 \times 0.150\,mol\,dm^{-3}$$

$$= 3.75 \times 10^{-3}\,mol$$

3. Now work out the concentration of base.

$$\text{Amount (mol) of KOH} = 2 \times \text{Amount (mol) of } H_2SO_4$$
$$= 7.50 \times 10^{-3}\,\text{mol}$$

Also

$$\text{Amount (mol) of KOH} = \text{Volume of KOH(aq)} \times \text{Concentration of KOH(aq)}$$

Therefore, if $c\,\text{mol dm}^{-3}$ is the concentration of KOH,

$$c\,\text{mol dm}^{-3} = 7.50 \times 10^{-3}\,\text{mol}/31.2 \times 10^{-3}\,\text{dm}^3$$
$$= 0.240\,\text{mol dm}^{-3}$$

The concentration of potassium hydroxide is $0.240\,\text{mol dm}^{-3}$.

CHECKPOINT 3K: TITRATION

1. $25.0\,\text{cm}^3$ of sodium hydroxide solution are neutralised by $15.0\,\text{cm}^3$ of a solution of hydrochloric acid of concentration $0.25\,\text{mol dm}^{-3}$. Find the concentration of the sodium hydroxide solution.

2. A solution of sodium hydroxide contains $10\,\text{g dm}^{-3}$.
(*a*) What is the concentration of the solution in mol dm^{-3}?
(*b*) What volume of this solution would be needed to neutralise $25.0\,\text{cm}^3$ of $0.10\,\text{mol dm}^{-3}$ hydrochloric acid?

3. $25.0\,\text{cm}^3$ of hydrochloric acid are neutralised by $20.0\,\text{cm}^3$ of a solution of $0.15\,\text{mol dm}^{-3}$ sodium carbonate solution.
(*a*) How many moles of sodium carbonate are neutralised by 1 mol HCl?
(*b*) What is the concentration of the hydrochloric acid?

4. The sixth form decide to test some antacid indigestion tablets. They obtain the results shown in the table by dissolving tablets and titrating the alkali in them against a standard acid.

Brand	*Price (£) of 100 tablets*	*Volume (cm^3) of $0.01\,mol\,dm^{-3}$ acid required to neutralise 1 tablet*
Stopit	0.91	2.8
Setlit	1.04	3.0
Mendit	1.30	3.3
Basit	1.56	3.6

(*a*) Which antacid tablets offer the best value for money?
(*b*) What other factors would you consider before choosing a brand?

5. A tanker of acid is emptied into a water supply by mistake. A water company chemist titrates the water and finds that $10.0\,\text{dm}^3$ of water are needed to neutralise $10.0\,\text{cm}^3$ of a $0.010\,\text{mol dm}^{-3}$ solution of sodium hydroxide. What is the concentration of acid in the water?

3.14.1 BACK-TITRATION

In back-titration the amount of reactant unused at the end of the reaction is found

In the technique known as **back-titration**, a known excess of one reagent **A** is allowed to react with an unknown amount of **B**. At the end of the reaction, the amount of **A** that remains is found by titration. A simple calculation gives the amount of **A** that has been used and the amount of **B** that has reacted.

Example A sample containing ammonium chloride was warmed with $100\,\text{cm}^3$ of $1.00\,\text{mol dm}^{-3}$ sodium hydroxide solution. After all the ammonia had been driven off, the excess of sodium hydroxide required $50.0\,\text{cm}^3$ of $0.250\,\text{mol dm}^{-3}$ sulphuric acid for neutralisation. What mass of ammonium chloride did the sample contain?

Method The two reactions which have taken place are

A worked example of back-titration

(*a*) The reaction between the ammonium salt and alkali:

$$NH_4^+(aq) + OH^-(aq) \rightarrow NH_3(g) + H_2O(l) \qquad [1]$$

(*b*) The neutralisation of the excess alkali:

$$2NaOH(aq) + H_2SO_4(aq) \rightarrow Na_2SO_4(aq) + 2H_2O(l) \quad [2]$$

Amount of NaOH initially present = $100 \times 10^{-3} \times 1.00$ = 0.100 mol

The amount of NaOH left unused is titrated against sulphuric acid.

Amount of H_2SO_4 = $50.0 \times 10^{-3} \times 0.250$ mol = 0.0125 mol

From equation [2]

Amount of NaOH = 2 × Amount of H_2SO_4 = 0.0250 mol

Amount of NaOH used in reaction = Initial amount − Amount left over

= 0.100 − 0.0250 mol = 0.0750 mol

From equation [1]

Amount of NH_4Cl = Amount of NaOH = 0.0750 mol

Mass of ammonium chloride = 0.0750×53.5 = 4.01 g

CHECKPOINT 3L: TITRATIONS

1. A solution is made by dissolving 5.00 g of impure sodium hydroxide in water and making it up to 1.00 dm^3 of solution. 25.0 cm^3 of this solution is neutralised by 30.3 cm^3 of hydrochloric acid, of concentration 0.102 mol dm^{-3}. Calculate the percentage purity of the sodium hydroxide.

2. Sodium carbonate crystals (27.8230 g) were dissolved in water and made up to 1.00 dm^3. 25.0 cm^3 of the solution were neutralised by 48.8 cm^3 of hydrochloric acid of concentration 0.100 mol dm^{-3}. Find *n* in the formula $Na_2CO_3 \cdot nH_2O$.

3. A fertiliser contains ammonium sulphate. A sample of 0.500 g of fertiliser was warmed with sodium hydroxide solution. The ammonia evolved was absorbed in 100 cm^3 of 0.100 mol dm^{-3} hydrochloric acid. The excess of hydrochloric acid required 55.9 cm^3 of 0.100 mol dm^{-3} sodium hydroxide for neutralisation. Calculate the percentage of ammonium sulphate in the sample.

4. When sodium hydroxide is added to copper (II) sulphate solution, a precipitate of $Cu_a(OH)_b(SO_4)_c$ is obtained. A 25.0 cm^3 portion of a 0.100 mol dm^{-3} solution of copper(II) sulphate required 3.75 cm^3 of a 1.00 mol dm^{-3} solution of sodium hydroxide to precipitate all the copper ions.

(*a*) Find the ratio, moles Cu^{2+} : moles OH^- in the precipitate,

(*b*) By considering the charges on the ions, find the simplest formula of $Cu_a(OH)_b(SO_4)_c$,

(*c*) Write an equation for the reaction between copper(II) sulphate and sodium hydroxide.

*See Footnote.

3.15 EQUATIONS FOR OXIDATION–REDUCTION REACTIONS

3.15.1 REDOX REACTIONS

Oxidising agents accept electrons . . .

. . . reducing agents donate electrons

Oxidising agents are substances which can accept electrons from other substances. **Reducing agents** are substances which can give electrons to other substances. **Oxidation** and **reduction** occur together. In an **oxidation–reduction** reaction or **redox** reaction, electrons pass from the reducing agent to the oxidising agent.

Iron(II) ions are reducing agents, losing electrons to form iron(III) ions:

$$Fe^{2+}(aq) \rightarrow Fe^{3+}(aq) + e^- \quad [1]$$

*For further practice, see E N Ramsden, *Calculations for A-Level Chemistry* (Stanley Thornes)

The half-reaction equations for the oxidising agent and the reducing agent combine to give the equation for the redox reaction

Chlorine is an oxidising agent, accepting electrons to form chloride ions:

$$Cl_2(aq) + 2e^- \rightarrow 2Cl^-(aq) \qquad [2]$$

Equations [1] and [2] are described as **half-reaction equations**. Free electrons never occur under ordinary laboratory conditions. To represent the real process, the half-reaction equations must be combined. If equation [1] is multiplied by 2 and added to equation [2], the result is

$$2Fe^{2+}(aq) + Cl_2(aq) + 2e^- \rightarrow 2Fe^{3+}(aq) + 2Cl^-(aq) + 2e^-$$

or, as the electrons cancel out

$$2Fe^{2+}(aq) + Cl_2(aq) \rightarrow 2Fe^{3+}(aq) + 2Cl^-(aq)$$

The technique of combining the half-reaction equations for the **oxidant** and the **reductant** is useful because often the equations for redox are more complicated than this example.

REACTION BETWEEN MANGANATE(VII) AND IRON(II)

Obtaining the equation for the redox reaction between acidic MnO_4^- and Fe^{2+}

Potassium manganate(VII), $KMnO_4$, in acidic solution is a powerful oxidising agent, widely used in titrimetric analysis. The sudden change from purple MnO_4^- ions to pale pink Mn^{2+} ions at the end-point means that no indicator is needed. In balancing the half-reaction equation, $8H^+(aq)$ are needed to combine with 4O in MnO_4^-:

$$MnO_4^-(aq) + 8H^+(aq) \rightarrow Mn^{2+}(aq) + 4H_2O(l)$$

The charge on the left-hand side (LHS) = −1 + 8 = +7 units.
The charge on the RHS = +2 units.
To equalise the charge on both sides of the equation, 5 electrons are needed on the LHS:

$$MnO_4^-(aq) + 8H^+(aq) + 5e^- \rightarrow Mn^{2+}(aq) + 4H_2O(l) \qquad [3]$$

Potassium manganate(VII) in acidic solution oxidises iron(II) ions. The equation for the reaction is obtained by combining half-reaction equations [1] and [3]. Since Fe^{2+} gives one electron and MnO_4^- needs five, equation [1] is multiplied by 5 and then added to equation [3]:

$$MnO_4^-(aq) + 8H^+(aq) + 5Fe^{2+}(aq) \rightarrow Mn^{2+}(aq) + 4H_2O(l) + 5Fe^{3+}(aq)$$

REACTION BETWEEN DICHROMATE(VI) AND ETHANEDIOATE

Potassium dichromate(VI), $K_2Cr_2O_7$, is an oxidising agent which is most effective in acidic solution. It is reduced to a chromium(III), Cr^{3+}, salt. Balancing the half-reaction equation with respect to mass gives

The redox reaction between acidic $Cr_2O_7^{2-}$ and $C_2O_4^{2-}$

$$Cr_2O_7^{2-}(aq) + 14H^+(aq) \rightarrow 2Cr^{3+}(aq) + 7H_2O(l)$$

Balancing the equation with respect to charge gives

$$Cr_2O_7^{2-}(aq) + 14H^+(aq) + 6e^- \rightarrow 2Cr^{3+}(aq) + 7H_2O(l) \qquad [4]$$

A check shows that the charge on the LHS = −2 + 14 − 6 = +6, and the charge on the RHS = +6 units.

Sodium ethanedioate and ethanedioic acid are oxidised to carbon dioxide. The half-reaction equation is

$$\begin{matrix} CO_2^- \\ | \\ CO_2^- \end{matrix}(aq) \rightarrow 2CO_2(g) + 2e^- \qquad [5]$$

To obtain the equation for the redox reaction between acidified potassium dichromate and sodium ethanedioate, equation [5] is multiplied by 3 and added to equation [4]:

$$Cr_2O_7^{2-}(aq) + 14H^+(aq) + 3C_2O_4^{2-}(aq) \rightarrow 2Cr^{3+}(aq) + 7H_2O(l) + 6CO_2(g)$$

REACTION BETWEEN IODINE AND SODIUM THIOSULPHATE

Obtaining the equation for the reaction between I_2 and $S_2O_3^{2-}$

Sodium thiosulphate(VI), $Na_2S_2O_3$, is a reducing agent. It is most often used in titrimetric analysis for reducing iodine to iodide ions, being oxidised in the process to sodium tetrathionate, $Na_2S_4O_6$. When the brown colour of iodine fades as the end-point approaches, a little starch solution is added. This gives an intense blue colour with even a trace of iodine. At the end-point the blue colour vanishes. The two half-reaction equations are

$$2S_2O_3^{2-}(aq) \rightarrow S_4O_6^{2-}(aq) + 2e^- \quad [6]$$

$$I_2(aq) + 2e^- \rightarrow 2I^-(aq) \quad [7]$$

Combining the two half-reaction equations gives

$$2S_2O_3^{2-}(aq) + I_2(aq) \rightarrow S_4O_6^{2-}(aq) + 2I^-(aq)$$

CHECKPOINT 3M: REDOX REACTIONS

1. Write balanced half-reaction equations for the oxidation of each of the following:

(*a*) $Sn^{2+}(aq) \rightarrow Sn^{4+}(aq)$

(*b*) $Cl^-(aq) \rightarrow Cl_2(aq)$

(*c*) $H_2S(aq) \rightarrow S(s) + H^+(aq)$

(*d*) $SO_3^{2-}(aq) \rightarrow SO_4^{2-}(aq) + H^+(aq)$

(*e*) $H_2O_2(aq) \rightarrow O_2(g) + H^+(aq)$

(*f*) $NO_2^-(aq) + H_2O(l) \rightarrow NO_3^-(aq) +$

(*g*) $MnO_2(s) \rightarrow MnO_4^-(aq) + H^+(aq)$

Check that the equations are balanced with respect to charge as well as mass. Remember that H_2O is present in all solutions; you will need it to balance some of the equations.

2. Write balanced half-reaction equations for the following reductions:

(*a*) $Br_2(aq) \rightarrow Br^-(aq)$

(*b*) $MnO_2(s) + H^+(aq) \rightarrow Mn^{2+}(aq)$

(*c*) $NO_3^-(aq) + 10H^+(aq) \rightarrow NH_4^+(aq)$

(*d*) $PbO_2(s) + H^+(aq) \rightarrow Pb^{2+}(aq)$

(*e*) $IO^-(aq) + H^+(aq) \rightarrow I_2(aq)$

(*f*) $ClO_3^-(aq) + H^+(aq) \rightarrow Cl_2(aq)$

Balance the equations for mass, using H_2O from the solution if needed, and then balance with respect to charge.

3. By combining half-reaction equations, write balanced equations for the following reactions:

(*a*) $Fe^{3+}(aq) + I^-(aq) \rightarrow$

(*b*) $Fe^{3+}(aq) + Sn^{2+}(aq) \rightarrow$

(*c*) $MnO_4^-(aq) + Cl^-(aq) + H^+(aq) \rightarrow$

(*d*) $MnO_4^-(aq) + H_2O_2(aq) + H^+(aq) \rightarrow$

(*e*) $MnO_4^-(aq) + H^+(aq) + I^-(aq) \rightarrow$

(*f*) $Cr_2O_7^{2-}(aq) + H^+(aq) + I^-(aq) \rightarrow$

(*g*) $Cr_2O_7^{2-}(aq) + H^+(aq) + Fe^{2+}(aq) \rightarrow$

(*h*) $Cr_2O_7^{2-}(aq) + H^+(aq) + NO_2^-(aq) \rightarrow$

(*i*) $MnO_4^-(aq) + H^+(aq) + Sn^{2+}(aq) \rightarrow$

(*j*) $Cr_2O_7^{2-}(aq) + H^+(aq) + SO_3^{2-}(aq) \rightarrow$

(*k*) $MnO_4^-(aq) + H^+(aq) + SO_3^{2-}(aq) \rightarrow$

(*l*) $MnO_4^-(aq) + H^+(aq) + Fe^{2+}(aq) \rightarrow$

(*m*) $Cr_2O_7^{2-}(aq) + H^+(aq) + Sn^{2+}(aq) \rightarrow$

(*n*) $PbO_2(s) + H^+(aq) + Cl^-(aq) \rightarrow$

(*o*) $ClO_3^-(aq) + H^+(aq) + I^-(aq) \rightarrow$

(*p*) $Br_2(aq) + I^-(aq) \rightarrow$

(*q*) $Cl_2(g) + NO_2^-(aq) + H_2O \rightarrow$

(*r*) $Cl_2(g) + H^+(aq) + IO^-(aq) \rightarrow$

(*s*) $MnO_2(s) + H^+(aq) + Cl^-(aq) \rightarrow$

(*t*) $Br_2(g) + H_2S(g) \rightarrow$

3.16 OXIDATION NUMBER

A method of expressing the combining power of elements is the idea of **oxidation number** or **oxidation state**.

Examples are:

- The oxidation number of sodium in Na^+ is +1.
- The oxidation number of aluminium in Al^{3+} is +3.
- The oxidation number of iodine in I^- is −1.
- The oxidation number of oxygen in O^{2-} is −2.

The use of oxidation numbers is extended to covalent compounds. Some elements are assigned positive oxidation numbers and others are assigned negative oxidation numbers in accordance with certain rules.

3.16.1 RULES FOR ASSIGNING OXIDATION NUMBERS

The oxidation number of an element in the uncombined state is zero . . .

1. The oxidation numbers of elements in their uncombined states, such as Na, Ca, Al, are zero. Similarly the oxidation numbers of iodine in I_2, oxygen in O_2 and sulphur in S_8 are zero.

. . . The oxidation number of an element in an ionic compound is equal to the charge on its ions, e.g. +1, +2, −1, −2 . . .

2. In ionic compounds the oxidation number is equal to the charge on the ion. The oxidation number of an element is not always the same. Iron has an oxidation number of +2 in Fe^{2+} and an oxidation number of +3 in Fe^{3+}.

. . . The oxidation numbers of the elements in a compound add up to zero

3. The sum of the oxidation numbers of all the atoms or ions in a compound is zero.

In NaCl,

$$(\text{Ox. No. of Na}) + (\text{Ox. No. of Cl}) = 0$$

$$(+1) + (-1) = 0$$

In Na_2O,

$$2(\text{Ox. No. of Na}) + (\text{Ox. No. of O}) = 0$$

$$2(+1) + (-2) = 0$$

In CuS,

$$(\text{Ox. No. of Cu}) + (\text{Ox. No of S}) = 0$$

$$(+2) + (-2) = 0$$

In $CaBr_2$,

$$(\text{Ox. No. of Ca}) + 2(\text{Ox. No. of Br}) = 0$$

$$(+2) + 2(-1) = 0$$

4. The sum of the oxidation numbers of all the atoms in an ion is equal to the charge on the ion. In SO_4^{2-}, the sum of the oxidation numbers (S = +6, O = −2) is

$$+6 + 4(-2) = -2$$

which is the charge on the ion.

Reference elements...

5. Some elements nearly always employ the same oxidation number in their compounds. They are used as reference points in assigning oxidation numbers to other elements. The reference elements are

K	Na	+1	H	+1	except in metal hydrides
Mg	Ca	+2	F	−1	
Al		+3	Cl	−1	except in compounds with O and F
			O	−2	except in peroxides, superoxides, fluorides

...Some worked examples

Example What is the oxidation number of thallium in $TlCl_3$?

Method Chlorine always has the oxidation number −1.

Therefore (Ox. No. of Tl) + 3(−1) = 0

and the oxidation number of thallium is +3.

Example What is the oxidation number of Cl in Cl_2O_7?

Method The exceptions to the rule that the oxidation number of Cl equals −1 are compounds with O and F. Oxygen is the reference point with the oxidation number −2.

Therefore 2 (Ox. No. of Cl) + 7(−2) = 0

and the oxidation number of chlorine is +7.

Example What is the oxidation number of Cr in $Cr(CN)_6^{3-}$?

Method The cyanide ion, CN^-, has a charge of −1.

Therefore (Ox. No. of Cr) + 6(−1) = −3

and the oxidation number of chromium is +3.

CHECKPOINT 3N: OXIDATION NUMBER

1. State the oxidation numbers of the elements in following atoms or ions:

Na, Na^+, Ba, Ba^{2+}, Rb^+, Rb, Ga, As, As^{3-}, Br^-, H_2, H^+, F_2, F^-

2. Give the oxidation numbers of the first element in each of the following compounds. Remember the oxidation numbers of the elements in a compound add up to zero. Take oxidation numbers for hydrogen (+1), oxygen (−2), fluorine (−1) and chlorine (−1) as reference points.

CuO, Cu_2O, H_2S, SO_2, SO_3, PbO, PbO_2, $AlCl_3$, SF_6, SCl_2, $TiCl_4$, V_2O_5

3. What is the oxidation number of the named element in the following species (ions or molecules)?

(a) N in NO, NO_2, N_2O_4, N_2O, NO_2^-, NO_3^-, N_2O_5

(b) Mn in $MnSO_4$, Mn_2O_3, MnO_2, MnO_4^-, MnO_4^{2-}

(c) As in As_2O_3, AsO_2^-, AsO_4^{3-}, AsH_3

(d) Cr in CrO_4^{2-}, $Cr_2O_7^{2-}$, CrO_3

(e) I in I^-, IO^-, IO_3^-, I_2, ICl_3, ICl_2^-

3.16.2 CHANGES IN OXIDATION NUMBER

Oxidation–reduction reactions are often discussed in terms of the change in oxidation number of each reactant. In the redox reaction,

$$2Fe^{2+}(aq) + I_2(aq) \rightarrow 2Fe^{3+}(aq) + 2I^-(aq)$$

When Fe^{2+} is converted into Fe^{3+}, the oxidation number increases from +2 to +3, and we say that Fe^{2+} has been oxidised to Fe^{3+}. When I_2 is converted into I^-, the oxidation number decreases from 0 in I_2 to -1 in I^-, and we say that I_2 has been reduced to I^-.

Change in Ox. No. of iron = No. of atoms × Change in Ox. No.

$= 2(+1) = +2$

Change in Ox. No. of iodine = No. of atoms × Change in Ox. No.

$= 2(-1) = -2$

Sum of changes in Ox. No. $= +2 - 2 = 0$

Oxidation numbers increase on oxidation and decrease on reduction...

In general, when an element is **oxidised**, its oxidation number increases; when an element is **reduced**, its oxidation number decreases. In a redox reaction

$$x\mathbf{A} + y\mathbf{B} \rightarrow$$

if the oxidation number of **A** changes by $+a$ units, and the oxidation number of **B** changes by $-b$ units

then $x(+a) + y(-b) = 0$

...Some worked examples

Example Consider the reduction of iron(III) ions by a tin(II) salt:

$$Sn^{2+}(aq) + 2Fe^{3+}(aq) \rightarrow Sn^{4+}(aq) + 2Fe^{2+}(aq)$$

For tin, change in Ox. No. $= +2$

For iron, change in Ox. No. $= -1$

And $1(+2) + 2(-1) = 0$

Example

$$3I_2(aq) + 3OH^-(aq) \rightarrow IO_3^-(aq) + 5I^-(aq) + 3H^+(aq)$$

The only element which changes its oxidation number is iodine.

On the LHS, in I_2 Ox. No. of I $= 0$

On the RHS, in IO_3^- (Ox. No. of I) $+ 3(-2) = -1$

and the Ox. No. of I $= +5$

In I^-, Ox. No. of I $= -1$

Iodine has changed from oxidation number zero on the LHS to a combination of Ox. No. +5 and Ox. No. −1 on the RHS. Part of the iodine has been oxidised and part has been reduced. A reaction of this kind is termed a **disproportionation reaction**.

3.16.3 BALANCING EQUATIONS BY THE OXIDATION NUMBER METHOD

Balancing equations...

The oxidation number method of balancing equations is best explained through an example.

Example Balance the equation

$$aKIO_3(aq) + bNa_2SO_3(aq) \rightarrow cKIO(aq) + dNa_2SO_4(aq)$$

Iodine changes from Ox. No. +5 in KIO_3 to +1 in KIO.

Change in Ox. No. of I = −4

Sulphur changes from Ox. No. +4 in Na_2SO_3 to +6 in Na_2SO_4.

...A worked example

Change in Ox. No. of S = +2

Therefore $a(-4) + b(+2) = 0$

If $a = 1$, $b = 2$ and the equation becomes

$$KIO_3(aq) + 2Na_2SO_3(aq) \rightarrow cKIO(aq) + dNa_2SO_4(aq)$$

By stoichiometry, it follows that $c = 1$ and $d = 2$, giving

$$KIO_3(aq) + 2Na_2SO_3(aq) \rightarrow KIO(aq) + 2Na_2SO_4(aq)$$

CHECKPOINT 3O: EQUATIONS AND OXIDATION NUMBERS

1. Use the oxidation number method to balance the equations:

(*a*) $MnO_4^-(aq) + H^+(aq) + Fe^{2+}(aq) \rightarrow Mn^{2+}(aq) + Fe^{3+}(aq) + H_2O(l)$

(*b*) $Mn^{2+}(aq) + BiO_3^-(aq) + H^+(aq) \rightarrow MnO_4^-(aq) + Bi^{3+}(aq) + H_2O(l)$

(*c*) $As_2O_3(s) + MnO_4^-(aq) + H^+(aq) \rightarrow As_2O_5(s) + Mn^{2+}(aq) + H_2O(l)$

(*d*) $Sn(s) + HNO_3(aq) \rightarrow SnO_2(s) + NO_2(g) + H_2O(l)$

(*e*) $Cu^{2+}(aq) + I^-(aq) \rightarrow CuI(s) + I_2(aq)$

(*f*) $Cl_2(g) + OH^-(aq) \rightarrow Cl^-(aq) + ClO^-(aq) + H_2O(l)$

(*g*) $Zn(s) + Fe^{3+}(aq) \rightarrow Zn^{2+}(aq) + Fe^{2+}(aq)$

(*h*) $I_2(aq) + S_2O_3^{2-}(aq) \rightarrow I^-(aq) + S_4O_6^{2-}(aq)$

2. One mole of the compound ICl_x reacts with an excess of potassium iodide solution to give two moles of I_2. Write an equation for the reaction, and state the oxidation number of I in ICl_x.

3. State the oxidation number of the species which is underlined in the following equations. Say whether the species is oxidised or reduced during the reaction. Complete and balance the equations:

(*a*) $H_2\underline{O}_2 + \underline{I}^- + H^+(aq) \rightarrow H_2\underline{O} + \underline{I}_2$

(*b*) $\underline{Cu} + \underline{N}O_3^- + H^+(aq) \rightarrow \underline{Cu}^{2+} + \underline{N}O + H_2O$

(*c*) $\underline{Fe}^{2+} + \underline{Cr}_2O_7^{2-} + H^+(aq) \rightarrow \underline{Fe}^{3+} + \underline{Cr}^{3+} + H_2O$

(*d*) $\underline{S}_2O_3^{2-} + \underline{I}_2 \rightarrow \underline{S}_4O_6^{2-} + \underline{I}^-$

(*e*) $K\underline{I}O_3 + K\underline{I} + H\underline{Cl} \rightarrow K\underline{Cl} + \underline{I}\underline{Cl} + 3H_2O$

(*f*) $\underline{Cr}O_4^{2-} + H^+(aq) \rightarrow \underline{Cr}_2O_7^{2-}$

3.17 OXIDATION NUMBERS AND NOMENCLATURE

3.17.1 SYSTEMATIC NOMENCLATURE

Oxidation numbers are used in the naming of compounds. Systematic nomenclature is set out by IUPAC (the International Union of Pure and Applied Chemistry) in their *Manual of Symbols and Terminology for Physiochemical Quantities and Units* and also by ASE (the Association for Science Education) in *Chemical Nomenclature, Symbols and Terminology* (3rd edition, 1985). The systems are not quite the same. Although the IUPAC system, coming from an international body, is more widely used, the examination boards follow the ASE system. This book, since it aims to prepare the readers for examinations, also follows the ASE system.

Systems for naming compounds

3.17.2 CATIONS

The oxidation state of the element is specified if it is variable

Cations (positive ions) are given the name of the element together with the oxidation number. This system of naming was devised by A Stock:

e.g. Fe^{2+} iron(II) ion Fe^{3+} iron(III) ion

Names of cations

When there is no doubt about the oxidation state because an element assumes one only, then it is omitted:

e.g. Na^+ sodium ion Al^{3+} aluminium ion

3.17.3 ANIONS

Elemental **anions** (negative ions) are named after the element, with the ending *-ide*:

e.g. H^- hydride N^{3-} nitride

Compound anions have names ending in *-ide*, *-ite* or *-ate*:

e.g. OH^- hydroxide NO_2^- nitrite NO_3^- nitrate

Many elements form more than one **oxoanion**, using more than one oxidation state (e.g., NO_2^-, NO_3^-). The names are derived from the name of the element which is combined with oxygen and the ending *-ate*:

e.g. SO_4^{2-} sulphate ion HCO_3^- hydrogencarbonate ion

Both ClO^- and ClO_3^- are chlorate ions. To distinguish between them, the oxidation number of chlorine is added:

Names of anions

e.g.	ClO^-	chlorate(I)	ClO_3^-	chlorate(V)
also	CrO_4^{2-}	chromate(VI)	$Cr_2O_7^{2-}$	dichromate(VI)
	MnO_4^{2-}	manganate(VI)	MnO_4^-	manganate(VII)
	NO_3^-	nitrate(V) or nitrate	NO_2^-	nitrate(III) or nitrite
	SO_4^{2-}	sulphate(VI) or sulphate	SO_3^{2-}	sulphate(IV) or sulphite

The Stock names for the last four examples have not been widely adopted, and people prefer to use nitrate, nitrite, sulphate and sulphite. These names date back to their usage before the changes in nomenclature of 1970.

3.17.4 ACIDS

Acids are named after their anions:

Names of acids

e.g. HClO chloric(I) acid

$HClO_2$ chloric(III) acid

$HClO_3$ chloric(V) acid

Again, the names nitrous acid and sulphurous acid are preferred to the Stock names (nitric(III) and sulphuric(IV)) for the acids HNO_2 and H_2SO_3.

3.17.5 SALTS

Salts are named by combining the names of the cation, with its oxidation number if that is variable, and the anion:

Names of salts

e.g. $FeSO_4$ iron(II) sulphate NaClO sodium chlorate(I)

When a salt is hydrated, the number of water molecules per formula unit is stated:

e.g. $CuSO_4 \cdot 5H_2O$ copper(II) sulphate-5-water

3.17.6 STOICHIOMETRIC FORMULAE

Some compounds are named by stoichiometry

The oxides, sulphides and halides of non-metallic elements are usually named, not by the Stock system but according to their stoichiometry:

e.g. NO	nitrogen oxide	CS_2	carbon disulphide
N_2O	dinitrogen oxide	$SiCl_4$	silicon tetrachloride
NO_2	nitrogen dioxide	$POCl_3$	phosphorus trichloride oxide
N_2O_4	dinitrogen tetraoxide	$SOCl_2$	sulphur dichloride oxide

Phosphorus compounds are sometimes named by stoichiometry, as above, and sometimes by the Stock system:

e.g. PCl_5 phosphorus pentachloride or phosphorus(V) chloride.

3.18 TITRIMETRIC ANALYSIS, USING REDOX REACTIONS

Titrimetric analysis using redox reactions...

Redox reactions are used in titrimetric analysis. For example, a solution of unknown concentration of a reductant is titrated against a standard solution of an oxidant. From the volumes of the two solutions and the equation for the reaction, the concentration of the unknown solution can be found.

Example (a) Find the concentration of an iron(II) sulphate solution, given that 25.0 cm^3 of the solution, when acidified, required 19.8 cm^3 of 0.0200 mol dm^{-3} potassium manganate(VII) for oxidation.

Method The equation for the reaction comes first. A combination of the two half-reaction equations, as described in § 3.5.11, gives

...Some worked examples

$$MnO_4^-(aq) + 5Fe^{2+}(aq) + 8H^+(aq) \rightarrow Mn^{2+}(aq) + 5Fe^{3+}(aq) + 4H_2O(l)$$

The equation indicates that 1 mol of MnO_4^- oxidises 5 mol of Fe^{2+}.

$$\text{Amount of } MnO_4^- \text{ in } 19.8\,cm^3 = 19.8 \times 10^{-3} \times 0.0200\,mol = 0.396 \times 10^{-3}\,mol$$

$$\text{Amount of } Fe^{2+} \text{ in } 25.0\,cm^3 = 5 \times \text{amount of } MnO_4^- = 1.98 \times 10^{-3}\,mol$$

$$\text{Concentration of } Fe^{2+} = (1.98 \times 10^{-3})/(25.0 \times 10^{-3})\,mol\,dm^{-3}$$

$$\text{Concentration of } FeSO_4 = 7.92 \times 10^{-2}\,mol\,dm^{-3}$$

Example (b) A standard solution is prepared by dissolving 1.185 g of 'AnalaR' potassium dichromate(VI) and making up to 250 cm^3 of solution. This solution is used to find the concentration of a sodium thiosulphate solution. A 25.0 cm^3 portion of the oxidant was acidified and added to an excess of potassium iodide to liberate iodine.

$$Cr_2O_7^{2-}(aq) + 6I^-(aq) + 14H^+(aq) \rightarrow 3I_2(aq) + 2Cr^{3+}(aq) + 7H_2O(l) \quad [1]$$

When the solution was titrated against sodium thiosulphate solution, 17.5 cm^3 of 'thio' were required. Find the concentration of the thiosulphate solution.

Method Combining the half-reaction equations

$$I_2(aq) + 2e^- \rightarrow 2I^-(aq)$$
$$2S_2O_3^{2-}(aq) \rightarrow S_4O_6^{2-}(aq) + 2e^-$$

gives

$$2S_2O_3^{2-}(aq) + I_2(aq) \rightarrow S_4O_6^{2-}(aq) + 2I^-(aq) \quad [2]$$

Concentration of $K_2Cr_2O_7$ = $(1.185/294) \times 4$ = 0.0161 mol dm^{-3}

Amount of I_2 in 25.0 cm^3 = $0.0161 \times 25.0 \times 10^{-3} \times 3$ mol
(from equation [1]) = 1.208×10^{-3} mol

Amount of thio in 17.5 cm^3 = $1.208 \times 10^{-3} \times 2$ mol
(from equation [2]) = 2.416×10^{-3} mol

Concentration of thiosulphate = $(2.416 \times 10^{-3})/(17.5 \times 10^{-3})$ mol dm^{-3}
= 0.138 mol dm^{-3}

CHECKPOINT 3P: REDOX TITRATIONS

1. A 0.1576 g piece of iron wire was converted into Fe^{2+} ions and then titrated against potassium dichromate solution of concentration 1.64×10^{-2} mol dm^{-3}. From the fact that 27.3 cm^3 of the oxidant were required, calculate the percentage purity of the iron wire.

2. A volume of 27.5 cm^3 of a 0.0200 mol dm^{-3} solution of potassium manganate(VII) was required to oxidise 25.0 cm^3 of a solution of hydrogen peroxide. Calculate the concentration of hydrogen peroxide and the volume of oxygen (at stp) evolved during the titration.

3. Calculate the percentage purity of an impure sample of sodium thiosulphate from the following data. A 0.2368 g sample of the sodium thiosulphate was added to 25.0 cm^3 of 0.0400 mol dm^{-3} iodine solution. The excess of iodine that remained after reaction needed 27.8 cm^3 of 0.0400 mol dm^{-3} thiosulphate solution in a titration.

4. What volume of potassium manganate(VII) solution of concentration 0.0100 mol dm^{-3} will oxidise 50.0 cm^3 of iron(II) ethanedioate solution of concentration 0.0200 mol dm^{-3} in acid conditions?

5. Mercury is oxidised by potassium manganate(VII) solution, with the formation of manganese(IV) oxide, MnO_2, and potassium hydroxide and an oxide of mercury. 50.0 cm^3 of 0.0200 mol dm^{-3} potassium manganate(VII) solution oxidise 0.600 g of mercury. Work out the equation for the reaction.

*See Footnote.

3.19 EQUILIBRIUM

Imagine that you are looking through the window of a popular restaurant during a busy lunchtime. You can see that all of the restaurant's 200 seats are taken. You come back 30 minutes later, and you see that all the seats are still occupied. However, you can see that people are entering the restaurant and other people are leaving the restaurant. The same situation continues over the next 2 hours. The population remains constant at 200 people while all the time people are entering and leaving the restaurant. There is a balance between the number leaving and the number arriving. This state of balance can be described as a state of **equilibrium**. If the same 200 people sat at the tables all the time, one would say that the situation was **static** (unchanging).

*For further practice, see E N Ramsden, *Calculations for A-Level Chemistry* (Stanley Thornes)

However, in the restaurant you are observing, there is motion as some customers arrive and others leave. The thing that remains constant is the balance between the number arriving and the number leaving. This is a **dynamic** (moving) equilibrium.

The restaurant you have been observing can be described as a system. The word **system** is used to describe a part of the universe which one wants to study in isolation from the rest of the universe. There are two kinds of systems: systems in a state of change and systems at equilibrium. A system in which a change in the properties of the system is occurring is described as 'a system in a state of change'. A system in which no change in its properties is occurring is described as 'a system at equilibrium'. An equilibrium may be a **static equilibrium** or a **dynamic equilibrium**.

Consider a system in which a physical change, vaporisation, occurs. Consider what happens when you drop 5 cm³ of the brown liquid, bromine, into a gas jar and replace the lid [see Figure 3.8].

FIGURE 3.8 Vaporisation of Liquid Bromine

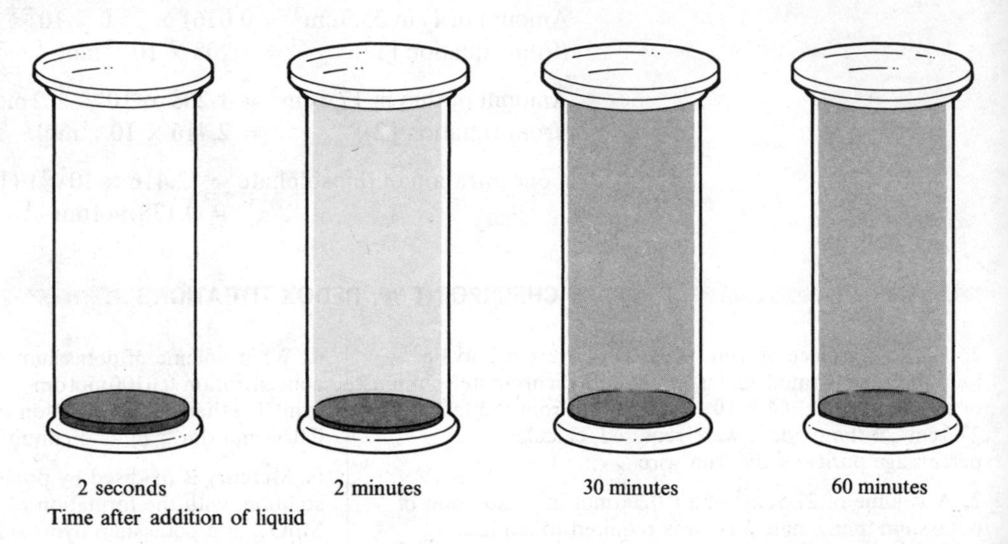

A phase is a physically distinct part of a system

As soon as the liquid enters the gas jar, it begins to **vaporise**: some molecules leave the liquid phase and enter the vapour phase. A **phase** is a part of a system which is physically distinct from other parts of the system. The system we are considering is the contents of the closed gas jar. Two phases are present: liquid (bromine) and gas (bromine vapour and air).

After 2 minutes, the gas in the gas jar is brown because it contains bromine molecules as well as air. Vaporisation (or **evaporation**) continues, and after 30 minutes the brown colour of bromine vapour is even more intense. The colour does not continue to deepen for ever. After 60 minutes, it is no more intense than after 30 minutes. It looks as though vaporisation has ceased, and the system is at equilibrium.

$Br_2(l)$ and $Br_2(g)$ reach equilibrium in a closed system...

If you could see individual molecules of bromine, however, you would see that the population of bromine molecules in the gas phase is constantly changing. Molecules of bromine are still passing from the liquid to the gas phase but, as fast as they do this, molecules of bromine pass from the gas phase to the liquid phase, that is, they **condense**. The system is at equilibrium because

$$\text{Rate of vaporisation} = \text{Rate of condensation}$$

...and the equilibrium is dynamic

This kind of system is described as being in **dynamic equilibrium**. Dynamic means *moving*, and, at a molecular level, the system is in motion. The properties of the system in bulk are unchanging; the volume of liquid bromine and the concentration of bromine in the gas phase are no longer changing:

$$Br_2(l) \rightleftharpoons Br_2(g)$$

If the system were not closed, it would not come to equilibrium. If the gas jar were open, bromine would continue to vaporise until there was no liquid bromine left.

Another physical change is dissolution (dissolving). Consider what happens when you stir a scoopful of copper(II) sulphate crystals in a beaker of water. As the salt dissolves, the solution becomes a more and more intense blue colour. [See Figure 3.9.]

FIGURE 3.9 Dissolution

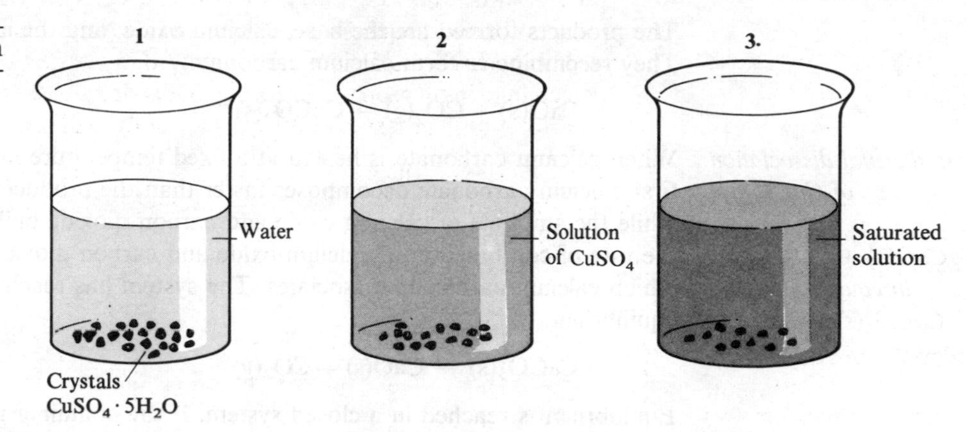
1
2
3.
Water
Solution of $CuSO_4$
Saturated solution
Crystals $CuSO_4 \cdot 5H_2O$

A saturated solution is in a state of dynamic equilibrium

After a while, the intensity of the blue colour remains constant, although (provided you have used an excess of crystals) undissolved copper(II) sulphate remains at the bottom of the beaker. The saturated solution is a system at equilibrium. Although nothing more seems to be happening, in fact copper(II) sulphate is still dissolving but, as fast as it does so, copper(II) sulphate is crystallising from solution:

$$CuSO_4 \cdot 5H_2O(s) + aq \rightleftharpoons Cu^{2+}(aq) + SO_4^{2-}(aq) + 5H_2O(l)$$

A radioactive tracer can be used to demonstrate the dynamic nature of the equilibrium

There is a way of demonstrating that this system is in dynamic equilibrium. It involves the use of a radioactive **tracer**. If some crystals of $Cu^{35}SO_4 \cdot 5H_2O$,which contain radioactive ^{35}S, are added, you might expect that none would dissolve because the solution is already saturated. After a time, however, it is found that the radioactivity is divided between the solution and the undissolved crystals. The reason is that undissolved solid is constantly dissolving, while solute crystallises from the solution at the same rate. [See Figure 3.10.]

FIGURE 3.10 An Experiment using a Radioactive Tracer

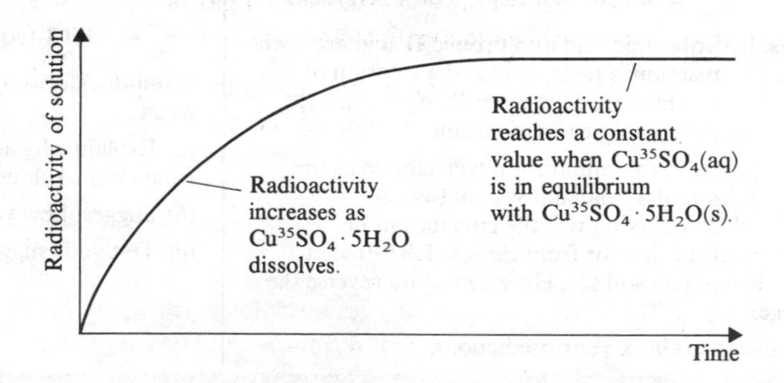
Radioactivity of solution
Time
Radioactivity increases as $Cu^{35}SO_4 \cdot 5H_2O$ dissolves.
Radioactivity reaches a constant value when $Cu^{35}SO_4(aq)$ is in equilibrium with $Cu^{35}SO_4 \cdot 5H_2O(s)$.

3.20 CHEMICAL EQUILIBRIA

The dynamic equilibria described above are physical changes. Chemical reactions can also come to equilibrium.

Chemical reactions, like physical changes, can reach a state of equilibrium

Some chemical reactions take place in one direction almost exclusively. For example, magnesium burns to form magnesium oxide:

$$2Mg(s) + O_2(g) \rightarrow 2MgO(s)$$

The tendency for magnesium oxide to split up to form magnesium and oxygen is negligible at normal temperatures.

Other chemical reactions take place in both directions at comparable rates. For example, when calcium carbonate is heated strongly, it decomposes:

$$CaCO_3(s) \rightarrow CaO(s) + CO_2(g)$$

The products formed are the base, calcium oxide, and the acid gas, carbon dioxide. They recombine to form calcium carbonate:

$$CaO(s) + CO_2(g) \rightarrow CaCO_3(s)$$

In the thermal dissociation of $CaCO_3$...

... $CaO(s) + CO_2(g)$ are in equilibrium with $CaCO_3(s)$ in a closed system...

When calcium carbonate is heated at a fixed temperature in a closed container, at first calcium carbonate decomposes faster than the products recombine. After a while the amounts of calcium oxide and carbon dioxide build up to a level at which the rate of combination of calcium oxide and carbon dioxide is equal to the rate at which calcium carbonate dissociates. The system has reached a state of dynamic equilibrium:

$$CaCO_3(s) \rightleftharpoons CaO(s) + CO_2(g)$$

... if one of the products is removed the equilibrium is disturbed

Equilibrium is reached in a closed system. If the container is open, carbon dioxide can escape. The equilibrium is disturbed, and more calcium carbonate dissociates to try to restore the equilibrium. When limestone is heated in a lime kiln, as the aim is to make plenty of quicklime, the carbon dioxide formed is removed by a powerful through draft of air in order to stop the system coming to equilibrium.

The pressure, the temperature and other external factors affect systems in equilibrium. H L Le Chatelier made a study of the way in which systems at equilibrium adjust when external factors are changed. His work is covered in Chapter 11.

CHECKPOINT 3Q: EQUILIBRIUM

1. An aqueous solution of bromine is called 'bromine water'. Some bromine molecules react with water molecules:

$$\underset{}{Br_2(aq)} + H_2O(l) \rightleftharpoons \underset{\text{Hydrobromic acid}}{HBr(aq)} + \underset{\text{Bromic(I) acid}}{HBrO(aq)}$$

The products, hydrobromic acid and bromic(I) acid are both strong acids. The reaction is reversible, and a solution of bromine in water reaches an equilibrium state in which the concentrations of all the species are constant.

Predict what change in the equilibrium will happen as the result of the addition of a small amount of sodium hydroxide. In which direction will the equilibrium be displaced, from left to right or from right to left? Predict what colour change you will see. How could you reverse the colour change?

Do an experiment to check your predictions.

2. A solution of bismuth trichloride in concentrated hydrochloric acid contains four substances: bismuth trichloride, $BiCl_3$, bismuth chloride oxide, $BiOCl$, hydrochloric acid and water. All four substances are in equilibrium:

$$BiCl_3(aq) + H_2O(l) \rightleftharpoons BiOCl(s) + 2HCl(aq)$$

Bismuth chloride oxide is a white solid which is insoluble in water.

(*a*) Explain why adding water makes the solution change from clear to cloudy.

(*b*) Suggest how you could make the solution clear again.

(*c*) Test your suggestions in practice.

3.21 EQUILIBRIUM CONSTANTS

The equilibrium constant is a measure of the extent of reaction

The extent to which the reactants are converted into the products before equilibrium is reached is measured by the **equilibrium constant** for the reaction. An example is the reaction between ethanoic acid and ethanol:

$$CH_3CO_2H(l) + C_2H_5OH(l) \rightleftharpoons CH_3CO_2C_2H_5(l) + H_2O(l)$$

The equilibrium constant K_c is given by

$$K_c = \frac{[CH_3CO_2C_2H_5][H_2O]}{[CH_3CO_2H][C_2H_5OH]}$$

where $[C_2H_5OH]$ is the concentration of ethanol in mol dm^{-3}. In general, in an equilibrium

$$m\mathbf{A} + n\mathbf{B} \rightleftharpoons p\mathbf{C} + q\mathbf{D}$$

$$K_c = \frac{[\mathbf{C}]^p[\mathbf{D}]^q}{[\mathbf{A}]^m[\mathbf{B}]^n}$$

The convention is to put the products in the numerator and the reactants in the denominator.

$$\text{Equilibrium constant} = \frac{\text{Product of the concentrations of the products raised to the appropriate powers}}{\text{Product of the concentrations of the reactants raised to the appropriate powers}}$$

The appropriate power is the coefficient before that species in the stoichiometric equation for the reaction.

3.21.1 CATALYSTS AND EQUILIBRIUM CONSTANTS

Catalysts increase the speed with which the equilibrium conditions are reached

Catalysts alter the rates of chemical reactions. In the case of a reaction which reaches a state of dynamic equilibrium, a catalyst increases the rates of both the forward reaction and the reverse reaction by the same ratio. The position of equilibrium is therefore unchanged. The equilibrium constant remains the same. What the catalyst does is to decrease the time needed for the system to reach a state of equilibrium. In industrial processes, a catalyst can make a valuable contribution to the economy of the process. In the Haber process, the percentage conversion of nitrogen and hydrogen to ammonia at the temperatures at which plants operate is small. The use of a catalyst to achieve the same percentage conversion in a shorter time increases the productivity of the plant.

3.22 OXIDATION–REDUCTION EQUILIBRIA

Like other reactions, oxidation–reduction reactions do not always go nearly to completion. In many cases, a state of equilibrium is reached. The reactants and products may be present at equilibrium in comparable amounts. If solutions of the reducing agent, iron(II) sulphate, and the oxidising agent, iodine, are mixed, the resulting solution will contain Fe^{3+}, Fe^{2+}, I_2 and I^-:

Redox reactions reach a position of equilibrium...

...for example

$$I_2(aq) + 2Fe^{2+}(aq) \rightleftharpoons 2I^-(aq) + 2Fe^{3+}(aq) \qquad [1]$$

$Fe^{2+} + I_2$...

When solutions of the oxidising reagent, Fe^{3+}, and the reducing agent, I^-, are mixed, the resulting solution will again contain all four species:

...$Fe^{3+} + I^-$

$$2Fe^{3+}(aq) + 2I^-(aq) \rightleftharpoons 2Fe^{2+}(aq) + I_2(aq) \qquad [2]$$

If the same amounts of iron and iodine have been used to make the two solutions, the compositions of solutions [1] and [2] will be the same. The reason is that equilibrium is established, and the position of equilibrium is the same, whether it is reached by route [1] or route [2]. The concentrations of the four species at equilibrium are related by the equilibrium constant. For reaction [2], the equilibrium constant is

The equilibrium constant

$$K_c = \frac{[Fe^{2+}]^2[I_2]}{[Fe^{3+}]^2[I^-]^2} = 6 \times 10^7 \, mol^{-1} \, dm^3$$

The concentration of each reactant and product is raised to the appropriate power, i.e., the coefficient before that reactant in the stoichiometric equation. The value of K_c is high because the position of equilibrium lies far over to the right hand side of equation [2]. In 1 dm^3 of solution made by adding 0.1 mol of Fe^{3+} and 0.05 mol of I_2, the concentration of Fe^{3+} remaining at equilibrium is only 5×10^{-5} mol dm^{-3}.

With other redox reactions, the **position of equilibrium** (the extent to which the reaction reaches completion) lies closer to the reactants side. The study of equilibria will be continued in Chapter 11.

QUESTIONS ON CHAPTER 3

1. Explain what is meant by
(*a*) stoichiometric equation
(*b*) oxidation
(*c*) reduction
(*d*) disproportionation

2. State what has been oxidised and what has been reduced in the following reactions:

(*a*) $Zn(s) + 2HCl(aq) \rightarrow ZnCl_2(aq) + H_2(g)$

(*b*) $CH_4(g) + 4Cl_2(g) \rightarrow CCl_4(l) + 4HCl(g)$

(*c*) $NH_4^+NO_3^-(s) \rightarrow N_2O(g) + 2H_2O(l)$

(*d*) $IO_3^-(aq) + 5I^-(aq) + 6H^+(aq) \rightarrow 3I_2(aq) + 3H_2O(l)$

(*e*) $2Fe(CN)_6^{4-}(aq) + Cl_2(g) \rightarrow 2Fe(CN)_6^{3-}(aq) + 2Cl^-(aq)$

(*f*) $2CrO_4^{2-}(aq) + 2H^+(aq) \rightarrow Cr_2O_7^{2-}(aq) + H_2O(l)$

(*g*) $2CuCl(aq) \rightarrow Cu(s) + CuCl_2(aq)$

3. What volume of 0.250 mol dm^{-3} sodium hydroxide solution is required to neutralise 25.0 cm^3 of 0.150 mol dm^{-3} sulphuric acid?

4. Explain what is meant by the terms
(*a*) equilibrium
(*b*) dynamic equilibrium
(*c*) equilibrium constant

5. A 25.0 g measure of household ammonia was dissolved in water and made up to 500 cm^3. A 25.0 cm^3 portion of this solution required 29.4 cm^3 of 0.250 mol dm^{-3} sulphuric acid for neutralisation. What is the percentage by mass of ammonia in the cleaning fluid?

6. Arsenic can be oxidised to arsenic(V) acid, H_3AsO_4. This acid oxidises I^- ions to I_2, which can be estimated by titration against a standard thiosulphate solution:

$$As + 5HNO_3 \rightarrow H_3AsO_4 + 5NO_2 + H_2O$$

$$H_3AsO_4 + 2HI \rightarrow H_3AsO_3 + I_2 + H_2O$$

If 0.1058 g of a sample containing arsenic required 28.7 cm^3 of a 0.0198 mol dm^{-3} solution of sodium thiosulphate in the final titration, what is the percentage of arsenic in the sample?

7. (*a*) Define oxidation–reduction in terms of electron transfer.

(*b*) Consider the reactions represented by the following equations (all substances are in the state in which they occur at room temperature unless otherwise stated).

A $LiH + H_2O \rightarrow LiOH + H_2$
B $Ca + H_2SO_4 \rightarrow CaSO_4 + H_2$
C $[Fe(H_2O_6]^{3+} + H_2O \rightarrow H_3O^+ + [Fe(H_2O)_5(OH)]^{2+}$
D $NH_3(g) + HCl(g) \rightarrow NH_4Cl(s)$
E $MnO_4^- + 8H^+ + 5Fe^{2+} \rightarrow Mn^{2+} + 4H_2O + 5Fe^{3+}$
F $BaO_2 + H_2SO_4 \rightarrow BaSO_4 + H_2O_2$
G $Cr_2O_7^{2-} + H_2O \rightarrow 2CrO_4^{2-} + 2H^+$
H $PbO_2(s) + 4HCl(aq) \rightarrow PbCl_2(s) + 2H_2O(l) + Cl_2(g)$

(i) *Some* of the reactions **A** to **H** are redox (oxidation–reduction) reactions. Use the letter key **A** to **H** to list *all* the redox reactions.

(ii) From the reactions listed in (*b*) above, select *three* of the redox reactions. For *each* reaction, complete the following table to show the element being oxidised and its initial and final oxidation state, and the element being reduced and its initial and final oxidation state.

	1	2	3
Reaction chosen (letter key)			
Element oxidised			
Initial oxidation state			
Final oxidation state			
Element reduced			
Initial oxidation state			
Final oxidation state			

(iii) For any *one* of the reactions listed in (*b*) re-write the equation as a pair of ion/electron half-reactions, one to represent the oxidation, and the other to represent the corresponding reduction.

Reaction chosen (letter key)
Oxidation half-reaction: *Reduction half-reaction:*

(WJEC 90)

8. This question is about sulphur dioxide. In the laboratory, the adsorption of sulphur dioxide may be demonstrated by passing sulphur dioxide through active charcoal, but this is not a practical method in industry. One way of tackling the acid rain problem, used in West German power stations, is to pass the waste gases containing sulphur dioxide through an aqueous suspension of limestone.

The overall reaction is

$$2SO_2(g) + 2CaCO_3(s) + O_2(g) \rightarrow 2CaSO_4(aq) + 2CO_2(aq)$$

Gypsum is then crystallised out as $CaSO_4{\cdot}2H_2O(s)$.
1.2 million tonnes of gypsum are produced per year in West Germany by this method.

(*a*) Describe the changes you would expect to *see* when sulphur dioxide is passed through an aqueous suspension of limestone.

(*b*) Sulphur dioxide can be detected by the reduction of $Cr_2O_7^{2-}$ ions to Cr^{3+} ions.

(i) Describe the changes you would expect to see when sulphur dioxide is passed through a solution of $Cr_2O_7^{2-}$ ions.

(ii) What is the oxidation number of chromium in the ion, $Cr_2O_7^{2-}$?

(iii) In the reaction, the oxidation number of sulphur increases from +4 to +6. Suggest the likely product of the oxidation of sulphur dioxide and hence deduce the equation for this reaction.

(*c*) (i) What does the term *adsorption* mean?

(ii) You are provided with a small cylinder containing sulphur dioxide. Draw a labelled diagram to show how you could measure out 100 cm³ of sulphur dioxide and then find the proportion adsorbed by 10 g of active charcoal.

(iii) How would you know when adsorption was complete?

(*d*) The amount of sulphur dioxide adsorbed by the active charcoal can also be determined by a titration method.

The 10 g of active charcoal containing the sulphur dioxide was added to 1000 cm³ of iodine solution of concentration 0.00500 mol of I_2 per dm³.

20.0 cm³ portions of this solution were then titrated with 0.0100 mol dm⁻³ sodium thiosulphate solution: 11.6 cm³ were required for complete reaction.

The relevant equations are:

$$SO_2(g) + I_2(aq) + 2H_2O(l) \rightarrow 2I^-(aq) + SO_4^{2-}(aq) + 4H^+(aq)$$

$$I_2(aq) + 2S_2O_3^{2-}(aq) \rightarrow 2I^-(aq) + S_4O_6^{2-}(aq)$$

(i) Calculate the number of moles of sodium thiosulphate in 11.6 cm³ of its solution.

(ii) Calculate the total number of moles of iodine, I_2, which reacted with the sodium thiosulphate.

(iii) Deduce the total number of moles of iodine which reacted with the sulphur dioxide.

(iv) Hence calculate the number of moles of sulphur dioxide present in 10 g of active charcoal.

(*e*) A power station produces 55 000 tonnes of gypsum per year (1 tonne = 10^3 kg).

(i) How many moles of gypsum are produced per year?
(Relative atomic masses: H = 1, O = 16, S = 32, Ca = 40)

(ii) What volume of sulphur dioxide was absorbed in the production of 55 000 tonnes of gypsum?

(1 mol of sulphur dioxide at this temperature has a volume of 24 dm³.)

(*f*) Suggest a use for the gypsum produced by this method.

(L(N) 90)

9. A 'copper' coin of mass 2.05 g consisted of copper with small amounts of tin and zinc. It was dissolved in moderately concentrated nitric acid and the solution was boiled free of oxides of nitrogen.

$$3Cu(s) + 8HNO_3(aq) \rightarrow 3Cu(NO_3)_2(aq) + 2NO(g) + 4H_2O(l) \quad [1]$$

The solution was diluted to 250 cm³ and 25.0 cm³ portions were neutralised and mixed with excess potassium iodide solution. Only the Cu^{2+} ions react to produce iodine.

$$2Cu^{2+}(aq) + 4I^-(aq) \rightarrow 2CuI(s) + I_2(aq) \quad [2]$$

The liberated iodine was titrated with sodium thiosulphate solution.

$$I_2(aq) + 2S_2O_3^{2-}(aq) \rightarrow 2I^-(aq) + S_4O_6^{2-}(aq) \quad [3]$$

(*a*) Which of the above reactions (equations [1]–[3]) are redox reactions?

(*b*) What indicator would you use in the titration?

(*c*) 25.0 cm³ of the solution liberated iodine which reacted with 30.0 cm³ of 0.100 mol dm⁻³ $Na_2S_2O_3$, sodium thiosulphate solution. Calculate:

(i) the number of moles of sodium thiosulphate $Na_2S_2O_3$ used in the titration

(ii) the number of moles of iodine, I_2, liberated by 25.0 cm³ of the solution in equation [2]

(iii) the percentage by mass of copper in the coin.

(*d*) This method of volumetric (titrimetric) analysis is rather slow.

(i) Suggest a more rapid technique for the regular analysis of Cu^{2+} ions in solution.

(ii) Would your suggested method be suitable for use with solutions of copper ions containing chromium? Explain your answer.

(O 91, AS)

10. Bronze is an alloy containing copper and tin (Sn). The percentage by mass of tin in the alloy was determined by the method described.

15.0 g of finely powdered bronze was warmed with excess dilute sulphuric acid, to convert all the tin to tin(II) sulphate. The mixture was then filtered, and the colourless filtrate was made up to 250 cm³ with distilled water.

25.0 cm³ portions of the tin(II) sulphate solution were titrated against 0.0200 mol dm⁻³ potassium manganate(VII) solution. 28.0 cm^3 of potassium manganate(VII) was needed to oxidise the tin(II) sulphate to colourless tin(IV) sulphate.

(*a*) What was removed by filtration after the reaction with the dilute sulphuric acid?

(*b*) Under what conditions would the potassium manganate(VII) titration be carried out?

(*c*) What colour change marks the end-point of the titration?

(*d*) Work out the ionic half-equations (ion-electron equations) for:
(i) the oxidation of the tin(II) ion
(ii) the reduction of MnO_4^- to Mn^{2+}.

(*e*) The ionic equation for the oxidation of Sn^{2+} ions by MnO_4^- ions is

$$5Sn^{2+} + 2MnO_4^- + 16H^+ \rightarrow 5Sn^{4+} + 2Mn^{2+} + 8H_2O$$

Calculate the percentage by mass of tin in the alloy.
(O 91, AS)

11. This question concerns the determination of the amount of preservative, sodium sulphite (Na_2SO_3), in a sample of beefburgers. In an experiment 1 kg of meat was boiled with an excess of dilute hydrochloric acid (Step 1). The sulphur dioxide gas released was completely absorbed in an excess of dilute aqueous sodium hydroxide (Step 2). The resulting solution was then acidified with dilute sulphuric acid and titrated with 0.02 M $KMnO_4$ solution (Step 3); 30.00 cm^3 were required to reach the end-point.

Use the following equations to answer the questions below:

Step 1 $Na_2SO_3 + 2HCl \rightarrow 2NaCl + SO_2 + H_2O$

Step 2 $SO_2 + 2OH^- \rightarrow H_2O + SO_3^{2-}$

Step 3 $5SO_3^{2-} + 2MnO_4^- + 6H^+$
$\rightarrow 5SO_4^{2-} + 2Mn^{2+} + 3H_2O$

(*a*) (i) How many moles of Na_2SO_3 are equivalent to 1 mol of MnO_4^-?
(ii) How many moles of MnO_4^- were used in the titration?
(iii) How many moles of Na_2SO_3 were present in 1 kg of the meat?
(iv) Government chemists often express the amount of Na_2SO_3 in meat as parts per million (1 ppm = 1 g of Na_2SO_3 in 10^6 g of meat). Express the amount of Na_2SO_3 in the meat as ppm.

(*b*) (i) In Step 1, why is it necessary to use an excess of dilute hydrochloric acid and to boil the solution?
(ii) In Step 3, why is it essential not to use dilute hydrochloric acid to acidify the solution?
(iii) In Step 3, what colour change would you observe at the end-point?
(NEAB 92)

12. The water-soluble mineral *schonite* has the formula

$$xMgSO_4 \cdot yK_2SO_4 \cdot zH_2O$$

(*a*) Give one simple test you could carry out which would provide evidence that schonite is a *double* salt, rather than a *complex* salt, of magnesium.

An analysis of the salt was carried out as follows. 8.04 g of the salt was dissolved in water and made up to a volume of 500 cm^3.

To 50.0 cm^3 of the solution, excess barium chloride solution was added. The precipitate of barium sulphate ($BaSO_4$, $M_r = 233$) was filtered off, washed and dried, and it weighed 0.932 g.

(*b*) Calculate the number of moles of barium sulphate formed.

(*c*) Calculate the total number of moles of potassium and magnesium sulphates in 8.04 g of schonite.

To a second 50.0 cm^3 of solution, ammonium chloride solution, ammonia and sodium dihydrogenphosphate were added. The magnesium was all precipitated as magnesium ammonium phosphate. This was filtered off, washed, dried and ignited to give 0.222 g of magnesium pyrophosphate, $Mg_2P_2O_7$.

$$2MgNH_4PO_4 \rightarrow Mg_2P_2O_7 + 2NH_3 + H_2O$$

(*d*) Calculate the number of moles of magnesium pyrophosphate formed.

(*e*) Calculate the number of moles of magnesium sulphate in 8.04 g of schonite.

(*f*) Calculate the number of moles of potassium sulphate in 8.04 g of schonite.

(*g*) Calculate the total mass of magnesium and potassium sulphate in 8.04 g of schonite.

(*h*) Use these figures to calculate the mass and number of moles of water in 8.04 g of schonite. Hence write the formula of the double salt.
(O 92, AS)

13. Answer *each* of the following.

(*a*) A compound **Z** has formula $Fe(C_2O_4) \cdot xH_2O$. When heated it decomposes according to

$$Fe(C_2O_4) \cdot xH_2O \rightarrow FeO + CO_2 + CO + xH_2O$$

When 3.596 g of **Z** were decomposed completely, the combined volume of CO_2 and CO formed was 896 cm^3 measured at stp. Calculate the value of x and the volume of acidified potassium manganate(VII) of concentration 0.020 mol dm^{-3} required to oxidise completely the same mass of **Z**.

(*b*) When 1.00 g of lithium is burnt in oxygen the product has a mass of 2.152 g and dissolves in water to give an alkaline solution. When 1.00 g of sodium is burnt in oxygen the product weighs 1.696 g. This product dissolves in water to form a solution which decolourises acidified aqueous potassium manganate(VII). Comment on these statements.

†(*c*) Suggest identities for **A** to **F** below and write appropriate equations.

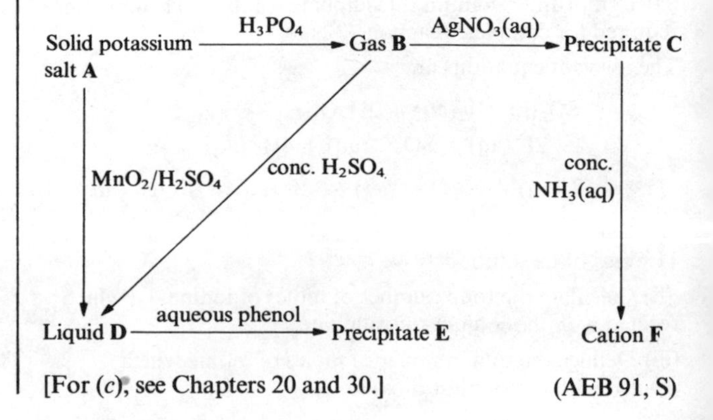

[For (*c*), see Chapters 20 and 30.] (AEB 91, S)

14. When a solution containing a high concentration of sulphate ions is electrolysed under carefully controlled conditions, the ion $S_2O_8{}^{2-}$ is produced at the anode. Give a balanced half-equation for this reaction.

In this reaction either the sulphur atoms or the oxygen atoms might have been oxidised. Calculate the average oxidation numbers of sulphur and oxygen in the ion $S_2O_8{}^{2-}$ for both of these possibilities.

Suggest the displayed formula for the ion $S_2O_8{}^{2-}$, given that the ion contains one O—O bond. On the basis of your displayed formula, explain the average oxidation numbers of sulphur and oxygen calculated previously.

The oxidising properties of the ion, $S_2O_8{}^{2-}$, were investigated. 25.0 cm³ of a solution containing 0.0500 mol dm⁻³ of $S_2O_8{}^{2-}$ was added to 10.0 cm³ of a solution of potassium iodide, KI, of concentration 1.00 mol dm⁻³. The iodine released required 20.0 cm³ of a solution of sodium thiosulphate, $Na_2S_2O_3$, of concentration 0.125 mol dm⁻³ for complete reaction. Calculate whether $S_2O_8{}^{2-}$ or iodide ions were present in excess and hence deduce the equation for the reaction of $S_2O_8{}^{2-}$ with iodide ions.

Would you expect $S_2O_8{}^{2-}$ ions to be able to oxidise I_2 to $IO_3{}^-(aq)$ or to H_5IO_6? Justify your answer.

(L(N) 91, S)

4

THE CHEMICAL BOND

4.1 IONS

4.1.1 SOME QUESTIONS

The background to the ionic theory

- Why are all copper(II) compounds blue, regardless of the non-metallic part of the compound?
- Why are all dichromates orange, regardless of the metallic part of the compound?
- Why do aqueous solutions of some substances conduct electricity?
- Why does a layer of copper appear on the negative electrode when a direct electric current is passed through a solution of a copper salt [see Figure 4.1, § 4.1.3]?
- Why is chlorine evolved at the positive electrode when a direct electric current passes through a concentrated aqueous solution of a chloride?
- Why do X ray diffraction patterns of salts show a regular pattern of dots [see Figure 6.2, § 6.1]?

To answer these questions and others, the **ionic theory** was developed. According to the ionic theory, some compounds consist of tiny particles which carry an electric charge, either positive or negative, and are called **ions**. The ionic theory was developed to explain the behaviour of compounds when a direct electric current is passed through the molten compound or an aqueous solution of the compound.

4.1.2 WHICH SUBSTANCES CONDUCT ELECTRICITY?

Substances can be divided into four groups according to their ability to conduct a direct current of electricity [see Table 4.1].

Solids: Metals and alloys and graphite conduct electricity

Liquids: Solutions of acids, alkalis and salts conduct electricity

TABLE 4.1
Electrical Conductors and Non-Conductors

Solids	
Electrical conductors	*Non-conductors, i.e. insulators*
All metallic elements All alloys One non-metallic element, the graphite allotrope of carbon	Non-metallic elements, e.g. sulphur Many compounds, e.g. polyethene Crystalline salts, e.g. sodium chloride, copper(II) sulphate
Liquids	
Electrolytes	*Non-electrolytes*
Solutions of acids and alkalis and salts; such liquids are called **electrolytes**. Chemical changes occur at the electrodes. For example, copper(II) chloride solution changes into copper and chlorine and water.	The liquids which do not conduct electricity are water and organic compounds, such as ethanol. They are called **non-electrolytes**.

4.1.3 WHAT HAPPENS WHEN COMPOUNDS CONDUCT ELECTRICITY?

Important terms: cell, electrode, anode, cathode, electrolyte

Crystals of copper(II) chloride do not conduct electricity, and distilled water does not conduct electricity. Yet when the two substances are mixed to make a solution, the solution conducts. Figure 4.1 shows what happens. The objects which conduct electricity into and out of the cell are called **electrodes**. The electrodes are usually made of elements such as platinum and graphite, which do not react with electrolytes. The electrode connected to the positive terminal of the battery is called the **anode**. The electrode connected to the negative terminal is called the **cathode**.

FIGURE 4.1 Electrolysis of Copper(II) Chloride

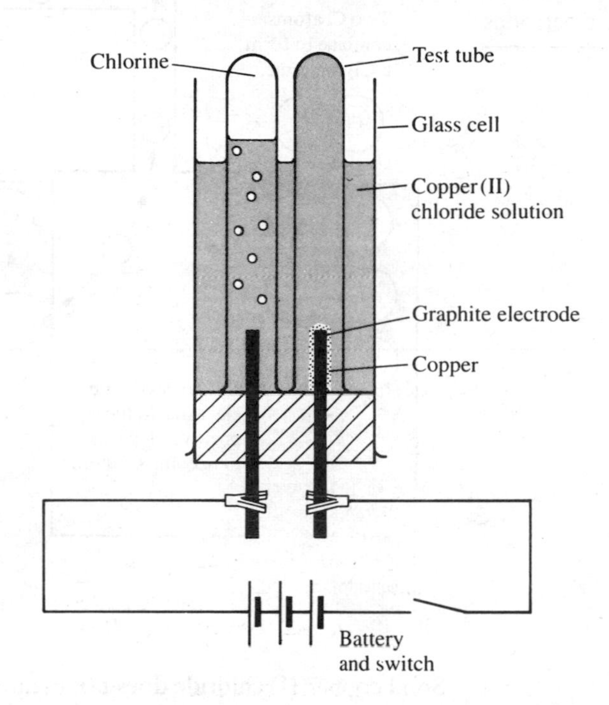

Solutions of some compounds are electrolysed . . .

. . . split up by an electric current . . .

. . . to form new substances

In the process, the solute, copper(II) chloride, is split up. Copper appears as a layer on the cathode. Chlorine is evolved at the anode. Copper(II) chloride has been **electrolysed**, that is, split up by an electric current. The process is called **electrolysis**. Copper(II) chloride is an **electrolyte**. An electrolyte is a compound which conducts electricity when molten or in solution. The compound is split up in the process. Any container in which electricity produces a chemical change (or in which a chemical change produces electricity) is called a **cell**.

Such compounds consist of ions . . .

In explaining electrolysis, we must take into account that, when copper salts are electrolysed, copper always appears at the negative electrode, never at the positive electrode. This must show that there is a positive charge associated with copper in all its salts. We accept that copper, the element, is composed of copper atoms. Does the combined copper in copper(II) chloride and other salts consist of a different kind of copper atoms with positive charges? Chlorine appears always at the positive electrode, never at the negative electrode. This indicates that there is a negative charge associated with the chlorine combined with copper in copper(II) chloride. Chlorine in its salts must consist of a type of chlorine atom with negative charge.

. . . positive ions (cations) . . .

. . . and negative ions (anions)

In 1834, Michael Faraday suggested a theory to explain electrolysis. He suggested the existence of tiny particles of matter carrying positive or negative electric charges. We now call them **ions**. Positively charged ions are called **cations**. Negatively charged ions are called **anions**.

In electrolysis ions are discharged at the electrodes

According to the ionic theory, the compound copper(II) chloride consists of positively charged copper ions and negatively charged chloride ions. During electrolysis, positively charged copper ions travel to the negative electrode. When they reach the negative electrode, the ions lose their charge: they are discharged to form copper atoms. Negatively charged chloride ions travel to the positive electrode. When they meet the positive electrode, they lose their charge: they are discharged, and molecules of chlorine are formed [see Figure 4.2].

FIGURE 4.2
What happens at the electrodes

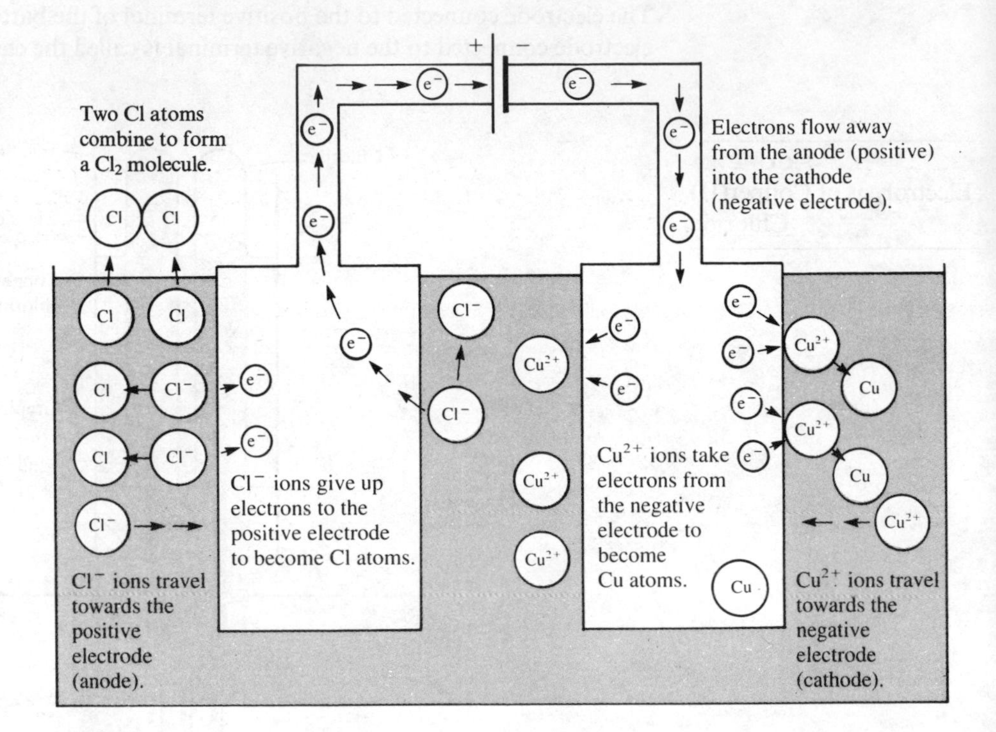

Solid copper(II) chloride does not conduct electrifcity, and water is a very poor electrical conductor. However, an aqueous solution of the salt is a good conductor and is an electrolyte. To explain this difference, scientists suggested that, in a solid, the ions are fixed in position, held together by strong attractive forces between positive and negative ions. In a solution of a salt, the ions are free to move [see Figure 4.3].

FIGURE 4.3
The Ions in a Solution are Free to Move

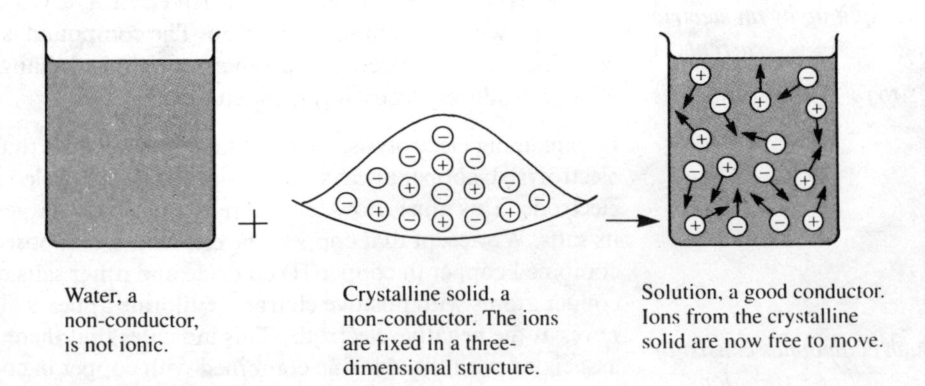

Water, a non-conductor, is not ionic.
Crystalline solid, a non-conductor. The ions are fixed in a three-dimensional structure.
Solution, a good conductor. Ions from the crystalline solid are now free to move.

The scientist who did the first work on electrolysis was Michael Faraday (1791–1867). He was the son of a poor blacksmith, and at the age of 14 he became an apprentice bookbinder. He educated himself by reading the books he was asked to bind and by joining various self-improvement groups. Faraday became especially interested in

Profile: Michael Faraday

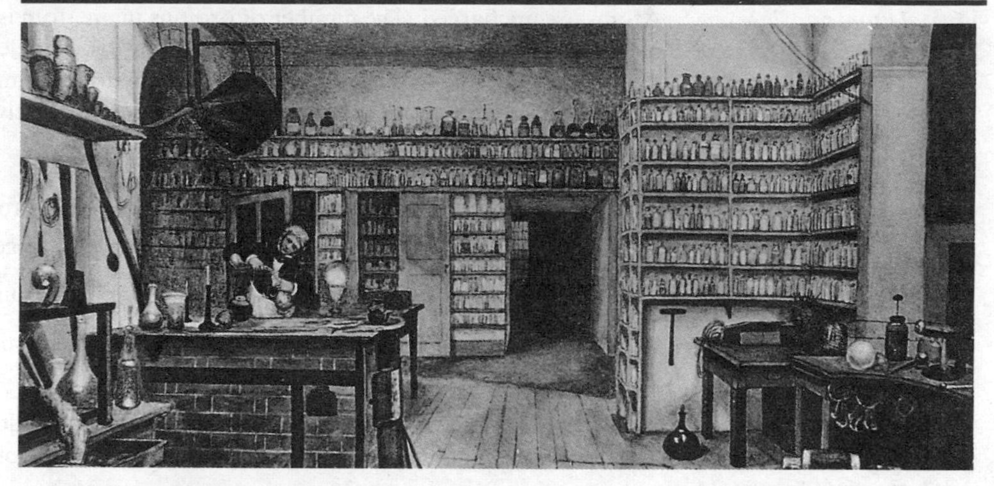

FIGURE 4.4 Michael Faraday in His Laboratory

chemistry and electricity. When he attended a course of lectures by Humphry Davy at the Royal Institution, Faraday made detailed notes, illustrated then with excellent diagrams, and bound them in a book. Davy was impressed with the young man's enthusiasm and offered him a job as a laboratory assistant.

Faraday was soon contributing actively to the research. The variety of his achievements is amazing. He made the first electric motor, the first transformer and the first dynamo. He formulated the First and Second Laws of Electrolysis. He discovered benzene and did research on steel, optical glass and the liquefaction of gases. Faraday became a superb scientific lecturer. He gave weekly evening lectures at the Royal Institution to popularise science and started a tradition of Christmas lectures for children which is still continued.

4.1.4 MOLTEN SALTS

Some compounds are electrolysed when molten These compounds are composed of ions

In a solution of a salt, the ions are free to move, and the solution can be electrolysed. Another way of enabling the ions to move is to melt the salt. You have probably seen the experiment shown in Figure 4.5. Lead(II) bromide is a convenient salt to use because it has a fairly low melting point.

FIGURE 4.5 Electrolysis of Molten Lead(II) Bromide

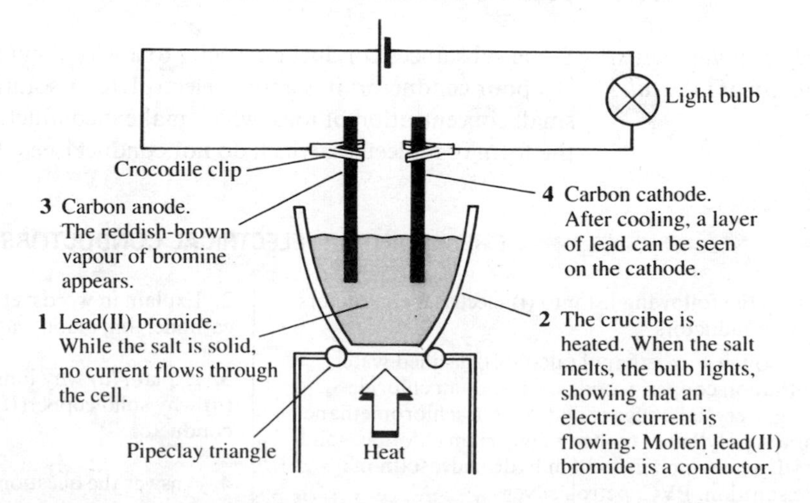
Light bulb
Crocodile clip
3 Carbon anode. The reddish-brown vapour of bromine appears.
4 Carbon cathode. After cooling, a layer of lead can be seen on the cathode.
1 Lead(II) bromide. While the salt is solid, no current flows through the cell.
2 The crucible is heated. When the salt melts, the bulb lights, showing that an electric current is flowing. Molten lead(II) bromide is a conductor.
Pipeclay triangle
Heat

The experiment shows that molten lead(II) bromide is electrolysed. In the molten salt, ions are free to move.

4.1.5 HOW DOES AN ATOM BECOME AN ION?

A metal atom may lose one or more electrons to become a cation

Atoms are uncharged. The number of protons in an atom is the same as the number of electrons [see § 1.5]. If an atom either gains or loses an electron, it will become electrically charged. Metal atoms and hydrogen atoms form positive ions. For example, a sodium atom loses one electron to form a positively charged sodium ion.

Sodium atom → Electron + Sodium ion

Na (11 protons, 11 electrons) → e^- + Na^+ (11 protons, 10 electrons)
uncharged charge = +1

A magnesium atom loses 2 electrons to become a magnesium ion, Mg^{2+}.

An aluminium atom loses 3 electrons to become an aluminium ion, Al^{3+}. The charge on a cation may be +1, +2 or +3.

An atom of a non-metal may gain one or more electrons to become an anion

Non-metallic elements form negative ions (anions). They do this by gaining electrons. A chlorine atom gains one electron to become a chloride ion, Cl^-.

Chlorine atom + Electron → Chloride ion

Cl (17 protons, 17 electrons) + e^- → Cl^- (17 protons, 18 electrons)
uncharged charge = −1

An oxygen atom gains 2 electrons to become an oxide ion, O^{2-}. Some anions contain oxygen combined with another element. Examples are: hydroxide ion, OH^-; nitrate ion, NO_3^-; sulphate ion, SO_4^{2-}. An anion may have a charge of −1, −2 or −3. Table 4.2, § 4.2.7, lists the symbols and formulae of some common ions.

4.1.6 NON-ELECTROLYTES

Non-electrolytes consist entirely of molecules

Some liquids do not conduct electricity. It follows that these substances do not contain ions. They consist of uncharged particles called molecules [see § 4.3]. Compounds formed between metallic and non-metallic elements are usually electrolytes, and compounds formed between non-metallic elements are generally non-electrolytes, e.g. ethanol (alcohol), or weak electrolytes [see § 4.1.7].

4.1.7 WEAK ELECTROLYTES

Weak electrolytes consist mainly of molecules

Some substances conduct electricity to a very slight extent. For example, ethanoic acid is a poor conductor; it is a weak electrolyte. A solution of ethanoic acid contains a small concentration of ions, which make it conduct. The compound exists mainly in the form of molecules, which do not conduct [see § 4.3].

CHECKPOINT 4A: ELECTRICAL CONDUCTORS

1. (*a*) Divide the following list into (i) electrical conductors and (ii) non-conductors:

solid wax, molten wax, ethanol (alcohol), distilled water, aqueous ethanol, copper, wood, steel, sodium chloride crystals, sugar crystals, sugar solution, tetrachloromethane (CCl_4), brass, polythene, molten magnesium chloride, solid sodium hydroxide, molten sodium hydroxide, sodium hydroxide solution, PVC, petrol, silver

(*b*) Say which of the conductors in the list are electrolytes.

2. Explain in words: electrolysis, electrolyte, cell, anode, cathode, ion, anion, cation.

3. Explain (*a*) why ions move towards electrodes and (*b*) why solid copper(II) sulphate is not an electrical conductor.

4. Answer the questions at the beginning of § 4.1.1.

4.2 THE IONIC BOND

Let us look in more detail at how atoms are able to form ions. How can one atom be able to give up an electron to form a cation? How can another atom be able to accept an electron to form an anion?

4.2.1 SODIUM CHLORIDE

Look at the electron configuration in a sodium atom [Figure 4.6]. There is just one electron more than there is in an atom of the noble gas, neon.

FIGURE 4.6
Atoms of Sodium and Neon

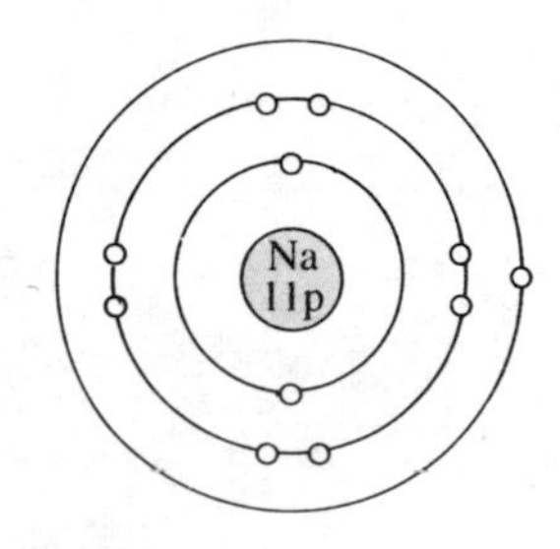

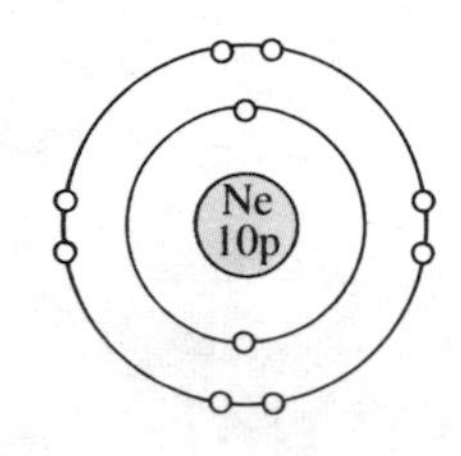

The noble gases have a full outer shell of electrons. This electron configuration is very stable

The noble gases [see § 2.5] are helium, neon, argon, krypton, xenon and radon. They take part in hardly any chemical reactions. With the electron configuration ns^2np^6 ($1s^2$ for He) the noble gases have a full outer shell of electrons [Figure 4.7] and it seems probable that it is this electron configuration that makes them stable, that is, chemically unreactive.

FIGURE 4.7
The Electron Configurations in Helium, Neon and Argon

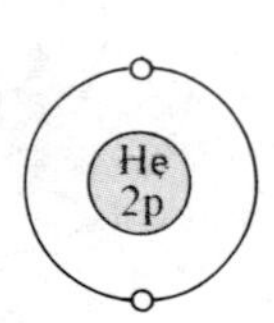

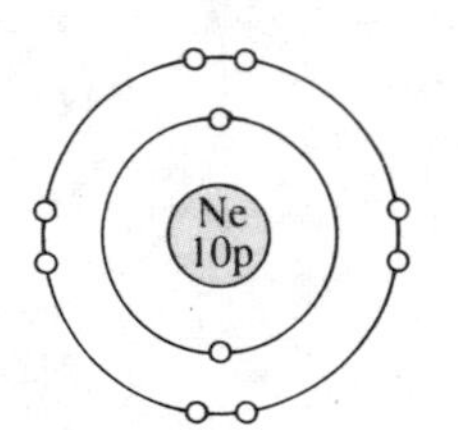

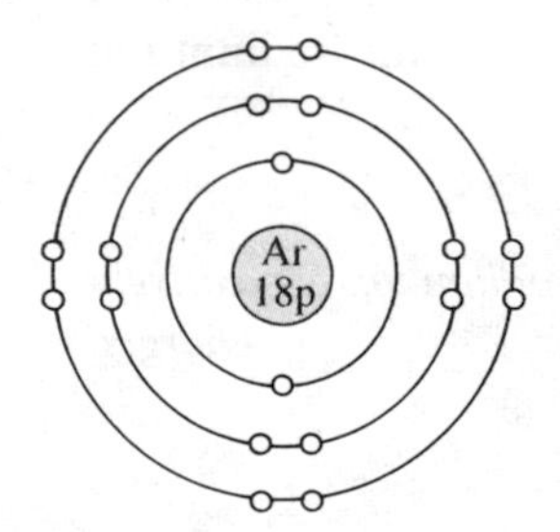

To achieve a full outer shell of electrons, a sodium atom must lose one electron . . .

When a sodium atom Na(2.8.1) loses the lone electron from its outermost shell, the outer shell that remains contains 8 electrons, the same as the noble gas neon Ne(2.8). This electron configuration is associated with stability. A sodium atom cannot lose an electron unless another atom will accept it. It can, however, give an electron to a chlorine atom. In fact, sodium burns vigorously in chlorine to form sodium chloride.

. . . and a chlorine atom must gain one electron

How can a chlorine atom accept an electron? With the electron configuration Cl(2.8.7), one more electron gives chlorine the same electron arrangement as the noble gas argon Ar(2.8.8). A full outer shell brings with it stability [Figure 4.8].

FIGURE 4.8
The Configuration of Electrons in Chlorine and Argon

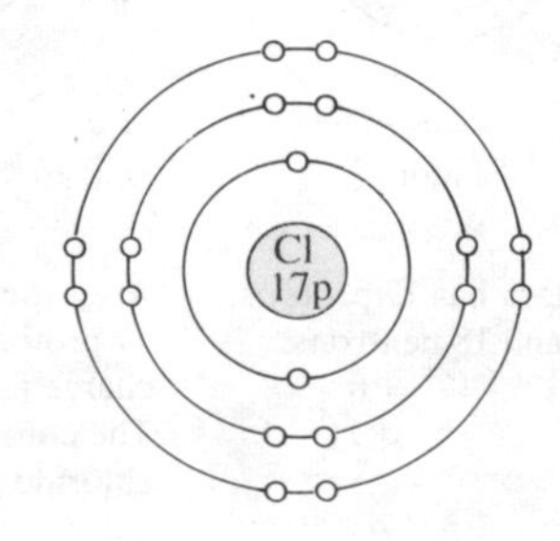

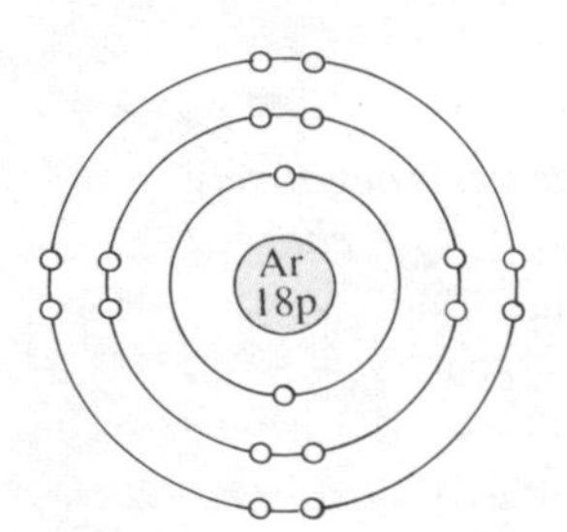

Figure 4.9 shows what happens when an atom of sodium donates an electron to an atom of chlorine. A full outer shell is left behind in sodium, and a full outer shell is created in chlorine.

FIGURE 4.9
The Formation of Sodium Chloride. (The sodium electrons have been shown as × and the chlorine electrons as ○.)

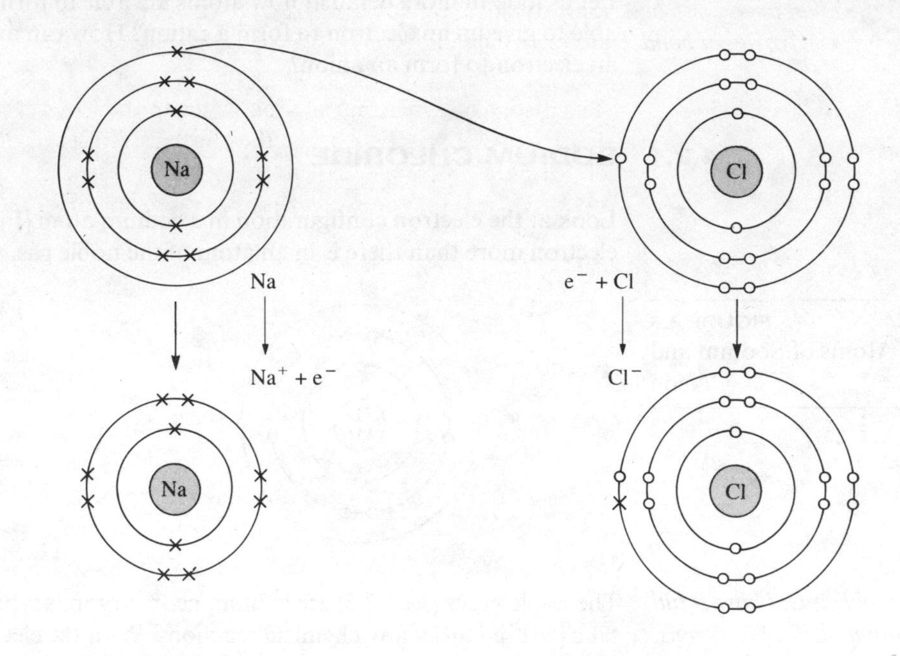

FIGURE 4.10
The Sodium Atom and the Sodium Ion

A sodium ion is positively charged

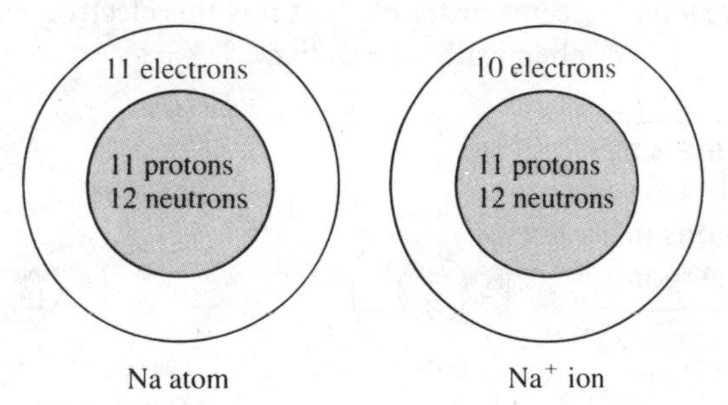

A sodium atom has 11 protons, 11 electrons and 12 neutrons. Charge = +11 − 11 = 0 charge unit

After it loses one electron, it has 11 protons and 10 electrons. Charge = +11 − 10 = +1 charge unit. The sodium atom, Na, has become a sodium ion, Na^+

FIGURE 4.11
A Chlorine Atom and a Chloride Ion

A chloride ion is negatively charged

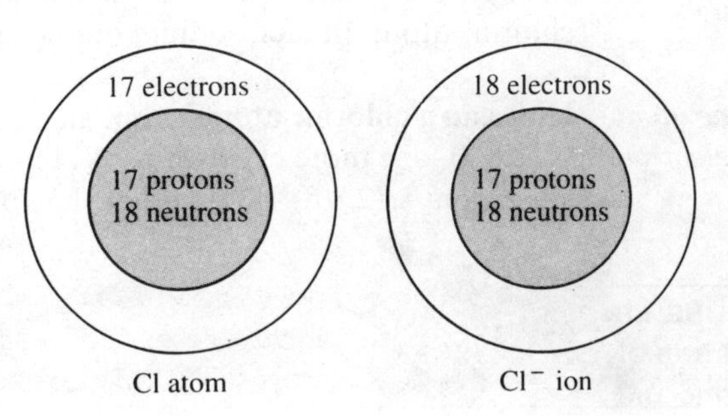

A chlorine atom has 17 protons, 17 electrons and 18 neutrons. Charge = + 17 – 17 = 0 charge unit

After it gains an electron, making 17 protons and 18 electrons, the charge is +17 − 18 = −1 charge unit. The chlorine atom, Cl, has become a chloride ion, Cl^-.

The oppositely charged ions are held together by an electrostatic attraction called an ionic bond or electrovalent bond

The electrostatic attraction between oppositely charged ions holds the ions Na^+ and Cl^- together. This electrostatic attraction is the chemical bond in the compound, sodium chloride. It is called an **ionic bond** or **electrovalent bond**. Sodium chloride is an ionic or electrovalent compound. The compounds which conduct electricity when they are melted or dissolved are electrovalent compounds.

The theory of the chemical bond is due to the work of W Kossel and G N Lewis. In 1916, working independently, they both put forward the theory that the formation of chemical bonds can be explained by the tendency of atoms to give or receive electrons in order to attain a noble gas type of electron configuration.

4.2.2 CRYSTALS

Ionic compounds form crystals

A feature of ionic compounds is that they form **crystals**. The crystals of sodium chloride are perfect cubes. In a dilute solution of sodium chloride, sodium ions and chloride ions are moving about independently of other ions. When the solution is evaporated to the point of crystallisation, the ions are much closer together. A sodium ion attracts chloride ions, as shown in Figure 4.12(a). Each chloride ion attracts other sodium ions, and a three-dimensional arrangement of ions called a **crystal structure** is built up [see Figure 4.12(b)]. There is no pair of Na^+ and Cl^- ions that could be regarded as a molecule of sodium chloride. The formula NaCl represents the ratio in which ions are present in the crystal structure. A pair of ions Na^+Cl^- is called a **formula unit** of sodium chloride.

FIGURE 4.12
The Arrangement of Ions in a Sodium Chloride Crystal

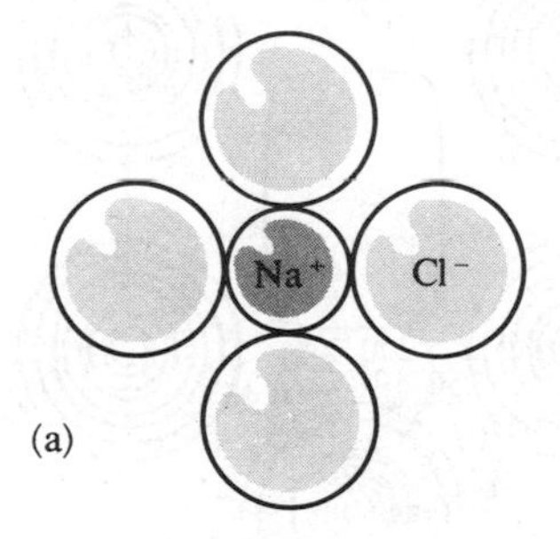

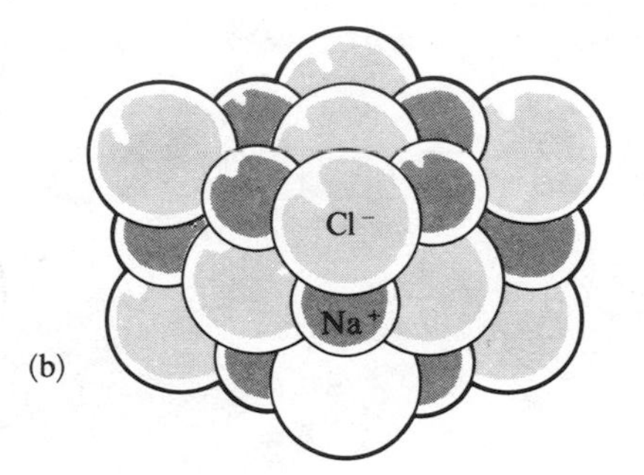

A crystal of sodium chloride is a three-dimensional structure of sodium ions and chloride ions

The crystal is uncharged because the number of sodium ions is equal to the number of chloride ions. The bonds between positive and negative ions are strong. This is why solid sodium chloride does not conduct electricity and is not electrolysed. In the solid, the ions cannot move out of their positions in the three-dimensional structure. When the salt is melted or dissolved, the ions are free to move and can travel towards the electrodes [see Figure 4.3].

X ray analysis demonstrates the existence of ions

The evidence for the existence of ions is summarised in § 12.6.1. An impressive demonstration of the existence of ions is the use of X ray analysis [§ 6.1] to obtain electron density maps. The one for sodium chloride is shown in Figure 4.13. It consists of regions of charge which are isolated from other regions of charge. This is the picture one would expect for a structure consisting of separate Na^+ and Cl^- ions. The technique is now sufficiently accurate to show that there are ten electrons, not eleven, associated with each sodium nucleus: the species present is Na^+, not Na. Figure 4.14 shows the electron density map for calcium fluoride.

FIGURE 4.13 Electron Density Map for Sodium Chloride

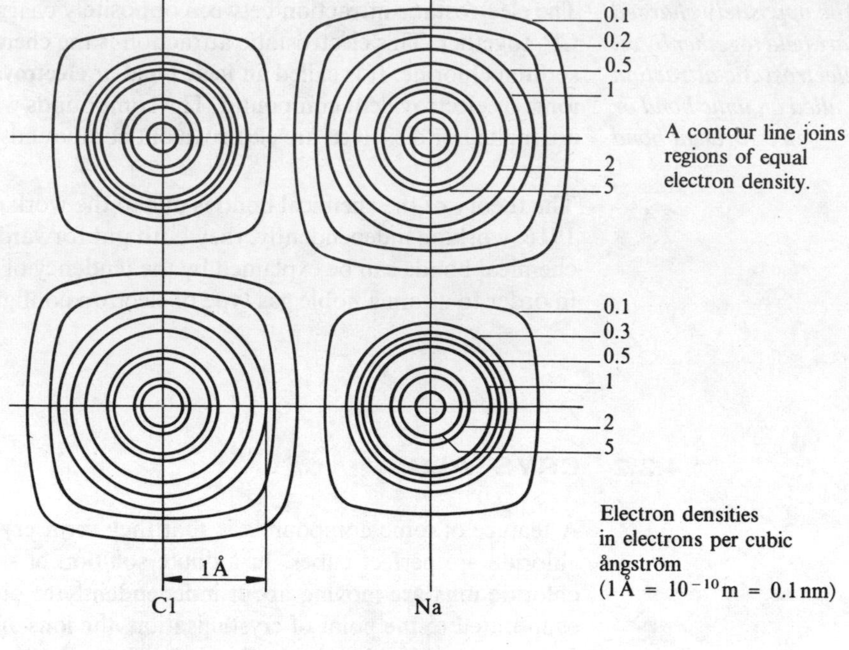

FIGURE 4.14 Electron Density Map for Calcium Fluoride

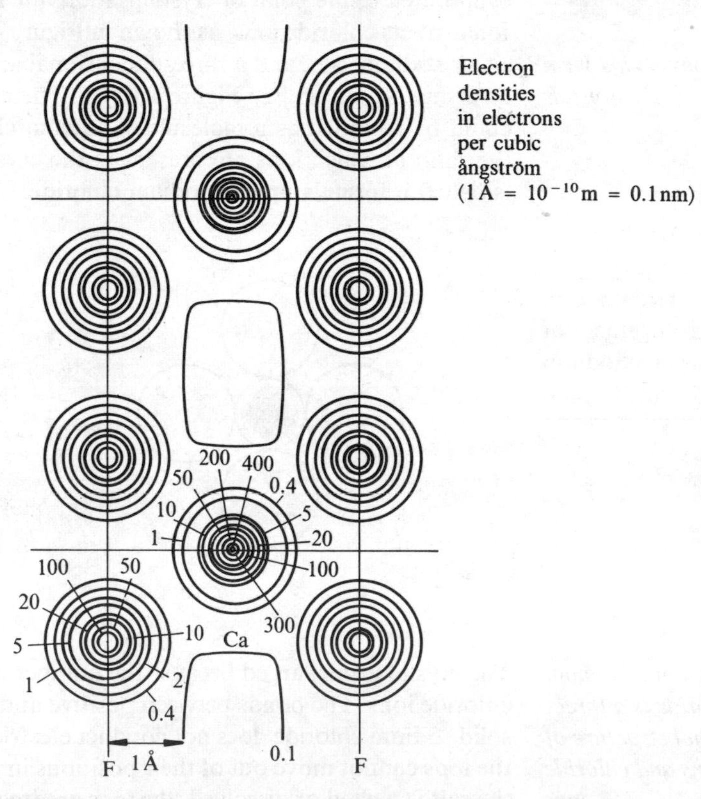

4.2.3 Magnesium fluoride

Magnesium fluoride contains two F^- ions for every Mg^{2+} ion

Magnesium (2.8.2) has two electrons in the outermost shell. It has to lose these to attain a stable, full outer shell of electrons. Fluorine (2.7) needs to gain an electron. For a magnesium atom to lose two electrons and a fluorine atom to gain only one electron, one magnesium atom has to combine with two fluorine atoms.

$$Mg\ (2.8.2) \rightarrow Mg^{2+}\ (2.8) + 2e^-$$

$$F\ (2.7) + e^- \rightarrow F^-\ (2.8)$$

$$Mg + F_2 + Mg^{2+}\ 2F^-$$

4.2.4 MAGNESIUM OXIDE

Magnesium oxide consists of Mg^{2+} ions and O^{2-} ions in equal numbers

The product formed when magnesium burns in air is magnesium oxide. One atom of magnesium, Mg (2.8.2), gives two electrons to one atom of oxygen, O (2.6). The ions Mg^{2+} (2.8) and O^{2-} (2.8) are formed. The electrostatic attraction between them is an ionic bond [see Figure 4.14].

FIGURE 4.15 The Formation of an Ionic Bond between Atoms of Magnesium and Oxygen

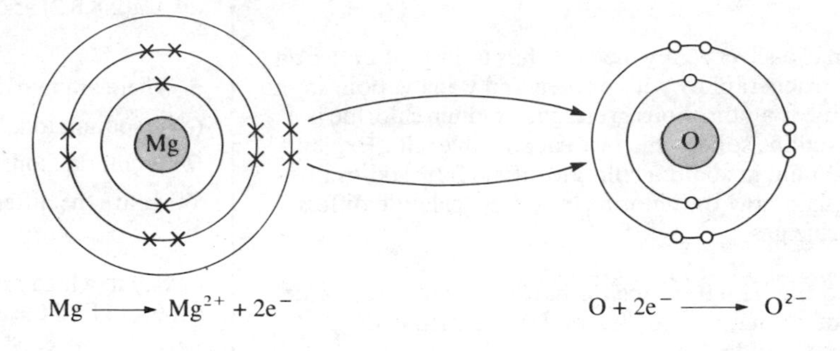

$Mg \longrightarrow Mg^{2+} + 2e^-$ $O + 2e^- \longrightarrow O^{2-}$

4.2.5 THE CHARGE ON THE IONS OF AN ELEMENT

Metals form ions with charge +1, +2 or +3

Non-metals form ions with charge −1, −2 or −3

Electrovalent compounds (ionic compounds) are formed when a metallic element combines with a non-metallic element. Metallic elements form positive ions, e.g. sodium ions, Na^+, magnesium ions, Mg^{2+} and aluminium ions, Al^{3+}. Non-metallic elements form negative ions, e.g. chloride ions, Cl^-, and oxide ions, O^{2-}.

Refer to Table 2.7, which lists the electron configurations of the atoms of some elements.

- Can you see from the electron configurations why potassium and sodium form ions K^+ and Na^+, while magnesium and calcium form ions Mg^{2+} and Ca^{2+}? What do the electron configurations of potassium and sodium have in common? What do the electron configurations of magnesium and calcium have in common?
- Can you see from the electron configurations why fluorine and chlorine form ions F^- and Cl^-, while oxygen and sulphur form ions O^{2-} and S^{2-} and nitrogen forms ions N^{3-}?
- In general,

The charge on the ions of a metallic element	=	Number of electrons in the outermost shell of an atom of that element
The charge on the ions of a non-metallic element	=	8 − (Number of electrons in the outermost shell of an atom of that element)

Non-metallic elements often form oxo-ions (which contain oxygen), e.g. sulphate, SO_4^{2-}, and nitrate, NO_3^-.

4.2.6 IONS AND THE PERIODIC TABLE

The charge depends on the number of electrons in the outermost shell . . .

- Compare the charges on the ions of the following elements with the group in which you find them in the Periodic Table [p. 792].

Ions: Na^+, K^+, Ca^{2+}, Mg^{2+}, Ba^{2+}, Sr^{2+}, Al^{3+}, N^{3-}, O^{2-}, S^{2-}, F^-, Cl^-, Br^-, I^-

- Can you see the relationship?

. . . and therefore on the position of the element in the Periodic Table

Charge on ions of metallic element = Group number of element
Charge on ions of non-metallic element = 8 − Group number of element

CHECKPOINT 4B: BONDING IN IONIC COMPOUNDS

1. Lithium is an alkali metal with the electron configuration Li(2.1). Fluorine is a halogen with the electron configuration F(2.7). Draw the configuration of electrons (*a*) in the atoms of Li and F and (*b*) in a pair of ions in the compound lithium fluoride.

2. Sodium is a silvery-grey metal. It has to be kept under oil because it reacts rapidly with oxygen and water vapour in the air. Chlorine is a poisonous green gas. Sodium chloride is a white, crystalline solid, which we eat as 'table salt'. Explain how the sodium in sodium chloride differs from sodium metal. Explain how the chlorine in sodium chloride differs from chlorine gas.

3. Refer to Figure 4.9 for the formation of sodium chloride. Draw a similar diagram to illustrate the formation of magnesium fluoride, MgF_2.

4. Use 'dot and cross' diagrams to show how the following atoms combine:

(*a*) Li(2.1) and O(2.6), (*b*) Na(2.8.1) and F(2.7),
(*c*) Be(2.2) and F(2.7), (*d*) Al(2.8.3) and F(2.7),
(*e*) Ca(2.8.8.2) and O(2.6).

5. There are two differences between

(*a*) a sodium ion, Na^+(2.8) and a neon atom, Ne(2.8)
(*b*) a chloride ion, Cl^-(2.8.8) and an argon atom, Ar(2.8.8).
What are the differences?

6. Say in which group of the Periodic Table you would expect to find the elements which form the following ions:
Rb^{2+}, Ga^{3+}, Sr^{2+}, At^-, P^{3-}, Se^{2-}, Cs^+, Ra^{2+}

4.2.7 FORMULAE OF IONIC COMPOUNDS

Electrovalent compounds consist of positive and negative ions. A compound is neutral because the charge on the positive ion (or ions) is equal to the charge on the negative ion (or ions). In zinc chloride, $ZnCl_2$, one zinc ion, Zn^{2+}, is balanced in charge by two chloride ions, $2Cl^-$.

You can work out the formula of an ionic compound by balancing the charges on the ions

Let us use this principle of balancing the charges to work out the formulae of electrovalent compounds.

Compound:	*Zinc chloride*
Which ions are present?	Zn^{2+}, Cl^-
How can you balance the charges?	One Zn^{2+} ion needs two Cl^- ions.
How many ions are needed?	Zn^{2+} and $2Cl^-$ ions
What is the formula?	$ZnCl_2$

Compound:	*Sodium carbonate*
Which ions are present?	Na^+, CO_3^{2-}
How can you balance the charges?	Two Na^+ are needed to balance one CO_3^{2-}.
How many ions are needed?	$2Na^+$ and CO_3^{2-}
What is the formula?	Na_2CO_3

Compound:	*Calcium hydroxide*
Which ions are present?	Ca^{2+}, OH^-
How can you balance the charges?	Two OH^- ions balance one Ca^{2+} ion.
How many ions are needed?	Ca^{2+} and $2OH^-$
What is the formula?	$Ca(OH)_2$ The brackets tell you that the 2 multiplies everything inside them. There are 2 O atoms and 2 H atoms, in addition to the 1 Ca ion.

Compound:	*Iron(II) sulphate*
Which ions are present?	Fe^{2+}, SO_4^{2-}
How can you balance the charges?	One Fe^{2+} ion balances one SO_4^{2-} ion.
How many ions are needed?	Fe^{2+} and SO_4^{2-}
What is the formula?	$FeSO_4$

Compound:	*Iron(III) sulphate*
Which ions are present?	Fe^{3+}, SO_4^{2-}
How can you balance the charges?	Two Fe^{3+} would balance three SO_4^{2-} ions.
How many ions are needed?	$2Fe^{3+}$ and $3SO_4^{2-}$
What is the formula?	$Fe_2(SO_4)_3$ The brackets tell you that all the atoms inside them must be multiplied by the 3 that follows them. The formula contains 2 Fe, 3 S and 12 O.

You need to learn the symbols and charges of the ions in Table 4.2. Then you can work out the formula of any electrovalent compound containing these ions.

You will notice that the sulphates of iron arc named iron(II) sulphate and iron(III) sulphate. The Roman numerals, II and III, show which type of ion, Fe^{2+} or Fe^{3+}, is present. This is always done with the compounds of elements of variable oxidation number.

TABLE 4.2 The Symbols and Formulae of some Common Ions

Ion	*Symbol*	*Ion*	*Symbol*
Aluminium	Al^{3+}	Bromide	Br^-
Ammonium	NH_4^+	Carbonate	CO_3^{2-}
Barium	Ba^{2+}	Chloride	Cl^-
Calcium	Ca^{2+}	Dichromate(VI)	$Cr_2O_7^{2-}$
Copper(I)	Cu^+	Hydrogencarbonate	HCO_3^-
Copper(II)	Cu^{2+}	Hydroxide	OH^-
Hydrogen	H^+	Iodide	I^-
Iron(II)	Fe^{2+}	Manganate(VII)	MnO_4^-
Iron(III)	Fe^{3+}	Nitrate	NO_3^-
Lead(II)	Pb^{2+}	Nitrite	NO_2^-
Magnesium	Mg^{2+}	Oxide	O^{2-}
Mercury(II)	Hg^{2+}	Phosphate	PO_4^{3-}
Potassium	K^+	Silicate	SiO_3^{2-}
Silver	Ag^+	Sulphate	SO_4^{2-}
Sodium	Na^+	Sulphide	S^{2-}
Zinc	Zn^{2+}	Sulphite	SO_3^{2-}

CHECKPOINT 4C: FORMULAE OF IONIC COMPOUNDS

1. Write the formulae of the following ionic compounds:

(*a*) silver chloride, (*b*) potassium nitrate, (*c*) silver nitrate, (*d*) zinc bromide, (*e*) magnesium iodide, (*f*) copper(II) bromide, (*g*) ammonium chloride, (*h*) ammonium sulphate, (*i*) calcium hydroxide, (*j*) aluminium chloride, (*k*) sodium hydrogencarbonate, (*l*) sodium sulphite, (*m*) iron(II) hydroxide, (*n*) iron(III) hydroxide, (*o*) aluminium oxide.

2. Name the following compounds:

(*a*) AlI_3, (*b*) $CuCO_3$, (*c*) $Zn(OH)_2$, (*d*) AgBr, (*e*) $Cu(NO_3)_2$, (*f*) $FeBr_2$, (*g*) $FeBr_3$, (*h*) Al_2O_3, (*i*) $KMnO_4$, (*j*) Na_2SiO_3, (*k*) Na_3PO_4, (*l*) KNO_2, (*m*) $K_2Cr_2O_7$, (*n*) $Ca_3(PO_4)_2$, (*o*) Na_2SO_3, (*p*) $BaSO_3$, (*q*) $Ca(HCO_3)_2$.

4.2.8 IONIC RADII

Ionic radii can be assigned such that the sum of the cationic and anionic radii is equal to the interionic distance in a crystal

The distance between the centres of the ions is the sum of the **cationic radius** and the **anionic radius**. Linus Pauling was able to apportion the **interionic distance** in potassium chloride (which equals 0.314 nm) between the K^+ and Cl^- ions. These ions are **isoelectronic** (have the same number of electrons). He assumed that the radius of each of the ions is inversely proportional to its **effective nuclear charge**. This is the nuclear charge modified by the **shielding effect** which the inner shells of electrons exert on the attraction of the nucleus for the valence electrons [§ 15.1].

Pauling obtained values for the ionic radii: K^+, 0.133 nm and Cl^-, 0.181 nm. Ionic radii appear to be *additive*. Pauling could subtract the radius of K^+ from the interionic distance in another potassium salt $K^+\mathbf{B}^-$, to obtain the radius of the anion $\mathbf{B}^-$. He used the radius of Cl^- and interionic distances in $\mathbf{M}^+Cl^-$ crystals to obtain values for the ionic radii of a number of cations. Some values of ionic radii are shown in Figure 15.4 [§ 15.1].

4.2.9 ENERGY CHANGES IN COMPOUND FORMATION

Electrovalent bonds are formed if the reaction between elements to give an ionic solid is exothermic

A detailed picture of the energy changes involved in the formation of an ionic compound was drawn by the theoreticians M Born and F Haber. They considered, for example, the formation of sodium chloride from its elements:

$$Na(s) + \tfrac{1}{2}Cl_2(g) \rightarrow NaCl(s)$$

They analysed the reaction as the sum of five steps. These are

1. Vaporisation of sodium $Na(s) \rightarrow Na(g)$ *Endothermic*
2. Ionisation of sodium $Na(g) \rightarrow Na^+(g) + e^-$ *Endothermic*
3. Dissociation of chlorine $\tfrac{1}{2}Cl_2(g) \rightarrow Cl(g)$ *Endothermic*
4. Ionisation of chlorine $Cl(g) + e^- \rightarrow Cl^-(g)$ *Exothermic*
5. Combination of ions to form a crystalline solid $Na^+(g) + Cl^-(g) \rightarrow NaCl(s)$ *Exothermic*

The sum of the five energy changes is exothermic. [See § 10.7 for a fuller treatment.]

The driving force behind the reaction is the fact that sodium metal and chlorine molecules can pass to a lower energy level by forming ionic bonds. The formation of sodium chloride is exothermic. This picture of electrovalency proves to be more fruitful than the simple picture of attaining a noble gas electron configuration. Elements will not form an ionic compound if it is at a higher energy level than the elements. They may combine by the formation of covalent bonds.

CHECKPOINT 4D: IONS

1. Write the formulae of the electrovalent compounds which contain the following pairs of ions:

(*a*) Mg^{2+} and N^{3-}

(*b*) Al^{3+} and F^-

(*c*) Al^{3+} and S^{2-}

(*d*) Fe^{2+} and O^{2-}

(*e*) Fe^{3+} and O^{2-}

(*f*) Co^{3+} and SO_4^{2-}

(*g*) Ni^{2+} and NO_3^-

2. Write the electron configuration of each of the following ions, and give the name of the *isoelectronic* noble gas (which has the same electron configuration):

Li^+, N^{3-}, Be^{2+}, K^+, S^{2-}

(Atomic numbers are: Li = 3, N = 7, Be = 4, K = 19, S = 16.)

3. Write the electron configurations for the following cations:

Mn^{2+}, Cu^+, Cu^{2+}, Zn^{2+}

(Atomic numbers are: Mn = 25, Cu = 29, Zn = 30.) With which noble gases are they isoelectronic?

4.3 THE COVALENT BOND

Some compounds are non-electrolytes. Since these compounds do not conduct electricity, they cannot consist of ions. Non-electrolytes contain a type of chemical bond which differs from the ionic bond.

In the single covalent bond, two atoms share one pair of electrons. By sharing, the bonded atoms both gain a full outer shell of electrons

Two atoms of chlorine combine to form a molecule, Cl_2. Both chlorine atoms have the electron configuration Cl(2.8.7). G N Lewis suggested that the bond involves each of the two chlorine atoms sharing one of its outermost electrons – **valence electrons** as they are termed – with the other chlorine atom. The two atoms have to approach sufficiently closely for their atomic orbitals to overlap. The shared pair of electrons is called a **covalent bond**. They occupy the same orbital with opposing spins [§ 2.4]. The Cl_2 molecule can be represented as shown in Figure 4.16:

FIGURE 4.16 Ways of Representing the Chlorine Molecule

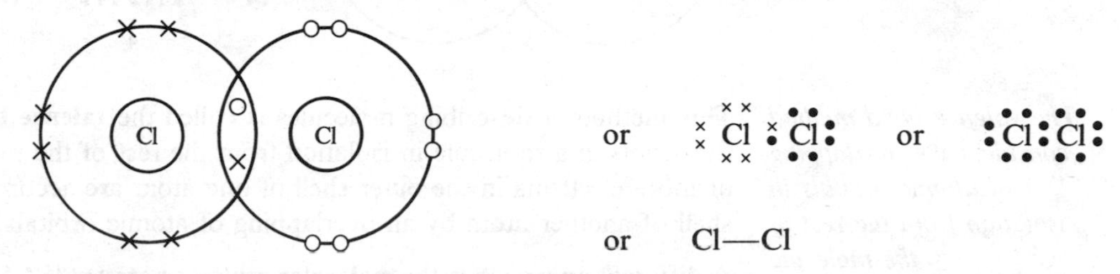

There are not really two types of electron, but it makes it easier to count the electrons if those from one atom are represented as crosses and those from the other as dots. By sharing a pair of electrons, each of the chlorine atoms has obtained eight electrons in its outer shell: it has 'completed its octet'. Electrons are shared when half-filled atomic orbitals of adjacent atoms overlap in space.

Covalent bonding is important in carbon compounds. The carbon atom, with four valence electrons, can attain a full octet by sharing one electron with each of four hydrogen atoms. The bonding in methane, CH_4, can be shown by a 'dot-and-cross' diagram [see Figure 4.17]. The hydrogen electrons are shown as dots and the carbon electrons as crosses. Carbon has completed its octet, and hydrogen has attained the noble gas configuration of helium by completing its 1s shell.

A double bond or a triple bond is formed when two atoms share two or three pairs of electrons

In carbon dioxide, the carbon atom shares two electrons with each of two oxygen atoms, in order to give all three atoms a full octet of valence electrons. [See Figure 4.18.]

FIGURE 4.17 Ways of Representing the Bonding in Methane

FIGURE 4.18 Ways of Representing the Carbon Dioxide Molecule

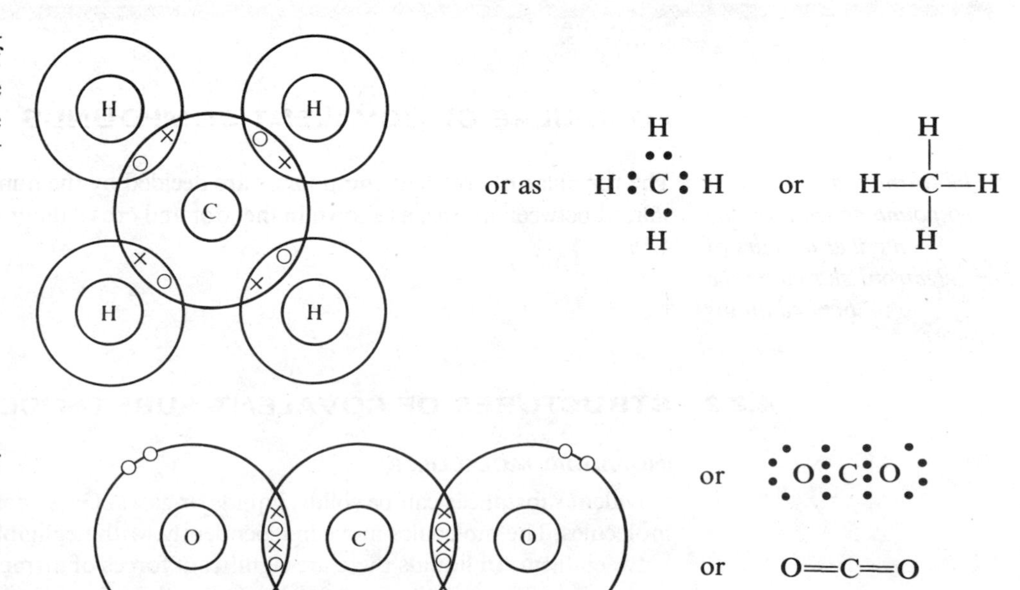

As each shared pair of electrons is a covalent bond, the two pairs of shared electrons between carbon and oxygen constitute a double bond. The pairs of electrons on the oxygen atoms which are not shared are described as **'lone pairs' of electrons**.

In a molecule of nitrogen, each nitrogen atom, with five electrons, needs to share three of its electrons with the other atom of nitrogen in order to complete its octet. The bonding can be written as shown in Figure 4.19.

FIGURE 4.19 Ways of Representing the N_2 Molecule

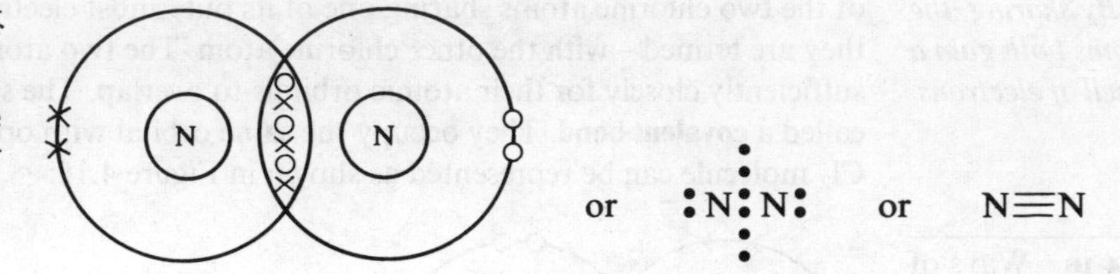

The valence bond method considers the overlapping of atomic orbitals in isolation from the rest of the molecule

This method of describing molecules is called the **valence bond method**. It considers the atoms in a molecule in isolation from the rest of the molecule, except that one or more electrons in the outer shell of one atom are accommodated in the outer shell of another atom by an overlapping of atomic orbitals.

A different approach is the **molecular orbital treatment** [§ 4.5.1].

CHECKPOINT 4E: BONDING IN COVALENT COMPOUNDS

1. The electron configurations of hydrogen and fluorine are H(1) and F(2.7). Show by means of a sketch (like those in Figures 4.16 to 4.19) what happens when H and F combine to form a covalent bond in HF.

2. Sketch the configuration of electrons in a molecule of the covalent compound CH_3Cl. (Electron configurations are C(2.4), H(l) and Cl(2.8.7).)

3. Sketch the configuration of electrons in a molecule of the covalent compound, NH_3. (N has the electron configuration 2.5.)

4. By means of a 'dot and cross' diagram, such those in Figures 4.6 to 4.19, sketch the arrangement of electrons in the molcule O_2.

5. Sketch the arrangement of electrons in a molecule of ethene, $H_2C{=}CH_2$.

4.3.1 FORMULAE OF COVALENT COMPOUNDS

The formula of a covalent compound depends on the number of pairs of electrons shared by the bonded atoms

The formulae of covalent compounds are decided by the number of pairs of electrons shared between atoms, as shown in the 'dot and cross' diagrams in Figures 4.16 to 4.19.

4.3.2 STRUCTURES OF COVALENT SUBSTANCES

INDIVIDUAL MOLECULES

Covalent substances can be solids, liquids or gases. Gases consist of individual molecules. The molecules move independently, with negligible forces of attraction between them. In liquids there are significant forces of attraction between the molecules. However, these are weaker forces than those in solids, and allow molecules to change their relative positions easily, so the substance can flow.

FIGURE 4.20
Models of Molecules of (a) CO_2, (b) CH_4, (c) H_2O, (d) $H_2C{=}CH_2$

MOLECULAR STRUCTURES

Many covalent substances are composed of individual molecules. In some covalent elements and compounds molecular structures form as weak forces which act between molecules

In some covalent substances, the attractive forces between molecules are strong enough to make the substances solids. For example, iodine is a shiny black crystalline solid. Forces of attraction operate between the molecules of iodine to hold them in a three-dimensional structure [see Figure 4.21]. There are two kinds of chemical bonds in it: the **intramolecular** bonds (inside the molecules), which are strong covalent I—I bonds, and the **intermolecular forces** (between individual molecules), which are weak forces of attraction between I_2 molecules. The origin of these weak forces of attraction is discussed in § 4.5.2. A structure of this kind, which consists of individual molecules bonded together in a regular arrangement by weak intermolecular bonds, is called a **molecular structure**. Other examples are described in § 6.4.

In a molecular structure, the intermolecular bonds are easily broken to allow molecules to move independently, that is, to enter the liquid phase. Such covalent substances have melting temperatures lower than those of ionic substances. Many are liquid or gaseous at room temperature. The boiling temperatures of many covalent substances are low for the same reason.

FIGURE 4.21
The Structure of an Iodine Crystal

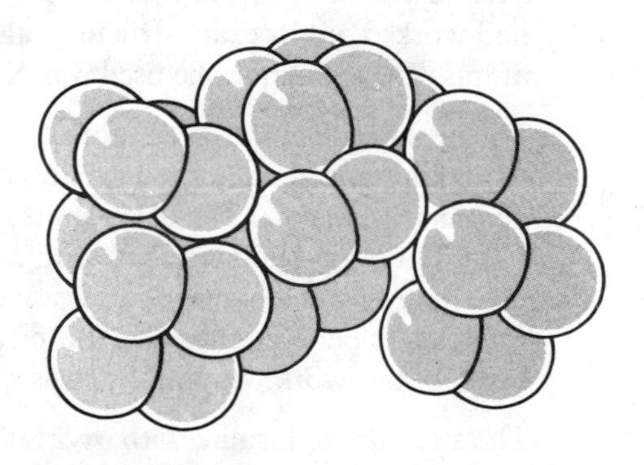

MACROMOLECULAR STRUCTURES

Other covalent substances are composed of macromolecules

Some covalent substances do not consist of individual molecules. They are **macromolecular structures**, or **giant molecular structures**. One example is diamond. Diamond is the hardest naturally occurring substance. The extraordinary properties of diamond arise from its structure. A crystal of diamond contains millions of atoms. Figure 4.22 shows how every carbon atom is joined by covalent bonds to four other carbon atoms. It is very difficult to break this macromolecular structure, and this is why diamond is so hard.

Other macromolecular structures, e.g. graphite [Figure 6.15], boron nitride (BN), silicon(IV) oxide [Figure 6.14] and silicon carbide (SiC) are described in § 6.5. The strong covalent bonds holding these giant molecules together give the substances high melting temperatures (higher than most ionic solids) and make them **involatile** (i.e. they have high boiling temperatures).

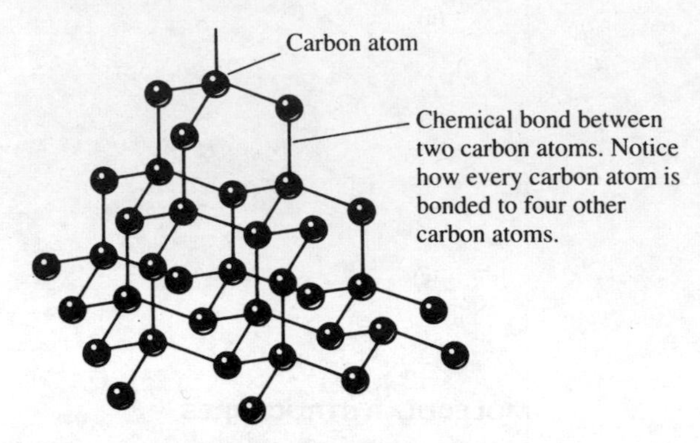
Carbon atom
Chemical bond between two carbon atoms. Notice how every carbon atom is bonded to four other carbon atoms.

FIGURE 4.22 The Bonding in Diamond

Profile: Dorothy Crowfoot Hodgkin (born 1910)

X ray crystallography has unravelled the molecular structures of three life-saving compounds: penicillin, vitamin B_{12} and insulin. This feat was accomplished by Dorothy Crowfoot Hodgkin. Dorothy Crowfoot was the daughter of archaeologists who worked in Egypt and the Sudan. Dorothy studied at the Universities of Oxford and Cambridge before becoming a fellow of Somerville College, Oxford. She decided to continue the work in X ray crystallography which she had started in Cambridge. Early in her scientific career, she married Thomas Hodgkin who gave his wife unfailing support and encouragement in all the challenges she undertook.

One of her first problems was penicillin. Although penicillin had been discovered by Sir Alexander Fleming in 1929, it was not isolated as a pure compound until 1941. At that time, Howard Florey and Ernst Chain were working to develop a method of culturing penicillin so that it could be used as a bactericide to treat casualties in the Second World War. They gave Dorothy Hodgkin 10 mg of the substance. By 1945 she had worked out its exact structure, allocating a position in space to each of the 41 atoms. The technique she used was X ray diffraction.

The next problem was even more taxing. In 1948 Glaxo laboratories gave Dorothy Hodgkin some beautiful red crystals of a substance which was effective against pernicious anaemia. The discoverers of this compound called it vitamin B_{12}. At first sight, Dorothy estimated that, with over 90 atoms, the structure of vitamin B_{12} would take many years to work out. However, she was able to use an advanced computer in the USA to perform the calculations, and the structure was completed in 1956. As a result of this work, vitamin B_{12} was synthesised.

The structure of insulin, with over 800 atoms, was a formidable task. With the new computing power at her disposal, Dorothy Hodgkin completed the work in 1969. In 1964 she received the Nobel Prize for Chemistry. Two other women have won this prize: Marie Curie in 1911 and her daughter Irene Joliot-Curie in 1935.

4.4 PROPERTIES OF IONIC AND COVALENT SUBSTANCES

The physical characteristics of substances, for example, whether they are solids, liquids or gases, depend on the types of chemical bond in the substances. Chemical behaviour also depends on the type of bond present.

TABLE 4.3
Characteristics of Ionic and Covalent Substances

The properties of an element or a compound depend on the type of chemical bond present. The physical state (s, l, g), the melting and boiling temperatures, the solubility and the electrolytic conductivity depend on whether ionic bonds or covalent bonds or weak intermolecular bonds are present

Ionic bonding	*Covalent bonding*
Ionic compounds are formed when metallic elements combine with non-metallic elements.	Atoms of non-metallic elements combine by forming covalent bonds.
(a) Atoms of metallic elements form positive ions. Elements in Groups 1, 2 and 3 of the Periodic Table form ions with charges +1, +2 and +3, e.g. Na^+, Ca^{2+}, Al^{3+}. (b) Elements of non-metallic elements form negative ions. Elements of Groups 6 and 7 of the Periodic Table form negative ions with charges −2 and −1, e.g. O^{2-} and Br^-.	The maximum number of covalent bonds that an atom can form is equal to the number of electrons in the outer shell. Often, an atom does not use all its electrons in covalent bond formation.
The ionic bond is a strong electrostatic attraction between ions of opposite charge. An ionic compound is composed of a giant regular structure consisting of millions of ions. Ionic compounds are crystalline because of this structure. The strong bond between ions of opposite charge makes it difficult to separate the ions. This is why ionic compounds have high melting temperatures and boiling temperatures.	There are three types of covalent substances: (a) Individual molecules. The bonds which hold the atoms together are strong, but there are negligible forces of attraction between molecules. Such covalent substances are gases, e.g. HCl, SO_2, CO_2, CH_4. (b) Molecular structures. Between individual molecules there are weak forces of attraction. Such covalent substances are low boiling temperature liquids, e.g. ethanol, and low melting temperature solids, e.g. iodine [Figure 4.21] and solid carbon dioxide, which are molecular structures. (c) Macromolecular structures or giant molecular structures. Large numbers of atoms link in chains or sheets, e.g. graphite [Figure 6.15] or in 3-dimensional structures, e.g. diamond [Figures 4.22 and 6.11] and quartz [Figure 6.14]. Substances with giant molecular structures have high melting and boiling temperatures.
Ionic compounds are electrolytes.	Covalent compounds are non-electrolytes.
Many ionic compounds dissolve in water. They are insoluble in organic solvents.	Covalent compounds are often insoluble in water. Many dissolve in organic solvents such as ethanol or propanone.

CHECKPOINT 4F: CHEMICAL BONDS

1. Say what kind of bonding you would expect between the following pairs of elements:

(*a*) Li and Br, (*b*) Sr and Cl, (*c*) Ba and S, (*d*) Ca and O, (*e*) S and O, (*f*) F and Cl, (*g*) Cl and O. [See the Periodic Table on p. 792 for help.]

2. From the information in Table 4.4, say what you can about the chemical bonds in **A**, **B**, **C**, **D** and **E**.

TABLE 4.4

Substance	*State*	*Melting temperature (°C)*	*Does it conduct electricity?*
A	Liquid	−60	Does not conduct electricity.
B	Solid	890	Conducts electricity when molten.
C	Solid	720	Does not conduct electricity when molten.
D	Solid	85	Does not conduct electricity when molten.
E	Gas	−100	Does not conduct electricity

4.5 COVALENT COMPOUNDS

4.5.1 THE MOLECULAR ORBITAL TREATMENT

The molecular orbital method considers the molecule as a whole ...

... and calculates the electron density over the whole molecule

We studied the valence bond method of describing molecules in §4.3. Another approach is to consider the *entire* molecule as a unit. Each electron is under the influence of all the nuclei and the electrons in the molecule. The atomic orbitals are replaced by molecular orbitals. The molecular orbital method of viewing the covalent bond uses **quantum mechanics** to calculate the distribution of electron density over the molecule. Figure 4.23 shows the results of the calculation for the hydrogen molecule. The contour lines join regions of the same electron density.

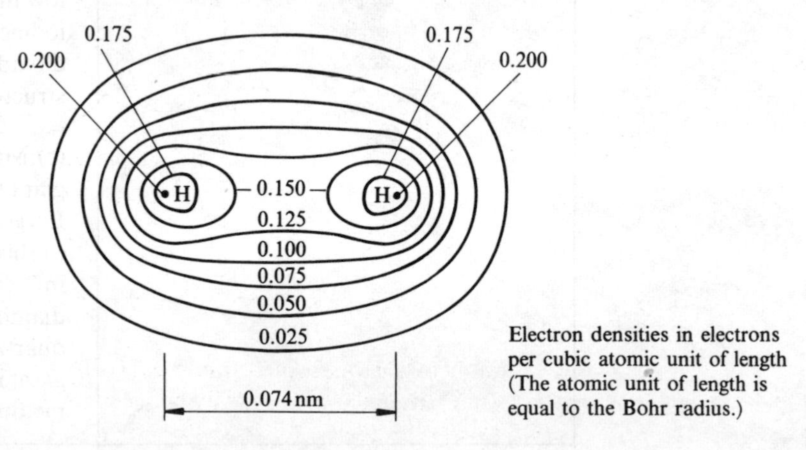

FIGURE 4.23 Electron Density Map for the Hydrogen Molecule

In the molecular orbital model the 2H in H_2 are held together by the attraction of the nuclei for the electron cloud between the two nuclei

The highest electron density is near each nucleus. There is also a region of high electron density between the two nuclei called the **electron cloud**. The electron density between the nuclei screens the nuclei from one another and prevents repulsion between the two positive nuclei from driving the atoms apart. Although there is a force of repulsion between the positively charged nuclei, the force of attraction between each nucleus and the electron cloud between the two nuclei is greater. It is this attraction that holds the atoms together in the molecule. [See Figure 4.24]

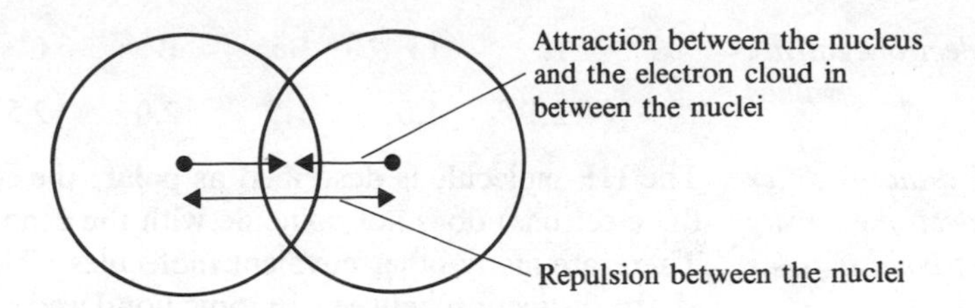

FIGURE 4.24 Attraction and Repulsion in the H_2 Molecule

4.5.2 COVALENT RADII

Covalent radii (atomic radii)

The distance between the nuclei of covalently bonded atoms is the sum of their **covalent radii**. Covalent radii, which are also called **atomic radii**, are *additive*. The sum of the covalent radii of chlorine and hydrogen gives the length of the covalent bond in hydrogen chloride [see Figure 4.25]. This figure also shows the **van der Waals radii**, which will be covered in § 4.7.2.

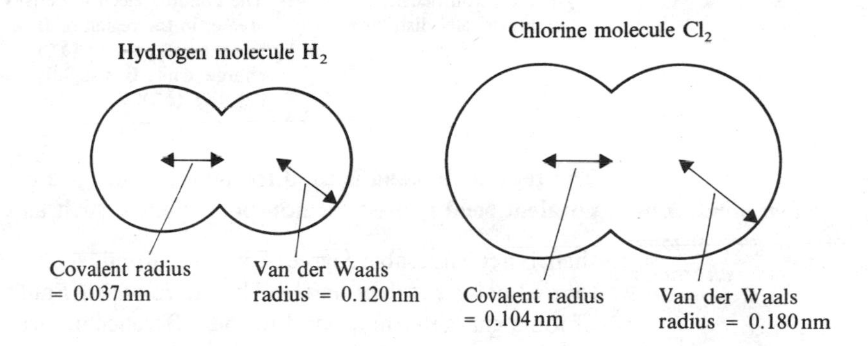

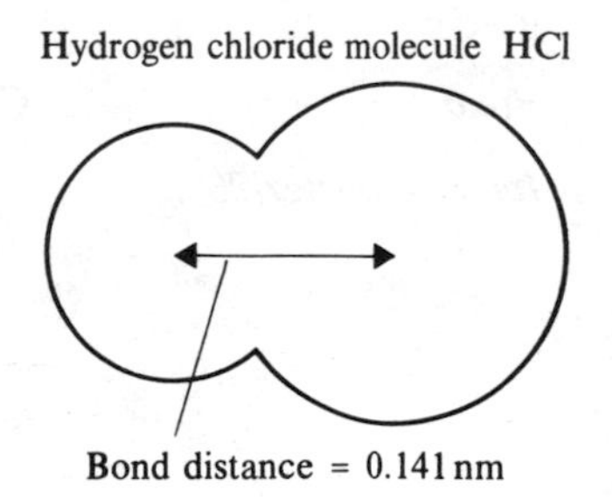

FIGURE 4.25 Covalent Radii and Van der Waals Radii for Hydrogen and Chlorine

4.5.3 ELECTRONEGATIVITY AND THE COVALENT BOND

Electronegativity is the ability of an atom in a covalent bond to attract electrons

Fluorine is the most electronegative of elements

In a bond between identical atoms, e.g. H—H, the electron density of the bonding orbital is distributed symmetrically between the bonded atoms [see Figure 4.23]. In a bond between different atoms, the bonding electrons may be more attracted to one of the bonded atoms than to the other. In the molecule HF, for example, the electron density of the bonding electrons lies more towards the fluorine atom than towards the hydrogen atom. The ability of an atom in a covalent bond to attract the bonding electrons is called **electronegativity**. Thus, fluorine is more **electronegative** than hydrogen. Electronegativity is not a quantity which can be measured or to which a unit can be assigned. Pauling derived a scale of relative electronegativity values. He assigned a value of 4.00 to fluorine, the most electronegative of elements. Some of his values are:

Various electronegativity values

H	Li	Be	B	C	N	O	F
2.1	1.0	1.5	2.0	2.5	3.0	3.5	4.0

If the bonded atoms differ in electronegativity a covalent bond is polar

The HF molecule is described as **polar**: the centre of the negative charge (due to the electrons) does not coincide with the centre of positive charge (due to the nuclei). There are many other covalent molecules which, like HF, are polar. There is no sharp distinction between an ionic bond and a covalent bond [see Figure 4.26].

FIGURE 4.26 Bond Types

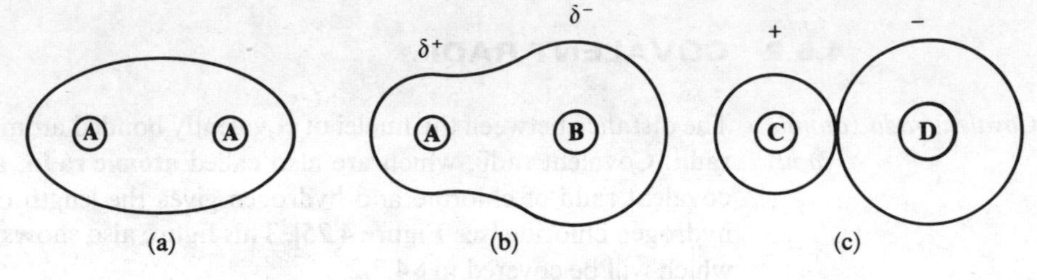

(a) Covalent bond. The electron density is symmetrically distributed.

(b) Polar covalent bond. The bonding electron density is greater in the region of **B**. **A** has a small positive (δ^+) charge, while **B** is slightly negative (δ^-).

(c) Ionic bond. The electron cloud of $\mathbf{C}^+$ does not come under the influence of $\mathbf{D}^-$. The electron cloud of $\mathbf{D}^-$ is not distorted by $\mathbf{C}^+$

Many covalent bonds have some degree of ionic character

The term **ionic bond** is used for bonds which are predominantly ionic. The term **covalent bond** is used for non-polar bonds, such as C—I, and also bonds in which there is a considerable degree of polarity. Bonds such as $\overset{\delta_+}{C}$—$\overset{\delta_-}{Cl}$ and $\overset{\delta_+}{C}$—$\overset{\delta_-}{O}$, are termed **polar covalent bonds**. The curve which Pauling drew to relate the percentage of ionic character in a bond to the difference in electronegativity between the bonded atoms is shown in Figure 4.27. Approximate values for some covalent bonds are given:

Bond	C—I	C—H	C—Cl	C—F
Ionic character/%	0	4	6	40

FIGURE 4.27 Curve relating Percentage of Ionic Character in a Bond to the Difference in Electronegativity between the Bonded Atoms

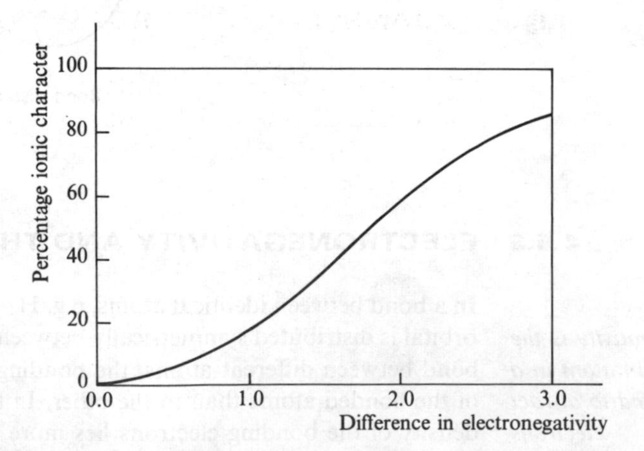

Pauling's electronegativity values can be applied to the consideration of a bond between two atoms, either in a diatomic molecule or between two atoms considered in isolation from other atoms and bonds. They should *not* be applied to a formula unit in a crystal structure as they will give an erroneous impression. If applied to $CaCl_2$, NaCl or LiF, electronegativity values will predict the nature of the bonding

Some ionic compounds show a degree of sharing of electron clouds

in an ion pair considered in isolation, e.g. $Na^+(g)\ Cl^-(g)$. In reality, we are interested in crystalline sodium chloride, in which each Cl^- ion is part of a crystal structure and its electron cloud is influenced symmetrically by six Na^+ ions around it. The net result is that the electron cloud of the Na^+ ion becomes rather cubical in shape, as shown in Figure 4.13, § 4.2.2. The electron cloud is drawn into the spaces between the ions in the structure, but sodium chloride is still overwhelmingly ionic in nature. The electron clouds associated with the cation and the anion influence each other to a more pronounced extent in lithium fluoride [see Figure 4.28]. In contrast, the electron density map of calcium fluoride [see Figure 4.14, § 4.2.2] shows an electron distribution which is, in cross-section, perfectly circular around each ion. Despite the deformation of the electron clouds in sodium chloride and lithium fluoride, these compounds are still ionic in character.

A cation may attract and deform the electron cloud of an anion

FIGURE 4.28 Electron Density Map for Lithium Fluoride

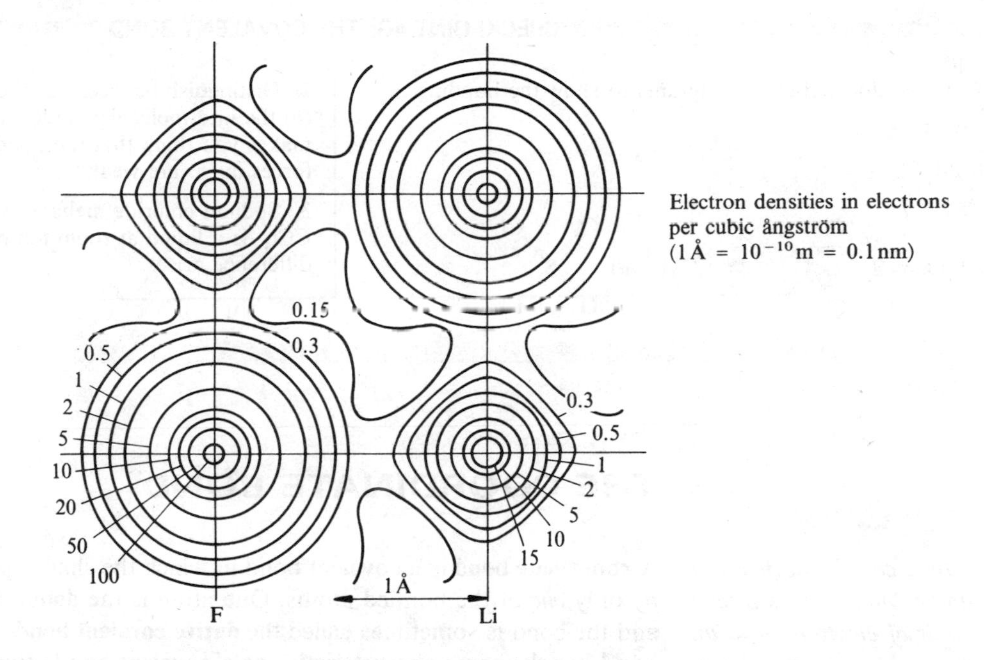

4.5.4 FAJANS' RULES OF BOND TYPE

The distortion of an electron cloud by a neighbouring charged ion or **dipole** is called **polarisation**. Comparing ions with similar nuclear charges, large ions are more easily polarised than small ions. The electrons in small ions are more closely controlled by the positively charged nucleus.

K Fajans formulated two rules to predict the proportions of ionic and covalent character in the bond formed between two atoms.

Fajans' rules predict the ionic and covalent character of bonds

1. Bonds will tend to be ionic if the ions formed are small in charge. For example, sodium chloride is likely to be ionic because Na^+ and Cl^- bear unit charges, whereas aluminium chloride is likely to be covalently bonded because Al^{3+} ions are highly charged.

2. Bonds will tend to be ionic if the radius of the possible cation is large (e.g. the alkali metals) and the radius of the possible anion is small (e.g. the smaller halogens).

A cation with a small radius and a high charge will polarise anions

The sizes of some ions are shown in Figure 15.4, § 15.1. Compare the radii of the ions Na^+, Mg^{2+} and Al^{3+}. The high charge on the Al^{3+} ion results in a small radius because the remaining electrons are drawn in close to the nucleus. The combination of a high charge and a small radius gives Al^{3+} a high **charge density** (i.e. charge/volume ratio), and this enables the cation to attract the electron cloud of a neighbouring anion (or of a molecule). The electron cloud of the anion will be distorted in such a way as to increase the electron density near the cation: the anion has become **polarised**.

A large anion is easily polarised

An anion is larger than the atom from which it was formed, and has one or two more electrons. The nucleus is less able to attract the electrons closely than it is in the parent atom. In a large anion, the electrons are further from the nucleus and less under its control than in a small anion, making the larger anion easier to polarise. If the cation is small and the anion is large, the cation will be able to polarise the anion, and there will be some sharing of the electron cloud of the anion: i.e. the bond will have some covalent character.

CHECKPOINT 4G: THE COVALENT BOND

1. Draw dot-and-cross diagrams to show the bonding in

(*a*) H—C≡N

(*b*) CCl_4

(*c*) O=S(—Cl)—Cl

(*d*) 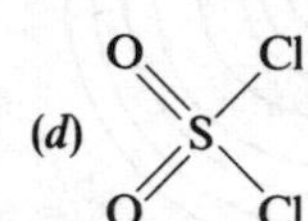

2. Distinguish between (*a*) the intramolecular bonds, and (*b*) the intermolecular forces in chloroform, $CHCl_3$. What makes you think that the bonds in (*a*) are strong and the forces in (*b*) are weak?

3. Sodium chloride melts at 800 °C. Tetrachloromethane, CCl_4, is a liquid at room temperature. Explain how this difference arises.

4.6 THE COORDINATE BOND

In a coordinate bond, a donor atom shares a lone pair of electrons with an acceptor atom

Once formed, a coordinate bond is like a covalent bond

A **coordinate bond** is a covalent bond in which the shared pair of electrons is provided by only *one* of the bonded atoms. One atom is the **donor**, the other is the **acceptor**, and the bond is sometimes called the **dative covalent bond**. Once formed, a coordinate bond has the same characteristics as a covalent bond. For an atom to act as a donor, it must have at least one pair of unshared electrons (a lone pair) in its outermost shell (the valence shell). The acceptor has at least one vacant orbital in its outer shell. It may be a metal cation or a transition metal atom or an atom in a molecule.

In water, H—$\overline{\underline{O}}$—H, the oxygen atom has two lone pairs of electrons. They can be shared with an atom which needs them to complete its valence shell. A proton H^+ has an empty 1s orbital. By accepting a pair of electrons from oxygen, it achieves a full s shell:

The oxonium ion, H_3O^+, is formed by coordination

$$\begin{array}{c}H\\ \cdot\cdot\\ H:\!O\!: \\ \cdot\cdot\end{array} + H^+ \longrightarrow \left[\begin{array}{c}H\\ \cdot\cdot\\ H:\!O\!:H \\ \cdot\cdot\end{array}\right]^+$$

The species formed is an **oxonium** ion, H_3O^+. The positive charge contributed by the proton is spread over the whole ion. The proton has one unit of positive charge spread over a surface area which is minute compared with other ions. The high charge density makes it extremely reactive: the proton cannot exist by itself. It is stabilised by the coordination of water molecules.

Water coordinates to metal ions...

Water also coordinates to metal ions. The fact that bonds are formed between metal ions and water is responsible for the solubility of many salts. Energy is required to break the bonds holding ions together in a crystal structure. If energy is given out

... an example is $[Ca(H_2O)_6]^{2+}$

when coordinate bonds form between metal ions and water, this may swing the balance in favour of solution. Metal ions are hydrated, e.g. $[Ca(H_2O)_6]^{2+}$:

$$Ca^{2+} + 6H:\ddot{\underset{..}{O}}:H \longrightarrow \left[Ca\left(:\overset{H}{\underset{..}{\ddot{O}}}:H\right)_6\right]^{2+}$$

The lone pair on N in NH_3 is responsible for the formation of coordination compounds by NH_3

The nitrogen atom in ammonia, $H\text{—}N(\text{—}H)_2|$, has an unshared pair of electrons which it is able to share with an atom that needs two electrons to complete its octet. When ammonia meets the vapour of aluminium fluoride, which consists of covalent AlF_3 molecules, a white solid of formula NH_3AlF_3 is formed. It is formed by the coordination of the lone pair of the nitrogen atom into the valence shell of the aluminium atom as shown below [see also Question 3, Checkpoint 4H].

$$H:\underset{H}{\overset{H}{\ddot{N}}}: + \underset{F}{\overset{F}{Al}}:F \longrightarrow H:\underset{H}{\overset{H}{N}}:\underset{F}{\overset{F}{Al}}:F$$

or

$$H\text{—}\underset{H}{\overset{H}{N}}: + \underset{F}{\overset{F}{Al}}\text{—}F \longrightarrow H\text{—}\underset{H}{\overset{H}{N}}\rightarrow\underset{F}{\overset{F}{Al}}\text{—}F$$

The symbol $\rightarrow$ is used for a coordinate bond, an arrow pointing from the donor towards the acceptor.

Ions such as $CuCl_4^{2-}$ are formed by coordination...

Copper(II) chloride is blue in solution. In the presence of a high concentration of chloride ions, the solution turns a very deep green. The colour is due to the formation of $CuCl_4^{2-}$ ions. Coordinate bonds form between Cl^- ions and Cu^{2+} ions:

$$4:\ddot{\underset{..}{Cl}}:^- + Cu^{2+}(aq) \longrightarrow \left[:\ddot{Cl}:\underset{:\ddot{Cl}:}{\overset{:\ddot{Cl}:}{Cu}}:\ddot{Cl}:\right]^{2-}$$

or

$$4Cl^- + Cu^{2+} \rightarrow \left[Cl\rightarrow \underset{\uparrow \atop Cl}{\overset{Cl \atop \downarrow}{Cu}} \leftarrow Cl\right]^{2-}$$

... they are called complex ions

The copper atom now has eight electrons in its valence shell. Ions such as $CuCl_4^{2-}$ and $Cu(NH_3)_4^{2+}$, which are formed by the combination of an ion with an oppositely charged ion or a molecule are called **complex ions**.

CHECKPOINT 4H: THE COORDINATE BOND

1. Draw a dot-and-cross diagram to show the arrangement of valence electrons in the complex ion $Cu(H_2O)_4^{2+}$. It is formed by coordination of H_2O molecules on to a Cu^{2+} ion.

2. Draw a dot-and-cross diagram to show the bonding in the complex ion $Fe(H_2O)_6^{3+}$. How many electrons has the iron atom in its valence shell? How can it accommodate this number?

3. Explain the bonding in the compound $NH_3 \cdot BF_3$.

4. The formula for carbon monoxide can be written $|C \overset{\leftarrow}{=} O|$. Draw a dot-and-cross diagram for this molecule.

The lone pair on the oxygen atom can be used to form a coordinate bond to a nickel atom. By means of a diagram, show the bonding in nickel carbonyl, $Ni(CO)_4$.

5. Explain the bonding in the tetraammine copper ion, $Cu(NH_3)_4^{2+}$.

6. The slightly soluble compound lead(II) chloride dissolves in concentrated hydrochloric acid to form a soluble complex ion. What do you think the formula of this ion might be? Explain the bonding in the ion.

4.7 INTERMOLECULAR FORCES

Dipole–dipole interactions, van der Waals forces and hydrogen bonds are all intermolecular forces

Intermolecular forces and bonds are of a number of types: dipole–dipole interactions, van der Waals forces and the hydrogen bond.

4.7.1 DIPOLE–DIPOLE INTERACTIONS

Attractions between dipoles result in an ordered arrangement of molecules

In the solid state, polar molecules interact to form an ordered arrangement. Dipole–dipole interactions between the molecules lead the molecules to pack in such a way that partial positive charges will be adjacent to partial negative charges [see Figure 4.29].

Solvation helps solids to dissolve

Polar solids dissolve in polar solvents. The energy required to break up the crystal is recouped by the energy released when polar solute molecules interact with polar solvent molecules. [See Figure 4.29.] This interaction is called **solvation**; if the solvent is water, it is called **hydration** [§ 17.9].

FIGURE 4.29 The Process of Dissolution

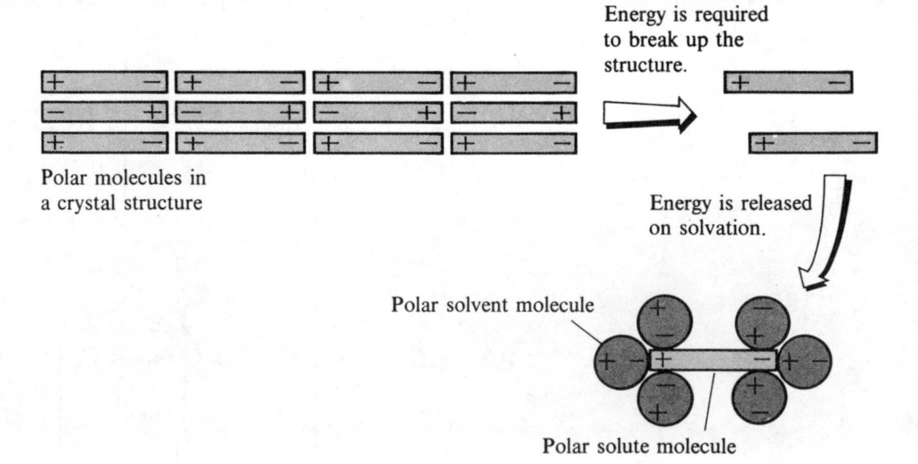
Energy is required to break up the structure.
Polar molecules in a crystal structure
Energy is released on solvation.
Polar solvent molecule
Polar solute molecule

4.7.2 VAN DER WAALS FORCES

There are different types of van der Waals forces...

When molecules pack together in the liquid or solid state, there must be forces of attraction between them. J D van der Waals postulated the existence of forces of attraction and repulsion that are neither ionic nor covalent. Such forces arise in a number of ways and are collectively called **van der Waals forces**. Dipole–dipole interactions [§4.7.1] between polar molecules are one type of van der Waals force.

...Attractive forces exist between non-polar molecules...

Attractive forces exist also between non-polar molecules. Even atoms of the noble gases are attracted to one another to a slight degree; this is why the noble gases can be liquefied. Consider two non-polar molecules which are very close together. Since they are non-polar, the arrangement of electrons is *on the average* symmetrical. Yet, *at any given instant*, the electron distribution in one molecule may be unsymmetrical. There may be a dipole in the molecule *for an instant*. Figure 4.30 shows how a temporary dipole in one molecule **A** can attract the electron cloud of a neighbouring molecule **B**. This means that both molecules will have dipoles, and the direction of the dipoles will be such that they attract one another. Since the electrons are moving about at high speed, the attraction has a fleeting existence. In the next instant, the dipole in **A** may be in the opposite direction. Again, the dipole which it **induces** in **B** will result in an attraction. The dipoles are temporary, but the net attraction which they produce is permanent.

...These forces are thought to be due to the momentary polarisation of molecules...

...leading to a permanent attraction

FIGURE 4.30 Attraction between Momentary Dipoles

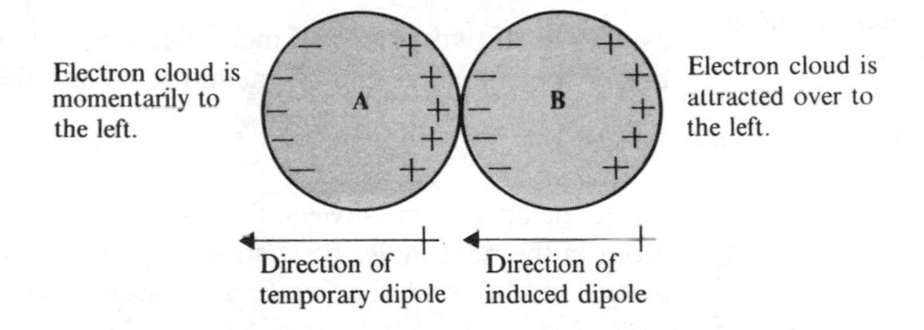
Electron cloud is momentarily to the left.
A
B
Electron cloud is attracted over to the left.
Direction of temporary dipole
Direction of induced dipole

The ease with which an electron cloud is distorted is called polarisability

The ease with which an electron cloud is distorted, and therefore the ease with which a dipole is induced, is called **polarisability**. Polarisability increases with the number of electrons in a molecule, and the strength of the forces of attraction therefore increases with molar mass. The shape of molecules is another factor: elongated molecules are more easily polarised than compact, symmetrical molecules.

Van der Waals forces determine distances between atoms in condensed phases

If a pair of molecules are far apart, there will be no induction of dipoles and no attractive forces between them. Should the molecules move too close together, repulsion between electron shells will predominate over the induction effect and drive the molecules apart. Figure 4.31 shows how closely argon atoms can approach in the liquid state. Half the distance between argon atoms at their closest distance of approach is the **van der Waals radius** of the atom.

FIGURE 4.31 Van der Waals Forces determine the Distance between Atoms in the Liquid State

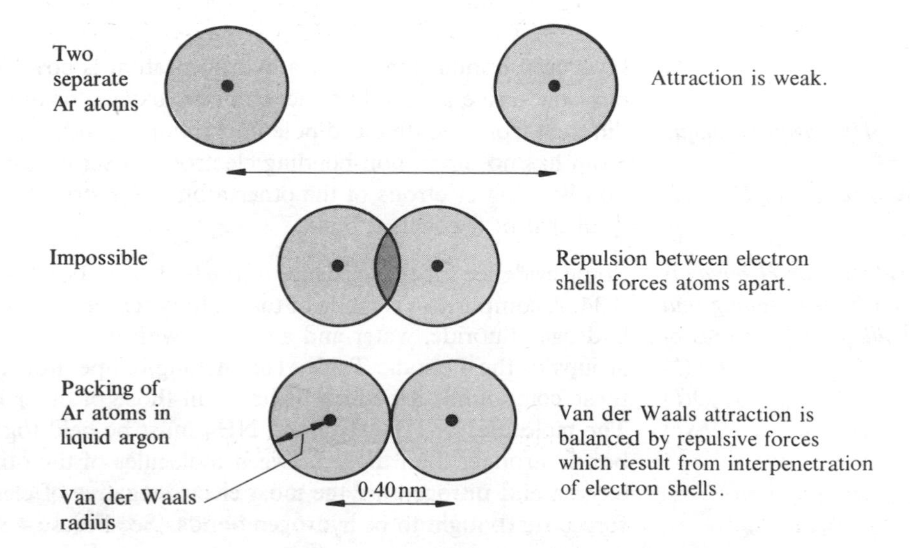
Two separate Ar atoms
Attraction is weak.
Impossible
Repulsion between electron shells forces atoms apart.
Packing of Ar atoms in liquid argon
Van der Waals radius
0.40 nm
Van der Waals attraction is balanced by repulsive forces which result from interpenetration of electron shells.

Van der Waals forces are stronger between large linear molecules

All types of van der Waals forces between small molecules are weak. Between molecules with long chains of atoms, giving many points of contact, van der Waals forces are stronger. This is why in the series of alkanes [§ 26.2] ethane, C_2H_6, is a gas at stp, hexane, C_6H_{14}, is a liquid, and octadecane, $C_{18}H_{38}$, is a solid. Branched-chain hydrocarbons are more **volatile** (have lower boiling temperatures) than unbranched-chain hydrocarbons. Hydrocarbon polymers have molecules which are continuous chains, containing thousands of repeating units. The long, strand-like molecules of poly(ethene), $-(CH_2\text{-}CH_2)_n-$ [§ 27.7.10] can align themselves to give thousands of contacts between atoms and set up very strong van der Waals forces. Poly(ethene) is an extremely tough material, which is used for the manufacture of laboratory and kitchen ware.

4.7.3 THE HYDROGEN BOND

THE NATURE OF THE BOND

When hydrogen is combined with an electronegative atom intermolecular hydrogen bonds can be formed

The bond in hydrogen fluoride is a polar covalent bond [§ 4.5.3] which can be written as $\overset{\delta+}{H}-\overset{\delta-}{F}$. If two polar HF molecules are close enough, there is an attraction between the positive end of one molecule and the negative end of the other molecule:

$$\overset{\delta+}{H}-\overset{\delta-}{F}\cdots\cdots\overset{\delta+}{H}-\overset{\delta-}{F}$$

Since the attraction between the hydrogen atom in one molecule and the fluorine atom in the next molecule is stronger than the repulsion between the two hydrogen atoms and between the two fluorine atoms, the two molecules are bonded together.

The attraction between a hydrogen atom in one molecule and an electronegative atom, such as fluorine, in another molecule is called a **hydrogen bond**. Hydrogen bonding holds a number of molecules of hydrogen fluoride together in liquid hydrogen fluoride [see Figure 4.32]. A molecule of HF forms an even stronger hydrogen bond to a fluoride ion, F^-. The ion $[F\cdots\cdots H\cdots\cdots F]^-$ is formed. Hydrogen fluoride forms acid salts, e.g. KHF_2, containing this anion.

FIGURE 4.32 Hydrogen Bonding in HF(l)

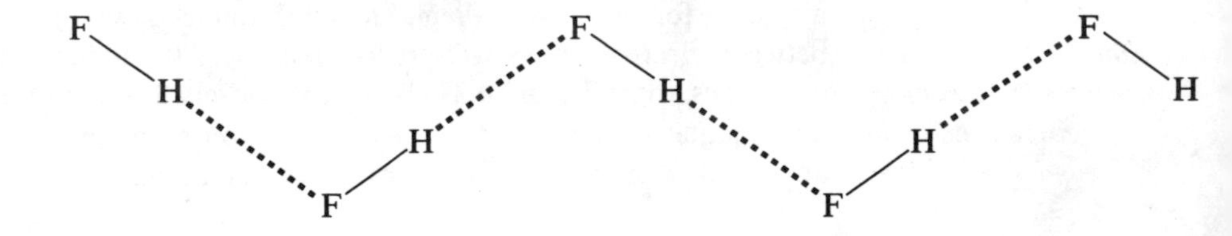

Hydrogen bonds form when a hydrogen atom is covalently bonded to one of the electronegative atoms, fluorine, chlorine, oxygen or nitrogen. Hydrogen bonding is different from the dipole–dipole interactions in other polar molecules. A hydrogen atom has no inner, non-bonding electrons to set up forces of repulsion with the non-bonding electrons of the other atom. The strength of a hydrogen bond is $\frac{1}{10}$ to $\frac{1}{20}$ of that of a covalent bond.

Hydrogen is unique

The strength of H bonds

Evidence for H bonds is given by the melting and boiling temperatures of HF(l) $H_2O(l)$ $NH_3(l)$

Some evidence for the existence of the hydrogen bond is shown in Figures 4.33 and 4.34. A comparison is made of the melting temperatures and boiling temperatures of hydrogen fluoride, water and ammonia with those of other hydrides in the same groups of the Periodic Table. The melting temperatures and boiling temperatures of these compounds are much higher than those of other hydrides in the same groups. The molecules of HF, H_2O and NH_3 must be held together by intermolecular bonds stronger than those between molecules of the other hydrides. Since fluorine, oxygen and nitrogen are the most electronegative of elements, the intermolecular forces are thought to be hydrogen bonds. [See Figure 4.35.]

FIGURE 4.33 Graph of Melting Temperature against Period Number

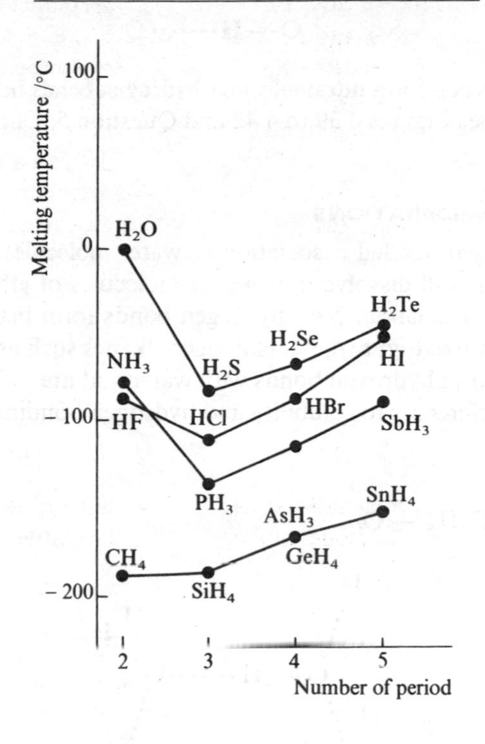

FIGURE 4.34 Graph of Boiling Temperature against Period Number

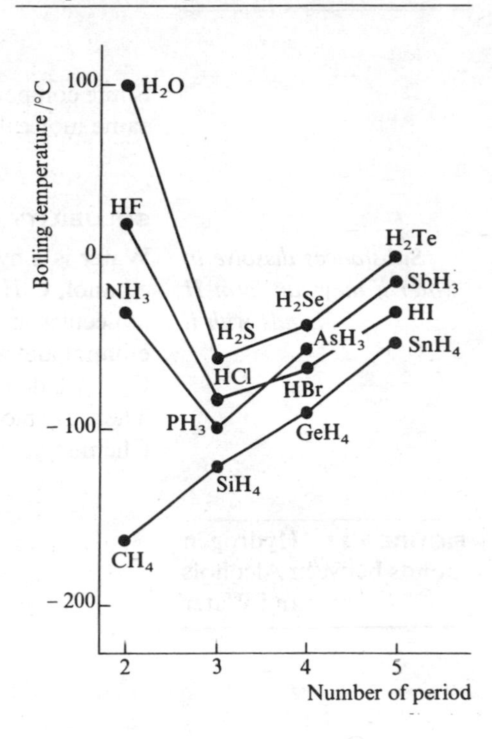

FIGURE 4.35 Hydrogen Bonding in $H_2O(l)$ and $NH_3(l)$

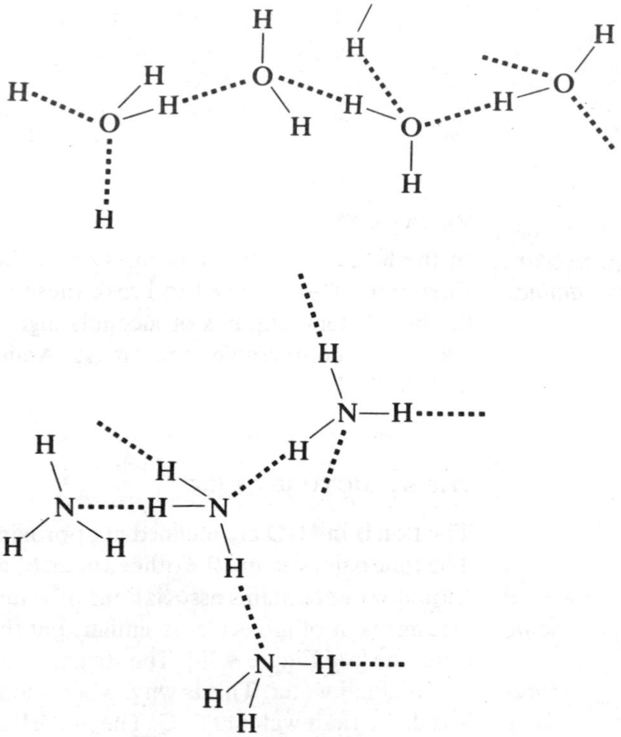

When the molar masses of carboxylic acids are found from measurements in the vapour phase and from solutions in organic solvents, the values are often up to twice the values calculated from the formulae. The molecules are thought to dimerise through the formation of hydrogen bonds [see Figure 4.36].

FIGURE 4.36 H Bonding in RCO_2H

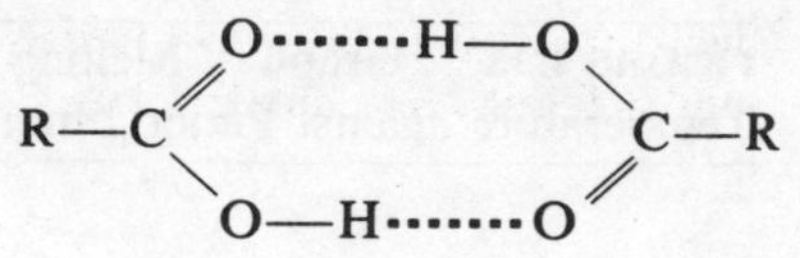

$$R-C\overset{\displaystyle O\cdots\cdots H-O}{\underset{\displaystyle O-H\cdots\cdots O}{}}C-R$$

Some compounds can form **intramolecular hydrogen bonds** between two groups in the same molecule. [See Figures 4.39 to 4.42 and Question 5, Checkpoint 4I.]

SOLUBILITY CONSIDERATIONS

Substances dissolve in water if they can form H bonds with it

Water is a hydrogen-bonded association of water molecules. A substance such as ethanol, C_2H_5OH, will dissolve in water as molecules of ethanol can displace water molecules in the association. New hydrogen bonds form between molecules of ethanol and water [see Figure 4.37]. Halogenoalkanes such as chloroethane, C_2H_5Cl, do not form hydrogen bonds with water and are only slightly soluble. There are more references to solubility and hydrogen bonding in Part 4: Organic Chemistry.

FIGURE 4.37 Hydrogen Bonds between Alcohols and Water

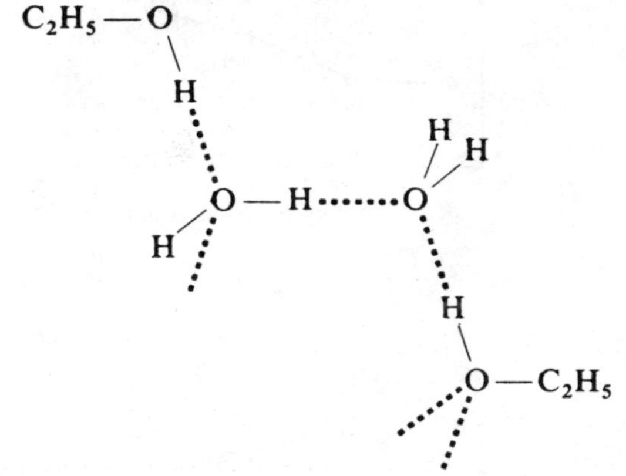

C_2H_5-O ; H ; H H ; O—H······O ; H ; H ; $O-C_2H_5$

VOLATILITY

Hydrogen bonding occurs in alcohols and amines

In the liquid state, the molecules of alcohols are associated by hydrogen bonding. Energy must be supplied to break these bonds when the liquid is vaporised, making the boiling temperatures of alcohols higher than those of non-associated liquids, e.g. alkanes, of comparable molar mass. Amines also are hydrogen-bonded in the liquid state [§ 32.3].

THE STRUCTURE OF ICE

Liquid water is associated by H bonding

Hydrogen bonding extends throughout the whole structure in ice...

The bonds in H_2O are inclined at approximately the tetrahedral angle of 109.5°. The lone pairs occupy the other apices of the tetrahedron [see Figure 5.6, § 5.1.3]. Liquid water contains associations of water molecules [see Figure 4.35]. In ice, the arrangement of molecules is similar, but the regularity extends throughout the whole structure [see Figure 4.38]. The structure spaces the molecules further apart than they are in liquid water. This is why, when water freezes, it expands by 9%, and why ice is less dense than water at 0 °C. The underlying structure of ice resembles that of diamond [see Figure 6.11, § 6.5].

...This explains why ponds freeze from the surface downwards

The fact that ice is less dense than water at 0 °C explains why ponds and lakes freeze from the surface downwards. Water reaches its maximum density at 4 °C. As it cools further, the water at the surface becomes less dense and therefore stays on top of the slightly warmer water until it freezes. The layer of ice on the surface helps to insulate the water underneath from further heat loss. Fish and plants survive under the ice in Canadian lakes and rivers for months.

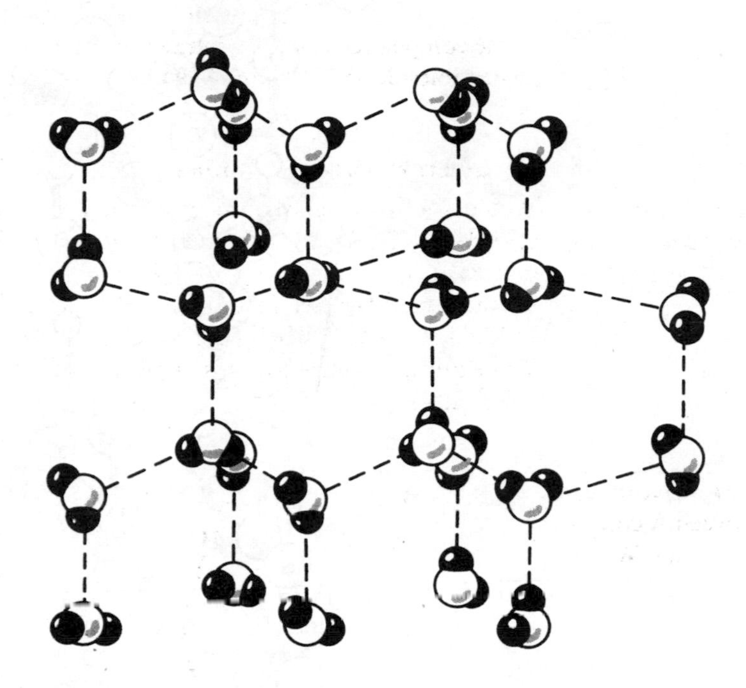

FIGURE 4.38 Hydrogen Bonding in Ice
Notes Each H_2O molecule uses both its H atoms to form hydrogen bonds and is also bonded to two other H_2O molecules by means of their H atoms. The arrangement of bonds about the O atoms is tetrahedral

THE STRUCTURE OF PROTEINS

Protein molecules have H bonds . . .

Hydrogen bonding is important in protein molecules. Proteins consist of long chains of formula

$$\left(\begin{array}{c} \text{R} \\ | \\ -\text{C}-\text{C}-\text{N}- \\ | \quad \| \quad | \\ \text{H} \quad \text{O} \quad \text{H} \end{array} \right)_n$$

. . . between $\backslash C{=}O /$ *and* $\backslash N{-}H /$ *groups*

R can be a number of groups. Since both the $\overset{\delta+}{C}{=}\overset{\delta-}{O}$ group and the $\overset{\delta-}{N}{-}\overset{\delta+}{H}$ groups are polar, hydrogen bonding can occur between them:

$$\overset{\delta+}{C}{=}\overset{\delta-}{O}\cdots\cdots\overset{\delta+}{H}{-}\overset{\delta-}{N}$$

The helical structure of proteins is sustained by H bonds

A single protein molecule contains many hydrogen bonds. They are one of the forms of intramolecular attraction which hold the protein in a three-dimensional arrangement described as the *secondary structure* of the protein. Figure 4.39 shows the α helical structure proposed by Pauling and co-workers as a result of their X ray diffraction studies on protein molecules. An **α helix** is a spiral, which, looking away from you, is spiralling in a clockwise direction.

FIGURE 4.39 A Protein Chain with the α Helical Structure

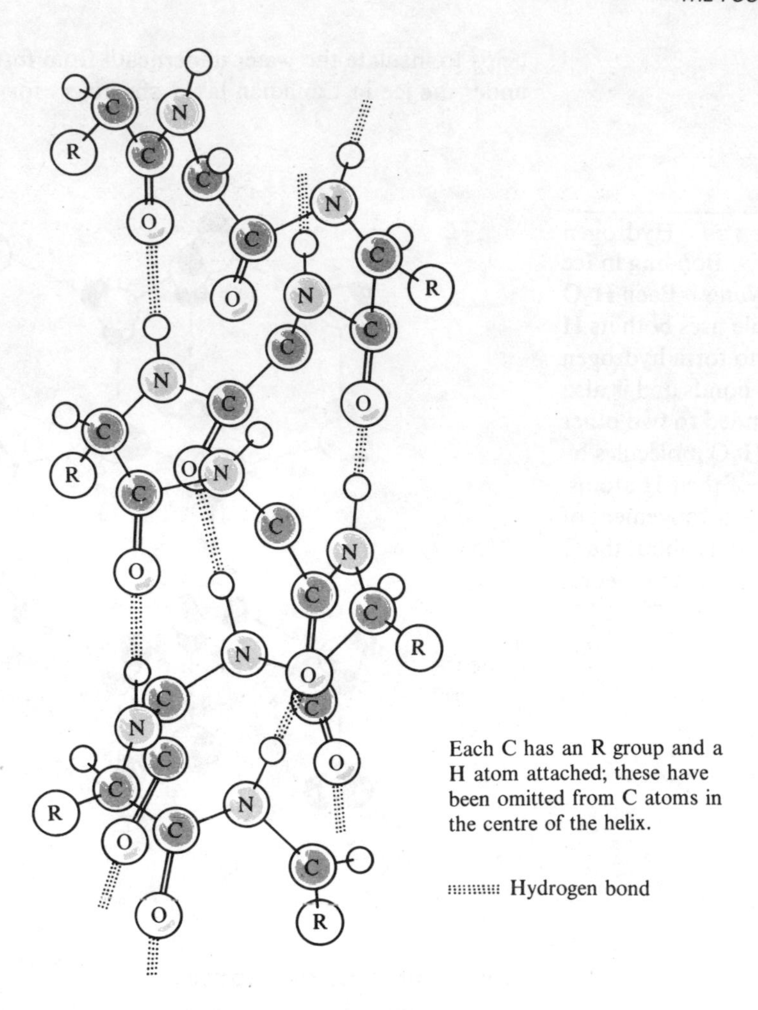

Each C has an R group and a H atom attached; these have been omitted from C atoms in the centre of the helix.
Hydrogen bond

THE DOUBLE HELIX

Nucleic acids contain phosphate groups, sugar groups and bases

Hydrogen bonding is important in the famous **double helix** of DNA. **Chromosomes** are the bodies in the nuclei of the cells of living organisms which carry genetic information. They contain macromolecular substances called **nucleic acids**. These are of two types: ribonucleic acid, RNA, and deoxyribonucleic acid, DNA. The macromolecular chains in DNA are of the type

```
—P—S—P—S—P—S—P—
   |     |     |
   B     B     B
```

where P is a phosphate group and S is the sugar deoxyribose:

P represents

```
   O     O⁻
    \\  /
      P
    /  \
 ⁻O     O⁻
```

S represents

```
        \
        H₂C
         |      O       B
         C  /      \  C /
    H /  \ H      H /  \ H
          C ——— C
          |       |
          O      OH
         /
```

The double helix of DNA is held in this configuration by H bonds

B is one of the four bases: adenine, thymine, cytosine and guanine. DNA consists of two macromolecular strands spiralling round each other in a double helix, as shown in Figure 4.40. The strands are held together by hydrogen bonding between the bases, as shown in Figure 4.41.

FIGURE 4.40 The Double Helix

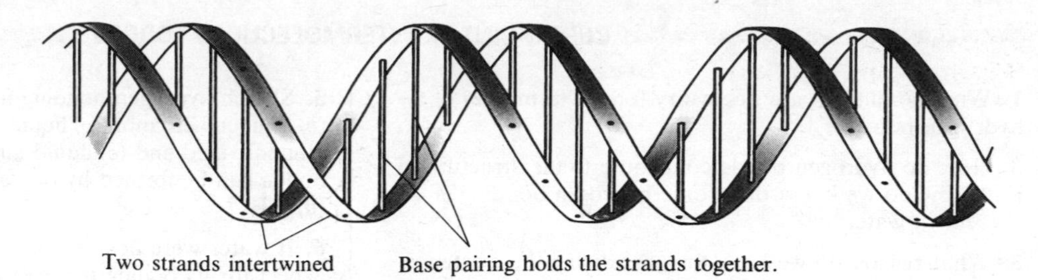

FIGURE 4.41 Hydrogen Bonding between Strands in DNA

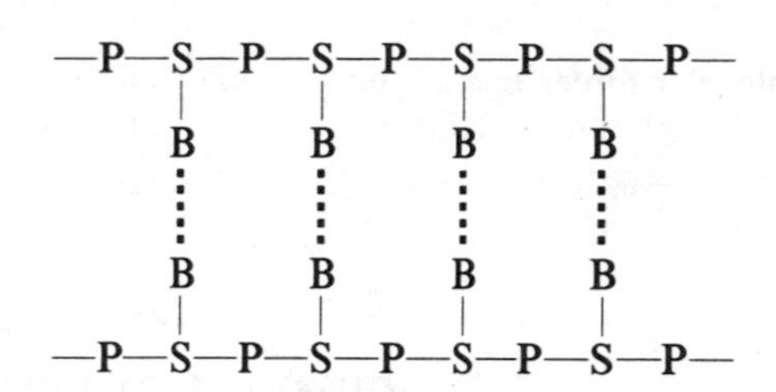

Of the four bases, *thymine* can pair up with *adenine* by hydrogen bonding and *cytosine* can form hydrogen bonds with *guanine*. The double helix brings these base pairs into contact so that they can form the bonds that keep the structure intact. Figure 4.42 shows the details of the hydrogen bonding.

FIGURE 4.42 Hydrogen Bonding of Base Pairs in DNA

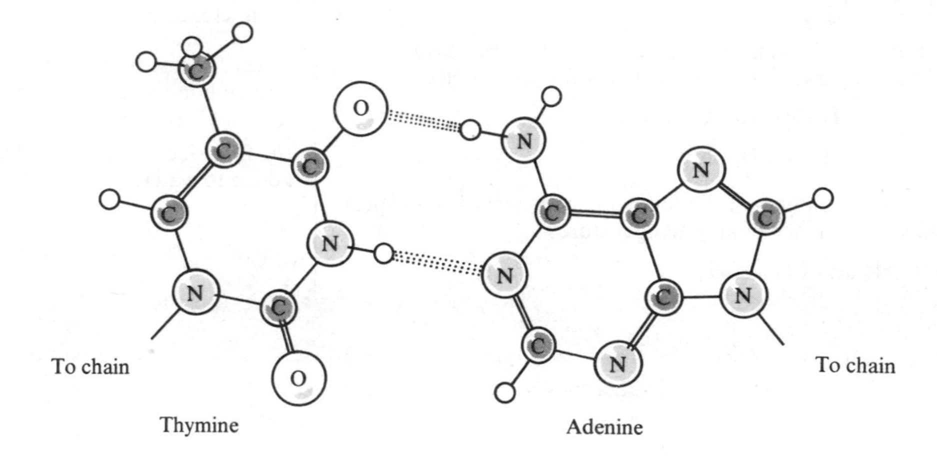

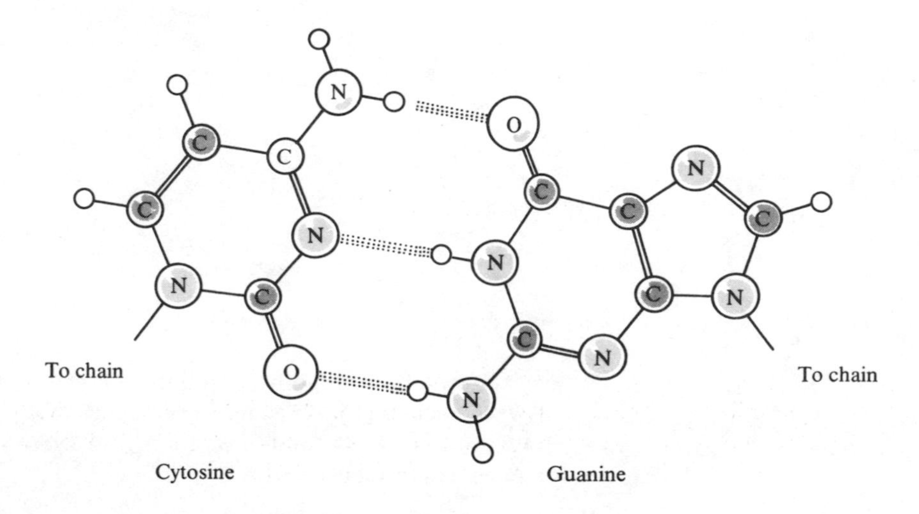

CHECKPOINT 4I: INTERMOLECULAR FORCES

1. What conditions are necessary for the formation of a hydrogen bond?

2. How do hydrogen bonds contribute to the structure of ice? How do we know that some hydrogen bonding is present in water?

3. What reason do we have for postulating the existence of van der Waals forces? What is responsible for the van der Waals attractive forces, and what gives rise to repulsion?

4. What type of intermolecular forces operate in (*a*) HBr(g), (*b*) Br_2(g), (*c*) ICl(g) and (*d*) HF(l)?

5. Sketch hydrogen bonding in (*a*) liquid ethanol, (*b*) aqueous ethanol, (*c*) liquid ethanoic acid, (*d*) aqueous ethanoic acid and (*e*) liquid ammonia. How many hydrogen bonds can be formed by one molecule of (i) H_2O, and (ii) NH_3?

6. If water were not hydrogen-bonded, what would you expect for its boiling temperature and melting temperature and the relative densities of the liquid and solid states?

QUESTIONS ON CHAPTER 4

1. Explain the 'octet theory' of valency. Point out examples of the success of the theory in explaining the formation of chemical bonds. Discuss cases in which the octet theory is inadequate.

2. Discuss the bonding in (*a*) NaH, (*b*) NH_4^+, (*c*) $BeCl_2$, (*d*) HF(l), (*e*) CCl_4 and (*f*) Cu.
[For (*f*), see § 6.2.1.]

†**3.** Put the following substances in order of increasing boiling temperatures, giving reasons for your choice:

C_4H_9OH, $CH_3CH_2CH_2CH_2CH_3$,

$(CH_3)_3CCH_3$, N_2

†**4.** Which members of the following pairs would you expect to have the higher boiling temperature?

(*a*) C_3H_8 and CH_3OCH_3

(*b*) $CH_3CH_2NH_2$ and CH_3CH_2OH

(*c*) CH_3CH_2OH and C_2H_6

(*d*) C_3H_8 and $(CH_3)_2C{=}O$

Give reasons for your choice.

5. Briefly explain the importance of the hydrogen bond to living creatures.

6. (*a*) Discuss the relation between physical properties and type of bonding found in the solid state for (i) Xe, (ii) Cu, (iii) NaCl, (iv) ice.

(*b*) Describe and explain what happens when solid NaCl is added to water.

(O 92)

5

THE SHAPES OF MOLECULES

5.1 THE ARRANGEMENT IN SPACE OF COVALENT BONDS

Ionic bonds are not directed in space whereas...

Electrovalent bonds are the electrostatic attractions that exist between oppositely charged ions. Since ions radiate a spherically symmetrical positive or negative field, ionic bonds are *non-directional.*

...covalent bonds have a preferred direction in space

When atoms approach one another closely, their atomic orbitals overlap and molecular orbitals are formed. Covalent bonds are formed when a bonding pair of electrons enters a molecular orbital of low energy. The bonding electrons must have opposing spins, in accordance with the Pauli Exclusion Principle [§ 2.3.2]. The more the atomic orbitals overlap, the more stable will be the molecular orbital formed. The strongest bonds will be formed if the atoms approach in such a way that there is maximum overlap between atomic orbitals. It follows that a covalent bond will have a *preferred direction.* A covalent molecule will have a shape which is determined by the angles between the bonds joining the atoms together.

$BeCl_2$ and $SnCl_2$ differ in shape...

...BCl_3 and NH_3 differ in shape

There must be a reason why beryllium chloride, $BeCl_2$, is a linear molecule without a dipole moment, while tin(II) chloride, $SnCl_2$, is a bent molecule with a dipole moment. There must also be a reason why the four atoms in boron(III) chloride, BCl_3, are coplanar, whereas in ammonia the nitrogen atom lies above three coplanar hydrogen atoms. [See Figure 5.1.]

FIGURE 5.1 $BeCl_2$ Compared with $SnCl_2$ and BCl_3 Compared with NH_3

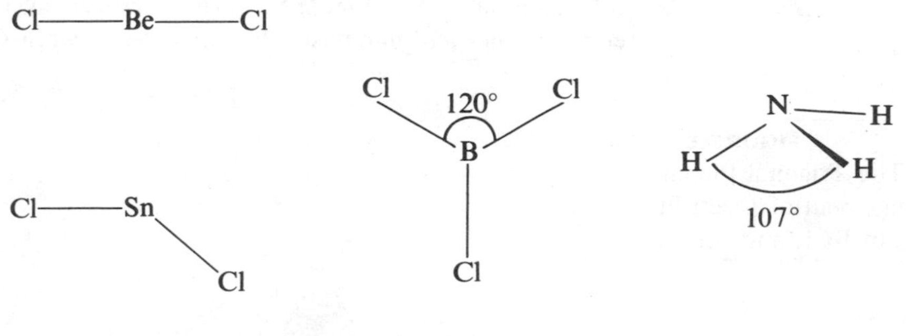

The differences can be explained by the atomic orbital approach or...

The explanation lies in the difference in the atomic orbitals used by beryllium and tin, and by boron and nitrogen. The atomic orbital approach to the question of molecular geometry is covered in § 5.2.

... the Sidgwick–Powell theory, which is the Valence Shell Electron Pair Repulsion Theory

A simpler but less precise theory to account for the shapes of molecules was put forward by Sidgwick and Powell in 1940. It is known as the **Valence Shell Electron Pair Repulsion Theory**. Sidgwick and Powell considered the shapes of small molecules and molecular ions, such as $BeCl_2$, BCl_3, NH_3, NH_4^+ and CH_4. They pointed out that the arrangement of electron pairs around the central atom in a molecule depends on the number of electron pairs. Between each electron pair and any other electron pair there is a force of electrostatic repulsion, which forces the orbitals as far apart as possible. Any lone pairs of electrons on the central atom occupy atomic orbitals, and they too repel the bonding pairs of electrons and influence the geometry of the molecule.

5.1.1 LINEAR MOLECULES

The molecules of gaseous beryllium chloride, $BeCl_2$, are linear. Beryllium, in Group 2 of the Periodic Table, has two electrons in its valence shell, and forms two covalent bonds. A **linear** arrangement of the atoms (a bond angle of 180°) puts the two electron clouds as far apart as possible:

Cl—Be—Cl

Other linear molecules are

Examples of linear molecules are $BeCl_2$, HC≡CH, H—C≡N and O=C=O

H—C≡C—H H—C≡N O=C=O

The electron pairs in a multiple bond are assumed on the Sidgwick–Powell theory to occupy the position of one electron pair in a single bond.

5.1.2 TRIGONAL PLANAR MOLECULES

The arrangement of 3 pairs of valence electrons is trigonal planar

Lone pairs of electrons can determine the shape of the molecule

When there are three pairs of electrons around the central atom, the bonds lie in the same plane at an angle of 120° to one another. Three atoms form a triangle about the central atom, and the arrangement is described as **trigonal planar**. An example is boron trichloride, BCl_3. Boron, in Group 3 of the Periodic Table, has three valence electrons and forms three covalent bonds. Gaseous tin(II) chloride, $SnCl_2$, has a dipole moment, proving that the molecule is not linear. The reason is that tin, in Group 4, is using only two of its four electrons for bond formation. The lone pair of electrons repel the bonding pairs and a trigonal planar arrangement of orbitals results. [See Figure 5.2.] This arrangement maximises the angle between the electron pairs and minimises the repulsion between them.

FIGURE 5.2 The Trigonal Planar Arrangement of Electron Pairs in BCl_3 and $SnCl_2$

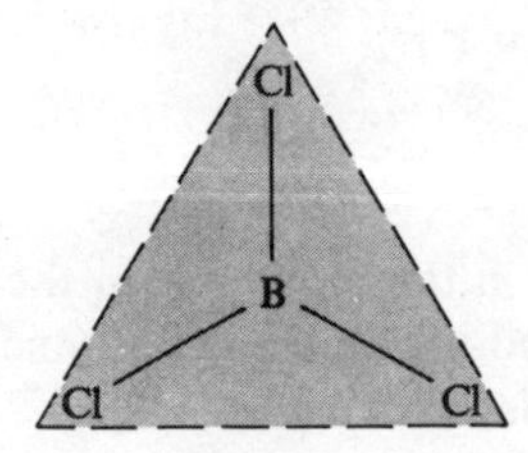

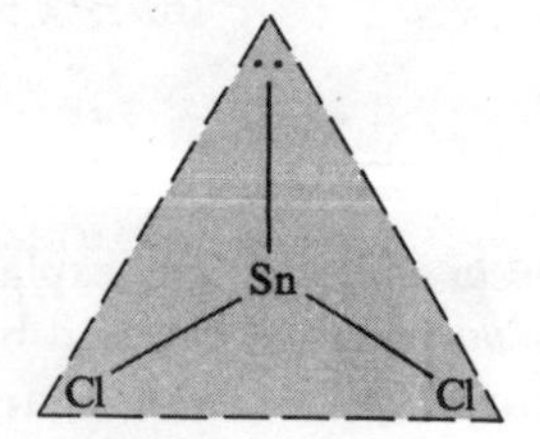

Other structures based on a trigonal planar arrangement are ethene, the nitrate ion and sulphur dioxide [see Figure 5.3].

FIGURE 5.3 The Arrangement of Electron Pairs in $CH_2{=}CH_2$, NO_3^-, and SO_2

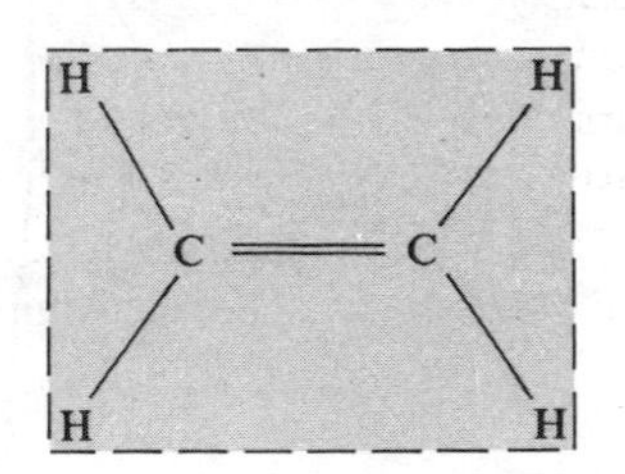

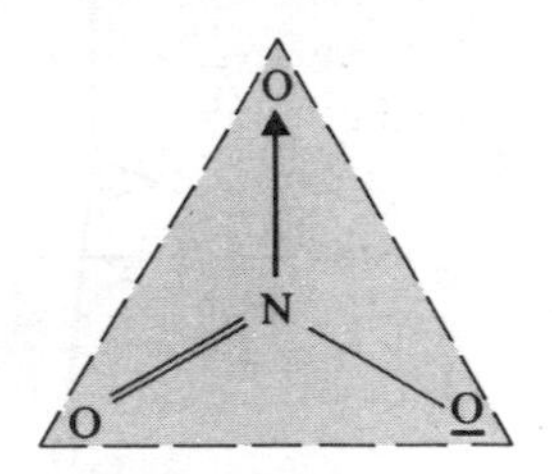

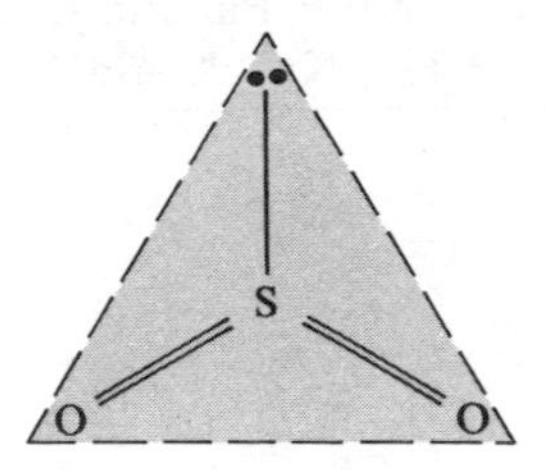

There is an interesting consequence of the coplanar arrangement. The formula of but-2-ene is $CH_3CH{=}CHCH_3$. There are two structures with this formula:

(a) CH_3—C—H ‖ CH_3—C—H (b) CH_3—C—H ‖ H—C—CH_3

cis- and trans- forms of but-2-ene

Structure (a), in which the hydrogen atoms are on the same side of the double bond, is called *cis*-but-2-ene, and structure (b), with the hydrogen atoms on opposite sides of the double bond, is called *trans*-but-2-ene. The existence of *cis*- and *trans*-forms of compounds is covered in § 25.9.2. [See Figure 5.4.]

FIGURE 5.4 *cis*- and *trans*- But-2-ene

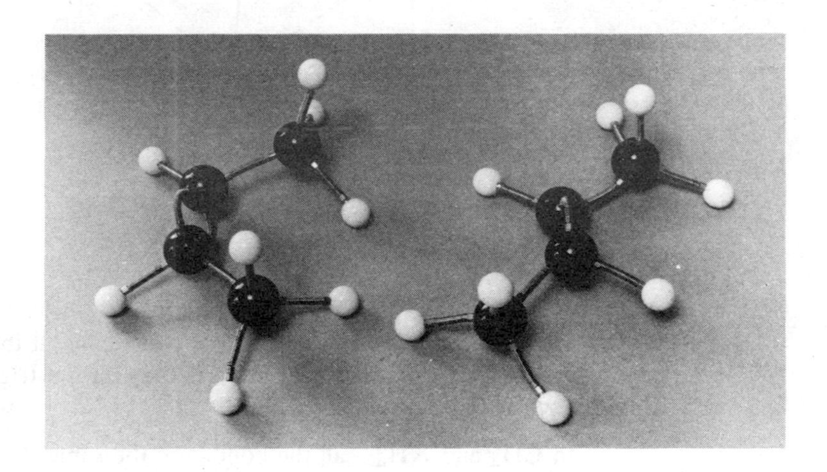

5.1.3 TETRAHEDRAL MOLECULES

4 electron pairs adopt a tetrahedral configuration

The molecules CH_4, NH_3, NH_4^+ and H_2O all have four pairs of electrons around the central atom. Whether they are bonding pairs or lone pairs of electrons, they experience mutual repulsion. To minimise this repulsion, the four electron orbitals adopt the spatial arrangement that maximises the angle between the orbitals. This is the **tetrahedral** arrangement. [See Figures 5.5 to 5.7.]

FIGURE 5.5 The Bonding in CH_4, NH_3, NH_4^+, H_2O

H—C—H (with H above and below C)   :N—H (with H above and below) $[H{\leftarrow}N—H]^+$ (with H above and below N)

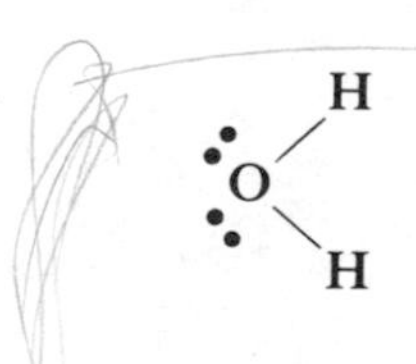

FIGURE 5.6
The Tetrahedral Arrangement of Valence Electron Pairs in CH_4, NH_3, NH_4^+ and H_2O

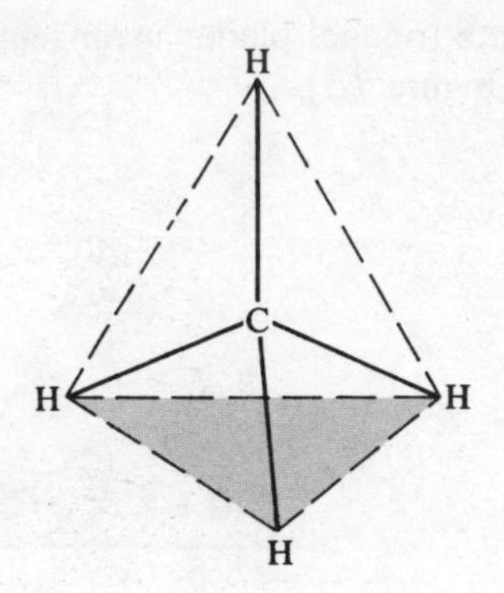

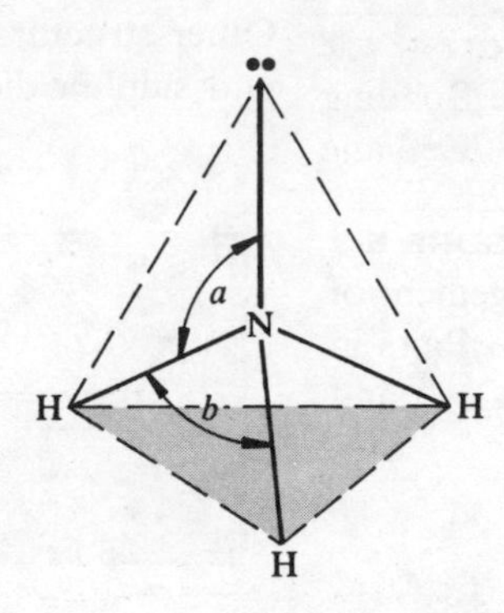

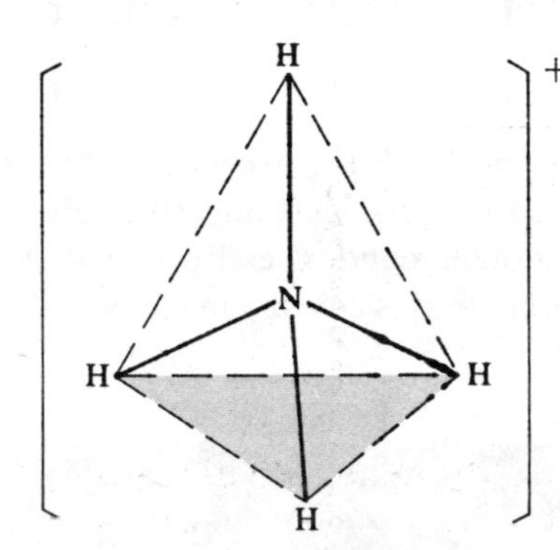

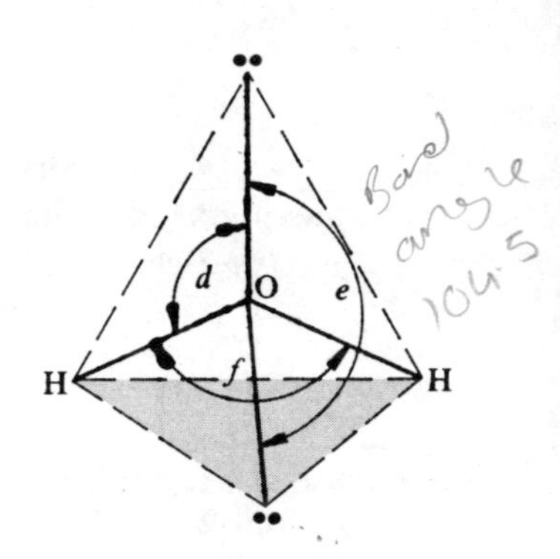

The arrangements of atoms are of course *not* all the same. In CH_4 and NH_4^+, the atoms form a tetrahedron, in NH_3 they form a trigonal pyramid, and in H_2O they form a bent line.

In CH_4 and NH_4^+ all the bonds are the same. Once formed, a coordinate bond is the same as a covalent bond. The structures are perfect tetrahedra with the tetrahedral angle of 109.5° between each pair of bonds. In NH_3 the bond angle is 107°, and in H_2O it is 104.5°.

Lone pairs are closer to the nucleus than bonding pairs and exert a greater repulsive force

To account for such departures from the expected bond angle, Gillespie and Nyholm suggested a refinement of the theory of valence shell electron pair repulsion. They suggested that, since lone pairs are closer to the nucleus than bonding pairs, they will exercise a greater force of repulsion. Repulsion between electron pairs decreases in the order

Lone pair : lone pair repulsion > Lone pair : bonding pair repulsion > Bonding pair : bonding pair repulsion

Repulsion between the lone pair and the bonding pairs in NH_3 makes the angle *a* in Figure 5.6 greater than the tetrahedral angle (109.5°) and consequently the angle *b* less than 109.5°. Similarly in H_2O, angles *d* and *e* are greater than 109.5°, and the angle *f* between the H—O—H bonds is 104.5°. Other structures based on the tetrahedron are the sulphate and sulphite ions [see Figure 5.8].

FIGURE 5.7
A Model of the Bonding Orbitals in Methane

FIGURE 5.8
The Shapes of SO_4^{2-} and SO_3^{2-}

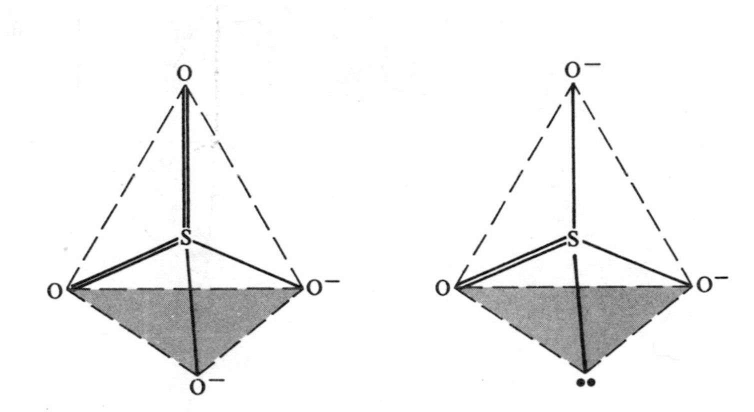

5.1.4 CHIRALITY

Six planes of symmetry run through the tetrahedron of hydrogen atoms in CH_4, and the centres of positive charge and negative charge coincide in the carbon atom at the centre of the tetrahedron. [See Figure 5.9.]

FIGURE 5.9
The Symmetry of the Tetrahedron

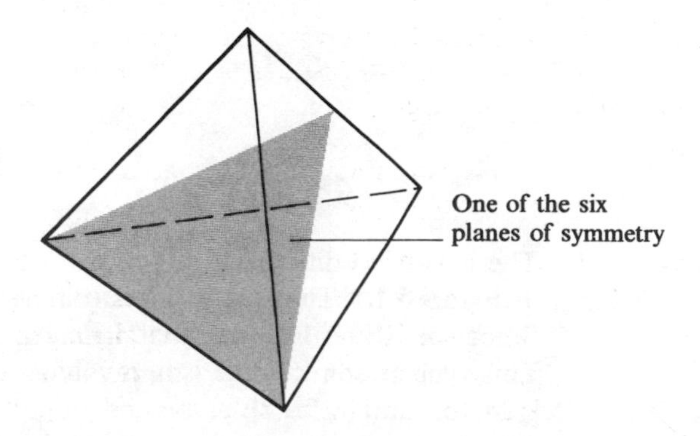

Four different groups bonded to carbon destroy the symmetry of the tetrahedron...

If four different atoms or groups are attached to a carbon atom, there is no longer a plane of symmetry in the molecule. Nor is there a centre of symmetry or an axis of symmetry. The carbon atom in CHClBrF is an **asymmetric** carbon atom. There are two ways of drawing a tetrahedral arrangement for this formula. You will see from Figure 5.10 that one molecule is the mirror image of the other; they are called **enantiomers**. The molecules cannot be superimposed, being related in the same way as a left hand and a right hand [see Figure 5.11]. They show the geometric property of **chirality**.

FIGURE 5.10 Enantiomers of Bromochlorofluoromethane

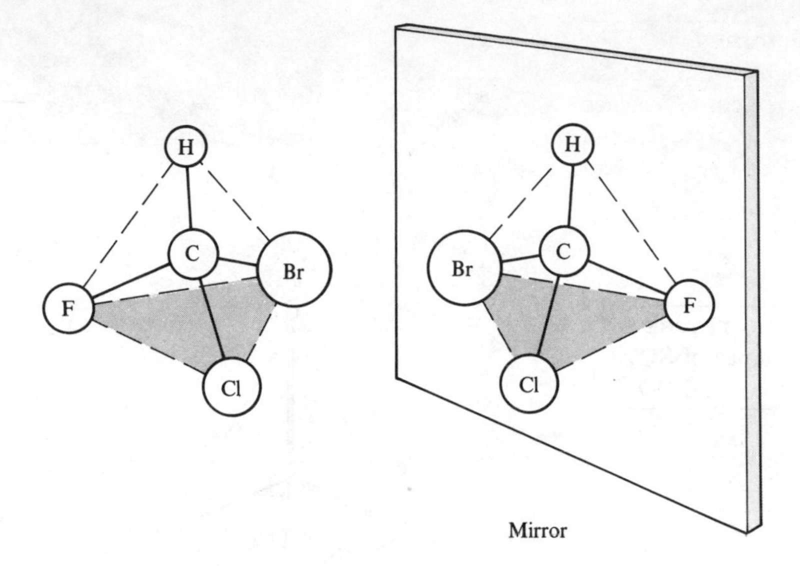

FIGURE 5.11 Left Hand and its Mirror Image

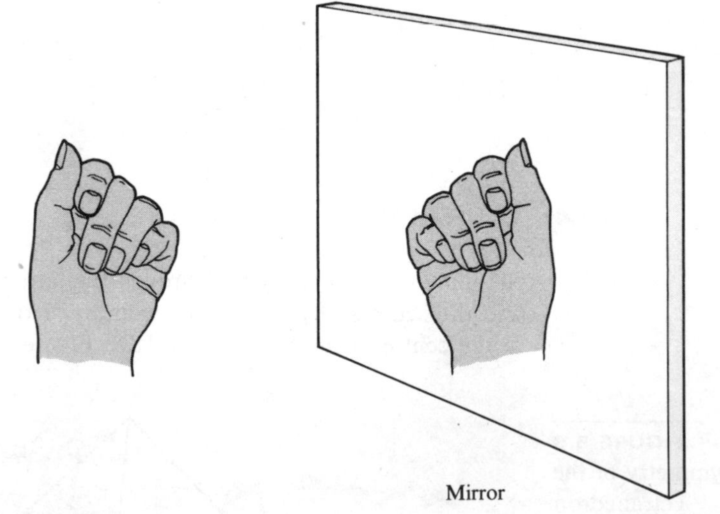

...such compounds exist in stereoisomeric forms

There are two different compounds with the two different kinds of molecules shown in Figure 5.10. They are **isomers** (different compounds with the same formula). Since they differ in the spatial arrangement of atoms, they are called **stereoisomers**. This type of stereoisomerism, involving enantiomers, is called **optical isomerism** [§24.13.6 and §25.9.2].

5.1.5 STRUCTURES WITH 5, 6 OR 7 PAIRS OF VALENCE ELECTRONS

Some atoms have more than 8 electrons in the valence shell

Structures with more than four pairs of electrons about the central atom may occur if the element is in the second short period or a later period. This is known as **expansion of the octet**. A molecule of phosphorus(V) chloride, PCl_5, with *five* bonding pairs of electrons, has the shape of a **trigonal bipyramid**. The angles between the bonds are 90° or 120°. Two Cl atoms occupy **axial** positions in the bipyramid, and three occupy **equatorial** positions [see Figure 5.12].

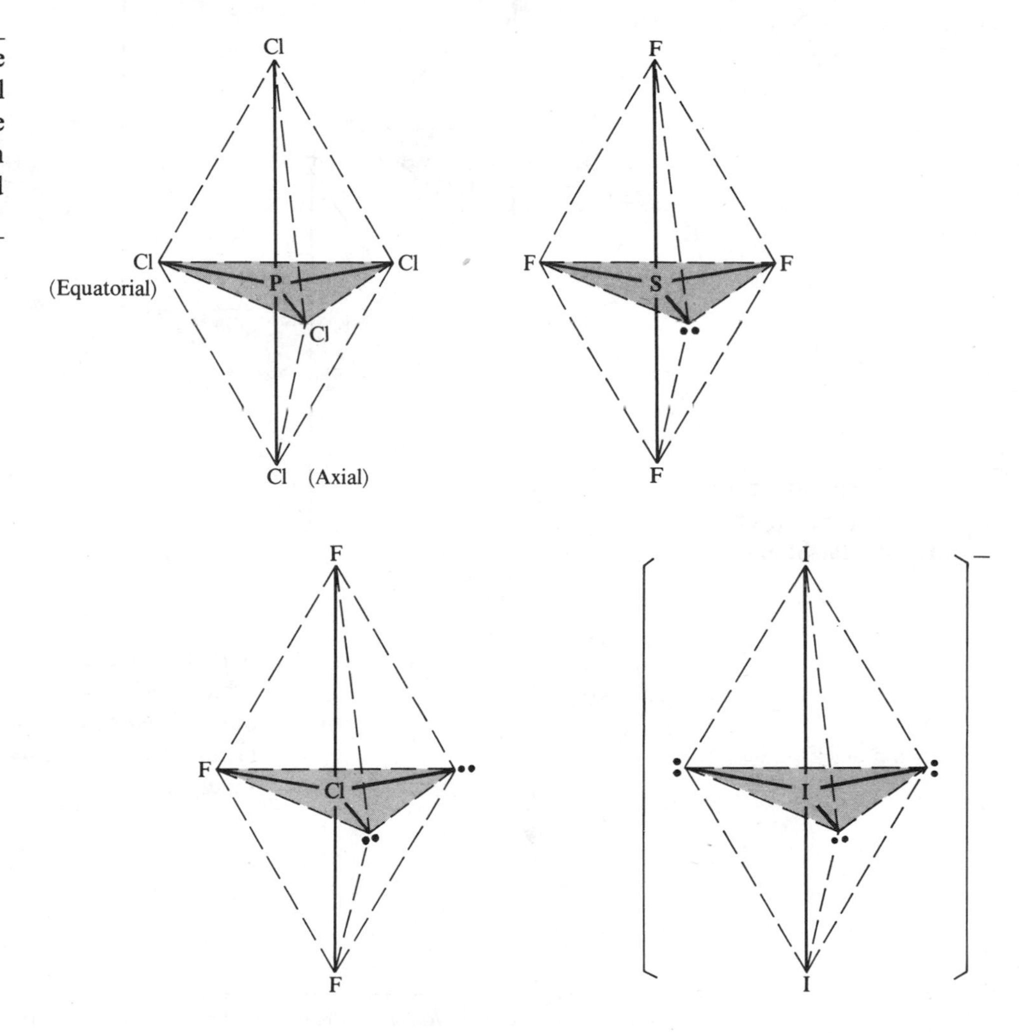

FIGURE 5.12 The Trigonal Bipyramidal Arrangement of Valence Shell Electron Pairs in PCl_5, SF_4, ClF_3, and I_3^-

The electron pairs take up positions so as to maximise the angle between electron pairs and minimise the repulsion between them

The same type of bond distribution is adopted by SF_4. Sulphur, in Group 6, uses four of its six electrons to form the bonds and has a lone pair of electrons. The lone pair could be in an axial position or, as shown in Figure 5.12, in an equatorial position. The choice of an equatorial position fits in with the Gillespie–Nyholm idea of the lone pair orbital being concentrated in a volume closer to the nucleus than the bond pairs. In this position it can have a bond angle of 120° with two orbitals and a bond angle of 90° with two others. The molecule of ClF_3 has the shape illustrated in Figure 5.12, with a trigonal bipyramidal arrangement of bonds. In the I_3^- ion, iodine (in Group 7) uses only two of its seven electrons in bond formation. The remaining five plus the electron that gives the ion its negative charge make up three lone pairs. The total five pairs of electrons are distributed in space as shown in Figure 5.12. The arrangement of bonds is trigonal bipyramidal; the arrangement of atoms is linear.

Structures with *six* pairs of electrons around the central atom are sulphur(VI) fluoride, SF_6, iodine(V) fluoride, IF_5, and the ICl_4^- ion. The **octahedral arrangement** of electron pairs is shown in Figure 5.13.

The arrangement of *atoms* in IF_5 is **square pyramidal**, a lone pair occupying the sixth position in the octahedron. In ICl_4^- the four chlorine atoms are in a **square planar** configuration, with lone pairs occupying the axial positions of the octahedron.

FIGURE 5.13 The Octahedral Shape of the Electron Pairs in SF_6, IF_5, ICl_4^-

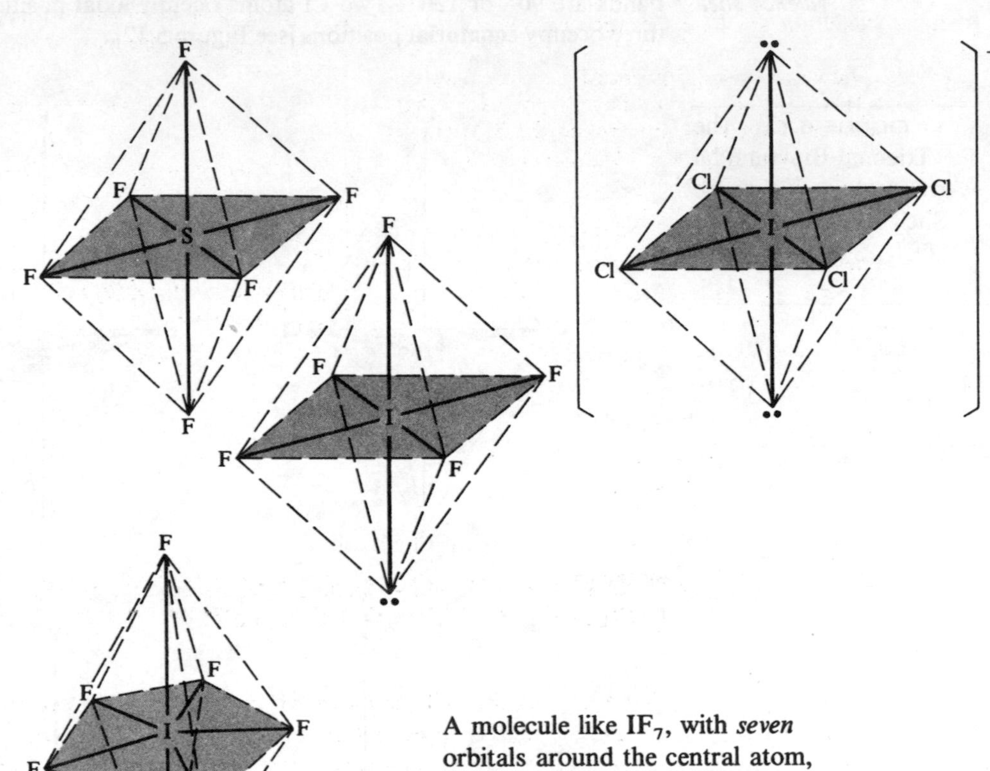

FIGURE 5.14 The Pentagonal Bipyramidal Shape of IF_7

A molecule like IF_7, with *seven* orbitals around the central atom, has the **pentagonal bipyramidal** arrangement of bonds shown in Figure 5.14.

5.1.6 SUMMARY

TABLE 5.1 A Summary of the Shapes of Molecules

No. of valence electrons	*No. of bond pairs*	*No. of lone pairs*	*Total electron pairs*	*Arrangement of orbitals*	*Arrangement of atoms*	*Example*
4	2	0	2	Linear	Linear	$BeCl_2$
6	3	0	3	Trigonal planar	Trigonal planar	BF_3
8	4	0	4		Tetrahedral	CH_4
8	3	1	4	Tetrahedral	Trigonal bipyramidal	NH_3
8	2	2	4		Bent line	H_2O
10	5	0	5		Trigonal pyramidal	PF_5
10	4	1	5	Trigonal		SF_4
10	3	2	5	bipyramidal		ClF_3
10	2	3	5		Linear	I_3^-
12	6	0	6		Octahedral	SF_6
12	5	1	6	Octahedral	Square pyramidal	IF_5
12	4	2	6		Square planar	ICl_4^-

CHECKPOINT 5A: SHAPES OF MOLECULES

1. (*a*) Take three long balloons, blow them up and tie the ends. Hold the three tied ends between your finger and thumb. What positions do the three balloons adopt?
(*b*) Add a fourth balloon and notice the positions which the balloons take up.

2. In BF_3, how many electron pairs are there around the B atom? Sketch the spatial distribution of bonds.

3. In BrF_3, how many electrons does Br use for bond formation? How many lone pairs does Br possess? What is the total number of electron pairs around the Br atom? Sketch the spatial distribution of bonds. How is this arrangement described?

4. Sketch the spatial distribution of bonds in HOBr. How would you describe the shape of (*a*) the electron orbitals, (*b*) the molecule?

5. Sketch the arrangement of bonds in CCl_4. If the $\overset{\delta+}{C}—\overset{\delta-}{Cl}$ bond has a dipole moment of *x* debyes, what is the dipole moment of the CCl_4 molecule?

6. (*a*) In the compound XeF_4, how many electrons is the noble gas xenon using for bond formation? How many lone pairs does xenon have? What is the total number of electron pairs around the central Xe atom? Sketch the arrangement of bonds. What shape is the molecule?
(*b*) Sketch the arrangement of bonds in XeF_2, XeF_6 and XeO_3.

7. Explain how the Sidgwick–Powell theory predicts the shape of the following molecules: (*a*) $SnCl_4$, (*b*) PH_3, (*c*) PF_5, (*d*) BH_3 and (*e*) BeH_2.

8. Sketch the spatial arrangement of bonds in the following: (*a*) F_2O, (*b*) $SeCl_4$, (*c*) SO_3, (*d*) ICl_3, (*e*) PF_6^- and (*f*) $COCl_2$.

5.2 *MOLECULAR GEOMETRY: A MOLECULAR ORBITAL TREATMENT

An alternative to the Sidgwick–Powell treatment is the molecular orbital approach

The Sidgwick–Powell theory provides a simple treatment of the shapes of covalent molecules. A more precise treatment of the spatial distribution of covalent bonds about a central atom involves a consideration of the atomic orbitals used in bond formation. The shapes of atomic orbitals are described in § 2.3.3 and shown in Figures 2.11 to 2.13. When **atomic orbitals** overlap, **molecular orbitals** are formed. The **molecular orbital approach** is illustrated by the following compounds of elements in the first short period of the Periodic Table:

$$HF \quad BeCl_2 \quad BF_3 \quad CH_4 \quad H_2O \text{ and } NH_3$$

5.2.1 HYDROGEN FLUORIDE

The electronic configurations of the fluorine atom and the hydrogen atom are

	F	H
2p	↑↓ ↑↓ ↑	
2s	↑↓	
1s	↑↓	↑
	F $1s^22s^22p^5$	H 1s

In HF, an s orbital and a p orbital overlap

When a molecule of hydrogen fluoride is formed, the s orbital of the hydrogen atom overlaps with the fluorine p orbital containing the unpaired electron. Figure 5.15 shows two of the three p orbitals at right angles to one another and shows how the bonding orbital of the fluorine atom becomes concentrated between the F and H nuclei when the bond is formed.

FIGURE 5.15
Hydrogen Fluoride

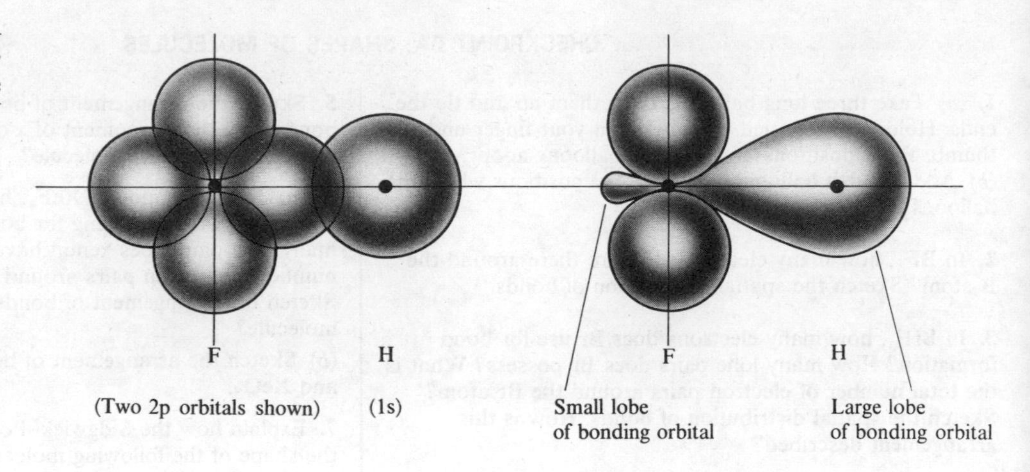

5.2.2 BERYLLIUM CHLORIDE

Beryllium has the electron configuration $1s^2 2s^2$, and chlorine has $1s^2 2s^2 2p^6 3s^2 3p^5$.

Be	Be*	Cl
2s ↑↓	2p ↑ ☐ ☐	3p ↑↓ ↑↓ ↑
1s ↑↓	2s ↑	3s ↑↓
Be $1s^2 2s^2$	1s ↑↓	2p ↑↓ ↑↓ ↑↓
	Be* $1s^2 2s 2p$	2s ↑↓
		1s ↑↓
		Cl $1s^2 2s^2 2p^6 3s^2 3p^5$

Beryllium uses two orbitals, an s and a p orbital . . .

. . . The bonds formed are identical . . .

. . . Two sp hybrid orbitals are used

In order to combine, beryllium needs unpaired electrons: otherwise it would be as unreactive as helium. If one 2s electron is promoted into the 2p shell, then the atom will have two unpaired electrons. The atom must absorb energy in order to promote the electron and is described as an 'excited' atom. The bonding orbitals formed by beryllium are not of two different kinds, derived simply from the s and p atomic orbitals, they are two identical **hybrid** orbitals. The electron densities in the s and p atomic orbitals combine in compound formation to give sp hybrid orbitals of the shape shown in Figure 5.16.

FIGURE 5.16
sp Hybrid Bonds

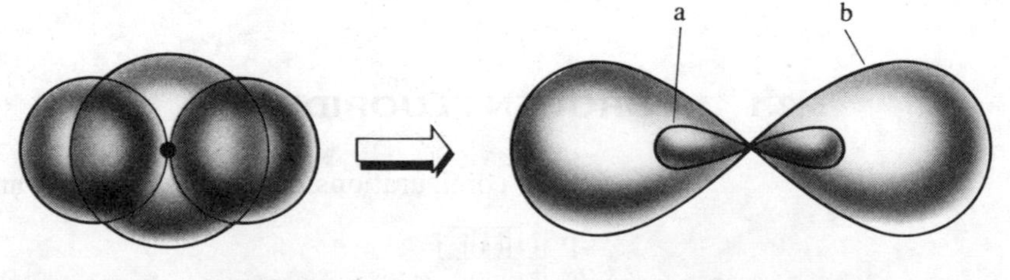

In future diagrams, the small lobe *a* will be omitted and only the main lobe *b* in each sp hybrid bond will be shown.

When the two beryllium sp hybrid orbitals overlap with p orbitals of chlorine atoms, the bonding is as shown in Figure 5.17.

FIGURE 5.17 $BeCl_2$, a Linear Molecule

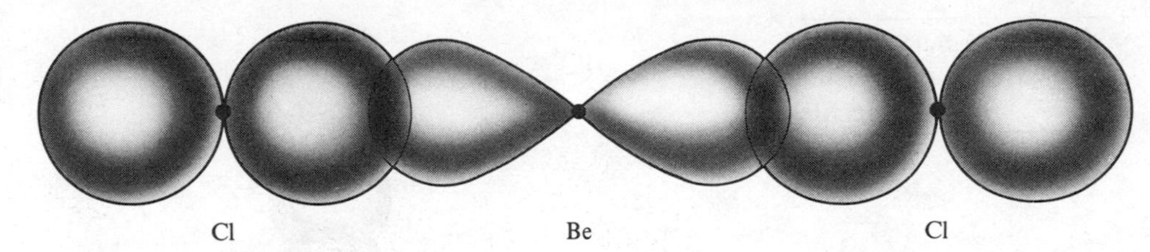

The non-bonding p orbitals of Cl are not shown.

This linear arrangement of atoms is as predicted by the Sidgwick–Powell theory.

5.2.3 BORON TRIFLUORIDE

In an atom of boron ($1s^22s^22p$), an s electron can be promoted to a p orbital.

	B	B*	F
2p	[↑][ ][ ]	[↑][↑][ ]	[↑↓][↑↓][↑]
2s	[↑↓]	[↑]	[↑↓]
1s	[↑↓]	[↑↓]	[↑↓]
	B $1s^22s^22p$	B* $1s^22s2p^2$	F $1s^22s^22p^5$

In boron, two p orbitals hybridise with one s orbital to form three sp^2 hybrid bonds . . .

The three orbitals which an excited atom of boron, B*, uses in bond formation are identical orbitals formed by a combination of the electron densities in the s and two p orbitals and called sp^2 hybrid orbitals. Quantum mechanical calculations give them the shape shown in Figure 5.18.

FIGURE 5.18 sp^2 Hybrid Orbitals

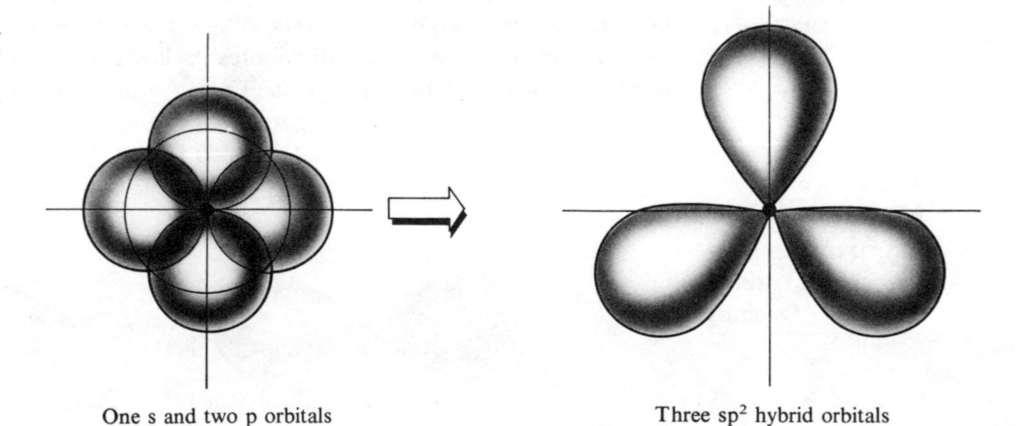

. . . The arrangement of orbitals is trigonal planar

When three sp^2 hybrid orbitals overlap with p orbitals from fluorine atoms, the bonding formed is as shown in Figure 5.19. The molecule is described as trigonal planar. Although the B—F bonds are polar, the centre of negative charge coincides with the centre of positive charge, the boron atom, and the molecule has no dipole moment.

FIGURE 5.19 BF_3, a Trigonal Planar Molecule

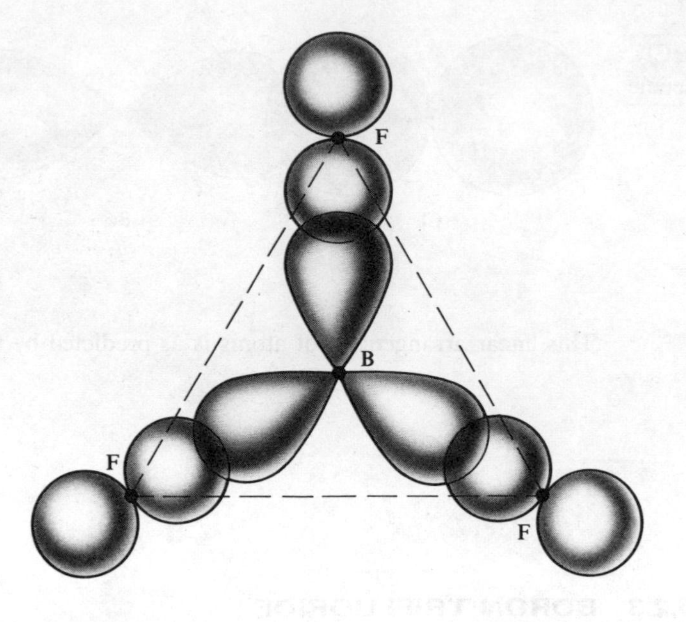

coplanar.
veen the bonds
non-bonding p
re not shown.)

5.2.4 METHANE

The electronic configurations of carbon in its normal, C, and excited, C*, states and hydrogen, H, are shown below:

	C	C*	H
2p	↑ ↑ ☐	↑ ↑ ↑	
2s	↑↓	↑	
1s	↑↓	↑↓	↑
	C $1s^2 2s^2 2p^2$	C* $1s^2 2s 2p^3$	H 1s

In carbon, an s electron is promoted to a p orbital...

...four sp^3 hybrid orbitals are formed

Each carbon atom, C, has two unpaired electrons, and one might expect carbon to form two bonds. It would not then attain a neon-like structure: it needs to share four electrons to do this. A sharing of four electrons can be achieved by promoting one of the 2s electrons into the 2p level. The excited carbon atom, C*, might be expected to form two different kinds of bond, using one s orbital and three p orbitals. Actually, the electron density distributes itself evenly through four bonding orbitals, which are called sp^3 hybrid orbitals. The sp^3 atomic orbital is more concentrated in direction than a p orbital [see Figure 5.20]. An sp^3 orbital is therefore able to overlap more extensively and form stronger bonds than a p orbital.

FIGURE 5.20 Comparison of Atomic Orbitals

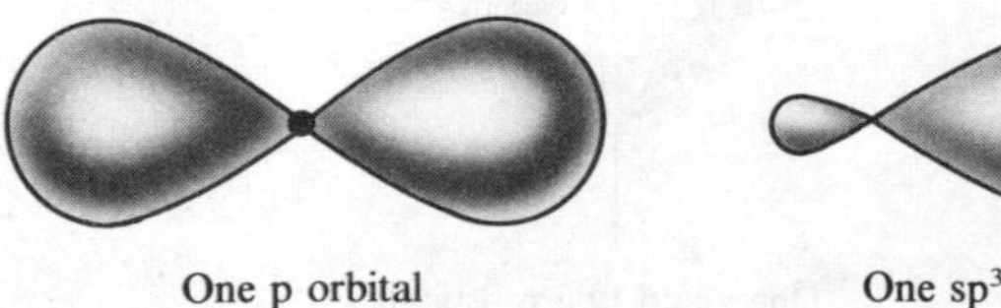

The tetrahedral arrangement of the four sp^3 orbitals of carbon is shown in Figure 5.21. The Sidgwick–Powell theory predicts this arrangement of bonds. Quantum mechanical calculations arrive at the same picture. Experimental evidence for the tetrahedral arrangement is gained from X ray diffraction studies of diamond. The angle between the bonds is shown to be 109.5° [see Figure 6.11, § 6.5].

FIGURE 5.21
Overlapping of Atomic Orbitals in Methane

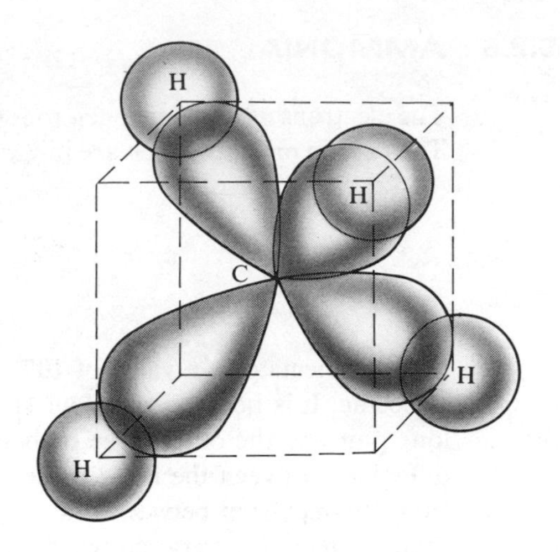

. . . they are tetrahedrally arranged in space

5.2.5 WATER

H_2O is not a linear molecule

The H_2O molecule has a dipole moment. If the molecule were linear, the dipoles in the two H—O bonds would cancel out: $\overset{\delta+}{H}—\overset{\delta-}{O}—\overset{\delta+}{H}$, and the molecule would have no dipole moment. Thc reason for its dipole moment lies in the atomic orbitals used for bonding:

	O $1s^2 2s^2 2p^4$	H 1s	O in H_2O
2p	↑↓ ↑ ↑		↑↓ ↑↓ ↑↓
2s	↑↓		↑↓
1s	↑↓	↓	↑↓

If O uses two p orbitals for bonding, the bond angle should be 90° . . .

If oxygen uses two p orbitals for bonding, the bond angle will be 90° [Figure 2.12(b), § 2.3.3]. In fact, X ray studies show the bond angle to be 105°. This is close to the tetrahedral angle of 109.5°. It is believed that hybridisation does occur between the s orbital and the three p orbitals of the oxygen atom. Of the four sp^3 hybrid orbitals, two are occupied by bonding pairs of electrons and the other two by lone pairs [see Figure 5.22]. The difference between the bond angle of 105° and the tetrahedral angle is explained by the greater repulsion between lone pair orbitals than between bonding orbitals [§ 5.1.3].

. . . In fact it uses sp^3 hybrid orbitals with bond angles of 105°

FIGURE 5.22
sp^3 Bonds in H_2O

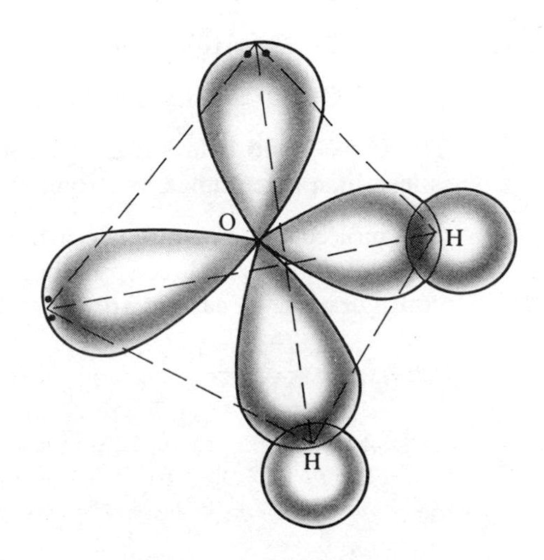

5.2.6 AMMONIA

The electrons used by nitrogen to form bonds are the three unpaired 2p electrons. They are in orbitals which are mutually perpendicular [see Figure 2.12(b), § 2.3.3].

2p	↑ ↑ ↑		↑↓ ↑↓ ↑↓
2s	↑↓		↑↓
1s	↑↓	↓	↑↓
	N $1s^2 2s^2 2p^3$	H 1s	N in NH_3

In NH_3, N uses four sp^3 hybrid orbitals...

... one is occupied by a lone pair of electrons

Measurements give a value of 107° for the angle between the bonds in the NH_3 molecule. It is believed that four sp^3 hybrid orbitals are formed, one occupied by a lone pair and the other three by bonding pairs of electrons [see Figure 5.23]. The difference between the bond angle, 107°, and the tetrahedral angle, 109.5°, arises from the repulsion between a lone pair and a bonding pair of electrons being greater than the repulsion between two bonding pairs [§ 5.1.3].

FIGURE 5.23
sp^3 Hybrid Bonds in NH_3

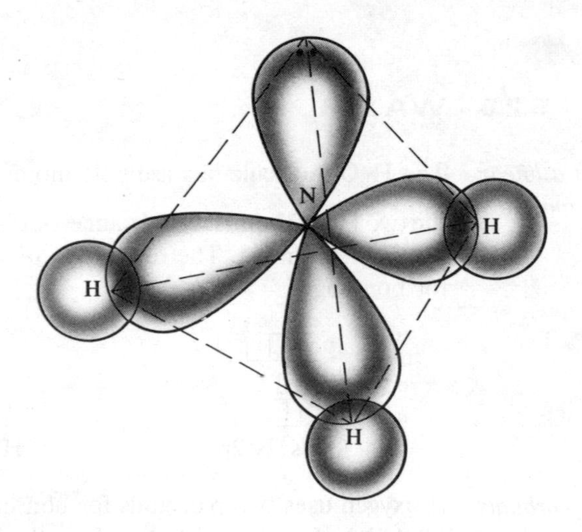

5.2.7 MULTIPLE BONDS

A double bond is less than twice as strong as a single bond...

... and a triple bond has less than three times the strength of a single bond

Carbon forms double bonds in compounds such as carbon dioxide, O=C=O, and ethene, $H_2C{=}CH_2$. The double bond is not simply two single bonds. The amount of energy required to break a certain bond in a mole of molecules is called the **standard bond enthalpy** [§ 10.6]. Standard bond enthalpies of carbon–carbon bonds are

C—C 346 kJ mol^{-1}

C=C 610 kJ mol^{-1}

C≡C 837 kJ mol^{-1}

The C=C bond is less than twice as strong as a C—C bond, and the C≡C bond is less than three times as strong as a C—C bond.

In a molecule of ethene, each carbon atom uses a 2s orbital and two of the three 2p orbitals to form three sp^2 hybrid bonds [see Figure 5.18, § 5.2.3]. The electronic configurations of carbon are shown below:

2p	↑ ↑ □	↑ ↑ ↑	↑↓ ↑↓ ↑
2s	↑↓	↑	↑↓
1s	↑↓	↑↓	↑↓
	C $1s^2 2s^2 2p^2$	C* $1s^2 2s 2p^3$	C in >C—

Each C has an unhybridised p orbital...

The carbon–carbon bond formed when the sp^2 orbitals of neighbouring carbon atoms overlap is called a σ, **sigma**, bond. In σ bonds, e.g. any single bond, overlap of atomic orbitals occurs along the line joining the two bonded atoms. There is an unhybridised p orbital at right angles to the plane of the two sp^2 orbitals, and the p orbitals on adjacent carbon atoms are close enough to overlap. The overlapping occurs at the sides of the orbitals [see Figure 5.24].

FIGURE 5.24
The Ethene Molecule

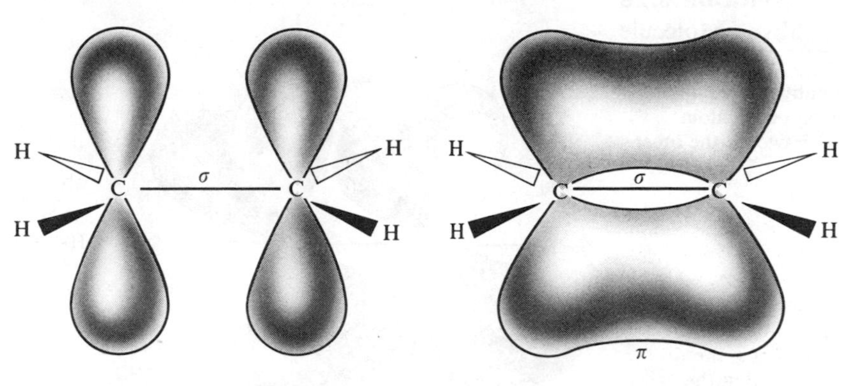

(a) The C atoms have unhybridised p orbitals.

(b) Sideways overlap between the two parallel p orbitals produces one π orbital.

...Sideways overlap between p orbitals is called a π bond...

This type of bond, produced by sideways overlapping of p orbitals above and below the plane of the sp^2 bonds, is called a π, **pi**, bond. It is not as strong as a σ bond since there is less overlapping of orbitals [see Figure 5.25]. This is why the C=C bond is less than twice as strong as a C—C bond.

...π bonds are less strong than σ bonds

Since overlapping of p orbitals on adjacent carbon atoms can occur only when the p orbitals are parallel, the two H_2C— structures must be coplanar, i.e., lie in the same plane. If one CH_2 group twists with respect to the other, the amount of overlapping of p orbitals will decrease, and the π bond will be partially broken. Since it requires energy to break a bond, the most stable arrangement of the molecule is the one in which all six atoms lie in the same plane [see Figure 27.2(a), § 27.4.3].

For π bonds to be formed the atoms in $H_2C{=}CH_2$ must be coplanar

FIGURE 5.25
The Difference between σ and π bonds

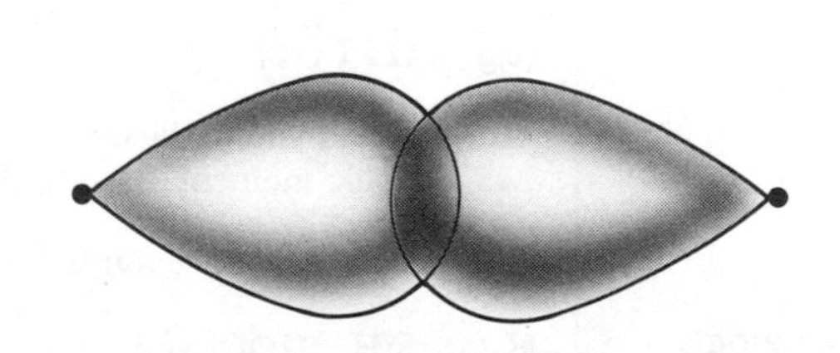

(a) σ Bonding. The orbitals point towards each other.

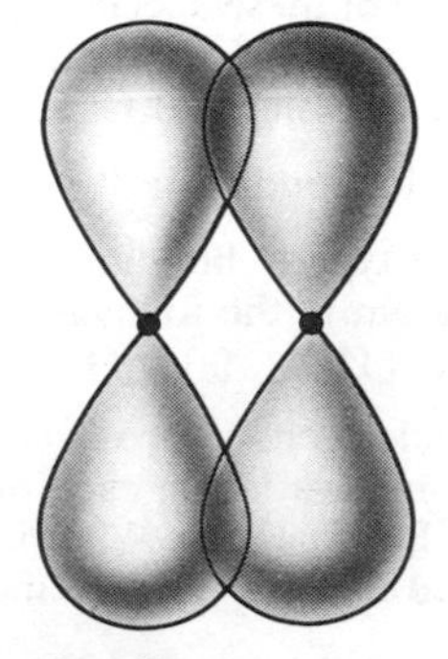

(b) π Bonding. The orbitals are parallel and overlap sideways.

In the —C≡C— bond each C uses two sp hybrid orbitals...

When a carbon atom is bonded to only two other atoms, as in HC≡CH or O=C=O the two σ bonds formed use sp hybrid orbitals [§ 5.2.2]. In H—C≡C—H, σ bonds are formed by the overlapping of sp hybrid orbitals of the two carbon atoms and by the overlapping of sp orbitals of each carbon atom

... the two unhybridised p orbitals overlap with those of the other C atom to form two π bonds

with the 1s orbital of a hydrogen atom. For maximum overlapping, the four atoms must lie in a straight line. Two π bonds are formed by overlapping of the remaining unhybridised p orbitals. There are two of these on each carbon atom at right angles to the sp hybrid orbitals [see Figure 5.26].

FIGURE 5.26
The Ethyne Molecule

(a) Two unhybridised p orbitals on one C atom overlap with two on the other C atom.

(b) Two π bonds are formed, one above and below the line of the σ bonds, the other in front and at the back.

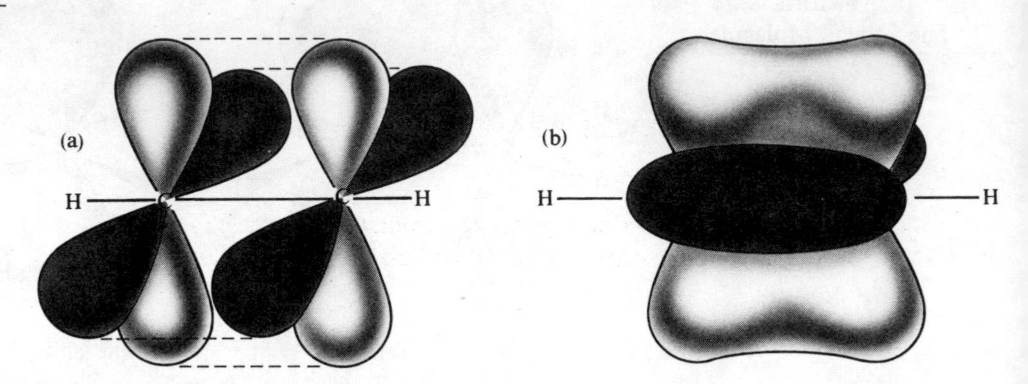

There is a π bond in CO_2

In carbon dioxide, there are sp hybrid bonds between the carbon atom and each of the oxygen atoms. Since two sp hybrid bonds are colinear, the CO_2 molecule is linear. Sideways overlapping of p orbitals produces single π bonds between each oxygen atom and the carbon atom.

π bond formation is restricted to small atoms

The formation of strong π bonds is restricted to members of the first short period: carbon, nitrogen and oxygen. In larger atoms, strong π bonds are not formed because, being removed from the line between the centres of the atoms, the π bond becomes rapidly weaker as the size of the atom increases.

CHECKPOINT 5B: BONDING

1. Match each of these species with one of the hybridisation schemes sp, sp^2, sp^3: (*a*) PH_3, (*b*) PH_4^+, (*c*) $BeCl_2$, (*d*) $SiCl_4$, (*e*) BrF_3, (*f*) Al_2Cl_6.

2. Describe the bonding in the molecules CO_2 and CO.

3. State what type of bonding orbitals are employed by the central atom in the following: (*a*) PH_3, (*b*) SCl_2, (*c*) HCHO, (*d*) HCN, (*e*) F_2O.

4. Draw an electrons-in-boxes diagram to show the electronic configuration in each of the underlined atoms: $\underline{Be}H_2$, $\underline{B}F_3$, $\underline{B}F_4^-$, H$\underline{C}$N. What is the nature of the hybrid bonds formed in each species, and what is the shape of each species?

5. Deduce the shapes of the following species: AsH_3, PH_4^+, H_3O^+, CS_2, $CH_2{=}C{=}CH_2$, $HC{\equiv}N$.

6. What is the arrangement of bonds around the central atom in each of the following species: CH_4, BF_3, NF_3, ICl_4^-, BrO_3^-, ClO_4^-, $CHCl_3$? State the nature of the hybridisation of the atomic orbitals on the central atom.

7. Write structures which show the arrangement of electrons in the bonding orbitals of O_2, CO_2, CO, NO_3^- and CN^-

(e.g., $:\ddot{O}::\ddot{O}:$)

8. What can you deduce from the fact that, whereas water has a dipole moment, carbon dioxide has none?

9. Write electron structures for PH_3, NH_3, NH_4Cl, H_2O, H_2O_2, SiH_4, HOCl and NO_2^- (e.g., $H:\ddot{\underset{\cdot\cdot}{P}}:H$ with H below P)

10. The ammonium ion and methane are said to be *isoelectronic*. What does this mean? Why do the compounds have different chemical properties?

11. (*a*) Sketch the arrangement of bonds in $CF_2{=}CF_2$.

(*b*) Explain why there are two isomers with the formula CFCl=CFCl.

5.3 DELOCALISED ORBITALS

Localised electrons are to be found between the nuclei of two bonded atoms...

In the compounds discussed so far, the electrons in the σ and π bonds have been located in the region *between* the nuclei of the bonded atoms. They are **localised** electrons. In some molecules, some of the electrons are **delocalised**: they do not remain between a pair of atoms.

5.3.1 BENZENE

Benzene, C_6H_6, is an aromatic hydrocarbon [§25.6]. Kekulé [§28.1] proposed the structural formula for benzene which is shown here.

An alternating system of single and double bonds is called a **conjugated double bond system**. Between each pair of adjacent carbon atoms is a σ bond, formed by overlapping of sp^2 hybrid orbitals. Since sp^2 bonds are coplanar, all the carbon atoms lie in the same plane and form a regular hexagon. The unhybridised p orbitals of the carbon atoms are perpendicular to the plane of the hexagonal benzene 'ring'.

FIGURE 5.27
The Benzene Ring

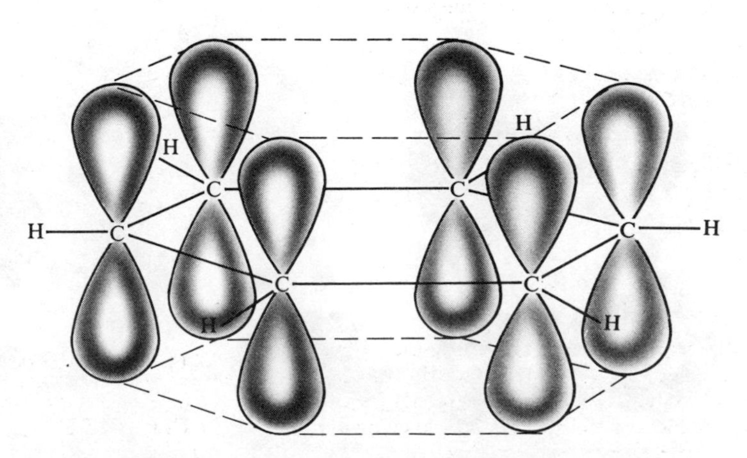

(a) Each of the C atoms has an unhybridised p orbital. These overlap sideways.

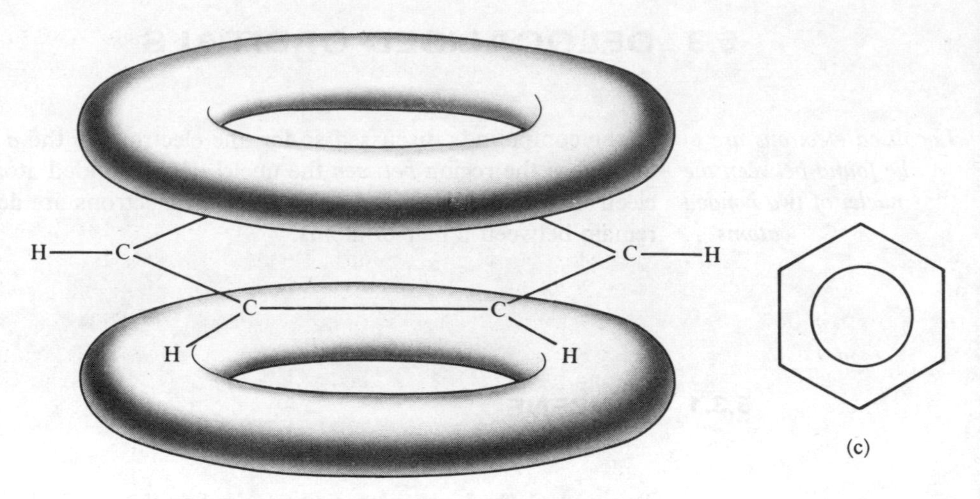

(b) The resulting distribution of π electron charge can be represented by two doughnut-shaped regions, one above and one below the hexagon of carbon atoms.

... Delocalised electrons do not remain between a pair of bonded atoms

As in ethene [see Figure 5.24], overlapping of p orbitals on adjacent carbon atoms gives rise to π bonds. In benzene, the p orbitals are able to overlap all round the ring [see Figure 5.27]. The electrons in the p orbitals cannot be regarded as located between any two carbon atoms: they are free to move between all the carbon atoms in the ring. They are described as **delocalised** and are represented as an annular cloud of electron density above and below the plane of the molecule.

Electrons in the π bonds in benzene are delocalised

The formula of benzene is often written as shown in Figure 5.27(c) to represent the delocalisation of π electrons.

FIGURE 5.28 Models of Ethene and Benzene

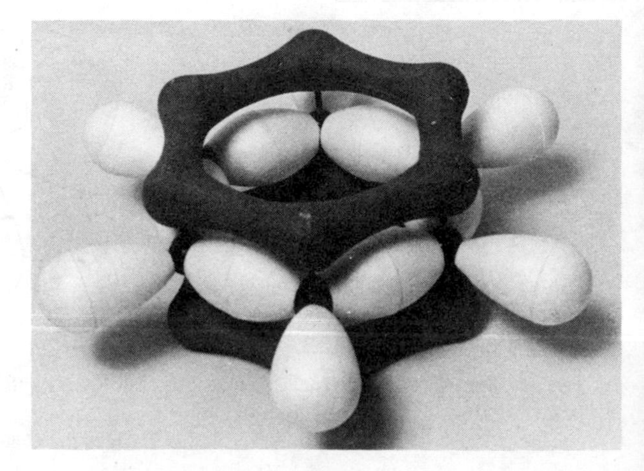

The delocalisation of π electrons confers stability on benzene. One can calculate the standard enthalpy content of benzene on the basis of the structure

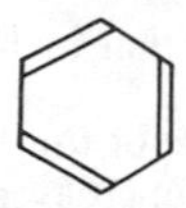

by adding the average standard bond enthalpies for the bonds [§ 5.2.7 and § 10.6]. The sum of 3(C—C) bonds plus 3(C=C) bonds plus 6(C—H) bonds is $-5350\,\text{kJ mol}^{-1}$. The standard enthalpy content can also be found from measurement of the standard enthalpy of combustion [§ 10.2]. The experimental value of $-5550\,\text{kJ mol}^{-1}$ is more negative than the theoretical value. This means that benzene is *more* stable than one would expect it to be on the basis of the formula

Delocalisation of electrons confers stability on benzene

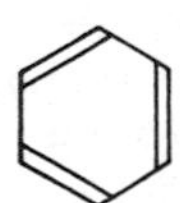

The *difference*, $-200\,\text{kJ mol}^{-1}$ is called the **delocalisation energy** of benzene.

CHECKPOINT 5C: HYBRID BONDS

1. How is it known that the four C—H bonds in methane are equivalent?

***2.** What is meant by the terms: (*a*) localised molecular orbital and (*b*) delocalised molecular orbital?

***3.** Describe the formation of the second bond between the two carbon atoms in ethene. Explain why the ethene molecule, $H_2C{=}CH_2$, is planar.

†**4.** Why was the Kekulé structure for benzene adopted? Why was it superseded? [Refer to Chapter 28.]

QUESTIONS ON CHAPTER 5

1. State the Sidgwick–Powell theory of electron pair repulsion. Sketch the arrangement of atoms in (*a*) $BeCl_2$, (*b*) BCl_3, (*c*) CCl_4, (*d*) NH_3, (*e*) H_2O and (*f*) IF_5. Explain how the arrangement of bonds is predicted on the Sidgwick–Powell theory.

2. Sketch the shapes of the following species:

CO_3^{2-}, NO_2^-, NO_2^+, NO_3^-, PCl_6^-,

PCl_4^+, ICl_4^-

3. Give the formula of a molecule whose atoms occupy each of the following shapes: (*a*) linear, (*b*) planar trigonal, (*c*) tetrahedral and (*d*) octahedral. State the angle between the bonds in each structure.

Some molecules which are based on a tetrahedral structure have bond angles different from the regular tetrahedral angle. Give an example, and explain the difference.

4. 'The shapes of simple molecules can be deduced from a consideration of the bonding electrons employed.' Discuss this statement. Apply the principle to the shapes of (*a*) NH_3, (*b*) H_2O, (*c*) CO_3^{2-}, (*d*) $H_2C{=}CH_2$, (*e*) ICl_3, (*f*) I_3^- and (*g*) SF_6.

5. (*a*) What are the main features of the electron pair repulsion model for accounting for the shapes of molecules?

(*b*) By considering the numbers of lone and bonding pairs of electrons, predict the general shapes of the following molecules or ions: F_2O, H_3O^+, ClF_4^-.

(*c*) Antimony, Sb, is in Group 5 of the Periodic Table. It forms a series of salts which contain the SbF_5^{n-} anion, the structure of which is a square-based pyramid:

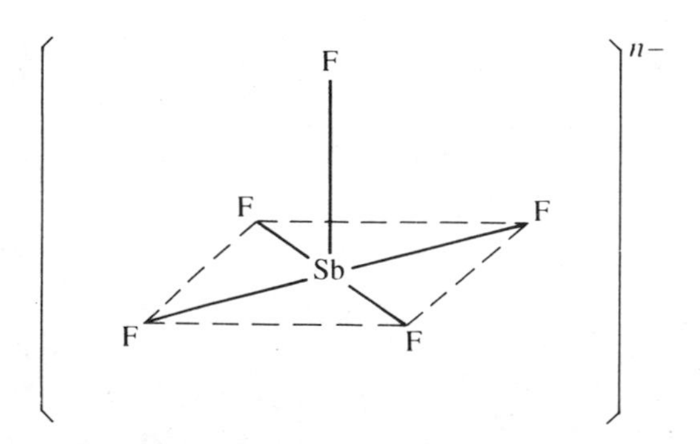

Deduce the total number of electrons around the antimony atom, the value of n and the oxidation number of Sb in this ion.

(C 91)

6. Explain the following statements:

(*a*) Two different compounds have the molecular formula

$$\begin{array}{ccc} H & & OH \\ & C & \\ CH_3 & & Cl \end{array}$$

(*b*) The spatial arrangement of atoms in BF_3 is different from that in NH_3.

(*c*) XeF_4 is planar, whereas CCl_4 is tetrahedral.

(*d*) The bond angle in NH_3 is greater than that in H_2O.

7. State the nature and spatial arrangement of the bonds in each of the following species: (*a*) CO_2, (*b*) HCN and (*c*) NO_3^-.

8. (*a*) Use the electron pair repulsion theory to explain the shapes of the following species: (i) SF_6 and ICl_4^-, (ii) PCl_5 and SF_4, (iii) NH_4^+ and PF_3.

(*b*) Predict the shapes of the molecules:

FNO, NSF

where the atoms are bonded in the orders shown, and the oxidation number of F is −1, of O is −2, and of S is +4.

(*c*) Give the shape of and the bonding in the ion NO_2^+.

(O 91)

9. (*a*) What are the essential ideas in the electron pair repulsion theory which can account for the shapes of simple molecules and ions?

(*b*) Use these ideas to account for the shapes of the following species. In *each* case state the number and the type of electron pairs involved, and sketch the full structure of the molecule.

(i) Methane, CH_4

(ii) Sulphur hexafluoride, SF_6

(iii) Water, H_2O

(iv) Phosphorus pentafluoride, PF_5

(v) Ethene, C_2H_4

(*c*) Predict the shapes of the following ions, explaining your reasoning:

(i) NH_4^+, (ii) H_3O^+, (iii) $[Fe(CN)_6]^{4-}$.

(*d*) Ammonia, NH_3, and boron trifluoride, BF_3, react together to form an addition compound, $NH_3{\cdot}BF_3$.

Draw diagrams to show the shapes of the two reactant molecules and that of the product.

(*e*) Why, and in what way, is the H—N—H bond angle in NH_3 different from that in NH_4^+?

(AEB 92)

6

CHEMICAL BONDING AND THE STRUCTURE OF SOLIDS

6.1 X RAY DIFFRACTION

A crystal is a regular three-dimensional arrangement of particles

A solid in which the arrangement of atoms or ions or molecules follows a regular three-dimensional design has a crystalline form. The surfaces of the crystal are planes, called **faces**. They intersect at angles that are characteristic for the substance. The ordered arrangement of particles in the crystal is called the **crystal structure**. It is based on a **lattice**, i.e., a geometrical arrangement of points in space. There are fourteen lattices on which crystal structures are based. The crystal structure gives rise to many series of equally spaced planes of particles [see Figure 6.1]. Since the wavelengths of X rays are comparable with the distances between the planes of particles, a crystalline solid acts as a three-dimensional diffraction grating for X rays.*

FIGURE 6.1 Some of the Planes in a Crystal (in two dimensions)

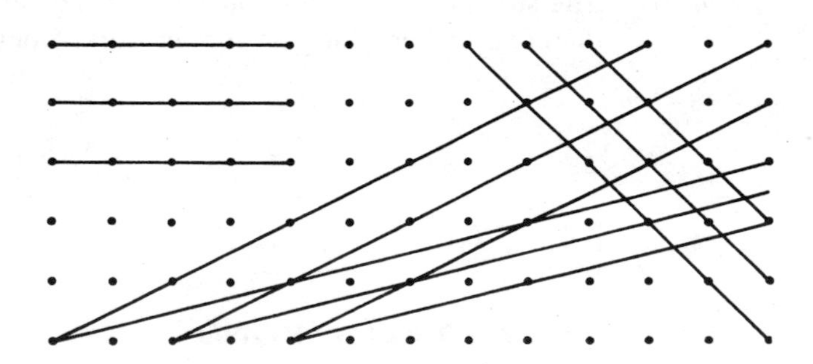

A crystal acts as a three-dimensional diffraction grating for X rays

When a beam of X rays meets the solid, X rays interact with electrons, and the beam is scattered. The scattered X rays must be made to produce a visible pattern, e.g., on a photographic film. From the pattern of scattering [see Figure 6.2] one can infer the pattern of distribution of electronic charge in the crystal and hence the nature of the crystal structure.

*See R Muncaster, *A-Level Physics* (Stanley Thornes)

Figure 6.2 shows a developed X ray film. The central spot has been produced by undeflected X rays, and the circles of spots are produced by X rays which have been diffracted through various angles by the planes of atoms or ions in the crystal.

FIGURE 6.2 An X ray Diffraction Pattern

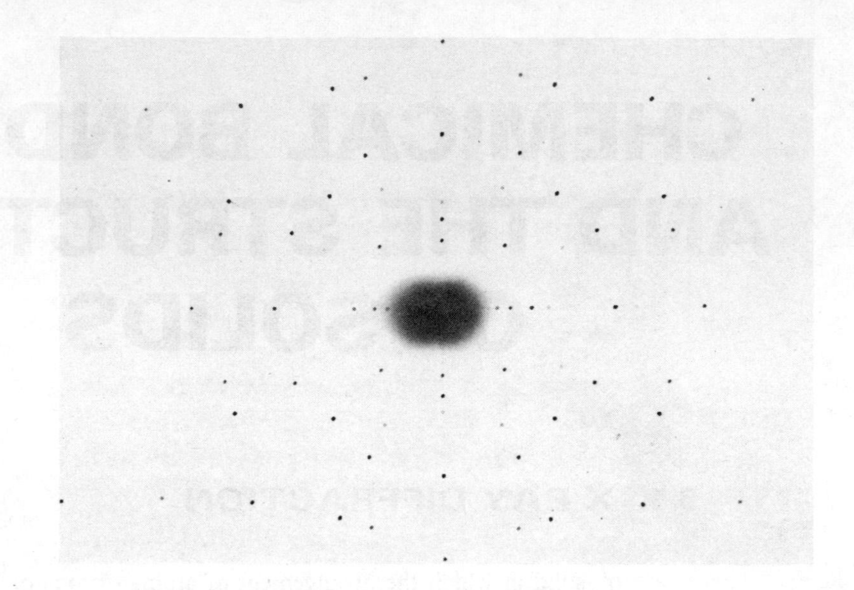

Crystallographers compare results of measurements with results calculated for models

Arriving at a crystal structure from an X ray photograph is not easy. A method which crystallographers employ is to construct a model of the crystal structure and calculate the angles and intensities of diffracted X rays which the model would produce. If the calculations do not agree with the measured results, a different model must be constructed and fresh calculations performed. Eventually, with the aid of a computer, agreement is reached between the model and the measurements. The model is then held to represent the structure of the crystal.

Hydrogen atoms do not show on X ray patterns

Since it is the *electrons* of atoms that scatter X rays, the small atoms in a compound, which have few electrons, especially hydrogen atoms, are difficult to detect. The structures of metals, ionic compounds and macromolecular substances (such as diamond and graphite) have been worked out from X ray diffraction measurements.

6.2 METALLIC SOLIDS

6.2.1 THE METALLIC BOND

The nature of the metallic bond must be responsible for the properties of metals

Modern technology is based on the use of metals. Most of our machines and most of our forms of transport are made of metal. There must be some feature of the bond between metal atoms that gives metals their special properties. Many metals are strong and can be deformed without breaking; many are **malleable** (can be hammered) and **ductile** (can be drawn out under tension). They are shiny when freshly cut and good conductors of heat and electricity. Any theory of the metallic bond must account for all these physical properties.

In a metal...

...Atomic orbitals overlap to form molecular orbitals...

The outer shell electrons of a metal (the valence electrons) are relatively easily removed, with the formation of metal cations. When two metal atoms approach closely, as in a metal structure, their outer shell orbitals overlap to form molecular orbitals. If a third atom approaches, its atomic orbitals can overlap with those of the first two atoms to form another molecular orbital. For a large number of atoms, a large number

... The valence electrons become delocalised,...

... Metal cations are formed...

... These are attracted to the electron cloud

of molecular orbitals are formed, extending over three dimensions. As a consequence of the multiple overlapping of atomic orbitals, the outer electrons from each atom come under the influence of a very large number of atoms. They are free to move through the structure and are no longer located in the outer shell of any one atom: they are **delocalised**. The removal of the electrons leaves behind metal cations. The reason why the cations are not pushed apart by the repulsion between them is that, in a pair of cations, each cation is attracted to the delocalised electron cloud between them [see Figure 6.3].

FIGURE 6.3
Deformation of Metal Structure

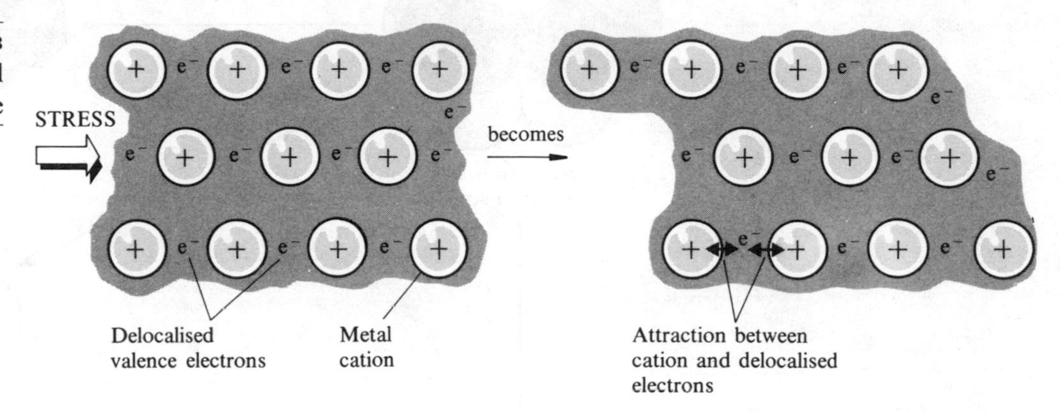

The metallic bond explains the strength of metals...

This theory of the metallic bond explains the physical properties of metals. If a stress is applied to the metal, the structure can change in shape without fracturing [see Figure 6.3]. This contrasts with the effect of stress on an ionic structure [see Figure 6.8, § 6.3].

... also their thermal conductivity...

The high thermal conductivity of metals is accounted for. When heat is supplied to one end of a piece of metal, the kinetic energy of the electrons is increased. The increase is transmitted through the system of delocalised electrons to other parts of the metal.

... and electrical conductivity...

Electrical conductivity can also be explained. If a potential difference is applied between the ends of a metal, the delocalised electron cloud will flow towards the positive potential.

... and the shiny appearance of metals

The shiny appearance of metals fits in which the theory of the nature of the metallic bond. The metal contains a large number of molecular orbitals at a large number of different energy levels. When light falls on to the metal, electrons are excited. A large number of transitions between energy levels is possible, with a whole range of frequencies being absorbed. As electrons return to lower energy levels, light is emitted and makes the metal shine.

6.2.2 THE STRUCTURE OF METALS

Types of metal structures

When identical spheres pack together so as to minimise the space between them, a close-packed structure is formed...

Metal atoms pack closely together in a regular structure. There is no way of packing spheres to fill a space completely without leaving gaps between them. Arrangements in which the gaps are kept to a minimum are called **close-packed** arrangements. X ray studies have revealed three main types of metallic structures. In the **hexagonal close-packed structure**, and the **face-centred-cubic close-packed structure**, the metal atoms pack to occupy 74% of the space. In the **body-centred-cubic structure**, the atoms occupy 68% of the total volume.

FIGURE 6.4
Face-centred-cubic Close-packed Structures

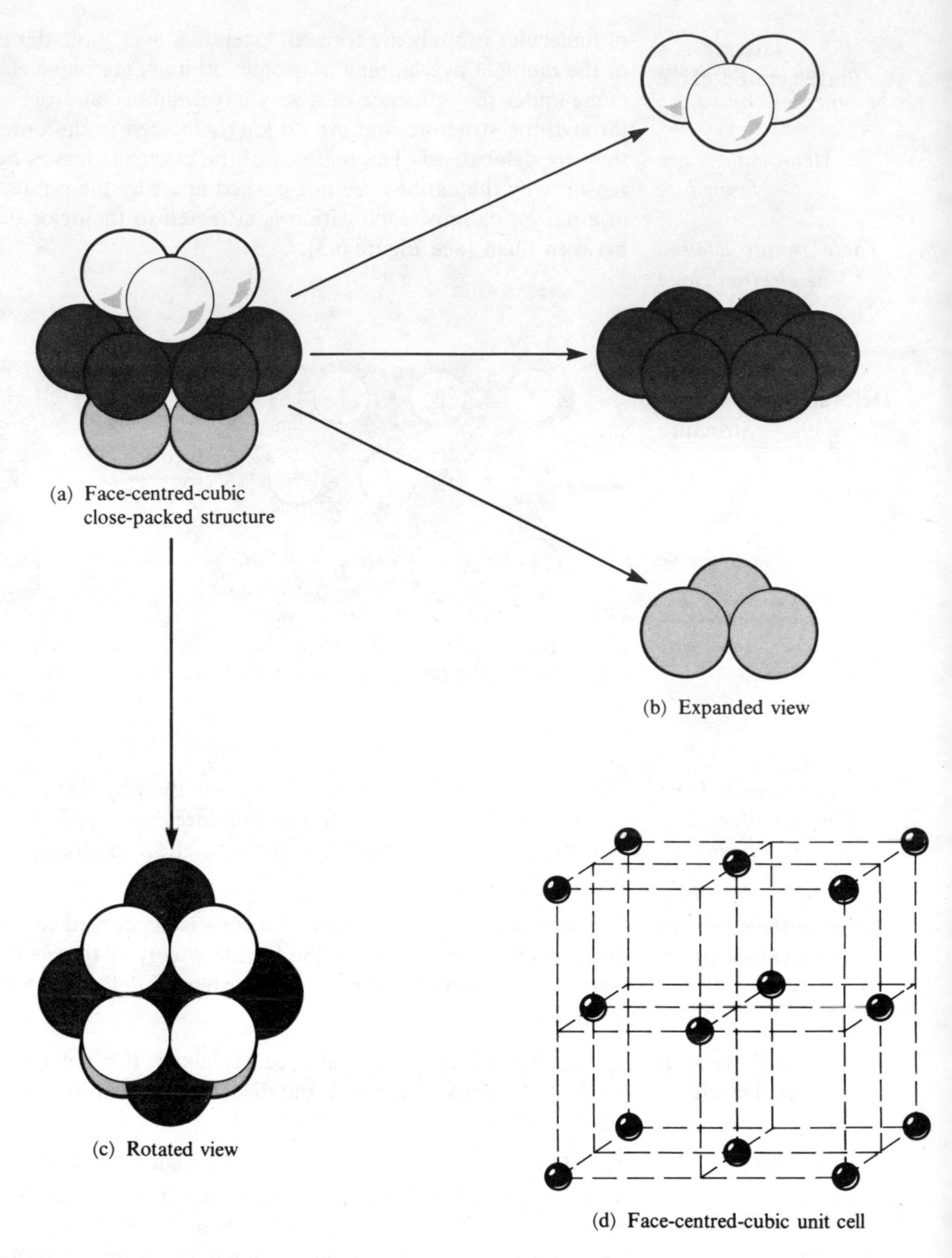
(a) Face-centred-cubic close-packed structure

(b) Expanded view

(c) Rotated view

(d) Face-centred-cubic unit cell

Figure 6.4 shows a face-centred-cubic close-packed structure. Since every atom is in contact with 12 others (6 in the same layer, 3 in the layer above and 3 in the layer below, it is said to have a **coordination number** of 12. The high coordination numbers in these structures arise from the non-directed nature of the metallic bond.

A unit cell

Also shown in Figure 6.4 is the **unit cell**. A unit cell is the smallest part of the crystal that contains all the characteristics of the structure. The whole structure can be generated by repeating the unit cell in three directions.

A body-centred-cubic structure is less close-packed

The less closely packed body-centred-cubic structure is shown in Figure 6.5(a). With one atom at each of the eight corners of a cube and one in the centre touching these eight, the coordination number is 8. Figure 6.5(b) shows an expanded view, and (c) shows the unit cell with tie-lines to show that the coordination number is 8.

FIGURE 6.5
Body-centred-cubic Structure

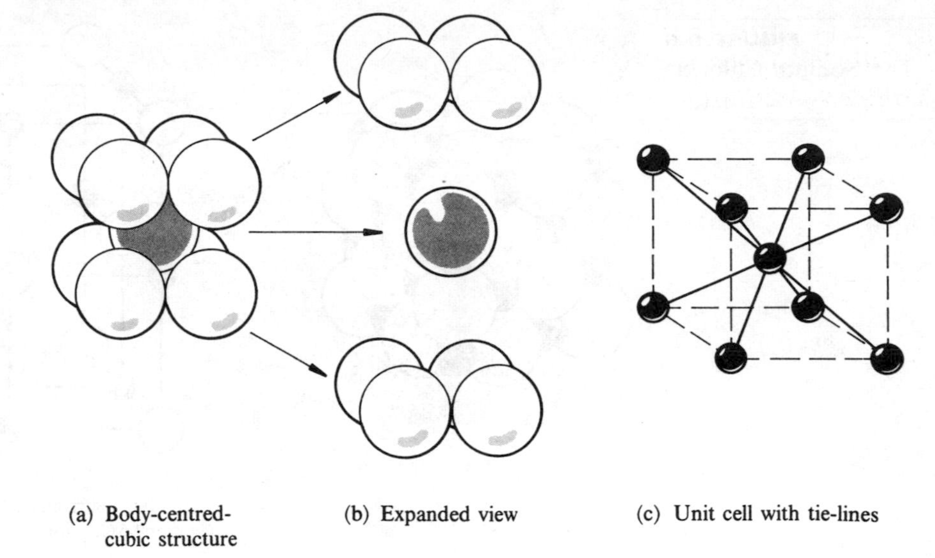

(a) Body-centred-cubic structure (b) Expanded view (c) Unit cell with tie-lines

6.2.3 METALLIC RADIUS

One half the distance between the nuclei of adjacent metal atoms is called the **metallic radius**.

CHECKPOINT 6A: METALLIC STRUCTURES

1. Why might someone describe an entire piece of metal as a large molecule?

2. How does the nature of the metallic bond account for the properties of metals?

3. Explain what is meant by (*a*) a close-packed structure, (*b*) a coordination number of 8.

6.3 IONIC STRUCTURES

Ionic structures are regular three-dimensional arrangements of ions

The alkali metal halides are ionic compounds. The ions are arranged in a regular three-dimensional structure. The melting and boiling temperatures of ionic compounds are high, owing to the strong forces of electrostatic attraction between the ions in the crystal. When the salts are melted or dissolved, the ions become free to move, and the salts conduct electricity. The crystals of alkali metal halides are *cubic* in shape, and X ray analysis shows two kinds of structures. The sodium chloride structure is illustrated in Figure 6.6(a).

In NaCl the anions form a face-centred-cubic lattice with cations occupying holes in the lattice...

... and 6:6 coordination

The best arrangement of ions in a structure, being the one with the lowest energy, is that which allows the greatest number of contacts between oppositely charged ions without pushing together ions with the same charge. Many structures are close-packed arrangements of anions, with the smaller cations occupying holes in the structure. Sodium chloride has a face-centred-cubic close-packed lattice of chloride ions (radius, 0.181 nm), which is expanded to accommodate sodium ions (radius 0.098 nm) in the lattice of anions [Figure 6.6(a)]. In Figure 6.6(b), only the centres of the ions are drawn in order to show the features of the structure clearly. There are six sodium ions surrounding each chloride ion: the coordination number of chlorine is 6. Similarly, there are six chloride ions surrounding each sodium ion: the coordination number of sodium is 6. The structure shows 6 : 6 coordination.

FIGURE 6.6
The Sodium Chloride Structure

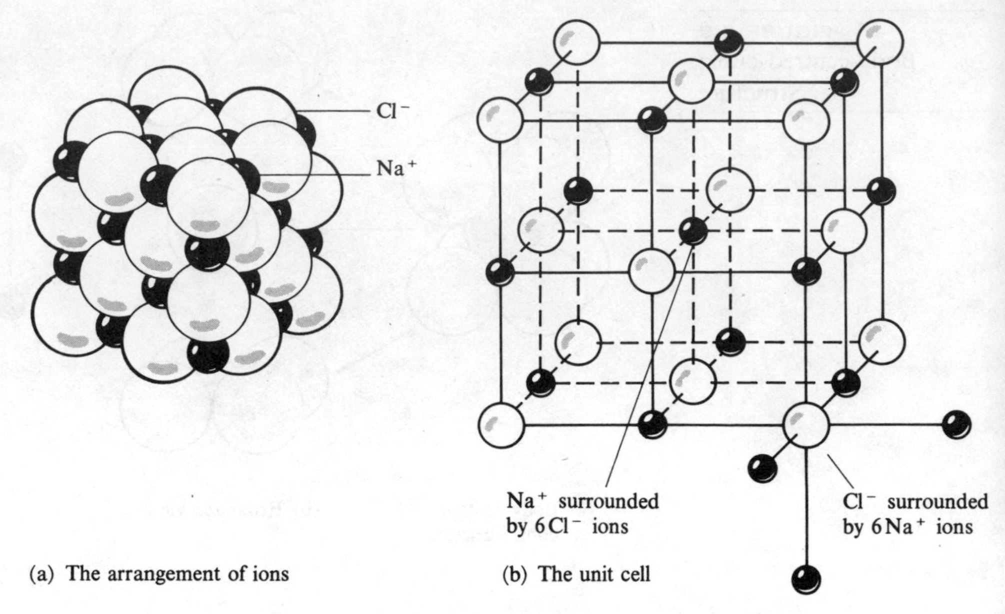

(a) The arrangement of ions (b) The unit cell

In CsCl the Cs^+ ions and Cl^- ions both adopt simple cubic structures with 8:8 coordination

The caesium chloride structure [see Figure 6.7(a)] is different. Since the caesium ion (radius 0.168 nm) is larger than the sodium ion, a larger number of chloride ions can surround it. The structure shows 8 : 8 coordination. Both the caesium ions and the chloride ions adopt simple cubic lattices, which interpenetrate [see Figure 6.7(b)] so that each cube of chloride ions has a caesium ion at its centre and vice versa.

FIGURE 6.7
The Caesium Chloride Structure

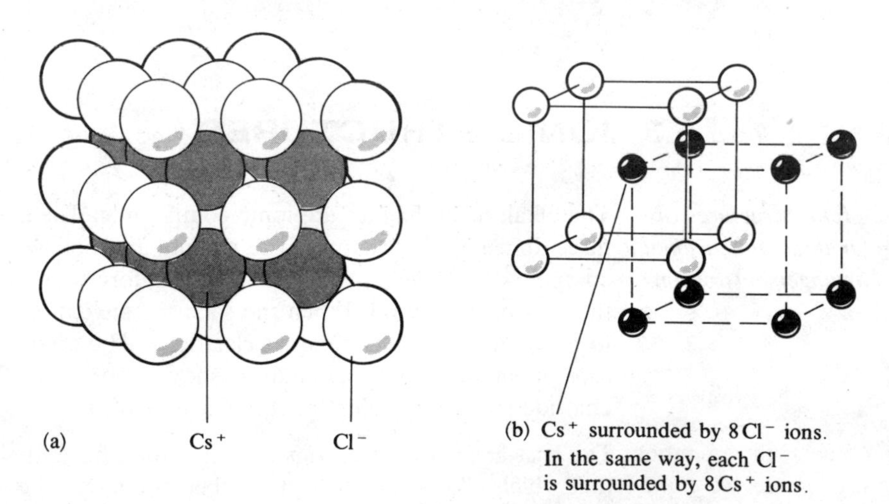

(a)

(b) Cs^+ surrounded by 8 Cl^- ions. In the same way, each Cl^- is surrounded by 8 Cs^+ ions.

Unlike metals, ionic crystals are brittle

Ionic crystals are brittle. Figure 6.8 shows what happens when an ionic crystal is subjected to stress. A slight dislocation in the crystal structure brings similarly charged ions together. Repulsion between the like charges fractures the crystal. This is a different picture from the effect of stress on a metallic crystal, where deformation of the structure does not result in fracture [see Figure 6.3, § 6.2.1].

FIGURE 6.8 An Ionic Structure is Easily Fractured

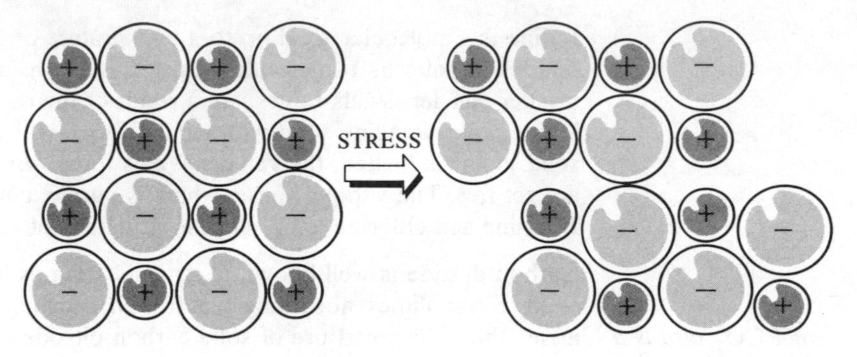

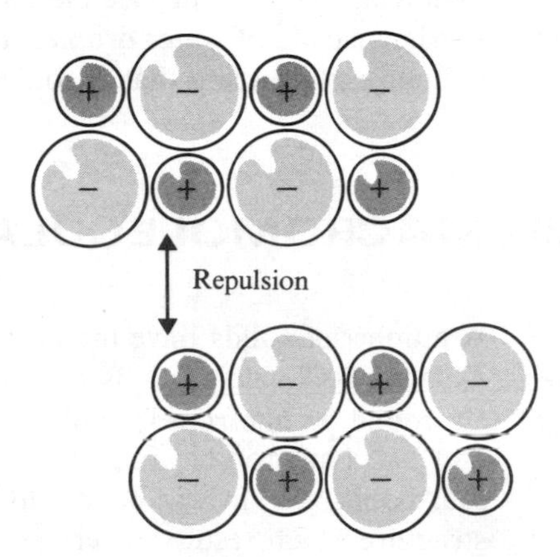

6.4 MOLECULAR SOLIDS

Van der Waals forces hold the molecules together in solid argon and in solid iodine...

Some solids are held together by weak attractions between individual molecules. They are described as **molecular solids** and said to have a **molecular structure**. At very low temperatures, even the noble gases can be solidified. Figure 6.9 shows the cubic close-packed structure of atoms in solid argon. The van der Waals forces between the atoms are very weak, and if the temperature rises above − 170 °C the solid melts. Liquid argon consists of separate atoms. Figure 6.10 shows the structure of solid iodine.

FIGURE 6.9 Solid Argon (face-centred cube)

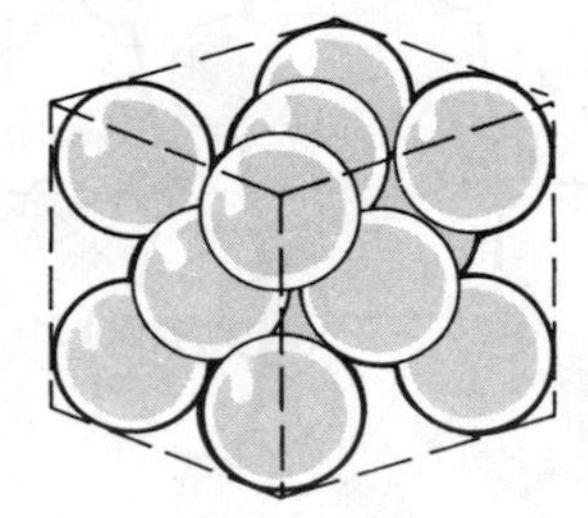

FIGURE 6.10 Solid Iodine (face-centred cube)

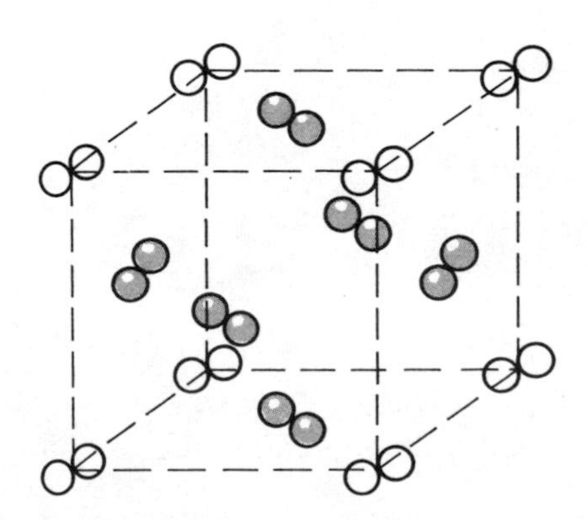

The I_2 molecules in the centre of each face are shaded.

Iodine is a molecular solid up to a temperature of 30 °C. The atoms are covalently bonded in pairs as I_2 molecules. Operating between the molecules are the much weaker van der Waals forces. As a result of the regular arrangement of molecules, iodine is a crystalline solid with regular faces, which give a shiny appearance. When solid iodine is heated, the van der Waals forces are broken and individual molecules are set free. The vapour phase, which is purple, consists of individual I_2 molecules. Bromine and chlorine adopt similar structures at lower temperatures.

...solid CO_2 also is a molecular structure

Carbon dioxide is well known in its solid form as 'dry ice' or 'drikold'. Above − 78 °C it sublimes, absorbing heat from its surroundings to do so. From this property arises the widespread use of solid carbon dioxide as a refrigerant, both in laboratory work and in the food industry. Pop singers sometimes like to enhance their performance by having lumps of dry ice on stage. As it sublimes, it cools the moist air, and swirling clouds of water droplets form. Solid carbon dioxide has a face-centred-cubic structure, resembling that of iodine, which is shown in Figure 6.10.

6.5 MACROMOLECULAR STRUCTURES

A macromolecular structure is held together by covalent bonds, e.g. diamond...

A number of solids have the kind of structure described as **macromolecular** or **giant molecular**. Covalent bonds between atoms bind all the atoms into a giant molecule. Diamond, an **allotrope** [§ 21.4] of carbon, is one of the hardest substances known and has a macromolecular structure [see Figure 6.11]. Each carbon atom forms four bonds (sp^3 hybrid bonds) to four other carbon atoms. The giant molecular structure which results is very strong. It is different from a molecular structure, in which, although the bonds between the atoms in a molecule are strong covalent bonds, the intermolecular forces of attraction are weak. Diamond remains a solid up to a temperature of 3500 °C, at which it sublimes.

...in which the strong covalent bonds result in a hard, abrasive character

The hard, abrasive character of diamond finds it many uses. Diamond-tipped tools are used for cutting and engraving, and diamond-tipped tools are used by oil prospectors for boring through rock [see Figure 6.12]. The high **refractive index*** of diamond gives it the sparkle that makes it the most prized of jewels [see Figure 6.13].

FIGURE 6.11 The Structure of Diamond

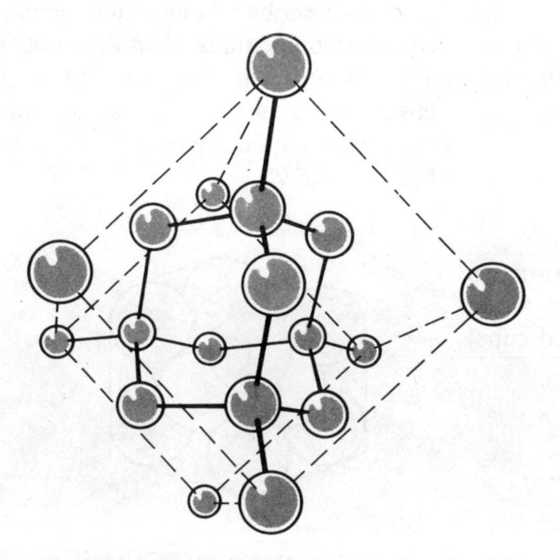

Each C atom is surrounded by 4 others: the coordination number is 4.

*See R Muncaster, *A-Level Physics* (Stanley Thornes)

FIGURE 6.12
Diamond-studded Drill Bits

FIGURE 6.13
A Selection of Diamonds

Other macromolecular structures are SiC, BN and SiO_2

Other solids with a diamond-like structure are silicon carbide $(SiC)_n$, and boron nitride $(BN)_n$. The formula unit, BN, is isoelectronic with the unit CC. Silicon(IV) oxide, SiO_2 (**silica**), also forms a three-dimensional structure. The Si—O bonds about each silicon atom are tetrahedrally distributed and each oxygen atom is bonded to two other silicon atoms [see Figure 6.14]. This structure occurs in **quartz** and other crystalline forms of silica. Quartz remains solid up to a temperature of 1700 °C.

FIGURE 6.14
Silicon(IV) Oxide Structure

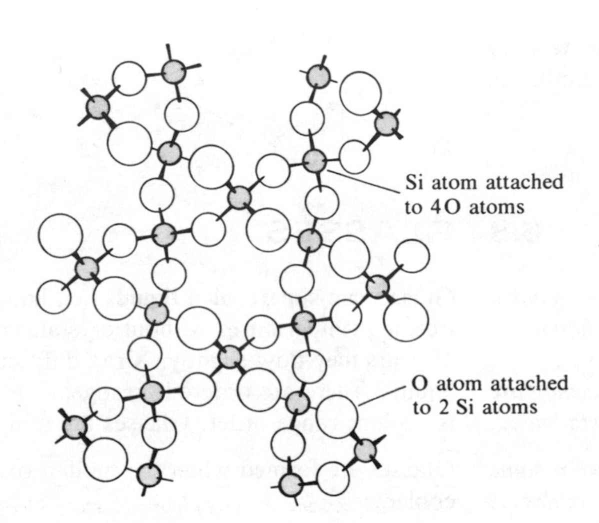

6.6 LAYER STRUCTURES

Layer structures have covalent bonds within each layer and weak van der Waals forces between layers e.g. graphite

Graphite, the other allotrope of carbon, has a **layered** structure. Within each layer, every carbon atom uses three coplanar sp^2 hybrid orbitals to bond to three other carbon atoms. A network of coplanar hexagons is formed, with a C—C bond distance of 0.142 nm. Between layers, the distance is 0.335 nm. The weak van der Waals forces of attraction between the layers allow one layer of bonded atoms to slide over another layer. The structure, which is shown in Figure 6.15, accounts for the properties of graphite. It is a lubricant, whereas diamond is abrasive. The unhybridised p electrons form a delocalised cloud of electrons similar to the metallic bond. They enable graphite to conduct electricity and are responsible for its shiny appearance.

FIGURE 6.15 The Structure of Graphite

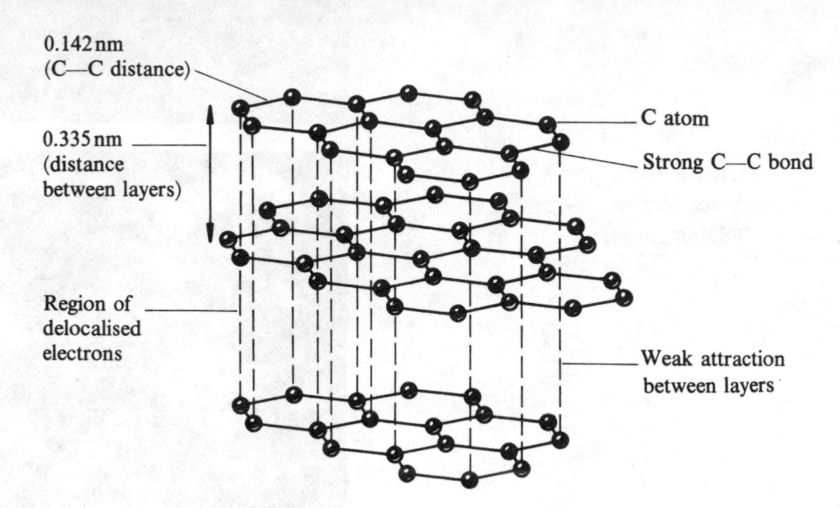

6.7 CHAIN STRUCTURES

Some substances exist in chain-like structures. One example is sulphur(VI) oxide, which crystallises in long shiny needles with the structure shown in Figure 6.16. Another is beryllium chloride which, in the solid state, has the structure shown in Figure 6.17.

FIGURE 6.16 The Structure of $(SO_3)_n$

FIGURE 6.17 The Structure of $(BeCl_2)_n$

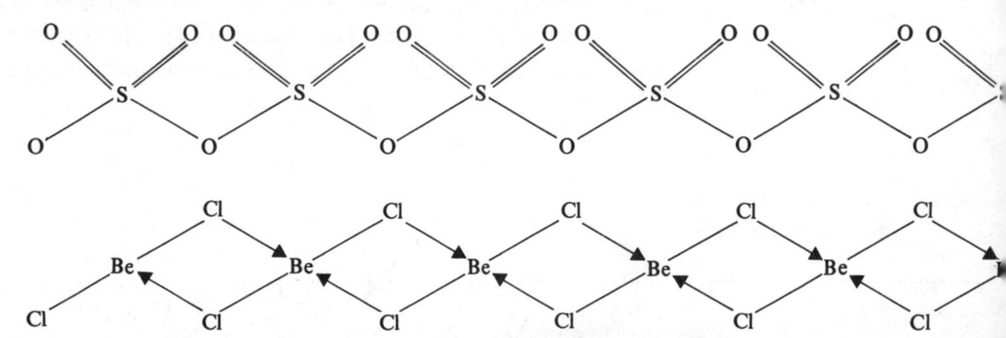

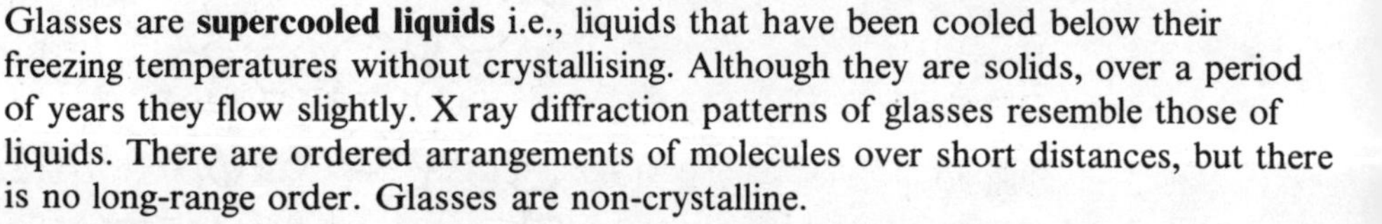

6.8 GLASSES

Glasses are supercooled liquids . . .

. . . they are non-crystalline . . .

. . . are formed from some molten oxides . . .

Glasses are **supercooled liquids** i.e., liquids that have been cooled below their freezing temperatures without crystallising. Although they are solids, over a period of years they flow slightly. X ray diffraction patterns of glasses resemble those of liquids. There are ordered arrangements of molecules over short distances, but there is no long-range order. Glasses are non-crystalline.

Glasses are formed when the molten oxides, SiO_2, B_2O_3, GeO_2 and P_4O_{10}, are cooled.

...and have many uses in industry and in the home...

Glasses are transparent to visible light. They have a multitude of uses in the construction industry, in the home and in the manufacture of optical instruments containing lenses, prisms and mirrors.

The manufacture of optical fibres is a new use for glass

Glass has an expanding use in the production of **optical fibres**. Glass can be melted and drawn out to form hair-thin fibres, which are as strong as steel. If a fibre is clad in a material of lower refractive index, light is guided along the fibre by total internal reflection at the surface of the fibre. With a source of light at one end and a light sensor at the other, optical fibres are used for communication. Lasers and light-emitting diodes are used as light sources. The big advantage of fibre optics is that, because the frequency of light is so high, a single fibre can carry a much larger number of channels of information than a coaxial cable.

The chemical inertness of glass has led to its use as a guardian of radioactive waste

The long term storage of liquid radioactive waste from nuclear power stations [§ 1.9.11] is a serious problem. Some of the waste needs to be stored for thousands of years. France has tackled the problem by **vitrification** (glass making). Concentrated radioactive waste is combined with glass-forming oxides. The resulting glass blocks are enclosed in steel canisters and stored behind thick concrete walls, which absorb radiation. The chemical inertness of glass should ensure that there is no natural mechanism by which the contents could contaminate the environment. Stores of this kind will have to be supervised for thousands of years. Some people feel that we have no right to impose this task on future generations. Others are concerned that the stores will present dangers from accidents, terrorists and war.

CHECKPOINT 6B: COVALENT STRUCTURES

1. (*a*) Why is it easy to rub away carbon atoms from graphite?

(*b*) Why is graphite used as a lubricant?

(*c*) Why is it impossible to rub away carbon atoms from a diamond?

(*d*) What characteristics of (i) diamond and (ii) graphite are useful in industry?

2. Solid iodine consists of shiny black crystals. Iodine vapour is purple. What is the difference in chemical bonding between solid and gaseous iodine?

3. 'Dry ice' is a name given to solid carbon dioxide.

(*a*) In what way does solid carbon dioxide resemble ice?

(*b*) In what way is solid carbon dioxide (i) better than ice as a refrigerant and (ii) more convenient than ice?

(*c*) Pop singers sometimes use dry ice to produce swirling clouds on stage. These clouds are clouds of condensed water vapour. Explain (i) where the water vapour comes from and (ii) why it is cool enough on stage to condense water vapour.

4. From Figures 6.11, § 6.5, and 6.15, § 6.6, which would you expect to have the higher density, diamond or graphite? Why is diamond used in cutting tools but not graphite?

5. What is meant by the statement that the electrons in diamond are *localised*, whereas graphite has *delocalised* electrons? Which electrons in graphite are delocalised? How do they affect the properties of graphite?

6. Sketch the structure of silicon carbide, SiC, which resembles diamond. Why do you think that **carborundum** (SiC) is used as an abrasive? Why does silicon carbide not exist in a graphite-type of structure?

***7.** Boron nitride, BN, has a structure like that of graphite. Explain the bonding in BN.

6.9 LIQUID CRYSTALS

TOPIC

You will read in Chapter 34 that pure solids melt sharply, the temperature remaining constant at the melting temperature until all the solid has melted. There are, however, many crystalline solids which pass at a sharp **transition temperature** to a turbid liquid phase before finally melting to form a clear liquid. Some compounds of this type are listed in Figure 6.18. The turbid liquid phases are liquids in that they can flow as liquids do and in possessing surface tension. Their molecules, however, possess some degree of order, with the result that these turbid liquids resemble crystals in certain optical properties. They are known as **liquid crystals**.

*See R. Muncaster, *A-Level Physics* (Stanley Thornes).

In many crystals, the velocity of light and therefore the refractive index is the same in every direction. Such crystals are described as **isotropic**. (Greek: *isos*, equal; *tropos*, a turn.) Other crystals (e.g., Iceland spar, $CaCO_3$) are **anisotropic**; the velocity of light is *not* the same in all directions. Such crystals are **birefringent**, i.e., an object viewed through the crystal is seen as a double refracted image. Besides this characteristic of birefringence*, anisotropic crystals give rise to interference patterns* in plane polarised light. Liquid crystals are anisotropic.

The liquid crystalline state exists between two temperatures, the melting temperature and the **clearing temperature**.

$$\text{Crystal} \underset{\text{melting temperature}}{\rightleftharpoons} \text{liquid crystal} \underset{\text{clearing temperature}}{\rightleftharpoons} \text{Isotropic liquid}$$

All liquid crystals are organic and have elongated molecules which end in a polar group,
e.g., —CN, —OR, $—NO_2$ and $—NH_2$. Many have flat portions, such as a benzene ring. The molecules are linear in conformation with doubly-bonded linking groups, e.g. >C=C< and —N=N(→O)—, to inhibit rotation and hold the molecule rigid about its long axis. The molecules possess strong dipoles and also easily polarisable groups. You can easily imagine how dipole–dipole interactions might give rise to intermolecular forces of attraction in compounds of this type. Such forces align the molecules side by side with their long axes parallel.

FIGURE 6.18 Some Liquid Crystalline Materials

(a) n-C_4H_9O—C_6H_4—N=N—C_6H_4—$COCH_3$

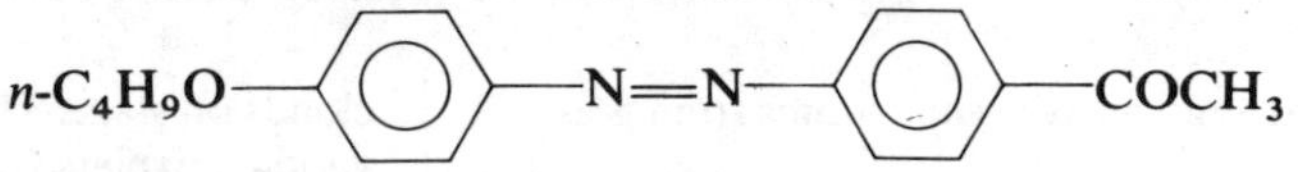

4-Butyloxy-4′-ethanoylazobenzene

(b) CH_3O—C_6H_4—N=N(→O)—C_6H_4—OCH_3

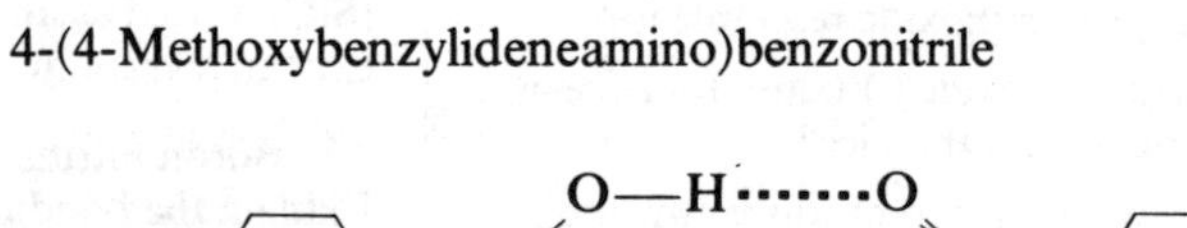

4,4′-Dimethoxyazoxybenzene

(c) CH_3O—C_6H_4—CH=N—C_6H_4—CN

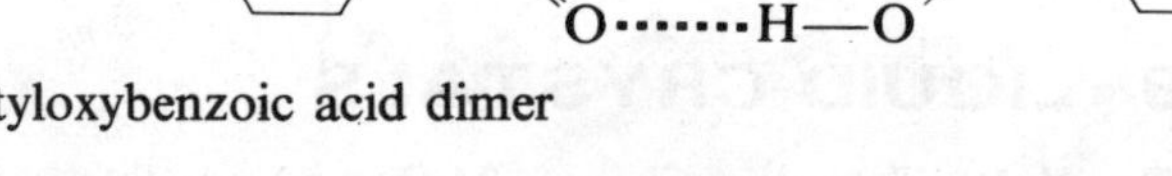

4-(4-Methoxybenzylideneamino)benzonitrile

(d) n-$C_8H_{17}O$—C_6H_4—C(O—H······O)(=O······H—O)C—C_6H_4—OC_8H_{17}-n

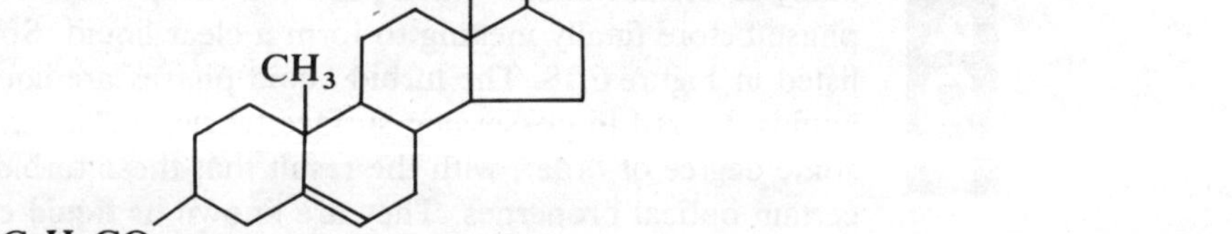

n-Octyloxybenzoic acid dimer

(e) $CH(CH_3)(CH_2)_3CH(CH_3)_2$; CH_3; CH_3; $C_6H_5CO_2$

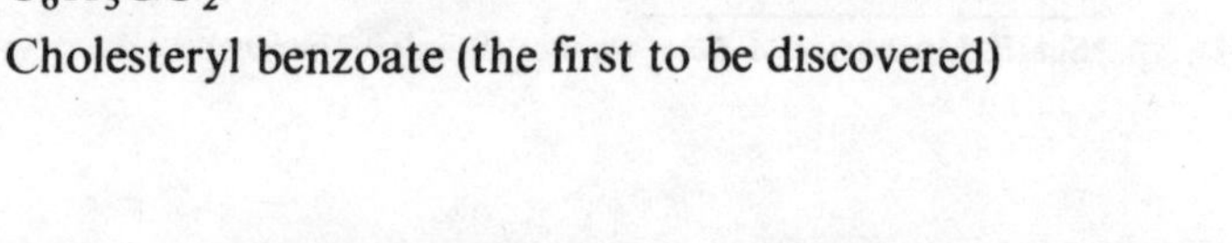

Cholesteryl benzoate (the first to be discovered)

(f) $CH_3(CH_2)_4$—⟨○⟩—⟨○⟩—CN

4′-Pentylbiphenyl-4-carbonitrile (a liquid crystal at room temperature)

There are three types of liquid crystals: **smectic**, **nematic** and **cholesteric**. Figure 6.19 shows the relationship between the smectic and nematic phases. Smectic liquids (Greek: *smektikos*: soaplike) do not flow freely: they glide along in one plane. X ray diffraction patterns suggest a structure consisting of a series of planes, which are further apart than the molecules in a crystal. A smectic phase may melt to form an isotropic liquid or may pass at a transition temperature into a nematic phase. Nematic phases flow readily, and their X ray diffraction patterns resemble those of true liquids. Viewed in plane-polarised light under a microscope, liquid crystals show characteristic coloured patterns. Nematic phases show threadlike patterns, which give the phase its name. (Greek: *nema*: thread.)

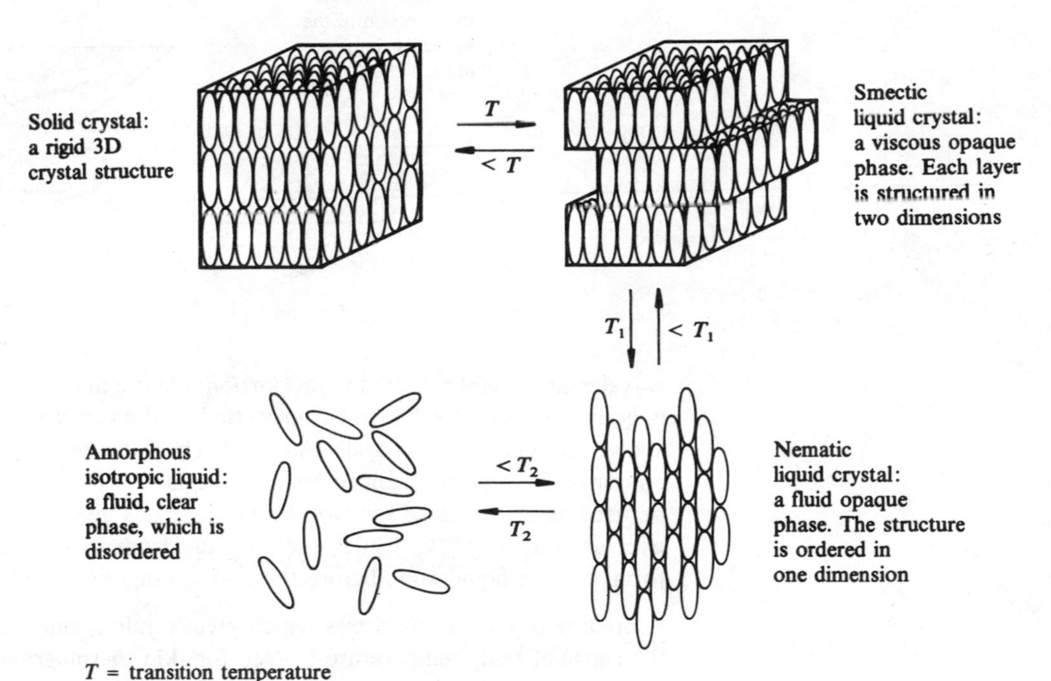
Solid crystal:
a rigid 3D
crystal structure
T
< T
Smectic
liquid crystal:
a viscous opaque
phase. Each layer
is structured in
two dimensions
T_1
< T_1
Amorphous
isotropic liquid:
a fluid, clear
phase, which is
disordered
< T_2
T_2
Nematic
liquid crystal:
a fluid opaque
phase. The structure
is ordered in
one dimension
T = transition temperature

FIGURE 6.19 Order in Liquid Crystals. (The elongated molecules are shown as ellipses. Except in the isotropic liquid they lie with their long axes parallel. T = transition temperature.)

A third type of liquid crystal phase (a modified nematic phase) is described as **cholesteric**. All the compounds which show cholesteric behaviour are **optically active** [§ 25.9.2]. Examples are the cholesteryl esters [e.g. Figure 6.18(e)]. Cholesteric liquid crystals may reflect different colours when viewed in ordinary light. The phenomenon is associated with a structure of layers 500 to 5000 molecules thick. Each layer is a stack of sheets of molecules. The long axes of the molecules in the same sheet point in the same direction. Figure 6.20 shows how, passing from one sheet to the next, the long axes of the molecules are displaced through a small angle. The displacement progresses in a clockwise direction (or, in some compounds, an anticlockwise direction), giving rise to a helical structure. The distance between one sheet in which the molecular long axes point in a certain direction and the next sheet in which the axes point in the same direction is the **pitch** of the helix. This distance is also the thickness of the layers mentioned above.

The structure reflects light of different wavelengths to different extents. If the reflected light is in the visible region, the cholesteric liquid crystal appears coloured. The

FIGURE 6.20
The Structure of a Cholesteric Liquid Crystal

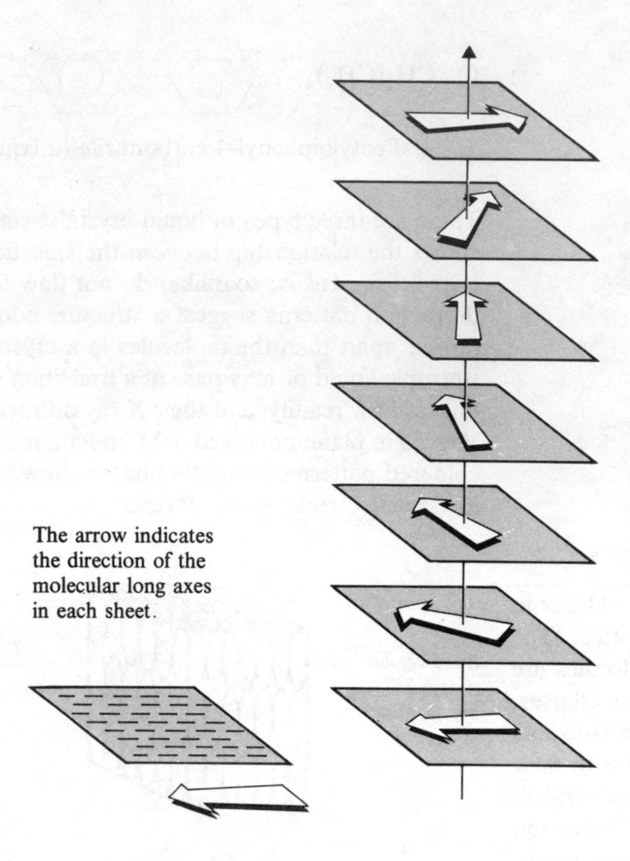
The arrow indicates the direction of the molecular long axes in each sheet.

wavelength of light reflected is proportional to the pitch of the helix. The light reflected is complementary to the light transmitted, and gives rise to the beautiful iridescence shown by such liquid crystals. As the pitch of the helix alters with temperature, the colour of the reflected light alters; this explains why cholesteric liquid crystals can be used as temperature sensors. Some materials give a full range of colour change over as little as 0.1 °C, while others change from violet to red over 40 °C, and some form the true liquid state before the colour range is completed.

A mixture of cholesteric esters which gives a full colour response over about 3 °C in the range of body temperature is used for **skin thermography**. Skin overlying veins and arteries is slightly warmer than in other areas, and the difference in temperature can be detected by cholesteric liquid crystals. Specialists can use the technique of skin thermography to detect blockages in veins and arteries. The technique has been successful in the early diagnosis of breast cancer. When a layer of cholesteric material is painted or sprayed on to the surface of the breast, a tumour shows up as a 'hot area', which is coloured blue.

The electronics industry uses cholesteric liquid crystals to find points of potential failure in circuits by detecting them as 'hot spots'. In the manufacture of aeroplanes, lightness is achieved by the use of laminates with a honeycomb interior. The quality of the bonding can be investigated by applying a film of cholesteric liquid, which will show up any defect in the bonding as a difference in thermal conductivity. Faults in welding and cracks due to metal fatigue can be detected in the same way.

Room thermometers contain cholesteric liquid crystals with a suitable temperature range. Figures show up in different colours as the temperature changes.

The digital displays you see in watches and calculators contain nematic liquid crystals. The nematic phase is fluid, and the liquid crystal molecules are polar. For these two reasons, the orientation of the molecules can be changed by the use of a very small electric field. If the change in orientation results in a change in optical properties, the liquid crystal can be used to display information, e.g., the time, or date. The timing of a watch display is controlled by a quartz crystal. A small electric current induces quartz to resonate* at 32 768 oscillations per second. A quartz watch has no mechanical moving parts, giving it a big advantage over traditional watches.

CHECKPOINT 6C: LIQUID CRYSTALS

1. Explain why the following groups are flat:

—(benzene ring)— —C(N—N)(N=N)C— (tetrazine ring)

2. Explain why the following groups have a dipole:

—OCH_3, —NO_2, —NH_2 and —N=N(→O)— .

3. Explain why these linking groups inhibit rotation and hold the molecules in which they are present in linear conformations:

—N=N—, —N=N(→O)— , —CH=CH—,

—C≡C— and —CH=N—.

4. Sketch a pair of molecules of the liquid crystal

$$4\text{-}4'\text{-}(CH_3)_2NC_6H_4N{=}NC_6H_4NO_2$$

(*a*) end to end and (*b*) side by side, in order to show how dipole–dipole interactions could hold the molecules in an ordered, parallel arrangement.

†**5.** Suggest methods of preparation for compounds (a), (b) and (d) in Figure 6.18.
[For help, see Part 4.]

QUESTIONS ON CHAPTER 6

1. What type of intramolecular and intermolecular bonds exist in (*a*) solid argon, (*b*) solid bromine, (*c*) diamond, (*d*) graphite and (*e*) silica?

2. Crystals of salts fracture easily, but metals are deformed under stress without fracturing. Explain the difference.

***3.** What is the coordination number of an ion? What is the coordination number of the cation in (*a*) the NaCl structure and (*b*) the CsCl structure? What is the reason for the difference?

4. Say what properties you would expect of substances which are (*a*) metals, (*b*) ionic compounds, (*c*) composed of individual covalent molecules, and (*d*) macromolecular covalent compounds.

5. Explain the bonding present in the solids sodium chloride, sodium, phosphorus(V) chloride (PCl_5) graphite and ice. Point out how the type of bonding determines the physical properties of the solids.

6. Give two examples of (*a*) ionic solids, (*b*) molecular solids and (*c*) covalent macromolecular solids. What are the factors that determine whether each of these types of solid will dissolve in water?

7. Describe the structure and bonding for each of the solids below, illustrating your answers with suitable diagrams. For each solid select *one* physical property and explain its dependence on structure and/or bonding.

caesium chloride, ice, poly(propene), aluminium, iodine

(O & C 91)

*See R Muncaster, *A-Level Physics* (Stanley Thornes)

8. (*a*) By means of a sketch graph showing how the number of neutrons varies with the number of protons present in stable nuclides, explain what is meant by the *stability range* of naturally occurring nuclides. Show the value of the initial slope of your graph and indicate where, if at all, the graph ends.

On your sketch label four points which lie outside the stability range, two points above it (one at low atomic number and one at high atomic number) and two points below (again one at low atomic number and one at high atomic number). Suggest the different ways in which nuclides such as those you have labelled can decay in order to become stable.

(*b*) (i) The crystal structures of sodium chloride and caesium chloride are different. The radii of the Na^+, Cs^+ and Cl^- ions are 95 pm, 169 pm and 181 pm respectively. Suggest reasons, in terms of these ionic sizes, for the differences between the structures of these crystals.

(ii) The rubidium ion has a radius of 148 pm. Suggest, with reasons, which of the two structures you discussed above is the more likely for rubidium chloride.

(iii) The co-ordination number of the Ca^{2+} ion in calcium fluoride is the same as that for Cs^+ in caesium chloride. Account for this observation and state the co-ordination number of F^- in this crystal.

(*c*) In terms of the concept of *order*, outline the major features of the *liquid* state, stressing its similarities to and differences from both the *crystalline* and the *gaseous* states.

It is a useful but crude simplification to link high boiling points in liquids with large relative molecular masses of the constituent molecules. Explain why this view is simplistic, indicating in particular why liquids whose molecules have similar relative molecular mass do not all share similar boiling points.

(NEAB 91, S)

9. (*a*) Describe, with the aid of diagrams, the structure of, and the nature of the forces present in the following solids: (i) iodine, (ii) diamond, (iii) sodium chloride and (iv) copper.

(*b*) Explain the following observations:

(i) Sodium is softer than copper but both are very good electrical conductors.

(ii) Diamond is hard and an electrical insulator, graphite is soft and an excellent electrical conductor.

(iii) Sodium chloride and caesium chloride have different structures.

(iv) Iodine dissolves readily in tetrachloromethane.

(AEB 90)

Part 2

PHYSICAL CHEMISTRY

7

GASES

7.1 STATES OF MATTER

The three main states of matter are the solid, liquid and gaseous states

There are three important **states of matter**: the gaseous, liquid and solid states. In addition, there are the **liquid crystalline state** [§ 6.9] and the **plasma state** [§ 7.4.3]. In solids and liquids, the molecules are close together, and considerable forces of attraction exist between them. When a solid melts, it expands slightly (with some exceptions) whereas, when a liquid evaporates, it forms many times its own volume of vapour. The gaseous state of a substance at temperatures below its critical temperature [§ 7.4.2] is called a **vapour**. Gases have much lower densities and are much more compressible than solids and liquids. Many scientists became interested in the behaviour of gases, and the results of the experimental studies which were made in the seventeenth, eighteenth and nineteenth centuries are embodied in the **gas laws**. A **law** is a concise statement of experimental results; it need not include an explanation of them. A **hypothesis** is an explanation of experimental findings by means of a concept or model. A well-established hypothesis becomes known as a **theory**.

The gas laws state the results of experimental studies on gases

7.2 THE GAS LAWS

7.2.1 BOYLE'S LAW

Boyle's Law deals with the effect of pressure on the volume of a gas

In 1662, Robert Boyle published his work on the **compressibility** of gases. He found that, at a constant temperature, the volume of a fixed mass of a given gas is inversely proportional to its pressure. Boyle's Law can be expressed as

$$PV = \text{Constant}$$

where P = pressure, and V = volume.

Figure 7.1 shows three ways of representing Boyle's Law graphically.

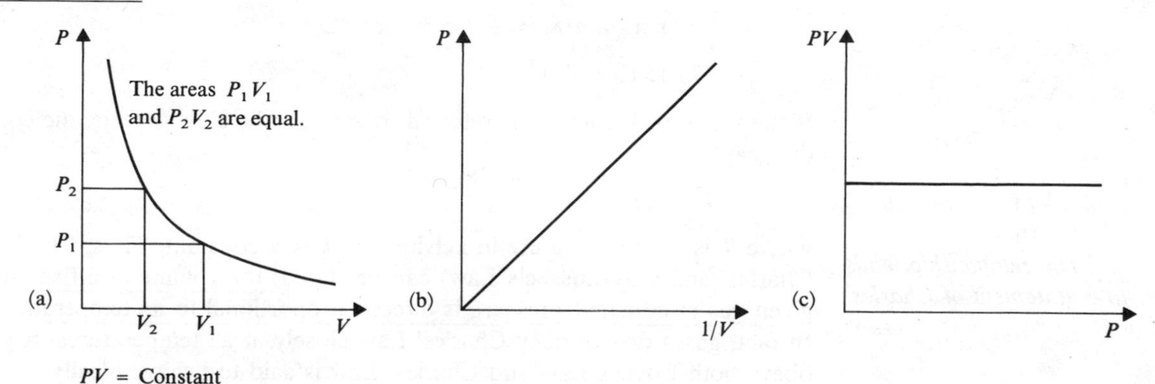

FIGURE 7.1 Graphs illustrating Boyle's Law (Plots of (*a*) P against V, (*b*) P against $1/V$ and (*c*) PV against P)

Further work has shown that gases do not obey Boyle's Law accurately under all conditions. All gases come closer to obeying the law at low density.

7.2.2 CHARLES' LAW

Charles' Law deals with the effect of temperature on the volume of a gas

Two French scientists, J L Gay-Lussac and J A C Charles, independently measured the expansion of gases that occurs when the temperature is raised. They found that the variation of volume with temperature was linear at constant pressure [see Figure 7.2].

FIGURE 7.2 Graph illustrating Charles' Law (Plot of volume against temperature (°C) for a gas at constant pressure)

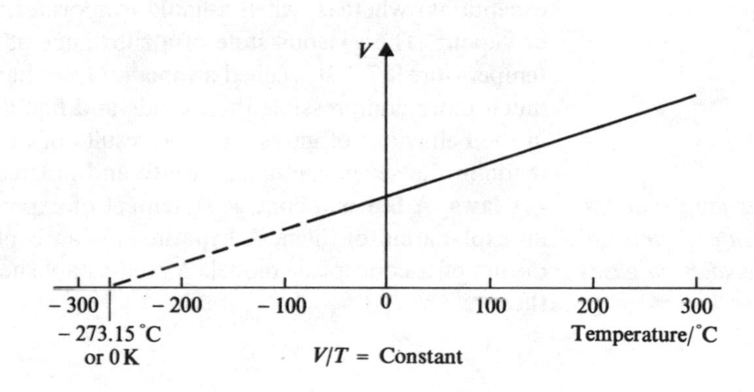

On extrapolation, the line cuts the temperature axis at – 273.15 °C. When similar plots are made for different gases, the extrapolated linear plots always cut the temperature axis at this point. It looks as though all gas volumes would become zero at – 273.15 °C, but in fact gases liquefy or solidify long before this temperature is reached. The temperature – 273.15 °C was adopted as the zero on a new temperature scale called the **absolute temperature scale** or the **Kelvin scale**. Temperatures on this scale are measured in kelvins and are obtained by adding 273.15 to temperatures on the Celsius scale:

Absolute zero and the Kelvin scale of temperature...

...273 K = 0 °C

$$\text{Temperature/K} = \text{Temperature/°C} + 273.15$$

$$273.15\,\text{K} = 0\,°\text{C}$$

If the graph in Figure 7.2 is referred to zero on the Kelvin scale, the equation for the graph is

V/T = constant...

$$V = kT$$

...The relationship leads to a statement of Charles' Law

where T is the temperature in kelvins and k is a constant. Thus, Charles' Law (or Charles' and Gay-Lussac's Law) can be stated: the volume of a fixed mass of a given gas at constant pressure is directly proportional to its temperature in kelvins. In fact, gases do not obey Charles' Law closely at all temperatures. A gas which obeys both Boyle's Law and Charles' Law is said to behave **ideally**.

7.2.3 THE EQUATION OF STATE FOR AN IDEAL GAS

By combining Boyle's Law (PV = constant) and Charles' Law (V/T = constant), one finds that PV/T = constant. This relationship is often written as

Combining Boyle's Law and Charles' Law gives the equation of state for an ideal gas...

$$\frac{P_1V_1}{T_1} = \frac{P_2V_2}{T_2}$$

... This can be used to calculate the volume of gas at stp

This equation is called **the equation of state for an ideal gas**. Real gases do not show ideal behaviour at all temperatures and pressures. The equation of state enables one to calculate the effect of a change in temperature and pressure on the volume of a gas. If a given mass of gas has a volume V_1 at a temperature T_1 and pressure P_1, then the volume, V_2, which the gas would occupy at a temperature T_2 and pressure P_2, can be found. Gas volumes are usually compared at 0 °C and 1 atmosphere, conditions which are referred to as **standard temperature and pressure** (stp). Sometimes the comparison is made at rtp, a room temperature of 20 °C and a pressure of 1 atm. The SI unit of pressure is the newton per square metre ($N\,m^{-2}$), called the pascal (Pa):

Pressure units

$$1 \text{ atmosphere} = 1.0132 \times 10^5\,N\,m^{-2} = 1.0132 \times 10^5\,Pa$$
$$= 760\,mm \text{ mercury}$$

The SI unit of volume is the cubic metre, m^3, but cubic decimetres, dm^3, cubic centimetres, cm^3, and litres, l, are also used.

Volume units

$$1\,m^3 = 10^3\,dm^3 = 10^6\,cm^3$$
$$1\,dm^3 = 1 \text{ litre}$$

Temperatures must be in kelvins in the ideal gas equation of state.

Example If the volume of a gas collected at 60 °C and $1.05 \times 10^5\,N\,m^{-2}$ is $60\,cm^3$, what would be the volume of gas at stp?

Method The experimental conditions are

A sample calculation of the volume of a gas at stp...

$$P_1 = 1.05 \times 10^5\,N\,m^{-2}$$
$$T_1 = 273 + 60 = 333\,K$$
$$V_1 = 60\,cm^3$$

Standard conditions are

$$P_2 = 1.01 \times 10^5\,N\,m^{-2}$$
$$T_2 = 273\,K$$

Since

$$\frac{P_1V_1}{T_1} = \frac{P_2V_2}{T_2}$$

The volume of gas at stp

$$V_2 = \frac{1.05 \times 10^5 \times 60 \times 273}{1.01 \times 10^5 \times 333}\,cm^3$$
$$= 51\,cm^3$$

CHECKPOINT 7A: CORRECTING GAS VOLUMES

1. Correct the following gas volumes to stp:

(*a*) 400 cm^3 of an ideal gas measured at 200 °C and $9.80 \times 10^4 N m^{-2}$

(*b*) 64.0 cm^3 of an ideal gas measured at 35 °C and $1.25 \times 10^5 N m^{-2}$

(*c*) 20.0 dm^3 of an ideal gas measured at 200 K and 201 $kN m^{-2}$

(*d*) 3.15 dm^3 of an ideal gas measured at 250 °C and 1.95 atm

2. (*a*) A volume 24.0 dm^3 of an ideal gas is collected at 1.00 atm and 25 °C. It is subjected to a pressure of 2.05 atm at 75 °C. What is its new volume?

(*b*) After 625 cm^3 of an ideal gas were collected at 90 °C and $9.67 \times 10^4 N m^{-2}$, conditions were changed to 25 °C and $1.13 \times 10^5 N m^{-2}$. What did the volume become?

7.2.4 GRAHAM'S LAW OF GASEOUS DIFFUSION AND EFFUSION

Gases diffuse . . .

. . . the rate of diffusion depend on the density of the gas . . .

Gases mix when brought into contact with one another. All gases spontaneously **diffuse** into one another to form a homogeneous mixture. The relative rates at which two gases diffuse into a third gas (e.g., air) depend on their densities. In 1829, T Graham stated that the relative rates of diffusion of gases, under the same conditions, are inversely proportional to the square roots of their densities. Comparing the rates of diffusion of gases **A** and **B**

. . . the rate is proportional to $\sqrt{1/\rho}$ . . .

$$\frac{r_A}{r_B} = \sqrt{\frac{\rho_B}{\rho_A}}$$

(where r = rate of diffusion and ρ = density). Graham's Law is now written as

. . . The rate is therefore proportional to $\sqrt{1/M}$ . . .

$$\frac{r_A}{r_B} = \sqrt{\frac{M_B}{M_A}}$$

(where M = molar mass). (It will be shown in § 7.2.7 that $\rho = M/V_m$ where V_m = molar volume.)

FIGURE 7.3 Gaseous Effusion

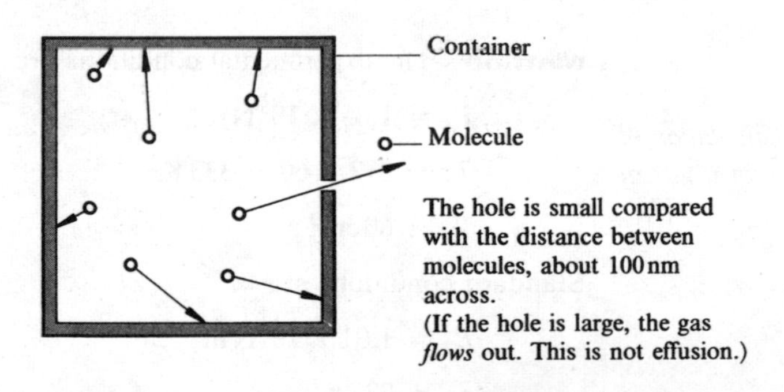
Container
Molecule
The hole is small compared with the distance between molecules, about 100 nm across.
(If the hole is large, the gas *flows* out. This is not effusion.)

. . . The same relationships are true of effusion

Graham's Law applies also to gaseous **effusion**. Effusion is the passage of a gas through a very small hole into a vacuum [see Figure 7.3].

From a practical point of view, it is easier to compare rates of effusion than rates of diffusion. The rate of effusion is found by measuring the fall in pressure that occurs when a gas effuses out of a container.

Applications of gaseous effusion

The application of gaseous effusion to the separation of uranium isotopes is illustrated in Figure 1.25, § 1.9.10. Another application is molar mass determination. The rate of effusion of the gas of unknown molar mass must be compared with that of a gas of known molar mass.

A worked example on molar mass determination from rates of effusion

Example 50.0 cm^3 of gas **A** effuse through a tiny aperture in 146 s. The same volume of carbon dioxide effuses under the same conditions in 115 s. Calculate the molar mass of **A**.

Method

$$\frac{\text{Rate } (CO_2)}{\text{Rate } (\mathbf{A})} = \frac{50.0/115}{50.0/146} = \sqrt{\frac{M_A}{M_{CO_2}}}$$

$$(1.27)^2 = M_A/44.0$$

The molar mass of **A**, $M_A = 71\,g\,mol^{-1}$.

CHECKPOINT 7B: DIFFUSION AND EFFUSION

1. In 5.00 minutes, 15.00 cm^3 of argon effuse through a pinhole. What volume of xenon will effuse through the same pinhole under the same conditions?

2. If it takes 2 minutes 50 seconds for 25.0 cm^3 of hydrogen to effuse through a pinhole, what time will be required for the same volume of bromine vapour to effuse through the same pinhole under the same conditions?

†**3.** A gaseous alkane diffuses through a porous partition at a rate of 2.56 $cm^3\,s^{-1}$. Helium diffuses through the same partition under the same conditions at a rate of 8.49 $cm^3\,s^{-1}$. What is the molar mass of the alkane? What is its molecular formula? [For alkanes, see Chapter 26.]

7.2.5 GAY-LUSSAC'S LAW AND AVOGADRO'S HYPOTHESIS

Gay-Lussac observed that gas volumes combine in a simple ratio

Gay-Lussac studied chemical reactions between gases. Among the observations he made were the facts that one volume of oxygen reacts with exactly twice its volume of hydrogen to form two volumes of steam, and one volume of hydrogen reacts with an equal volume of chlorine to form two volumes of hydrogen chloride. In his **Law of Combining Volumes** in 1809, Gay-Lussac stated that, in a reaction between gases, the volumes of the reacting gases, measured at the same temperature and pressure, are in simple ratio to one another and to the volumes of any gaseous products.

Avogadro's explanation of Gay-Lussac's results

In order to explain the simple relationship which Gay-Lussac had found, Avogadro in 1811 suggested that equal volumes of gases, measured at the same temperature and pressure, contain the same number of molecules. This suggestion is called **Avogadro's Hypothesis**. Since it has been used successfully for nearly two centuries, it is often called **Avogadro's Law**. (This is an imprecise use of the term *law*, which should be applied to a statement of experimental observations and not to a theoretical explanation.) On the basis of Avogadro's Hypothesis, one can interpret the observation

Avogadro's Hypothesis can be used to find the formula of a gaseous compound

1 volume hydrogen + 1 volume chlorine → 2 volumes hydrogen chloride

as 1 molecule hydrogen + 1 molecule chlorine → 2 molecules hydrogen chloride

As it must be possible to make 1 molecule of hydrogen chloride

$\frac{1}{2}$ molecule hydrogen + $\frac{1}{2}$ molecule chlorine → 1 molecule hydrogen chloride

The molecules of hydrogen and chlorine must both contain an even number of atoms. The simplest hypothesis is that each molecule contains two atoms. In this case

$$H_2 + Cl_2 \rightarrow 2HCl$$

It has never been found necessary to suggest a number greater than two.
It follows from Avogadro's Hypothesis that if equal volumes of gases contain equal numbers of molecules then the volume occupied by one mole of molecules must be

Gas molar volume

the same for all gases. It is called the **gas molar volume** and measures $22.414 \, dm^3$ at stp (0 °C and 1 atm) or $24.056 \, dm^3$ at rtp (20 °C and 1 atm). This information is used in calculations on the volumes of gases formed in chemical reactions. [See §3.12.]

7.2.6 THE IDEAL GAS EQUATION

For ideal gases which obey Boyle's Law and Charles' Law the dependence of volume upon external conditions is given by the equation

The ideal gas equation is PV = nRT

$$\frac{P \times V}{T} = \text{Constant (for a given mass of gas)}$$

R *is the Universal Gas Constant...*

It follows from Avogadro's Hypothesis that, if one mole of gas is considered, the constant will be the same for all gases. It is called the Universal Gas Constant, and given the symbol R, so that the equation becomes, when $V = V_m$, the gas molar volume

$$PV_m = RT$$

This equation is called the **ideal gas equation**. For n moles of gas, the equation becomes

$$PV = nRT$$

The value of the constant R can be calculated. Consider one mole of an ideal gas at stp. Its volume V_m is $22.414 \, dm^3$. Inserting values of P, V_m and T in SI units into the ideal gas equation gives

...Finding the value of R...

$$P = 1.0132 \times 10^5 \, N \, m^{-2}$$

$$T = 273.16 \, K$$

$$V_m = 22.414 \times 10^{-3} \, m^3$$

$$n = 1 \, mol$$

$$1.0132 \times 10^5 \times 22.414 \times 10^{-3} = 1 \times 273.16 \times R$$

$$R = 8.314$$

...and the unit of R

The unit of R is PV/nT i.e., $\dfrac{N \, m^{-2} \, m^3}{mol \, K} = N \, m \, mol^{-1} \, K^{-1} = J \, K^{-1} \, mol^{-1}$.

Thus $R = 8.314 \, J \, K^{-1} \, mol^{-1}$ (joules per kelvin per mole).

Real gases depart from ideal behaviour at high pressures and at low temperatures.

7.2.7 *THE DENSITY AND MOLAR MASS OF A GAS

The density of a gas can be found and used to give its molar mass

The density of a gas is found by weighing a known volume of gas. Since

$$PV = nRT = \frac{m}{M} RT$$

(where m = mass and M = molar mass)
and density

$$\rho = m/V$$

$$M = \rho RT/P$$

CHECKPOINT 7C: GAS MOLAR VOLUME

($R = 8.31\,J\,K^{-1}\,mol^{-1}$,
Gas molar volume = $22.4\,dm^3$ at stp)

1. Calculate the molar mass of a gas which has a density of $2.615\,g\,dm^{-3}$ at 298 K and $101\,kN\,m^{-2}$.

2. At 273 K and $1.01 \times 10^5\,N\,m^{-2}$, 6.319 g of a gas occupy $2.00\,dm^3$. Calculate the molar mass of the gas.

3. An ideal gas occupies $225\,cm^3$ at 280 K and $4.80 \times 10^5\,N\,m^{-2}$. What amount of gas (in moles) is present?

4. Calculate the volume occupied by 1.15 mol of an ideal gas at $1.01 \times 10^5\,N\,m^{-2}$ and 20 °C.

7.2.8 DALTON'S LAW OF PARTIAL PRESSURES

Dalton considered the pressures exerted by the constituents of a mixture of gases. Since air is approximately $\frac{4}{5}$ nitrogen and $\frac{1}{5}$ oxygen by volume, $\frac{4}{5}$ of the air pressure is due to nitrogen and $\frac{1}{5}$ is due to oxygen. The contribution that each gas makes to the total pressure is called the **partial pressure**. The partial pressure of a gas is the pressure that each gas would exert if it alone occupied the container. Dalton's Law of partial pressures states that, in a mixture of gases which do not react chemically, the total pressure of the mixture is the sum of the partial pressures of the constituent gases. It is an ideal gas law.

Definition of partial pressure

Dalton's Law

The partial pressure of each gas is equal to the product of the total pressure of the mixture and the **mole fraction** of the gas. Mole fractions are one method of expressing the composition of a mixture. The mole fraction of **A** in a mixture of **A** and **B** is

Definition of mole fraction

$$\text{Mole fraction of A} = \frac{\text{Moles of A}}{\text{Moles of A} + \text{Moles of B}} = \frac{n_A}{n_A + n_B}$$

From Avogadro's Hypothesis, it follows that

$$\frac{n_A}{n_A + n_B} = \frac{\text{Volume of A}}{\text{Total volume}}$$

Calculation of partial pressure

Example $4.00\,dm^3$ of oxygen at a pressure of 400 kPa and $1.00\,dm^3$ of nitrogen at a pressure of 200 kPa are introduced into a $2.00\,dm^3$ vessel. What is the total pressure in the vessel?

Method When oxygen contracts from $4.00\,dm^3$ to $2.00\,dm^3$, the pressure increases from 400 kPa to

$$400 \times \frac{4.00}{2.00} \quad \text{i.e. } 800\,kPa.$$

The partial pressure of oxygen in the vessel is 800 kPa.
When nitrogen expands from $1.00\,dm^3$ to $2.00\,dm^3$, the pressure decreases from 200 to

$$200 \times \frac{1.00}{2.00} = 100\,kPa.$$

The partial pressure of nitrogen is 100 kPa.

$$\text{The total pressure} = p(O_2) + p(N_2) = 800 + 100 = 900\,kPa$$

The total pressure in the vessel is 900 kPa.

*7.2.9 HENRY'S LAW

The solubility of a gas is proportional to its partial pressure

Henry's Law is concerned with the solubility of gases in liquids. It states that the mass of gas dissolved at constant temperature per unit volume of solvent is directly proportional to the partial pressure of the gas. This law may be expressed as

$$m_s = kp$$

where m_s is the mass of gas dissolved p is its partial pressure, and k is a constant.

Definition of solubility...

The **solubility** of a gas is the volume which will dissolve in unit volume of the solvent under stated conditions of temperature and pressure. For example, the solubility of carbon dioxide at 30 °C and 1 atm is 30 cm³ in 1.00 dm³ of water.

CHECKPOINT 7D: PARTIAL PRESSURES

1. 50.0 cm³ of carbon dioxide at 10^5 N m^{-2} are mixed with 150 cm³ of hydrogen at the same pressure. If the pressure of the mixture is 1.00×10^5 N m^{-2}, what is the partial pressure of carbon dioxide?

2. A mixture of gases at a pressure 1.01×10^5 N m^{-2} has the volume composition of 30% CO, 50% O_2, 20% CO_2.

(*a*) What is the partial pressure of each gas?

(*b*) If the carbon dioxide is removed by the addition of some pellets of sodium hydroxide, what will be the partial pressures of O_2 and CO?

3. Into a 10.0 dm³ vessel are introduced 4.00 dm³ of methane at a pressure of 2.02×10^5 N m^{-2}, 12.5 dm³ of ethane at a pressure of 3.50×10^5 N m^{-2} and 1.50 dm³ of propane at a pressure of 1.01×10^5 N m^{-2}. What is the pressure of the resulting mixture?

4. A mixture of 20% NH_3, 55% H_2 and 25% N_2 by volume has a pressure of 9.80×10^4 N m^{-2}.

(*a*) What is the partial pressure of each gas?

(*b*) What changes will take place in the partial pressures of hydrogen and nitrogen if the ammonia is removed by the addition of solid phosphorus(V) oxide?

5. At room temperature, the ratio of the solubility of oxygen in water to nitrogen in water is 1.90. The percentage composition by volume of air is taken as 20% oxygen : 80% nitrogen. What is the percentage composition by volume of the mixture of oxygen and nitrogen that is obtained by degassing water saturated with air at room temperature?

7.3 THE KINETIC THEORY OF GASES

7.3.1 THE BASIC ASSUMPTIONS

The gas laws need an explanation...

... The kinetic theory of gases offers one

All these laws describe the behaviour of gases, and they call for an explanation. Many scientists made a contribution to the development of a theoretical model to represent the behaviour of gases. Even as far back as the seventeenth century, many realised that in a gas the molecules themselves occupy only a tiny fraction of the volume in which the gas is contained. This is evident from the large volume of vapour formed from one volume of liquid, and it explains the low densities and high compressibilities of gases. A number of scientists envisaged the molecules of a gas as being in motion, but it was R J Clausius (1857) and J C Maxwell (1859) who presented the **kinetic theory of gases** in its present form. (Greek: *kinein*, to move.) The theory rests on a number of *basic assumptions*.

Molecules of gas are in constant motion...

According to the kinetic theory of gases, the molecules of a gas are in constant motion, moving in straight lines unless they collide with the walls of the container or with another molecule. As the molecules collide with the walls of the container, they exert a pressure on the container. If the volume of the container is decreased, the molecules collide more frequently with the container, and the pressure increases.

... This explains gas pressure

If the temperature of the gas is raised, the molecules gain more kinetic energy and move faster and collide more often with the container, thus increasing the pressure.

Molecular collisions are elastic

The collisions are held to be perfectly *elastic*; the molecules bouncing back off the walls with, on average, no loss of kinetic energy. If this were not so, if the molecules rebounded with less energy, they would be giving up energy to the container, and the temperature of the container would rise while the temperature of the gas fell.

The volume of the molecules is negligible compared with that of the container

The distances between molecules are great

The volume of the molecules is held to be negligible compared with the space occupied by the gas. The distances between molecules are so great compared with the sizes of the molecules (10^3 or 10^4 times the diameter of a molecule) that the molecules exert no force upon one another except during collisions. The kinetic energy of the molecules is assumed to be proportional to the temperature of the gas on the Kelvin scale.

The kinetic theory equation

From these assumptions, R J Clausius deduced the **kinetic theory equation***:

$$PV = \tfrac{1}{3}mN\overline{c^2}$$

where P = pressure, V = volume, m = mass of one molecule, N = number of molecules, $\overline{c^2}$ = mean square velocity. If there are n_1 molecules with velocity c_1, n_2 with velocity c_2, etc., $\overline{c^2}$ is given by

$$\overline{c^2} = \frac{n_1c_1{}^2 + n_2c_2{}^2 + n_3c_3{}^2 + \ldots}{n_1 + n_2 + n_3 + \ldots}$$

The root mean square velocity

The square root of $\overline{c^2}$, $\sqrt{\overline{c^2}}$, is called the **root mean square velocity** (rms velocity).

The kinetic theory makes a number of assumptions about the behaviour of gaseous molecules. The ideal gas laws describe the behaviour of gases. If the assumptions of the kinetic theory are correct, it should be possible to derive the ideal gas laws from the kinetic theory.

7.3.2 CALCULATION OF ROOT MEAN SQUARE VELOCITY

Calculation of rms velocity

The kinetic theory can be used to calculate the root mean square velocity of gas molecules, for example, that of oxygen molecules at stp.

Method Molar mass of oxygen = $32.0\,\text{g}\,\text{mol}^{-1}$

In the equation $PV = \tfrac{1}{3}mN\overline{c^2}$, substitute $PV = RT$ for 1 mole of gas, which gives

$$\tfrac{1}{3}mN\overline{c^2} = RT$$

Then, $N = L$, the Avogadro constant, and $mL = M$, the molar mass of gas in $\text{kg}\,\text{mol}^{-1}$, which gives

$$\overline{c^2} = \frac{3RT}{M}$$

$$\overline{c^2} = 3 \times 8.314 \times 273/(32.0 \times 10^{-3})$$

$$\sqrt{\overline{c^2}} = 461\,\text{m}\,\text{s}^{-1}\text{, as before}$$

The rms velocity of oxygen molecules at stp is $461\,\text{m}\,\text{s}^{-1}$.

*For a derivation, see R Muncaster, *A-level Physics* (Stanley Thornes)

7.3.3 CALCULATION OF MOLECULAR ENERGY

The **translational kinetic energy** of a particle is given by $\frac{1}{2}mc^2$. The translational kinetic energy of a number of molecules of gas, E_t, is given by

The value of molecular energy...

$$E_t = \tfrac{1}{2}mc_1^2 + \tfrac{1}{2}mc_2^2 + \tfrac{1}{2}mc_3^2 + \ldots + \tfrac{1}{2}mc_n^2$$

If N = number of molecules and $\overline{c^2}$ = mean square velocity

$$E_t = \tfrac{1}{2}mN\overline{c^2}$$

Since $\frac{1}{3}mN\overline{c^2} = RT$

$$E_t = \tfrac{3}{2}RT$$

...and a calculation

For example, the translational kinetic energy of one mole of a gas at 273 K is given by

$$E_t = \tfrac{3}{2} \times 8.314 \times 273 = 3.405\,\text{kJ mol}^{-1}$$

CHECKPOINT 7E: THE KINETIC THEORY AND THE IDEAL GAS EQUATION

(Take Avogadro constant = $6.022 \times 10^{23}\,\text{mol}^{-1}$, $R = 8.314\,\text{J K}^{-1}\,\text{mol}^{-1}$, 1 atm = $1.013 \times 10^5\,\text{N m}^{-2}$)

1. State the main postulates of the Kinetic Theory of Gases.

2. How does the motion of molecules in a gas differ from that in a liquid? When a liquid is evaporated, heat energy must be supplied. What happens to this energy?

3. Find the ratio of the root mean square velocities of oxygen and hydrogen at the same temperature.

4. Find the ratio of the root mean square velocities of $^{235}UF_6$ and $^{238}UF_6$ at the same temperature.

5. Calculate the number of molecules of oxygen in a 0.500 dm^3 flask at a pressure of 101 N m^{-2} at 273 K.

7.3.4 THE DISTRIBUTION OF MOLECULAR ENERGIES

Molecular energy has a spread of values at any temperature

J C Maxwell and L Boltzmann plotted distribution curves to show how the spread in the values of molecular energy (translational kinetic energy) alters with temperature. They calculated the fraction of the total number of molecules with energy x, and plotted this against the value of molecular energy x, as in Figure 7.4.

FIGURE 7.4 Distribution of Molecular Energies

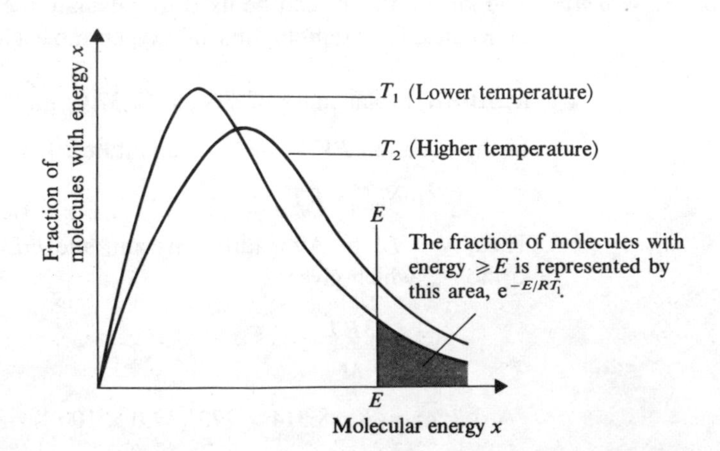

At high temperature...

Only a very small fraction of the molecules have very low energies and only a very small fraction have very high energies. The figure shows how the molecules are distributed between different energy levels at two temperatures T_1 and T_2, where T_2 is higher than T_1. In both cases, the area under the curve is the same as all the

... the average energy increases...

fractions add up to unity. As the temperature is raised, the average energy of the molecules increases proportionately [§ 7.3.3].

... and the fraction of molecules with high energy increases

In addition, the shape of the distribution curve alters as the fraction of molecules with low energies decreases and the fraction of molecules with high energies increases. Maxwell and Boltzmann derived an expression for the fraction of molecules possessing high energy. The most important part of this expression is the exponential function $e^{-E/RT}$. For a value of molecular energy E, the fraction of molecules with energy greater than or equal to E is given approximately by $e^{-E/RT}$.

7.4 REAL GASES

7.4.1 NON-IDEAL BEHAVIOUR

Real gases do not show ideal behaviour at high P *and low* T...

Real gases depart from the ideal behaviour illustrated in Figure 7.1, § 7.2.1 at *high pressures* and *low temperatures*. Figure 7.5 shows how the product PV departs from a constant value as pressure increases. You will see that the departure from ideal behaviour occurs at very high pressures; many gases approximate to ideal behaviour at pressures in the neighbourhood of 1 atm. The deviation of PV from constancy is greatest for easily liquefied gases such as ammonia, and least for gases which are far above their boiling temperatures. At low temperatures, all gases approach their boiling temperatures and deviate from ideal behaviour.

FIGURE 7.5 Plot of PV against P at a Constant Temperature for one Mole of Gas at 273 K

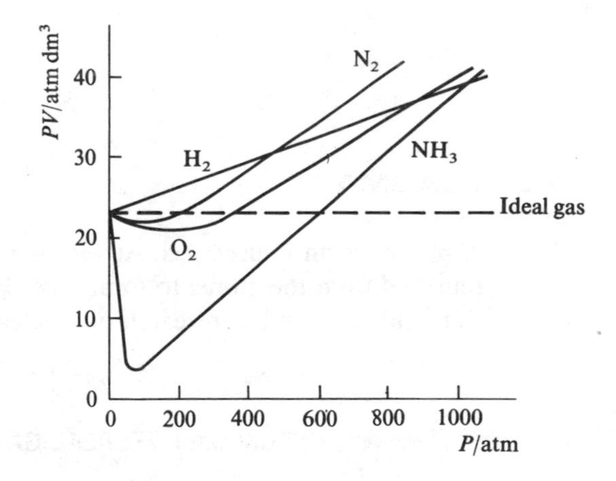

... Van der Waals offered an explanation of this deviation from ideal behaviour

Two of the assumptions of the kinetic theory have to be modified in gases at high pressures or low temperatures. One is that there are no forces acting between molecules. When gases are compressed, the molecules come close enough together for van der Waals forces of attraction [§ 4.7.2] to operate between them. These attractions make the gas more compressible than it is at low pressure, and for one mole of gas $PV < RT$. This is shown in the medium pressure range of Figure 7.5.

At high pressure, there are forces between the gas molecules...

... and the volume of the molecules cannot be neglected

At still higher pressures, the molecules are pushed so close together that repulsive forces operate between them [see Figure 4.31, § 4.7.2]. These repulsions make the gas less compressible than it is at low pressure, and $PV > RT$. This is shown in the high pressure region of Figure 7.5. The second assumption is that the volume of the molecules is negligible compared with the volume occupied by the gas. This is no longer true in a highly compressed gas.

7.4.2 LIQUEFACTION OF GASES

Gases cannot be liquefied above their critical temperatures

For every gas there is a **critical temperature** above which the gas cannot be liquefied. This is why many early attempts to liquefy gases by compression alone failed. For gases like hydrogen, oxygen and nitrogen, with low critical temperatures (around – 200 °C), the problem of cooling them sufficiently to be liquefied was a difficult one. It was solved by using the Joule–Thomson effect.

Sudden expansion results in cooling...

... This cooling, the Joule–Thomson effect, is used to liquefy air

J P Joule and Sir William Thomson noticed that when a gas was allowed to expand through a porous plug its temperature fell. The expanding gas has to do work: it has to push back the gas in front of it, and also the gas molecules become more widely separated against intermolecular forces of attraction. If the expansion takes place too rapidly for energy to be absorbed from the surroundings, the energy must come from the gas itself, and the gas cools.

The industrial liquefaction of air is illustrated in Figure 7.6. After several cycles of compression followed by sudden expansion, the temperature is low enough to allow air to be liquefied at – 200 °C.

FIGURE 7.6
Liquefaction of Air

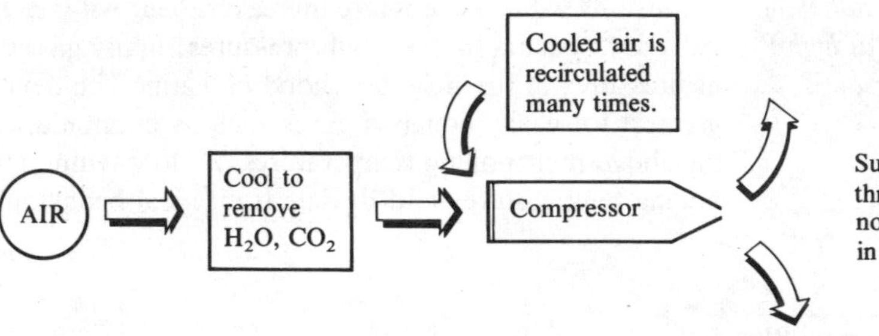

7.4.3 PLASMAS

A **plasma** is an ionised gas. At very high temperatures, 10^4 to 10^5 K, electrons are removed from the atoms to form ions. Plasmas are studied in the atmosphere of the Sun and stars and in research on nuclear fusion.

CHECKPOINT 7F: REAL GASES

1. State two ways in which a real gas differs from an ideal gas.

2. Which of the gases: hydrogen, propane and hydrogen chloride, would you expect to depart the furthest from ideal gas behaviour? Explain your choice.

QUESTIONS ON CHAPTER 7

1. (*a*) How does the pressure of a given mass of gas in a fixed volume change as the temperature rises? How does the kinetic theory explain this change?

(*b*) At stp, a certain mass of gas has a volume of 1.00 dm^3. At 30 atm pressure, the volume is 31.2 cm^3. At 60 atm, the volume is 14.9 cm^3. Does the gas show ideal behaviour? Explain your answer.

2. How does the kinetic theory of gases explain (*a*) the interdiffusion of gases, (*b*) gas pressure and (*c*) the compressibility of gases?

3. From the equation

$$pV = \frac{1}{3}mN\overline{c^2}$$

calculate the kinetic energy of 1 mole of an ideal gas at stp (R = 8.314 J K^{-1} mol^{-1}).

†4. 80 cm^3 of oxygen effused from an apparatus in 70 s, while 80 cm^3 of nitrogen dioxide took 100 s. Calculate a value for the molar mass of nitrogen dioxide. After studying § 11.5.5 explain why the value is different from that corresponding to the formula NO_2.

5. Given $R = 8.314\,J\,K^{-1}\,mol^{-1}$, $L = 6.022 \times 10^{23}\,mol^{-1}$, $1\,atm = 1.01 \times 10^5\,N\,m^{-2}$, calculate

(*a*) the number of molecules of hydrogen in a bulb of capacity 0.500 dm^3 which has been evacuated down to a pressure of 1.00 $N\,m^{-2}$ at 298 K,

(*b*) the kinetic energy of the molecules in 1.00 mol hydrogen at 298 K,

(*c*) the ratio of the root mean square velocities of hydrogen and hydrogen bromide at 298 K.

6. Hydrogen diffused 7.94 times faster than a gaseous fluoride of phosphorus (under the same conditions). Calculate the molar mass of the fluoride, and suggest its formula.

7. (*a*) The ideal gas equation can be written as $pV = nRT$. Use this equation to calculate the volume occupied by one mole of an ideal gas at 300 K and 100 kPa pressure.

(*b*) An organic compound, **X**, contains carbon, hydrogen and oxygen only. When vaporised at 101 kPa and 373 K, 0.100 g of **X** occupied a volume of 66.7 cm^3. Calculate the relative molecular mass of **X**.

(*c*) On combustion in excess oxygen, 1 mol of **X** produced 2 mol of carbon dioxide and 3 mol of water.

(i) What is the molecular formula of **X**?

(ii) Write structures for *two* compounds with this molecular formula.

(iii) Write a balanced equation for the complete combustion of **X** in oxygen.

(*d*) **X** is a liquid at room temperature. When **X** is treated with metallic sodium, hydrogen is evolved.

(i) Use this information to deduce the structure of **X**.

(ii) Write a balanced equation for the reaction of **X** with metallic sodium.

(AEB 91)

8. (*a*) The table below shows the effect of increasing pressure, *P*, on the product of pressure and volume, *PV*, for 1 mol of ammonia gas at 25 °C.

P/atm	1.0	2.0	5.0	9.8	10.0	20.0
PV/atm dm^3	24.3	24.2	23.8	23.1	0.2	0.4

(i) Plot these data as a curve of *PV* against *P* on graph paper, labelling and numbering the axes.

(ii) Also plot *PV* against *P* for 1 mol of an ideal gas at 25 °C, given that it occupies a volume of 22.4 dm^3 at 1 atm (1.01×10^5 Pa) pressure and a temperature of 0 °C (273 K).

(iii) Use your knowledge of kinetic-molecular theory and of the intermolecular forces present in ammonia:

(*1*) to account for the shape of the ammonia graph

(*2*) to explain what happens when the pressure on the ammonia gas is increased at 25 °C.

(*b*) Sketch a curve showing how the vapour pressure of a pure liquid changes with temperature. Mark on your sketch the temperature at which the liquid will boil if heated in an open flask.

(*c*) The boiling points and relative molecular masses of hydrides from two groups of the Periodic Table are listed below.

Hydride	HF	HCl	HI	CH_4	SiH_4	SnH_4
Boiling point/°C	19	−85	−35	−161	−112	−52
Relative molecular mass	20	36.5	128	16	32	123

(i) Comment, without explanation, on trends and any anomalies in the two sets, and on any boiling point differences between the groups of hydrides.

(ii) State and discuss the types of intermolecular force which are present in these hydrides and thus explain the trends, anomalies and differences mentioned in (*c*)(i).

(WJEC 90)

9. (*a*) Draw a sketch graph with labelled axes to show the distribution of kinetic energies of the molecules in an ideal gas. Describe how the distribution changes when the temperature of the gas is increased.

(*b*) State Graham's Law of diffusion, and give an industrial application which is based upon it.

(*c*) Figure 7.7 illustrates the behaviour of one mole each of the three gases, argon, oxygen and carbon dioxide at 0 °C; *p* is the pressure and *V* the volume.

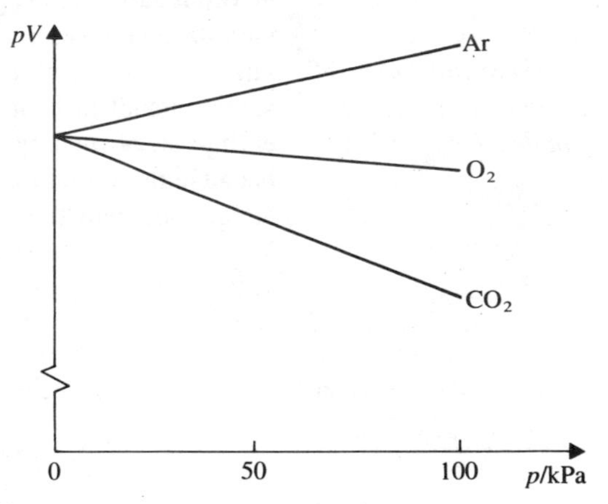

FIGURE 7.7

(i) Make a copy of this diagram in your answer book and draw on this copy a dotted line which applies to an ideal gas at 0 °C. Explain why the line you have drawn applies to an ideal gas.

(ii) Why do the three lines for argon, oxygen and carbon dioxide intersect at zero pressure?

(iii) Give *two* reasons for the non-ideality of real gases.

(iv) Use these reasons to explain why *pV* decreases with increasing pressure for oxygen and carbon dioxide but increases for argon.

(v) An ideal gas occupies a volume of 50.0 dm^3 at 300 K and 1.52×10^6 Pa. Calculate the amount in moles of this gas present in this volume.

(O 91)

8

LIQUIDS

8.1 THE LIQUID STATE

In a liquid, the molecules are closer together than in a gas and less ordered than in a solid

In a liquid, the molecules are closer together than in a gas. This is evident from the fact that one volume of water is formed by the condensation of 1300 volumes of steam. The formation of a liquid from a gas is due to the attractive forces between molecules. These must be strong enough to overcome the kinetic energy of the molecules. When a gas is cooled, the molecules lose kinetic energy until they reach a temperature at which their kinetic energy will no longer overcome the forces of intermolecular attraction, and the gas liquefies.

There are forces of attraction between the molecules in a liquid

The liquid state bears more resemblance to the solid state than to the gaseous state. The particles in a liquid are further apart than in a solid and less ordered. When solids melt, there is usually an expansion of the order of 10%. The amount of heat required to convert 1 mole of a liquid to vapour at 1 atm is called the **standard enthalpy of vaporisation, $\Delta H^{\ominus}_{\text{Vaporisation}}$**. The same amount of heat is released when 1 mole of vapour condenses at 1 atm. The amount of heat required to convert 1 mole of a solid into a liquid at 1 atm is called the **standard enthalpy of melting, $\Delta H^{\ominus}_{\text{Melting}}$**. The same amount of heat is released when 1 mole of a liquid solidifies at 1 atm. Standard enthalpies of melting are much smaller than standard enthalpies of vaporisation because the transition from solid to liquid involves less disruption of intermolecular attractions than the transition from liquid to gas.

FIGURE 8.1 The Distance between Molecules of a Liquid

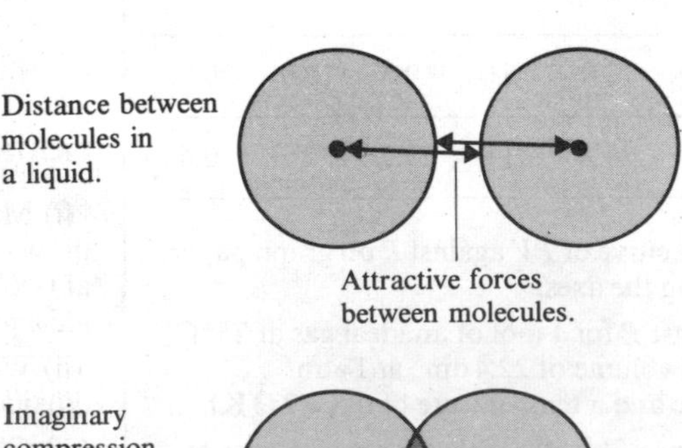
Distance between molecules in a liquid.
Attractive forces between molecules.

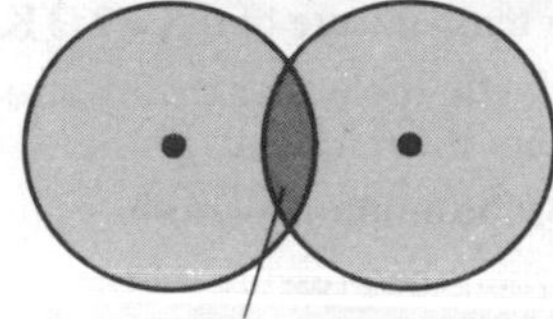
Imaginary compression of liquid forces molecules closer together.
Interpenetration of electron clouds causes repulsion, and molecules move further apart.

Liquids are slightly compressible...

Liquids have a definite volume and are only slightly compressible. The distance between molecules in a liquid is governed by a balance between forces of attraction between molecules and forces of repulsion between neighbouring electron clouds [see Figure 8.1]. The forces of attraction between the molecules, being less strong than those in solids, are not strong enough to hold the liquid in a definite shape. A liquid flows to fit the shape of its container.

...and liquids flow

Diffusion takes place more slowly in liquids than in gases

Gases are able to diffuse because the molecules are in rapid motion. The rates of diffusion of liquids are much less than those of gases, and those of solids are extremely low. Although the molecules of a liquid are in rapid motion, they are not able to move freely from one part of a liquid to another: they are restricted by the attractive forces which bind them into a loose network with neighbouring molecules. Occasionally, a molecule breaks free from the network of molecules that surround it, and moves to another region of the liquid: it **diffuses**.

Molecules with high energy escape from the liquid to the vapour phase

Since molecules of liquid are in constant motion, some of them will have enough energy to escape from the liquid into the vapour state. (The term **vapour** is applied to a gas below its critical temperature.) Those that escape will be molecules with energy considerably above average, sufficient to remove them against the attraction of other molecules of liquid. The molecules that remain in the liquid will be of lower energy than those that escape, and the temperature of the liquid will fall. Since the fraction of molecules with high energy increases as the temperature rises, the rate of evaporation increases with temperature.

At equilibrium, the rate of evaporation equals the rate of condensation

Evaporation will continue until no liquid remains. If the liquid is in a closed container, however, the molecules in the vapour state will collide with the walls of the container, and some will be directed back towards the liquid. Some of these will re-enter the liquid, i.e. **condense**. [See Figure 8.2.] **Equilibrium** will be reached when the rate at which molecules of liquid evaporate is equal to the rate at which molecules of vapour condense.

FIGURE 8.2 A Liquid and its Vapour reach Equilibrium in a Closed Container. The rate of evaporation is equal to the rate of condensation.

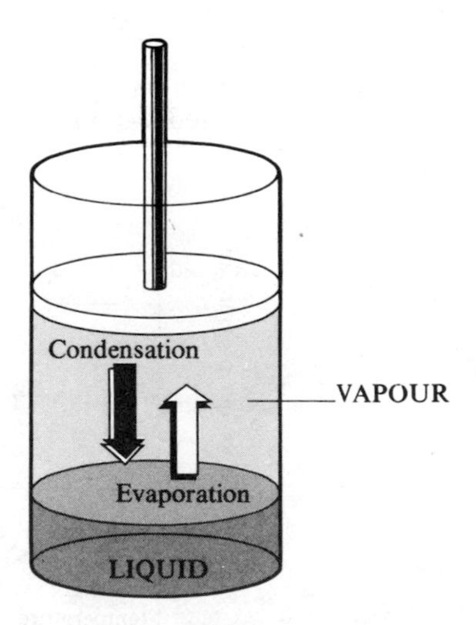

8.2 SATURATED VAPOUR PRESSURE

The maximum vapour pressure developed by a liquid is its saturated vapour pressure at that temperature

As the molecules of vapour collide with the walls of the container they exert a pressure. The maximum vapour pressure that can be developed by a liquid is called the **saturated vapour pressure** of that liquid. Since the fraction of molecules with high energy increases with temperature, evaporation increases as the temperature rises and the saturated vapour pressure increases with temperature. The magnitude of the saturated vapour pressure depends on the *identity* of the liquid and the *temperature*; it does not depend on the amount of liquid present. Solids have vapour pressures too, but, at room temperature, the vapour pressures of most solids are low.

The saturated vapour pressure of a liquid can be measured as shown in Figure 8.3.

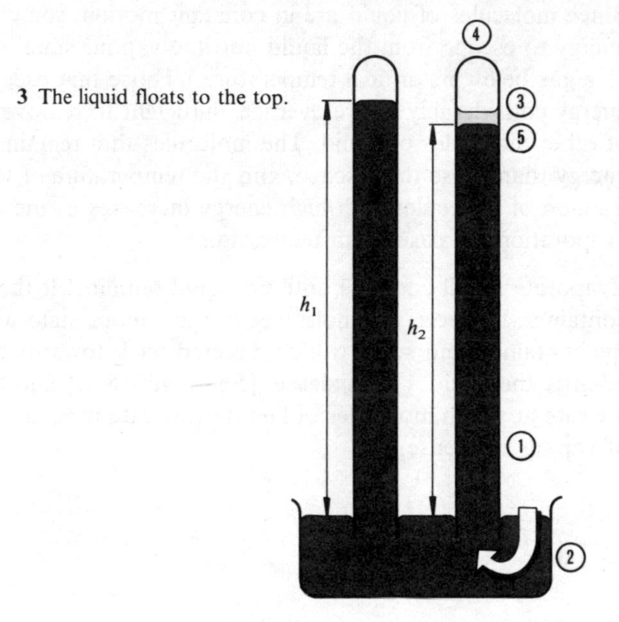
3 The liquid floats to the top.
4 Vapour is formed. The vapour exerts a pressure on the column of mercury.
5 The level of mercury drops. The difference, $(h_1 - h_2)$ mm, is the saturated vapour pressure of the liquid at this temperature.
h_1
h_2
1 Mercury barometer
2 Some liquid is introduced into the barometer.

FIGURE 8.3 Measuring the Saturated Vapour Pressure of a Liquid

A plot of saturated vapour pressure against temperature is shown in Figure 8.4.

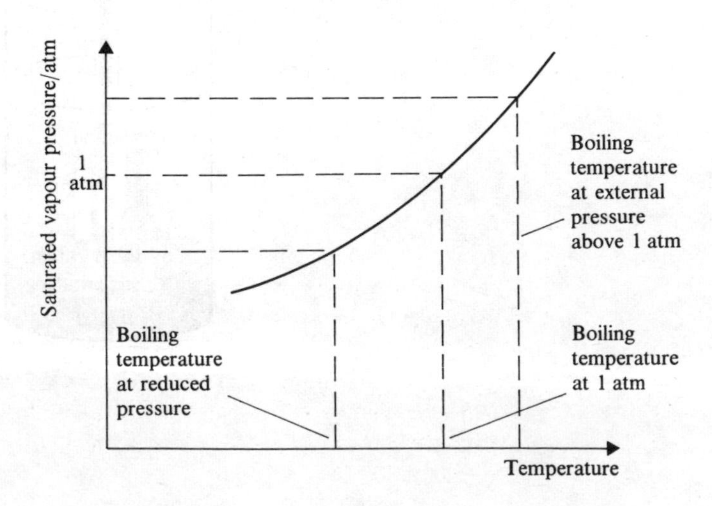
Saturated vapour pressure/atm
1 atm
Boiling temperature at external pressure above 1 atm
Boiling temperature at reduced pressure
Boiling temperature at 1 atm
Temperature

FIGURE 8.4 Variation of Saturated Vapour Pressure with Temperature

Saturated vapour pressure, svp, increases with temperature

If a liquid is heated in an open container, its saturated vapour pressure increases until it becomes equal to atmospheric pressure. When this happens, bubbles of vapour form in the interior of the liquid and escape into the atmosphere because the vapour pressure is high enough to push the air aside: the liquid boils. The **boiling temperature** of a liquid is defined as the temperature at which its saturated vapour pressure, svp, is equal to the external pressure, normally 1 atm. While a liquid is boiling, the heat taken in is used to produce molecules with enough energy to escape into the vapour phase. The average kinetic energy of the molecules remaining in the liquid does not increase, and its temperature remains constant. Figure 8.4 shows how the boiling temperature increases at pressures greater than 1 atm and decreases at pressures less than 1 atm. Distillation under reduced pressure (sometimes called **vacuum distillation**) is used in the purification of substances which decompose at their boiling temperatures. The apparatus shown in Figure 8.5 could be used. By means of vacuum pumps, pressures down to 10^3 or $10^2\,N\,m^{-2}$ can be obtained.

When the svp equals the external pressure, the liquid boils

Distillation under reduced pressure is used for the purification of some substances

The measurement of boiling temperature

A pure liquid can be identified by its boiling temperature at a certain pressure (normally 1 atm). When a boiling temperature is measured, the thermometer must measure the temperature of the vapour in equilibrium with the boiling liquid [see Figure 8.5]. It should not be immersed in the liquid because **superheating** (heating above the boiling temperature) can occur if a liquid is heated rapidly.

FIGURE 8.5 Distillation Apparatus

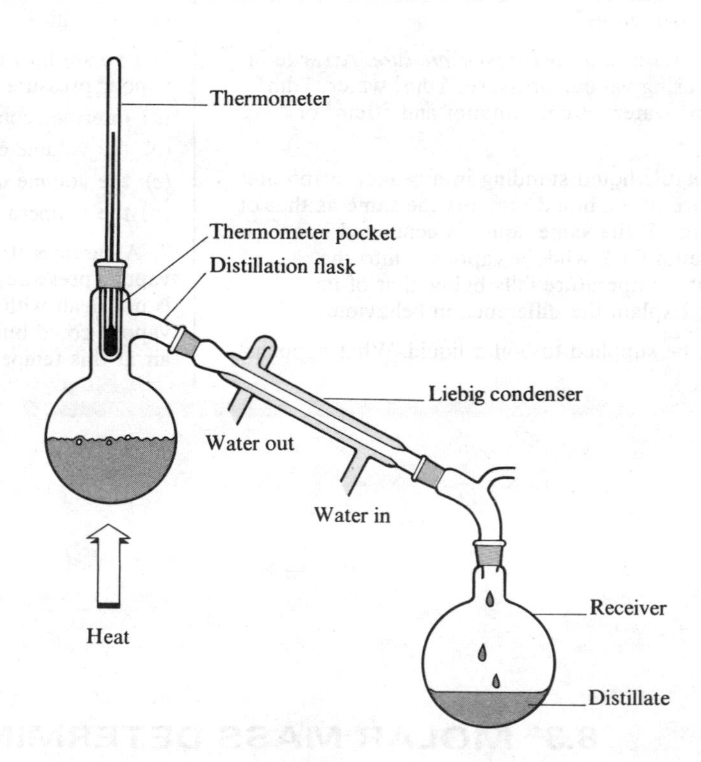

The saturated vapour pressure of a liquid can be used to calculate the mass of the liquid that is present in the vapour state in a given volume at a stated temperature.

The mass of a liquid that vaporises can be calculated

Example A 1.000 dm^3 vessel contains air at $1.01 \times 10^5\,N\,m^{-2}$ and 0 °C. After 1.000 g of water is introduced into the vessel, the temperature is raised to 90 °C. Calculate the mass of water that will vaporise. The saturated vapour pressure of water at 90 °C is $6.99 \times 10^4\,N\,m^{-2}$.

Method Use the Ideal Gas Equation

$PV = nRT$

Substitute $P = 6.99 \times 10^4\,\text{N m}^{-2}$

$R = 8.314\,\text{J K}^{-1}\,\text{mol}^{-1}$

$T = 363\,\text{K}$

$V = 1.000\,\text{dm}^3 = 1.000 \times 10^{-3}\,\text{m}^3$

$6.99 \times 10^4 \times 1.000 \times 10^{-3} = n \times 8.314 \times 363$

Amount of water $n = 2.32 \times 10^{-2}\,\text{mol}$

Mass of water $= 18.0 \times 2.32 \times 10^{-2}\,\text{g}$

Mass of water vapour $= 0.417\,\text{g}$

CHECKPOINT 8A: VAPOUR PRESSURE

1. Why are liquids more compressible than solids and less compressible than gases?

2. Explain the term *saturated vapour pressure.* Arrange in order of increasing vapour pressure: 1 dm³ water, 1 dm³ ethanol, 50 cm³ water, 50 cm³ ethanol and 50 cm³ of ethoxyethane.

3. While a volatile liquid standing in a beaker evaporates, the temperature of the liquid remains the same as that of its surroundings. If the same liquid is contained in an insulated vacuum flask while it vaporises into the atmosphere, its temperature falls below that of its surroundings. Explain the difference in behaviour.

4. Heat must be supplied to boil a liquid. What happens to this heat?

5. Why is distillation under reduced pressure often employed in the purification of chemicals?

6. Explain how each of the following factors influences the vapour pressure of a liquid:
(*a*) intermolecular forces in the liquid
(*b*) the volume of the liquid in the system
(*c*) the volume of vapour in the system
(*d*) the temperature of the liquid..

7. A careless student breaks a thermometer. The saturated vapour pressure of mercury at 20 °C is 0.160 Pa. If the spill is not dealt with immediately, what mass of mercury vapour could build up in each cubic metre of laboratory air at this temperature?

8.3 MOLAR MASS DETERMINATION

The volume of vapour formed by a known mass of liquid is measured...

The gas syringe method of determining the molar mass of a volatile liquid is shown in Figure 8.6. A weighed quantity of liquid is injected into a gas syringe, where it vaporises, and the volume occupied by the vapour is measured.

In the method that uses the apparatus designed by Victor Meyer, a known mass of liquid is vaporised, and the volume of air which it displaces is measured. Neither method gives a very accurate result. These methods are used in conjunction with accurately known empirical formulae to establish the molecular formulae of compounds.

FIGURE 8.6 Gas Syringe Method of Molar Mass Determination

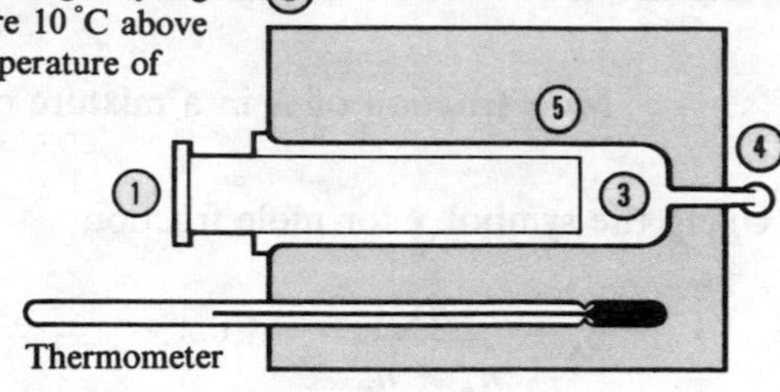

2 Furnace keeps the gas syringe at a temperature 10 °C above the boiling temperature of the liquid
1 Gas syringe
Thermometer
3 The gas syringe contains air. The volume v_1 cm³ is noted.
4 Self-sealing rubber cap. Liquid is injected through a hypodermic needle from a weighed syringe (m_1 g). The syringe is reweighed (m_2 g). The mass of liquid injected = $(m_1 - m_2)$ g.
5 When the liquid is injected, the plunger moves backwards. The volume of gas in the barrel is read (v_2 cm³). Volume of vapour = $(v_2 - v_1)$ cm³.

...and its molar mass is calculated...

...A sample calculation

Method of calculation

$$PV = nRT = \frac{m}{M}RT$$

where $V = v_2 - v_1$, and $m = m_1 - m_2$
T = temperature recorded by thermometer,
P = the measured atmospheric pressure
R = the gas constant = $8.314\,\text{J K}^{-1}\text{mol}^{-1}$
$\therefore$ M, the molar mass of the gas, can be found.

Example $m_1 = 20.255\,\text{g}$, $m_2 = 20.120\,\text{g}$, $v_1 = 9.80\,\text{cm}^3$, $v_2 = 65.8\,\text{cm}^3$, $T = 363\,\text{K}$, $P = 1.01 \times 10^5\,\text{N m}^{-2}$

When these experimental results are inserted into the equation

$$PV = \frac{m}{M}RT$$

$$M = \frac{0.135 \times 8.314 \times 363}{1.01 \times 10^5 \times 56.0 \times 10^{-6}}$$

The molar mass, $M = 72\,\text{g mol}^{-1}$

CHECKPOINT 8B: MOLAR MASS OF VOLATILE LIQUIDS

($R = 8.314\,\text{J K}^{-1}\text{mol}^{-1}$, $1\,\text{atm} = 1.01 \times 10^5\,\text{N m}^{-2}$)

1. When 0.184 g of a liquid was injected into a gas syringe at 45 °C, the volume of gas formed was 55.8 cm³ at 1 atm. What value does this indicate for the molar mass of the liquid? By referring to the empirical formula CH_2 obtain an accurate value for the molar mass of the liquid.

2. When 0.125 g of a liquid was injected into a gas syringe at 50 °C, it formed 36.8 cm³ of vapour at atmospheric pressure. The empirical formula of the compound is CH_2Cl. What value of molar mass do the experimental data give? What is the accurate value?

3. A liquid of empirical formula CH_2 was vaporised. From 0.108 g of liquid, the volume of vapour was 50.8 cm³, measured at 22 °C and $9.8 \times 10^4\,\text{N m}^{-2}$. Calculate the molar mass of the liquid.

4. By vaporising 0.100 g of a liquid at 100 °C and $1.01 \times 10^5\,\text{N m}^{-2}$, 20.0 cm³ of vapour are obtained. Find the molar mass of the liquid.

8.4 SOLUTIONS OF LIQUIDS IN LIQUIDS

8.4.1 RAOULT'S LAW

A **solution** is a homogeneous (i.e., the same all through) mixture of two substances. When one liquid dissolves in another, the saturated vapour pressure of the solution depends on the saturated vapour pressures of the components and on the composition

Definition of mole fraction

of the mixture. One way of expressing the composition of a mixture of liquids is to state the **mole fraction** of each constituent. By definition

$$\text{Mole fraction of } \mathbf{A} \text{ in a mixture of } \mathbf{A} \text{ and } \mathbf{B} = \frac{\text{Number of moles of } \mathbf{A}}{\text{Total number of moles}}$$

Using the symbol x for mole fraction

$$x_A = \frac{n_A}{n_A + n_B}$$

Raoult's Law . . .

. . . gives the vapour pressure of a solution

The vapour above a mixture of liquids **A** and **B** will contain both **A** and **B**. **Raoult's Law** (after F Raoult) states that the saturated vapour pressure of each component in the mixture is equal to the product of the mole fraction of that component and the saturated vapour pressure of that component when pure. This can be expressed

$$p_A = x_{A(l)} p_A^0$$

where p_A = saturated vapour pressure of **A**, p_A^0 = saturated vapour pressure of pure **A**, and $x_{A(l)}$ = mole fraction of **A** in the solution. Similarly for **B**

$$p_B = x_{B(l)} p_B^0$$

There are conditions for obeying Raoult's Law

Raoult's Law is obeyed by mixtures of similar compounds. They are said to form **ideal solutions**. The substances **A** and **B** form an ideal solution if the intermolecular forces **A**-----**A**, **A**-----**B** and **B**-----**B** are all equal. There is neither a volume change nor an enthalpy (heat) change on mixing.

The vapour above a mixture of liquids does not have the same composition as the liquid. If $x_{A(v)}$ and $x_{B(v)}$ are the mole fractions of **A** and **B** in the vapour phase, then

The vapour is richer than the liquid in the more volatile component

$$\frac{x_{A(v)}}{x_{B(v)}} = \frac{p_A}{p_B} = \frac{x_{A(l)} p_A^0}{x_{B(l)} p_B^0}$$

If **A** is more volatile than **B**

$$p_A^0 > p_B^0$$

and

$$\frac{x_{A(v)}}{x_{B(v)}} > \frac{x_{A(l)}}{x_{B(l)}}$$

The vapour is richer than the liquid in **A**, the more volatile component.

Example: how to calculate the composition of the vapour above a solution of two liquids

Example At 80 °C, the vapour pressure of benzene is $10.0 \times 10^4\,\text{N m}^{-2}$; and that of methylbenzene is $4.00 \times 10^4\,\text{N m}^{-2}$. Estimate the mole fraction of methylbenzene in the vapour in equilibrium with a liquid mixture of benzene and methylbenzene at 80 °C in which the mole fraction of methylbenzene is 0.60.

Method

$$\frac{\text{Amount of methylbenzene in vapour}}{\text{Amount of benzene in vapour}} = \frac{0.60}{0.40} \times \frac{4.00 \times 10^4}{10.0 \times 10^4} = 0.60$$

$$\text{Mole fraction of methylbenzene} = \frac{\text{Moles of methylbenzene}}{\text{Total number of moles}}$$

$$= \frac{0.60}{1.60} = 0.375$$

Figure 8.7 shows the results of a number of calculations of this kind on an ideal mixture of A and B. The composition of vapour in equilibrium with liquid mixtures of different compositions is shown.

FIGURE 8.7 Vapour Pressure–Composition Curve for an Ideal Mixture of Liquids

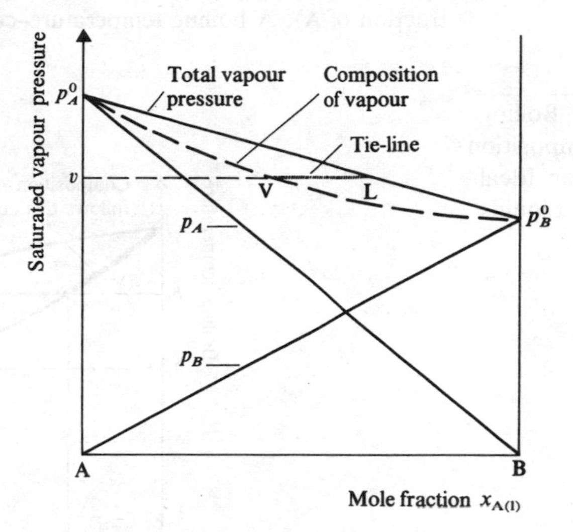

The total saturated vapour pressure is the sum of the partial vapour pressures, $p_A + p_B$. When the total vapour pressure is v, the composition of the liquid is represented by the point L, and the composition of the vapour by the point V. The line V-----L is called a **tie-line**. The vapour is richer than the liquid in the more volatile component, A. The difference in composition between the liquid and the vapour is exaggerated in the figure.

Figure 8.8 shows the vapour pressure–composition curves at two temperatures, T_1 and T_2.

FIGURE 8.8 Saturated Vapour Pressure–Composition Curves at Two Temperatures

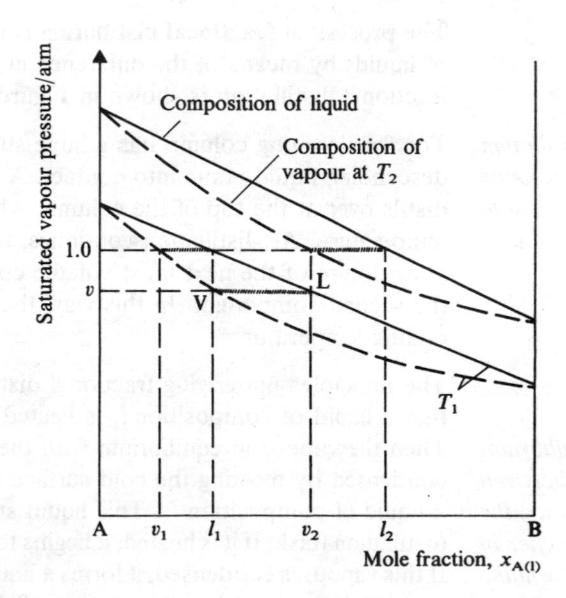

Boiling temperature depends on composition...

... The relationship can be plotted

When the vapour pressure of a liquid reaches atmospheric pressure, the liquid boils. T_1 is the boiling temperature of a liquid with composition l_1 in equilibrium with vapour of composition v_1. T_2 is the boiling temperature of a liquid which has composition l_2, in equilibrium with a vapour of composition v_2. These values can be plotted in the form of a boiling temperature–composition curve. There is a gradual variation in boiling temperature with composition. Neither the liquid curve nor the vapour curve is linear, even for an ideal mixture. The same shape is obtained when boiling temperature is plotted against the mass fraction or percentage composition by mass, instead of the mole fraction (since percentage of **A** by mass = 100 × mass fraction of **A**). A boiling temperature–composition curve is shown in Figure 8.9.

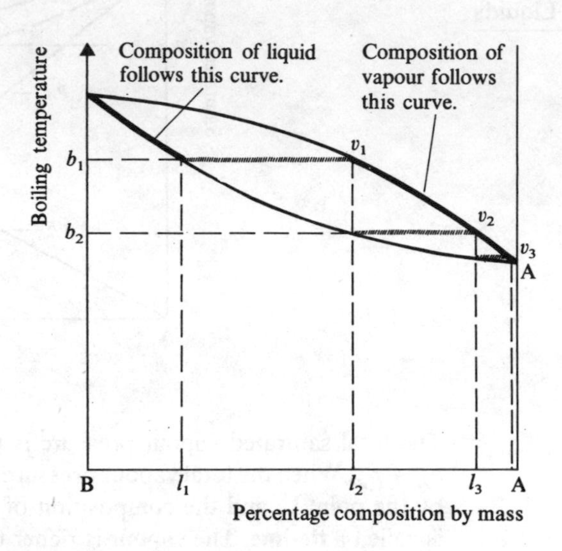

FIGURE 8.9 Boiling Temperature–Composition Curve for an Ideal Mixture of Liquids

8.4.2 FRACTIONAL DISTILLATION

The process of **fractional distillation** is used to separate the components of a mixture of liquids by means of the difference in their boiling temperatures. An apparatus for fractional distillation is shown in Figure 8.10.

Fractional distillation separates the components of a mixture according to their boiling temperatures

The fractionating column has a large surface area on which ascending vapour and descending liquid come into contact. A mixture rich in the most volatile component distils over at the top of the column, where the thermometer registers its boiling temperature. As distillation continues, the temperature rises towards the boiling temperature of the next most volatile component. The receiver is changed to collect the second component. In this way, the components are distilled over at their boiling temperatures.

Each time that equilibrium is established between liquid and vapour, the vapour becomes richer in the more volatile component

The principles underlying fractional distillation are illustrated in Figure 8.9. Imagine that a liquid of composition l_1 is heated until it begins to boil at a temperature b_1. Then the vapour in equilibrium with the liquid has composition v_1. If the vapour is condensed by meeting the cold surface of a distillation column, it condenses to form a liquid of composition l_2. This liquid starts to trickle down the column towards the distillation flask. If it is heated, it begins to boil at b_2, to form a vapour of composition v_2. If this vapour is condensed, it forms a liquid of composition l_3. By repeated vaporisation and condensation, the composition of the vapour is made to follow the curve $v_1v_2v_3$, becoming richer and richer in **A**, the more volatile component. The liquid is becoming richer in the less volatile component, **B**, and its composition follows the curve from l_1 towards **B**. The longer the column, the more vaporisation followed by condensation steps will be achieved, and the closer to pure **A** and pure **B** will the **distillate** and **residue** become.

FIGURE 8.10 Fractional Distillation

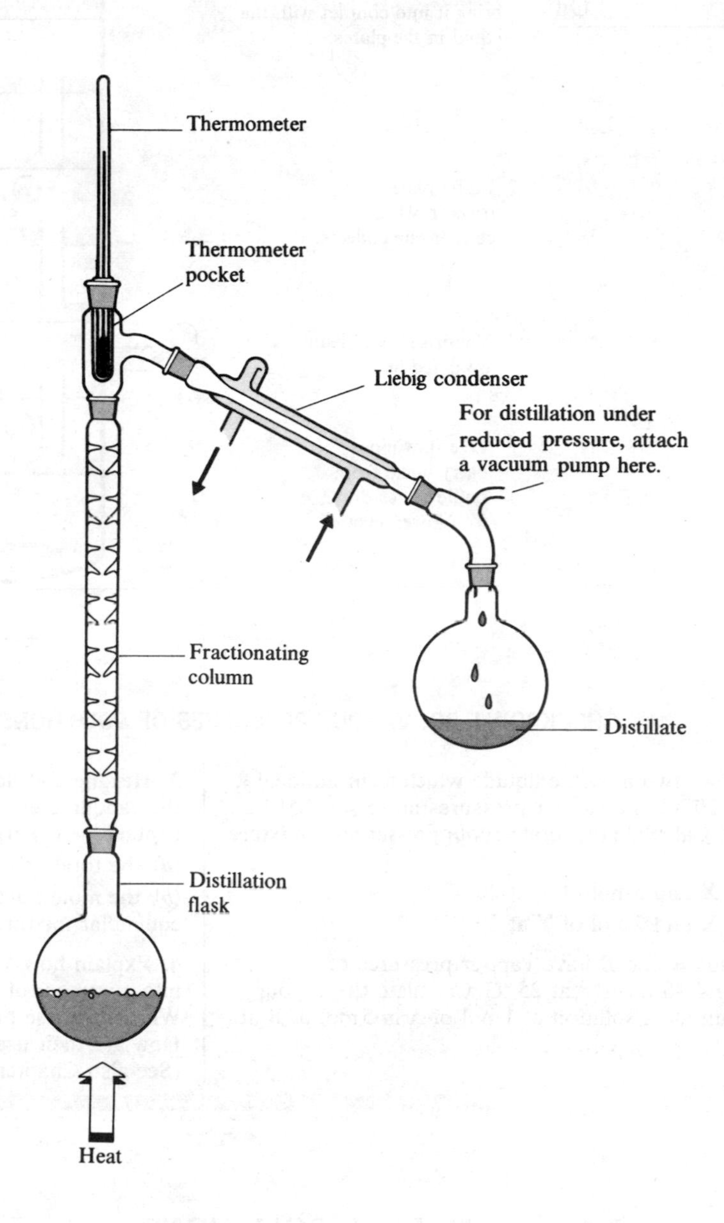

8.4.3 CONTINUOUS FRACTIONAL DISTILLATION

Fuels are obtained from crude oil by fractional distillation

The oil industry uses fractional distillation. Crude petroleum oil is vaporised and fed into a massive fractionating column, which may be 30 to 60 m high and 3 to 6 m in diameter. Different fractions, such as gasoline and kerosene, are drawn off continuously from the column at different levels. Figure 8.11 shows how liquid and vapour attain equilibrium at each level, so that low boiling temperature fractions rise to the top of the column and high boiling temperature fractions are drawn off from the bottom of the column.

FIGURE 8.11
Continuous Fractional Distillation of Petroleum Oil

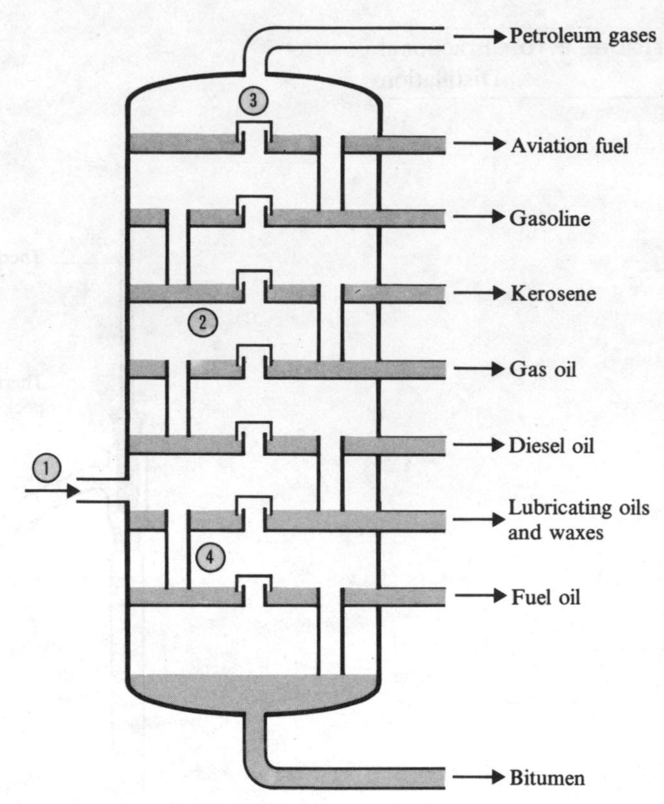
3 Ascending vapour must pass through *bubble caps*, which bring it into contact with the liquid in the plates.
2 Baffle plates, trays in which condensate collects.
1 Vaporised petroleum oil is fed in.
4 Overflow pipes carry liquid from baffle plates down to a lower level.
Petroleum gases
Aviation fuel
Gasoline
Kerosene
Gas oil
Diesel oil
Lubricating oils and waxes
Fuel oil
Bitumen

CHECKPOINT 8C: VAPOUR PRESSURES OF SOLUTIONS OF TWO LIQUIDS

1. **X** and **Y** are two miscible liquids which form an ideal solution. At 20 °C, the vapour pressures are: p_X = 25 kPa; p_Y = 45 kPa. Calculate the total vapour pressure of a mixture of:

(*a*) 1 mol of **X** and 5 mol of **Y** at 20 °C

(*b*) 5 mol of **X** and 2 mol of **Y** at 20 °C

2. Two liquids **A** and **B** have vapour pressures of 15 kN m^{-2} and 40 kN m^{-2} at 25 °C. Calculate the vapour pressure of an ideal solution of 1 mol of **A** in 5 mol of **B** at 25 °C.

3. Hexane and heptane form an ideal solution. At 30 °C, the vapour pressures are hexane = 30 kN m^{-2}; heptane = 12 kN m^{-2}. Calculate

(*a*) the total vapour pressure

(*b*) the mole fraction of heptane in the vapour above an equimolar mixture of hexane and heptane

4. Explain how fractional distillation separates crude oil into a number of products.
What are these products used for?
How are their uses related to their boiling temperatures?
(See also Chapter 26)

8.4.4 NON-IDEAL SOLUTIONS

Raoult's Law is obeyed by ideal solutions . . .

. . . Non-ideal solutions have vapour pressures greater or less than those predicted

Solutions which have a vapour pressure greater than that predicted from Raoult's Law are said to show a **positive deviation** from the law. Those with a vapour pressure lower than the calculated value are said to show a **negative deviation**. [See Figure 8.12.] Typical of a pair of liquids showing a slight positive deviation are hexane and ethanol. The molecules are very different, and molecules of ethanol can more easily escape from a mixture of ethanol and hexane than from pure ethanol, where hydrogen bonding holds molecules together. A slight negative deviation from Raoult's Law is shown by a mixture of trichloromethane, $CHCl_3$, and ethoxyethane, $C_2H_5OC_2H_5$. Forces of attraction between the two kinds of molecules tend to prevent the molecules escaping into the vapour phase. When a pair of liquids shows a very large positive deviation from Raoult's Law, the vapour pressure–composition curve has a maximum. A system with a very large negative deviation shows a minimum [see Figure 8.12].

FIGURE 8.12 Saturated Vapour Pressure–Composition Curves for Non-ideal Solutions

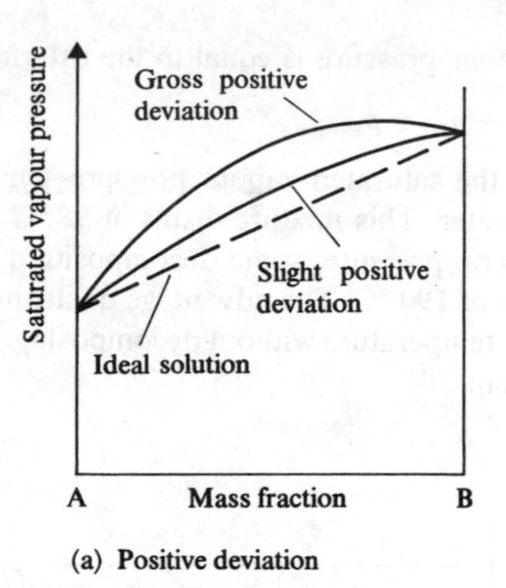

(a) Positive deviation

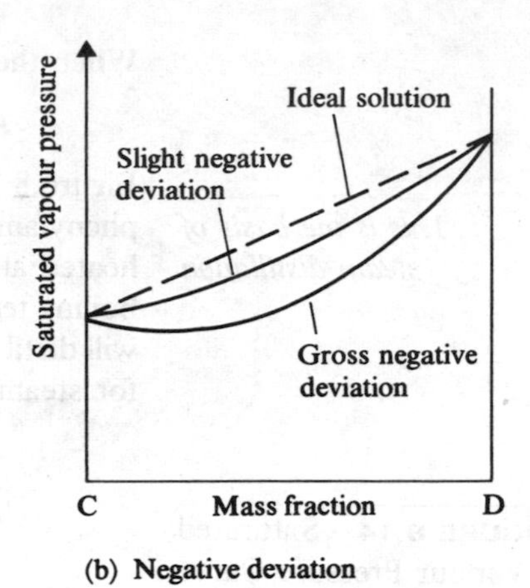

(b) Negative deviation

Boiling temperature–composition curves for systems which show gross deviations from Raoult's Law are shown in Figure 8.13.

FIGURE 8.13 Boiling Temperature–Composition Curves for Non-ideal Solutions

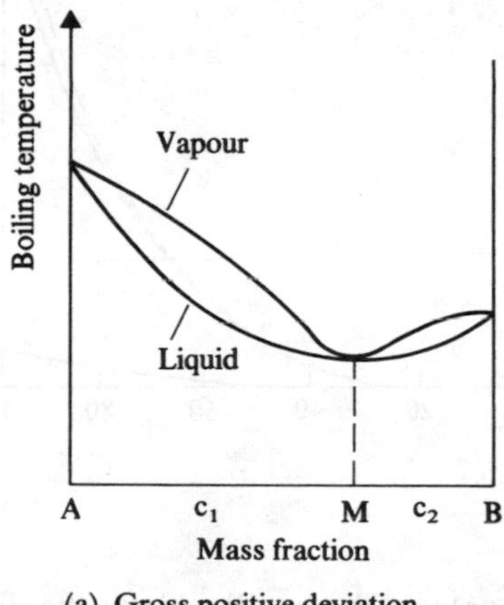

(a) Gross positive deviation

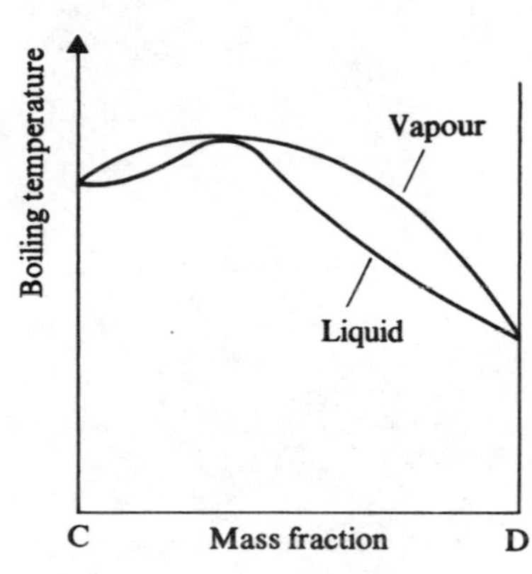

(b) Gross negative deviation

If there is a maximum or minimum in the svp–composition curve, an azeotrope distils

A maximum in the saturated vapour pressure curve results in a minimum in the boiling temperature–composition curve and vice versa. An application of the principles of fractional distillation [§ 8.4.2] shows that, in Figure 8.13(a), a liquid of composition c_1 can be separated by distillation into almost pure **A** and a mixture of composition **M**. A liquid of composition c_2 can be separated by distillation into almost pure **B** and the mixture M. If a liquid of composition M is boiled, the vapour has the same composition as the liquid. This means that the composition, and therefore the boiling temperature, T_b, do not change as distillation is continued. The mixture is therefore called an **azeotropic** or **constant-boiling mixture**. Examples are nitric acid/water (T_b is a maximum) and benzene/ethanol (T_b is a minimum). Azeotropes are *not* classified as compounds because their compositions vary with pressure.

8.5 IMMISCIBLE LIQUIDS

8.5.1 STEAM DISTILLATION

A mixture boils when the combined svp equals the external pressure...

In a system of immiscible liquids, each liquid exerts its own vapour pressure, independently of the other. The saturated vapour pressure of the mixture is equal to the sum of the saturated vapour pressures of the pure components. For a mixture of **A** and **B**,

$$p_{\text{total}} = p_A^0 + p_B^0$$

When the total vapour pressure is equal to the external pressure, the mixture distils.

$$\text{At } T_b,\ p_A^0 + p_B^0 = p_{\text{external}}$$

... This is the basis of steam distillation

Figure 8.14 shows the saturated vapour pressure–temperature curve for a mixture of phenylamine and water. This mixture distils at 98 °C at 1 atm. When phenylamine is heated at atmospheric pressure, some decomposition occurs before it reaches its boiling temperature of 190 °C. The advantage of steam distillation is that phenylamine will distil at a lower temperature without decomposing. Figure 8.15 shows an apparatus for steam distillation.

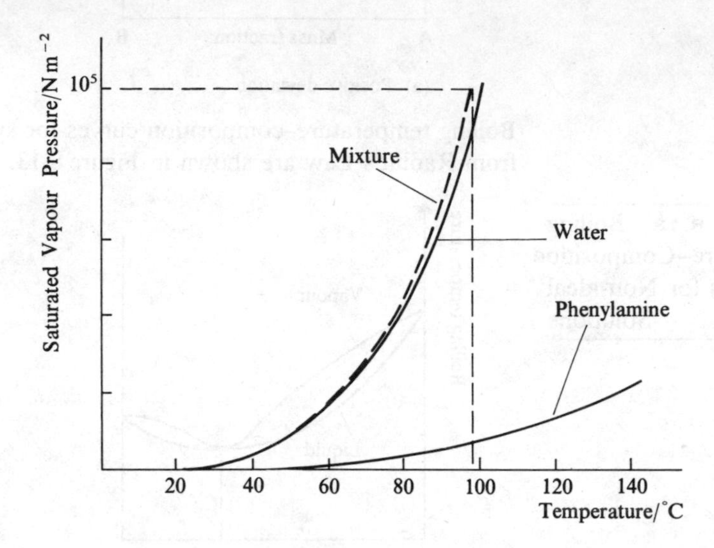

FIGURE 8.14 Saturated Vapour Pressure Curve for a Pair of Immiscible Liquids

The composition of the steam distillate can be calculated...

The ratio of the amounts of the two liquids in the distillate is equal to the ratio of their vapour pressures. If p_A, p_W are the saturated vapour pressures of phenylamine and water (where $p_A = 3.99 \times 10^3\ \text{N m}^{-2}$, $p_W = 9.70 \times 10^4\ \text{N m}^{-2}$ at 98 °C) and n_A, n_W are the amounts of phenylamine and water in a volume V of gaseous distillate, then

$$p_A V = n_A RT$$

$$p_W V = n_W RT$$

$$\text{and } \frac{n_A}{n_W} = \frac{p_A}{p_W}$$

$$= \frac{3.99 \times 10^3}{9.70 \times 10^4} = 4.11 \times 10^{-2}$$

... A sample calculation

If m_A and m_W are the masses and M_A and M_W are the molar masses of phenylamine and water respectively (where $M_A = 93\ \text{g mol}^{-1}$ and $M_w = 18\ \text{g mol}^{-1}$)

$$\frac{m_A/M_A}{m_W/M_W} = 4.11 \times 10^{-2}$$

$$\frac{m_A}{m_W} = 4.11 \times 10^{-2} \times 93/18 = 0.212$$

The ratio of phenylamine to water by mass in the distillate is 0.212; the percentage by mass of phenylamine is 17.5%.

FIGURE 8.15 Steam Distillation

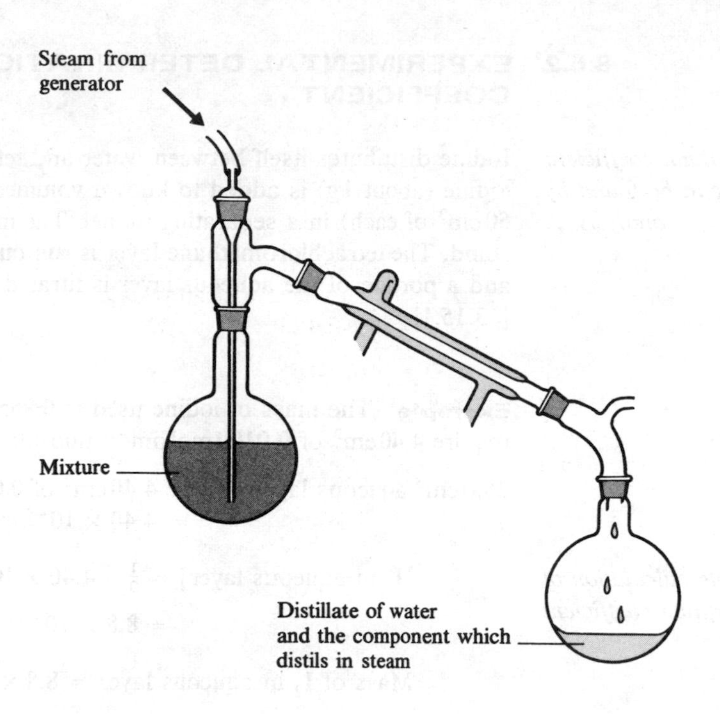

CHECKPOINT 8D: STEAM DISTILLATION

1. An organic liquid distils in steam. The partial pressures of the two liquids at the boiling point are: organic liquid 5.3 kPa; water 96.0 kPa. The distillate contains the liquids in a ratio of 0.48 g of organic liquid : 1.00 g of water. What is the molar mass of the organic liquid?

2. Chlorobenzene distils in steam at 91 °C at $1.01 \times 10^5\,N\,m^{-2}$. At this temperature, the vapour pressure of chlorobenzene is $2.90 \times 10^4\,N\,m^{-2}$. Calculate the percentage by mass of chlorobenzene in the distillate.

8.6 DISTRIBUTION OF A SOLUTE BETWEEN TWO SOLVENTS

8.6.1 THE DISTRIBUTION LAW

A solute will distribute itself between a pair of immiscible solvents...

...in accordance with its distribution coefficient...

When a solute is added to a pair of immiscible liquids, it may dissolve in both liquids. In this case the solute will distribute itself between the two solvents. It may well be more soluble in one solvent than the other. The way in which a solute is partitioned between two solvents can be studied by shaking the solute with a pair of immiscible liquids in a separating funnel. The two layers can be run off separately and analysed. Provided that there is insufficient solute to saturate either of the two solvents, and provided that the solute is in the same molecular state in both solvents, then it is found that the ratio of the solute concentrations in the two solvents is always the same. If c_U and c_L are the concentrations in the upper and lower layers, then

$$c_U/c_L = k$$

The constant, k, is called the **partition coefficient** or **distribution coefficient**, and is constant for a particular temperature. Since a dynamic equilibrium exists between

...which is temperature-dependent

Solute in upper layer $\rightleftharpoons$ Solute in lower layer

k is a type of equilibrium constant. Like other equilibrium constants it is temperature-dependent.

8.6.2 EXPERIMENTAL DETERMINATION OF PARTITION COEFFICIENT

The partition coefficient can be found by analysis...

Iodine distributes itself between water and tetrachloromethane. A known mass of iodine (about 1 g) is added to known volumes of water and tetrachloromethane (say 50 cm^3 of each) in a separating funnel. The mixture is shaken well and allowed to stand. The tetrachloromethane layer is run out of the bottom of the separating funnel and a portion of the aqueous layer is titrated against standard thiosulphate solution [§3.15.1].

Example The mass of iodine used is 0.9656 g and 25.0 cm^3 of the aqueous layer require 4.40 cm^3 of 0.0100 $mol\,dm^{-3}$ thiosulphate. $A_r(I) = 127$.

25.0 cm^3 aqueous layer require 4.40 cm^3 of 0.0100 $mol\,dm^{-3}$ thiosulphate
$= 4.40 \times 10^{-5}$ mol thiosulphate

...A sample calculation of partition coefficient

$$[I_2 \text{ in aqueous layer}] = \tfrac{1}{2} \times 4.40 \times 10^{-5}/(25.0 \times 10^{-3})$$

$$= 8.8 \times 10^{-4}\,mol\,dm^{-3}$$

$$\text{Mass of } I_2 \text{ in aqueous layer} = 8.8 \times 10^{-4} \times 254 \times 50.0 \times 10^{-3} = 0.0112\,g$$

$$\text{Mass of } I_2 \text{ in } CCl_4 \text{ layer} = 0.9656 - 0.0112 = 0.9544\,g$$

$$[I_2 \text{ in water}]/[I_2 \text{ in } CCl_4] = 0.0112/0.9544 = 1.17 \times 10^{-2}$$

The partition coefficient between water and tetrachloromethane is 1.17×10^{-2}.

Ethoxyethane extraction is used in the purification of organic products...

Ethoxyethane (ether) extraction is often used in organic preparations. Ethoxyethane is a good solvent for many organic compounds and is immiscible with water. When an aqueous solution of a product is shaken with ethoxyethane in a separating funnel, the product will distribute itself between the two layers. If the partition coefficient is high, a large fraction of product will pass into the ether layer. The ether layer is separated from the aqueous layer and dried. With its low boiling temperature, ethoxyethane is easily distilled off to leave the product. It is more efficient to use the ether in portions for repeated extractions than to use it all in one operation. The following calculation illustrates this point.

Example The product of an organic synthesis, 5.00 g of **X**, is obtained in solution in 1.00 dm^3 of water. Calculate the mass of **X** that can be extracted from the aqueous solution by (*a*) 50.0 cm^3 of ethoxyethane and (*b*) two successive portions of 25.0 cm^3 of ethoxyethane. The partition coefficient of **X** between ethoxyethane and water is 40.0 at room temperature.

Method

(*a*) Let the mass of **X** extracted by 50.0 cm^3 of ether $= a_1$ g

$$[\mathbf{X} \text{ in ether}]/[\mathbf{X} \text{ in water}] = 40.0$$

$$\frac{a_1/50.0}{(5.00 - a_1)/1000} = 40.0$$

$$a_1 = 3.33\,g$$

The mass of **X** that can be extracted by using all the ether at once is 3.33 g.

It is more efficient to use the solvent in portions

(*b*) Let a_2 g = mass of **X** extracted by the first 25.0 cm^3 of ether, and
a_3 g = mass of **X** extracted by the second 25.0 cm^3 of ether

$$\frac{a_2/25.0}{(5.00 - a_2)/1000} = 40.0$$

$$a_2 = 2.50\,\text{g}$$

If 2.50 g of **X** are extracted by ether, 2.50 g remain in the aqueous solution. Therefore

$$\frac{a_3/25.0}{(2.50 - a_3)/1000} = 40.0$$

$$a_3 = 1.25\,\text{g}$$

Total mass of **X** extracted by ether in two portions is

$$2.50 + 1.25 = 3.75\,\text{g}$$

This is greater than the value of 3.33 g calculated for the mass of **X** extracted by using all the ether at once.

CHECKPOINT 8E: PARTITION

1. The partition coefficient of **S** between ethoxyethane and water is 5.0. A solution containing 10.0 g of **S** in 500 cm^3 of water is extracted with 100 cm^3 of ether. Which of the following gives the mass of **S** extracted from the water?

A 2.5 g B 5.0 g C 7.5 g D 10.0 g E 12.5 g

2. A substance **T** of $M_r = 60.0\,\text{g mol}^{-1}$ is present at a concentration of $0.0800\,\text{mol dm}^{-3}$ in 500 cm^3 of aqueous solution. The distribution coefficient of **T** between water and tetrachloromethane is 2.78×10^{-2}. What mass of **T** will be extracted from the aqueous solution by shaking it with 50.0 cm^3 of tetrachloromethane?

3. An aqueous solution contains 5.0 g of **X** in 100 cm^3 of solution. The partition coefficient of **X** between water and tetrachloromethane is 0.200. Calculate the mass of **X** extracted by shaking 100 cm^3 of the aqueous solution with

(*a*) 50 cm^3 of tetrachloromethane

(*b*) two successive 25 cm^3 portions of tetrachloromethane.

8.7 PARTITION CHROMATOGRAPHY

8.7.1 THE THEORY

Chromatographic separations use repeated partition of solutes between two solvents

It has been shown in an example in § 8.6.2 and in Question 3, Checkpoint 8E, that repeated extractions with organic solvents are effective in removing a considerable quantity of solute from an aqueous solution. The technique of repeated extractions can be applied to the separation of a number of solutes in a solution, provided that the solutes differ in their solubility in a second solvent.

In column chromatography...

... the stationary phase is adsorbed on a solid which packs the column...

In partition chromatography, many extractions are performed in succession in one operation. The solutes are partitioned between a **stationary phase** and a **mobile phase**. The stationary phase is a solvent (often water) **adsorbed** (bonded to the surface) on a solid. This may be paper or a solid such as alumina or silica gel, which has been packed into a column or spread on a glass plate. The mobile phase is a second solvent which trickles through the stationary phase.

8.7.2 COLUMN CHROMATOGRAPHY

... The moving phase is a second solvent...

The *column* is a glass tube packed with an **adsorbent** (e.g., alumina, silica, starch, magnesium silicate). The solid adsorbent is made into a slurry with water, poured in and allowed to settle. [See Figure 8.16.] A solution of the mixture to be analysed is poured on to the top of the column so that the components can be adsorbed on the column. The second solvent, called the **eluant**, is allowed to trickle slowly through the column.

FIGURE 8.16 Column Chromatography

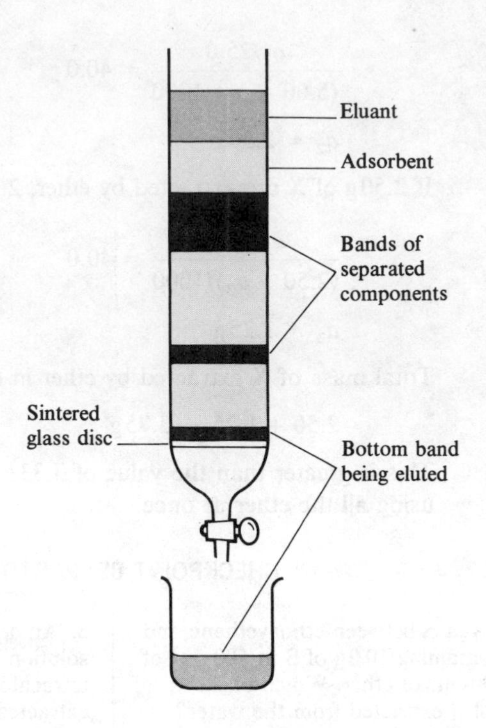

... Solutes partition between the solvents ...

... The rate at which a solute travels down the column depends on its partition coefficient

Partition of the solutes takes place between the moving solvent and the stationary adsorbed water, and the eluant carries the solutes with it, to be readsorbed further down the column. The rates at which the solutes travel down the column depend on their partition coefficients. If these are sufficiently different then, provided the eluant is kept flowing continuously, a complete separation is achieved, and the components become spread out along the column in order of their partition coefficients. The separated bands are called a **chromatogram**. The components can be recovered by dissolving them separately out of the column – **eluting** them – with more solvent, and then evaporating off the solvent.

8.7.3 PAPER CHROMATOGRAPHY

In paper chromatography...

... the stationary phase is water adsorbed on paper...

... The solvent ascends...

In paper chromatography, the principle is the same as for column chromatography, but the adsorbent material is *paper*. A solution of the mixture to be separated is applied to a strip of chromatography paper. This is hung in a glass tank so that the end dips into solvent in the bottom of the tank [see Figure 8.17]. The solvent may be water, ethanol, glacial ethanoic acid, butanol, etc., or a mixture of any of these. As the solvent rises through the paper, it meets the sample, and the component bands spread out. The separation is stopped when the solvent has travelled nearly to the top of the paper. The distance travelled by the **solvent front** is measured. Then for each component, the $\mathbf{R_F}$ **value** is calculated, by

$$R_F = \frac{\text{Distance travelled by the component}}{\text{Distance travelled by the solvent front}}$$

The R_F value can be used to assist in identifying the components of a mixture. The separated components can be obtained by cutting the paper into strips and dissolving out each compound.

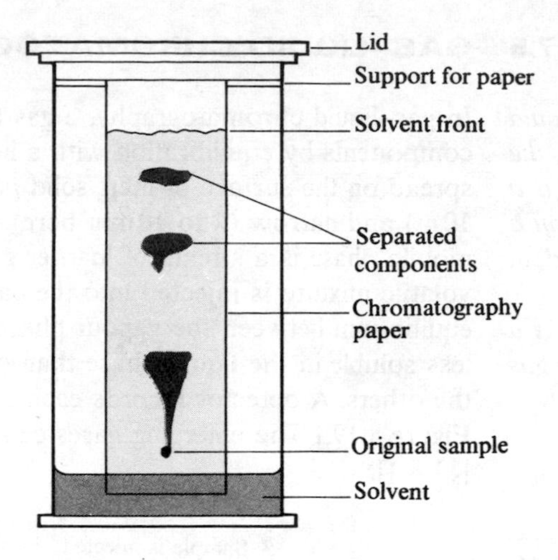

FIGURE 8.17 Paper Chromatography

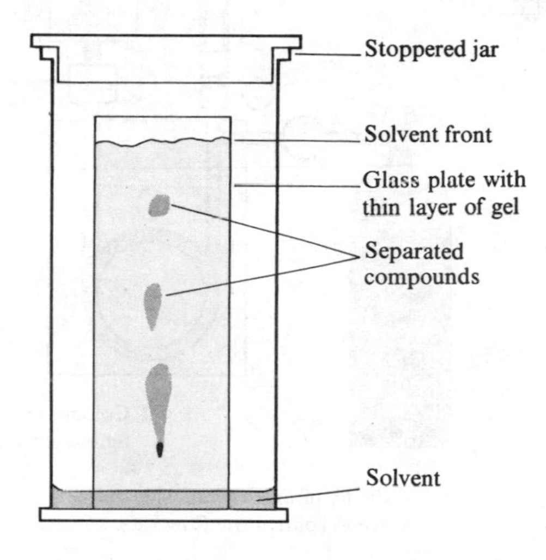

FIGURE 8.18 Thin Layer Chromatography

... or descends through the paper

There is a different design of chromatography tank, with a trough at the top of the tank to hold the solvent. The paper hangs down from this trough, and solvent descends through the paper. The two techniques are naturally described as **ascending** and **descending paper chromatography**.

8.7.4 THIN LAYER CHROMATOGRAPHY

A thin layer of adsorbent solid can be used for chromatography

The solid adsorbent may be in the form of a *thin layer* on the surface of a glass plate. The adsorbent (e.g., silica gel or calcium sulphate) is made into a thick paste and spread evenly over a glass plate. The thin layer of paste is allowed to dry and baked in an oven. Spots of the mixture are applied and a chromatogram is developed in the same manner as for paper chromatography. [See Figure 8.18.] Thin layer chromatography has the advantage that a variety of different adsorbents can be used. Since the thin layer is more compact than paper, more equilibrations take place while the solvent front travels over the same distance. Often separations can be achieved in a few centimetres, and coated microscope slides are frequently used for thin layer chromatography.

8.7.5 GAS–LIQUID CHROMATOGRAPHY

In gas–liquid chromatography, the stationary phase is a liquid adsorbed on a solid…

…The mobile phase is a stream of carrier gas

In gas–liquid chromatography, a gas (or volatile liquid or solid) is separated into its components by equilibration with a liquid. The liquid is the stationary phase. It is spread on the surface of inert solid particles which pack the column. Long (5 to 10 m) and narrow (2 to 10 mm bore), the column is often wound into a coil. The mobile phase is a stream of 'carrier gas', nitrogen or a noble gas. When a sample of volatile mixture is injected into the carrier gas, each component sets up a partition equilibrium between the vapour phase and the liquid phase. Some components are less soluble in the liquid phase than others and emerge from the column ahead of the others. A detector records each component as it leaves the column. [See Figure 8.19.] The emerging gases can also be analysed by a mass spectrometer [§ 1.8.1].

FIGURE 8.19 Gas–Liquid Chromatography

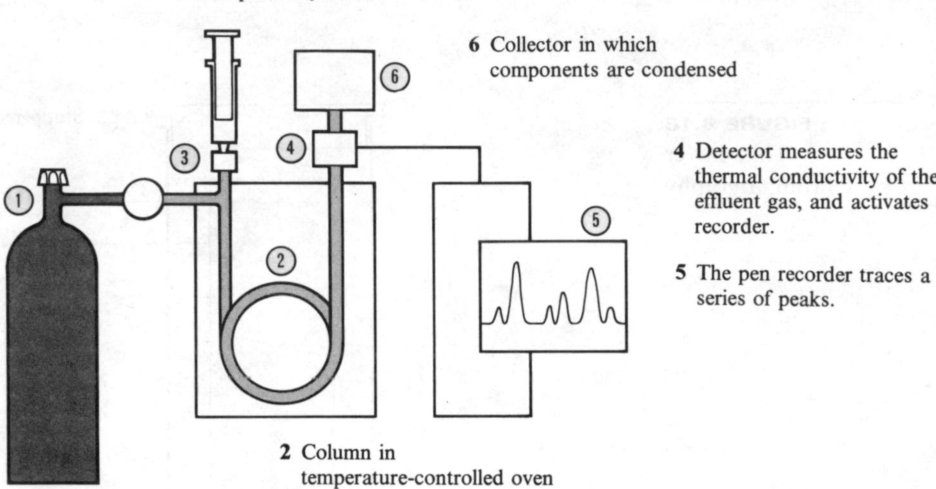
3 Sample is injected.
6 Collector in which components are condensed
4 Detector measures the thermal conductivity of the effluent gas, and activates a recorder.
5 The pen recorder traces a series of peaks.
2 Column in temperature-controlled oven
1 Cylinder of carrier gas with valve to control the flow rate
1
2
3
4
5
6

8.7.6 ION EXCHANGE

Ion exchange resins replace cations by $H^+(aq)$…

…and anions by $OH^-(aq)$

Ion exchange is a type of partition of ionic compounds. The **ion exchange resin** is a polymer which contains at intervals polar groups which can remove undesirable cations (or anions) and replace them with other cations (or anions). The water softener Permutit®, is sodium aluminium silicate, which replaces calcium and magnesium ions in hard water by sodium ions. To purify water, a resin must replace all the cations and anions present by hydrogen ions and hydroxide ions. A combination of a **cation exchanger** and an **anion exchanger** is needed. Cation exchange resins often contain sulphonic acid groups, $—SO_3^-H^+$, and anion exchangers often contain quaternary ammonium groups, e.g., $—\overset{+}{N}(CH_3)_3OH^-$. If water passes slowly through a **deioniser**, equilibrium is set up at each level between hydrogen ions attached to the resin and hydrogen ions in solution:

$$—SO_3^-H^+(\text{resin}) + Na^+(aq) \rightleftharpoons —SO_3^-Na^+(\text{resin}) + H^+(aq)$$

As the water moves on, two of the components in the equilibrium ($H^+(aq)$ and $Na^+(aq)$) are removed and a fresh equilibrium is established. Each successive equilibration increases the replacement of metal cations by hydrogen ions. At the same time, hydroxide ions are replacing other anions. If tap water is run slowly through an ion exchange resin, 'deionised' water of high purity can be obtained.

QUESTIONS ON CHAPTER 8

1. Nitric acid (T_b = 87 °C) and water form a constant boiling mixture, of T_b = 122 °C and composition 65% by mass nitric acid.

(*a*) Draw the boiling temperature–composition curve for nitric acid and water.

(*b*) State what is meant by a constant boiling mixture.

(*c*) State Raoult's Law. Does this mixture show a positive or negative deviation from this law?

(*d*) What interaction between nitric acid and water could give rise to this type of deviation?

(*e*) What happens when a mixture containing 30% by mass of nitric acid is distilled?

2. Explain the principles underlying steam distillation. For what types of compound is it useful?

3. Mixture A = water + nitrobenzene
Mixture B = benzene + methylbenzene

State the relationship between the total vapour pressure of the mixture and the vapour pressures of the components in both mixtures.

4. Explain

(*a*) how liquids mix

(*b*) why some of a liquid evaporates below its boiling temperature

(*c*) how the vapour pressure of a system of two immiscible liquids varies with (i) composition and (ii) temperature.

5. Describe practical methods which you have used in the laboratory for column chromatography, paper chromatography and thin-layer chromatography. Give examples of mixtures which can be separated by these methods.

6. Discuss the physical principles involved in

(*a*) the ethoxyethane extraction of a product from aqueous solution

(*b*) column chromatography

(*c*) the separation of two immiscible liquids.

7. Calculate the percentage by volume composition of the vapour over a mixture of 64 g of methanol and 46 g of ethanol at 330 K. The vapour pressures/N m^{-2} at this temperature are CH_3OH = 8.1 × 10^4; C_2H_5OH = 4.5 × 10^4.

8. The saturated vapour pressures of phenylamine and water at a number of temperatures are shown below.

Temperature/°C	70	80	90	100
Phenylamine: vapour pressure/10^5 N m^{-2}	0.0141	0.0242	0.0384	0.0606
Water: vapour pressure/10^5 N m^{-2}	0.311	0.473	0.679	1.010

(*a*) By means of a graph, find the temperature at which a mixture of phenylamine and water boils at 1.01 × 10^5 N m^{-2} (1 atm).

(*b*) Calculate the percentage of phenylamine in the distillate.

9. Compare the fractional distillation of an ideal mixture, such as hexane (T_b = 69 °C) and heptane (T_b = 98.5 °C) with that of a non-ideal mixture, such as ethanol (T_b = 78 °C) and water. Draw boiling temperature–composition curves for the mixtures.

10. Define the term *partition coefficient*. A compound **Z** has a partition coefficient of 4.00 between ethoxyethane and water. Calculate the mass of **Z** extracted from 100 cm^3 of an aqueous solution of 4.00 g of **Z** by two successive extractions with 50 cm^3 of ethoxyethane.

11. A hot gas syringe was injected with a small quantity of a liquid **X** through a gas tight seal. **X** has a molecular formula which may be represented as $C_aH_bO_c$ where *a*, *b* and *c* are whole numbers. The liquid vaporised immediately and the volume of gas produced was measured and corrected to room temperature and pressure:

Mass of liquid **X** injected = 0.30 g
Volume of vapour produced = 82.7 cm^3

A second sample of **X** was completely burned in oxygen when it produced only water and carbon dioxide, whose volume was measured at room temperature and pressure:

Mass of liquid **X** burnt in oxygen = 0.40 g
Mass of water produced = 0.455 g
Volume of carbon dioxide produced = 551 cm^3

(*a*) Calculate:

(i) the relative molecular mass of **X**

(ii) the number of moles of water produced by burning one mole of **X**

(iii) the number of moles of carbon dioxide produced by burning one mole of **X**.

(*b*) Write a balanced equation for the combustion of **X** and use it to find the molecular formula for **X**.

(*c*) Draw structural formulae for *two* possible isomers of **X**.

(O & C 92, AS)

12. (*a*) State Raoult's Law as applied to mixtures of miscible liquids.

(*b*) Two liquids, **A** and **B**, have vapour pressures p_A and p_B at 25 °C. **A** is the more volatile liquid.

Assuming that **A** and **B** behave ideally,

(i) Draw fully labelled graphs, on the same axes, showing how p_A and p_B and total vapour pressure p vary with the mole fraction for mixtures of **A** and **B**.

(ii) Draw a fully labelled temperature–composition diagram for the liquid and vapour phases of **A** and **B**.

(*c*) At 50 °C, the vapour pressure of hexane is 54.0 kPa and that of heptane is 22.0 kPa. Calculate the partial vapour pressures of hexane and heptane above a mixture of 8.60 g of hexane and 10.0 g of heptane at this temperature. Assume that this mixture obeys Raoult's Law.

(*d*) (i) State the law governing the distribution of a solute, **X**, between two immiscible solvents, **P** and **Q**.

(ii) By application of this law, show that, at constant temperature,

$$\frac{m_1 V_2}{m_2 V_1} = \frac{p_1}{p_2}$$

where m_1 and m_2 are the masses of a gas X dissolved respectively in volume V_1 of a liquid at pressure p_1 and volume V_2 of the same liquid at pressure p_2.

(iii) The solubility of carbon dioxide in water at 25 °C is 0.0338 mol dm^{-3} at 1.00 atm pressure. Calculate the pressure required to dissolve 1.00 mol of carbon dioxide in 10.0 dm^3 water at the same temperature, assuming that the solution contains only molecular carbon dioxide.

(O 92)

13. (*a*) Make a careful drawing of the pressure–temperature phase diagram for carbon dioxide. Label the axes and indicate the approximate positions of 1 atmosphere pressure and 298 K on these axes.

Explain the significance of:

(i) an area

(ii) a line on your diagram

illustrating your answer with *one* area and line of your choice.

(iii) Mark the triple point and explain its significance.

Explain why it follows from your diagram that:

(iv) carbon dioxide expands on melting

(v) solid carbon dioxide sublimes at atmospheric pressure.

†(*b*) Illustrate *monotropy* and *enantiotropy* by considering carbon and tin.

[For (*b*) see § 23.2.]

(O 91)

14. (*a*) Draw diagrams to show the variation of *boiling point* with composition for mixtures of

(i) methanol and ethanol

(ii) cyclohexane (b.p. 81 °C) and ethanol (b.p. 78.5 °C)

Your diagrams should have clearly labelled axes and show both the liquid and vapour compositions.

(*b*) (i) State Raoult's Law.

(ii) Explain why the mixture in (*a*)(i) above is described as 'nearly ideal' and that in (*a*)(ii) above is described as 'showing a positive deviation'.

(*c*) Explain how some liquid mixtures which show a positive deviation from Raoult's Law can form a constant-boiling (azeotropic) mixture. Your explanation should include a boiling point/composition diagram.

(*d*) A mixture of trichloromethane and ethyl ethanoate is said to 'show a negative deviation' from Raoult's Law'. Explain the temperature change observed when samples of these two liquids, each at the same starting temperature, are mixed.

(*e*) A text-book describes some of the final stages of the preparation of phenylamine as follows:

'After the phenylamine has been freed from the salt using excess of the concentrated aqueous sodium hydroxide, the mixture is *steam-distilled* until no more oily drops pass down the condenser. The distillate is placed in a separating funnel and *extracted using ethoxyethane as solvent*. Use 90 cm^3 of ethoxyethane in three approximately equal portions.'

(i) In what circumstances is *steam distillation* a useful method for recovering an organic product from a reaction mixture?

(ii) Briefly explain the term *solvent extraction*. Why are *successive* small amounts of solvent used, rather than one large amount? (No quantitative treatment is required).

(AEB 92)

15. The partition of iodine between water and 1,1,1-trichloroethane (TCE) may be determined by the following procedure:

100 cm^3 of water and 100 cm^3 of TCE were placed, together with 4 g of finely powdered iodine, in a 250 cm^3 stoppered bottle. The contents were shaken for one hour in a water bath at 30 °C. 5.0 cm^3 of the TCE layer were titrated with 0.1 M sodium thiosulphate solution. The bottle was re-stoppered and shaken for further periods of 30 minutes and a further 5.0 cm^3 of the TCE layer were titrated. When two titration values were the same the water layer was titrated with 0.01 M thiosulphate solution. The whole experiment was repeated using 2 g and 8 g of iodine.

(*a*) (i) Why was finely ground iodine used?

(ii) Why was the mixture shaken?

(iii) What may be deduced when two successive titration values for the TCE layer are the same?

(iv) What indicator may be used in the titration of iodine with sodium thiosulphate?

(v) Write a balanced equation for the reaction of iodine with sodium thiosulphate.

(*b*) The following results were obtained from the experiment:

Mass of iodine/g	V_1 titration value for TCE layer/cm^3 of 0.1 M thiosulphate solution	V_2 titration value for water layer/cm^3 of 0.01 M thiosulphate solution	$\frac{V_1}{V_2}$
4	15.5	6.1	
2	7.8	3.0	
8	32.0	12.3	

(i) Complete the third column in the table.

(ii) Comment on the values obtained for V_1/V_2.

(iii) The ratio V_1/V_2 was obtained for the partition of iodine between the solvents at 30 °C. Explain why you would expect the ratio to change with temperature.

(*c*) In a test for bromide, chlorine is passed into aqueous sodium bromide and the mixture shaken with a small amount of TCE. The organic layer becomes yellow-red.

(i) Write a balanced equation for the reaction of chlorine with aqueous sodium bromide solution.

(ii) Why does the organic layer become coloured?

(iii) Suggest why a *small* amount of TCE is used.

(iv) Describe and explain another test for bromide ions.

(NI 91)

9

SOLUTIONS

9.1 SOLUTIONS OF SOLIDS IN LIQUIDS

Definition of a saturated solution

A solution that contains as much solute as can be dissolved at that temperature in the presence of undissolved solute is called a **saturated** solution. In a saturated solution, a state of dynamic equilibrium exists, with particles of solid constantly dissolving, and solute constantly being deposited. The quantity of solid that dissolves in a certain quantity of liquid to form a saturated solution at a stated temperature is the **solubility** of the solid. There are various ways of expressing solubility, e.g., g solute dm^{-3} solution or mol dm^{-3}; g solute kg^{-1} solvent or mol kg^{-1} and mole fraction : mol solute/(mol solute + mol solvent). The solubility of a solute can be found by (a) titrimetric analysis, (b) evaporating a known mass of solution to dryness and weighing the solute that remains, and (c) other physical techniques.

Solubility can be expressed in various units...

...There are a number of methods of measurement

Figure 9.1 shows a **solubility curve**, a graph of solubility against temperature.

FIGURE 9.1
A Solubility Curve

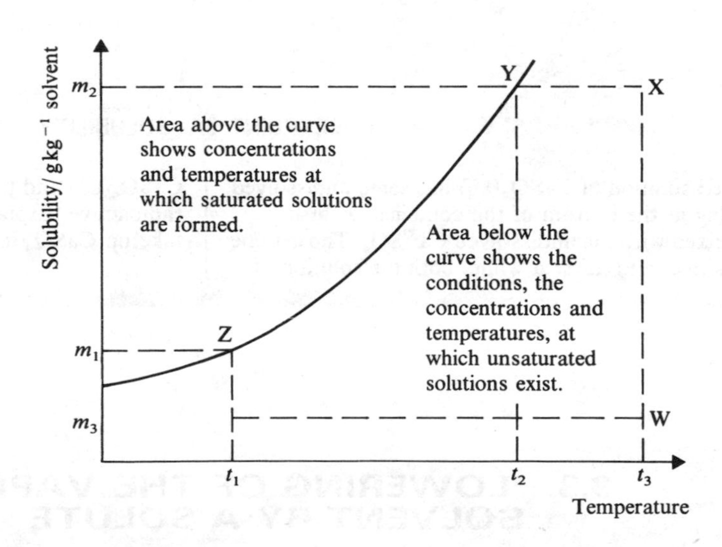

A solubility curve shows the effect of temperature on solubility...

...and enables one to predict the quantity of solute that will crystallise when a solution is cooled

A solution at point **X** contains m_2 g of solute in 1 kg of solvent at a temperature t_3. It is an unsaturated solution, and when it is cooled it does not deposit solute until it reaches a temperature t_2. At t_2, solute is deposited, and the concentration of the solution decreases. On further cooling, more solute is deposited, and the concentration of the solution again decreases. As the temperature falls, the concentration follows the curve **YZ**. At the temperature t_1, the mass of solute that can remain in solution is m_1 g. The mass of solute that has crystallised between t_2 and t_1 is $(m_2 - m_1)$ g. A solution **W**, containing m_3 g of solute in 1 kg of solvent at t_3, can be cooled from t_3 to t_1 without depositing any solute as the solubility is never exceeded.

9.2 RECRYSTALLISATION

Recrystallisation of a solute is used as a method of purification

Solids are purified by **recrystallisation** from a suitable solvent, e.g., water, ethanol, propanone. The impure material is dissolved in the minimum quantity of hot solvent. The hot solution is filtered to remove insoluble impurities. A heated Buchner funnel and flask [see Figure 9.2] are used to prevent cooling of the solution and crystallisation of the solute. The filtered solution is allowed to cool so that the solute crystallises out. The impurities are present in smaller quantities and remain in solution at the lower temperature. The crystals are filtered, washed in the funnel with a little cold solvent, and dried. The purity of a solid is assessed by finding its melting temperature. [See § 34.2.1 and Figure 34.3.]

FIGURE 9.2 Filtration under Reduced Pressure

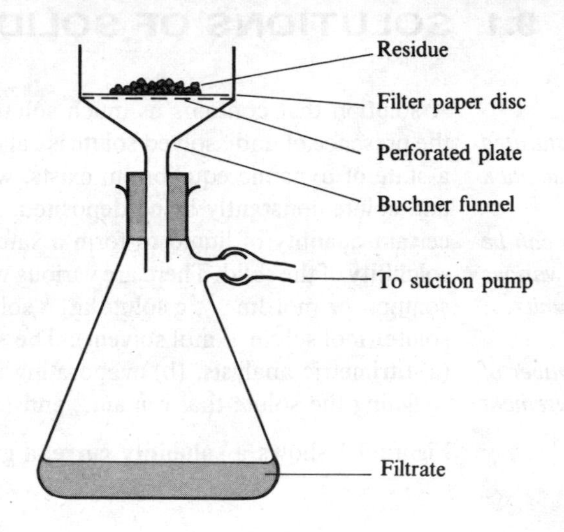

CHECKPOINT 9A: SOLUBILITY

1. A saturated solution of $CaSO_4(aq)$ has some undissolved $CaSO_4(s)$ lying at the bottom of the container. A little $Ca^{35}SO_4$ is mixed with the undissolved $Ca^{32}SO_4$. The isotope ^{35}S is radioactive, and, after a while, both the solution of $CaSO_4(aq)$ and the undissolved $CaSO_4(s)$ are found to be radioactive. Explain how the solution has been able to take up $CaSO_4$ in spite of being saturated.

9.3 *LOWERING OF THE VAPOUR PRESSURE OF A SOLVENT BY A SOLUTE

For a solution of two liquids **A** and **B**, Raoult's Law [§ 8.4.1] states

$$p = x_A p_A^0 + x_B p_B^0$$

where x_A = mole fraction and p_A^0 = saturated vapour pressure of pure **A** and so on. In a solution of a solid **B** in a liquid **A**, if the solid is involatile, i.e., p_B^0 is negligible, then

A solute lowers the saturated vapour pressure of a solvent . . .

$$p = x_A p_A^0$$

Since x_A is a fraction, $p < p_A^0$.

The presence of a solute has lowered the saturated vapour pressure of the solvent. It follows that the lowering of the vapour pressure, $p_A^0 - p$, is equal to the mole fraction of the solute. This conclusion holds for ideal solutions, and can be expressed by

$$\frac{p^0 - p}{p^0} = \frac{n_1}{n_1 + n_2}$$

p^0 = vapour pressure of pure solvent
p = vapour pressure of solution
n_1 = amount of solute
n_2 = amount of solvent

In dilute solutions $n_2 \gg n_1$ and the expression becomes

$$\frac{p^0 - p}{p^0} = \frac{n_1}{n_2}$$

Notice that the relative lowering of the saturated vapour pressure depends on the molar concentration of the solute, not on its identity. Properties such as vapour pressure lowering, which depend on the concentration of dissolved particles and not on their nature, are called *colligative properties*.

9.4 OSMOTIC PRESSURE

Osmosis is the passage of solvent from a dilute solution to a more concentrated solution

A **semipermeable membrane** is a film of material which can be penetrated by a solvent but not by a solute. When two solutions are separated by a semipermeable membrane, solvent passes from the more dilute to the more concentrated. This phenomenon is called **osmosis**. Examples of semipermeable membranes are animal and plant cell walls. Artificial semipermeable membranes are formed by allowing two solutions to meet in the pores of a porous material.

Copper(II) salts and potassium hexacyanoferrate(II) react in the pores of a porous pot to produce a semipermeable membrane of copper(II) hexacyanoferrate(II), $Cu_2Fe(CN)_6$.

Definition of osmotic pressure

When a solution and its solvent are separated by a semipermeable membrane, the pressure which must be applied to the solution to prevent the solvent from entering is called the **osmotic pressure** of the solution. There is an analogy with gas pressures. One mole of a solid **A**, when vaporised, occupies a volume of 22.4 dm^3 at 0 °C and $1.01 \times 10^5\,N\,m^{-2}$. One mole of **A** dissolved in 22.4 dm^3 of solvent at 0 °C exerts an osmotic pressure of $1.01 \times 10^5\,N\,m^{-2}$.

The osmotic pressure equation resembles the Ideal Gas Equation

The expression which relates osmotic pressure to concentration and temperature is similar to the Ideal Gas Equation:

$$\pi V = nRT$$

where π = osmotic pressure, V = volume, T = temperature/K, n = amount of solute/mol, and R = a constant which has the same value as the gas constant, $8.314\,J\,K^{-1}\,mol^{-1}$. This equation, the van 't Hoff equation, is obeyed by ideal solutions.

9.4.1 MEASUREMENT OF OSMOTIC PRESSURE

The osmotic pressure of a solution depends on the molar concentration of solute present: it is a colligative property. Measurements of osmotic pressure can be used to give the molar masses of solutes.

Figure 9.3 shows the Berkeley and Hartley method for the measurement of osmotic pressure.

Example Calculate the molar mass of Z, given that a solution of 60.0 g of Z in 1.00 dm^3 of water exerts an osmotic pressure of $4.31 \times 10^5 \text{ N m}^{-2}$ at 25 °C.

FIGURE 9.3 Berkeley and Hartley Method for the Measurement of Osmotic Pressure

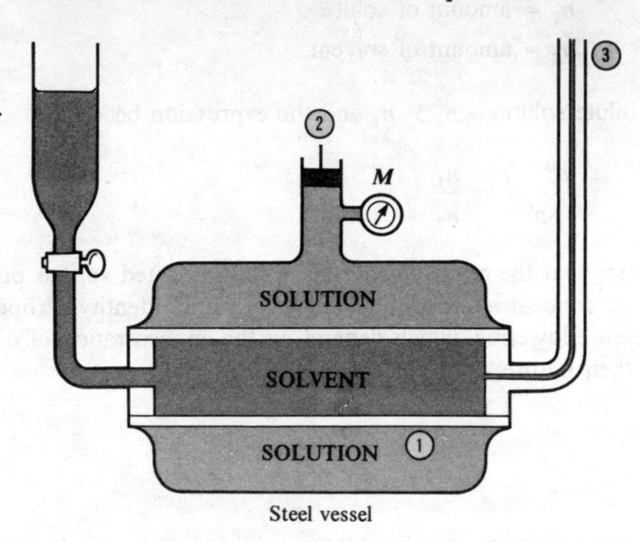

3 Capillary tube indicates movement of solvent into the solution.

2 Pressure is applied to the solution. It is adjusted until there is no flow of solvent into the solution.

1 Cylindrical tube is porous. In the pores, a semipermeable membrane has been laid down, e.g. $Cu_2Fe(CN)_6$.

Method

A calculation of molar mass from osmotic pressure

$$\pi V = nRT$$

$$\pi = 4.31 \times 10^5 \text{ N m}^{-2}$$

$$V = 1.00 \times 10^{-3} \text{ m}^3$$

$$R = 8.314 \text{ J K}^{-1} \text{ mol}^{-1}$$

$$T = 298 \text{ K}$$

$$\therefore \quad 4.31 \times 10^5 \times 1.00 \times 10^{-3} = \frac{60.0}{M} \times 8.314 \times 298$$

The molar mass $M = 345 \text{ g mol}^{-1}$

... and on temperature-sensitive substances

Osmotic pressure measurements are useful for substances of biological interest. There is the advantage that measurements can be made at room temperature. Many naturally occurring substances, such as proteins, are temperature-sensitive and undergo changes at 0 °C and 100 °C. Mass spectrometry [§ 1.8.1] is the most convenient method for measuring molar masses. Some large molecules, however, fragment and do not give molecular ions. For such compounds, osmotic pressure measurement offers a valuable alternative.

9.4.2 OSMOSIS AND PLANT CELLS

The plasma membrane in a plant cell is a semipermeable membrane. The fluid inside the cell exerts an osmotic pressure. Figure 9.4 shows what happens when a plant cell is placed in solutions of different osmotic pressures. Osmosis is of vital importance in the life of plants, being the means by which the plant roots take in water.

FIGURE 9.4
A Plant Cell

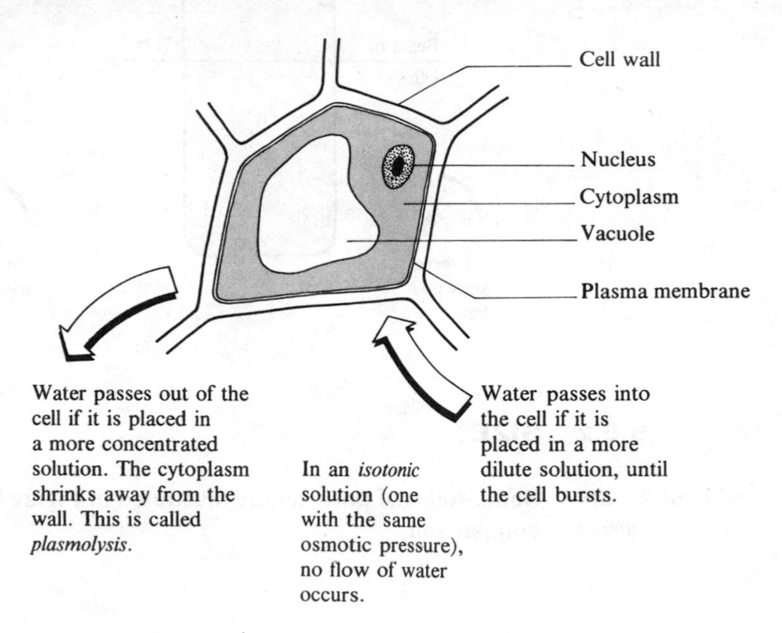
Cell wall
Nucleus
Cytoplasm
Vacuole
Plasma membrane
Water passes out of the cell if it is placed in a more concentrated solution. The cytoplasm shrinks away from the wall. This is called *plasmolysis*.
In an *isotonic* solution (one with the same osmotic pressure), no flow of water occurs.
Water passes into the cell if it is placed in a more dilute solution, until the cell bursts.

9.4.3 OSMOSIS AND BLOOD

Red blood corpuscles behave in the same way. Any solution which is injected into the bloodstream must be **isotonic** with blood, i.e., have the same osmotic pressure.

CHECKPOINT 9B: OSMOTIC PRESSURE

1. Calculate the osmotic pressure at 20 °C of a solution of sucrose, $C_{12}H_{22}O_{11}$, containing 20.0 g dm^{-3}.

2. A solution of 2.00 g of a polymer in 1.00 dm^3 of water has an osmotic pressure of 300 N m^{-2} at 20 °C. Calculate the molar mass of the polymer.

3. A solution containing 3.47 g of **X** in 250 cm^3 of solution at 20 °C has osmotic pressure 2.06×10^3 N m^{-2}. Calculate the molar mass of **X**.

4. The osmotic pressure of a solution of 1.00 g of chymotrypsin in 100 cm^3 of water at 300 K is 994 N m^{-2}. Calculate the molar mass of chrymotrypsin.

9.5 COLLOIDS

A colloidal dispersion is a type of solution containing large particles of solute . . .

Silica, SiO_2, is insoluble in water: the sand on the seashore is evidence of this fact. Yet when an aqueous solution of a silicate is acidified, silica is not precipitated. According to the pH, silica is obtained as a **colloidal dispersion** or a gelatinous precipitate or a solid-like **gel**, in which liquid is trapped. A colloidal dispersion resembles a solution in being clear, apart from a slight opalescence. A colloidal dispersion may contain up to 30% by mass of silica. Measurements of colligative properties indicate that the silica is present as a small concentration of large particles.

9.5.1 OPTICAL PROPERTIES

. . . The solute particles are large enough to scatter light

The particles of a colloid are not big enough to be visible, but they are big enough to scatter light. This effect, the **Tyndall effect**, was named after its discoverer. It resembles the scattering of light in dust-laden air, as shown in Figure 9.5.

FIGURE 9.5 Scattering of Light by a Colloid

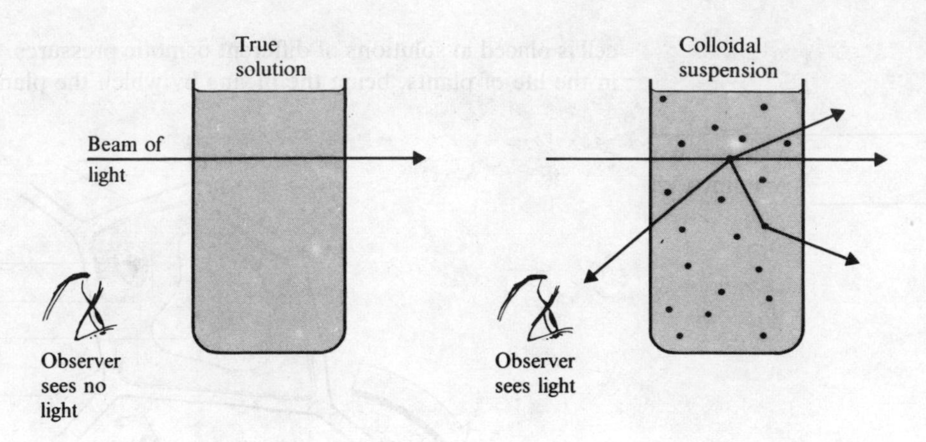

9.5.2 SIZE

Colloidal particles are 100 nm across

Some colloidal particles are illustrated in Figure 9.6, with ions and molecules for comparison.

FIGURE 9.6 Size of Colloidal Particles

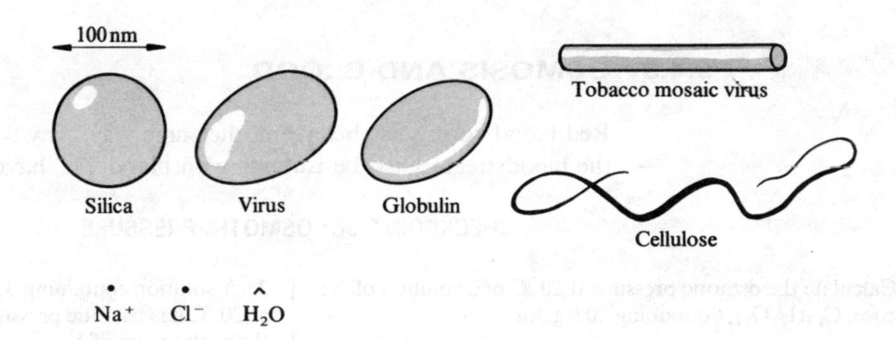

9.5.3 CHARGE

The reason why colloidal particles do not coalesce is that their surfaces are charged. Colloidal silica particles adsorb hydroxide ions on their surfaces and are negatively charged. Repulsion between like charges keeps the particles in suspension.

Charged colloidal particles migrate in an electric field, as ions do.

9.5.4 PRECIPITATION OF COLLOIDS

Colloids can be precipitated by neutralising their charges

Colloids are precipitated by neutralising the charge on the particles, often by the addition of ions of opposite charge. This is done at one stage in the purification of drinking water, when clay particles and other colloidal suspensions must be removed. Impure water is treated with, e.g., aluminium sulphate. When the negative charges on the clay particles are neutralised by Al^{3+} ions, clay coagulates and settles out of solution.

Colloidal proteins in blood are negatively charged, and styptic pencils, which help blood to clot, contain Al^{3+} ions or Fe^{3+} ions to neutralise the charges on the surfaces of the colloidal particles.

9.5.5 CLASSIFICATION

Colloids are classified as **lyophilic** (solvent-loving; **hydrophilic** if the solvent is water) and **lyophobic** (solvent-fearing; **hydrophobic** if the solvent is water). Table 9.1 lists some types of colloids.

TABLE 9.1 Some Types of Colloids

Dispersed phase	*Dispersion medium*	*Type*	*Example*
Liquid	Gas	Aerosol	Fog, mist from aerosol spray can
Solid	Gas	Aerosol	Smoke, dust-laden air
Gas	Liquid	Foam	Soap suds, whipped cream
Liquid	Liquid	Emulsion	Oil in water, milk, mayonnaise
Solid	Liquid	Sol	Clay, colloidal gold
Gas	Solid	Solid foam	Lava, pumice
Liquid	Solid	Solid emulsion	Pearl, opal
Solid	Solid	Solid sol	Some gems, e.g., black diamond, ruby glass

QUESTIONS ON CHAPTER 9

1. What is a saturated solution? Explain how you could measure the solubility of the crystalline solid, ethanedioic acid-2-water, $(CO_2H)_2 \cdot 2H_2O$, in water at room temperature.

2. Why does the vapour pressure of a liquid increase with temperature? Why does the vapour pressure of a liquid decrease when an involatile solute is dissolved in it?

3. Explain why osmotic pressure offers a practical method for finding the molar masses of polymers with molar masses of the order of $10^5\ \text{g mol}^{-1}$.

There is a method for finding molar masses which is very much more economical in material and time. What is it? How much material does this method use? Is it suitable for polymers?

4. (*a*) Give the names and the formulae of the repeating units of *each* of the following:

(i) *two* addition polymers

(ii) *two* condensation polymers

(*b*) Give a specific use for *each* of the four polymers named in (*a*) and state what properties make it suitable for the use given.

(*c*) The osmotic pressure of solutions of different concentrations of a polymer were measured in a suitable osmometer. The results are given below.

Temperature 298 K:

Osmotic pressure/Pa	*Concentration*/g dm^{-3}
118	2.0
480	6.0
1000	10.0
1680	14.0

At these concentrations, the osmotic pressure, π, is given by the equation

$$\pi/c = RT/M + kc$$

where c is the concentration in g m^{-3}, k is a constant, R is the gas constant, T is the temperature (in K) and M is the mean relative molecular mass of the polymer.

Use the above data and a graphical method to determine the value of M for the polymer.

(AEB 92, S)

10

THERMOCHEMISTRY

10.1 ENERGY

Energy is the ability to do work

Work = force × distance

This chapter is about energy. An object which is capable of doing work is said to possess **energy**. The trouble with equating energy with the capacity for doing work is that one then has to explain what work is. When a force is applied to a stationary object, the object moves, and, when this happens, the force is said to be doing **work**. The amount of work done is the product of the force and the distance which the object moves in the direction of the force:

$$\text{Work} = \text{Force} \times \text{Distance}$$

The unit of force is the **newton**. (When 1 newton acts on a mass of 1 kg, it causes the mass to move with an acceleration of $1\,\text{m s}^{-2}$.) The unit of work is the **joule**. (1 joule = 1 newton metre, i.e., 1 J = 1 Nm.) Energy and work are measured in the same unit.

10.1.1 ENERGY CHANGES

The energy changes in chemical reactions are important

Energy comes from the Sun

The energy changes that accompany chemical reactions are of vital importance to us. Our own life processes depend on the energy content of the food we eat. The energy of the chemical bonds in the compounds present in food has come from the Sun. In the process of photosynthesis, plants which contain the catalyst chlorophyll convert carbon dioxide and water and the energy of sunlight into sugars:

$$6CO_2(g) + 6H_2O(l) \xrightarrow[\text{chlorophyll}]{\text{sunlight}} C_6H_{12}O_6(aq) + 6O_2(g)$$

A reaction during which energy is absorbed is termed an **endothermic** reaction. Photosynthesis takes in 15 MJ of energy for every kilogram of glucose synthesised.

The energy in fossil fuels can be harnessed

The kind of life we lead depends on harnessing energy from different sources. We harness the fossil fuels, coal, oil and natural gas. These are formed by the slow conversion of plant material over millions of years, and thus derive their energy ultimately from the Sun. In the combustion of octane in a plentiful supply of air, 40 MJ of energy are obtained for every litre of octane burnt.

There are many forms of energy...

...All are either kinetic or potential energy...

Heat is kinetic energy

There are many forms of energy: heat, light, chemical energy, nuclear energy, etc., but basically there are only two kinds of energy, **kinetic energy** and **potential energy**. The energy which an object possesses because it is moving is called kinetic energy. The amount of kinetic energy that an object possesses is the amount of work that it can do before it comes to rest, having used up all its kinetic energy. The energy which an object possesses because of its position or because of the arrangement of its component parts is called potential energy. **Heat** is a form of kinetic energy: it is the kinetic energy associated with the motion of atoms and molecules. The energy

The energy of chemical bonds is potential energy

of chemical bonds is a form of potential energy, arising from the positions of atoms and molecules with respect to one another.

The study of the energy changes that accompany chemical reactions is called **thermochemistry** or **chemical thermodynamics**. It has its origins in the study of heat engines, in which the ideas of heat and movement combined to give the name **thermodynamics**.

The First Law of Thermodynamics states that energy is conserved

Energy can be converted from one form into another. Electrical energy can be converted into heat energy. Our bodies can convert the energy of the chemical bonds in food into other kinds of energy. Calculations on energy conversions show that energy is never created and never destroyed. Observations on physical changes and chemical reactions are summarised in the **First Law of Thermodynamics**. This law states that *energy can be changed from one form into another, but it can neither be created nor destroyed.*

10.1.2 HEAT AND TEMPERATURE

When an object receives heat, its temperature rises. The rise in temperature depends on the **heat capacity** of the object:

$$\Delta T = \Delta Q/C \qquad [1]$$

Heat capacity and specific heat capacity

The relationship between heat absorbed and rise in temperature

ΔT = rise in temperature, ΔQ = heat absorbed and C = heat capacity. You can see that the heat capacity of a mass of substance is the quantity of heat required to raise its temperature by 1 K. The heat capacity of unit mass of a substance is the specific heat capacity of that substance.

$$c = C/m \qquad [2]$$

where c = specific heat capacity and m = mass

Combining equations [1] and [2] gives

$$\Delta Q = mc\Delta T \qquad [3]$$

From equation [3], the rise in temperature that results from the absorption of a certain quantity of heat by a mass of substance can be calculated. For an example, see § 10.3.1.

10.2 INTERNAL ENERGY AND ENTHALPY

Matter possesses energy in the form of ...

... kinetic energy and

... potential energy ...

... the sum of which = internal energy

Matter contains energy. This is in the form of kinetic energy and potential energy. The kinetic energy of matter is the energy of motion at a molecular level. The atoms or ions or molecules in a solid are vibrating and rotating and translating (moving from one place to another). The potential energy of matter arises from the positions of the atoms relative to one another. Bond-breaking and bond-making involve changes in potential energy. When two solids react, as in the reaction

$$Ca(s) + I_2(s) \rightarrow CaI_2(s)$$

there is little change in kinetic energy, but there is a big change in potential energy as the bonding in the product is different from the bonding in the reactants. The kinetic energy and potential energy together make up the **internal energy** of matter.

Heat may be given out or taken in during a chemical reaction

Frequently, during the course of a chemical reaction, heat is either given out or taken in from the surroundings. The heat absorbed during a reaction is equal to the internal energy of the products minus the internal energy of the reactants, provided that no work is done by the system on the surroundings [see Figure 10.1].

FIGURE 10.1 Heat Changes in Chemical Reactions

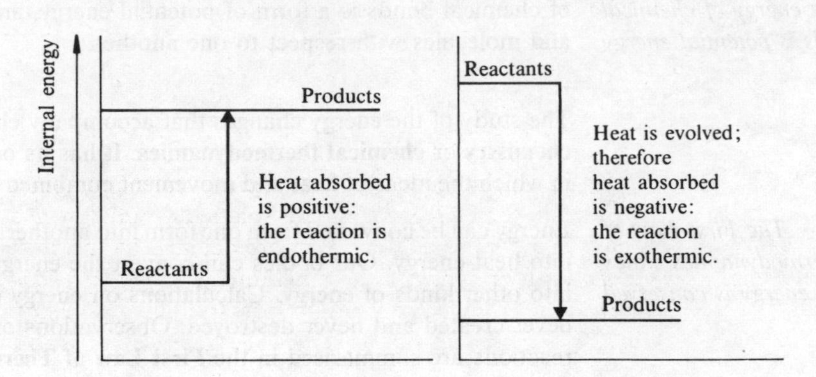

If the reaction takes place at constant pressure, any gases formed are allowed to escape, and they do work in forcing back the atmosphere. The heat absorbed during the reaction is equal to the increase in internal energy plus the work done in expansion.

$$\begin{pmatrix}\text{Heat absorbed at}\\ \text{constant pressure}\end{pmatrix} = \begin{pmatrix}\text{Change in}\\ \text{internal energy}\end{pmatrix} + \begin{pmatrix}\text{Work done on}\\ \text{surroundings}\end{pmatrix} \quad [a]$$

Since most laboratory work is carried out at constant pressure, it is convenient to define a quantity enthalpy, such that

$$H = U + PV$$

Heat absorbed at constant pressure = change in enthalpy

where H = enthalpy, U = internal energy, P = pressure, and V = volume. Then, at constant pressure, $\Delta P = 0$, and

$$\Delta H = \Delta U + P\Delta V$$

where ΔV is the change in volume that occurs during reaction. Since $P\Delta V$ is the work done by a gas when it expands by a volume ΔV at a pressure P*, you can see that

$$\begin{pmatrix}\text{Change in}\\ \text{enthalpy}\end{pmatrix} = \begin{pmatrix}\text{Change in}\\ \text{internal energy}\end{pmatrix} + \begin{pmatrix}\text{Work done on}\\ \text{surroundings}\end{pmatrix} \quad [b]$$

Comparing equations [a] and [b] shows that the change in enthalpy is equal to the heat absorbed at constant pressure.

If ΔV is positive, as in an expansion, $\Delta H > \Delta U$

If ΔV is negative, as in a contraction, $\Delta H < \Delta U$

If ΔV is zero, i.e. in reaction at constant volume, $\Delta H = \Delta U$

For reactions of gases, ΔH differs from ΔU

Reactions of solids and liquids do not involve large changes in volume, and ΔH is close to ΔU. Reactions in which there is a considerable change in volume are those which involve gases, and ΔH can be calculated from the ideal gas equation. Since

$$PV = nRT$$

then $$P\Delta V = \Delta nRT$$

Δn, the increase in the number of moles of gas, is indicated by the equation for the reaction. For example, in the reaction

$$NH_4NO_3(s) \rightarrow N_2O(g) + 2H_2O(g)$$

$\Delta n = 3$.

*See, e.g., R Muncaster, *A-level Physics* (Stanley Thornes)

The standard state of a substance

When you are talking about the enthalpy of a substance, you must state the temperature, pressure and physical state of the substance. It is usual to compare the enthalpies of substances in their **standard states**. The standard state of a substance is the pure substance in a specified state (solid, liquid or gas) at 1 atmosphere pressure.

Standard enthalpy changes under standard conditions

The value of an enthalpy change depends on the temperature, the physical states (s, l, g) of the reactants and products, the pressures of gaseous reactants and products and the concentrations of solutions. Enthalpy changes are therefore stated under standard conditions, and are denoted by the symbol, $\Delta H^{\ominus}_{T}$, meaning the standard enthalpy change at temperature, T. **Standard conditions** are: gases at a pressure of 1 atm, solutions at unit concentration, substances in their normal physical states at the specified temperature. The temperature must be specified. If the temperature is not quoted, as in $\Delta H^{\ominus}$, you can assume that it is 298 K (25 °C). Thus ΔH refers to heat absorbed at constant pressure and $\Delta H^{\ominus}$ refers to heat absorbed under standard conditions.

Definitions of some standard enthalpy changes

$\Delta H^{\ominus}_{F}$ for an element is zero

Associated with any physical or chemical change is a standard enthalpy change. Some standard enthalpy changes are defined below. **Standard enthalpy of formation,** $\Delta H^{\ominus}_{F}$, is the heat absorbed when 1 mole of a substance is formed from its elements in their standard states. For example, the standard enthalpy of formation of sodium chloride is calculated for the reaction between solid sodium and gaseous chlorine molecules, Na(s) and Cl_2(g). It follows from this definition that all elements in their standard states are assigned a value of zero for their standard enthalpies of formation. Figure 10.2 illustrates how the value of $\Delta H^{\ominus}_{F}$ may be negative (as for sodium chloride) or positive (as for ethyne). If the enthalpy absorbed is negative, enthalpy is released, and the reaction is **exothermic**.

FIGURE 10.2 Standard Enthalpy of Formation

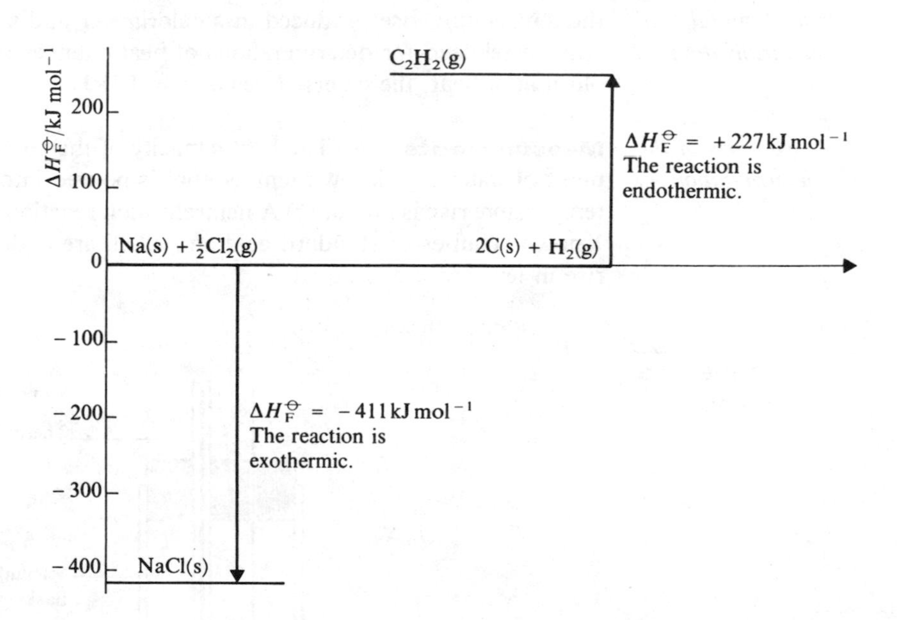

More definitions...

Standard enthalpy of combustion, $\Delta H^{\ominus}_{C}$, is the heat absorbed when 1 mole of a substance is completely burned in oxygen at 1 atm. Since heat is usually evolved in such a reaction, $\Delta H^{\ominus}_{C}$ will be negative.

Standard enthalpy of neutralisation is the heat absorbed when an acid and a base react to form 1 mole of water under standard conditions.

...$\Delta H^{\ominus}$ = molar enthalpy change at 1 atm

Standard enthalpy of reaction, $\Delta H^{\ominus}_{R}$, is the heat absorbed in a reaction at 1 atm between the number of moles of reactants shown in the equation for the reaction.

In the reaction

$$4H_2O(l) + 3Fe(s) \rightarrow Fe_3O_4(s) + 4H_2(g)$$

there is no reason why the standard enthalpy of reaction should be related to 1 mole of iron or 1 mole of steam or 1 mole of iron oxide. Instead, it is related to the whole reaction, as written, between 4 moles of steam and 3 moles of iron.

Standard enthalpy of dissolution is the heat absorbed when 1 mole of a substance is dissolved at 1 atm in a stated amount of solvent. This may be 100 g or 1000 g of solvent or it may be an 'infinite' amount of solvent, i.e., a volume so large that on further dilution there is no further heat exchange.

Standard enthalpy of atomisation is the enthalpy absorbed when a substance decomposes to form 1 mole of gaseous atoms.

10.3 EXPERIMENTAL METHODS FOR FINDING THE STANDARD ENTHALPY OF REACTION

10.3.1 STANDARD ENTHALPY OF NEUTRALISATION

The reaction to be studied is the formation of water from oxonium (hydrogen) ions and hydroxide ions:

$$H_3O^+(aq) + OH^-(aq) \rightarrow 2H_2O(l)$$

Method of finding standard enthalpy of neutralisation...

The heat released when a known amount of water is formed is found by measuring the temperature rise produced in a calorimeter and its contents [see Figure 10.3]. Any vessel used for determinations of heat changes is called a **calorimeter**, after the old unit of heat, the **calorie** (1 calorie = 4.18 J).

... The two parts of the determination...

Measurements (*a*) The heat capacity of the calorimeter is found. A known mass of water at a known temperature is poured into the calorimeter, and the temperature rise is noted. (*b*) A neutralisation reaction is carried out in the calorimeter. Known volumes of standard acid and alkali are added to the calorimeter, and the rise in temperature is noted.

FIGURE 10.3
A Calorimeter

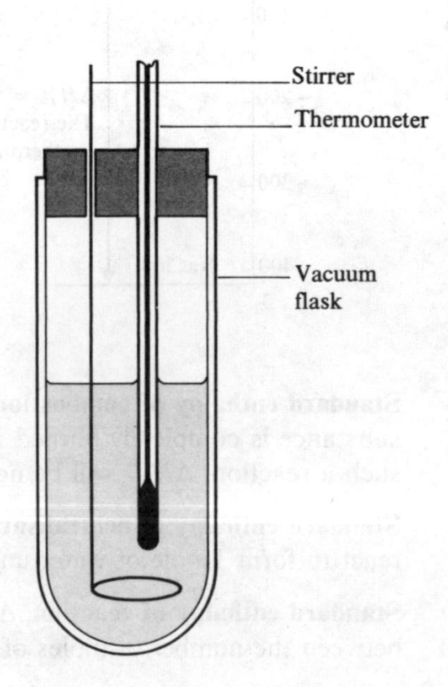

... the results ...

A Typical Result (*a*) When 100 g of water at 94.0 °C were added to a calorimeter at 17.5 °C, the temperature rose to 80.5 °C. (*b*) 250 cm³ of sodium hydroxide (0.400 mol dm⁻³) were added to 250 cm³ of hydrochloric acid (0.400 mol dm⁻³) in the calorimeter. The temperature of the two solutions was 17.5 °C initially and rose to 20.1 °C.

... and method of calculation

Calculation of $\Delta H^{\ominus}$ (Neutralisation) The assumption is made that the specific heat capacities of the solutions are the same as that of water, 4180 J kg⁻¹ K⁻¹.
In (*a*)

Heat given out by water = Heat received by calorimeter of heat capacity C

$$0.100 \times 4180(94.0 - 80.5) = C(80.5 - 17.5)$$

$$C = 90\,\text{J K}^{-1}$$

In (*b*)

$$\text{Heat from neutralisation} = \text{Heat received by calorimeter + solutions}$$
$$= 90(20.1 - 17.5) + (0.500 \times 4180(20.1 - 17.5))$$
$$= 5670\,\text{J}$$

$$\text{Amount of water formed} = 250 \times 10^{-3} \times 0.400 = 0.100\,\text{mol}$$

$$\text{Heat evolved per mole} = 5670/0.100 = 56\,700\,\text{J mol}^{-1}$$

The standard enthalpy of neutralisation is −56.7 kJ mol⁻¹.

10.3.2 EXPERIMENTAL METHOD FOR FINDING ENTHALPY OF COMBUSTION

Figure 10.4 shows a simple method for obtaining an approximate value for the enthalpy of combustion of a fuel.

FIGURE 10.4
Apparatus for Finding Enthalpy of Combustion

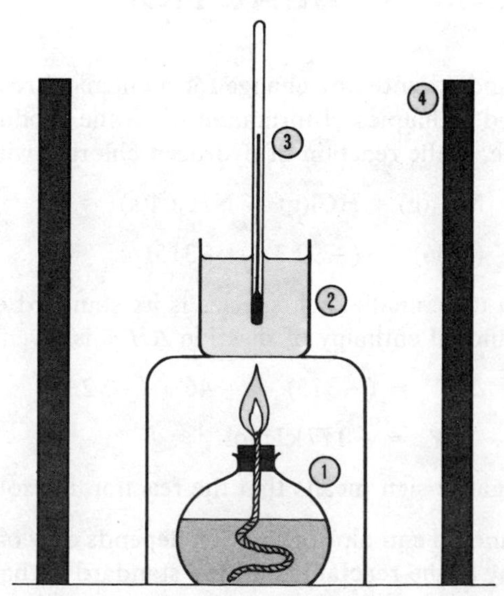
4 Shield reduces heat loss to surroundings.
3 Thermometer records rise in temperature, t °C.
2 Metal calorimeter contains a known mass of water, m_2 g.
1 Spirit burner contains fuel. Weighing before and after burning gives mass of fuel burnt, m_1 g.

A worked example of enthalpy of combustion

Example When ethanol was burnt in the apparatus shown in Figure 10.4, the results were: $m_1 = 1.50\,g$, $m_2 = 500\,g$, $t = 19.5\,°C$. Find the enthalpy of combustion of ethanol. Compare the experimental value with the listed value of $-1368\,kJ\,mol^{-1}$.

Method

$$\text{Heat evolved} = 500 \times 4.18 \times 19.5 = 40.8\,kJ$$

$$\text{Amount of ethanol burnt} = 1.50/46 = 0.0326\,mol$$

$$\text{Molar enthalpy of combustion} = 40.8/0.0326 = 1250\,kJ\,mol^{-1}$$

The enthalpy of combustion is $-1250\,kJ\,mol^{-1}$. This is lower than the listed value because some of the heat evolved is lost to the surroundings.

CHECKPOINT 10A: COMBUSTION

1. Calculate the amount of energy a woman takes in during a year on a 2200 kcal a day diet (1 calorie = 4.18 J). Calculate the amount of energy she spends in driving 5000 miles a year at 40 miles to the gallon of octane, given 1 gallon of octane weighs 3.12 kg, and $\Delta H^{\ominus}_{c}$ (octane) = $-5512\,kJ\,mol^{-1}$.

2. A cocktail contains 33 g of ethanol. If $\Delta H^{\ominus}_{c}$ (ethanol) is $-1370\,kJ\,mol^{-1}$, calculate the standard enthalpy released when this amount of ethanol is combusted inside the tissues of the drinker. Your result will be slightly in error because the value of $\Delta H^{\ominus}_{c}$ which you have been given is not exactly appropriate. What value of $\Delta H^{\ominus}_{c}$ should be used in the calculation?

3. Compare the liquid fuels propane, butane and iso-octane, and decide which gives the best calorific value for money.

Fuel	*Propane (l)*	*Butane (l)*	*Iso-octane (l)*
Price/pence kg^{-1}	52.5	61.0	57.6
$\Delta H^{\ominus}_{c}/kJ\,mol^{-1}$	2220	2880	5510

4. You want to boil a kettle. The kettle contains 2.00 kg of water at 20 °C. What mass of natural gas (methane) must be burned to raise this quantity of water to 100 °C? (Assume no heat is lost.) (Specific heat capacity of water = $4.18\,J\,g^{-1}\,K^{-1}$; $\Delta H^{\ominus}_{c}(CH_4) = -890\,kJ\,mol^{-1}$)

10.4 STANDARD ENTHALPY CHANGE FOR A CHEMICAL REACTION

The standard enthalpy change for a chemical reaction can be calculated from the standard enthalpies of formation of all the products and reactants involved. For example, in the reaction of hydrogen chloride with ammonia

Finding $\Delta H^{\ominus}$ of reaction

$$NH_3(g) + HCl(g) \rightarrow NH_4Cl(s)$$
$$(-46) \quad (-92.3) \quad (-315)$$

Written underneath each species is its standard enthalpy of formation in $kJ\,mol^{-1}$. The standard enthalpy of reaction $\Delta H^{\ominus}$ is given by

$$\Delta H^{\ominus} = (-315) - (-46 + (-92.3))$$
$$= -177\,kJ\,mol^{-1}$$

The negative sign means that the reaction is exothermic.

The standard enthalpy of reaction depends only on the difference between the standard enthalpy of the reactants and the standard enthalpy of the products and not on the

Hess's Law

route by which the reaction occurs. This idea is embodied in **Hess's Law**, which states that, if a reaction can take place by more than one route, the overall change in enthalpy is the same, whichever route is followed.

Hess's Law is a special case of the First Law of Thermodynamics. In the set of reactions shown in the **enthalpy diagram** in Figure 10.5, the standard enthalpy change in going from **A** to **B** by the direct route 1 is $\Delta H_1^\ominus$. The standard enthalpy change in going from **A** to **B** by the indirect route 2 is $\Delta H_2^\ominus + \Delta H_3^\ominus$.

FIGURE 10.5 An Enthalpy Diagram

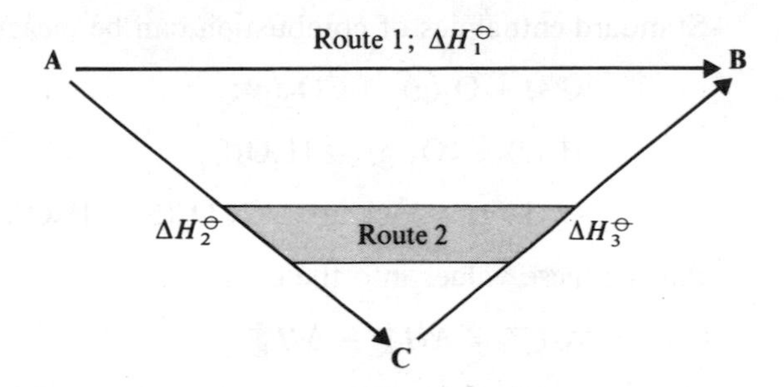

If the sum $(\Delta H_2^\ominus + \Delta H_3^\ominus)$ were less than $\Delta H_1^\ominus$, it would be possible to create energy by making **A** from **B** by route 1 and then converting **A** into **B** by route 2. This would be contrary to the First Law of Thermodynamics. It follows that

$$\Delta H_1^\ominus = \Delta H_2^\ominus + \Delta H_3^\ominus$$

and the standard enthalpy change is the same for the different reaction routes. It follows also that the standard enthalpy change for the reaction $\mathbf{B} \rightarrow \mathbf{A}$ is $-\Delta H_1^\ominus$.

10.4.1 FINDING THE STANDARD ENTHALPY OF FORMATION OF A COMPOUND INDIRECTLY

$\Delta H_F^\ominus$ of a compound may be found directly by experiment or indirectly, by applying Hess's Law...

Sometimes, the standard enthalpy of formation of a compound can be measured directly by allowing known amounts of elements to combine and measuring the heat evolved. Other reactions are difficult to study, and the standard enthalpy of reaction must be found indirectly.

To find the standard enthalpy of formation of ethyne from practical measurements is impossible as attempts to make ethyne from carbon and hydrogen

$$2C(s) + H_2(g) \rightarrow C_2H_2(g)$$

...that is, by using an enthalpy diagram

will result in the formation of a mixture of hydrocarbons. This is where Hess's Law comes to the rescue. In Figure 10.6, it follows from Hess's Law that the standard enthalpy change via route A is the same as the standard enthalpy change via route B.

FIGURE 10.6 Enthalpy Diagram for Ethyne

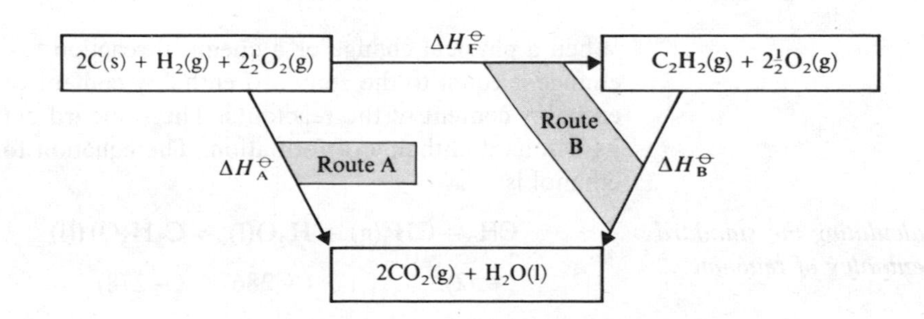

The change in standard enthalpy when carbon and hydrogen burn to form carbon dioxide and water is the same as the sum of the standard enthalpy changes when carbon and hydrogen combine to form ethyne and then ethyne burns to form carbon dioxide and water. Thus

$$\Delta H_A^\ominus = \Delta H_F^\ominus + \Delta H_B^\ominus$$

$$\Delta H_A^\ominus = 2(\Delta H^\ominus \text{ of combustion of C}) + (\Delta H^\ominus \text{ of combustion of } H_2)$$

$$\Delta H_B^\ominus = \Delta H^\ominus \text{ of combustion of } C_2H_2$$

Standard enthalpies of combustion can be measured accurately. They are

$$C(s) + O_2(g) \rightarrow CO_2(g); \quad \Delta H_1^\ominus = -394\,\text{kJ mol}^{-1} \quad [a]$$

$$H_2(g) + \tfrac{1}{2}O_2(g) \rightarrow H_2O(l); \quad \Delta H_2^\ominus = -286\,\text{kJ mol}^{-1} \quad [b]$$

$$C_2H_2(g) + 2\tfrac{1}{2}O_2(g) \rightarrow 2CO_2(g) + H_2O(l); \quad \Delta H_3^\ominus = -1300\,\text{kJ mol}^{-1} \quad [c]$$

Putting these values into the equation

$$\Delta H_F^\ominus = \Delta H_A^\ominus - \Delta H_B^\ominus$$

gives $\Delta H_F^\ominus = 2(-394) + (-286) - (-1300) = +226\,\text{kJ mol}^{-1}$

A value of $226\,\text{kJ mol}^{-1}$ is obtained for the standard enthalpy of formation of ethyne. Ethyne is described as an **endothermic compound** since $\Delta H_F^\ominus$ is positive.

A simple method of calculating $\Delta H_F^\ominus$

Alternative Method of Calculation There is a simpler method of calculation. It involves three steps.

1. Write the equation for the combustion of ethyne, since this is the reaction for which $\Delta H^\ominus$ can be measured.

2. Under each species, write $\Delta H_F^\ominus$. You can see from equations [a] and [b] that the standard enthalpies of formation of carbon dioxide and water are -394 and $-286\,\text{kJ mol}^{-1}$ respectively.

$$C_2H_2(g) + 2\tfrac{1}{2}O_2(g) \rightarrow 2CO_2(g) + H_2O(l); \quad \Delta H_3^\ominus = -1300\,\text{kJ mol}^{-1}$$

$$\Delta H_F^\ominus(C_2H_2) \quad 0 \quad 2(-394) \quad (-286)$$

3. Since

$$\Delta H^\ominus = \text{Total } H^\ominus \text{ of products} - \text{Total } H^\ominus \text{ of reactants}$$

$$\Delta H_3^\ominus = -1300 = 2(-394) + (-286) - \Delta H_F^\ominus(C_2H_2)$$

$$\Delta H_F^\ominus(C_2H_2) = +226\,\text{kJ mol}^{-1}$$

The standard enthalpy of formation of ethyne is $226\,\text{kJ mol}^{-1}$, as before.

10.4.2 FINDING THE STANDARD ENTHALPY OF REACTION

When a physical change or a chemical reaction takes place, the standard enthalpy change is equal to the standard enthalpy content of the products minus the standard enthalpy content of the reactants. The standard enthalpy content of a substance is its standard enthalpy of formation. The equation for the hydration of ethene to form ethanol is

Calculating the standard enthalpy of reaction...

$$CH_2{=}CH_2(g) + H_2O(l) \rightarrow C_2H_5OH(l)$$

$$(+52) \quad (-286) \quad (-278)$$

Written beneath each species is the value of $\Delta H_F^\ominus$ in kJ mol^{-1}. The **standard enthalpy of reaction** is given by

$$\Delta H_R^\ominus = -278 - (52 - 286)$$
$$= -44\,\text{kJ mol}^{-1}.$$

The reaction is exothermic by 44 kJ mol^{-1}. In short

...The method

$$\Delta H^\ominus_{\text{Reaction}} = \sum \Delta H^\ominus_{\text{Formation}} \text{ of products} - \sum \Delta H^\ominus_{\text{Formation}} \text{ of reactants}$$

Alternative Method of Calculation A different method of finding the standard enthalpy of reaction, $\Delta H_R^\ominus$, is illustrated by the physical change from one allotrope of tin to the other:

$$\text{Sn(s, white)} \rightarrow \text{Sn(s, grey)};\ \Delta H_R^\ominus = ?$$

FIGURE 10.7 Enthalpy Diagram for Tin

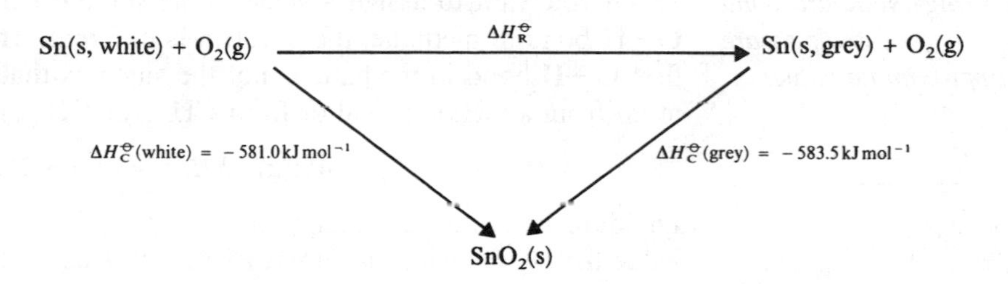

$\Delta H^\ominus_{Reaction}$ can also be found from an enthalpy diagram...

The best way of finding $\Delta H_R^\ominus$ for the change is to measure the standard enthalpy of combustion of each allotrope. These are shown in the enthalpy diagram, Figure 10.7.

From Hess's Law

$$\Delta H_C^\ominus(\text{white}) = \Delta H_R^\ominus + \Delta H_C^\ominus(\text{grey})$$
$$\Delta H_R^\ominus = -581.0 - (-583.5)$$
$$= +2.5\,\text{kJ mol}^{-1}$$

The method of calculation in this example is simply

...Alternative method

$$\Delta H^\ominus_{\text{Reaction}} = \sum \Delta H^\ominus_{\text{Combustion}} \text{ of reactants} - \sum \Delta H^\ominus_{\text{Combustion}} \text{ of products}$$

CHECKPOINT 10B: ENTHALPY CHANGES

1. State the sign of the standard enthalpy change in the following:
(*a*) the combustion of octane
(*b*) the condensation of steam
(*c*) the freezing of water
(*d*) the electrolysis of water
(*e*) the combustion of sodium in chlorine.

2. Why is it necessary to compare the enthalpy content of substances under standard conditions?

3. Arrange the following in order of increasing standard enthalpy:
(*a*) 1 mol H_2(g) + $\frac{1}{2}$mol O_2(g) at 25 °C and 1 atm
(*b*) 1 mol H_2O(g) at 100 °C and 1 atm
(*c*) 1 mol H_2O(l) at 25 °C and 1 atm
(*d*) 1 mol H_2O(s) at 0 °C and 1 atm
(*e*) 1 mol H_2O(l) at 0 °C and 1 atm
(*f*) 1 mol H_2O(l) at 100 °C and 1 atm

10.5 STANDARD BOND DISSOCIATION ENTHALPY

Standard bond dissociation enthalpies can be measured accurately

The **standard bond dissociation enthalpy** is the energy that must be absorbed to separate the two atoms in a bond. When hydrogen chloride dissociates

$$HCl(g) \rightarrow H(g) + Cl(g); \quad \Delta H^{\ominus} = 429.7\,kJ\,mol^{-1}$$

The standard bond dissociation enthalpy of the H—Cl bond in HCl is $429.7\,kJ\,mol^{-1}$. The value is accurately known as it is found by precise spectroscopic measurements.

10.6 AVERAGE STANDARD BOND ENTHALPY

Average standard bond enthalpies are approximate values...

When you want to assign a value to the standard enthalpy of dissociation of the C—H bond in methane, the problem is different. The energy required to break the first C—H bond in methane is not the same as that required to remove a hydrogen atom from a $CH_3\cdot$ radical or from $CH_2\colon$ or $CH\vdots$. In the complete dissociation

$$CH_4(g) \rightarrow C(g) + 4H(g); \Delta H^{\ominus} = +1662\,kJ\,mol^{-1}$$

Dividing the standard enthalpy change equally between the four bonds gives an average value for the C—H bond of $416\,kJ\,mol^{-1}$. This value is called the **average standard bond enthalpy** of the C—H bond.

... which give approximate values for $\Delta H^{\ominus}_{Reaction}$...

Tables of average standard bond enthalpies make the assumption that the standard enthalpy of a bond is the same in different molecules. This is only roughly true. Since standard bond enthalpies vary from one compound to another, the use of average standard bond enthalpies gives only approximate values for standard enthalpies of reaction calculated from them. Experimental methods are used to obtain standard enthalpies of reaction whenever possible. Calculations based on average standard bond enthalpies are used only for reactions which have not been studied experimentally, for example, the reactions of a substance which has not been isolated in a pure state.

... and are often called bond energy terms

Average standard bond enthalpy is often called the **bond energy term**. One can say that the bond energy term for the C—H bond is $416\,kJ\,mol^{-1}$. The sum of all the bond energy terms for a compound is the standard enthalpy absorbed in atomising that compound in the *gaseous state*. The standard enthalpy of formation of a hydrocarbon includes the sum of the bond energy terms and also the standard enthalpy of atomisation of the carbon atoms and the standard enthalpy of atomisation of the hydrogen atoms.

Method of calculation of $\Delta H^{\ominus}_F$ from bond energy terms

Example Calculate the standard enthalpy of formation of ethane, using the following information. The average standard bond enthalpies are: C—C = 348, C—H = $416\,kJ\,mol^{-1}$. The standard enthalpies of atomisation are: C(s) = $718\,kJ\,mol^{-1}$, and $\frac{1}{2}H_2(g)$ = 218 kJ per mole of H atoms formed.

Method In the C_2H_6 molecule are

one C—C bond of standard enthalpy = $348\,kJ\,mol^{-1}$

six C—H bonds of standard enthalpy = $2496\,kJ\,mol^{-1}$

The total standard enthalpy of the bonds = $2844\,kJ\,mol^{-1}$

This is the energy given out when the atoms combine:

$$2C(g) + 6H(g) \rightarrow C_2H_6(g);\ \Delta H^{\ominus} = -2844\,kJ\,mol^{-1}$$

The standard enthalpy of formation of C(g) from C(s) is $718\,kJ\,mol^{-1}$.
The standard enthalpy of formation of H(g) from $\frac{1}{2}H_2(g)$ is $218\,kJ\,mol^{-1}$.
These values of the standard enthalpy content of each species can be put into the equation:

$$2C(g) + 6H(g) \rightarrow C_2H_6(g);\ \ \Delta H^{\ominus} = -2844\,kJ\,mol^{-1}$$

$$2(718)\quad 6(218)\quad \Delta H^{\ominus}_F$$

$$\Delta H^{\ominus} = \Delta H^{\ominus}_F(\text{product}) - \Delta H^{\ominus}_F(\text{reactants})$$

$$-2844 = \Delta H^{\ominus}_F - 2(718) - 6(218)$$

$$\Delta H^{\ominus}_F = -100\,kJ\,mol^{-1}$$

The standard enthalpy of formation of ethane is $-100\,kJ\,mol^{-1}$.

10.6.1 STANDARD ENTHALPY OF REACTION FROM AVERAGE STANDARD BOND ENTHALPIES

When the standard enthalpy change for a reaction cannot be measured, an approximate value can be obtained by using average standard bond enthalpies. During a reaction, energy must be supplied to break bonds in the reactants, and energy is given out when the bonds in the products form. The standard enthalpy of reaction is the difference between the sum of the average standard bond enthalpies of the products and the sum of the average standard bond enthalpies of the reactants.

Method of calculation of $\Delta H^{\ominus}_R$ from bond energy terms

Example Calculate the standard enthalpy change of the reaction

$$(CH_3)_2C{=}O(g) + HCN(g) \rightarrow (CH_3)_2C\begin{matrix} / OH \\ \backslash CN \end{matrix}\ (g)$$

Mean standard bond enthalpies/$kJ\,mol^{-1}$ are C=O, 743; C—H, 412; C—O 360; C—C, 348; H—O, 463.

Method

Bonds broken are one C=O of $\Delta H^{\ominus} = 743\,kJ\,mol^{-1}$
one C—H of $\Delta H^{\ominus} = 412\,kJ\,mol^{-1}$

Total standard enthalpy absorbed $= 1155\,kJ\,mol^{-1}$

Bonds created are one C—O of $\Delta H^{\ominus} = 360\,kJ\,mol^{-1}$
one O—H of $\Delta H^{\ominus} = 463\,kJ\,mol^{-1}$
one C—C of $\Delta H^{\ominus} = 348\,kJ\,mol^{-1}$

Total standard enthalpy absorbed $= -1171\,kJ\,mol^{-1}$

Standard enthalpy change of reaction $= -1171 + 1155$
$= -16\,kJ\,mol^{-1}$

10.7 *THE BORN–HABER CYCLE

The Born–Haber cycle is a technique for applying Hess's Law to the standard enthalpy changes which occur when an ionic compound is formed. Think of the reaction between sodium and chlorine to form sodium chloride. The reaction can be considered to occur by means of the following steps, even though the reaction itself may not follow this route.

The formation of an ionic compound can be treated as a sequence of separate steps

Vaporisation of sodium
$Na(s) \rightarrow Na(g)$; $\Delta H_S^\ominus$ = standard enthalpy of sublimation or vaporisation of sodium [1]

Ionisation of sodium
$Na(g) \rightarrow Na^+(g) + e^-$; $\Delta H_I^\ominus$ = standard ionisation enthalpy of sodium [2]

Dissociation of chlorine molecules
$\frac{1}{2}Cl_2(g) \rightarrow Cl(g)$; $\frac{1}{2}\Delta H_D^\ominus$ = $\frac{1}{2}$ standard bond dissociation enthalpy of chlorine molecules [3]

Ionisation of chlorine atoms
$Cl(g) + e^- \rightarrow Cl^-(g)$; $\Delta H_E^\ominus$ = electron affinity of chlorine [4]

Reaction between ions
$Na^+(g) + Cl^-(g) \rightarrow NaCl(s)$; $\Delta H_L^\ominus$ = standard lattice enthalpy [5]

Definitions of $\Delta H^\ominus$ terms involved in the steps of the Born–Haber cycle

The standard enthalpy changes can be defined as follows:

[1] The standard enthalpy of sublimation or vaporisation is the enthalpy absorbed when 1 mole of sodium atoms is vaporised.

[2] The standard enthalpy of ionisation of sodium is the enthalpy required to remove 1 mole of electrons from 1 mole of gaseous sodium atoms.

[3] The standard bond dissociation enthalpy of chlorine is the enthalpy required to dissociate 1 mole of chlorine molecules into atoms.

FIGURE 10.8
Born–Haber Cycle for Sodium Chloride

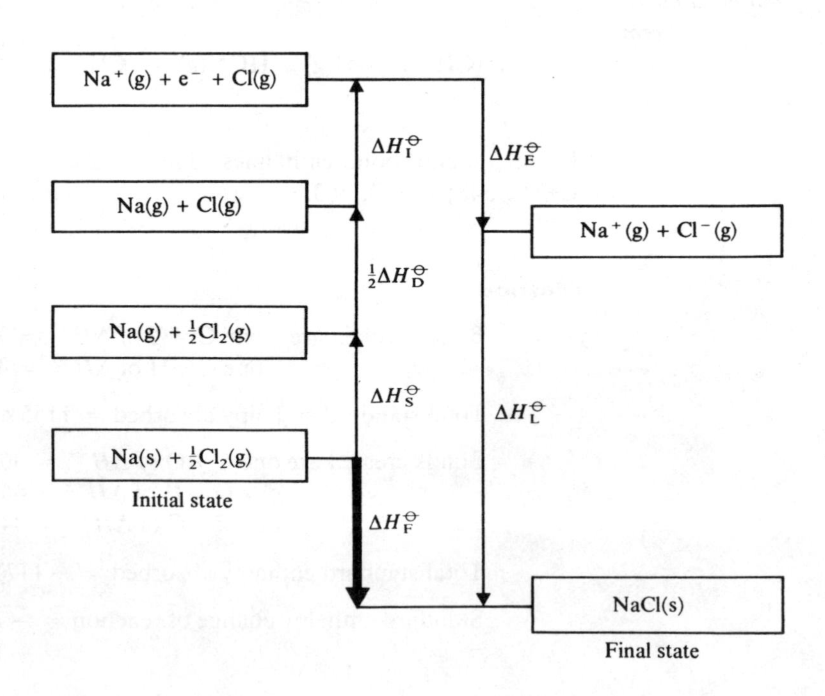

[4] The electron affinity of chlorine is the enthalpy absorbed when 1 mole of chlorine atoms accept 1 mole of electrons to become chloride ions. It has a negative value, showing that this reaction is exothermic.

[5] The standard lattice enthalpy is the enthalpy absorbed when 1 mole of sodium chloride is formed from its gaseous ions; its value is negative. (The standard lattice dissociation enthalpy is the enthalpy absorbed when 1 mole of sodium chloride is separated into its gaseous ions; it has a positive value.)

The standard enthalpy changes in Steps [1] to [5] are represented in Figure 10.8. The steps in a Born–Haber cycle are represented as going upwards if they absorb energy and downwards if they give out energy.

Application of Hess's Law to the Born–Haber cycle

Applying Hess's Law to this cycle, it follows that the sum of the standard enthalpy terms [1] to [5] is equal to the difference in standard enthalpy between the product, sodium chloride, and the reactants, solid sodium and gaseous chlorine molecules, that is the standard enthalpy of formation of sodium chloride. Inserting numerical values gives

$$\Delta H_F^\ominus = \Delta H_S^\ominus + \tfrac{1}{2}\Delta H_D^\ominus + \Delta H_I^\ominus + \Delta H_E^\ominus + \Delta H_L^\ominus$$

$$= +109 + 121 + 494 - 380 - 755 = -411\,\text{kJ mol}^{-1}$$

In practice, it is easier to measure standard enthalpies of formation than to measure some of the other steps. The electron affinity is the hardest term to measure experimentally: Born–Haber cycles are often used to calculate electron affinities.

The Born–Haber cycle for magnesium oxide is shown in Figure 10.9.

You will be able to identify $\Delta H_1^\ominus$, $\Delta H_2^\ominus$, $\Delta H_3^\ominus$ and $\Delta H_5^\ominus$ by comparison with the sodium chloride cycle. In the case of oxygen, $\Delta H_4^\ominus$ is an endothermic term. The reason is that, although energy is given out in the process

$$O(g) + e^- \rightarrow O^-(g)$$

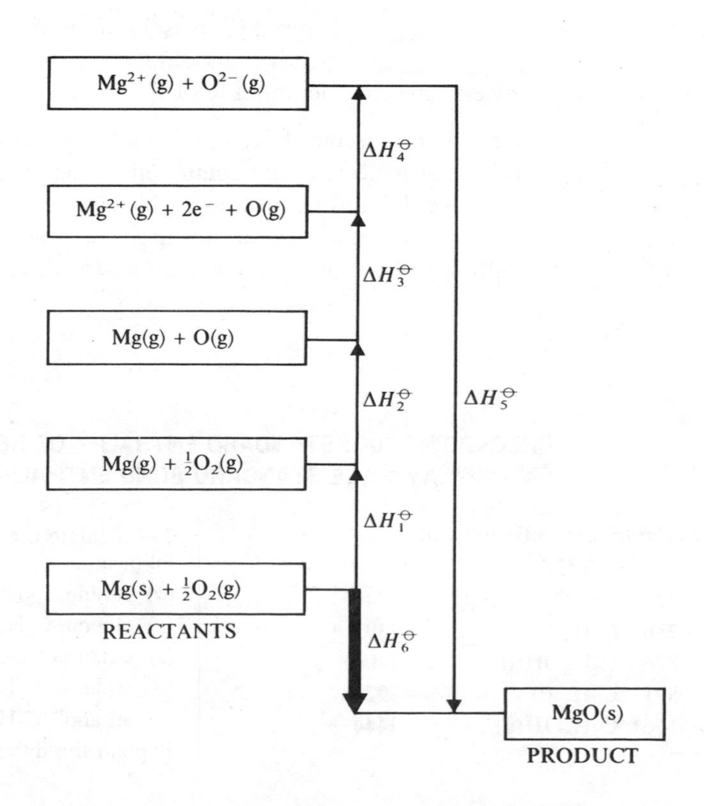

FIGURE 10.9 Born–Haber Cycle for Magnesium Oxide

If the lattice enthalpy is strongly exothermic, this may be the term that decides whether an ionic compound is formed

energy is required for the introduction of a second electron against the repulsion of O^-, in the process

$$O^-(g) + e^- \rightarrow O^{2-}(g)$$

making $\Delta H_4^\ominus$ endothermic. The only exothermic term is the lattice enthalpy. The reason why the formation of the ionic lattice is strongly exothermic is that both ions Mg^{2+} and O^{2-}, are small, and both carry two units of charge. When these opposite charges are brought close together, there is a big release of energy. This example shows the importance of lattice enthalpy in determining whether ionic compounds are formed [see Question 3, Checkpoint 10C].

10.8 ENTHALPY CHANGES INVOLVED WHEN IONIC COMPOUNDS DISSOLVE

When an ionic solid dissolves in a solvent, two enthalpy terms are involved.

When salts dissolve, energy is required to separate the ions...

1. The ions must be separated from the ionic lattice. The energy required is the lattice dissociation enthalpy.

...Solvation of the ions releases energy...

2. The separate ions interact with the molecules of solvent. If the solvent is polar, a charged ion can be attracted to one end of a polar solvent molecule. The energy released as these attractive forces come into play is compensation for the energy required to dissociate the lattice:

$$\Delta H^\ominus_{\text{Dissolution}} = \Delta H^\ominus_{\text{Lattice dissociation}} + \Delta H^\ominus_{\text{Solvation}}$$

(a positive quantity) (a negative quantity)

...On balance $\Delta H^\ominus$ of dissolution can be positive or negative

If $\Delta H^\ominus_{\text{Dissolution}}$ is negative, dissolution is favoured by enthalpy considerations. If $\Delta H^\ominus_{\text{Dissolution}}$ is positive, dissolution may still occur endothermically if it is favoured by entropy considerations [§ 10.9].

Highly negative enthalpies of solvation make water a good solvent

As water molecules have large dipoles, powerful interactions are possible between the polar molecules and solute ions. These powerful interactions result in a high negative enthalpy of solvation, which makes water a good solvent. Non-polar solvents, such as hydrocarbons, do not dissolve ionic substances because there is no negative enthalpy of solvation to compensate for the positive lattice dissociation enthalpy.

CHECKPOINT 10C: STANDARD ENTHALPY OF REACTION AND AVERAGE STANDARD BOND ENTHALPIES

The following are standard enthalpies of combustion/kJ mol^{-1} at 298 K:

C(graphite)	− 394	$CH_3CO_2C_2H_5(l)$	− 2246
$H_2(g)$	− 286	C_2H_4	− 1393
$CH_3CO_2H(l)$	− 876	$C_2H_5OH(l)$	− 1400
$CH_4(g)$	− 891	$C_6H_{12}(l)$	− 3924
$C_2H_6(g)$	− 1561	$C_2H_5OH(g)$	− 1444

1. Calculate the standard enthalpy of formation of the following:

(*a*) ethane, $C_2H_6(g)$
(*b*) ethene, $C_2H_4(g)$
(*c*) ethanoic acid, $CH_3CO_2H(l)$
(*d*) ethanol, $C_2H_5OH(l)$
(*e*) ethanol, $C_2H_5OH(g)$
Explain the difference between the values of (*d*) and (*e*).

2. Find the standard enthalpy of formation of ethylethanoate(l), ethanol(l), ethanoic acid(l), water(l). Calculate the standard enthalpy change in the reaction

$$CH_3CO_2H(l) + C_2H_5OH(l) \rightarrow CH_3CO_2C_2H_5(l) + H_2O(l)$$

***3.** Refer to Figure 10.9. To what do the symbols $\Delta H_1^\ominus$ to $\Delta H_5^\ominus$ refer? What is the significance of an arrow pointing upwards? Given that the values, in $kJ\,mol^{-1}$ are: $\Delta H_1^\ominus = +153$, $\Delta H_2^\ominus = +248$, $\Delta H_3^\ominus = +2180$, $\Delta H_4^\ominus = +745$, and $\Delta H_5^\ominus = -3930$, calculate $\Delta H_6^\ominus$, the standard enthalpy of formation of magnesium oxide.

4. The following values are $\Delta H_F^\ominus/kJ\,mol^{-1}$ at 298 K:

$CH_4(g)$ −76; $CO_2(g)$ −394; $H_2O(l)$ −286;
$H_2O(g)$ −242; $NH_3(g)$ −46.2; $C_2H_5OH(l)$ −278;
$C_8H_{18}(l)$ −210; $C_2H_6(g)$ −85

Calculate the standard enthalpy changes at 298 K for the reactions:

(*a*) $C_2H_6(g) + 3\frac{1}{2}O_2(g) \rightarrow 2CO_2(g) + 3H_2O(l)$
(*b*) $C_2H_5OH(l) + 3O_2(g) \rightarrow 2CO_2(g) + 3H_2O(l)$
(*c*) $H_2(g) + \frac{1}{2}O_2(g) \rightarrow H_2O(l)$
(*d*) $C_8H_{18}(l) + 12\frac{1}{2}O_2(g) \rightarrow 8CO_2(g) + 9H_2O(l)$

***5.** The Born–Haber cycle for rubidium chloride is shown in Figure 10.10. The letters **A** to **F** represent standard enthalpy changes. Give the names of these quantities. Their values in $kJ\,mol^{-1}$ are: **A** = −431, **B** = +86, **C** = +122, **D** = +408, **F** = −675. Calculate the value of **E**.

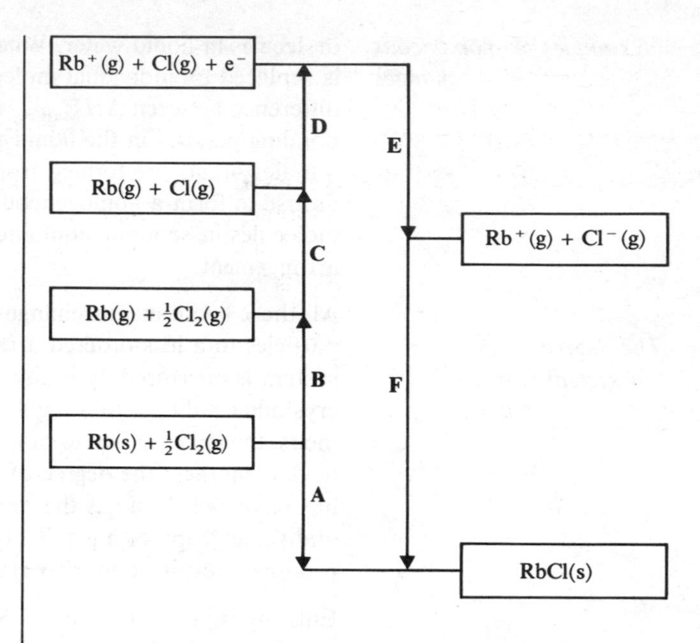

FIGURE 10.10
Born–Haber Cycle for Rubidium Chloride (not to scale)

10.9 *FREE ENERGY AND ENTROPY

Some reactions occur spontaneously...

A reaction which happens of its own accord, without any external help, is said to occur **spontaneously**. Some spontaneous reactions are endothermic. Examples are [1] the dissolution of potassium chloride in water, [2] the melting of ice, [3] the evaporation of water, [4] the dissociation of ammonium carbonate:

$$KCl(s) + aq \rightarrow KCl(aq);\ \Delta H^\ominus = 19\,kJ\,mol^{-1} \quad [1]$$

$$H_2O(s) \rightarrow H_2O(l);\ \Delta H^\ominus = 6.0\,kJ\,mol^{-1} \quad [2]$$

$$H_2O(l) \rightarrow H_2O(g);\ \Delta H^\ominus = 44\,kJ\,mol^{-1} \quad [3]$$

$$(NH_4)_2CO_3(s) \rightarrow 2NH_3(g) + CO_2(g) + H_2O(g);\ \Delta H^\ominus = 68\,kJ\,mol^{-1} \quad [4]$$

...Such reactions increase the 'disorder' of the system

The difference in enthalpy between the products and the reactants cannot be the only factor which decides whether a chemical reaction takes place. There must be an additional factor involved.

All these four physical or chemical changes have a factor in common. When potassium chloride dissolves, the regular arrangement of the crystal structure is replaced by a random distribution of ions in solution. When ice melts, the regular hydrogen-bonded structure of ice [see Figure 4.38, §4.7.3] is replaced by the fluid association of water

Examples of spontaneous changes

molecules in liquid water. When water evaporates, the association of water molecules is replaced by individual molecules moving independently in the gas phase. (The big difference between $\Delta H^{\ominus}_{\text{Melting}}$ and $\Delta H^{\ominus}_{\text{Evaporation}}$ shows that extensive hydrogen bonding persists in the liquid phase.) In the dissociation of ammonium carbonate, 4 moles of gas are formed from 1 mole of solid. When gases come into contact, they diffuse to form a homogeneous mixture: instead of a tidy arrangement of different molecules in separate containers, the molecules are now mixed up in a random arrangement.

The degree of 'disorder' of a system is measured by its entropy...

All these spontaneous changes involve a transition from an ordered arrangement of particles to a less ordered arrangement. The degree of disorder or randomness in a system is measured by a physical quantity termed the **entropy** of the system. A crystalline solid, with a regular arrangement of ions, has a low entropy. When it melts, the ions are free to move, and it is impossible to describe their positions relative to one another: the degree of order in the system has *decreased*, and the entropy has *increased*. In a gas the molecules move completely independently of one another, and the entropy of a gas is high. Physical and chemical changes in which gases are produced result in an increase in entropy.

...Estimating whether a change involves an increase or a decrease in entropy, S...

Entropy is given the symbol S, standard entropy, $S^{\ominus}$, and change in entropy, ΔS. An increase in the degree of disorder of a system shows in a positive value of ΔS. It is often possible to tell whether a reaction has a positive or negative value of ΔS by an inspection of the equation. The equation

$$NH_4NO_3(s) \rightarrow N_2O(g) + 2H_2O(g)$$

...and predicting the sign of $\Delta S^{\ominus}$

shows that 1 mole of the crystalline solid, ammonium nitrate, forms 1 mole of a gas, dinitrogen oxide, and 2 moles of steam, a total of 3 moles of gas. The value of ΔS is positive. Under standard conditions, the water formed is a liquid, and the increase in entropy is less: $\Delta S^{\ominus}$ has a smaller positive value. Other reactions for which it is easy to predict the sign of ΔS are

$$CaO(s) + H_2O(l) \rightarrow Ca(OH)_2(s);\ \Delta S \text{ negative}$$

$$CaCO_3(s) \rightarrow CaO(s) + CO_2(g);\ \Delta S \text{ positive}$$

Both factors, the change in enthalpy and the change in entropy, are important in deciding whether a physical or chemical change will occur. They are combined in the equation

The change in free energy, ΔG

Free energy, G = Enthalpy, H − (Temperature/K × Entropy, S)

$$G = H - TS$$

from which it follows that, at a constant temperature

$$\Delta G = \Delta H - T\Delta S$$

...must be negative if a change is to occur...

All spontaneous physical and chemical changes take place in the direction of a *decrease* in free energy. They may involve an increase or a decrease in enthalpy; they may involve an increase or a decrease in entropy; they all involve a decrease in free energy. The sign of ΔG for a spontaneous change must be negative. A change is therefore assisted by a decrease in enthalpy (ΔH negative) and by an increase in entropy (ΔS positive).

...If ΔG is negative, the reaction is feasible...

A reaction with a negative value of ΔG is said to be **feasible**. This means that, if the reaction takes place, it will go in the direction of the reactants forming the products and not in the reverse direction. To say that a reaction is feasible does not imply anything about the rate of the reaction. The reacting species may have to surmount an energy barrier, **the activation energy**, before reaction can occur [§ 14.9].

... The value of $\Delta G^{\ominus}$ can be used to predict the feasibility of a change

If the change takes place under standard conditions, i.e., with each reactant and product at unit concentration (or pressure), then the free energy change is equal to the standard free energy change, $\Delta G^{\ominus}$. When reaction takes place under non-standard conditions, the free energy change, ΔG, differs from $\Delta G^{\ominus}$ as ΔG depends on the concentrations (or pressures) of the reactants and products. It is easy to obtain $\Delta G^{\ominus}$ from tables of standard enthalpies of formation and standard entropies, but one really wants to know the value of ΔG for the real conditions, and this is not easy to compute. However, if $\Delta G^{\ominus}$ has a sufficiently large positive or negative value, $\Delta G^{\ominus}$ may determine the feasibility of reaction over a large range of concentrations (or pressures).

CHECKPOINT 10D: *ENTROPY

1. Give examples of spontaneous processes which are
(*a*) exothermic
(*b*) endothermic
(*c*) accompanied by an increase in entropy
(*d*) accompanied by a decrease in entropy.

2. State the sign of the entropy change in the following reactions:
(*a*) $NH_3(g) + HCl(g) \rightarrow NH_4Cl(s)$
(*b*) $COCl_2(g) \rightarrow CO(g) + Cl_2(g)$
(*c*) $PCl_3(g) + Cl_2(g) \rightarrow PCl_5(g)$
(*d*) $N_2(g) + 3H_2(g) \rightarrow 2NH_3(g)$
(*e*) $C_6H_{12}(l) + 9O_2(g) \rightarrow 6CO_2(g) + 6H_2O(g)$

3. Arrange the following in order of increasing entropy:
(*a*) 1 mol $H_2O(l)$ 100 °C, 1 atm
(*b*) 1 mol $H_2O(s)$ 0 °C, 1 atm
(*c*) 1 mol $H_2O(l)$ 0 °C, 1 atm
(*d*) 1 mol $H_2O(g)$ 100 °C, 1 atm
(*e*) 1 mol $H_2O(l)$ 25 °C, 1 atm
(*f*) 1 mol $H_2O(g)$ 100 °C, $\frac{1}{2}$ atm

4. When sodium hydroxide is dissolved in water, the temperature of the solution rises. When ammonium nitrate dissolves in water, the temperature of the solution falls. Explain the difference in behaviour.

5. Why is the entropy of a solid less than that of the gaseous form of the same substance?

***6.** Do you agree with the statement: 'In a spontaneous chemical process, the system goes to a state of lower energy'? If not, explain why and formulate a correct statement.

7. Is the change in standard entropy for each of the following processes positive or negative?
(*a*) 1 mol solid ethanoic acid → 1 mol liquid ethanoic acid
(*b*) 1 mol liquid ethanol → 1 mol gaseous ethanol
(*c*) 1 mol $N_2O_4(g) \rightarrow$ 2 mol $NO_2(g)$
(*d*) 1 mol $O_2(g)$ + 1 mol $N_2(g) \rightarrow$ 2 mol $NO(g)$
(*e*) $\frac{1}{2}$ mol $O_2(g)$ + 1 mol $Cu(s) \rightarrow$ 1 mol $CuO(s)$

10.9.1 *ELLINGHAM DIAGRAMS

Ellingham plotted $\Delta G^{\ominus}$ against T

In 1944, T Ellingham published plots of $\Delta G^{\ominus}$ against T for a number of reactions. Figure 10.11 shows an Ellingham diagram for some oxidation reactions. Values of $\Delta G^{\ominus}$ are compared for reactions written with 1 mole of oxygen on the left-hand side.

For the oxidation of the element, E

$$2E(s) + O_2(g) \rightarrow 2EO(s)$$

Reactions of this type involve a decrease in entropy as 1 mole of gas is converted into a solid. It follows from § 10.9 that, if $\Delta S^{\ominus}$ is negative, $\Delta G^{\ominus}$ increases (becomes less negative) with temperature.

FIGURE 10.11 An Ellingham Diagram for the Formation of Some Oxides (Values of $\Delta G^{\ominus}$ relate to the reaction of 1 mole of O_2)

The diagram shows why Ag_2O decomposes when heated, but MgO is stable...

...A glance at the diagram also shows that $\Delta G^{\ominus}$ for the reduction of Cr_2O_3 by Al is always negative

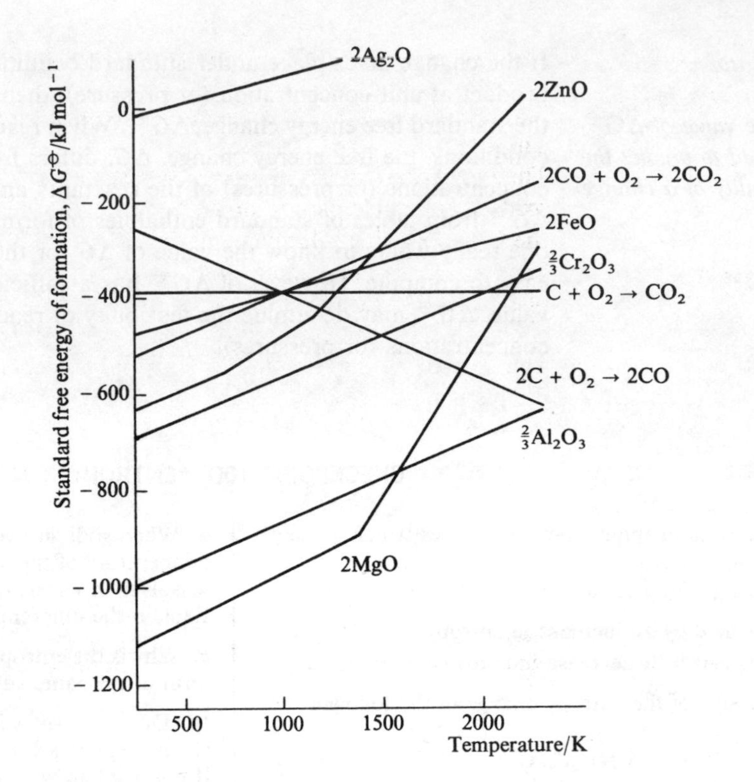

For the formation of silver oxide, $\Delta G^{\ominus}$ is negative up to 440 K. Above this temperature, $\Delta G^{\ominus}$ is positive, and the reverse reaction occurs: silver oxide decomposes into silver and oxygen. For the oxidation of magnesium, $\Delta G^{\ominus}$ is negative over the whole range of temperature shown: magnesium oxide does not decompose when heated to 2500 K. The break in the MgO line occurs at the boiling temperature of magnesium.

The reaction

$$2Mg(g) + O_2(g) \rightarrow 2MgO(s)$$

involves a bigger decrease in entropy than the reaction

$$2Mg(s) + O_2(g) \rightarrow 2MgO(s)$$

Above the boiling temperature of magnesium, therefore, ΔS is more negative.

Since $\Delta G^{\ominus} = \underset{\text{Negative}}{\Delta H^{\ominus}} - \underset{\text{Negative}}{T\Delta S^{\ominus}}$

when $\Delta S^{\ominus}$ is more negative, the term $-T\Delta S^{\ominus}$ is more positive, and $\Delta G^{\ominus}$ is less negative.

Compare the aluminium oxide line and the chromium oxide line. At all temperatures, the line for the oxidation of aluminium lies below that for chromium. This shows that the value of $\Delta G^{\ominus}$ for the reaction

$$2Al(s) + Cr_2O_3(s) \rightarrow 2Cr(s) + Al_2O_3(s)$$

is negative over the complete temperature range shown. Aluminium will reduce chromium oxide at any temperature in the range shown [§ 19.2.3]. In general, the height of a line on the diagram is a measure of the instability of the oxide.

Ellingham diagrams are useful in explaining the reduction of some metal oxides by carbon and carbon monoxide. In the reaction

The conditions employed in the blast furnace allow CO to be formed and to reduce iron oxides

$$2C(s) + O_2(g) \rightarrow 2CO(g)$$

$\Delta S^\ominus$ is positive, and $\Delta G^\ominus$ therefore decreases (becomes more negative) as T increases. The $CO \rightarrow CO_2$ line lies below the $Fe \rightarrow FeO$ line at temperatures below 1000 K. The temperature must be *below* 1000 K to allow carbon monoxide to reduce iron oxides. To enable carbon monoxide to be formed, the $C \rightarrow CO$ line must lie below the $C \rightarrow CO_2$ line. This happens *above* 1000 K. In the blast furnace [see Figure 24.18, §24.14.1] the combustion of coke at the bottom of the furnace produces a temperature of 2000 K, and carbon monoxide is formed. As it rises up the furnace, carbon monoxide cools until, in the centre of the furnace, a temperature of 1000 K allows it to reduce iron oxides. At the bottom of the furnace, carbon is able to act as a reductant because above 1000 K the lines $C \rightarrow CO$ and $C \rightarrow CO_2$ both lie below the $Fe \rightarrow FeO$ line.

CHECKPOINT 10E: *FREE ENERGY

1. Explain the difference between a spontaneous reaction and a feasible reaction.

2. Magnesium is mined as magnesium carbonate. Explain what is wrong with the following suggested method for extracting magnesium:

$$MgCO_3(s) \xrightarrow{\text{heat}} MgO(s) + CO_2(g)$$

$$MgO(s) + CO(g) \xrightarrow{\text{heat}} Mg(s) + CO_2(g)$$

3. Refer to Figure 10.11.

(*a*) Explain why specks of carbon appear when magnesium is burnt in a jar of carbon dioxide.

(*b*) Explain the change in gradient in the MgO graph at 1400 K.

(*c*) Give the temperature range over which magnesium oxide can be reduced by coke.

(*d*) When zinc nitrate is heated, the solid product is zinc oxide. When silver nitrate is heated, the solid product is silver.
Explain the reason for this difference is behaviour.

(*e*) Name (i) an oxide which you predict will not be reduced by aluminium,
(ii) an oxide (other than chromium(III) oxide) which you predict will be reduced by aluminium.

10.10 LINK WITH KINETICS

10.10.1 WHICH DECIDES WHETHER REACTIONS HAPPEN, THERMODYNAMICS OR KINETICS?

Thermodynamic studies tell us $\Delta G^\ominus$ and therefore whether a change is feasible...

If $\Delta G^\ominus$ for a change is negative, we can state that the reactants are thermodynamically unstable relative to that change. Methane is thermodynamically stable with respect to dissociation into its elements:

$$CH_4(g) \rightarrow C(g) + 2H_2(g);\ \Delta G^\ominus = +51\,\text{kJ mol}^{-1}$$

Methane is thermodynamically unstable relative to oxidation:

$$CH_4(g) + 2O_2(g) \rightarrow CO_2(g) + 2H_2O(l);\ \Delta G^\ominus = -580\,\text{kJ mol}^{-1}$$

...Kinetic studies tell us the energy of activation and therefore how fast a change will happen

Thermodynamic instability ($\Delta G^\ominus$ is negative) relative to a certain change is a necessary condition for that change to occur, but it is not the only necessary condition. Methane does not react with air at a measurable rate at room temperature (in the absence of UV light). The reason is that a large amount of energy, called the **activation energy**, must be supplied to enable the molecules to react. The concept of activation energy is covered in Chapter 14. Briefly, the idea is that, although the product molecules have a lower free energy content than the reactant molecules, there is an energy barrier in between the reactant and the product molecules. The reactant molecules must be given enough energy to surmount this barrier in order to react. Very, very few molecules of reactant acquire this activation energy at room temperature. In consequence, the overall rate of reaction is so slow as to be unobservable. The reaction is described as being **under kinetic control**. If the temperature is raised sufficiently, the proportion of energetic molecules increases, and the reaction takes place at a measurable rate. Substances which react slowly because the activation energy of the reaction is high and the temperature is too low are described as **unreactive** or **non-labile** or **inert**.

QUESTIONS ON CHAPTER 10

1. (*a*) Describe how you would find the standard enthalpy of neutralisation of a strong acid by a strong base.

†(*b*) When you have studied § 12.7, explain why the same value is obtained for the neutralisation by sodium hydroxide of nitric, hydrochloric and sulphuric acids, but a different value is obtained with ethanoic acid [§ 12.7.4].

2. (*a*) State the First Law of Thermodynamics.

(*b*) The heat change in a reaction can be expressed as ΔH or ΔU. What is the distinction between these quantities? Write an equation which relates the two quantities.

(*c*) Which of the two, ΔH or ΔU, is the more useful in the study of chemical reactions? Explain your answer.

***3.** Construct a Born–Haber cycle for the formation of solid potassium chloride from its elements in their standard states. Use the data below, and calculate the standard enthalpy of formation of KCl(s).

$K(s) \rightarrow K(g)$; $\Delta H^{\ominus} = 90\,kJ\,mol^{-1}$

$K(g) \rightarrow K^+(g) + e^-$; $\Delta H^{\ominus} = 418\,kJ\,mol^{-1}$

$\frac{1}{2}Cl_2(g) \rightarrow Cl(g)$; $\Delta H^{\ominus} = 122\,kJ\,mol^{-1}$

$Cl(g) + e^- \rightarrow Cl^-(g)$; $\Delta H^{\ominus} = -348\,kJ\,mol^{-1}$

$Cl^-(g) + K^+(g) \rightarrow KCl(s)$; $\Delta H^{\ominus} = -718\,kJ\,mol^{-1}$

4. Use the data below to find (*a*) the C—C bond enthalpy in ethane and (*b*) the value of $\Delta H^{\ominus}$ for $3C(g) + 8H(g) \rightarrow C_3H_8(g)$

$C(g) + 4H(g) \rightarrow CH_4(g)$; $\Delta H^{\ominus} = -1664\,kJ\,mol^{-1}$

$2C(g) + 6H(g) \rightarrow C_2H_6(g)$; $\Delta H^{\ominus} = -2827\,kJ\,mol^{-1}$

***5.** Construct a Born–Haber cycle, and use it to find the standard lattice enthalpy of cadmium(II) iodide.

$Cd(s) \rightarrow Cd(g)$; $\Delta H^{\ominus} = +113\,kJ\,mol^{-1}$

$Cd(g) \rightarrow Cd^{2+}(g) + 2e^-$; $\Delta H^{\ominus} = +2490\,kJ\,mol^{-1}$

$I_2(s) \rightarrow I_2(g)$; $\Delta H^{\ominus} = +19.4\,kJ\,mol^{-1}$

$I_2(g) \rightarrow 2I(g)$; $\Delta H^{\ominus} = +151\,kJ\,mol^{-1}$

$I(g) + e^- \rightarrow I^-(g)$; $\Delta H^{\ominus} = -314\,kJ\,mol^{-1}$

$Cd(s) + I_2(s) \rightarrow CdI_2(s)$; $\Delta H^{\ominus} = -201\,kJ\,mol^{-1}$

6. (*a*) State Hess's Law.

(*b*) Find the standard enthalpy change for the reaction

$CO(g) + 2H_2(g) \rightarrow CH_3OH(l)$

Use the data

$CO(g) + \frac{1}{2}O_2(g) \rightarrow CO_2(g)$; $\Delta H^{\ominus} = -283\,kJ\,mol^{-1}$

$H_2(g) + \frac{1}{2}O_2(g) \rightarrow H_2O(l)$; $\Delta H^{\ominus} = -286\,kJ\,mol^{-1}$

$CH_3OH(l) + 1\frac{1}{2}O_2(g) \rightarrow CO_2(g) + 2H_2O(l)$; $\Delta H^{\ominus} = -715\,kJ\,mol^{-1}$

(*c*) Explain how Hess's Law underwrites the method of calculation.

7. For benzene and cyclohexane, the standard enthalpies of combustion/kJ mol^{-1} are $C_6H_6(l)$ − 3280; $C_6H_{12}(l)$ − 3920. For C(s) and $H_2(g)$ the values of $\Delta H^{\ominus}_C$ are − 393 and − 286 kJ mol^{-1} respectively. (*a*) Calculate the standard enthalpies of formation of benzene and cyclohexane.
(*b*) Comment on the relative stability of these compounds.

8. Find the standard enthalpy of formation of carbon disulphide, CS_2. It burns in air to form CO_2 and SO_2. Standard enthalpies of combustion/kJ mol^{-1} are $CS_2(l)$ − 1075; S(s) − 297; C(s), − 394.

9. The standard enthalpies of hydrogenation of cyclohexene and benzene are − 120 and − 208 kJ mol^{-1} respectively.

(*a*) Explain why the value for benzene is not three times that for cyclohexene.

(*b*) Estimate the standard enthalpies of hydrogenation of cyclohexa-1,3-diene and cyclohexa-1,4-diene.

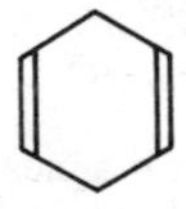
Cyclohexene

Cyclohexa-1,3-diene

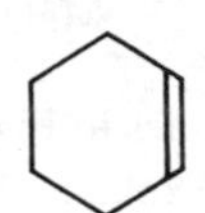
Cyclohexa-1,4-diene

10. A pellet of potassium hydroxide weighing 0.166 g is added to 50.0 g of water in a styrofoam cup. The temperature of the water rises from 19.4 to 20.2 °C. Find the standard enthalpy of solution of KOH.

11. The industrial process for the gasification of coal can be represented approximately by the equation

$$2C(s) + 2H_2O(g) \rightarrow CH_4(g) + CO_2(g)$$

From the values of $\Delta H^{\ominus}_{Combustion}$ listed in Checkpoint 10C find $\Delta H^{\ominus}$ for this reaction.

12. (*a*) Describe the dissolution of an ionic solid in water, discussing the energetics of the process.

(*b*) Suggest why, in general, $A^{2+}B^{2-}$ compounds are less soluble than A^+B^- compounds.

(*c*) Suggest why (i) LiF is less soluble than NaF,
(ii) LiCl is more soluble than NaCl,
(iii) NaF is less soluble than NaCl.

	LiCl	LiF	NaCl	NaF
Lattice enthalpies/kJ mol^{-1}	− 843	− 1029	− 775	− 968
Hydration enthalpies/kJ mol^{-1}	− 883	− 1023	− 778	− 965

13. (i) What do you understand by the term *entropy*?
(ii) For each of the following reactions, say, with reasons, whether the entropy is likely to increase or decrease or stay approximately the same:

(*a*) $N_2(g) + 3H_2(g) \rightarrow 2NH_3(g)$

(*b*) $SO_2(g) + Cl_2(g) \rightarrow SO_2Cl_2(g)$

(*c*) $H_2NCO_2NH_4(s) \rightarrow CO_2(g) + 2NH_3(g)$
(*d*) $CO(g) + H_2O(g) \rightarrow CO_2(g) + H_2B(g)$

14. (*a*) (i) Define *standard enthalpy change* of a reaction.
(ii) What is the value of the enthalpy of formation of an element in its standard state?
(*b*) Calculate the standard enthalpy change for the reaction

$$2HN_3 + 2NO \rightarrow H_2O_2 + 4N_2$$

given that standard enthalpies of formation are

for HN_3 $\Delta H_F^\ominus = +264.0\ kJ\ mol^{-1}$

for NO $\Delta H_F^\ominus = +90.3\ kJ\ mol^{-1}$

and for H_2O_2 $\Delta H_F^\ominus = -187.8\ kJ\ mol^{-1}$

Suggest why this reaction has been considered for use as a rocket fuel.

(O 91)

15. (*a*) Set out, in the form of a diagram, the complete Born–Haber cycle for the formation of sodium chloride from metallic sodium and gaseous chlorine. Clearly show all the species involved and all the energy/enthalpy changes.
(*b*) Use the following data to calculate the lattice energy of calcium oxide.

Standard molar enthalpy change	$kJ\ mol^{-1}$
Formation of calcium oxide	−635
Atomisation (sublimation) of calcium	+193
Sum of first two ionisation energies of calcium	+1740
Atomisation (dissociation) of oxygen, per mole of atoms formed	+250
Sum of first two electron affinities of oxygen	+702

(*c*) Although the standard molar enthalpy change of formation of magnesium oxide is almost the same as that of calcium oxide, its lattice energy is about 300 $kJ\ mol^{-1}$ greater in magnitude.
Which enthalpy changes in the Born–Haber cycle for magnesium oxide will differ significantly from those in the corresponding cycle for calcium oxide? Briefly account for any differences.
(*d*) The compound $CaCl_2$ is well known but neither $CaCl_3$ nor CaCl has yet been made as a stable compound. Calculations show that
(i) $CaCl_3$ would have a *large positive* standard molar enthalpy change of formation
(ii) CaCl would have a *small negative* standard molar enthalpy change of formation.
Account for each of (i) and (ii) in terms of enthalpy change of atomisation, ionisation energy, electron affinity, and lattice energy.

(AEB 92)

16. Sodium fluoride and caesium fluoride have different interionic distances, lattice energies and enthalpy changes of hydration.

Ionic radius/nm	Na^+ F^-	0.102 0.133	Cs^+ F^-	0.170 0.133
Lattice energy/$kJ\ mol^{-1}$	NaF	−918	CsF	−747
Enthalpy change of hydration/$kJ\ mol^{-1}$	Na^+ F^-	−390 −457	Cs^+ F^-	−248 −457

(*a*) In terms of the electronic structure of the elements concerned, comment on the relative sizes of the three ions.
(*b*) Comment on the differences in lattice energies and enthalpy changes of hydration.
(*c*) At room temperature, the solubility in water of NaF is 0.987 $mol\ dm^{-3}$ and that of CsF is 38.4 $mol\ dm^{-3}$. Comment on these values in terms of the relevant energy changes.
(*d*) At room temperature, the solubility in water of $NaClO_4$ is 17.0 $mol\ dm^{-3}$ and that of $CsClO_4$ is 0.086 $mol\ dm^{-3}$. Comment on the difference in the solubilities of these two compounds compared with that of the fluorides.
(*e*) Suggest, with a reason, which of $NaClO_4$ and $CsClO_4$ would be the more stable to heat.

(C 92, S)

17. (*a*) By using the following data, draw an appropriate energy cycle and calculate the enthalpy change of hydration of (i) the chloride ion, (ii) the iodide ion. Comment on the difference in their values.

Enthalpy change of solution of NaCl(s) = $-2\ kJ\ mol^{-1}$
Enthalpy change of solution of NaI(s) = $+2\ kJ\ mol^{-1}$
Enthalpy change of hydration of $Na^+(g)$ = $-390\ kJ\ mol^{-1}$
Lattice energy of sodium chloride = $-772\ kJ\ mol^{-1}$
Lattice energy of sodium iodide = $-699\ kJ\ mol^{-1}$

(*b*) The taste of solutions of alkali metal halides depends on the sum of the ionic radii of the ions: sodium iodide and potassium chloride are salty; rubidium chloride is both salty and bitter; caesium chloride and rubidium bromide are bitter.
Use the following data to predict the tastes of sodium bromide and potassium iodide, showing your reasoning.

Ion	*Radius*/nm	*Ion*	*Radius*/nm
Na^+	0.098	Cl^-	0.181
K^+	0.133	Br^-	0.196
Rb^+	0.148	I^-	0.219
Cs^+	0.167		

(C 90)

18. (*a*) The combustion of some fuels produces large amounts of carbon dioxide, which may modify the Earth's climate (the 'greenhouse effect').

Substance	$CH_4(g)$	$CO_2(g)$	$H_2O(l)$	$C(s)$	$O_2(g)$
Standard enthalpy of formation ($\Delta H_F^\ominus$ (298 K)/$kJ\ mol^{-1}$)	−75.0	−394.0	−286.0	0	0

($A_r(C) = 12.0$, $A_r(H) = 1.0$)

(i) Use the data given to calculate the standard enthalpy change of combustion of coal (essentially carbon) *and* of natural gas (essentially methane).

(ii) Using these results, calculate which fuel produces more energy on combustion

(*1*) per gram of fuel

(*2*) per mole of carbon dioxide formed.

(iii) State which of these two fuels is likely to have the smaller effect on the climate per kilojoule of energy produced, and give a reason.

(*b*) (i) State the principle which may be used to predict the shapes of simple molecules.

(ii) Use this principle to predict, giving your reasons, the shapes of the methane, water, and carbon dioxide molecules.

(*c*) (i) Name the type of bond occurring within the molecules in (*b*)(ii) (CH_4, H_2O and CO_2).

(ii) State and explain the difference (*not* the difference in bond dissociation enthalpies) between the nature of the single bonds within the hydrides CH_4 and H_2O.

(*d*) Explain the fact that the boiling points of methane and water differ by over 250 °C, despite the fact that their relative molecular masses are nearly equal (16 and 18 respectively).

(WJEC 91)

19. (*a*) Describe the *Born–Haber cycle*, stating the principle or law on which it is based, and giving *one* example of its use.

(*b*) State and discuss the factors which govern the magnitude of

(i) the enthalpies of atomisation of elements, e.g. C, He, H and Pb

(ii) the lattice energy of an ionic crystal.

(*c*) In 1962 Bartlett prepared the ionic crystal $O_2^+PtF_6^-$ by reacting oxygen with PtF_6. He then reasoned that the noble gas compound $Xe^+PtF_6^-(s)$ might also be stable and proceeded to make it by reacting xenon (Xe) with platinum hexafluoride (PtF_6).

Using the data given and simplified cycles of the Born–Haber type, explain Bartlett's reasoning and also predict which other members of the noble gas family might form similar stable compounds.

[N.B. Assume for this purpose that O_2^+ and all the noble gas mono-cations have the same ionic radius.]

Substance	O_2	He	Ne	Ar	Kr	Xe	Rn
*First ionisation energy/*kJ mol^{-1}	1164	2378	2087	1527	1357	1177	1043

$$Xe^+(g) + PtF_6^-(g) \rightarrow Xe^+PtF_6^-(s);$$
$$\Delta H^\ominus(298\text{ K}) = -500\text{ kJ mol}^{-1};$$

$$PtF_6(g) + e^- \rightarrow PtF_6^-(g);$$
$$\Delta H^\ominus(298\text{ K}) = -770\text{ kJ mol}^{-1}$$

(WJEC 91, S)

20. The industrial preparation of the polymer, poly(tetrafluoroethene) or PTFE, is based on the synthesis of the monomer tetrafluoroethene, $CF_2{=}CF_2$, which is produced by thermal cracking of chlorodifluoromethane, $CHClF_2$, according to reaction [1] below.

$$2CHClF_2(g) \rightleftharpoons CF_2{=}CF_2(g) + 2HCl(g) \qquad [1]$$

Here the $CHClF_2$ is diluted by superheated steam, which also acts as the heat source. The monomer $CF_2{=}CF_2$ is also obtained via reaction [2].

$$2CHF_3(g) \rightleftharpoons CF_2{=}CF_2(g) + 2HF(g); \quad \Delta H^\ominus = +198.1\text{ kJ mol}^{-1} \qquad [2]$$

Consider this information, together with the data in the table below, and answer the following questions.

Compound	$\Delta H_F^\ominus$/ kJ mol^{-1}	*Molecule* X—X	D(X—X)/ kJ mol^{-1}
HCl(g)	−92.3	F—F(g)	154.7
$CHClF_2$(g)	−485.2	Cl—Cl(g)	246.7
$CF_2{=}CF_2$(g)	−658.3		
CF_4(g)	−679.6		
CCl_4(g)	−106.6		

(*a*) (i) Calculate the value of the enthalpy change, $\Delta H^\ominus$, for reaction [1]. State, giving your reasons, how you would expect the yield of the tetrafluoroethene monomer to be affected by: (*1*) increase of temperature and (*2*) increase of pressure. In the latter case explain how your conclusion is compatible with the experimental conditions described.

(ii) Indicate and explain whether there are any drawbacks to the use of reaction [2] which would make reaction [1] preferable.

(*b*) (i) Use the expressions

$$CX_4(g) \rightarrow C(s) + 2X_2(g); \quad \Delta H^\ominus = -\Delta H_F^\ominus$$
$$C(s) \rightarrow C(g); \quad \Delta H^\ominus = +718.0\text{ kJ mol}^{-1}$$
$$\text{and} \quad 2X_2(g) \rightarrow 4X(g); \quad \Delta H^\ominus = 2D(X{-}X)$$

where **X** = F, Cl, to calculate $\Delta H^\ominus$ for the *two* processes

$$CX_4(g) \rightarrow C(g) + 4X(g)$$

Hence find the average C—**X** bond energies for the species CX_4(g) (where **X** = F and **X** = Cl). Given that the average C—H bond energy is 416.1 kJ mol^{-1}, explain the implications of your results for the relative chemical reactivities of C—H, C—F and C—Cl bonds.

(ii) Chlorofluorocarbons (CFCs) are widely used as propellent gases for aerosols. In the upper atmosphere, photochemically induced homolytic fission of one of the carbon–halogen bonds of CFCs produces halogen radicals which then attack the ozone layer. Use your results from (*b*) (i) above to suggest *which* halogen is likely to be the dominant cause of such damage.

(WJEC 92)

21. (*a*) State the First Law of Thermodynamics and discuss the relationship between this law and Hess's Law.

(*b*) Describe how you could measure the molar enthalpy of combustion (ΔH_C) of ethanol by a simple laboratory experiment. Discuss the practical precautions which would be necessary to minimise experimental error. Explain how a value for ΔH_C could be calculated from the experimental results.

(*c*) Methanol can be produced from methane by a two-step process.

Step 1 $CH_4(g) + H_2O(g) \rightleftharpoons CO(g) + 3H_2(g)$

Step 2 $CO(g) + 2H_2(g) \rightleftharpoons CH_3OH(g)$

(i) Use the following enthalpies of combustion to calculate the enthalpy change, ΔH, for each of the two steps.

	$CH_4(g)$	$CO(g)$	$H_2(g)$	$CH_3OH(g)$
ΔH_C/kJ mol^{-1}	−808	−283	−245	−671

[*Note* Where water is a product of combustion the figures refer to the formation of $H_2O(g)$.]

(ii) Discuss how changes in temperature and pressure will affect the yield of products in each step.

(iii) Discuss *two* economic advantages of operating these two steps in reaction vessels close to each other in an industrial plant.

(NEAB 92)

22. (*a*) State Hess's Law, and relate it to the conservation of energy.

(*b*) Hydrogen chloride gas dissolves in water to form hydrochloric acid. Use the standard enthalpies of formation shown to calculate $\Delta H^{\ominus}$ for this reaction at 298 K.

Standard enthalpies of formation at 298 K/kJ mol^{-1}	
$HCl(g)$	−92.3
$H^+(aq)$	0
$Cl^-(aq)$	−167.2

(*c*) 25 cm^3 of 1.00 mol dm^{-3} hydrochloric acid, at 298 K, was added to 50 cm^3 of 1.00 mol dm^{-3} sodium hydroxide, also initially at 298 K, in a polystyrene cup calorimeter. The temperature of the mixture rose to 302.4 K. The specific heat capacity of water is 4.20 J g^{-1} K^{-1} and the density of water is 1.00 g cm^{-3}.

(i) Calculate a value for the standard enthalpy change for the reaction.

(ii) What assumptions have to be made in order to carry out this calculation?

(iii) Within experimental error, an identical value is obtained if the reactants are aqueous nitric acid and aqueous potassium hydroxide. Explain the significance of this.

(O 92, S)

11

CHEMICAL EQUILIBRIUM

An introduction to this topic has been made in Chapter 3.

11.1 REVERSIBLE REACTIONS

Some reactions take place in both directions...

Chemical reactions which take place in both directions are called reversible reactions. An example of a reversible reaction between gases is the reaction between hydrogen and iodine to form hydrogen iodide:

...e.g., the reaction between hydrogen and iodine

$$H_2(g) + I_2(g) \rightleftharpoons 2HI(g)$$

M Bodenstein made a detailed study of this reaction from 1890 to 1900. His method was to seal in glass bulbs either a mixture of hydrogen and iodine or pure hydrogen iodide, and place the bulbs in a thermostat bath. After a time interval, he cooled the bulbs rapidly so that chemical reaction stopped. Then he analysed the contents of each bulb to find the amounts of hydrogen, iodine and hydrogen iodide present.

FIGURE 11.1 Bodenstein's Results for a Set of Experiments at 448 °C

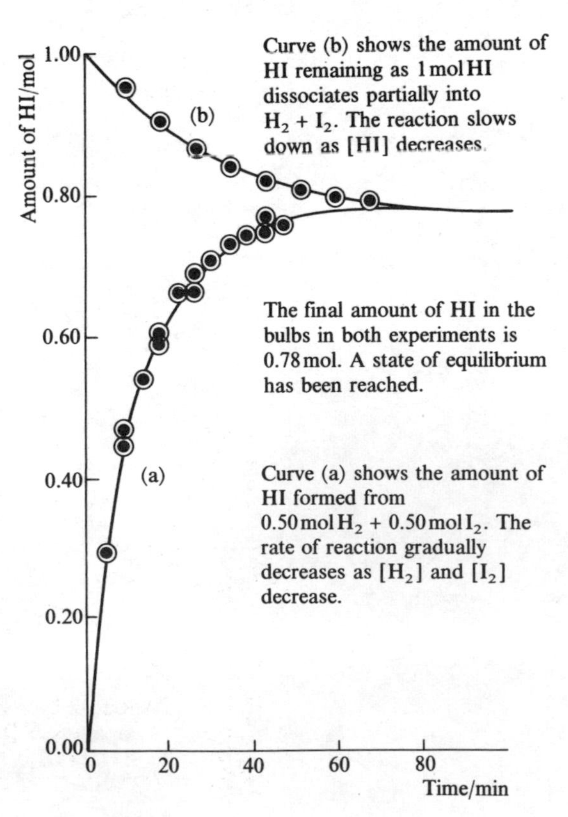

Amount of HI/mol
1.00
0.80
0.60
0.40
0.20
0.00
0
20
40
60
80
Time/min
(a)
(b)
Curve (b) shows the amount of HI remaining as 1 mol HI dissociates partially into $H_2 + I_2$. The reaction slows down as [HI] decreases.
The final amount of HI in the bulbs in both experiments is 0.78 mol. A state of equilibrium has been reached.
Curve (a) shows the amount of HI formed from 0.50 mol H_2 + 0.50 mol I_2. The rate of reaction gradually decreases as $[H_2]$ and $[I_2]$ decrease.

Figure 11.1 shows a set of results at a certain temperature and pressure. If the reaction between 0.50 mol H_2 and 0.50 mol I_2 went to completion, 1.00 mol HI would be formed. It appears that, with 0.78 mol HI formed, no further reaction takes place. The system has reached equilibrium. The same amount of HI is present in the bulbs when equilibrium is reached from the other direction, by the dissociation of hydrogen iodide.

A state of dynamic equilibrium is reached...

...when the forward and reverse reactions occur at the same rate

In an equilibrium mixture of hydrogen, iodine and hydrogen iodide, it is not obvious that chemical reactions are still occurring. In fact, both the forward and the reverse reactions are still taking place. Since the rates of the forward and reverse reactions are equal, the concentration of each species remains constant. The system is said to be in **dynamic equilibrium**. It is possible to prove this by injecting iodine containing a small quantity of radioactive iodine-131 into an equilibrium mixture. Radioactive iodine appears in the hydrogen iodide, showing that the synthesis of hydrogen iodide and the decomposition of hydrogen iodide are still occurring.

The equilibrium concentrations of HI, I_2 and H_2 fit the simple law

$$\frac{[HI]^2}{[H_2][I_2]} = K_c$$

Square brackets represent concentrations in mol dm^{-3}. The ratio, K_c, has a constant value at a particular temperature, no matter what amounts of HI, I_2 and H_2 are taken initially. K_c is the **equilibrium constant** for the reaction in terms of concentration.

The esterification–hydrolysis reaction comes to equilibrium

An example of a reversible reaction in solution is the esterification of ethanoic acid by ethanol to make ethyl ethanoate. This ester is hydrolysed by water to ethanoic acid and ethanol:

$$CH_3CO_2H(l) + C_2H_5OH(l) \rightleftharpoons CH_3CO_2C_2H_5(l) + H_2O(l)$$

Figure 11.2 shows what happens when ethanoic acid and ethanol are mixed and allowed to come to equilibrium.

FIGURE 11.2
An Esterification reaches Equilibrium

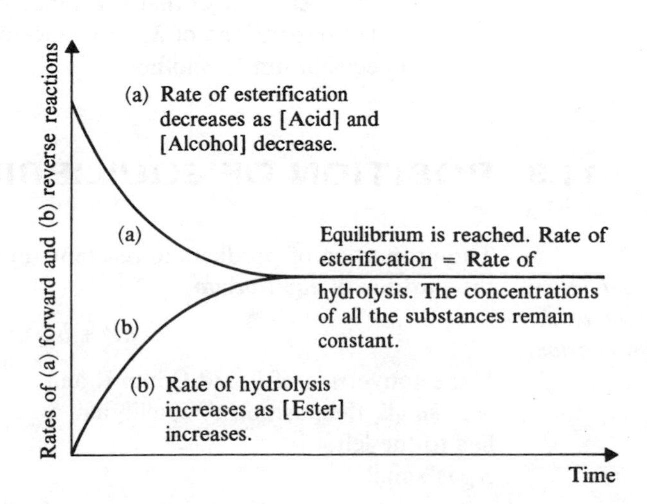
Rates of (a) forward and (b) reverse reactions
(a) Rate of esterification decreases as [Acid] and [Alcohol] decrease.
(a)
Equilibrium is reached. Rate of esterification = Rate of hydrolysis. The concentrations of all the substances remain constant.
(b)
(b) Rate of hydrolysis increases as [Ester] increases.
Time

The concentrations of the reactants and products at equilibrium fit the equilibrium law

The equilibrium concentrations of acid, alcohol, ester and water obey the law

$$\frac{[CH_3CO_2C_2H_5][H_2O]}{[CH_3CO_2H][C_2H_5OH]} = K_c$$

K_c is the equilibrium constant for the esterification.
Esterification is catalysed by inorganic acids. The presence of a catalyst does not alter the equilibrium constant. Its effect is to decrease the time needed for the system to reach a state of equilibrium.

These two reactions are examples of **homogeneous** equilibria, the first in the gas phase and the second in the liquid phase. If the reactants are in different phases, the equilibrium is described as **heterogeneous**. An example of a heterogeneous equilibrium is the reaction between steam and heated iron to form iron(II) iron(III) oxide Fe_3O_4, and hydrogen:

The reaction between iron and steam reaches equilibrium provided that it takes place in a closed container

$$3Fe(s) + 4H_2O(g) \rightleftharpoons Fe_3O_4(s) + 4H_2(g)$$

If the reaction takes place in a closed container, the system reaches equilibrium. In an open container, hydrogen escapes. In an effort to restore equilibrium, more iron reacts with steam to make more hydrogen. When this reaction was used industrially to make hydrogen, the yield was increased by passing a stream of steam over heated iron, and not allowing the system to come to equilibrium [see also § 11.5.4].

11.2 THE EQUILIBRIUM LAW

The equilibrium law summarises the results of a vast amount of research on reactions such as the three mentioned above. For a reaction

$$a\mathbf{P} + b\mathbf{Q} \rightleftharpoons c\mathbf{R} + d\mathbf{S}$$

the equilibrium constant, expressed in terms of concentrations, K_c, is given by the **Equilibrium Law**:

$$K_c = \frac{[\mathbf{R}]^c[\mathbf{S}]^d}{[\mathbf{P}]^a[\mathbf{Q}]^b}$$

A statement of the equilibrium law

The law can be stated as follows. If a reversible reaction is allowed to reach equilibrium, then the product of the concentrations of the products (raised to the appropriate powers) divided by the product of the concentrations of the reactants (raised to the appropriate powers) has a constant value at a particular temperature. The 'appropriate power' is the coefficient of that substance in the stoichiometric equation for the reaction. The dimensions of K_c are concentration$^{(c+d-a-b)}$, and the units vary from one equilibrium to another.

11.3 POSITION OF EQUILIBRIUM

The proportion of products to reactants in the equilibrium mixture is described as the **position of equilibrium**:

The position of equilibrium is not the same as the equilibrium constant

$$a\mathbf{P} + b\mathbf{Q} \rightleftharpoons c\mathbf{R} + d\mathbf{S}$$

If the conversion of **P** and **Q** into **R** and **S** is small, the position of equilibrium lies to the left.
K_c is small.

If the equilibrium mixture is largely composed of **R** and **S**, the position of equilibrium lies to the right.
K_c is large.

A change in conditions, e.g., pressure, will affect the position of equilibrium

The equilibrium constant K_c is not the same as the position of equilibrium. While K_c is constant at a particular temperature, a change in external conditions can alter the position of equilibrium. Chemists are often interested in finding the best conditions for operating a manufacturing process. They want to shift equilibrium reactions in the direction of forming the products. The study of the factors which alter the position of equilibrium is commercially important.

11.4 THE EFFECT OF CONDITIONS ON THE POSITION OF EQUILIBRIUM

11.4.1 LE CHATELIER'S PRINCIPLE

Le Chatelier's Principle describes the effects of external factors on equilibria

Le Chatelier studied the influence of pressure, temperature and concentration on equilibria. His views are known as **Le Chatelier's Principle**. This states that in any equilibrium, when a change is made to some external factor (such as temperature or pressure), the change in the position of equilibrium is such as to tend to change the external factor in the opposite direction. The system cannot completely cancel the change in the external factor, but it moves in the direction that will minimise the change.

11.4.2 CHANGES IN CONCENTRATION

When there is a change in concentration...

An aqueous solution of bismuth(III) chloride is cloudy because of the hydrolysis

$$BiCl_3(aq) + H_2O(l) \rightleftharpoons BiOCl(s) + 2HCl(aq)$$

... the position of equilibrium changes; the equilibrium constant is unchanged

If a little concentrated hydrochloric acid is added, the position of equilibrium shifts in the direction that will absorb acid, i.e., from right to left. The solution cannot absorb *all* the acid added. The hydrolysis of bismuth(III) chloride is much reduced, and a clear solution results.

11.4.3 CHANGES IN PRESSURE

When there is a change in pressure...

It is mainly with gaseous reactions that changes in pressure are important (except for geological reactions). Le Chatelier's Principle can be applied to the formation of ammonia:

... the position of equilibrium changes; the equilibrium constant is unchanged

$$N_2(g) + 3H_2(g) \rightleftharpoons 2NH_3(g); \quad \Delta H^\ominus = -92\,\text{kJ mol}^{-1}$$

If the pressure of an equilibrium mixture of nitrogen, hydrogen and ammonia is increased, the equilibrium shifts in the direction that tends to decrease the pressure. It does this by decreasing the total number of molecules present, i.e., by moving from left to right of the equation. Although the position of equilibrium has moved towards the ammonia side of the equation, the equilibrium constant has not changed.

11.4.4 CHANGES IN TEMPERATURE

The equilibrium constant of an exothermic reaction decreases with temperature...

The formation of ammonia is exothermic. If the temperature is raised, the system can absorb heat by the dissociation of ammonia into nitrogen and hydrogen. The equilibrium constant for the formation of ammonia is decreased. This is a different kind of effect from the influence of pressure, where the position of equilibrium is altered but the equilibrium constant remains the same.

In the case of an endothermic reaction, such as

$$N_2(g) + O_2(g) \rightleftharpoons 2NO(g);\ \Delta H^\ominus = 180\,kJ\,mol^{-1}$$

...and that of an endothermic reaction increases with temperature

increasing the temperature increases the equilibrium constant.

There is further discussion of the importance of choosing the most favourable conditions for carrying out reversible reactions in the Haber process [§ 22.4.1] and the Contact process [§ 21.11]. Catalysts do not alter the position of equilibrium [§ 14.10]. They change only the rate at which equilibrium is attained.

11.4.5 PHYSICAL CHANGES

Le Chatelier's Principle applies to physical changes as well as chemical changes. In the reversible change of state

$$\text{Ice} \rightleftharpoons \text{Water};\ \Delta H^\ominus = 6.00\,kJ\,mol^{-1}$$

Physical changes also obey Le Chatelier's Principle

When pressure is exerted on a mixture of ice and water in equilibrium, the system adjusts in such a manner as to tend to decrease its volume, i.e., ice melts. (The density of water increases from 0 °C to 4 °C.) When the temperature rises, the system adjusts itself in such a manner as to tend to resist an increase in temperature. This can be accomplished by the melting of ice, which is an endothermic process. As long as there is any ice left, the temperature will remain constant at 0 °C (at 1 atm).

CHECKPOINT 11A: LE CHATELIER'S PRINCIPLE

1. Consider the equilibrium

$$N_2(g) + 3H_2(g) \rightleftharpoons 2NH_3(g);\ \Delta H^\ominus = -90\,kJ\,mol^{-1}$$

At 25 °C and 10 atm, the percentage conversion to ammonia in the equilibrium mixture is 5%. If the pressure is increased to 40 atm, while the temperature remains the same, will the percentage conversion to ammonia be greater or less? If the temperature is increased to 250 °C, while the pressure remains at 10 atm, will the equilibrium amount of ammonia be greater or less? What will happen if some iron and molybdenum are added to the equilibrium mixture?

2. Consider the equilibrium

$$2NO_2(g) \rightleftharpoons N_2O_4(g);\ \Delta H^\ominus = -54\,kJ\,mol^{-1}$$

How are (i) the extent of conversion of NO_2 to N_2O_4 and (ii) the equilibrium constant affected by (*a*) an increase in temperature, (*b*) an increase in pressure and (*c*) an increase in volume at constant temperature? Explain your answers.

3. In the equilibrium

$$PCl_5(g) \rightleftharpoons PCl_3(g) + Cl_2(g);\ \Delta H^\ominus = 90\,kJ\,mol^{-1}$$

what is the effect on (i) the position of equilibrium and (ii) the equilibrium constant of (*a*) increasing the temperature, (*b*) decreasing the volume of the container, (*c*) adding a catalyst, (*d*) adding $Cl_2(g)$ and (*e*) adding a noble gas? Explain your answers.

4. The freezing of water at 0 °C is represented

$$H_2O(l, \rho = 1.00\,kg\,dm^{-3}) \rightleftharpoons H_2O(s, \rho = 0.92\,kg\,dm^{-3})$$

Explain why the application of pressure to ice at 0 °C makes it melt. Why do most other solids not show the same behaviour?

11.5 EXAMPLES OF REVERSIBLE REACTIONS

11.5.1 REACTION 1: ESTERIFICATION

Example Calculate the amount of ethyl ethanoate formed when 1.0 mol ethanoic acid and 1.0 mol ethanol reach equilibrium. At room temperature, $K_c = 4.0$.

Esterification of ethanoic acid and ethanol...

Method Let the amount of ethyl ethanoate at equilibrium = x mol and the volume of the solution = $V\,dm^3$. The equilibrium amounts and concentrations are as follows:

$$CH_3CO_2H(l) + C_2H_5OH(l) \rightleftharpoons CH_3CO_2C_2H_5(l) + H_2O(l)$$

Amount/mol	$1 - x$	$1 - x$	x	x
Concentration/mol dm^{-3}	$(1 - x)/V$	$(1 - x)/V$	x/V	x/V

Since

$$K_c = \frac{[CH_3CO_2C_2H_5][H_2O]}{[CH_3CO_2H][C_2H_5OH]} = 4.0$$

$$x^2/(1 - x)^2 = 4.0$$

$$3x^2 - 8x + 4 = 0$$

Solving this quadratic equation gives $x = 2$ or $\frac{2}{3}$. The value 2 can be excluded because it is higher than the amount of ethanoic acid present initially. The amount of ester formed = $\frac{2}{3}$mol.

AN EXPERIMENTAL DETERMINATION OF K_c

A method of finding K_c for the esterification reaction between ethanol and ethanoic acid

The reaction to be studied is the esterification discussed above. It is catalysed by acid. The stages of the experiment are as follows.

1. A number of mixtures are made up, containing ethanol, ethanoic acid, ethyl ethanoate and water. Each mixture is different. Every mixture contains a small amount of hydrochloric acid to catalyse the reaction.

2. The mixtures are put into stoppered bottles and left for a week in a thermostat bath.

3. At the end of the week, the contents of each flask are titrated against standard sodium hydroxide solution, using phenolphthalein as indicator. The titration gives the amount of CH_3CO_2H + the amount of HCl. The amount of HCl is still the same as that present initially. The amount of CH_3CO_2H is found by subtraction.

4. The amounts of the other substances present are calculated.

A sample calculation A $10.0\,cm^3$ mixture contains the initial amounts/mol ethanol 0.0515; ethanoic acid 0.0525; water 0.0167; ester 0.0314; $H^+(aq)$ 1.00×10^{-3}.

The equilibrium amount of ethanoic acid = 0.0255 mol.
Since the amount of ethanoic acid has decreased by 0.0270 mol, ethanol has decreased by the same amount, and ester and water have both increased by this amount.

Species	CH_3CO_2H +	$C_2H_5OH \rightleftharpoons$	$CH_3CO_2C_2H_5$ +	H_2O
Initial amount/mol	0.0525	0.0515	0.0314	0.0167
Equilibrium amount/mol	0.0255	0.0245	0.0584	0.0437

Since

$$K_c = \frac{[CH_3CO_2C_2H_5][H_2O]}{[CH_3CO_2H][C_2H_5OH]}$$

All the substances are present in the same volume of solution, therefore

$$K_c = \frac{0.0584 \times 0.0437}{0.0255 \times 0.0245} = 4.1$$

From calculations on all the mixtures, an average value of K_c at the chosen temperature is obtained. K_c is dimensionless.

11.5.2 REACTION 2: THE REACTION BETWEEN HYDROGEN AND IODINE

For gaseous reactions, it is usual to employ K_p, the equilibrium constant in terms of partial pressures

In the reaction between hydrogen and iodine

$$H_2(g) + I_2(g) \rightleftharpoons 2HI(g)$$

$$K_c = \frac{[HI]^2}{[H_2][I_2]}$$

Since this is a reaction between gases, the concentration of each gas can be expressed as a partial pressure [§ 7.2.8]. Then

$$K_p = \frac{p_{HI}^{\ 2}}{p_{H_2} \times p_{I_2}}$$

K_p is the equilibrium constant in terms of partial pressures.

A detailed treatment of the H_2 and I_2 reaction and...

You can see that, since the pressure units cancel out in the expression, K_p is dimensionless. K_p predicts the same equilibrium composition, no matter what the pressure may be. This is not the case for all gaseous equilibria [contrast Reaction 3, below. (If the gases are not ideal and do not obey Dalton's Law of Partial Pressures accurately, there will be some discrepancies at high pressures.)

...a worked example

Example When 1.00 mol hydrogen and 1.00 mol iodine are allowed to reach equilibrium in a 1.00 dm³ flask at 450 °C and $1.01 \times 10^5\,N\,m^{-2}$, the amount of hydrogen iodide at equilibrium is 1.56 mol. Calculate K_p at 450 °C.

Method

P = total pressure	$H_2(g)$	$+ I_2(g)$	$\rightleftharpoons 2HI(g)$	*Total*
Initial amount/mol	1.00	1.00	0	2.00
Equilibrium amount/mol	$1 - a$	$1 - a$	$2a$	2

Since $2a = 1.56 \quad a = 0.78$ mol

Equilibrium amount/mol	0.22	0.22	1.56	2.00
Equilibrium partial pressure	$\left(\frac{0.22P}{2.00}\right)$	$\left(\frac{0.22P}{2.00}\right)$	$\left(\frac{1.56P}{2.00}\right)$	P

$$K_p = \frac{p_{HI}^{\ 2}}{p_{H_2}p_{I_2}} = \left(\frac{1.56P}{2.00}\right)^2 \Big/ \left(\frac{0.22P}{2.00}\right)^2 = 50$$

11.5.3 REACTION 3: THE HABER PROCESS

In the **Haber process** for making ammonia

$$N_2(g) + 3H_2(g) \rightleftharpoons 2NH_3(g)$$

$$K_p = \frac{p_{NH_3}^{\ 2}}{p_{N_2} \times p_{H_2}^{\ 3}}$$

A detailed treatment of the Haber Process

Let the mole fractions of the components in the equilibrium mixture be x_{NH_3}, x_{N_2} and x_{H_2}. The total pressure = P. The partial pressures are

$$p_{N_2} = x_{N_2}P; \; p_{H_2} = x_{H_2}P; \; p_{NH_3} = x_{NH_3}P$$

$$K_p = \frac{(x_{NH_3}P)^2}{x_{N_2}P(x_{H_2}P)^3}$$

$$= \frac{{x_{NH_3}}^2}{x_{N_2}{x_{H_2}}^3 P^2}$$

The unit of K_p is P^{-2}, e.g., atm^{-2} or $N^{-2}\,m^4$.
When P increases, K_p stays constant; therefore x_{NH_3} must increase, and x_{N_2} and x_{H_2} must decrease. The higher the pressure, the greater will be the percentage conversion into ammonia.

...and a worked example

Example Nitrogen and hydrogen are mixed in a molar ratio 1 : 3. At equilibrium, at 600 °C and 10 atm, the percentage of ammonia in the mixture of gases is 15%. Find K_p at 600 °C.

Method Let a be the fraction of each mole of N_2 that has reacted at equilibrium. Then the equilibrium amounts of the gases are as listed below.
P = total pressure = 10 atm.

Species	$N_2(g)$	$+ 3H_2(g)$	$\rightleftharpoons 2NH_3(g)$	*Total*
Initial amount/mol	1	3	0	4
Equilibrium amount/mol	$1 - a$	$3(1 - a)$	$2a$	$4 - 2a$
Mole fraction	$\frac{1-a}{4-2a}$	$\frac{3(1-a)}{4-2a}$	$\frac{2a}{4-2a}$	

The percentage of NH_3 = $\frac{2a}{4-2a} \times 100 = 15 \quad \therefore a = 0.26$

Mole fraction	$\frac{0.74}{3.48}$	$\frac{2.22}{3.48}$	$\frac{0.52}{3.48}$	1
Equilibrium partial pressure	$\frac{0.74P}{3.48}$	$\frac{2.22P}{3.48}$	$\frac{0.52P}{3.48}$	P

$$K_p = {p_{NH_3}}^2 / p_{N_2}{p_{H_2}}^3$$

$$= \left(\frac{0.52P}{3.48}\right)^2 \Big/ \left(\frac{0.74P}{3.48}\right)\left(\frac{2.22P}{3.48}\right)^3$$

$$= 0.40P^{-2} = 4.0 \times 10^{-3}\,atm^{-2}$$

11.5.4 REACTION 4: THE REACTION BETWEEN IRON AND STEAM

The heterogeneous reaction between iron and steam...

In the reaction between steam and heated iron

$$3Fe(s) + 4H_2O(g) \rightleftharpoons Fe_3O_4(s) + 4H_2(g)$$

$$K_p = \frac{{p_{H_2}}^4}{{p_{H_2O}}^4}$$

The solids do not appear in the expression. Their vapour pressures remain constant (at a constant temperature) as long as there is some of each solid present. These constant vapour pressures are incorporated into the value of the constant K_p.

Example A mixture of iron and steam was allowed to reach equilibrium at 600 °C. The equilibrium pressures of hydrogen and steam were 3.2 kPa and 2.4 kPa respectively. Calculate the value of the equilibrium constant in terms of partial pressures.

Method

$$K_p = \frac{p_{H_2}{}^4}{p_{H_2O}{}^4} = \left(\frac{3.2}{2.4}\right)^4 = 3.16$$

The equilibrium constant, K_p is 3.2. It is dimensionless.

CHECKPOINT 11B: EQUILIBRIUM CONSTANTS

1. Write an expression for the equilibrium constant, K_p, for each of the following reactions:

(*a*) $CS_2(g) + 4H_2(g) \rightleftharpoons CH_4(g) + 2H_2S(g)$

(*b*) $4NH_3(g) + 5O_2(g) \rightleftharpoons 4NO(g) + 6H_2O(g)$

(*c*) $2NO_2(g) + 7H_2(g) \rightleftharpoons 2NH_3(g) + 4H_2O(g)$

2. Write an expression for the equilibrium constant, K_c, for each of the following reactions:

(*a*) $Sn^{2+}(aq) + 2Fe^{3+}(aq) \rightleftharpoons Sn^{4+}(aq) + 2Fe^{2+}(aq)$

(*b*) $Ag^+(aq) + Fe^{2+}(aq) \rightleftharpoons Fe^{3+}(aq) + Ag(s)$

(*c*) $2Cr^{3+}(aq) + Fe(s) \rightleftharpoons 2Cr^{2+}(aq) + Fe^{2+}(aq)$

3. Equilibrium is established in the reaction

$$\mathbf{A}(aq) + \mathbf{B}(aq) \rightleftharpoons 2\mathbf{C}(aq)$$

If equilibrium concentrations are [**A**] = 0.25, [**B**] = 0.40, [**C**] = 0.50 mol dm^{-3}, what is the value of K_c?

4. The following reaction was allowed to reach equilibrium:

$$2\mathbf{D}(aq) + \mathbf{E}(aq) \rightleftharpoons \mathbf{F}(aq)$$

The initial amounts of the reactants present in 1.00 dm^3 of solution were 1.00 mol **D** and 0.75 mol **E**. At equilibrium, the amounts were 0.70 mol **D** and 0.60 mol **E**. Calculate the equilibrium constant, K_c.

5. The gases SO_2, O_2 and SO_3 are allowed to reach equilibrium. The partial pressures of the gases are p_{SO_2} = 0.050 atm, p_{O_2} = 0.025 atm, p_{SO_3} = 1.00 atm. Find the values of K_p for the equilibria

(*a*) $SO_2(g) + \frac{1}{2}O_2(g) \rightleftharpoons SO_3(g)$

(*b*) $2SO_2(g) + O_2(g) \rightleftharpoons 2SO_3(g)$

6. A mixture contained 1.00 mol of ethanoic acid and 5.00 mol of ethanol. After the system had come to equilibrium, a portion of the mixture was titrated against 0.200 mol dm^{-3} sodium hydroxide solution. The titration showed that the whole of the equilibrium mixture would require 289 cm^3 of the standard alkali for neutralisation. Find the value of K_c for the esterification reaction.

11.5.5 REACTION 5: THERMAL DISSOCIATION

Another type of reaction which reaches an equilibrium position is thermal dissociation. For example, when phosphorus(V) chloride is heated, it dissociates partially into phosphorus(III) chloride and chlorine:

$$PCl_5(g) \rightleftharpoons PCl_3(g) + Cl_2(g)$$

As a result, the pressure of the gas (at constant volume) or, alternatively, the volume of the gas (at constant pressure) is greater than expected.

The fraction of molecules that dissociate is called the **degree of dissociation**, and is represented by the letter α. The amounts of substances in the equilibrium mixture formed from 1 mole of PCl_5 are as follows:

When substances dissociate in the gas phase, the actual pressure is greater than expected...

Species	$PCl_5(g)$	$PCl_3(g)$	$Cl_2(g)$	*Total*
Amount	$1 - \alpha$	α	α	$1 + \alpha$

$$\therefore \quad \frac{\text{Actual number of particles}}{\text{Expected number of particles}} = \frac{1 + \alpha}{1}$$

Since the pressure of a gas is proportional to the concentration of particles present, at constant volume

$$\frac{\text{Actual pressure of gas}}{\text{Expected pressure of gas}} = 1 + \alpha$$

For a substance in which one molecule dissociates into n molecules

...From the increase in pressure, the degree of dissociation can be found...

$$\frac{\text{Actual number of particles}}{\text{Expected number of particles}} = 1 + (n - 1)\alpha$$

The equilibrium constant for the thermal dissociation of PCl_5 is

...and also the equilibrium constant for the dissociation

$$K_c = \frac{[Cl_2][PCl_3]}{[PCl_5]}$$

If V = the volume of the container

$$K_c = \frac{(\alpha/V) \times (\alpha/V)}{(1 - \alpha)/V} = \frac{\alpha^2}{(1 - \alpha)V}$$

If c = the initial concentration of PCl_5, then $c = \dfrac{1}{V}$ and

$$K_c = \frac{\alpha^2 c}{1 - \alpha}$$

The equilibrium constant K_p, in terms of partial pressures, is

$$K_p = \frac{p_{Cl_2} p_{PCl_3}}{p_{PCl_5}}$$

If $p_{Cl_2} = p_{PCl_3} = \left(\dfrac{\alpha}{1 + \alpha}\right)P$, and $p_{PCl_5} = \left(\dfrac{1 - \alpha}{1 + \alpha}\right)P$, where P = total pressure,

$$K_p = \frac{\left(\dfrac{\alpha}{1 + \alpha}\right)^2 P^2}{\left(\dfrac{1 - \alpha}{1 + \alpha}\right)P} = \frac{\alpha^2 P}{1 - \alpha^2}$$

The unit of K_p is a pressure unit.
If α is small, $1 - \alpha^2 \approx 1$, and $\alpha^2 \approx K_p/P$, i.e., the higher the pressure, the lower is the degree of dissociation.

Example When 0.100 mol of phosphorus(V) chloride is heated at 150 °C in a 1.00 dm^3 vessel, a pressure of $4.38 \times 10^5\,N\,m^{-2}$ is measured. Calculate the degree of dissociation and the value of K_p at this temperature.

Method (a) Find α from a comparison of real and expected pressures. The expected pressure is calculated from $PV = nRT$ [§ 7.2.6]. Putting $V = 1.00 \times 10^{-3}\,m^3$, $T = 423\,K$, $n = 0.100$ and $R = 8.314\,J\,K^{-1}\,mol^{-1}$ gives

$$P = 3.51 \times 10^5\,N\,m^{-2}$$

Since

$$\text{Measured pressure/Expected pressure} = 1 + \alpha$$

$$4.38 \times 10^5/(3.51 \times 10^5) = 1 + \alpha$$

The degree of dissociation, $\alpha = 0.25$

(b) Find K_p.

Since $K_p = \alpha^2 P/(1 - \alpha^2)$

and $\alpha = 0.25$, and $P = 4.38 \times 10^5\,N\,m^{-2}$

$$K_p = 2.92 \times 10^4\,N\,m^{-2}$$

Thermal dissociation results in a low measured value for the molar mass...

Another sign of thermal dissociation is an unexpectedly low value for the molar mass. [See method, § 8.2.] From the equation

$$PV = \frac{m}{M}RT$$

you can see that, if the volume, V, of a mass of gas, m, is greater than expected, then the value obtained for the molar mass, M, will be less than expected. Since

$$\text{Actual volume of gas/Expected volume of gas} = 1 + \alpha$$

$$\text{Expected molar mass/Measured molar mass} = 1 + \alpha$$

If one molecule dissociates into n particles

$$\text{Expected molar mass/Measured molar mass} = 1 + (n - 1)\alpha$$

Example A molar mass determination gave a value of 82.5 g mol^{-1} for dinitrogen tetraoxide, N_2O_4, at 25 °C. Calculate the degree of dissociation at this temperature.

Method The dissociation that occurs is

...An example is the dissociation of N_2O_4

$$N_2O_4(g) \rightleftharpoons 2NO_2(g)$$

Thus, $n = 2$ and

$$\text{Calculated molar mass/Measured molar mass} = 1 + \alpha$$

$$\therefore \quad 92.0/82.5 = 1 + \alpha$$

$$\text{Degree of dissociation } \alpha = 0.115$$

CHECKPOINT 11C: DISSOCIATION

1. When 0.100 mol of carbon dichloride oxide, $COCl_2$, was vaporised at 400 °C in a 1.00 dm^3 vessel, a pressure of 8.00×10^5 N m^{-2} developed. Calculate the degree of dissociation of $COCl_2$(g) into CO(g) + Cl_2(g).

2. The vapour formed from 0.200 g of antimony(V) chloride, $SbCl_5$, occupied 40.4 cm^3 at 250 °C and 1.01×10^5 N m^{-2}. Calculate the degree of dissociation into $SbCl_3$(g) + Cl_2(g) and the value of K_p at 250 °C.

3. The dissociation of nitrogen chloride oxide is represented by the equation

$$2NOCl(g) \rightleftharpoons 2NO(g) + Cl_2(g)$$

If the degree of dissociation is α, how many moles of substance are formed from 1 mole of NOCl? At 200 °C, a value of 54.3 g mol^{-1} was obtained by a molar mass determination. Calculate the value of α at this temperature, and state what fraction of the molecules in the equilibrium mixture are chlorine molecules.

4. Phosphorus(V) chloride is 30.0% dissociated at a certain temperature and a pressure of 1.01×10^5 N m^{-2}. Find the value of K_p at this temperature for the dissociation,

$$PCl_5(g) \rightleftharpoons PCl_3(g) + Cl_2(g)$$

11.6 PHASE EQUILIBRIUM DIAGRAMS

Definition of a phase

A **phase** is a homogeneous part of a system which is physically distinct from other parts of the system. A mixture of gases is one phase. A solution is one phase. A saturated solution in the presence of an excess of solute is a two-phase system. A mixture of solids consists of as many phases as there are solids present.

In a phase diagram...

When a substance exists in different physical states, the conditions under which each state exists can be represented by a **phase diagram**. An **area** on a phase diagram represents one phase. A **line** represents the conditions under which two phases can exist in equilibrium. A **triple point** describes the conditions under which three phases can coexist.

11.6.1 WATER

Figure 11.3 shows the phase diagram for water.

...an area = one phase...

...a line = two phases

...a point = three phases, and is named a triple point

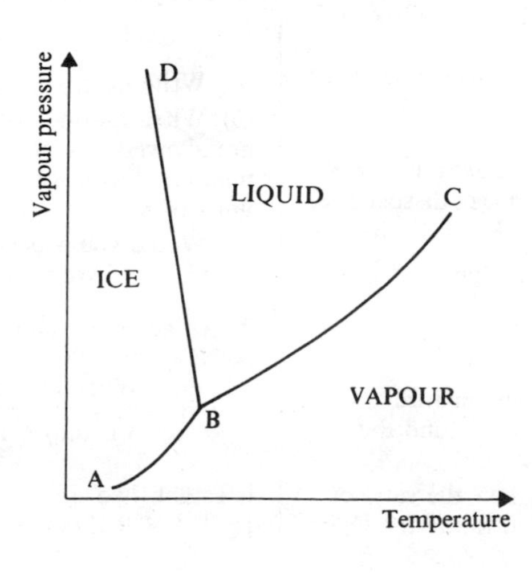

BD = Effect of pressure on freezing temperature of water

BC = Effect of temperature on saturated vapour pressure of water

AB = Effect of temperature on saturated vapour pressure of ice

B = Conditions under which solid, liquid and gaseous forms can coexist — a triple point

FIGURE 11.3 Phase Equilibrium Diagram for Water (not to scale)

11.6.2 CARBON DIOXIDE

Figure 11.4 shows the phase diagram for carbon dioxide [see Question 8 at the end of this chapter].

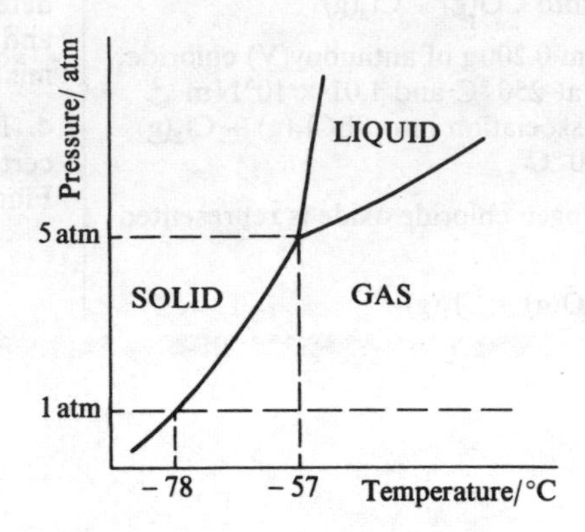

FIGURE 11.4 Phase Equilibrium Diagram for Carbon Dioxide (not to scale)

CHECKPOINT 11D: PHASE EQUILIBRIA

1. Figure 11.5 shows the pressure–temperature phase diagram for substance **X**.

(*a*) Trace the figure. On your copy, mark the areas S (solid), L (liquid) and G (gas).

(*b*) Name the point A, and explain its significance.

(*c*) Draw in the vapour pressure–temperature curve for a dilute solution in **X** of a non-volatile solute.

(*d*) Mark the freezing temperature of the solution from (*c*).

(*e*) Say, with an explanation, how the freezing temperature of pure **X** varies with an increase in pressure.

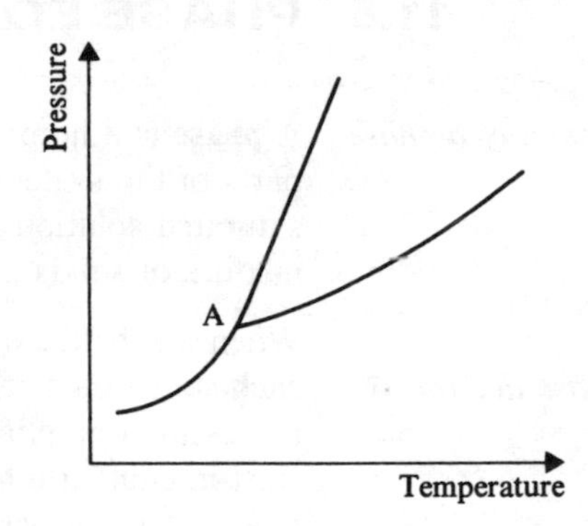

FIGURE 11.5

QUESTIONS ON CHAPTER 11

1. Study the gaseous equilibrium

$$N_2O_4(g) \rightleftharpoons 2NO_2(g);$$

$$\Delta H^{\ominus} = 54\,kJ\,mol^{-1} \textit{ Endothermic}$$

(*a*) Write an expression for the equilibrium constant, K_p, in terms of partial pressures.

(*b*) State, with reasons, the effect on the equilibrium position of (i) an increase in total pressure and (ii) an increase in temperature.

(*c*) At 60 °C, 1.00 dm^3 of the gas weighed 2.585 g under a pressure of $1.01 \times 10^5\,N\,m^{-2}$. Find the degree of dissociation of N_2O_4 and the value of K_p. ($R = 8.314\,J\,K\,mol^{-1}$)

2. The **Contact process** involves the equilibrium

$$2SO_2(g) + O_2(g) \rightleftharpoons 2SO_3(g)$$

(*a*) Write an expression for K_p.

(*b*) At 1000 K and 1.00 atm, the equilibrium mixture contains 27.0% SO_2 and 40.0% O_2 by volume. Find the value of K_p.

(*c*) Use the expression from (*a*) to show how the ratio of SO_3/SO_2 in equilibrium mixtures is related to the total pressure in the presence of a fixed excess of oxygen.

(*d*) Explain why industrial plants operate at pressures of around 2 atm.

3. Study the equilibrium

$$H_2O(g) + CO(g) \rightleftharpoons H_2(g) + CO_2(g)$$

(*a*) Write an expression for K_p.

(*b*) When 1.00 mol of steam and 1.00 mol of carbon monoxide are allowed to reach equilibrium, 33.3% of the equilibrium mixture is hydrogen. Calculate the value of K_p. State the units of K_p.

(*c*) Would you expect an increase in total pressure to affect the yield of hydrogen? Explain your answer.

4. An aqueous solution is made by dissolving 1.00 mol of $AgNO_3$ and 1.00 mol of $FeSO_4$ in water and making up to 1.00 dm^3. When the equilibrium

$$Ag^+(aq) + Fe^{2+}(aq) \rightleftharpoons Fe^{3+}(aq) + Ag(s)$$

is established, $[Ag^+] = [Fe^{2+}] = 0.44\,mol\,dm^{-2}$, and $[Fe^{3+}] = 0.56\,mol\,dm^{-3}$. Find K_c for the equilibrium.

5. Iron is added to a solution containing the ions, Cr^{3+}, Cr^{2+} and Fe^{2+}. The equilibrium concentrations of the ions present in solution are $[Cr^{3+}] = 0.030\,mol\,dm^{-3}$, $[Cr^{2+}] = 0.27\,mol\,dm^{-3}$, $[Fe^{2+}] = 0.11\,mol\,dm^{-3}$. Find K_c for the equilibrium

$$2Cr^{3+}(aq) + Fe(s) \rightleftharpoons 2Cr^{2+}(aq) + Fe^{2+}(aq)$$

6. For the equilibrium

$$2\mathbf{P}(g) + \mathbf{Q}(g) \rightleftharpoons 2\mathbf{R}(g)$$

K_c is numerically equal to 6.0. Into a $1.00\,dm^3$ flask are introduced 3.0 mol of **P**, 3.0 mol of **Q** and 3.0 mol of **R**.

(*a*) State the unit in which K_c is expressed.

(*b*) Is the mixture at equilibrium?

(*c*) If not, what must the volume of the flask be in order for such a mixture to exist in equilibrium at the temperature for which K_c is given?

7. Into a $1.00\,dm^3$ vessel at 1000 K are introduced simultaneously 0.500 mol of SO_2, 0.100 mol of O_2 and 0.700 mol of SO_3.
For the equilibrium

$$2SO_2(g) + O_2(g) \rightleftharpoons 2SO_3(g)$$

the value of K_c at 1000 K is $2.8 \times 10^2\,mol\,dm^{-3}$.

(*a*) Is the mixture at equilibrium?

(*b*) If it is not at equilibrium, in which direction will the reaction proceed?

8. Figure 11.4, § 11.6.2 shows the phase equilibrium diagram for carbon dioxide.

(*a*) Using the diagram, predict what will happen when solid CO_2 is warmed above $-78\,°C$ at 1 atm.

(*b*) Give the conditions under which gaseous CO_2 can be liquefied.

(*c*) At what temperature is liquid CO_2 stable when the pressure is below 5 atm?

(*d*) The phase diagram for CO_2 differs in two ways from that for water [see Figure 11.3, § 11.6.1]. What are they?

***9.** The following values of K_p were obtained at the stated temperatures in a study of the reaction

$$2SO_2(g) + O_2(g) \rightleftharpoons 2SO_3(g)$$

Use them to obtain a value for the standard enthalpy change of the reaction.

Temperature/°C	856	780	732	684	636
K_p/kPa	21.1	96.5	287	921	3343

10. Consider the reversible reaction used in the Contact process for the manufacture of sulphuric acid:

$$2SO_2(g) + O_2(g) \rightleftharpoons 2SO_3(g);$$

$$\Delta H^{\ominus} = -196\,kJ\,mol^{-1}$$

and assume that 2 mol of sulphur dioxide is mixed with 1 mol of oxygen and the mixture is then allowed to reach equilibrium at 250 °C and 10^5 Pa pressure.

(*a*) (i) Write down an expression for the equilibrium constant for this reaction in terms of the concentrations of the species present.

(ii) What would be the effect on the position of equilibrium if the vessel were now reduced in volume?

(iii) What would be the effect on the value of the equilibrium constant of raising the temperature?

(*b*) The usual conditions used in industry for this process are that it operates at about 400 °C, at pressures just above 10^5 Pa and in the presence of a catalyst.

(i) Bearing in mind your answer to part (ii) above, why do you think that higher pressures are not used?

(ii) Bearing in mind your answer to part (iii) above, why do you think that lower temperatures are not used?

(iii) Pure oxygen is not used in industry. Air is cheaper but contains mostly nitrogen, which is unreactive in this process. Forecast the effect that the presence of the nitrogen will have on the position of equilibrium and give your reasoning.

(C 91, AS)

11. (*a*) What is meant by the term *dynamic equilibrium*?

(*b*) (i) Write the expression for K_p for the equilibrium

$$N_2O_4(g) \rightleftharpoons 2NO_2(g)$$

(ii) Explain, in outline, how a value for K_p may be determined at 323 K.

(iii) At 298 K the value for the equilibrium constant, K_p, for the above reaction is 12.26 kPa.

Calculate the density, in $g\,dm^{-3}$, of dinitrogen tetraoxide vapour at 298 K and 101.3 kPa pressure.

(iv) At 305 K the value of K_p for the reaction is 20.57 kPa. If $\ln K_p = -\Delta H^{\ominus}/RT +$ constant, calculate $\Delta H^{\ominus}$ for the reaction.

(AEB 92, S)

12. (*a*) State Le Chatelier's Principle.

(*b*) Many commercially important processes involve equilibrium reactions. Give *two* examples of such processes, naming each process and giving the appropriate equation(s).

(*c*) In the Mond process for the purification of nickel, carbon monoxide is passed over impure nickel at 50 °C to form, at this temperature, gaseous nickel tetracarbonyl, $Ni(CO)_4$. This vapour is then passed over pure nickel pellets at 230 °C, when the nickel tetracarbonyl decomposes, depositing pure nickel.

(i) Write one equation that summarises both processes.

(ii) Give the expression for K_c for the reaction in (*c*)(i).

(iii) If the concentration of CO(g) in an equilibrium mixture is doubled, calculate the change in the nickel tetracarbonyl concentration.

(iv) Hence suggest a reason why a new Canadian plant operates at an elevated pressure of 20 atmospheres in the first stage, compared with the atmospheric pressure of older plants.

(L 91)

13. *The Production of Ammonia.*
(Adapted from *The Essential Chemical Industry* – The Polytechnic of North London, 1985)

Raw materials
Nearly all UK production is from natural gas.

Manufacture
This involves two main stages:

(1) The manufacture and purification of synthesis gas (a mixture of nitrogen and hydrogen).

(2) Ammonia synthesis

$$N_2(g) + 3H_2(g) \rightleftharpoons 2NH_3(g); \quad \Delta H = -92\,\text{kJ mol}^{-1}$$

The effects of temperature and pressure on the synthesis are shown in Figure 11.6.

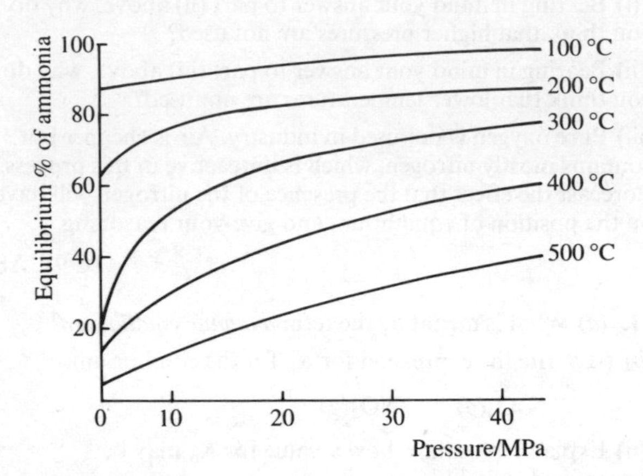

FIGURE 11.6

The synthesis gas is compressed to 20 MPa and passed over a catalyst at 380–450 °C. About 15% ammonia is produced per pass.

Uses
About 80% of ammonia is used in fertiliser production, 7% for conversion to nitric acid and about 5% for nylon manufacture.

(*a*) (i) Explain, by reference to the equation, why high equilibrium percentages of ammonia are obtained at low temperatures.

(ii) Suggest a reason why a low temperature is not used in commercial ammonia synthesis.

(iii) Explain, by reference to the equation, why high pressures increase the equilibrium percentage of ammonia.

(iv) Suggest a reason why very high pressures are not used in commercial ammonia synthesis.

(*b*) For the large-scale manufacture of ammonia give

(i) *two* fixed costs for the process

(ii) *two* variable costs for the process.

(*c*) Suggest *four* factors that should be considered when deciding upon the location of a plant for the production of ammonia.

(*d*) The data in the table below refers to ammonia synthesis at a pressure of 20 MPa and at a temperature between 380–450 °C.

Gas	N_2	H_2	NH_3
Equilibrium partial pressure/MPa	4.3	12.8	3.1

(i) Write an expression for the equilibrium constant, K_p, for ammonia synthesis.

(ii) Calculate a value for K_p using the data in the table. Include units in your answer.

(AEB 90, AS)

14. At room temperature, gaseous dinitrogen tetraoxide and nitrogen dioxide are in dynamic equilibrium according to the following equation:

$$N_2O_4(g) \rightleftharpoons 2NO_2(g); \quad \Delta H = +58\,\text{kJ mol}^{-1}$$

(*a*) Explain what is meant by the term *dynamic equilibrium*, and write the expression for the equilibrium constant, K_p, for this reaction.

(*b*) At a temperature of 25 °C (298 K), 1.00 g of a mixture of these two gases takes up a volume of $3.17 \times 10^{-4}\,\text{m}^3$ at a pressure of 101 kPa ($1.01 \times 10^5\,\text{N m}^{-2}$). Calculate the average relative molecular mass of the mixture.

(*c*) State *Le Chatelier's Principle*, and use it to deduce qualitatively the effect on the average relative molecular mass of this gaseous mixture of increasing

(i) the pressure

(ii) the temperature.

(*d*) Nitrogen dioxide (from car exhaust fumes) can react with sulphur dioxide (from the burning of fossil fuels) in the presence of water vapour in the atmosphere to produce sulphuric acid (acid rain) and nitrogen monoxide, NO. The nitrogen monoxide is rapidly re-oxidised to nitrogen dioxide by oxygen.

Construct balanced equations for these two reactions and hence suggest the role played by nitrogen dioxide in the overall process.

(C 92)

15. The equation for the reaction of ethanol and ethanoic acid is given below.

$$CH_3CO_2H(l) + C_2H_5OH(l) \rightleftharpoons CH_3CO_2C_2H_5(l) + H_2O(l)$$

3.0 g of ethanoic acid and 2.3 g of ethanol were equilibrated at 100 °C for about one hour and then quickly cooled in an ice-bath. 50 cm³ of aqueous sodium hydroxide of concentration $1.0\,\text{mol dm}^{-3}$ were then added and the mixture titrated with hydrochloric acid of the same concentration. 33.3 cm³ of acid were required.

(*a*) What is the meaning of the term 'equilibrate'? Why is the equilibrium mixture cooled rapidly in an ice-bath? Suggest a suitable indicator for use in the titration and give a reason for its choice.

(*b*) State Le Chatelier's Principle and predict the effect of adding ethanol to the equilibrium mixture.

(*c*) Give an expression for K_c and calculate its value using the data provided.

†(*d*) By using average bond enthalpies, show how you could calculate a value for ΔH for the forward reaction.

†(*e*) What would be the effect on the reaction of adding hydrogen ions as a catalyst?

†(*f*) Describe with essential experimental details how you would prepare a pure sample of ethyl ethanoate using this reaction.

[For (*d*) see Chapter 10.]
[For (*e*) see Chapter 14.]
[For (*f*) see Chapter 33.]

(L 91)

16. *(a)* (i) State Le Chatelier's principle.

(ii) List *two* factors that can shift the position of a chemical equilibrium between species in aqueous solution. Explain what effect changes in these factors have on the position of equilibrium.

(iii) Describe the application of Le Chatelier's principle to the optimising of the economic yield in the ammonia synthesis (the Haber process).

(b) The equilibrium constant K_p for the reaction

$$PCl_5(g) \rightleftharpoons PCl_3(g) + Cl_2(g)$$

is 1.06×10^6 N m^{-2} at 250 °C.

A sample of PCl_5 at an initial pressure of 1.01×10^6 N m^{-2} dissociates in the vapour phase at 250 °C in a vessel of fixed volume.

(i) Calculate the partial pressure of each species present at equilibrium and the final total pressure.

(ii) Will the same partial pressures be obtained if equal amounts (i.e. equal numbers of moles) of PCl_3 and Cl_2 are mixed at 250 °C and the *final* pressure is the same as in part (i)? Explain your reasoning.

[The solution to a quadratic equation of the general form $ax^2 + bx + c = 0$ is $x = (-b \pm \sqrt{b^2 - 4ac})/2a$.]

(U & C 91)

17. *(a)* (i) Explain the concept of *chemical equilibrium*.

(ii) Define the term *equilibrium constant* (*K*) and explain why it is useful.

(iii) Write down the *simplest* possible expressions for equilibrium constants (K_p or K_c as appropriate) for the following equilibria and give the units of *K* in each case:

(1) $H_2(g) + \frac{1}{2}O_2(g) \rightleftharpoons H_2O(g)$

(2) $CaCO_3(s) \rightleftharpoons CaO(s) + CO_2(g)$

(3) $2Fe^{3+}(aq) + Sn^{2+}(aq) \rightleftharpoons 2Fe^{2+}(aq) + Sn^{4+}(aq)$

(b) In a closed-vessel experiment on the Haber process, nitrogen at 50 atm pressure and hydrogen at 150 atm pressure were reacted together at constant temperature. After a certain time interval it was found that the ammonia formed had a pressure of 40 atm. Given that the equilibrium constant, K_p, at the reaction temperature is 7.316×10^{-5} atm^{-2}, calculate whether or not the system had reached equilibrium.

(c) State and discuss the factors which govern the rates at which chemical reactions occur.

(WJEC 91)

12 ELECTROCHEMISTRY

12.1 ELECTROLYTIC CONDUCTION

Electrolytic conduction differs from electronic conduction . . .

. . . Faraday pioneered this subject

An electric current in a metallic conductor is a stream of electrons. The nature of the metallic bond [§ 6.2.1] allows metals to conduct electricity. This process is called **electronic conduction**. It is a physical change: no permanent change occurs in the metal.

Some compounds conduct electricity when in the molten state or in solution. This subject was introduced in Chapter 4. When compounds conduct electricity, chemical changes occur, and new substances are formed. Michael Faraday, who did much of the early work on this phenomenon, described it in 1834 as **electrolytic conduction**. He called the compounds which conduct electricity and are decomposed by it **electrolytes**. The chemical changes which take place are called **electrochemical reactions**. The vessel in which an electrochemical change takes place is called a **cell**, and the conductors which carry electricity into and out of the cell are called **electrodes**.

12.1.1 FARADAY'S LAWS OF ELECTROLYSIS

The electrolysis of a solution of a silver salt to deposit a layer of silver on the cathode can be carried out as shown in Figure 12.1.

FIGURE 12.1 Electrolysis of Silver Nitrate Solution

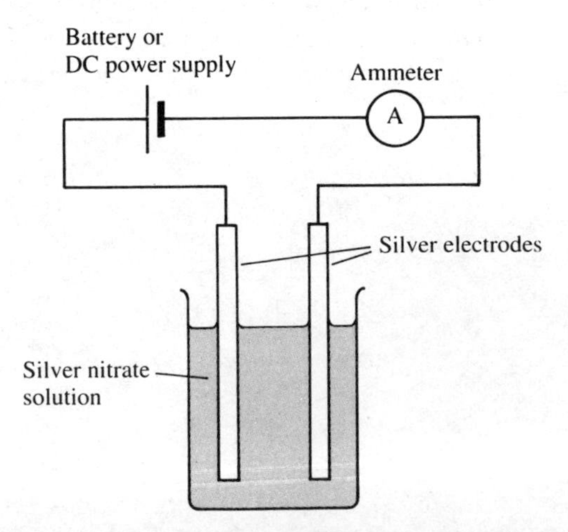

With this apparatus, it is possible to find out how much electric charge is needed to deposit a certain mass of silver. If the cathode is weighed before and after the passage of the current, the difference gives the mass of silver deposited. If the current is measured with a milliammeter and the electrolysis is timed, one can work out the quantity of electric charge which has passed. Electric charge is measured in coulombs (C).

One coulomb of charge = One ampere of current flowing for one second

Charge in coulombs (C) = Current in amperes (A) × Time in seconds (s)

The results of many experiments of this kind led Faraday to state his **First Law of Electrolysis**: The mass of a substance liberated at or dissolved from an electrode is proportional to the quantity of electricity passed. If we look at the mole concept [see Chapter 3] we can understand why this is so.

The cathode process is

$$Ag^+(aq) + e^- \rightarrow Ag(s)$$

According to this equation, the amount of silver discharged is proportional to the number of electrons which flow through the circuit – which is the same as the quantity of electricity that passes. We can take this further. According to the equation,

1 mol silver ions + 1 mol electrons → 1 mol silver atoms

one can measure the mass of silver discharged by a certain quantity of electric charge. For example, in an experiment in which 40.0 mA flowed for 2.00 hours,

Mass of silver deposited on the cathode = 0.332 g

Quantity of electricity = Current × Time

$= 0.040\,A \times 2.00 \times 60 \times 60\,s$

$= 288\,C$

Quantity of electricity needed to discharge 1 mol silver = $288\,C \times 108/0.322$

$= 96.5 \times 10^3\,C$

This quantity is 96 500 coulombs. This quantity of electric charge must be the charge on one mole of electrons. The ratio 96 500 coulombs per mole ($C\,mol^{-1}$) is called the **Faraday constant**.

Faraday conducted experiments to find the masses of different elements discharged by the same quantity of electricity. These experiments led him to state his **Second Law of Electrolysis**: When the same quantity of electricity passes through different electrolytes, equivalent masses of the elements are formed at the electrodes. What Faraday meant by 'equivalent masses' is explained below.

12.1.2 COULOMETRY

Coulometry relates electric charge to mass or volume of product

The apparatus shown in Figure 12.2 enables one to repeat Faraday's work on coulometry. A **coulometer** is a cell in which measurements can be made of the quantity of electricity passed and the masses of substances liberated at or dissolved from the electrodes. Also part of the apparatus is a **voltameter**, the name given to a cell which is constructed to allow any gases evolved to be collected.

EXAMPLE OF RESULTS OBTAINED IN COULOMETRY

In an experiment in which 40.0 mA flowed for 2.00 hours, the masses deposited on the cathodes were the same as the masses dissolved from the anodes: 0.322 g silver, 0.0948 g copper, and 0.0518 g chromium. The volumes of gases at room temperature and pressure were 35.8 cm^3 hydrogen and 17.9 cm^3 oxygen. (Gas molar volume = 24.0 dm^3 at rtp.)

FIGURE 12.2
An Apparatus for Coulometry

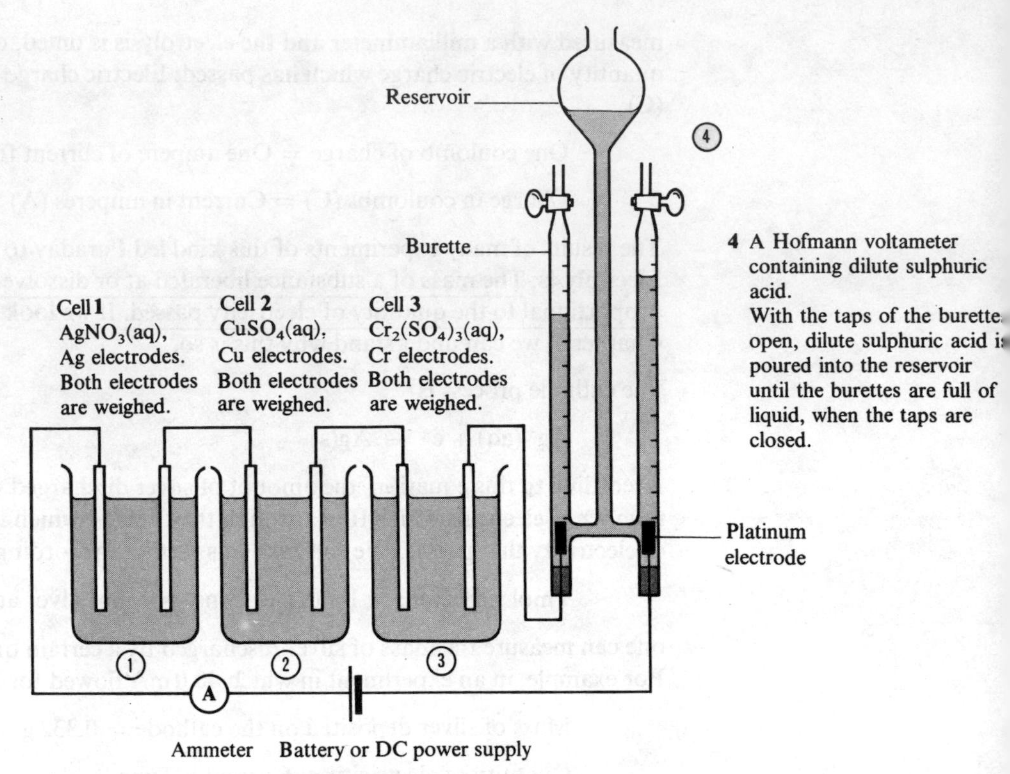
Reservoir
Burette
Cell 1
$AgNO_3(aq)$,
Ag electrodes.
Both electrodes are weighed.
Cell 2
$CuSO_4(aq)$,
Cu electrodes.
Both electrodes are weighed.
Cell 3
$Cr_2(SO_4)_3(aq)$,
Cr electrodes.
Both electrodes are weighed.
4 A Hofmann voltameter containing dilute sulphuric acid.
With the taps of the burette open, dilute sulphuric acid is poured into the reservoir until the burettes are full of liquid, when the taps are closed.
Platinum electrode
1
2
3
4
A
Ammeter
Battery or DC power supply

Treatment of results Quantity of electricity/coulombs

$$= \text{Current/amperes} \times \text{Time/seconds}$$
$$= 0.040 \times 2.00 \times 60 \times 60 = 288\,\text{C}$$

The quantity of electricity required to discharge 1 mole of each element can be calculated.

Results obtained by coulometry

Quantity of electricity required for 1 mol Ag $= 288 \times 108/0.322 = 96.5 \times 10^3\,\text{C}$
Quantity of electricity required for 1 mol Cu $= 288 \times 63.5/0.0948 = 193 \times 10^3\,\text{C}$
Quantity of electricity required for 1 mol Cr $= 288 \times 52.0/0.0518 = 289 \times 10^3\,\text{C}$
Quantity of electricity required for 1 mol H_2 $= 288 \times 24\,000/35.8 = 193 \times 10^3\,\text{C}$
Quantity of electricity required for 1 mol O_2 $= 288 \times 24\,000/17.9 = 386 \times 10^3\,\text{C}$

When 96.5×10^3 C of electricity pass through the cells, the amounts of metals deposited are Ag, 1 mol; Cu, $\frac{1}{2}$mol; Cr, $\frac{1}{3}$mol; i.e., 1 mol/the valency of the element. These amounts are what Faraday referred to as the **equivalents** of the metals in his second law.

Interpretation of Faraday's work

Faraday did his work before the structure of the atom and the nature of the chemical bond were elucidated. He postulated that an electric current was conducted through a cell by charged particles, which he called **ions**. Looking at electrolytic conduction from today's viewpoint, we can see that the cathode processes can be represented as

$$Ag^+(aq) + e^- \rightarrow Ag(s)$$
$$Cu^{2+}(aq) + 2e^- \rightarrow Cu(s)$$
$$Cr^{3+}(aq) + 3e^- \rightarrow Cr(s)$$

You can see from these equations that 1 mole of electrons will discharge 1 mole of Ag^+ ions, $\frac{1}{2}$ mole of Cu^{2+} ions or $\frac{1}{3}$ mole of Cr^{3+} ions.

In general

$$\text{Amount of element discharged/mole} = \frac{\text{Number of moles of electrons}}{\text{Number of charges on one ion of the element}}$$

[For the formation of hydrogen and oxygen see § 12.3.]

A worked example

Example Find how long it will take to deposit 1.00 g of chromium when a current of 0.120 A flows through a solution of chromium(III) sulphate solution.

Method

Since

$$Cr^{3+}(aq) + 3e^- \rightarrow Cr(s)$$

3 mol electrons are needed to deposit 1 mol chromium (i.e., 52 g chromium)
3/52 mol electrons deposit 1 g chromium
Number of coulombs required = 96 500 × 3/52 C

Since

Number of coulombs = Amperes × Seconds = 0.120 × t (t = time/s)

$0.120 \times t = 96\,500 \times 3/52$

$t = 46\,400$ s (approximately)

The current passes for 46 400 s (12.9 h).

12.2 EXAMPLES OF ELECTROLYSIS

When a molten salt is electrolysed, the products are predictable. When an aqueous solution is electrolysed, hydrogen and oxygen sometimes appear at the cathode and anode. The products formed from a few electrolytes are shown in Table 12.1.

TABLE 12.1 Products of Electrolysis

(*a*) Using inert electrodes (platinum or graphite)

Electrolyte	*Cathode*	*Anode*
$PbBr_2$(l)	Pb(s)	Br_2(g)
NaCl(l)	Na(s)	Cl_2(g)
$CuCl_2$(aq)	Cu(s)	Cl_2(g)
NaCl(aq)	H_2(g)	Cl_2(g)
KNO_3(aq)	H_2(g)	O_2(g)
$CuSO_4$(aq)	Cu(s)	O_2(g)
H_2SO_4(aq)	H_2(g)	O_2(g)
NaOH(aq)	H_2(g)	O_2(g)

(b) When the electrodes take part in the reactions		
Electrolyte	*Copper cathode*	*Copper anode*
$CuSO_4(aq)$	Cu(s) deposited	Cu(s) dissolves to form Cu^{2+} ions

12.3 EXPLANATION OF ELECTROLYSIS

When a molten salt is electrolysed, the metal ions arrive at the cathode. This negatively charged electrode supplies electrons, which discharge the cations. In the case of lead(II) chloride

Equations for electrode processes

$$Pb^{2+}(l) + 2e^- \rightarrow Pb(l)$$

The anions travel to the anode, which, being positively charged, takes electrons from the anions, thus discharging them.

$$Cl^-(l) \rightarrow Cl(g) + e^-$$

The electrode process is followed by the formation of Cl_2 molecules:

$$2Cl(g) \rightarrow Cl_2(g)$$

The flow of current through the external circuit

The anode takes electrons from the anions. The electrons flow through the external circuit from anode to cathode. At the cathode, the electrons are available to discharge cations. The electric current is conducted through the cell by ions and through the external circuit by electrons.

In aqueous solutions, hydrogen ions may be discharged at the cathode...

When an aqueous solution is electrolysed, the products of electrolysis are not as predictable. Some metal cations are not discharged from aqueous solution. Sodium ions, for example, are present at the cathode during the electrolysis of sodium nitrate solution, but are not discharged. A small concentration of hydrogen ions arises from the dissociation of water:

$$2H_2O(l) \rightleftharpoons H_3O^+(aq) + OH^-(aq)$$

The hydrogen ions accept electrons from the cathode to form hydrogen atoms:

$$H_3O^+(aq) + e^- \rightarrow H(g) + H_2O(l)$$

Subsequently, hydrogen atoms combine rapidly to form hydrogen molecules:

$$2H(g) \rightarrow H_2(g)$$

Although the concentration of hydrogen ions is only 10^{-7} mol dm^{-3} in water, when these hydrogen ions are discharged, more are formed by the dissociation of more water molecules. This gives a vast supply of hydrogen ions, and sodium ions remain in solution while hydrogen is evolved.

...and hydroxide ions may be discharged at the anode

At the anode, both nitrate ions and hydroxide ions are present. Hydroxide ions are easier to discharge than nitrate ions. Nitrate ions remain in solution, while the electrode process is

$$OH^-(aq) \rightarrow OH(aq) + e^-$$

followed by

$$4OH(aq) \rightarrow O_2(g) + 2H_2O(l)$$

Profile: Sir Humphry Davy (1778–1829)

Sir Humphry Davy is famous for his invention of the miner's safety lamp. He refused to patent his invention, saying 'My sole object was to serve the cause of humanity'.

Davy was born in Penzance, Cornwall, in 1778, the eldest son of a woodcarver. On leaving school at 16, he was employed by a surgeon, who encouraged Davy to educate himself. At the age of 19, Davy went to work in an institute for medical research in Bristol. Within a few months, he had discovered the anaesthetic properties of laughing gas, N_2O. Davy was quick to take advantage of Volta's invention of the galvanic pile, which gave a reliable source of electricity that could be used to investigate electrolysis. It enabled Davy to isolate by electrolysis six new elements: sodium, potassium, magnesium, calcium, strontium and barium. He demonstrated that chlorine is an element and identified iodine. Davy was appointed to a lectureship in the Royal Institution. He gave spectacular lectures which were so popular that traffic jams formed in the street outside whenever he lectured. At the age of 34, Davy was knighted.

12.3.1 THE ELECTROCHEMICAL SERIES

An order can be drawn up for cations to show the relative ease of discharge...

... The same can be done for anions...

If a solution contains copper(II) ions and zinc ions, during electrolysis zinc ions remain in solution, while copper(II) ions are discharged. If a solution contains bromide ions and iodide ions, iodide ions are discharged, while bromide ions remain in solution. Cations can be arranged in order, according to their relative ease of discharge at a cathode. Anions can be arranged in order of their relative ease of discharge at an anode. The list of ions in order of ease of discharge is called the **Electrochemical Series.** A shortened version of it is shown in Table 12.2 [see also Table 13.1, § 13.1.2].

... The result is the Electrochemical Series

TABLE 12.2 The Electrochemical Series

Cations				*Anions*
K^+				NO_3^-
Ca^{2+}				SO_4^{2-}
Na^+			Ease of discharge $A^{n-} \rightarrow A + ne^-$ increases ↓	Cl^-
Mg^{2+}	Ease of discharge $M^{n+} + ne^- \rightarrow M$ increases ↓			Br^-
Al^{3+}				I^-
Zn^{2+}				OH^-
Fe^{2+}		Chemical reactivity $M \rightarrow M^{n+} + ne^-$ decreases ↓		
Sn^{2+}				
Pb^{2+}				
H_3O^+				
Cu^{2+}				
Ag^+				
Au^{3+}				

If the concentration of one ion is very much greater than the concentration of another, this factor may interfere with the expected order of discharge...

In the electrolysis of a solution containing a mixture of ions, to find out which of the ions will be discharged, you read from the bottom of the Electrochemical Series. When all the ions of that species have been discharged, then the ions next higher up in the series are discharged.

Sometimes, this rule does not predict the observed result. In solutions of halides, the relative concentrations of halide ions and hydroxide ions affect the result. The concentration of hydroxide ions in water is only 10^{-7} mol dm^{-3}. In an aqueous solution of a halide of concentration 0.1 mol dm^{-3}, the concentration of halide ions is 10^6 times greater than the concentration of hydroxide ions. Although, according to the

Electrochemical Series, hydroxide ions should be discharged, the concentration of halide ions is so much greater than the concentration of hydroxide ions that halide ions are discharged.

Similarly, inspection of the Electrochemical Series would lead you to expect that hydrogen would be evolved in the electrolysis of a lead(II) salt in aqueous solution. In fact, lead is deposited at the cathode. Since the concentration of lead ions is many powers of ten greater than the concentration of hydrogen ions in an aqueous solution of a lead(II) salt, the effect of concentration overrides the positions of the ions in the Electrochemical Series.

... The nature of the cathode can affect the order of discharge of cations

Another factor which can lead to a departure from the expected order of discharge of ions is the nature of the cathode. At a mercury cathode, the voltage needed to discharge hydrogen ions is much greater than at a platinum or graphite electrode. Sodium ions are discharged in preference to hydrogen ions at a mercury cathode, with the formation of a sodium mercury amalgam. A mercury cathode is the basis of the industrial electrolysis of aqueous sodium chloride [see Figure 18.4, § 18.5.5]. The Electrochemical Series reflects the reactivity of metals in reactions in which they form ions:

$$\mathbf{M} \rightarrow \mathbf{M}^{n+} + ne^-$$

The Electrochemical Series reflects the reactivity of the metals...

This reaction is the opposite to that which occurs when metal ions are discharged:

$$\mathbf{M}^{n+}(\text{aq}) + ne^- \rightarrow \mathbf{M}(\text{s})$$

... with some exceptions

For this reason, the order of discharge of ions is the reverse of the order of reactivity of the metals. Here again, there are some exceptions. Although sodium is below calcium in the Electrochemical Series, it reacts more vigorously with water than calcium does. The reason is probably that calcium compounds are less soluble than sodium compounds, and the rates at which metals react with water may be governed by the rates at which the compounds formed can dissolve. Aluminium does not appear to be as reactive in practice as its position in the Electrochemical Series would indicate. This is because a film of aluminium oxide forms on the surface of the metal, and this film is much less reactive than the metal itself. If the oxide layer is removed, then aluminium shows its true reactivity [see Chapter 19].

12.4 APPLICATIONS OF ELECTROLYSIS

There are many applications of electrolysis

1. Extraction of sodium by the electrolysis of molten sodium chloride n the Downs process [§ 18.3].

2. Manufacture of sodium hydroxide by the electrolysis of aqueous sodium chloride in a mercury cathode cell [§ 18.5.5] and a diaphragm cell [§ 18.8.2].

3. Manufacture of sodium chlorate(I) and sodium chlorate(V) [§ 20.10] by the electrolysis of aqueous sodium chloride.

4. Manufacture of chlorine by the electrolysis of sodium chloride as in 1 and 2 above.

5. Manufacture of hydrogen by the electrolysis of brine (aqueous sodium chloride) as in 2 above.

6. Extraction of magnesium and calcium by electrolysis of the fused (molten) chlorides [§ 18.3].

7. Extraction of aluminium by the electrolysis of fused bauxite [§ 19.2.2].

8. Anodisation and dyeing of aluminium [§ 19.2.1].

9. Purification of copper by electrolysis, using a lump of impure copper as the anode [§ 24.16].

10. Electroplating, e.g., chromium plating [§ 24.14.5].

FIGURE 12.3 Manufacture of Sodium Hydroxide: Diaphragm Cells at ICI, Lostock

FIGURE 12.4 Purification of Copper by Electrolysis

12.4.1 THE WRECK OF THE *TITANIC*

The liner *Titanic* was thought to be unsinkable, yet she sank on her maiden voyage in 1912 after colliding with an iceberg. The wreck was discovered by divers in 1985, and 1800 objects have been recovered from the sea bed. Electricité de France, EDF, has a laboratory which specialises in the restoration of objects recovered from shipwrecks and archaeological digs. Jewellery, watches, coins and many other objects have been restored.

EDF used an electrochemical process to remove the crust-like deposit of salts which had formed on the objects. This avoided the use of abrasive cleaning products. EDA wired the metal objects as cathodes in electrolysis cells. The chloride ions and other anions in the crust travelled towards the anode. Non-conducting objects, e.g. porcelain and wood, were placed between two electrodes so that ions passed through the object and flushed out the deposit. The electrochemical treatment of each object took between 200 and 1000 hours. The objects were rinsed with de-ionised water. To avoid cracking, they were dried slowly at low temperature and pressure.

CHECKPOINT 12A: ELECTROLYSIS

1. The same current was passed through molten sodium chloride and through molten cryolite containing aluminium oxide. If 4.60 g of sodium were liberated in one cell, what mass of aluminium was liberated in the other?

2. What mass of each of the following substances will be liberated by the passage of 0.200 mol of electrons?
(*a*) Mg(s) from $MgCl_2$(l), (*b*) Cl_2(g) from NaCl(aq),
(*c*) Cu(s) from $CuSO_4$(aq), (*d*) Pb(s) from $Pb(NO_3)_2$(aq),
(*e*) H_2(g) from NaCl(aq).

3. During the electrolysis of a 1.00 mol dm^{-3} solution of copper(II) sulphate at 20 °C, a current of 0.100 A was passed for 1.00 h. The mass of the copper cathode increased by 0.118 g. State how the change in mass would be affected if the experiment were repeated

(*a*) at a current of 0.200 A

(*b*) at 30 °C

(*c*) with a 2.00 mol dm^{-3} solution of copper(II) sulphate

(*d*) for a time of 2.00 h

4. When a metal of relative atomic mass 207 is deposited by electrolysis, a current of 0.0600 A flowing for 66 min increases the mass of the cathode by 0.254 g.

Find the number of moles of metal deposited and the number of moles of electrons that have passed. Deduce the number of units of charge on the cations of this metal.

12.5 ELECTROLYTIC CONDUCTIVITY

Ohm's Law

The conduction of current through an electrolytic cell obeys **Ohm's Law**. This law states that

$$I \propto V$$

where I = current and V = potential difference between electrodes.
It follows from Ohm's Law that

$$R = V/I$$

where R = resistance.

Resistivity, ρ (rho), is defined by the equation

Definition of resistivity...

$$R = \rho l/A \qquad [1]$$

Definition of conductivity...

where l = length and A = cross-sectional area of conductor. The reciprocal of resistance is conductance; the reciprocal of resistivity is **conductivity**, κ (kappa):

$$1/\rho = \kappa \qquad [2]$$

Combining equations [1] and [2] gives an equation for **conductivity**:

$$\kappa = l/AR \qquad [3]$$

Molar conductivity, Λ (lambda), is defined by the equation

Definition of molar conductivity

$$\Lambda = \kappa/c$$

where Λ (lambda) = molar conductivity, κ = conductivity, and c = concentration. If κ is in $\Omega^{-1}\,m^{-1}$ and c is in $mol\,m^{-3}$, then Λ has the unit $\Omega^{-1}\,m^2\,mol^{-1}$, and Λ is numerically equal to the conductivity of 1 mole of the electrolyte.

12.5.1 MOLAR CONDUCTIVITY AND CONCENTRATION

Molar conductivity increases with dilution. For strong electrolytes Λ soon reaches a maximum...

The molar conductivity of a solute depends on its concentration. When values of molar conductivity, Λ (lambda), are plotted against the volume of solution containing 1 mole of solute, that is the **dilution**, the graphs obtained for different electrolytes fall into two categories. These are shown in Figure 12.5. **Strong electrolytes** are those which are completely ionised in solution. They give graphs of shape **A**, swiftly rising to a maximum. The maximum value of Λ is called the **molar conductivity at infinite dilution**, as further dilution has no effect on its value. It is represented as Λ_∞ or Λ_0.

FIGURE 12.5 Plots of Molar Conductivity against Dilution

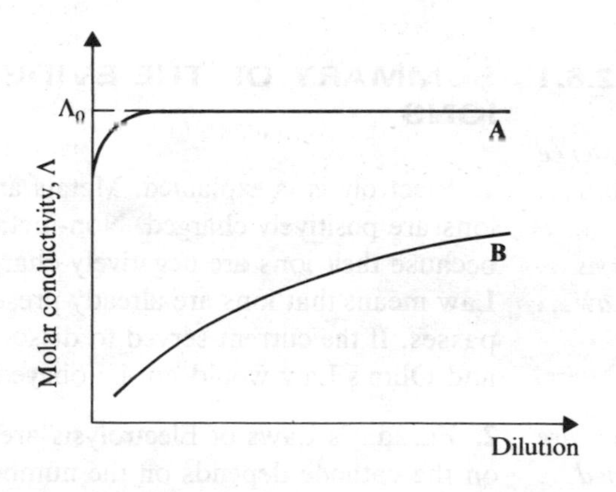

...For weak electrolytes, Λ cannot be followed to a maximum

Weak electrolytes, which are incompletely ionised, give graphs similar in shape to **B**. The value of Λ increases as the solution becomes more and more dilute. There is a limit to the range of concentration over which measurements can be made as, in very dilute solutions, the conductivity of the solution is very little greater than the conductivity of water, and no worthwhile values can be obtained.

The degree of ionisation of a weak electrolyte increases with dilution...

S Arrhenius, in 1873, explained the shape of curve **B** as arising from an increase in the ionisation of the solute with dilution. He suggested that the fraction α of the solute which is ionised gradually increases with dilution. He postulated that, if measurements could be extended to even more dilute solutions, α would approach a value of 1, and Λ would level off at a maximum value, Λ_0. The value of the degree of ionisation at any concentration is given by

$$\alpha = \Lambda/\Lambda_0$$

The ratio Λ/Λ_0 is called the **conductivity ratio**.

The shapc of curve A suggested to Arrhenius that strong electrolytes are completely ionised in solution, except at very high concentrations. In the light of the twentieth century view of the nature of the chemical bond, this idea is unacceptable. We cannot imagine that an electrolyte such as copper(II) sulphate can be other than completely ionised: there is no possibility of covalent bonding.

Strong electrolytes show low Λ at high concentration

According to the modified ionic theory put forward by P Debye and E Hückel and improved by L Onsager, strong electrolytes are always completely ionised in solution. The decrease in molar conductivity in concentrated solutions must have another explanation. This is discussed in § 12.6.2. The value of α obtained from Λ/Λ_0 is called the **apparent degree of ionisation** or the **conductivity ratio**.

α, the conductivity ratio

CHECKPOINT 12B: CONDUCTIVITY

1. Explain what is meant by (*a*) the electrolytic conductivity and (*b*) the molar conductivity of an electrolyte. Sketch curves to show how (*a*) and (*b*) vary with dilution for (i) a strong electrolyte and (ii) a weak electrolyte.

12.6 THE IONIC THEORY

Arrhenius put forward the theory in 1887 that, when an electrolyte dissolves, a certain fraction of it dissociates into positively and negatively charged particles called ions. His theory explained an enormous volumc of work on electrolysis and electrolytic conduction. Evidence for the existence of ions is summarised below.

12.6.1 SUMMARY OF THE EVIDENCE FOR THE EXISTENCE OF IONS

Evidence for the existence of ions...

...from electrolysis...

...and Ohm's Law...

1. Electrolysis is explained. Metals are deposited only at the cathode because their ions are positively charged. Non-metallic elements are liberated only at the anode because their ions are negatively charged. The fact that electrolytes obey Ohm's Law means that ions are already present in the solution or the melt before the current passes. If the current served to dissociate the electrolyte, energy would be consumed, and Ohm's Law would not be obeyed.

...Faraday's Laws are explained...

2. Faraday's Laws of Electrolysis are explained. The amount of a metal deposited on the cathode depends on the number of moles of electrons which flow through the circuit and on the number of charges on one ion of the metal.

Evidence from $\Delta H^\ominus$ of neutralisation...

3. The standard enthalpy of neutralisation has a constant value of approximately $-58\,\text{kJ mol}^{-1}$, regardless of which strong acid and base are used for its measurement. This is because the neutralisation reaction is in each case the reaction between hydrogen ions and hydroxide ions:

$$H_3O^+(aq) + OH^-(aq) \rightarrow 2H_2O(l)$$

...from the increase of Λ with dilution...

4. The increase in molar conductivity with dilution can be explained. In the case of weak electrolytes, it is attributed to an increase in the degree of ionisation with increasing dilution. The variation of molar conductivity with dilution for strong electrolytes is explained by the Debye–Hückel and Onsager Theory see below.

...from chemical properties...

5. The chemical properties of an electrolyte are the sum of the properties of its ions. All chlorides give a white precipitate with silver nitrate solution. All iron(II) salts give a green precipitate with sodium hydroxide solution.

...and from X ray crystallography

6. X ray crystallography has shown that salts consist of ions, even in the solid state. X ray diffraction patterns of sodium chloride show a three-dimensional structure of alternate sodium and chloride ions [see § 6.3, Figure 6.6 and Figure 4.13, § 4.2.2].

The last piece of evidence was not available to Arrhenius. Since his day, the structure of the atom has been elucidated [see Chapters 1 and 2]. The nature of the chemical bond in electrolytes is believed to be an electrostatic attraction between oppositely charged ions. These developments led to a modification of the Arrhenius Theory by Debye and Hückel and by Onsager.

12.6.2 THE DEBYE–HÜCKEL AND ONSAGER THEORY: INTERIONIC ATTRACTIONS

The low values of molar conductivity for strong electrolytes at high concentrations are explained

Debye and Hückel (1923) and Onsager (1927) modified the Ionic Theory as presented by Arrhenius. They said that strong electrolytes are always completely ionised. They explained the low molar conductivities of ions in concentrated solutions in a different way. They postulated that ions move more slowly in concentrated solutions. Since ions are closer together in concentrated solutions than in dilute solutions, the electrical forces between ions are greater. As unlike charges attract, a cation will have more anions than cations in its immediate neighbourhood. The attractive forces between these anions and the cation reduce the velocity of the cation [see Figure 12.6]. The anions are similarly retarded in concentrated solutions.

FIGURE 12.6 Ions are slowed down in Concentrated Solutions

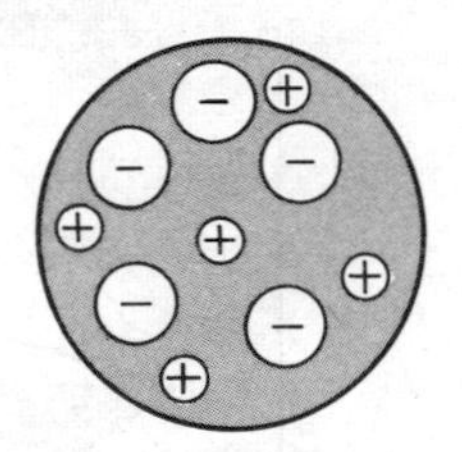

(a) In a concentrated solution, the distribution of ions is not completely random. A cation has more anions than cations surrounding it.

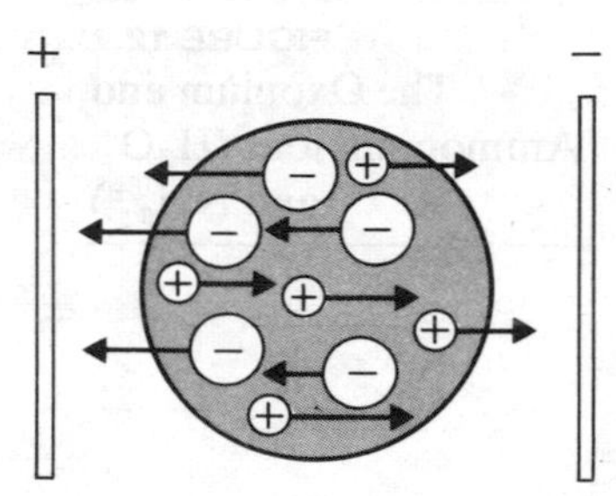

(b) As a cation moves towards the cathode, it is slowed down by the forces of attraction between it and the anions.

12.7 IONIC EQUILIBRIA

12.7.1 ACIDS AND BASES

Early descriptions of acids and alkalis

Before the work of W Ostwald and S Arrhenius on the dissociation of electrolytes, there were attempts to define acids and bases. The sour taste and the effect on vegetable colourings, such as litmus, characterised acids. The soapy feel and detergent power characterised alkalis. It was recognised that acids react with alkalis and also with some other compounds to give salts. The term **base** superseded the term **alkali** as meaning the opposite of an acid. A base was defined as a substance which would react with an acid to form a salt. The study of acids progressed from Boyle (1663) through Lavoisier (1780) and Davy (1810) to Liebig (1838). Liebig stated that acids are compounds which contain hydrogen that can be displaced by metals. The explanation of how the possession of hydrogen confers acidic properties was not forthcoming until the Ostwald–Arrhenius **Theory of Electrolytic Dissociation** in 1880. They identified the hydrogen atoms which give rise to acidic properties as being those which form hydrogen ions in solution. Bases were said to produce hydroxide ions in solution and to neutralise acids by the reaction

In the dissociation theory...

...acids give H^+ ions and bases give OH^- ions

$$H^+ + OH^- \rightarrow H_2O$$

Some difficulties arose with these definitions

The Dissociation Theory definitions of acids and bases ran into difficulties. As pure hydrogen chloride does not conduct electricity, should it be classified as an acid, or does it only become an acid in contact with water? A solution of sodium ethoxide in ethanol has strongly basic properties. It contains no OH^- ions, but it does contain $C_2H_5O^-$ ions. Bases such as ammonia neutralise acids by picking up a proton, rather than by providing hydroxide ions:

$$NH_3 + H^+ \rightarrow NH_4^+$$

H^+ ions cannot exist in solution: they are so small that the charge density is high, and they must be solvated...

As work proceeded, doubt was cast on the existence of the hydrogen ion, H^+, in solution. The proton, H^+, is very small (10^{-15} m diameter) compared with other cations (around 10^{-10} m diameter). The electric field in its neighbourhood is so intense that it attracts any molecule with unshared electrons, such as H_2O. The reaction

$$H^+ + H_2O \rightarrow H_3O^+$$

was shown (by spectroscopic measurements) to liberate 1300 kJ mol^{-1}. As the reaction is so exothermic, unhydrated protons do not exist in solution. (They are produced in gaseous reactions and in nuclear reactions.) The hydrated proton, H_3O^+, is called the **oxonium ion**, and is also referred to as the hydrogen ion. In this text, H_3O^+ will be called a hydrogen ion. Its structure is shown in Figure 12.7 with that of the ammonium ion for comparison.

...In water, H_3O^+ ions are formed

FIGURE 12.7 The Oxonium and Ammonium ions (H_3O^+ and NH_4^+)

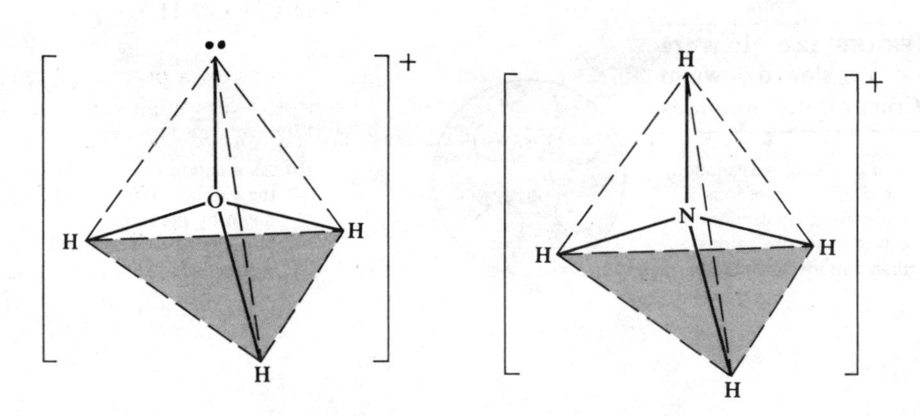

The H_3O^+ ion, like other cations, is solvated by water molecules, and is sometimes written as $H_9O_4^+$. In solution in ethanol, the hydrogen ion is $C_2H_5OH_2^+$, and in liquid ammonia it is NH_4^+.

THE BRÖNSTED–LOWRY DEFINITION OF ACIDS AND BASES

Brönsted and Lowry described acids as proton-donors and bases as proton-acceptors

Acknowledging that the proton does not exist in solution made it necessary to review the definition of acids. The best definition of acids and bases is that proposed by T M Lowry and also, independently, by J N Brönsted in 1923. They said that an acid is a substance which can donate a proton to another substance. A base is a substance which can accept a proton from another substance. The relationship is

$$\text{Acid} \rightleftharpoons \text{Base} + H^+$$

This equation does not represent an actual reaction in solution since the proton, H^+, cannot exist in solution. The acid and base which are related in this way, by the exchange of a proton, are called a **conjugate acid–base pair**. Since

$$\text{Acid 1} \rightleftharpoons \text{Conjugate base 1} + H^+$$

and $$\text{Base 2} + H^+ \rightleftharpoons \text{Conjugate acid 2}$$

a reaction between an acid and a base is

Conjugate acid–base pairs

$$\text{Acid 1} + \text{Base 2} \rightleftharpoons \text{Conjugate base 1} + \text{Conjugate acid 2}$$

Reactions between acids and bases

Acid 1 is transformed into its conjugate base, and Base 2 is transformed into its conjugate acid. No substance can act as an acid in solution unless a base is present to accept a proton: the reactions of acids are reactions between acids and bases. Similarly, all reactions of bases in solution are acid–base reactions.

Examples of Lowry–Brönsted acids and bases

Examples of **Brönsted–Lowry acids** are

$$HCl + H_2O \rightarrow H_3O^+ + Cl^- \quad [1]$$

$$HSO_4^- + H_2O \rightleftharpoons H_3O^+ + SO_4^{2-} \quad [2]$$

$$CH_3CO_2H + H_2O \rightleftharpoons H_3O^+ + CH_3CO_2^- \quad [3]$$

Examples of **Brönsted–Lowry bases** are:

$$NH_3 + H_2O \rightleftharpoons NH_4^+ + OH^- \quad [4]$$

$$RNH_2 + H_2O \rightleftharpoons RNH_3^+ + OH^- \quad [5]$$

$$HSO_4^- + H_3O^+ \rightleftharpoons H_2SO_4 + H_2O \quad [6]$$

$$CH_3CO_2^- + H_2O \rightleftharpoons CH_3CO_2H + OH^- \quad [7]$$

In [1], [2] and [3], water is acting as a proton-acceptor, a base; in [4], [5], [6] and [7], water is acting as a proton-donor, an acid. Water is described as an **amphoteric** or **amphiprotic** solvent. The HSO_4^- ion is an amphoteric or amphiprotic species.

THE LEWIS DEFINITION OF ACIDS AND BASES

There are reactions which appear to us, on common-sense grounds, to be acid–base reactions, and which do not come within the scope of the Brönsted–Lowry definition. Such reactions are

$$CaO + SO_3 \rightarrow CaSO_4$$

$$NH_3 + BF_3 \rightarrow NH_3BF_3$$

A Lewis base gives a lone pair of electrons to a Lewis acid... ...with the formation of a covalent bond

To accommodate reactions of this type, G N Lewis (during 1930 to 1940) proposed a fresh definition of acids and bases. He described as acid–base reactions those reactions in which an unshared electron pair in the base molecule is accepted by the acid molecule, with the formation of a covalent bond. For example, in the reaction

$$H_3N\!: + BF_3 \rightarrow H_3N\!\rightarrow\!BF_3$$

A comparison of the two definitions

ammonia is the base, and boron trifluoride is the acid. The Lewis definition of a base includes the Brönsted–Lowry bases because a species with a lone pair of electrons will accept a proton from a Brönsted–Lowry acid. Lewis acids, such as BF_3 and SO_3, are not acids in the Brönsted–Lowry sense, and acids such as HCl, H_2SO_4 and CH_3CO_2H are not acids according to the Lewis definition.

The Brönsted–Lowry description of acids and bases lends itself readily to a quantitative treatment of the strengths of acids and bases [§ 12.7.3 and § 12.7.6]. No such quantitative treatment is possible for Lewis acids and bases.

12.7.2 THE IONIC PRODUCT FOR WATER

The hydrogen ion concentration in a solution can be denoted by means of the **hydrogen ion exponent** or **pH** of the solution. The pH of a solution is the negative logarithm to the base ten of the hydrogen ion concentration in $mol\,dm^{-3}$. **pOH** is related in the same way to the hydroxide ion concentration:

pH and pOH...

$$pH = -\lg[H_3O^+/mol\,dm^{-3}]$$

$$pOH = -\lg[OH^-/mol\,dm^{-3}]$$

Water is only slightly ionised:

$$2H_2O \rightleftharpoons H_3O^+ + OH^-$$

The product of the concentrations of hydrogen ions and hydroxide ions is equal to $1.00 \times 10^{-14} mol^2 dm^{-6}$ at 25 °C. This product is called the **ionic product** for water, K_w.

... The ionic product K_w and pK_w

$$[H_3O^+][OH^-] = K_w = 1.00 \times 10^{-14} mol^2 dm^{-6}$$

$$pH + pOH = pK_w = 14 \text{ at } 25\,°C$$

This relationship is true of aqueous solutions as well as water.

12.7.3 CALCULATION OF pH AND pOH FOR STRONG ACIDS AND BASES

The pH of a solution of a strong acid or strong base is simply calculated.

Example What is the pH of a solution of hydrochloric acid of concentration $0.1\,mol\,dm^{-3}$?

Method

Since

Calculation of pH...

$$[H_3O^+] = 0.1\,mol\,dm^{-3} = 10^{-1} mol\,dm^{-3}$$

$$\lg[H_3O^+] = -1 \text{ and } pH = 1$$

Example What is the pH of a solution of sodium hydroxide of concentration $0.01\,mol\,dm^{-3}$ at 25 °C?

Method

Since

... Calculation of pOH

$$[OH^-] = 0.01 = 10^{-2} mol\,dm^{-3}$$

$$pOH = 2$$

$$pH = 14 - pOH = 12 \text{ at } 25\,°C$$

At 25 °C, solutions with a pH of 7 are neutral, solutions with a pH less than 7 are acidic, and solutions with a pH greater than 7 are alkaline. The relationship between pH and $[H_3O^+]$ and $[OH^-]$ is illustrated in Table 12.3.

TABLE 12.3
pH Values of Acidic and Alkaline Solutions

$[H_3O^+]$/ mol dm^{-3}	1	10^{-1}	10^{-2}	10^{-3}	10^{-4}	10^{-5}	10^{-6}	10^{-7}	10^{-8}	10^{-9}	10^{-10}	10^{-11}	10^{-12}	10^{-13}	10^{-14}
$[OH^-]$/ mol dm^{-3}	10^{-14}	10^{-13}	10^{-12}	10^{-11}	10^{-10}	10^{-9}	10^{-8}	10^{-7}	10^{-6}	10^{-5}	10^{-4}	10^{-3}	10^{-2}	10^{-1}	1
pH	0	1	2	3	4	5	6	7	8	9	10	11	12	13	14
	Strongly acidic			Weakly acidic				Neutral		Weakly alkaline			Strongly alkaline		

CHECKPOINT 12C: ACIDS AND BASES

1. Define the terms *Brönsted acid* and *Brönsted base*. Give two examples of each, explaining how they fit the definitions. Define the terms *Lewis acid* and *Lewis base*, and give one example of each.

†**2.** Explain the following statements:
(*a*) $[Al(H_2O)_6]^{3+}$ is classified as an acid [§ 19.2.4].
(*b*) $ClCH_2CO_2H$ is a stronger acid than CH_3CO_2H [§ 12.7.7].
(*c*) The Friedel–Crafts catalyst, $AlCl_3$, is classified as a Lewis acid [§ 28.7.6].

3. How are Lewis acids and bases different from Brönsted–Lowry acids and bases and from Arrhenius acids and bases? In the equilibrium

$$NH_3 + H_2O \rightleftharpoons NH_4^+ + OH^-$$

which species is the acid and which the base, according to the definitions of (*a*) Arrhenius and (*b*) Lowry and Brönsted?

4. According to the Brönsted–Lowry theory, water can function as an acid and as a base. Quote one reaction in which water acts as an acid and one in which it acts as a base.

5. At 0 °C, $K_w = 1.14 \times 10^{-15}\ mol^2\ dm^{-6}$. Find:
(*a*) pK_w at 0 °C
(*b*) the pH at 0 °C of a 0.01 mol dm^{-3} solution of sodium hydroxide

6. Name the Lewis acid and the Lewis base in the following reactions:
(*a*) $RCOBr + FeBr_3 \rightleftharpoons RCO^+FeBr_4^-$
(*b*) $Ag^+ + 2NH_3 \rightarrow Ag(NH_3)_2^+$
(*c*) $CH_3CO_2^- + 2HF \rightarrow CH_3CO_2H + HF_2^-$

12.7.4 WEAK ELECTROLYTES

The degree of ionisation of a weak electrolyte

Weak electrolytes consist of molecules, some of which dissociate to form ions. The fraction of molecules which dissociate is called the **degree of ionisation** or **degree of dissociation**. As the concentration of a solution of a weak electrolyte decreases, the degree of ionisation increases. **Ostwald's Dilution Law** gives a relationship between the degree of ionisation α of a weak electrolyte and its concentration c. Consider a weak acid HA. An equilibrium is set up between undissociated molecues HA and the ions H_3O^+ and A^-:

$$HA + H_2O \rightleftharpoons H_3O^+ + A^-$$

The equilibrium constant [§ 11.1] is called the **acid dissociation constant**, K_a, and is given by

The dissociation constant of a weak acid

$$K_a = \frac{[H_3O^+][A^-]}{[HA]}$$

The square brackets represent concentration in mol dm^{-3}, e.g., $[H_3O^+]$ is the concentration of hydrogen ions. The expression for K_a does not include $[H_2O]$ because the concentration of water remains constant in all reasonably dilute solutions. If 1 mole of acid is dissolved in V dm^3 of solution, the amounts and concentrations of each species at equilibrium are as follows:

	HA	+ $H_2O \rightleftharpoons$	H_3O^+	+ A^-
Amount/mol	$1 - \alpha$		α	α
Concentration/mol dm^{-3}	$(1 - \alpha)/V$		α/V	α/V

Putting these concentrations into the expression for K_a gives

$$K_a = \frac{(\alpha/V)^2}{(1-\alpha)/V} = \frac{\alpha^2}{(1-\alpha)V}$$

Since V = volume containing 1 mol of solute, $1/V = c$, and

Ostwald's Dilution Law

$$K_a = \frac{\alpha^2 c}{(1-\alpha)}$$

This expression embodies Ostwald's Dilution Law.

For many weak electrolytes, α is so small that the error involved in putting $(1 - \alpha) = 1$ is negligible. Then

$$K_a = c\alpha^2$$

$$\alpha = \sqrt{K_a/c}$$

Ostwald's Dilution Law can be applied to weak bases. In the case of a weak base **B**, which is partially ionised in solution

$$\mathbf{B} + H_2O \rightleftharpoons \mathbf{BH}^+ + OH^-$$

The base dissociation constant, K_b, is given by the equation

$$K_b = \frac{[\mathbf{BH}^+][OH^-]}{[\mathbf{B}]}$$

$$K_b = c\alpha^2/(1-\alpha)$$

If $\alpha \ll 1$ $K_b = c\alpha^2$

The value of K_a for an acid is a quantitative measure of the strength of the acid in all its typical reactions...

The value of its dissociation constant tells you how strong an acid is and how vigorously it will take part in the reactions which are typical of acids. It enables you to calculate the conductivity and the pH of a solution of the acid [§ 12.7.3]. It is a measure of the effectiveness of an acid in acid-catalysed reactions. All these aspects of acid behaviour are covered by one physical constant. From the values of K_a, you can say that chloroethanoic acid is 80 times stronger than ethanoic acid. The value of the dissociation constant of a base is equally important. From the values of K_b, you can say that methylamine is 23 times as strong a base as ammonia. The quantitative measure of the strengths of acids and bases provided by the K_a and K_b values is a splendid feature of the Brönsted–Lowry treatment of acids and bases. No such quantitative treatment can be made of Lewis acids and bases.

...K_b for a base is equally important

Tables often list pK values of acids and bases. These are defined as

$$pK_a = -\lg(K_a/\text{mol dm}^{-3})$$

$$pK_b = -\lg(K_b/\text{mol dm}^{-3})$$

The higher the value of K_a (or K_b), the lower the value of pK_a (or pK_b), and the stronger is the acid (or base).

12.7.5 CONJUGATE ACID–BASE PAIRS

The relationship between an acid and its conjugate base

Some tables list the base dissociation constants K_b of bases such as amines. Others list the acid dissociation constants K_a of their conjugate acids. The conjugate acid $\mathbf{BH^+}$ of a base **B** dissociates according to the equilibrium

$$\mathbf{BH^+} + H_2O \rightleftharpoons \mathbf{B} + H_3O^+$$

The acid dissociation constant of $\mathbf{BH^+}$ is

$$K_a = \frac{[\mathbf{B}][H_3O^+]}{[\mathbf{BH^+}]}$$

Multiplying K_a by K_b [§ 12.7.4] gives $K_aK_b = [H_3O^+][OH^-]$ that is

$$K_aK_b = K_w$$

This is a useful relationship between the dissociation constants of acid–base pairs. It can also be expressed as

$$pK_a + pK_b = pK_w$$

12.7.6 HOW TO CALCULATE THE pH OF A SOLUTION OF A WEAK ACID OR A WEAK BASE

In order to calculate the pH of a solution of a weak acid or base, you must know the concentration of the solution and the dissociation constant of the acid or base. The converse is also true: if the pH of a solution is measured, you can use it to find the dissociation constant of the weak electrolyte.

Example Calculate the pH of a 1.00×10^{-2} mol dm^{-3} solution of butanoic acid, for which $K_a = 1.51 \times 10^{-5}$ mol dm^{-3}.

Method Use the equation

A sample calculation of the pH of a solution of a weak acid...

$$K_a = \frac{[H_3O^+][C_3H_7CO_2^-]}{[C_3H_7CO_2H]}$$

The concentrations, $[H_3O^+]$ and $[C_3H_7CO_2^-]$ are equal.

The concentration $[C_3H_7CO_2H]$ is very little less than 1.00×10^{-2} mol dm^{-3}. Since the degree of ionisation is small, the approximation $[C_3H_7CO_2H] = 1.00 \times 10^{-2}$ mol dm^{-3} is made to simplify the calculation. Then

$$[H_3O^+]^2 = 1.51 \times 10^{-5} \times 1.00 \times 10^{-2}$$

$$[H_3O^+] = 3.89 \times 10^{-4} \text{ mol dm}^{-3}$$

$$pH = 3.42$$

How valid was the approximation $[C_3H_7CO_2H] = 1.00 \times 10^{-2}\,mol\,dm^{-3}$?
Since

$$[H_3O^+] = 3.89 \times 10^{-4}\,mol\,dm^{-3} \text{ (approximately)}$$

$$[C_3H_7CO_2H] = (1.00 \times 10^{-2}) - (3.89 \times 10^{-4}) = 0.96 \times 10^{-2}\,mol\,dm^{-3}$$

If this approximate value is used in a new calculation, a new value of pH = 3.42 is obtained. This is the same as in the first case, and shows that the approximation was justified*. For most weak acids and bases, the approximation can be safely made.

...and a calculation of a dissociation constant from the pH of a solution of a weak base

Example A solution of dimethylamine of concentration $1.00 \times 10^{-2}\,mol\,dm^{-3}$ has a pH of 7.64 at 25 °C. Calculate (*a*) the dissociation constant of the base and (*b*) the degree of dissociation.

Method The dissociation of the base can be represented by

$$(CH_3)_2NH + H_2O \rightleftharpoons (CH_3)_2NH_2^+ + OH^-$$

Thus $$K_b = \frac{[(CH_3)_2NH_2^+][OH^-]}{[(CH_3)_2NH]}$$

$$= [OH^-]^2/(1.00 \times 10^{-2})$$

Since

$$pH = 7.64,\ pOH = 14.0 - 7.64 = 6.36 \text{ at } 25\,°C$$

$$[OH^-] = \text{antilg}(-6.36) = 4.37 \times 10^{-7}\,mol\,dm^{-3}$$

$$K_b = (4.37 \times 10^{-7})^2/(1.00 \times 10^{-2}) = 1.91 \times 10^{-11}\,mol\,dm^{-3}$$

From the Ostwald Dilution Law, if α = degree of dissociation

$$K_b = \frac{\alpha^2 c}{1 - \alpha}$$

For a weak base

$$K_b = \alpha^2 c$$

$$1.91 \times 10^{-11} = \alpha^2 \times 1.00 \times 10^{-2}$$

Degree of dissociation

$$\alpha = 4.37 \times 10^{-5}$$

12.7.7 *HOW SUBSTITUENTS AFFECT THE STRENGTH OF ACIDS AND BASES

If X is more electronegative [§4.5.3] than carbon, the acid XCH_2CO_2H is a stronger acid than CH_3CO_2H. An example is chloroethanoic acid, $ClCH_2CO_2H$, which is 80 times stronger than ethanoic acid. The chlorine nucleus in the anion $ClCH_2CO_2^-$ attracts the electrons in the Cl—C bond, enabling the charge to be spread through the anion more than it is in $CH_3CO_2^-$:

$$Cl{-}CH_2{-}C\left\{\begin{matrix}O\\O\end{matrix}\right\}^- \qquad H{-}CH_2{-}C\left\{\begin{matrix}O\\O\end{matrix}\right\}^-$$

*See E N Ramsden, *Calculations for A-level Chemistry* (Stanley Thornes)

Electron withdrawing substituents make acids stronger, bases weaker...

The reduction of the charge located on the oxygen atoms makes $ClCH_2CO_2^-$ a weaker proton acceptor (a weaker base) than $CH_3CO_2^-$, and therefore makes $ClCH_2CO_2H$ a stronger acid than CH_3CO_2H. Dichloroethanoic acid is stronger still, and trichloroethanoic acid is as strong as some mineral acids. Values of K_a are given below:

CH_3CO_2H 1.8×10^{-5} mol dm^{-3} Cl_3CCO_2H 2.2×10^{-1} mol dm^{-3}

$ClCH_2CO_2H$ 1.4×10^{-3} mol dm^{-3} HCO_2H 1.7×10^{-4} mol dm^{-3}

...Electron donating substituents make acids weaker, bases stronger

Methanoic acid HCO_2H, is a stronger acid than ethanoic acid. It is inferred from this that the group —CH_3 is less electronegative than hydrogen:

$$\mathrm{H{-}C}\left\{\begin{matrix}\mathrm{O}\\ \\ \mathrm{O}\end{matrix}\right\}^- \qquad \mathrm{H_3C{-}C}\left\{\begin{matrix}\mathrm{O}\\ \\ \mathrm{O}\end{matrix}\right\}^-$$

By donating electrons to the —CO_2^- group, —CH_3 makes it a stronger base (a better proton acceptor) and makes CH_3CO_2H a weaker acid than HCO_2H.

Bases are proton acceptors. If a group X is substituted for hydrogen in methylamine, then XCH_2NH_2 will have a different value of K_b from CH_3NH_2. Groups such as —CH_3, which increase the availability of electrons at the nitrogen atom, increase its power to attract protons, i.e., its basicity. Thus dimethylamine, $(CH_3)_2NH$ is a stronger base than methylamine. Values of basic dissociation constants are

CH_3NH_2 4.4×10^{-4} mol dm^{-3} $(CH_3)_2NH$ 5.9×10^{-4} mol dm^{-3}

A group (e.g. Cl—) which is more electronegative than carbon is described as having a negative inductive effect (a −I effect). A group which is more electropositive than carbon is said to have a positive inductive (+I) effect. Inductive effects are permanent polarisations of molecules. They affect the physical properties of the compounds [see §4.5.3].

CHECKPOINT 12D: pH AND DISSOCIATION CONSTANTS

1. The value of K_w, the ionic product for water, increases with temperature. Will the pH of pure water be greater or less than 7 at 100 °C?

2. The pH values of 0.100 mol dm^{-3} solutions of hydrochloric acid and ethanoic acid are 1 and approximately 3 respectively. How does this difference arise?

3. Give an expression for the dissociation constant, K_a, of a weak acid. How is pK_a related to K_a? If two acids, HA and HB, have pK_a values of 3.4 and 4.4, what can you say about the relative strengths of the two acids?

4. Explain what is meant by a conjugate acid–base pair. Give two examples.

5. Find the pH of the following solutions at 25 °C:
(*a*) 0.00100 mol dm^{-3} HCl(aq)
(*b*) 2.50×10^{-2} mol dm^{-3} $HClO_4$(aq) (a strong acid)
(*c*) 3.60×10^{-5} mol dm^{-3} HNO_3(aq)
(*d*) 2.50×10^{-2} mol dm^{-3} NaOH(aq)
(*e*) 3.00×10^{-3} mol dm^{-3} $Ca(OH)_2$(aq)

6. Find the pH of the following solutions:
(*a*) 1.00×10^{-2} mol dm^{-3} ethanoic acid (pK_a = 4.76)
(*b*) 0.100 mol dm^{-3} methanoic acid (pK_a = 3.75)

7. Find the dissociation constants of the following acids from the data:
(*a*) A solution of 2.00 mol dm^{-3} hydrogen cyanide has a pH of 4.55.
(*b*) A solution of 1.00 mol dm^{-3} iodic(I) acid has a pH of 5.26.

8. Calculate pK_b for the following weak bases at 25 °C from the data:
(*a*) A solution of 1.00×10^{-2} mol dm^{-3} $C_6H_5CH_2NH_2$ has a pH of 10.7.
(*b*) A solution of 1.00×10^{-2} mol dm^{-3} $(C_2H_5)_2NH$ has a pH of 11.5.

12.7.8 INDICATORS

Indicators are weak acids or weak bases...

The indicators used in acid–base titrations are weak acids or weak bases. The ions are of a different colour from the undissociated molecules. Litmus is a weak acid, which can be represented by the formula HL. In solution

$$HL + H_2O \rightleftharpoons H_3O^+ + L^-$$

...Litmus

The molecules HL are red, and the anions L^- are blue. The dissociation constant of the indicator is

$$K_a = \frac{[H_3O^+][L^-]}{[HL]}$$

In acid solution, $[H_3O^+]$ is high, and H_3O^+ ions combine with L^- ions to form HL molecules, which are red. In alkaline solution, H_3O^+ ions are removed to form molecules of water, and HL molecules react with OH^- ions to form L^- ions, which are blue:

Their molecules and ions differ in colour...

$$HL + OH^- \rightleftharpoons L^- + H_2O$$

If $[HL] = [L^-]$, the indicator appears purple. Since

$$[H_3O^+] = K_a\frac{[HL]}{[L^-]}$$

this happens when

$$[H_3O^+] = K_a$$

and $\quad pH = pK_a$

...They change colour over 2 units of pH...

If the ratio $[HL]/[L^-] \geqslant 10/1$, the solution appears red. If the ratio $[HL]/[L^-] \leqslant 1/10$, the solution appears blue. Thus litmus changes from red to blue over a range of hydrogen ion concentration from

$$[H_3O^+] = K_a \times 10/1 \text{ to } [H_3O^+] = K_a \times 1/10$$

which gives a range of pH from $pH = pK_a + 1$ to $pH = pK_a - 1$. The colour change takes place over a range of about 2 pH units.

The indicator methyl orange is a weak base, which can be represented as B [Figure 12.10]:

...Methyl orange

$$B + H_3O^+ \rightleftharpoons BH^+ + H_2O$$

The molecules B are yellow, and the cations BH^+ are red. If the ratio $[B]/[BH^+] \geqslant 10/1$, the indicator appears yellow; if the ratio $\leqslant 1/10$, the indicator appears red. The indicator changes from yellow to red over the range of pH given by $pK_b + 1$ to $pK_b - 1$, i.e., 2 pH units. At a pH equal to pK_b, the ratio $[B]/[BH^+] = 1$, and the indicator is orange.

12.7.9 ACID–BASE TITRATIONS

To understand how indicators show the end-point in acid–base titrations, it is necessary to calculate the way in which pH changes during the course of a titration.

TITRATION OF A STRONG BASE INTO A STRONG ACID

Consider the titration of 25.0 cm³ of 0.100 mol dm⁻³ hydrochloric acid with 0.100 mol dm⁻³ sodium hydroxide solution. At the beginning of the titration

$$[H_3O^+] = 1.00 \times 10^{-1}\,\text{mol dm}^{-3};\ pH = 1.00$$

When $24.0\,cm^3$ of sodium hydroxide have been added

$$[H_3O^+] = (1.0\,cm^3 \text{ of } 0.100\,mol\,dm^{-3} \text{ acid})/(49.0\,cm^3 \text{ of solution})$$
$$[H_3O^+] = (1.0 \times 10^{-3} \times 0.100)/(49.0 \times 10^{-3}) = 2.04 \times 10^{-3}\,mol\,dm^{-3}$$
$$pH = 2.69$$

Calculation of pH changes during titration of a strong acid by a strong base . . .

Similar calculations have been employed to give all the pH values plotted in Figure 12.8(a). The figure shows how pH varies during the course of the titration. [See Question 4, Checkpoint 12E.]

FIGURE 12.8 Changes in pH during Titration (of base ($0.100\,mol\,dm^{-3}$) into $25.0\,cm^3$ of acid ($0.100\,mol\,dm^{-3}$))

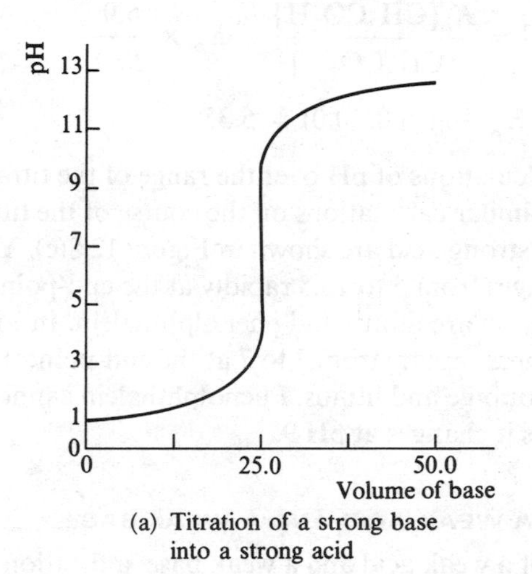

(a) Titration of a strong base into a strong acid

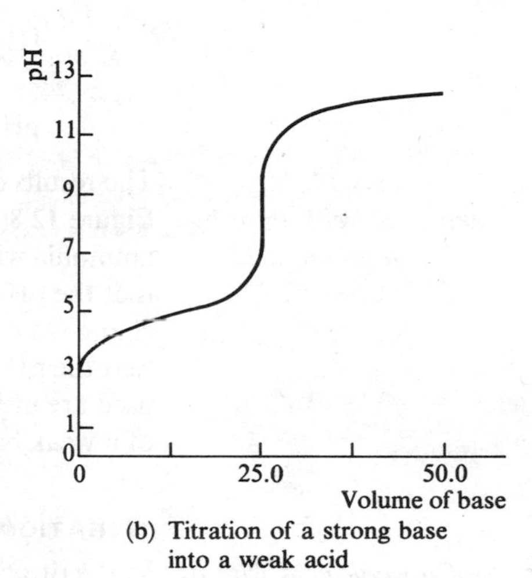

(b) Titration of a strong base into a weak acid

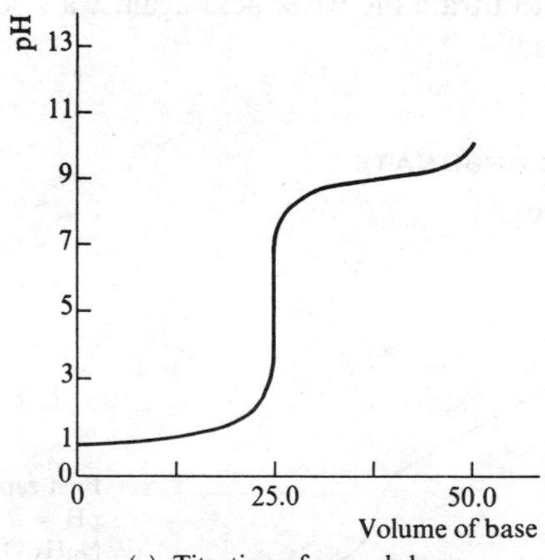

(c) Titration of a weak base into a strong acid

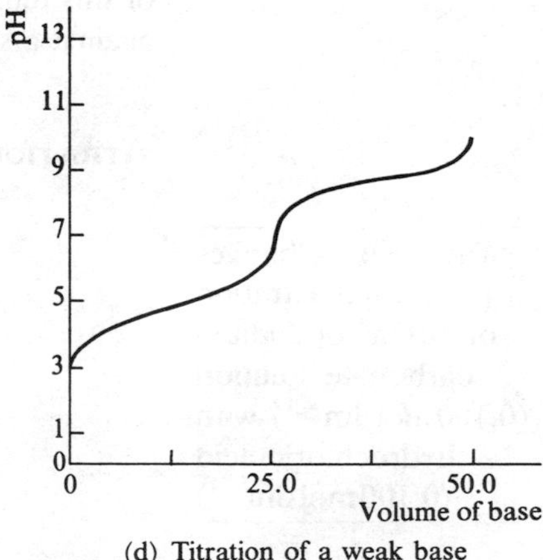

(d) Titration of a weak base into a weak acid

. . . pH changes rapidly at the end-point

The end-point of the titration occurs when $25.0\,cm^3$ of alkali have been added. The pH changes rapidly from 3.5 at $24.9\,cm^3$ of alkali to 10.5 at $25.1\,cm^3$ of alkali. Any indicator which changes colour over the range of pH 3.5 to 10.5 can be used to show the end-point of the titration.

TITRATION OF A STRONG BASE INTO A WEAK ACID AND A WEAK BASE INTO A STRONG ACID

The weak acid ethanoic acid can be titrated with sodium hydroxide solution. At the beginning of the titration, 25.0 cm^3 of 0.100 mol dm^{-3} ethanoic acid have a pH given by

$$[H_3O^+]^2 = K_a[CH_3CO_2H]$$

$$\therefore \quad pH = 2.88$$

Calculation of pH changes during the titration of a weak acid by a strong base...

When 20.0 cm^3 of alkali have been added

$$[CH_3CO_2H] = 5.0\,cm^3 \times 0.100\,mol\,dm^{-3}/45.0\,cm^3$$

$$[CH_3CO_2^-] = 20.0\,cm^3 \times 0.100\,mol\,dm^{-3}/45.0\,cm^3$$

$$[H_3O^+] = \frac{K_a[CH_3CO_2H]}{[CH_3CO_2^-]} = K_a \times \frac{5.0}{20.0}$$

$$pH = pK_a - \lg(5.0/20.0) = 5.35$$

and of a weak base by a strong acid...

The results of calculations of pH over the range of the titration are shown in Figure 12.8(b). Similar calculations on the course of the titration of the weak base ammonia with a strong acid are shown in Figure 12.8(c). You can see in Figure 12.8(b) that the pH changes from 5 to 10.5 rapidly at the end-point. Indicators which change in this region are litmus and phenolphthalein. In Figure 12.8(c), you can see that the pH changes rapidly from 3 to 7 at the end-point. Indicators which can be used are methyl orange and litmus. Phenolphthalein cannot be used in the titration of a weak base as it changes at pH 9.

TITRATION OF A WEAK ACID AND A WEAK BASE

...of a weak acid with a weak base

In the titration of a weak acid and a weak base, a titration curve like that in Figure 12.8(d) is obtained. The change in pH at the end-point is gradual, and indicators will change colour gradually. No indicator will give a sharp end-point. The way out of this difficulty is to titrate the weak acid against a strong base and the weak base against a strong acid.

TITRATION OF A CARBONATE

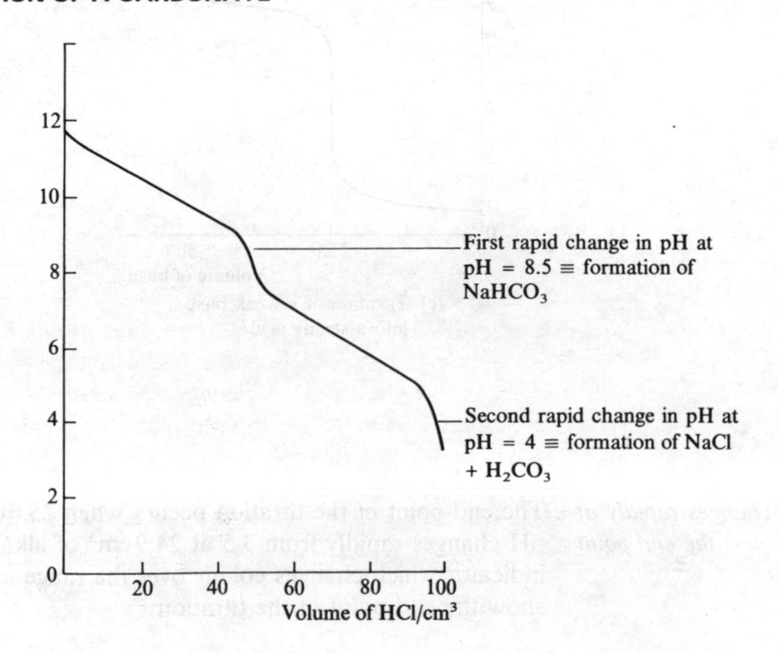

FIGURE 12.9 Changes in pH during Titration of 50 cm^3 of sodium carbonate solution (0.100 mol dm^{-3}) with hydrochloric acid (0.100 mol dm^{-3})

...and of a carbonate... Figure 12.9 shows the pH changes during the titration of sodium carbonate with hydrochloric acid. The two stages in the titration are

$$Na_2CO_3(aq) + HCl(aq) \rightarrow NaHCO_3(aq) + NaCl(aq)$$
$$NaHCO_3(aq) + HCl(aq) \rightarrow NaCl(aq) + H_2O(l) + CO_2(g)$$

Figure 12.10 shows indicators which will change at the two equivalence-points. Use can be made of this two-stage titration to estimate sodium carbonate in a mixture of sodium carbonate and sodium hydroxide or sodium carbonate and sodium hydrogencarbonate.

Example A 25.0 cm^3 portion of a solution containing sodium carbonate and sodium hydrogencarbonate needed 22.5 cm^3 of a solution of hydrochloric acid of concentration 0.100 mol dm^{-3} to decolourise phenolphthalein. On addition of methyl orange, a further 28.5 cm^3 of the acid were needed to turn this indicator to its neutral colour. Calculate the concentrations of sodium carbonate and sodium hydrogencarbonate in the solution.

Method In the first stage

...Worked examples

Amount of HCl needed to convert CO_3^{2-} to HCO_3^-

$= 22.5 \times 10^{-3} \times 0.100 = 2.25 \times 10^{-3}$ mol

Amount of CO_3^{2-} $= 2.25 \times 10^{-3}$ mol

Concentration of CO_3^{2-} $= (2.25 \times 10^{-3})/(25.0 \times 10^{-3})$

$= 9.00 \times 10^{-2}$ mol dm^{-3}

The volume of acid needed to neutralise the total $HCO_3^-(aq)$ is 28.5 cm^3. Of this, 22.5 cm^3 are needed to neutralise the $HCO_3^-(aq)$ formed from $CO_3^{2-}(aq)$ in the first stage. The remaining 6.00 cm^3 neutralise the $HCO_3^-(aq)$ present in the original solution.

Amount of HCl needed for HCO_3^- $= 6.00 \times 10^{-3} \times 0.100$ mol

$= 6.00 \times 10^{-4}$ mol

Amount of HCO_3^- $= 6.00 \times 10^{-4}$ mol

Concentration of HCO_3^- $= (6.00 \times 10^{-4})/(25.0 \times 10^{-3})$ mol dm^{-3}

$= 2.40 \times 10^{-2}$ mol dm^{-3}

The concentrations are sodium carbonate, 9.00×10^{-2} mol dm^{-3}; sodium hydrogencarbonate, 2.40×10^{-2} mol dm^{-3}.

Example 25.0 cm^3 of a solution containing sodium hydroxide and sodium carbonate was titrated against 0.100 mol dm^{-3} hydrochloric acid, using phenolphthalein as indicator. After 30.0 cm^3 of acid had been used, the indicator was decolorised. Methyl orange was added, and a further 12.5 cm^3 of hydrochloric acid were needed to turn the indicator orange. Calculate the concentrations of sodium hydroxide and sodium carbonate in the solution.

Method (*a*) 30.0 cm^3 of acid were needed for the first stage:

$$OH^-(aq) + H^+(aq) \rightarrow H_2O(l)$$
$$CO_3^{2-}(aq) + H^+(aq) = HCO_3^-(aq)$$

(*b*) $12.5\,cm^3$ of acid were needed for the second stage:

$$HCO_3^-(aq) + H^+(aq) \rightarrow CO_2(g) + H_2O(l)$$

$$\text{Amount of HCl used} = 12.5 \times 10^{-3} \times 0.100 = 1.25 \times 10^{-3}\,mol$$

This is equal to the amount of HCO_3^- formed in the first stage, which is equal to the amount of CO_3^{2-} in $25.0\,cm^3$ of solution.
Therefore

$$[CO_3^{2-}] = 1.25 \times 10^{-3}/(25.0 \times 10^{-3}) = 0.0500\,mol\,dm^{-3}$$

$$\text{Volume of HCl needed by NaOH} = 30.00 - 12.50 = 17.50\,cm^3$$

$$\text{Amount of NaOH in } 25.0\,cm^3 = 17.50 \times 10^{-3} \times 0.100\,mol$$

FIGURE 12.10 Changes in pH for some Common Indicators

pH / *Indicator*	1	2	3	4	5	6	7	8	9	10	11
Thymol blue	Red	Change			Yellow			Change		Blue	
Methyl orange		Red		Change	Yellow						
Methyl red				Red	Change		Yellow				
Litmus					Red		Change		Blue		
Bromothymol blue					Yellow		Change	Blue			
Phenolphthalein						Colourless			Change		Red
Universal Indicator		Red		Orange		Yellow	Green	Blue		Violet	

Therefore

$$[NaOH] = 1.75 \times 10^{-3}/(25.0 \times 10^{-3}) = 0.0700\,mol\,dm^{-3}$$

The concentrations are $[Na_2CO_3] = 0.050\,mol\,dm^{-3}$

$[NaOH] = 0.070\,mol\,dm^{-3}$

THE pH RANGE OF INDICATORS

The range over which different indicators can be used is shown in Figure 12.10. Universal Indicator is a mixture of several indicators.

12.7.10 CONDUCTIMETRIC TITRATION

The conductance of an acid changes during the course of a titration

Indicators are not the only means of following a change in pH. During the course of a titration of hydrochloric acid by sodium hydroxide, hydrogen ions are removed from solution to form molecules of water and are replaced by sodium ions. After the end-point, if addition of alkali continues, hydroxide ions accumulate. Since hydrogen ions have a higher value of molar conductivity than any other ions, the graph of conductivity against the volume of alkali added has the form shown in Figure 12.11(b). An apparatus which could be used for **conductimetric titration** is shown in Figure 12.11(a). [For potentiometric titration, see § 13.2.2.]

FIGURE 12.11
Conductimetric Titration

(a)

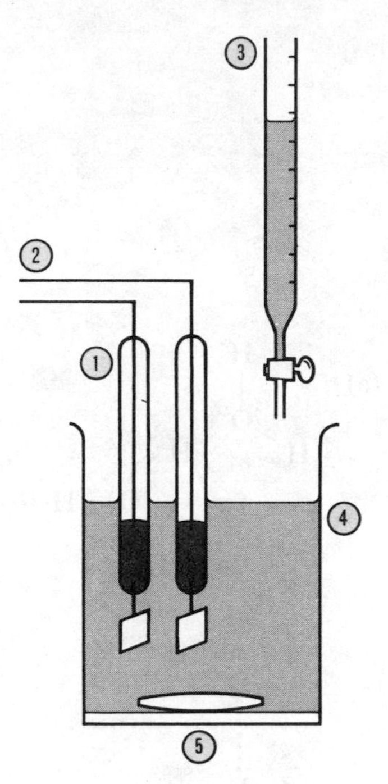
3 Burette measures volume of alkali (1 mol dm^{-3}) added.
2 Leads to conductance meter. This measures the conductance of the solution between the electrodes.
4 Measured volume of 0.1 mol dm^{-3} acid contains H_3O^+ ions of high molar conductivity.
1 Two electrodes. The current flowing between these electrodes depends on the concentration of ions present in the solution.
5 Magnet and magnetic stirrer mix the solution after each addition of alkali.
6 The change in the conductance of the solution as the titration proceeds is shown in (b).

(b)

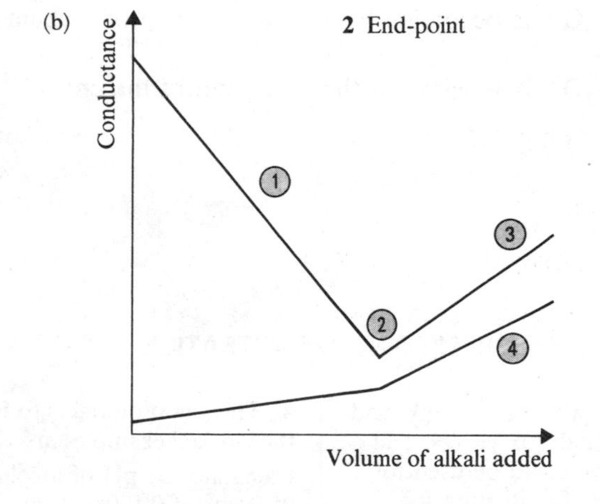
Conductance
2 End-point
Volume of alkali added
1 Conductance falls as H_3O^+ ions are replaced by cations from the alkali, e.g. Na^+, of lower conductivity.
3 Conductance rises as OH^- ions are added.
4 Curve for weak acid and strong base. If a weak base is used, the increase in conductance after the end-point is much smaller.

12.7.11 MOLAR CONDUCTIVITY OF THE HYDROGEN ION

Λ_0 for $H^+(aq)$ is high . . .

The value of molar conductivity at infinite dilution, Λ_0, for the hydrogen ion is much greater than for other ions. The explanation lies in the hydrogen bonding of water molecules [§4.7.3]. Since long chains of water molecules are linked by hydrogen bonds [see Figure 12.12] if a hydrogen ion enters into the bonding arrangement at one end of a chain, a rearrangement of bonds can take place to release a hydrogen ion at the other end of the chain. These changes can take place very much more quickly than the movement of an ion.

. . . an explanation in terms of hydrogen bonding

FIGURE 12.12 Water Molecules linked by Hydrogen Bonding

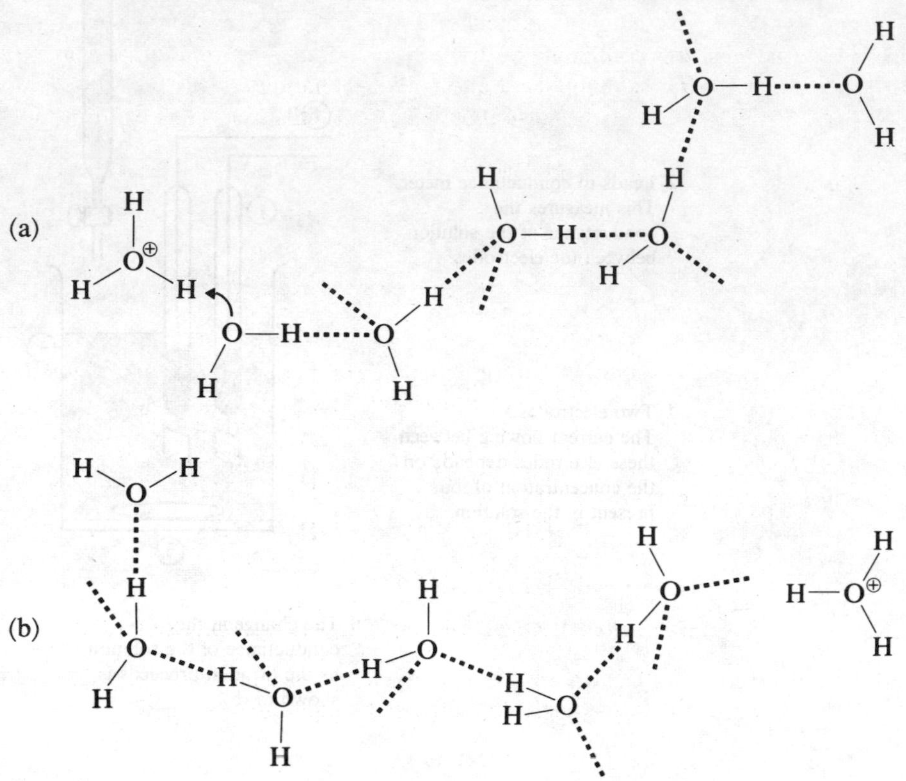

(a) H_3O^+ is becoming bonded to one end of a chain of water molecules

(b) H_3O^+ is released at the other end of the chain

CHECKPOINT 12E: TITRATION

1. A vegetable extract contains a dye which is a weak acid with $K_a = 10^{-5}\,mol\,dm^{-3}$. At pH 2 the dye is yellow, and at pH 8 it is blue. Explain how this dye could be used for acid–base titrations. Give an example of a titration for which it could be used, and one for which the indicator would be unsuitable.

2. Explain why it is possible to use one indicator for a number of different titrations which reach equivalence-point at different values of pH.
By means of a diagram, explain why phenolphthalein is not used in the titration of a weak base into a strong acid. What kind of indicator should be used for such a purpose?

3. Sketch the change in pH that occurs during the titration of aqueous ammonia into hydrochloric acid of the same concentration.
Suggest an indicator which could be used for the titration, and name one that should be avoided.

4. This question refers to Figure 12.8(a), § 12.7.9. Refer to the worked examples in § 12.7.9 if necessary.
Calculate the pH of a solution produced by adding to $25.0\,cm^3$ of $0.100\,mol\,dm^{-3}$ hydrochloric acid the following volumes of $0.100\,mol\,dm^{-3}$ sodium hydroxide solution:
(*a*) $24.5\,cm^3$, (*b*) $24.7\,cm^3$, (*c*) $24.9\,cm^3$, (*d*) $25.1\,cm^3$, (*e*) $25.5\,cm^3$. Check to see whether your results fit the shape of Figure 12.8(a).

5. A 1.000 g sample of a mixture of Na_2CO_3(anhydrous) and $NaHCO_3$ was dissolved in water. With phenolphthalein as indicator, $17.5\,cm^3$ of $0.200\,mol\,dm^{-3}$ hydrochloric acid were required to neutralise the solution. Calculate the percentage by mass composition of the mixture.

6. Explain the high value of Λ_0 of the OH^- ion. Refer to Figure 12.12 for help.

12.7.12 BUFFER SOLUTIONS

A buffer solution absorbs small amounts of $H^+(aq)$ and $OH^-(aq)$ without a change in pH

A **buffer solution** is one which will resist changes in pH due to the addition of small amounts of acid and alkali. An effective buffer can be made by preparing a solution containing both a weak acid and also one of its salts with a strong base, e.g., ethanoic acid and sodium ethanoate. This will absorb small amounts of hydrogen ions because they react with ethanoate ions to form molecules of ethanoic acid:

$$CH_3CO_2^- + H_3O^+ \rightleftharpoons CH_3CO_2H + H_2O$$

Hydroxide ions are absorbed (in small amounts) by combining with ethanoic acid molecules to form ethanoate ions and water:

$$OH^- + CH_3CO_2H \rightleftharpoons CH_3CO_2^- + H_2O$$

The combination of a weak acid and its salt with a strong base acts as a buffer...

A solution of a weak base and one of its salts formed with a strong acid, e.g., ammonia solution and ammonium chloride, will act as a buffer. If hydrogen ions are added, they largely combine with ammonia, and, if hydroxide ions are added, they largely combine with ammonium ions:

$$NH_3 + H_3O^+ \rightleftharpoons NH_4^+ + H_2O$$

$$OH^- + NH_4^+ \rightleftharpoons NH_3 + H_2O$$

*CALCULATION OF THE pH OF A BUFFER SOLUTION

The pH of a buffer solution consisting of a weak acid HA and its salt with a strong base is calculated from the equation

$$K_a = \frac{[H_3O^+][A^-]}{[HA]}$$

$$[H_3O^+] = K_a\frac{[HA]}{[A^-]}$$

$$pH = pK_a + \lg\frac{[A^-]}{[HA]}$$

Since the salt is completely ionised and the acid only slightly ionised, we can assume that all the anions come from the salt, and put

$$[A^-] = [Salt]$$

$$[HA] = [Acid]$$

$$\therefore \quad pH = pK_a + \lg\frac{[Salt]}{[Acid]}$$

An effective buffering action is obtained at pH values fairly close to pK_a.

...A weak base and its salt with a strong acid act as a buffer...

For a buffer made from a weak base **B** and its salt with a strong acid BH^+X^-

$$K_b = \frac{[BH^+][OH^-]}{[B]}$$

$$pOH = pK_b + \lg\frac{[BH^+]}{[B]}$$

$$pH = pK_w - pK_b + \lg\frac{[B]}{[BH^+]}$$

Since the weak base is only slightly ionised, we can put

$$[B] = [\text{Base added}]$$
$$[BH^+] = [\text{Salt added}]$$
$$\therefore \quad pH = pK_w - pK_b + \lg\frac{[\text{Base}]}{[\text{Salt}]}$$

Example (a) Three solutions contain ethanoic acid ($K_a = 1.80 \times 10^{-5}\,\text{mol dm}^{-3}$) at a concentration $0.10\,\text{mol dm}^{-3}$ and sodium ethanoate at a concentration (*a*) $0.10\,\text{mol dm}^{-3}$, (*b*) $0.20\,\text{mol dm}^{-3}$, (*c*) $0.50\,\text{mol dm}^{-3}$. Calculate the pH values of the three solutions.

Method

$$pH = pK_a + \lg\frac{[\text{Salt}]}{[\text{Acid}]}$$

... Calculation of pH values of buffers

In solution (*a*) $pH = 4.75 + \lg(0.10/0.10)$
$= 4.75 + \lg 1.0$
$= 4.75$

In solution (*b*) $pH = 4.75 + \lg(0.20/0.10)$
$= 4.75 + \lg 2.0$
$= 5.05$

In solution (*c*) $pH = 4.75 + \lg(0.50/0.10)$
$= 5.45$

The pH values are (*a*) 4.75, (*b*) 5.05, (*c*) 5.45

CHECKPOINT 12F: BUFFERS

1. How does a buffer maintain an almost constant pH, even when small amounts of acid or alkali are added to a solution? What two components are present in a buffer?

2. Solution (*a*) contains $0.100\,\text{mol dm}^{-3}$ potassium chloride solution. Solution (*b*) contains $0.100\,\text{mol dm}^{-3}$ ammonium ethanoate. Both solutions are neutral. Explain why the addition of $1.00\,\text{cm}^3$ of $1.00\,\text{mol dm}^{-3}$ hydrochloric acid to $1.00\,\text{dm}^3$ of solution (*a*) changes its pH to 3, but the same treatment has very little effect on the pH of solution (*b*).

3. Given a $0.100\,\text{mol dm}^{-3}$ solution of dimethylamine, $(CH_3)_2NH$, what would you need in order to prepare a buffer solution? Explain how the buffer would react to the addition of (*a*) hydrogen ions and (*b*) hydroxide ions.

***4.** A solution of $0.100\,\text{mol dm}^{-3}$ ethanoic acid and $0.400\,\text{mol dm}^{-3}$ sodium ethanoate has a pH of 5.35. Find the dissociation constant of the acid.

***5.** Ethanoic acid, CH_3CO_2H, has $pK_a = 4.76$ at room temperature. Find the pH of the solutions which result from the addition of $50\,\text{cm}^3$ of $0.040\,\text{mol dm}^{-3}$ NaOH(aq) to (*a*) $50\,\text{cm}^3$ of $0.080\,\text{mol dm}^{-3}$ CH_3CO_2H(aq), (*b*) $50\,\text{cm}^3$ of $0.060\,\text{mol dm}^{-3}$ CH_3CO_2H(aq).
Explain why the pH values of the solutions do not change much when small amounts of acid or alkali are added.

***6.** What is the pH of a solution that has been prepared by the addition of $50.0\,\text{cm}^3$ of $0.200\,\text{mol dm}^{-3}$ sodium hydroxide to $50.0\,\text{cm}^3$ of $0.400\,\text{mol dm}^{-3}$ ethanoic acid ($K_a = 1.75 \times 10^{-5}\,\text{mol dm}^{-3}$)?

12.7.13 SALT HYDROLYSIS

Many salts dissolve in water to give neutral solutions. Some salts, however, react with water to form acidic or alkaline solutions. These reactions are described as **salt hydrolysis.**

Some salts react with water...

Sodium carbonate is the salt of a strong base and a weak acid. The carbonate ion, CO_3^{2-}, is a base, a proton abstractor. The reaction

... Their solutions are acidic or alkaline...

$$CO_3^{2-}(aq) + H_2O(l) \rightleftharpoons OH^-(aq) + HCO_3^-(aq)$$

makes a solution of sodium carbonate alkaline. Solutions of sodium hydrogencarbonate are less strongly alkaline because HCO_3^- is a weaker base than CO_3^{2-}:

$$HCO_3^-(aq) + H_2O(l) \rightleftharpoons H_2CO_3(aq) + OH^-(aq)$$

... Such reactions are called salt hydrolysis

Other basic anions are S^{2-}, HS^-, CN^-, $CH_3CO_2^-$ and the anions of many other organic acids.

The hydrolysis of a salt of a weak base and a strong acid, e.g., ammonium chloride, results in an increase in the concentration of hydrogen ions:

$$NH_4^+(aq) + H_2O(l) \rightleftharpoons NH_3(aq) + H_3O^+(aq)$$

The hydrolysis of aluminium salts is covered in § 19.2.4, and transition metal salts in § 24.14.6.

CHECKPOINT 12G: SALT HYDROLYSIS

1. Predict whether the pH of the following solutions will be 7 or > 7 or < 7. Give your reasons.

(*a*) 0.10 mol dm^{-3} ammonium chloride
(*b*) 0.010 mol dm^{-3} methylammonium chloride
(*c*) 0.10 mol dm^{-3} potassium cyanide
(*d*) 0.10 mol dm^{-3} sodium methanoate

2. The pH readings below refer to the titration of sodium hydroxide solution into 25.0 cm^3 of 0.100 mol dm^{-3} ethanoic acid.

V/cm^3	0	4.0	6.0	8.0	10.0	12.0	14.0	14.4	14.6	14.8	15.0	15.2	15.4	16.0
pH	2.8	3.8	4.2	4.6	5.1	5.5	6.2	6.5	6.8	7.6	9.0	9.8	10.5	11.4

Plot V, the volume of sodium hydroxide solution, against pH. Use your graph to find the following quantities:

(*a*) the pH at the equivalence point
(*b*) the concentration in mol dm^{-3} of sodium hydroxide.
Name an indicator which could be used for this titration.

12.7.14 *SOLUBILITY AND SOLUBILITY PRODUCT

Definition of solubility product...

Many salts which we refer to as insoluble do in fact dissolve to a small extent. In a saturated solution, dissolved ions and undissolved salt are in equilibrium. When a saturated solution of nickel(II) sulphide is in contact with solid nickel(II) sulphide, an equilibrium is established:

$$NiS(s) \rightleftharpoons Ni^{2+}(aq) + S^{2-}(aq)$$

The product of the concentrations of nickel ions and sulphide ions is called the solubility product (K_{sp}) of nickel(II) sulphide.

$$K_{sp} = [Ni^{2+}][S^{2-}]$$

The **solubility product** of a salt is the product of the concentrations of the ions in a saturated solution of the salt, raised to the appropriate powers. It applies only to sparingly soluble salts. For lead(II) hydroxide, $Pb(OH)_2$

$$K_{sp} = [Pb^{2+}][OH^-]^2$$

Like other equilibrium constants, solubility products vary with temperature.

... and solubility

The solubility of a salt is expressed in various ways [§ 9.1]. They include the mass of solute dissolved in 1 dm^3 of solution and the amount/mol of solute in 1 dm^3 of solution at a stated temperature. The relationship between solubility and solubility product can be seen in the following examples.

Example (a) Given that the solubility product for calcium sulphate at 25 °C = $2.4 \times 10^{-5}\,mol^2\,dm^{-6}$, calculate the solubility at this temperature.

Method

$$K_{sp} = [Ca^{2+}][SO_4^{2-}] = [Ca^{2+}]^2$$

$$[Ca^{2+}] = \sqrt{K_{sp}} = \sqrt{2.4 \times 10^{-5}}$$

$$= 4.9 \times 10^{-3}\,mol\,dm^{-3}$$

Since each mole of $CaSO_4$ that dissolves puts 1 mole of Ca^{2+} into solution

$$[CaSO_4] = [Ca^{2+}] = 4.9 \times 10^{-3}\,mol\,dm^{-3}$$

The solubility of $CaSO_4$ is $4.9 \times 10^{-3}\,mol\,dm^{-3} = 0.67\,g\,dm^{-3}$.

Example (b) If the solubility of PbF_2 is $0.64\,g\,dm^{-3}$, what is the value of the solubility product?

Method

$$\text{Solubility of } PbF_2 = 0.64\,g\,dm^{-3} = 0.64/245\,mol\,dm^{-3}$$

$$= 2.61 \times 10^{-3}\,mol\,dm^{-3}$$

$$\therefore \quad [Pb^{2+}] = 2.61 \times 10^{-3}\,mol\,dm^{-3}$$

$$\text{and} \quad [F^-] = 2 \times 2.61 \times 10^{-3}\,mol\,dm^{-3}$$

$$K_{sp} = [Pb^{2+}][F^-]^2 = 4(2.61 \times 10^{-3})^3 = 7.1 \times 10^{-8}\,mol^3\,dm^{-9}$$

You will see that the unit in which the solubility product is expressed is the result of multiplying three concentrations together.

12.7.15 *THE COMMON ION EFFECT

The solubility of MA is reduced by the addition of $M^{2+}(aq)$ or $A^{2-}(aq)$

A saturated solution of a salt **MA** is in contact with solid **MA**.

$$MA(s) \rightleftharpoons M^{2+}(aq) + A^{2-}(aq)$$

$$K_{sp} = [M^{2+}][A^{2-}]$$

If a solution containing M^{2+} ions is added, $[M^{2+}]$ is increased. Even when the ions are not present in equimolar concentrations, the solubility product, K_{sp}, remains the same. So that the product $[M^{2+}][A^{2-}]$ shall not exceed K_{sp}, M^{2+} ions will be removed from solution as solid **MA**. The addition of a solution containing M^{2+} ions will therefore result in the precipitation of solid **MA**. The addition of a solution containing A^{2-} ions will have the same effect. The precipitation of a solute from solution on addition of an electrolyte solution which has an ion in common with the solute is an example of the **common ion effect**.

Another example of the common ion effect is the change in the concentration of ions produced by the dissociation of a weak acid in the presence of a solution of one of its ions. [See buffer solutions, § 12.7.12.]

Example (c) Calculate the solubility of manganese(II) sulphide (*a*) in water and (*b*) in a 1.0×10^{-2} mol dm^{-3} solution of sulphide ion. K_{sp}(MnS) at 25 °C = 2.5×10^{-13} mol^2 dm^{-6}.

Method

(*a*)

$$K_{sp} = [Mn^{2+}][S^{2-}]$$

$$[Mn^{2+}]^2 = 2.5 \times 10^{-13} \text{ mol}^2 \text{dm}^{-6}$$

$$[Mn^{2+}] = 5.0 \times 10^{-7} \text{ mol dm}^{-3}$$

$$[MnS] = 5.0 \times 10^{-7} \text{ mol dm}^{-3}$$

(*b*) The concentration of S^{2-} ions due to dissolved MnS can be neglected in comparison with 1.0×10^{-2} mol dm^{-3}.

$$[S^{2-}] = 1.0 \times 10^{-2} \text{ mol dm}^{-3}$$

$$K_{sp} = [Mn^{2+}](1.0 \times 10^{-2})$$

$$[Mn^{2+}] = 2.5 \times 10^{-13}(1.0 \times 10^{-2})$$

$$= 2.5 \times 10^{-11} \text{ mol dm}^{-3}$$

$$[MnS] = 2.5 \times 10^{-11} \text{ mol dm}^{-3}$$

Comparison of the results of (*a*) and (*b*) show that the solubility of MnS has been reduced by a factor of 5.0×10^3 by the presence of 1.0×10^{-2} mol dm^{-3} sulphide ions.

12.7.16 *APPLICATIONS OF SOLUBILITY PRODUCTS

PRECIPITATION TITRATIONS

Chloride solutions are estimated by titration against silver nitrate solution...

Chlorides are titrated by running silver nitrate solution into a solution of a chloride. Silver chloride is precipitated. The indicator potassium chromate gives a red precipitate of silver chromate at the end-point. It is essential that silver chromate does not precipitate until all the chloride ions have precipitated as silver chloride.

Solubility products are $K_{sp}(AgCl) = 1.2 \times 10^{-10}$ mol^2 dm^{-6}; $K_{sp}(Ag_2CrO_4) = 2.4 \times 10^{-12}$ mol^3 dm^{-9}.

$[Ag^+]$ in equilibrium with solid AgCl = 1.1×10^{-5} mol dm^{-3} (See Example (a))

...Potassium chromate is used as the indicator

$[Ag^+]$ in equilibrium with solid Ag_2CrO_4 = 1.7×10^{-4} mol dm^{-3} (See Example (b))

Silver chloride therefore precipitates in preference to silver chromate.

QUALITATIVE ANALYSIS

Schemes for the qualitative analysis of mixtures of ions employ methods of separating cations on the basis of the differences between the solubility products of their salts. For example, when hydrochloric acid is added to a solution containing a number of metal cations, only metal chlorides with low solubility products are precipitated:

$$K_{sp}(AgCl) = 1.2 \times 10^{-10} \text{ mol}^2 \text{dm}^{-6}$$

The precipitation of insoluble salts is a useful method for identifying cations

If the concentration of hydrochloric acid is 0.10 mol dm^{-3},

$$[Ag^+](0.10) = 1.2 \times 10^{-10}$$

$$[Ag^+] = 1.2 \times 10^{-9} \text{ mol dm}^{-3}$$

Only this tiny concentration of silver ions can remain in solution, and if silver ions are present in the mixture, a precipitate of silver chloride is obtained.

CHECKPOINT 12H: *SOLUBILITY PRODUCTS

1. Distinguish between the terms *solubility* and *solubility product*. Name three substances to which the term *solubility product* can be applied.

2. Write expressions for the solubility products of the following salts, given their solubilities:

$BaCO_3$, *a* mol dm^{-3}; BaI_2, *b* mol dm^{-3}; Ag_3PO_4, *c* mol dm^{-3}

3. Explain why, when hydrogen sulphide is passed into an acidic solution containing Cu^{2+} ions and Fe^{2+} ions, CuS is precipitated but FeS is not precipitated. ($K_{sp}(CuS) = 6 \times 10^{-36}$, $K_{sp}(FeS) = 6 \times 10^{-18}$ mol^2 dm^{-6}).

***4.** $K_{sp}(CaSO_4) = 2.0 \times 10^{-5}$ mol^2 dm^{-6} at 25 °C. Calculate the solubility of $CaSO_4$ in mol dm^{-3}
(*a*) in water,
(*b*) in 0.100 mol dm^{-3} aqueous sodium sulphate,
(*c*) in 0.200 mol dm^{-3} aqueous calcium nitrate.

5. $K_{sp}(AgCl) = 1.2 \times 10^{-10}$ mol^2 dm^{-6}, $K_{sp}(AgI) = 8.3 \times 10^{-17}$ mol^2 dm^{-6}
A solution of potassium iodide is shaken with solid silver chloride. Describe what you would expect to see. Test your prediction. Explain the changes that occur.

12.7.17 COMPLEX IONS

Complex ions are formed by the combination of a cation with a neutral molecule (or molecules) or an oppositely charged ion (or ions). Coordinate bonding is involved [§ 4.6]. Examples are $[Al(H_2O)_6]^{3+}$, $[Ag(NH_3)_2]^+$, $[CuCl_4]^{2-}$ and $[Al(OH)_4]^-$.

Complex ions have dissociation constants...

The formation of a complex ion is an equilibrium reaction. In the formation of tetracyanozincate ions, $[Zn(CN)_4]^{2-}$, there is set up an equilibrium:

$$Zn^{2+}(aq) + 4CN^-(aq) \rightleftharpoons Zn(CN)_4{}^{2-}(aq)$$

The dissociation constant of the complex ion is

$$K_d = \frac{[Zn^{2+}][CN^-]^4}{[Zn(CN)_4{}^{2-}]} = 2.0 \times 10^{-15} \text{ mol}^4 \text{ dm}^{-12}$$

The low value of the dissociation constant shows that the complex ion is very stable. The units are (concentration)^5/(concentration), i.e. mol^4 dm^{-12}.

... The reciprocal is the stability constant

The reciprocal of the dissociation constant is called the **stability constant** of the complex.

Edta can be used in complexometric titration...

A number of metal ions form complexes with edta [§ 24.13]. These complexes are so stable that they can be used for the estimation of the metal ions by **complexometric titration**. The disodium salt of edta is a primary standard. It is used in alkaline solution so that all the carboxyl groups are ionised as

$$(^-O_2CCH_2)_2NCH_2CH_2N(CH_2CO_2{}^-)_2$$

... with an indicator

The end-point in the titration is shown by an indicator which forms a coloured complex with the metal ion which is being titrated. If Eriochrome Black T is used as indicator, the metal-indicator complex colour of red is seen at the beginning of the titration. As the edta solution is added, metal ions are removed from the indicator and complex with edta. At the end-point, the blue colour of the free indicator is seen:

$$\text{Metal–indicator (red)} + \text{edta} \rightarrow \text{Metal–edta} + \text{Indicator (blue)}$$

A worked example of complexometric titration

Example Hardness in water is caused by calcium and magnesium ions. Both these ions complex strongly with edta. To a 200 cm^3 sample of tap water were added an alkaline buffer and a few drops of Eriochrome Black T. A volume 3.50 cm^3 of 0.100 mol dm^{-3} edta was used in titration. Find the concentration of calcium and magnesium ion in the water. The complexes have the formulae Ca(edta) and Mg(edta).

Method

$$\text{Amount of edta in titration} = 3.50 \times 0.100 \times 10^{-3} = 3.50 \times 10^{-4}\,\text{mol}$$

$$\text{Amount of } Ca^{2+} + Mg^{2+} = 3.5 \times 10^{-4}\,\text{mol}$$

$$[Ca^{2+}] + [Mg^{2+}] = 3.50 \times 10^{-4}/(200 \times 10^{-3})$$
$$= 1.75 \times 10^{-3}\,\text{mol dm}^{-3}$$

CHECKPOINT 12I: COMPLEX IONS

*1. Explain these observations, with the help of your knowledge of inorganic chemistry:

(*a*) The solubility of $PbCl_2$ in water decreases on the addition of dilute hydrochloric acid and increases on the addition of concentrated hydrochloric acid.

(*b*) The addition of aqueous ammonia to aqueous magnesium sulphate gives a white precipitate which is soluble in aqueous ammonium chloride.

(*c*) The solubility products of CdS and CuS are 1.6×10^{-28} and 6.3×10^{-36} mol^2 dm^{-6} respectively. Explain why, when hydrogen sulphide is bubbled through solutions of (i) Cd^{2+} ions and an excess of CN^- ions, (ii) Cu^{2+} ions and an excess of CN^- ions, CdS is precipitated but CuS is not.

(*d*) A solution of $FeCl_3$ is yellow. Addition of ammonium thiocyanate, NH_4CNS, produces a red colour, which can be discharged by the addition of sodium fluoride.

QUESTIONS ON CHAPTER 12

1. Write equations for the reactions that occur when each of the following is separately dissolved in water:
(*a*) HCl(g), (*b*) NH_3(g), (*c*) CH_3CO_2H(l), (*d*) $C_2H_5NH_2$(g), (*e*) $H_2NCH_2CO_2H$(s).

2. Classify the following species as oxidising agent/reducing agent/Brönsted acid/Brönsted base. Quote one reaction (with its equation) for each species to illustrate its typical behaviour. Some species fall into more than one category:
(*a*) NH_3(aq), (*b*) NH_4^+(aq), (*c*) HSO_3^-(aq), (*d*) $(CO_2H)_2$, (*e*) $Fe(H_2O)_6^{3+}$(aq).

3. Explain the terms *conductivity* and *molar conductivity*. Explain why the molar conductivity of potassium nitrate remains constant over a wide range of dilution. How do the slightly lower values of molar conductivity at high concentrations arise?

4. What is an ion exchange resin? Give an example of the use of such a resin, and explain the principles underlying its use [see also § 8.7.6].

5. Explain the significance of the Faraday constant, 96 478 C mol^{-1}. Write an equation for the electrode process that results in the evolution of oxygen at the anode during the electrolysis of an aqueous solution. Calculate the time in minutes for which a current of 0.500 A would need to be passed in order to yield 500 cm^3 (at stp) of oxygen.

6. Sketch titration curves showing the change in pH which occurs during the addition of an excess of sodium hydroxide of concentration 1.0 mol dm^{-3} to 25 cm^3 of (*a*) 1.0 mol dm^{-3} hydrochloric acid, (*b*) 1.0 mol dm^{-3} sulphuric acid, (*c*) 1.0 mol dm^{-3} ethanoic acid ($pK_a = 4.75$).

Explain why the pH at the end-point in (*c*) is different from that in (*a*).

7. (*a*) Explain why an aqueous solution of ammonium chloride is acidic.

(*b*) Explain why a mixture of ammonium chloride and ammonia in solution has a buffering action.

8. Describe an experiment which you could carry out to find the number of units of charge on a metal cation.

9. A current of 0.200 A is passed for 2.00 h through 200 cm^3 of 0.0500 mol dm^{-3} aqueous silver nitrate. Calculate the volume (at stp) of hydrogen evolved.

10. Three voltameters were connected in series. They contain silver nitrate solution, copper(II) sulphate solution and acidified chromium(III) sulphate solution. After a current was passed for 30.0 min, 0.216 g of silver had been deposited on the cathode of the first cell. Calculate:

(*a*) the current

(*b*) the mass of copper deposited in the second cell

(*c*) the volume (at stp) of oxygen evolved in the second cell

(*d*) the mass of chromium deposited in the third voltameter

11. A current of 1.75 A passed for 1.00 h through a solution of copper(II) sulphate deposited 1.245 g of copper, and some hydrogen was evolved. Calculate the volume of hydrogen at stp.

12. Silver chloride is sparingly soluble in water. It is even less soluble in dilute hydrochloric acid, more soluble in concentrated hydrochloric acid and much more soluble in aqueous ammonia. Explain the effects of (*a*) dilute hydrochloric acid, (*b*) concentrated hydrochloric acid and (*c*) aqueous ammonia on the solubility.

13. A solution of 0.100 mol dm^{-3} sodium hydroxide was added to 25.0 cm^3 of 0.100 mol dm^{-3} hydrochloric acid:

(*a*) In one experiment, the pH was measured at intervals during the titration.

(*b*) In a second experiment, the electrical conductivity was measured throughout the titration.

(*c*) In the third experiment, the temperature of the acid was recorded as titration proceeded.

Sketch the changes that occur in (*a*), (*b*) and (*c*) as alkali is added to the acid.

14. A 50 cm^3 portion of aqueous ammonia (1.00 mol dm^{-3}) is titrated with 1.00 mol dm^{-3} hydrochloric acid. Draw a graph to show roughly (without doing detailed calculations) the change in pH that occurs as titration proceeds, until 60 cm^3 of the acid have been added.

What kind of indicator would be suitable for use in this titration ($K_b(NH_3) = 1.8 \times 10^{-5}$ mol dm^{-3})?

15. Apple juice has a pH of 3.5.

(*a*) (i) Define pH.

(ii) Calculate the molar concentration of hydrogen ions in apple juice.

Apple juice can be titrated with standard alkali. A 25.0 cm^3 sample of apple juice was exactly neutralised by 27.5 cm^3 of 0.10 mol dm^{-3} sodium hydroxide using phenolphthalein as indicator.

(*b*) Assuming that apple juice contains a single acid which is monobasic, calculate the molar concentration of the acid in the juice.

(*c*) (i) How can you explain the difference between the two results you have obtained in (*a*) (ii) and (*b*)?

(ii) What constant can be determined from these two results?

(iii) Calculate a numerical value of this constant.

(*d*) Suggest *two* reasons why phenolphthalein is a suitable indicator for this titration.

(C 90)

16. (*a*) State whether the pHs of 0.1 mol dm^{-3} solutions of the following salts are greater than, equal to, or less than 7. Explain your answer in each case.

(i) NH_4Cl, (ii) NaCl, (iii) CH_3CO_2Na.

(*b*) (i) Define and explain the concept of *solubility product* (K_s).

(ii) The solubility product of silver chloride (AgCl) is 2×10^{-10} mol^2 dm^{-6} at 25 °C. Calculate the solubility of solid AgCl in water at this temperature and explain the basis of your calculation.

(iii) State, giving a reason, whether the solubility of AgCl in water would be altered by addition of KCl in relatively small amounts to the saturated solution.

(*c*) Describe the structure and bonding in a diamond crystal and relate these to its hardness and volatility.

(*d*) (i) State the main oxidation states shown by the elements of Group 4 in their compounds.

(ii) State, giving examples, how the relative stability of these oxidation states in (*d*)(i) changes as the group is descended.

(iii) Give a reason for the trend you describe in (*d*)(ii) above.

(WJEC 91)

17. (*a*) (i) Summarise the Brönsted–Lowry theory of acids and bases, and define the terms *acid, base, neutralisation, conjugate acid,* and *conjugate base* on the basis of this theory.

(ii) State which of the following in aqueous solution normally act as Brönsted–Lowry

(*1*) acids

(*2*) bases.

$[Al(H_2O)_6]^{3+}$, $CH_3CO_2^-$, CH_3NH_2, NH_4^+, CO_3^{2-}, NaBr, C_6H_5OH

(iii) Give *one* example in *each* case of *neutralisation, a conjugate acid* and *a conjugate base.*

(*b*) Write down the two *new* equilibria which might be set up if anhydrous ethanoic acid and anhydrous hydrogen fluoride were mixed together in equal molar proportions.

State which of these equilibria would have the larger equilibrium constant, if the K_a values in aqueous solution below are a good guide to the relative strengths of the two acids under anhydrous conditions.

($K_a(CH_3CO_2H) = 1.8 \times 10^{-5}$ mol dm^{-3};
$K_a(HF) = 5.6 \times 10^{-4}$ mol dm^{-3})

(*c*) Calculate the pH of an aqueous solution containing 0.1 mol dm^{-3} of ammonia, given that K_b for ammonia is 1.8×10^{-5} mol dm^{-3} and that the ionic product of water, K_w, is 1.0×10^{-14} mol^2 dm^{-6}.

(*d*) Streams made acidic by rain or the run-off of water from old mine tips may dissolve normally insoluble compounds, such as lead(II) hydroxide, $Pb(OH)_2$, from the soil. Some of these are potentially poisonous.

Use your knowledge of solubility products and pH to explain why this process of solution of $Pb(OH)_2$ occurs. Thus derive an equation to relate the maximum concentration of Pb^{2+} in the stream to its hydrogen ion concentration, assuming that no interfering anions are present, i.e. that the only solid present is $Pb(OH)_2$.

(WJEC 90)

18. (*a*) (i) Describe how you would measure the pH of an aqueous solution of ethanoic acid ($0.1\ mol\ dm^{-3}$) using a pH meter.

(ii) Describe the principles underlying this method of pH measurement.

(*b*) $30\ cm^3$ of aqueous ethanoic acid ($0.1\ mol\ dm^{-3}$) are added in small amounts to $15\ cm^3$ of aqueous sodium hydroxide ($0.1\ mol\ dm^{-3}$). Without calculating the exact pH at the end-point, draw a sketch graph to show how the pH of the mixture would change as the ethanoic acid was added.

(*c*) $10.0\ cm^3$ of aqueous ethanoic acid ($0.100\ mol\ dm^{-3}$) are mixed with $10.0\ cm^3$ of aqueous sodium ethanoate ($0.200\ mol\ dm^{-3}$) at 298 K. The pH of the resulting solution was found to be 5.06.

(i) Write down an expression for the acid dissociation constant, K_a, of ethanoic acid.

(ii) Calculate the concentrations of the ethanoate ion and the ethanoic acid in the mixture.

(iii) Calculate the value of K_a for ethanoic acid at 298 K.

(*d*) Explain the behaviour of a *buffer* solution when an acid is added.

(*e*) For most weak acids, ionisation is an endothermic process. State how you would expect K_a to change with an increase in (i) temperature and (ii) concentration of the acid.

(O 92)

19. The ingredients of a typical lemonade are

INGREDIENTS: CARBONATED WATER, CITRIC ACID, FLAVOURING, ACIDITY REGULATOR: SODIUM CITRATE; ARTIFICIAL SWEETENER: ASPARTAME, PRESERVATIVE: SODIUM BENZOATE; STABILIZER: CMC

(*a*) Carbonated water is an aqueous solution of carbon dioxide. Some of the dissolved carbon dioxide reacts to form carbonic acid, H_2CO_3.

The carbonic acid then ionises

$$H_2CO_3(aq) + H_2O(l) \rightleftharpoons H_3O^+(aq) + HCO_3^-(aq)$$

$K_a = 2.0 \times 10^{-4}\ mol\ dm^{-3}$

$pK_a = 3.7$

(i) Explain why carbonic acid is classified as a Lowry–Brönsted acid.

(ii) Comment on the strength of carbonic acid as an acid.

(iii) Write an expression for the ionisation (dissociation) constant for carbonic acid.

(iv) Calculate the pH of $0.1\ mol\ dm^{-3}$ carbonic acid. State any assumptions you make.

(v) Give the name for the mixture of a solution of a weak acid and the salt of the acid, like citric acid and sodium citrate.

(vi) Discuss the behaviour of a solution of citric acid and sodium citrate when an acid is added.

(*b*) Suggest the signs of the entropy change in the system, $\Delta S^\ominus$, and the total entropy change, $\Delta S^\ominus_{total}$, for the escape of carbon dioxide from lemonade. Justify your suggestions.

†(*c*) When a lemonade bottle is left open, the lemonade goes 'flat' as carbon dioxide escapes from the carbonated water. Laboratory tests suggest that this is a first-order process. Describe how you would attempt to verify this by an experiment. You should include in your answer

(i) a description of the apparatus

(ii) details of any chemicals used

(iii) the measurements you would make

(iv) an outline of how you would process the results

(v) the way you would deduce the order of the reaction.

[For (*c*) see Chapter 14.]

(L(N) 92, AS)

20. (*a*) The pH of acid rain is 5. What is the hydrogen ion concentration of such rain?

(*b*) The hydrogen ion concentration of a solution can be measured using an electrochemical cell, and this is the basis of the pH meter. Describe the electrode sensitive to hydrogen ions which is used, and state the other essential parts of the meter.

(*c*) Such a pH meter is usually calibrated by using a buffer solution.

$1.00\ dm^3$ of a buffer solution, pH 5.00, was prepared using ethanoic acid and sodium ethanoate. The solution contained 50.0 g of anhydrous sodium ethanoate. What mass of ethanoic acid did it contain?

The K_a for ethanoic acid is $1.8 \times 10^{-5}\ mol\ dm^{-3}$.

(*d*) (i) If $1.00\ cm^3$ of $10.0\ mol\ dm^{-3}$ hydrochloric acid is added to $1.00\ dm^3$ of the buffer solution, what will be the new pH of the solution?

(ii) What would have been the pH of a solution obtained by adding $1.00\ cm^3$ of $10.0\ mol\ dm^{-3}$ hydrochloric acid to $1.00\ dm^3$ of aqueous acid of pH 5.00?

(iii) Explain the difference between these two results.

(O 91, S)

13 OXIDATION–REDUCTION EQUILIBRIA

13.1 ELECTRODE POTENTIALS

A piece of metal consists of metal cations and a cloud of delocalised valence electrons [§ 6.2.1]. If a strip of metal is placed in a solution of its ions, some of the cations in the metal may dissolve leaving a build-up of electrons on the metal.

$$M(s) \rightarrow M^{2+}(aq) + 2e^-$$

The metal will become negatively charged. Alternatively, metal ions may take electrons from the strip of metal and be discharged as metal atoms:

$$M^{2+}(aq) + 2e^- \rightarrow M(s)$$

The electrode potential of a metal may be positive or negative

In this case, the metal will become positively charged. The potential difference between the strip of metal and the solution depends on the nature of the metal and on the concentration of the ions involved in the equilibrium at the metal surface. Zinc acquires a more negative potential than copper, since it has a greater tendency to dissolve as ions and a smaller tendency to be deposited as metal.

The two metals, zinc and copper in solutions of their ions, may be combined as shown in Figure 13.1 to make an electrochemical cell. The solutions of zinc sulphate and copper(II) sulphate are separated by a porous partition. The metals form the electrodes of the cell, and are connected through a voltmeter. Since the electrode reactions are

$$Zn(s) \rightarrow Zn^{2+}(aq) + 2e^-$$

$$Cu^{2+}(aq) + 2e^- \rightarrow Cu(s)$$

Two metals and solutions of their ions combine to make a cell which can produce a current...

zinc is negatively charged, and copper is positively charged, and electrons flow through the external circuit from zinc to copper. Zinc dissolves from the zinc electrode, and copper is deposited on the copper electrode. Ions flow through the porous partition. The overall cell reaction is

$$Zn(s) + Cu^{2+}(aq) \rightarrow Zn^{2+}(aq) + Cu(s)$$

...and is called a galvanic or voltaic cell

The type of cell in which a chemical reaction results in the production of an electric current is called a **galvanic** or **voltaic** cell. The electromotive force (emf) of the cell is a measure of the tendency of electrons to flow through the external circuit. Under **reversible conditions**, the emf is equal to the difference between the potentials of the two electrodes. To achieve reversible conditions, an electronic voltmeter is used to measure emf. This has a resistance so high that the current which it takes from the cell is negligible. When no current flows, the cell is operating reversibly.

FIGURE 13.1
A Zinc–Copper Electrochemical Cell

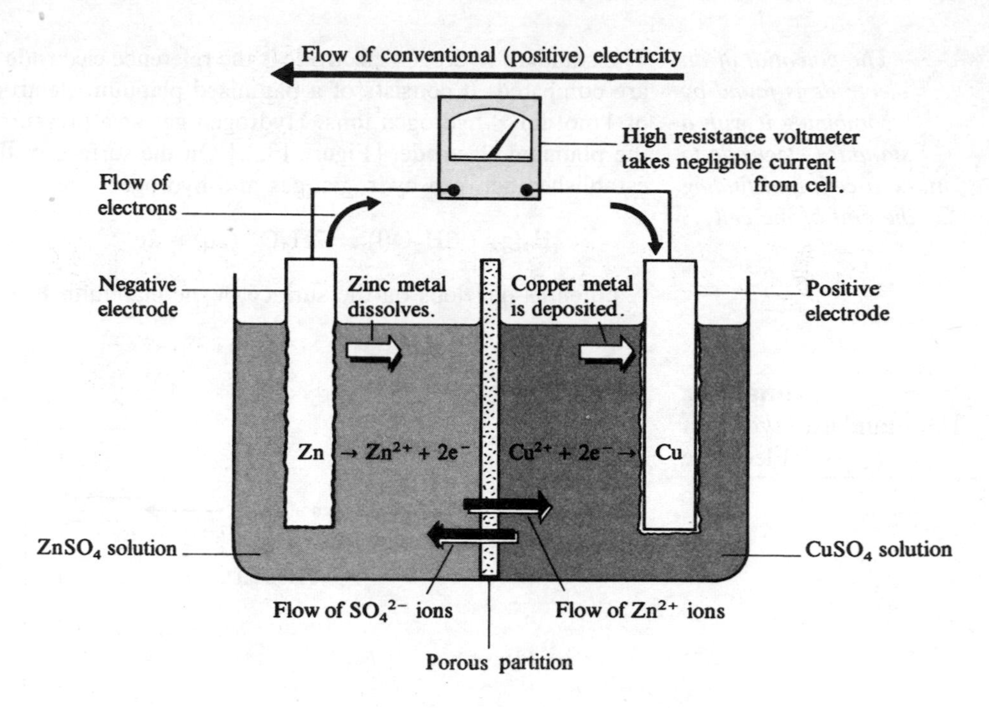

Convention for the sign of the emf of a cell

By convention, the emf, E, is taken to act from left to right through the cell. Thus

$$E_{\text{cell}} = E_{\text{RHS electrode}} - E_{\text{LHS electrode}}$$

Then a positive emf indicates that electrons are flowing from left to right through the external circuit [in Figure 13.1, from Zn to Cu]. The flow of conventional electricity (positive electricity) is therefore from right to left through the external circuit (from Cu to Zn in Figure 13.1).

The cell in Figure 13.1 can be represented as

$$\text{Zn(s)} \mid \text{ZnSO}_4\text{(aq)} \,\vdots\vdots\, \text{CuSO}_4\text{(aq)} \mid \text{Cu(s)}$$

Its emf is given by

$$E = E_{\text{Cu}} - E_{\text{Zn}}$$

13.1.1 STANDARD ELECTRODE POTENTIAL

Definition of the standard electrode potential, $E^{\ominus}$, of a metal

If the metal is immersed in a solution of its ions of concentration 1 mol dm^{-3} at 25 °C, then the potential acquired under these standard conditions is the **standard electrode potential** of that metal, $E^{\ominus}$. In the cell shown in Figure 13.1, if each electrode is immersed in a 1 mol dm^{-3} solution of its ions at 25 °C, then its potential will be its standard electrode potential. The values are $E^{\ominus}_{\text{Cu}} = +0.34\,\text{V}$, $E^{\ominus}_{\text{Zn}} = -0.76\,\text{V}$. The voltmeter will then register an emf of 1.1 V. This emf is the difference, $E^{\ominus}_{\text{Cu}} - E^{\ominus}_{\text{Zn}}$.

13.1.2 FINDING ELECTRODE POTENTIALS

If a cell is constructed from a standard electrode (i.e., one of known potential) and an electrode of unknown potential, the emf of the cell can be measured and used to find the unknown electrode potential. For a cell with the standard electrode on the left-hand side

$$E_{\text{cell}} = E_{\text{electrode of unknown potential}} - E_{\text{standard electrode}}$$

The potential of an electrode is found by combining it with a standard electrode to make a cell, and finding the emf of the cell...

The standard hydrogen electrode is the reference electrode with which other electrodes are compared. It consists of a platinised platinum electrode immersed in a solution of 1 mol dm^{-3} hydrogen ions. Hydrogen gas at a pressure of 1 atm is bubbled over the platinum electrode. [Figure 13.2.] On the surface of the platinum, equilibrium is established between hydrogen gas and hydrogen ions.

$$H_2(g) + 2H_2O(l) \rightleftharpoons 2H_3O^+(aq) + 2e^-$$

A potential develops on the surface of the platinum. It is assigned a value of zero volts.

FIGURE 13.2 The Standard Hydrogen Electrode

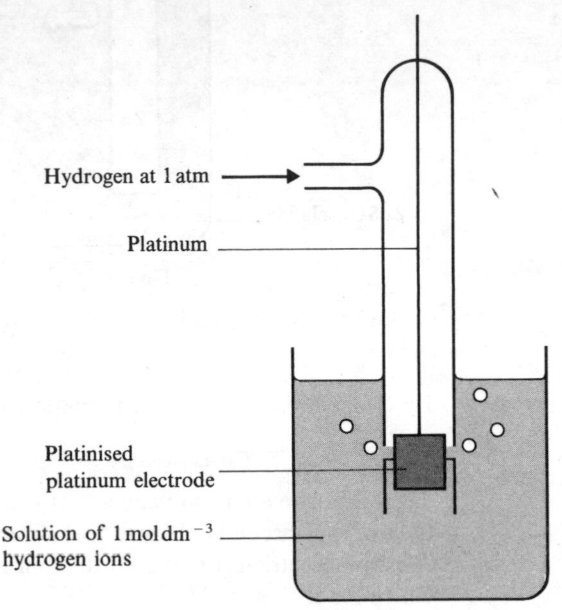

... E$^{\ominus}$ for another electrode can be found by combination with a standard hydrogen electrode...

The standard electrode potentials of other systems can be found by combining them with a standard hydrogen electrode and measuring the emf of the cell formed. A voltaic cell which combines a standard zinc electrode and a standard hydrogen electrode is shown in Figure 13.3.

FIGURE 13.3 Apparatus for finding the Standard Electrode Potential of the $Zn(s) \rightarrow Zn^{2+} + 2e^-$ system

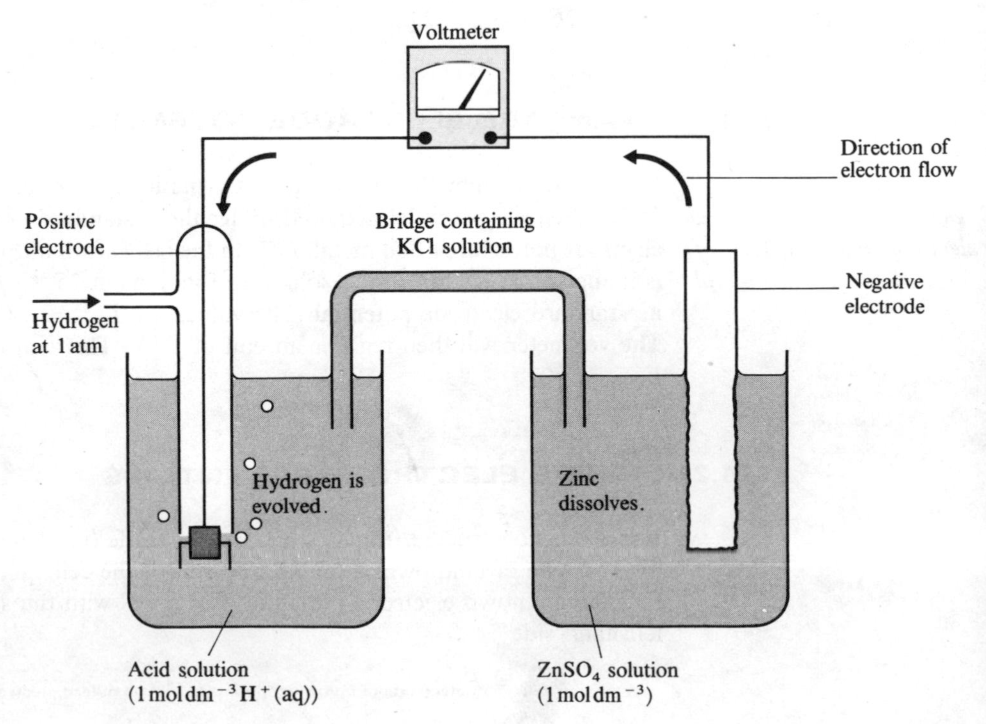

The two compartments in the figure are connected by a **salt bridge**. This contains an electrolyte such as potassium chloride, which conducts electricity but does not allow mixing of the two solutions in the half-cells. The emf of this cell is −0.76 V. The voltmeter shows that electrons flow through the external circuit from zinc to the hydrogen electrode, showing that zinc has a standard electrode potential of −0.76 V.

... but the hydrogen electrode is difficult to operate

The hydrogen electrode is not a convenient reference electrode to use in measurements: maintaining a stream of hydrogen at 1 atm takes careful management. Another standard electrode is the calomel electrode. Calomel is a name for mercury(I) chloride. When mercury and solid mercury(I) chloride are both in contact with a solution of chloride ions, the equilibrium

$$Hg_2Cl_2(s) + 2e^- \rightleftharpoons 2Hg(l) + 2Cl^-(aq)$$

More convenient is the calomel electrode

is set up. The electrode potential of this system depends on the concentration of chloride ions. The easiest way of keeping the concentration constant is to keep the solution saturated by having some undissolved potassium chloride present. The potential of a saturated calomel electrode at 25 °C is +0.244 V. Once the electrode potential has been determined by measuring the emf of a cell composed of a saturated calomel electrode and a standard hydrogen electrode, the calomel electrode can be used as a reference electrode in other measurements. [See Figure 13.4.]

FIGURE 13.4 Two Forms of the Calomel Electrode (in (a) contact is made through the side arm; in (b) contact is made through the sintered glass disc)

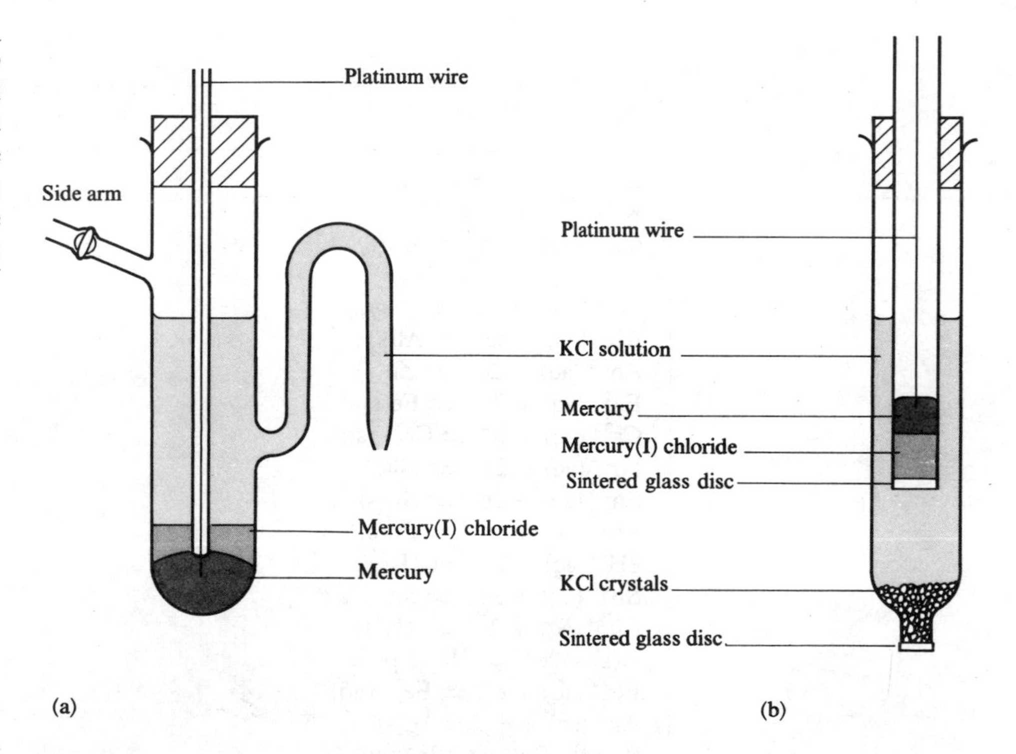

In the electrode process

$$Zn(s) \rightarrow Zn^{2+}(aq) + 2e^-$$

Redox systems have electrode potentials

zinc is acting as a reducing agent, supplying electrons. Oxidation–reduction systems other than metals in equilibrium with their ions also have electrode potentials. If a piece of platinum wire is immersed in an acidic solution of manganate(VII) ions, the equilibrium

$$MnO_4^-(aq) + 8H^+(aq) + 5e^- \rightleftharpoons Mn^{2+}(aq) + 4H_2O(l)$$

Definition of standard electrode potential for a redox half-cell

takes electrons from the piece of platinum, which therefore becomes positively charged. If each of the species involved in the equilibrium has a concentration of 1 mol dm^{-3}, the potential acquired by the platinum is the **standard electrode potential** or **standard reduction potential** of the redox system.

The convention is to write each half-cell reaction as a reduction process:

$$\text{Oxidant} + ne^- \rightleftharpoons \text{Reductant}$$

The standard electrode potentials quoted in tables refer to the reaction written in this way, i.e., as a reduction reaction. They are therefore reduction potentials. Reactions which proceed from left to right more readily than the reduction of hydrogen ions

$$H^+(aq) + e^- \rightleftharpoons \tfrac{1}{2}H_2(g)$$

are given a positive standard electrode potential. Oxidising agents have high positive standard electrode potentials. Reducing agents, such as metals, have highly negative standard electrode potentials.

When all the redox systems are arranged in order of their standard electrode potentials, the Electrochemical Series is obtained. Table 13.1 shows some common redox systems:

Reaction	$E^{\ominus}$/V (at 298 K)
$Li^+(aq) + e^- \rightleftharpoons Li(s)$	−3.04
$K^+(aq) + e^- \rightleftharpoons K(s)$	−2.92
$Ca^{2+}(aq) + 2e^- \rightleftharpoons Ca(s)$	−2.87
$Na^+(aq) + e^- \rightleftharpoons Na(s)$	−2.71
$Mg^{2+}(aq) + 2e^- \rightleftharpoons Mg(s)$	−2.38
$Al^{3+}(aq) + 3e^- \rightleftharpoons Al(s)$	−1.66
$Zn^{2+}(aq) + 2e^- \rightleftharpoons Zn(s)$	−0.76
$Fe^{2+}(aq) + 2e^- \rightleftharpoons Fe(s)$	−0.44
$Cr^{3+}(aq) + e^- \rightleftharpoons Cr^{2+}(aq)$	−0.41
$Ni^{2+}(aq) + 2e^- \rightleftharpoons Ni(s)$	−0.25
$Sn^{2+}(aq) + 2e^- \rightleftharpoons Sn(s)$	−0.14
$Pb^{2+}(aq) + 2e^- \rightleftharpoons Pb(s)$	−0.13
$2H^+(aq) + 2e^- \rightleftharpoons H_2(g)$	0.00
$Sn^{4+}(aq) + 2e^- \rightleftharpoons Sn^{2+}(aq)$	0.15
$Cu^{2+}(aq) + 2e^- \rightleftharpoons Cu(s)$	0.34
$I_2(s) + 2e^- \rightleftharpoons 2I^-(aq)$	0.54
$Fe^{3+}(aq) + e^- \rightleftharpoons Fe^{2+}(aq)$	0.77
$Ag^+(aq) + e^- \rightleftharpoons Ag(s)$	0.80
$Br_2(l) + 2e^- \rightleftharpoons 2Br^-(aq)$	1.07
$MnO_2(s) + 4H^+(aq) + 2e^- \rightleftharpoons Mn^{2+}(aq) + 2H_2O(l)$	1.23
$Cr_2O_7^{2-}(aq) + 14H^+(aq) + 6e^- \rightleftharpoons 2Cr^{3+}(aq) + 7H_2O(l)$	1.33
$Cl_2(g) + 2e^- \rightleftharpoons 2Cl^-(aq)$	1.36
$Ce^{4+}(aq) + e^- \rightleftharpoons Ce^{3+}(aq)$ (in $H_2SO_4(aq)$)	1.45
$PbO_2(s) + 4H^+(aq) + 2e^- \rightleftharpoons Pb^{2+}(aq) + 2H_2O(l)$	1.47
$MnO_4^-(aq) + 8H^+(aq) + 5e^- \rightleftharpoons Mn^{2+}(aq) + 4H_2O(l)$	1.52
$Ce^{4+}(aq) + e^- \rightleftharpoons Ce^{3+}(aq)$ (in $HNO_3(aq)$)	1.61
$H_2O_2(aq) + 2H^+(aq) + 2e^- \rightleftharpoons 2H_2O(l)$	1.77
$F_2(g) + 2e^- \rightleftharpoons 2F^-(aq)$	2.87

TABLE 13.1 Standard Electrode Potentials

The criterion for a spontaneous cell reaction is that $E^{\ominus}_{cell}$ is positive

When two electrodes combine to form a cell, the value of $E^{\ominus}$ for the cell must be positive if the cell reaction is to happen spontaneously. For example, when copper and silver are in contact with solutions of their ions, two equilibria are set up:

$$Cu^{2+}(aq) + 2e^- \rightleftharpoons Cu(s); \quad E^{\ominus} = +0.34\,V$$

$$Ag^+(aq) + e^- \rightleftharpoons Ag(s); \quad E^{\ominus} = +0.80\,V$$

So that $E^{\ominus}$ shall have a positive value, the reactions that take place are

$$Ag^+(aq) + e^- \rightarrow Ag(s); \quad E^{\ominus} = +0.80\,V$$

$$Cu(s) \rightarrow Cu^{2+}(aq) + 2e^-; \quad E^{\ominus} = -0.34\,V$$

$$\text{Total}: Cu(s) + 2Ag^+(aq) \rightarrow Cu^{2+}(aq) + 2Ag(s); \quad E^{\ominus} = +0.46\,V$$

If solutions containing Ce^{4+}, Ce^{3+}, Fe^{3+} and Fe^{2+} are mixed, the redox equilibria in the solution are

$$Fe^{3+}(aq) + e^- \rightleftharpoons Fe^{2+}(aq); \quad E^{\ominus} = +0.77\,V$$

$$Ce^{4+}(aq) + e^- \rightleftharpoons Ce^{3+}(aq); \quad E^{\ominus} = +1.45\,V$$

The redox reaction that takes place is that for which $E^{\ominus}$ is positive, i.e.

$$Ce^{4+}(aq) + Fe^{2+}(aq) \rightarrow Ce^{3+}(aq) + Fe^{3+}(aq); \quad E^{\ominus} = +0.68\,V$$

A redox reaction will go almost to completion between two redox systems which differ by 0.3 V or more in their electrode potentials.

STANDARD ELECTRODE POTENTIAL AND STANDARD FREE ENERGY CHANGE

There is a relationship between electrochemistry and thermochemistry. In the electrode reaction,

$$M^{n+}(aq) + ne^- \rightleftharpoons M(s)$$

if the standard free energy change is $\Delta G^{\ominus}$ then the value of the standard electrode potential, $E^{\ominus}$ at the same temperature is given by

$$\Delta G^{\ominus} = -nFE^{\ominus}$$

where n is the number of electrons transferred in the electrode reaction and F is the Faraday constant.

For example, $E^{\ominus} = +0.34$ V at 298 K for the electrode reaction

$$Cu^{2+}(aq) + 2e^- \rightleftharpoons Cu(s)$$

The value of $\Delta G^{\ominus}$ is given by

$$\Delta G^{\ominus} = -2 \times 96\,500 \times (-0.34)$$

$$= 65.6\,kJ\,mol^{-1}$$

CHECKPOINT 13A: ELECTRODE POTENTIALS

1. Does a high positive standard electrode potential for a redox system indicate that the system acts as an oxidising agent or a reducing agent?

2. Explain the terms (*a*) electrode potential, (*b*) standard electrode potential.

Explain how you could find the standard electrode potential for the system

$$Fe^{3+}(aq)|Fe^{2+}(aq)|Pt$$

3. This cell is set up:

$$Ag(s)|Ag^+(aq, 1\,mol\,dm^{-3})\,\vdots\vdots\,Cu^{2+}(aq, 1\,mol\,dm^{-3})|Cu(s)$$

(*a*) State the emf of the cell. Refer to Table 13.1.

(*b*) Write the equation for the chemical reaction that takes place in the cell when the copper and silver electrodes are connected by an external circuit.

(*c*) State the direction in which electrons flow through the external circuit.

4. Which electrode, anode or cathode, is associated in an electrochemical cell with oxidation?

5. Refer to Table 13.1. Which of the following reactions will occur spontaneously? (Assume all concentrations are 1 mol dm^{-3}.)

(*a*) $Fe(s) + Zn^{2+}(aq) \rightarrow Fe^{2+}(aq) + Zn(s)$

(*b*) $Fe(s) + Sn^{2+}(aq) \rightarrow Fe^{2+}(aq) + Sn(s)$

(*c*) $Sn^{4+}(aq) + 2I^-(aq) \rightarrow Sn^{2+}(aq) + I_2(s)$

(*d*) $Zn(s) + Mg^{2+}(aq) \rightarrow Zn^{2+}(aq) + Mg(s)$

(*e*) $Zn(s) + Sn^{2+}(aq) \rightarrow Zn^{2+}(aq) + Sn(s)$

(*f*) $Sn^{4+}(aq) + 2Fe^{2+}(aq) \rightarrow Sn^{2+}(aq) + 2Fe^{3+}(aq)$

(*g*) $Cr_2O_7^{2-}(aq) + 14H^+(aq) + 6Cl^-(aq) \rightarrow 2Cr^{3+}(aq) + 7H_2O(l) + 3Cl_2(g)$

(*h*) $2Ce^{4+}(aq) + 2Br^-(aq) \rightarrow 2Ce^{3+}(aq) + Br_2(l)$

13.1.3 THE GLASS ELECTRODE

The potential of the glass electrode depends on pH

There is another electrode which has a potential that depends on pH. The potential of the glass electrode varies in a regular way with the pH of the solution in which it is immersed.

For pH measurement, a glass electrode is combined with a reference electrode

The glass electrode is coupled with a calomel electrode to complete the cell. As the glass electrode has a high resistance, a high resistance voltmeter is used to measure the emf of the cell.

The potential of the glass electrode at 25 °C is given by

$$E = K - 0.0592\,\text{pH}$$

where K = a constant.

FIGURE 13.5 The Glass Electrode (Hanna Instruments)

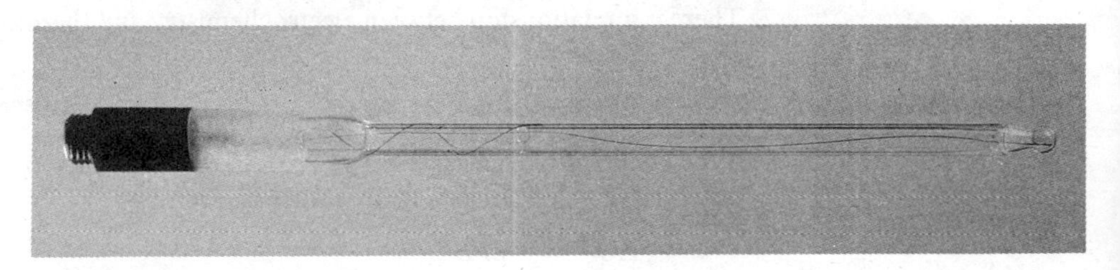

13.2 POTENTIOMETRY

13.2.1 ACID–BASE TITRATIONS

The potential of a glass electrode changes during the course of a titration

A potentiometric method can be used to follow the course of a titration. To perform an acid–base titration, a combination of a glass electrode and a calomel electrode is used. The potential of the glass electrode depends on the pH of the solution. As the pH changes during the titration, the potential of the glass electrode changes, and the emf of the cell changes. An apparatus for potentiometric titration is shown in Figure 13.6.

A plot of emf against the volume of titrant gives the end-point

A plot of emf against the volume of titrant added is plotted. Such plots have the same shapes as the titration curves in Figure 12.8, § 12.7.9: at the end-point, there is a sharp change in emf. Automatic titrators have been developed to add titrant and record the course of the titration automatically.

FIGURE 13.6
A Potentiometric Method for Acid–Base Titration

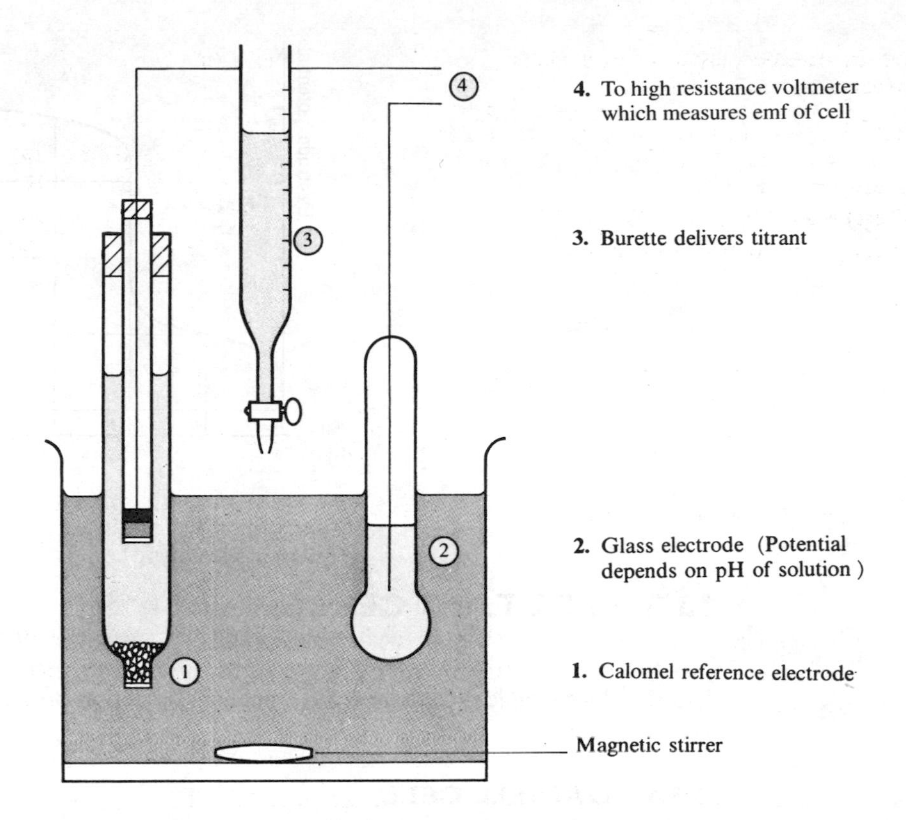

13.2.2 OXIDATION–REDUCTION TITRATIONS

The potential of a platinum electrode varies during the course of a redox titration

A platinum electrode dipping into the redox solution measures the potential of the system. The usual reference electrode is a calomel electrode. Figure 13.7(a) shows apparatus which could be used. The variation of electrode potential during the course of the titration follows the curve shown in Figure 13.7(b).

FIGURE 13.7
A Potentiometric Method for A Redox Titration

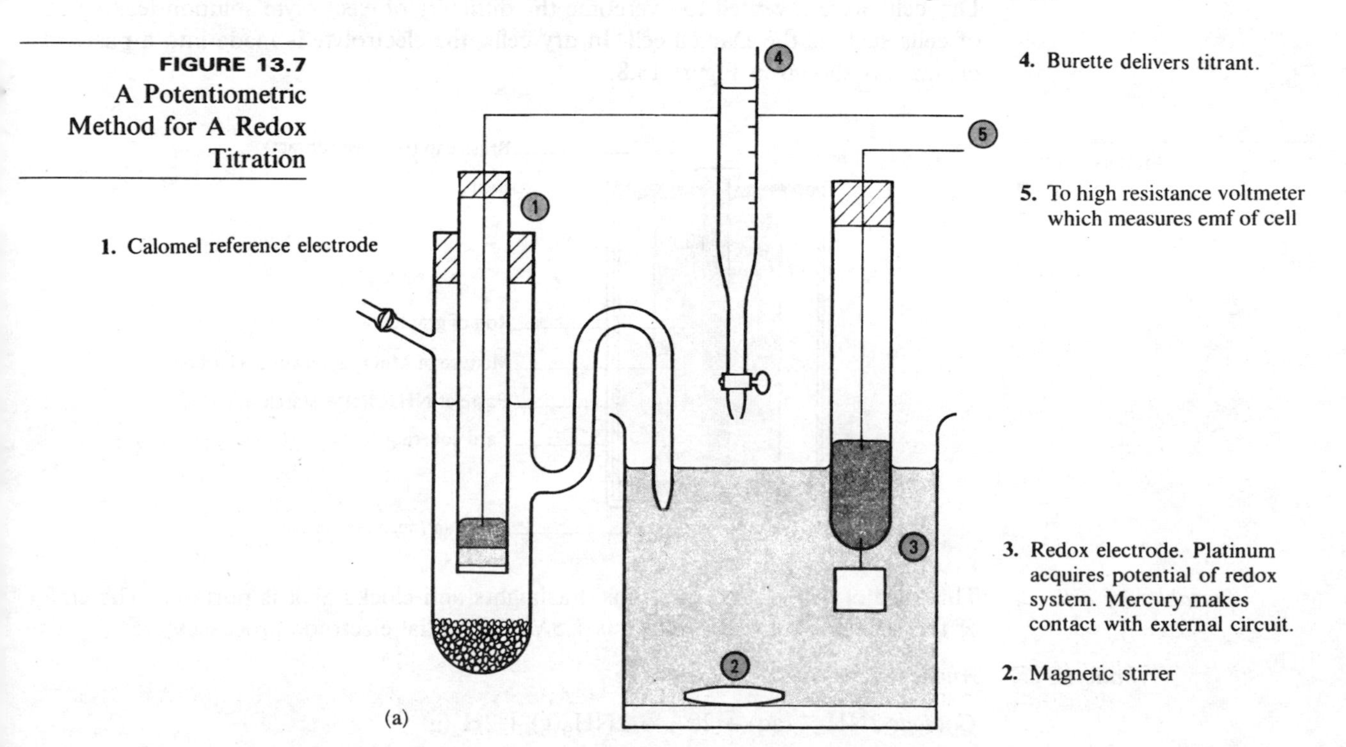

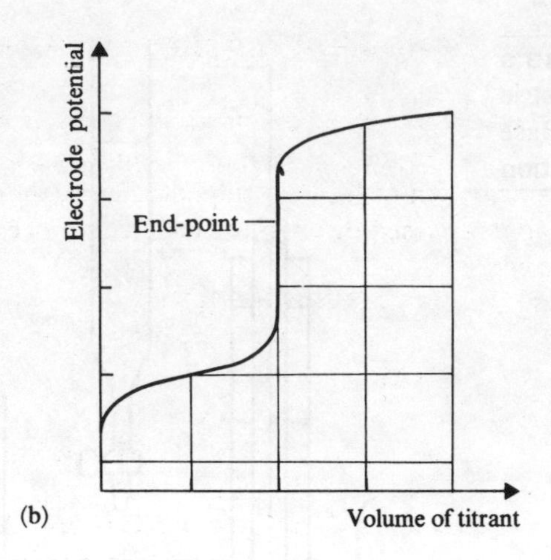

13.3 VOLTAIC CELLS

In galvanic or voltaic cells, a chemical reaction produces an electric current.

13.3.1 DANIELL CELL

The Daniell cell utilises zinc and copper in solutions of their ions as shown in Figure 13.1, § 13.1. The emf of the cell is 1.1 V.

13.3.2 DRY CELLS

Dry cells were invented to overcome the difficulty of electrolyte solution leaking out of cells such as the Daniell cell. In dry cells, the electrolyte is made into a paste. An example is shown in Figure 13.8.

FIGURE 13.8 A Dry Cell

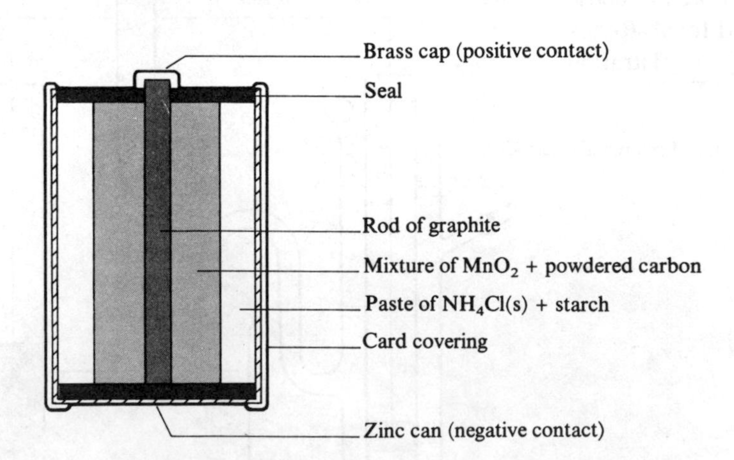

Portable dry cells...

This type of cell is used in radios, flashlights and clocks as it is portable. The emf of the cell shown in Figure 13.8 is 1.5 V. The initial electrode processes are

...batteries...

Anode $Zn(s) \rightarrow Zn^{2+}(aq) + 2e^-$

Cathode $2NH_4^+(aq) + 2e^- \rightarrow 2NH_3(g) + 2H_2(g)$

13.3.3 THE LEAD–ACID ACCUMULATOR

...accumulators

This cell stores or accumulates electric charge. It consists of two lead plates dipping into a 30% solution of sulphuric acid. Both plates become covered with an insoluble film of lead(II) sulphate. First, the cell must be charged. A direct electric current is passed through the cell. The processes which take place are

The lead-acid accumulator is charged by passing a direct electric current through it...

CHARGE

Positive plate

$$PbSO_4(s) + 2e^- \rightarrow Pb(s) + SO_4^{2-}(aq)$$

Negative plate

$$PbSO_4(s) + 2H_2O(l) \rightarrow PbO_2(s) + 4H^+(aq) + SO_4^{2-}(aq) + 2e^-$$

The plates are now different and therefore have different potentials, so that, when they are connected, an electric current will flow between them. When the cell supplies electric current, i.e., discharges, the processes which take place are

...During discharge, it supplies an electric current

DISCHARGE

Negative plate

$$Pb(s) + SO_4^{2-}(aq) \rightarrow PbSO_4(s) + 2e^-$$

Positive plate

$$2PbO_2(s) + 4H^+(aq) + SO_4^{2-}(aq) + 2e^- \rightarrow PbSO_4(s) + 2H_2O(l)$$

When all the PbO_2 and Pb have been converted to $PbSO_4$, there is no difference between the plates, and the cell can no longer give a current.

The polarity of the plates reverses between charge and discharge

You will notice that the Pb plate is the positive plate during charge, and the negative plate during discharge. The polarity of the PbO_2 plate is also reversed, from negative during charge to positive during discharge. The reactions can be summarised as

$$Pb(s) + PbO_2(s) + 2H_2SO_4(aq) \underset{\text{charge}}{\overset{\text{discharge}}{\rightleftharpoons}} 2PbSO_4(s) + 2H_2O(l)$$

Car batteries

Car batteries consist of six lead accumulator cells joined in series to give an emf of 12 V. When the car is in motion, it drives a generator which charges the battery. If there is too much stopping and starting, the battery loses its charge and becomes 'flat', until it is recharged by the passage of a direct current from a transformer. During discharge, sulphuric acid is used up, and the density of the liquid in the cells drops. A **hydrometer** can be used to measure the density of the liquid and assess the state of the battery.

13.3.4 FUEL CELLS

Fuel cells are a promising source of energy for the future

A fuel cell is a galvanic cell which converts the chemical energy of a continuous supply of reactants into electrical energy. Fuel is supplied to one electrode and an oxidant, usually oxygen, to the other [see Figure 13.9]. A great deal of research is being done on fuel cells as they are a promising source of energy for the future. The American Gemini space probes and Apollo moon probes used hydrogen–oxygen fuel cells. The astronauts used the product of the reaction to supplement their drinking water.

FIGURE 13.9
A Hydrogen–Oxygen Fuel Cell

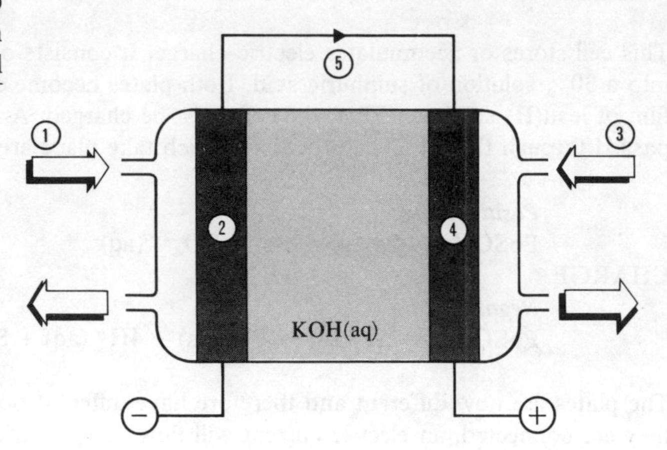

1 Stream of hydrogen

2 Hydrogen diffuses through the porous cathode (e.g., of Ni). When it comes into contact with the electrolyte, KOH(aq), adsorbed H_2 is oxidised:

$\frac{1}{2}H_2 + OH^- \rightarrow H_2O + e^-$

3 Stream of oxygen

4 Oxygen diffuses through a porous anode (e.g., of nickel). Adsorbed oxygen is reduced to OH^- ions:

$\frac{1}{2}O_2 + H_2O + 2e^- \rightarrow 2OH^-(aq)$

The overall reaction is

$2H_2(g) + O_2(g) \rightarrow 2H_2O(l)$

5 Electrons flow through an external circuit from cathode to anode.

QUESTIONS ON CHAPTER 13

1. Refer to Table 13.1, § 13.1.2.
Which of the following species are oxidised by manganese(IV) oxide?
Br^-, Ag, I^-, Cl^-

2. Which of the following species are reduced by Sn^{2+}?
I_2, Ni^{2+}, Cu^{2+}, Fe^{2+}

3. The rusting of iron is prevented by (*a*) a coating of paint or (*b*) a layer of zinc or (*c*) a layer of tin. How do these different methods work? What is meant by 'sacrificial protection'? Give examples of its use [see § 24.14.5].

4. Refer to Table 13.1, § 13.1.2. Relate the differences between the chemical properties of the elements mentioned below to the differences between the standard electrode potentials of the following systems:

(*a*) $K^+(aq)|K(s)$ (*d*) $Fe^{2+}(aq)|Fe(s)$
(*b*) $Mg^{2+}(aq)|Mg(s)$ (*e*) $Cu^{2+}(aq)|Cu(s)$
(*c*) $Ca^{2+}(aq)|Ca(s)$ (*f*) $Cl_2(g)|Cl^-(aq)$

5. Two half-cells are
(*a*) $Co^{2+}(aq, 1\,mol\,dm^{-3})|Co(s)$
(*b*) $Cl_2(g, 1\,atm)|Cl^-(aq, 1\,mol\,dm^{-3})|Pt$
State which will be the positive and which the negative electrode when the two half-cells are connected. Write the equation for the cell reaction. If the emf of the cell is 1.63 V, what is $E^\ominus$ for the cobalt half-cell? [See Table 13.1, § 13.1.2. for $E^\ominus(Cl_2/Cl^-)$.]

6. Calculate the standard emf's of the following cells at 298 K:
(*a*) $Ni(s)|Ni^{2+}(aq)\vdots\vdots Sn^{2+}(aq), Sn^{4+}(aq)|Pt(s)$
(*b*) $Pt(s)|I_2(s), I^-(aq)\vdots\vdots Ag^+(aq)|Ag(s)$
(*c*) $Pt(s)|Cl_2(g), Cl^-(aq)\vdots\vdots Br_2(l), Br^-(aq)|Pt(s)$
(*d*) $Sn(s)|Sn^{2+}(aq)\vdots\vdots Ag^+(aq)|Ag(s)$
(*e*) $Ag(s)|Ag^+(aq)\vdots\vdots Cu^{2+}(aq)|Cu(s)$
(*f*) $Fe(s)|Fe^{2+}(aq)\vdots\vdots Cu^{2+}(aq)|Cu(s)$
(*g*) $Zn(s)|Zn^{2+}(aq)\vdots\vdots Pb^{2+}(aq)|Pb(s)$

7. Refer to Table 13.1, § 13.1.2 and to the standard reduction potentials listed here:

$$E^\ominus(VO_2^+(aq)|VO^{2+}(aq)) = +1.00\,V$$
$$E^\ominus(Cd^{2+}(aq)|Cd(s)) = -0.40\,V$$
$$E^\ominus(BrO_3^-(aq)|Br_2(g)) = +1.52\,V$$
$$E^\ominus(S_4O_6^{2-}(aq)|S_2O_3^{2-}(aq)) = 0.090\,V$$

(*a*) State which of the species MnO_4^-, Ce^{4+}, $Cr_2O_7^{2-}$, VO_2^+, Fe^{3+} are able to liberate chlorine from an acidic solution of sodium chloride.

(*b*) Write a balanced equation for the reaction between MnO_4^- and VO^{2+} in acid solution.

(*c*) Find $E^\ominus$ for the cell

$Zn(s)|ZnSO_4(1\,mol\,dm^{-3})\vdots\vdots CdSO_4(1\,mol\,dm^{-3})|Cd(s)$

(*d*) Put the following into order of their power as oxidising agents in acid solution: $Cr_2O_7^{2-}$, Cl_2, MnO_4^-, I_2, BrO_3^-, $S_4O_6^{2-}$

8. (*a*) Define *oxidation*.

(*b*) Some standard electrode potentials are given below.

	$E^\ominus$/V
$H^+ + e^- \rightleftharpoons \frac{1}{2}H_2$	0.00
$Fe^{3+} + 3e^- \rightleftharpoons Fe$	−0.04
$Cr^{3+} + 3e^- \rightleftharpoons Cr$	−0.74
$Ti^{2+} + 2e^- \rightleftharpoons Ti$	−1.63
$Al^{3+} + 3e^- \rightleftharpoons Al$	−1.66

(i) Why are these sometimes referred to as *redox potentials*?

(ii) Write the names of the four metals in order of decreasing reactivity with acids, giving the *most reactive first*.

(iii) The hydrogen electrode potential is *defined* as *zero*. Why do you think that hydrogen was chosen as the standard and what are its disadvantages?

(iv) What would you expect to *see* if a piece of titanium (Ti) metal were dropped into a solution of iron(III) chloride?

(*c*) If a large block of chromium (Cr) metal is immersed in cold, concentrated ethanoic acid (CH_3CO_2H) then the evolution of hydrogen gas is very slow. Give *three* reasons why the reaction should be slow.

(O & C 92, AS)

9. (*a*) The electrode potential of zinc is the maximum potential difference between a zinc rod and a solution of its ions with respect to a standard hydrogen electrode, all under standard conditions.

(i) Name the instrument used to measure the potential difference and explain why it must be used.

(ii) Describe the essential features of a standard hydrogen electrode.

(*b*) Using the information below, measured under standard conditions, calculate the enthalpy change for the reaction

$$H_2(g) + Cu^{2+}(aq) \rightarrow Cu(s) + 2H^+(aq)$$

Reaction	*Enthalpy change*/kJ mol^{-1}
$Cu(s) \rightarrow Cu(g)$	339 ($\Delta H^{\ominus}_{\text{atomisation}}$)
$Cu(g) - 2e^- \rightarrow Cu^{2+}(g)$	2711 ($\Delta H^{\ominus}_{\text{ionisation 1st + 2nd}}$)
$Cu^{2+}(g) + aq \rightarrow Cu^{2+}(aq)$	−2100 ($\Delta H^{\ominus}_{\text{hydration}}$)
$\frac{1}{2}H_2(g) - e^- + aq \rightarrow H^+(aq)$	446 ($\Delta H^{\ominus}_{\text{formation}}$ [$H^+(aq)$])

(*c*) Given that the standard entropy change, $\Delta S^{\ominus}$, for the reaction

$$H_2(g) + Cu^{2+}(aq) \rightarrow Cu(s) + 2H^+(aq)$$

is 6.70×10^{-3} kJ K^{-1} mol^{-1} at 298 K and that one Faraday equals 9.648×10^4 C mol^{-1}, calculate

(i) the value of the standard free energy change, $\Delta G^{\ominus}$

(ii) the value of the standard electrode potential, $E^{\ominus}$ (Cu^{2+}/Cu), at this temperature.

(iii) In the light of your value for $\Delta G^{\ominus}$, explain the behaviour of copper metal when placed in dilute sulphuric acid.

(AEB 90)

10. (*a*) (i) Define *oxidation* and *reduction* in terms of electron transfer.

(ii) Define *standard electrode potential.*

(iii) Give the equation for the process which is arbitrarily assigned a standard electrode potential of zero.

(*b*) The table below gives the standard electrode potentials of various processes.

Process	$E^{\ominus}$/V
$2HCO_2^- + 2CO_2 + 6H^+(aq) + 6e^- \rightleftharpoons (C_4H_4O_6)^{2-} + 2H_2O$	+0.20
$O_2 + 2H^+(aq) + 2e^- \rightleftharpoons H_2O_2$	+0.68
$H_2O_2 + 2H^+(aq) + 2e^- \rightleftharpoons 2H_2O$	+1.77

(i) Hydrogen peroxide can act as an oxidant or a reductant. Hydrogen peroxide solution is added to an acid solution containing HCO_2^- and $(C_4H_4O_6)^{2-}$ ions, under standard conditions. Predict and explain the reaction which is likely to occur between hydrogen peroxide and these species.

(ii) Write the equation for this reaction.

(iii) State what is meant by *disproportionation.*

(iv) By considering the standard electrode potentials given, explain the disproportionation of hydrogen peroxide when treated with a manganese(IV) oxide catalyst.

†(*c*) The rusting of iron is an electrolytic process in which different regions of the iron act as anodes and as cathodes.

(i) Write the ionic equation for the reaction involving oxygen which occurs at a cathodic region.

(ii) Write the equation for the reaction occurring at an anodic region.

(iii) Which regions on the surface of the iron are most likely to behave as anodes?

(iv) Why does the presence of sodium chloride accelerate the rusting process?

[For (*c*) see Chapter 24.]

(O 92)

11. This question is based on electrode potentials and on phase equilibria.

(*a*) (i) Define the term *standard (reduction) electrode potential.*

The following standard electrode potentials will be needed to answer the remainder of this section.

Electrode	$E^{\ominus}$ *(298* K)/V
$Mg^{2+}(aq) \mid Mg(s)$	−2.34
$Zn^{2+}(aq) \mid Zn(s)$	−0.76
$Cu^{2+}(aq) \mid Cu(s)$	+0.34
$Fe^{3+}(aq) \mid Fe^{2+}(aq)$	+0.77
$Ag^+(aq) \mid Ag(s)$	+0.80
$\frac{1}{2}Br_2(l) \mid Br^-(aq)$	+1.06

(ii) Draw a labelled sketch of the cell formed by combining the $Zn^{2+}(aq) \mid Zn(s)$ and $Cu^{2+}(aq) \mid Cu(s)$ half cells.

Mark on the sketch

(*1*) the positive electrode

(*2*) the direction of electron flow in a metal wire which joins the electrodes

(*3*) the electrode at which oxidation occurs.

(iii) Calculate the standard emf of this cell (in (*a*)(ii)).

(iv) State, giving a reason in *each* case, what you would expect to happen when

(*1*) copper powder is shaken with aqueous silver nitrate solution

(*2*) copper powder is shaken with aqueous magnesium chloride solution

(*3*) an aqueous solution of Fe^{2+} ions is shaken with bromine.

(*b*) (i) Draw a boiling point–composition diagram for an ideal system of two miscible liquids of different boiling points, labelling the axes and explaining the significance of the curves drawn.

(ii) Use the diagram that you have drawn in (*b*)(i) to explain how the two components may be completely separated by fractional distillation.

(iii) Give *one* example of the use of fractional distillation in industry.

(*c*) (i) Explain the principle involved in the process of *solvent extracton*. State why the process is useful.

(ii) During nuclear fuel reprocessing, solvent extraction with an organic solvent is used to recover uranium salts from aqueous solution. If the partition coefficient of a uranium salt between the organic solvent and acidified water has a value of 50, calculate the ratio of the total number of moles of uranium in the organic solvent layer to that in the water layer after 500 dm^3 of an aqueous solution of uranium salt has been extracted with 200 dm^3 of organic solvent.

[*Hint* Number of moles = Concentration × Volume.]

(WJEC 92)

12. (*a*) Explain the origin of the potential difference, *E*, that exists between a metal and a solution of its ions. Explain why different metal/metal ion systems have different values of *E*.

(*b*) For the electrochemical cell represented by the cell diagram

$$Pt \,|\, H_2(g), |\, H^+(aq) \,||\, Cu^{2+}(aq) \,|\, Cu(s)$$

describe how you would investigate the variation of electrode potential with change in copper ion concentration. You may assume that you have access to standard hydrogen and copper electrode systems. You should draw a fully labelled diagram and describe the essential steps in the method including details of how you would make careful and accurate measurements.

(*c*) The following data, with copper as the positive electrode, were obtained from such an experiment using a standard hydrogen electrode.

$[Cu^{2+}(aq)]$/mol dm^{-3}	*E*/V
1.0	0.34
1.0×10^{-2}	0.28
1.0×10^{-4}	0.22
1.0×10^{-6}	0.16

(i) The emf, *E*, of the cell varies with log $[Cu^{2+}(aq)]$. Plot a graph of *E* against log $[Cu^{2+}(aq)]$, choosing your axes to allow extrapolation to log $[Cu^{2+}(aq)] = -12$.

(ii) What is the copper ion concentration when the copper electrode potential is equal to the standard hydrogen electrode potential?

(iii) Write the equation of the reaction that occurs in the cell when it is short circuited.

(iv) When the cell emf is zero the cell reaction has reached equilibrium. Write the equilibrium constant expression and determine the equilibrium constant, K_c, of the reaction. Comment on the accuracy of the value that you obtain.

(v) Comment on the likelihood of this reaction occurring when hydrogen is bubbled into an aqueous solution of copper sulphate.

(O & C 92, S)

14

REACTION KINETICS

14.1 THE RATE OF REACTION

The rate of a chemical reaction depends on a number of factors . . .

Many people are interested in knowing how to alter the rates of chemical reactions. Fertiliser manufacturers are interested in speeding up the formation of ammonia from nitrogen and hydrogen. Car manufacturers are interested in slowing down the rate at which iron rusts. A number of factors can be changed to alter the speed of a chemical reaction. These are:

1. the size of the particles of a solid reactant
2. the concentrations of reactants in solution
3. the pressures of gaseous reactants
4. the temperature
5. the presence of light
6. the addition of a catalyst

In § 14.1, we will take a look at these factors in a qualitative manner before going on to a fuller, quantitative treatment in later sections of this chapter.

14.1.1 PARTICLE SIZE

The reaction between calcium carbonate and dilute hydrochloric acid is used to prepare carbon dioxide [see Figure 23.12, § 23.6.3].

$$CaCO_3(s) + 2HCl(aq) \rightarrow CO_2(g) + CaCl_2(aq) + H_2O(l)$$

. . the size of particles of a solid reactant . . .

The calcium carbonate used is in the form of marble chips. The reaction can be used to find out whether large lumps of a solid react at the same speed as small lumps of the same solid. Figure 14.1 shows an apparatus which could be used.

The mass of the flask and contents is noted at various times after the start of the reaction. When the mass is plotted against time, results such as those in Figure 14.2 are obtained. The results show that the smaller the size of the particles of calcium carbonate, the faster the reaction takes place [see Question 3, Checkpoint 14A].

FIGURE 14.1
An Apparatus for following the Loss in Mass during a Reaction

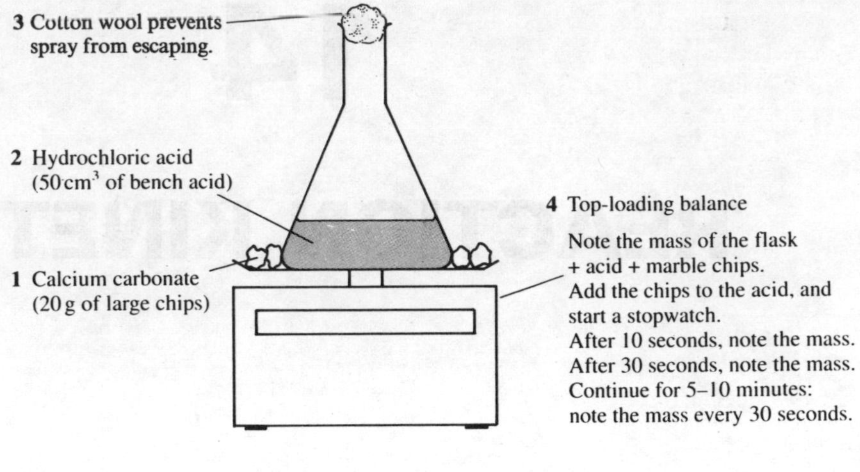

3 Cotton wool prevents spray from escaping.

4 Top-loading balance

Add the chips to the acid, and start a stopwatch.
After 10 seconds, note the mass.
After 30 seconds, note the mass.
Continue for 5–10 minutes: note the mass every 30 seconds.

FIGURE 14.2
Results obtained with Different Sizes of Marble Chips

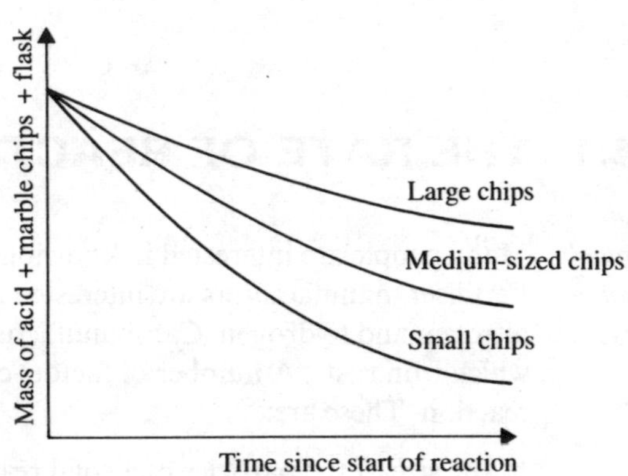

Mass of acid + marble chips + flask

Medium-sized chips

Time since start of reaction

14.1.2 CONCENTRATION

... the concentration of a reactant in solution ...

Many chemical reactions take place in solution. The concentration of a reactant may affect the speed of the reaction. In the reaction between sodium thiosulphate and acid,

$$Na_2S_2O_3(aq) + 2HCl(aq) \rightarrow S(s) + SO_2(g) + 2NaCl(aq) + H_2O(l)$$

Sodium thiosulphate Sulphur

Sulphur appears as very small particles of solid suspended in the solution. A method of studying the speed at which sulphur is formed is shown in Figure 14.3. Graphs of readings of time and 1/time against the concentration of thiosulphate are shown in Figures 14.4 and 14.5.

The faster the reaction takes place, the shorter the time before the deposit of sulphur is dense enough to hide the cross from view. That is, the speed of the reaction is inversely proportional to the time taken for the reaction to finish.

$$\text{Speed of reaction} \propto 1/\text{Time}$$

FIGURE 14.3
Measuring the Time needed for Reaction

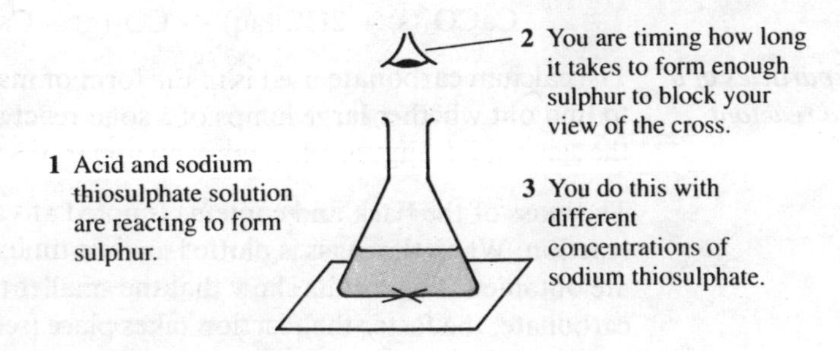

2 You are timing how long it takes to form enough sulphur to block your view of the cross.
1 Acid and sodium thiosulphate solution are reacting to form sulphur.
3 You do this with different concentrations of sodium thiosulphate.

FIGURE 14.4
A Graph of Time against Concentration

The results show that the cross disappears soonest when the solution is most concentrated.

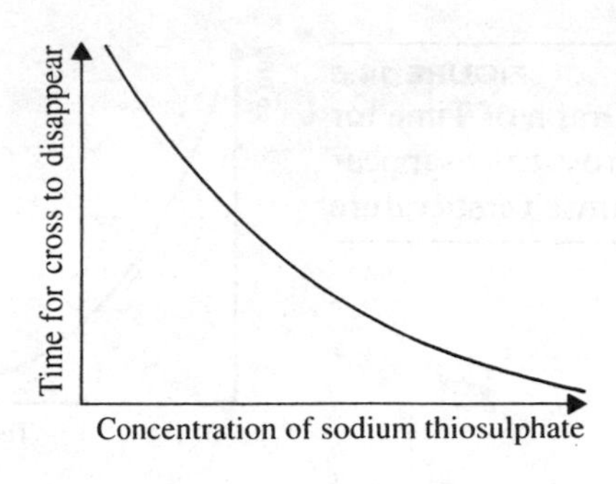

FIGURE 14.5
A Graph of 1/Time against Concentration

When 1/(time for cross to disappear) is plotted against concentration, a straight line is obtained. This shows that for this reaction,

1/Time $\propto$ Concentration

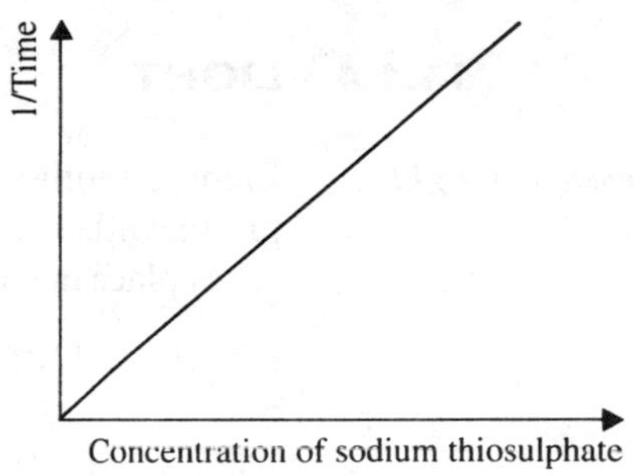

You can see from the graph in Figure 14.5 that

$$\text{Speed of reaction} \propto \text{Concentration of thiosulphate}$$

If you repeat the experiment, keeping the concentration of thiosulphate constant and varying the concentration of acid, you will find that

$$\text{Speed of reaction} \propto \text{Concentration of acid}$$

The reason why the speed increases with concentration is that the ions are closer together in a concentrated solution. The closer together they are, the more frequently do the ions collide. The more often they collide, the greater is their chance of reacting.

14.1.2 THE PRESSURES OF GASEOUS REACTANTS

... the pressure of a gaseous reactant ...

When gases react, molecules have first to collide before they can react. If the pressure is increased, molecules of gas are pushed closer together. As a result, they collide more frequently and react more rapidly.

14.1.3 TEMPERATURE

... the temperature ...

To study the effect of altering the temperature on the rate of a chemical reaction, one reaction that could be chosen is the thiosulphate–acid reaction [see Figure 14.3]. Figure 14.6 shows a typical plot of values of the time taken for the cross to be obscured by sulphur against the temperature. A plot of 1/time against temperature shows a steep increase in the speed of the reaction as the temperature rises [see Figure 14.7]. This reaction goes approximately twice as fast at 30 °C as it does at 20 °C.

At the higher temperature, the ions or molecules of solute have more energy. They move with a higher velocity and collide more frequently and with more force. Once the ions have collided, there is a chance that they will react. The increased collision frequency results in a higher rate of reaction.

FIGURE 14.6
A Graph of Time for Cross to Disappear against Temperature

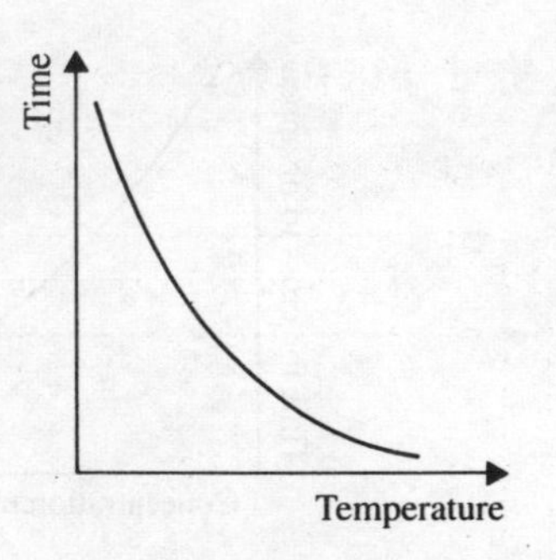

FIGURE 14.7
A Graph of 1/Time against Temperature for the Reaction

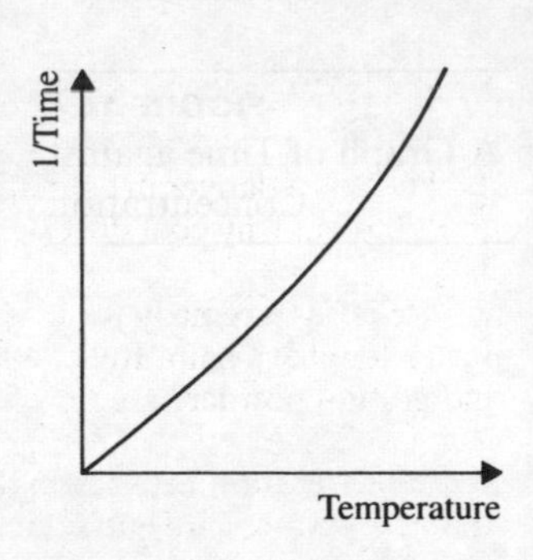

14.1.4 LIGHT

... the presence of light ...

Light is another form of energy that will speed up chemical reactions. One example is photosynthesis, the reaction by which green plants synthesise sugars. The reaction takes place in sunlight and in the presence of chlorophyll, the green pigment in plants.

$$6CO_2(g) + 6H_2O(l) \rightarrow C_6H_{12}O_6(aq) + 6O_2(g)$$

Sucrose

Another example is the formation of silver from silver salts that takes place when a photographic film is exposed to light.

14.1.5 CATALYSTS

... and the presence of a catalyst

Photosynthesis takes place only in green plants. This is because chlorophyll, the pigment in green plants, must be present. Chlorophyll is a catalyst for this reaction. A **catalyst** is a substance which increases the speed of a reaction without being used up in the reaction.

Catalysts are important in industry. A manufacturer tries to find a catalyst which will enable a reaction to give a good yield at a low temperature. Then fuel bills will be lower and profits will be higher. When plastics are manufactured, often the monomer is polymerised under high pressure. If a catalyst can be found to enable the reaction to give a good yield at low pressure, then the industrial plant will not have to withstand high pressure; it can be constructed of less robust materials at lower cost. Table 14.1 lists some important industrial catalysts.

All the processes which take place in plants and animals need catalysts. The catalysts which occur in living things are called **enzymes**. They catalyse a specific reaction, such as the digestion of a certain protein.

TABLE 14.1
Some Important Industrial Catalysts

Catalyst	*Reaction*
Platinum	Oxidation of ammonia to give nitrogen monoxide, a step in the manufacture of nitric acid [see § 22.7.3]
Vanadium(V) oxide	Oxidation of sulphur dioxide to sulphur trioxide, a step in the manufacture of sulphuric acid [see § 21.11]
Nickel	Hydrogenation of unsaturated compounds to form saturated compounds; used in the manufacture of margarine [see § 17.3]
Iron	The combination of nitrogen and hydrogen to form ammonia in the Haber process [see § 22.4.1]

CHECKPOINT 14A: REACTION RATES

1. There is a danger in coal mines that coal dust may catch fire. Explain why coal dust is more dangerous than coal.

2. 'Relief' is a remedy for acid indigestion. Which will work faster to relieve pain, Relief indigestion tablets or Relief indigestion powder?

3. In the reaction between marble chips and acid, assume that the particles are cubes and the length of the side is a.

(*a*) What is (i) the surface area of a cube of side a, (ii) the volume of a cube of side a, (iii) the ratio: Surface area/Volume?

(*b*) As a decreases, does the ratio Surface area/Volume increase or decrease? How does your answer explain the change in the rate of reaction when large chips are replaced with small chips?

4.

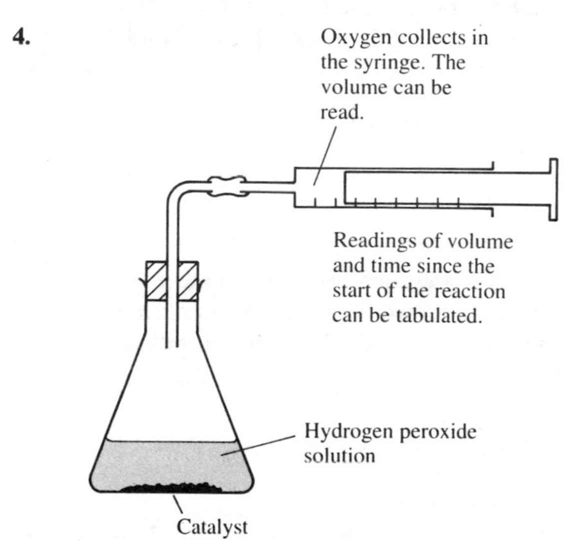
Oxygen collects in the syringe. The volume can be read.
Readings of volume and time since the start of the reaction can be tabulated.
Hydrogen peroxide solution
Catalyst

FIGURE 14.8

Some manganese(IV) oxide (a catalyst) was added to a solution of hydrogen peroxide in the apparatus shown in Figure 14.8. Oxygen was liberated in the reaction

$$2H_2O_2(aq) \rightarrow O_2(g) + 2H_2O(l)$$

Hydrogen peroxide — Oxygen — Water

Table 14.2 shows readings of the volume of oxygen (measured at rtp) collected at various times after the start of the reaction.

TABLE 14.2

Volume of oxygen/cm^3	*Time*/min
37	1
67	2
87	3
107	4
114	5
120	6
120	7
120	8

(*a*) Plot a graph of the volume of oxygen against the time after the start of the experiment.

(*b*) What amount of oxygen (in mol) was formed in the reaction?

(*c*) What mass of hydrogen peroxide was present?

(*d*) Find the time taken for the decomposition of half the mass of hydrogen peroxide.

(*e*) On your graph, sketch the curve that you would predict if the reaction were carried out at 30 °C.

(*f*) Suggest another way of altering the speed of the reaction, and sketch the curve that would result from the change that you suggest.

5. An ester [see Chapter 33] is hydrolysed to form an alcohol and an acid. The table shows the percentage of ester remaining after certain intervals of time.

Time/s	0	100	200	300	400	600
% of ester remaining	100	70.5	55.8	44.5	37.8	29.7

(*a*) Plot a graph of the percentage of ester remaining against time. Find the percentage of ester which reacted (i) in the first 100 s, (ii) between 100 s and 200 s and (iii) between 400 s and 500 s.

(*b*) Draw a tangent to the graph at Time = 0. The gradient of this tangent gives the value of (% ester/Time) at the start of the reaction. Calculate this value. This is the initial rate of reaction.

(*c*) Explain why there is a difference between the values of % ester/Time) at different times.

14.1.6 THE IMPORTANCE OF REACTION RATES

The rates of chemical reactions vary greatly

Chemical reactions vary enormously in speed. Some proceed slowly over a period of months (e.g., the rusting of iron); others take weeks to reach completion (e.g., the fermentation of ethanol). Some reactions are fast (e.g., the precipitation of insoluble salts); others are so fast as to be explosive (e.g., the reaction between hydrogen and oxygen). In Chapter 10, we saw how the sign of $\Delta G^{\ominus}$, the standard free energy

The rate of a reaction cannot be predicted from the standard free energy change

change, indicates whether it is possible for a reaction to occur. If $\Delta G^{\ominus}$ is negative, the reaction is feasible, but the study of thermodynamics does not tell us how fast a reaction will occur. This is something that must be found out by experiment. Industrial chemists are interested in knowing how fast a reaction takes place as the speed is a factor in deciding whether a manufacturing process can be carried out profitably.

Many factors affect the rate of a reaction…

Many factors influence the rate of a chemical reaction, and it is important to discover the conditions under which a reaction will proceed most economically. The rate at which the product is formed is only one factor. The cost of the energy consumed if a high temperature is needed must be computed. If the process requires high pressure, the cost of a plant which is robust enough to withstand the conditions will be high.

…Reaction kinetics is the study of these factors

The study of the factors that affect the rates of chemical reactions is called **reaction kinetics**. Such studies throw light on the **mechanisms** of reactions. All reactions take place in one or more simple steps, and the sequence of steps is called the mechanism. The number of reacting species (molecules, atoms, ions or free radicals) that take part in a reaction step is the **molecularity** of that step. Reaction steps are described as **unimolecular**, **bimolecular** or **trimolecular** [§ 14.9.2]. Examples of reaction mechanisms are covered in §§ 27.7.5, 27.7.6, 28.7.2, 28.7.7, 29.8.1, 31.7.2 and 33.13.1.

14.2 AVERAGE RATE

The rate of a reaction is the rate of change of concentration of a reactant or a product

The average rate of a chemical reaction over a certain interval of time is equal to the change in the concentration of a reactant or product that occurs during that time divided by the time. When an ester is hydrolysed, if the concentration of the ester decreases from $1.00\,\text{mol}\,\text{dm}^{-3}$ to $0.50\,\text{mol}\,\text{dm}^{-3}$ in 1.00 hour, then the average rate of reaction over this time interval can be given:

$$\text{Average rate} = (1.00 - 0.50)\,\text{mol}\,\text{dm}^{-3}/(1.00 \times 60 \times 60)\,\text{s}$$
$$= 1.39 \times 10^{-4}\,\text{mol}\,\text{dm}^{-3}\,\text{s}^{-1}.$$

In the example in Figure 14.9, the rate of reaction slows down as the concentration of reactant decreases

The dimension of reaction rate is concentration time^{-1}. Consider a reaction of the type

$$\mathbf{A} \rightarrow \mathbf{B}$$

where 1 mole of the reactant produces 1 mole of the product. Figure 14.9 shows how the concentration of product increases and the concentration of reactant decreases as the time which has passed since the start of the reaction increases.

FIGURE 14.9 Variation of Concentrations of Reactant and Product with Time

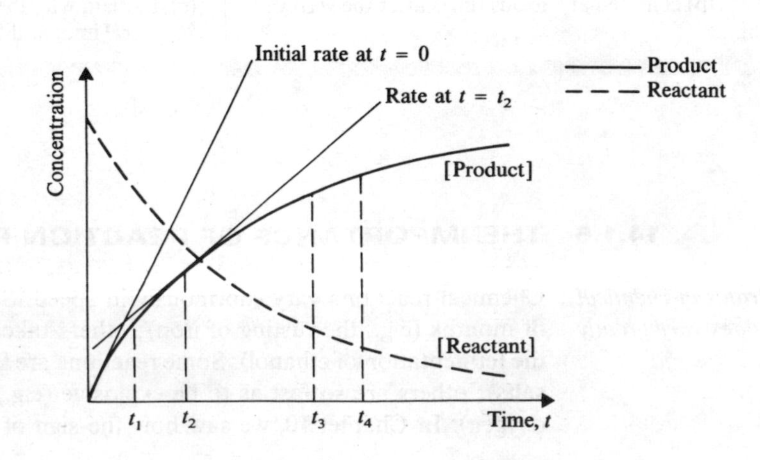

The rate at the very beginning of the reaction is called the initial rate

You can see that less product has been formed between t_3 and t_4 than in an equal time interval between t_1 and t_2. The rate of reaction decreases as the reaction proceeds and the reactant is gradually used up. One can only state the rate of reaction at a certain time. At time t_2 the rate of reaction is the gradient of the tangent to the curve at this point [see Figure 14.9]. The rate at the start of the reaction, when an infinitesimally small amount of the reactant has been used up, is called the **initial rate** of the reaction. In Figure 14.9, the gradient of the tangent to the curve at $t = 0$ gives the initial rate.

The rate of the reaction

$$\mathbf{A} \rightarrow \mathbf{B}$$

is the rate of decrease in concentration of **A** or the rate of increase in concentration of **B**:

$$\text{Rate} = -\frac{d[\mathbf{A}]}{dt} = \frac{d[\mathbf{B}]}{dt}$$

where $[\mathbf{A}]$ = concentration/mol dm^{-3} of **A**. In the reaction

$$BrO_3^-(aq) + 5Br^-(aq) + 6H^+(aq) \rightarrow 3Br_2(aq) + 3H_2O(l)$$

$$\text{Rate} = -\frac{d[BrO_3^-]}{dt} = -\frac{1}{5}\frac{d[Br^-]}{dt} = \frac{1}{3}\frac{d[Br_2]}{dt} \text{ and so on}$$

The rate of a reaction is found by measuring some property of a reactant or a product at various times after the start of the reaction. Some of the methods of 'following' the reaction in this way will now be described.

14.3 METHODS OF FINDING THE RATES OF CHEMICAL REACTIONS

14.3.1 CHEMICAL METHODS

Titration can be used to follow the change in the concentration of the reactant or the product

The progress of a reaction can often be followed by chemical analysis. The reaction is carried out in a thermostatically controlled water bath. Solutions of the reactants of known concentrations are mixed, and a stop clock is started. A sample of the reacting mixture is withdrawn with a pipette, and the reaction is stopped. This may be done by removing one of the reactants by a chemical reaction. Alternatively, the reaction may be suddenly slowed down by cooling or by dilution. This is done by pipetting a sample into a freezing mixture or into an excess of the solvent. A titration is then performed to find the concentration of one of the reactants or one of the products.

Example (a) The alkaline hydrolysis of an ester

$$CH_3CO_2C_2H_5(aq) + NaOH(aq) \rightarrow CH_3CO_2Na(aq) + C_2H_5OH(aq)$$

The course of an alkaline hydrolysis of an ester is followed by measuring the concentration of alkali at various times after the start of the reaction

Solutions of ester and alkali of known concentrations are allowed to reach the temperature of a thermostat bath. The solutions are mixed, and the time of mixing is noted. A sample of the reaction solution is withdrawn by pipette, and run into about four times its volume of ice-cold water. The dilution and cooling reduce the rate of reaction almost to zero. The alkali that remains is titrated against standard acid, using phenolphthalein as indicator. Ethanoate ions do not affect the colour change of this indicator. The analysis is repeated at various intervals of time after the start of the reaction.

Example (b) The chlorination of aromatic compounds, e.g., 4-chlorophenoxyethanoic acid

$$\text{OCH}_2\text{CO}_2\text{H-C}_6\text{H}_4\text{-Cl} + \text{Cl}_2 \xrightarrow[\text{ethanoic acid}]{\text{in solution in glacial}} \text{OCH}_2\text{CO}_2\text{H-C}_6\text{H}_3\text{Cl-Cl} + \text{HCl}$$

4-Chlorophenoxyethanoic acid — 2,4-Dichlorophenoxyethanoic acid

A chlorination reaction is followed by means of thiosulphate titrations

Solutions of known concentrations of ether and chlorine are mixed at the required temperature, and the time is noted. A sample of the reacting mixture is withdrawn and run into an excess of potassium iodide solution. All the chlorine remaining reacts very rapidly with potassium iodide to form iodine, and the chlorination stops. The amount of iodine formed can be found by titration against a standard solution of sodium thiosulphate.

14.3.2 PHYSICAL METHODS

A CHANGE IN GAS VOLUME

The volume of gas evolved can be measured after various time intervals

In a reaction in which a gas is formed, the volume of gas can be recorded at various times [see Figure 14.10]. Examples are the reaction of a metal with an acid and the decomposition of hydrogen peroxide:

$$\text{Mg(s)} + 2\text{HCl(aq)} \rightarrow \text{H}_2\text{(g)} + \text{MgCl}_2\text{(aq)}$$

$$2\text{H}_2\text{O}_2\text{(aq)} \rightarrow 2\text{H}_2\text{O(l)} + \text{O}_2\text{(g)}$$

FIGURE 14.10
Measuring the Evolution of a Gas in a Reaction

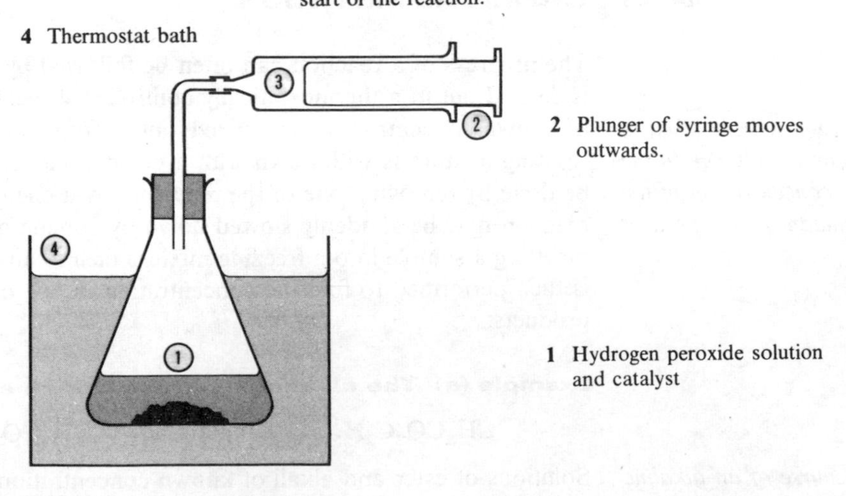
3 Oxygen. Volume is recorded at certain times after the start of the reaction.
4 Thermostat bath
2 Plunger of syringe moves outwards.
1 Hydrogen peroxide solution and catalyst

A CHANGE IN GAS PRESSURE

An increase or decrease in gaseous pressure can be used to follow many gaseous reactions

Some reactions between gases involve an increase in the number of moles of gas, e.g.

$$2\text{N}_2\text{O}_5\text{(g)} \rightarrow 2\text{N}_2\text{O}_4\text{(g)} + \text{O}_2\text{(g)}$$

If the reaction takes place at constant volume, the resulting increase in pressure can be followed. The reaction

$$2H_2(g) + O_2(g) \rightarrow 2H_2O(g)$$

is accompanied by a decrease in the number of moles of gas and by a decrease in pressure at constant volume.

Such reactions can be followed by the method shown in Figure 14.11.

FIGURE 14.11 Apparatus for Following Changes in Gas Pressure

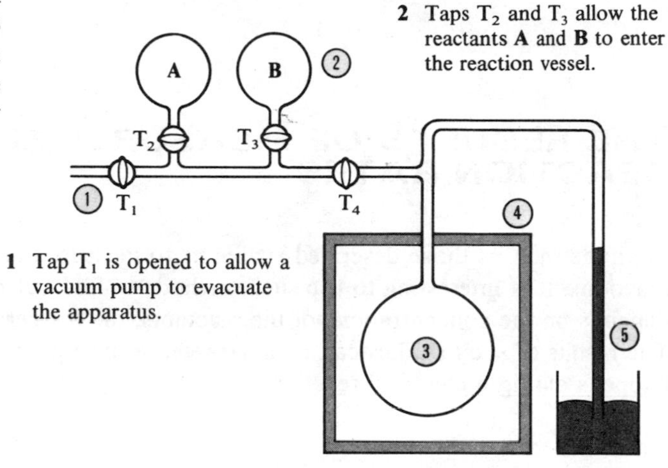

2 Taps T_2 and T_3 allow the reactants **A** and **B** to enter the reaction vessel.
A
B
T_2
T_3
T_1
T_4
3 Reaction vessel. Tap T_4 traps gases **A** and **B** inside.
4 Thermostatically controlled furnace
1 Tap T_1 is opened to allow a vacuum pump to evacuate the apparatus.
5 Manometer records pressure inside reaction vessel.

CHANGES IN THE ABSORPTION SPECTRUM

The use of a spectrophotometer enables light absorption to be measured

Many substances absorb light, either in the visible region or, more frequently, in the ultraviolet region. A **spectrophotometer** is an instrument for measuring the absorption of light at various wavelengths. If a reaction is carried out in a cell inside a spectrophotometer, the change in the absorption spectrum can be used to follow the course of the reaction. Figure 14.12 shows the change in the absorption spectrum during the reaction

$$A \rightarrow C$$

FIGURE 14.12 Change in Absorption Spectrum during Reaction

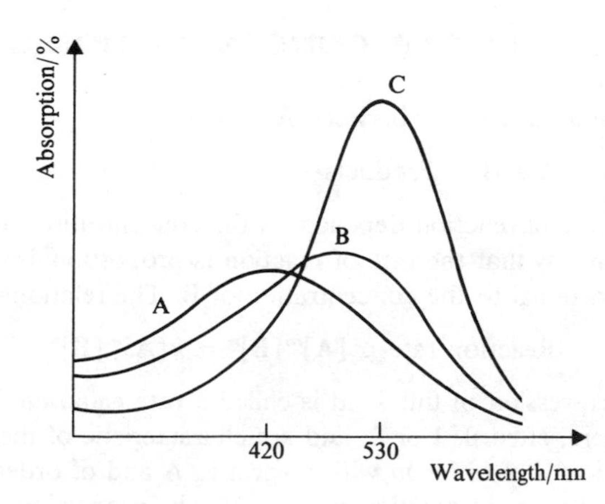

Absorption/%
C
B
A
420
530
Wavelength/nm
Curve A = absorption spectrum of **A**
Curve C = absorption spectrum of **C**
Curve B = absorption spectrum at an intermediate stage in the reaction
The height of the peak at 530 nm could be used to follow the reaction, because **A** does not absorb at this wavelength.

ELECTRICAL PROPERTIES

A change in conductance indicates a change in the concentration of ions...

A change in the conductance of a solution occurs if ions are used up or created during a reaction.

. . . as does a change in electrode potential

The potential of an electrode in contact with a solution of its ions changes if the concentration of ions changes. Potentiometric methods [see § 13.2] can often be used to follow the course of a reaction which involves ions.

THERMAL CONDUCTIVITY

Thermal conductivity

A gaseous reaction can be followed by measuring the thermal conductivity of the mixture of reacting gases.

14.4 THE RESULTS OF MEASUREMENTS OF REACTION RATES

Methods such as those described enable us to measure the rates of chemical reactions. It is interesting to use such methods to find out how the rate of reaction depends on the concentrations of the reactants, the temperature and other factors. The results of such studies can be interpreted to give a detailed picture of what happens during a chemical reaction.

14.4.1 THE EFFECT OF PARTICLE SIZE ON REACTION RATE

Solids react faster in finely divided form

Reactions of solids take place faster when the solids are in a finely divided state. This is because the ratio of surface area to mass is greater in small particles than in large particles, and the area over which the solid can come into contact with liquid or gaseous reactants is greater. Examples are the reaction of powdered zinc and granulated zinc with acids and the reaction of powdered calcium carbonate and marble chips with acid.

14.4.2 THE EFFECT OF CONCENTRATION ON REACTION RATE

Consider a reaction between **A** and **B**:

$$\mathbf{A} + \mathbf{B} \rightarrow \text{Products}$$

The way in which rate is related to concentration is governed by the order of the reaction

The rate of reaction depends on the concentrations of **A** and **B**, but one cannot simply say that the rate of reaction is proportional to the concentration of **A** and proportional to the concentration of **B**. The relationship is

$$\text{Reaction rate} \propto [\mathbf{A}]^m[\mathbf{B}]^n = k[\mathbf{A}]^m[\mathbf{B}]^n$$

An expression of this kind is called a **rate equation**. The indices m and n are usually integers, often 0, 1 or 2, and are characteristic of the reaction. One says that the reaction is of **order** m with respect to **A** and of order n with respect to **B**. The overall order of reaction is $(m + n)$. The proportionality constant k is called the **rate constant** or **velocity constant** for the reaction.

The rate constant for a reaction relates the rate to the concentrations of the reactants

Example (a) A solution of **Q**, of concentration 0.20 mol dm^{-3} undergoes a **first-order** reaction at an initial rate of 3.0×10^{-4} mol dm^{-3} s^{-1}. Calculate the rate constant.

Method Since, for a first-order reaction

$$\text{Initial rate} = k[\mathbf{Q}]_0$$

$$3.0 \times 10^{-4}\,\text{mol dm}^{-3}\,\text{s}^{-1} = k \times 0.20\,\text{mol dm}^{-3}$$

$$\text{The rate constant } k = 1.5 \times 10^{-3}\,\text{s}^{-1}$$

The dimension of a first-order rate constant is time^{-1}.

First-order and second-order rate constants have different dimensions

Example (b) A **second-order** reaction takes place between the reactants **P** and **Q**, which are both initially present at concentration 0.20 mol dm^{-3}. If the initial rate of reaction is 1.6×10^{-4} mol dm^{-3} s^{-1}, what is the rate constant?

Method

$$\text{Initial rate} = k[\mathbf{P}]_0[\mathbf{Q}]_0$$

$$\therefore \quad 1.6 \times 10^{-4}\,\text{mol dm}^{-3}\,\text{s}^{-1} = k \times (0.20\,\text{mol dm}^{-3})^2$$

$$\text{The rate constant, } k = 4.0 \times 10^{-3}\,\text{dm}^3\,\text{mol}^{-1}\,\text{s}^{-1}$$

A second-order rate constant has the dimensions concentration^{-1} time^{-1}.

14.5 ORDER OF REACTION

The stoichiometric equation for a reaction does not reveal the order of the reaction

The order of a reaction does not follow from its stoichiometric equation. The reaction between bromate(V) ions, bromide ions and hydrogen ions to give bromine is represented by the equation

$$BrO_3^-(aq) + 5Br^-(aq) + 6H^+(aq) \rightarrow 3Br_2(aq) + 3H_2O(l)$$

The results of kinetic measurements give the following rate equation:

$$-\frac{d[BrO_3^-]}{dt} \propto [BrO_3^-][Br^-][H^+(aq)]^2$$

The reaction is first-order with respect to bromate(V), first-order with respect to bromide, second-order with respect to hydrogen ion and fourth-order overall. The negative sign means that $[BrO_3^-]$ decreases with time.

14.5.1 ORDER OF REACTION FROM INITIAL RATE

In a reaction

$$\mathbf{A} \rightarrow \mathbf{X}$$

The order of a reaction can be found by comparing the initial rates of two reactions at known initial concentrations

the rate of reaction $= k[\mathbf{A}]^n$, where n = the order of reaction. Two experiments to find the rate of reaction, two 'runs', are done at different concentrations of **A**. $[\mathbf{A}_0]_1$ = initial concentration of **A** in Run 1, and $(v_0)_1$ = the initial rate in Run 1.

$$(v_0)_1 = k[\mathbf{A}_0]_1{}^n$$

$$(v_0)_2 = k[\mathbf{A}_0]_2{}^n$$

The ratio

$$\frac{(v_0)_1}{(v_0)_2} = \left(\frac{[\mathbf{A}_0]_1}{[\mathbf{A}_0]_2}\right)^n$$

The order of reaction may be found by comparing the initial rates of reactions at different concentrations. This can be done by inspection, as in the example below, or by taking logarithms:

$$\lg\left(\frac{(v_0)_1}{(v_0)_2}\right) = n \lg\left(\frac{[A_0]_1}{[A_0]_2}\right)$$

A worked example: how to find the order by comparing initial rates

Example The following results were obtained for a reaction between **A** and **B**:

Run	*Concentrations*/mol dm^{-3}		*Initial rate*/mol dm^{-3} s^{-1}
	[**A**]	[**B**]	
(a)	0.50	1.0	2.0
(b)	0.50	2.0	8.0
(c)	0.50	3.0	18
(d)	1.0	3.0	36
(e)	2.0	3.0	72

What is the order of reaction with respect to **A** and with respect to **B**? What is the rate equation for the reaction? Calculate the rate constant.

Method Let the rate equation be

$$\text{Rate} = k[\mathbf{A}]^m[\mathbf{B}]^n$$

Compare runs (d) and (e), in which [**B**] is constant:

$$\frac{\text{Rate (e)}}{\text{Rate (d)}} = \left(\frac{2.0}{1.0}\right)^m = \frac{72}{36} \qquad \text{therefore } m = 1$$

Compare runs (b) and (a), in which [**A**] is constant

$$\frac{\text{Rate (b)}}{\text{Rate (a)}} = \left(\frac{2.0}{1.0}\right)^n = \frac{8}{2} \qquad \text{therefore } n = 2$$

The reaction is first-order with respect to **A** and second-order with respect to **B**. The rate equation is

Once the order has been found, the rate constant can be found

$$\text{Rate} = k[\mathbf{A}][\mathbf{B}]^2$$

The rate constant can now be calculated from the results of any run, for example (c)

$$18 = k \times 0.5 \times (3.0)^2$$

$$k = 4.0\,\text{dm}^6\,\text{mol}^{-2}\,\text{s}^{-1}$$

This third-order rate constant has the dimension concentration^{-2}time^{-1}.

CHECKPOINT 14B: INITIAL RATES

1. Explain the terms: *rate of reaction, order of reaction, stoichiometry of reaction* and *rate constant.*

2. Describe three physical methods which can be used to follow the course of a chemical reaction. What are the advantages of physical methods over chemical methods?

3. For the reaction

$$A + B \rightarrow C$$

the following results were obtained for kinetic 'runs' at the same temperature:

$[A]_0$/mol dm^{-3}	$[B]_0$/mol dm^{-3}	Initial rate/mol dm^{-3} s^{-1}
0.20	0.10	0.20
0.40	0.10	0.80
0.40	0.20	0.80

Find

(*a*) the rate equation for the reaction

(*b*) the rate constant

(*c*) the initial rate of a reaction, when $[A]_0 = 0.60\,\text{mol dm}^{-3}$ and $[B]_0 = 0.30\,\text{mol dm}^{-3}$

4. Tabulated are values of initial rates measured for the reaction

$$2A + B \rightarrow C + D$$

Experiment	*[A]*/mol dm^{-3}	*[B]*/mol dm^{-3}	*Initial rate*/ mol dm^{-3} min^{-1}
1	0.150	0.25	1.4×10^{-5}
2	0.150	0.50	5.6×10^{-5}
3	0.075	0.50	2.8×10^{-5}
4	0.075	0.25	7.0×10^{-6}

(*a*) Find the order with respect to **A**, the order with respect to **B** and the overall order of the reaction.

(*b*) Find the value of the rate constant.

(*c*) Find the initial rate of reaction when $[A]_0 = 0.120\,\text{mol dm}^{-3}$ and $[B]_0 = 0.220\,\text{mol dm}^{-3}$.

14.5.2 ZERO-ORDER REACTIONS

If the rate is independent of concentration, the reaction is zero order...

In a **zero-order** reaction, the rate is independent of the concentration of the reactant. A plot of the concentration of the reactant, [A], against time has the form shown in Figure 14.13.

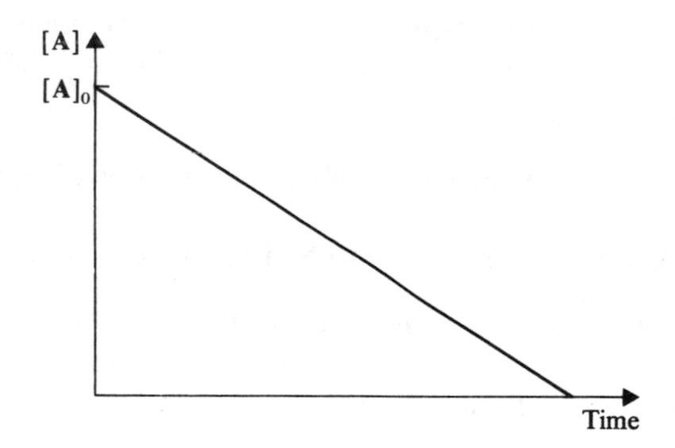

FIGURE 14.13 A Plot of Concentration against Time for a Zero-Order Reaction

The rate equation for a zero-order reaction is

$$\text{Rate} = k[A]^0$$

and the rate constant has the dimension concentration time^{-1}.

...One example is the reaction of iodine and propanone...

In the iodination of propanone [§ 14.11], the reaction rate does not change if the concentration of iodine is changed:

$$CH_3COCH_3(aq) + I_2(aq) \xrightarrow[\text{buffer}]{\text{acid}} CH_3COCH_2I(aq) + HI(aq)$$

The reaction is said to be zero-order with respect to iodine.

...Another example is the adsorption of reactants in gaseous reactions

Sometimes, reactions between gases are zero-order with respect to one of the reactants. This often indicates that this reactant has been adsorbed on the surface of the vessel. The rate of reaction then depends on the frequency with which molecules of the non-adsorbed gas collide with the inside of the vessel. This frequency is proportional to the concentration of the non-adsorbed reactant.

14.5.3 FIRST-ORDER REACTIONS

If the reaction

$$A \rightarrow \text{Products}$$

is a first-order reaction, the rate equation will be

$$\text{Rate} = k[A] \quad \text{i.e.} \ -d[A]/dt = k[A]$$

If $[A]_0$ = initial concentration of A, the integrated form of this equation is

$$kt = \ln\frac{[A]_0}{[A]} = 2.303\lg\frac{[A]_0}{[A]}$$

HALF LIFE

Definition of half-life, $t_{1/2}$...

The time taken for the reaction to go to half-completion is called the **half-life** of the reaction, $t_{1/2}$. At time $t_{1/2}$

$$[A] = [A]_0/2$$

$$\therefore \quad kt_{1/2} = \ln 2 = 0.693$$

This shows that the half-life of a first-order reaction is independent of the initial concentration. Radioactive decay is an example of a reaction showing first-order kinetics [§ 1.9.6].

Example The half-life of radium is 1590 years. How long will it take for a sample of radium to decay to 10% of its original radioactivity?

Method This problem is solved in two steps:

A worked example on radioactive decay

(*a*) Use $t_{1/2}$ to find k, the rate constant, from the equation

$$kt_{1/2} = 0.693$$

Then

$$k = 0.693/1590 = 4.36 \times 10^{-4}\,\text{year}^{-1}$$

(*b*) Insert this value for k into the equation

$$kt = \ln\frac{[A]_0}{[A]}$$

$$4.36 \times 10^{-4} \times t = \ln(100\%/10\%)$$

$$t = 5280 \text{ years}$$

After 5280 years, 10% of the original radioactivity remains.

Example Carbon-14 dating shows that a piece of ancient wood gives 10 counts per minute per gram of carbon, compared with $15\,\text{cpm}\,\text{g}^{-1}$ of carbon from a sample of new wood. The half-life of ^{14}C is 5600 years. What is the age of the ancient wood?

Method Again, there are two steps in the calculation:

A worked example on C-14 dating

(*a*) Use $t_{1/2}$ to find k

$$k = 0.693/5600 = 1.24 \times 10^{-4}\,\text{year}^{-1}$$

(*b*) Insert this value of k into the equation:

$$kt = \ln\frac{[\mathbf{A}]_0}{[\mathbf{A}]}$$

Since $$\frac{[\mathbf{A}]_0}{[\mathbf{A}]} = \frac{^{14}\text{C content in new wood}}{^{14}\text{C content in ancient wood}} = \frac{15\,\text{cpm}}{10\,\text{cpm}}$$

$$t = \ln\frac{15}{10}\Big/(1.24 \times 10^{-4}) = 3270\text{ years}$$

The wood is 3270 years old.

14.5.4 PSEUDO-FIRST-ORDER REACTIONS

The acid-catalysed hydrolysis of an ester, e.g., ethyl ethanoate

$$CH_3CO_2C_2H_5(l) + H_2O(l) \xrightarrow{\text{acid}} CH_3CO_2H(l) + C_2H_5OH(l)$$

If the concentration of one reactant is very large, the reaction appears to be zero order with respect to that reactant

is first-order with respect to ester and first-order with respect to water. If water is present in large excess, only a small fraction of the water will be used up in the reaction. The concentration of water is practically constant, and the rate depends on the concentration of ester alone:

$$-\text{d}[CH_3CO_2C_2H_5]/\text{d}t = k'[CH_3CO_2C_2H_5]$$

k' = a first-order rate constant. The reaction appears to be zero-order with respect to water.

CHECKPOINT 14C: FIRST-ORDER REACTIONS

1. A radioactive element decays with a rate constant of $2.0 \times 10^{-4}\,\text{s}^{-1}$. How long will it take for 0.50 g of the substance to decay to 0.10 g?

2. If the half-life of a radioactive element is 150 s, what percentage of the isotope will remain after 600 seconds?

3. The results listed were obtained for a 'run' on the reaction

$$\mathbf{A} \rightarrow \mathbf{B} + \mathbf{C}$$

Plot [**A**] against t.

(*a*) From the graph, find the order of the reaction with respect to **A**.

(*b*) By drawing a tangent, find the initial rate of reaction.

(*c*) Calculate the rate constant for the reaction.

Time/s	*[A]*/mol dm^{-3}
0	0.800
400	0.580
800	0.400
1200	0.280
1600	0.200
2000	0.140
2400	0.100

4. If a radioisotope loses 95% of its activity in 110 minutes, what is its half-life?

5. The decomposition of benzene diazonium chloride is first-order:

$$C_6H_5N_2Cl(aq) \rightarrow C_6H_5Cl(aq) + N_2(g)$$

The following results give the volume of nitrogen at 50 °C, measured at various time intervals after the start of the reaction, obtained in the decomposition of 500 cm^3 of a 1.10×10^{-3} mol dm^{-3} solution at 50 °C:

Time/min	2.0	4.0	6.0	9.0	12.0	16.0	22.0	28.0
Volume of N_2/cm^3	1.7	3.4	4.9	6.6	8.1	9.5	11.2	12.2

(*a*) Plot the volume of nitrogen evolved against the time interval.

(*b*) Calculate the volume of nitrogen at 50 °C that will be formed at $t = \infty$ from the amount of benzene diazonium chloride specified. Enter this on the graph.

(*c*) From the graph, estimate the half-life of the reaction, $t_{1/2}$.

(*d*) What evidence have you that the reaction is first-order?

(*e*) Use the value of $t_{1/2}$ to find the rate constant k.

(*f*) Obtain the initial rate of reaction.

(*g*) From it, calculate the rate constant.

14.5.5 SECOND-ORDER REACTIONS

In the second order reaction

$$\mathbf{A} + \mathbf{B} \rightarrow \text{Products}$$

the rate equation is

$$\text{Rate} = -\mathrm{d}[\mathbf{A}]/\mathrm{d}t = -\mathrm{d}[\mathbf{B}]/\mathrm{d}t = k[\mathbf{A}][\mathbf{B}]$$

14.6 SUMMARY OF THE DEPENDENCE OF THE CONCENTRATIONS OF REACTANT AND PRODUCT ON TIME

The manner in which the rate of reaction depends on the concentration of the reactant is shown in Figure 14.14. For a zero-order reaction, the rate remains constant as the concentration of reactant changes. For a first-order reaction, the rate of reaction is directly proportional to the concentration of the reactant. For a second-order reaction, the rate of reaction increases with the concentration as shown in the figure. It is therefore sometimes possible to tell the order of a reaction from an inspection of a graph of rate against concentration over a wide range of concentration.

FIGURE 14.14 Graphs of Rate against concentration

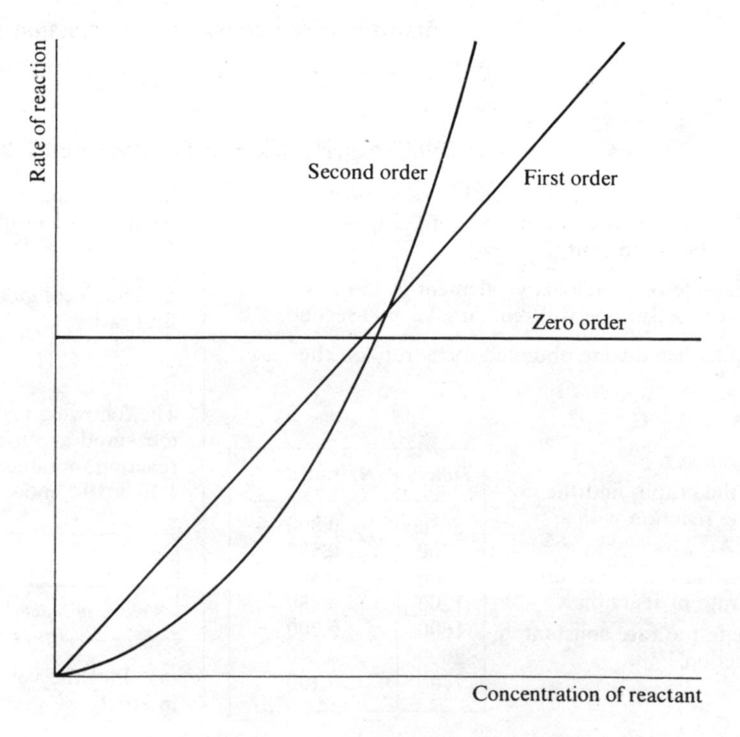

CHECKPOINT 14D: ORDER OF REACTION

1. In a reaction

$$R \rightarrow S$$

the initial rate is $1.7 \times 10^{-4}\,mol\,dm^{-3}\,s^{-1}$ when $[R]_0 = 0.25\,mol\,dm^{-3}$. Predict the initial rate when $[R]_0 = 0.75\,mol\,dm^{-3}$ if the reaction is (*a*) zero-order, (*b*) first-order and (*c*) second-order.

2. The results tabulated refer to the isomerisation

trans-CHCl=CHCl → *cis*-CHCl=CHCl

Time/s	0	600	900	1200	1500	1800
Trans-*isomer*/mol	1.00	0.90	0.85	0.81	0.77	0.73

From a suitable plot, find the order of reaction and the rate constant.

3. The following results were obtained for the acid-catalysed hydrolysis of methyl ethanoate:

Time/s	0	1150	2050	3600	5050	8000
[Ester]/mol dm^{-3}	0.500	0.375	0.300	0.216	0.150	0.071

From a graph of [Ester] against time, find the order of the reaction and the rate constant.

14.7 THE EFFECT OF LIGHT ON REACTION RATES: PHOTOCHEMICAL REACTIONS

Some reactions take place faster in the presence of light

Reactions with very high rates often involve free radicals. When a covalent bond splits **homolytically**, each of the bonded atoms or groups takes one of the bonding pair of electrons, and each product is termed a **free radical**:

$$A:B \rightarrow A\cdot + \cdot B$$

(In **heterolytic** fission

$$A:B \rightarrow A^+ + :B^-$$

one of the bonded atoms takes both bonding electrons, and ions are formed).

Light energy may split bonds to form free radicals

Evidence for the existence of free radicals is obtained from mass spectrometry and other methods.

Energy must be supplied to break the bonds and produce free radicals. In thermal reactions, this energy comes from collisions with other molecules. Those reactions which are started by the absorption of light energy are called photochemical reactions.

A photochemical reaction takes place between H_2 and Cl_2

Hydrogen and chlorine react slowly in the dark, unless heated above 200 °C, to form hydrogen chloride. In the presence of sunlight, however, reaction takes place rapidly at room temperature:

$$H_2(g) + Cl_2(g) \rightarrow 2HCl(g)$$

It is believed that the first step in the photochemical reaction is the absorption of light, leading to the dissociation of chlorine molecules:

$$Cl_2 + h\nu \rightarrow 2Cl\cdot \quad [1]\ \textit{Initiation reaction}$$

The chlorine atoms (or radicals) formed react with hydrogen molecules:

$$Cl\cdot + H_2 \rightarrow HCl + H\cdot \quad [2]\ \textit{Propagation reaction}$$

The hydrogen atoms formed in [2] react with chlorine molecules:

$$H\cdot + Cl_2 \rightarrow HCl + Cl\cdot \qquad [3] \textit{ Propagation reaction}$$

These two steps are repeated many times, setting up a **chain reaction**. For the absorption of one quantum of light by one chlorine molecule, many thousands of molecules of hydrogen chloride are formed. The chain does not go on for ever; it is brought to an end by the combination of free radicals:

$$\left.\begin{array}{ll} 2Cl\cdot \rightarrow Cl_2 & [4] \\ 2H\cdot \rightarrow H_2 & [5] \\ H\cdot + Cl\cdot \rightarrow HCl & [6] \end{array}\right\} \textit{Chain termination reactions}$$

The chlorination of methane is a photochemical reaction

Another photochemical reaction is the chlorination of methane in sunlight to chloromethane and other derivatives [§ 26.3.8].

The absorption of radiation from radioactive elements, for example X rays and γ rays, is sometimes used to initiate chemical reactions.

14.8 THE EFFECT OF TEMPERATURE ON REACTION RATES

Reaction rates increase with temperature . . .

An increase in temperature increases the rate of a reaction by increasing the rate constant. Figure 14.15 shows plots of rate constant k against temperature T and of $\ln k$ against $1/T$.

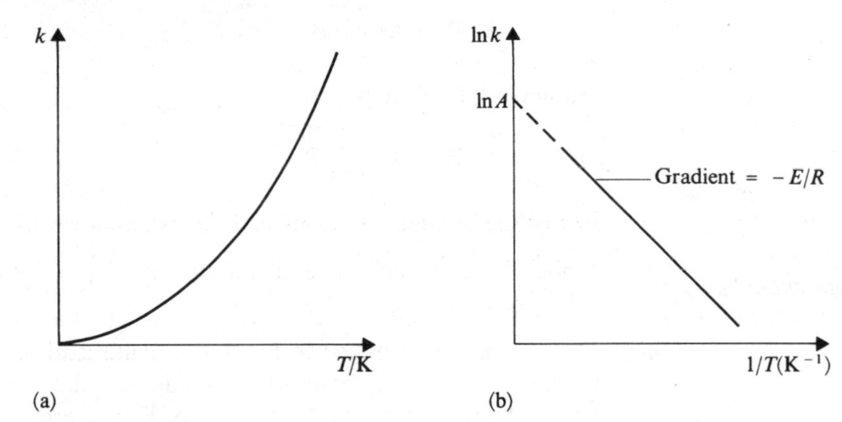

FIGURE 14.15
(a) Plot of k against T,
(b) Plot of $\ln k$ against $1/T$

The variation of rate constant with temperature was studied by Arrhenius and found to fit the equation

$$k = Ae^{-E/RT}$$

. . . The dependence of rate on temperature fits the Arrhenius equation

This equation is known as the **Arrhenius equation**. R is the gas constant [§ 7.2.6] and A and E are constant for a given reaction. The constants can be found by using the equation in logarithmic form:

$$\ln k = \ln A - E/RT$$

ln k against 1/T gives a linear plot

A plot of $\ln k$ against $1/T$ is a straight line of gradient $-E/R$ and an intercept on the y-axis of $\ln A$. If $\lg k$ is plotted against $1/T$, the gradient of the line is $-E/2.303R$, and the intercept is $\lg A$.

The significance of the temperature dependence is discussed below, and the method of determining E and A is exemplified in § 14.9.1.

14.9 THEORIES OF REACTION RATES

14.9.1 THE COLLISION THEORY

Gas molecules must collide in order to react...

... The rate of reaction is much less than the rate of collision

From the Kinetic Theory of Gases, was developed the **Collision Theory of bimolecular reactions in the gas phase**. In a reaction between two gaseous substances **A** and **B**, a molecule of **A** must collide with a molecule of **B** before reaction can occur. It has been shown that the collision frequency Z is proportional to the product [**A**][**B**]. If every collision results in reaction, the rate of reaction will equal the collision frequency. Theoreticians have calculated the collision frequency, and have found that the rate constant for a bimolecular gaseous reaction should be of the order of $10^{11}\,dm^3\,mol^{-1}\,s^{-1}$. Although there are reactions which have rate constants of this magnitude, e.g.

$$H\cdot + Br_2 \rightarrow Br\cdot + HBr$$

such reactions are few. The reason why most reactions are slower than this is that only a small fraction of collisions results in reaction.

In the Arrhenius equation

$$k = Ae^{-E/RT}$$

Molecules must not only collide; they must have enough energy to react

E is termed the **energy of activation**, A is the rate of collision between molecules, and E is the energy which the colliding molecules must possess before a collision will result in reaction.

The Arrhenius equation ties in with the Maxwell–Boltzmann equation

The fraction $e^{-E/RT}$ appears in the Maxwell–Boltzmann calculations [§ 7.3.4]. They calculated that for a value of molecular energy E/J mol^{-1}, the fraction of molecules at temperature T that possess energy $\geqslant E$ is given approximately by the value of $e^{-E/RT}$. Calculation of k from the Arrhenius equation shows that a reaction with an activation energy of 50 kJ mol^{-1} should approximately double in rate over 10 K. This behaviour is observed for many such reactions.

The term A *in the Arrhenius equation*

In the Arrhenius equation, it was suggested that the term A might be the collision frequency. Calculated values of collision frequencies, however, are often higher than values obtained from plots such as Figure 14.15(b), § 14.8. The explanation must be that molecules must collide not only with sufficient energy for reaction to occur, but also in a favourable orientation in space. The term A is therefore thought to be the rate at which molecules collide in an orientation which is favourable to reaction.

The Collision Theory applies to reactions in solution as well as to reactions in the gas phase

The Collision Theory was developed for bimolecular reactions in the gas phase, but it has been found to apply to reactions in solution. Although solvent molecules prevent molecules of reactant coming together as frequently as they would in the gas phase, once they have come together, reactant molecules are less able to escape, and will collide repeatedly. These repeated collisions between a pair of molecules may well lead to reaction. Some reactions which have been studied both in the gas phase and in solution are found to have similar rate constants in the two phases.

To summarise: the Collision Theory of bimolecular reactions states that, in order to react, molecules must (*a*) collide, (*b*) collide in a favourable orientation and (*c*) collide with enough energy to react.

*THE METHOD OF FINDING THE ACTIVATION ENERGY

The activation energy and the pre-exponential factor for a reaction can be found from measurements of the rate constant at different temperatures.

A worked example...

Example Find the activation energy and the pre-exponential factor for the reaction

$$2HI(g) \rightarrow H_2(g) + I_2(g)$$

...finding the activation energy from rate constants at different temperatures

Values of the rate constant k at various temperatures are given below:

T/K	556	629	700	781
$k/dm^3\,mol^{-1}\,s^{-1}$	7.04×10^{-7}	6.04×10^{-5}	2.32×10^{-3}	7.90×10^{-2}
$10^3/T$	1.80	1.59	1.43	1.28
$\ln k$	− 14.2	− 9.71	− 6.07	− 2.54

Method Figure 14.16 shows the linear plot of $\ln k$ against $10^3/T$.

$$\text{Gradient} = -2.35 \times 10^4\,K = -E/R$$

$$\text{Activation energy } E = 195\,kJ\,mol^{-1}$$

From the graph when

$$\ln k = -4.00,\ 1/T = 1.34 \times 10^{-3}$$

Since $\ln A = \ln k + E/RT$

$$\ln A = -4.00 + \frac{195 \times 10^3 \times 1.34 \times 10^{-3}}{8.314}$$

Pre-exponential factor $A = 7.9 \times 10^{11}\,dm^3\,mol^{-1}\,s^{-1}$.

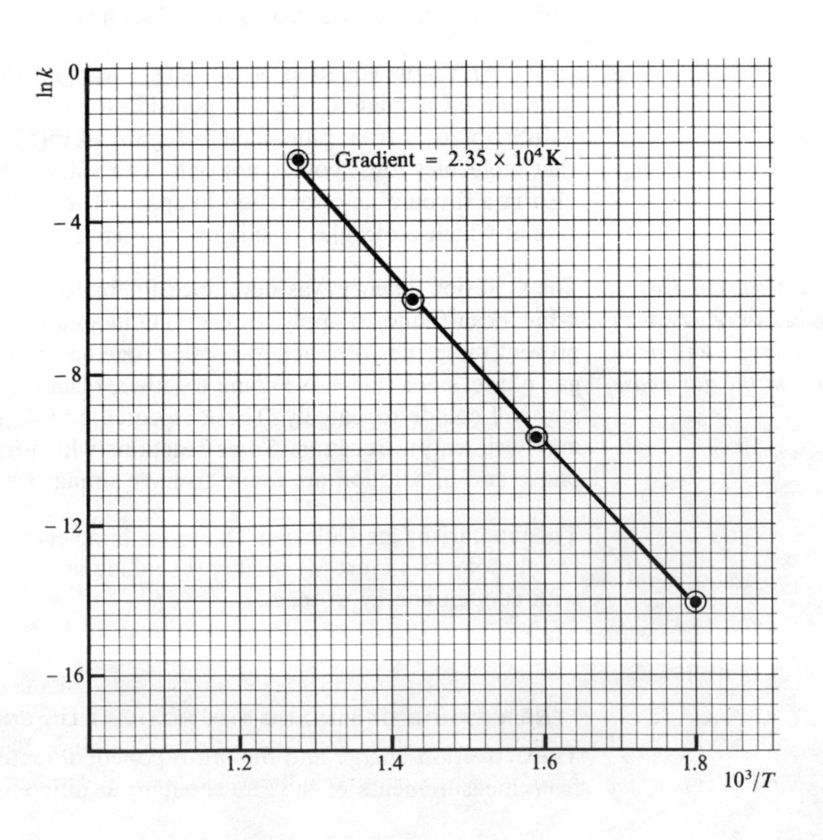

FIGURE 14.16 Graph of $\ln k$ against $10^3/T$

14.9.2 THE TRANSITION STATE THEORY

The Transition State Theory looks at the course of a collision in detail

The **Transition State Theory** is concerned with what actually happens during a collision. It follows the energy and orientation of the reactant molecules as they collide and seeks an explanation of why such a small fraction of collisions results in reaction.

When two molecules approach each other in a collision, the electron clouds experience a gradual increase in their mutual repulsions, and the molecules begin to slow down. While this is happening, the kinetic energy of the molecules is being converted into potential energy. If the molecules had little kinetic energy to begin with, i.e., if they were not moving very fast, they will come to a stop before their electron clouds have interpenetrated very much, and then fly apart again without reacting [see Figure 14.17(a)].

FIGURE 14.17 Collisions between molecules

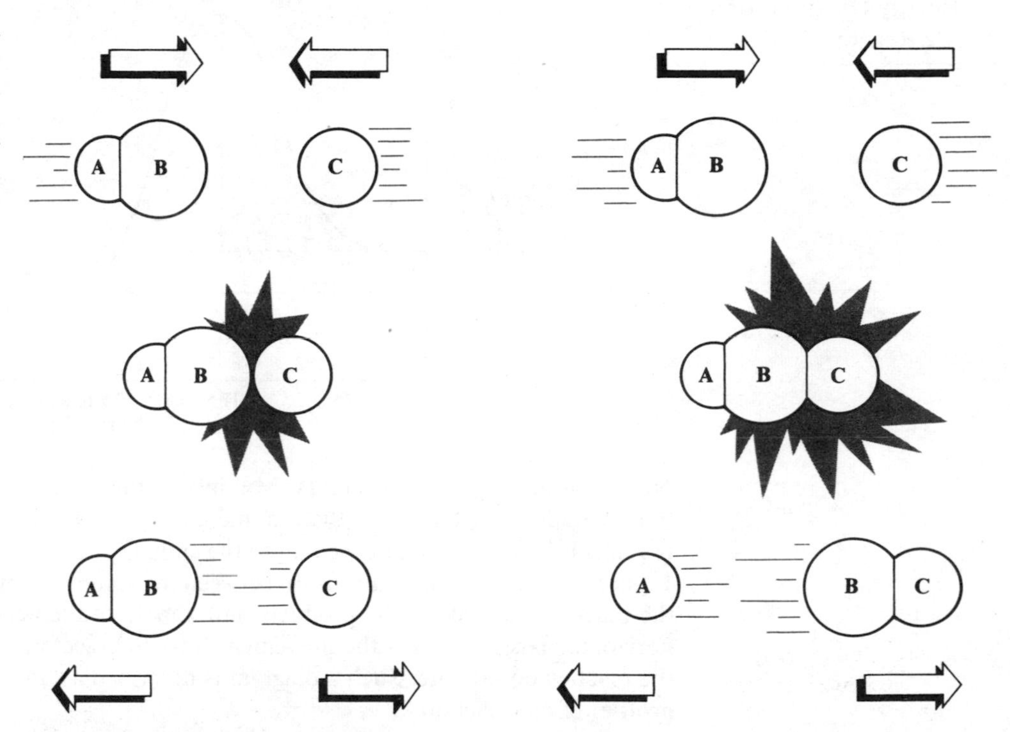

(a) Two slow-moving molecules collide. The electron clouds do not interpenetrate.

(b) Two fast-moving molecules collide. Atoms approach closely, and electron clouds interpenetrate. This leads to reaction.

Kinetic energy is converted into potential energy as reacting molecules collide

When two fast-moving molecules collide, they have a lot of kinetic energy that can be converted into potential energy. They are able to overcome the forces of repulsion between their electron clouds and to approach each other closely. The interpenetration of the electron clouds that occurs permits a rearrangement of valence electrons, with the breaking of old bonds and the formation of new bonds, i.e., a chemical reaction [see Figure 14.17(b)].

A methyl radical and a molecule of hydrogen chloride may react if they collide:

$$CH_3\cdot + HCl \rightarrow CH_4 + Cl\cdot$$

As the reactant molecules move along the reaction coordinate towards becoming the products, the potential energy passes through a peak...

The change in potential energy that takes place during the course of one reactive collision (i.e., one that changes the reactant molecules into the product molecules) is shown in Figure 14.18. The horizontal axis is called the **reaction coordinate**. Positions along the reaction coordinate represent the distance that the reacting species have moved towards forming the products. As $CH_3\cdot$ and HCl approach one another, their potential energy increases to a maximum. The arrangement of atomic nuclei and bonding electrons at the potential energy maximum is called the **activated complex**.

It can be represented as

... The arrangement of atomic nuclei and bonding electrons at this peak is called the activated complex

$$H_3C\text{-----}H\text{-----}Cl$$

As the new bond, $H_3C—H$, forms, it assists the breaking of the old bond, $H—Cl$. The activated complex exists in a **transition state** along the reaction coordinate. Once formed, the transition state is transformed into the products. The difference in potential energy between the activated complex and the reactants is called the activation energy E_A. The number of reacting species that take part in the formation of the transition state is the **molecularity** of the reaction step [§ 14.1.6]. This reaction step is **bimolecular**.

FIGURE 14.18 Potential Energy Diagram for a Single Reactive Collision

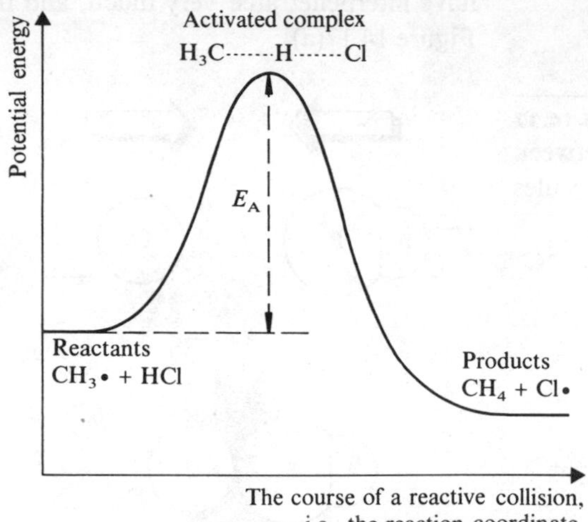

Now consider 1 mole of reactants. Multiplying the potential energy of a molecule by the Avogadro constant gives the internal energy of 1 mole of such molecules [§ 10.2]. This is approximately equal to the enthalpy of 1 mole of molecules. Figure 14.19 shows the relationship between the standard enthalpy $H^{\ominus}$ of a mole of the reactants, a mole of the products and a mole of the activated complex. The horizontal axis represents the movement from the reactants to the products along the reaction coordinate. Such a diagram is described as an **enthalpy profile** (or **energy profile**) of the reaction.

FIGURE 14.19 An Enthalpy Profile for a Reaction

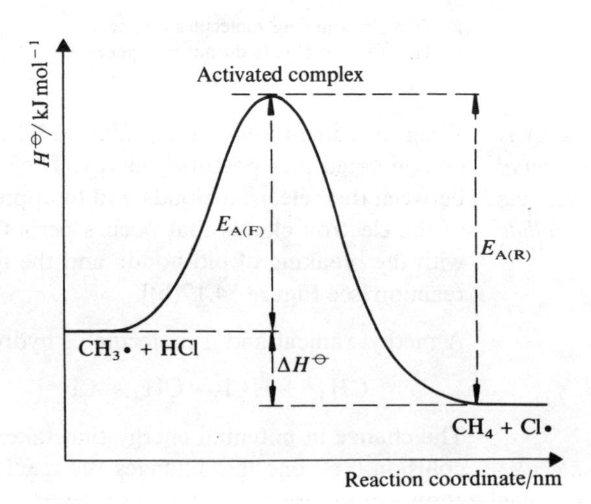

$E_{A(F)}$ and $E_{A(R)}$ in Figure 14.19 represent the activation energies of the forward and reverse reactions. $\Delta H^{\ominus}$ is the standard enthalpy of reaction:

$$\Delta H^{\ominus} = H^{\ominus}_{\text{Products}} - H^{\ominus}_{\text{Reactants}}$$

As can be seen from Figure 14.11

$$\Delta H^{\ominus} = E_{A(F)} - E_{A(R)}$$

The Transition State Theory applies to reactions in solution as well as to gaseous reactions. The alkaline hydrolysis of primary halogenoalkanes (e.g., C_2H_5Br) is discussed in § 29.8.1.

Some reactions take place via a reactive intermediate

Some reactions take place via a **reactive intermediate**. One example is the nitration of benzene [§ 28.7.1]. Another is the hydrolysis of tertiary halogenoalkanes [§ 29.8.2]:

$$(CH_3)_3CBr \rightarrow (CH_3)_3C^+ + Br^-$$

$$(CH_3)_3C^+ + H_2O \rightarrow (CH_3)_3COH + H^+(aq)$$

The reactive intermediate is the carbocation, $(CH_3)_3C^+$. It is preceded by and followed by a transition state. The enthalpy profile for a reaction of this kind is shown in Figure 14.20.

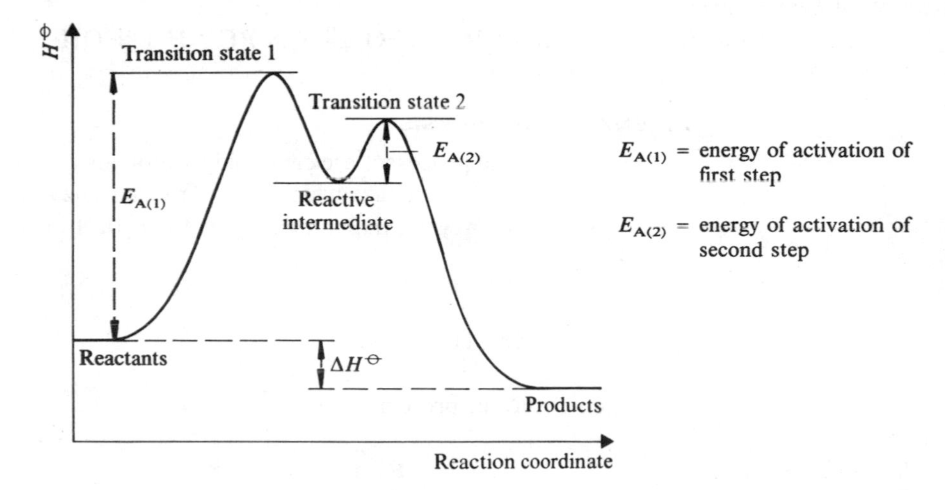

FIGURE 14.20 Enthalpy Profile of a Reaction which involves a Reactive Intermediate

14.10 CATALYSIS

A catalyst lowers the activation energy of a reaction by providing a different route from reactants to products

A **catalyst** is a substance which alters the rate of a chemical reaction without being consumed in the reaction. Thus, a small amount of catalyst is able to catalyse the reaction of a large amount of reactant. A catalysed reaction has a lower activation energy than an uncatalysed reaction. It is believed that a catalyst provides a different mechanism for the reaction, with a lower activation energy [see Figure 14.21].

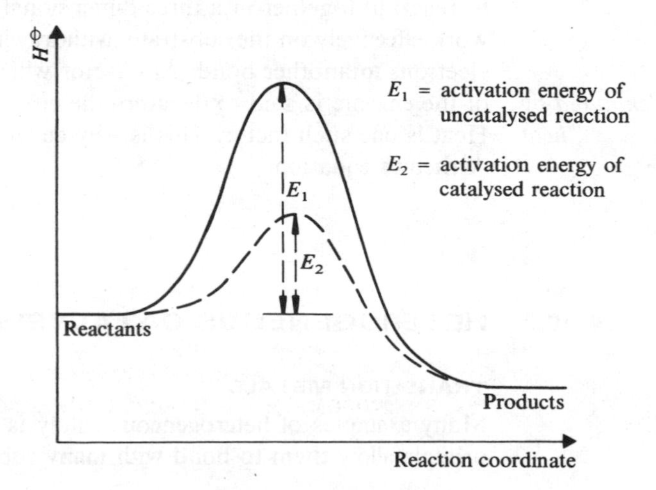

FIGURE 14.21 Energy Profiles for a Catalysed and an Uncatalysed Reaction

Catalysis may be homogeneous or heterogeneous

If the catalyst and the reactants are in the same phase, the process is described as **homogeneous catalysis. Heterogeneous catalysis** takes place at the surface of a catalyst which is in a different phase from the reactants (e.g., a solid catalysing a reaction between gases). The reactants are adsorbed on to the surface of the catalyst, where bonds are broken and new bonds are formed. The products are then desorbed from the surface. Catalysis is an extremely widespread phenomenon, and only a few examples are given here.

14.10.1 HOMOGENEOUS CATALYSIS

ACID–BASE CATALYSIS

Ester hydrolysis is an example of homogeneous catalysis

Many reactions are catalysed by acids and bases. An example is the acid-catalysed hydrolysis of esters to give a carboxylic acid and an alcohol or a phenol:

$$RCO_2R' + H_2O \xrightleftharpoons{H^+(aq)} RCO_2H + R'OH$$

ENZYME CATALYSIS

Most biological processes are catalysed by proteins called **enzymes**. For many enzymes there is a specific reaction which that particular enzyme is ideally suited to catalyse. The digestive enzyme, papain, catalyses the hydrolysis of the peptide bond

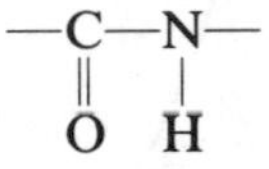

$$\begin{array}{cc} —C— & N— \\ \| & | \\ O & H \end{array}$$

which occurs in proteins:

Enzymes are proteins which act as catalysts for specific reactions

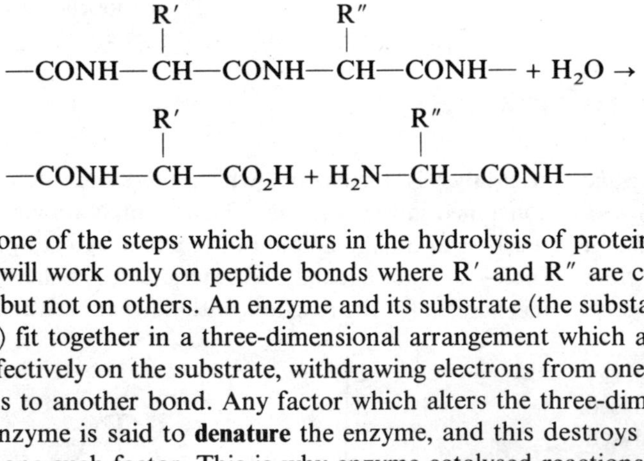

$$\begin{array}{l} \qquad\quad R' \qquad\qquad\qquad R'' \\ \qquad\quad | \qquad\qquad\qquad\; | \\ —CONH—CH—CONH—CH—CONH— + H_2O \rightarrow \end{array}$$

$$\begin{array}{l} \qquad\quad R' \qquad\qquad\qquad\qquad R'' \\ \qquad\quad | \qquad\qquad\qquad\qquad\; | \\ —CONH—CH—CO_2H + H_2N—CH—CONH— \end{array}$$

Enzymes are denatured by heat

This is one of the steps which occurs in the hydrolysis of proteins to amino acids. Papain will work only on peptide bonds where R′ and R″ are certain amino acid groups, but not on others. An enzyme and its substrate (the substance which it enables to react) fit together in a three-dimensional arrangement which allows the enzyme to work effectively on the substrate, withdrawing electrons from one bond and supplying electrons to another bond. Any factor which alters the three-dimensional structure of the enzyme is said to **denature** the enzyme, and this destroys its catalytic activity. Heat is one such factor. This is why enzyme-catalysed reactions do not obey the Arrhenius equation.

14.10.2 HETEROGENEOUS CATALYSIS

TRANSITION METALS

Many examples of heterogeneous catalysis involve transition metals. Their empty d orbitals allow them to bond with many substances to form reactive intermediates.

In the Haber Process for the manufacture of ammonia, the catalyst is a mixture of iron and vanadium:

Many transition metals act as heterogeneous catalysts...

$$N_2(g) + 3H_2(g) \xrightarrow{Fe/V} 2NH_3(g)$$

In the hydrogenation of alkenes, a process which is used in the conversion of liquid oils into solid fats, nickel is used. It is finely divided to increase the area of surface over which the reactants can come into contact.

...Reaction takes place at the surface

$$R_2C{=}CR_2(g) + H_2(g) \xrightarrow{Ni} R_2CH{-}CHR_2(g)$$

The oxidation of sulphur dioxide in the Contact Process is catalysed by platinum or by vanadium(V) oxide. If platinum is used, it readily absorbs impurities present in the sulphur dioxide and loses its activity: it is easily 'poisoned'.

'CRACKING'

Catalysis is important in the petroleum industry

The 'cracking' of hydrocarbons in the petroleum industry is catalysed by a mixture of silica and alumina:

$$\underset{\text{Octane}}{C_8H_{18}(g)} \xrightarrow{Al_2O_3/SiO_2} \underset{\text{Butane}}{C_4H_{10}(g)} + \underset{\text{Butene}}{C_4H_8(g)}$$

CHECKPOINT 14E: REACTION KINETICS

1. Explain the following statements:
(*a*) The rate of a bimolecular reaction cannot be calculated from the collision frequency alone.
(*b*) The increase in the rate of a chemical reaction with an increase in temperature is much greater than the corresponding increase in the collision frequency.
(*c*) The presence of a catalyst can make a big change in the rate of a chemical reaction.

***2.** The Collision Theory postulates that collisions between molecules are necessary before reaction can occur. A unimolecular reaction involves the dissociation or isomerisation of a single molecule.
How does the Collision Theory apply to unimolecular reactions?

3. What is an activated complex? Give an example to illustrate your answer. On the Transition State Theory, what is it assumed that an activated complex will do?

4. What is a spontaneous reaction? Do spontaneous reactions always take place rapidly? Give examples. In what way does a catalyst affect the spontaneous reaction?

5. Describe what is meant by a *chain reaction*. By referring to a chosen example, explain what are chain-initiating steps, chain-propagating steps and chain-terminating steps.

6. The root mean square speed of gaseous molecules [§ 7.3.1] is given by

$$\tfrac{1}{3}mL\overline{c^2} = RT$$

Calculate the ratio

rms speed at 308 K/rms speed at 298 K

For a certain reaction, the ratio

reaction rate at 308 K/reaction rate at 298 K = 2.0

How can you explain the difference between the two ratios?

7. According to thermodynamics, an exothermic reaction is *feasible*. Why do some exothermic reactions proceed very slowly?

8. Why does the probability that a collision will result in reaction depend on the orientation of the colliding molecules? Illustrate your answer with reference to the reactions

$$2HI(g) \rightarrow H_2(g) + I_2(g)$$

$$CH_3Br + I^- \xrightarrow{\text{in propanone}} CH_3I + Br^-$$

$$(CH_3)_3N + C_2H_5I \xrightarrow[\text{solvent}]{\text{in an organic}} (CH_3)_3\overset{+}{N}C_2H_5 + I^-$$

*9. (a) The reaction

$$P \rightarrow Q$$

doubles its rate constant between 25 °C and 35 °C. What is its activation energy?

(b) Calculate the activation energy of the reaction

$$R \rightarrow S$$

which triples its rate constant over the same temperature range.

*10. By means of a suitable graph, use the following results of kinetic 'runs' on the reaction

$$H_2(g) + I_2(g) \rightarrow 2HI(g)$$

to obtain a value for the activation energy:

T/K	555	606	645	714	769
k/mol^{-1}dm^{3}s^{-1}	3.72×10^{-5}	7.32×10^{-4}	5.41×10^{-3}	0.111	0.819

14.11 A DETAILED KINETIC STUDY

In a kinetic study of the iodination of propanone...

The iodination of propanone

$$CH_3COCH_3(aq) + I_2(aq) \rightarrow CH_3COCH_2I(aq) + HI(aq)$$

is a reaction with interesting kinetics. The reaction is acid-catalysed. From the equation, one might postulate that the rate equation would be

$$-\frac{d[I_2]}{dt} = k[CH_3COCH_3]^a[I_2]^b$$

A study of the reaction will involve finding a and b.

First: Find the order with respect to iodine

... the concentration of propanone is kept approximately constant

It is arranged that the propanone concentration is much greater than the iodine concentration, e.g., $[CH_3COCH_3] = 1.00\,mol\,dm^{-3}$ and $[I_2] = 0.00500\,mol\,dm^{-3}$ so that, at the end of a run, $[CH_3COCH_3] = 0.995\,mol\,dm^{-3}$, a decrease of 0.5%. One can say that $[CH_3COCH_3]$ is effectively constant, and

$$-\frac{d[I_2]}{dt} = k_1[I_2]^b \text{ so that } b \text{ can be found.}$$

The reaction is started by mixing the reagents...

...A sample of the solution is pipetted...

... The reaction is stopped...

... Titration gives the iodine concentration

Solutions of known concentration of (a) propanone, (b) iodine in potassium iodide, and (c) an acid buffer of known pH are prepared and brought to the required temperature in a thermostat bath. The reaction is started by pipetting volumes of the three solutions into a flask, and a stop watch is started. After a few minutes, a sample of the reacting mixture is pipetted from the solution into a sodium hydrogencarbonate solution. This stops the reaction instantly by neutralising the acid. The time at which the reaction stops is recorded. The iodine that remains is determined by titration against a standard solution of sodium thiosulphate. The analysis is repeated at intervals of a few minutes. The volume of thiosulphate required is plotted against the time elapsed since the start of the reaction. Figure 14.22 shows the plot obtained. It is a straight line: the gradient of the graph does not change as the concentration of iodine decreases. This shows that the rate of reaction remains constant as the iodine concentration decreases:

It is found that the rate of the reaction remains constant as the iodine concentration decreases

$$-\frac{d[I_2]}{dt} = \text{Constant}$$

The reaction is zero-order with respect to iodine, and the rate equation becomes

$$-\frac{d[I_2]}{dt} = k[CH_3COCH_3]^a$$

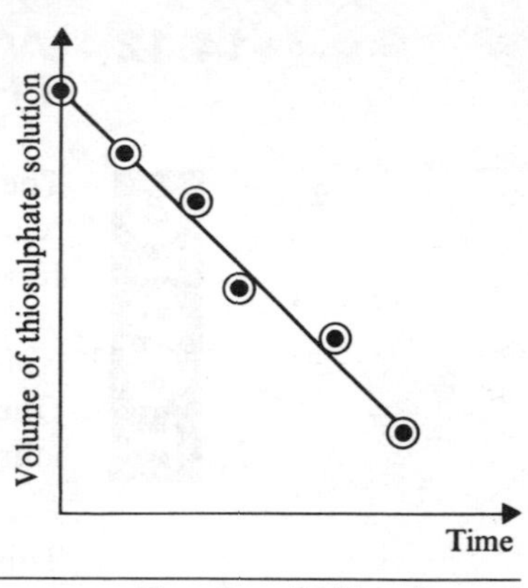

FIGURE 14.22 Graph of Results of Iodine Titration

Second: Find the order with respect to propanone

Runs at different propanone concentrations

The procedure is repeated with different concentrations of propanone. It is found that doubling the concentration of propanone doubles the rate of the reaction: the reaction is first-order with respect to propanone:

$$-\frac{d[I_2]}{dt} = k[CH_3COCH_3]$$

Third: Find the effect of acid concentration

Runs at different acid concentrations

A set of runs in buffers with different values of pH and fixed concentrations of propanone and iodine shows that the rate of reaction is proportional to the hydrogen ion concentration. The rate equation is

$$-\frac{d[I_2]}{dt} = k[CH_3COCH_3][H^+(aq)]$$

Interpretation of the rate equation

From the results of kinetic studies, a mechanism for the reaction can be worked out

Iodine must be present for iodination to occur, but its concentration does not appear in the rate equation. To explain this, it is suggested that the reaction takes place in steps. It is the rate of the slowest step that determines the rate of the overall reaction. If iodine is involved in a step which is too fast to be rate-determining, it will not appear in the rate equation. The suggested mechanism for the reaction is

$$CH_3{-}\underset{\substack{\|\\O}}{C}{-}CH_3 + H^+(aq) \underset{}{\overset{\text{slow}}{\rightleftharpoons}} CH_3{-}\underset{\substack{\|\\O{-}H\\+}}{C}{-}CH_3 \overset{\text{fast}}{\rightleftharpoons} CH_3{-}\underset{\substack{|\\O{-}H}}{C}{=}CH_2 + H^+(aq) \qquad [1]$$

This is the slow, rate-determining step. An enol is formed. Since halogens are electrophiles [§25.8.3], iodine reacts rapidly with the $>C{=}C<$ bond through its π-electrons:

$$CH_3{-}\underset{\substack{|\\O{-}H}}{C}{=}CH_2 + I{-}I \xrightarrow{\text{fast}} CH_3{-}\underset{\substack{|\\O{-}H}}{\overset{+}{C}}{-}CH_2{-}I + I^- \qquad [2]$$

The intermediate formed has a positive charge on a carbon atom, and fast loses a proton to form the iodoketone:

$$CH_3{-}\underset{\substack{|\\O{-}H}}{\overset{+}{C}}{-}CH_2{-}I \xrightarrow{\text{fast}} CH_3{-}\underset{\substack{\|\\O}}{C}{-}CH_2{-}I + H^+(aq) \qquad [3]$$

[For curly arrows see §27.7.5.]

14.12 WORK CAN BE AS MUCH FUN AS GOING TO THE RACES

TOPIC

The great team of Harcourt and Esson devoted their working lives to research on reaction kinetics. They published the results of their work in the *Proceedings of the Royal Society* between 1866 and 1912. Augustus Vernon Harcourt worked in the laboratory in Christ Church College in Oxford. His co-worker, William Esson, worked out the mathematical interpretations of the results of Harcourt's experiments. One of the reactions they studied was that between hydrogen peroxide and iodide ions, which is still known as the Harcourt–Esson reaction.

This book does not include instructions for practical work. The reason is lack of space, not a lack of respect for the importance of practical work. The author assumes that you, the student, are spending a good deal of time in the laboratory. The Harcourt–Esson reaction is an interesting investigation for you to make during the course of a practical session.

The overall reaction is

$$H_2O_2(aq) + 2H_3O^+(aq) + 2I^-(aq) \rightarrow I_2(aq) + 4H_2O(l)$$

Kinetic measurements showed that the reaction is second-order. The rate is proportional to the iodide ion concentration and to the hydrogen ion concentration. To explain the order of reaction, Harcourt and Esson suggested that the reaction takes place in stages.

1. The first step is a slow reaction between H_2O_2 and I^-:

$$H_2O_2(aq) + I^-(aq) \xrightarrow{\text{slow}} IO^-(aq) + H_2O(l) \qquad [1]$$

2. The second step is the establishment of equilibrium by the weak acid, iodic(I) acid, HIO:

$$IO^-(aq) + H_3O^+(aq) \overset{\text{fast}}{\rightleftharpoons} HIO(aq) + H_2O(l) \qquad [2]$$

This happens rapidly.

3. Lastly, iodine is formed in a fast reaction between iodic(I) acid, hydrogen ions and iodide ions:

$$HIO(aq) + H_3O^+(aq) + I^-(aq) \xrightarrow{\text{fast}} I_2(aq) + 2H_2O(l) \qquad [3]$$

This mechanism agrees with the stoichiometry of the overall reaction. You can check up by adding [1] + [2] + [3]. The kinetics and overall order of reaction are those of the slowest step, [1], which is the rate-determining step for the whole reaction. This research work of Harcourt and Esson illustrates one of the important applications of reaction kinetics, which is to throw light on the mechanisms of chemical reactions.

These workers also studied the effect of temperature on the rates of chemical reactions. By extrapolation from measured rates, they predicted the temperature at which, in theory, reaction rates would become zero. They called this temperature the 'zero of chemical action'. It is intriguing to note that their zero was close to the absolute zero of temperature based on the gas laws and on thermodynamics!

After an innings of 46 years, the partnership came at last to a close, and in 1912 Harcourt wrote the last paper of the series. It was published in the *Philosophical Transactions of the Royal Society* in 1912. He ended his last paper in this way.

'Having carried thus far an experimental inquiry which is suggestive of much further work, the author ventures to express a hope that it may attract the attention and pass into the hands of some younger chemist. The mode of working adds to the usual interest of research the particular excitement which attaches to all observations and predictions of time, sporting and scientific, whether it be of the time of a race or of the moment of an occultation.'

Occultation is the eclipsing of one heavenly body by another. People become very excited, waiting to watch an eclipse of the Sun or Moon. As for timing a race, you know how excited people become over watching their horse pass the finishing post or waiting for the first marathon runner to finish the course. This is how Harcourt felt about reaction kinetics. He got the same thrill out of timing reactions, plotting his experimental points and seeing them fall on a straight line. As well as the intellectual satisfaction of thinking out explanations for his results and formulating theories, he got a real thrill out of being in the laboratory. After 46 years of research, his enthusiasm still comes through in this passage. Perhaps you would like to consider his recommendation of the scientific life to 'some younger chemist'?

QUESTIONS ON CHAPTER 14

1. Explain what is meant by the terms *rate constant, order of reaction, molecularity of reaction*. Under what conditions are order and molecularity different? What are the differences between zero-, first- and second-order reactions?

2. In the reaction

$$A \rightarrow B + C$$

what are the units of the rate constant for (*a*) a zero-order reaction, (*b*) a first-order reaction and (c) a second-order reaction?

3. The results of experiments on the hydrolysis of ethyl ethanoate in aqueous solution show first-order kinetics. Explain why.

4. What is a chain reaction? Identify the steps described as chain-initiating, chain-propagating and chain-terminating in (*a*) the chlorination of methane and (*b*) the fission of ^{235}U.

5. The radioactive isotope $^{24}Na^+(aq)$ is injected into an animal. If the half-life is 15 h, how many hours will elapse before the radioactivity has fallen to 10% of the original dose?

6. A freshly-felled piece of wood gives 15.0 cpm (g of carbon)$^{-1}$. If a piece of wood from an Egyptian mummy case gives 9.5 cpm g^{-1}, how old is the case? The half-life of ^{14}C is 5600 years.

7. Ethanal decomposes thermally to form methane and carbon monoxide:

$$CH_3CHO(g) \rightarrow CH_4(g) + CO(g)$$

Standard enthalpies of formation/kJ mol^{-1} [$\Delta H^{\ominus}_{F}$, § 10.2] are: $CH_3CHO(g) = -166$, $CH_4(g) = -75$, $CO(g) = -110$. The energy of activation, $E_A = 190\,kJ\,mol^{-1}$. When the decomposition is catalysed by iodine, $E_A = 136\,kJ\,mol^{-1}$. Draw the reaction profiles for the catalysed and uncatalysed reactions, indicating the values of E_A and the values of $\Delta H^{\ominus}_{Reaction}$.

8. The following results were obtained in a study of the reaction

$$S_2O_8^{2-}(aq) + 2I^-(aq) \rightarrow 2SO_4^{2-}(aq) + I_2(aq)$$

between peroxodisulphate and iodide ions:

Experiment	*[$S_2O_8^{2-}$]* /mol dm^{-3}	*[I^-]* /mol dm^{-3}	*Initial rate,* $\frac{d[S_2O_8^{2-}]}{dt}$ / mol dm^{-3} s^{-1}
1	0.040	0.040	9.6×10^{-6}
2	0.080	0.040	1.92×10^{-5}
3	0.080	0.020	9.6×10^{-6}

Find the order of reaction with respect to (*a*) $S_2O_8^{2-}$, (*b*) I^- and (*c*) find the rate constant. What is the initial rate of the reaction when $[S_2O_8^{2-}]_0 = 0.12\,mol\,dm^{-3}$, and $[I^-]_0 = 0.015\,mol\,dm^{-3}$?

9. The nitration of the aromatic compound ArH is zero-order with respect to ArH. Suggest an explanation for this observation.

10. Results are given below for three reactions of a substance **A**:

Reaction 1

Time/min	*[A]*/mol dm^{-3}
0	1.00
2.0	0.82
4.0	0.67
7.0	0.49
10.0	0.37
14.0	0.24
20.0	0.14
25.0	0.08

Reaction 2

Time/min	*[A]*/mol dm^{-3}
0	1.00
2.0	0.79
4.0	0.59
7.0	0.30
10.0	0.00

Reaction 3

Time/min	*[A]*/mol dm^{-3}
0	1.00
2.0	0.84
4.0	0.72
7.0	0.58
10.0	0.50
15.0	0.40
20.0	0.33
25.0	0.29

(*a*) Construct plots of **[A]** against time. From an inspection of the plots, say which reaction is (i) first-order, (ii) second-order and (iii) zero-order.

(*b*) Read off the half-life for each reaction.

(*c*) Give the concentration of **A** remaining after 9.5 min in (i) the first-order, (ii) the second-order and (iii) the zero-order reaction.

11. Give examples of reactions which you could *follow* by means of (*a*) a conductimetric method, (*b*) a colorimetric method, (*c*) pressure measurements and (*d*) volume measurements.

12. Describe how you would attempt to follow the decomposition of nitrogen(V) oxide (*a*) in the gas phase and (*b*) in solution in tetrachloromethane, in which oxygen is insoluble:

(*a*) $2N_2O_5(g) \rightarrow 4NO_2(g) + O_2(g)$

(*b*) $2N_2O_5(CCl_4) \rightarrow 4NO_2(CCl_4) + O_2(g)$

13. Ethanal decomposes at 800 K according to the equation

$$CH_3CHO(g) \rightarrow CH_4(g) + CO(g)$$

Explain what is meant by these statements:

(*a*) The reaction is second-order.

(*b*) Iodine catalyses the reaction.

(*c*) The reaction in the presence of iodine is first-order with respect to ethanal and first-order with respect to iodine.

(*d*) The reaction that takes place in the presence of iodine has a lower activation energy than the reaction of ethanal alone.

14. What is meant by the *mechanism* of a chemical reaction? Explain why, when chemical reactions take place in more than one step, the overall rate of the reaction is determined by the rate of the slowest step. Give an example of a reaction of this kind.

***15.** Explain the following statements:

(*a*) At room temperature, a mixture of hydrogen and chlorine will not react until irradiated with ultraviolet light.

(*b*) The calculated frequency of collisions between molecules in the gas phase is often 10^6 to 10^{10} times greater than the measured rate at which molecules react (expressed in the same unit).

(*c*) The collision frequency between molecules in the gas phase is proportional to the square root of the temperature in kelvins, but a rise in temperature of 10 K often doubles the rate of reaction.

(*d*) It is not possible to tell the order of a reaction from its stoichiometric equation.

***16.** Chemists use two models to interpret the results of measurements of reaction rates. These are the Collision Theory and the Transition State Theory. Describe the theories briefly. Point out the similarities between the two theories. What are the differences?

***17.** Describe the effect of temperature on the Maxwell–Boltzmann distribution of molecular speeds. How does a knowledge of this effect help us to understand the effect of change of temperature on the rate constants of reactions in the gas phase?

***18.** The Arrhenius equation states

$$k = A\,e^{-E/RT}$$

State what the letters *k*, *R*, e and *T* represent. Explain what quantities are represented by *A* and *E*. The following results were obtained for the decomposition of N_2O_5 at different temperatures:

Temperature/K	k/s^{-1}
298	1.74×10^{-5}
308	6.61×10^{-5}
318	2.51×10^{-4}
328	7.59×10^{-4}
338	2.40×10^{-3}

From a suitable plot, find the value of E. Calculate a value for A.

19. The curve represents the variation of energy with reaction coordinate for the reaction

$$A(g) + B(g) \rightleftharpoons C(g) + D(g)$$

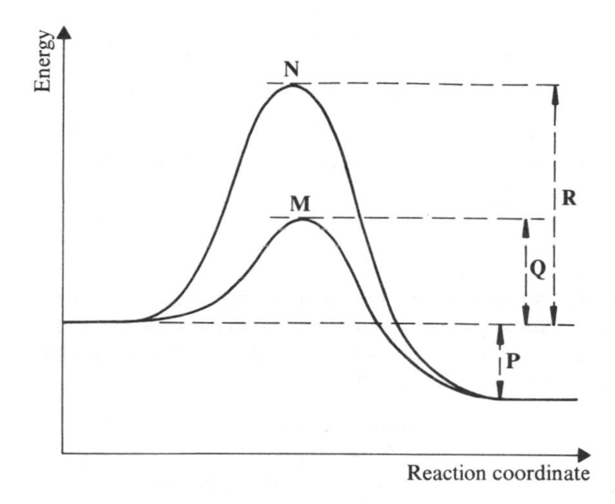

FIGURE 14.23

(*a*) Of the curves, **M** and **N**, one represents the uncatalysed reaction, and one the reaction in the presence of a catalyst. Which is which?

(*b*) What are the quantities marked **P, Q** and **R**?

(*c*) How does a catalyst work?

20. (*a*) Write an equation for the decomposition of hydrogen peroxide.

(*b*) Explain how you could find the rate of decomposition of hydrogen peroxide in aqueous solution, in the presence of a catalyst, using (i) a gas syringe, (ii) a standard solution of potassium manganate(VII).

(*c*) Tabulated below are values of the initial rate of decomposition at different concentrations of hydrogen peroxide. Plot a graph of initial rate against concentration.

$[H_2O_2]$/mol dm^{-3}	0.100	0.175	0.250	0.300
Rate/10^{-4} mol dm^{-3} s^{-1}	0.593	1.04	1.48	1.82

(*d*) From the graph, find the order of reaction and the rate constant.

21. Sucrose is hydrolysed in dilute acid solution to give a mixture of glucose and fructose

$$\underset{\text{Sucrose}}{C_{12}H_{22}O_{11}(aq)} + H_2O(l) \rightarrow \underset{\text{Glucose}}{C_6H_{12}O_6(aq)} + \underset{\text{Fructose}}{C_6H_{12}O_6(aq)}$$

In a study of this reaction, the molar concentrations [S] of sucrose remaining at intervals of time were determined. A graph (Figure 14.24) of the logarithm of the molar concentration ($\ln[S] = \ln[C_{12}H_{22}O_{11}]$) as a function of time t was drawn.

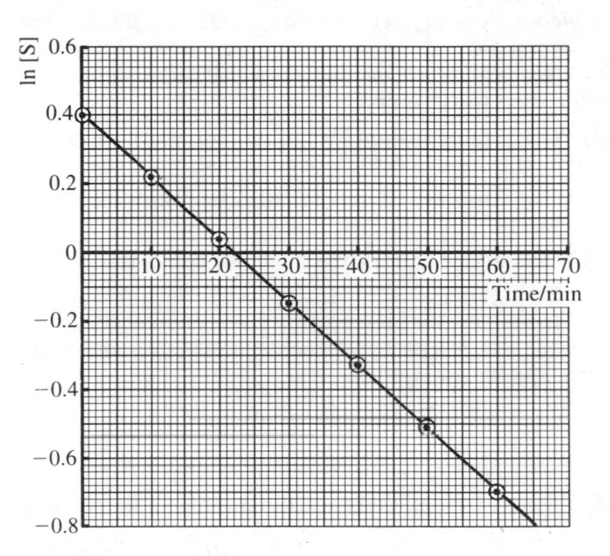

FIGURE 14.24

(*a*) (i) What was the initial concentration of the sucrose?

(ii) What is the order of the reaction (with respect to sucrose)? Explain your answer.

(iii) Calculate the rate constant for the reaction and give its units.

(*b*) (i) Why do you suppose that even though the reaction is a hydrolysis the concentration of water does not appear to be involved in the reaction rate?

(ii) If the acid concentration is doubled then so is the reaction rate. What does this tell you about the kinetics of the reaction?

(O 90, AS)

22. (*a*) The stoichiometric equation for the reaction between bromate(V) and bromide ions in acid solution is

$$5Br^- + BrO_3^- + 6H^+ \rightarrow 3Br_2 + 3H_2O$$

A possible mechanism for this reaction is given below.

Step 1	$H^+ + Br^-$	$\rightarrow HBr$	Fast
Step 2	$H^+ + BrO_3^-$	$\rightarrow HBrO_3$	Fast
Step 3	$HBr + HBrO_3$	$\rightarrow HBrO + HBrO_2$	Slow
Step 4	$HBrO_2 + HBr$	$\rightarrow 2HBrO$	Fast
Step 5	$HBrO + HBr$	$\rightarrow H_2O + Br_2$	Fast

(i) Distinguish between *molecularity* and *order of reaction*.

(ii) What is the molecularity of this reaction?

(*b*) The reaction is found to be first-order with respect to Br^- and BrO_3^-. A series of experiments was performed at 300 K to determine the order with respect to H^+. The initial concentrations of Br^- and BrO_3^- were constant, but the initial concentration of H^+ varied. The time taken to form 1.50×10^{-6} mol Br_2 was recorded.

The following table gives the results of such an experiment.

Initial concentration of H^+/mol dm^{-3}	3.00×10^{-3}	2.50×10^{-3}	1.80×10^{-3}	1.25×10^{-3}	0.800×10^{-3}
Time/s	11	16	32	63	158

(i) Using a graphical method, show that the order of reaction with respect to H^+ is two.

(ii) Write a rate equation for this reaction which shows how the concentration of BrO_3^- ions decreases with time.

(iii) What symbol in the rate equation represents the rate constant?

(iv) What are the units in which the rate constant is expressed?

(*c*) Explain how the presence of a *homogeneous* catalyst alters the rate of a chemical reaction. Illustrate your answer with a specific example.

(O 92)

23. Figure 14.25 shows the reaction profile for the reaction

$$H_2(g) + \tfrac{1}{2}O_2(g) \rightarrow H_2O(g); \quad \Delta H^{\ominus} = -242\ \text{kJ mol}^{-1}$$

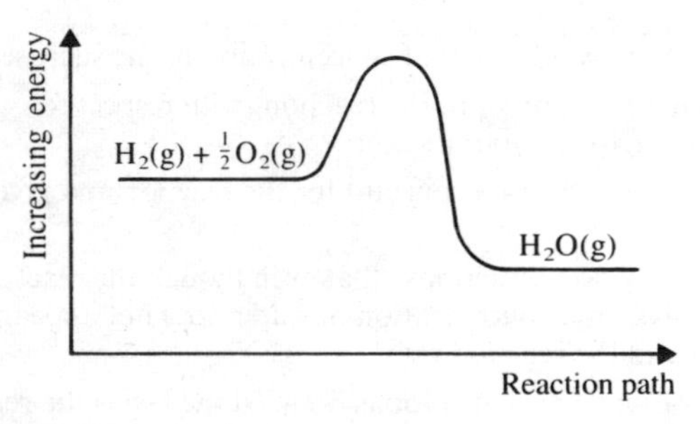

FIGURE 14.25

(*a*) On a copy of the figure, label clearly:

(i) the activation energy of the reaction

(ii) the overall enthalpy change.

(*b*) Platinum can be used to speed up this reaction. On your copy of Figure 14.25 indicate an alternative energy path which the catalysed reaction might take.

(*c*) Given the following bond energies and using the value for the enthalpy of the above reaction, calculate a value for the O=O bond energy.

$$\text{H—H}; \quad \Delta H^{\ominus} = +436\ \text{kJ mol}^{-1}$$
$$\text{O—H}; \quad \Delta H^{\ominus} = +464\ \text{kJ mol}^{-1}$$

(*d*) Explain why a mixture of hydrogen and oxygen will explode if ignited, but will only react very slowly by itself.

(*e*) Given the following information:

$$H_2(g) + O_2(g) \rightarrow H_2O_2(l); \quad \Delta H^{\ominus} = -188\ \text{kJ mol}^{-1}$$
$$H_2O_2(l) \rightarrow H_2O(l) + \tfrac{1}{2}O_2(g); \quad \Delta H^{\ominus} = -98\ \text{kJ mol}^{-1}$$

calculate the heat of formation of water, and use it to explain why hydrogen peroxide is not likely to be formed by direct reaction of hydrogen and oxygen.

(O 92, AS)

24. (*a*) What general methods are available for following the rate of a reaction? How can the order of a reaction be obtained from experimental data?

(*b*) Benzoyl chloride reacts with phenylamine according to the equation:

$$C_6H_5COCl + C_6H_5NH_2 \rightarrow C_6H_5C(=O)NHC_6H_5 + HCl$$

In an experiment at 298 KK, equal numbers of moles of the two reactants were mixed together in an inert solvent and the concentration of the phenylamine was measured at different times. The following results were obtained:

Time/s	$[C_6H_5NH_2]$/mol dm^{-3}
0	0.0200
100	0.0173
200	0.0151
300	0.0132
500	0.0109
700	0.0093
1000	0.0077
1500	0.0058
2000	0.0048

(i) Determine the order of this reaction.

(ii) Suggest three possible rate equations.

(iii) Calculate the rate constant for this reaction.

(iv) How would you expect the rate constant to vary with temperature?

(L(N) 91, S)

25. (*a*) Explain clearly what is meant by the terms *order of reaction* and *rate constant*.

(*b*) Dinitrogen pentoxide, N_2O_5, decomposes according to the equation:

$$2N_2O_5(g) \rightarrow 2N_2O_4(g) + O_2(g)$$

The following experimental results were obtained at 60 °C:

$[N_2O_5]$/mol dm^{-3}	*Rate of disappearance of* N_2O_5/mol dm^{-3} s^{-1}
22.70×10^{-3}	11.80×10^{-6}
19.10×10^{-3}	9.90×10^{-6}
13.50×10^{-3}	7.00×10^{-6}
8.00×10^{-3}	4.15×10^{-6}
3.00×10^{-3}	1.50×10^{-6}

(i) Plot these results on a suitable graph. Use your graph to calculate the rate when the concentration is 20.0×10^{-3} mol dm^{-3}.

(ii) What is the order of the reaction? Explain how you arrived at your answer.

(iii) Calculate the value of the rate constant at 60 °C.

(*c*) (i) What is meant by the term *half-life*?

(ii) At 60 °C the half-life of N_2O_5 is 6 minutes. How long would it take for the concentration of N_2O_5 to decrease from 20.0×10^{-3} to 5.0×10^{-3} mol dm^{-3}?

(*d*) (i) Briefly explain how the change in concentration of N_2O_5 with time might have been followed in such an experiment.

(ii) Sketch and label a suitable apparatus to do so.

(*e*) Suggest a suitable mechanism for the reaction, explaining your reasoning.

(AEB 90)

26. Reaction of aqueous hydrogen peroxide with iodide ions in aqueous acid leads to the formation of the triiodide ion (I_3^-).

(*a*) Write a balanced equation for this reaction.

(*b*) In a series of experiments at 293 K, run with different concentrations of the reactants, the initial rates of formation of the triiodide ion were determined. The results are shown in the table.

Experiment	*Initial concentrations/* (mol dm^{-3} × 10^{-3})			*Initial rate/* (mol dm^{-3} s^{-1} × 10^{-7})
	$[H_2O_2]$	$[I^-]$	$[H^+]$	
1	5.0	5.0	5.0	2.8
2	10.0	5.0	5.0	5.6
3	5.0	10.0	5.0	5.6
4	5.0	5.0	10.0	2.8

(i) Explain what is meant by the expression *order of reaction*.

(ii) Use the information to establish the order of reaction with respect to each of the three reactants and the overall order of the reaction.

(iii) Explain the term *rate constant*, and calculate its value for this reaction at 293 K.

(iv) In a fifth experiment, using an initial hydrogen peroxide concentration of 3.0×10^{-3} mol dm^{-3} and aqueous acid of pH 2.1, the initial rate was shown to be 4.5×10^{-7} mol dm^{-3} s^{-1}. What was the initial concentration of iodide used in the experiment?

(v) Explain why it is important that all the experimental measurements were made at a constant temperature.

(O 92, S)

27. Define carefully what is meant by (i) the *order* of a chemical reaction, and (ii) the *molecularity* of the reaction, distinguishing clearly between them.

Explain also what you understand by the terms *elementary process, rate determining step* and *mechanism*, and use these concepts to explain why the stoichiometry and the kinetic rate law for any given process may or may not coincide.

(*b*) Consider the reaction $H_2 + Br_2 \rightarrow 2HBr$. Neglecting the reverse reaction, the gas-phase process is thought to occur by the following sequence of reactions.

$$Br_2 \underset{k_{-1}}{\overset{k_1}{\rightleftharpoons}} 2Br \quad \left(\text{rapid equilibrium, } K = \frac{k_1}{k_{-1}}\right)$$

$$Br + H_2 \xrightarrow{k_2} HBr + H \quad \text{(slow)}$$

$$H + Br_2 \xrightarrow{k_3} HBr + Br \quad \text{(fast)}$$

(i) Deduce an expression for the kinetic rate law for this process whereby the initial rate of production of HBr, d[HBr]/d*t*, is related to the initial concentrations of hydrogen and bromine, $[H_2]$ and $[Br_2]$.

(ii) State, with an explanation, which (if any) of the following sets of kinetic data is consistent with the reaction sequence shown above and, if appropriate, demonstrate that it is consistent with the kinetic rate law deduced in (*b*)(i) above.

Data set	*Initial* $[H_2]$/ mol dm^{-3}	*Initial* $[Br_2]$/ mol dm^{-3}	*Relative initial rate, d[HBr]/dt*
1	*a*	*b*	1
	3*a*	*b*	3
	3*a*	4*b*	6
	a	9*b*	3
2	*a*	*b*	1
	2*a*	*b*	2
	2*a*	2*b*	8
	a	2*b*	4
3	*a*	*b*	1
	4*a*	*b*	2
	4*a*	2*b*	4
	9*a*	2*b*	6
4	*a*	*b*	1
	2*a*	*b*	2
	a	2*b*	2
	2*a*	2*b*	4

(*c*) Consider the following two possible schemes for the gas-phase reaction $H_2 + I_2 \rightarrow 2HI$, again disregarding the reverse reaction:

$$(1) \quad H_2 + I_2 \xrightarrow{k_1} 2HI \quad \text{(slow)}$$

$$(2) \quad I_2 \underset{k_{-1}}{\overset{k_1}{\rightleftharpoons}} 2I \quad \left(\text{rapid equilibrium, } K = \frac{k_1}{k_{-1}}\right)$$

$$\text{Then } H_2 + I + I \xrightarrow{k_2} 2HI \quad \text{(slow)}$$

Discuss whether or not it would be possible, on the basis of the resulting kinetic rate laws, to distinguish between these two possible routes. Explain also whether the enhanced dissociation of I_2 into 2I brought about by irradiation with ultraviolet light could be used to elucidate the course of the reaction.

(WJEC 90, S)

28. Peroxodisulphate ions react with iodide ions according to the equation

$$S_2O_8^{2-} + 2I^- \rightarrow 2SO_4^{2-} + I_2$$

The rate of the reaction may be found by mixing the two solutions together, adding water to keep the total volume constant and measuring the time for the iodine concentration to reach a given value (using a colorimeter or a titration method).

In an experiment, the peroxodisulphate and iodide solutions were both 0.050 mol dm^{-3}. The time taken for the iodine concentration to become 0.0010 mol dm^{-3} was measured. The following results were obtained:

Experiment number	*Vol. of $S_2O_8^{2-}(aq)$/cm^3*	*Vol. of $I^-(aq)$/cm^3*	*Vol. of $H_2O(l)$/cm^3*	*Time/s*
1	5	10	10	56
2	3	10	12	95
3	2	10	13	145
4	5	5	15	120
5	5	3	17	185

(*a*) Assuming that the rate equation has the form

$$\text{Rate} = k[S_2O_8^{2-}]^a[I^-]^b$$

(i) deduce the values of *a* and *b*

(ii) find the value of *k* and state its units.

(*b*) In the light of this rate equation, what can be said about the mechanism of the reaction between peroxodisulphate ions and iodide ions?

(*c*) The experiment was repeated in the presence of iron(II) ions which act as a catalyst.

Experiment number	*Vol. of $S_2O_8^{2-}(aq)$/cm^3*	*Vol. of $I^-(aq)$/cm^3*	*Vol. of $Fe^{2+}(aq)$/cm^3*	*Vol. of $H_2O(l)$/cm^3*	*Time/s*
6	5	10	1	9	19
7	5	10	2	8	10
8	5	10	3	7	6

(i) How does the concentration of the catalyst affect the rate of the reaction?

(ii) Suggest a possible mechanism for the action of the catalyst.

(iii) By using the table of $E^{\ominus}$ values in Table 13.1, § 13.1.2, together with a value for the $S_2O_8^{2-}/2SO_4^{2-}$ redox potential of +2.01 V, justify the mechanism you have proposed in (ii).

(iv) State, with a reason, whether you think tin(II) ions would act as a catalyst for this reaction.

(C 92, S)

29. When investigating rates of reaction, experiments are designed to take advantage of some change in the reaction as time proceeds. For example, there may be a change in colour, pH or conductivity, and this could be monitored in following the reaction. In this question, the reaction involved is the decomposition of the reactive organic intermediate benzenediazonium chloride. It decomposes according to the equation:

$$C_6H_5N_2^+Cl^-(aq) + H_2O(l)$$
$$\rightarrow C_6H_5OH(aq) + N_2(g) + HCl(aq)$$

The reaction can therefore be followed by monitoring the evolution of gas with time. It can be shown that the concentration of benzenediazonium chloride at time *t* is proportional to $V - V_t$, where V is the volume of gas collected when the reaction is complete, and V_t is the volume of gas collected at time *t*. Plotting $V - V_t$ against time is essentially the same as plotting the concentration of benzenediazonium chloride against time.

(*a*) Design an experiment, including the preparation of the benzenediazonium chloride solution, that would enable you to collect volume/time data for the reaction above. The experiment should use apparatus normally available in a school laboratory.

The following points may be useful:

(i) Benzenediazonium chloride decomposes at a convenient rate at about 40 °C.

(ii) The V reading requires approximately two hours at this temperature.

(iii) There is no need to monitor the reaction from $t = 0$, i.e. as soon as the solution of benzenediazonium chloride has been produced.

(*b*) Typical data of V_t against time are given below. Use the data as fully as possible in describing the kinetics of the decomposition of benzenediazonium chloride.

Time/min	*Volume of gas/cm^3*
10	35
20	53
30	65
40	72
50	76
60	78
Complete reaction	80

(*c*) Explain how you would modify the experiment to obtain sufficient data to deduce the activation energy, E_a, for the reaction. You should indicate how you would use the data that you collect to obtain your value for E_a.

(*d*) Give *two* other reactions which rely on different physical techniques to monitor them, outlining the technique briefly (details are not expected).

(L 92)

APPENDIX 1

A SELECTION OF QUESTIONS ON PHYSICAL CHEMISTRY

1. Coal is an important resource which can be burned to provide energy or converted into other chemicals. Gasoline, which contains octane, can be manufactured on an industrial scale from coal by reactions similar to the following.

$C(s) + H_2O(g) \rightleftharpoons CO(g) + H_2(g)$ *Reaction 1*

$8CO(g) + 17H_2(g) \xrightleftharpoons[\text{catalyst}]{\text{Fe/Co}} C_8H_{18}(g) + 8H_2O(g)$ *Reaction 2*

(*a*) Use the following data to estimate the enthalpy change for *each* reaction:

	$H_2O(g)$	*$CO(g)$*	*$C_8H_{18}(g)$*
Enthalpy of formation/ ($kJ\,mol^{-1}$)	−242	−111	−169

(*b*) Predict and explain the optimum industrial conditions for

(i) the production of carbon monoxide and hydrogen by Reaction 1

(ii) the production of octane by Reaction 2.

(*c*) Under what circumstances is the production of octane from coal likely to make a profit? Discuss any hazards and pollution problems which may be associated with the process.

(NEAB 91, AS)

2. (*a*) (i) Draw a Born–Haber cycle to illustrate the formation of solid $BaCl_2$ from its elements in their standard states. Name the enthalpy change involved at each stage.

(ii) Using the following data, calculate the enthalpy of formation of $BaCl_2(s)$:

Process	$\Delta H^{\ominus}/kJ\,mol^{-1}$
$Ba(s) \rightarrow Ba(g)$	176
$Ba(g) \rightarrow Ba^{+}(g) + e^{-}$	508
$Ba^{+}(g) \rightarrow Ba^{2+}(g) + e^{-}$	972
$\frac{1}{2}Cl_2(g) \rightarrow Cl(g)$	121
$Cl(g) + e^{-} \rightarrow Cl^{-}(g)$	−364
$Ba^{2+}(g) + 2Cl^{-}(g) \rightarrow BaCl_2(s)$	−2018

(*b*) The enthalpy of solution of barium chloride is the enthalpy change of the process

$$BaCl_2(s) \xrightarrow{aq} Ba^{2+}(aq) + 2Cl^{-}(aq)$$

Describe an experiment which you could perform to determine the enthalpy of solution of barium chloride, indicating how you would calculate the results.

(*c*) (i) The *mass* concentration of sulphate ions in a saturated solution of barium sulphate is $9.55 \times 10^{-4}\,g\,dm^{-3}$.

Calculate the solubility product of barium sulphate.

(ii) A concentrated solution of barium chloride is added slowly to aqueous sodium sulphate ($0.25\,mol\,dm^{-3}$).

Calculate the concentration of barium ions when barium sulphate just begins to precipitate.

(O 92)

3. In the laboratory carbon dioxide is made when a carbonate reacts with an acid.

(*a*) (i) Write a balanced ionic equation for the reaction of an acid with a carbonate.

(ii) Explain why a mixture of calcium carbonate and dilute sulphuric acid is not used to prepare carbon dioxide in the laboratory.

(*b*) Explain, giving balanced equations, how it is possible to distinguish between sodium carbonate and sodium hydrogencarbonate using magnesium ions.

(*c*) Carbon dioxide is not an 'ideal gas'. Argon is closer to being 'ideal'. Some of the assumptions made about ideal gases do not hold at very high pressures or very low temperatures.

(i) State two of these assumptions.

(ii) Explain the limitations of one assumption for a real gas at very high pressure and very low temperature.

(iii) The distribution of velocities for carbon dioxide molecules at 20 °C may be represented by the graph below.

Copy the graph and draw another graph on the same axes to show how the distribution would change when the temperature is raised.

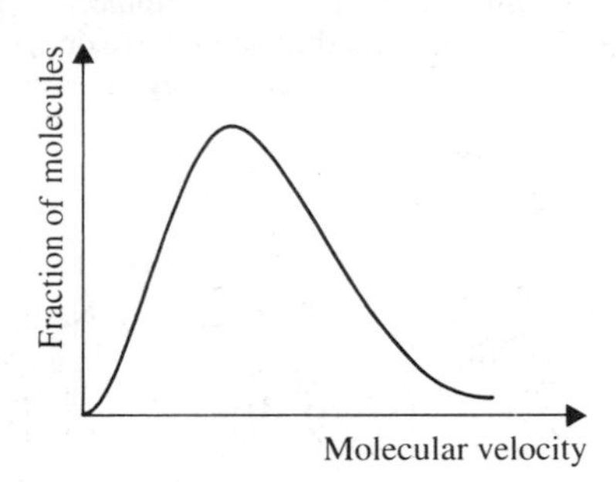

(*d*) Liquid carbon dioxide is increasingly being used as a solvent. It is used to extract caffeine from coffee and hop oil from hops.

Although carbon and oxygen have different electronegativities liquid carbon dioxide is non-polar.

(i) Explain why carbon dioxide is non-polar.

(ii) What type of compound would liquid carbon dioxide be expected to dissolve?

(*e*) A solution of carbon dioxide in water at 25 °C may be regarded as a 0.0037 M solution of carbonic acid.

$$CO_2 + H_2O \rightleftharpoons H_2CO_3$$

Assume that the H^+ in solution arises from the first dissociation of carbonic acid.

$$K_a = \frac{[H^+][HCO_3^-]}{[H_2CO_3]} = 4.3 \times 10^{-7}$$

Calculate the pH of the solution to the first decimal place.

(NI 91)

4. (*a*) (i) Define the term *standard first molar ionisation energy*.

(ii) State the fundamental factors which govern the values of first and successive ionisation energies (IE) and, using the values listed as examples or otherwise, discuss and explain how electronic structure (or position in the Periodic Table) affects these values.

IE/ $kJ\,mol^{-1}$	Element							
	K	*Ca*	*Ba*	*Ge*	*Pb*	*Br*	*I*	*Kr*
First	419	590	502	760	715	1142	1009	1351
Second	3069	1146	967	1537	—	2080	—	—
Third	—	4942	3500	—	—	—	—	—

(iii) Relate the value of an element's first ionisation energy to whether that element

(*1*) tends to form ionic or covalent compounds

(*2*) is very reactive, e.g. with water, or not

(*3*) is a reducing agent or not.

(iv) Explain why the first ionisation energy of sodium is $496\,kJ\,mol^{-1}$ whereas the energy required to remove 1 mol of electrons from sodium metal is only $210\,kJ\,mol^{-1}$.

(*b*) Diamond and graphite are both technologically important materials. Compare the structure of and the bonding in the two allotropes and relate these to their useful properties.

(*c*) Give an account of the types of intermolecular force which exist and of the effects that these have on the physical properties of the substances in which they are present.

(WJEC 92)

5.

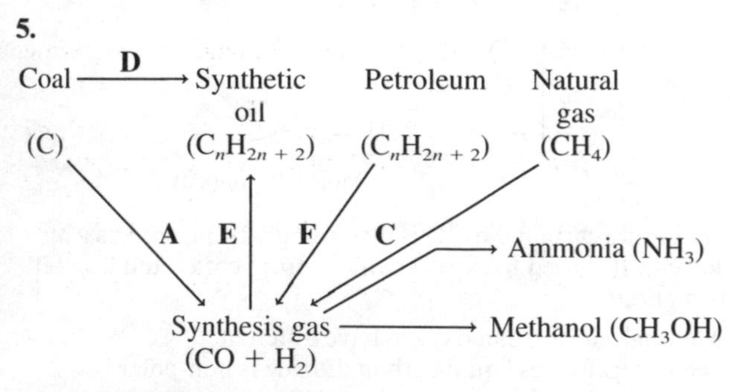

The figure above shows basic fuels (coal, synthetic oil, petroleum and natural gas) which are also sources of organic chemicals, synthesis gas, and the routes to ammonia and methanol, the production of each of which exceeds ten million tonnes per annum worldwide and which are important industrial products.

(*a*) Route **D** was important in Germany during World War II, route (**A** + **E**) is still important in South Africa, and, in Britain, route **A** was replaced by **F** which was itself replaced by **C**.

Discuss the economic, technical and other factors which are, or have been, responsible for the above facts: make reasonable inferences if you are not sure of the factual background.

(*b*) (i) Explain why ammonia and methanol plants are often sited together.

(ii) The gas-phase enthalpy changes for the two syntheses are

$$\tfrac{1}{2}N_2 + \tfrac{3}{2}H_2 \rightarrow NH_3; \quad \Delta H^{\ominus}(298\,K) = -45\,kJ\,mol^{-1}$$

$$CO + 2H_2 \rightarrow CH_3OH; \quad \Delta H^{\ominus}(298\,K) = -100\,kJ\,mol^{-1}$$

In *each* case state what general conditions of temperature and pressure should be used to obtain a high equilibrium yield of product.

(iii) State what other physico-chemical factor is important in a successful process and explain how it is affected by reaction conditions.

(iv) The equilibrium constant (K_p) for the methanol synthesis as written has a value of $2.85 \times 10^{-4}\,atm^{-2}$ at 300 °C. Calculate the equilibrium pressure of methanol at this temperature if the CO/H_2 ratio is stoichiometric (i.e. 1 : 2) and the combined pressure of CO plus H_2 *at equilibrium* is 60 atm.

(*c*) Originally the methanol process operated at 400 °C and 200 atm pressure but the current ICI process employs 225 °C and 50 atm.

(i) State why these are improved conditions and suggest why they can now be used.

(ii) Relative rate constants (*k*) for the methanol synthesis at various temperatures are as follows:

Temperature/K	500	525	552	579
Relative rate constant, k	1.0	3.0	9.0	27.0

Given that the Arrhenius equation, which relates rate constant to temperature, may be written as

$$\ln k = \text{constant} - \frac{E}{R}\left(\frac{1}{T}\right)$$

where *E* is the activation energy, *R* is the gas constant ($8.31\,J\,K^{-1}\,mol^{-1}$) and *T* is the temperature in kelvin, calculate graphically the activation energy for the process.

(iii) From this value and the data given in (*b*), calculate also the activation energy of the back reaction

$$CH_3OH \rightarrow CO + 2H_2$$

(*d*) The industrial synthesis of ammonia requires severe conditions of temperature and pressure and yet ammonia is synthesised from nitrogen on the roots of leguminous plants at 20 °C and 1 atm pressure. Comment on this anomaly and suggest how progress might be made towards easier manufacture of ammonia.

(WJEC 92, S)

6. (*a*) (i) Draw a graph with curves showing the distribution of speed or energy of gas molecules at *two* different temperatures, labelling both axes and the higher temperature curve.

(ii) Using this graph (from (*a*)(i)) explain the effect of increasing temperature on the rates of gas reactions.

(*b*) The table lists some relative pressure–volume (pV) values for 1 mol of CO_2 and of H_2 at 298 K.

Pressure (p)/atm		1	50	100	200
Relative pV *value*	CO_2	1.0	0.74	0.27	0.41
	H_2	1.0	1.03	1.06	1.13

(i) Plot pV against p for both gases on graph paper and label the axes.

(ii) On your graph show the variation of pV against p for an ideal gas over the same pressure range.

(iii) Comment on the shape of the curves. Use arguments based on the kinetic theory of gases to account for these shapes.

(*c*) (i) For an aqueous solution of an acid, explain clearly the difference between the meaning of the terms *weak acid* and *dilute acid.*

(ii) Some acid/base indicators are listed below along with the pH range over which they change colour.

Indicator	*Name*	*pH range*
A	Cresol red	0.3–1.7
B	Methyl red	4.2–6.2
C	Phenolphthalein	8.3–10.0
D	Nitramine	10.8–13.0

Give the correct letter *or letters* to show which indicator or indicators would be suitable for the following titrations at 0.1 mol dm^{-3} concentrations:

(*1*) $CH_3COOH/NaOH$, (*2*) $HCl/NH_3(aq)$ (*3*) $HCl/NaOH$.

(iii) State the properties which a substance must possess in order to be useful as an acid/base indicator and explain how such indicators work.

(iv) For an acid of pK_a 6.0, calculate the pH given by an aqueous solution of it at 0.1 mol dm^{-3} concentration.

(WJEC 92)

7. In an experiment to test the effectiveness of a new genetically engineered micro-organism a sample of sewage was treated with it. Half of the carbohydrate contamination, $C_6H_{12}O_6$, of the sewage was oxidised to carbon dioxide and water and 10% underwent anaerobic decomposition into methane and carbon dioxide only. 5 m^3 of methane gas, measured at 298 K and 10^5 Pa, were produced by the plant each day.

(*a*) (i) Write the balanced equations for the two processes for the removal of carbohydrate contamination (oxidation and anaerobic decomposition).

(ii) Calculate the total volume of gas produced per day.

(iii) Given that the enthalpy of combustion of methane is −890 kJ mol^{-1}, calculate the amount of heat energy that could be produced each day by burning the methane produced from the sewage.

(*b*) The pH of the solution was maintained at 3.0 by the addition of ethanoic acid. Calculate the approximate concentration (in mol dm^{-3}) of the ethanoic acid in the solution at that pH, given that the acid dissociation constant, K_a, for ethanoic acid is 1.8×10^{-5}.

(C 90, AS)

8. (*a*) Equal masses of three different substances are introduced into three identical evacuated vessels. The temperature of each vessel is then adjusted until the contents of each is entirely gaseous and the pressure in all three is identical.

(*b*) The three substances are given below. Temperatures needed for identical pressures are shown in brackets.

$$\text{Ar}\,(133.3\,\text{K}) \quad CH_3COOH\,(400.0\,\text{K}) \quad PCl_5\,(500.0\,\text{K})$$

Assuming that argon is an ideal gas ($M_r = 40$), estimate the apparent relative molecular mass of CH_3COOH and of PCl_5. Compare these values with the expected relative molecular masses, explain any differences and, if the pressure in each vessel is 250 kPa, calculate the value(s) of any equilibrium constants involved.

(*b*) Write the conventional representation of a cell whose emf is that of the standard nickel half-cell, justifying your choice. State *two* precautions needed (other than the choice of temperature and pressure) in order to make a reliable measurement of the emf of such a cell.

At 298 K, the equilibrium constant (K_c) of a cell reaction is related to the standard emf of the cell ($E^\ominus$/V) by the expression

$$\log_{10} K_c = \frac{nE^\ominus}{0.059}$$

where n is the number of electrons transferred in the cell reaction.

What is the concentration of cobalt(II) ions present if a solution made by mixing equal volumes of two solutions, one of which is 2.0 M in cobalt(II) ions and the other 2.0 M in nickel(II) ions, is shaken with some finely divided cobalt and nickel metal?

$$Co^{2+}(aq) + 2e^- \rightarrow Co(s); \quad E^\ominus = -0.277\,\text{V}$$
$$Ni^{2+}(aq) + 2e^- \rightarrow Ni(s); \quad E^\ominus = -0.250\,\text{V}$$

(*c*) The equilibrium in the system

Ethanol + Ethyl ethanoate + Ethanoic acid + Water
(*acetate*) (*acetic*)

was investigated by taking 10.19 g of ethyl ethanoate, adding a solution of concentrated hydrochloric acid, and heating in a sealed tube at 25 °C. The equilibrium was then frozen by adding an excess of a solution of exactly neutralised sodium ethanoate. The resulting mixture was made up to 1000 cm^3 with distilled water and aliquots of 25 cm^3 were titrated as follows:

(i) with 0.100 M $Ba(OH)_2$: average titre = 10.90 cm^3

(ii) with 0.100 M $AgNO_3$: average titre = 10.20 cm^3.

If the number of moles of water present at equilibrium (*before* adding sodium ethanoate and diluting) is 0.230, determine the equilibrium constant for the esterification reaction at this temperature.

(NEAB 91, S)

9. (*a*) When phenol dissolves in water, the solution is slightly acidic owing to the equilibrium

$$C_6H_5OH(aq) \rightleftharpoons C_6H_5O^-(aq) + H^+(aq)$$

pK_a at 298 K for this equilibrium may be taken to have a value of 10.0.

A saturated aqueous solution of phenol at 298 K contains 6.70 g of phenol in 100 cm^3 of water. Calculate the pH of this solution.

(*b*) When phenol is dissolved in sodium hydroxide solution, an equilibrium is obtained involving the formation of sodium phenoxide:

$$C_6H_5OH(aq) + NaOH(aq) \rightleftharpoons C_6H_5ONa(aq) + H_2O(l)$$

0.100 mol of phenol was dissolved in 100.0 cm^3 of 1.00 mol dm^{-3} sodium hydroxide. This solution was shaken with 25.0 cm^3 of hexane until the phenol had distributed itself between the two solvents:

$$C_6H_5OH(aq) \rightleftharpoons C_6H_5OH(hexane)$$

The hexane layer was carefully separated from the aqueous layer and the hexane allowed to evaporate to dryness. The solid residue of phenol weighed 1.25 g.

The equilibrium constant for the distribution of phenol between hexane and water is

$$K_c = \frac{[C_6H_5OH(hexane)]_{eqm}}{[C_6H_5OH(aq)]_{eqm}} = 2.50$$

Calculate a value for the equilibrium constant for the reaction between phenol and sodium hydroxide in aqueous solution.

(*c*) Why does a saturated aqueous solution of phenol not cause evolution of carbon dioxide when added to 1.0 mol dm^{-3} sodium carbonate solution? Carry out short calculations to aid your explanation, using your answer to (*a*) and given $K_a : H_2O + CO_2 \rightleftharpoons H^+ + HCO_3^- = 4.5 \times 10^{-7}$ mol dm^{-3}; $K_a : HCO_3^- \rightleftharpoons H^+ + CO_3^{2-} = 2 \times 10^{-4}$ mol dm^{-3}. (It is not necessary to attempt any lengthy or involved calculations.)

(L(N) 92, S)

Part 3

INORGANIC CHEMISTRY

15

PATTERNS OF CHANGE IN THE PERIODIC TABLE

15.1 PHYSICAL PROPERTIES

The Periodic Table

The Periodic Table was introduced in § 2.5. The reason why this arrangement of the elements was proposed is that the elements show **periodicity** in their physical and chemical properties. **Periodic** means repeating after a regular interval. The way in which alkali metals, halogens and noble gases occur at regular intervals was described in § 2.8. The regular intervals are a consequence of the way in which electron shells are filled [§ 2.7].

The physical properties of elements show periodicity, i.e., they repeat after an interval of 8 or 18 elements

Some of the physical properties which show periodicity are illustrated in Figures 15.1 to 15.6. The melting temperature [see Figure 15.1] of an element depends on the strength of the bonds and also on the structure of the element. When metals melt, some metallic bonding remains in the liquid phase. When macromolecular substances, such as carbon, melt, nearly all the bonds have to be broken to melt the solid. The standard enthalpy of melting varies in a similar way with the proton number (atomic number) of the element.

FIGURE 15.1 Periodicity of Melting Temperatures of the Elements

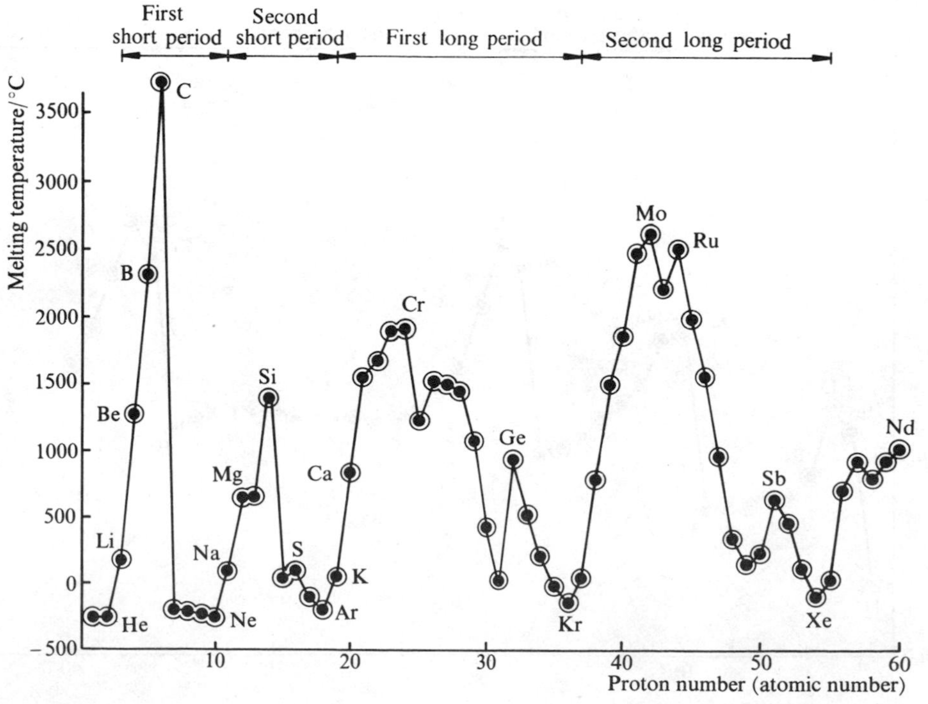

The variation of boiling temperature with proton number is shown in Figure 15.2. The standard enthalpies of vaporisation [§ 8.1] vary in a similar manner.

FIGURE 15.2 Periodicity of the Boiling Temperatures of the Elements

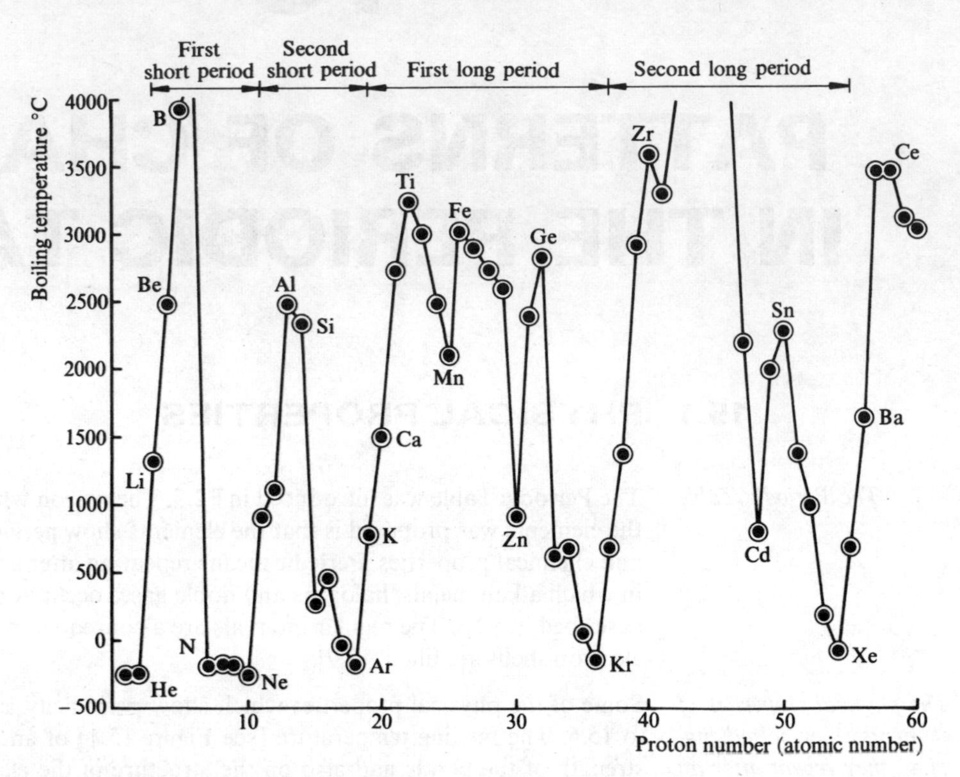

Atomic radius is plotted against proton (atomic) number in Figure 15.3. The sizes of some atoms and ions are compared in Figure 15.4.

FIGURE 15.3 Variation of Atomic Radius with Proton Number. (The values for noble gases are van der Waals radii. The values for other elements are single covalent bond radii.)

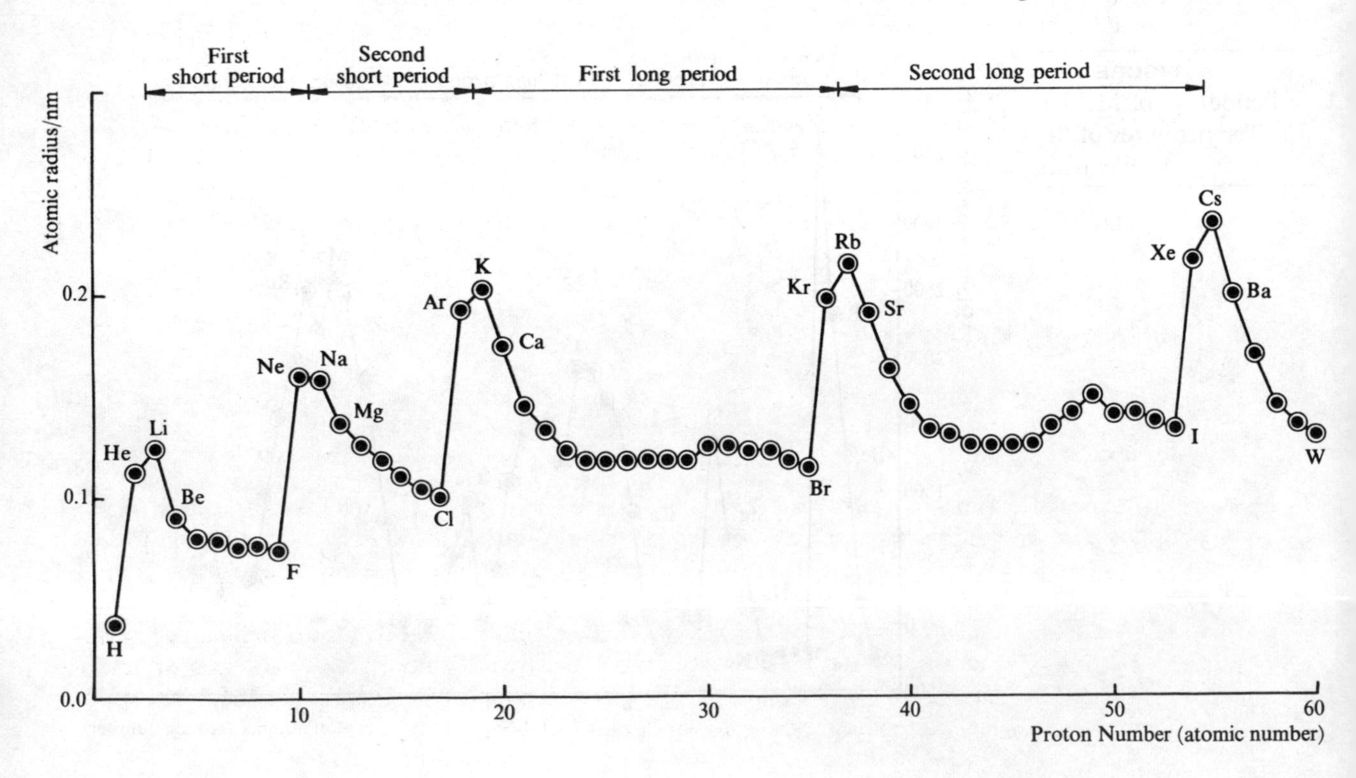

FIGURE 15.4 Atomic and Ionic Radii (○ = 0.10 nm = 100 pm)

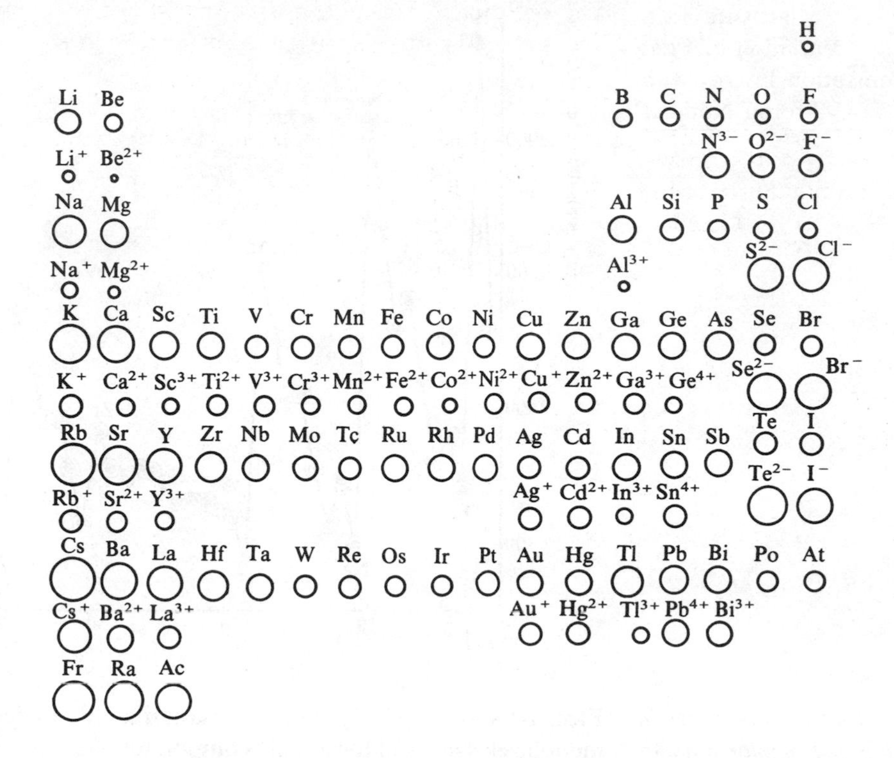

The radii of atoms and ions increase from top to bottom of a group...

From Figure 15.4, you can see that ionic and atomic radii increase from top to bottom of any group. Both the nuclear charge and the number of electron shells increase down a group. Although the increasing nuclear charge decreases the sizes of individual shells, the effect of adding a shell is the dominating effect.

...and decrease from left to right across a period as the effective nuclear charge increases

Covalent and ionic radii both decrease from left to right across any period of the Periodic Table. In the first short period, (Li–F), the nuclear charge increases from 3 to 9. As the nuclear charge increases, it pulls the K electrons closer to the nucleus, and the radius of the K shell decreases. The effect on the L electrons is complicated by the fact that they are *screened* or shielded from the nucleus by the K shell, so that the effective nuclear charge is less than the actual nuclear charge. For example, in lithium the outermost electron is attracted by a nucleus with a charge of +3 screened by two electrons. The net nuclear charge is closer to +1 than +3. In beryllium, the L electrons are attracted by a nucleus which has a charge of +4, and is screened by 2 electrons, which make its effective charge close to +2. Nevertheless, reading from left to right across a period, the effective nuclear charge increases and causes a steady decrease in atomic radius across a period [see Figure 15.4]. A comparison of ions with the same numerical charge, e.g., M^{2+}, shows that ionic radii follow the same pattern.

A series of transition metals differ little in atomic radius

Transition metals show a very small change in atomic radius across a series. In the first transition series, beginning with scandium, the size of the atoms is governed by the N shell, while the additional electrons are entering the M shell.

Ionisation energy varies in a periodic manner

The first ionisation energy [see Figure 15.5] shows periodic variation. It reaches a peak at each noble gas. From helium to lithium and from neon to sodium, the ionisation energy decreases sharply. The additional electrons (a 2s electron in Li and a 3s electron in Na) are much more easily removed than the electrons in the noble gases. Across the periods from lithium to neon and from sodium to argon, the increasing nuclear charge makes it more difficult to remove an electron, and the ionisation energy increases. The increasing ionisation energies for the removal of successive electrons are shown in Figure 2.17, §2.4.

FIGURE 15.5 Variation of First Ionisation Energy with Proton Number

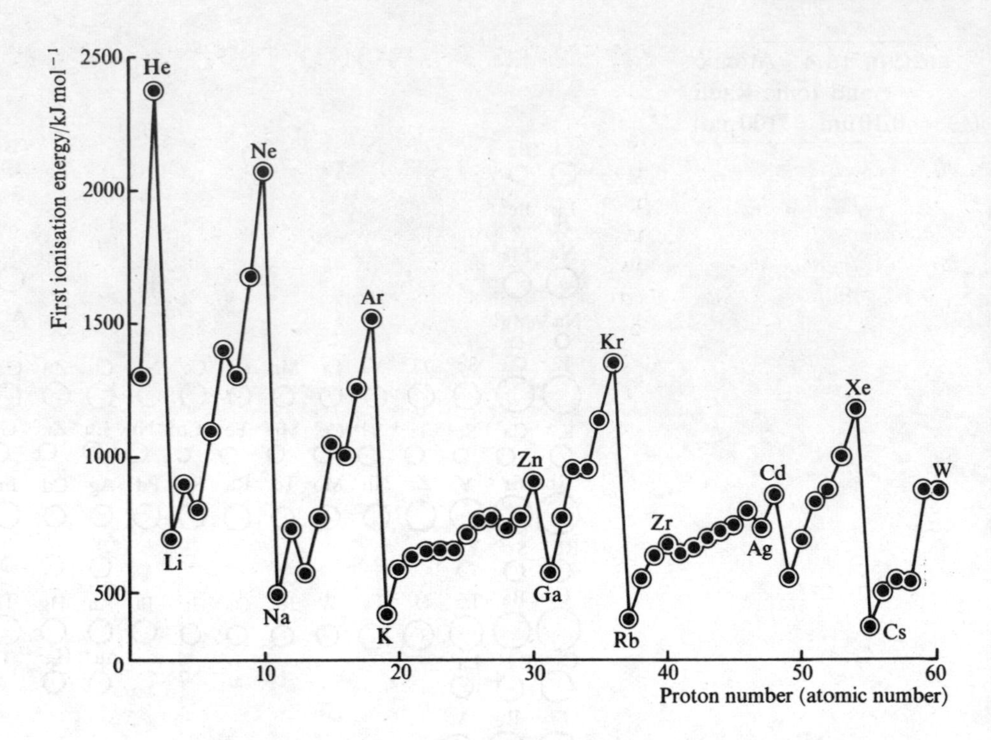

Electron affinity varies in a regular manner

Electron affinity [§ 10.7] is an important quantity in determining whether a non-metallic element will form ionic compounds. Across a period, as nuclear charge increases, electron affinity increases (becomes more negative, more exothermic). If an electron enters a shell close to the nucleus it becomes more tightly bound, with the release of more energy, than if it enters a shell distant from the nucleus. Electron affinity therefore decreases (becomes less negative, less exothermic) from top to bottom of a group.

Electronegativity increases across a period from left to right and decreases down a group

The property of elements called electronegativity was covered in § 4.5.3. The difference in electronegativity between bonded atoms determines the percentage of ionic character in a covalent bond [see Figure 4.27, § 4.5.3]. Across a period, as nuclear charge increases, electronegativity increases. It decreases down a group as the number of electron shells increases. An element with a very low electronegativity is described as **electropositive**. Figure 15.6 shows how the electronegativity of an element is related to its position in the Periodic Table.

FIGURE 15.6 The Electronegativity Scale of the Elements, showing Relation to the Periodic Table (after L Pauling)

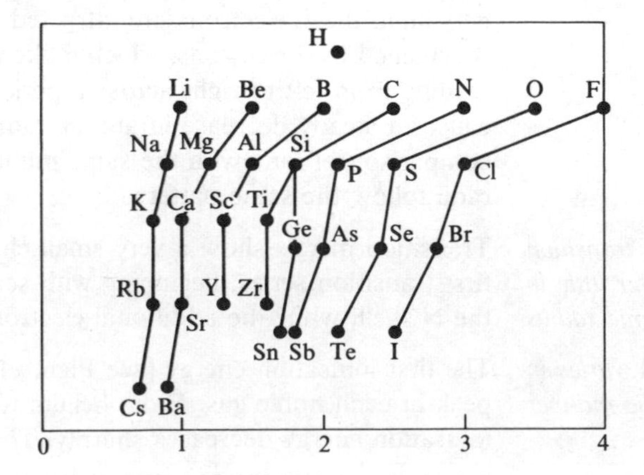

Other physical properties which illustrate periodicity are the standard enthalpy of atomisation, the standard enthalpy of hydration of ions and atomic volume. The last-mentioned was the first property to be studied. Lothar Meyer plotted atomic volume against proton (atomic) number in 1867.

15.2 TRENDS ACROSS THE PERIODIC TABLE

The polarising power of cations and the polarisability of anions vary in a regular manner across the periods and down the groups of the Periodic Table

The trends in ionic radius, first ionisation energy, first electron affinity and electronegativity are summarised in Figure 15.7. These quantities determine the type of bonds formed by an element. Fajans' Rules, which predict the ratio of ionic/covalent character in a bond between two elements are listed in § 4.5.4. They describe the power of small cations and highly charged cations to **polarise** anions, introducing a degree of covalent character into an ionic bond. It follows from Fajans' Rules that the degree of electrovalent character in the compounds formed by cationic elements and anionic elements will vary as shown in Figure 15.7. Electropositive elements will show their most electrovalent behaviour in the top right-hand corner of the table, where oxygen, fluorine and chlorine are to be found.

FIGURE 15.7 Gradations in Bond Type across the s and p Blocks

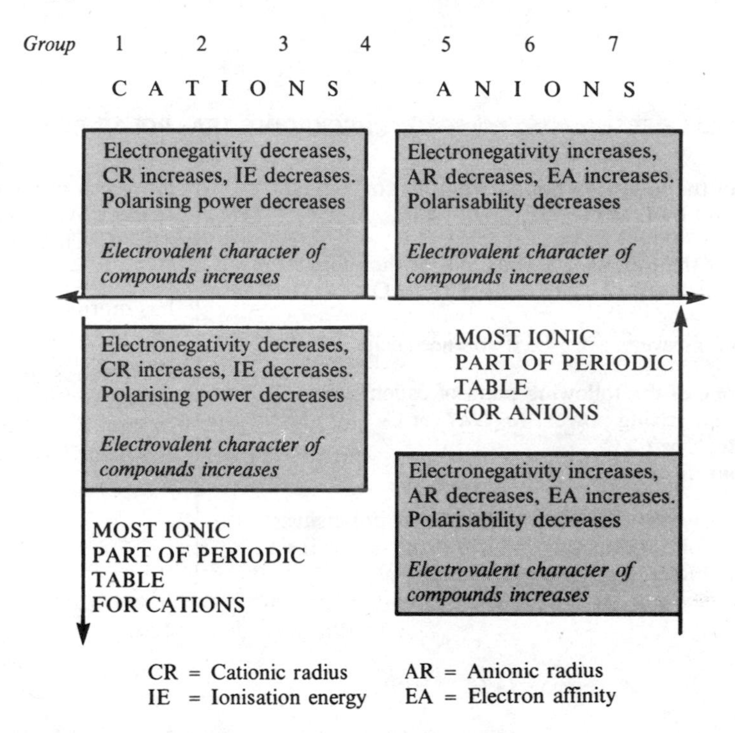

Trends in chemical properties can be observed across sections of the Periodic Table

In practice, one cannot consider a set of cations across a whole period; for example, the cations from Li^+ to F^{7+} do not all exist. Nor can one follow anionic behaviour across a whole period. The gradations in bond type are applied to sections of periods. One can say that

Period 2:	Li Be	B C N	O F
	form cations	form covalent bonds	form anions

One can predict that from Li^+ to Be^{2+}, there will be an increase in the covalent character of the bonds, and from O^{2-} to F^-, there will be an increase in the electrovalent character of the bonds.

Diagonal relationships are explained in terms of the polarising power of cations and the polarisability of anions

Figure 15.8 explains the existence of diagonal relationships. The first element in Group 1, lithium, is diagonally related to the second element in Group 2, magnesium. Since the polarising power of the cations decreases from lithium down to sodium and increases across from sodium to magnesium, lithium and magnesium form compounds with a similar degree of covalent character. In fact, the two elements are similar in their chemistry [§ 18.7]. There is a diagonal relationship between beryllium in Group 2 and aluminium in Group 3 and between boron in Group 3 and silicon in Group 4 [§ 19.4].

FIGURE 15.8 Diagonal Relationships in the Periodic Table

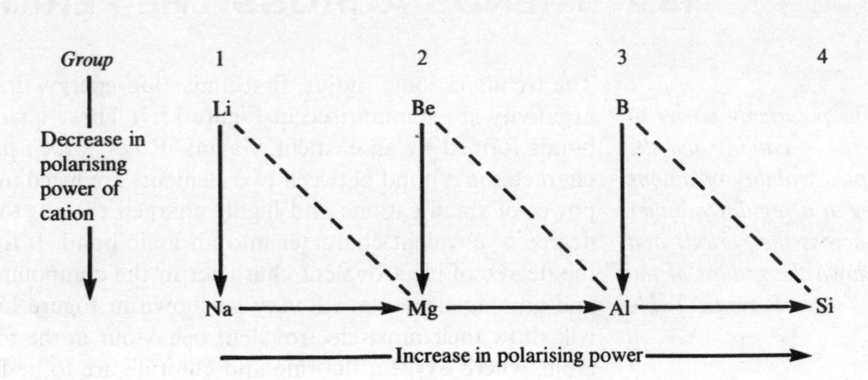

CHECKPOINT 15A: POLAR BONDS

1. Which of the following bonds are polar: (*a*) S—O, (*b*) Cl—Cl, (*c*) I—I, (*d*) C—H, (*e*) C—Cl.

†2. Which of the following anions will be the more easily polarised? (*a*) Cl^- or I^-, (*b*) Cl^- or O^{2-}, (*c*) O^{2-} or S^{2-}.
Explain your answers. See §4.5 if you need help.

3. Which one of the following pairs of cations has the greater polarising power? (*a*) Na^+ or Cs^+, (*b*) Mg^{2+} or Al^{3+}, (*c*) Be^{2+} or Ca^{2+}.
Explain your answers.

4. Explain why the cation Fe^{3+} is strongly polarising. Arrange the following compounds in order of increasing covalent character: $FeBr_3$, $FeCl_3$, FeF_3, FeI_3.

5. Which of the following salts has the highest percentage of electrovalent character?
LiCl, CsCl, CsF, CsI, LiF, LiI.

6. Which one of the following pairs of compounds has more covalent character? (*a*) $AlCl_3$ or Al_2O_3, (*b*) SnF_4 or $SnCl_4$.

7. Which is the more polarising cation, Li^+ or Cs^+? Which of the chlorides, LiCl or CsCl, has the larger degree of covalent character? One of these chlorides is soluble in organic solvents. Which do you suppose it is?

15.3 CHEMICAL PROPERTIES AND BOND TYPE

The chemical behaviour of compounds is determined by the character of the bonds in the compound.

15.3.1 OXIDES

Ionic oxides are basic

Electrovalent oxides are basic. The high negative charge/volume ratio of the O^{2-} ion makes it act as a proton-acceptor, e.g.

$$2Na^+O^{2-}(s) + H_2O(l) \rightarrow 2Na^+(aq) + 2OH^-(aq)$$

Covalent oxides are acidic or neutral

Covalent oxides are either acidic or neutral. An oxide like SO_3 cannot act as a proton-donor, but it can give rise to oxonium ions by acting as an electron-acceptor (a Lewis acid, see §12.7.1).

$$SO_3(s) + 2H_2O(l) \rightarrow H_3O^+(aq) + HSO_4^-(aq)$$

Oxides of intermediate bond character are amphoteric.

15.3.2 HALIDES

Many ionic chlorides dissolve in water

Electrovalent chlorides simply dissolve in water, with the exception of a few sparingly soluble chlorides, e.g.

$$Na^+Cl^-(s) + aq \rightarrow Na^+(aq) + Cl^-(aq)$$

Many covalent chlorides are hydrolysed

Covalent chlorides may, like CCl_4, neither dissolve in nor react with water or they may, like PCl_3, be hydrolysed to an acid and hydrogen chloride:

$$PCl_3(l) + 3H_2O(l) \rightarrow H_3PO_3(aq) + 3HCl(aq)$$

The phosphorus atom has empty d orbitals, into which electrons from the oxygen atoms in a water molecule can coordinate. Coordination is followed by hydrolysis. Chlorides of polarising cations may be partially hydrolysed:

Chlorides of polarising cations are acidic in solution

$$BiCl_3(aq) + H_2O(l) \rightleftharpoons BiOCl(s) + 2HCl(aq)$$

The solutions of chlorides and other salts of highly polarising cations are acidic on account of salt hydrolysis [§ 12.7.13].

$$[Al(H_2O)_6]^{3+}(aq) + H_2O(l) \rightleftharpoons [Al(OH)(H_2O)_5]^{2+}(aq) + H_3O^+(aq)$$

15.3.3 HYDRIDES

Electrovalent hydrides have a H^- ion, which acts as a proton-acceptor, and they therefore react readily with water to form a base:

$$Na^+H^-(s) + H_2O(l) \rightarrow H_2(g) + Na^+(aq) + OH^-(aq)$$

Covalent compounds of hydrogen can be proton-donors (HCl, H_2S), proton-acceptors (NH_3), amphoteric (H_2O) or neutral (CH_4).

15.3.4 OXIDES, HALIDES AND HYDRIDES IN RELATION TO THE PERIODIC TABLE

Figures 15.9 and 15.10 relate the properties of oxides, chlorides and hydrides to the positions of elements in the Periodic Table. [See § 17.7 for hydrides, §§ 20.8 and 20.9 for halides and § 21.8 for oxides.]

FIGURE 15.9 Bonding in Oxides, Chlorides and Hydrides

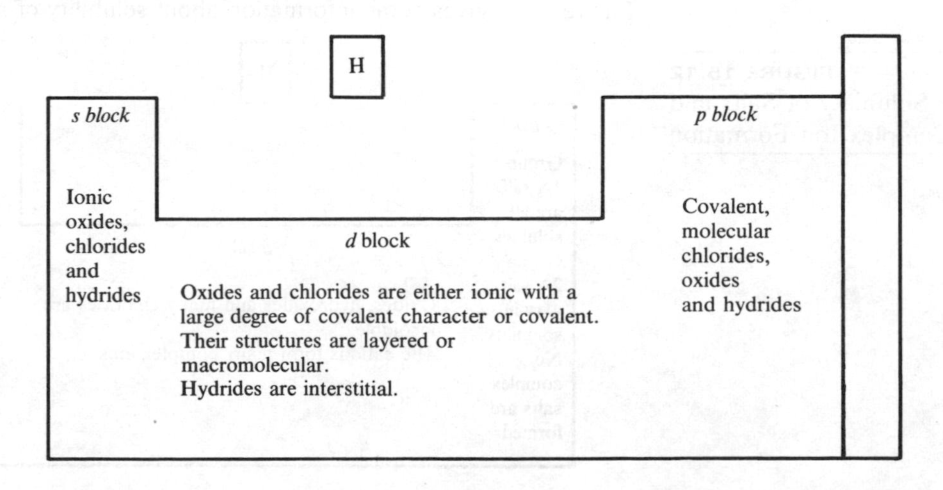

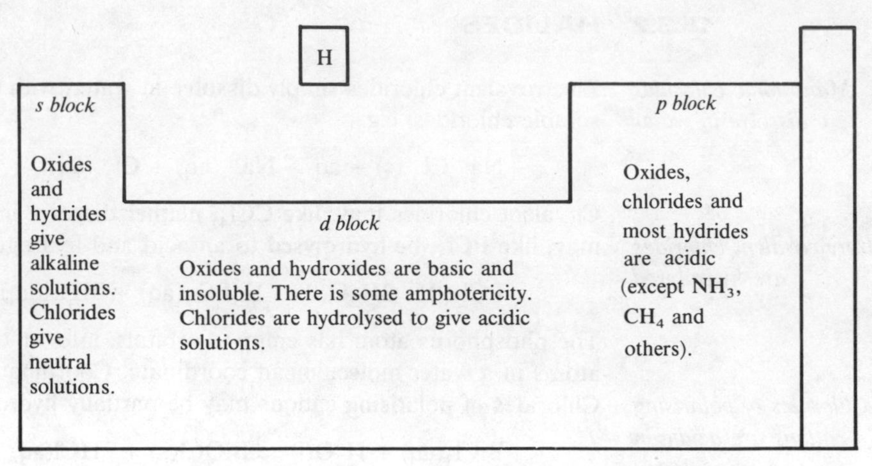

FIGURE 15.10 Acidic and Basic Compounds

The relationship of electronegativity to the Periodic Table is shown in Figure 15.6, § 15.1. The oxidising or reducing properties of the elements are related to their electronegativity, as shown in Figure 15.11.

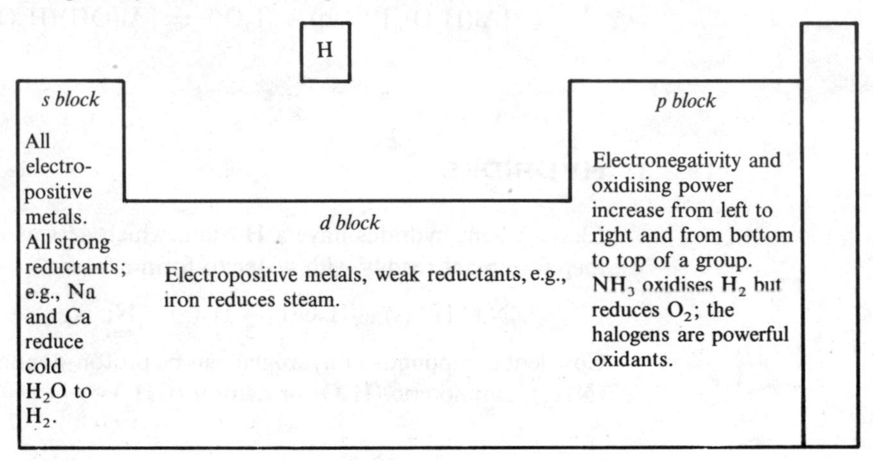

FIGURE 15.11 Redox Properties and the Periodic Table

15.3.5 SALTS

Thermal stability is related to bond character

Nitrates, sulphates and carbonates in which the electrovalent character of the bonds is high are stable to heat. Salts with more polarising cations are more easily decomposed by heat: calcium carbonate is decomposed, whereas sodium carbonate is stable. Hydrogencarbonates do not exist after Group 2, and solid hydrogencarbonates are obtained only in Group 1.

Figure 15.12 gives some information about solubility of salts and complex formation.

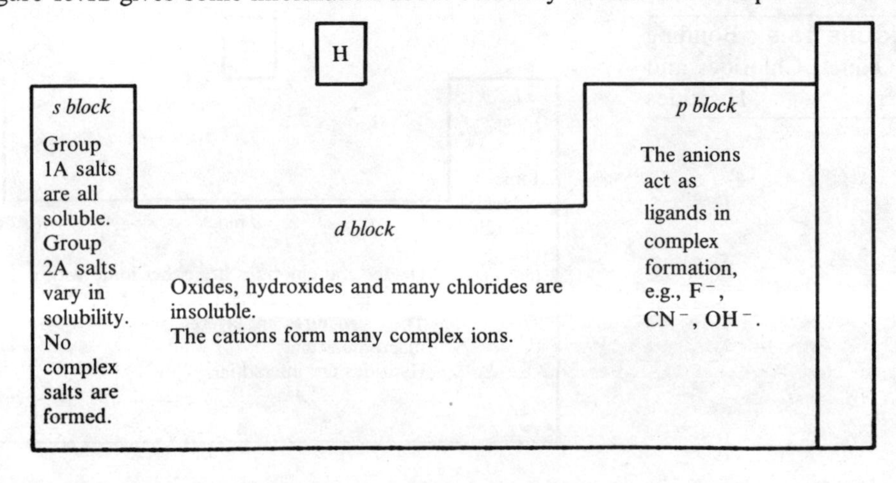

FIGURE 15.12 Solubility of Salts and Complex Ion Formation

15.3.6 VARIABLE OXIDATION STATE

The compounds of a metal in a high oxidation state show more covalent character than compounds of the metal in a lower oxidation state

In this section an attempt has been made to relate the properties of the compounds of an element to the position of the element in the Periodic Table. The oxidation state of the element also influences the nature of its compounds. Since the polarising power of a cation M^{4+} is greater than that of M^{2+}, M(IV) compounds will show more covalent character than M(II) compounds. For example, lead(II) chloride is a crystalline solid which dissolves sparingly in water to form a solution of Pb^{2+}(aq) and Cl^-(aq) ions. Lead(IV) chloride is a liquid, which is rapidly hydrolysed by water to lead(IV) dichloride oxide and hydrogen chloride:

$$PbCl_4(l) + H_2O(l) \rightarrow PbOCl_2(aq) + 2HCl(aq)$$

Lead(II) oxide, PbO, is basic, while lead(IV) oxide is amphoteric, reacting with alkalis to form plumbates(IV), containing the anion, $Pb(OH)_6^{2-}$.

QUESTIONS ON CHAPTER 15

1. This question is about ionisation energies, atomic structure and their importance in chemistry.
(*a*) Define the term, *first ionisation energy* of an element.
(*b*) State any two factors which determine the first ionisation energies of the elements in a given group of the Periodic Table:
(c) Three elements, A, B and C, have atomic numbers 8, 10 and 11 respectively.
(i) Write down the electronic structures of the three elements.
(ii) What is the order of increasing first ionisation energy of the three elements?
(iii) Explain your reasoning in (ii).
(*d*) The following Table shows the first ionisation energies, I, in kJ mol^{-1} and atomic numbers, Z, of some Group II and Group III elements:

Group II			Group III		
	Z	I		Z	I
Be	4	900	B	5	799
Mg	12	736	Al	13	577
Ca	20	590	Ga	31	577

(i) Explain why the value of the first energy of Be is greater than that of B.
(ii) Explain the change in relative magnitudes of the first ionisation energies of the Group II elements from Be to Ca.
(iii) Explain the change in relative magnitudes of the first ionisation energies of the Group III elements from B to Ga.
(*e*) Two elements have atomic numbers of 19 and 9. What type of compound will they form when they react? Explain your reasoning.

(O&C88)

2. (*a*) Which of the cations, Ca^{2+} or Al^{3+}, has the greater polarising power?
(*b*) Which of the chlorides, $CaCl_2$ or $AlCl_3$, is the more covalent in character?
One of these chlorides sublimes when it is heated. Which one do you suppose it is?

3. (*a*) Compare the polarising power of Be^{2+} and Ba^{2+}.
(*b*) What would you predict about the degree of covalent character in $BeCl_2$ and $BaCl_2$?
One of the chlorides is hydrolysed in aqueous solution. Which one do you suppose it is?

4. Of the compounds $SnCl_2$ and $SnCl_4$, which one has the more covalent character? One of the chlorides is a crystalline solid, and one is a fuming liquid. Which is which?

5. Below is a representation of two periods of the Periodic Table:

Group	1	2	3	4	5	6	7
	Li	Be	B	C	N	O	F
	Na	Mg	Al	Si	P	S	Cl

Indicate, by the symbol, which of these elements you consider to be
(i) the most reactive non-metal
(ii) the most reactive metal
(iii) the one with the smallest atom
(iv) the one which has a first ionisation energy of 736 kJ mol^{-1} when the previous two elements had corresponding values of 2080 and 494 kJ mol^{-1} respectively
(v) the one which forms the largest anion
(vi) the one which has the highest electronegativity
(vii) a p-block metal
(viii) likely to form an amphoteric oxide of formula X_2O_3
(ix) in the same group as an element of atomic number 31
(x) the one which has the highest boiling point

(xi) likely to combine with nitrogen to form only single bonds with a bond energy of approximately 350 kJ mol^{-1}

(xii) able to conduct electricity and to form a dioxide.

(C 90, AS)

6. (*a*) Successive ionisation energies for the element boron are given below.

Electron removed	First	Second	Third	Fourth	Fifth
Iionisation energy/ kJ mol^{-1}	800	2400	3600	25 000	33 000

(i) Write the equation which represents the second ionisation energy of boron.

(ii) State the electronic configuration of boron, and explain how the above values provide evidence for the existence of electronic shells.

(*b*) First ionisation energies for the elements lithium to nitrogen are given below.

Element	Li	Be	B	C	N
Ionisation energy/ kJ mol^{-1}	520	900	800	1100	1400

Explain

(i) the overall increase in first ionisation energy across the period,

(ii) how the values provide evidence for the existence of electron sub-shells.

(*c*) (i) State, and explain in terms of sub-shells, the maximum number of electrons which can be accommodated in the third quantum shell.

(ii) Explain why Period 3 (Na to Ar) consists of only eight elements.

(*d*) Describe how the melting points of the elements change across Period 3 (Na to Ar) and account for these changes in terms of structure and bonding.

(NEAB 91, AS)

7. (*a*) The atomic radii and melting points of elements in Period 3 are given below.

	Na	*Mg*	*Al*	*Si*	*P*	*S*	*Cl*
Atomic radius/nm	0.157	0.136	0.125	0.117	0.110	0.104	0.099
Melting point/°C	98	651	660	1410	44	114	−101

(i) State and explain the trend shown by the values for the atomic radii across Period 3.

(ii) Account for the melting points of these elements in terms of structure and bonding.

(*b*) Discuss, writing equations where appropriate, how the following properties vary across Period 3:

(i) the acid–base character of the oxides

(ii) the reaction with water of the chlorides of the elements sodium to phosphorus.

(NEAB 92, AS)

8. (*a*) What is meant by *periodicity* in the Periodic Table? Illustrate your answer with two examples, taken from different groups.

(*b*) How are the oxidation states of the elements of the period sodium to chlorine determined by their electronic structures?

(*c*) Describe how the chemistry of elements and compounds changes across a short period of the Periodic Table by considering: (i) the elements sodium to chlorine, and (ii) their hydrides.

In (i) and (ii), discuss one physical and one chemical property.

(*d*) State the two principal methods used for the extraction of metals. What factors affect the choice between them?

(O 91)

9. (*a*) Consider the following oxides:

Al_2O_3, CaO, CO, MgO, Na_2O, P_4O_{10}, SO_2, SiO_2

State which of these oxides are:

(i) *basic* and give *two* further examples

(ii) *acidic* and give *two* further examples

(iii) *neutral* and give *one* further example

(iv) *amphoteric* and give *one* further example.

(*b*) (i) Describe, *in outline*, tests which you would carry out to demonstrate whether a given oxide was basic, acidic or neutral.

(ii) Describe tests which you would carry out to show how amphoteric character is exhibited chemically, *stating clearly the experimental observations expected.* Write equations for the reactions which occur.

(*c*) Consider the following chlorides:

CCl_4, HCl, $MgCl_2$, NaCl, PCl_5, $PbCl_2$, $PbCl_4$, $SiCl_4$

(i) State, with *brief* explanation, which of the above chlorides contain dominantly (*1*) ionic bonding, and (*2*) covalent bonding. For compounds which you describe as ionic, state whether this is for the anhydrous compound or for aqueous solution or both.

(ii) Explain why you may be unable to assign clearly certain of these chlorides to either category.

(iii) Discuss with particular (but not necessarily exclusive) reference to the above listed chlorides, the characteristic differences in behaviour (e.g. physical properties, interaction with water, electrolytic properties) between ionic and covalent compounds.

(WJEC 90)

10. (*a*) Define the terms *ionisation energy* and *electron affinity*.

Explain the difference between the first and second ionisation energies of an element.

(*b*) Explain briefly how the value of the first electron affinity for fluorine can be obtained using a Born–Haber cycle.

(*c*) The first electron affinities *E* of some of the elements in the second short period of the Periodic Table are:

Element	Li	B	C	N	O	F	Ne
E/kJ mol^{-1}	−60	−28	−122	+7	−142	−328	+29

(i) Discuss the differences between the values for carbon and nitrogen in terms of the electronic configurations of these elements.

(ii) By considering the electronic configurations, predict whether the electron affinity of beryllium will be more negative or less negative than that of lithium, giving your reasons.

(iii) The second electron affinity for oxygen has a value of $+844\,kJ\,mol^{-1}$. Comment on this value in comparison with the values for the first electron affinities of oxygen and fluorine.

(iv) Predict the sign of the second electron affinity for fluorine, giving your reasons.

(O 91, S)

11. (*a*) What is meant by *periodicity* in the Periodic Table? Illustrate your answer with *two* examples, taken from different groups.

(*b*) State, and as far as possible explain, the changes which occur in atomic radius and first ionisation energy of an element (i) in a group and (ii) in a short period of the Periodic Table.

(*c*) Suggest reasons why gaseous Na_2 and Cl_2 are covalent, but solid NaCl is ionic.

(*d*) Compare and contrast the reactions (and products) of sodium and chlorine with water.

(O 92)

12. (*a*) In older versions of the Periodic Table the *s*-block metals were often written in groups containing transition elements:

Group 1	*Group 2*
Li	Be
Na	Mg
K	Ca
Cu	Zn

(i) Outline *two* similarities in the chemistries of magnesium and zinc which might justify their being grouped together.

(ii) Choosing *either* copper *or* zinc, give two reasons for not including it in these groups.

(*b*) Most metals can be produced by the reduction of their oxides but the circumstances can vary. Iron is normally produced by using carbon monoxide as the reducing agent although in certain situations the oxide may be reduced by aluminium powder. Titanium oxide is first converted to the chloride and it is then reduced to titanium by magnesium metal.

(i) Using the standard enthalpies of formation for iron(III) oxide and aluminium oxide as -822 and $-1669\,kJ\,mol^{-1}$ respectively, calculate the enthalpy change for the reduction of iron(III) oxide by aluminium.

(ii) Suggest a situation in which iron might be produced by this method and give a reason for doing so.

(iii) Why do you think that iron is not normally produced by this method?

(iv) Why do you think that aluminium is not produced by the reduction of its oxide by carbon monoxide?

(*c*) (i) Titanium dioxide is converted to the tetrachloride by passing chlorine over a hot mixture with carbon. Write an equation for this reaction.

(ii) Suggest why titanium is not produced by reducing the oxide with magnesium.

(C 90, AS)

16

GROUP 0: THE NOBLE GASES

16.1 MEMBERS OF THE GROUP

The outermost shell of electrons is full with two electrons in the case of helium and eight for the other elements [see Table 16.1]. There is stability associated with a full valence shell, and the noble gases are very unreactive.

TABLE 16.1 The Noble Gases

Name	*Symbol*	*Z*	*Electron Configuration*	*Concentration in air/ppm*
Helium	He	2	$1s^2$	5.0
Neon	Ne	10	$(He)2s^22p^6$	20
Argon	Ar	18	$(Ne)3s^23p^6$	10 000
Krypton	Kr	36	$(Ar)3d^{10}4s^24p^6$	1.0
Xenon	Xe	54	$(Kr)4d^{10}5s^25p^6$	0.08
Radon	Rn	86	$(Xe)5d^{10}6s^26p^6$	

16.1.1 HELIUM

Uses of helium

Helium is obtained from some natural gas wells where the gas may contain up to 5% of helium. It is used to inflate airships, weather balloons and aeroplane tyres, being safer than hydrogen for these purposes. A mixture of oxygen and helium is used by divers instead of air. If they use air, nitrogen dissolves in the blood under pressure and then comes out of the blood when they surface, causing painful spasms which they call 'the bends'.

16.1.2 ARGON

Uses of argon

Argon forms 1% of air and is obtained during the fractional distillation of liquid air. It is used to provide an inert atmosphere for gas–liquid chromatography, for risky welding jobs and for some chemical reactions.

16.1.3 NEON, KRYPTON AND XENON

Neon and krypton lights

Neon, krypton and xenon are also obtained from air. Neon gives out a red glow when an electrical discharge is passed through the gas at low pressure, and finds widespread use in neon lights. Krypton also is used in discharge tubes.

16.1.4 RADON

Radon is a radioactive gas which is formed when radium decays.

16.2 COMPOUNDS OF THE NOBLE GASES

Compounds of xenon...

...XeF_2, XeF_4, XeF_6...

The first noble gas compound ever made was the ionic solid xenon hexafluoroplatinate, $XePtF_6$, which resulted from a reaction between xenon and the powerful oxidising agent platinum(VI) fluoride. This discovery, in 1962, was followed by the synthesis of xenon(II) fluoride, XeF_2; xenon(IV) fluoride, XeF_4; and xenon(VI) fluoride, XeF_6. The bonding is covalent, with one, two and three of xenon's 5p electrons being promoted to 5d orbitals. The energy required for promotion is great and can be compensated for only by reaction with the most electronegative of elements, fluorine.

Compounds of krypton...

...KrF_2, KrF_4

The krypton compounds krypton(II) fluoride, KrF_2, and krypton(IV) fluoride, KrF_4, have been synthesised. These require more severe conditions for their synthesis than the xenon fluorides.

This spatial arrangement of the bonds and the lone pairs of electrons about the xenon atom in xenon compounds is shown in Figure 16.1.

FIGURE 16.1 Structures of Some Xenon Compounds

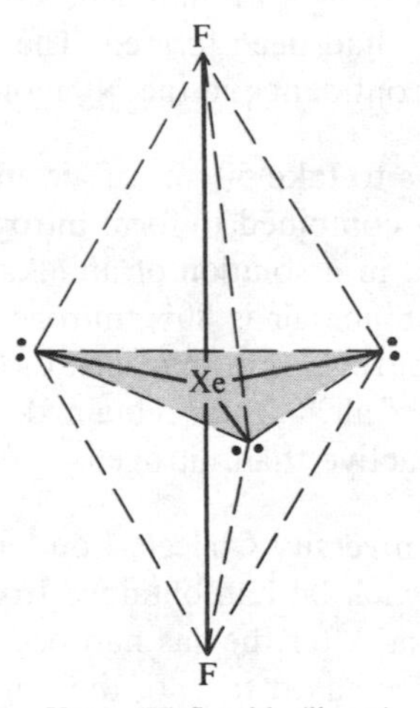

Xenon(II) fluoride (linear)

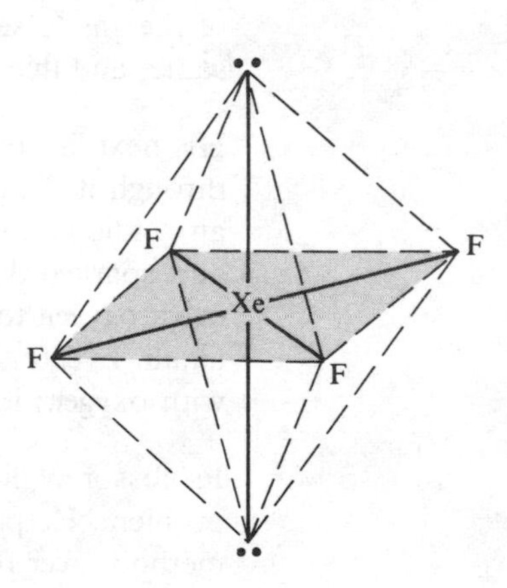

Xenon(IV) fluoride (square planar)

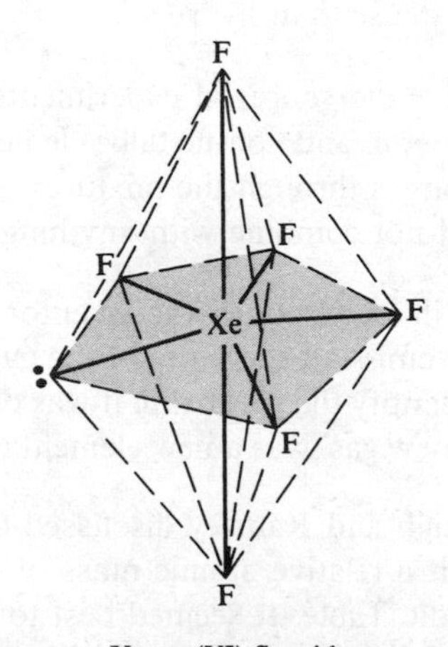

Xenon(VI) fluoride
(The bonds form a pentagonal bipyramid.)

16.3 THE 'NEW GAS'

TOPIC

The relative atomic masses of the elements were found by accurate, painstaking work done by scientists in the nineteenth century. Lord Rayleigh, professor at the Cavendish Laboratory in Cambridge from 1880 to 1900, was one of these meticulous workers. One of his methods of preparing nitrogen was the oxidation of ammonia. Another method was to start with air and remove from it oxygen, carbon dioxide and water vapour. Rayleigh found that nitrogen from air was always 0.5% denser than nitrogen from ammonia. He put forward four possible explanations:

1. Nitrogen from air might contain oxygen, which is denser than nitrogen.

2. Nitrogen from ammonia might contain the less dense gas, hydrogen, as impurity.

3. Nitrogen from air might be in the molecular form N_3 as well as the usual form of N_2 molecules. He favoured this theory because he knew that dioxygen, O_2, could be converted into trioxygen or ozone, O_3.

4. Nitrogen from ammonia might contain some N atoms as well as N_2 molecules.

Rayleigh tested these theories. He could not detect oxygen or hydrogen as impurities. Since he knew that ozone, O_3, could be made by passing a silent electric discharge through oxygen, he did the same thing with nitrogen, and then measured the density of the gas to see whether any N_3 had been formed. The density was exactly the same, and this undermined his confidence in the N_3 molecules theory.

His next line of investigation was to take 50 cm³ of air and pass an electric discharge through it. Nitrogen and oxygen combined to form nitrogen dioxide, which, being an acidic gas, could be absorbed in a solution of an alkali. He added more oxygen, and sparked the mixture again. Since air is 80% nitrogen, he had to keep on adding extra oxygen to convert all the nitrogen into nitrogen dioxide, which he absorbed in alkali. Eventually, of the 50 cm³ of air, 0.3 cm³ remained. This gas would not combine with oxygen: it was even less reactive than nitrogen.

Professor William Ramsey of University College, London, became interested in the problem. He passed nitrogen, which he had obtained from the air by the Rayleigh method, over red-hot magnesium. After the gas had been passed to and fro over the metal many times, the volume decreased to $\frac{1}{80}$ of the original volume. This residue of gas would not combine with magnesium as nitrogen did. It was 19.075 times more dense than hydrogen.

Ramsey did scores of experiments with the 'new gas' as he called it. He heated it with metals and non-metallic elements, he mixed it with other gases and passed electrical discharges through the mixtures. All his efforts had the same result: the 'new gas' would not combine with anything.

Sir William Crookes, the inventor of the discharge tube [§ 1.3], was asked to look at the emission spectrum of the gas [§ 2.1]. He was amazed to find that he could not identify the spectrum: it was different from the spectra of all the known elements. The 'new gas' was a new element!

Rayleigh and Ramsey discussed the new element. The density they had found for it gave it a relative atomic mass of 40. It was difficult to fit the new element into the Periodic Table. It seemed best to put it in between Cl = 35.5 and K = 39. This was one of the anomalies that led to the arrangement of elements in order of atomic number (proton number) rather than relative atomic mass.

Group	1	2	3	4	5	6	7	
Element A_r	Li 7	Be 9	B 11	C 12	N 14	O 16	F 19	
Element A_r	Na 23	Mg 24	Al 27	Si 28	P 31	S 32	Cl 35.5	? 40
Element A_r	K 39	Ca 40						

TABLE 16.2 Part of the Periodic Table

Rayleigh, who had started the work, and Ramsey, who had finished it, announced their discovery together at the British Association for the Advancement of Science in 1894. They claimed to have discovered a new element, which did not fit into any group of the Periodic Table. When the chairman of the meeting suggested calling the gas *argon* (Greek: *argos*, lazy) this was the name they adopted.

Ramsey went on to discover helium, neon, krypton and xenon. Their relative atomic masses and their lack of chemical reactivity placed them in a group with argon, and they formed a new Group 0 of the Periodic Table. They were called the **inert gases**, but are now generally called the **noble gases**.

CHECKPOINT 16A: ARGON

1. Why was Rayleigh looking for impurities in nitrogen?
2. A sequence of reactions which Ramsey did was

$$\text{Air} \rightarrow \text{Nitrogen} \xrightarrow{Mg} \text{Magnesium nitride}$$
$$\downarrow H_2O$$
$$\text{Nitrogen} \xleftarrow{CuO} \text{Ammonia}$$

Write equations for the last three reactions. Would the density of the nitrogen obtained at the end of this sequence of reactions be the same as Rayleigh's value for nitrogen from the air or nitrogen from ammonia?

3. What made Ramsey believe that he had discovered a new element?

QUESTIONS ON CHAPTER 16

1. The electron configurations of the noble gases were, until recently, thought to be perfectly stable. Explain how this belief was the starting-point in the first explanations of the nature of the chemical bond.
2. How did the discovery of the noble gases confirm people's acceptance of the Periodic Table? [See § 16.3.]
3. What shapes would you predict for the molecules KrF_2 and KrF_4?
4. Why do you think that it is more difficult for krypton to form compounds than for xenon? Would you expect argon to be more reactive or less reactive than krypton?
5. Write an account of the noble gases (inert gases) with particular reference to their discovery, their importance in the Periodic Classification, and their part in the development of ideas on bonding.

(NI 82)

17

HYDROGEN

17.1 OCCURRENCE

Hydrogen in the Sun

Hydrogen is the most abundant element in the universe. The Sun and the stars derive their energy from the nuclear fusion reaction

$$4\,^{1}_{1}H \rightarrow \,^{4}_{2}He + 2\,^{0}_{1}e \text{ (positron)} + \gamma \text{(radiation)}$$

There is very little free hydrogen in the Earth's atmosphere, but there is plenty of hydrogen on the Earth, combined as water and organic compounds.

17.2 MANUFACTURE

1. Hydrogen is manufactured from natural gas, which is largely methane. The process is illustrated in Figure 17.1.

FIGURE 17.1 Hydrogen from Natural Gas

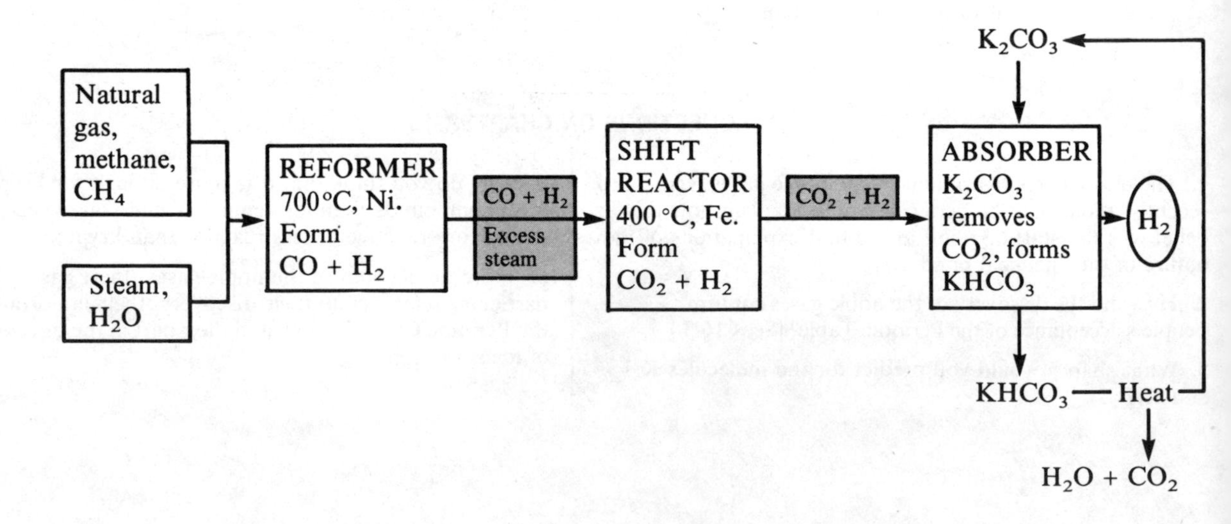

The reactions that take place are

Hydrogen is manufactured from natural gas...

Reforming: $CH_4(g) + H_2O(g) \rightarrow CO(g) + 3H_2(g)$

Shift Reaction: $CO(g) + H_2O(g) \rightarrow CO_2(g) + H_2(g)$

Absorption: $K_2CO_3(aq) + CO_2(g) + H_2O(l) \rightleftarrows 2KHCO_3(aq)$

...from brine...

2. Hydrogen is produced during the electrolysis of brine for the manufacture of sodium hydroxide [see Figure 18.4, § 18.5.5].

...from cracking of hydrocarbons

3. It is also a by-product of the cracking of hydrocarbons in a refinery [§ 26.3.2].

17.3 INDUSTRIAL USES

Figure 17.2 shows the main industrial uses of hydrogen.

FIGURE 17.2 Some Industrial Uses of Hydrogen

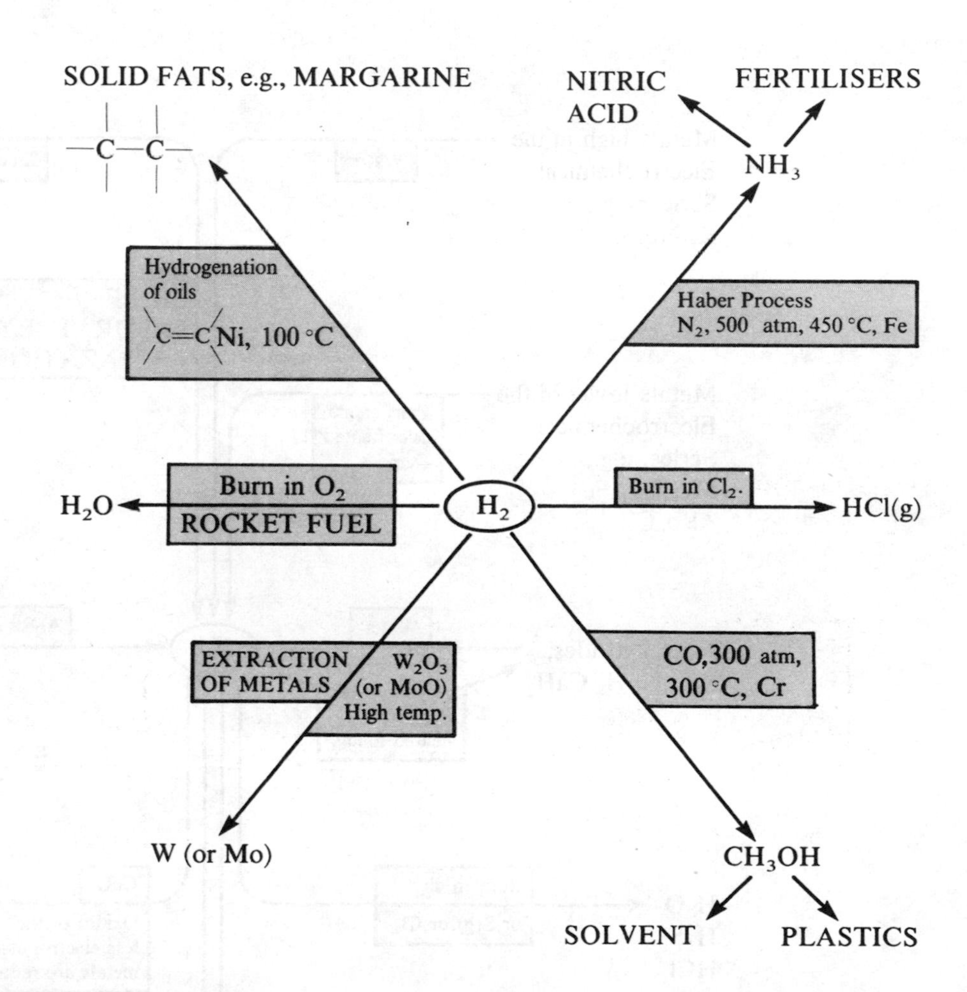

17.4 LABORATORY PREPARATION AND REACTIONS

Hydrogen is prepared in the laboratory from a metal + HCl(aq) or $H_2SO_4(aq)$

The most convenient laboratory preparation is the action of a dilute acid (not nitric acid) on zinc or another metal which is fairly high in the Electrochemical Series. Made in this way, hydrogen contains some hydrogen sulphide, which comes from the zinc sulphide present as an impurity in zinc.

Other laboratory preparations are shown in Figure 17.3. Also shown are some of the reactions of hydrogen. Equations for some of the reactions shown are:

$$2Na(s) + 2H_2O(l) \rightarrow 2NaOH(aq) + H_2(g)$$

$$Mg(s) + H_2O(g) \rightarrow MgO(s) + H_2(g)$$

$$CaH_2(s) + 2H_2O(l) \rightarrow Ca(OH)_2(aq) + 2H_2(g)$$

$$2Al(s) + 2OH^-(aq) + 6H_2O \rightarrow 2Al(OH)_4^-(aq) + 3H_2(g)$$

FIGURE 17.3 Some Reactions of Hydrogen

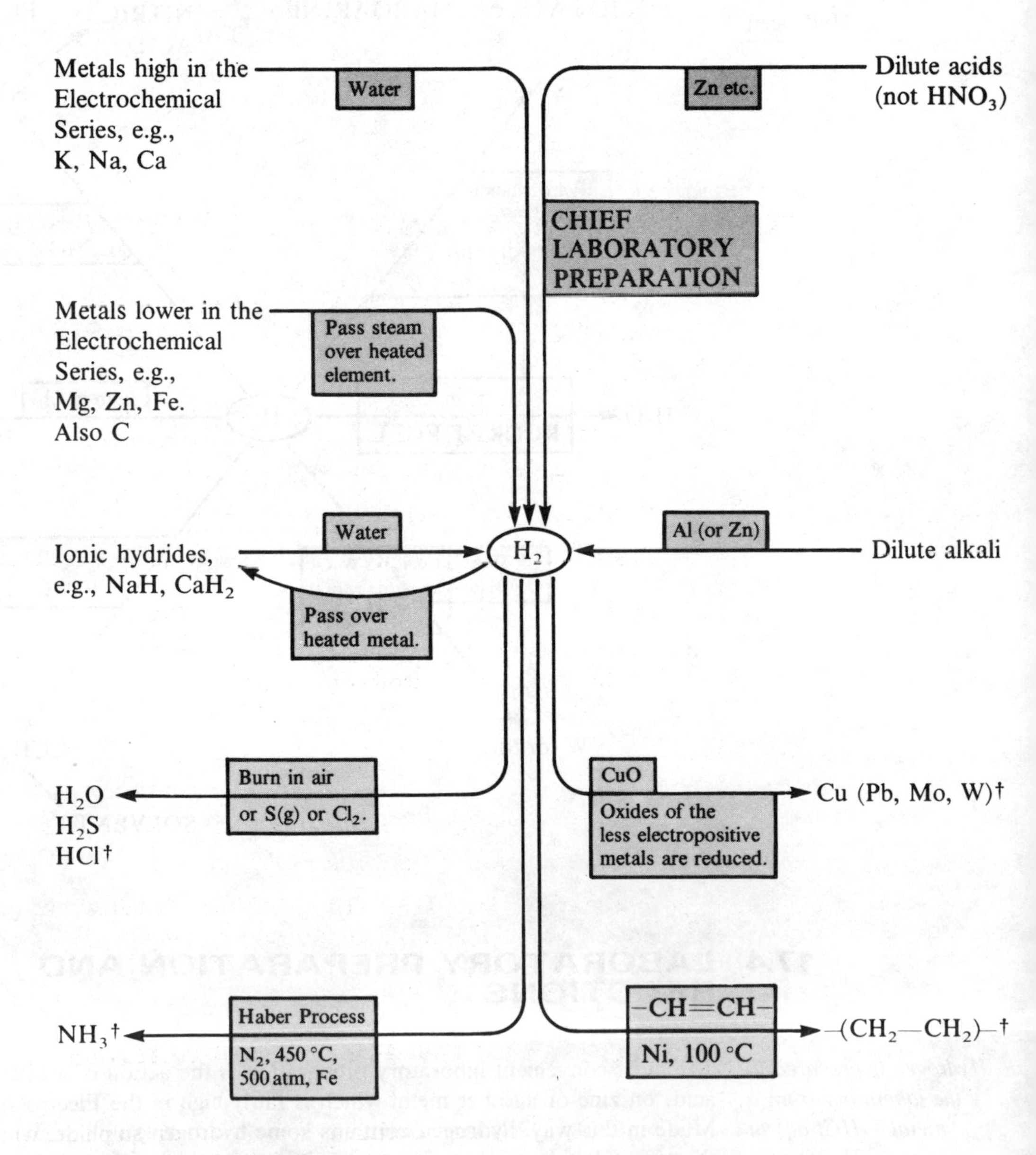

CHECKPOINT 17A: REACTIONS OF HYDROGEN

1. Write equations for the reactions between
(*a*) calcium and water
(*b*) zinc and dilute sulphuric acid
(*c*) magnesium and dilute hydrochloric acid
(*d*) zinc and sodium hydroxide solution to give the zincate ion, $Zn(OH)_4^{2-}$

2. Write equations for the reactions of hydrogen with
(*a*) ethene
(*b*) tungsten(III) oxide
(*c*) oxygen
(*d*) sulphur
(*e*) chlorine
(*f*) nitrogen

Which of these reactions are important in industry? What conditions are employed to give good yields?

3. 'Water gas' was an important fuel twenty years ago. It is a mixture of $CO + H_2$ formed by the reaction of steam and red-hot coke. Write an equation for its formation. Why do you think it is no longer produced? How might you obtain hydrogen from this mixture in the laboratory?

4. State the conditions under which hydrogen will react with the following, and name the products:
(*a*) AgO
(*b*) Pb_3O_4
(*c*) W_2O_3
(*d*) Al_2O_3
(*e*) CaO

17.5 HYDROGEN IONS AND HYDRIDE IONS

The formation of the oxonium ion, H_3O^+, is endothermic

The hydrogen atom, H ($1s^1$), can lose one electron to form a proton, H^+, or gain one electron to fill the outer shell in H^- ($1s^2$), or it can fill the outer shell by the formation of a covalent bond.

In aqueous solution, the proton exists as the hydrated oxonium ion, H_3O^+ (aq) [§ 12.7.1]. The oxonium ion is often referred to as a hydrogen ion, and its formula is often written as H^+(aq). The formation of H_3O^+(aq) is endothermic:

$$\tfrac{1}{2}H_2(g) + H_2O \rightarrow H_3O^+(aq); \quad \Delta H^\ominus = +320\,kJ\,mol^{-1}$$

The cation H_3O^+ is formed in combination with an anion, e.g.

$$HCl(aq) + H_2O(l) \rightarrow H_3O^+(aq) + Cl^-(aq)$$

The formation of an anion is usually exothermic. If this is the case, then the formation of $H_3O^+A^-$ may well be exothermic overall.

The formation of the hydride ion, H^-, is exothermic

The formation of H^- is exothermic:

$$\tfrac{1}{2}H_2(g) \rightarrow H^-(g); \quad \Delta H^\ominus = -150\,kJ\,mol^{-1}$$

The anion H^- is formed in combination with cations, e.g.

$$Ca(s) + H_2(g) \rightarrow CaH_2(s)$$

The formation of a hydride M^+H^- or $M^{2+}2H^-$ is exothermic only if the ionisation enthalpy of the cation is small. Only with metals in Groups 1 and 2 is this true.

17.6 REACTIONS OF HYDROGEN IONS

The reactions which characterise acids are those of the hydrogen ion, H_3O^+(aq) or H^+(aq). They are summarised in Figure 17.4.

FIGURE 17.4 Some Reactions of the Hydrogen Ion

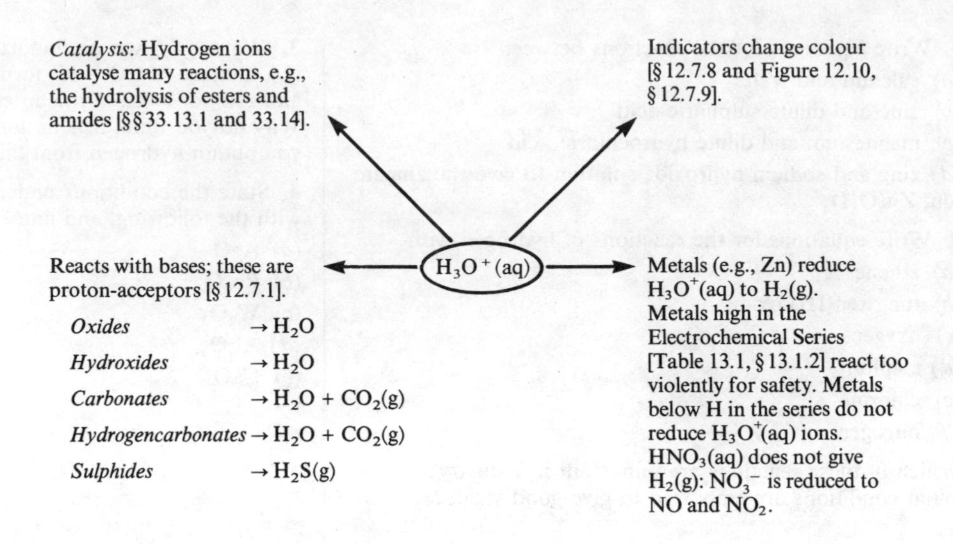

Catalysis: Hydrogen ions catalyse many reactions, e.g., the hydrolysis of esters and amides [§§ 33.13.1 and 33.14].
Indicators change colour [§ 12.7.8 and Figure 12.10, § 12.7.9].
$H_3O^+(aq)$
Reacts with bases; these are proton-acceptors [§ 12.7.1].
Oxides → H_2O
Hydroxides → H_2O
Carbonates → $H_2O + CO_2(g)$
Hydrogencarbonates → $H_2O + CO_2(g)$
Sulphides → $H_2S(g)$
Metals (e.g., Zn) reduce $H_3O^+(aq)$ to $H_2(g)$. Metals high in the Electrochemical Series [Table 13.1, § 13.1.2] react too violently for safety. Metals below H in the series do not reduce $H_3O^+(aq)$ ions. $HNO_3(aq)$ does not give $H_2(g)$: NO_3^- is reduced to NO and NO_2.

17.7 HYDRIDES

17.7.1 IONIC HYDRIDES

Ionic hydrides are formed by Group 1 and 2 metals

Ionic hydrides are formed by the metals of Groups 1 and 2. They are made by passing hydrogen over the heated metal. These hydrides are crystalline solids with structures similar to the corresponding halides. They react vigorously with water to give hydrogen, and burn vigorously in air:

The hydrides react with water and burn in air

$$H^-(s) + H_2O(l) \rightarrow H_2(g) + OH^-(aq)$$

$$2H^-(s) + O_2(g) \rightarrow O^{2-}(s) + H_2O(g)$$

Ionic hydrides, including complex hydrides, are reducing agents

In these reactions the hydride is acting as a reducing agent. Complex hydrides, such as lithium tetrahydridoaluminate, $Li^+[AlH_4]^-$, and sodium tetrahydridoborate, $Na^+[BH_4]^-$, are preferred as reducing agents because they are more stable. Both are valued reducing agents in organic chemistry. Sodium tetrahydridoborate is stable in cold water, but lithium tetrahydridoaluminate must be used in solution in dry ethoxyethane because it reacts with water.

Figure 17.5 shows where in the Periodic Table ionic hydrides are formed.

17.7.2 COVALENT HYDRIDES

Covalent hydrides are molecular and often gaseous

Covalent hydrides are molecular. Except for water and hydrogen fluoride, which are associated by hydrogen bonding, covalent hydrides are gaseous.

Hydrogen combines directly with all other non-metallic elements, but the yield of hydride is sometimes low (e.g., NH_3). Hydrolysis is often used to prepare hydrides, for example

Methods of preparing covalent hydrides

(*a*) Zinc sulphide + dilute acid → hydrogen sulphide, H_2S

(*b*) Calcium nitride + water or dilute acid → ammonia, NH_3

(*c*) Calcium carbide + water or dilute acid → ethyne, C_2H_2

(*d*) Magnesium silicide + dilute acid → silane, SiH_4

FIGURE 17.5 Hydrides

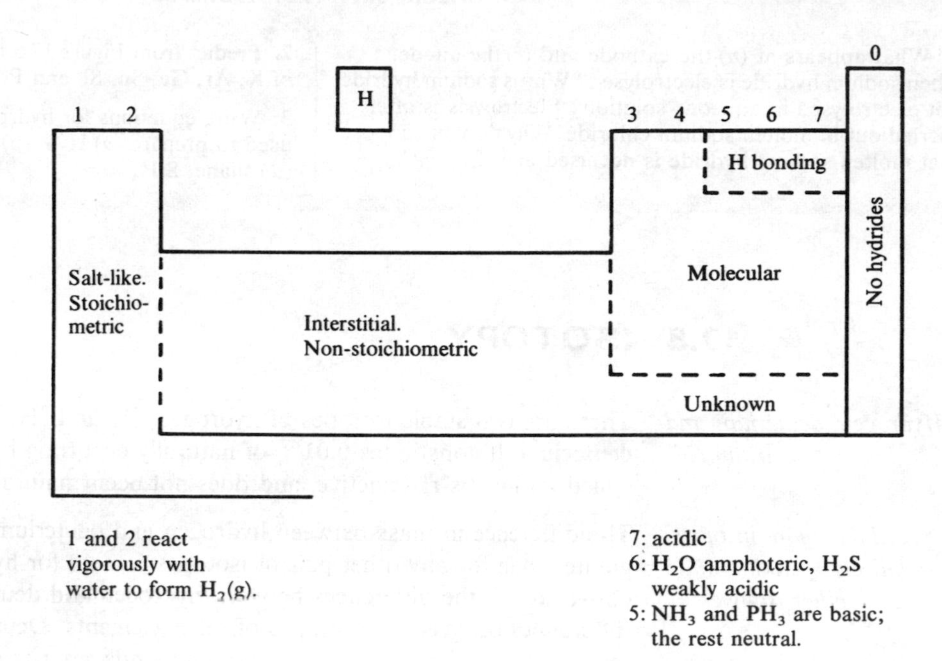

Table 17.1 summarises the hydrides of the elements in Periods 2 and 3 of the Periodic Table.

TABLE 17.1 The Properties of Hydrides

	Hydride	*Physical state*	*Reaction with air*	*Reaction with water*
Gp 1	LiH NaH	Solid Solid	Burn vigorously	React violently → $H_2(g) + OH^-(aq)$
Gp 2	BeH_2 MgH_2	Solid Solid	Burn vigorously	React → $H_2(g) + OH^-(aq)$
Gp 3	BH_3 AlH_3	Gas Gas	Burns Burns	Neutral solution Reacts → $H_2(g)$
Gp 4	CH_4 SiH_4	Gas Gas	Burns Ignites	Do not dissolve or react
Gp 5	NH_3 PH_3	Gas Gas	Burns in O_2 Ignites in air	Forms weak alkali Insoluble
Gp 6	H_2O H_2S	Liquid Gas	None Burns	— Forms weak acid
Gp 7	HF HCl	Liquid Gas	None None	Forms acid Forms strong acid

CHECKPOINT 17B: HYDRIDES

1. What appears at (*a*) the cathode and (*b*) the anode, when sodium hydride is electrolysed? Why is sodium hydride not electrolysed in aqueous solution? Electrolysis is often carried out in molten sodium chloride. Why do you suppose that molten sodium hydride is not used alone?

2. Predict from Figure 17.5 the properties of the hydrides of K, Ar, Ge, Sn, Se and Pd.

3. Write equations for hydrolysis reactions which can be used to prepare (*a*) H_2S, (b) NH_3, (*c*) ethyne, C_2H_2, and (*d*) silane, SiH_4.

17.8 ISOTOPY

Hydrogen, deuterium and tritium . . .

There are two stable isotopes of hydrogen, ${}^1_1H$ and ${}^2_1H$. The heavier isotope is called **deuterium**. It constitutes 0.01% of naturally occurring hydrogen. The isotope ${}^3_1H$, called **tritium**, is radioactive, and does not occur naturally.

. . . differ more in mass, relatively speaking, than other isotopes . . .

. . . There are some chemical differences between them

The difference in mass between hydrogen and deuterium is, relatively speaking, greater than for any other pair of isotopes (except for hydrogen and tritium). In consequence, the differences between hydrogen and deuterium are greater than the differences between the isotopes of other elements. Deuterium is usually less reactive than hydrogen. Deuterium compounds react at different rates from hydrogen compounds, sometimes differing by a factor of ten in rate. During the electrolysis of water, deuterium ions are discharged more slowly than hydrogen ions. As electrolysis proceeds, the water gradually becomes richer in deuterium. Pure D_2O, **heavy water**, can eventually be obtained. This is how deuterium oxide is obtained commercially. Other deuterium compounds are made from it [see Figure 17.6]. Deuterium oxide has T_m = 3.8 °C, T_b = 101.8 °C and density = 1.10 g cm^{-3}.

FIGURE 17.6 Routes to some Deuterium Compounds

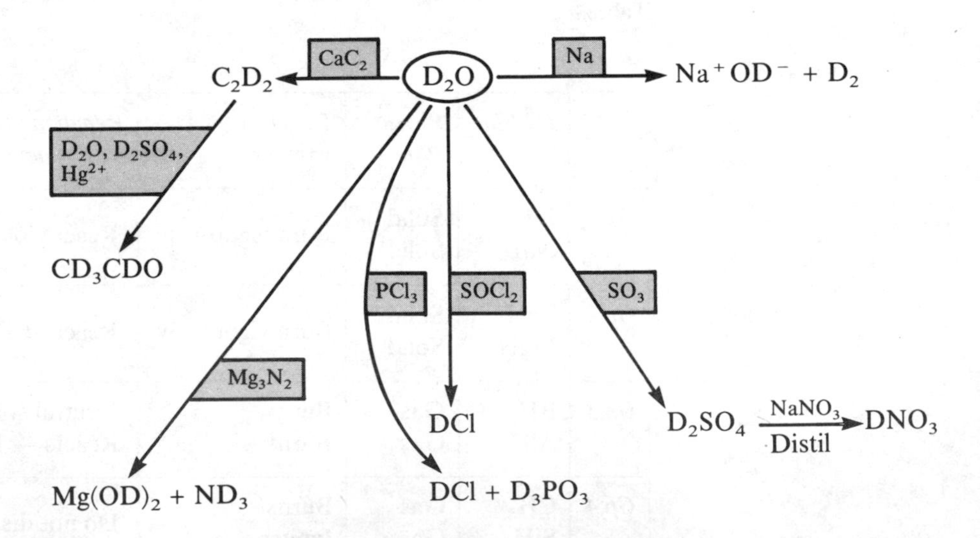

CHECKPOINT 17C: HYDROGEN

1. Explain the following statements:

1*a*) There is very little hydrogen in the Earth's atmosphere.

(*b*) Methane, CH_4, burns in air at 500 °C, whereas silane, SiH_4, ignites spontaneously in air.

2. In what ways is the chemistry of hydrogen similar to that of Group 1 and in what ways does it differ?

3. How does hydrogen (*a*) resemble and (*b*) differ from the halogens?

4. State how the following could be prepared from D_2O:

D_2, NaD, ND_3, DCl, C_2D_2

17.9 WATER

The physical properties of water

Some of the physical properties of water have been mentioned. These are: bond angle in H_2O [§ 5.2.5 and Figure 5.6 § 5.1.3], hydrogen bonding [§ 4.7.3], melting and boiling temperatures [§ 4.7.3], ice [§ 4.7.3] and the dissolution of organic solutes [§ 4.7.3].

The polar nature of water molecules

$$\overset{\delta_+}{H}-\overset{\delta_-}{O}-\overset{\delta_+}{H}$$

explains why water is a good solvent for ionic compounds. Figure 17.7 shows how a salt dissolves in water. More detail is given in § 10.8.

FIGURE 17.7 Dissolution of a Salt in Water

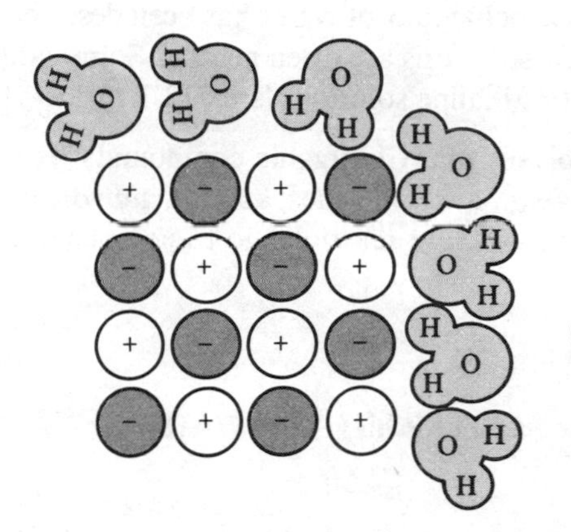

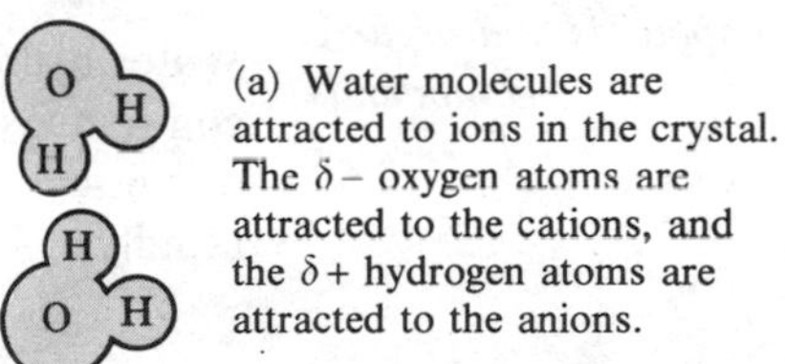

(a) Water molecules are attracted to ions in the crystal. The $\delta-$ oxygen atoms are attracted to the cations, and the $\delta+$ hydrogen atoms are attracted to the anions.

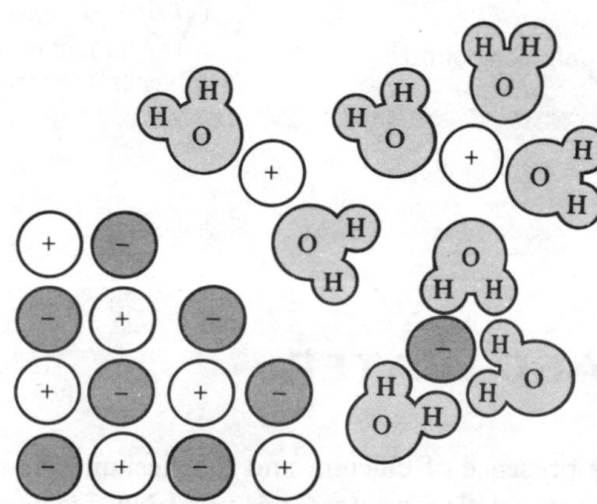

(b) As forces of attraction come into play, energy is given out. This compensates for the energy required to break up the crystal structure. Ions leave the structure.

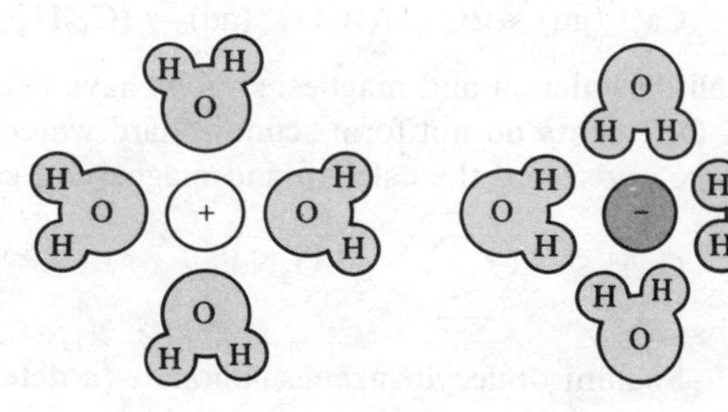

(c) These hydrated ions are called *aqua* ions. Many are surrounded by six water molecules.

FIGURE 17.8
The Aquasodium Ion, $[Na(H_2O)_6]^+$ or $Na^+(aq)$

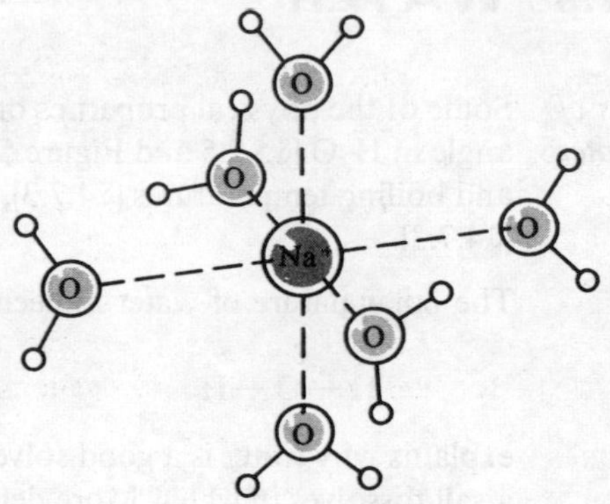

Aqua ions

Hydrated ions (ions surrounded by water molecules) are called **aqua ions**. Figure 17.8 shows the aquasodium ion, $[Na(H_2O)_6]^+$ or $Na^+(aq)$ for short.

Water hydrolyses some salts, covalent inorganic compounds and organic compounds

The acid–base behaviour of water has been described in § 12.7.1. When salts dissolve in water, their solutions are often neutral. Some salts, however, react with water to form acidic or alkaline solutions [see § 12.7.13].

Water hydrolyses many inorganic compounds (e.g., $SiCl_4$, PCl_3) and organic compounds (e.g., acid chlorides, amides, anhydrides and esters). Unlike the hydrolysis of salts, many of these reactions go to completion, sometimes slowly and sometimes rapidly.

CHECKPOINT 17D: WATER

1. Explain the following statements:
(*a*) Ponds freeze from the top downwards.
(*b*) Oil and water do not mix.
(*c*) Whisky and water do mix.
(*d*) Water is a good solvent for ionic compounds.
(*e*) Many classes of organic compounds dissolve in water.
(*f*) A solution of sodium sulphide has an unpleasant smell.
(*g*) It is dangerous to dissolve potassium cyanide in water.
(*h*) Aluminium sulphate solution is used in fire-extinguishers to generate carbon dioxide.

17.10 HARD WATER

Calcium and magnesium ions form 'scum' with soap

The presence of calcium and magnesium ions makes water hard. A soap such as sodium octadecanoate, $C_{17}H_{35}CO_2Na$, reacts with calcium and magnesium ions to form an insoluble 'scum' of calcium and magnesium octadecanoate:

$$Ca^{2+}(aq) + 2C_{17}H_{35}CO_2^-(aq) \rightarrow (C_{17}H_{35}CO_2)_2Ca(s)$$

Detergents lather, even in hard water

when all the calcium and magnesium ions have been precipitated, then the soap will lather. Detergents do not form scum in hard water. They are the sodium salts of sulphonic acids, and the calcium and magnesium salts are soluble:

$$C_{12}H_{25}-C_6H_4-SO_3Na$$

Sodium dodecylbenzenesulphonate (a detergent)

17.10.1 TEMPORARY HARDNESS

Temporary hardness is caused by the hydrogen carbonates of calcium and magnesium...

Limestone reacts with rain water that contains dissolved carbon dioxide to form the soluble salt calcium hydrogencarbonate:

$$CaCO_3(s) + H_2O(l) + CO_2(aq) \underset{\text{boil hard water}}{\overset{\text{limestone and rainwater}}{\rightleftharpoons}} Ca(HCO_3)_2(aq)$$

...and is removed by boiling

The reaction is reversed by boiling. When hard water containing calcium hydrogencarbonate is boiled, calcium carbonate is deposited, and the water becomes soft. Hardness caused by hydrogencarbonates is therefore called **temporary hardness**.

17.10.2 PERMANENT HARDNESS

Permanent hardness is caused by calcium and magnesium sulphates...

When water trickles over rocks containing calcium or magnesium sulphates, these minerals dissolve. The resulting hard water cannot be softened by boiling, and is described as **permanently hard water**. It can be softened by the following methods, which also remove temporary hardness.

WASHING SODA

...and is removed by washing soda...

Washing soda, $Na_2CO_3 \cdot 10H_2O$, is added to precipitate calcium and magnesium ions as insoluble carbonates.

ION EXCHANGE

Hard water can be softened by passing it slowly through a column containing sodium aluminium silicate (Permutit®). An exchange of ions takes place:

$$Ca^{2+}(aq) + 2NaAlSilicate(s) \underset{\text{Regenerating spent Permutit®}}{\overset{\text{Permutit® softening hard water}}{\rightleftharpoons}} 2Na^+(aq) + Ca(AlSilicate)_2(s)$$

When all the sodium ions in the Permutit® have been replaced by calcium and magnesium ions, the Permutit® must be regenerated. This is done by passing through it a concentrated solution of sodium chloride. After this solution has been washed out of the column, the Permutit® is ready for use. [See § 8.7.6 for the theory of ion exchange.]

CALGON

...and by complexing agents

Calgon® and similar products are the sodium salts of polyphosphate ions. They are able to form insoluble complexes with calcium ions and magnesium ions, releasing sodium ions as they do so.

17.10.3 ESTIMATION OF TOTAL HARDNESS

See § 12.7.17.

17.10.4 PURE WATER

Pure water

Pure water is made by distillation [see Figures 8.5, § 8.2 and 8.10, § 8.4.2]. Many laboratories find it more convenient to purify water by running it through a column containing ion-exchange resins [§ 8.7.6].

CHECKPOINT 17E: HARD WATER

1. Explain the following statements:
(*a*) Washing soda softens water.
(*b*) Washing soda makes water alkaline.
(*c*) Stalactites and stalagmites and the 'scale' in kettles are all formed by the same chemical reaction.
(*d*) Mixed ion-exchange resins do not neutralise each other. (I.e., $\mathbf{R}^+OH^- + H^+\mathbf{P}^- \rightarrow \mathbf{R}^+\mathbf{P}^- + H_2O$ does not happen.)
(*e*) Biochemists prefer to use distilled water in their work, rather than de-ionised water.

17.11 FLUORIDATION OF WATER

TOPIC

Tooth enamel consists of calcium hydroxide phosphate, $Ca_5(PO_4)_3OH$, also called calcium hydroxyapatite. In this ionic structure, an equilbrium exists:

$$Ca_5(PO_4)_3OH(s) + aq \underset{\text{Remineralisation}}{\overset{\text{Demineralisation}}{\rightleftharpoons}} 5Ca^{2+}(aq) + 3PO_4^{3-}(aq) + OH^-(aq)$$

In the environment of the mouth, calcium hydroxyapatite can be dissolved and reformed. The equilibrium lies to the left. If acids are present in the mouth, however, they react with hydroxide ions and favour demineralisation. Normally the pH of the mouth is 6.8. Within the plaque (a gelatinous mass of microorganisms) which coats teeth, the pH may be much lower. Sugar is the chief culprit because bacteria in the plaque convert sugars into acids.

Calcium hydroxide phosphate (calcium hydroxyapatite) in tooth enamel is attacked by acids. Fluoride ion can replace some of the hydroxide ion in hydroxyapatite. The fluoridated compound is less susceptible to attack by acids.

A relationship has been found between the extent of dental caries (decay) and the concentration of fluoride ion in the drinking water. It has been observed that in cities where the concentration of natural fluoride in the water is high the incidence of tooth decay is lower than average. Fluoride inhibits certain enzymes, such as those that catalyse the fermentation of sugars to lactic acid. Fluoride ion also substitutes for some of the hydroxide ion in calcium hydroxyapatite to form $Ca_5(PO_4)_3(OH)_{1-x}F_x$. The fluoridated hydroxyapatite is less easily attacked by acidic solutions than is hydroxyapatite.

Many people have objected to fluoridation as a kind of compulsory mass medication. Very high levels of fluoride cause damage to teeth. Long term exposure to high levels of fluoride can cause damage to bone, kidney and thyroid. A study by the Royal College of Physicians has concluded, however, that there is no risk to the individual or the environment from levels of fluoride up to 1 ppm. Many water authorities now add sodium fluoride to bring the level of fluoride in drinking water up to 1 ppm. The alternative of taking fluoride tablets is less effective because a large dose of fluoride is rapidly excreted.

CHECKPOINT 17F: FLUORIDATION OF WATER

1. Which can do the more damage, a chewy caramel which sticks between your teeth or a chocolate cream? Explain your answer.

2. Why do many people use toothpastes containing calcium fluoride?

3. How does fluoridation of drinking water benefit (*a*) the individual and (*b*) the Government?

4. Give two reasons why fluoridation of drinking water is more effective than issuing fluoride tablets.

QUESTIONS ON CHAPTER 17

***1.** Deuterium is an isotope of hydrogen. From your knowledge of the chemistry of hydrogen compounds, suggest methods for the preparation from D_2O of (*a*) D_2SO_4, (*b*) D_2S, (*c*) ND_3, (*d*) CD_4, (*e*) C_2D_2, (*f*) D_2, (*g*) DCl, (*h*) $Ca(OD)_2$.

2. 'In some ways, water is an abnormal liquid.' Give three properties which support this statement, and explain how these properties arise.

3. Distinguish between *hydration* and *hydrolysis* with examples. If you have started your study of organic chemistry, include some examples from this area.

†**4.** Discuss the acid–base reactions of water. (Do this question after studying Chapter 12.)

5. Describe with the aid of diagrams the bonding in (*a*) NaH, (*b*) CaH_2, (*c*) CH_4, (*d*) NH_3, (*e*) HCl, (*f*) HF. What reactions take place when these hydrides are added to water?

6. The enthalpies of formation of methane, octane, carbon dioxide and water are −75, −250, −394 and $-286\ \text{kJ mol}^{-1}$ respectively.

(*a*) Calculate the energy produced by burning 1 kg of each of hydrogen, methane and octane.

(*b*) Hydrogen is often suggested as a replacement fuel for either methane or octane on the grounds of being more environmentally friendly.

(i) Why should hydrogen be seen as a better fuel than methane or octane?

(ii) What environmental problems are likely to arise from the massive use of fuels such as methane and octane over the last century?

(iii) What disadvantages might there be in using hydrogen as the fuel for homes?

(iv) How might hydrogen be produced in an environmentally acceptable manner?

(Relative atomic masses: H = 1, C = 12, O = 16)

(O & C 90, AS)

7. (*a*) What is an isotope?

(*b*) How is deuterium oxide ^2H_2O (D_2O) obtained from water, and why is it not feasible to prepare tritium oxide, ^3H_2O (T_2O), in the same way?

(*c*) Tritium is made by the bombardment of lithium-6 with neutrons.

(i) From what other product must it be separated?

(ii) Suggest a method by which the separation might be achieved.

(*d*) When phosphinic acid, H_3PO_2, is dissolved in a large excess of deuterium oxide, the compound H_2DPO_2 is formed.

(i) How could the incorporation of one deuterium atom per molecule be shown to have taken place?

(ii) What deductions can be made about the structure of phosphinic acid?

(iii) If ammonium chloride had been treated in the same way with excess D_2O, what would be the formula of the product which could be isolated by evaporation to dryness?

Suggest a mechanism for what happens.

†(*e*) Starting from deuterium oxide, D_2O, and any appropriate organic compound, how could a sample of 1,2-dideuterioethane, $CH_2D—CH_2D$, be prepared?

[For (*e*) see Chapter 27.]

(O 91, S)

8. There are three isotopes of hydrogen.

Isotope	*Symbol*	*Stability*
Hydrogen	^1_1H	Stable
Deuterium	^2_1D	Stable
Tritium	^3_1T	Radioactive

Tritium emits low energy β-radiation.

(*a*) (i) Sketch the structure of a tritium atom showing subatomic particles.

(ii) Explain what is meant by β-radiation.

(*b*) The half-life of tritium is 12.35 years. How long will it take for 16 μg (microgram) of tritium to decay to 1 μg?

(*c*) Tritium is used to place a radioactive label in molecules. For instance, Cornforth has investigated the action of the enzyme *ethanol dehydrogenase* on ethanol labelled with tritium i.e. CH_3CHTOH.

(i) Explain why CH_3CHTOH is considered to be chiral.

(ii) Draw the two stereoisomers of CH_3CHTOH showing their three-dimensional nature.

(iii) Explain whether both of the stereoisomers drawn in (ii) will react with the enzyme.

(*d*) Tritium gas (T_2), or tritium oxide (T_2O), is used to introduce tritium into compounds. Write balanced equations to show how you could prepare the following compounds using T_2 or T_2O: (i) TCl, (ii) NT_3, (iii) $Ca(OT)_2$.

(NI 91)

9. There was considerable interest last year in the reported invention of a hydrogen-powered car engine which would seem to be considerably 'greener' than the petrol engine.

(*a*) Using the following enthalpies of combustion:

Hydrogen: $H_2 + \frac{1}{2}O_2 \rightarrow H_2O$; $\Delta H_c = -286\ \text{kJ mol}^{-1}$

Carbon: $C + O_2 \rightarrow CO_2$; $\Delta H_c = -394\ \text{kJ mol}^{-1}$

and the enthalpy of formation, ΔH_F, of octane (petrol) as $-250\ \text{kJ mol}^{-1}$, calculate the energy produced by burning 1 kg of octane, C_8H_{18}, according to the equation

$$C_8H_{18} + 17\tfrac{1}{2}O_2 \rightarrow 8CO_2 + 9H_2O$$

(*b*) Assuming that engines for motor cars could be designed to run smoothly and efficiently on either hydrogen or octane as fuels, give *one* possible advantage and *one* possible disadvantage for each of these fuels.

(*c*) (i) If we assume that the hydrogen to be used as fuel had to be produced by the electrolysis of water and the electricity was to be made by burning coke (to produce steam to drive turbines) how might this affect the plausibility of the argument for the hydrogen engine?

(ii) How might hydrogen be produced without the combustion of fossil fuels?

(*d*) (i) Give *two* possible advantages of making motor cars entirely out of plastics (polymers).

(ii) What developments would you consider necessary in the properties of polymers for it to be possible for the hydrogen-powered car to have an *engine* made entirely from plastics?

(O & C 92, AS)

18

THE s BLOCK METALS: GROUPS 1 AND 2

18.1 THE MEMBERS OF THE GROUPS

s block:
*Group 1; (core)*ns
*Group 2; (core)*ns^2

The s block metals are the metals in Group 1 and Group 2 of the Periodic Table. They are called the s block elements because they occupy an area of the Periodic Table following the noble gases, an area in which the s orbitals are being filled. The s block metals have much in common, and aluminium in Group 3 shares many of their properties. The s block metals are listed in Table 18.1.

TABLE 18.1 Physical Properties of s block Metals

Group 1				T_m/°C	T_b/°C	*AR*/nm	*IR*/nm	*IE*/kJ mol^{-1}	$E^{\ominus}$/V
Lithium	Li	3	1s^{2}2s	180	1330	0.15	0.06	519	−3.05
Sodium	Na	11	(Ne)3s	98	892	0.19	0.10	494	−2.71
Potassium	K	19	(Ar)4s	64	760	0.23	0.13	418	−2.93
Rubidium	Rb	37	(Kr)5s	39	688	0.24	0.15	402	−2.92
Caesium	Cs	55	(Xe)6s	39	690	0.26	0.17	376	−2.92
Francium	Fr	87	(Rn)7s	A radioactive element which is artificially made					
Group 2									
Beryllium	Be	4	1s^{2}2s^2	1280	2770	0.11	0.03	2660	−1.85
Magnesium	Mg	12	(Ne)3s^2	650	1110	0.16	0.07	2186	−2.37
Calcium	Ca	20	(Ar)4s^2	840	1440	0.20	0.10	1740	−2.87
Strontium	Sr	38	(Kr)5s^2	768	1380	0.21	0.11	1608	−2.89
Barium	Ba	56	(Xe)6s^2	714	1640	0.22	0.13	1468	−2.91
Radium	Ra	88	(Rn)7s^2	A radioactive element					

(Where T_m/T_b = melting/boiling temperature, AR = atomic radius, IR = ionic radius, IE = ionisation energy; for Group 2, the sum of the first and second ionisation energies, $E^{\ominus}$ = standard electrode potential.)

In these metals, the metallic bond is relatively weak

The members of Groups 1 and 2 are all metals. They are silvery coloured and tarnish rapidly in air. They show relatively weak metallic bonding because they have only one or two valence electrons. They differ in a number of ways from metals later in the Periodic Table:

1. They are soft: they can be cut with a knife.

2. Their melting and boiling temperatures are low.

3. They have low standard enthalpies (heats) of melting and vaporisation.

4. They have low densities. (Li, Na, K are less dense than water.) Group 2, with two valence electrons, show stronger metallic bonding, which is reflected in their physical properties.

The flame colours

The outer electron or electrons can be excited to a higher energy level. When they fall to a lower energy level, energy is emitted. For these metals, the energy is sufficiently low to have a wavelength in the visible spectrum [§ 2.1]. These elements therefore colour flames: Li—red, Na—yellow, K—lilac, Rb—red, Cs—blue, (Be—colourless), Mg—brilliant white, Ca—brick red, Sr—crimson, Ba—apple green.

Ionisation energies are low

The elements show constant oxidation numbers of +1 in Group 1 and +2 in Group 2. The ionisation energy required for the process

$$M(g) \rightarrow M^{n+}(g) + ne^-$$

is low. The s electrons are shielded from the attraction of the nucleus by the noble gas core and are easily removed. As the size of the atoms increases down the groups, the electrons to be removed become more distant from the nuclear charge, and the ionisation energy decreases.

Metals are reducing agents. The highly negative $E^{\ominus}$ values show that s block metals are powerful reducing agents

Metals are reducing agents. The power of s block metals as reducing agents is shown by the vigour with which they reduce water to hydrogen. The standard electrode potential, $E^{\ominus}$ [§ 13.1.1], measures the tendency for the reduction process

$$M^{n+}(aq) + ne^- \rightarrow M(s)$$

to occur. A highly negative value for $E^{\ominus}$ indicates that the reverse process will take place: the metal atoms will form ions and electrons.

FIGURE 18.1 Trends in Ionisation Energy and Standard Electrode Potential

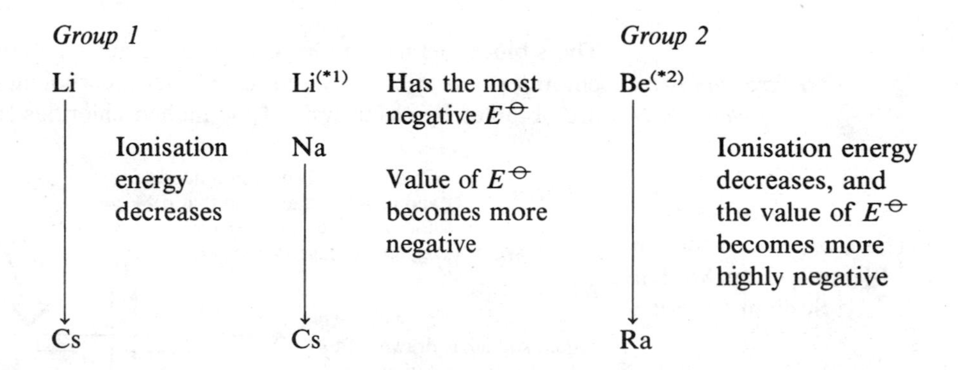

(*1) The small size of Li^+ gives it a high standard enthalpy of hydration. This is why Li^+ has the most negative value of $E^{\ominus}$ in the group. The small size of Li^+ enables it to polarise anions, and its compounds have some covalent character.

(*2) Be^{2+} is small and highly charged. It polarises anions, and beryllium compounds are mainly covalent.

18.2 USES

SODIUM

Sodium is used in some nuclear reactors

1. Molten sodium is used as a coolant in some types of nuclear reactor. Its high thermal conductivity and low melting temperature and the fact that its boiling temperature is much higher than that of water make sodium suitable for this purpose.

. . . in electrical circuits . . .

2. Sodium wire is used in electrical circuits for special applications. It is very flexible and has a high electrical conductivity. The wire is coated with plastics to exclude moisture.

. . . in lamps . . .

3. Sodium vapour lamps are used for street lighting.

. . . as a reducing agent . . .

4. Sodium amalgam and sodium tetrahydridoborate, $NaBH_4$ [§ 35.4], are used as reducing agents.

5. Sodium cyanide is used in the extraction of silver and gold.

BERYLLIUM

Beryllium is used for making containers for uranium-238 as it does not absorb neutrons, and therefore does not become radioactive.

MAGNESIUM

Magnesium is used in alloys and in flares

1. Magnesium is alloyed with aluminium to make Duralumin® [§ 19.2.1].

2. Magnesium is used as a sacrificial anode to prevent iron from rusting [§ 24.14.5].

3. The intense white light of burning magnesium is used in flares and distress signals.

18.3 OCCURRENCE AND EXTRACTION

Extraction is by electrolysis

The s block metals are too reactive to occur uncombined. Group 1 are found as chlorides. Group 2 are found as chlorides, carbonates and sulphates. The metals are obtained by electrolysis of the molten chlorides [see Figure 18.2].

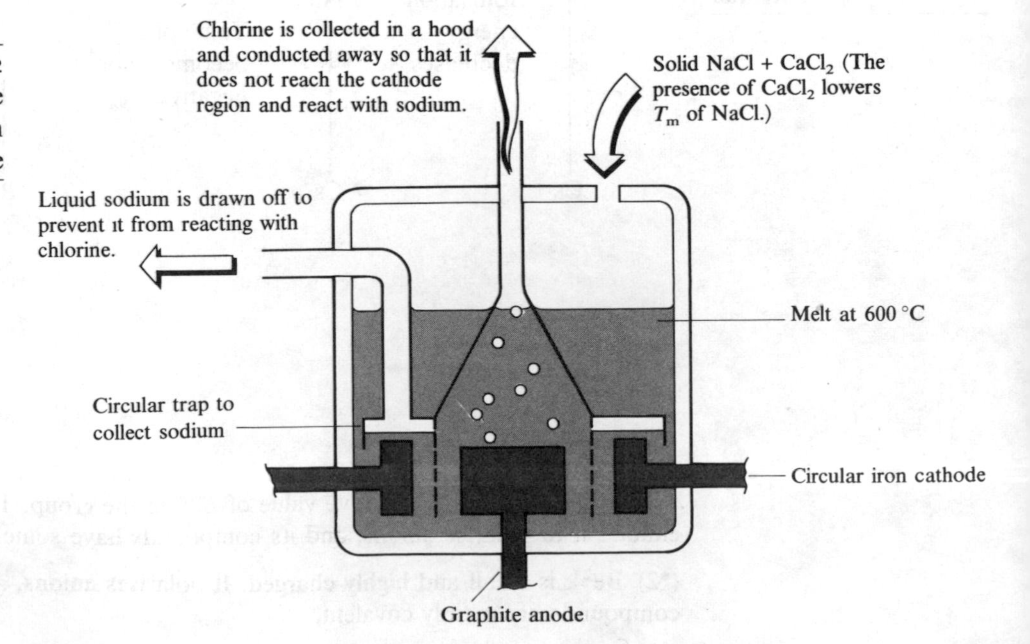

Chlorine is collected in a hood and conducted away so that it does not reach the cathode region and react with sodium.
Solid NaCl + CaCl₂ (The presence of CaCl₂ lowers Tₘ of NaCl.)
Liquid sodium is drawn off to prevent it from reacting with chlorine.
Melt at 600 °C
Circular trap to collect sodium
Circular iron cathode
Graphite anode

FIGURE 18.2 The Downs Cell for the Electrolysis of Molten Sodium Chloride

Chemical reducing agents cannot reduce the oxides of s block metals

Electrolysis is an expensive way of obtaining metals. When possible, a reducing agent, such as carbon monoxide or carbon, is employed to reduce a metal oxide to the metal. The metals of Groups 1 and 2 are themselves such powerful reducing agents that their oxides cannot be reduced by chemical reducing agents. The metallurgist must resort to electrolysis.

CHECKPOINT 18A: THE METALS

1. Refer to Table 18.1, §18.1.

(*a*) Considering Group 1, explain why the radius of each ion is smaller than that of the atom.

(*b*) Explain why the ionic radius increases from Li^+ to Cs^+.

(*c*) Explain why, although Na^+ and Mg^{2+} have the same electron configuration, the radius of Mg^{2+} is only 0.07 nm.

(*d*) Which of the ions listed in Table 18.1 is likely to have the highest enthalpy (energy) of hydration? Explain your choice.

2. Explain why Group 2 metals have higher boiling temperatures and melting temperatures than Group 1 metals. What other physical properties differ?

3. For the process

$$M(s) \rightarrow M^{n+}(aq) + ne^-$$

list the individual steps by which this process can be considered to occur [see Born–Haber cycle, § 10.7]. Name the enthalpy changes associated with each step.

4. Calcium chloride solution is the waste-product of the ammonia–soda process [§ 18.5.6]. Explain why calcium cannot be made by the electrolysis of this solution.

18.4 REACTIONS

18.4.1 REACTION WITH HYDROGEN

Hydrides contain H^-, except for those of Be and Mg

The highly electropositive metals in Groups 1 and 2 react with hydrogen to form **hydrides**, in which hydrogen is the anion, H^- [§ 17.7]. Beryllium and magnesium in Group 2 are less electropositive than the rest of the s block and form covalent hydrides.

18.4.2 REACTION WITH WATER

All (except Be and Mg) react with cold water...

Except for beryllium and magnesium, Groups 1 and 2 react with cold water to give hydrogen and the metal hydroxide. Group 1 metals are kept under oil to prevent them from reacting with water vapour in the air:

$$2Na(s) + 2H_2O(l) \rightarrow 2NaOH(aq) + H_2(g)$$

Magnesium reacts only slowly with cold water but with steam rapidly forms the oxide and hydrogen:

$$Mg(s) + H_2O(g) \rightarrow H_2(g) + MgO(s)$$

The vigour of the reaction with water is a good illustration of the increase in reactivity in passing down the groups. In short:

...Reactivity increases down each group

Lithium reacts slowly with water; violently with acids.

Sodium reacts vigorously with water; violently with acids.

Potassium reacts so vigorously with water that the hydrogen formed catches fire.

Rubidium and caesium react even more violently.

Beryllium does not react with water, but reacts rapidly with acids.

Magnesium reacts slowly with water, fast with steam and fast with acids.

Calcium undergoes a steady reaction with water and a vigorous reaction with acids.

Strontium and barium are more reactive than calcium and about equal in reactivity to lithium.

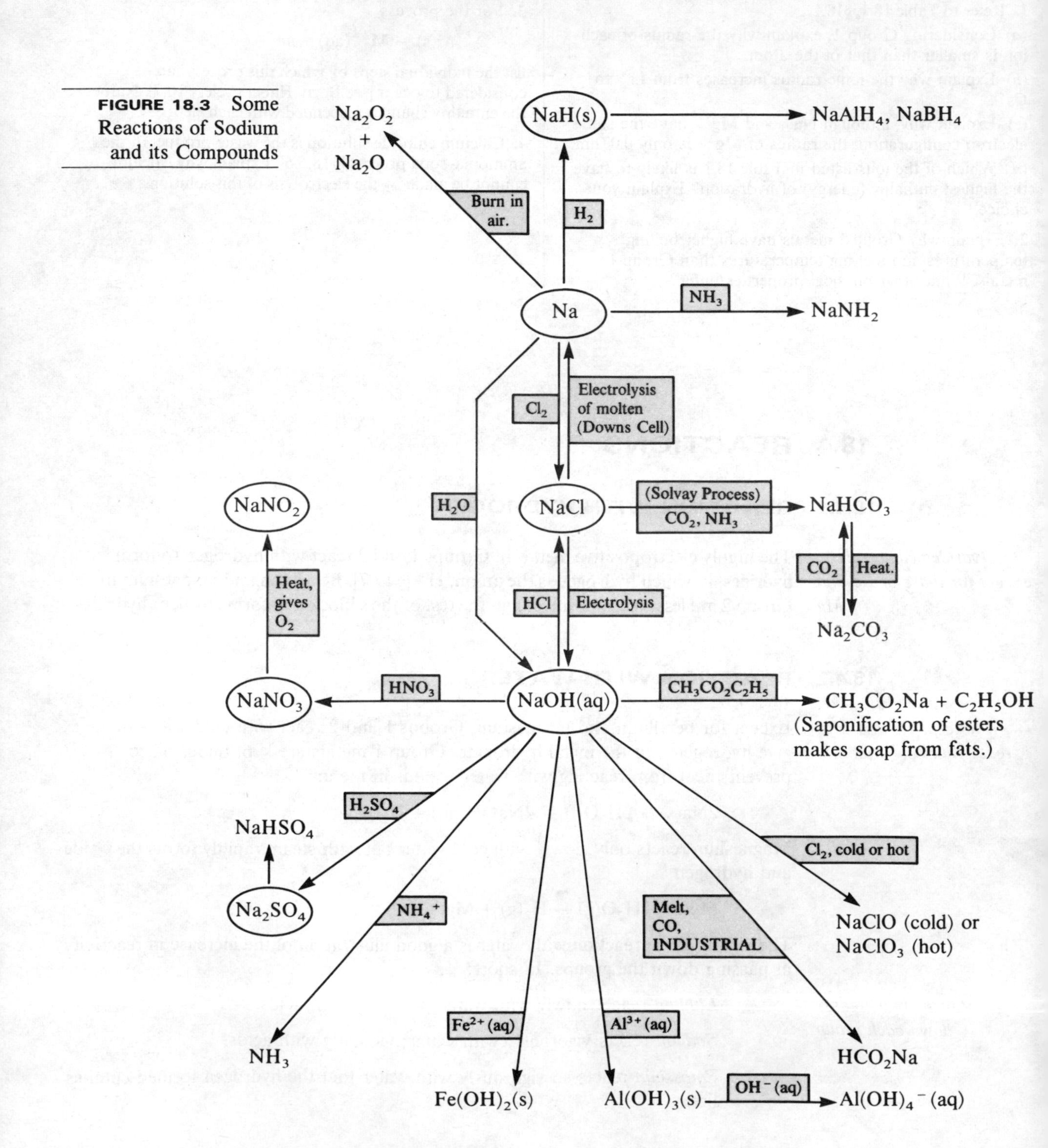

FIGURE 18.3 Some Reactions of Sodium and its Compounds

18.4.3 REACTION WITH NON-METALLIC ELEMENTS

All the s block metals react with oxygen. They tarnish in air as they soon become coated with a film of oxide. They burn readily in air and caesium and rubidium inflame spontaneously in air.

All react readily with oxygen and with the halogens...

They all react with halogens on heating to form halides.

...some react with nitrogen...

When heated in nitrogen, the metals that react are lithium in Group 1 and all of Group 2 except beryllium. Those nitrides are formed that have sufficiently high lattice enthalpies to compensate for the high ionisation enthalpies of the metal ions:

$$6Li(s) + N_2(g) \rightarrow 2Li_3N(s)$$

On addition of water, nitrides give ammonia:

$$Mg_3N_2(s) + 6H_2O(l) \rightarrow 3Mg(OH)_2(s) + 2NH_3(g)$$

...Group 2 react with carbon

Group 2 metals react with carbon at high temperatures to form carbides; Group 1 do not react. Calcium dicarbide, CaC_2, reacts with water to give ethyne [Figure 27.7, §27.10].

18.5 COMPOUNDS

18.5.1 IONIC CHARACTER

The ions are hydrated, not hydrolysed, and form few complex ions

Li and Be compounds have some covalent character

The salts of Group 1 have the highest amount of electrovalent character of any salts [§15.2]. The metal ions are hydrated in solution. The ions of Group 1 form very complex compounds. Being small with high charge densities, the ions Li^+ and Be^{2+} polarise anions, and form compounds with some degree of covalent character [§15.2]. The cations formed by the rest of the elements in the s block, being larger, have smaller charge densities and do not polarise anions appreciably. The Born–Haber treatment [§10.7] explains why these metals prefer to form electrovalent compounds.

18.5.2 SOLUBILITY

Solubilities are determined by two factors

Solubilities are not easy to explain because they are determined by two factors. Differences in ionic size have opposing effects on the two factors:

Small ions:
- *Factor 1* Crystal lattices are hard to break up. (The lattice dissociation enthalpy is highly endothermic.)
- *Factor 2* Much energy is released when the ions are hydrated. (The enthalpy of hydration is highly exothermic.)

GROUP 1 SALTS AND HYDROXIDES

The solubilities of salts and hydroxides in Group 1...

The salts and hydroxides of Group 1 are soluble because, since the cations are singly charged, the lattice dissociation enthalpies are not high. The only insoluble salts are $NaZn(UO_2)_3(CH_3CO_2)_9$ and $K_2Na[Co(NO_2)_6]$.

GROUP 2 SALTS AND HYDROXIDES

Compared with Group 1

(*a*) Lattice enthalpies are higher.

(*b*) Enthalpies of solvation are higher.

These two factors have opposing effects on solubility.

... and in Group 2

TABLE 18.2 Solubility of Group 2 Compounds

Sparingly soluble compounds			*Soluble salts*	
MSO_4 and MCO_3		*$M(OH)_2$*	*$M(NO_3)_2$*	
Mg Ca Sr Ba	↓ *Decrease* in solubility is due to the decrease in the enthalpy of hydration (*Note 1*).	↓ *Increase* in solubility is due to the decrease in the lattice dissociation enthalpy (*Note 2*).	Mg Ca Sr Ba	↓ *Increase* (Mg → Ca) ↓ *Decrease* (Sr → Ba)

Note 1 The lattice enthalpies do not vary much. Since in size

Anion ≫ Cation

the differences in cation size do not greatly affect the lattice enthalpy.

Note 2 This outweighs the change in the enthalpy of hydration.

18.5.3 THERMAL STABILITY OF COMPOUNDS

Lattice enthalpies depend on two factors

The thermal stability of an ionic solid is measured by its standard lattice enthalpy [§ 10.7]. This depends on two factors:

Standard lattice enthalpy
- *Factor 1* The greater the charges on the ions, the greater is the attraction between them, and the greater is the lattice enthalpy.
- *Factor 2* The smaller the ions, the more closely they can approach in the lattice, and the greater is the lattice enthalpy.

NITRATES

Nitrates...

Nitrates decompose when heated. Those of Group 1 (except Li) decompose to form nitrites (nitrates(III)) and oxygen:

$$2KNO_3(s) \rightarrow 2KNO_2(s) + O_2(g)$$

Nitrates of Group 2 and lithium decompose further to form the metal oxide, nitrogen dioxide and oxygen:

$$2Mg(NO_3)_2(s) \rightarrow 2MgO(s) + 4NO_2(g) + O_2(g)$$

Since the NO_2^- ion is smaller than the NO_3^- ion, the solid lattice of $Na^+NO_2^-$ is more stable than that of $Na^+NO_3^-$. The nitrite lattice is sufficiently stable to avoid further decomposition to the oxide. This is not the case for Group 2 nitrates. The

O^{2-} ion is smaller and more highly charged than the NO_3^- ion. Oxides $M^{2+}O^{2-}$ have high lattice enthalpies, and nitrates of Group 2 metals therefore decompose to form oxides.

CARBONATES AND HYDROGENCARBONATES

...carbonates and hydrogencarbonates...

The carbonates of Group 1 (except Li) are thermally stable. Those of Group 2 (and Li) decompose when heated to form the oxides:

$$CaCO_3(s) \rightarrow CaO(s) + CO_2(g)$$

The hydrogencarbonates of Group 1 are solids which decompose at 100 °C:

$$2NaHCO_3(s) \xrightarrow{100\,^\circ C} Na_2CO_3(s) + CO_2(g) + H_2O(g)$$

This is why sodium hydrogencarbonate is used in baking powder.

The hydrogencarbonates of Group 2 are less stable. They exist only in solution. When their solutions are boiled, they decompose to form the carbonates [§ 17.10.1].

HYDROXIDES

...hydroxides

Hydroxides follow the same pattern as carbonates. Those of Group 1 (except Li) are stable. Those of Group 2 and lithium are decomposed by heat to form oxides:

$$Mg(OH)_2(s) \xrightarrow{heat} MgO(s) + H_2O(l)$$

18.5.4 OXIDES

Normal oxides, peroxides and superoxides...

...all react with water...

...and all give $OH^-(aq)$

The s block metals form three kinds of oxides. These are (a) **normal oxides** containing the anion O^{2-}, (b) **peroxides** containing the anion O_2^{2-} and (c) **superoxides** (or **hyperoxides**) containing the anion O_2^-. All the oxides react with water to give solutions of hydroxide ions. Peroxides give hydrogen peroxide in addition.

$$O^{2-}(s) + H_2O(l) \rightarrow 2OH^-(aq)$$

$$O_2^{2-}(s) + 2H_2O(l) \rightarrow 2OH^-(aq) + H_2O_2(aq)$$

18.5.5 HYDROXIDES

Group 1 hydroxides are strongly basic and are soluble...

All Group 1 hydroxides are soluble [§ 18.5.2]. All (except LiOH) are **deliquescent**, i.e., they absorb water vapour from the air and dissolve in it. The strongly alkaline solutions give Group 1 their name of the **alkali metals**.

...Group 2 hydroxides are less soluble

The hydroxides of Ca, Sr and Ba dissolve in water [§ 18.5.2], but their lower solubilities make them weaker alkalis. The hydroxides of Be and Mg are insoluble in water. Beryllium hydroxide is amphoteric.

The manufacture of NaOH

The manufacture of sodium hydroxide by the electrolysis of brine is illustrated in Figure 18.4. With most electrodes, the electrolysis of brine yields hydrogen at the cathode and chlorine at the anode [§ 12.2]. With a mercury cathode, hydrogen has a high overvoltage which raises the negative potential required for the discharge of hydrogen ions. At the same time, the discharge potential of sodium ions is lowered because discharged sodium atoms combine with mercury to form an amalgam. The amalgam passes to a chamber where it can react with water to form sodium hydroxide and hydrogen, with the regeneration of mercury, which is returned to the electrolysis cell:

$$Na^+(aq) + e^- + Hg(l) \rightarrow Na \cdot Hg(l)$$

$$2Na \cdot Hg(l) + 2H_2O(l) \rightarrow 2NaOH(aq) + H_2(g) + 2Hg(l)$$

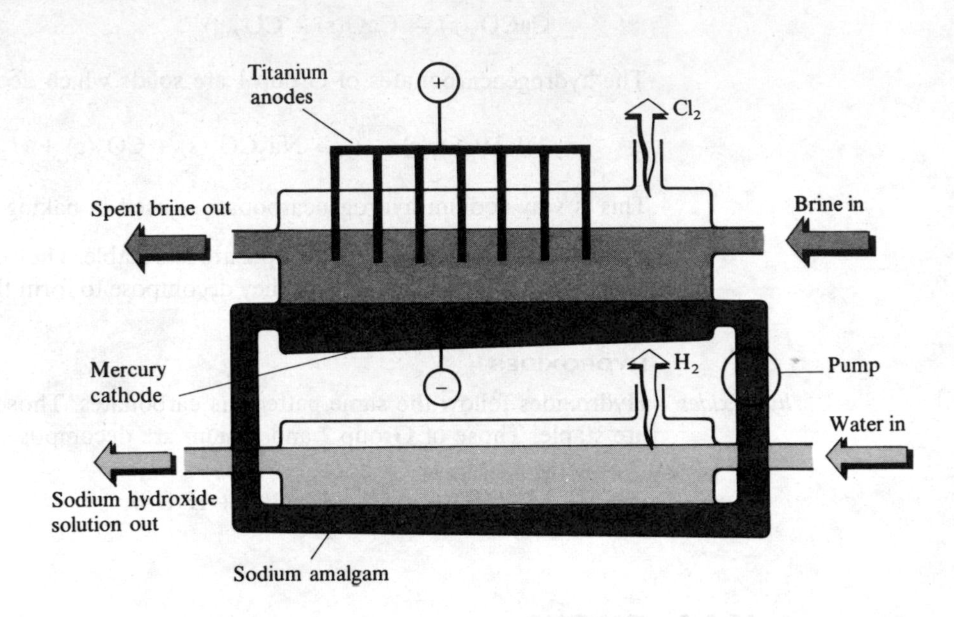

FIGURE 18.4(a) Manufacture of Sodium Hydroxide in the Mercury Cathode Cell (top) and Decomposer (bottom)

The reactions of alkalis are:

Alkalis react with acids...

1. They neutralise acids to form salts:

$$Na^+(aq)\,OH^-(aq) + H^+(aq)\,Cl^-(aq) \rightarrow Na^+(aq)\,Cl^-(aq) + H_2O(l)$$

...metal cations...

2. They precipitate insoluble metal hydroxides from solution:

$$Cu^{2+}(aq) + 2OH^-(aq) \rightarrow Cu(OH)_2(s)$$

3. They precipitate and then dissolve amphoteric hydroxides:

$$Al^{3+}(aq) + 3OH^-(aq) \rightarrow \underset{\text{Aluminium hydroxide}}{Al(OH)_3(s)} \xrightarrow{OH^-(aq)} \underset{\text{Aluminate ion}}{Al(OH)_4^-(aq)}$$

...with salts of weak bases...

4. They displace weak bases (ammonia and amines) from their salts:

$$NH_4^+(s) + OH^-(aq) \rightarrow NH_3(g) + H_2O(l)$$

...organic compounds, e.g., esters...

5. They hydrolyse esters and other organic compounds:

$$\underset{\text{Ethyl ethanoate}}{CH_3CO_2C_2H_5(aq)} + OH^-(aq) \rightarrow \underset{\text{Ethanoate ion}}{CH_3CO_2^-(aq)} + \underset{\text{Ethanol}}{C_2H_5OH(aq)}$$

...and with halogens

6. Reactions with halogens are covered in § 20.7; reactions with phosphorus in § 22.3.

FIGURE 18.4(b) Mercury Cathode Cells at ICI, Runcorn

TABLE 18.3 Uses of Hydroxides

Compound	*Uses*
Sodium hydroxide (Potassium hydroxide is similar.)	(*1*) Titrimetric analysis of acids [§ 3.14]. (*2*) Hydrolysis of organic compounds, e.g., esters [§ 33.13.1], amides [§ 33.14] and nitriles [§ 33.16]. **Saponification**, the hydrolysis of the esters of glycerol to give soaps [§ 33.13.3]. (*3*) Manufacture of sodium methanoate, HCO_2Na [§ 33.5.1].
Calcium hydroxide (Limestone is heated in a lime kiln to give calcium oxide, *quicklime*, which is 'slaked' with water to give *slaked lime*, $Ca(OH)_2$.)	Called *slaked lime*; the aqueous solution is called *limewater*. (*1*) Treatment of fields which have become too acidic for healthy plant growth. (*2*) Mortar = Slaked lime + Sand + Water The paste sets and then hardens as it reacts with carbon dioxide in the air to form calcium carbonate. (*3*) Manufacture of calcium hydrogensulphite. The paper industry needs this to remove lignin from wood and leave cellulose, ready to be made into paper. (*4*) Reaction with chlorine to form bleaching powder, $Ca(OCl)_2 \cdot CaCl_2$. This is a useful source of chlorine, which it liberates readily on treatment with acid.

18.5.6 CARBONATES

Thermal stability has been mentioned in § 18.5.3.

Hydrates of sodium carbonate

When sodium carbonate crystallises from solution, it appears as crystals of sodium carbonate-10-water, $Na_2CO_3 \cdot 10H_2O$. These crystals **effloresce** (give off water vapour) to form the monohydrate, $Na_2CO_3 \cdot H_2O$. When heated at 100 °C, the hydrates are converted into the anhydrous salt.

Sodium carbonate is used as a primary standard

Anhydrous sodium carbonate is used as a primary standard in titrimetric analysis [§§ 3.14 and 12.7.9]. Since it is deliquescent, anhydrous potassium carbonate cannot be used as a primary standard.

Commercial uses of sodium carbonate
Manufacture of sodium carbonate by the ammonia–soda process

Sodium carbonate crystals (washing soda) are used in softening hard water [§ 17.10.2]. Anhydrous sodium carbonate is required in great quantities by the glass and paper industries and for the manufacture of soaps and detergents. The manufacture of sodium carbonate by the ammonia–soda process or Solvay process is illustrated in Figure 18.6. The process uses the materials coke, limestone, sodium chloride and ammonia. It employs a clever recycling of materials so that the only by-product is calcium chloride. Although some of this is used as a drying agent, it is largely a waste-product.

The Solvay process begins with heating limestone in a kiln. The energy required is supplied by burning coke. Limestone and coke are mixed in a kiln, and the coke is ignited:

The raw materials are plentiful...

$$CaCO_3(s) \rightarrow CaO(s) + CO_2(g) \qquad [1]$$

The carbon dioxide produced is passed up the Solvay tower, down which trickles ammoniacal brine (a solution of ammonia in brine). Cooling coils in the tower remove the heat of reaction. This reaction occurs:

$$NaCl(aq) + CO_2(g) + NH_3(g) + H_2O(l) \rightarrow NaHCO_3(s) + NH_4Cl(s) \qquad [2]$$

...Clever recycling of materials reduces costs

The precipitated sodium hydrogencarbonate is filtered off and washed free of ammonium chloride. It is heated to give sodium carbonate and carbon dioxide, which is sent to the Solvay tower:

$$2NaHCO_3(s) \rightarrow Na_2CO_3(s) + CO_2(g) + H_2O(g) \qquad [3]$$

The calcium oxide produced in [1] is slaked:

$$CaO(s) + H_2O(l) \rightarrow Ca(OH)_2(s) \qquad [4]$$

Calcium hydroxide produced in [4] and ammonium chloride from [2] react to give ammonia, which is dissolved in brine, and calcium chloride:

$$2NH_4Cl(aq) + Ca(OH)_2(s) \rightarrow 2NH_3(g) + CaCl_2(aq) + 2H_2O(l) \qquad [5]$$

$KHCO_3$ is more soluble than $NaHCO_3$

§ 18.8.1 tells you more about the ammonia–soda process. Potassium carbonate is not made in the same way because potassium hydrogencarbonate is too soluble to crystallise at the bottom of the Solvay tower.

$MgCO_3$...
...$CaCO_3$ and its uses

Calcium and magnesium are mined as the carbonates, $MgCO_3$, *magnesite*; $MgCO_3 \cdot CaCO_3$, *dolomite*; and $CaCO_3$, *calcite* (Iceland spar); *marble*; *limestone*; *chalk* and *aragonite* in coral shells. Calcium carbonate is used in the ammonia–soda process, in the iron and steel industry, in the glass industry and in the manufacture of cement.

FIGURE 18.5 Reactions of Calcium and Magnesium

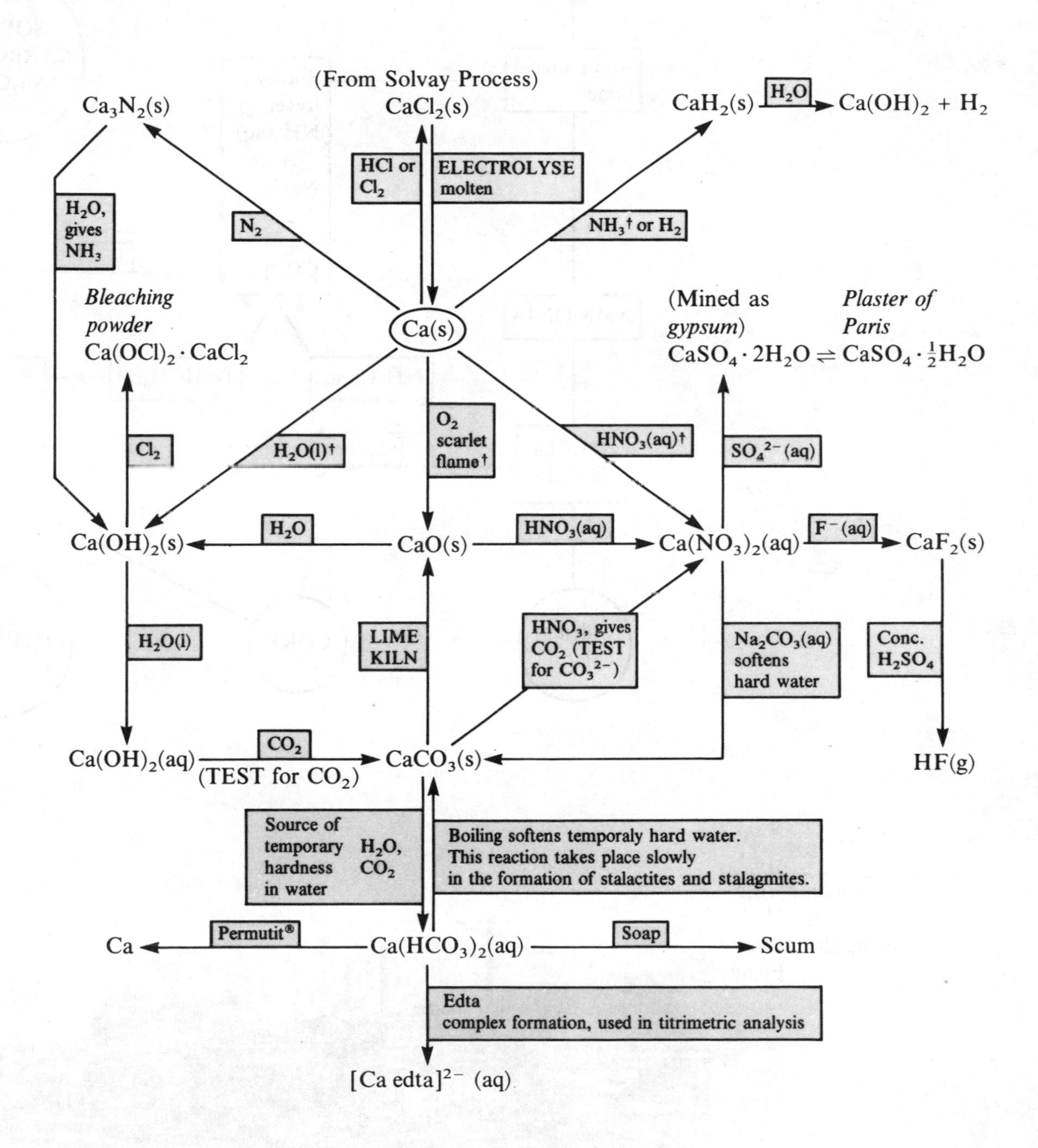

The reactions of magnesium are similar, but

† $Mg + NH_3 \rightarrow Mg_3N_2$

† Mg burns with a bright white flame.

† Mg reacts with steam to form MgO, but only slowly with water.

† Mg + cold, dilute $HNO_3(aq)$ gives $H_2(g)$, not $NO_2(g)$.

FIGURE 18.6
The Solvay Process for the Manufacture of Sodium Carbonate

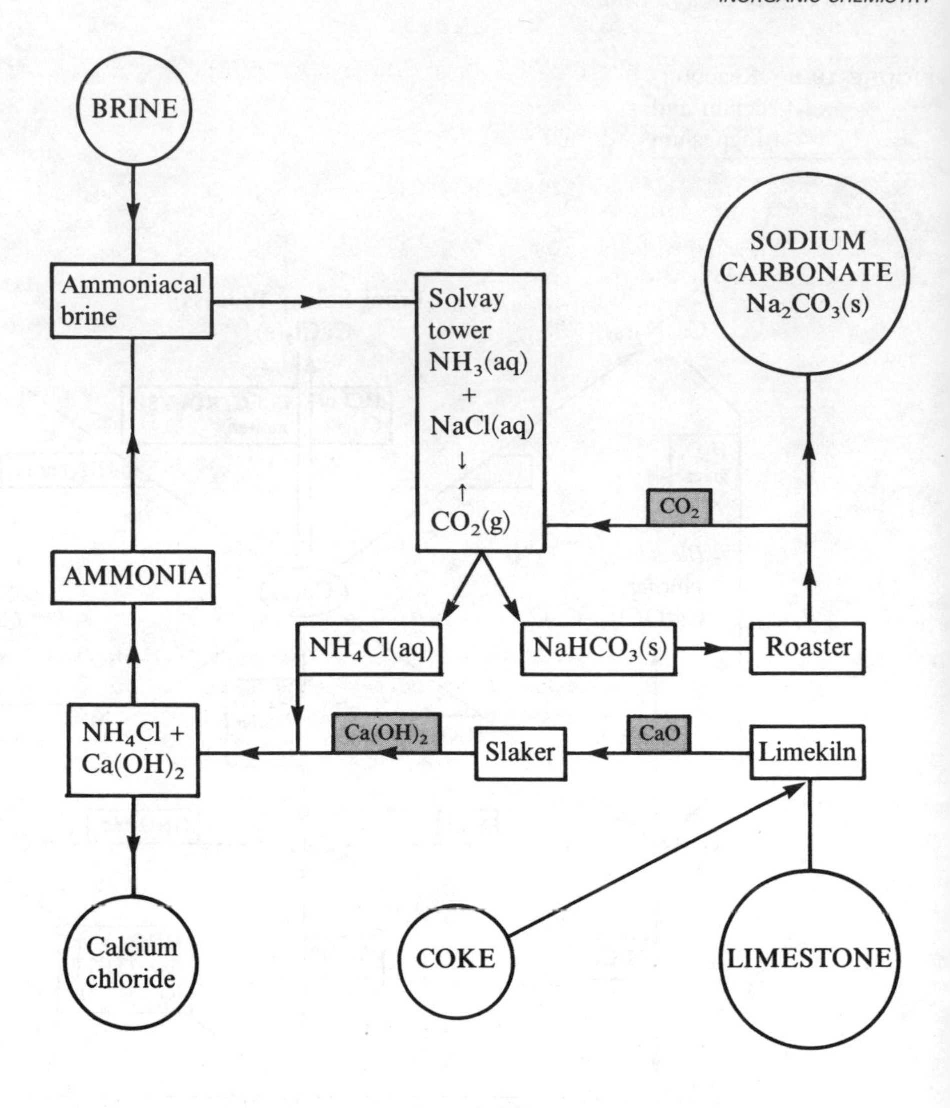

FIGURE 18.7
Winnington ICI, including the Solvay Plant

CHECKPOINT 18B: SOME COMPOUNDS OF s BLOCK METALS

1. (*a*) Explain why CaO(s) has a higher standard lattice enthalpy than $CaCO_3$(s).

(*b*) Explain why $MgCO_3$(s) is more easily decomposed by heat than $BaCO_3$(s).

2. A pellet of sodium hydroxide weighing 0.254 g was left to stand in the air. It changed to a colourless liquid and then to colourless, transparent crystals. After some days it formed a white solid, weighing 0.394 g. Explain the changes that have occurred.

3. Explain why $Mg(HCO_3)_2$ and $Ca(HCO_3)_2$ make water hard, and how the water can be softened.

4. What is the net reaction in the Solvay process? Comment on the cost and availability of the raw materials used. What techniques are used to keep running costs to a minimum?

Explain why the presence of NaCl reduces the solubility of $NaHCO_3$ in the Solvay tower [see § 12.7.15].

18.5.7 HALIDES

Halides...

Common salt, NaCl, is mined, e.g., in Cheshire. It is used in the manufacture of sodium [§ 18.3], sodium carbonate [§ 18.5.6], sodium hydroxide [§ 18.5.5] and sodium chlorate(I) and sodium chlorate(V) [§ 20.7].

Calcium chloride is the by-product of the ammonia–soda process. It is deliquescent, and is used as a drying agent except for ammonia and ethanol, with which it forms complexes. It crystallises as $CaCl_2 \cdot 6H_2O$.

Potassium bromide is used as a sedative and also in the production of silver bromide for photographic films.

18.5.8 NITRATES

...nitrates...

Sodium nitrate and potassium nitrate are mined. They are used as fertilisers.

The nitrates of Group 2 are made by the reaction of dilute nitric acid on the metal, metal oxide, hydroxide or carbonate.

The thermal stability of nitrates has been covered in § 18.5.3.

18.5.9 SULPHATES

...sulphates

SODIUM SULPHATE
Na_2SO_4

Prepared by titrating aqueous sodium hydroxide with sulphuric acid to pH 7. Crystallises as $Na_2SO_4 \cdot 10H_2O$.

$NaHSO_4$

If a second, equal volume of sulphuric acid is added, the hydrogensulphate is formed. It is acidic, dissociating partially to give H^+(aq) and SO_4^{2-} ions.

MAGNESIUM SULPHATE
$MgSO_4$

Mined as $MgSO_4 \cdot 7H_2O$.
Present in tap water, causing permanent hardness.
Used as the laxative Epsom salts.

CALCIUM SULPHATE
$CaSO_4$
$CaSO_4 \cdot 2H_2O$
$(CaSO_4)_2 \cdot H_2O$

Mined as *anhydrite*, $CaSO_4$, and as *gypsum*, $CaSO_4 \cdot 2H_2O$.
The latter gives plaster of Paris when heated at 100 °C:

$$2CaSO_4 \cdot 2H_2O(s) \underset{\text{Add water}}{\overset{100\,^{\circ}\text{C}}{\rightleftarrows}} (CaSO_4)_2 \cdot H_2O(s) + H_2O(l)$$

Calcium sulphate-2-water   Plaster of Paris

On the addition of water, plaster of Paris expands slightly and sets to form calcium sulphate-2-water.
Calcium sulphate is used to make the fertiliser ammonium sulphate.
Present in tap water, it is a cause of permanent hardness [§ 17.10.2].

18.6 HYDROLYSIS OF SALTS

Sodium carbonate solutions are strongly alkaline

The larger cations of Groups 1 and 2 are hydrated in solution. Their salts do not undergo hydrolysis unless they are the salts of weak acids, e.g., H_2S, HCN, H_2CO_3 [§§ 12.7.13 and 17.9]. The strongly alkaline nature of sodium carbonate solutions gives them their detergent power. When a solution of sodium carbonate is added to a solution of copper(II) sulphate, a precipitate of basic copper(II) carbonate, $CuCO_3 \cdot Cu(OH)_2 \cdot xH_2O$, is obtained. A less strongly alkaline sodium hydrogencarbonate solution must be used to give a precipitate of copper(II) carbonate, $CuCO_3$.

Mg^{2+} salts are hydrolysed

When magnesium chloride solutions are evaporated, they do not give anhydrous magnesium chloride. Hydrolysis occurs, with the formation of the basic chloride:

$$MgCl_2(aq) + H_2O(l) \rightleftharpoons Mg(OH)Cl(s) + HCl(aq)$$

If evaporation is carried out in a stream of hydrogen chloride, the equilibrium is reversed, and magnesium chloride can be obtained.

The hydrolysis of aluminium salts is mentioned in § 19.2.4 and that of transition metal salts in § 24.14.6.

18.7 COMPARISON OF LITHIUM WITH MAGNESIUM

The diagonal relationship between lithium and magnesium

The reasons for this diagonal relationship are outlined in § 15.2. Lithium ions, Li^+, being extremely small, are able to polarise anions and give compounds a high degree of covalent character. The features in which lithium resembles magnesium and differs from other alkali metals are

1. the action of heat on the hydroxide, carbonate and nitrate

2. the formation of a nitride

3. the absence of a peroxide.

CHECKPOINT 18C: SALTS

1. Explain the following statements:

(*a*) Sodium carbonate solutions are alkaline.

(*b*) Calcium chloride is the waste product of the Solvay Process.

(*c*) Calcium sulphate is a compound with (i) medical and (ii) agricultural importance.

(*d*) In limestone regions, tap water is hard.

(*e*) Sodium forms two sulphates.

(*f*) Beryllium chloride solutions are acidic.

(*g*) Beryllium chloride is a covalent substance in the vapour state.

18.8 TWO INDUSTRIES BASED ON SALT

18.8.1 THE AMMONIA–SODA PROCESS

TOPIC

The ammonia–soda process [§ 18.5.6] was established in the UK in 1872. John Brunner and Ludwig Mond acquired the right to make sodium carbonate by the process which had been patented by Alfred and Ernest Solvay in Belgium in 1861. The Brunner–Mond plant, built at Winnington in Cheshire [see Figure 18.8], is now a part of ICI. At nearby Northwich, Middlewich and Nantwich, there are vast salt deposits underground. Salt was mined to such an extent that Northwich was suffering from subsidence by 1890. Another method of extracting salt is employed now. A hole is drilled in the ground, and water is pumped in. After the water has been left underground long enough to become saturated with salt, brine is pumped out. Pillars of salt are left intact at intervals. The holes are left full of saturated brine so that the ground above will not subside.

The brine that is pumped out of the ground contains sulphates. These need to be removed before the brine goes to the ammonia–soda plant. 'Milk of lime', a suspension of calcium hydroxide, is added to precipitate sulphate ion as calcium sulphate:

$$Na_2SO_4(aq) + Ca(OH)_2(aq) \rightarrow CaSO_4(s) + 2NaOH(aq)$$

The solid calcium sulphate is dumped down a bore hole into one of the underground brine-filled caverns.

The ammonia–soda process discharges no pollutant gases into the air. The only by-product is calcium chloride. Some of this is sold for use as a drying agent. Most of it is waste. It is discharged into a short stretch of river which flows into the sea. The amount of chloride ion dumped in the sea annually is negligible compared with that present in sea water. Calcium ion is constantly added to sea water by the action of rainwater on limestone [§ 17.10.1]. Marine creatures remove calcium ion from the sea to build their shells. The contribution which the ammonia–soda process makes to the calcium content of the sea is negligible.

In Japan, the ammonia–soda process is operated to produce sodium carbonate, ammonium chloride and calcium oxide. Reaction [5] in § 18.5.6 is eliminated.

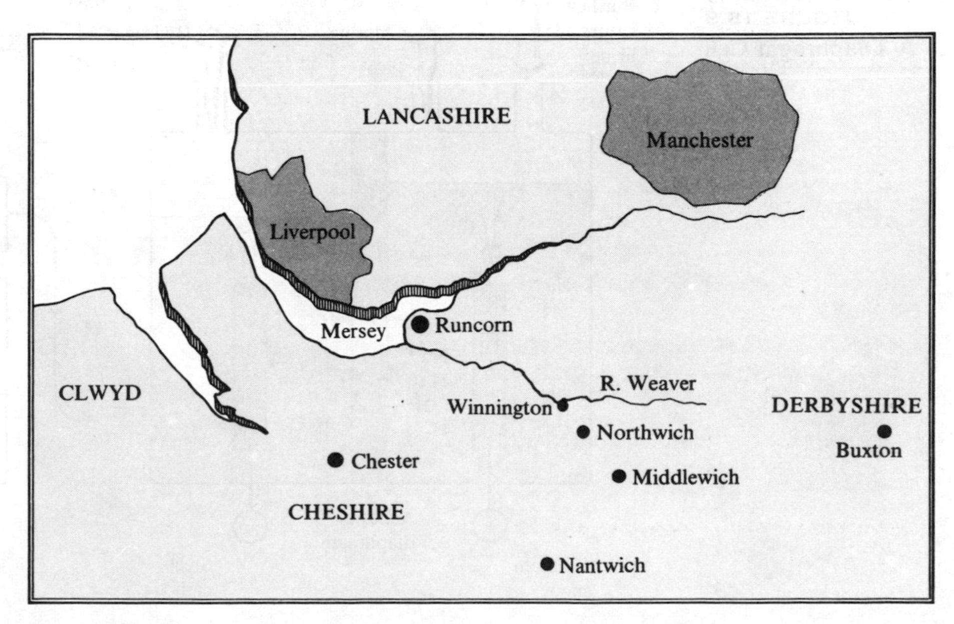

FIGURE 18.8 Winnington, Runcorn and Environment

Ammonium chloride is a good fertiliser for rice. It is not used as a fertiliser on dry land crops because it leads to a build-up of chloride ion, which crops will not tolerate. Rice is grown in paddy fields which are continuously irrigated by large volumes of water. Ammonium ions are absorbed by the plants, and chloride ions are carried away by the flow of water. The calcium oxide produced is sold for use in the iron and steel industry and in the cement industry.

Japan has no salt deposits. It obtains salt from Australia. With plenty of sunshine and large areas of land available, Australia uses the method of solar evaporation to obtain salt from sea water. Japan also has access to sea water and has a sunny climate, but the rainfall is too high for solar evaporation to yield salt from sea water.

18.8.2 THE CHLOR–ALKALI INDUSTRY

Mercury cathode cells [see Figure 18.4, § 18.5.5] produce a large fraction of the chlorine and caustic soda that industry uses. ICI has a chlor-alkali plant at Runcorn [see Figure 18.8]. There are disadvantages to the process. Mercury is expensive, and it is poisonous. Although mercury is recirculated, some mercury escapes with the spent brine. This converts it into the soluble salt, mercury(II) chloride. It has been the practice to discharge the effluent into lakes and rivers. High levels of mercury build up in fish, which take in the mercury compounds but cannot excrete them. In 1970, 700 lakes in Canada and the USA were closed to fishermen because of the high mercury content of fish caught there. It was feared that the Minamata tragedy might be repeated. In 1958, 100 people died and thousands were maimed in Minamata in Japan. They had eaten fish with a very high mercury content caught in Minamata Bay. The source of mercury there was the effluent from a PVC plant, which used mercury(II) chloride as a catalyst in the polymerisation of ethene [§ 27.7.10]. The mercury in the North American lakes was traced to the chlor-alkali industry.

The industry immediately took steps to reclaim mercury from the effluent by precipitation as mercury(II) sulphide. Since this date, a larger share in the production of chlorine and sodium hydroxide and hydrogen has been taken by two types of electrolytic cell which do not use mercury. These are the diaphragm cell and the membrane cell. The USA chlor-alkali industry employs 2 diaphragm cells for every mercury cell. The UK employs 1 diaphragm cell for every 20 mercury cells.

FIGURE 18.9
A Diaphragm Cell

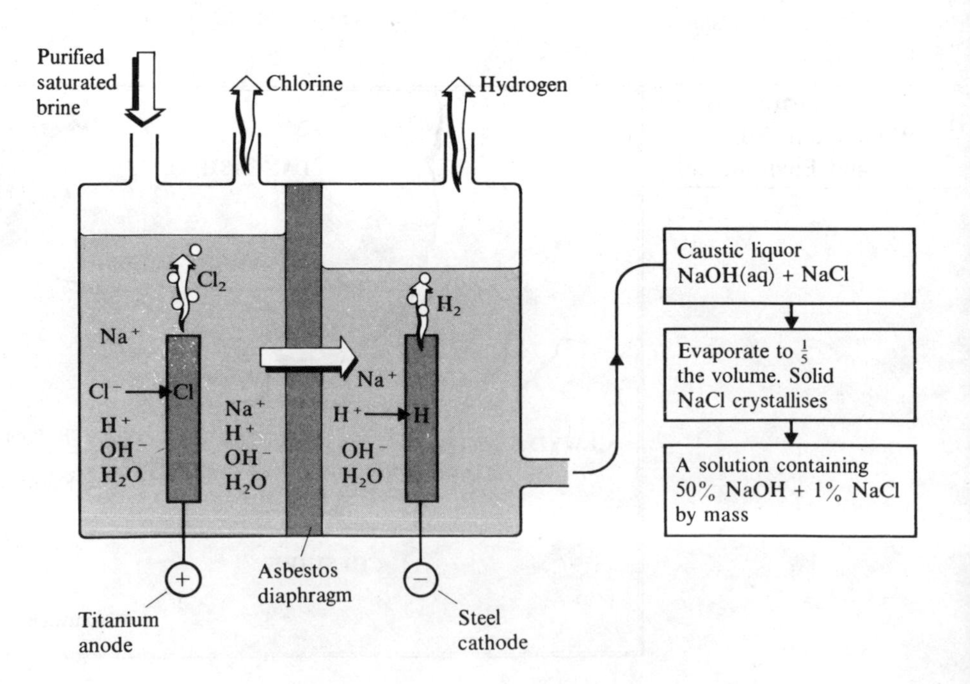

In the **diaphragm cell**, a porous asbestos diaphragm separates the cathode and anode. Purified saturated brine is fed into the anode compartment. Purification is necessary to remove calcium and magnesium ions which would precipitate as insoluble hydroxides and block the pores of the diaphragm. The level of liquid in the anode compartment is higher than that in the cathode compartment so that brine will seep through the diaphragm. The cell reactions are shown in Figure 18.9.

A newer type of cell is the **membrane cell**. The electrode reactions are the same as those in the diaphragm cell. The electrodes are separated by an ion-exchange membrane. Since chloride ions cannot pass through this membrane, the sodium hydroxide formed in the cathode compartment is free from sodium chloride. It is sold as a 35% by mass solution of NaOH or evaporated to give a 50% solution or solid sodium hydroxide.

Installations contain 50 to 100 single cells joined in series. The advantages of each type of cell are summarised in Table 18.4.

TABLE 18.4 The Cells used for the Electrolysis of Brine

Type	*Construction*	*Operation*	*Product*
Mercury	Expensive to construct and fill	Mercury must be reclaimed from any effluent to avoid pollution.	High-purity NaOH(aq) is 50% by mass NaOH.
Diaphragm	Simpler and lower in cost	The diaphragm is replaced frequently. No pollution	Must be evaporated to remove NaCl and to concentrate the product
Membrane	Simple, low cost. Single cells are small and easily transported.	Needs brine of high purity. The membrane lasts for 2–3 years. No pollution	High-purity product. Evaporation increases the concentration.

CHECKPOINT 18D: THE CHLOR-ALKALI INDUSTRY

1. Where do you think Winnington ICI obtains (*a*) limestone and (*b*) coal? [See Figure 18.8.] Why is their product important?

2. What effluent is produced by the ammonia–soda process? Is it an environmental hazard? How well is the Winnington plant situated for effluent disposal?

3. (*a*) Why does Australia not mine salt?

(*b*) Why does Japan not produce calcium chloride as a by-product?

4. How does the electrolysis of brine in the mercury cathode cell compare with the ammonia–soda process with respect to pollution?

5. What are the two strategies that are being adopted by the chlor-alkali industry to reduce pollution by mercury?

6. What are the three important products of the chlor-alkali industry? What are they used for?

QUESTIONS ON CHAPTER 18

1. From your knowledge of the chemistry of Group 2, predict what you can of the properties of radium. In particular, comment on

(*a*) the reaction of radium with water

(*b*) the solubility of its hydroxide and a likely value for the pH of its solution

(*c*) the solubility of the sulphate, chloride and carbonate

(*d*) the action of heat on the nitrate and carbonate.

2. Francium, the last member of Group 1, is a short-lived radioactive element. From what you know of the chemistry of Group 1, deduce what the properties of francium are likely to be, with respect to

(*a*) the nature of its hydride and the reaction of the hydride with water

(*b*) combination with nitrogen

(*c*) combination with oxygen

(*d*) the action of heat on the carbonate, hydrogencarbonate and nitrate

(*e*) the solubility of its salts in water and in organic solvents.

3. Comment on the statement, 'The metals of Group 1 are a similar set of elements, yet some gradation in properties can be observed from top to bottom of the group.'

4. Comment on the statement. 'In some ways, the members of Group 2 are a very similar set of elements; yet one can also look at them as a pair of similar elements (Be, Mg) and a trio of elements (Ca, Sr, Ba).'

5. Give an account of Groups 1 and 2. Point out the similarities and differences between the two groups.

6. 'In some ways hydrogen resembles the alkali metals.' Discuss this statement.

7. The table shows some data relating to members of Group 2 in the Periodic Table.

Element **M**	*Mg*	*Ca*	*Sr*	*Ba*
Enthalpy of hydration of $\mathbf{M}^{2+}$/kJ mol^{-1} (enthalpy change for $\mathbf{M}^{2+}(g) + aq \rightarrow \mathbf{M}^{2+}(aq)$)	−1920	−1650	−1480	−1360
Standard electrode potential $E^{\ominus}$/V	−2.37	−2.87	−2.89	−2.91

(*a*) (i) Which of the elements listed is the most powerful reducing agent? Give a reason for your answer.

(ii) Write an equation showing the reduction of hydrogen gas by magnesium.

(iii) Explain, by reference to the $E^{\ominus}$ values, how you would expect the elements of Group 2 to be manufactured from their chlorides.

(iv) Suggest a reason for the trend in hydration enthalpies shown in the table.

(*b*) Which of the elements listed forms

(i) the sulphate which is least soluble in water

(ii) the carbonate which is most stable to heat?

(AEB 90)

8. Element **X** belongs to Group 2 of the Periodic Table (Mg, Ca, Sr, Ba) and undergoes the reactions shown in the scheme below.

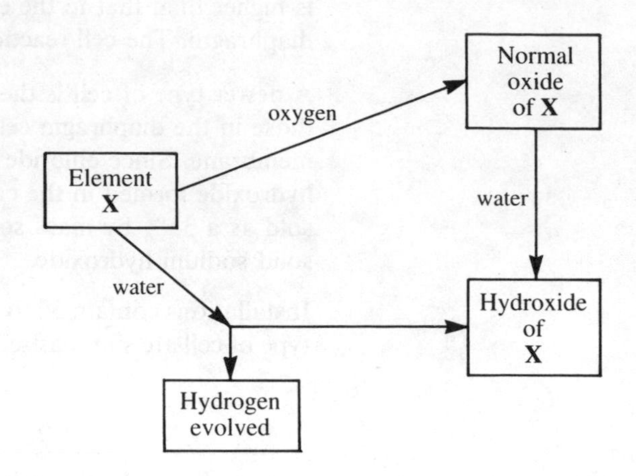

(*a*) (i) Element **X** reacts readily with oxygen to form a normal oxide: write down a balanced chemical equation for the reaction of one atom of **X** in this way.

(ii) The normal oxide of **X** thus formed reacts vigorously with water to form the hydroxide of **X**; write down a balanced chemical equation for this reaction.

(iii) State what you can deduce from the information above concerning the identity of **X**.

(iv) The hydroxide of **X** thus formed is sparingly soluble in cold water; state what you can deduce about the identity of **X** from this information.

(v) State how you would expect the solubilities in water of the hydroxides of the Group 2 elements to vary with increasing atomic number.

(*b*) (i) Element **X** also reacts rapidly, but not violently, with cold water to form the hydroxide; explain what can be deduced about the identity of **X** from this information.

(ii) Write a balanced chemical equation, based on one atom of **X**, for the reaction in (*b*)(i) above.

(*c*) (i) When y g of element **X** reacts to form the normal oxide, 1.3992y g of the latter is produced. Calculate the relative atomic mass of element **X**.

($A_r(O) = 16.00$)

(ii) When y g of element **X** reacts with water as in (*b*) above, the resulting alkaline solution requires 49.90 cm^3 of 0.1000 mol dm^{-3} hydrochloric acid for complete reaction. Calculate (*1*) the number of moles of hydroxide ion produced by the reaction of y g of element **X** with water and (*2*) the mass, y, of element **X** used.

(*d*) (i) Use the Periodic Table on p. 792 to identify element **X** from the value of its relative atomic mass.

(ii) State what colour you would expect the salts of element **X** to exhibit in a flame test.

(WJEC 92)

9. (*a*) Give an account of the properties of calcium by referring to the metal, calcium hydrogencarbonate, and at least four other compounds of commercial or environmental importance.

(*b*) What advantage is there in using barium peroxide rather than sodium peroxide for the preparation of an aqueous solution of hydrogen peroxide?

(*c*) Account for the fact that beryllium fluoride is water soluble, whereas calcium fluoride is insoluble in water.

(*d*) Anhydrous beryllium chloride shows properties which are not typical of other Group 2 metal chlorides. List these anomalous properties, and provide an explanation for them.

(O 92, S)

10. (*a*) (i) What are the main factors responsible for the observed trends in the physical and chemical properties of the Group 2 elements?

(ii) Give reasons for the atypical properties of beryllium.

(*b*) In the light of the trends demonstrated in the sequence magnesium – calcium – strontium – barium, predict the following behaviour of radium and its compounds. Give brief reasons for your prediction.

(i) the reaction of radium with water

(ii) the solubility in water of radium sulphate

(iii) the solubility in water of radium hydroxide

(iv) the type of bonding in radium compounds, and the typical physical properties of these compounds

(v) a method for obtaining radium from its compounds

(*c*) (i) Part of the decay series of naturally occurring thorium is as follows. The half-lives are shown under the arrows.

$$^{228}\text{Ra} \xrightarrow[6.67\text{ years}]{\beta} \text{W} \xrightarrow[6.13\text{ hours}]{\beta} \text{X} \xrightarrow[1.91\text{ years}]{\alpha} \text{Y} \xrightarrow[3.64\text{ days}]{\alpha} \text{Z}$$

Use the Periodic Table on p. 792 to identify **W**, **X**, **Y** and **Z**, and give the correct mass number and atomic number for each.

(ii) Why is the decay of a radioisotope described as a *first-order* process?

(iii) A sample of ^{228}Ra with an initial mass of 100 mg is left in a safe container. What mass of ^{228}Ra would remain after 20 years?

(AEB 92)

11. (*a*) (i) State the conditions under which magnesium and calcium will react with water, and write balanced equations for the reactions.

(ii) Explain any differences between the two reactions in terms of the atomic properties of the two metals.

(*b*) Compare the chemistries of magnesium and calcium with reference to the following:

(i) the solubilities of their sulphates in water

(ii) the thermal stabilities of their carbonates

(iii) the reaction of their oxides with water

(*c*) A mineral, which can be represented by the formula $Mg_xBa_y(CO_3)_z$, was analysed as described below. From the results, calculate the formula of the mineral.

A sample of the mineral was dissolved in excess hydrochloric acid and the solution made up to 100 cm^3 with water. During the process 48 cm^3 of carbon dioxide, measured at 25 °C and 1 atmosphere pressure, were evolved.

A 25.0 cm^3 portion of the resulting solution required 25.0 cm^3 of EDTA solution of concentration 0.02 mol dm^{-3} to reach an end-point. A further 25.0 cm^3 portion gave a precipitate of barium sulphate of mass 0.058 g on treatment with excess dilute sulphuric acid. You may assume that Group 2 metal ions form 1 : 1 complexes with EDTA.

(Molar volume of any gas at 25 °C and 1 atmosphere pressure = 24 dm^3.)

(AEB 90)

12. (*a*) (i) Explain why the molecule of boron trifluoride, BF_3, is planar, but that of ammonia, NH_3, shows a distorted tetrahedral structure. Indicate, with your reasoning, the geometry which you would expect for the tetrahydridoborate(III) anion, BH_4^-.

(ii) Explain why silver chloride, AgCl (obtained as a white precipitate by the reaction of chloride ions with silver nitrate) is soluble in aqueous ammonia.

(*b*) An element, **X**, has standard first, second and third molar ionisation energies of 738, 1449 and 7728 kJ mol^{-1} respectively. The elements immediately preceding and following **X** in the Periodic Table have standard first molar ionisation energies of 496 and 578 kJ mol^{-1} respectively. Halides of element **X** show no characteristic flame test.

Aqueous solutions of the nitrate of element **X** show the following properties:

1. No precipitate is formed with dilute sulphuric acid.
2. A white precipitate is produced with sodium carbonate solution.
3. A white precipitate is formed with sodium hydroxide solution which is not soluble in excess.

(i) Interpret all the above information. Suggest, with your reasoning, an identity for **X**. Give balanced chemical equations wherever appropriate.

(ii) State your expectations concerning the thermal stabilities of the hydroxide and carbonate of **X**, giving your reasons.

(iii) Indicate and explain how you would expect the thermal stabilities of the hydroxides and carbonates of other members of the group to compare with those of the hydroxide and carbonate of **X**.

(WJEC 92)

19

GROUP 3

19.1 THE MEMBERS OF THE GROUP

The elements in Group 3 of the Periodic Table are listed in Table 19.1.

TABLE 19.1 The Elements of Group 3

Name	*Symbol*	Z	*Electron Configuration*
Boron	B	5	$1s^22s^22p$
Aluminium	Al	13	$(Ne)3s^23p$
Gallium	Ga	31	$(Ar)3d^{10}4s^24p$
Indium	In	49	$(Kr)4d^{10}5s^25p$
Thallium	Tl	81	$(Xe)5d^{10}6s^26p$

The elements of Group 3

With the exception of boron, they are metals. With the exception of boron, they form ionic compounds by losing the s and p electrons to form an ion $\mathbf{M^{3+}}$. They also form covalent compounds, through the promotion of an s electron to an unoccupied p orbital and the formation of three sp^2 hybrid bonds [§ 5.2.3]. Thallium forms some Tl^+ compounds. Aluminium is by far the most important of the metals.

19.2 ALUMINIUM

19.2.1 USES

Aluminium has a low density and it is not corroded

It is a good thermal conductor and also a reflector of heat

It is used in headlights…

…in electrical cables…

Aluminium is the most abundant metal in the surface of the earth; yet it has been extracted in quantity only since the end of the nineteenth century. Every month new uses are being found for this metal which resists corrosion and which has a low density. Being completely resistant to corrosion, it is ideal for packaging food [see Figure 19.2]. Aluminium is amazing in being a good thermal conductor which can also be used as a thermal insulator. As a thermal conductor, it is used for the manufacture of saucepans and cooking foil. The insulating property of aluminium arises from its ability to reflect radiant heat (i.e., infrared rays). Prematurely born babies are sometimes wrapped in aluminium foil, which keeps them warm by reflecting heat lost from the body. Firefighters in the USA wear suits which are coated with aluminium to reflect the heat from the fire and keep them cool [see Figure 19.3]. The polished surface of aluminium finds it a use in the reflectors of car headlights. Aluminium is a good electrical conductor and is replacing copper in overhead cables: to support aluminium cables, which are lighter, the pylons can be spaced at longer intervals.

. . . and for the construction of boats and planes

Some parts of cars are made of aluminium

Since it has a low density and is not corroded, aluminium has obvious advantages over iron as a manufacturing material. Pure aluminium is too soft for construction purposes, but alloys (e.g., Al/Mg and Duralumin®, Al/Mg/Cu) have a higher tensile strength and are used for the construction of aeroplanes and small boats [see Figure 19.4]. More and more parts of cars are being made of aluminium: engine blocks can be cast from aluminium; piston heads are made of aluminium and encircled by steel rings, and rocker covers are made of aluminium. Some vehicles have an aluminium body, but the chassis needs the strength of steel, and the engine is of cast iron. When a car is scrapped because the iron in it has rusted, the aluminium parts are as good as new and can be recycled. Another advantage of incorporating aluminium parts is that the vehicle becomes lighter and consumes less petrol.

The metal is coated with a film of aluminium oxide, which is unreactive

The use of aluminium alloys in construction is possible because the metal is coated with a thin film of aluminium oxide, which resists attack by corrosive reagents.

FIGURE 19.1
The Tanker is made of Aluminium with an Inner Tank of Stainless Steel. The Door of the Inner Tank can be seen

FIGURE 19.2
The Can is made of Mild Steel or Aluminium. The Ring-pull Cap is of Aluminium

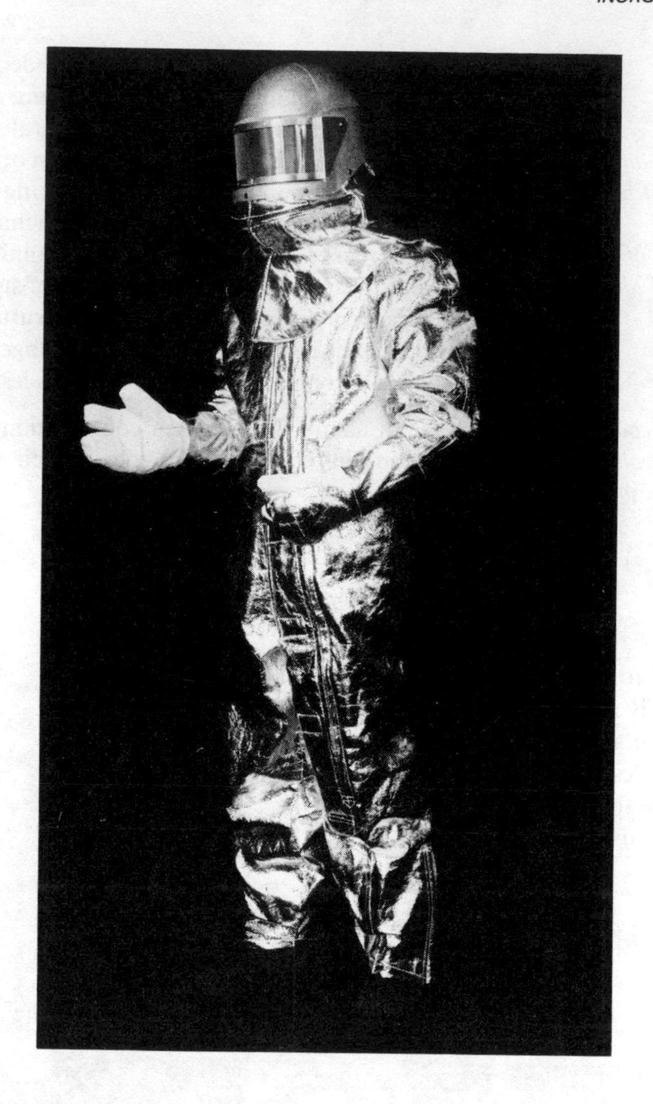

FIGURE 19.3 Firefighter in an Aluminium Fabric Suit

FIGURE 19.4 The A300 Airbus made from Aluminium Alloys

Anodised aluminium can be dyed, and is used for construction purposes

When aluminium is **anodised**, that is, made the anode in an electrolytic cell of sulphuric acid or chromic acid, the layer of oxide is thickened. When formed in this way, the oxide is hydrated and can absorb dyes. Dyed anodised aluminium is used for door frames and window frames, which are decorative as well as weatherproof.

FIGURE 19.5 Aluminium can be used together with Gallium and Arsenic, to make a Semi-conductor Laser-chip. (The temperature inside this gold-lined furnace is 1000 °C.)

19.2.2 EXTRACTION OF ALUMINIUM

Purification of Al_2O_3 requires separation from Fe_2O_3 and SiO_2

Aluminium is mined as the ore bauxite, aluminium oxide-2-water, $Al_2O_3 \cdot 2H_2O$, which contains silicon(IV) oxide and iron(III) oxide as impurities. Pure aluminium oxide is obtained from the ore by utilising the fact that it is amphoteric, whereas, of the impurities, silicon(IV) oxide is acidic and iron(III) oxide is basic. After being ground, the ore is treated as shown in Figure 19.6.

FIGURE 19.6 Purification of Aluminium Oxide

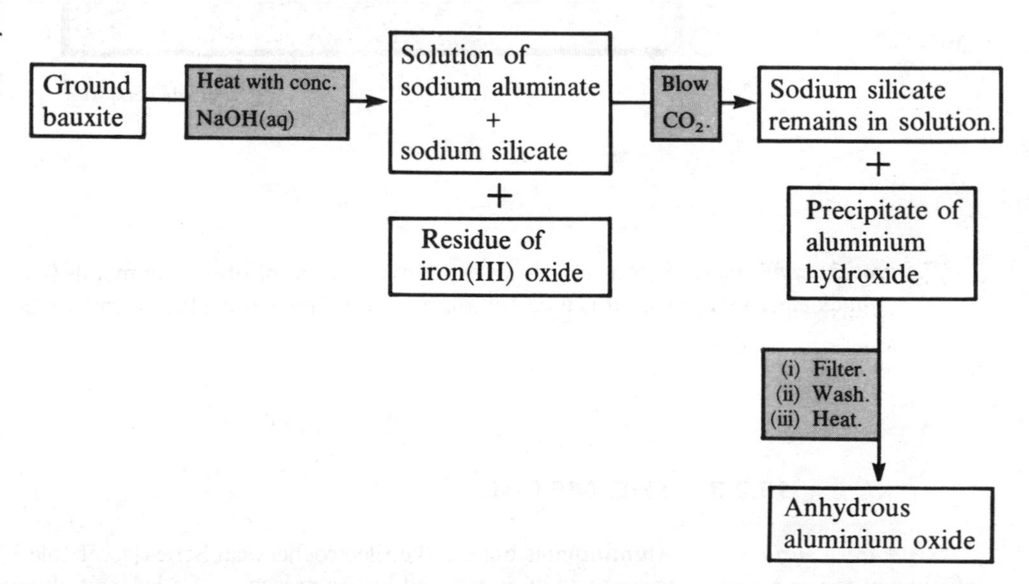

Electrolysis of Al_2O_3 in molten Na_3AlF_6 is used to extract aluminium

Aluminium is obtained from its oxide by electrolysis. Since the melting temperature of aluminium oxide (2050 °C) is so high that electrolysis of the molten oxide cannot be accomplished, a solvent must be used. The cell shown in Figure 19.7 contains a molten mixture of the ore cryolite, sodium hexafluoroaluminate, Na_3AlF_6, with calcium fluoride and aluminium fluoride added to lower its melting temperature.
Aluminium oxide is dissolved in this melt, and electrolysed at 850 °C to give aluminium and oxygen.

The electrode processes

It is postulated that the equilibrium

$$Al_2O_3 \rightleftharpoons Al^{3+} + AlO_3^{3-}$$

gives rise to the electrode processes

Cathode $Al^{3+} + 3e^- \rightarrow Al$

Anode $4AlO_3^{3-} \rightarrow 2Al_2O_3 + 3O_2 + 12e^-$

FIGURE 19.7
A Hall–Héroult Cell
(5 m × 3 m × 1 m,
30 000 A, 5 V)

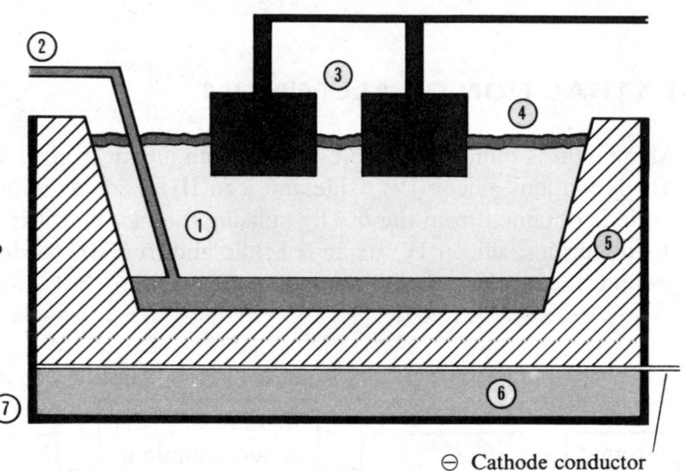

3 Carbon anode blocks replaced often because of oxidation to CO_2 by the O_2 evolved.

2 Molten Al is syphoned off.

1 Electrolyte: molten cryolite, Na_3AlF_6 ($+ CaF_2 + AlF_3$ to lower T_m) $+ Al_2O_3$. More Al_2O_3 is added periodically.

7 Steel case

4 Crust of solid Al_2O_3 protects molten Al from oxidation.

5 Carbon cathode

6 Insulation

The process consumes much electricity

Electrolysis is an expensive method of obtaining metals from their ores. The reasons why it is used for aluminium follow from § 10.9.1 and are described in § 19.5.

19.2.3 THE METAL

A fresh surface of aluminium reacts rapidly with water vapour in the air

Aluminium is high in the Electrochemical Series [see Table 12.2, § 12.3.1], but its true reactivity is masked by the presence of a layer of aluminium oxide on its surface. This can be removed by reaction with mercury or mercury(II) chloride. When a fresh aluminium surface is exposed, it reacts immediately with water vapour in the air to form strands of aluminium hydroxide.

Aluminium reacts with non-metallic elements...

Aluminium reacts directly with the non-metallic elements oxygen, sulphur, nitrogen, carbon (at a high temperature) and the halogens to form compounds Al_2O_3, Al_2S_3, AlN, Al_4C_3 and AlX_3. The high electropositivity of aluminium leads to its use in the thermit process for extracting other metals (e.g., chromium) from their oxides:

... with metal oxides...

$$2Al + Cr_2O_3 \rightarrow 2Cr + Al_2O_3$$

Nitric acid makes aluminium 'passive' by increasing the thickness of the oxide film. Hydrochloric acid and sulphuric acid in fairly concentrated solutions react with aluminium to form salts:

... with HCl(aq) and $H_2SO_4(aq)$...

$$2Al(s) + 6HCl(aq) \rightarrow 2AlCl_3(aq) + 3H_2(g)$$

$$2Al(s) + 6H_2SO_4(aq) \rightarrow Al_2(SO_4)_3(aq) + 3SO_2(g) + 6H_2O(l)$$

Alkalis react with aluminium to form an aluminate with the evolution of hydrogen:

... and with alkalis

$$2Al(s) + 2OH^-(aq) + 6H_2O(l) \rightarrow 2Al(OH)_4{}^-(aq) + 3H_2(g)$$

The reaction starts slowly and speeds up as the metal is stripped of its protective film.

19.2.4 THE Al^{3+} ION

The small size of the Al^{3+} ion gives its compounds a high degree of covalent character

The charge/radius ratio of the Al^{3+} ion is 3 units of charge/0.05 nm = 60 units of charge nm^{-1}. This ratio is high compared with the ratios Na^+ = 10 and Mg^{2+} = 30 units of charge nm^{-1}. It is similar to that of Be^{2+} = 66 units of charge nm^{-1}. The Al^{3+} ion is able to polarise the electron cloud of an anion [§§ 4.5.4 and 15.2] to form a bond with a high degree of covalent character. Aluminium fluoride is ionic, the oxide is largely ionic, with some covalent character, and the anhydrous chloride, bromide and iodide have polar covalent bonds.

In aqueous solution, H_2O molecules coordinate to the Al^{3+} ion

In aqueous solution, the Al^{3+} ion is stabilised by the coordination of water molecules to form the complex hexaaquaaluminium(III) ion, $[Al(H_2O)_6]^{3+}$. The high ionisation energy required to form Al^{3+} ions is largely balanced by the energy released when Al—OH_2 bonds are formed. Six water molecules are distributed octahedrally about an Al^{3+} ion [see Figure 19.8(a)]. The coordination of water molecules occurs through the donation by the oxygen atom of a lone pair of electrons [see Figure 19.8(b)].

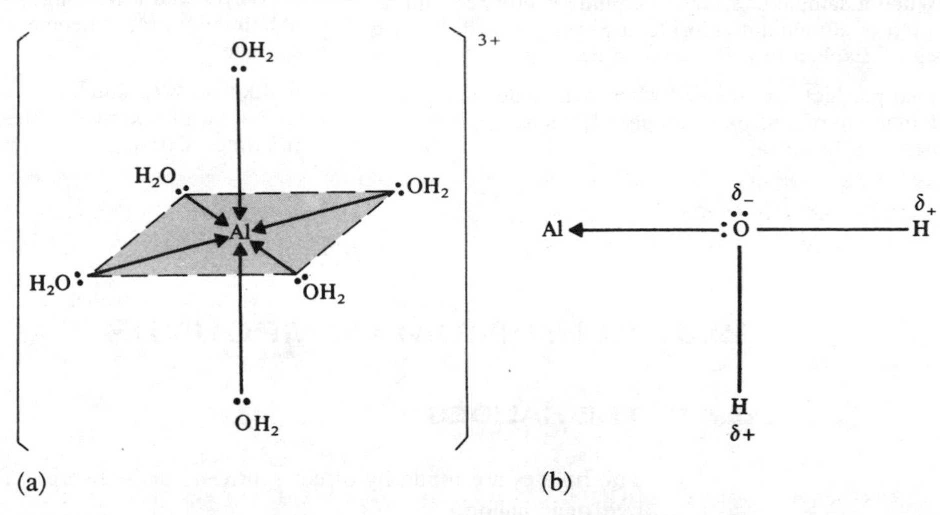

FIGURE 19.8
(a) $[Al(H_2O)_6]^{3+}$
(b) Polarity of O—H Bonds

Coordinated water molecules tend to donate protons to free water molecules

In consequence, the electrons in the O—H bond move closer to the oxygen atom, and the hydrogen atoms have a greater degree of positive charge than in free water molecules. The partial positive charges on the hydrogen atoms attract bases, which may abstract protons from the coordinated water molecules. Water molecules of the solvent itself may function as bases:

This makes aluminium salts acidic in solution

$$[Al(H_2O)_6]^{3+} + H_2O \rightleftharpoons [Al(OH)(H_2O)_5]^{2+} + H_3O^+$$

$$[Al(OH)(H_2O)_5]^{2+} + H_2O \rightleftharpoons [Al(OH)_2(H_2O)_4]^+ + H_3O^+$$

As they do so, oxonium ions are formed, and the solution becomes acidic.

The process is called salt hydrolysis

These equilibria are examples of salt hydrolysis [§ 12.7.13]. If stronger bases are present (e.g., OH^-, CO_3^{2-}, S^{2-}), H_3O^+ ions are removed, and the hydrolysis equilibria move to the right. The stronger base can also remove the third proton:

$$[Al(OH)_2(H_2O)_4]^+ + OH^- \rightleftharpoons [Al(OH)_3(H_2O)_3](s) + H_2O$$

A precipitate of hydrated aluminium hydroxide appears. If an excess of hydroxide ions is added, protons are removed from the precipitate:

$$[Al(OH)_3(H_2O)_3](s) + OH^- \rightleftharpoons [Al(OH)_4(H_2O)_2]^-(aq) + H_2O$$

The solution formed contains diaquatetrahydroxoaluminate(III) ions:

$$[Al(OH)_4(H_2O)_2]^-$$

usually written $Al(OH)_4^-(aq)$ and called tetrahydroxoaluminate(III) ions or simply aluminate ions. If an acid is added to the solution, the above equilibrium is reversed, and hydrated aluminium hydroxide is precipitated.

Al^{3+}(aq) ions are used as coagulating agents

The high charge/radius ratio of aluminium ions makes them useful coagulating agents. Aluminium ions are adsorbed on to the surface of negatively charged colloidal particles. The charges on the surfaces of the colloidal particles are reduced, and they are able to join to form a solid precipitate. Aluminium sulphate is used in water treatment plants to remove colloidal organic material from water [§ 9.5.4].

CHECKPOINT 19A: ALUMINIUM

1. (*a*) What are the advantages and disadvantages of iron and aluminium for use in car engines, car bodies and car bumpers?
(*b*) Why have aluminium saucepans become more popular than iron cooking pans?

2. When a sample of sodium sulphide is dropped into a solution of aluminium chloride, a gas is detected. What is this gas? Explain how it comes to be formed.

3. Two products are formed when hydroxide ions react with diaquatetrahydroxoaluminate(III) ions. Give their names and formulae.

4. Which of the following solutions do you think will be acidic?

$$FeSO_4,\ Fe_2(SO_4)_3,\ ZnSO_4,\ Cr_2(SO_4)_3$$

Explain your answer.

5. Why would it be dangerous to allow a solution of aluminium nitrate to come into contact with sodium cyanide?

6. Explain why aluminium finds use in (*a*) mirrors, (*b*) overhead electrical cables, (*c*) milk-bottle tops and (*d*) window frames.

19.3 ALUMINIUM COMPOUNDS

19.3.1 THE HALIDES

The halides are made by direct synthesis or by heating aluminium in a stream of hydrogen halide:

The synthesis of halides

$$2Al + 3X_2 \rightarrow 2AlX_3$$

$$2Al + 6HX \rightarrow 2AlX_3 + 3H_2$$

An apparatus which could be used for the preparation of anhydrous aluminium chloride is shown in Figure 19.9.

The dimerisation of AlX_3

Aluminium chloride sublimes as the dimer Al_2Cl_6. This reacts readily with water to give Al^{3+}(aq) and Cl^-(aq). The bromide and iodide also exist as dimers in the gaseous state. They dissociate into the monomers on further heating. Figure 19.10 shows the structures of the monomers and the dimers.

FIGURE 19.9 Preparation of Aluminium Chloride

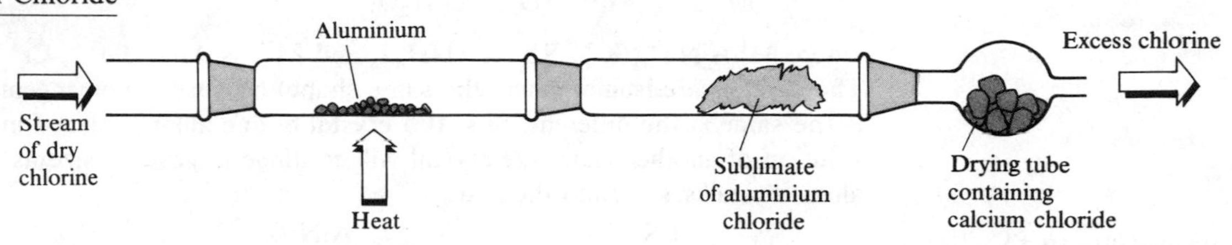

The chloride, bromide and iodide dissolve in covalent solvents like benzene. They are important catalysts in the Friedel–Crafts reactions [§§28.7.4 to 28.7.5], because of their ability to act as Lewis acids (electron-acceptors) [§§12.7.1 and 28.7.7].

FIGURE 19.10 The Trigonal Planar Structure of $AlBr_3$ and the Tetrahedral Arrangement of Bonds in Al_2Br_6

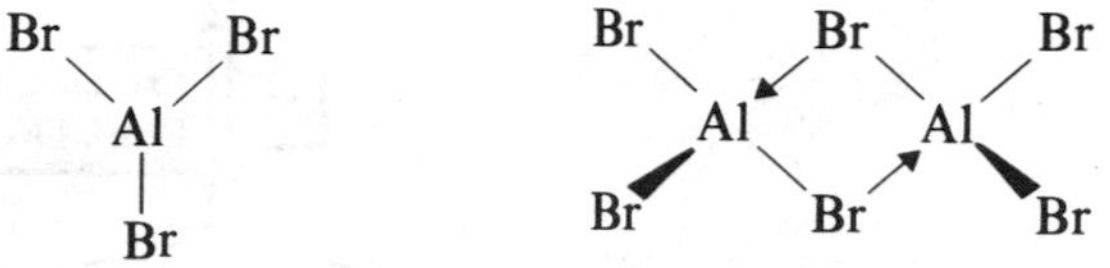

19.3.2 THE OXIDE AND HYDROXIDE

Aluminium oxide is amphoteric...

Aluminium oxide is mined as bauxite. It is amphoteric. With acids, it reacts to form salts of the $[Al(H_2O)_6]^{3+}$ ion (usually written $Al^{3+}(aq)$), and with alkalis it forms salts of the tetrahydroxoaluminate ion (the aluminate ion), $Al(OH)_4^-$.

...It is used as a catalyst...

...and in chromatography

Aluminium oxide (often called alumina) catalyses the cracking of alkanes in the petrochemical industry [§26.3.2] and the dehydration of ethanol to ethene [§30.7.2]. It is frequently used as the stationary phase in column chromatography [§8.7.2].

The hydroxide also is amphoteric...

Aluminium hydroxide is a white, gelatinous precipitate formed when ammonia solution is added to a solution of an aluminium salt. It is amphoteric. On standing, it loses water to become hydrated aluminium oxide.

...It is used as a mordant

Aluminium hydroxide is used in the dyeing industry. It is called a *mordant* (Latin: *mordere*, to bite) as it helps the dye to 'bite' the cloth. The cloth is soaked in a solution of aluminium sulphate, and alkali is added so that a gelatinous precipitate of the hydroxide is deposited in the fibres of the cloth. When the cloth is dipped into a vat of dye, the dye is adsorbed by the water occluded in the gelatinous precipitate and held by the charge on the $Al^{3+}(aq)$ ion.

19.3.3 ALUMS

Alums are double salts

Aluminium sulphate crystallises as the hydrate, $Al_2(SO_4)_3 \cdot 18H_2O$. If equimolar amounts of aluminium sulphate and potassium sulphate are allowed to crystallise together, a **double salt**, aluminium potassium sulphate-24-water

$$K_2SO_4 \cdot Al_2(SO_4)_3 \cdot 24H_2O$$

crystallises. It is not a complex salt: in solution, it behaves as a mixture of the two salts. The double salt has a high lattice enthalpy (energy), making it less soluble than the simple salts from which it forms.

Crystals of alums are isomorphous

Crystals of the same octahedral shape are obtained with other double salts. They are called **alums** and have the general formula

$$M_2{}^{I}SO_4 \cdot M_2{}^{III}(SO_4)_3 \cdot 24H_2O$$

where M^I is Na^+, K^+, Rb^+ or $NH_4{}^+$ and M^{III} is Al^{3+}, Fe^{3+}, Cr^{3+} or Mn^{3+}. The crystals are **isomorphous** (the same shape) because the arrangement of the ions is the same in the different salts. If a crystal of one alum is placed in a saturated solution of another alum, the crystal will continue to grow in size as the second alum crystallises around the first.

FIGURE 19.11 Some Reactions of Aluminium and its Compounds

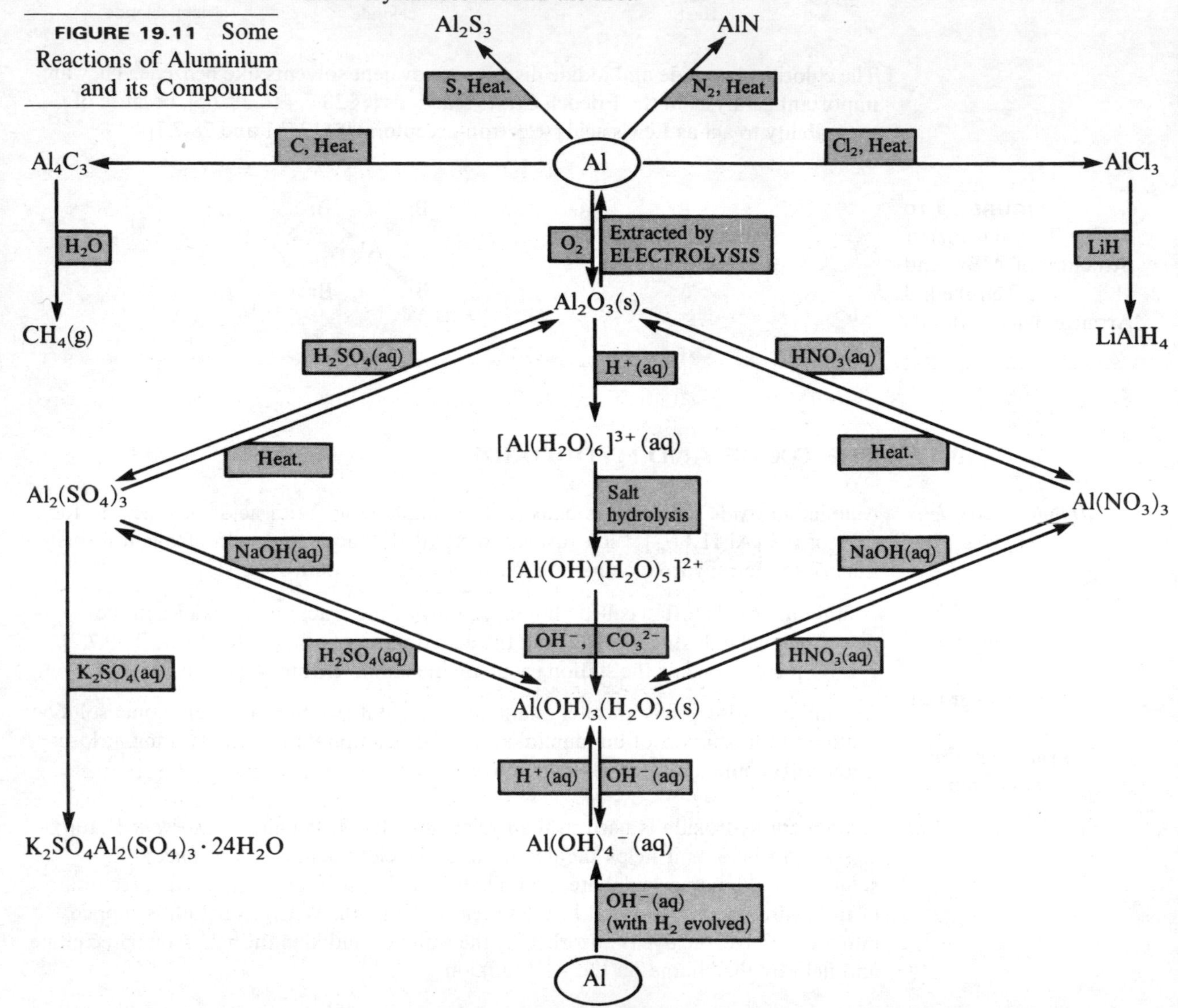

CHECKPOINT 19B: ALUMINIUM COMPOUNDS

1. From the reaction of aluminium with hydrochloric acid can be obtained crystals of $AlCl_3 \cdot 6H_2O$. Why is it not possible to obtain anhydrous aluminium chloride by heating these crystals?

2. Why do aluminium halides act as Lewis acids? The ion $AlBr_4{}^-$ is an intermediate in a Friedel–Crafts reaction. Name the ion and describe the arrangement of bonds about the central atom.

3. If you need to prepare aluminium hydroxide, why is it better to add a solution of ammonia to a solution of an aluminium salt, rather than to add a solution of sodium hydroxide?

4. What do you think would happen if a crystal of aluminium ammonium sulphate-24-water were suspended in a saturated solution containing equimolar amounts of chromium(III) sulphate and sodium sulphate? Explain your answer.

5. *Potash alum*, $K_2SO_4 \cdot Al_2(SO_4)_3 \cdot 24H_2O$, can be made by allowing an excess of aluminium to react with warm potassium hydroxide solution until the evolution of hydrogen ceases, decanting, acidifying with dilute sulphuric acid solution and setting the solution aside to crystallise. Explain what reactions take place and write equations for them.

6. Why do you think aluminium salts are used in 'styptic pencils' to assist blood to clot?

19.4 *DIAGONAL RELATIONSHIPS

Beryllium, at the beginning of Group 2, resembles aluminium, the second element in Group 3:

1. The oxides and hydroxides are amphoteric.

With alkali they form the complex ions, beryllate ion, $[Be(OH)_4]^{2-}$ and aluminate ion, $[Al(OH)_4]^-$.

2. The halides are covalent when anhydrous (except AlF_3 which is ionic) and dimerise as e.g. Al_2Cl_6.

The ionise in water. They form complex ions, e.g. $[BeF_4]^{2-}$, $[AlF_6]^{3-}$.

Boron, at the head of Group 3, resembles silicon, the second element in Group 4:

1. Both are non-metallic elements. Boron is a poor electrical conductor. Silicon is well known as a semiconductor.

2. The oxides, B_2O_3 and SiO_2, are acidic, covalent and macromolecular.

3. The halides are covalent, and completely hydrolysed by water.

4. The hydrides are flammable gases, e.g. B_2H_6, SiH_4, Si_2H_6.

19.5 THE ALUMINIUM PROBLEM

TOPIC

Aluminium is the most abundant metal in the surface of the earth, yet the metal was extracted from its ore only 150 years ago and remained a rarity for another 60 years. The Tsar of Russia gave his baby son an aluminium rattle to play with as it was more expensive than gold and he wanted the little fellow to have only the best.

Two compounds of aluminium have been known for hundreds of years. One is a beautiful crystalline substance which chemists called *alum* and which we know as aluminium potassium sulphate. The other compound known to chemists of old was a basic substance from which they could make alum. They called it *alumina* and deduced that it was the oxide of a metal, the metal aluminium, which no one had ever seen.

Sir Humphrey Davy tried in 1807 to obtain aluminium by electrolysing alumina, but he failed. In 1825, the Danish scientist, H C Oersted, decided to try a chemical method of isolating aluminium. He made aluminium chloride from alumina, charcoal and chlorine. Then he allowed aluminium chloride to react with potassium amalgam. Potassium was the most reactive metal he knew. He hoped that it would displace aluminium from aluminium chloride and leave him with potassium chloride and aluminium amalgam. He distilled what he hoped was aluminium amalgam under reduced pressure. Mercury distilled over. Left in the distillation flask was 'a lump of metal which in colour and lustre somewhat resembles tin'.

Wöhler was a German chemistry professor. He was unable to get Oersted's method to work and in 1827 he tried an extraction from alum. He obtained aluminium hydroxide by adding alkali to a solution of alum, and then converted it into aluminium chloride. Wöhler put aluminium chloride and potassium into a crucible, and, when he just touched it with a flame, a violent reaction occurred. To get rid of the excess of potassium, he plunged the crucible into a trough of water. To his delight, a grey powder floated to the surface, and after several experiments he had enough grey powder to melt it and get a lump of metal. This was the first certain extraction of aluminium.

A French chemist called Henri Sainte-Claire Déville decided to try to improve on Wöhler's extraction of aluminium, and turn it into a commercial process. Déville took aluminium oxide, charcoal and salt and heated the mixture in a stream of chlorine. The compound sodium hexachloroaluminate, Na_3AlCl_6, was formed. When he melted this compound with an excess of sodium, he obtained sodium chloride and, as he had hoped, molten aluminium, which he was able to run off and cast into ingots. Iron had been known for 5000 years, but this was the first time, in 1860, that aluminium had been obtained in quantity.

Now that there was a commercial process for making aluminium on a large scale, people began to discover what a useful metal aluminium is and to invent new uses for it. It was still an expensive metal because of the high cost of the large quantities of sodium used in its extraction. Since the reactive metals (Na, K, Ca, Mg) are extracted from their ores by electrolysis, it was natural that people should return to the possibility of using an electrolytic method for the extraction of aluminium. The big problem was that the oxide could not be melted to give a conducting liquid.

The problem was solved by a young American student called Charles Martin Hall. Hall's professor had worked under Wöhler, and his accounts of the early research work had made Hall impatient to contribute his chapter to the aluminium story. Hall bought some batteries and found an outhouse where he could experiment. He discovered that he could melt the ore *cryolite*, sodium hexafluoroaluminate, Na_3AlF_6, at 1000 °C, and then dissolve aluminium oxide in the molten cryolite to form a conducting solution. Hall electrolysed the melt, using graphite electrodes and, to his joy, he obtained aluminium at the cathode. The year was 1886, and Hall was just 21 years old! The Aluminium Company of America was founded to develop the process he had discovered.

Across the Atlantic, another young man was obsessed with the same problem. In France, Paul Héroult (23 years old) was working away in another makeshift laboratory. He arrived at the same conclusion as Hall. The electrolytic cell used in aluminium plants is called the Hall–Héroult cell [see Figure 19.7, § 19.2.2].

Aluminium is mined in many parts of the world. The ore is called *bauxite*, as it has been mined for a long time in Baux in France. The production of aluminium uses a great deal of electricity. It takes 15 000 kilowatt hours to make 1 tonne (10^3 kg) of aluminium. Hydroelectric power is the most economical form of electricity, and aluminium plants are built close to a waterfall or dam which can be used as a source of power. Hydroelectric power stations are often in regions of great natural beauty; aluminium plants are unsightly. There can be a conflict of interest between conservationists, who want to preserve the landscape, and industrialists, who want to provide us with more and more of this amazingly useful metal.

CHECKPOINT 19C: THE ALUMINIUM PROBLEM

1. Write equations for (*a*) the sequence of reactions by which Oersted obtained aluminium from alumina and (*b*) the sequence of reactions by which Wöhler obtained aluminium from alum.

2. How did the Hall–Héroult method improve on Déville's method of extracting aluminium?

3. In the Hall–Héroult cell, cryolite, Na_3AlF_6, is present. Why is sodium not formed at the cathode? Why does aluminium not form at the cathode when molten cryolite is electrolysed without added alumina?

4. Calculate how many coulombs of electricity are required to produce 1 tonne of aluminium (1 tonne = 10^3 kg, A_r(Al) = 27, Faraday constant = 96 500 C mol^{-1}).

QUESTIONS ON CHAPTER 19

1. Describe the manufacture of aluminium, explaining the reasons for the conditions employed. What advantages does aluminium have over steel?
Give examples of the uses to which aluminium is put.

2. Aluminium fluoride boils at 1270 °C and aluminium bromide at 265 °C. Can you explain the large difference in boiling temperatures?

3. Describe the preparation of $Al_2(SO_4)_3 \cdot 12H_2O$ from aluminium. Why are solutions of this salt acidic?

4. How does aluminium react with (*a*) chlorine, (*b*) hydrochloric acid, (*c*) sodium hydroxide solution?

5. What is 'potash alum'? How can this salt be made, using aluminium as a starting material?

6. This question is concerned with the preparation and analysis of a sample of aluminium chloride.

(*a*) Aluminium chloride can be prepared in the laboratory by the reaction between aluminium and chlorine using the apparatus shown in Figure 19.12.

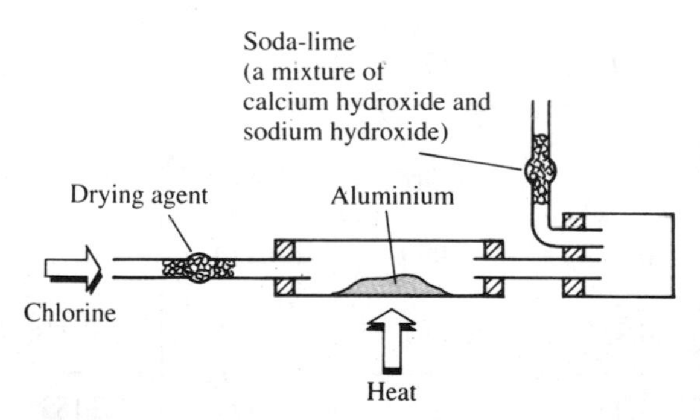

FIGURE 19.12

(i) Why is it necessary to dry the chlorine?

(ii) What is the purpose of the soda-lime in this preparation?

(iii) The equation of the reaction between aluminium and chlorine to form aluminium chloride can be represented as
$2Al(s) + 3Cl_2(g) \rightarrow 2AlCl_3(s)$

In a typical experiment 1.00 g of aluminium produced 2.30 g of aluminium chloride. Calculate the % yield of aluminium chloride.

$$(A_r(\text{Al}) = 27.0, \quad A_r(\text{Cl}) = 35.5)$$

(*b*) The formula of a pure sample of aluminium chloride was determined as follows.

A solution **X** was prepared by adding 0.120 g of the sample to 100 cm^3 of water in a conical flask. This solution needed 52.0 cm^3 of 0.0500 mol dm^{-3} aqueous silver nitrate to react according to the equation

$$Ag^+(aq) + Cl^-(aq) \rightarrow AgCl(s)$$

The equation of the reaction between aluminium chloride and water can be represented by

$$AlCl_3 + 3H_2O \rightarrow Al(OH)_3 + 3HCl$$

In 100 cm^3 of solution **X** calculate

(i) the number of moles of chloride ions

(ii) the mass of chloride ions

(iii) the number of moles of aluminium.

(iv) What is the apparent formula of the aluminium chloride based on these results?

(v) Suggest a reason why this apparent formula is not $AlCl_3$.

(*c*) Under certain conditions aluminium can form the ion $AlCl_4^-$.

(i) Draw the electronic structure (i.e. the dot–cross diagram) showing only the outermost electrons of each atom for $AlCl_4^-$.

(ii) Name the shape which you would expect for this ion and draw a sketch showing its three-dimensional structure. Give reasons for your prediction.

(O & C 91)

7. To produce 1 tonne of aluminium metal, 6.61 tonnes of bauxite, 0.081 tonnes of sodium hydroxide, 0.041 tonnes of aluminium fluoride, 0.031 tonnes of cryolite (Na_3AlF_6), 0.61 tonnes of carbon and 2×10^4 kW h of electricity are needed.

(*a*) (i) What are the functions of the cryolite?

(ii) What is the role of the carbon?

(*b*) Typically the bauxite is mined in Canada, purified in Ireland and turned into aluminium in Wales.

(i) Suggest *two* reasons why the purification of the bauxite should be carried out in Ireland and not in either Canada or Wales.

(ii) The aluminium producing plant is on Anglesey, at the far end of the island, away from the mainland.

Suggest *two* reasons for its location and *one* disadvantage for it being there.

(*c*) A new process for the manufacture of aluminium is under consideration. This involves the conversion of bauxite to aluminium chloride which is then mixed with sodium chloride and lithium chloride, melted at 600 °C and electrolysed. Approximately 1×10^4 kW h of electricity are needed per tonne of aluminium. Chlorine is given off at the anodes.

(i) Give *four* advantages of the new process over the old process.

(ii) Suggest why, despite these advantages, it is unlikely that the Anglesey plant will be converted to the new process.

(*d*) (i) Why are aqueous solutions of aluminium chloride acidic?

(ii) Describe the bonding in anhydrous aluminium chloride.

(C 91, AS)

20

GROUP 7: THE HALOGENS

20.1 THE MEMBERS OF THE GROUP

The appearance and physical state of the halogens

The elements of Group 7 are fluorine, chlorine, bromine, iodine and astatine. Fluorine is a poisonous pale yellow gas, chlorine is a poisonous dense green gas (Greek: *chloros*, green), bromine is a caustic and toxic brown volatile liquid (Greek: *bromos*, stench) and iodine is a shiny black solid which sublimes to form a violet vapour on gentle heating. Astatine is radioactive and does not occur naturally. Some properties of the elements are shown in Table 20.1.

Property	*Fluorine* F	*Chlorine* Cl	*Bromine* Br	*Iodine* I
Proton number (Atomic number)	9	17	35	53
Outer electron configuration	$2s^22p^5$	$3s^23p^5$	$3d^{10}4s^24p^5$	$4d^{10}5s^25p^5$
Atomic radius/nm	0.072	0.099	0.114	0.133
Ionic radius/nm	0.136	0.181	0.195	0.216
Boiling temperature/°C	− 187	− 35	59	183
Standard enthalpy of dissociation/kJ mol^{-1} *of X*	79.1	122	111	106
Electron affinity/kJ mol^{-1}	− 333	− 348	− 340	− 297
Standard electrode potential/V	+ 2.87	+ 1.36	+ 1.07	+ 0.54
Electronegativity	4.00	2.85	2.75	2.20
Oxidation states	− 1	− 1,1,3,5,7	− 1,1,3,5,7	− 1,1,3,5,7
Standard enthalpy of formation of NaX/kJ mol^{-1}	− 573	− 414	− 361	− 288
Standard lattice enthalpy of NaX/kJ mol^{-1}	− 902	− 771	− 733	− 684

TABLE 20.1 Physical Properties of the Halogens

Van der Waals forces between halogen molecules

The halogens are a very similar set of non-metallic elements. Their name 'halogens' is derived from the Greek for 'salt formers'. They exist as diatomic molecules, X_2. The strength of the van der Waals forces between X_2 molecules increases as the number of electrons in the molecule of X_2 increases, i.e.

$$F_2 < Cl_2 < Br_2 < I_2$$

This explains the order of melting and boiling temperatures. In iodine, the van der Waals forces are strong enough to sustain a solid structure of iodine molecules [see Figure 6.10, § 6.4].

20.2 IONIC BOND FORMATION

Halogens react to form X^- ions ...

... in the formation of metal halides

The halogens form salts by accepting one electron to complete an octet of valence electrons, with the formation of a halide ion X^-. The steps involved in the formation of ionic halides are described in the Born–Haber cycle in § 10.7. The relative ease with which the halogens form ionic halides is determined by three factors. They are

Factor (a) The standard bond dissociation enthalpy of $X_2(g)$:

Energy is absorbed in $X_2(g) \rightarrow 2X(g)$

Factor (b) The first electron affinity of X:

Energy is released in $X(g) + e^- \rightarrow X^-(g)$

Factor (c) The standard lattice enthalpy of the halide formed:

Energy is released in $\mathbf{M}^+(g) + X^-(g) \rightarrow \mathbf{M}X(s)$

Values of *(a)*, *(b)* and *(c)* are listed in Table 20.1.

The X—X bond strength decreases down the group ...

Factor (a) One would expect the X—X bond to decrease in strength as the size of X increases [see Figure 20.1]. This happens for the bond strength in

$$Cl_2 > Br_2 > I_2$$

[See Figure 20.2.]

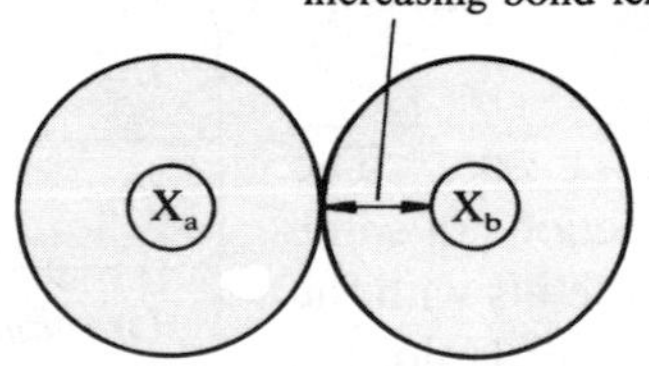

FIGURE 20.1 The Bonding in a Halogen Molecule

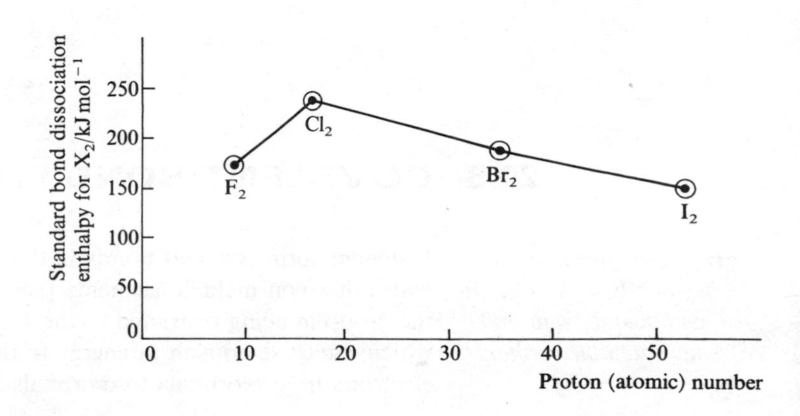

FIGURE 20.2 Standard Bond Dissociation Enthalpy for $X_2(g)$

... with the exception that the F—F bond is weaker than expected

The F—F bond is unexpectedly weak [see Figure 20.3]. The reason for this is thought to be that the small size of the fluorine atom brings the lone pairs of electrons in F_2 closer together than in other halogen molecules.

FIGURE 20.3 The F—F Bond

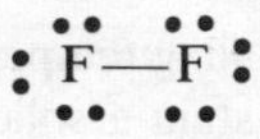

$$:\ddot{F}-\ddot{F}:$$

Repulsion between lone pairs of electrons weakens the F—F bond.

Electron affinity

Factor (*b*) Values of first electron affinities are shown in Table 20.1. You will find it interesting to plot electron affinity against atomic number.

Standard lattice enthalpy of halides

Factor (*c*) The standard lattice enthalpies of NaX are shown in Table 20.1. The F^- ion is so small that ionic fluorides have highly negative standard lattice enthalpies: much energy is given out when metal cations and fluoride ions form ionic lattices.

Fluorine brings out the highest oxidation states of metals [see Table 20.2]. Factor (*a*) is less endothermic for fluorine than for other halogens. Factors (*b*) and (*c*) are highly exothermic for fluorine and compensate for the ionisation enthalpy of the metal. Metals can therefore use their highest oxidation states in combination with fluorine. People working with fluorine use containers of copper or nickel steel. Although fluorine attacks these metals, the coating of metal fluoride which is formed resists further attack.

TABLE 20.2 The Reactions of some Metals with the Halogens

Metal	*Fluorine*	*Chlorine*	*Bromine*	*Iodine*
Most metals	Catch fire	Combine when heated		
Gold	Reacts	Does not react		
Platinum	PtF_6	$PtCl_4$	Does not react	
Silver	AgF_2	AgCl	Does not react	
Iron	FeF_3	$FeCl_3$	$FeBr_3$	FeI_2
Tin	SnF_4	$SnCl_4$	$SnBr_2$	SnI_2
Uranium	UF_6	UCl_4	UBr_4	UI_4

20.3 COVALENT BOND FORMATION

Halogens form covalent bonds in X_2 and in compounds with other non-metallic elements

Halogens form covalent bonds in the diatomic molecules X_2 and in compounds with other non-metallic elements [see Table 20.3]. Fluorine differs from the rest of the group in being restricted to the L shell. The other elements have empty d orbitals which are close enough in energy to the occupied p orbitals to allow promotion of electrons from p orbitals to d orbitals [see Figure 20.4].

FIGURE 20.4 Electron Configurations in F, Cl and Cl*

F	Cl	Cl*
	3d [ ][ ][ ][ ][ ]	3d [↑][ ][ ][ ][ ]
2p [↑↓][↑↓][↑]	3p [↑↓][↑↓][↑]	3p [↑↓][↑][↑]
2s [↑↓]	3s [↑↓]	3s [↑↓]
$1s^2$	$1s^22s^22p^6$	$1s^22s^22p^6$
F ground state can form one covalent bond.	Cl ground state can form one covalent bond.	Cl* first excited state can form three covalent bonds.

TABLE 20.3 The Reactions of some Non-metallic Elements with the Halogens

Element	*Fluorine*	*Chlorine*	*Bromine*	*Iodine*
He, Ne, Ar, N_2	Do not react			
Kr, Xe	React when heated	Do not react		
S	Reacts	Reacts when heated		
C	Reacts	Does not react		
O_2	Reacts	Does not react		
H_2	Reacts explosively, even in the dark at − 200 °C	Reacts explosively in sunlight; slowly in the dark below 200 °C	Reacts above 200 °C and at lower temperatures with Pt catalyst	Reacts to form an equilibrium mixture of H_2, I_2, HI
Other non-metallic elements	React readily. Fluorine brings out the highest oxidation state. Examples of fluorides which have no corresponding chlorides are SF_6, SiF_6, IF_7, XeF_6.			

Elements employ their highest oxidation states in combination with fluorine

Fluorine brings out the highest oxidation states of elements with which it combines. The reason is the high standard bond enthalpies of covalent bonds between fluorine and other elements. Much energy is given out when these bonds are formed. Atoms can promote electrons from shared orbitals to unoccupied orbitals because the energy required for promotion will be repaid when covalent bonds are formed.

CHECKPOINT 20A: BONDING

1. Why is fluorine more reactive than the other halogens?

2. Why do metals show their highest oxidation states in combination with fluorine? Explain why uranium forms UF_6 but the highest chloride is UCl_4.

3. Explain why sulphur combines with fluorine to form SF_6 but with chlorine forms SCl_4; why iodine forms IF_7 with fluorine but forms ICl_3 with chlorine; why bromine forms BrF_5 but the highest chloride is $BrCl_3$.

4. For iodine, draw electrons-in-boxes diagrams for the ground state and the first, second and third excited states. Say how many covalent bonds can be formed by iodine in each state. Explain why electron promotion occurs more readily in iodine than in chlorine.

5. For the halogens, referring to Table 20.1, plot:

(*a*) covalent bond radius against the proton (atomic) number of the halogen,

(*b*) ionic radius against proton number.

20.4 OXIDISING REACTIONS

Fluorine is the most powerful oxidant of the halogens

The oxidising power of the halogens is measured by the value of the standard reduction potential for the process

$$X_2(aq) + 2e^- \rightleftharpoons 2X^-(aq)$$

In most of their oxidising reactions, the halogens are reacting as X_2 molecules and forming hydrated ions, $X^-(aq)$.

From the values in Table 20.1, § 20.1 and Figure 20.5, you can see that fluorine is the most powerful oxidant of the group.

TABLE 20.4 Some Oxidising Reactions of the Halogens

Oxidant	*Reaction*
All halogens	Sulphite, SO_3^{2-} → Sulphate, SO_4^{2-}
All halogens	Hydrogen sulphide, H_2S → Sulphur, S
Cl_2, Br_2	Thiosulphate, $S_2O_3^{2-}$ → Sulphate, SO_4^{2-}
I_2	Thiosulphate, $S_2O_3^{2-}$ → Tetrathionate, $S_4O_6^{2-}$ (This reaction is used for titrimetric analysis of iodine [§ 3.18].)
Cl_2, Br_2	Organic compounds are oxidised, e.g., methane, CH_4, ethyne, C_2H_2.
F_2	Reacts explosively with organic compounds
I_2	Does not oxidise organic compounds

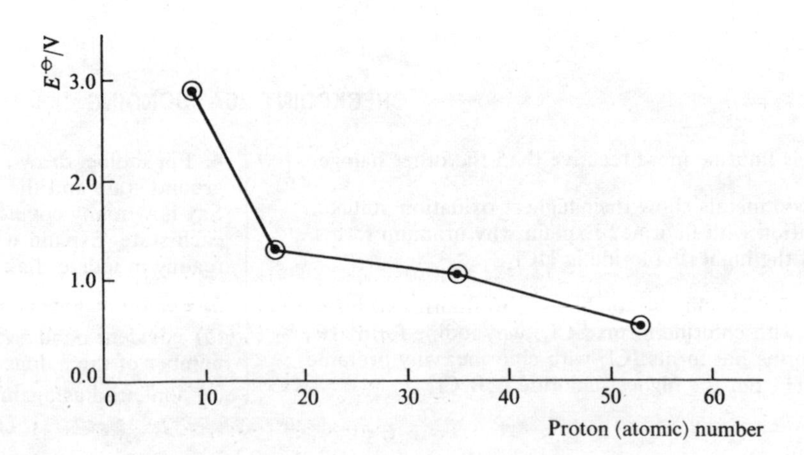

FIGURE 20.5 Standard Electrode Potentials for the Halogens

Standard electrode potentials measure oxidising power

By combining the values for $E^{\ominus}$ shown in Table 20.1, you can answer these questions:

Will fluorine oxidise chloride ions?
Will fluorine oxidise bromides?
Will chlorine oxidise (a) bromides, (b) iodides and (c) fluorides?
Which halogens will be oxidised by bromine?

[Remember that a reaction will happen if the value of $E^{\ominus}$ is positive.]

Chlorine water is often used as an oxidising agent

Fluorine is rarely used as an oxidising reagent as it is difficult to handle. Chlorine and its aqueous solution, 'chlorine water', are often used as oxidising agents. In chlorine water, there are two oxidising agents: chlorine and chloric(I) acid, HClO [§ 20.7].

CHECKPOINT 20B: REACTIVITY

1. Explain why bromine oxidises iodides, but bromides are oxidised by chlorine.

2. Some half-reaction equations are listed below:

(*a*) $SO_3^{2-}(aq) + H_2O(l) \rightarrow SO_4^{2-}(aq) + 2H^+(aq) + 2e^-$
(*b*) $H_2S(aq) \rightarrow S(s) + 2H^+(aq) + 2e^-$
(*c*) $X_2(aq) + 2e^- \rightarrow 2X^-(aq)$
(*d*) $S_2O_3^{2-}(aq) + 5H_2O(l) \rightarrow 2SO_4^{2-}(aq) + 10H^+(aq) + 8e^-$
(*e*) $2S_2O_3^{2-}(aq) \rightarrow S_4O_6^{2-}(aq) + 2e^-$

By combining half-reaction equations, obtain equations for the reactions of chlorine with (i) sulphites, (ii) hydrogen sulphide, (iii) a thiosulphate and (iv) obtain an equation for the reaction of iodine with a thiosulphate.

3. Choose two reactions which illustrate the gradation in reactivity down Group 7.

20.5 OCCURRENCE AND EXTRACTION

The halogens are too reactive to occur free. The methods used for the commercial extraction of the halogens and the laboratory preparation of the halogens are related to their oxidising power. There is no oxidant that will oxidise fluorides to fluorine, and electrolysis is used. Chlorine also is made by electrolysis industrially. Sufficiently powerful oxidants will oxidise bromides and iodides to the halogens.

20.5.1 COMMERCIAL EXTRACTION

Fluorine is mined as fluorspar, CaF_2, and as cryolite, Na_3AlF_6. It is obtained by an electrolytic method.

Chlorine is obtained by the electrolysis of molten sodium chloride [§ 18.3] and the electrolysis of brine [§ 18.5.5].

Bromine is obtained from seawater as shown in Figure 20.6.

Iodine is mined as sodium iodate(V), $NaIO_3$, which is present in Chile saltpetre, $NaNO_3$. Sodium hydrogensulphite is employed to reduce iodate(V) ions to iodide ions. A reaction between iodide ions and iodate(V) ions produces iodine:

$$IO_3^-(aq) + 3HSO_3^-(aq) \rightarrow I^-(aq) + 3HSO_4^-(aq)$$

$$IO_3^-(aq) + 5I^-(aq) + 6H^+(aq) \rightarrow 3I_2(s) + 3H_2O(l)$$

FIGURE 20.6 Extraction of Bromine from Sea Water

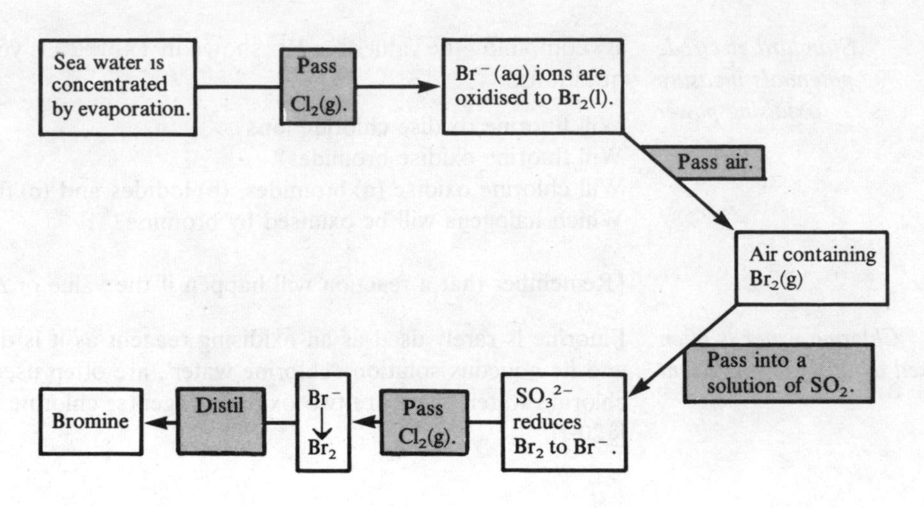

20.5.2 LABORATORY PREPARATION

In the laboratory, chlorine, bromine and iodine are obtained by oxidation of the halides:

$$2X^- \rightleftharpoons X_2 + 2e^-$$

Figure 20.7 shows the preparation of chlorine by the action of concentrated sulphuric acid and manganese(IV) oxide on sodium chloride.

FIGURE 20.7 Laboratory Preparation of dry Chlorine (in fume cupboard)

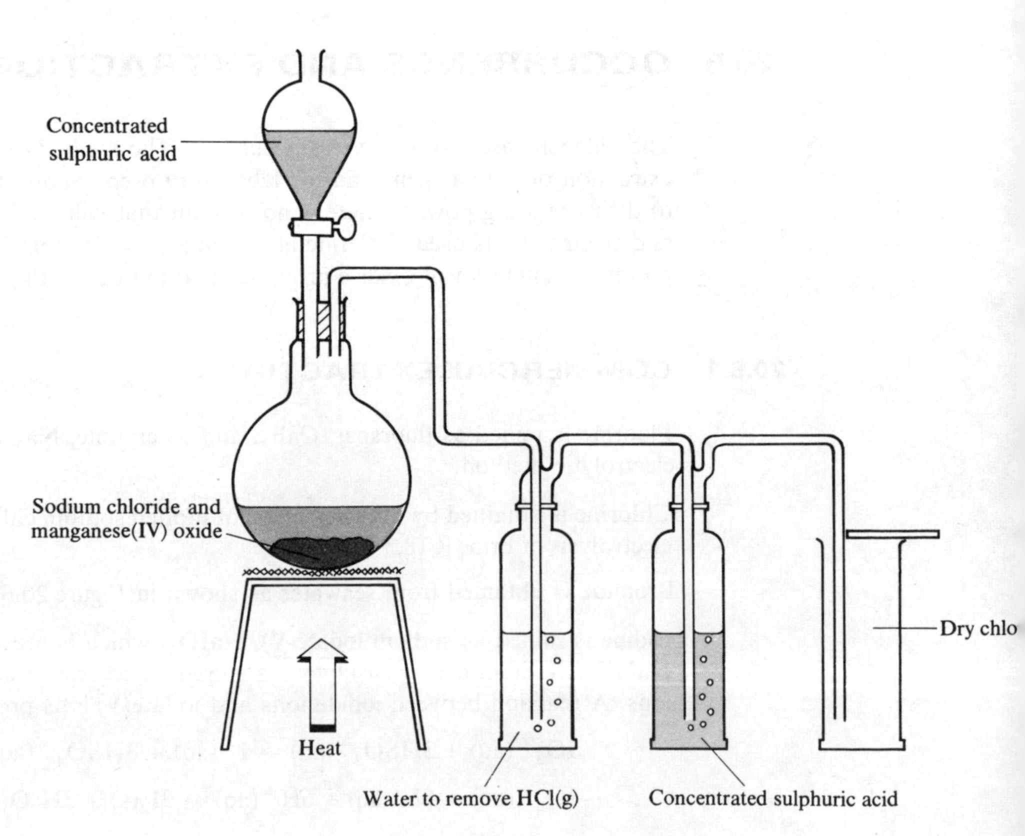

Alternatively, concentrated hydrochloric acid can be used as a source of chlorine. It can be run from a tap funnel on to potassium manganate(VII) or warmed with manganese(IV) oxide or lead(IV) oxide. If the chlorine is not required dry, it can be collected over water, in which it is only slightly soluble.

FIGURE 20.8
Laboratory Preparation of Bromine

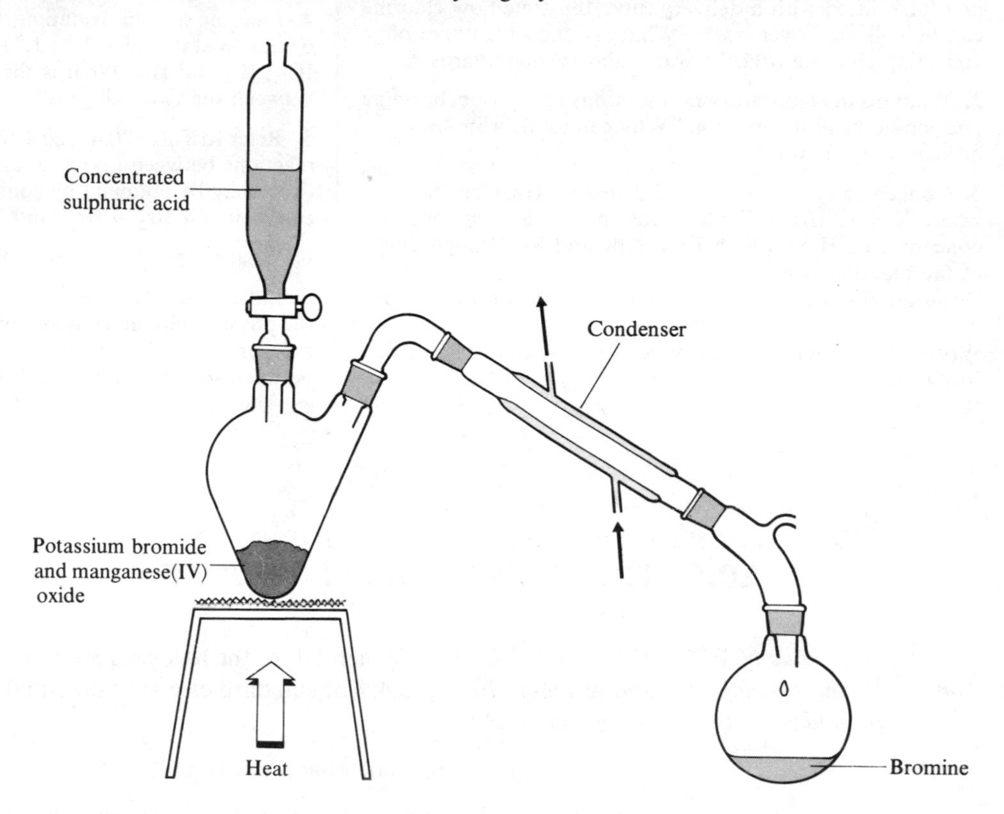

The laboratory preparation of bromine is shown in Figure 20.8 and that of iodine in Figure 20.9. Concentrated sulphuric acid displaces the hydrogen halide from its salt, and manganese(IV) oxide oxidises it to the halogen [see Question 3, Checkpoint 20C].

A solution of chlorine in water is readily made by dissolving bleaching powder, $CaCl_2 \cdot Ca(ClO)_2$, in water and adding dilute hydrochloric acid:

$$Ca(ClO)_2 + 4H^+(aq) + 2Cl^-(aq) \rightarrow Ca^{2+}(aq) + 2H_2O(l) + 2Cl_2(aq)$$

FIGURE 20.9
Laboratory Preparation of Iodine

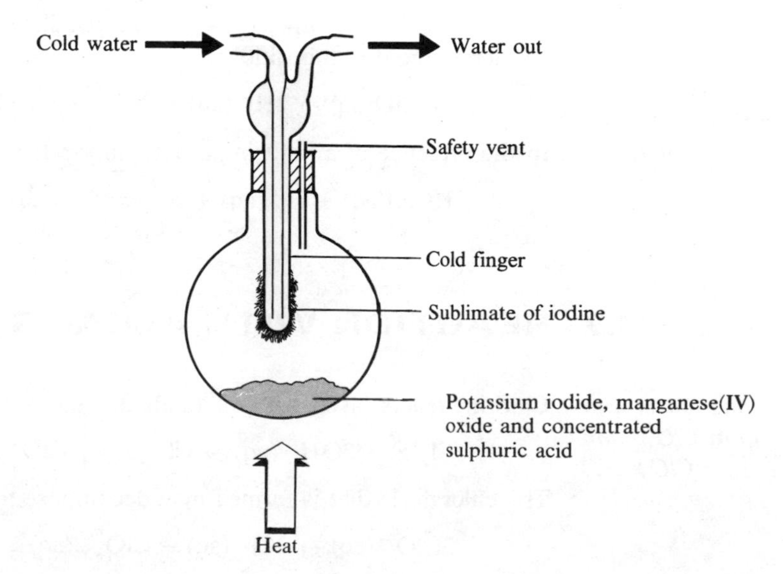

CHECKPOINT 20C: PREPARATIONS

1. Draw an apparatus in which concentrated hydrochloric acid is run from a tap funnel on to potassium manganate(VII) in a flask fitted with a delivery tube. Illustrate how chlorine can be collected over water. What are the advantages of collecting chlorine (*a*) over water and (*b*) downwards?

2. What do the laboratory preparations of chlorine, bromine and iodine have in common? Why cannot fluorine be prepared in this way?

3. Concentrated sulphuric acid displaces HX from KX, where X = Cl, Br, I. Write equations for the reactions of concentrated H_2SO_4 with KCl, KBr and KI. One product of the reaction is $KHSO_4$.
Write a half-reaction equation for the oxidation of X^- to X_2.
Write a half-reaction equation for the reduction of acidified MnO_2 to Mn^{2+}.
Combine the half-reaction equations to give the equation for the oxidation of X^- by MnO_2 and acid.

4. Fluorine has an oxidation state of −1 only. Iodine has oxidation states of −1, +1, +3, +5 and +7 in I^-, ICl, ICl_3, IF_5 and IF_7. What is the reason for this difference between the two halogens?

5. Refer to Table 20.4, § 20.4. Write equations for the reactions between (*a*) Cl_2 and Br^-, (*b*) SO_3^{2-} and Br_2. (This may be obtained by combining the half-reaction equations for $Br_2 \rightarrow Br^-$ and $SO_3^{2-} \rightarrow SO_4^{2-}$.)

6. What is the quickest way of preparing chlorine water?

7. Little is known of the chemistry of astatine. Describe the physical characteristics which you would expect for this halogen. What do you think would be formed in a reaction between sodium astatide and concentrated sulphuric acid?

20.6 REACTION WITH WATER

The value of $E^\ominus$ measures the strength of an oxidising agent

The standard electrode potentials of the halogens are shown in Table 20.1, § 20.1, and in Figure 20.5, § 20.4. The standard electrode potential (reduction potential) for oxygen is +1.23 V.

$$O_2(g) + 4H^+(aq) + 4e^- \rightleftharpoons 2H_2O(l);\ E^\ominus = +1.23\,V$$

Fluorine and chlorine are thus capable of oxidising water, while bromine and iodine are not. Fluorine oxidises water to oxygen (and some hydrogen peroxide and ozone):

$$2F_2(g) + 2H_2O(l) \rightarrow 4HF(aq) + O_2(g)$$

Chlorine reacts slowly with water to form hydrochloric acid and chloric(I) acid, HClO:

$$Cl_2(g) + H_2O(l) \rightarrow HClO(aq) + H^+(aq) + Cl^-(aq)$$

Fluorine oxidises water

Chlorine slowly oxidises water, with the formation of HClO

Chloric(I) acid decomposes to give oxygen slowly, unless sunlight is present to accelerate the decomposition:

$$2HClO(aq) \rightarrow 2H^+(aq) + 2Cl^-(aq) + O_2(g)$$

In the presence of a reducing agent, chloric(I) acid acts as an oxidising agent:

$$HClO(aq) + H^+(aq) + 2e^- \rightleftharpoons Cl^-(aq) + H_2O(l)$$

20.7 REACTION WITH ALKALIS

Chlorine reacts with alkalis, to form ClO^- and ClO_3^- ...

Chlorine reacts faster with dilute alkalis than with water:

$$Cl_2(g) + 2OH^-(aq) \rightarrow Cl^-(aq) + ClO^-(aq) + H_2O(l)$$

The chlorate(I) that is formed may decompose to form a chloride and a chlorate(V):

$$3ClO^-(aq) \rightarrow 2Cl^-(aq) + ClO_3^-(aq)$$

... The reaction has commercial use

The decomposition is slow at room temperature but fast at 70 °C. A reaction like this, in which a species is simultaneously oxidised and reduced, is called a **disproportionation** reaction. Part of the ClO^- is oxidised to ClO_3^-, while the rest is reduced to Cl^-. These reactions are used commercially in the manufacture of sodium chlorate(I), NaClO, a widely used mild antiseptic (Milton®), and sodium chlorate(V), $NaClO_3$, a powerful weedkiller (Tandar®). Both chlorine and sodium hydroxide are products of the electrolysis of brine. If they are allowed to come into contact, sodium chlorate(I) is produced; if the temperature is raised, they can be made to form sodium chlorate(V).

Br_2 and I_2 react with alkalis to form the halate(V) ions

Bromine and iodine react with dilute alkalis, either cold or warm, to give a mixture of halide and halate(V):

$$3Br_2(aq) + 6OH^-(aq) \rightarrow BrO_3^-(aq) + 5Br^-(aq) + 3H_2O(l)$$

Alkalis + F_2 give OF_2 or O_2

Fluorine reacts with cold, dilute alkalis to give oxygen difluoride, OF_2, and with warm concentrated alkalis to give oxygen:

$$2F_2(g) + 2OH^-(aq) \xrightarrow{\text{cold, dil.}} OF_2(g) + 2F^-(aq) + H_2O(l)$$

$$2F_2(g) + 4OH^-(aq) \xrightarrow{\text{warm, conc.}} O_2(g) + 4F^-(aq) + 2H_2O(l)$$

FIGURE 20.10 Some Reactions of Chlorine

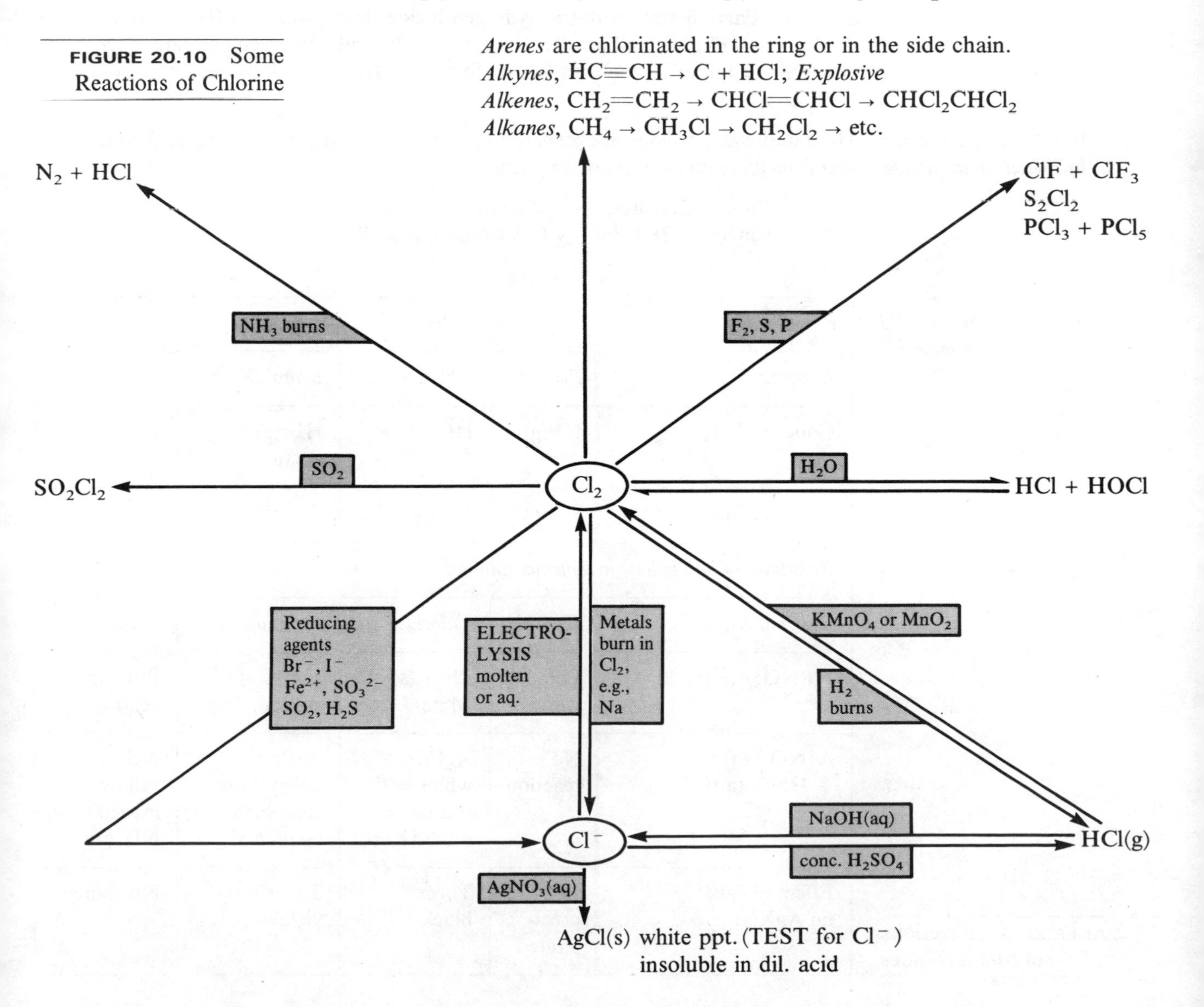

CHECKPOINT 20D: REACTIONS

1. Write the equation for the reaction of chlorine with water. Explain how the addition of an alkali speeds up the reaction.

2. Write the oxidation number of chlorine in each of the species

$$3\underline{Cl}O^-(aq) \rightarrow 2\underline{Cl}^-(aq) + \underline{Cl}O_3^-(aq)$$

Why is this reaction described as *disproportionation*?

3. Write the oxidation numbers of oxygen and fluorine in each of the species

(*a*) $2\underline{F}_2(g) + 2\underline{O}H^-(aq) \rightarrow \underline{O}\underline{F}_2(g) + 2\underline{F}^-(aq) + H_2\underline{O}(l)$

(*b*) $2\underline{F}_2(g) + 4\underline{O}H^-(aq) \rightarrow \underline{O}_2(g) + 4\underline{F}^-(aq) + 2H_2\underline{O}(l)$

4. How do the standard electrode potentials of the X_2/X^- half-cells determine (*a*) the reactions of the halogens with water and (*b*) the methods of preparation of the elements?

20.8 METAL HALIDES

Metal + X_2 or HX(g) gives metal halide

Anhydrous metal halides are prepared by heating the metal in a stream of dry halogen or hydrogen halide [see Figure 19.9, § 19.3.1]. Metals with variable oxidation states usually give the halide of a higher oxidation state with the halogen and the halide of a lower oxidation state with the hydrogen halide. Iron gives iron(III) chloride, $FeCl_3$, with chlorine, and iron(II) chloride, $FeCl_2$, with hydrogen chloride. Exceptions are the reaction of iodine with iron to form iron(II) iodide, FeI_2, and with copper to form copper(I) iodide, CuI.

Metal + HX(aq) gives hydrated metal halide

Hydrated metal halides are prepared by the reaction of a hydrohalic acid with a metal or its oxide, hydroxide or carbonate:

$$Zn(s) + 2HCl(aq) \rightarrow ZnCl_2(aq) + H_2(g)$$
$$CuO(s) + 2HCl(aq) \rightarrow CuCl_2(aq) + H_2O(l)$$

Test reactions of solid halides...

...and of solutions

TABLE 20.5 Reactions of Metal Halides

Reactions of solid halides				
Reagent	*Fluoride*	*Chloride*	*Bromide*	*Iodide*
Conc. H_2SO_4	HF(g)	HCl(g)	HBr(g) $Br_2(g)$	$I_2(g)$
Conc. H_2SO_4 + MnO_2	HF(g)	$Cl_2(g)$	$Br_2(g)$	$I_2(g)$
Reactions of halide ions in aqueous solution				
Reagent	*Fluoride*	*Chloride*	*Bromide*	*Iodide*
$Pb(NO_3)_2(aq)$	$PbF_2(s)$ white	$PbCl_2(s)$ white	$PbBr_2(s)$ cream	$PbI_2(s)$ yellow
$AgNO_3(aq)$ + $HNO_3(aq)$	No reaction	AgCl(s) white soluble in dil. $NH_3(aq)$	AgBr(s) pale yellow soluble in conc. $NH_3(aq)$	AgI(s) yellow insoluble in $NH_3(aq)$
Effect of light on AgX(s)	Not preci- pitated	Turns black	Turns yellow	No change

Many halides crystallise with water of crystallisation, e.g., $MgCl_2 \cdot 6H_2O$, $AlCl_3 \cdot 6H_2O$. When the hydrates are heated in an attempt to obtain the anhydrous salt, they are often hydrolysed to give a basic chloride or hydroxide:

$$MgCl_2 \cdot 6H_2O(s) \rightleftharpoons MgCl(OH)(s) + HCl(g) + 5H_2O(g)$$

$$AlCl_3 \cdot 6H_2O(s) \rightleftharpoons Al(OH)_3(s) + 3HCl(g) + 3H_2O(g)$$

Some hydrates are hydrolysed on heating

The degree of hydrolysis depends on the degree of covalent character in the bonds [§ 15.3.4]. Anhydrous magnesium chloride can be obtained by heating the crystals of the hydrate in a stream of hydrogen chloride gas. This moves the equilibrium over to the left-hand side.

Most metal halides are soluble

Metal halides are soluble, except for the lead halides and the chloride, bromide and iodide of mercury(I) and silver. Some ionic fluorides differ in solubility from the corresponding chlorides. Calcium fluoride is insoluble, and silver fluoride is soluble. Some tests for metal halides are summarised in Table 20.5.

20.9 NON-METAL HALIDES

Combination with non-metallic elements

Halogens combine with many non-metallic elements. Fluorine brings out the highest oxidation states of elements with which it combines. In the case of other halides, the use of an excess of the halogen often results in the formation of the halide of the element in its highest oxidation state. For example, phosphorus reacts with chlorine to form phosphorus(III) chloride PCl_3, or phosphorus(V) chloride, PCl_5, depending on the proportion of chlorine used.

The halides of non-metallic elements are covalent and are hydrolysed by water, with the exception of tetrachloromethane, CCl_4 [§ 23.7.2]:

Halides of non-metallic elements are hydrolysed with the formation of HX(g)

$$SiCl_4(l) + 2H_2O(l) \rightarrow SiO_2(aq) + 4HCl(aq)$$

Silicon(IV) chloride   Silicon(IV) oxide

$$PCl_3(l) + 3H_2O(l) \rightarrow H_3PO_3(aq) + 3HCl(g)$$

Phosphorus(III) chloride   Phosphonic acid

20.9.1 HYDROGEN HALIDES

The hydrogen halides, hydrogen fluoride, hydrogen chloride, hydrogen bromide and hydrogen iodide, can be made by direct synthesis. The method works well for hydrogen chloride, which is manufactured industrially by burning a stream of hydrogen in chlorine:

$$H_2(g) + Cl_2(g) \rightarrow 2HCl(g)$$

Industrial manufacture of HCl(g)...

The reaction between hydrogen and fluorine is dangerously fast, and the reactions between hydrogen and bromine or iodine do not give a good yield.

Hydrogen fluoride is toxic and caustic. It is made by the action of concentrated sulphuric acid on calcium fluoride:

...and HF(g)

$$CaF_2(s) + 2H_2SO_4(l) \rightarrow 2HF(g) + Ca(HSO_4)_2\,(\text{in solution})$$

Laboratory preparation of hydrogen chloride...

A similar displacement reaction between sodium chloride and concentrated sulphuric acid can be used for the laboratory preparation of hydrogen chloride [see Figure 20.11].

FIGURE 20.11
Laboratory Preparation of Hydrogen Chloride

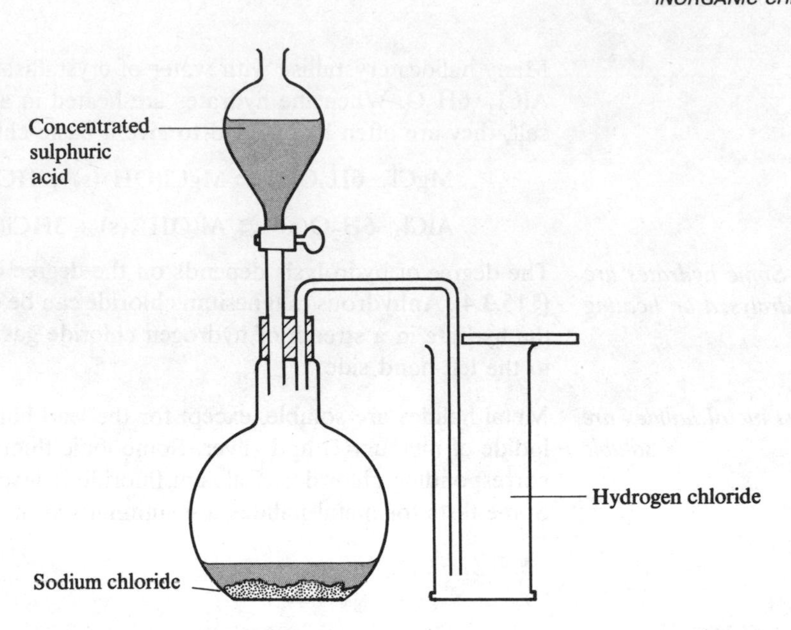

... hydrogen bromide and hydrogen iodide

Hydrogen bromide and iodide cannot be made in this way because concentrated sulphuric acid oxidises them to the halogens. They are made by hydrolysis of phosphorus tribromide or phosphorus triiodide.

TABLE 20.6
The Physical Properties of the Hydrogen Halides

	HF	*HCl*	*HBr*	*HI*
$\Delta H^{\ominus}_{Formation}$/kJ mol^{-1}	−270	−92	−36	+26
$\Delta H^{\ominus}_{Bond\ dissociation}$/kJ mol^{-1}	+560	+430	+370	+300
Boiling temperature/°C	20	−85	−67	−35
pK_a	3.25	−7.4	−9.5	−10

From Table 20.6, you can see that the standard bond dissociation enthalpy decreases in the order

$$HF > HCl > HBr > HI$$

Thermal stability of HX

This is why the thermal stability of the compounds decreases in the same order. Another example of the effect of standard bond dissociation enthalpies is the ability of hydrogen halides in aqueous solution to act as proton-donors:

$$HX(aq) + H_2O(l) \rightleftharpoons H_3O^+(aq) + X^-(aq)$$

Acid strength of HX

The acidic strength increases in the order

$$HF \ll HCl < HBr < HI$$

Hydrogen fluoride is a much weaker acid than the rest.

HX(aq) show typical acid reactions

The hydrohalic acids react with metals above hydrogen in the Electrochemical Series and with metal oxides, hydroxides and carbonates. Hydrofluoric acid attacks glass, and glass apparatus cannot be used for work with hydrogen fluoride or fluorine.

HF reacts with glass

Fluorine does not attack perfectly dry glass, but since fluorine reacts with water to form hydrogen fluoride, it will attack damp glassware.

The characteristics of chlorides are summarised in Figure 20.12.

FIGURE 20.12 The Nature of Chlorides

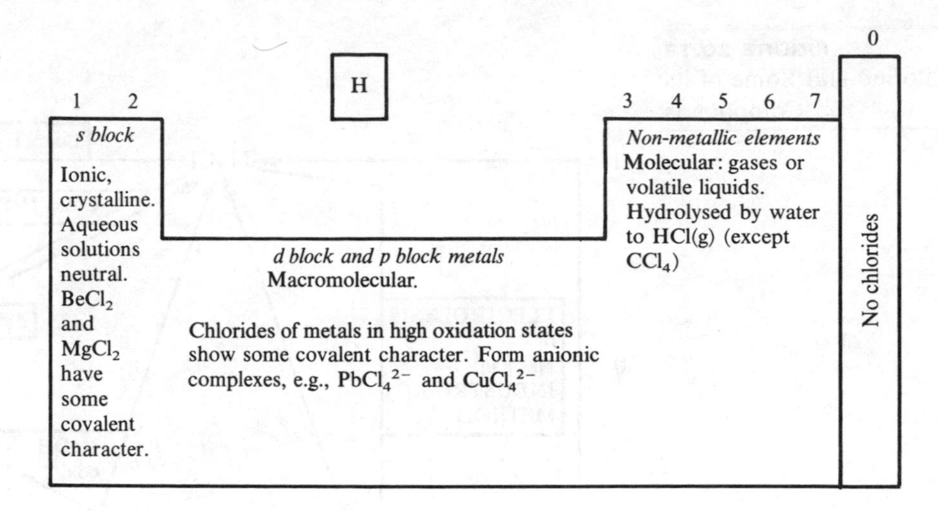

H
0
1 2
3 4 5 6 7
s block
Ionic, crystalline. Aqueous solutions neutral. BeCl2 and MgCl2 have some covalent character.
d block and p block metals
Macromolecular.
Chlorides of metals in high oxidation states show some covalent character. Form anionic complexes, e.g., PbCl4 2− and CuCl4 2−
Non-metallic elements
Molecular: gases or volatile liquids. Hydrolysed by water to HCl(g) (except CCl4)
No chlorides

CHECKPOINT 20E: HALIDES

1. Which of the halogens has (*a*) the smallest electron affinity, (*b*) the greatest oxidising power, (*c*) the greatest ability to form hydrogen bonds and (*d*) the weakest acid HX?

2. What is meant by describing the hydrohalic acids as 'fuming' gases? Why do they behave in this way?

†**3.** The boiling temperatures of the hydrogen halides depend on intermolecular forces.

(*a*) Explain the order of T_b:

$$HCl < HBr < HI$$

(*b*) Explain why T_b of HF is so much greater than those of the other hydrogen halides [see § 4.7.3].

4. Would you expect a solution of HAt to be a strong acid or a weak acid? What would you expect to see when a solution of silver nitrate was added to a solution of HAt(aq)?

5. Tetrachloromethane, CCl_4, does not react with water, but silicon(IV) chloride, $SiCl_4$, is rapidly hydrolysed. Explain this difference.

6. (*a*) In standard bond dissociation enthalpy

$$HF > HCl > HBr > HI$$

(*b*) In thermal stability

$$HF > HCl > HBr > HI$$

(*c*) In acid strength

$$HF \ll HCl < HBr < HI$$

Explain how statements (*b*) and (*c*) follow from statement (*a*).

7. Discuss the hydrogen halides with respect to their power as reducing agents.

20.10 OXO-ACIDS AND THEIR SALTS

The oxo-acids of chlorine

The oxo-acids of chlorine are:

$HClO$	chloric(I) acid	Thermal stability increases ↓
$HClO_2$	chloric(III) acid	Acid strength increases ↓
$HClO_3$	chloric(V) acid	Oxidising power increases ↓
$HClO_4$	chloric(VII) acid	

Some salts . . .

. . . $NaClO$ $NaClO_3$ $KClO_3$ $KBrO_3$ KIO_3

Some of the salts of these acids have been mentioned, e.g., $NaClO$ and $NaClO_3$. Potassium chlorate(V), $KClO_3$, is used in matches and fireworks. Potassium bromate(V) is used as a source of bromine in volumetric analysis. Potassium iodate(V) is used as a primary standard in volumetric analysis [§ 3.13.1].

FIGURE 20.13
Chlorine and Some of its Compounds

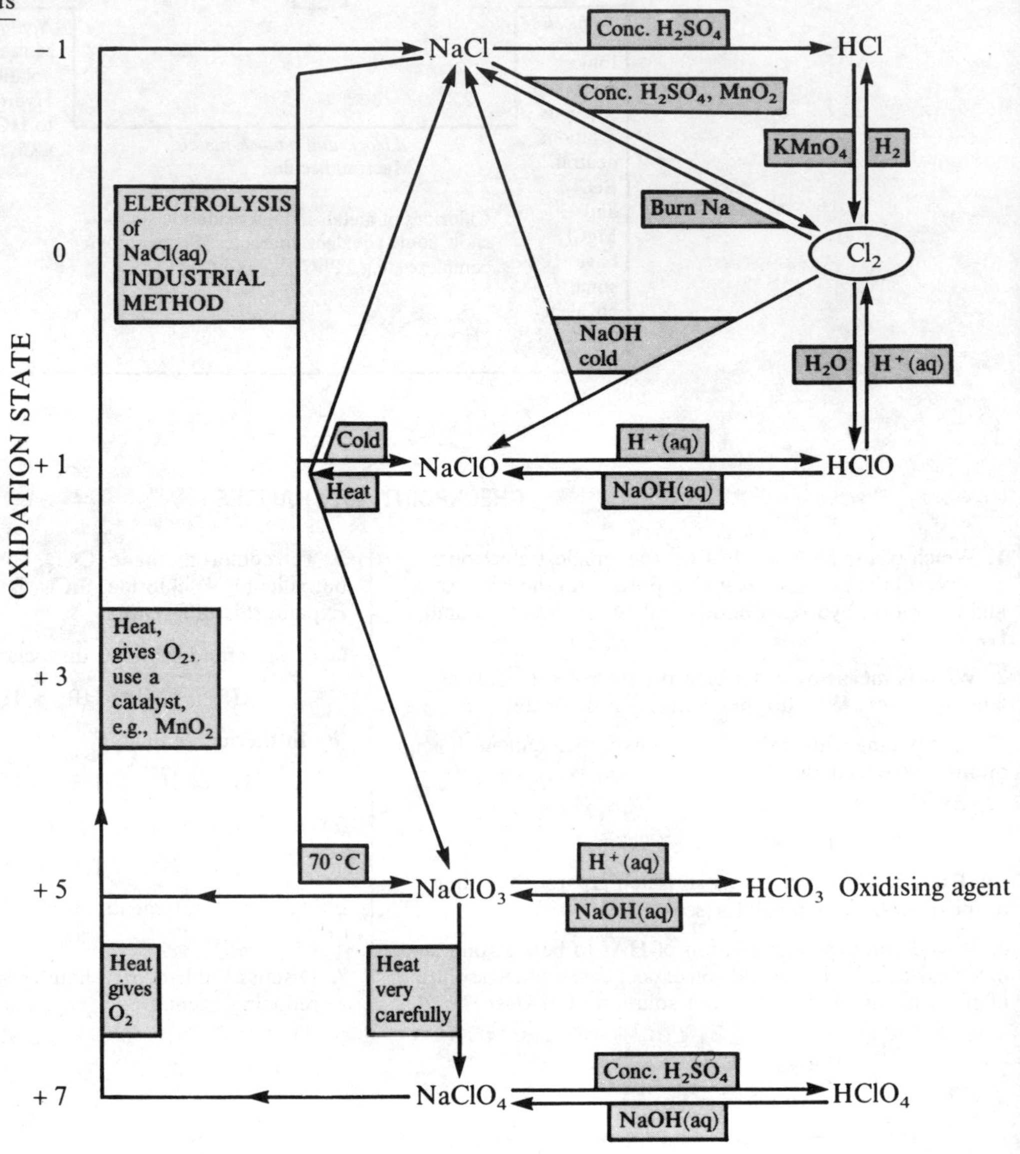

20.11 THYROXINE

TOPIC

Your body contains 1/2 500 000 by mass of iodine (a mass about the size of a pinhead). The absence of this mass of iodine would be fatal as it is needed for the production of thyroxine.

$$\text{HO}-\text{C}_6\text{H}_2\text{I}_2-\text{O}-\text{C}_6\text{H}_2\text{I}_2-\text{CH}_2-\text{CH}(\text{NH}_2)-\text{C}(=\text{O})-\text{OH}$$

Thyroxine

The hormone thyroxine is a compound of iodine

Thyroxine is a hormone that regulates the body's chemical activity. It is produced by the thyroid gland in the throat. A lack of iodine in the diet causes an enlargement of the thyroid known as goitre. The body attempts to increase the production of thyroxine by increasing the number of cells in the thyroid. This attempt cannot succeed if the concentration of iodine remains low.

CHECKPOINT 20F: SALTS OF OXO-ACIDS

1. Explain why it is more accurate to weigh out potassium iodate(V) as a source of iodine than to weigh out iodine.

2. Write an equation for the formation of bromine from potassium bromate(V).

3. Draw the spatial arrangement of bonds in ClF, ClF_3, BrF_5, IF_7 and ICl_4^-, ICl_2^-, HOI.

4. Write the formulae of three ions containing astatine.

5. Classify the following compounds as ionic or molecular or macromolecular: KBr, $AlBr_3$, BCl_3, CCl_4, $PbCl_2$, $PbCl_4$, $SiCl_4$, PCl_3, PCl_5. What tests would you make on the substances to substantiate your classification?

20.12 SUMMARY OF GROUP 7

Summary of Group 7B

1. The ions X^- are readily formed. Electron affinity decreases down the group. All the halogens react vigorously with metals. With the s block metals, they form ionic compounds.

2. Covalent compounds and complex ions are formed by the halogens in combination with p block and d block metals. Ionic radii increase down the group, and polarisation of X^- therefore increases down the group. Fluorides are usually ionic.

3. The ions XO^-, XO_2^-, XO_3^- and XO_4^- are formed, except by fluorine.

4. The halogens form molecular compounds with non-metallic elements.

5. The halogens are all oxidising agents. Oxidising power decreases in the order

$$F_2 > Cl_2 > Br_2 > I_2$$

Reducing power of X^- decreases in the order

$$I^- > Br^-$$

Cl^- and F^- are not reductants.

6. Intermolecular forces increase down the group, and lead to a transition in physical state from fluorine (pale yellow gas) and chlorine (green gas), to bromine (brown volatile liquid) and iodine (shiny black solid which sublimes to a violet vapour).

20.13 *DIFFERENCES BETWEEN FLUORINE AND OTHER HALOGENS

1. Fluorine is exceptionally reactive because of the low dissociation enthalpy (energy) of the F—F bond.

Fluorine differs from the rest of the group...

...in its high reactivity...

2. Fluorine compounds are very stable. There are two reasons for this:

(*a*) The small size of the fluorine atom allows it to form covalent bonds which are stronger than those of other halogens. Since covalent fluorides have highly negative standard enthalpies of formation, they are stable with respect to dissociation either into their elements or into compounds with lower oxidation numbers. This is why fluorine brings out higher oxidation numbers of other atoms than do the rest of the halogens.

...in the stability of its compounds...

(*b*) The small size of the fluoride ion gives ionic fluorides higher lattice enthalpies than the corresponding compounds of other halogens. Fluorides are often more ionic than other halides. AlF_3 has an ionic structure; $AlCl_3$ has a layer structure, and Al_2Br_6 and Al_2I_6 are molecular compounds. Fluorides sometimes differ in solubility from other halides. The high lattice enthalpy makes calcium fluoride insoluble, in contrast to other calcium halides. Silver fluoride differs from other silver halides in being soluble, owing to the high enthalpy of hydration of the fluoride ion [§ 10.8].

...in its single oxidation state...

3. Fluorine is restricted to an L shell of eight electrons. This means that its only oxidation state is -1, and it explains the inertness of many fluorine compounds, e.g., fluorocarbons.

...and in the strength of the hydrogen bonds which it forms

4. Being the most electronegative of elements, fluorine is able to form strong hydrogen bonds. Hydrogen bonding and the strength of the H—F bond are responsible for the weakly acidic nature of hydrogen fluoride.

5. In its reactions with water and alkalis, fluorine differs from other halogens. It forms no oxo-ions.

20.14 USES OF HALOGENS

Uses of fluorine include the manufacture of 'freon', PTFE

Fluorine is used in the manufacture of fluorohydrocarbons. The commercial refrigerant gas *freon* contains CCl_2F_2, $CClF_3$ and other chlorofluoromethanes. It is extremely unreactive. Tetrafluoroethene, $CF_2{=}CF_2$, can be polymerised to give poly(tetrafluoroethene), $-\!(CF_2-CF_2)\!-_n$, PTFE. This material is a plastic which resists attack by most chemicals and is used in the chemical industry for the manufacture of corrosion-proof valves and seals. Its insulating property finds it a use for coating electrical wiring, and its low coefficient of friction finds it a use as a non-stick coating for cooking pans and for skis. Because of its volatility and its stability, it is used as the volatile component in aerosol cans.

Uses of chlorine include the manufacture of disinfectants...

Chlorine is used as a domestic bleach, and as a disinfectant in swimming baths. It is used in the manufacture of the gentle antiseptic, sodium chlorate(I), and the powerful weedkiller sodium chlorate(V).

...solvents...

...antiseptics...

...and insecticides

Chlorinated organic compounds find many uses. Tetrachloromethane, CCl_4, and trichloroethene, $CHCl{=}CCl_2$, are used as solvents for grease removal. Many antiseptics are chloro-compounds. TCP contains 2,4,6-trichlorophenol. Chloro-compounds, for example, DDT, have proved to be valuable insecticides [§ 20.15].

Bromine compounds are used as petrol additives

Bromine compounds, e.g., $C_2H_4Br_2$, are used as petrol additives. When tetraethyllead is added to petrol as an antiknock, lead oxides would be deposited in the cylinders in the absence of these organic bromo-compounds. They convert lead into volatile compounds, which are discharged through the vehicle exhaust into the air. Rising concern about the level of lead compounds in the air has led a number of governments, to ban the use of lead compounds as antiknocks [§ 26.3.1]. Since 1990, all new motor vehicles in the UK have been designed to run on lead-free petrol.

20.15 DDT

TOPIC

A research student called Othmar Geidler made DDT, in 1874. Sixty years later, another chemist called Paul Mueller repeated the synthesis so that he could try out DDT in his work on insecticides. He found that it was extremely poisonous to houseflies and other insects.

1,1,1-Trichloro-2,2-bis(4-chlorophenyl)ethane
(The letters DDT come from its former name, dichlorodiphenyltrichloroethane.)

DDT, a chloro-compound with insecticidal properties, was used against mosquitoes and lice in the Second World War

This work was done in the 1930s, and when the Second World War started in 1939, chemists had found more uses for DDT. During a war, in addition to those killed in action, many people die of disease through lack of medical supplies, shortage of water, overcrowding and poor sanitation. The use of DDT in the Second World War helped to alleviate some of this misery. When the Allies landed in islands in the Pacific, they faced the danger of malaria as well as enemy forces. By spraying DDT from aeroplanes, they were able to wipe out the mosquito population, and remove the source of malaria. Dusting with DDT kept the troops free from the body lice which had plagued soldiers in earlier wars. After the Allies landed in Italy and occupied Naples, an epidemic of typhus broke out. To kill the lice which carry the disease, the whole area was sprayed with DDT and the population dusted themselves with DDT. Within days, the epidemic was over. In 1948, Mueller was awarded the Nobel prize, for discovering the life-saving properties of DDT.

2,4-Dichlorophenoxyethanoic acid
(2,4-D)

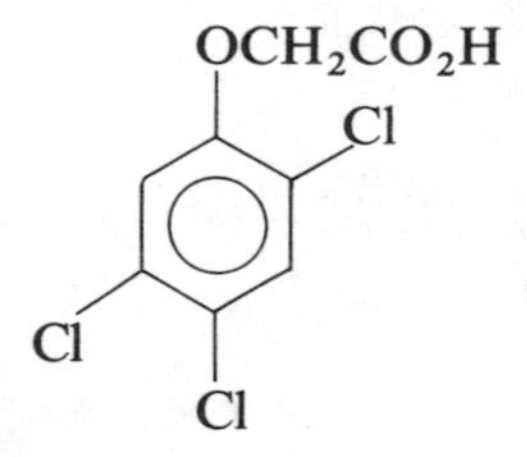

2,4,5-Trichlorophenoxyethanoic acid
(2,4,5-T)

DDT and related compounds were used as insecticides and herbicides by farmers

After the war, farmers welcomed DDT-related compounds to replace the non-selective insecticides and herbicides which they had been using. DDT killed insects but not farm animals. The herbicides 2,4-D and 2,4,5-T killed weeds and left grass to grow. Dieldrin and aldrin were found to be even more potent than DDT:

Dieldrin

Aldrin

Some insects soon became resistant

As early as 1946, however, it began to appear that the new chloro-compounds were not a perfect solution to the insect problem. Some species of housefly soon became resistant to DDT. In the USA, the cotton farmers had been delighted with the way DDT had attacked the cotton boll weevil, but, by 1960, they were having to spray more and more frequently with higher and higher doses as the weevil became resistant to the insecticide.

These compounds are very stable, and remain in a treated area to be eaten by microorganisms, fish and birds and small animals

In 1962, Rachel Carson, in her book *The Silent Spring*, called for a halt to the widespread and indiscriminate spraying of insecticides and herbicides. These compounds are so stable that they persist for a long time in areas where they have been sprayed. If they are eaten by birds and animals, they cannot be excreted because they are insoluble in water, but they can be stored in the body because they are soluble in fat. In the spring of 1956, large numbers of birds were found dead in cereal-growing regions and were found to have a high content of dieldrin and aldrin, with which the cereals had been treated. Scientists began to investigate the spreading of insecticides and herbicides. They found DDT in penguins in the Antarctic, where the spray had never been used. They reasoned that if DDT is used in a malarial region, it will be taken up by small organisms. If these are eaten by a fish and the fish is eaten by a bird, the bird may carry DDT for hundreds of miles. DDT can concentrate up a food chain. If the content of DDT in seawater is 1×10^{-6} ppm, plankton can have a content of 3×10^{-4} ppm, and fish eating the plankton may contain 0.5 ppm. Birds of prey concentrate DDT further until they may contain 10 ppm. It is possible that fish and birds which have high DDT contents may be eaten by human beings. If DDT is so toxic to insects, can it be completely harmless to human beings? In 1964, the Advisory Committee on Poisonous Substances used in Agriculture and Food Storage recommended that dieldrin and aldrin should be used only when drastic measures are needed. Restrictions have been placed on the use of DDT, and consumption of DDT has fallen to about half of what it was in 1964.

The use of DDT has declined

CHECKPOINT 20G: PESTICIDES

1. The subject for debate is:

'Modern synthetic pesticides do more harm than good.'

Some suggestions for students to put forward in debate are given below. You may need to do some library research before presenting your point. A chairperson is needed to call on the speakers and to take a vote.

Student 1: Define 'pesticide'. Describe what would happen if we had no pesticides.

Student 2: Say what chemical pesticides and other methods of control were used a century ago. Say what advantages and disadvantages they had compared with present day pesticides.

Student 3: Explain the advantages of modern pesticides.

Student 4: Explain why some people fear modern synthetic pesticides. Give examples.

Student 5: Say what alternatives there are to modern synthetic pesticides.

Chair: Take a vote.

2. An enzyme in mammals and resistant insects converts DDT into the substance shown here. This substance is biologically inactive. Explain how the shape of the molecule shown here differs from that of DDT. Suggest why this might make it unable to interact with the receptor site in insects.

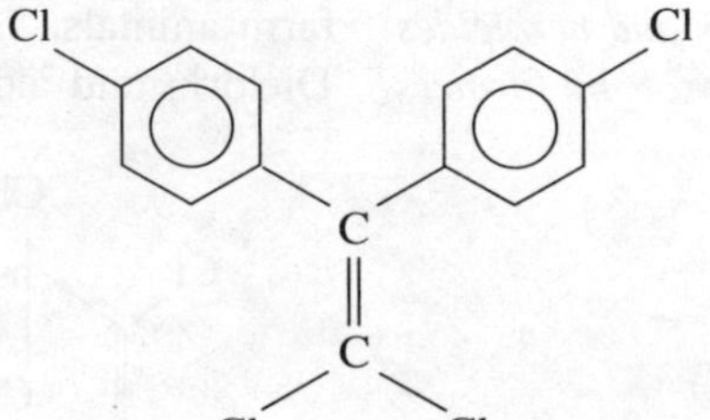

3. Would you use DDT if you were a farmer? If your answer is no, what would you do to protect your crops from weeds and pests? (Many people are trying to figure out an answer to this question!)

4. (*a*) Give *two* examples of catalysis in organic chemistry. For *each* example, give the name of the catalyst and of the reactants and write an equation for the reaction.

(*b*) Under appropriate conditions, trichloroethanal reacts with chlorobenzene to produce the pesticide DDT.

$$CCl_3CHO + 2\,C_6H_5Cl \rightarrow (ClC_6H_4)_2CH(CCl_3) + H_2O$$

(i) Classify this type of reaction.

(ii) If one mole of the product is refluxed with excess aqueous sodium hydroxide, acidified with aqueous nitric acid and then treated with excess aqueous silver nitrate, what mass of silver chloride would you expect to obtain?

(iii) What would you expect to observe if trichloroethanal was warmed with aqueous ammoniacal silver nitrate?

(iv) What type of reaction has occurred in (*b*)(iii)?

(*c*) The pesticide DDT has been restricted in its use.

(i) Give *two* reasons why this restriction has been imposed.

(ii) Suggest an alternative method to the use of DDT for the control of harmful insects.

(AEB 91)

QUESTIONS ON CHAPTER 20

1. Compare the compounds HF, HCl, HBr and HI with respect to (*a*) the strengths of the acids formed in aqueous solution and (*b*) thermal stability.

2. What trends are observed in Group 7 in the oxidising power of the halogens?

3. Give three general methods for preparing anhydrous halides of elements. Which of the halogens would you expect to stabilise the highest oxidation state of a particular element? Why?

4. Describe the manufacture of sodium hydroxide and chlorine from sodium chloride. Explain the principles involved. Why is there a big demand for these two substances?

5. Discuss the chemistry of Group 7. Suggest reasons for the trends in properties. Include the reactions of the halogens with (*a*) metals, (*b*) hydrogen, (*c*) water, (*d*) sodium hydroxide solution, (*e*) solutions of alkali metal halides. Include also the reaction of sodium halides with sulphuric acid and the strengths of the halogen hydracids.

6. Predict the chemistry of astatine, the last member of Group 7. How would you expect it to react with (*a*) hydrogen, (*b*) metals and (*c*) fluorine?

7. (*a*) Indicate , by stating the necessary reagents and observations, how you would distinguish between sodium bromide and sodium iodide using a chemical test.

(*b*) A number of oxoanions of chlorine are known; examples include ClO^-, ClO_3^- and ClO_4^-.

(i) What is the oxidation number of chlorine in (*1*) ClO_3^-, (*2*) ClO^-?

(ii) ClO^- is formed when chlorine reacts with aqueous alkali. Write an ionic equation for this reaction.

(iii) When $KClO_3$ is heated just above its melting point it forms KCl and $KClO_4$. Write an equation for this reaction.

(*c*) (i) What is meant by the term *electron affinity*?

(ii) How is the electron affinity of fluorine anomalous?

(iii) Suggest an explanation for this anomaly.

(*d*) The acid dissociation constant, K_a, for hydrofluoric acid is 5.6×10^{-4} whereas that for hydrochloric acid is 1×10^7.

(i) What difference in properties of the two acids do these figures indicate?

(ii) Give two factors which are responsible for the difference in these values.

(*e*) Bonds such as C—Cl are termed 'polar' covalent bonds.

(i) What is meant by the term *polar covalent bond*?

(ii) Explain why the H—F bond has greater ionic character than the H—Cl bond.

(L 91, AS)

8. This question is concerned with various aspects of the chemistry of the halogens, their hydrides, and the salts of their oxoacids.

(*a*) The standard molar enthalpies of formation, $\Delta H_F^\ominus$(298 K), for HF(g), HCl(g), HBr(g) and HI(g) are respectively −268.5, −92.4, −36.2 and +25.9 kJ mol^{-1}. Describe and explain how you would expect (i) the thermal stabilities and (ii) the acid strengths to vary within the hydrogen halide series.

†(*b*) (i) Describe the initial reaction which occurs when iodine is added to *cold dilute* sodium hydroxide. State the type of reaction which takes place and write an equation for it.

(ii) Describe how the solution in (*b*)(i) above reacts with propanone (acetone) to form triiodomethane (iodoform), CHI_3, indicating the appropriate conditions. State what you would expect to observe and write equation(s) for the reaction(s) which occur(s).

(iii) The reaction in (*b*)(ii) above is often referred to as the iodoform reaction: it is useful as a diagnostic test for identifying organic compounds which contain a particular group. State what this group is, giving a formula, and describe how you would use the iodoform reaction to distinguish between $CH_3CH_2COCH_2CH_3$ and $CH_3CH_2CH_2COCH_3$.

(*c*) (i) Describe the reaction which occurs when iodine is heated with excess concentrated sodium hydroxide solution. State the type of reaction which takes place and write an equation for it.

(ii) A quantity of iodine was treated as in (*c*)(i) above and the resulting colourless solution acidified with aqueous sulphuric acid. A brown mixture was obtained due to the re-formation of iodine. Determine the relationship between the amount of iodine originally used and the amount liberated on acidification, giving your reasoning.
[For (*b*) see Chapter 31.]

(WJEC 90)

9. (*a*) State *two* compounds of chlorine, other than sodium chloride, with large-scale uses and describe *one* such use for each.

(*b*) Explain the chemistry that is occurring during the following series of reactions that all take place under aqueous conditions:

When sodium chloride is added to silver nitrate, a white precipitate forms which dissolves in an excess of dilute ammonia. The subsequent addition of sodium bromide to this latter solution causes the precipitation of a cream-coloured solid.

Suggest why this latter precipitate dissolves when sodium cyanide is added.

(C 90)

10. An aqueous solution of chlorine can be used as a disinfectant, for example in swimming pools.

(*a*) The amount of chlorine in pool water can be determined by adding excess potassium iodide solution which reacts to form iodine:

$$Cl_2(aq) + 2I^-(aq) \rightarrow 2Cl^-(aq) + I_2(aq)$$

The amount of iodine formed is found by titration with sodium thiosulphate solution of known concentration.

A student carried out the determination of chlorine in a sample of pool water. The student's complete record of all the measurements taken is shown below:

Volume of water sample tested = $1000\,cm^3$

Initial reading of burette = $7\,cm^3$
Final reading of burette = $16.3\,cm^3$
Volume added from burette = $9.3\,cm^3$

Concentration of sodium thiosulphate = $0.00500\,mol\,dm^{-3}$

(i) Write a balanced equation, including state symbols, for the reaction between iodine and sodium thiosulphate in aqueous solution.

(ii) The record of measurements reveals faults in both the procedure and the recording of measurements. State one fault in each of these.

(iii) Calculate the concentration of chlorine molecules, Cl_2, in the sample of pool water using the results obtained by the student.

(iv) At these low concentrations, the end-point of an iodine/sodium thiosulphate titration, at which the last sign of the yellow-brown colour of iodine solution disappears, is very difficult to see.

What indicator would you use to make this end-point more visible?

Give the name of the indicator and the colour change at the end-point.

(*b*) The disinfecting action is due to the presence of chloric(I) acid, HClO, formed by the reaction of chlorine with water:

$$Cl_2(aq) + H_2O(l) \rightleftharpoons HClO(aq) + H^+(aq) + Cl^-(aq)$$

Chloric(I) acid ionises as a weak acid:

$$HClO(aq) \rightleftharpoons H^+(aq) + ClO^-(aq)$$

In many swimming pools, chemicals other than chlorine are used to form the chloric(I) acid. This is partly because the use of chlorine gas causes much more corrosion of metal parts in the pool than does chloric(I) acid.

Compounds used to chlorinate pool water in this way include calcium chlorate(I) and chlorine dioxide, ClO_2.

(i) Why should chlorine in pool water cause much more corrosion of metal parts than chloric(I) acid?

(ii) Suggest one other reason why the use of chlorine itself is undesirable.

(iii) Write down the formula for calcium chlorate(I).

(iv) Chlorine dioxide undergoes a disproportionation reaction when it reacts with water, one of the products being chloric(I) acid. By use of oxidation numbers or otherwise, predict a balanced equation for this reaction.

(*c*) The disinfecting action of chloric(I) acid involves two possible oxidation processes. The relevant electrode potentials for the two possible redox reactions are:

$$HClO(aq) + H^+(aq) + 2e^- \rightleftharpoons Cl^-(aq) + H_2O(l); \quad E^\ominus = +1.49\,V$$

$$ClO^-(aq) + H_2O(l) + 2e^- \rightleftharpoons Cl^-(aq) + 2OH^-(aq); \quad E^\ominus = +0.89\,V$$

The oxidation reaction involved depends on the pH of the pool water, as the oxidant in alkaline solution is $ClO^-(aq)$.

The pH of swimming pools is kept slightly acidic at a pH of about 6. Suggest why a pH of 6 is preferred to

(i) a more acidic pH, by considering the equilibria involved in (*b*),

(ii) an alkaline pH, by considering the above electrode potentials for the two possible redox reactions. (L(N) 91)

11. (*a*) State how aqueous hydrogen fluoride differs in its acidic behaviour from that of aqueous solutions of the other hydrogen halides.

(*b*) Hydrogen fluoride reacts with a mixture of sodium hydroxide and aluminium hydroxide to give a commercially important compound, **X**, which occurs naturally, but in limited quantities only. **X** contains only sodium, aluminium and fluorine. In an analysis for aluminium 0.2100 g of **X** gave 0.0510 g Al_2O_3. All the fluorine present in the same mass of sample was precipitated as the mixed lead halide, PbClF, and gave 1.569 g of this compound.

(i) What is the formula of **X**?

(ii) What is the structure of the anion present in **X**? Draw the structure.

(iii) How is **X** used commercially and why is it so used?

(*c*) (i) Explain why the boiling point of hydrogen fluoride is much higher than the boiling points of the other hydrogen halides.

(ii) Protein structures contain many hydrogen bonds. Suggest an explanation for the fact that liquid hydrogen fluoride is widely used in biochemical research for dissolving proteins which are insoluble in water. (O & C 91)

12. (*a*) Describe the industrial production of sodium hydroxide and chlorine by the electrolysis of aqueous sodium chloride. Your answer should include a labelled sketch of a cell and a careful explanation of what is taking place at each stage of the process.

(*b*) Describe how pure chlorine may be safely prepared in the laboratory and how it may be collected. Include a sketch of the apparatus.

(*c*) Give *two* large-scale uses of chlorine.

(O 92)

13. (*a*) Describe and explain the relative thermal stabilities of the halogen hydrides.

(*b*) One major use of chlorine is in the manufacture of sodium chlorate(I), used in solution as a household bleach. Describe the reaction you could use in the laboratory to make a solution that contains sodium chlorate(I).

(*c*) Acidified aqueous potassium bromate(V), $KBrO_3$, reacts with hydrogen sulphide, H_2S, to give a precipitate of sulphur and an orange-red solution. On shaking the solution with trichloroethane, the colour is transferred to the organic layer. Describe the type of reaction taking place and suggest an identity for the orange-red product.

(C 91)

21 GROUP 6

21.1 THE MEMBERS OF THE GROUP

The elements of Group 6 are listed in Table 21.1.

TABLE 21.1 The Elements of Group 6

Elements		Z	*Electron configuration*	*AR*/nm	*IR*/nm
Oxygen	O	8	2.6	0.074	0.140
Sulphur	S	16	2.8.6	0.104	0.184
Selenium	Se	34	2.8.18.6	0.117	0.198
Tellurium	Te	52	2.8.18.18.6	0.137	0.221
Polonium	Po	84	2.8.18.32.18.6	0.152	–

There is a transition down the group from non-metallic to metallic properties. All except polonium form E^{2-} ions. All form covalent hydrides, H_2E. Oxygen differs in many respects from the rest of the group [§ 21.15].

21.2 OXYGEN

Oxygen and respiration

Oxygen constitutes 21% by volume of air. It is the essential part of the air for animal respiration. Oxygen is sufficiently soluble in water to be able to support the respiration of marine animals. The Earth's crust is 89% by mass oxygen.

Its physical properties and chemical reactivity

Oxygen is a colourless diatomic gas, O_2. Properly speaking, we should refer to molecular oxygen as dioxygen. It boils at −183 °C and freezes at −219 °C. Oxygen reacts directly with most other elements, and forms compounds with all elements except helium, neon, argon and krypton.

21.2.1 MANUFACTURE

Industrial manufacture

Oxygen is obtained industrially by the fractional distillation of liquid air [§ 22.2.1]. It is stored under pressure in cylinders.

21.2.2 INDUSTRIAL AND OTHER USES

Oxygen is used in the steel industry...

The steel industry uses oxygen to burn off the carbon which is present as an impurity in cast iron and which makes iron brittle [§ 24.14.3]. Since so much oxygen is needed (one tonne of oxygen for every tonne of steel), steel works have their own oxygen

. . . and for cutting and welding metals . . .

plants on the same site. Oxygen is also used in 'oxy-acetylene' torches which burn ethyne in oxygen in an extremely exothermic reaction and are used for cutting and welding metals.

. . . to help patients with breathing difficulties . . .

. . . and for mountaineers, divers and space travel

Hospitals use oxygen to revive accident victims and patients with breathing difficulties due to complaints such as pneumonia and asthma. Deep-sea divers, mountaineers and high-altitude pilots use oxygen. Space flights consume an enormous amount of oxygen. The Saturn rockets which launched the American astronauts on their journey to the Moon carried 2200 tonnes of liquid oxygen. The first stage burned 450 tonnes of kerosene in 1800 tonnes of oxygen, stages two and three were powered by hydrogen burning in oxygen, and there was also oxygen for the astronauts to breathe.

21.2.3 LABORATORY PREPARATION

The methods by which oxygen can be obtained in the laboratory are summarised in Figure 21.1.

FIGURE 21.1 Methods of Preparing Oxygen

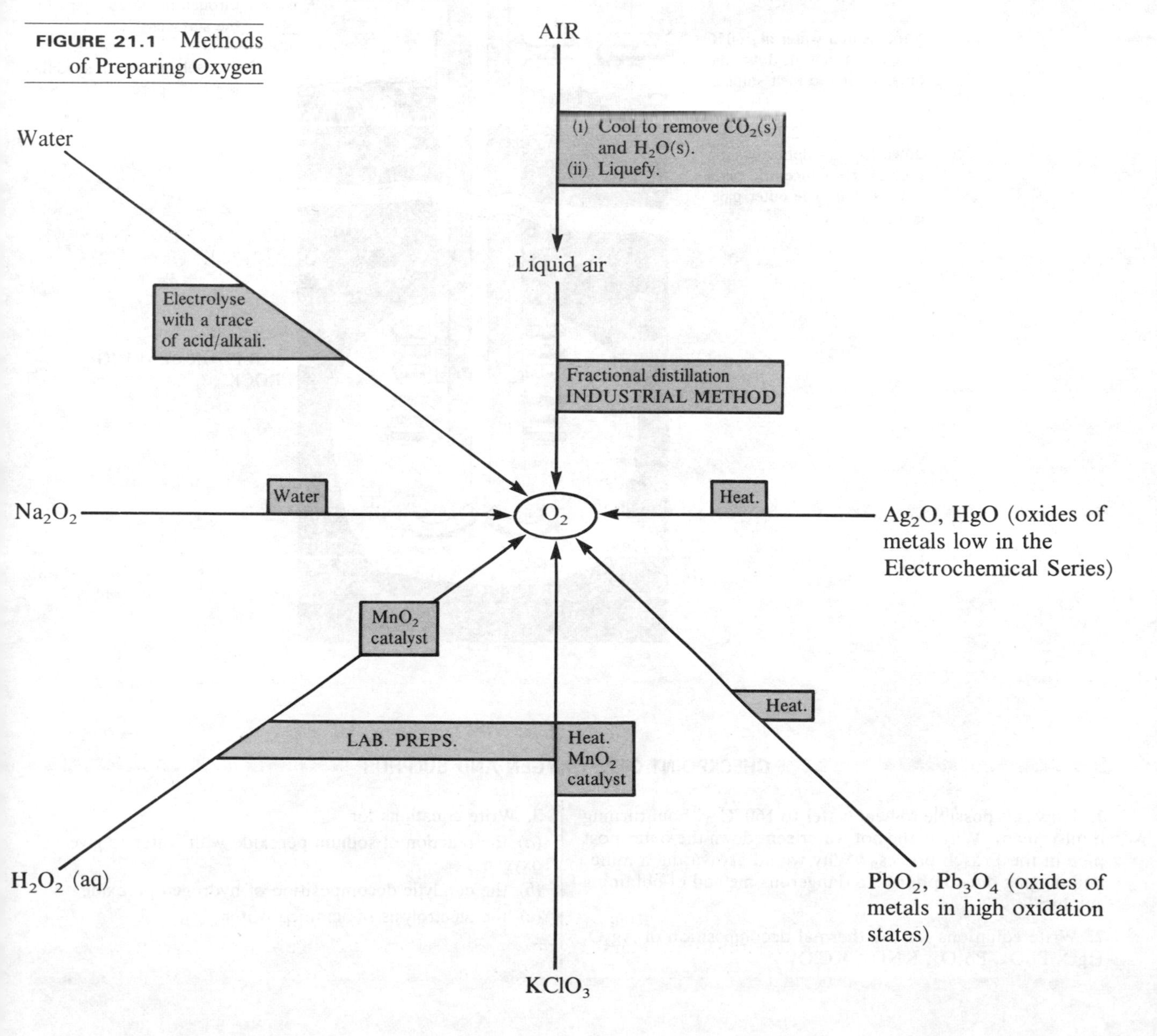

21.3 EXTRACTION OF SULPHUR

Sulphur is found as the element and as metal sulphide ores and as a number of sulphates, e.g., calcium and magnesium sulphates.

H Frasch invented a method of bringing sulphur to the surface from deposits of elemental sulphur below ground level [see Figure 21.2]. It is used in the USA, Poland and Japan.

FIGURE 21.2
The Frasch Process for Mining Sulphur

3 Compressed air is sent down the inside pipe.

2 Superheated water at 160 °C and 10 atm is sent down the outside pipe to melt sulphur.

1 A shaft is sunk 200–400 m down to the sulphur-bearing rock. Three concentric pipes are inserted. The outer pipe is 1 m across.

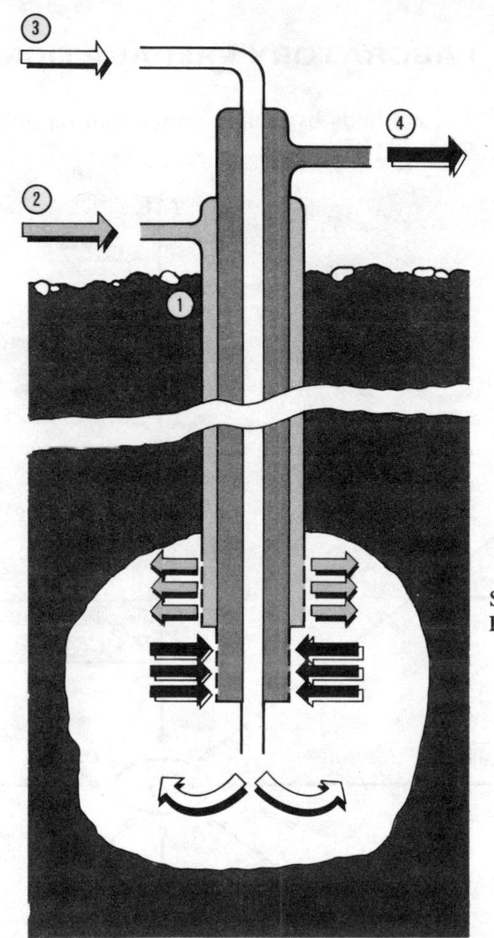

4 A foam of molten sulphur, water and air is forced up through the middle pipe to the surface.

5 The sulphur which solidifies is 99.5% pure.

SULPHUR-BEARING ROCK

CHECKPOINT 21A: OXYGEN AND SULPHUR

1. How is it possible to heat water to 160 °C without turning it into steam? Why is the hot water sent down the outermost pipe in the Frasch process? Why would excavating a mine and digging out sulphur be a dangerous method of obtaining sulphur?

2. Write equations for the thermal decomposition of Ag_2O, HgO, PbO_2, Pb_3O_4, KNO_3, $KClO_3$.

3. Write equations for

(*a*) the reaction of sodium peroxide with water to give oxygen

(*b*) the catalytic decomposition of hydrogen peroxide

(*c*) the electrolysis of acidified water.

21.4 ALLOTROPY

21.4.1 OXYGEN

The element oxygen exists in two forms, dioxygen, O_2, and trioxygen, O_3. The existence of an element in two forms is called **allotropy**. Trioxygen, ozone, is the less stable allotrope:

Allotropy of oxygen...

$$3O_2(g) \rightarrow 2O_3(g); \quad \Delta H^{\ominus} = 142\,kJ\,mol^{-1} \text{ of } O_3$$

...O_2 and O_3

Dioxygen must absorb energy in order to form trioxygen, but trioxygen decomposes spontaneously to form dioxygen.

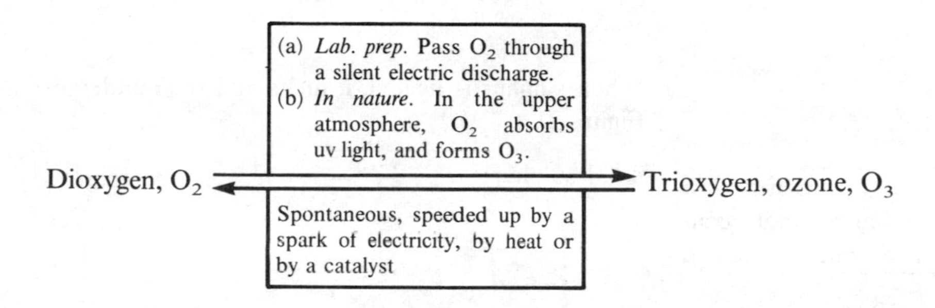

Monotropic allotropy

Since the direction of spontaneous change is always the same under all conditions, the allotropy is described as **monotropic** (moving in one direction).

Ozone has a characteristic smell. In concentrations above 1000 ppm, it is damaging to health.

The oxygen molecule $:\ddot{O}=\ddot{O}:$

has a σ bond and a π bond between the two atoms. In the ozone molecule, there is an angle of 117° between the bonds. The structure can be represented by the delocalised electron structure:

The structure of ozone

O (central), O, O — delocalised bonds

21.4.2 SULPHUR

Allotropy of sulphur...

...rhombic and monoclinic

There are two main allotropes of sulphur. Both **rhombic** and **monoclinic** sulphur consist of S_8 molecules. They differ in crystal structure [see Figure 21.3]. Above the transition temperature, 95.6 °C at 1 atm, monoclinic sulphur is the stable form; below 95.6 °C, rhombic sulphur is the more stable form. Transition between the two forms takes place in either direction, depending on the temperature. The allotropy is described as **enantiotropic** (moving in both directions).

FIGURE 21.3 Allotropes of Sulphur

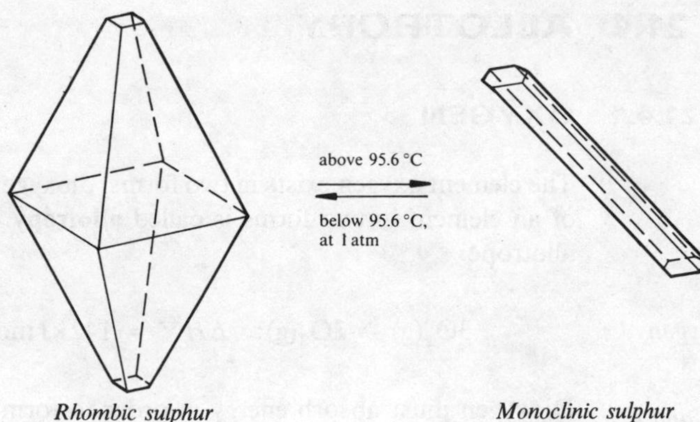

Rhombic sulphur

Yellow transparent crystals, T_m = 113 °C, obtained when sulphur crystallises from solution

Monoclinic sulphur

Amber coloured needles, T_m = 119 °C, obtained when sulphur solidifies above 95.6 °C

When sulphur is heated, it melts and then undergoes a set of changes, as shown in Figure 21.4.

FIGURE 21.4 The Changes that occur when Sulphur is Heated

Solid sulphur — S_8 rings

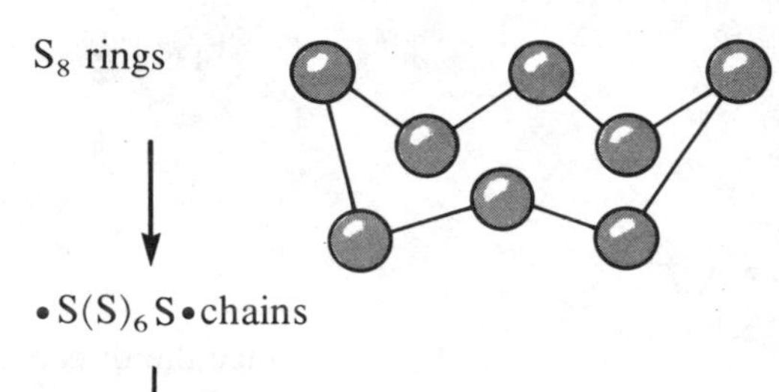

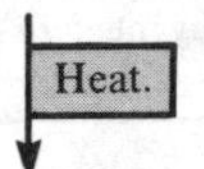

Transparent yellow liquid — $\bullet S(S)_6S\bullet$ chains

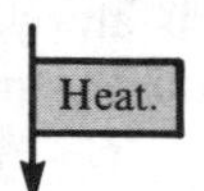

Colour darkens. Liquid reaches maximum viscosity at 200 °C. — Long chains of around 10^5 S atoms become entangled and make the liquid viscous.

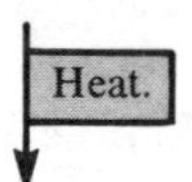

Liquid becomes mobile at 400 °C. — Chains break up to form smaller units, and the liquid becomes mobile.

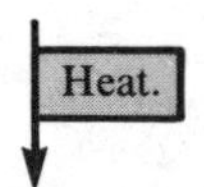

Liquid sulphur boils at 444 °C.

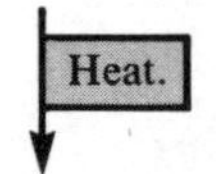

Vapour — Vapour contains S_8, S_4 and S_2 molecules

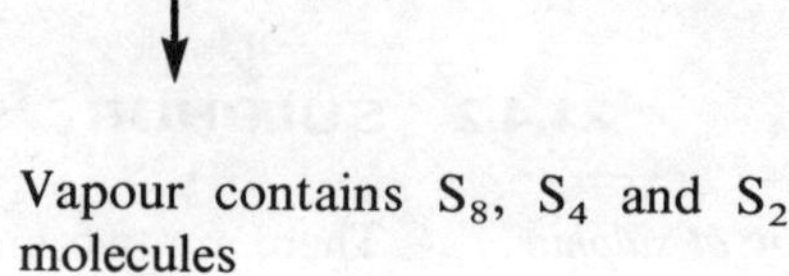

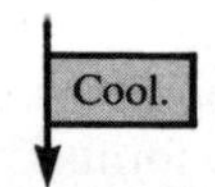

Solidifies as $S_8(s)$ — S_8 molecules

21.5 THE OZONE LAYER

TOPIC

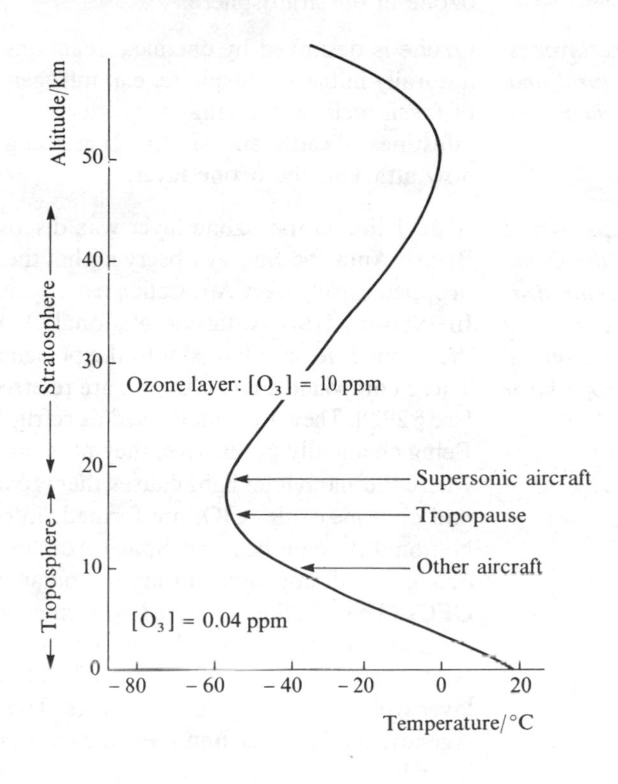

FIGURE 21.5
The Structure of the Atmosphere

There is a layer of ozone in the stratosphere (upper atmosphere). The first step in the formation of ozone in the stratosphere is the dissociation of oxygen molecules by UV light. The ozone layer protects oxygen in the troposphere (lower atmosphere) from UV light

The sunlight that falls on the upper layers of the atmosphere (the stratosphere; see Figure 21.5) contains much more ultraviolet (UV) light than the radiation which reaches the surface of the earth. Ultraviolet light has enough energy to bring about photochemical reactions that convert dioxygen, O_2 into ozone, O_3. As plant life evolved on earth, photosynthesis increased the concentration of oxygen in the atmosphere. In consequence, the concentration of ozone in the stratosphere increased until equilibrium was reached between the rate of formation of ozone from oxygen and the rate of decomposition of ozone. Ozone can absorb UV light. In doing so, ozone protects oxygen in the lower atmosphere from being dissociated and keeps most of these harmful rays from penetrating to the Earth's surface. The energy of the absorbed radiation heats up the ozone, creating warm layers high in the stratosphere. The maximum concentration of ozone, about 10 ppm, occurs 25–50 km from the surface of the Earth.

Some of the reactions involving dioxygen, O_2, and ozone (trioxygen, O_3) are:

(*a*) Dioxygen is dissociated by solar UV rays:

$$O_2 + h\nu(\text{UV}) \rightarrow \text{O}\cdot + \text{O}\cdot$$

(*b*) Some of the oxygen atoms formed combine with dioxygen molecules to form ozone:

$$\text{O}\cdot + O_2 \rightarrow O_3$$

(*c*) Ozone absorbs UV light and dissociates:

$$O_3 + h\nu(\text{UV}) \rightarrow O_2 + \text{O}\cdot$$

(*d*) Ozone reacts with oxygen atoms to form dioxygen

$$O_3 + \text{O}\cdot \rightarrow 2O_2$$

The rates of formation and destruction result in a steady state concentration of ozone in the stratosphere.

The ozone layer is attacked by natural and man-made chemicals

Ozone is destroyed by chemical reactions with a number of substances that occur naturally in the stratosphere, e.g. nitrogen oxides (from microbes and the combustion of fossil fuels) and methane (produced by microbes in swamps, rice paddies and the intestines of cattle and sheep). Increasing quantities of man-made substances are now attacking the ozone layer.

Scientists have discovered a decrease in the ozone layer over Antarctica. They relate it to the attack by CFCs on the ozone layer

A depletion of the ozone layer was discovered in 1983, when scientists with the British Antarctic Survey observed that the concentration of ozone in the stratosphere dropped rapidly over Antarctica each spring, to be replenished by the end of November. In 1986 the USA Antarctic National Ozone Expedition confirmed the decrease in the ozone layer and linked it to the presence in the stratosphere of chlorofluoroalkanes. These compounds, e.g. CCl_2F_2, are referred to as CFCs, short for chlorofluorocarbons [see § 29.9]. They are widely used as refrigerants, thermal insulation and aerosols. Being chemically unreactive, they pass through the troposphere to the stratosphere, where the ultraviolet light causes them to dissociate. As a result, atomic chlorine and chlorine oxide, ClO, are formed. Information gathered by satellites of the USA's National Aeronautics and Space Administration (NASA) showed that 50% of the ozone had disappeared during the polar thaw, and demonstrated a connection with CFCs. The satellites detected concentrations of chlorine oxide a hundred times as great as concentrations in latitudes where there is less diminution of the ozone layer. NASA predicted that, if the present rate of CFC production continues, the ozone layer will decrease by 3% in 20 years. The US Government Environmental Protection Agency, on the other hand, estimates a possible destruction of 50% of the ozone layer by the year 2050.

Reactions which take place after the photolysis of CFCs are:

(*a*) $O_3 + Cl\cdot \rightarrow O_2 + ClO$
(*b*) $ClO + O\cdot \rightarrow Cl + O_2$
(*c*) $O_3 + ClO \rightarrow O_2 + ClO_2$

Reactions with atomic chlorine, chlorine oxide and nitrogen oxide destroy ozone. Chain reactions are set up

Notice that the reactions (*a*) and (*b*) constitute a chain reaction. As a result, one Cl atom can destroy thousands of molecules of ozone. Nitrogen oxide from the exhausts of the supersonic aircraft which fly in the stratosphere take part in the reactions:

(*d*) $O_3 + NO \rightarrow O_2 + NO_2$
(*e*) $NO_2 + O\cdot \rightarrow NO + O_2$

Reactions (*d*) and (*e*) form a chain reaction. The effect of reactions (*a*)–(*e*) is to lower the steady state concentration of ozone.

In 1989 an international team of scientists working in northern Canada detected a decrease in the ozone layer over the Arctic. This is even more serious than the 'hole' over the Antarctic as it is closer to densely populated countries.

The result that is forecast for a decrease in the ozone layer is an increase in the amount of UV radiation reaching the earth, with a consequent increase in the incidence of skin cancer

The results that are forecast for a depletion of the ozone layer are alarming. There is a strong connection between UV radiation and the incidence of non-melanoma skin cancer in humans. This is generally a non-fatal cancer, but there is some evidence to support a connection between UV radiation and melanoma skin cancer which is a more frequently fatal form of the disease. The US National Academic of Sciences report in 1986 concluded that there would be an increase in malignant melanoma, a serious form of cancer which frequently causes death. The report predicted that a 10% reduction in stratospheric ozone would result in a 20% increase in both forms of skin cancer. This would mean 160 000 extra cases of non-melanoma skin cancers per year in the US and 8000 in the UK. Excessive UV radiation has also been linked to cataracts and lowered immunity to disease.

Since the ozone layer warms the stratosphere, any decrease in the ozone should cause a decrease in the stratospheric temperature. There is no way of predicting what the effect on the temperature of the troposphere would be, but it would be reasonable to expect some effect.

Ozone levels in the troposphere are rising. Often the levels exceed that recommended by the WHO

In the stratosphere ozone is beneficial, shielding the earth from UV radiation. In the troposphere, ozone is an undesirable chemical that causes damage to humans, vegetation and many materials, e.g. rubber and textiles. Ozone is suspected of triggering asthma attacks and bronchitis. In unpolluted air over Britain and the rest of Europe, the ozone concentration is 20–50 ppb. On still, sunny days, it can rise to over 60 ppb. World Health Organisation guidelines for human health state that the ozone concentration should not exceed a mean of 60 ppb over an 8 hour interval. This is often exceeded in southern England.

In the troposphere ozone is a dangerous substance, which damages plants, animals and materials. It causes respiratory diseases in humans, serious damage to trees and contributes to the greenhouse effect

Ozone in the troposphere has been linked to dying forests. Originally acid rain [see §21.14] was thought to be the decisive factor in damage to forests. Now data for North America and Europe suggest that ozone may be of primary importance, with acid rain ranking second in Europe and perhaps third in North America. In Sweden, a research project has been set up to find out how ozone damages pine and fir trees. They have a theory that ozone impedes water utilisation and photosynthesis, perhaps by disrupting mechanisms that regulate the opening of the stomata (holes in the leaves).

Ozone also contributes to the greenhouse effect [see §23.9].

Ozone is formed in the troposphere by photochemical reactions between oxygen, hydrocarbons and oxides of nitrogen. A reduction in the emission of oxides of nitrogen by motor vehicles would solve the problem

In the troposphere ozone is formed from oxygen by photochemical reactions with oxides of nitrogen and hydrocarbons. Both of these are emitted by motor vehicles [see §26.4]. Power stations are another source of nitrogen oxides. The use of excessive quantities of nitrogenous fertilisers releases oxides of nitrogen into the atmosphere. Hydrocarbons come from many sources: ruminating cattle, termites, marshes, rice paddies and leaking North Sea gas as well as vehicle exhausts. The concentration of ozone close to the ground has doubled in Europe over the past 30 years.

The production of ozone in the troposphere must be stemmed. It is difficult to remove hydrocarbons when they come from so many natural sources. The best hope is to reduce the emission of nitrogen oxides from vehicle exhausts [see §26.4].

CHECKPOINT 21B: THE OZONE LAYER

1. (*a*) What is meant by a *chain reaction*?
(*b*) What chain reaction results in the formation of ozone from oxygen?
(*c*) What is meant by *steady state concentration*?
(*d*) If the rate of destruction of ozone increases, will the ozone concentration decrease to a lower level or fall to zero? Explain your answer.
(*e*) Why are oxygen molecules dissociated in the stratosphere and not in the troposphere?

2. What results are predicted for the decrease in the ozone layer? Mention effects on the climate and on human health.

3. Ozone is something of an enigma. In the stratosphere it is beneficial to human health; in the troposphere it is detrimental to human health. Discuss this statement.

4. The temperature of the troposphere decreases with height. At the height labelled 'tropopause' in Figure 21.5 the temperature starts to increase again. Why does the temperature decrease and then increase with height?

21.6 REACTIONS OF OXYGEN AND SULPHUR

Oxygen combines directly with most elements [see Figure 21.6]. Sulphur combines with most metals when heated and with the non-metallic elements fluorine, chlorine, oxygen and carbon. It is oxidised by concentrated nitric acid and by concentrated sulphuric acid:

$$S(s) + 6HNO_3(aq) \rightarrow H_2SO_4(l) + 6NO_2(g) + 2H_2O(l)$$

$$S(s) + 2H_2SO_4(l) \rightarrow 3SO_2(g) + 2H_2O(l)$$

FIGURE 21.6 Some Reactions of Oxygen

Animal respiration. Complexes with haemoglobin in the blood, oxidises carbohydrates in the body tissues.

Air oxidises, e.g., alcohol to acid, Fe^{2+} to Fe^{3+}, SO_3^{2-} to SO_4^{2-},

H_2O_2

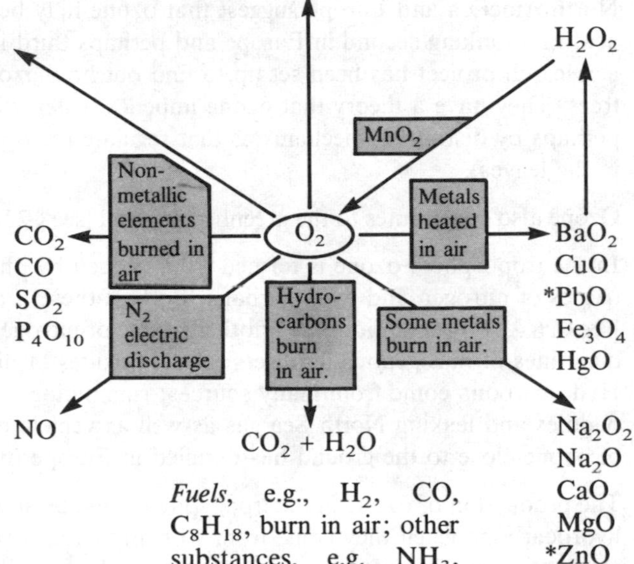

Acidic: Some of these oxides dissolve in water to form acids; all react with bases to form salts, with the exception of one or two neutral oxides, e.g., NO.

Basic: All these oxides are basic and react with acids to form salts. *Some are amphoteric.

Fuels, e.g., H_2, CO, C_8H_{18}, burn in air; other substances, e.g. NH_3, burn in pure O_2.

CaO
MgO
*ZnO
*Al_2O_3

CHECKPOINT 21C: BONDING

1. Why is oxygen a gas and sulphur a solid?

***2.** Calculate the standard enthalpy of formation of (*a*) O^{2-} and (*b*) S^{2-} by using the following data.

Element	$\Delta^{\ominus}H_{Atomisation}$/kJ mol^{-1}	*1st electron affinity*/kJ mol^{-1}	*2nd electron affinity*/kJ mol^{-1}
Oxygen	248	− 142	+ 844
Sulphur	223	− 200	+ 532

3. Use the standard bond enthalpies in kJ mol^{-1}: S=S, 431; S—S, 264; O=O, 498 and O—O, 142. Calculate the values of $\Delta H^{\ominus}$ for the changes

(*a*) $S_8(g) \rightarrow 4S_2(g)$

(*b*) $O_8(g) \rightarrow 4O_2(g)$

Comment on the values you obtain.

21.7 HYDRIDES OF OXYGEN AND SULPHUR

21.7.1 WATER

See § 17.9.

21.7.2 HYDROGEN SULPHIDE

Laboratory preparation of hydrogen sulphide...

Hydrogen sulphide, H_2S, is an unpleasant-smelling gas which is poisonous at a level of 1000 ppm. It occurs in natural gas and in bad eggs, where it is formed by the decay of proteins. In the laboratory, it is made by the action of dilute hydrochloric acid on iron(II) sulphide:

$$S^{2-}(s) + 2H^+(aq) \rightarrow H_2S(g)$$

Although it is slightly soluble, it can be collected over water.

IONISATION

... which is a weak, diprotic acid...

Hydrogen sulphide is a weak acid, ionising to give hydrogen ions, sulphide ions and hydrogensulphide ions. In its reducing reactions, it is oxidised to sulphur:

$$H_2S(aq) \rightleftharpoons S(s) + 2H^+(aq) + 2e^-; \quad E^\ominus = -0.51\,V$$

Hydrogen sulphide is a reducing agent

Hydrogen sulphide reduces acidified manganate(VII), acidified dichromate(VI), iron(III) ions, moist chlorine and moist sulphur dioxide [see Question 2, Checkpoint 21D].

21.7.3 HYDROGEN PEROXIDE

The arrangement of bonds in H_2O_2

Hydrogen peroxide is a liquid of formula H_2O_2. Its structure is illustrated in Figure 21.7. The molecules are associated by hydrogen bonding, and the liquid is viscous.

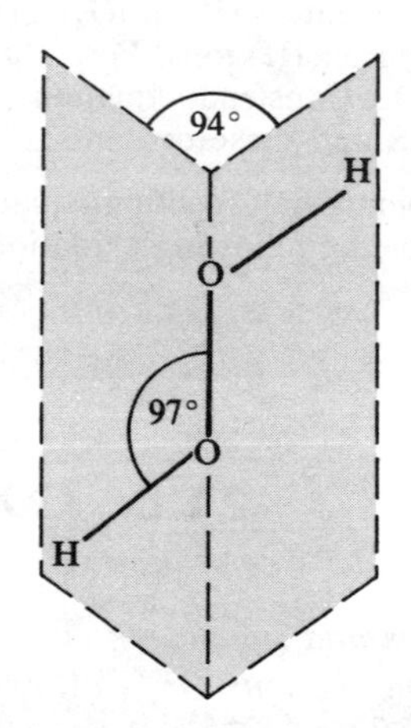

FIGURE 21.7
The Structure of H_2O_2

PREPARATION

The laboratory preparation of hydrogen peroxide

The action of dilute sulphuric acid on barium peroxide, BaO_2, at 0 °C is used for the preparation of hydrogen peroxide:

$$BaO_2(s) + H_2SO_4(aq) \rightarrow BaSO_4(s) + H_2O_2(aq)$$

The insoluble barium sulphate is removed by filtration. The aqueous solution of hydrogen peroxide can be concentrated by distillation under reduced pressure.

Hydrogen peroxide is used as a source of oxygen

Hydrogen peroxide readily decomposes to give oxygen:

$$2H_2O_2(aq) \rightarrow 2H_2O(l) + O_2(g)$$

The decomposition is catalysed by a number of metals and metal oxides, the most popular being manganese(IV) oxide, MnO_2. In the laboratory, hydrogen peroxide is used as aqueous solutions which are designated as, for example, '20 volume', meaning that 1 volume of the solution will provide 20 volumes of oxygen at stp.

OXIDISING AGENT/REDUCING AGENT

Hydrogen peroxide acts as an oxidising agent:

$$H_2O_2(aq) + 2H^+(aq) + 2e^- \rightarrow 2H_2O(l); \quad E^\ominus = +1.77\,V$$

and as a reducing agent:

$$H_2O_2(aq) \rightarrow O_2(g) + 2H^+(aq) + 2e^-; \quad E^\ominus = -0.68\,V$$

Hydrogen peroxide is an oxidising agent and a reducing agent

Acidic solutions favour the oxidising action, e.g., Fe^{2+} to Fe^{3+}; PbS to $PbSO_4$; I^- to I_2. Alkaline solutions favour the reducing reaction, e.g. ClO^- to Cl^-. Even in acid solution, hydrogen peroxide reduces manganate(VII), MnO_4^-, to Mn^{2+}, with the formation of oxygen [see Question 3, Checkpoint 21D].

USE

Hydrogen peroxide is used as a bleach for textiles, wood pulp and human hair. The bleaching action depends on its oxidising property. A convenient feature is that the only by-product is water.

CHECKPOINT 21D: HYDRIDES

1. Sketch an apparatus which could be used for the laboratory preparation of hydrogen peroxide.

†**2.** See § 21.7.2 for the half-reaction equation for the oxidation of hydrogen sulphide. Referring to Chapter 3 if you need to, write half-reaction equations for the reduction of (*a*) acid manganate(VII), MnO_4^-, (*b*) acid dichromate(VI), $Cr_2O_7^{2-}$, (*c*) iron(III) ions, Fe^{3+}, (*d*) moist Cl_2 and (*e*) moist SO_2. Construct equations for the reactions between H_2S and these oxidants.

3. By combining half-reaction equations, construct equations for the following oxidation reactions:

(*a*) I^- to I_2 by acidified H_2O_2
(*b*) PbS to $PbSO_4$ by H_2O_2
(*c*) H_2O_2 to O_2 by Cl_2
(*d*) H_2O_2 to O_2 by acidified $KMnO_4$
(*e*) Fe^{2+} to Fe^{3+} by acidified H_2O_2.

4. Suggest a method for the titrimetric analysis of a solution of hydrogen peroxide. What is the concentration/mol dm^{-3} of '20 volume' hydrogen peroxide?

21.8 OXIDES AND SULPHIDES

Ionic oxides and sulphides containing the ions O^{2-} and S^{2-} are formed

The formation of O^{2-} is an endothermic process. When it forms an ionic structure, however, the small size and double charge on the anion result in a highly negative (exothermic) value for the lattice enthalpy. When the lattice enthalpy compensates for the ionisation enthalpies of the cation and anion, an ionic compound may be formed [see MgO, § 10.7]. Many ionic oxides are formed. Since the S^{2-} ion is larger and more easily polarised than the O^{2-} ion, sulphides tend to be more covalent in character than the corresponding oxides. Only the sulphides of Groups 1 and 2 appear to be ionic in character.

The sulphides have more covalent character than the oxides

Oxides and sulphides that dissolve in water give alkaline solutions due to the basic nature of the O^{2-} and S^{2-} ions:

$$O^{2-}(aq) + H_2O(l) \rightarrow 2OH^-(aq)$$

$$S^{2-}(aq) + H_2O(l) \rightarrow HS^-(aq) + OH^-(aq)$$

21.8.1 CLASSIFICATION OF OXIDES

Oxides may be classified according to their formulae [see Table 21.2], or according to their structure [see Figure 21.8], or according to their acid–base properties [see Table 21.3 and Figure 21.9].

Oxide	*Types of bonds in the oxides* E_xO_y
Normal oxides	Bonds are between E and O only. Some are ionic, e.g., $Ca^{2+}O^{2-}$; others are covalent, e.g., CO_2, $(SiO_2)_n$
Peroxides	Bonds are between E and O and also between O atoms. Some are ionic, e.g., $2Na^+$ O—O^{2-}; others are covalent, e.g., H—O—O—H
Mixed oxides	e.g., Pb_3O_4, which reacts as a mixture $2PbO \cdot PbO_2$ and Fe_3O_4, which reacts as a mixture $FeO \cdot Fe_2O_3$
Non-stoichiometric oxides	Transition metals form oxides of formula $M_{0-1}O$, e.g., $Fe_{0.9}O$

TABLE 21.2 Classification of Oxides according to Formula

	Oxides of metallic elements	*Oxides of non-metallic elements*
(*a*)	Oxides of metals in lower oxidation states are basic. Some react with water to give $OH^-(aq)$, e.g., CaO, MgO.	Most are acidic. Some dissolve in water to form solutions with a high concentration of hydrogen ions, e.g., SO_3.
(*b*)	Others are insoluble in water, but react with acids and acidic oxides, e.g., Fe_2O_3, CuO.	Macromolecular oxides, e.g., $(SiO_2)_n$, $(B_2O_3)_n$, do not dissolve, but react with basic oxides and amphoteric oxides to give salts.
(*c*)	Strongly basic oxides, e.g., K_2O, CaO react with amphoteric oxides.	A few are neutral, e.g., N_2O, NO.
(*d*)	Some metal oxides are amphoteric, reacting with both basic oxides and acidic oxides, e.g., ZnO, SnO, SnO_2, PbO, PbO_2, Cr_2O_3, Al_2O_3.	

TABLE 21.3 Oxides of Metallic and Non-metallic Elements

FIGURE 21.8 Classification of Oxides according to Structure

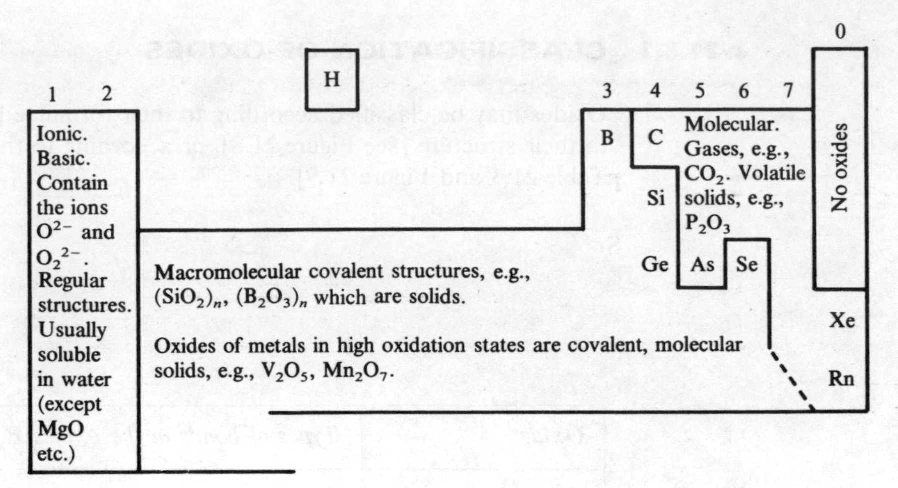

FIGURE 21.9 The Acid–Base Character of Oxides

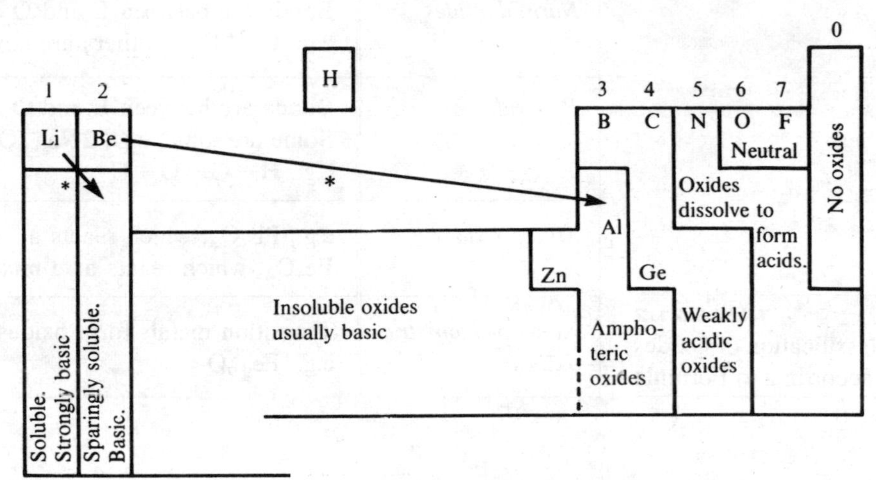

CHECKPOINT 21E: OXIDES

1. Classify the following (i) according to structure and (ii) according to acid–base character:
(*a*) Na_2O
(*b*) CaO
(*c*) SiO_2
(*d*) Fe_2O_3
(*e*) Fe_3O_4

2. Which member of each of the following pairs is the more acidic oxide?
(*a*) BaO and CO
(*b*) MnO and Mn_2O_7
(*c*) NO and N_2O_5
(*d*) Cr_2O_3 and CrO_3
(*e*) CaO and FeO
(*f*) SO_2 and SeO_2

3. Explain why oxygen and sulphur differ in
(*a*) the ionic character of MO and MS (where M is a metal)
(*b*) the boiling temperatures of H_2O and H_2S
(*c*) the formulae of their fluorides, F_2O and SF_6.

21.9 SULPHUR DIOXIDE

The laboratory preparation from sulphites

The industrial sources of sulphur dioxide, SO_2, are listed in §21.14. A convenient laboratory preparation is the action of a dilute acid on a sulphite. The gas is collected downwards because it is denser than air:

$$SO_3^{2-}(s) + 2H^+(aq) \rightarrow SO_2(g) + H_2O(l)$$

It is also made by the action of hot, concentrated sulphuric acid on copper [§21.11.2].

21.9.1 PHYSICAL PROPERTIES

Sulphur dioxide is a dense gas with a choking smell. It fumes in air and is extremely soluble in water. It boils at −10 °C, and liquefies at room temperature under a pressure of 3 atm.

The bonding in SO_2

The bonding in SO_2 is shown in Figure 21.10. The S=O bonds are double, and the sulphur atom has ten electrons in its valence shell. It has 'expanded its octet'.

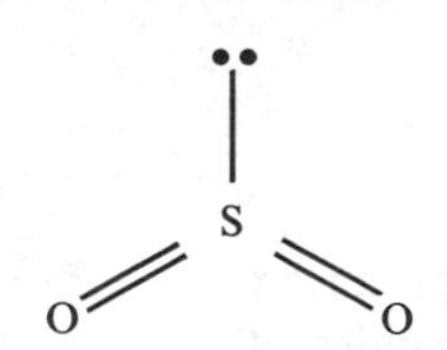

FIGURE 21.10
The Bonding in SO_2

21.9.2 SULPHUROUS ACID AND SULPHITES

Sulphur dioxide is the anhydride of sulphurous acid, which is a weak acid...

Sulphur dioxide dissolves in water to form a solution which contains sulphurous acid:

$$SO_2(aq) + H_2O(l) \rightleftharpoons H_2SO_3(aq)$$

The acid is a weak acid, ionising to give hydrogensulphite ions, HSO_3^-, and sulphite ions, SO_3^{2-}. Attempts to concentrate the solution result in the decomposition of sulphurous acid and the evolution of sulphur dioxide gas.

...It reacts with alkalis to form sulphites or hydrogensulphites

Sulphur dioxide reacts with aqueous sodium hydroxide to form sodium sulphite solution. If more sulphur dioxide is passed into this solution, sodium hydrogensulphite is formed:

$$SO_2(g) + 2NaOH(aq) \rightarrow Na_2SO_3(aq) + H_2O(l)$$

$$Na_2SO_3(aq) + SO_2(g) + H_2O(l) \rightarrow 2NaHSO_3(aq)$$

Moist sulphur dioxide, or sulphurous acid, and sulphites behave as reducing agents:

SO_2 and SO_3^{2-} are reducing agents

$$SO_3^{2-}(aq) + H_2O(l) \rightleftharpoons SO_4^{2-}(aq) + 2H^+(aq) + 2e^-; \quad E^\ominus = -0.17\,V$$

Sulphite ions are oxidised by air, chlorine, iron(III) ions, dichromate(VI) ions and manganate(VII) ions. The reaction with dichromate(VI), resulting in a change from orange to blue, and the decolorisation of manganate(VII) are used as tests for sulphur dioxide. Hydrogen sulphide is a stronger reducing agent than sulphur dioxide, and it reduces sulphur dioxide to sulphur.

CHECKPOINT 21F: SULPHUR DIOXIDE

1. Write equations for the ionisation of sulphurous acid.

2. Write half-reaction equations for the oxidants mentioned above.

Combine these with the half-reaction equation for $SO_3^{2-}(aq)$ to obtain equations for the reactions of $SO_3^{2-}(aq)$ with O_2, Cl_2, $Fe^{3+}(aq)$, $Cr_2O_7^{2-}(aq)$, $MnO_4^-(aq)$.

Obtain an equation for the reaction of H_2S and SO_3^{2-}.

21.10 SULPHUR(VI) OXIDE

Sulphur(VI) oxide exists as molecules of monomer, $SO_3(g)$, trimer, $(SO_3)_3(s)$ and polymer, $(SO_3)_n(s)$ [see Figure 21.11].

The laboratory preparation of sulphur(VI) oxide is illustrated in Figure 21.12. Dry sulphur dioxide and dry oxygen combine on the surface of platinised asbestos at 400 °C. The product solidifies as long needle-shaped crystals in the cooled receiver:

$$2SO_2(g) + O_2(g) \underset{400\,°C}{\overset{Pt}{\rightleftharpoons}} 2SO_3(s)$$

FIGURE 21.11 Molecules of Sulphur(VI) Oxide

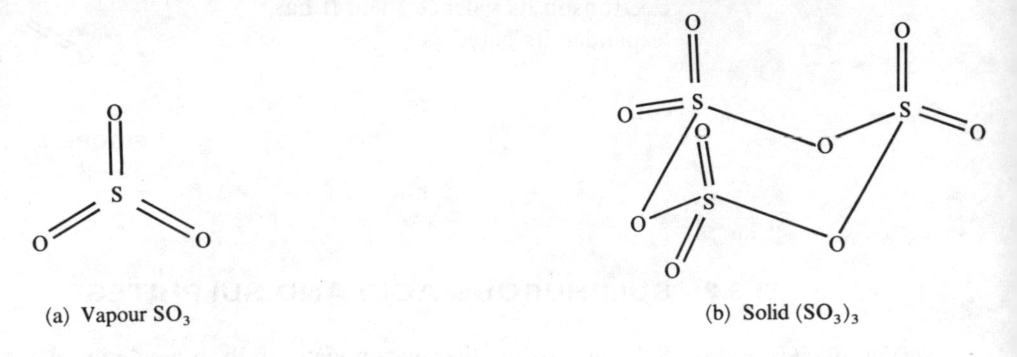

(a) Vapour SO_3

(b) Solid $(SO_3)_3$

(c) Solid $(SO_3)_n$

FIGURE 21.12 Laboratory Preparation of Sulphur(VI) Oxide

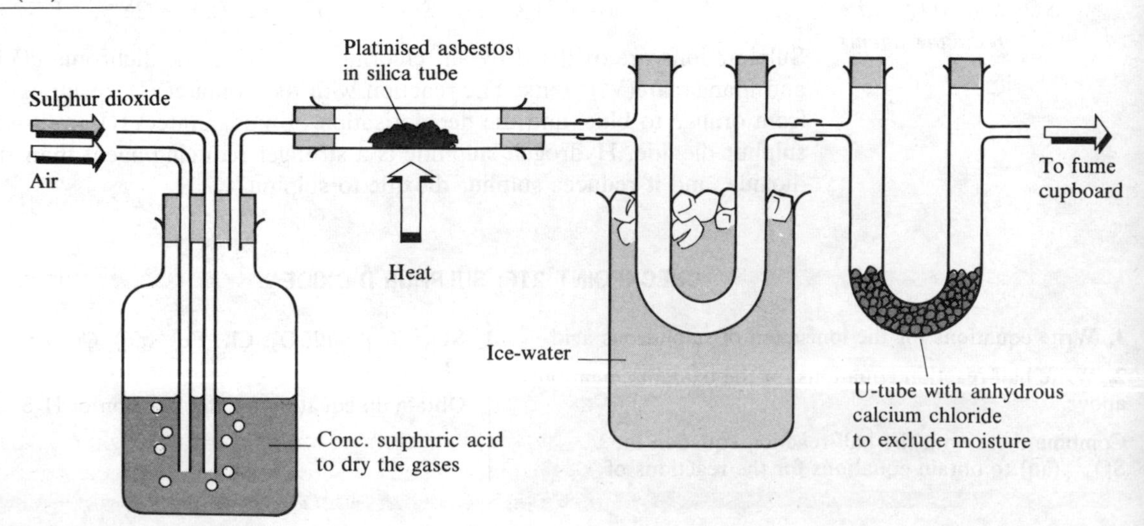

21.10.1 PROPERTIES

Sulphur(VI) oxide fumes in moist air, and has acidic reactions and oxidising reactions

Sulphur(VI) oxide is an acid oxide. It fumes in moist air as it reacts with moisture to form a mist of sulphuric acid:

$$SO_3(g) + H_2O(l) \rightarrow H_2SO_4(l)$$

It reacts with basic oxides to form sulphates:

$$CaO(s) + SO_3(g) \rightarrow CaSO_4(s)$$

Sulphur(VI) oxide is an oxidising agent. It takes part in the same oxidising reactions as sulphuric(VI) acid, of which it is the anhydride.

A comparison of SO_2 and SO_3

TABLE 21.4 A Comparison of the Oxides of Sulphur

Sulphur dioxide, SO_2	*Sulphur(VI) oxide, SO_3*
Colourless gas	Solid. Easily vaporised.
Fumes. Pungent smell	Fumes. Pungent smell
Very soluble	Very soluble
Solution is weakly acidic	Solution is strongly acidic
Reducing agent (e.g., $Fe^{3+} \rightarrow Fe^{2+}$). Combines with $O_2 \rightarrow SO_3$ and with $Cl_2 \rightarrow SO_2Cl_2$	Oxidising agent (e.g., $Br^- \rightarrow Br_2$)

21.11 SULPHURIC ACID

Manufacture of sulphuric acid needs SO_2

Sulphuric acid is made by the Contact process. The anhydride of sulphuric acid is sulphur(VI) oxide, SO_3. It is made by the oxidation of sulphur dioxide, which is obtained from the following sources:

Sources of sulphur dioxide

1. Sulphur is burned in air.

2. Hydrogen sulphide from crude oil is burned in air.

3. During the extraction of metals from sulphide ores, the ores are roasted in air, giving sulphur dioxide.

4. When anhydrite, $CaSO_4$, is heated with coke and sand in the manufacture of cement, $CaSiO_3$, sulphur dioxide is a by-product.

5. Flue-gas desulphurisation in power stations will in the future supply sulphur dioxide [see § 21.14].

Sulphur dioxide is passed through an electrostatic precipitator to remove dust and impurities.

$SO_2 + O_2$ combine to form SO_3 in a reaction which reaches equilibrium

When sulphur dioxide and oxygen combine

$$2SO_2(g) + O_2(g) \rightleftharpoons 2SO_3(g); \quad \Delta H^\ominus = -197\,\text{kJ mol}^{-1}$$

the reactants and product reach a state of equilibrium. According to Le Chatelier's Principle [§ 11.4.1] the reaction will be favoured by a low temperature and a high pressure. At high pressures, however, sulphur dioxide liquefies, and at low temperatures the rate of attainment of equilibrium is slow. The catalyst vanadium(V) oxide is so effective that a 95% conversion is achieved at 450 °C and 2 atm. The catalyst is easily 'poisoned' by absorbing impurities in the sulphur dioxide. These take up the surface of the catalyst and impair its efficiency. The temperature of the catalyst is maintained at 450 °C in spite of the exothermicity of the reaction by using the heat generated in the catalyst chamber to heat the incoming gases [see Figure 21.13].

The principles which govern the position of equilibrium are discussed in the light of Le Chatelier's Principle

The sulphur(VI) oxide formed is not absorbed in water because the reaction

$$SO_3(g) + H_2O(l) \rightarrow H_2SO_4(l)$$

is so intensely exothermic as to vaporise the sulphuric acid formed. It can be safely absorbed in 98% sulphuric acid to form fuming sulphuric acid, *oleum*, $H_2S_2O_7$, which reacts with water to form sulphuric acid:

SO_3 is absorbed in H_2SO_4

$$SO_3(g) + H_2SO_4(l) \rightarrow H_2S_2O_7(l)$$

$$H_2S_2O_7(l) + H_2O(l) \rightarrow 2H_2SO_4(l)$$

FIGURE 21.13
The Contact Process

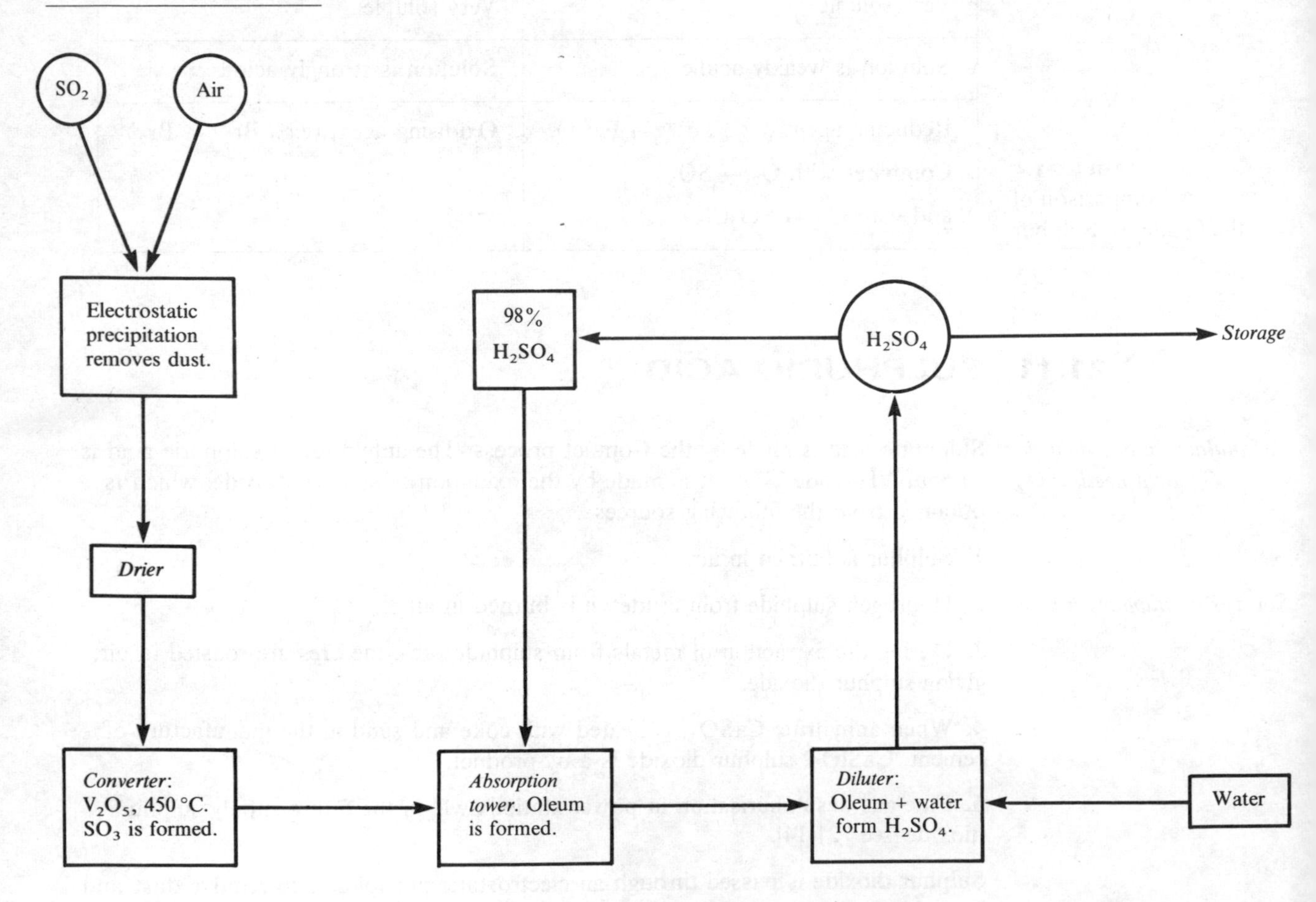

21.11.1 PHYSICAL PROPERTIES

Physical properties of H_2SO_4

Sulphuric acid is a viscous liquid. It is a covalent substance with the structure shown in Figure 21.14(a). The viscosity and the high boiling temperature (338 °C) are due to hydrogen bonding [see Figure 21.14(b)].

FIGURE 21.14
Sulphuric Acid
(*a*) the Structure
(*b*) the Intermolecular
Hydrogen Bonding

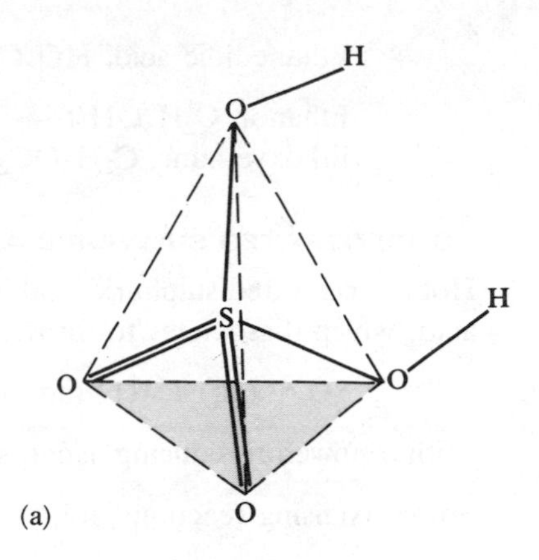

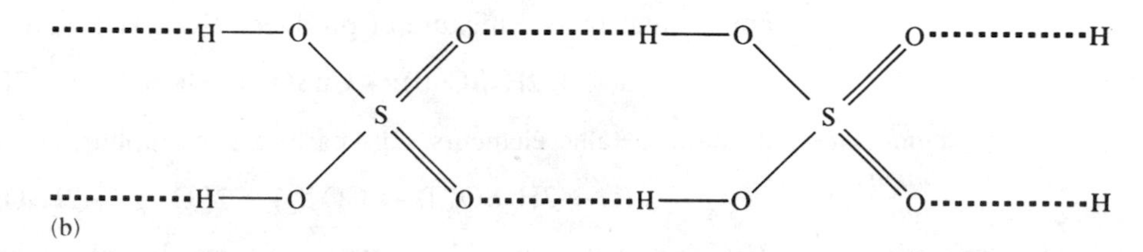

With water, sulphuric acid forms a constant-boiling [§ 8.4.4] mixture containing 98.3% acid.

21.11.2 CHEMICAL PROPERTIES

ACID PROPERTIES

Sulphuric acid is a diprotic acid, the second ionisation being incomplete . . .

In aqueous solution, sulphuric acid functions as an acid. The ionisation

$$H_2SO_4(aq) + H_2O(l) \rightarrow H_3O^+(aq) + HSO_4^-(aq)$$

is virtually complete. The ionisation

$$HSO_4^-(aq) + H_2O(l) \rightleftharpoons H_3O^+(aq) + SO_4^{2-}(aq)$$

is about 10% complete.

. . . It reacts as a typical acid

Sulphuric acid in solution reacts with metals and their oxides, hydroxides and carbonates in typical acid fashion.

CONCENTRATED SULPHURIC ACID AS A DRYING AGENT

Concentrated sulphuric acid must be diluted with care

Concentrated sulphuric acid reacts exothermically with water. When a solution is made it is essential to pour the acid into water, stirring to disperse the heat evolved. It is dangerous to add water to concentrated sulphuric acid as small pockets of water are likely to boil.

It is used to dry gases . . .

Gases are dried by bubbling them through concentrated sulphuric acid, as shown in Figure 21.12. For basic gases, another drying agent must be used.

CONCENTRATED SULPHURIC ACID AS A DEHYDRATING AGENT

When concentrated sulphuric acid removes water from a mixture, it is termed a **drying agent**. When it removes the elements of water from a compound, with the formation of a new compound, it is described as a **dehydrating agent**. Some dehydrating reactions of concentrated sulphuric acid are:

. . . and as a dehydrating agent

$$\text{Glucose, } C_6H_{12}O_6(s) \rightarrow \text{Sugar charcoal, } C(s) + H_2O$$

$$\text{Methanoic acid, } HCO_2H(l) \rightarrow CO(g) + H_2O$$

Ethanedioic acid, $HO_2CCO_2H(l) \rightarrow CO(g) + CO_2(g) + H_2O(l)$

Ethanol, $C_2H_5OH(l) \rightarrow$ Ethene, $CH_2{=}CH_2(g)$ or
Ethoxyethane, $C_2H_5OC_2H_5(g)$ depending on the conditions [§ 30.7.2]

CONCENTRATED SULPHURIC ACID AS AN OXIDISING AGENT

Hot concentrated sulphuric acid is an oxidising agent. It may be reduced to sulphurous acid, which dissociates to form sulphur dioxide and water:

As an oxidising agent, concentrated sulphuric acid oxidises metals...

$$SO_4^{2-}(aq) + 4H^+(aq) + 2e^- \rightleftharpoons SO_2(aq) + 2H_2O(l); \quad E^\ominus = +0.17\,V$$

With a powerful reducing agent, sulphuric acid may be reduced to hydrogen sulphide.

Some oxidising reactions are:

1. Metals are oxidised to sulphates. Equations represent these reactions only approximately as a mixture of products, SO_2, H_2S and S, is formed:

$$Cu(s) + 2H_2SO_4(l) \rightarrow CuSO_4(aq) + SO_2(g) + 2H_2O(l)$$

... non-metallic elements...

2. Non-metallic elements, e.g., carbon and sulphur, are converted into oxides:

$$C(s) + 2H_2SO_4(l) \rightarrow CO_2(g) + 2SO_2(g) + 2H_2O(l)$$

... and compounds, e.g., HBr, HI, H_2S

3. Compounds are oxidised, with the formation of sulphur dioxide or hydrogen sulphide. Hydrogen bromide, iodide and sulphide are oxidised:

$$2HBr(g) + H_2SO_4(l) \rightarrow Br_2(l) + SO_2(g) + 2H_2O(g)$$

$$3H_2S(g) + H_2SO_4(l) \rightarrow 4S(s) + 4H_2O(l)$$

FIGURE 21.15 Reactions of Concentrated Sulphuric Acid

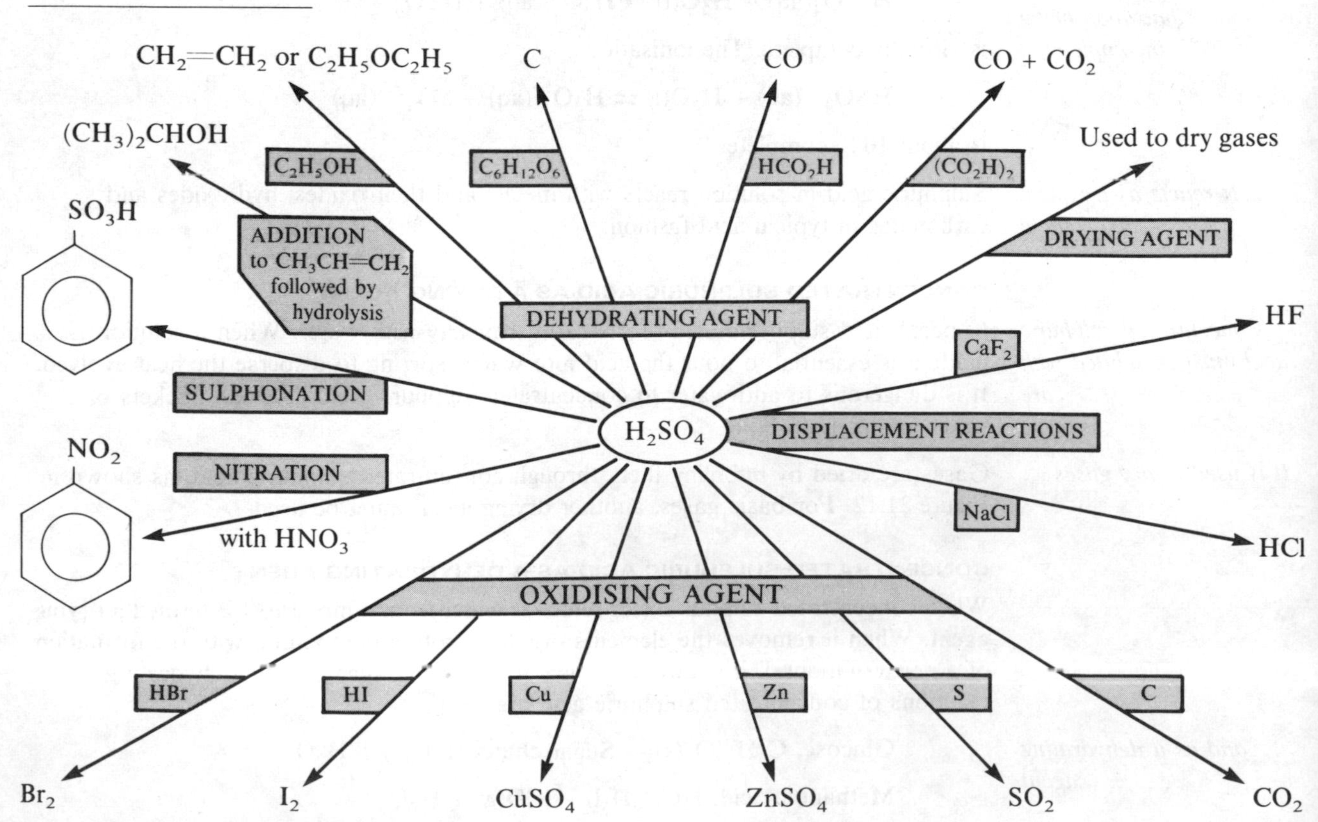

DISPLACEMENT REACTIONS OF CONCENTRATED SULPHURIC ACID

Concentrated sulphuric acid displaces other acids from their salts

When concentrated sulphuric acid is heated with a nitrate, it displaces the other acid from its salt, and nitric acid is formed. In the cold, an equilibrium is set up:

$$NO_3^-(s) + H_2SO_4(l) \rightleftharpoons HNO_3(l) + HSO_4^-(s)$$

If the mixture is heated, nitric acid distils at 120 °C, while sulphuric acid, with T_b 270 °C remains. The removal of nitric acid moves the equilibrium over to the right-hand side. Similarly, the more volatile acids, hydrogen fluoride, hydrogen chloride and phosphoric(V) acid, are displaced from fluorides, chlorides and phosphates(V).

REACTIONS WITH ORGANIC COMPOUNDS

Conc. H_2SO_4 in organic chemistry

Concentrated sulphuric acid is a sulphonating agent, able to introduce the —SO_3H group into an aromatic ring [§ 28.7.3]. It is used in nitration [§ 28.7.1] and in addition to alkenes [§ 27.7.7].

CHECKPOINT 21G: SULPHURIC ACID

1. Concentrated sulphuric acid reacts with sodium iodide in a two-stage reaction to give iodine. Explain how sulphuric acid is reacting, and write equations for the two steps.

2. Concentrated sulphuric acid reacts with potassium bromide to form a solid and three gases. Name these products, and explain how they come to be formed.

3. A convenient laboratory preparation for carbon monoxide is the action of concentrated sulphuric acid on sodium methanoate, HCO_2Na.
Explain how carbon monoxide is formed. A similar reaction occurs between concentrated sulphuric acid and sodium ethanedioate, $(CO_2Na)_2$. Write equations for the two reactions.

4. What would you expect to see when concentrated sulphuric acid is added to copper(II) sulphate crystals?

5. Phosphorus(V) oxide, P_2O_5, is a more powerful dehydrating agent than sulphuric acid. What is the fuming gas that is formed in a reaction between the two chemicals?

6. Give examples, with equations, of reactions in which sulphuric acid acts as *(a)* an acid, *(b)* a dehydrating agent, *(c)* an oxidising agent, *(d)* a sulphonating agent.

7. The Contact process achieves a 95% conversion of sulphur dioxide. What happens to the rest?

21.12 SULPHATES

The sulphate ion, SO_4^{2-}, can be represented [see Figure 21.16] as a delocalised structure. The distribution of bonds is tetrahedral.

FIGURE 21.16
The Structure of the Sulphate Ion

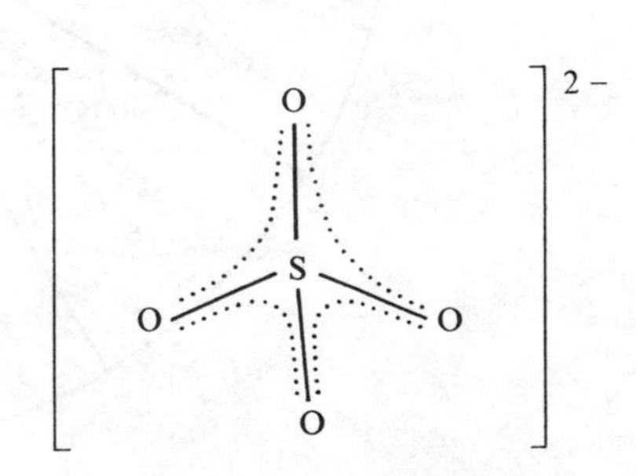

Methods of preparing sulphates

Soluble sulphates are prepared by the action of dilute sulphuric acid on the metal or its oxide, hydroxide or carbonate. Insoluble sulphates are made by precipitation.

The members of Groups 1 and 2 also form hydrogensulphates, which decompose when heated to form sulphates.

Thermal decomposition of $FeSO_4$ and $Fe_2(SO_4)_3$

Some sulphates also decompose on strong heating. Both iron(II) and iron(III) sulphates decompose to form iron(III) oxide:

$$Fe_2(SO_4)_3(s) \rightarrow Fe_2O_3(s) + 3SO_3(g)$$

$$2FeSO_4(s) \rightarrow Fe_2O_3(s) + SO_3(g) + SO_2(g)$$

How to test for sulphates and sulphites

A test for the sulphate ion in solution is the addition of barium chloride solution and dilute hydrochloric acid. A white precipitate of barium sulphate denotes the presence of a sulphate or a hydrogensulphate:

$$SO_4^{2-}(aq) + Ba^{2+}(aq) \rightarrow BaSO_4(s)$$

Sulphites give a white precipitate of barium sulphite on the addition of barium chloride solution. When dilute hydrochloric acid is added, the precipitate dissolves with an effervescence of sulphur dioxide.

21.13 THIOSULPHATES

Thiosulphates contain the ion $S_2O_3^{2-}$. The corresponding acid has never been isolated.

Laboratory preparation of sodium thiosulphate

Sodium thiosulphate is made by boiling a solution of sodium sulphite with sulphur. After filtration and evaporation, crystals of $Na_2S_2O_3 \cdot 5H_2O$ are obtained:

$$SO_3^{2-}(aq) + S(s) \rightarrow S_2O_3^{2-}(aq)$$

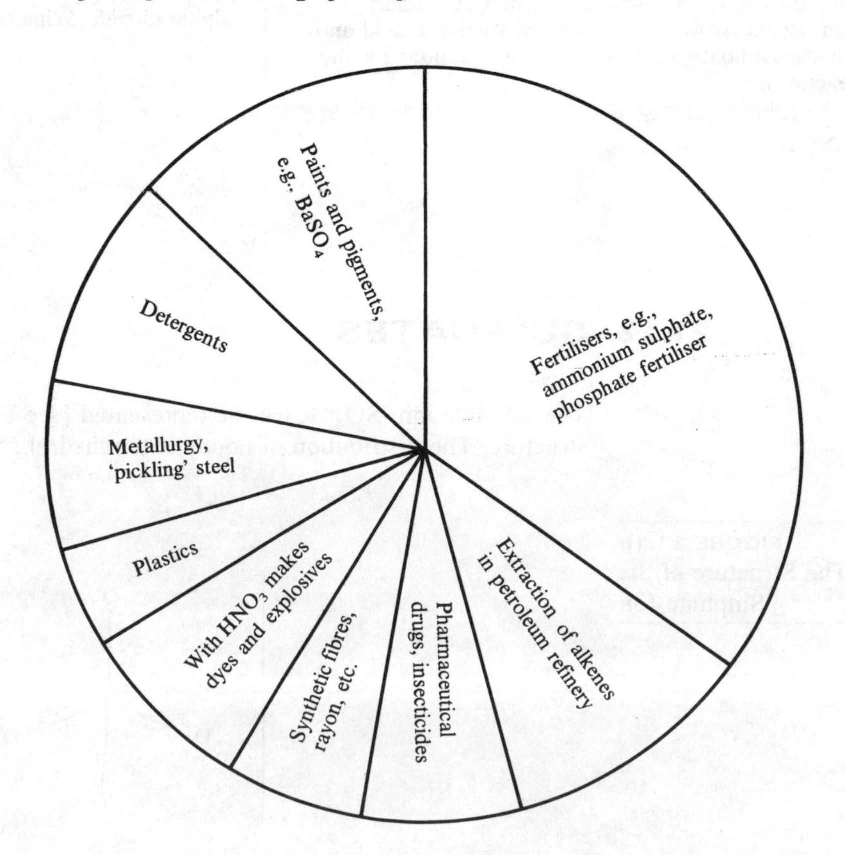

FIGURE 21.17 Some of the Industrial Uses of Sulphuric Acid (The world consumption totals 130 million tonnes per annum.)

The action of acid on thiosulphate

On the addition of acid to a thiosulphate solution, sulphur is precipitated and sulphur dioxide is evolved:

$$S_2O_3^{2-}(aq) + 2H^+(aq) \rightarrow SO_2(g) + S(s) + H_2O(l)$$

The reaction with I_2 and Cl_2

Sodium thiosulphate is a reducing agent. It is used for the titrimetric analysis of iodine [§ 3.15.1] which oxidises thiosulphate to tetrathionate, $S_4O_6^{2-}$. Industrially, thiosulphate is used to remove an excess of chlorine from fabrics after they have been bleached. Chlorine oxidises thiosulphates to sulphates.

CHECKPOINT 21H: SULPHATES

1. The substance $FeSO_4 \cdot 7H_2O$ occurs naturally and has been known for centuries as *green vitriol*. On heating, green vitriol gives a reddish-brown pigment called *jeweller's rouge* and some gaseous products. When these are condensed, another valuable chemical is formed. What is it? What is the formula of jeweller's rouge? Write the equation for the thermal decomposition.

2. Outline the steps you would take in the laboratory preparation of
(*a*) $MgSO_4 \cdot 7H_2O$ crystals
(*b*) $Na_2SO_4 \cdot 5H_2O$ crystals
(*c*) $NaHSO_4 \cdot 6H_2O$ crystals.

3. Copy the following table. In each box, write what you would observe, and explain why.

Solution	*Reagents added separately to solutions*		
	$BaCl_2(aq)$	$HCl(aq)$	$BaCl_2(aq) + HCl(aq)$
$Na_2SO_4(aq)$			
$Na_2SO_3(aq)$			
$NaHSO_4(aq)$			
$Na_2S_2O_3(aq)$			

4. Draw up a table comparing sulphurous acid, H_2SO_3, and sulphuric acid, H_2SO_4, with respect to
(*a*) the ease of isolation of the pure acids
(*b*) the extent of ionisation in aqueous solution
(*c*) their reactions as typical acids
(*d*) their power as oxidising agents
(*e*) their power as reducing agents
(*f*) any other points.

5. Draw the arrangement of bonds in the species:

$$O_3, S_8, SF_6, SO_2, SO_3, SO_4^{2-}, S_2O_3^{2-}$$

6. Write half-reaction equations for (*a*) the oxidation of thiosulphate to tetrathionate, (*b*) the oxidation of thiosulphate to sulphate, (*c*) the reduction of I_2 to I^-, (*d*) the reduction of Cl_2 to Cl^-. Combine equations (*a*) and (*c*). Combine equations (*b*) and (*d*).

21.14 ACID RAIN

TOPIC

Rain is naturally slightly acidic because it reacts with carbon dioxide in the air to form carbonic acid. Natural rain-water has a pH of about 5.6. In central Europe, rain-water is much more acidic, with a pH of about 4.1, and on the fringes of Europe, e.g. Ireland and Portugal, rain-water has a pH of about 4.9. Rain from individual storms can have a pH below 3, and water droplets in fogs can be even more acidic.

All rain water is slightly acidic. Acid rain has a pH below 5.0

Acid rain is now thought to be the cause of the extensive damage to Europe's trees and to the death of fish in the lakes of Canada, Norway, Sweden, Wales, Scotland and other countries. Europe shows the worst signs of damage by acid rain, but acid rain is becoming a global phenomenon.

Lakes in many countries of Europe and North America have become so acidic that their stocks of fish have died

Lakes in Scandinavia, Scotland, Canada and the USA have become much more acidic. Thousands of lakes which once stocked fish are now dead. The death of fish is usually attributed to poisoning by aluminium. Sulphate ions in acid rain can combine with aluminium in complex compounds to form soluble aluminium sulphate, which washes into streams. There it interferes with the operation of fish gills, so that they become clogged with mucus. The fish die from lack of oxygen.

Many of Europe's trees have lost much of their foliage. In some countries large areas of forest have died. The cause is thought to be acid rain

Forests have declined. In the mid-1970s Alpine forests started to lose their fir trees. Then in West Germany, including the famous Black Forest, spruce began to thin and their needles turned brown. After 1984, the decline stabilised in Germany. The Netherlands, Czechoslovakia, Switzerland and Britain recorded that 20–30% of their trees were severely defoliated. Acidic rain-water draining from soils washes out nutrients and liberates aluminium ion, which the roots of trees may take up. Without essential nutrients, e.g. calcium and magnesium, trees starve to death.

Building materials which are basic... and iron and steel... are attacked by acid rain

Building stone may be limestone or marble (both calcium carbonate) or a sandstone in which quartz grains are held together by a coating of calcium carbonate or iron oxide. These materials (apart from quartz) are attacked by acid rain. Metallic structures, e.g. bridges, ships and motor vehicles, are also attacked by acid rain.

FIGURE 21.18 The Formation of Acid Rain

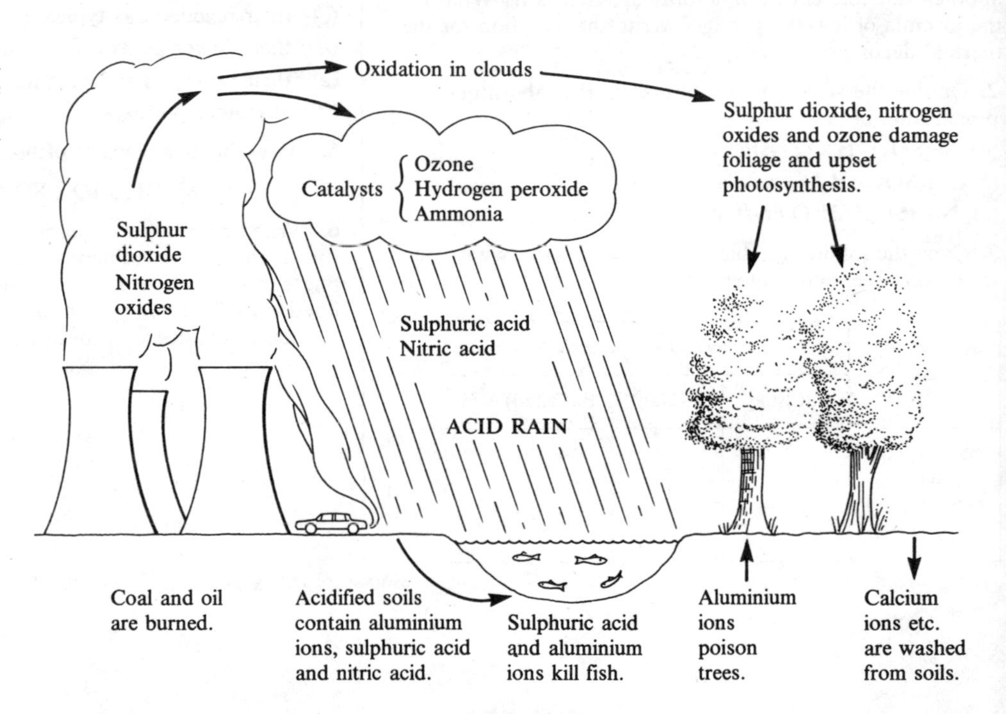

Sulphur dioxide is one of the causes of acid rain. It is formed in the combustion of fossil fuels. In the atmosphere, sulphur dioxide reacts with oxygen and water vapour to form sulphuric acid

One of the chief culprits in the formation of acid rain is sulphur dioxide. Natural sources, such as volcanoes, sea spray, rotting vegetation and plankton send sulphur dioxide into the atmosphere. Half the sulphur dioxide in the atmosphere, however, comes from the combustion of fossil fuels. Over Europe the proportion of sulphur dioxide in the air that comes from fuels is 85%. When sulphur dioxide reaches the atmosphere, it reacts with moisture and oxygen to form sulphuric acid. Not all the sulphur dioxide is converted into sulphuric acid: in dry air, sulphur dioxide can travel hundreds of kilometres with little conversion to acid, and can descend to ground level unconverted. When sulphur dioxide is incorporated into clouds, conversion into sulphuric acid takes place within two hours. In heavily polluted city air, tiny particles of metals catalyse the reactions. Ozone and hydrogen peroxide also seem to assist them, as does ammonia, which is given off in large quantities from slurry tanks.

FIGURE 21.19 Sources of Sulphur Dioxide

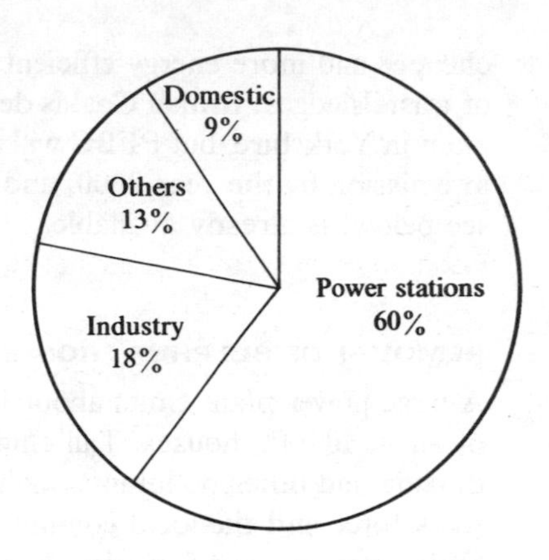

The 'Thirty Per Cent Club' of nations have agreed to reduce their emission of sulphur dioxide

Acid rain has become an important political issue, souring relationships between countries which export pollution and those which receive it; for example, between Britain and Norway and between the USA and Canada. In 1984 The UK joined the 'Thirty Percent Club', a group of nations committed to reducing sulphur dioxide emission by 30% by the year 2000. Figure 21.19 shows that the biggest source of sulphur dioxide is power stations. In 1984, the Central Electricity Generating Board (CEGB) estimated a cost of £120 million for treating the exhaust gases from each existing power station and an increase in running costs of 10–15% which would increase the price of electricity by 5%.
Several measures can be taken to reduce this pollution.

LOW SULPHUR FUELS

Some of the sulphur in oil and coal can be removed before the fuels are burnt

Oil power stations emit on average about 10% more sulphur dioxide than coal power stations. Modern oil refineries can, however, produce low-sulphur oil. The coal used in British power stations contains an average of 1.6% sulphur. About half the sulphur is combined as 'iron pyrites', FeS_2. Of this, 80% can be removed by grinding the coal and using various separation techniques. Removal of organically bound sulphur is more difficult. If the coal is converted into a gaseous fuel by one of the modern methods of *coal gasification*, sulphur can be removed as the gas is formed. Coal gasification has another advantage in that it may supplement our diminishing reserves of natural gas. Of the sulphur in bituminous coal, 25% can be removed by crushing and cleaning. A number of countries in Eastern Europe use lignite which cannot be cleaned by present technology.

NEW BURNERS

Pulverised fluidised bed combustion (PFBC) is a technique for removing sulphur from coal as it burns. PFBC can be built into new power stations

Pulverised fluidised bed combustion (PFBC) offers a new, more efficient way of burning coal in a bed of limestone, which removes sulphur as the coal burns. PFBC can cut the emission of nitrogen oxides as well as sulphur dioxide. Pilot tests suggest that such burners could remove up to 80% of sulphur dioxide. These systems may soon be widely used in pollution-conscious countries.

In 1984 the House of Lords published a report on air pollution. The Lords urged the Government to take preventive measures against air pollution. The report recommended the development and installation of PFBC plants. They recommended PFBC over other systems (for FGD, see later) on the grounds that PFBC would be cheaper and more energy-efficient and would create fewer problems in the disposal of waste sludges. British Coal is developing the technology in its Grimethorpe research plant in Yorkshire, but PFBC will not be ready in time to achieve the needed reduction in emission by the year 2000, and an alternative method (flue-gas desulphurisation; see below) is already available.

REMOVAL OF SULPHUR FROM EXHAUST GASES

A large power plant emits about 10^8 m^3 of gases daily — roughly equal to the volume of air in 100 000 houses. Tall chimneys take the exhaust gases, including sulphur dioxide and other pollutants, high up into the air before they are discharged. The work force and the local community are protected. At one time, the managers of power stations and factories thought that was the end of the problem. Now that we know acidic pollutants are to blame for the acid rain that falls perhaps thousands of kilometres away, we realise that tall chimneys do not solve the problem; they merely transfer it to another region.

Flue gas desulphurisation (FGD) is a technique for removing sulphur dioxide from the exhaust gases in the chimney stacks of power stations. FGD plants can be fitted on to existing power stations. The UK is investing in FGD

Sulphur dioxide can be removed from the exhaust gases before they leave the chimney stack of the power station. Systems that carry out such *flue gas desulphurisation* (FGD) have been developed and are increasingly used in power stations. The exhaust gases are washed by an alkaline solution, which converts the sulphur dioxide into a waste sludge. The systems can remove up to 95% of the sulphur in the flue gases. Sweden and West Germany have programmes of fitting FGD systems, and the US has installed them in some areas. The Central Electricity Generating Board (CEGB) of the UK has been very reluctant to spend money on desulphurisation. In 1988 Prince Charles said 'I would have thought that the CEGB was doing too little and too late. Our responsibilities do lie in not exporting our problems abroad.' He asked why Britain was so slow to respond to what he saw as growing public concern about the environment and why it took so long for legislation to be put into effect. He suggested that on attitudes to the environment many industrialists were out of step with the 'ordinary bloke'.

The concentration of sulphur dioxide in flue gases from power stations is about 0.3%. Some of the alkaline solutions and suspensions used to remove it are as follows:

(1) A slurry of limestone and lime is used to 'scrub' the flue gases.

(*a*) $CaCO_3 + SO_2(g) \rightarrow CaSO_3 + CO_2(g)$

(*b*) $CaO + SO_2 \rightarrow CaSO_3$

(*c*) $2CaSO_3 + O_2 + 4H_2O \rightarrow 2CaSO_4 \cdot 2H_2O$

Reactions (*a*), (*b*) and (*c*) produce an impure sludge which is dumped.

(2) A slurry of magnesium oxide is used as a 'scrubber'.

(*d*) $MgO + SO_2 \xrightarrow{H_2O} MgSO_3 \xrightarrow{heat} MgO + SO_2$

The magnesium sulphite produced is heated to give magnesium oxide, which is recycled, and sulphur dioxide at a concentration high enough to allow it to be used in the manufacture of sulphuric acid.

(3) A solution of sodium sulphite can be used for 'scrubbing'.

(*e*) $Na_2SO_3 + H_2O + SO_2 \rightarrow 2NaHSO_3$

The sodium hydrogensulphite produced can be heated to give sodium sulphite for recycling and sulphur dioxide for sale to sulphuric acid manufacturers.

In 1988 the CEGB announced that Drax power station in Yorkshire, Western Europe's largest coal-fired power station, would be fitted with an FGD plant. A flow diagram is shown in Figure 21.20. The £400 million FGD plant will remove 90% of the sulphur dioxide discharged from the power station. The power station at Fidler's Ferry in Cheshire will also be fitted with an FGD plant. The calcium sulphate (gypsum) produced in the process will be sold to the plaster board industry and cement manufacturers. Should there be a surplus, it will be used for land-filling and reclamation of quarries, gravel pits and open-cast workings. The Drax FGD plant will come into operation in stages between 1993 and 1995.

Flue gas desulphurisation is easier for the smelting industries, where the concentration of sulphur dioxide in the gases can be as high as 5–15%. Direct conversion to sulphuric acid is possible. This is a saleable commodity. If there is not a ready market, it can be converted into sulphur which is easier to store and to transport.

FIGURE 21.20 Plans for Drax Coal-fired Power Station FGD Plant

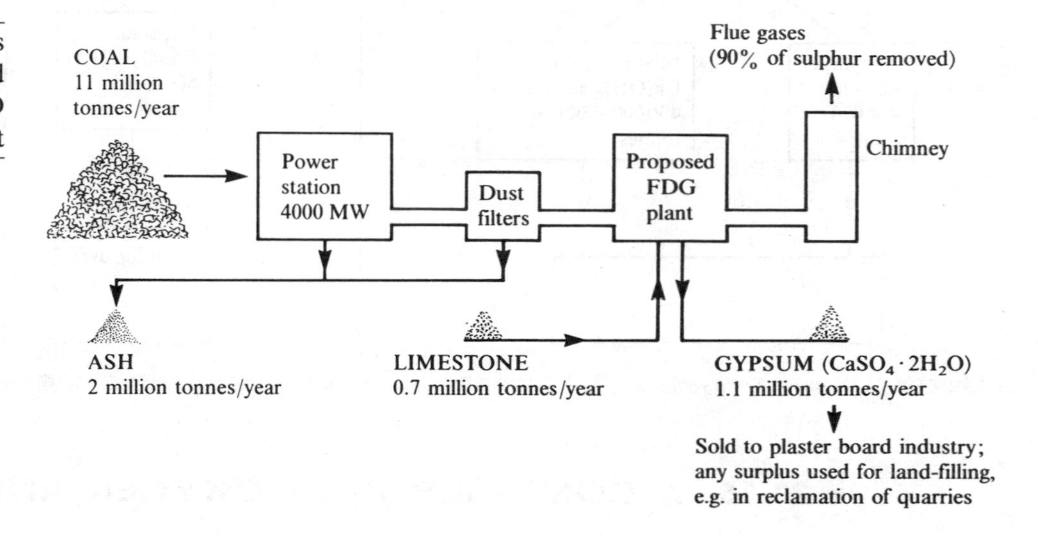

OTHER SOLUTIONS

Nuclear power stations and renewable energy sources are other options for reducing acid rain

Other solutions to the problem of pollution from power stations are a switch to nuclear power stations and the development of technology for harnessing 'renewable energy sources', e.g. solar energy, wind, waves, hydropower and geothermal energy.

CHECKPOINT 21I: ACID RAIN

1. (*a*) Compared with the rain in Scandinavia, is the rain in the UK
A more acidic **B** less acidic **C** the same?

(*b*) How much of Western Europe's sulphur dioxide emission comes from the UK?
A 1/4 **B** 1/6 **C** 1/10

(*c*) How many new car models sold in the UK would pass the US Clean Air Laws?
A 20 **B** 5 **C** 1

(*d*) Of the 87 lochs surveyed in the Galloway and Loch Ard areas of Scotland in 1979, how many had become acidic?
A almost all **B** half **C** very few

(*e*) What percentage of oaks in the UK are unhealthy, having lost more than 10 per cent of their leaves?
A 20% **B** 50% **C** 80%

2. The UK imports 750 000 tonnes of sulphur annually. The emission of sulphur from one power station is 1.3×10^6 tonnes of sulphur per year. Comment on these figures.

3. Refer to the comparison of PFBC and FGD in Figure 21.21.

(*a*) Which process is better from the point of view of using limestone more carefully?

(*b*) Which process will obtain more energy from coal?

(*c*) How do the processes compare with respect to the disposal of waste products?

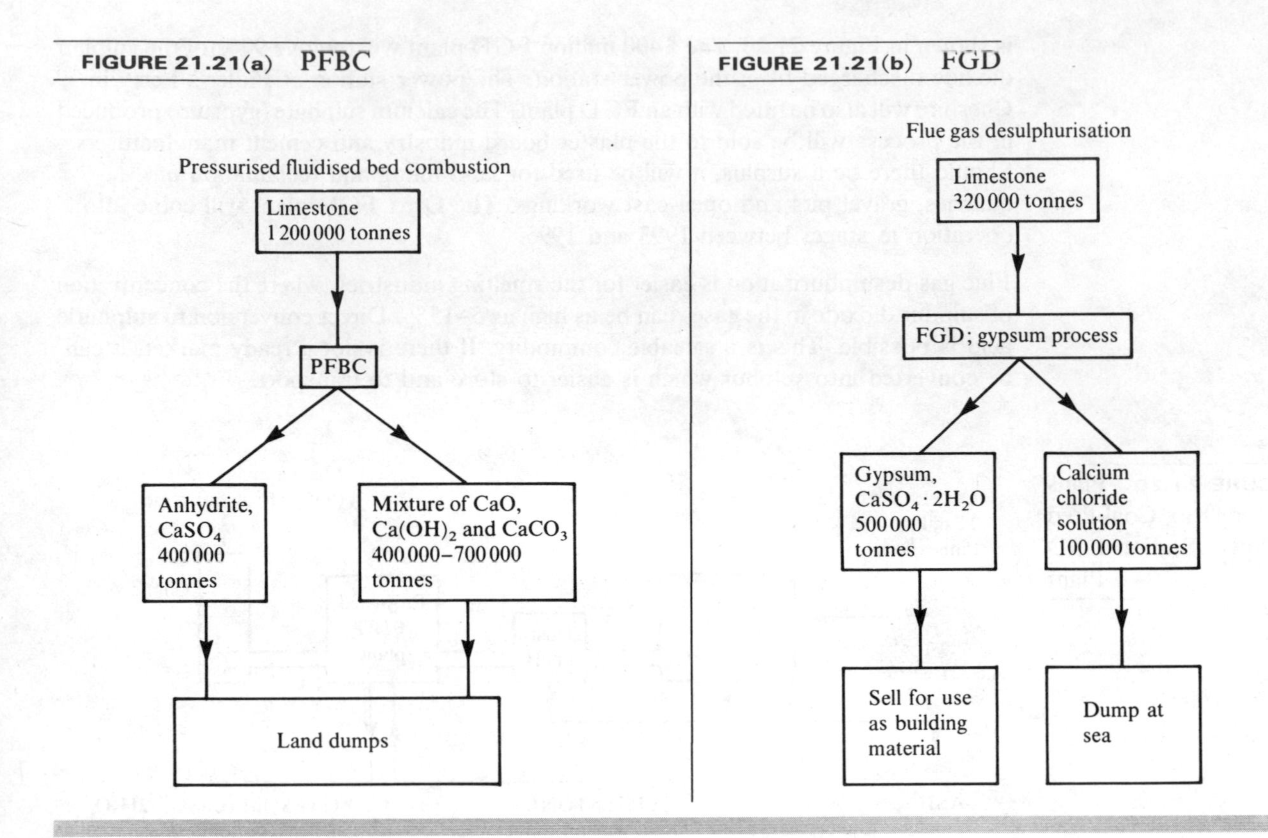

21.15 A COMPARISON OF OXYGEN AND SULPHUR

SIMILARITIES

1. Both have the electron configuration (core)ns^2np^4. They form O^{2-} and S^{2-} and also two covalent bonds.

2. They are non-metallic elements.

3. They are very reactive elements, combining directly with metals and with non-metallic elements.

Oxygen and sulphur are similar in many respects

4. Both show allotropy: oxygen as O_2 and O_3; sulphur as the rhombic and monoclinic forms.

5. The hydrides H_2O and H_2S are stable, while H_2O_2 and H_2S_2 are unstable.

6. The oxides and sulphides of Groups 1 and 2 are ionic.

7. The oxides and sulphides of Group 1 undergo similar hydrolysis reactions:

$$O^{2-}(aq) + H_2O(l) \rightarrow 2OH^-(aq)$$

$$S^{2-}(aq) + H_2O(l) \rightarrow HS^-(aq) + OH^-(aq)$$

DIFFERENCES

1. Oxygen exists as O_2 molecules (and as O_3 [§ 21.4.1]).
Sulphur is usually in the form of S_8 molecules. The reason for the difference is the standard bond enthalpies for the O=O, O—O, S=S and S—S bonds [see Question 3, Checkpoint 21C].

2. **Oxygen is restricted to an L shell of 8 electrons and an oxidation state of −2 (except in peroxides and fluorides). Sulphur can use its d orbitals and employ oxidation states of −2, +2, +4 and +6.**

3. **The diameters of the ions are O^{2-}, 0.280 nm and S^{2-}, 0.368 nm. The S^{2-} ion is therefore more polarisable than the O^{2-} ion. Oxides are more ionic in character than the corresponding sulphides. Ionic sulphides are formed by large cations, e.g., those of Group 1A.**

4. **Oxygen is more electronegative than sulphur. The difference leads to hydrogen bonding in water but not in hydrogen sulphide.**

5. **Hydrogen sulphide is a weak acid. Water can act both as a proton-donor and as a proton-acceptor.**

6. **The allotropy of oxygen is monotropic; that of sulphur is enantiotropic.**

QUESTIONS ON CHAPTER 21

1. Describe the Contact process for the manufacture of sulphuric acid. Explain the choice of conditions used for the process. Mention three uses for sulphuric acid.

2. How and under what conditions does sulphuric acid react with (*a*) potassium bromide, (*b*) ethanedioic acid, (*c*) sodium methanoate and (*d*) sucrose?

3. Illustrate the use of sulphuric acid (*a*) as an acid, (*b*) as an oxidising agent, (*c*) as a dehydrating agent, (*d*) in the displacement of other acids from their salts and (*e*) as a sulphonating agent.

4. Sulphur dioxide and chlorine are both used as bleaches. They act on different dyes. Why can sulphur dioxide bleach materials that are not bleached by chlorine?

5. Review the oxidation states of sulphur, giving examples of compounds of sulphur in its different oxidation states. Give examples, with equations, of redox reactions in which sulphur changes its oxidation number.

6. Describe the changes in appearance that occur when sulphur is heated to its boiling temperature. How can the changes be explained in terms of the molecular structure of sulphur?

7. Write the formulae of the two compounds of sulphur with chlorine and oxygen. Say how they can be made from sulphur dioxide. Say how they react with propanoic acid.

8. (*a*) When a hydrocarbon fuel is burned with the correct amount of air required for combustion, carbon monoxide is generally present in the exhaust gases.

(i) Write the equation for the complete combustion of a hydrocarbon of formula C_xH_y.

(ii) Suggest *two* reasons why the formation of carbon monoxide is undesirable.

(iii) Adding an excess of air successfully reduces the formation of carbon monoxide. Suggest an important disadvantage of doing this.

(*b*) Processes involving the roasting of a metal sulphide ore produce high concentrations of sulphur dioxide in the exhaust gas. Chemists have devised several ways of solving this pollution problem, and an extension to an existing plant can produce saleable by-products from the sulphur dioxide.

How might political or economic factors interfere with the implementation of these pollution remedies in certain countries?

(*c*) One solution to the pollution problems referred to in (*b*) is to oxidise the sulphur dioxide catalytically and to use the resulting sulphur trioxide to make sulphuric acid. What mass of sulphuric acid ($M_r = 98$) would be produced from one tonne of pyrites (FeS_2, $M_r = 120$) if all of the sulphur were converted into sulphuric acid?

(*d*) There are other solutions to these pollution problems. In the Resox process, sulphur dioxide ($M_r = 64$) is reduced to sulphur by pulverised coal in the presence of steam as a catalyst. The sulphur produced can be sold. The consumable needs of the process are stated to be 'about 0.2 kg coal and 0.05 kW h electricity per kilogram of inlet sulphur dioxide'.

(i) Write an equation for this chemical process.

(ii) From this equation, estimate the coal consumption of the process per kilogram of sulphur dioxide, thus verifying (or otherwise) the stated claim. Assume for simplicity that the coal consists only of carbon.

(iii) Suggest a reason why the electricity is needed.

(iv) The exhaust gases from roasting a metal sulphide ore contain 10% by volume of sulphur dioxide. What volume of exhaust gas before treatment will contain '1 kg of inlet sulphur dioxide'? The molar volume under the conditions used is 60 litres mol^{-1}.

(NEAB 91)

9. (*a*) Give an account of the preparation of hydrogen.

(*b*) Describe *two* properties which illustrate differences between H_2O and H_2S.

(*c*) Briefly outline a method of preparation of each of the following:

(i) H_2S, (ii) $Na_2S_2O_3$, (iii) Na_2CO_3 from Na, (iv) Ca from $CaCO_3$.

Confine your answers to essential conditions, equations and separations (or collections).

(O 91)

10. 'Freon 12' (CF_2Cl_2) is a liquid used as a refrigerant, and also (until relatively recently) as an aerosol propellant. Freon 12 is a very unreactive compound and, once released, the gas drifts slowly up into the stratosphere, 20–40 km above the Earth's surface.

In the stratosphere, freon 12 reacts with the ozone layer which protects the Earth from harmful solar radiation.

Part of the mechanism of the destruction of ozone is

(i) $CF_2Cl_2 + hf \rightarrow CF_2Cl\cdot + Cl\cdot$
(ii) $Cl\cdot + O_3 \rightarrow ClO\cdot + O_2$

where hf represents electromagnetic radiation of frequency f acting on the freon.

(*a*) Give the systematic name of freon 12.

(*b*) Suggest why the drift of freon 12 up into the stratosphere is fairly slow.

(*c*) Why is the inertness of freon 12 a big advantage in its use as an aerosol propellant?

(*d*) Suggest why this inertness is responsible for the damage it causes to the ozone layer.

(*e*) What *type* of process is stage (i)?

(*f*) What term describes particles like $CF_2Cl\cdot$ and $ClO\cdot$?

(*g*) Describe briefly a reaction involving *either* methane *or* ethene which involves particles of the same type as $CF_2Cl\cdot$ and $ClO\cdot$.

(*h*) Nitrogen monoxide ($NO\cdot$) is also present in the stratosphere, mainly from exhaust fumes of transport burning fossil fuels. Suggest an equation, similar to (ii), to show how nitrogen monoxide destroys ozone.

(*i*) Supersonic aircraft (such as Concorde) fly in the stratosphere. In view of the information in this question, outline an environmental implication of an increase in supersonic flight.

(O 92, AS)

11. (*a*) By reference to the F_2 molecule, explain the meaning of the term *bond energy*.

Sulphur hexafluoride can be made by reacting sulphur tetrafluoride with fluorine in the gas phase:

$$SF_4(g) + F_2(g) \rightarrow SF_6(g); \quad \Delta H = -434\,\text{kJ mol}^{-1}$$

By considering the bonds broken and bonds formed during this reaction, calculate an average value for the S—F bond energy. State any assumptions you have made.

(*b*) (i) Construct dot-and-cross diagrams to illustrate the bonding in the molecules of sulphur difluoride, SF_2, and sulphur hexafluoride, and predict their shapes.

(ii) Caesium fluoride has a similar formula mass to sulphur hexafluoride. Describe *three* major differences you would expect to find in the physical properties of the two compounds.

(C 92)

12. (*a*) Give equations and essential conditions for the chemical reactions which are involved in the manufacture of sulphuric acid from sulphur dioxide.

(*b*) Concentrated sulphuric acid may function as

(i) an oxidising agent
(ii) a dehydrating agent
(iii) a catalyst.

In each case, give one reaction, with an equation, which illustrates the property of sulphuric acid.

(*c*) Sulphur dioxide undergoes a redox reaction with hydrogen sulphide to form sulphur and water. Write a balanced equation for the reaction and indicate the initial and final oxidation states of the sulphur compounds concerned.

(*d*) (i) Old oil-paintings which contain lead carbonate as a white pigment may be darkened on exposure to air polluted with hydrogen sulphide. Suggest a reason for this effect.

(ii) The colour of the painting may be restored by treatment with hydrogen peroxide. Suggest what reaction is taking place and the nature of the product.

(O & C 91)

13. (*a*) Outline the chemical processes by means of which

(i) ozone is formed in the upper atmosphere
(ii) chlorofluorocarbons (CFCs) may lead to the destruction of ozone in the upper atmosphere
(iii) chloroalkanes are formed from alkanes and chlorine.

(*b*) The mechanisms of the three chemical processes in part (*a*) have certain features in common. Identify *three* of them.

(*c*) (i) Outline *two* uses to which halogenated hydrocarbons (including CFCs) have been put.

(ii) Which chemical features of halogenated hydrocarbons have led to their widespread use?

(iii) Indicate how some of these features have also led to environmental problems.

(O & C 92)

22

GROUP 5

22.1 THE MEMBERS OF THE GROUP

The elements in Group 5 are listed in Table 22.1.

TABLE 22.1 The Members of Group 5

Element	*Symbol*	Z	*Outer shell of electrons*	*Oxidation states*	$T_m/°C$	$T_b/°C$
Nitrogen	N	7	$2s^22p^3$	3,5	−210	−196
Phosphorus	P	15	$3s^23p^3$	3,5	44(white) 590(red)	280
Arsenic	As	33	$4s^24p^3$	3,5	–	613
Antimony	Sb	51	$5s^25p^3$	3,5	630	1380
Bismuth	Bi	83	$6s^26p^3$	3(5)	271	1560

22.2 OCCURRENCE, EXTRACTION AND USES

22.2.1 NITROGEN

Nitrogen in nature

Nitrogen constitutes 78% by volume of dry air. It is an essential element in all living things, being present in proteins and nucleic acids. Figure 22.1 illustrates the **nitrogen cycle**. This is the balance between the reactions which take nitrogen out of the air and out of the soil and the reactions which put nitrogen into the air and into the soil. Nitrogen is mined as nitrate ores, for example, *Chile nitre*, $NaNO_3$.

Nitrogen is obtained from liquid air and used in the manufacture of ammonia

For industrial use, nitrogen is obtained by the fractional distillation of liquid air. Nitrogen distils at −196 °C, while the liquid becomes richer in oxygen, which has a boiling temperature of −183 °C.

Nitrogen is used for the manufacture of ammonia and thence nitric acid. Gaseous nitrogen is used to provide an inert atmosphere for reactions which cannot be carried out in the presence of oxygen. It is used as a carrier gas in gas–liquid chromatography.

Dinitrogen consists of diatomic molecules

$$N\equiv N$$

The low reactivity of nitrogen can be attributed largely to the strength of the triple bond:

$$N_2(g) \rightarrow 2N(g); \quad \Delta H^\ominus = +940\,kJ\,mol^{-1}$$

FIGURE 22.1
The Nitrogen Cycle

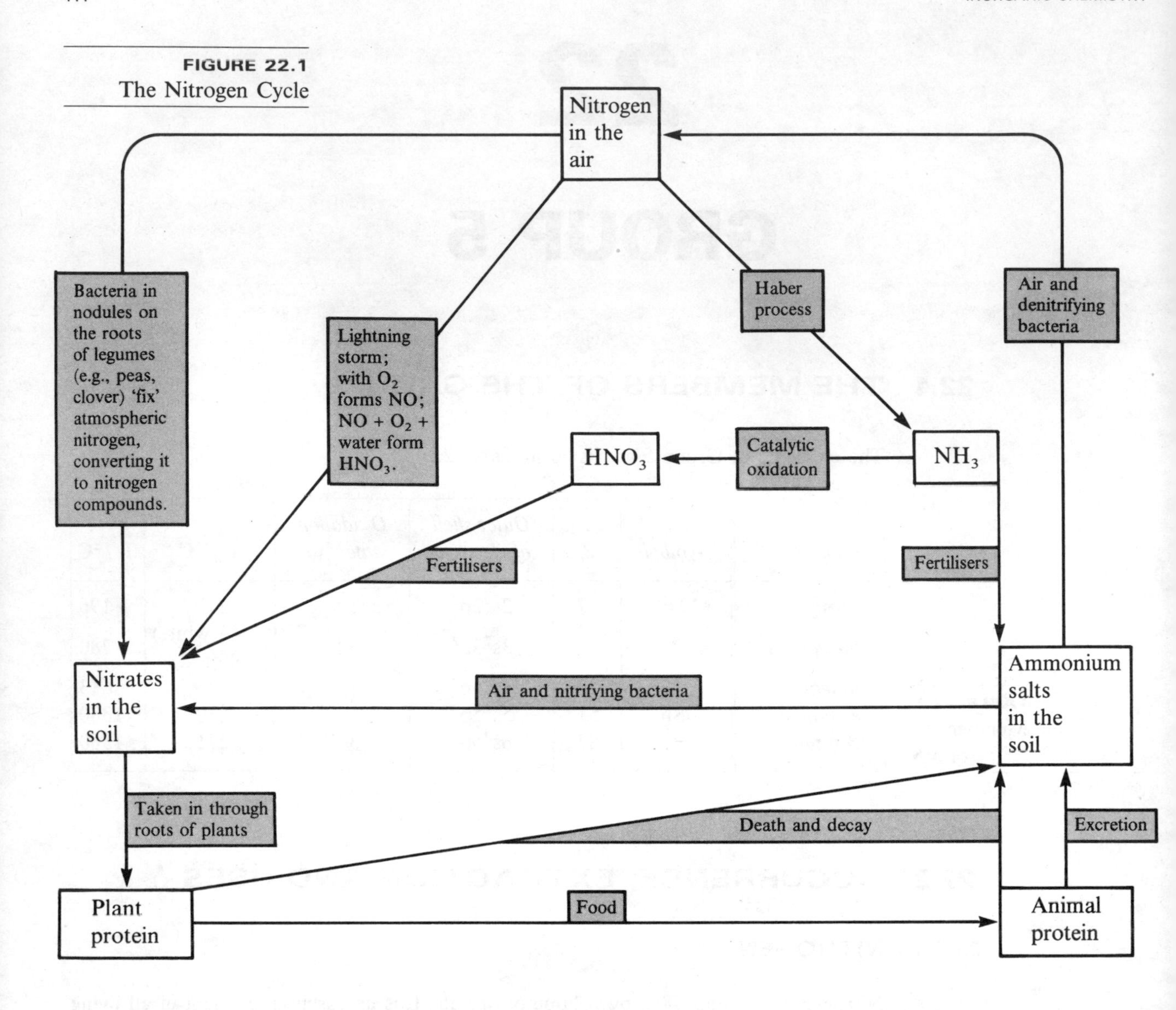

22.2.2 PHOSPHORUS

Phosphorus is extracted from calcium phosphate

Phosphorus is too reactive an element to occur free. It is mined as phosphate ores, e.g., calcium phosphate, $Ca_3(PO_4)_2$. When this is heated with silica and coke in an electric furnace, phosphorus sublimes over.

There are red and white allotropes

The two chief allotropes of phosphorus are the white and red forms. The allotropy is monotropic, red phosphorus being more stable than white under all conditions.

Phosphorus is used in match-heads and on the sides of boxes for safety matches.

22.3 BOND FORMATION

Nitrogen forms N^{3-} ions; other members of the group are 3-covalent...

The members of Group 5 all accept three electrons to attain a full outer octet. The formation of N^{3-} anions is highly endothermic. Ionic nitrides are formed only if their lattice enthalpies are highly exothermic, i.e., by Groups 1 and 2. The reason the rest of the Group do not form A^{3-} anions is that larger ions, e.g., P^{3-}, cannot approach cations sufficiently closely to form a lattice of high enthalpy.

All the elements of Group 5 show oxidation states of +3 and +5 in their oxo-anions:

+3 in NO_2^-, PO_2^-, AsO_3^{3-}

Some cations are formed, e.g. Sb^{3+}, Bi^{3+}, SbO^+, BiO^+

+5 in NO_3^-, PO_3^-, PO_4^{3-}, AsO_4^{3-}, SbO_3^- and BiO_3^-

Antimony and bismuth form the cations Sb^{3+}(aq) and Bi^{3+}(aq), which are readily hydrolysed to the oxo-cations, SbO^+(aq) and BiO^+(aq). They form complex ions, e.g., $BiCl_4^-$(aq).

All except nitrogen show a covalency of 5 as well as 3

All the elements in this group show a covalency of 3, using the three unpaired p electrons. All except nitrogen show a covalency of 5 by promoting one of the s electrons to a d orbital. This cannot happen with nitrogen because it has no d orbitals at energy levels comparable with the occupied L shell orbitals [see Figure 22.2].

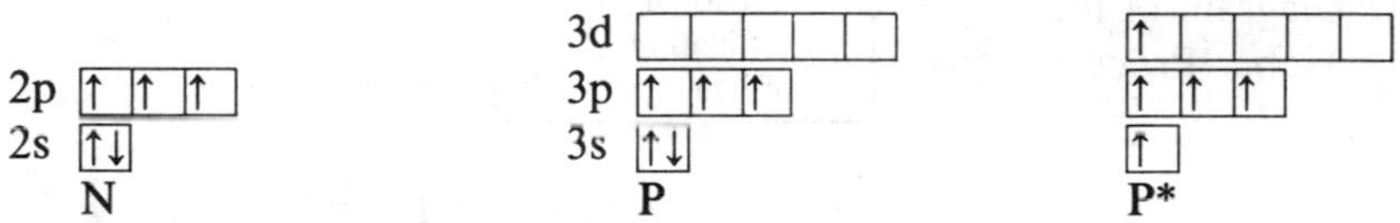

N: 2p ↑ ↑ ↑; 2s ↑↓

P: 3d (empty); 3p ↑ ↑ ↑; 3s ↑↓

P*: 3d ↑; 3p ↑ ↑ ↑; 3s ↑

FIGURE 22.2 Bonding Orbitals in N, P in its Ground State and P* in its Excited State

When nitrogen uses its lone pair of electrons to form a coordinate bond, as in NH_4^+ and NO_3^-, it has a covalency of 4.

Reactant	*Nitrogen*	*Phosphorus*
Metals	Ionic or interstitial nitrides formed	Phosphides formed
Oxygen	Some NO formed at high temperatures and in an electric discharge	Ignites, white at 35 °C, red at 260 °C, to form P_2O_3 + P_2O_5
Sulphur	No reaction	Mixture of sulphides formed
Halogens	No reaction	PX_3 + PX_5 formed
Hydrogen	Some NH_3 formed at high pressure	No reaction
Alkali	No reaction	White P (not red) reacts to form PH_3 + $H_2PO_2^-$
Conc. HNO_3	No reaction	H_3PO_4 formed

TABLE 22.2 Reactions of Nitrogen and Phosphorus

22.4 THE HYDRIDES

22.4.1 AMMONIA

The Haber process for ammonia

Ammonia is made by the Haber process from nitrogen and hydrogen:

$$N_2(g) + 3H_2(g) \rightleftharpoons 2NH_3(g); \quad \Delta H^\ominus = -92\,\text{kJ mol}^{-1}$$

The reaction is exothermic, and involves a decrease in the number of moles of gas. An application of Le Chatelier's Principle [§ 11.4.1] shows that the forward reaction should be assisted by a low temperature and a high pressure. At low temperature, the rate of attainment of equilibrium is low. At high temperature, the position of equilibrium is over to the left. A compromise temperature is adopted, and a catalyst is employed to speed up the attainment of equilibrium concentrations. The conditions employed in industrial plants are 200–1000 atm, 500 °C and, as catalyst, iron with aluminium oxide as a promoter. The yield is about 10%, and unreacted gases are recycled [see Figure 22.3].

FIGURE 22.3(a) A Flow Diagram of the Haber Process

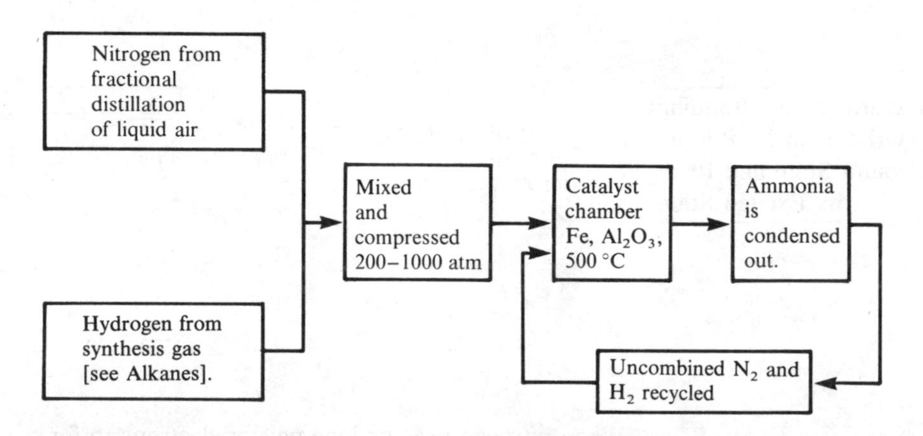

FIGURE 22.3(b) ICI Billingham. The Haber Process

The laboratory preparation of ammonia from an ammonium salt and a base

In the laboratory, ammonia is made by the action of a strong base on an ammonium salt:

$$NH_4^+(s) + OH^-(s) \rightarrow NH_3(g) + H_2O(g)$$

The reaction is accelerated by warming, and the gas is collected upwards after passing through a drying tower containing calcium oxide. Acidic drying agents cannot be used, and ammonia reacts with calcium chloride to form complex ions.

The bonds in NH_3 are tetrahedrally arranged

The arrangement of bonds in ammonia is tetrahedral, with the lone pair on the nitrogen atom occupying the fourth apex [see Figure 5.6, § 5.1.3]. The lone pair enables ammonia to act as a base, accepting a proton to form the ammonium ion, NH_4^+. This is what happens when ammonia and hydrogen chloride come into contact and form ammonium chloride. This reaction is used as a test for ammonia:

The reaction with hydrogen chloride is a test for ammonia

$$H_3N\!:(g) + H{-}Cl(g) \longrightarrow [NH_4]^+ \; Cl^-(s)$$

Ammonia is a Lewis base, able to use its lone pair to form a coordinate bond to another molecule [§ 12.7.1].

Ammonia forms hydrogen bonds with water

Ammonia is extremely soluble in water (1300 volumes per unit volume of water) because it can form hydrogen bonds with water. A saturated solution of ammonia has a density of $0.880\,g\,cm^{-3}$ and is often referred to as 880 ammonia.

An aqueous solution of ammonia is ionised to a small extent, to form NH_4^+ ions and OH^- ions. The basic dissociation constant, K_b, is $1.8 \times 10^{-5}\,mol\,dm^{-3}$ at 298 K [§ 12.7.4], which means that a solution of concentration $1.0\,mol\,dm^{-3}$ is 0.4% ionised.

REACTIONS OF AQUEOUS AMMONIA

Basic reactions of NH_3...

1. Ammonia neutralises acids to form ammonium salts.

...precipitation reactions...

2. It precipitates insoluble metal hydroxides from solutions of metal salts. Ammonia is useful for precipitating amphoteric metal hydroxides which dissolve in an excess of a strong alkali (e.g., $Al(OH)_3$, $Pb(OH)_2$).

...complex ion formation

3. Some metal hydroxides dissolve in an excess of ammonia to form soluble ammine complex ions, e.g., $[Ag(NH_3)_2]^+(aq)$ and $[Cu(NH_3)_4]^{2+}(aq)$.

AMMONIA AS A REDUCING AGENT

1. Ammonia is oxidised to nitrogen in reactions with chlorine, chlorate(I) ions and some heated metal oxides:

Ammonia is a reducing agent...

$$2NH_3(g) + 3Cl_2(g) \rightarrow N_2(g) + 6HCl(g)$$

$$2NH_3(aq) + 3ClO^-(aq) \rightarrow N_2(g) + 3Cl^-(aq) + 3H_2O(l)$$

$$2NH_3(g) + 3CuO(s) \rightarrow N_2(g) + 3Cu(s) + 3H_2O(g)$$

...and burns in oxygen to form N_2 or NO

2. Ammonia burns in oxygen to form nitrogen, but, in the presence of platinum, nitrogen monoxide is formed. This is a stage in the manufacture of nitric acid [see § 22.7.3].

FIGURE 22.4 Some Reactions of Ammonia

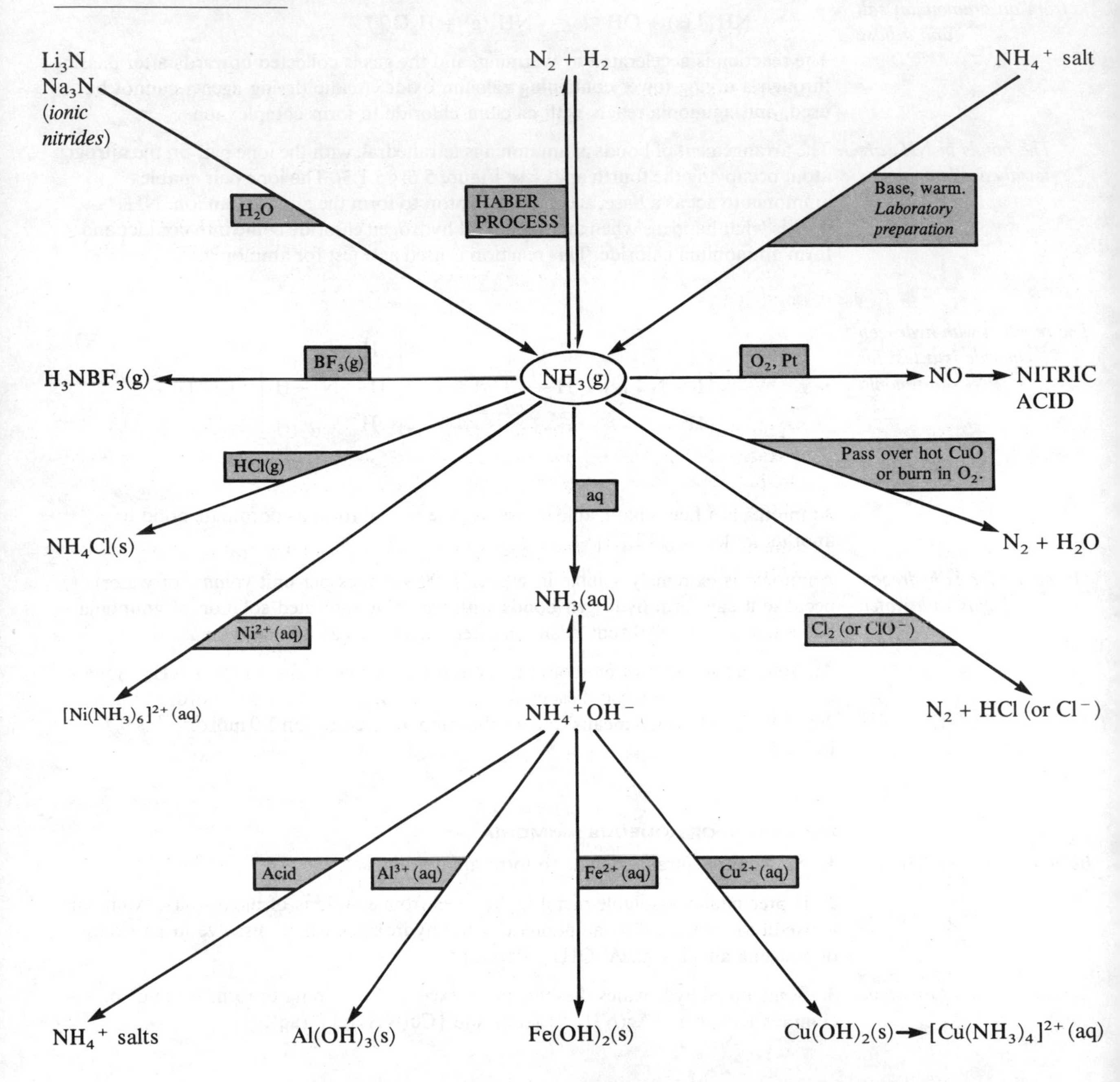

AMMONIUM SALTS

Ammonium salts are easily decomposed by heat. Ammonium carbonate and ammonium chloride dissociate:

$$(NH_4)_2CO_3(s) \rightleftharpoons 2NH_3(g) + CO_2(g) + H_2O(g)$$

$$NH_4Cl(s) \rightleftharpoons NH_3(g) + HCl(g)$$

Some ammonium salts undergo thermal dissociation...

When the ammonium salts of oxidising acids are decomposed on heating, the ammonia formed is oxidised to nitrogen or an oxide of nitrogen. The decomposition of ammonium nitrate(V) and ammonium nitrate(III) (nitrite) can be explosive:

... Some decompose explosively

$$NH_4NO_2(s) \xrightarrow[\text{decomposition}]{\text{explosive}} N_2(g) + 2H_2O(g)$$

$$NH_4NO_3(s) \xrightarrow[\text{decomposition}]{\text{explosive}} N_2O(g) + 2H_2O(g)$$

Ammonium dichromate decomposes spectacularly, orange crystals erupting to form a mound of dark green chromium(III) oxide.

$$(NH_4)_2Cr_2O_7(s) \xrightarrow{\text{heat}} N_2(g) + Cr_2O_3(s) + 4H_2O(g)$$

A test for ammonium salts

A test for ammonium salts is the evolution of ammonia when the salt is warmed with a strong base. The reaction can also be used for the estimation of ammonium salts since the amount of base required to decompose the salt can be found by titration [§ 3.14].

22.4.2 *PHOSPHINE

Phosphorus reacts with alkali to form phosphine, PH_3

Phosphine, PH_3, is made by heating white phosphorus in concentrated sodium hydroxide solution. It is a poisonous gas with an unpleasant smell. When pure, it is stable, but, when prepared by this method, it contains phosphorus vapour and it is spontaneously flammable in air.

Phosphides are hydrolysed to give phosphine

Phosphine is also made by the hydrolysis of ionic phosphides, just as ammonia is made from nitrides:

$$Ca_3P_2(s) + 6H_2O(l) \rightarrow 2PH_3(g) + 3Ca(OH)_2(s)$$

Phosphine is a weak base...

Phosphine is a weaker base than ammonia, and is only slightly soluble in water. Phosphonium salts are easily decomposed.

... it is also a reducing agent

Phosphine is a reducing agent.

CHECKPOINT 22A: HYDRIDES

1. Draw an apparatus which could be used for the laboratory preparation and collection of dry ammonia from ammonium sulphate and calcium hydroxide. Explain the choice of drying agent and method of collection.

2. What uses do calcium chlorate(I) and sodium chlorate(I) have in the home? Why should they not be used in conjunction with household ammonia?

3. Give two examples of reactions in which ammonia acts as a reducing agent.

4. Describe the industrial manufacture of ammonia, explaining the chemical principles involved.

22.5 HALIDES

22.5.1 HALIDES OF NITROGEN

Nitrogen forms halides of formula NX_3. The rest of the group form halides EX_3, and some form EX_5. Nitrogen trichloride is a covalent, highly explosive oil, which is made by a highly dangerous reaction between ammonia and chlorine. It is readily decomposed by water to form ammonia and chloric(I) acid:

NCl_3 is hydrolysed

NF_3 is stable to hydrolysis

$$NCl_3(l) + 3H_2O(l) \rightarrow NH_3(g) + 3HClO(aq)$$

followed by

$$NH_3(aq) + HClO(aq) \rightleftharpoons NH_4^+(aq) + ClO^-(aq)$$

The hydrolysis of a chloride of the first member of a group is unusual and contrasts with the stability to hydrolysis of tetrachloromethane, CCl_4. Nitrogen trifluoride is not hydrolysed.

22.5.2 HALIDES OF PHOSPHORUS

TABLE 22.3 Halides of Phosphorus

Halide	*Preparation*	*Reaction with water*	*Other reactions*
PCl_3 (PBr_3 and PI_3 are similar.)	Chlorine is passed over white phosphorus. P burns, and PCl_3 distils over.	Easily hydrolysed to H_3PO_3, phosphonic acid.	With $O_2 \rightarrow POCl_3$ With $Cl_2 \rightarrow PCl_5$
PCl_5 (PBr_5 is similar. No PI_5 exists.)	Chlorine is passed over PCl_3.	With cold water, gives $POCl_3$, phosphorus trichloride oxide. Boiling water gives H_3PO_4, phosphoric(V) acid.	PCl_5, PBr_3, PI_3 react with alcohols $ROH \rightarrow RX$ and with carboxylic acids $RCO_2H \rightarrow RCOX$

CHECKPOINT 22B: HALIDES

1. Draw 'electrons-in-boxes' diagrams to explain why phosphorus can form PCl_3 and PCl_5, but nitrogen can form only NCl_3.

2. Explain why NCl_3 is hydrolysed more slowly than PCl_3, to give different products. Why is NF_3 stable to hydrolysis? How does the ease of hydrolysis change from PCl_3 through $AsCl_3$ and $SbCl_3$ to $BiCl_3$?

22.6 OXIDES

The oxides formed by Group 5 elements are listed in Table 22.4.

TABLE 22.4 The Oxides of Group 5

	OXIDATION STATE				
Element	*+1*	*+2*	*+3*	*+4*	*+5*
Nitrogen	$N_2O(g)$	$NO(g)$	$N_2O_3(g)$	$NO_2(g)$, $N_2O_4(g)$	$N_2O_5(s)$
Phosphorus			$P_2O_3(s)$	$PO_2(s)$*	$P_2O_5(s)$
Arsenic			$As_2O_3(s)$		$As_2O_5(s)$
Antimony			$Sb_2O_3(s)$*	$SbO_2(s)$*	$Sb_2O_5(s)$*
Bismuth			$Bi_2O_3(s)$*		

*Macromolecular structure

The gradation in acidic character of the oxides can be represented

Acidic character of oxides

E_2O_3, E_2O_5:	*N*		*P*		*As*		*Sb*		*Bi*
Character	Strongly acidic	→			Weakly acidic	→	Amphoteric (Sb_2O_3 acidic)	→	Weakly basic

Dinitrogen oxide, N_2O, and nitrogen monoxide, NO, are neutral, and the other oxides increase in acidity

$$N_2O_3 < NO_2 < N_2O_5$$

22.6.1 OXIDES OF NITROGEN

DINITROGEN OXIDE

Dinitrogen oxide is made from an ammonium salt and a nitrate

Dinitrogen oxide (nitrogen(I) oxide), N_2O, is made by heating an ammonium salt with a nitrate:

$$NH_4^+(s) + NO_3^-(s) \rightarrow N_2O(g) + 2H_2O(g)$$

It rekindles a glowing splint...

...and is neutral...

The alternative method of heating ammonium nitrate is not employed because it is explosive. The gas is an anaesthetic called 'laughing gas' with a sickly-sweet smell. It rekindles a glowing splint because this is hot enough to decompose the gas into nitrogen and oxygen. Dinitrogen oxide is insoluble and neutral.

NITROGEN MONOXIDE

Nitrogen monoxide is formed in lightning flashes and by the oxidation of ammonia

Nitrogen monoxide or nitrogen oxide, NO, is an insoluble, neutral oxide. On meeting air, it reacts rapidly with oxygen to form nitrogen dioxide, NO_2. Nitrogen monoxide is formed during lightning flashes [see Figure 22.1, § 22.2.1], in vehicle engines [see § 26.4] and in the catalytic oxidation of ammonia [§ 22.4.1].

Nitrogen monoxide can be made in the laboratory by the following methods:

It is prepared from copper and nitric acid...

...and from a nitrite and an acid

1. The reaction of copper with a 50 : 50 solution of nitric acid and water [§ 22.7.3].

2. The reaction of sodium nitrite with a 50 : 50 solution of sulphuric acid in water. Nitrous acid is formed, and disproportionates with the formation of nitrogen monoxide [§ 22.7.1].

Nitrogen monoxide extinguishes a burning splint, but allows magnesium to burn in it.

NITROGEN DIOXIDE

Nitrogen dioxide dimerises

Nitrogen has an oxidation state of +4 in nitrogen dioxide, NO_2, and in its dimer, dinitrogen tetraoxide, N_2O_4. Nitrogen dioxide is a brown gas which dimerises on cooling to form a colourless liquid:

$$N_2O_4(l) \underset{\text{cool}}{\overset{\text{heat}}{\rightleftharpoons}} N_2O_4(g) \underset{\text{cool}}{\overset{\text{heat}}{\rightleftharpoons}} 2NO_2(g)$$

Colourless *Brown*

Nitrogen dioxide can be made in the laboratory by the following methods:

It is prepared from nitrates...

1. The action of heat on an anhydrous nitrate (other than those of Group 1):

$$2Pb(NO_3)_2(s) \rightarrow 2PbO(s) + 4NO_2(g) + O_2(g)$$

The gas can be collected by condensing it as liquid N_2O_4.

...and from nitric acid

2. The action of a metal on concentrated nitric acid (e.g., Cu):

$$Cu(s) + 4HNO_3(conc) \rightarrow Cu(NO_3)_2(aq) + 2NO_2(g) + 2H_2O(l)$$

NO_2 is the anhydride of nitrous and nitric acids

Nitrogen dioxide is a mixed anhydride. It dissolves in water to form a mixture of nitric and nitrous acids:

$$2NO_2(g) + H_2O(l) \rightarrow HNO_2(aq) + HNO_3(aq)$$

The nitrous acid in the solution is unstable, and disproportionates into nitric acid and nitrogen monoxide [§ 22.7.1].

A molecule of NO_2 possesses an unpaired electron. [See Figure 22.5.] The structure cannot be represented by any one valence bond structure, such as (a). It can be represented by the delocalised π bonding in structure (b). The unpaired electrons in two NO_2 molecules can pair up to form a N—N bond in N_2O_4, as shown in structure (c).

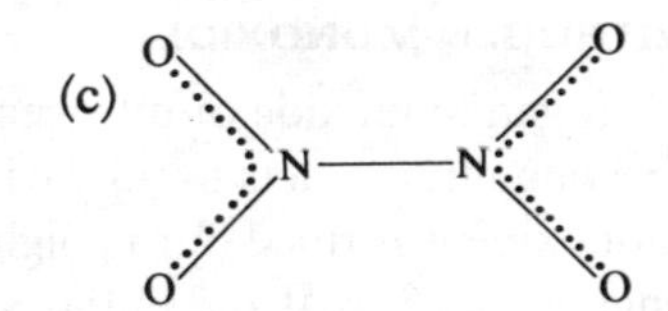

FIGURE 22.5
The Structure of NO_2 and N_2O_4

22.6.2 OXIDES OF PHOSPHORUS

Oxide	*Preparation*	*Reactions*
Phosphorus(III) oxide, P_2O_3	Burn phosphorus in a limited supply of air.	Is the anhydride of phosphonic acid, H_3PO_3. With $O_2 \rightarrow P_2O_5$; with $Cl_2 \rightarrow POCl_3$.
Phosphorus(V) oxide, P_2O_5	Burn phosphorus in an excess of air.	With a limited amount of water forms trioxophosphoric(V) acid, HPO_3; with more water forms tetraoxophosphoric(V) acid, H_3PO_4. Used as a drying agent and as a dehydrating agent, e.g., *amide*, $RCONH_2 \rightarrow$ *nitrile*, RCN [§ 33.14]

TABLE 22.5
The Oxides of Phosphorus

CHECKPOINT 22C: OXIDES

1. Describe the part played by nitrogen monoxide in the nitrogen cycle.

2. Given copper and concentrated nitric acid, how could you prepare (*a*) nitrogen dioxide, (*b*) nitrogen monoxide?

***3.** State the products of the reactions of phosphorus(III) oxide with (*a*) water, (*b*) oxygen, (*c*) chlorine. Give equations.

†4. Explain why nitrogen monoxide is formed in the internal combustion engine [see § 26.4].

5. How can you distinguish by a chemical test between (*a*) NO_2 and N_2O, (*b*) N_2O and O_2?

6. Discuss the gradation in acidic character of the oxides of Group 5.

22.7 THE OXO-ACIDS OF NITROGEN

22.7.1 NITROUS ACID (NITRIC(III) ACID, HNO_2)

Nitrous acid, HNO_2, is a weak acid of pK_a 3.34 at 25 °C. A solution of nitrous acid can be made by adding a mineral acid to a cold, dilute solution of a nitrite:

Nitrites react with mineral acid to give nitrous acid, which disproportionates at room temperature to form NO_2

$$NaNO_2(aq) + HCl(aq) \xrightarrow{5\,^\circ C} HNO_2(aq) + NaCl(aq)$$

The solution is fairly stable below 5 °C and is used in the preparation of diazonium compounds [§ 32.8.3]. At room temperature, nitrous acid disproportionates to give nitric acid and nitrogen monoxide, which immediately reacts with air to form nitrogen dioxide:

$$3HNO_2(aq) \rightarrow HNO_3(aq) + 2NO(g) + H_2O(l)$$

$$2NO(g) + O_2(g) \rightarrow 2NO_2(g)$$

This reaction is used to distinguish between nitrites and nitrates.

22.7.2 NITRITES (NITRATES(III))

Nitrites of Na and K can be made by heating the nitrates

The nitrites commonly used in the laboratory are those of sodium and potassium. They are made by heating the nitrates:

$$2NaNO_3(s) \rightarrow 2NaNO_2(s) + O_2(g)$$

The structure of NO_2^-

The structure of the nitrite ion can be represented as

$$\left[O \overset{N}{\diagup\;\diagdown} O \right]^-$$

22.7.3 NITRIC ACID (NITRIC(V) ACID, HNO_3)

The industrial method of making nitric acid is the catalytic oxidation of ammonia. A flow diagram for the process, which was invented by Ostwald, is shown in Figure 22.6. The reactions which take place are

The catalytic oxidation of ammonia gives nitric acid

$$4NH_3(g) + 5O_2(g) \xrightarrow[900\,^\circ C]{Pt,\,Rh} 4NO(g) + 6H_2O(l) \qquad [1]$$

$$2NO(g) + O_2(g) \rightarrow 2NO_2(g) \qquad [2]$$

$$4NO_2(g) + O_2(g) + 2H_2O(l) \rightarrow 4HNO_3(l) \qquad [3]$$

FIGURE 22.6 A Flow Diagram for the Manufacture of Nitric Acid

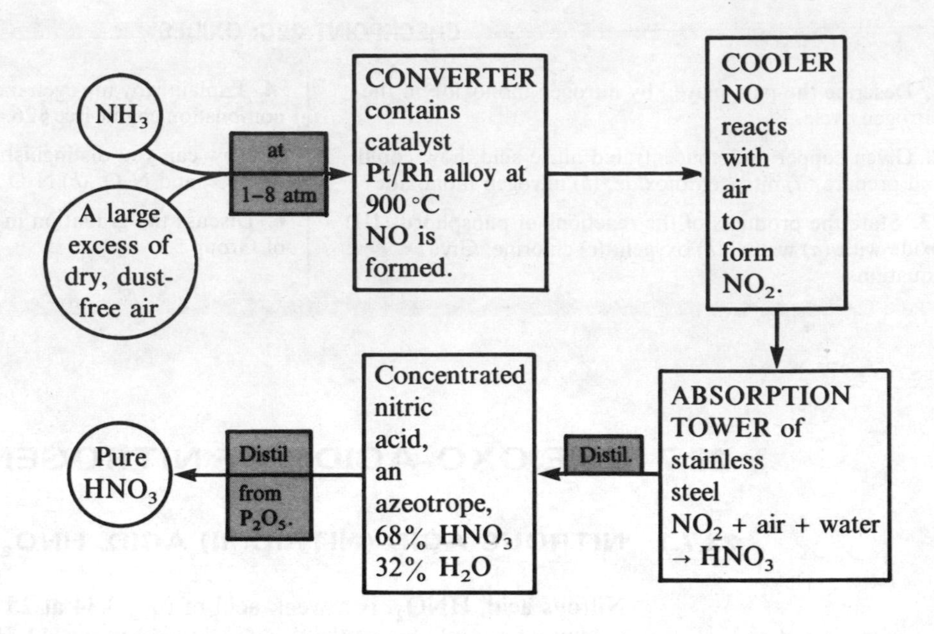

The azeotrope of HNO_3/H_2O

The laboratory acid called 'concentrated nitric acid' is a constant-boiling mixture of 68% nitric acid with water. It can be distilled from phosphorus(V) oxide to give pure nitric acid.

The laboratory preparation from a nitrate and conc. sulphuric acid

In the laboratory, nitric acid is made by distilling sodium nitrate and concentrated sulphuric acid:

$$NaNO_3(s) + H_2SO_4(l) \rightarrow HNO_3(l) + NaHSO_4(s)$$

Concentrated nitric acid is usually yellow. In the light, it decomposes to form nitrogen dioxide (the source of the yellow colour) and oxygen:

$$4HNO_3(aq) \rightarrow 4NO_2(g) + O_2(g) + 2H_2O(l)$$

The decomposition is accelerated by heat.

PROPERTIES OF NITRIC ACID

Pure nitric acid is a colourless liquid which boils at 86 °C. The vapour consists of molecules which have the structure

The structure of HNO_3

$$\text{H}-\text{O}-\text{N}(\text{O})_2$$

Ionisation in the presence of water

In aqueous solution nitric acid is a strong acid:

$$HNO_3(l) + H_2O(l) \rightleftharpoons H_3O^+(aq) + NO_3^-(aq)$$

REACTIONS OF NITRIC ACID

Aqueous nitric acid gives the reactions typical of mineral acids [§ 12.7.1]. Metals react to form nitrates. Since nitric acid is an oxidising agent, hydrogen is rarely formed. Only magnesium and calcium react with cold, dilute nitric acid to give hydrogen:

$$Mg(s) + 2HNO_3(aq) \xrightarrow[\text{dilute}]{\text{cold}} Mg(NO_3)_2(aq) + H_2(g)$$

Other metals reduce the nitrate ion in preference to the hydrogen ion. The reduction products formed are governed by the redox equilibria:

Nitric acid is an oxidising agent, being reduced to a number of different products

(a) $NO_3^-(aq) + 2H^+(aq) + e^- \rightleftharpoons H_2O + NO_2$

(b) $NO_3^-(aq) + 4H^+(aq) + 3e^- \rightleftharpoons 2H_2O + NO$

(c) $NO_3^-(aq) + 5H^+(aq) + 4e^- \rightleftharpoons 2\frac{1}{2}H_2O + \frac{1}{2}N_2O$

(d) $NO_3^-(aq) + 8H^+(aq) + 6e^- \rightleftharpoons 2H_2O + H_3\overset{+}{N}OH$

(e) $NO_3^-(aq) + 10H^+(aq) + 8e^- \rightleftharpoons 3H_2O + NH_4^+$

In these reactions, nitrogen is reduced from an oxidation state of +5 in NO_3^- to (a) +4, (b) +2, (c) +1, (d) −1 or (e) −3.

The reaction of conc. HNO_3 with metals...

...e.g., Cu, Pb...

...e.g., Mg, Zn...

Metals low in the Electrochemical Series (e.g., Cu, Pb) react with concentrated nitric acid according to equation (a) to give NO_2, and with dilute nitric acid according to equation (b) to give NO. Metals high in the Electrochemical Series (e.g., Mg, Zn) react with dilute nitric acid according to equation (c), (d) or (e), depending on the concentration and the temperature.

...and cations...

Cations of metals with variable oxidation states (e.g., Fe^{2+}, Sn^{2+}) are oxidised by concentrated nitric acid.

...non metals...

Non-metallic elements (e.g., S, P) are oxidised by concentrated nitric acid to the acids corresponding to their highest oxidation states (e.g., H_2SO_4, H_3PO_4).

and anions

Anions oxidised by concentrated nitric acid include Cl^-, Br^-, I^-, S^{2-}.

Weakly metallic elements are oxidised to hydrated oxides:

$$Sn(s) + 4HNO_3(conc.) \rightarrow SnO_2(s) + 2H_2O(l) + 4NO_2(g)$$

In reaction with conc. HNO_3 some metals are rendered 'passive'

With some metals (e.g., Al, Cr, Fe) concentrated nitric acid reacts to cover the surface with a film of oxide which will not react with any acid. The metal is then described as 'passive'.

Noble metals are attacked by 'aqua regia'

The few metals that are not attacked by concentrated nitric acid (e.g., Pt, Au) are called **noble metals**. They are attacked by *aqua regia*, a mixture of concentrated nitric acid and concentrated hydrochloric acid. Concentrated nitric acid reacts with gold to a slight extent to form Au^{3+} ions. Chloride ions remove these Au^{3+} ions to form soluble complex ions:

$$Au^{3+}(aq) + 4Cl^-(aq) \rightleftharpoons AuCl_4^-(aq)$$

Devarda's alloy (Al, Cu, Zn) in alkaline solution reduces nitric acid and nitrates and nitrites to ammonia:

$$4Zn(s) + NO_3^-(aq) + 7OH^-(aq) + 6H_2O(l) \rightarrow 4[Zn(OH)_4]^{2-}(aq) + NH_3(g)$$

A test for a nitrate is its reduction to ammonia

This reaction is used as a test for nitrates. Since the test will only be meaningful in the absence of ammonium salts, these are removed by heating the suspected nitrate with alkali before the alloy is added. A separate test must be used to distinguish a nitrate from a nitrite.

CHECKPOINT 22D: OXIDATION REACTIONS

The half-reaction equations for the oxidation reactions of nitric acid are given above. Write equations for the redox reactions mentioned above by combining the following half-reaction equations:

1. $Cu \rightarrow Cu^{2+} + 2e^-$ and equation (a)
2. $Cu \rightarrow Cu^{2+} + 2e^-$ and equation (b)
3. $Zn \rightarrow Zn^{2+} + 2e^-$ and equation (c)

4. $Fe^{2+} \rightarrow Fe^{3+} + e^-$ and equation (b)
5. $Au \rightarrow Au^{3+} + 3e^-$ and equation (a)
6. $S + 4H_2O \rightarrow SO_4^{2-} + 8H^+(aq) + 6e^-$ and equation (a)
7. $2I^- \rightarrow I_2 + 2e^-$ and equation (a)
8. $P + 4H_2O \rightarrow PO_4^{3-} + 8H^+(aq) + 5e^-$ and equation (a)
9. $2Cl^- \rightarrow Cl_2 + 2e^-$ and equation (a)
10. $Sn^{2+} \rightarrow Sn^{4+} + 2e^-$ and equation (b)

USES OF NITRIC ACID

Nitric acid is used in the manufacture of some explosives and dyes

Nitric acid is used in the manufacture of organic nitro-compounds [§ 28.7.1]. Some of these are useful explosives (e.g., dynamite and TNT). Others are useful intermediates in the dye industry. A large amount of nitric acid is used in the manufacture of nitrate fertilisers.

FIGURE 22.7 The Reactions of Nitric Acid

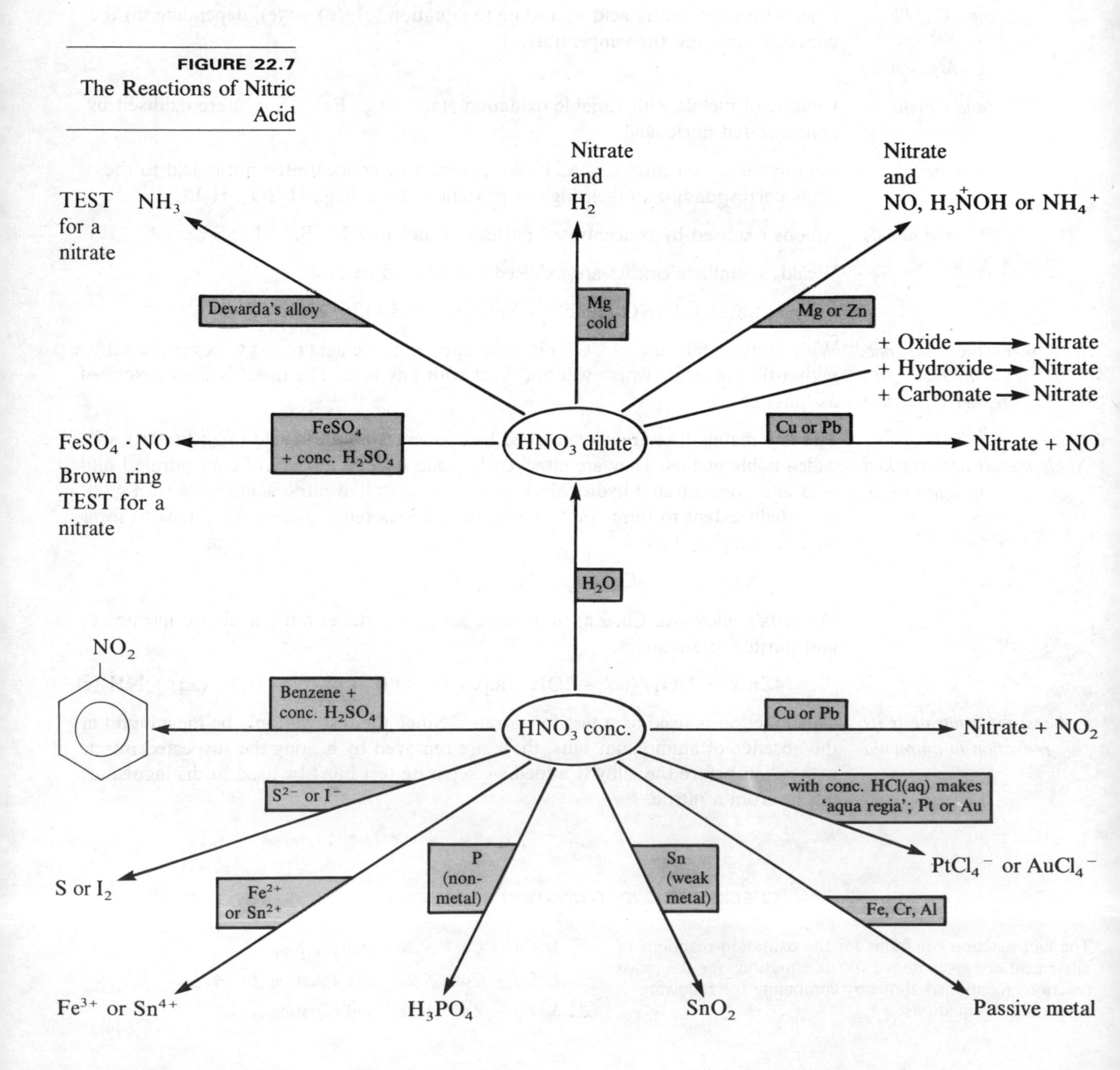

22.7.4 NITRATES

Nitrates are soluble and are decomposed by heat

Nitrates are prepared by the action of nitric acid on metals, metal oxides, hydroxides and carbonates. They all dissolve in water and decompose when heated [see Table 22.6].

FIGURE 22.8 Nitrogen Compounds Illustrating the Different Oxidation States of Nitrogen

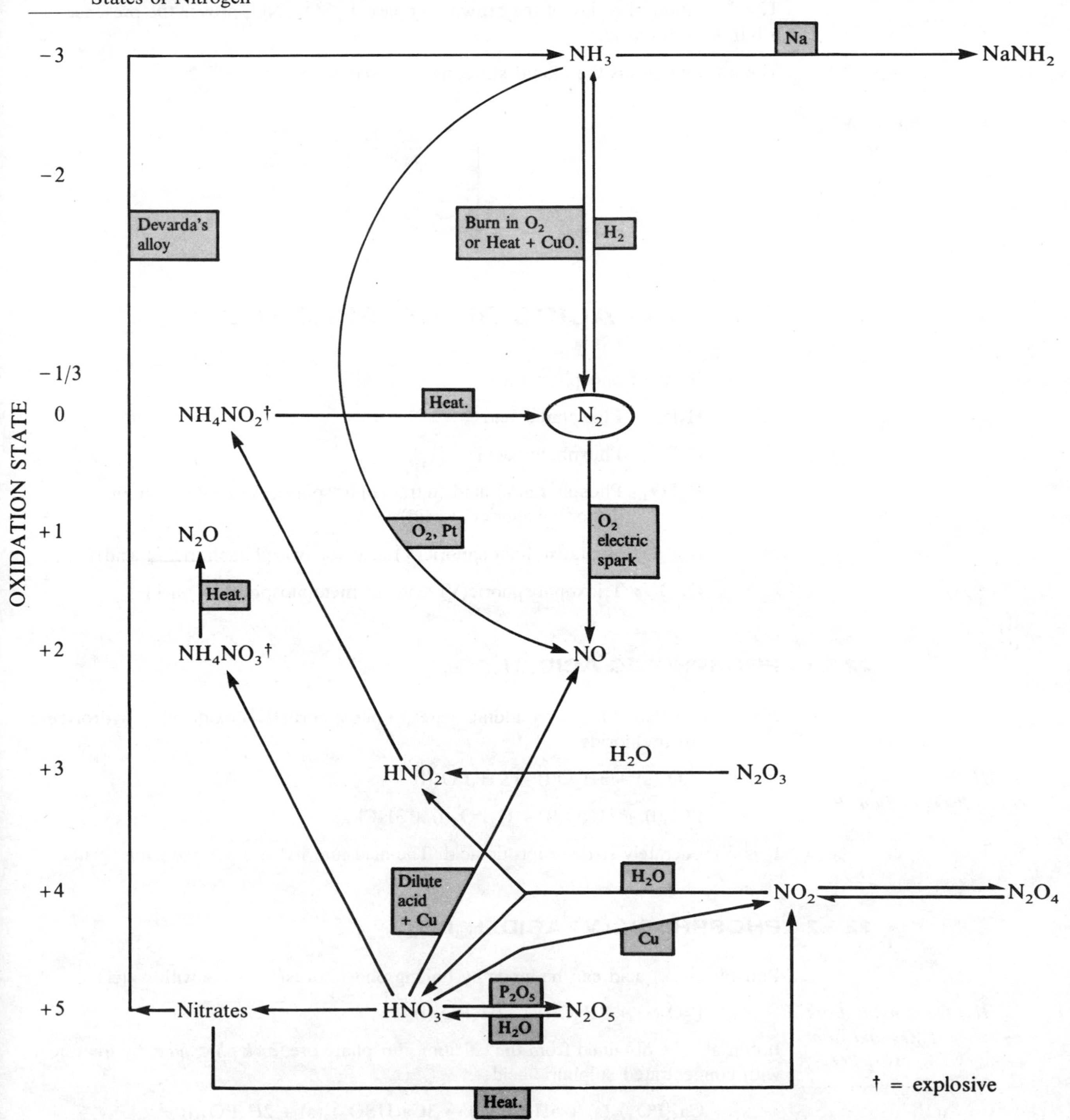

TABLE 22.6 Thermal Decomposition of Nitrates

Nitrate	*Products of thermal decomposition*
Group 1 (except Li)	Nitrite + Oxygen: $MNO_2 + O_2$
Most metals	Oxide + Oxygen + Nitrogen dioxide: $MO + O_2 + NO_2$
Unreactive metals (Ag, Hg)	Metal + Oxygen + Nitrogen dioxide: $M + O_2 + NO_2$
Ammonium	Explosive → Dinitrogen oxide, $N_2O + H_2O$

The brown ring test for a nitrate or a nitrite

One test for a nitrate, the reduction to ammonia by Devarda's alloy, has been described. An alternative, the **brown ring test**, requires the addition to a solution of the suspected nitrate of iron(II) sulphate solution and a few drops of concentrated sulphuric acid. The formation of a ring of the brown complex, $FeSO_4 \cdot NO$, proves the presence of a nitrate or a nitrite.

The nitrate ion has the planar structure represented:

The structure of NO_3^-

$$\left[\begin{array}{c} O \cdots N \cdots O \\ \| \\ O \end{array} \right]^-$$

22.8 THE OXO-ACIDS OF PHOSPHORUS

The oxo-acids of phosphorus are

- H_3PO_2 Phosphinic acid
- H_3PO_3 Phosphonic acid
- H_3PO_4 Phosphoric(V) acid (tetraoxophosphoric(V) acid in full, or orthophosphoric(V) acid)
- $H_4P_2O_7$ Heptaoxodiphosphoric(V) acid (or pyrophosphoric(V) acid)
- HPO_3 Trioxophosphoric(V) acid (or metaphosphoric(V) acid)

22.8.1 PHOSPHONIC ACID, H_3PO_3

Phosphonic acid is made by adding water to phosphorus(III) oxide or by hydrolysing phosphorus trichloride:

H_3PO_3 is made from P_2O_3 or from PCl_3

$$P_2O_3(s) + 3H_2O(l) \rightarrow 2H_3PO_3(l)$$

$$PCl_3(l) + 3H_2O(l) \rightarrow H_3PO_3(l) + 3HCl(g)$$

It is a moderately strong diprotic acid. The acid and its salts are reducing agents.

22.8.2 PHOSPHORIC(V) ACID, H_3PO_4

Phosphoric(V) acid can be made by boiling phosphorus(V) oxide with water:

H_3PO_4 is made from P_2O_5 and from $Ca_3(PO_4)_2$

$$P_2O_5(s) + 3H_2O(l) \rightarrow 2H_3PO_4(l)$$

It can also be obtained from the calcium phosphate ore, *rock phosphate*, by treatment with concentrated sulphuric acid:

$$Ca_3(PO_4)_2(s) + 6H_2SO_4(l) \rightarrow 3Ca(HSO_4)_2(s) + 2H_3PO_4(l)$$

Phosphoric(V) acid is a deliquescent crystalline solid. It is often sold as an 85% solution in water called 'syrupy phosphoric acid'. The high viscosity of this solution arises from intermolecular hydrogen-bonding.

Phosphoric(V) acid is a relatively weak triprotic acid. Most of its salts are insoluble in water.

22.8.3 PHOSPHATES

Test for a phosphate(V)

Phosphates(V), PO_4^{3-}, give a yellow precipitate of ammonium phosphomolybdate when warmed with nitric acid and ammonium molybdate solution. Heptaoxodiphosphates(V), $P_2O_7^{4-}$, and trioxophosphates(V), PO_3^-, also give positive results, but the precipitates are slower to appear.

22.8.4 USES OF PHOSPHORIC ACID AND PHOSPHATES

Uses of phosphates in fertilisers

... the food industry...

... and water softeners

Calcium phosphate occurs naturally and is used in the manufacture of fertilisers. Reaction with concentrated sulphuric or phosphoric or nitric acid yields the fertilisers *superphosphate of lime*, $Ca(H_2PO_4)_2 + CaSO_4$, or *triple superphosphate*, $Ca(H_2PO_4)_2$, or *nitrophos*, $Ca(H_2PO_4)_2 + Ca(NO_3)_2$. Sodium phosphate is used in the food industry. Calgon®, a water softener, is a polymeric sodium phosphate.

Phosphoric acid is used in rust-inhibition

Phosphoric(V) acid is used in the rustproofing of steel. If Fe^{3+} ions are formed by rusting, they react to form insoluble iron(III) phosphate(V) which protects the steel beneath from further attack.

22.9 A SUMMARY OF GROUP 5 TRENDS

1. The character of the elements changes from non-metallic nitrogen, which consists of N_2 molecules, to metallic bismuth, which has a close-packed metallic structure.

2. The 'inert pair effect' [§ 23.1] is shown by bismuth.

3. The stability to heat of the hydrides and their basic strength decrease in the order $NH_3 \gg PH_3 > AsH_3 > SbH_3 > BiH_3$.

4. The oxides of nitrogen are acidic (except N_2O and NO), as are the oxides of phosphorus and arsenic. While antimony(III) oxide is amphoteric, antimony(V) oxide is acidic, and bismuth(III) oxide is basic. The basicity of the oxides increases down the group, and the +3 oxides are more basic than the +5 oxides.

5. Antimony and bismuth show metallic behaviour in the formation of the cations, Sb^{3+} and Bi^{3+}.

22.10 DIFFERENCES BETWEEN NITROGEN AND OTHER MEMBERS

1. Nitrogen is the only member of the group that readily forms multiple covalent bonds.
The N≡N bond is very strong, and makes nitrogen unreactive.

2. Dinitrogen is a gas; other members are solids.

3. Nitrogen is the most electronegative member of the group and the only one to form hydrogen bonds.

4. It is the only group member that forms an ion E^{3-}.

5. Nitrogen is restricted to an L shell of electrons. Other members of the group can expand their octets by using their d orbitals.

6. Nitrogen forms oxides other than E_2O_3 and E_2O_5, which are formed by the rest of the group.

7. The properties of nitric acid and nitrous acid are not shared by the corresponding acids of the rest of the group.

QUESTIONS ON CHAPTER 22

1. Sketch the spatial arrangement of bonds in NH_3, NH_4^+, PCl_5.

2. Show the arrangement of valence electrons in NH_3, NH_4^+, NO_2, N_2O_4, NO_3^-.

***3.** List the differences between the chemical properties of nitrogen and phosphorus. What is the reason for the large difference in reactivity between them? What similarities justify the inclusion of the two elements in the same group of the Periodic Table?

†4. What is the importance of nitrogen monoxide in relation to (*a*) the ozone layer [§ 21.5] and (*b*) photochemical smog [§ 27.11]?

5. Write equations for the thermal decomposition of KNO_3, $Zn(NO_3)_2$, $Hg(NO_3)_2$, NH_4NO_3, NH_4NO_2 and NH_4Cl.

6. How could you find the concentration of (*a*) a solution of nitrate ions and (*b*) a solution of ammonium ions?

7. Describe the Haber process for the manufacture of ammonia. Explain the reasons for the conditions employed, and state the sources of the raw materials used. Mention three important uses of nitrogen compounds.

***8.** Outline preparations for (*a*) NO, (*b*) N_2O_4, (*c*) N_2O, (*d*) P_2O_5. State how each of the oxides reacts with water.

9. Compare the chemistry of nitrogen and phosphorus with respect to (*a*) the structure of the elements, (*b*) the hydrides (bonding, basicity, thermal stability and reaction with oxygen), and (*c*) the chlorides (preparation, bonding and reaction with water).

†10. What part do oxides of nitrogen play in the formation of acid rain? How can the emission of oxides of nitrogen be reduced? [See §§ 21.14 and 26.4.]

11. Discuss the application of the principles of kinetics and equilibrium to the manufacture of (*a*) ammonia, (*b*) nitric acid [see Chapters 11 and 14].

†12. Describe and explain what happens when aqueous ammonia is added slowly, until it is in excess, to solutions of (*a*) copper(II) sulphate, (*b*) copper(I) chloride and (*c*) silver nitrate. What use is made of the final products of (*b*) and (*c*) in organic chemistry?

***13.** Explain why the inclusion of nitrogen and phosphorus in the same group of the periodic table is justified. Consider the elements, their hydrides, chlorides and oxides.

14. Explain why nitrogen compounds are added to farmland. Find out, by calculation, which is the best source of nitrogen: sodium nitrate; ammonium nitrate; ammonium sulphate; or urea, CH_4N_2O.

15. (*a*) Write electron configurations for nitrogen, phosphorus and arsenic in their ground states.

(*b*) Explain why nitrogen can form N^{3-} ions, but arsenic cannot form As^{3-} ions. Name two elements which react with nitrogen to form ionic nitrides.

(*c*) Explain why phosphorus and arsenic can show a covalency of 5, but nitrogen cannot. Give examples of compounds in which phosphorus shows a covalency of 5.

(*d*) Which of the hydrides, NH_3 and PH_3, is the stronger base? Explain the reason for the differences between them.

(*e*) The boiling temperatures of the hydrides are: $NH_3 = -33\,°C$, $PH_3 = -87\,°C$. Offer an explanation of the difference in terms of structure and bonding.

16. (*a*) A laboratory method to find the percentage by mass of ammonium ions in a sample of fertiliser involves boiling the fertiliser with excess aqueous sodium hydroxide until no more ammonia is given off.

$$NH_4^+(aq) + OH^-(aq) \rightarrow NH_3(g) + H_2O(l)$$

The amount of sodium hydroxide in excess is found by titration with hydrochloric acid of known concentration.

(i) Give a simple test for ammonia gas.

(ii) Explain the reaction between ammonium ions and hydroxide ions in terms of the Brönsted–Lowry acid–base theory.

(iii) What indicator would be suitable to detect the end-point in the titration between sodium hydroxide and hydrochloric acid? Describe the colour change at the end-point.

(iv) The results of an experiment show that 0.0500 mol of sodium hydroxide is needed to react with the ammonium ions in 10.0 g of fertiliser. Calculate the percentage by mass of ammonium ions in the fertiliser.

(*b*) Growmore fertiliser is a mixture of the compounds KCl, NH_4NO_3 and $(NH_4)_3PO_4$. Describe a test, and give its expected result, to show the presence in the fertiliser of (i) potassium ions and (ii) chloride ions.

(*c*) (i) Give *two* reasons why a farmer might use Growmore fertiliser on his fields.

(ii) Give *two* environmental problems that might arise if Growmore fertiliser were used to excess on fields.

(*d*) Farmers spread calcium hydroxide (lime) on their fields to neutralise acidic soil.

(i) What would a farmer measure to help him decide whether *liming* is needed?

(ii) Write an ionic equation to show how hydrogen ions are neutralised by calcium hydroxide.

(AEB 92, AS)

17. (*a*) (i) Explain how nitrogen is fixed on an industrial scale.

(ii) What compounds containing nitrogen and phosphorus are commonly found in fertilisers?

(iii) Mention some environmental problems which may be associated with their large-scale use.

(*b*) Ammonia reacts with oxygen in two different ways. Give equations for both these reactions and explain how one is used industrially to produce nitric acid.

(*c*) Identify compound **A** and write an equation for each reaction involved.

A is a white solid which decomposes on heating to give a brown gas mixture capable of rekindling a glowing splint and a coloured solid residue. The brown gas can be partially condensed at −20 °C to a near colourless solid. An aqueous solution of **A** gives a white precipitate with acidified NaCl solution. The precipitate is soluble in hot water.

(O 91)

18. (*a*) Propose shapes for the molecules $B(CH_3)_3$, $C(CH_3)_4$, and $O(CH_3)_2$ and discuss their potential Lewis acid–base properties.

(*b*) Identify the compounds **A**, **B**, **C** and **D** in the following reaction sequence. Deduce probable shapes for the compound **A** and for the two phosphorus species present in **B**.

The reaction of PF_3 with chlorine gave a gas **A** with the percentage composition: Cl 44.7, F 35.8, P 19.5; $M_r = 159$. On standing for several hours, samples of **A** deposited a crystalline white solid **B** with the same empirical formula as **A** but with $M_r = 318$. **B** contains one cation and one anion per formula unit. Reaction of **B** with sodium fluoride gave a gas **C** and a solid **D**. Elemental analysis of **C** gave Cl 74.0%, F 9.9%, P 16.1%; $M_r = 192$. Elemental analysis of **D** gave F 67.9%, Na 13.7%, P, 18.4%; **D** contains one cation and one anion per formula unit.

(*c*) Discuss how you would expect the solid **B** to interact with $N(CH_3)_3$.

(NEAB 91, S)

19. (*a*) Nitric acid, HNO_3, is a strong acid. It is miscible in all proportions with water, and, unless very dilute, is a powerful oxidising agent. Why might it be very difficult to obtain a sample of iron(II) nitrate crystals by the action of nitric acid on iron?

(*b*) A possible method of obtaining the crystals would be to start with barium nitrate and iron(II) sulphate-7-water. Suggest a practical procedure by which this might be done. You should give an idea of the amounts you would use, and explain why your method would give hydrated iron(II) sulphate crystals free of other salts.

(*c*) Hydrated iron(II) ammonium sulphate, $FeSO_4 \cdot (NH_4)_2SO_4 \cdot 6H_2O$, is a much purer source of iron(II) ions than the simple sulphate. Would the double sulphate be a better choice than the simple sulphate for the preparation of iron(II) nitrate by your method? Explain your answer.

(O 92, AS)

23

GROUP 4

23.1 THE BONDS FORMED IN GROUP 4

There is a big gradation in properties down Group 4

Some of the physical properties of the members of Group 4 are listed in Table 23.1. The gradation in properties from carbon (a non-metallic element) to lead (a metallic element) is much more marked than that in groups on the extreme left or right of the Periodic Table.

Valencies 2 and 4 are met in the group

The elements show valencies of 2 and 4. All the Group 4 elements have the electron configuration (core)ns^2np^2:

	1s	2s	2p
C (ground state)	↑↓	↑↓	↑ ↑
C* (excited state)	↑↓	↑	↑ ↑ ↑

If an electron is promoted from a 2s orbital to a vacant p orbital, the element can show a covalency of 4. If the energy required to promote the s electron is compensated by the energy released when two additional covalent bonds are formed, the element will show a covalency of 4. Passing down the group, the strength of the covalent bonds formed with other elements decreases and there is an increasing tendency to show a covalency of 2. The behaviour of later members of a group, showing a valency of 2 less than the group valency through a failure to use their s electrons, is called the **'inert pair effect'**. Tin(II) compounds are reducing agents because tin(IV) is the more stable oxidation state of tin, but lead(IV) compounds are oxidising agents because lead(II) is the stable oxidation state for lead.

The 'inert pair effect' favours a valency of 2

Both covalent and ionic compounds are met in Group 4

Carbon, silicon and germanium form covalent compounds, with a valency of 4. A few compounds of tin and lead are sometimes described as containing the E^{4+} ion, but such compounds are predominantly covalent in character. Tin(II) compounds also are predominantly covalent, but lead(II) compounds are predominantly ionic.

Carbon is restricted to an octet of valence electrons

The covalency of carbon never exceeds 4 since the number of electrons in the valence shell cannot exceed 8. Later elements in the group can expand their octets. Carbon forms no complex ions as the later elements do, e.g., SiF_6^{2-}, $SnCl_6^{2-}$, $GeCl_6^{2-}$, $PbCl_6^{2-}$. The stability of these complex ions increases as the central atom becomes more electropositive.

Carbon is the only element in the group which forms strong π bonds [§ 5.2.7].

The behaviour of lead(II) compounds and some of the properties of tin(II) compounds are attributed to the ions Pb^{2+} (aq) and Sn^{2+} (aq).

Element, E	*Carbon*	*Silicon*	*Germanium*	*Tin*	*Lead*
Atomic number	6	14	32	50	82
Electron configuration	(core)$2s^2 2p^2$	(core)$3s^2 3p^2$	(core)$4s^2 4p^2$	(core)$5s^2 5p^2$	(core)$6s^2 6p^2$
Ionic radius of E^{2+}/nm	–	–	–	0.112	0.120
Covalent radius/nm	0.077	0.117	0.122	0.140	0.154
Standard enthalpy of vaporisation/kJ mol^{-1}	717	440	380	290	180
Density/g cm^{-3}	2.22 (graphite) 3.51 (diamond)	2.33	5.32	7.3	11.3
Electronegativity	2.50	1.75	2.00	1.70	1.55
Ionisation energy/MJ mol^{-1} (1)	1.09	0.79	0.76	0.71	0.72
(2)	2.40	1.60	1.50	1.40	1.50
(3)	4.60	3.20	2.30	2.90	3.10
(4)	6.20	4.40	4.40	3.90	4.10
Standard enthalpy					
of **E—E** *bond*/kJ mol^{-1}	348	176	188	150	–
of **E—O** *bond*/kJ mol^{-1}	360	374	360	–	–
of **E=E** *bond*/kJ mol^{-1}	612	–	–	–	–
of **E=O** *bond*/kJ mol^{-1}	743	640	–	–	–
of **E—H** *bond*/kJ mol^{-1}	412	338	285	251	–
Structure of element	Macromolecular (diamond, graphite)	Macromolecular (like diamond)		Metallic (and diamond lattice)	Metallic

TABLE 23.1 Group 4

The principal oxidation states

The principal oxidation states of the elements are:

Carbon	+4 (−4 in CH_4)
Silicon	+4
Germanium	+4 and +2 (a reducing state)
Tin	+4 and +2 (a reducing state)
Lead	+2 and +4 (an oxidising state)

23.2 THE STRUCTURES OF GROUP 4 ELEMENTS

The structure of carbon...

Carbon has two well-known allotropes, diamond and graphite (and also the allotropes known as 'bucky balls'; see § 34.10). The structures of diamond and graphite have been described in §§ 6.5 and 6.6 and shown in Figures 6.11 and 6.15. The allotropy is monotropic:

$$\text{C(diamond)} \rightleftharpoons \text{C(graphite)}; \quad \Delta H^{\ominus} = -2.1\,\text{kJ mol}^{-1}$$

... which shows monotropic allotropy

Industrial diamonds are made from graphite

Although graphite is the more stable allotrope, the activation energy for the change from diamond to graphite is high. As a result, diamond is stable under normal conditions. Since diamond is the denser of the allotropes, the effect of pressure, in accordance with Le Chatelier's Principle, is to increase the stability of diamond relative to graphite. Industrial diamonds are made by subjecting graphite to high temperatures and pressures (e.g., 2000 °C, 10^5 atm), with a catalyst. Graphite is more reactive than diamond: the closed structure of diamond gives rise to high activation energies.

The structures of charcoal...

The impure forms of carbon (charcoal, etc.) are microcrystalline forms of graphite. Being more finely divided, they are more reactive than graphite.

... silicon and germanium...

Silicon and germanium have structures of the diamond type. The decrease in melting temperature from carbon through silicon to germanium reflects the decreasing bond enthalpies.

... and tin which shows enantiotropic allotropy

Tin exists in allotropic forms, two metallic (cubic close-packed) and one of the diamond type. The allotropy is enantiotropic. At low temperatures, β tin changes to α tin, expanding and crumbling as it does so.

$$\alpha\ \text{tin} \xrightleftharpoons{13\,°\text{C}} \beta\ \text{tin} \xrightleftharpoons{161\,°\text{C}} \gamma\ \text{tin} \xrightleftharpoons{232\,°\text{C}} \text{Liquid}$$

	α tin	β tin	γ tin
	Grey	*White*	
	Diamond structure	Metallic structure	Metallic structure
Density/g cm^{-3}	5.8	7.3	

... and lead

Lead exists in one form only, which is metallic and cubic close-packed.

23.3 OCCURRENCE, EXTRACTION AND USES OF GROUP 4 ELEMENTS

23.3.1 CARBON

DIAMONDS

Diamonds are used for jewellery and in industry

Diamonds are mined in Brazil and South Africa. They are prized as jewellery because of their high reflectivity and high refractive index, which makes them sparkle, especially when they have been expertly cut. Small diamonds are used in industry to tip cutting tools and drills.

GRAPHITE

The manufacture of graphite and its uses

Some countries have natural supplies of graphite. Others manufacture it by heating coke and sand in an electric furnace. It is used for making electrodes, as a lubricant, as a mixture with clay in 'lead' pencils and for slowing down neutrons in nuclear reactors [§ 1.9.11].

Charcoal

Some of the impure forms of carbon are wood charcoal, animal charcoal, sugar charcoal, carbon black, coal and coke.

23.3.2 SILICON

Sources of silicon and purification

Silicon occurs naturally as silicon(IV) oxide, SiO_2, in sand and quartz and also as a number of silicates. It is extracted as shown in Figure 23.1. When it is required in a high state of purity, for use as a semiconductor in transistors [see § 23.4], it is purified by zone refining, as shown in Figure 23.2.

FIGURE 23.1
Extraction of Silicon

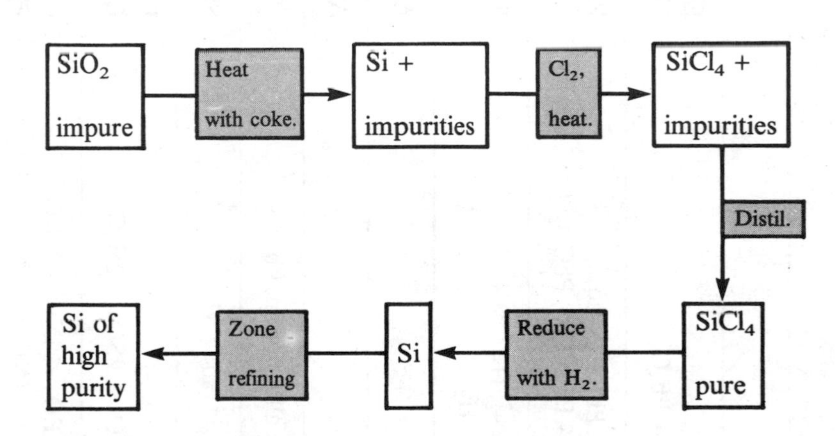

SiO_2 impure
Heat with coke.
Si + impurities
Cl_2, heat.
$SiCl_4$ + impurities
Distil.
$SiCl_4$ pure
Reduce with H_2.
Si
Zone refining
Si of high purity

FIGURE 23.2
Zone Refining

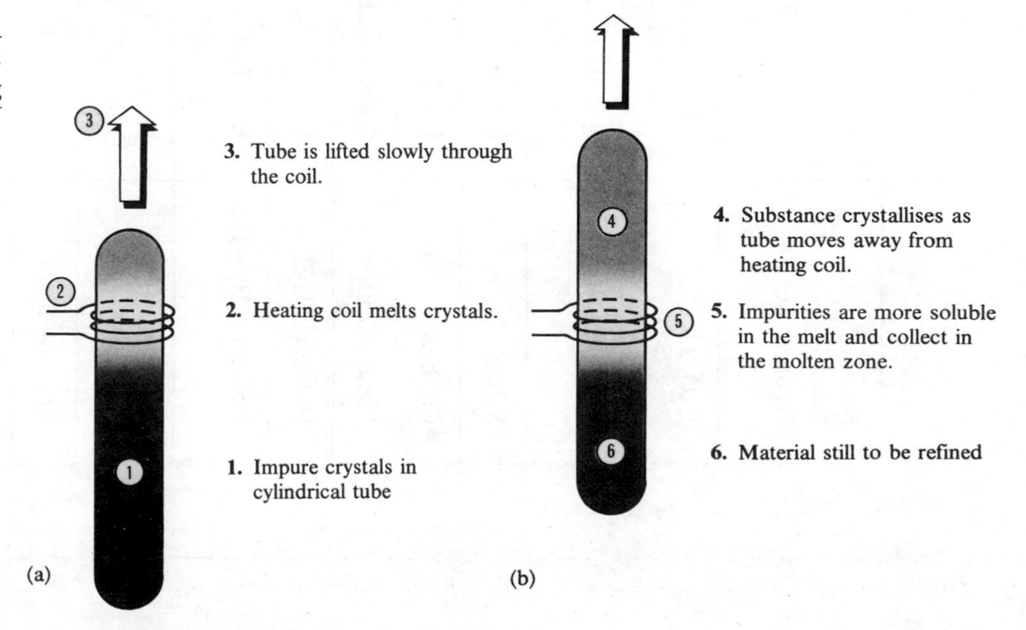

3. Tube is lifted slowly through the coil.
2. Heating coil melts crystals.
1. Impure crystals in cylindrical tube
4. Substance crystallises as tube moves away from heating coil.
5. Impurities are more soluble in the melt and collect in the molten zone.
6. Material still to be refined
(a)
(b)

Reagent	*Reaction*	*Comment*
O_2 or air, heat.	$E + O_2 \rightarrow EO_2$ $Pb + O_2 \rightarrow PbO \rightarrow Pb_3O_4$	E = C, Si, Ge, Sn Pb only ($C \rightarrow CO$ in limited air supply)
H_2O, 20 °C	Pb + soft water $\rightarrow Pb(OH)_2(aq)$ + hard water $\rightarrow PbSO_4(s), PbCO_3(s)$	Pb only
Steam, heat.	$E + 2H_2O \rightarrow EO_2 + 2H_2$ $C + H_2O \rightarrow CO + H_2$	E = Sn, Si
S, heat.	$E + 2S \rightarrow ES_2$	All except Pb; $Pb \rightarrow PbS$
Cl_2, heat.	$E + 2Cl_2 \rightarrow ECl_4$	All except Pb; $Pb \rightarrow PbCl_2$
Metals, heat.	Carbides, silicides, alloys of Sn, Pb	
Hot, conc. HCl	$E + 2H^+(aq) \rightarrow E^{2+}(aq) + H_2$	Sn and (slowly) Pb
Hot, conc. H_2SO_4	$E + H_2SO_4 \rightarrow E^{n+} + SO_2$ $C + 2H_2SO_4 \rightarrow CO_2 + 2SO_2 + 2H_2O$	$E^{n+} = Sn^{4+}, Pb^{2+}$ C only
Conc. HNO_3	$3E + 4HNO_3 \rightarrow 3EO_2 + 4NO + 2H_2O$ $3Pb + 8HNO_3 \rightarrow 3Pb(NO_3)_2 + 2NO + 4H_2O$	C; Ge and Sn $\rightarrow$ hydrated oxide. Not Si, Pb Pb only
Aqueous alkali	$Si + 2OH^- + H_2O \rightarrow SiO_3^{2-} + 2H_2$	Not C, Ge, Pb; Sn slowly $\rightarrow Sn(OH)_6^{2-}$
Molten base	Form SiO_4^{4-}, GeO_4^{4-}, $Sn(OH)_6^{2-}$, $Pb(OH)_4^{2-}$	Not C

TABLE 23.2 Reactions of Group 4

23.3.3 GERMANIUM

Semiconductor Germanium is a semiconductor, and for use in transistors it is, like silicon, purified by zone refining.

23.3.4 TIN

Extraction of tin... Tin is mined as *tinstone*, SnO_2. The ore is pulverised, washed by flotation, roasted to remove sulphur and reduced to the metal by heating with coke. Tin is used to

...uses plate iron to prevent it from rusting [§24.14.5] and as the alloys *type metal* (Sn, Sb, Pb), *bronze* (Sn, Cu) and *solder* (Pb, Sn).

23.3.5 LEAD

The extraction of lead... Lead is mined as *galena*, PbS. The ore is roasted to form lead(II) oxide, which can be reduced to the metal by coke.

...and the uses of lead Traditionally, lead has been used in plumbing, but for many purposes it has been replaced by copper (e.g., water pipes) and by plastics (e.g., gas pipes, guttering and sheathing electrical cables). Alloys of lead are *type metal* (Sn, Sb, Pb) and *solder* (Sn, Pb). Lead is used as a protective shield from radioactivity [§1.9.8], in accumulators [§13.3.3] and in the manufacture of the antiknock, tetraethyllead, $Pb(C_2H_5)_4$ [§26.4]. Lead compounds are used as pigments, e.g., white basic lead carbonate, $Pb(OH)_2 \cdot PbCO_3$, and the orange pigment 'red lead', Pb_3O_4.

23.4 SILICON THE SEMICONDUCTOR

TOPIC

Semiconductors are a special class of materials which have an electrical resistance between those of electrical conductors and electrical insulators. They all conduct electricity better as the temperature rises; that is, the resistance falls as the temperature rises.

Energy bands are used to explain the conduction of electricity by semiconductors. The electrons in the shell of an isolated atom have discrete energy levels [see §2.2.1]. Those in the highest energy level are the valence electrons, which are used in bond formation [see Chapter 4]. When atoms are present in a solid structure, the vibration of the atoms about their mean positions causes the energy levels to spread out into narrow bands. The band containing the valence electrons is called the **valence band**. In metals, electrons in the valence band can move easily to an unfilled energy level in a band of higher energy. This band is called the **conduction band**. Electrons in the conduction band are delocalised; they have broken free from individual atoms and are able to move through the material when a potential difference is applied across it.

In metals there is no energy gap between the valence band and the conduction band. In insulators the energy gap is too great (about 5 eV) to allow electrons to move between the bands. In semiconductors, the energy gap is smaller (about 1 eV), and the chance of electrons jumping the gap increases as the temperature rises [see Figure 23.3].

FIGURE 23.3
Valence Bands and Conduction Bands

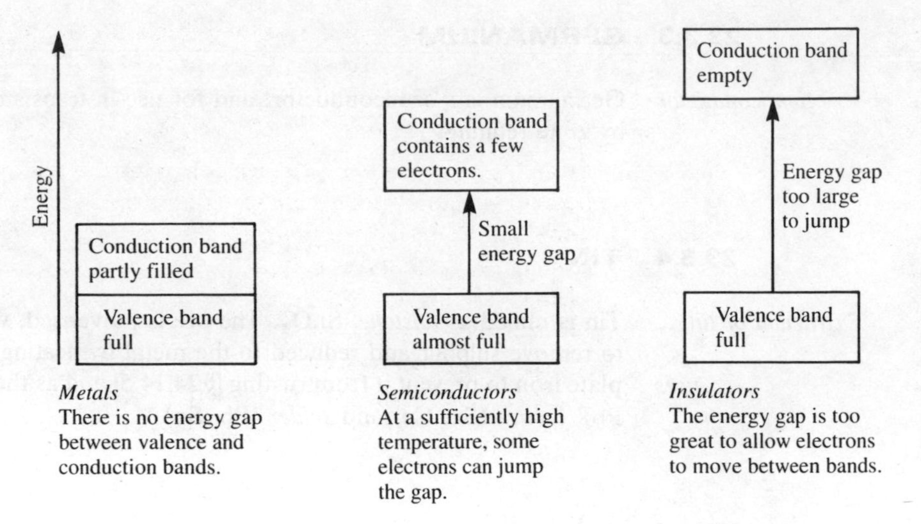

Semiconductors include both elements and compounds. The semiconductors which are pure elements and compounds are described as **intrinsic semiconductors**. Silicon is the intrinsic semiconductor which is used for the production of the integrated circuits known as **silicon chips**. The electron configuration of silicon is (2.8.4). As each atom forms four tetrahedrally directed covalent bonds to other silicon atoms, a macromolecular crystal structure in three dimensions is built up [see Figure 23.4].

FIGURE 23.4
Part of a Silicon Crystal (shown in two dimensions)

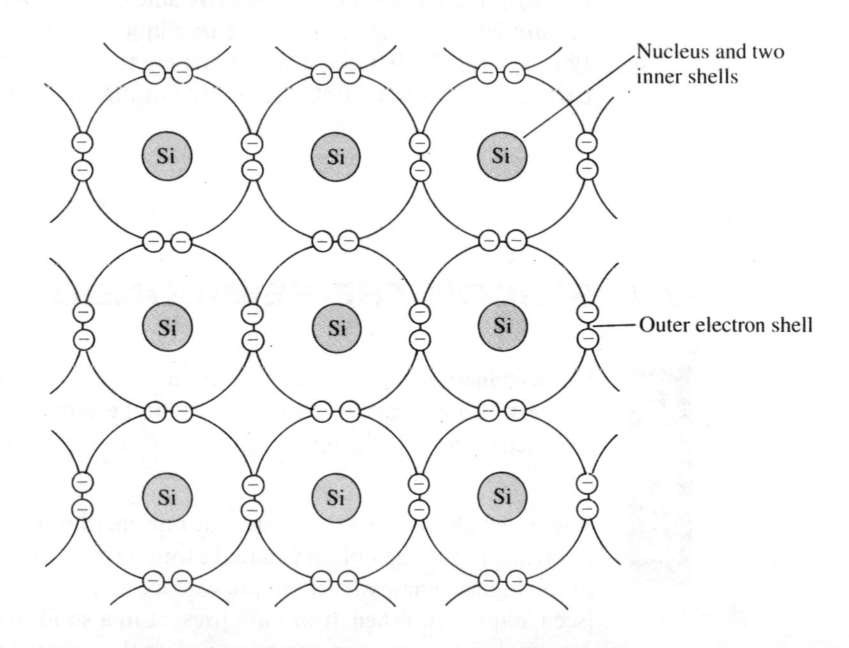

At absolute zero, all the electrons are localised in the valence band, and the material is an electrical insulator. When heat is supplied to the piece of silicon, some of the electrons gain enough energy to jump from the valence band into the conduction band. These delocalised electrons can carry charge through the material when a potential difference is applied across it. As the temperature rises, more electrons jump into the conduction band, and the resistance of the material falls.

Extrinsic semiconductors are made by adding certain substances to intrinsic semiconductors. This can be done by exposing the semiconductor to the vapour of the substance to be added in a furnace. The chosen substance is added in carefully controlled amounts to bring its content up to only a few parts per million. The atoms of the added substance are therefore well spaced out in the semiconductor so that its crystal structure is not weakened. The process is called **doping**, and the added substances are called **dopants**. The dopants are chosen to produce the required change in the elecrical properties.

There are two types of extrinsic semiconductors: **n-type** and **p-type**. In n-type semiconductors, the dopants are Group 5 elements, such as phosphorus and antimony, which have five electrons in the outer shell. When a dopant atom replaces an atom of silicon in the structure, it uses four of its five valence electrons to form covalent bonds with silicon atoms. The fifth electron is supplied to the material [see Figure 25.5], creating a negative charge. This is why such semiconductors are called n-type (n for negative). The added atoms are called **donor atoms** because they donate electrons to the material. The possession of delocalised electrons makes the extrinsic semiconductor a much better electrical conductor than the parent intrinsic conductor. The addition of donor atoms leaves the crystal uncharged overall because the additional electrons are associated with additional positive charges on the donor nuclei.

FIGURE 23.5 n-Type Extrinsic Semiconductor

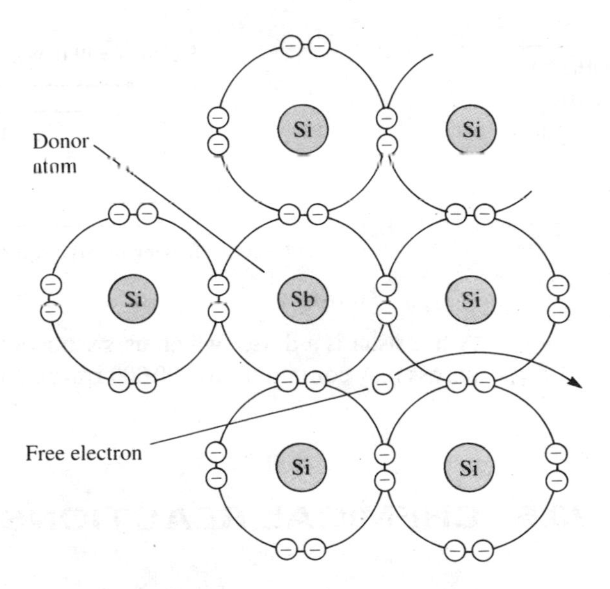

In p-type semiconductors, the dopants are Group 3 elements, such as boron and aluminium, which have three electrons in the outer shell. When a dopant atom replaces a silicon atom, it forms three electron-pair bonds with three silicon atoms, but the fourth bond is incomplete: it has only one electron. The vacancy is called an **electron hole** and is positively charged. This is why this type of semiconductor is called p-type (p for positive). The added atoms are called **acceptor atoms** because they can accept electrons to fill the 'holes' in the bonds.

In p-type semiconductors, conduction takes place by means of electrons jumping across from bonds into holes [see Figure 25.6]. The electron which jumps from a bond into a hole creates a hole in the bond which it left. Another electron jumps from a bond into this hole, and so it goes on across the structure.

What happens when an n-type and a p-type semiconductor are placed together is shown in Figure 25.7. Electrons flowing through the p-type electrode stop at the pn-junction. Electrons flowing through the n-type electrode pass through the junction because they can pass from a structure with surplus electrons into a structure with electron holes. A junction of this kind will transform alternating current into direct current. It is the basis of the **silicon-controlled rectifier**.

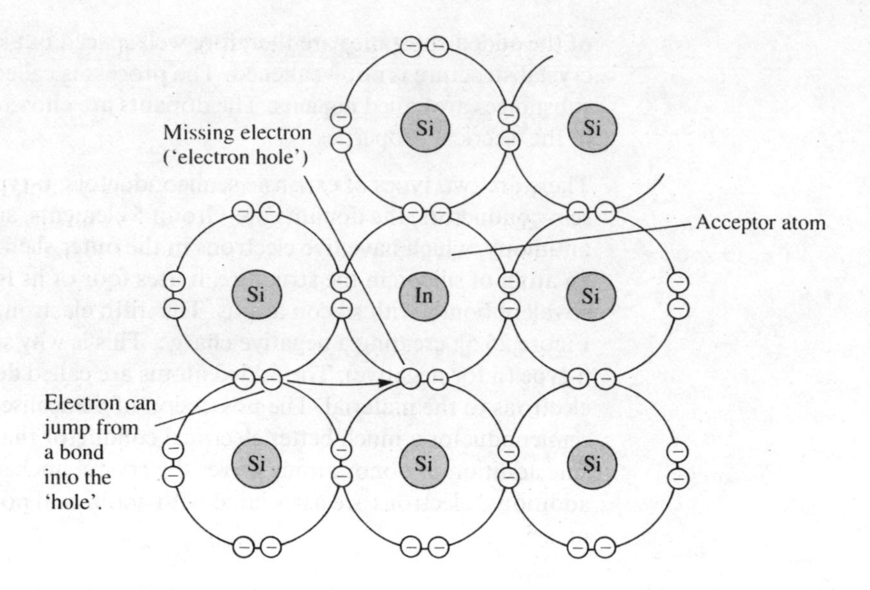

FIGURE 23.6
p-Type Extrinsic Semiconductor

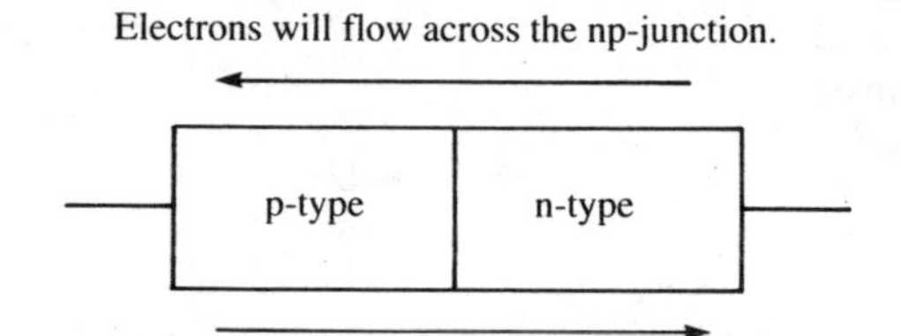

Electrons will flow across the np-junction.
p-type
n-type
Electrons flowing in this direction cannot cross the pn-junction.

FIGURE 23.7
A pn-Junction and an np-Junction

A **transistor** is a device which uses semiconductor electrodes. A silicon chip 0.5–1.0 cm across may contain up to 200 000 transistors in one integrated circuit*.

23.5 CHEMICAL REACTIONS OF GROUP 4

Reactions of Group 4

Although in Group 4 the difference between the first and last members of the group has reached a maximum, the elements do form a set. The reactions which they have in common are summarised in Table 23.2. As in the other groups, there are differences between the first member and the second. The demarcation in properties between the Period 2 element and the Period 3 element is maximal in Group 4. The special features of carbon chemistry are discussed in § 23.6.

CHECKPOINT 23A: STRUCTURE AND REACTIONS

1. Describe preparations of (*a*) diamond and (*b*) graphite. List the uses of the two allotropes, and explain the difference in properties between them.

***2.** Describe how extremely pure germanium is obtained.

3. Carbon and lead have proton (atomic) numbers 6 and 82. Write the electronic structure of each atom. Explain, on the basis of their electronic structures, why lead is metallic and carbon is non-metallic.

4. How does the structure of graphite explain (*a*) its ability to mark paper, (*b*) its use as a lubricant and (*c*) its use as an electrical conductor?

*For information about transistors, see J. Breithaupt, *Understanding Physics for Advanced Level* (Stanley Thornes)

23.6 SPECIAL FEATURES OF CARBON CHEMISTRY

23.6.1 CATENATION

Carbon catenates to form —C—C—C— chains

Catenation is the ability of an element to form bonds between atoms of the same element. Carbon catenates to form chains and rings, with single, double and triple bonds. To be able to catenate, an element **E** must have a valency ⩾2 and must form **E**—**E** bonds which are similar in strength to those of **E** to other elements, particularly **E**—O bonds. Since elements are exposed to the possibility of reaction with air and water, if **E**—O bonds are stronger than **E**—**E** bonds, then energy considerations will favour the formation of compounds containing **E**—O bonds rather than **E**—**E**—**E** chains [see Question 2, Checkpoint 23B]. For silicon, it is energetically more favourable to form a chain

Silicon forms —Si—O—Si— bonds in preference

```
  |       |       |
—Si—O—Si—O—Si—
  |       |       |
```

while for carbon, energy considerations favour a chain

```
 |  |  |  |
—C—C—C—C—
 |  |  |  |
```

While no compounds containing Si—Si bonds are found in nature, silica and a variety of silicates contain chains and networks of bonds:

```
     |      |
—O—Si—O—Si—O—
     |      |
```

SiH_4 is unstable

Alkanes are stable at room temperature

The silanes (SiH_4, etc.) and hydrides of other members of the group are thermodynamically unstable (i.e., $\Delta G^{\ominus}$ is negative) with respect to dissociation into the elements, oxidation and hydrolysis. The alkanes (CH_4, etc.) are thermodynamically stable relative to dissociation into the elements and relative to hydrolysis. Although thermodynamically unstable relative to oxidation to carbon dioxide and water, alkanes are inert at room temperature because the activation energy for oxidation is high.

23.6.2 MULTIPLE BONDS

Carbon forms >C=C< bonds and —C≡C— bonds

Carbon forms double and triple bonds between carbon atoms (in alkenes, alkynes and arenes); with nitrogen atoms (in nitriles) and with oxygen atoms (in aldehydes, ketones, carboxylic acids and their derivatives). The rest of Group 4 form no corresponding compounds. These compounds involve π bonds, and only elements in the first short period form strong π bonds [§ 5.2.7].

23.6.3 GASEOUS OXIDES

The oxides of carbon are gases

Carbon forms gaseous oxides, CO and CO_2, in contrast to the other members. In this group, it is only for carbon that the **E**=O bond is more than twice as strong as the **E**—O bond. Carbon forms two double bonds in O=C=O, whereas for silicon the formation of four Si—O bonds is preferred on energy grounds [see Question 2, Checkpoint 23B].

23.6.4 L SHELL RESTRICTION

Carbon forms no complexes

When carbon forms four covalent bonds, it has completed its octet. This explains the lack of reactivity of many carbon compounds. With no lone pairs of electrons and a complete octet, it cannot give or accept electrons and therefore forms no complexes.

Carbon compounds are not hydrolysed as are those of later members of the group. Comparing the hydrolysis of tetrachloromethane, CCl_4, and silicon tetrachloride, $SiCl_4$, it can be calculated that the hydrolysis

$$ECl_4(l) + 2H_2O(l) \rightarrow EO_2(g \text{ or } s) + 4HCl(g)$$

CCl_4 is stable to hydrolysis

is thermodynamically feasible for both compounds (i.e., $\Delta G^\ominus$ values are negative). Although silicon tetrachloride is hydrolysed with ease, tetrachloromethane is inert to hydrolysis. The reason for the difference in reactivity is that a different mechanism of hydrolysis operates in the two cases. It is believed that, in the hydrolysis of silicon tetrachloride, the first step is the attack by the negatively charged oxygen atom in a water molecule or hydroxide ion on the positively charged silicon atom. Silicon can use an unoccupied 3d orbital to accommodate the lone pair from the oxygen atom, forming a short-lived intermediate which dissociates to form hydrogen chloride and silicon trichloride hydroxide, $SiCl_3OH$. Repetition of these steps gives hydrated silicon(IV) oxide:

The hydrolysis of $SiCl_4$ employs 3d orbitals

$$\overset{\delta-}{Cl}-\overset{\delta+}{Si}(\overset{\delta-}{Cl})_3 + \overset{\delta-}{O}(\overset{\delta+}{H})_2 \longrightarrow \left[Cl_3Si \leftarrow OH_2 \right]$$

$$Si(OH)_4 \text{ or } SiO_2 \cdot xH_2O \longleftarrow Si(OH)Cl_3 + HCl\,(g)$$

Since carbon cannot expand its octet because its empty 3d orbitals are too different in energy from the 2p orbitals to come into play, a similar mechanism cannot operate for the hydrolysis of tetrachloromethane. This must proceed by a mechanism with a much higher activation energy and takes place extremely slowly.

23.6.5 ELECTRONEGATIVITY

Carbon is the most electronegative element in the group

Carbon is the most electronegative member of the group. Whereas the Si—H bond is polarised

$$\overset{\delta+}{Si}-\overset{\delta-}{H}$$

and the silanes behave similarly to ionic hydrides, the C—H bond is polarised

$$\overset{\delta-}{C}-\overset{\delta+}{H}$$

In the dicarbides (e.g., Cu_2C_2), carbon is present as the anion

$$^-C{\equiv}C^-$$

and the methanides (e.g., Al_4C_3) contain highly polar covalent bonds.

CHECKPOINT 23B: CARBON

1. Explain the following statements:
(*a*) Carbon forms a vast number of compounds.
(*b*) Methane boils at a lower temperature than silane, SiH_4, but in Group 6, H_2O boils at a higher temperature than H_2S.

2. To explain why CO_2 is molecular while SiO_2 exists as a macromolecular structure, calculate the standard enthalpy change for the polymerisation of one EO_2 unit.

(*a*) O=Si=O ⟶ [polymeric Si–O structure]

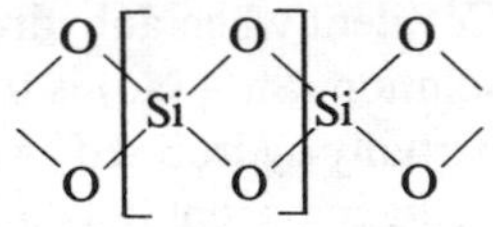

(*b*) O=C=O ⟶ [polymeric C–O structure]

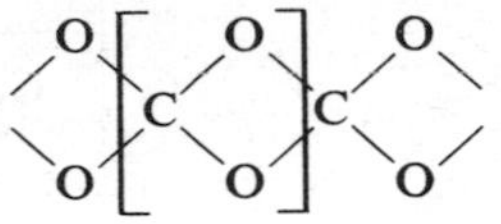

Use the data in Table 23.1, § 23.1.

3. Explain the nature of the bonding in
(*a*) >C=C<
(*b*) —C≡C—
(*c*) —C≡N

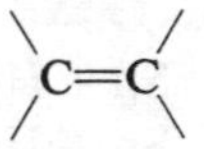

23.7 THE COMPOUNDS OF GROUP 4

23.7.1 THE HYDRIDES

Reactivity of the hydrides increases down Group 4

The hydrides of Group 4 are summarised in Table 23.3. The reactivity of the hydrides increases down the group. Silanes are spontaneously flammable in air, are readily hydrolysed by bases to hydrated silicon(IV) oxide, are strongly reducing and dissociate into the elements above 400 °C.

TABLE 23.3 Hydrides of Group 4

Element	*Hydrides*	*Preparation*
Carbon	Alkanes, alkenes, alkynes, arenes	Petroleum industry
Silicon	Silanes, Si_nH_{2n+2}; $n = 1–10$	Mg_2Si + Acid
Germanium	Ge_nH_{2n+2}; $n = 1–6$	Mg_2Ge + Acid
Tin	Stannane, SnH_4, only	Sn · Hg + Acid
Lead	Plumbane, PbH_4, only	Pb · Hg + Acid

23.7.2 THE HALIDES

Stability of EX_4 decreases down the group and decreases from F to I

All the elements of Group 4 form tetrahalides of formula EX_4, where X = F, Cl, Br or I, except for $PbBr_4$ and PbI_4. All the tetrahalides are volatile covalent compounds, except for the tetrafluorides of tin and lead, SnF_4 and PbF_4, which have some ionic character and form macromolecular structures. The tetrahalides are stable with respect to dissociation into the elements, except for lead(IV) chloride which dissociates at room temperature into lead(II) chloride and chlorine. Stability decreases down the group and decreases from fluorides to iodides. Except for tetrachloromethane and tetrafluoromethane, the tetrahalides are hydrolysed in solution:

$$SiCl_4(l) + 2H_2O(l) \rightarrow SiO_2(s) + 4HCl(g)$$

When silicon tetrafluoride is hydrolysed, the products react to form the complex hexafluorosilicate ion, SiF_6^{2-}.

Table 23.4 summarises the preparations of the halides.

TABLE 23.4 Halides of Group 4

Halide	*Preparation*	*Properties*
CCl_4	$Cl_2 + CS_2$, Fe catalyst $\rightarrow CCl_4 + S_2Cl_2$	Uses [§ 29.5]
ECl_4 (*except CCl_4*)	E + X_2, heat EO_2 + HX(g) EO_2 + HX(aq, conc.)	$PbCl_4$ must be kept below 5 °C.
$SnCl_2$	Sn + HCl(g), heat Sn + HCl(aq, conc.)	Covalent when anhydrous; in solution, Sn^{2+}(aq) is formed; partially hydrolysed to Sn(OH)Cl(s). Reducing agent [§ 32.7.2]
$PbCl_2$	Pb + Cl_2(g), heat Pb + HCl(aq, conc.)	Ionic [see Figure 23.17, § 23.7.4] The reaction is possible because the insoluble $PbCl_2$ formed is converted by conc. HCl into soluble $PbCl_4^{2-}$(aq).
PbX_2	Precipitation	Ionic [see Figure 23.17, § 23.7.4]

CHECKPOINT 23C: HALIDES

1. (*a*) Why is an aqueous solution of tin(II) chloride cloudy? How can a clear solution be made? (*b*) What happens when this solution reacts with (i) NaOH(aq) and (ii) $FeCl_3$(aq)?

2. Outline the methods used in the preparation of Group 4 chlorides. Discuss the trends in valency and bond type shown by the chlorides.

3. Suggest methods for the preparation of (*a*) $PbCl_4$, (*b*) $PbCl_2$. Why is $PbCl_2$ very soluble in concentrated hydrochloric acid?

4. How does SiF_4 differ from $SiCl_4$ in its hydrolysis?

23.7.3 THE OXIDES

The basic character of the oxides increases as the group is descended. The +4 oxides are more acidic than the +2 oxides.

CARBON MONOXIDE, CO

CO is produced by petrol engines

Carbon monoxide is formed when carbon and hydrocarbons burn incompletely. It is present in the exhaust fumes of petrol-driven vehicles [§ 26.3.1]. At concentrations above 0.1% carbon monoxide is poisonous [§ 24.13.8]. It is the more dangerous for being colourless and odourless.

It is a poisonous gas . . .

. . . and a reducing agent

Carbon monoxide is an important reducing agent. It is used for the reduction of iron ores [§ 24.14.1].

In the laboratory, carbon monoxide can be made by the action of concentrated sulphuric acid on sodium methanoate or sodium ethanedioate:

The laboratory preparation...

$$HCO_2Na(s) + H_2SO_4(l) \rightarrow NaHSO_4(s) + CO(g) + H_2O(l)$$

$$C_2O_4Na_2(s) + 2H_2SO_4(l) \rightarrow 2NaHSO_4(s) + CO(g) + CO_2(g) + H_2O(l)$$

In the second case, carbon dioxide is removed by bubbling the mixture of gases through alkali.

...electronic structure...

The electronic structure of carbon monoxide can be represented as

...and complex formation

$$:C \equiv O:$$

The lone pair of electrons on the carbon atom enable it to act as a ligand in the formation of carbonyl complexes, e.g., $[Cr(CO)_6]^{3+}$ and $Ni(CO)_4$ [§ 24.13].

FIGURE 23.8 Some Reactions of Carbon Monoxide

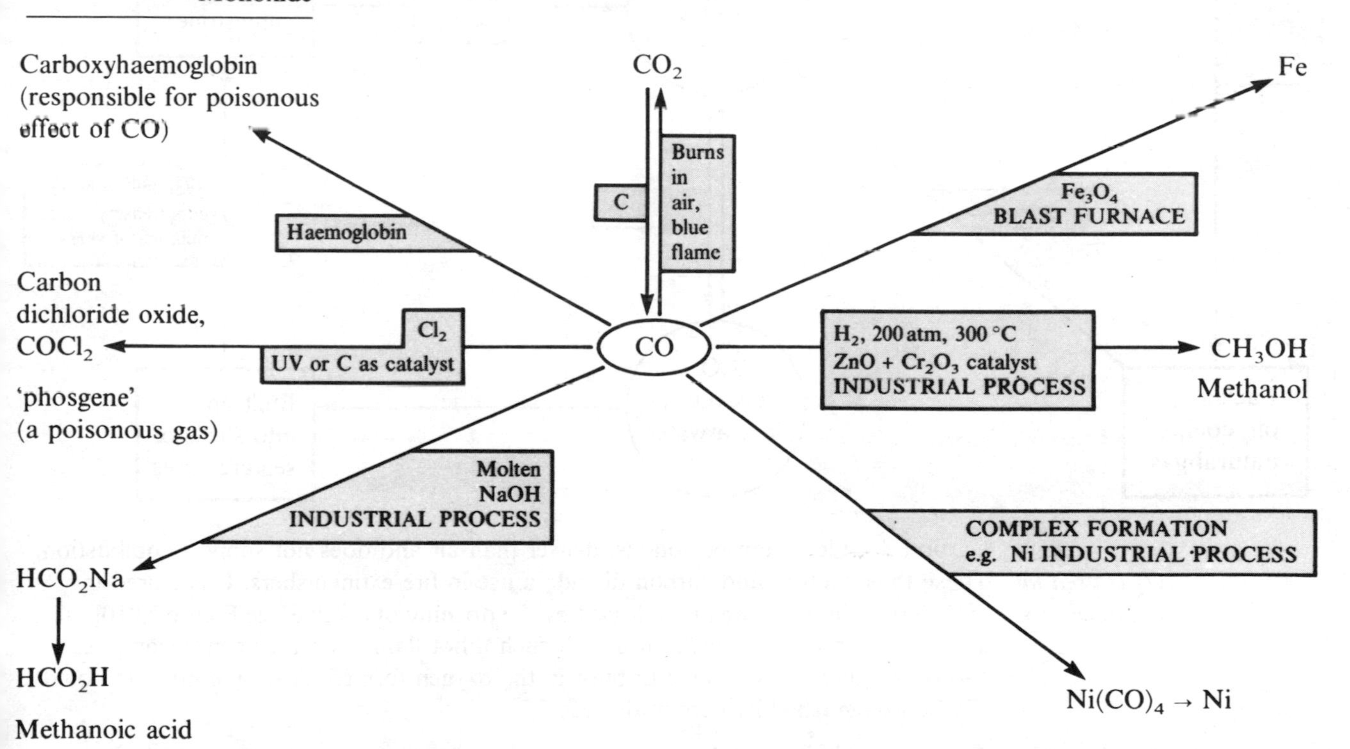

CARBON DIOXIDE, CO_2

Photosynthesis

Carbon dioxide is present in air at a level of 0.03% by volume. Plants use carbon dioxide in the process of photosynthesis [§ 10.1.1] and both plants and animals evolve carbon dioxide in respiration. The balance of processes which give out carbon dioxide and those which use carbon dioxide is called the carbon cycle and is illustrated in Figure 23.9.

CO_2 is used in soft drinks

Carbon dioxide is colourless and odourless, and is slightly soluble in water. Flavoured solutions of carbon dioxide under pressure are the basis of the soft drinks industry.

FIGURE 23.9 The Carbon Cycle

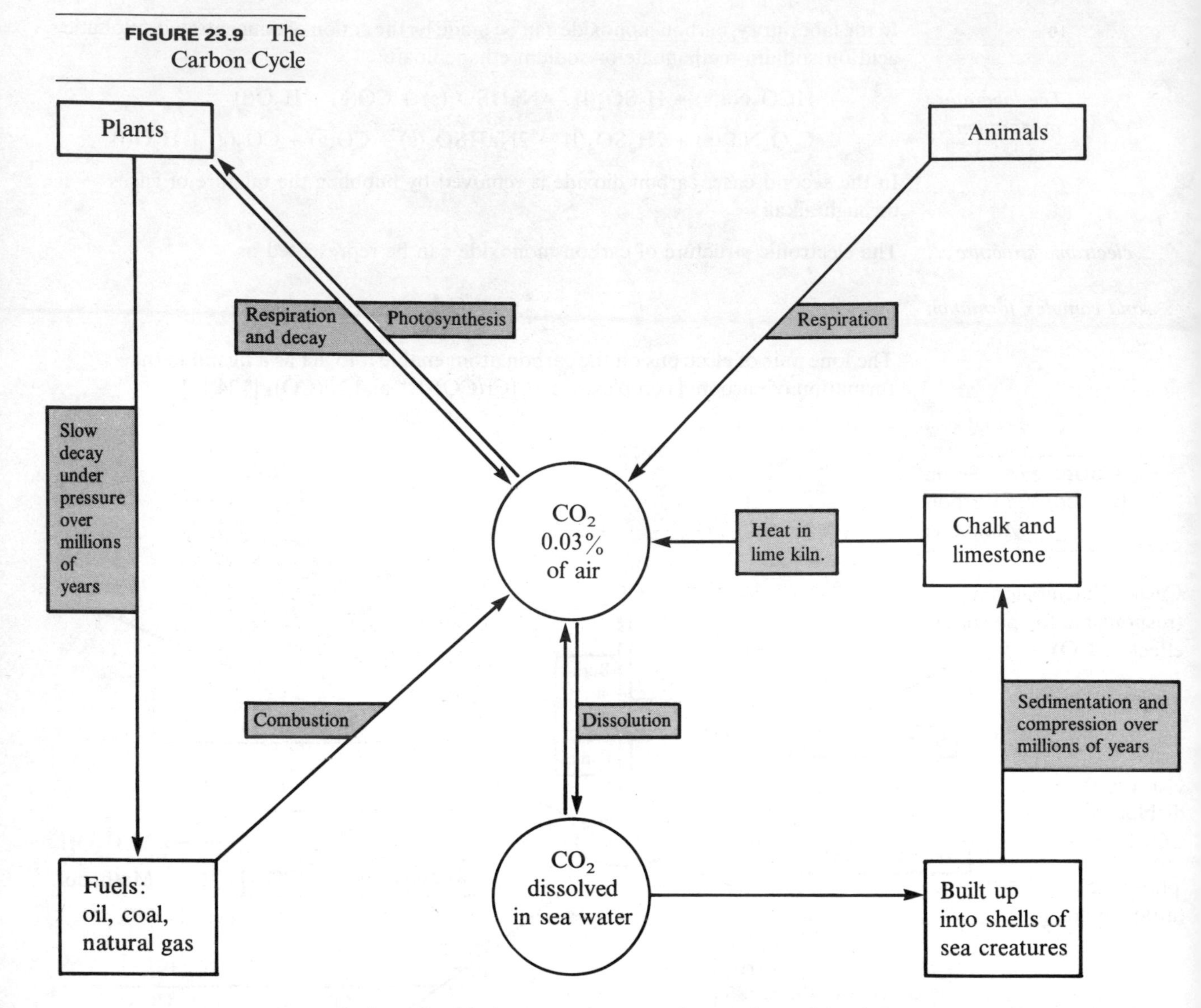

CO_2 is used in fire-extinguishers

Carbon dioxide is non-poisonous, denser than air and does not support combustion. These three factors find carbon dioxide a use in fire-extinguishers. It is stored in cylinders under pressure and released by the opening of a valve [see Figure 23.10]. There are substances which burn with such a hot flame that they can decompose carbon dioxide and continue to burn in the oxygen formed. A magnesium fire could not be extinguished by carbon dioxide.

CO_2 is easily liquefied

Solid CO_2 is called 'dry ice'

The gas can be liquefied at room temperature by a pressure of 60 atm. If the compressed gas is allowed to expand rapidly, solid carbon dioxide is obtained. It is a white solid, with the structure described in § 6.4. Its high standard enthalpy of vaporisation makes it a useful refrigerant, and, since it sublimes to form gaseous carbon dioxide, rather than melting, it is known as 'dry ice' or 'Drikold' [see Figure 23.11].

The laboratory preparation of CO_2

Industrially, carbon dioxide is formed as a by-product during the manufacture of quicklime from limestone [Table 18.3, § 18.5.5] and in the fermentation of sugars to ethanol [§ 30.3.3]. In the laboratory, it can be made by the action of dilute hydrochloric acid or dilute nitric acid on marble chips. The gas can be collected over water. If required pure and dry, it is bubbled through water (to remove any hydrogen chloride), through concentrated sulphuric acid and then collected downwards [see Figure 23.12].

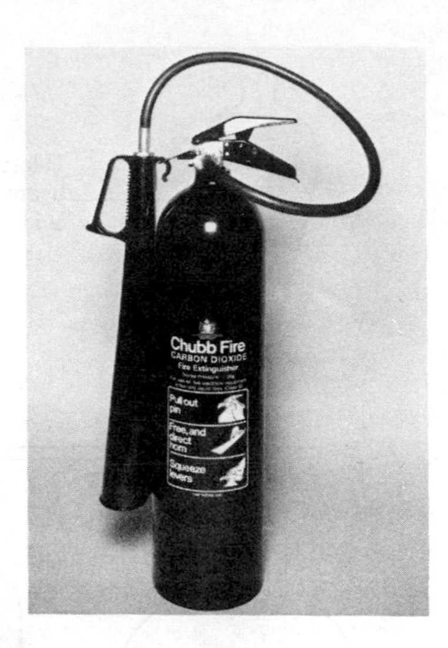

FIGURE 23.10 A Carbon Dioxide Fire-extinguisher

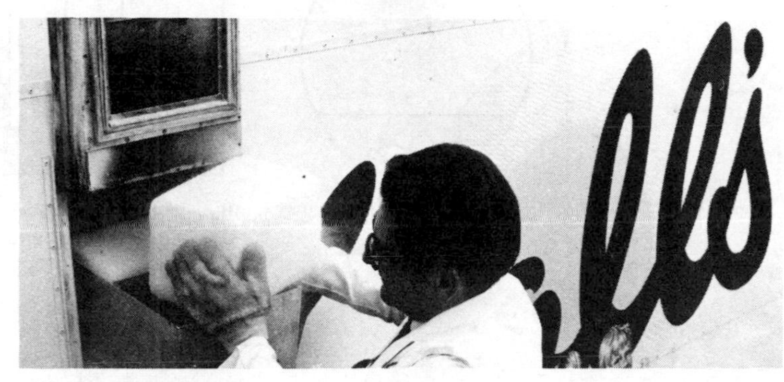

FIGURE 23.11 Drikold, ICI's Solid CO_2

CARBONIC ACID AND CARBONATES

The solution of CO_2 forms carbonic acid

Carbon dioxide is an acidic gas. It is the anhydride of carbonic acid, H_2CO_3, but only about 0.4% of dissolved carbon dioxide is converted into carbonic acid, and when a solution is boiled nearly all the dissolved carbon dioxide is expelled:

$$CO_2(g) + H_2O(l) \rightleftharpoons CO_2(aq) + H_2O(l) \rightleftharpoons H_2CO_3(aq)$$

Carbonic acid is a weak diprotic acid.

Carbon dioxide reacts with bases. When passed into sodium hydroxide solution, it forms first sodium carbonate and then sodium hydrogencarbonate:

$$CO_2(g) + 2NaOH(aq) \rightarrow Na_2CO_3(aq) + H_2O(l) \xrightarrow{CO_2(g)} 2NaHCO_3(aq)$$

Carbonates and hydrogencarbonates

A similar reaction with calcium hydroxide solution (limewater) is used as a test for carbon dioxide. A precipitate of calcium carbonate appears and then dissolves to form soluble calcium hydrogencarbonate:

The limewater test for CO_2

$$CO_2(g) + Ca(OH)_2(aq) \rightarrow CaCO_3(s) + H_2O(l) \xrightarrow{CO_2(g)} Ca(HCO_3)_2(aq)$$

The structure of CO_3^{2-}

The structure of the planar carbonate ion can be represented by a molecular orbital structure:

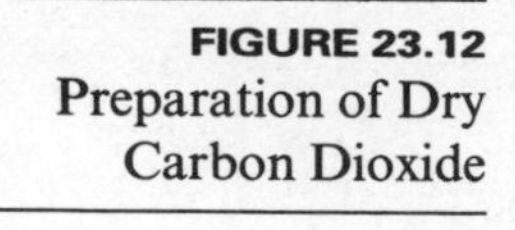

FIGURE 23.12 Preparation of Dry Carbon Dioxide

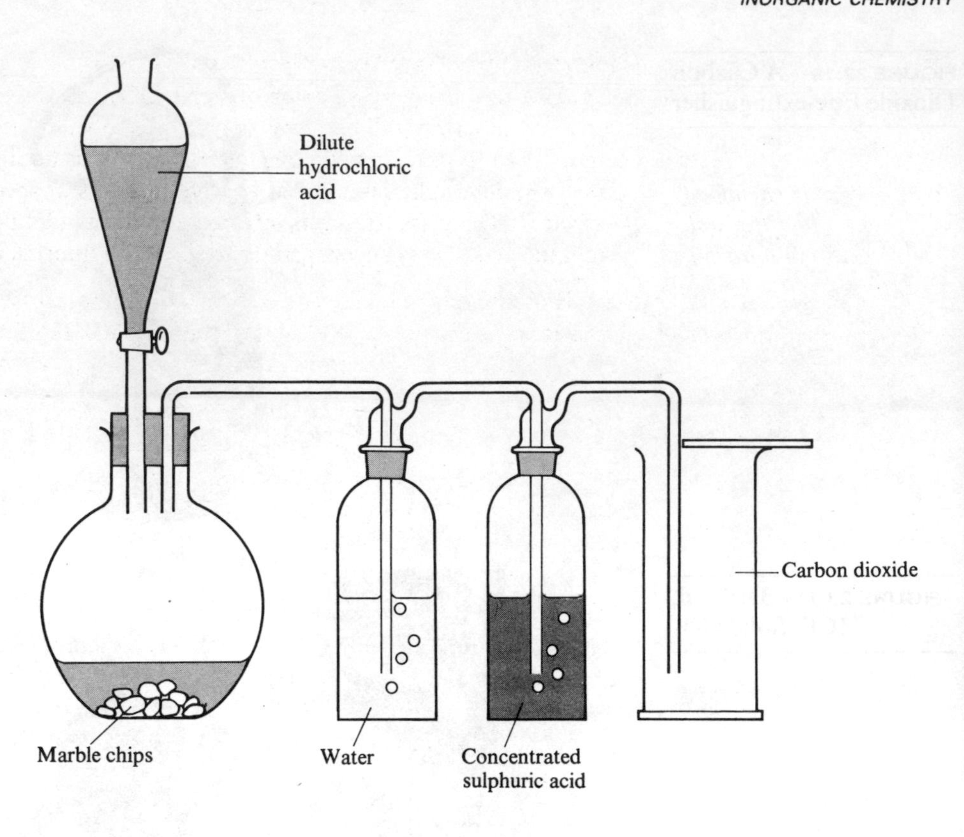

FIGURE 23.13 Some Reactions of Carbonates

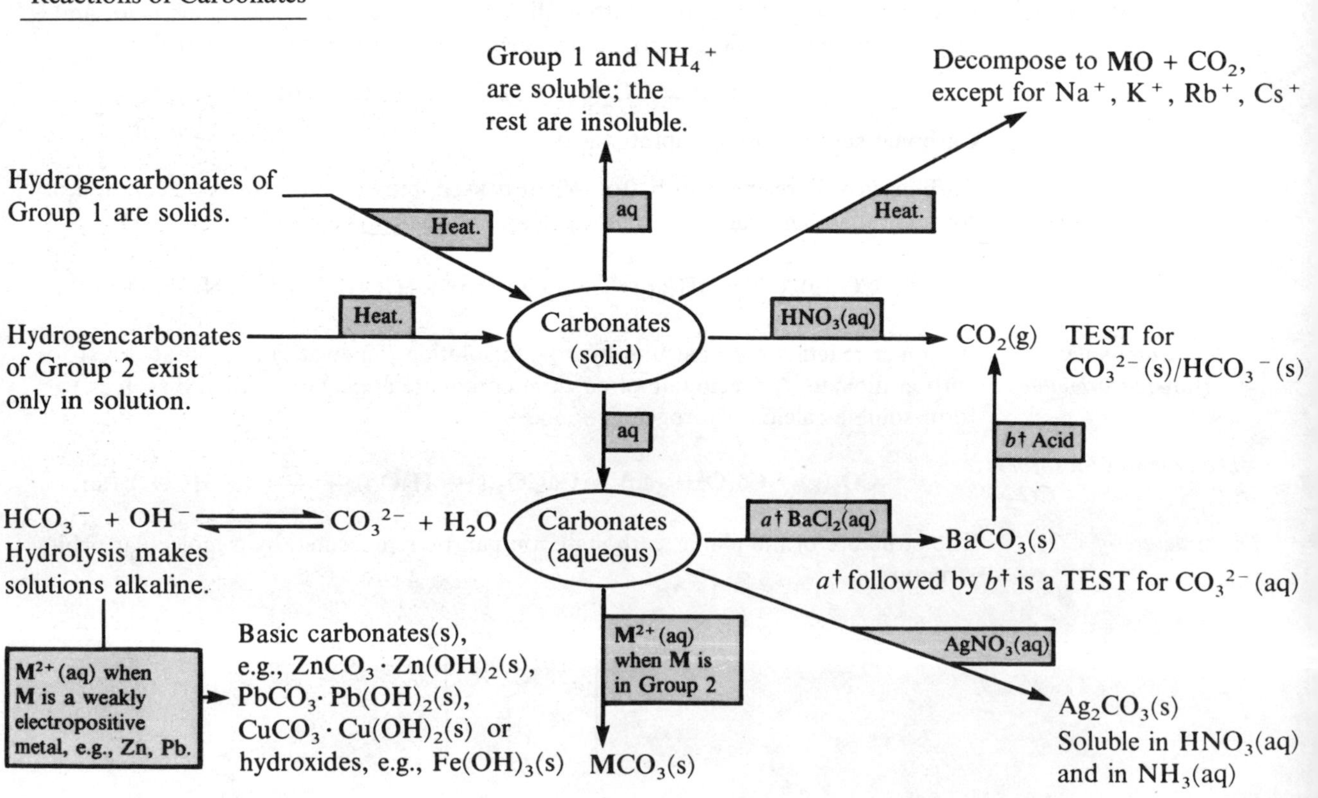

SILICON(IV) OXIDE, SILICA, SiO_2

Silica glass is chemically inert and transmits infrared and ultraviolet light

Several forms of silicon(IV) oxide or *silica*, SiO_2, are known. The structure of quartz and other crystalline forms of silica is shown in Figure 6.14, § 6.5. Silica melts at 1710 °C to form a viscous liquid. When this liquid is cooled, it forms a glass [§ 6.8]. Silica glass is used for the manufacture of specialised laboratory glassware. Silica glass transmits infrared and ultraviolet light. It is chemically inert, being attacked only by hydrogen fluoride, damp fluorine and fused bases.

Uses for silica

Silica is used in the manufacture of cement and mortar [see Table 18.3, § 18.5.5], the anhydrite process for making sulphur dioxide [§ 21.11], the extraction of phosphorus [§ 22.2.2] and the manufacture of the abrasive, silicon carbide, SiC.

Quartz watches

Quartz is used in the manufacture of electronic equipment and timing devices. Quartz watches have recently taken a large slice of the market away from watches with moving parts. The timing mechanism is controlled by a quartz crystal which is induced to vibrate at 32 768 oscillations per second by the application of a small electric field [see end of § 6.9].

SILICATES

When silicates are formed by the reaction of silicon(IV) oxide and a molten base, the silicate ion that is formed depends on the amounts of base and silica present. The silicate ion may be (a) SiO_4^{4-}, a tetrahedral structure, (b) $Si_2O_7^{6-}$, (c) $Si_3O_9^{6-}$ and (d) chains or sheets of extended $(SiO_3)_n^{2n-}$ and $(SiO_2)_n^{n-}$ anions [see Figure 23.9].

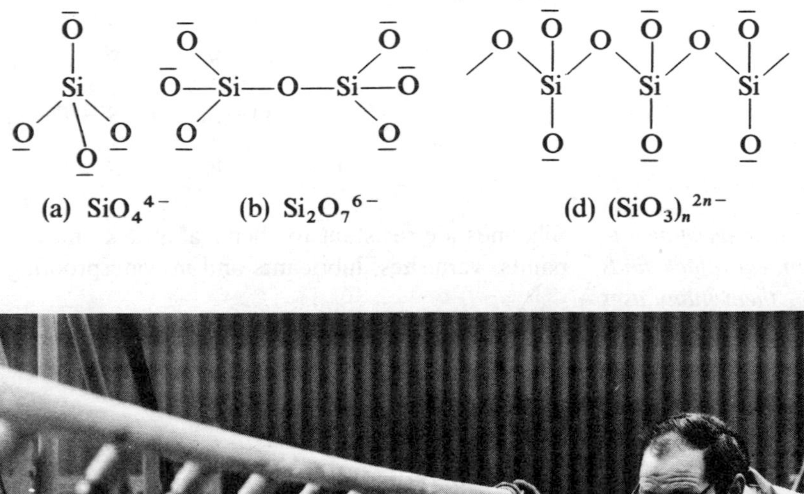

(a) SiO_4^{4-} (b) $Si_2O_7^{6-}$ (d) $(SiO_3)_n^{2n-}$

FIGURE 23.14 Silicate Ions

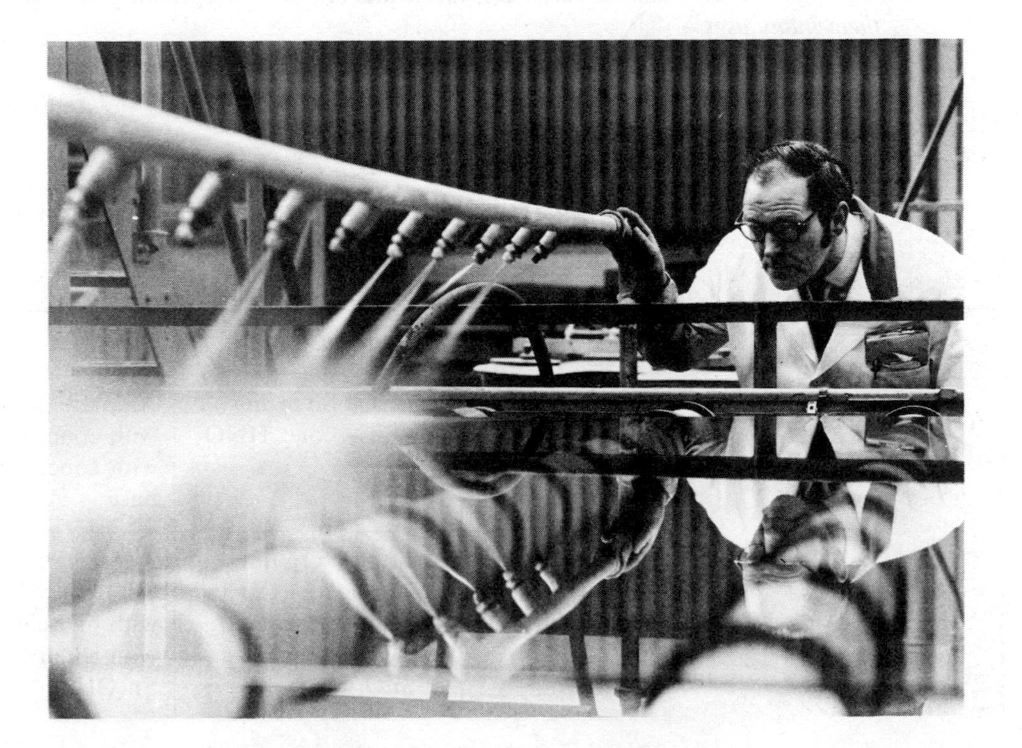

FIGURE 23.15 A Continuous Sheet of Glass leaving the Pilkington Float Glass Plant

Some aluminosilicates are zeolites, and act as molecular sieves

If some Si^{4+} ions are replaced by Al^{3+} ions, aluminosilicates are formed. To preserve electrical neutrality, another cation, e.g., Na^+ or Ca^{2+}, must be incorporated. Some aluminosilicates lose water on heating to form an open structure which has a large surface area and is porous. These aluminosilicates are called *zeolites* and are used as cation exchangers [§ 8.7.6] and as 'molecular sieves'. They retain molecules which are of a size to fit into the cavities in the structure and allow larger and smaller molecules to pass through.

Soda glass is made from silica...

Silica is used in the manufacture of glass. *Soda glass* is a mixture of sodium silicate and calcium silicate, which is made by melting the carbonates with silica at 1500 °C:

$$CaCO_3(s) + SiO_2(s) \rightarrow CaSiO_3(s) + CO_2(g)$$

$$Na_2CO_3(s) + SiO_2(s) \rightarrow Na_2SiO_3(s) + CO_2(g)$$

...as are cobalt glass...

Coloured glasses are made by adding metal oxides to the melt, e.g., cobalt(II) oxide for *blue glass*. Pyrex® is a borosilicate glass, made by adding boron oxide to the melt. It can withstand higher temperatures than soda glass. Sodium silicate,

...and water glass

Na_4SiO_4, is a water-soluble glass called *water glass*.

SILICONES

Silicones are polymers

Silicones are polymers. One type of structure (in which R is an alkyl group) is shown below.

```
      R       R       R       R
      |       |       |       |
—O—Si—O—Si—O—Si—O—Si—
      |       |       |       |
      R       R       R       R
```

Their resistance to chemical attack finds them many uses

Silicones are resistant to chemical attack and are water-repellent. They are used in paints, varnishes, lubricants and in waterproofing fabrics.

OXIDES OF TIN AND LEAD

The oxides of tin and lead are amphoteric. Their preparation and properties are summarised in Table 23.5.

TABLE 23.5 Oxides of Tin and Lead

Oxide	*Preparation*	*Properties*
Tin(IV) oxide, SnO_2, the more stable oxide	(a) Heat Sn in air. (b) Sn + conc. HNO_3	Amphoteric; with conc. $H_2SO_4 \rightarrow Sn(SO_4)_2$; with conc. alkali or fused base → stannate(IV), $Sn(OH)_6^{2-}$
Tin(II) oxide, SnO	Heat SnC_2O_4, tin(II) ethanedioate. (CO and CO_2 are formed and prevent oxidation of SnO.)	Amphoteric, but more basic than SnO_2; with dilute acid → Sn^{2+} salt; with alkali → stannate(II), $Sn(OH)_4^{2-}$

Oxide	*Preparation*	*Properties*
Lead(IV) oxide, PbO_2, brown	Warm Pb^{2+} salt with oxidising agent, e.g., $ClO^-(aq)$.	Lead–acid accumulator [§ 13.3.3] Powerful oxidising agent; Heat → $PbO + O_2$ with SO_2 → $PbSO_4$ with warm HCl → Cl_2 Amphoteric; with HCl(aq) below 20 °C → $PbCl_4(l)$; with alkali or fused base → plumbate(IV), $Pb(OH)_6^{2-}$
Lead(II) oxide, PbO, yellow and red polymorphs	(a) Heat $Pb(NO_3)_2$. (b) Heat $PbCO_3$.	Amphoteric; with HCl(aq) → $PbCl_2$; with alkali or fused base → plumbate(II), $Pb(OH)_4^{2-}$
Dilead(II) lead(IV) oxide, Pb_3O_4, 'red lead'	Heat PbO at 400 °C in air.	Behaves as a mixed oxide, $2PbO \cdot PbO_2$ With $HNO_3(aq)$ → $Pb(NO_3)_2 + PbO_2$ With conc. HCl(aq) → $PbCl_2 + Cl_2$

23.7.4 SALTS

Tin(II) and lead(II) compounds are ionic. Their reactions are shown in Figure 23.16 and Figure 23.17.

FIGURE 23.16 Some Reactions of $Sn^{2+}(aq)$

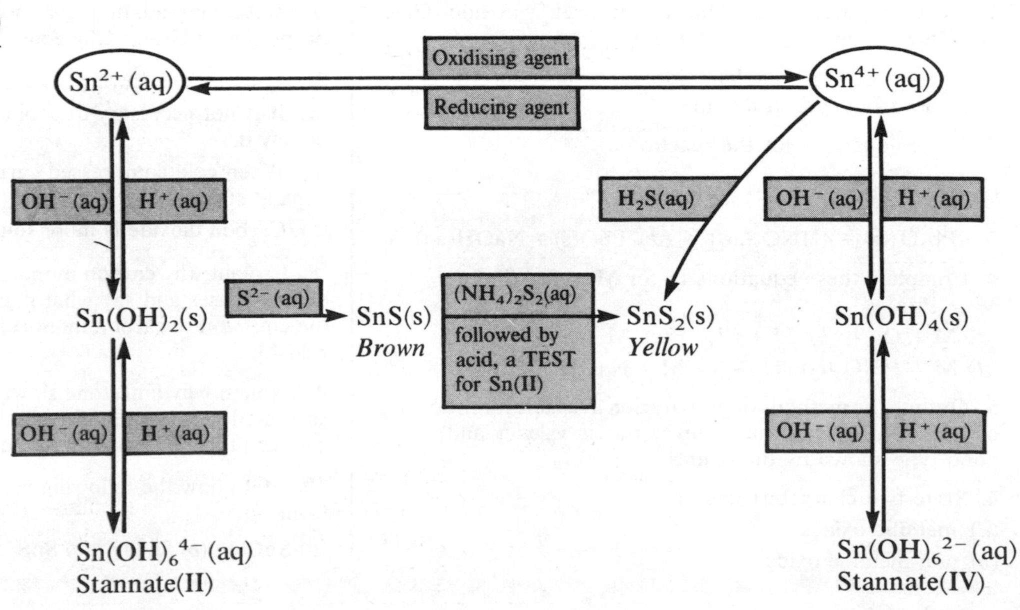

FIGURE 23.17 Some Reactions of $Pb^{2+}(aq)$

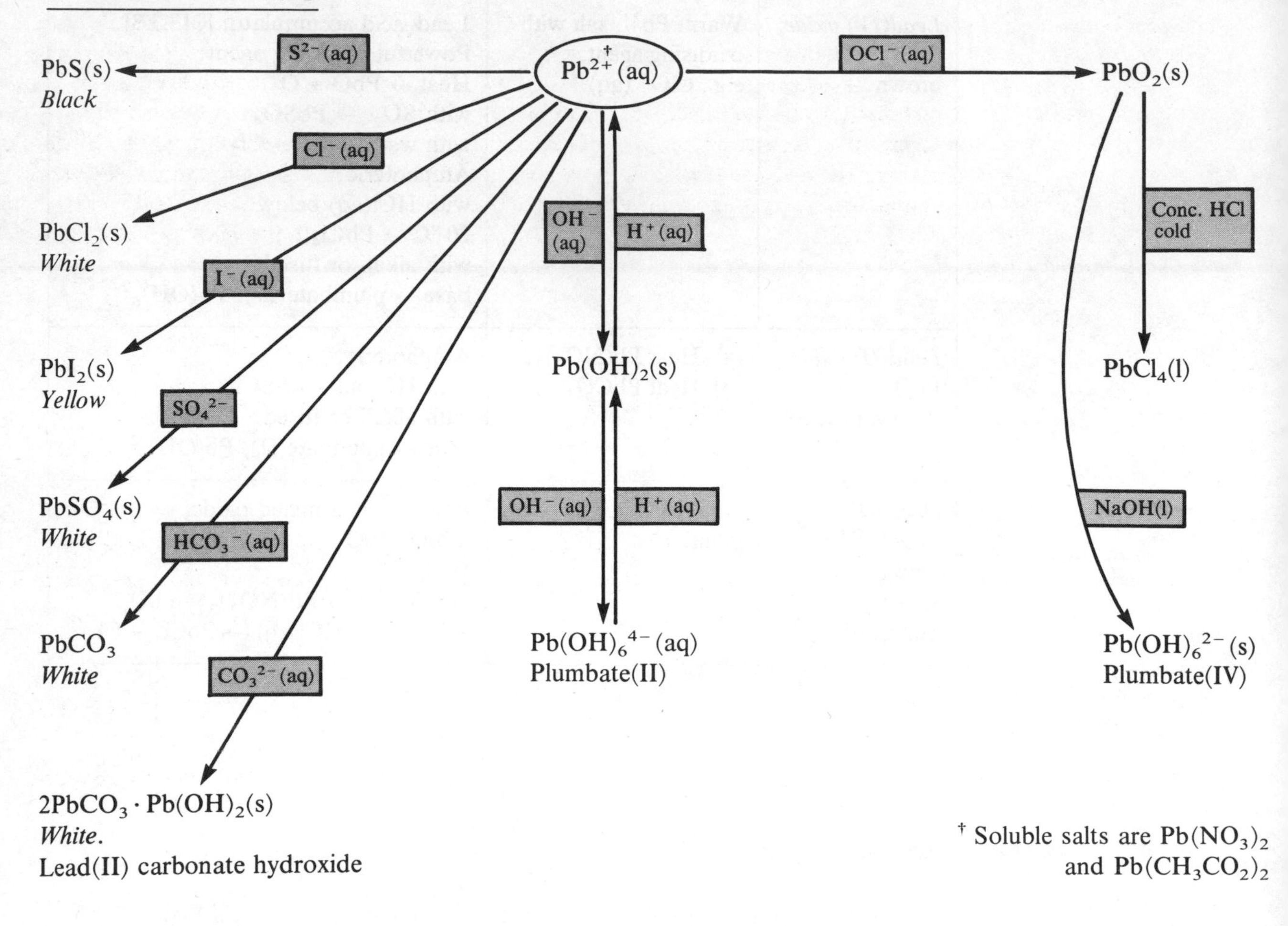

CHECKPOINT 23D: OXIDES

1. 'Lead(II) oxide is more basic than lead(IV) oxide'. Give evidence to support this statement.

2. Explain why silicon(IV) oxide has a different type of structure from carbon dioxide.

3. Write equations for the reactions:

(*a*) $Pb_3O_4(s) \xrightarrow{\text{heat}}$ (*b*) $PbO_2(s) \xrightarrow{\text{heat}}$

(*c*) $Pb_3O_4(s) + 4HNO_3(aq) \rightarrow$ (*d*) $PbO(s) + NaOH(aq) \rightarrow$

4. Complete these equations (i) for **M** = Sn, (ii) for **M** = Pb:

(*a*) $\mathbf{M} + O_2 \rightarrow$ (*b*) $\mathbf{M} + Cl_2 \rightarrow$ (*c*) $\mathbf{M} + S \rightarrow$

(*d*) $\mathbf{M} + HNO_3$(conc.) $\rightarrow$ (*e*) $\mathbf{M}$ + NaOH(conc.) $\rightarrow$

5. Outline the methods of preparation available for the oxides of Group 4. Discuss the group trends in valency and bond type shown by the oxides.

6. State two characteristics of

(*a*) metallic oxides

(*b*) non-metallic oxides.

Illustrate the gradation from non-metallic to metallic properties in Group 4 by considering the oxides.

7. Explain the following statements:

(*a*) It is not necessary to cool carbon dioxide in order to liquefy it.

(*b*) When cold compressed carbon dioxide is allowed to expand suddenly, it solidifies.

(*c*) Carbon dioxide is more soluble under pressure.

†**8.** Explain why carbon monoxide is present in vehicle exhaust gases and say what ill effects it has. Suggest how the emission of carbon monoxide can be reduced. [See § 26.4.]

9. Explain why limestone slowly dissolves in water that is saturated with carbon dioxide. What is the connection between this reaction and (*a*) hard water and (*b*) stalactites?

10. State how the following compounds can be obtained from tin.

(*a*) $SnCl_2$ (*b*) $SnCl_4$ (*c*) SnS (*d*) SnS_2 (*e*) $Na_2Sn(OH)_6$

23.8 A COMPARATIVE LOOK AT GROUP 4

There is a non-metal to metal transition down Group 4 ...

In Group 4, the differences between the first member of the group and the last have reached a maximum. There is a transition from carbon and silicon, which are non-metallic elements, through germanium, which is intermediate in character, a **metalloid**, to tin and lead which are metals. Carbon and silicon have covalent, macromolecular structures. Carbon (except for graphite), is a non-conductor. Silicon and germanium are semiconductors. Tin and lead have metallic structures and are conductors (except for α tin with its diamond-type structure). The reactions with dilute acids and with concentrated nitric acid illustrate the non-metal to metal transition:

... This is illustrated by the reactions with dilute acid and with conc. nitric acid

Dilute acid + C, Si, Ge – no reaction

Dilute acid + Sn – reacts

Dilute acid + Pb – too low in the Electrochemical Series to react

Concentrated nitric acid + C, Si, Ge, Sn → hydrated oxides

Concentrated nitric acid + Pb → metal nitrate

Valencies of 2 and 4 are met in the group

The members of the group show valencies of 4 or 2. Passing down the group, there is an increase in the tendency to use a valency of 2 and an increase in the electrovalent character of the bonds.

There is a transition from covalency to electrovalency down the group

C, Si, Ge: covalent compounds, almost exclusively 4-valent

Sn: covalent +4 and ionic +2 states are formed with almost equal ease, but Sn^{2+} is a reducing agent.

Sn and Pb show a valency of 2

Pb: mainly ionic +2 state; also covalent +4 state. Pb(IV) is an oxidising state.

Carbon differs in many respects from the rest of the group ...

In addition to the trends down the group, there is a sharp line of demarcation between carbon and silicon. In every group, the first member differs from the rest in having no d orbitals of energy comparable with the occupied p orbitals and in being therefore unable to 'expand its octet'. The main differences between carbon and silicon are:

... Hydrocarbons ...

1. Carbon forms a huge number of hydrocarbons. Silicon forms only a few silanes, Si_nH_{2n+2}, which are spontaneously flammable in air [§ 23.7.1].

... Oxides ...

2. Carbon forms monomeric oxides, CO and CO_2. Silicon forms the polymeric $(SiO_2)_n$ [§ 23.7.3].

... Halides ...

3. Carbon halides, CX_4, are not hydrolysed, whereas those of silicon, SiX_4, are readily hydrolysed [§ 23.7.2].

Carbon forms no complexes

4. Carbon forms no complexes. Silicon and other members of the group can expand their octets to form, e.g., SiF_6^{2-}, $SnCl_6^{2-}$, $PbCl_4^{2-}$.

Halides, EX_4

The +4 halides, $\mathbf{EX_4}$, are, with the exception of the carbon tetrahalides, CX_4, all hydrolysed readily. Of the halides excluding those of carbon, the silicon halides are the most covalent in character and the most readily hydrolysed, to give silicate ions. The halides SnX_4 and PbX_4 are hydrolysed to give basic salts or stannates(IV) or plumbates(IV).

SnX_2 PbX_2

Tin and lead form +2 halides, those of lead being more electrovalent than those of tin.

Oxides increase in basicity down the group

The oxides increase in basicity down the group, and the +2 oxides are more basic than the +4 oxides. The oxides of carbon and silicon are acidic; those of germanium, tin and lead are amphoteric:

CO_2 dissolves in water to form a weak acid

CO reacts with fused NaOH → sodium methanoate

SiO_2 reacts with a fused base → a silicate

SnO_2, *SnO* } react with dilute acid → Sn^{4+} and Sn^{2+} salts and react with a fused base or concentrated aqueous alkali to form a stannate(IV) or a stannate(II).

PbO_2, *PbO* } react with dilute acids → Pb(IV) compounds and Pb^{2+} salts and with a fused base or concentrated aqueous alkali to form a plumbate(IV) or a plumbate(II). PbO is more basic than PbO_2.

23.9 THE GREENHOUSE EFFECT

TOPIC

The mean temperature of our planet is fixed by a steady state balance between the energy received from the Sun and an equal quantity of heat energy radiated back into space by the Earth. If disturbances in either incoming or outgoing energy upset this balance, the average temperature of the Earth's surface will drift to a different steady state value. The resulting changes in the Earth's climate could upset food production, create deserts, raise the level of oceans or start a new ice age. One mechanism for regulating the Earth's temperature is the **greenhouse effect** [see Figure 23.18].

FIGURE 23.18 The Atmospheric Greenhouse Effect

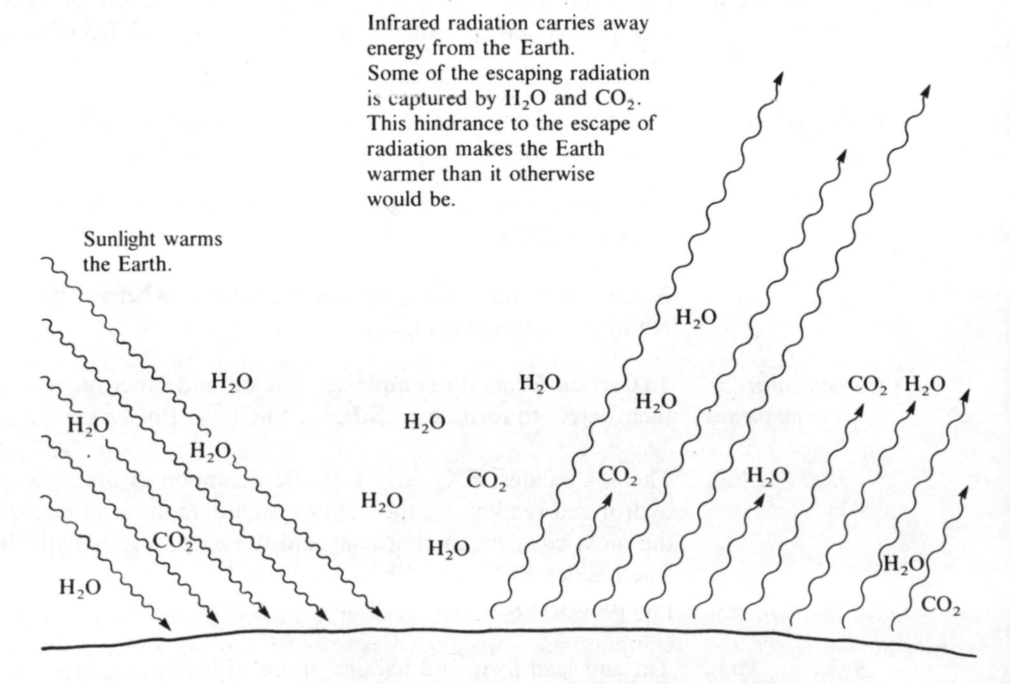

Infrared radiation carries away energy from the Earth. Some of the escaping radiation is captured by H₂O and CO₂. This hindrance to the escape of radiation makes the Earth warmer than it otherwise would be.
Sunlight warms the Earth.
H₂O
CO₂

The Earth receives energy from the Sun...

The Sun emits radiation in a band of wavelengths from ultraviolet to infrared (UV to IR, 200 nm to 3000 nm) with a maximum in the visible spectrum at 500 nm. This radiation passes through the atmosphere of the Earth with very little absorption. When the radiation reaches the Earth, it warms the ground or sea. The warm surface of the Earth radiates energy outwards at the longer infrared wavelengths. Unlike sunlight, infrared radiation cannot travel freely through air. Infrared radiation is absorbed by water vapour, carbon dioxide, ozone and other gases in the lower atmosphere, warming up the lower layers of the atmosphere, which radiate some heat back to the ground and some out into space.

...and radiates energy back into space at the IR wavelengths

Gases in the lower atmosphere, e.g. water vapour and carbon dioxide, absorb IR radiation from the Earth and radiate energy back to Earth. This warming effect is called the greenhouse effect

The warming effect of carbon dioxide and water vapour has been named the **greenhouse effect**. The effect is compared with the glass of a greenhouse, which lets sunlight enter and prevents infrared radiation from leaving. (Actually, in addition to reflecting infrared radiation, real greenhouses trap heat mainly by preventing warm air from escaping by convection.)

There is more water in the atmosphere than carbon dioxide, so most of the greenhouse heating of the Earth's surface is due to water vapour. However, there is a gap in the absorption by water, which is partly plugged by carbon dioxide [see Figure 23.19].

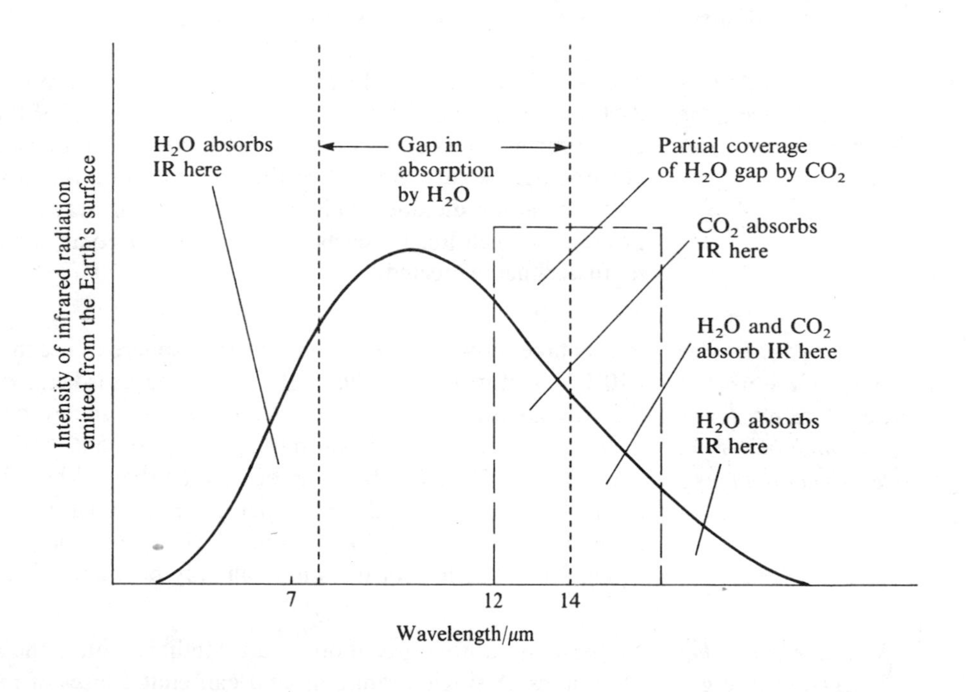

FIGURE 23.19 Carbon Dioxide Partially Plugs the Gap in the Water Cover

Carbon dioxide partially fills a gap in the absorption of IR radiation by water vapour

Increasing the concentration of carbon dioxide in the atmosphere increases the amount of infrared radiation absorbed and increases the warming effect of the lower atmosphere. A concentration could be reached at which the carbon dioxide would capture *all* of the radiation at the correct wavelengths for absorption by carbon dioxide; after this concentration is reached any further increase cannot lead to a further rise in temperature. It follows that there is no danger of a runaway effect which might result in desert conditions over all the Earth.

Without the warming greenhouse effect, the earth would be uninhabitable

The greenhouse effect is a natural phenomenon. Without it, the Earth would be uninhabitable. It is what keeps us from being a frozen planet. If gases such as carbon dioxide did not trap the Sun's energy, the Earth's mean temperature would be about −20 °C, rather than its current 15 °C.

The concentration of carbon dioxide in the atmosphere is increasing. It is forecast that by the year 2050 the level of carbon dioxide will be twice what it was before industrialisation. The predicted result is an increase in global temperature of 2–3 °C. The rise in temperature would be greater at the poles than at the tropics

As long ago as 1890, Svante Arrhenius [see §§ 12.5.1 and 12.7.1] began to worry about the unprecedented release of carbon dioxide into the air from the massive burning of coal during the Industrial Revolution. Arrhenius made the remarkable prediction that doubling the concentration of atmospheric carbon dioxide would eventually lead to a 5 °C warming of the earth. Before the Industrial Revolution, the carbon dioxide concentration was 280–90 ppm; now it is 340 ppm. By 2050, if the burning of fossil fuels continues at its present level, the concentration of carbon dioxide will have doubled. Climatologists estimate that doubling the carbon dioxide concentration would increase mean temperatures over the globe by 2–3 °C. They also predict that the increase would be greatest at high latitudes (near the poles) and that there would be shifts in rainfall patterns. They are able to make these predictions by analysing data from their records for warm years and cold years. Such analysis shows that in warm years the rise in temperature is greater near the poles than in the tropics. Warm years are associated with a 1–2% increase in rainfall. On the basis of such analysis, climatologists predict that an average warming of 2–3 °C would mean Russia warming up by, say 3 °C, North America by 1–2 °C and some countries, e.g. Japan, India and Spain showing little change in temperature. A less optimistic prediction from the UK Meteorological Office is that a doubling of the carbon dioxide concentration would lead to a mean rise of 5 °C in global temperature and a temperature rise in polar regions of 12 °C.

The combustion of fossil fuels is the source of most of the carbon dioxide that is enhancing the greenhouse effect

As Third World countries industrialise, they burn more fossil fuel to provide the energy they need. At present, the global consumption of fossil fuel is increasing at over 4% a year — doubling in 16 years. The increase in carbon dioxide levels is 2–4 ppm each year. This corresponds to about half of the carbon dioxide produced each year by the burning of fossil fuels. Most of the rest is dissolved in the oceans. No one can be sure that, when the concentration of carbon dioxide has doubled, half the carbon dioxide produced each year will stay in the atmosphere. Perhaps the processes which absorb carbon dioxide will become saturated, and the build-up will be faster than expected.

As a result of extensive felling of the world's forests, there are fewer trees to take in carbon dioxide in photosynthesis

Plants take carbon dioxide from the atmosphere to use in photosynthesis [see § 10.1.1]. Unfortunately, the world's forest cover is being drastically cut. In Brazil, the Amazon rain forest once covered 3 million square miles. As the region has been developed for agriculture and mining, 10% to 15% of the forest has been destroyed and a further 20% has been seriously disturbed. When the felled trees are burned or left to rot, carbon dioxide and other greenhouse gases are released. The same kind of deforestation is going on in some African countries, Indonesia, India and the Philippines. The loss of trees may already be making the Earth warmer.

Methane is another greenhouse gas

Methane is another greenhouse gas. Methane enters the atmosphere from a number of sources. A single termite mound can emit 5 litres of methane per minute. Swamps, rice fields, leaking North Sea gas pipes, landfill rubbish dumps and herds of cows all send methane into the atmosphere. In addition to acting as a greenhouse gas in the lower atmosphere, methane also destroys ozone in the upper atmosphere [see § 21.5].

Chlorofluorocarbons, CFCs, are powerful greenhouse gases

A host of other gases are now known to add to the greenhouse effect. Chlorofluoroalkanes (often called chlorofluorocarbons, CFCs), which are used in refrigerators and air conditioners and in aerosol cans, are powerful greenhouse gases. The international agreement signed at Montreal in 1987 will reduce the emission of CFCs by 50% by the year 2000 [see § 29.9]. At another international meeting in London in 1989, the UK and other countries pledged to reduce their use of CFCs by 85% by the year 2000.

An increase of 0.5 °C has been observed since the turn of the century. This may not seem alarming, but scientists expect any greenhouse warming to be masked for quite a time by the enormous capacity of the oceans to absorb heat.

A catastrophic effect of global warming would be the melting of ice at the north and south polar ice caps. It would lead to a rise in the level of the oceans and widespread flooding

If the global temperature rises by 3–4 °C, glaciers and ice sheets will melt and the massive polar ice caps will begin to melt. This melting would take centuries. The water in the oceans would expand as it warmed up. One estimate is that the ocean level would ultimately rise by about 70 m. Long before that, it would flood many of the coastal areas of the world. Some countries, e.g. Bangladesh, Thailand, Guyana, the Philippines and Britain, would find low-lying land flooded by rising seas. Other countries would suffer yearly droughts and some areas would become deserts. The US Environmental Protection Agency figures predict that by the year 2100, sea levels could have risen by 0.5–3.5 m.

If the temperature of the Earth rises, as predicted, it will rise more in the countries far removed from the equator. The rich countries of the North will be more affected than the Third World countries

There is a political dimension to the greenhouse effect. The forecast rise in global temperature would affect countries at different latitudes differently. In the corn belt of the US, any change in temperature and decline in rainfall could prove disastrous. A change of 1 °C would mean a reduction of 10% in yields. The wheat belt of the CIS would suffer a reduction of 20% for a 1 °C rise. On the other hand, for a global warming of 1 °C or more, rice yields would be increased, even if rainfall decreased. In regions where slight warming and increased rainfall are forecast, rice yields could increase by 10% or more. Although the Third World countries will intensify the greenhouse problem by their economic growth, they will suffer few adverse consequences themselves. The rich Northern countries would suffer the worst consequences. In the rich Northern countries, the problem of carbon dioxide emission can be solved by turning to nuclear power and alternative energy sources. The Third World cannot afford to invest in nuclear reactors. Rapid industrial growth is the priority of Third World countries. They can achieve it most easily by using coal as their main source of energy.

To avoid an increase in global temperature, it is essential to stop using CFCs ...

... to decrease the combustion of fossil fuels ...

... and to put a stop to deforestation

One move to reduce the greenhouse effect is to cut back on the production of 'chlorofluorocarbons' [CFCs, see § 29.9]. The most far-reaching step would be to cut back on the use of fossil fuels. This would be hard to do in industrialised countries without a tremendous effort in energy conservation and the development of alternative energy sources. In developing countries, reductions in the burning of fossil fuels would be difficult to impose because they would delay industrialisation of these countries.

A very effective step would be to protect the tropical rain forests. There is a scheme for achieving this aim. It is called **debt-for-nature exchange** or **debt-swap**. The scheme recognises that many of the countries which are cutting down forests are struggling to repay foreign debts and are desperately in need of the revenue they will get for timber. In debt-swap, a rich country agrees to cancel a debt from a poor country if that country will agree to conserve a certain area of forest in exchange. Cameroon and Madagascar in Africa and Costa Rica and Bolivia in South America are some of the countries which have benefited from the scheme. The whole world will benefit if the scheme expands.

CHECKPOINT 23E: THE GREENHOUSE EFFECT

1. (*a*) About 6×10^9 tonnes of carbon in fossil fuels are burned annually. If this mass of carbon were burned completely, what mass of carbon dioxide would it produce?
(*b*) The present mass of carbon dioxide in the atmosphere is 2.5×10^{12} tonnes. It is assumed that half of the additional carbon dioxide will enter the sea and other 'sinks'. In how many years will the carbon dioxide content double?

2. An unknown number of 'multiplier effects' could make the greenhouse effect worse than expected.

(*a*) If the polar ice caps began to melt, the land beneath them would be exposed. How would the exposure of this land add to global warming?

(*b*) An increase in temperature would lead to evaporation of water from the oceans. How would this add to global warming?

(*c*) An increase in temperature would decrease the solubility of carbon dioxide. How would this add to global warming?

(*d*) An increase in temperature might lead to increased volcanic acitivity. How would this affect the global temperature?

3. Suppose experts were to state that, beyond reasonable doubt, the present rate of carbon dioxide emission would have catastrophic repercussions (melting of the polar ice caps etc.) by the year 2100. Draw up a plan for world response to the threat. Discuss the political aspects of international cooperation versus national goals, the desire of underdeveloped nations to industrialise, the standard of living in developed countries, the role of electrical energy and nuclear power and other factors.

This is a problem which is baffling nations: you may want to form a group to discuss it and pool ideas!

4. How does deforestation add to the greenhouse problem? Which countries are cutting down forests to make arable land? Which countries are cutting down trees for use as fuel? If it were up to you to persuade such countries to sacrifice their short-term ends for the long-term future of the planet, what would you say? What resistance do you think you might meet?

Again, this is a tough problem, and a group discussion might be the best way to approach it.

QUESTIONS ON CHAPTER 23

1. Outline the preparations of CO_2, SiO_2, CCl_4 and $SiCl_4$. Explain why CO_2 is a gas, whereas SiO_2 is a solid of high melting temperature.

2. Give two properties of (*a*) metallic oxides, (*b*) non-metallic oxides, (*c*) metallic chlorides and (*d*) non-metallic chlorides. Illustrate the transition in properties down Group 4 by considering the properties of (i) their oxides and (ii) their chlorides.

3. Compare the elements of Group 4 with respect to

(*a*) the crystal structure of the element

(*b*) the thermal stability of the hydrides

(*c*) the stability to hydrolysis of the halides

(*d*) the oxidation states of the elements

(*e*) the basicity of the oxides EO_2 and EO.

4. Some chemists view Group 4B as a pair of similar elements, C and Si, followed by a trio of similar elements, Ge, Sn and Pb. What evidence can you offer to support this view?

5. Do you agree with these statements? Explain your answers.

(*a*) The chemistry of carbon is dominated by the ability of the carbon atom to bond to itself and to form multiple bonds to itself and to other atoms.

(*b*) The chemistry of silicon is dominated by the ability of silicon to form strong bonds to oxygen and the reluctance of silicon to bond to itself.

6. What are allotropes?
Sketch the structures of the two allotropes of carbon. How does the structure explain the physical properties of each allotrope?

How could you show that the allotropes are both pure carbon?

7. Outline how you could prepare (*a*) CO from C, (*b*) PbO from Pb, (*c*) SnI_4 from Sn and (*d*) PbI_2 from Pb.

8. Considering Group 4, discuss

(*a*) the variation of acid/base character of the oxides MO and MO_2 down the group

(*b*) the reactivity of the tetrahalides with water

(*c*) the thermal stability of the dichlorides and tetrachlorides.

In what ways is carbon atypical of the group?

9. Silicon is the second most abundant element in the surface crust of the Earth and occurs mainly as silica and silicates. It is element number 14 in the Periodic Table and has isotopes of mass number 28, 29 and 30. The relative atomic mass is 28.09.

Silicon occurs in Group 4 of the Periodic Table and has chemical properties similar to carbon.

(*a*) (i) Write the electronic structure of silicon in terms of s and p electrons.

(ii) Explain how you would deduce that $^{28}_{14}Si$ is the most commonly occurring isotope of silicon.

(*b*) Metals are good conductors and non-metals are insulators. Silicon is a semiconductor.

(i) Explain electronic conduction in metals in terms of delocalised bonding.

(ii) Name an element that is added to silicon to convert it into a p-type semiconductor.

(*c*) Silica (silicon dioxide) occurs mainly as quartz and sand. Heating a mixture of sodium carbonate, calcium carbonate and sand produces glass.

(i) Sketch the repeating three-dimensional structure of quartz showing two repeating units.

(ii) Write a balanced equation for the reaction of molten sodium hydroxide with silica.

(iii) Suggest and explain why concentrated alkalis should not be used in burettes which have ground glass joints.

(*d*) The silicon hydrides were extensively studied by Stock who treated magnesium silicide with acid to obtain a mixture of silanes. There are similarities with hydrocarbons, as silicon can form four covalent bonds.

	SiH_4	Si_2H_6	Si_3H_8	Si_4H_{10}	*Higher silanes*
% in mixture	40	30	15	10	5
M.pt.	−185	−132	−117	−84	—
B.pt.	−112	−15	53	107	—

(i) Name a physical technique that can be used to separate the silanes.

(ii) Si_4H_{10} exists as two isomers. Draw the structures of these two isomers.

(iii) Both CH_4 and SiH_4 burn in air. Write balanced equations for the *complete* combustion of SiH_4 and the *incomplete* combustion of methane.

(iv) Above 500 °C silanes are decomposed completely to hydrogen and silicon whereas hydrocarbons are stable. Suggest why this is so.

(NI 90)

10. (*a*) Copy and complete the table below to show how the chloride of each of the following reacts with water.

Element/oxidation state	*Reaction with water*
Carbon	
Silicon	
Tin(IV)	
Lead(II)	
Lead(IV)	

(*b*) (i) Choose two chlorides from the above list, whose reactions with water differ substantially, and give a reason for the difference.

(ii) Comment on the fact that two chlorides of lead, in different oxidation states, are listed compared with only one for silicon.

(iii) By analogy with carbon, suggest a formula for another chloride of silicon(IV) and describe its structure (a clear diagram will suffice).

(*c*) The following are two lists of bond enthalpies/kJ mol^{-1}.

A: Si—O 374; Si=O 638; C—O 360; C=O 743

B: Si—Si 176; C—C 348; Si—H 318; C—H 412

(i) Use list A to help explain why silicon(IV) oxide has a giant atomic structure but carbon dioxide has the structure O=C=O.

(ii) Use list B to help explain why extra-terrestrial life forms based on silicon are unlikely.

(L 91)

11. (*a*) Discuss the trends in the chemical properties of the elements of Group 1 by considering their reactions with water and oxygen.

(*b*) Illustrate the transition from non-metallic to metallic properties exhibited by Group 4 elements, C, Si, Sn, Pb, by considering the properties of their oxides and chlorides.

(*c*) Outline a method for the preparation of XeF_4; include the collection (or separation) of the product.

(O 92)

12. (*a*) Discuss with the help of diagrams the structures of diamond and graphite.

(*b*) Account for the fact that graphite is used as a lubricant, whereas diamond is used as an abrasive.

(*c*) Silicon carbide (SiC) is used industrially as a substitute for diamond. Suggest a structure for silicon carbide.

(*d*) Impure silicon tetrachloride is prepared from sand. Suggest, with your reasons, a suitable method for the purification of silicon tetrachloride.

(*e*) Silicon tetrachloride can be reduced to give either silicon or silane (SiH_4). What reagents can be employed to carry out these two different processes?

(*f*) Suggest how, and why methane and silane differ when treated with aqueous sodium hydroxide.

(O 92, S)

13. (*a*) The standard enthalpy change from graphite to diamond is +2.1 kJ mol^{-1}. Their densities are 2.25 g cm^{-3} and 3.53 g cm^{-3} respectively.

Industrial diamonds are made by heating graphite to 2000 °C under a pressure of 70 000 atm and in the presence of a catalyst.

Suggest why each of these conditions is necessary.

(*b*) Below 13 °C, metallic tin changes into a form with a diamond structure. The change is rapid at −30 °C.

(i) Suggest why this conversion takes place more readily than the conversion of graphite into diamond.

(ii) Why do you think that cans made from tinplate should not be used in very cold climates?

(*c*) Suggest a reason why silicon has a diamond structure, no form corresponding to graphite being known.

(*d*) (i) The melting points of the tetrafluorides of the Group 4 elements are as follows: carbon −128 °C, silicon −96 °C, germanium −36 °C, tin 705 °C.

Discuss these values.

(ii) The compound PbF_4 decomposes at 700 °C before it melts. Suggest why the behaviour of this fluoride differs from that of the tetrafluorides of the other Group 4 elements.

(C 92, S)

24

THE TRANSITION METALS

24.1 INTRODUCTION

The first transition series in Period 4 are a very similar set of elements

Across the second and third periods of the Periodic Table, there is a gradation in properties, from the alkali metals to the halogens. The fourth period begins in the same way, with an alkali metal (potassium) and an alkaline earth (calcium). The next ten elements do not show the gradation in properties of previous periods: they are remarkably similar to one another in their properties and are all metals. They are called the **first transition series**. Periods five and six also contain transition series. The reason for the similarity of the first transition series is that, considering the series from left to right, while each additional electron is entering the 3d shell, the chemistry of the elements continues to be determined largely by the 4s electrons.

The difference between transition metals is the number of d electrons. This affects their chemistry less than a difference in s or p electrons

From one transition element to the next, the nuclear charge increases by 1 unit, and the number of electrons also increases by 1. Since each additional electron enters the 3d shell, it helps to shield the 4s electrons from the increased nuclear charge, with the result that the effective nuclear charge remains fairly constant across the series of transition elements. The sizes of the atoms and the magnitudes of the first ionisation energies are therefore very similar and the elements have comparable electropositivities. The electron configurations of the first transition series are shown in Table 24.1, in which (Ar) = $1s^22s^22p^63s^23p^6$. You will notice that chromium completes occupying its d orbitals with unpaired electrons at the expense of its 4s electrons, and copper completes its full d^{10} shell at the expense of its 4s electrons. There appears to be a certain measure of stability associated with a full d^{10} shell and with a half-filled d^5 shell.

TABLE 24.1 Electron Configurations of the First Transition Series

	3d					4s
Sc (Ar)	↑					↑↓
Ti (Ar)	↑	↑				↑↓
V (Ar)	↑	↑	↑			↑↓
Cr (Ar)	↑	↑	↑	↑	↑	↑
Mn (Ar)	↑	↑	↑	↑	↑	↑↓
Fe (Ar)	↑↓	↑	↑	↑	↑	↑↓
Co (Ar)	↑↓	↑↓	↑	↑	↑	↑↓
Ni (Ar)	↑↓	↑↓	↑↓	↑	↑	↑↓
Cu (Ar)	↑↓	↑↓	↑↓	↑↓	↑↓	↑
Zn (Ar)	↑↓	↑↓	↑↓	↑↓	↑↓	↑↓

Transition metals are called d block metals

Scandium, zinc and copper

Transition metals are often referred to as d block metals. They are defined as elements which form some compounds in which there is an incomplete subshell of d electrons. Scandium ($3d^0$ in compounds) and zinc ($3d^{10}$ in compounds) are excluded by this definition, and copper is included only in copper(II) ($3d^9$) compounds. It is convenient to include these metals in a treatment of transition metals, however, on account of the chemical resemblance of their compounds to transition metal compounds.

24.2 PHYSICAL PROPERTIES OF TRANSITION METALS

The metallic bond is stronger than in s block metals

The first transition metals are all hard and dense, good conductors of heat and electricity and possessing useful mechanical properties. Their melting and boiling temperatures [see Figure 24.1] and standard enthalpies of melting are higher than those of s block metals. All these properties are a measure of the strength of the metallic bond. With d electrons as well as s electrons available to take part in delocalisation, the metallic bond is strong in transition metals. Figure 24.1 shows how both melting and boiling temperatures drop at manganese. In manganese, the d subshell is half full, with 5 electrons in 5 orbitals. This electron configuration appears to make the d electrons less available for delocalisation.

There is a measure of stability in a half-full d subshell

FIGURE 24.1 Boiling and Melting Temperatures of Transition Elements

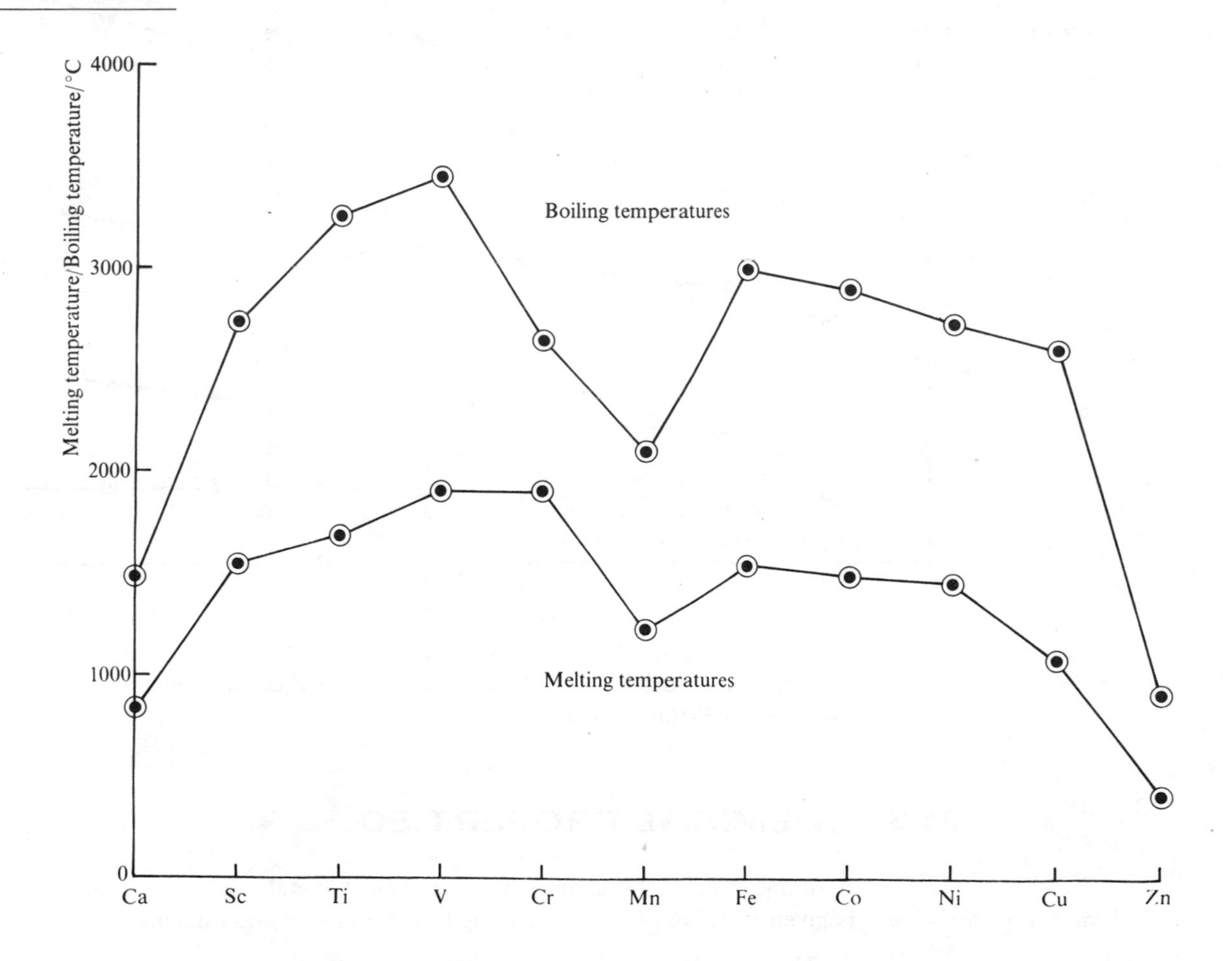

24.2.1 DENSITY

d block metals are denser than s block metals

While the size of the atom, measured by the metallic radius, increases slightly from scandium to zinc, the relative atomic mass increases considerably. The result is an increase in density from scandium to zinc. The d block metals are, in general, denser than the s block metals.

24.2.2 IONISATION ENERGY

Figure 24.2 shows how the first and second ionisation energies increase only slightly from scandium to zinc. The reason is that, as discussed in 24.1, the effective nuclear charge increases only slightly across the series. The increases in the third and fourth ionisation energies across the series are more rapid as d electrons are being removed, and the effective nuclear charge therefore increases by a significant amount from one element to the next.

FIGURE 24.2 Ionisation Energies of Transition Metals

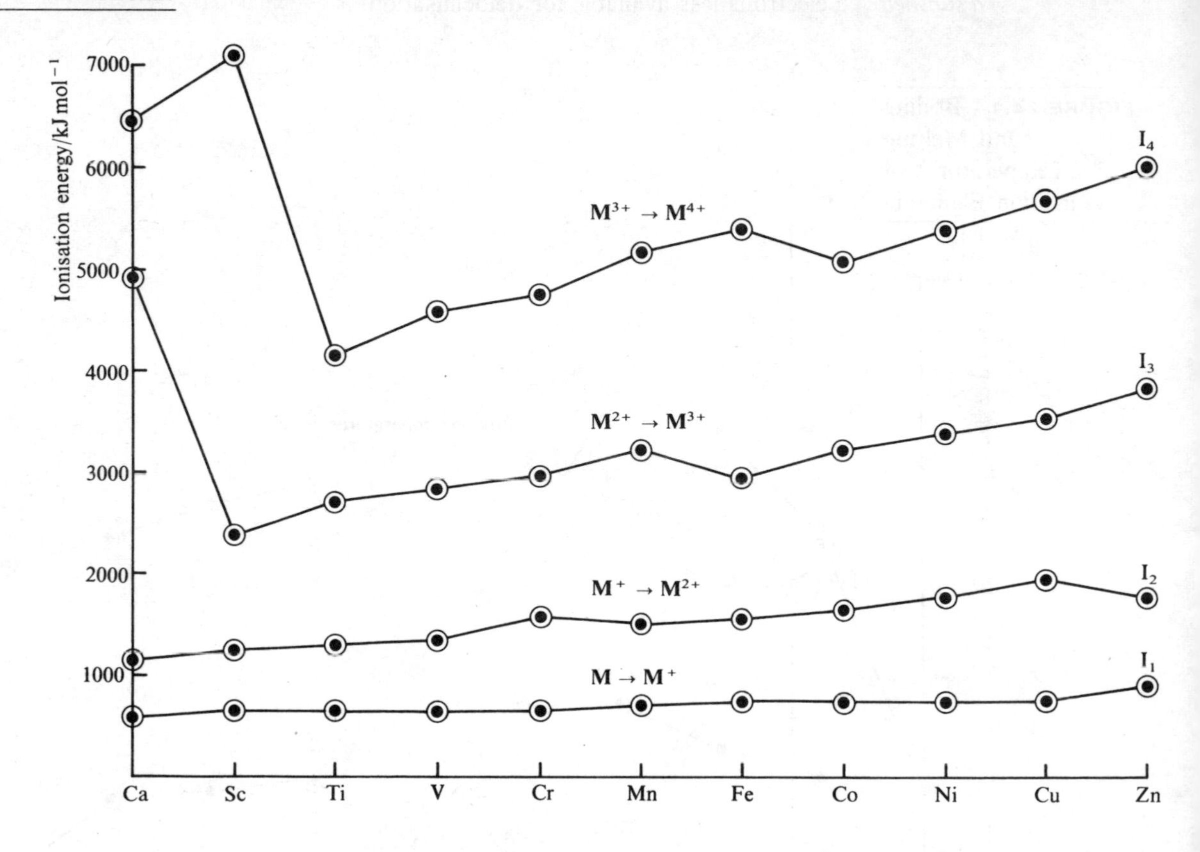

Ionisation energies increase across the period

Each of the curves shows a maximum; these are at Cr^{2+}, at Mn^{3+} and at Fe^{4+}. In each case, the peak in ionisation energy occurs when the removal of an electron disturbs a half-full d^5 shell.

24.3 CHEMICAL PROPERTIES

Features of transition metals

A summary of the differences between d block metals and s block and p block metals is given in Table 24.2. Outstanding features of transition metals are:

1. The variable oxidation state [§ 24.6].
2. Their use as catalysts [§ 24.7].
3. Paramagnetism [§ 24.8].
4. The formation of complex ions [§ 24.13].
5. The formation of coloured ions [§ 24.13.2].

TABLE 24.2
A Comparison of Metals

	s block	*p block*	*d block*
Physical properties	Soft, low melting temperature	Harder, with higher T_m than s block	Harder still, with higher T_m than p block
Reaction with water	React, often vigorously	React only slowly with cold water	
Reaction with non-metallic elements	React vigorously	React less vigorously than s block metals	
Reaction with hydrogen	Form ionic hydrides	Form no hydrides	Some form interstitial hydrides
Bonding	Usually ionic	Usually covalent or complex ions	
Properties of ions	Form simple ions with noble gas configuration	Simple ions have a completed d shell. Easily form complex ions	Some simple ions are formed. Many complex ions are formed.
Complex ions	Simple ions can be loosely hydrated to form colourless complex ions.	Colourless complex ions are formed, rather than simple ions.	Complex ions are formed readily. Usually coloured
Oxidation numbers	Oxidation No. = Group No.	Employ Ox. No. = Gp. No. and also Ox. No. = Gp. No. −2	Ox. No. varies usually by 1 unit at a time, +2, +3 are common.

24.4 METHODS OF EXTRACTION

Extraction by reduction of the oxide

Transition metals occur mainly as sulphides and oxides. The least reactive (e.g., Cu, Au, Pt) are also found **native** (i.e., uncombined). Reduction of the oxide by carbon or carbon monoxide is the usual method of extraction. The theory is covered in § 10.9.1. The following steps may be involved in the extraction:

1. The ore is concentrated. Sometimes flotation is employed: a stream of water carries away debris and leaves the denser ore behind.

2. Sulphide ores are then roasted to convert them to oxides.

3. Heating with coke reduces the oxide to the metal.

4. Carbon is present as a major impurity in the metal. It is removed by heating the metal in a stream of air. Further purification may be achieved by electrolysis (e.g., Cu, Ag, Cr).

24.5 USES OF TRANSITION METALS

Uses of iron, steel...

...and titanium

Iron is our most important metal. Steels are made by alloying iron with carbon and with other transition metals, such as vanadium, manganese, cobalt and nickel [§ 24.14.2]. Titanium is a metal with the same kind of mechanical strength as steel and two big advantages: it is less dense than steel, and it does not corrode. It is stronger than aluminium. The high cost of titanium has limited its use to applications where no expense is spared. It is used in the construction of space capsules. It is better able than steel to withstand the high temperatures that are experienced when a space capsule re-enters the Earth's atmosphere.

Titanium is being used as a twentieth-century remedy for a nineteenth-century mistake. When repairs were made to the Acropolis in 1896–1933, steel bolts were used to join broken pieces of marble, and steel girders were used to reinforce ancient architraves and porches. The Ancient Greeks, twenty-four centuries ago, had used iron bolts coated with lead. By 1970, the newer steel pins had rusted and, in doing so, had expanded, causing cracks in the marble. The only solution to this serious problem is to remove all the steel pins and replace them with bolts of titanium, which does not corrode under any natural conditions.

24.6 OXIDATION STATES

The electron configuration is (Ar) $4s^2 3d^n$. Once the 4s electrons have been removed, the 3d electrons may also be removed. The difference in energy between the 3d and the 4s electrons is much smaller than the difference between the 3s and the 3p electrons. The oxidation states employed by the elements are shown in Table 24.3.

Some of the oxidation states are uncommon and unstable. The important ones are underlined. The stability of the +2 oxidation state relative to +3 and higher oxidation states increases from left to right across the series. It reflects the increasing difficulty of removing a 3d electron as the nuclear charge increases.

Figure 24.3 shows both the complete range of oxidation states shown by each element and also the important oxidation states in the chlorides and oxides of transition elements. In their lower oxidation states, elements form ionic compounds, but in their higher oxidation states they form covalent compounds [§ 15.3.6]. Also included in Figure 24.3 are the carbonyl compounds, in which the metal has an oxidation state of zero.

TABLE 24.3 Oxidation States of Transition Metals

Sc	*Ti*	*V*	*Cr*	*Mn*	*Fe*	*Co*	*Ni*	*Cu*	*Zn*
								+1	
	+2	+2	+2	+2	+2	+2	+2	+2	+2
+3	+3	+3	+3	+3	+3	+3	+3	+3	
	+4	+4	+4	+4	+4	+4	+4		
		+5	+5	+5					
			+6	+6	+6				
				+7					

24.7 CATALYSIS BY TRANSITION METALS

Many important reactions are catalysed by transition metals . . .

Transition metals and their compounds are important catalysts. Some industrial reactions which are catalysed by transition metals are: the Contact process (vanadium(V) oxide), the Haber process (iron or iron(III) oxide), the hydrogenation of unsaturated oils (finely divided nickel) and the oxidation of ammonia (platinum or platinum–rhodium alloy). These examples all involve heterogeneous catalysis [§ 14.10.2], in which the reactant molecules are adsorbed on the surface of the catalyst. It is likely that the 3d electrons enable the transition metal catalyst to form temporary bonds with reactant molecules. In homogeneous catalysis [§ 14.10.1], which is usually found in reactions in solution, the variable oxidation number of the transition metal may enable it to take part in a sequence of reaction stages and emerge unchanged at the end. An example is the oxidation of iodide ions by peroxodisulphate ions, $S_2O_8^{2-}$, according to the equation

$$S_2O_8^{2-}(aq) + 2I^-(aq) \rightarrow 2SO_4^{2-}(aq) + I_2(aq) \qquad [1]$$

. . . The oxidation of iodide ions is an example of homogeneous catalysis

Iron(II) ions catalyse the reaction, and it is thought that they may provide an alternative route for the reaction via steps [2] and [3]:

$$Fe^{2+}(aq) + \tfrac{1}{2}S_2O_8^{2-}(aq) \rightarrow Fe^{3+}(aq) + SO_4^{2-}(aq) \qquad [2]$$

$$Fe^{3+}(aq) + I^-(aq) \rightarrow Fe^{2+}(aq) + \tfrac{1}{2}I_2(aq) \qquad [3]$$

The alternative route involves two reactions between oppositely charged ions and therefore has a lower activation energy [§ 14.10] than reaction [1] between ions of the same charge. The Fe^{2+} catalyst is regenerated in step [3].

24.8 PARAMAGNETISM

Transition metal ions with unpaired electron spins are paramagnetic

Paramagnetic substances are weakly attracted by a magnetic field. Any species with an unpaired electron is paramagnetic because there is a magnetic moment associated with the spinning electron. Transition metal ions that have unpaired d electrons are paramagnetic. The greater the number of unpaired electrons, the more paramagnetic is the ion. The metals, iron, cobalt and nickel are **ferromagnetic**, that is, they are strongly attracted to a magnetic field.

FIGURE 24.3 Important Oxides and Halides of the Transition Metals

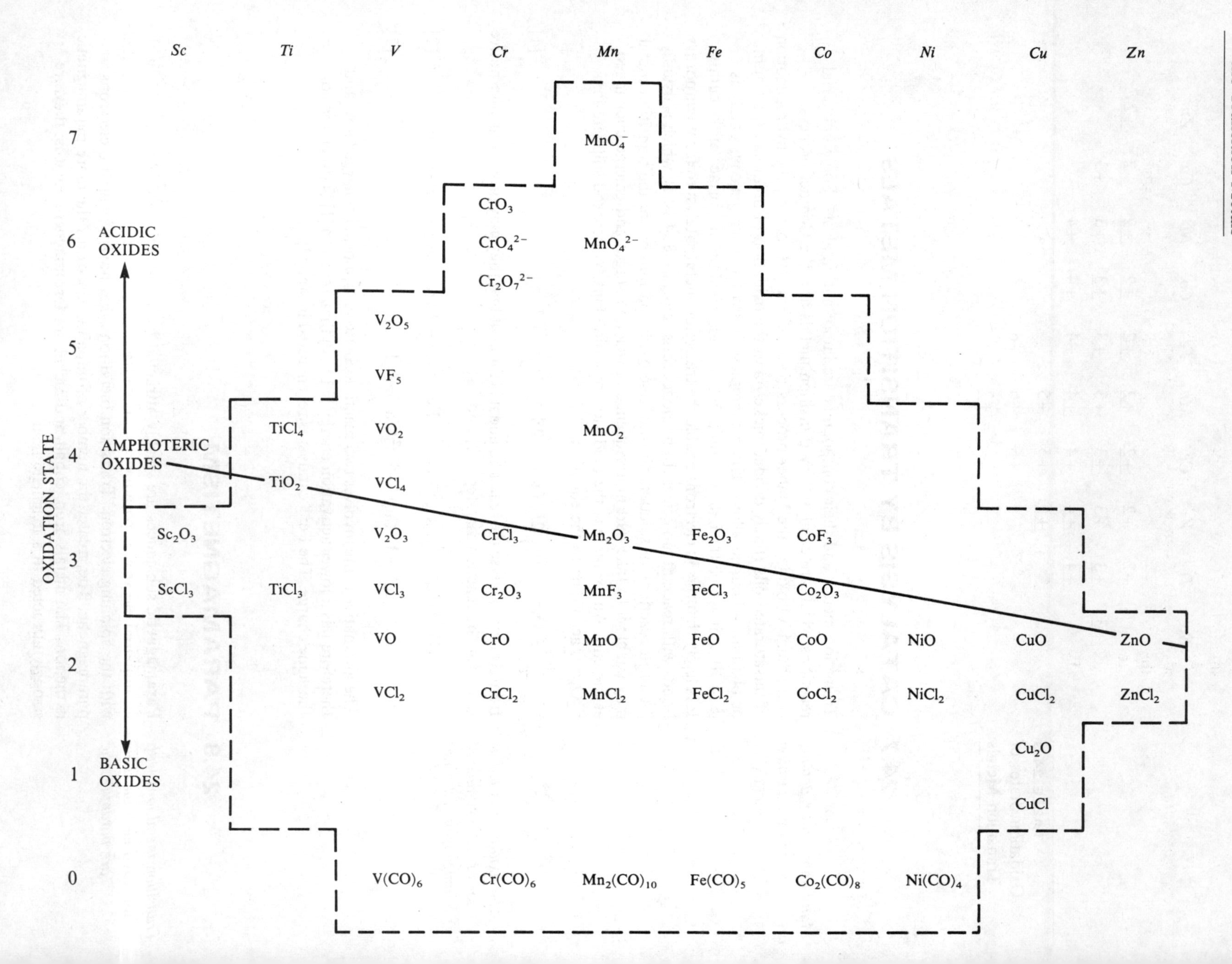

24.9 OXIDES AND HYDROXIDES OF TRANSITION METALS

The oxides are insoluble, black or coloured, with covalent character...

Transition metals react with oxygen to form oxides, with the exception of those (e.g., Ag, Au) that are low in the Electrochemical Series. The important oxides formed by the first transition series are shown in Figure 24.3. They are nearly all insoluble in water and either black or coloured. The covalent character of the bonds is appreciable.

24.9.1 BASICITY

...some are acidic, some basic, some amphoteric

The basicity of the oxides of the transition metals in an oxidation state of +2 decreases from left to right across the series. For any one metal, the basicity of the oxides decreases as the oxidation state of the metal increases. Figure 24.3 shows the rough dividing line between basic oxides and acidic oxides. Oxides below the line are basic.

24.9.2 REDUCTION OF OXIDES

The metals can be extracted from oxides

Iron

Transition metal oxides can be reduced to the metal. For the less electropositive metals (excluding Ti and V), carbon and carbon monoxide are often used as the reducing agents. In the blast furnace for the extraction of iron, carbon monoxide is the reducing agent [§§ 10.9.1 and 24.14.1]. The ore *chromite*, $FeO \cdot Cr_2O_3$, is reduced to an alloy of iron and chromium by heating it with carbon:

$$FeO \cdot Cr_2O_3(s) + 4C(s) \rightarrow Fe(s) + 2Cr(s) + 4CO(g)$$

Ferrochrome alloy

The ferrochrome alloy produced is used in the production of stainless steel.

Titanium

Titanium is mined as rutile, TiO_2 and other oxide ores which contain iron. Titanium is not extracted from titanium(IV) oxide because of the high value of the enthalpy of this reaction, 800 kJ mol^{-1}. Titanium(IV) oxide is converted into titanium(IV) chloride and this is reduced by magnesium to titanium. For the conversion,

$$TiO_2(s) + 2Cl_2(g) + 2C(s) \rightarrow TiCl_4(g) + 2CO(g)$$

the values of both ΔH and ΔS are favourable. The steps in the conversion are illustrated in Figure 24.4.

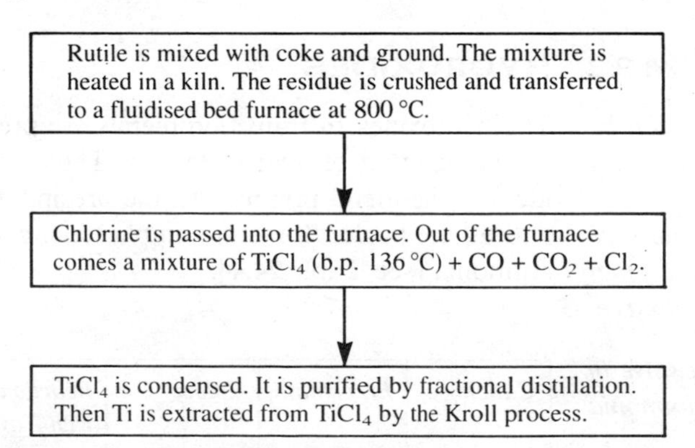

Rutile is mixed with coke and ground. The mixture is heated in a kiln. The residue is crushed and transferred to a fluidised bed furnace at 800 °C.
Chlorine is passed into the furnace. Out of the furnace comes a mixture of $TiCl_4$ (b.p. 136 °C) + CO + CO_2 + Cl_2.
$TiCl_4$ is condensed. It is purified by fractional distillation. Then Ti is extracted from $TiCl_4$ by the Kroll process.

FIGURE 24.4 The Conversion of Rutile, TiO_2, into Titanium(IV) Chloride, $TiCl_4$

The Kroll process (named after its inventor) for the reduction of titanium(IV) chloride is illustrated in Figure 24.5. The process is a **batch process**. Titanium forms inside the reactor, which must be allowed to cool so that the 'sponge' of titanium can be scraped out. The reactor is cleaned after each batch.

The titanium sponge from the reactor is purified either by leaching with acid to remove magnesium and magnesium chloride (some titanium reacts with the acid) or by distillation under reduced pressure. Then the metal is melted in an electric arc furnace and converted into ingots.

FIGURE 24.5
The Kroll Process for the Reduction of Titanium(IV) Chloride to Titanium

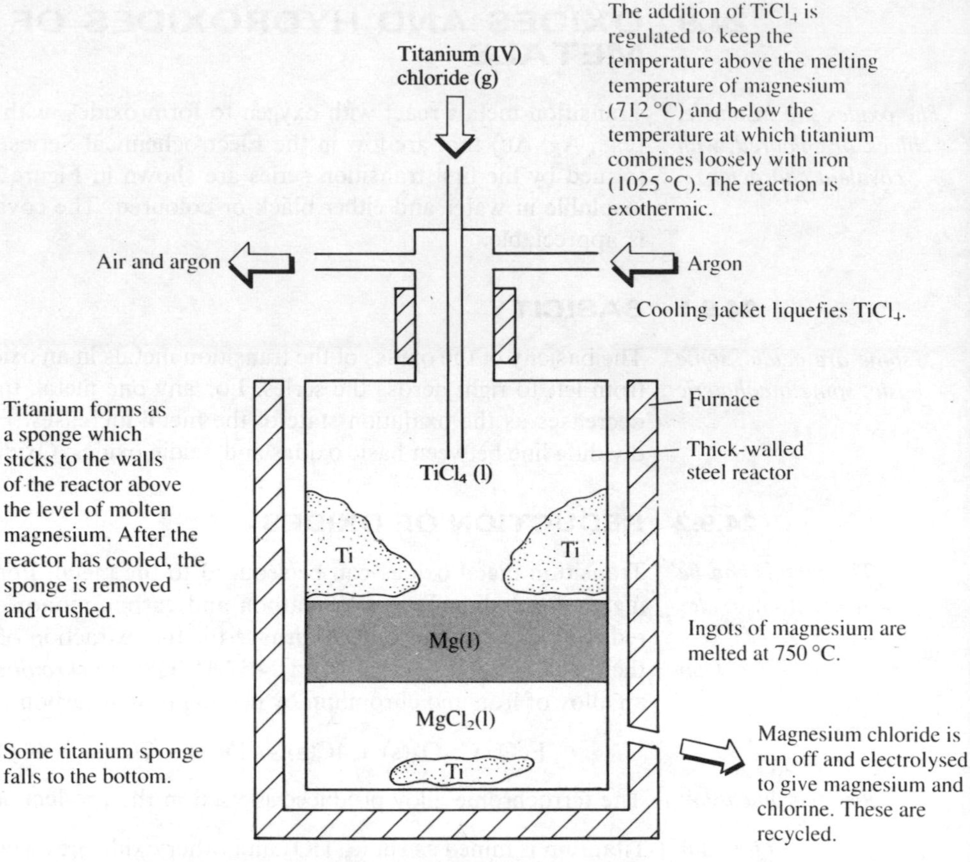
Titanium (IV) chloride (g)
The addition of $TiCl_4$ is regulated to keep the temperature above the melting temperature of magnesium (712 °C) and below the temperature at which titanium combines loosely with iron (1025 °C). The reaction is exothermic.
Air and argon
Argon
Cooling jacket liquefies $TiCl_4$.
Furnace
Thick-walled steel reactor
Titanium forms as a sponge which sticks to the walls of the reactor above the level of molten magnesium. After the reactor has cooled, the sponge is removed and crushed.
$TiCl_4$ (l)
Ti
Ti
Mg(l)
Ingots of magnesium are melted at 750 °C.
$MgCl_2$(l)
Some titanium sponge falls to the bottom.
Ti
Magnesium chloride is run off and electrolysed to give magnesium and chlorine. These are recycled.

Nickel

Nickel is obtained by reducing nickel(II) oxide with carbon.

Thermit reactions use Al to reduce metal oxides

The more electropositive elements cannot be obtained from their oxides by reduction with carbon. Aluminium is used. It reduces metal oxides with the evolution of heat in a set of reactions called thermit reactions. Chromium(III) oxide, Cr_2O_3, vanadium(V) oxide, V_2O_5 and cobalt(II) dicobalt(III) oxide, Co_3O_4, are reduced in this way:

$$Cr_2O_3(s) + 2Al(s) \rightarrow 2Cr(s) + Al_2O_3(s)$$

24.9.3 HYDROXIDES

The hydroxides can be made by precipitation

The precipitates are gelatinous and often coloured...

...Some dissolve in ammonia

The hydroxides of transition metals are precipitated from solutions of the metal ions by the addition of hydroxide ions. The colour of the precipitate can often be used to identify the metal present. All the precipitates are gelatinous, owing to hydration, and all are basic. Some are amphoteric, and some form soluble complex ions with ammonia [see Table 24.4].

TABLE 24.4
Transition Metal Hydroxides

Cation	*Precipitate*	*Colour*	*Reaction with NaOH(aq)*	*Reaction with NH_3(aq)*
Cr^{3+}(aq)	$Cr(OH)_3$	Green	Chromate(III) ion, CrO_3^{3-}(aq)	–
Mn^{2+}(aq)	$Mn(OH)_2$	Beige	–	–
Fe^{2+}(aq)	$Fe(OH)_2$	Green	–	–
Fe^{3+}(aq)	$Fe(OH)_3$	Rust	–	–
Co^{2+}(aq)	$Co(OH)_2$	Pink	Cobaltate(II) ion, $Co(OH)_4^{2-}$(aq)	$Co(NH_3)_6^{2+}$(aq)
Ni^{2+}(aq)	$Ni(OH)_2$	Green	–	$Ni(NH_3)_6^{2+}$(aq)
Cu^{2+}(aq)	$Cu(OH)_2$	Blue	–	$Cu(NH_3)_4^{2+}$(aq)
Zn^{2+}(aq)	$Zn(OH)_2$	White	Zincate ion, $Zn(OH)_4^{2-}$(aq)	$Zn(NH_3)_4^{2+}$(aq)

24.10 OXO-IONS OF TRANSITION METALS

Dichromate(VI) and manganate(VII) are powerful oxidising agents

In their higher oxidation states, the transition metals occur, not as cations M^{2+}, but combined with oxygen as anions MO_4^{3-} and MO_4^{2-}. The most important of these are the vanadate(V) ion, $V_3O_9^{3-}$; chromate(VI), CrO_4^{2-}; dichromate(VI), $Cr_2O_7^{2-}$; manganate(VI), MnO_4^{2-}; and manganate(VII), MnO_4^{-}. The sodium and potassium salts of these anions are soluble in water. Their power as oxidising agents enables them to be used in titrimetric analysis.

24.10.1 CHROMATES

FIGURE 24.6 Some Reactions of Chromium and its Compounds

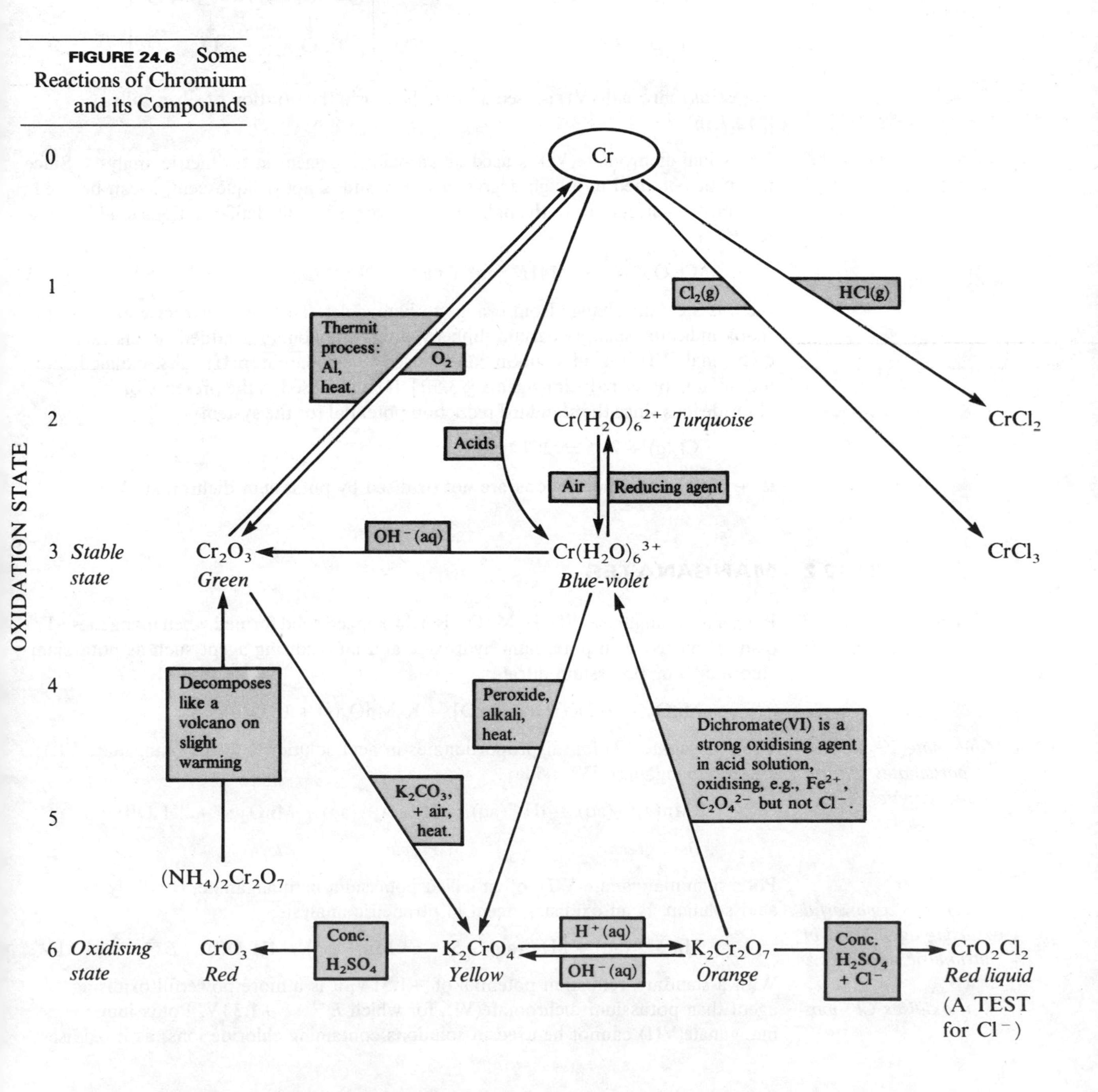

When a chromium(III) salt is heated in alkaline solution with a peroxide [see Figure 24.5], a chromate(VI) ion is formed:

The preparation of chromate(VI) and of dichromate(VI)

$$2Cr^{3+}(aq) + 4OH^-(aq) + 3O_2^{2-}(aq) \rightarrow 2CrO_4^{2-}(aq) + 2H_2O(l)$$

The Cr^{3+}(aq) ion is blue-violet; the CrO_4^{2-}(aq) ion is yellow. In acid solution, yellow chromate(VI) ions condense to form orange dichromate(VI) ions:

$$2CrO_4^{2-}(aq) + 2H^+(aq) \rightleftharpoons Cr_2O_7^{2-}(aq) + H_2O(l)$$

Yellow chromate(VI) ion Orange dichromate(VI) ion

The oxidation state of chromium is +6 in both ions, as can be seen from the structures:

$$\left[\begin{array}{c} O \\ \uparrow \\ O \leftarrow Cr \rightarrow O \\ \downarrow \\ O \end{array}\right]^{2-} \qquad \left[\begin{array}{c} O \quad\quad\; O \\ \uparrow \quad\quad\;\; \uparrow \\ O \leftarrow Cr - O - Cr \rightarrow O \\ \downarrow \quad\quad\;\; \downarrow \\ O \quad\quad\; O \end{array}\right]^{2-}$$

Potassium chromate(VI) is used as an indicator in the titration of silver salts [§ 12.7.16].

Potassium dichromate(VI) is an oxidising agent used in titrimetric analysis

Potassium dichromate(VI) is used as an oxidising agent in titrimetric analysis. Since it can be obtained in a high degree of purity and is not deliquescent, it can be used as a primary standard. In the half-reaction [see § 3.15.1 for half-reactions, and § 13.1.1 for $E^{\ominus}$]

$$Cr_2O_7^{2-}(aq) + 14H^+(aq) + 6e^- \rightarrow 2Cr^{3+}(aq) + 7H_2O(l); \quad E^{\ominus} = +1.33\,V$$

there is a colour change from orange to blue-violet. To give a sharper end-point, a redox indicator such as barium diphenylamine sulphonate is added. Potassium dichromate(VI) in acid solution can be used to estimate iron(II) salts, ethanedioates, iodides and other reducing agents [§ 3.5.1]. It can be used in the presence of chloride ions since the standard reduction potential for the system

$$Cl_2(g) + 2e^- \rightleftharpoons 2Cl^-(aq)$$

is +1.36 V, and chloride ions are not oxidised by potassium dichromate(VI).

24.10.2 MANGANATES

The preparation of manganate(VI)

Potassium manganate(VI), K_2MnO_4, is a dark green solid formed when manganese(IV) oxide is melted with potassium hydroxide and an oxidising agent such as potassium chlorate(V) or potassium nitrate:

$$MnO_2(s) + 2KOH(s) + [O] \rightarrow K_2MnO_4(s) + H_2O(l)$$

Manganate(VI) disproportionates to form manganate(VII)...

The manganate(VI) ion disproportionates in acid solution into the manganate(VII) ion and manganese(IV) oxide:

$$3MnO_4^{2-}(aq) + 4H^+(aq) \rightarrow 2MnO_4^-(aq) + MnO_2(s) + 2H_2O(l)$$

Dark green *Purple* *Black*

...this is a powerful oxidising agent used in titrimetric analysis...

Potassium manganate(VII), often called potassium permanganate, is widely used in acid solution as an oxidising agent in titrimetric analysis:

$$MnO_4^-(aq) + 8H^+(aq) + 5e^- \rightleftharpoons Mn^{2+}(aq) + 4H_2O(l); \quad E^{\ominus} = +1.51\,V$$

...it oxidises Cl^- ions

With a standard reduction potential of +1.51 V, it is a more powerful oxidising agent than potassium dichromate(VI), for which $E^{\ominus}$ = +1.33 V. Potassium manganate(VII) cannot be used in solutions containing chloride ions, as it oxidises them to chlorine. Solutions of potassium manganate(VII) are kept in brown bottles because in the presence of light they slowly oxidise water to oxygen. Substances which can be estimated by titration against acidified potassium manganate(VII) are iron(II) salts, hydrogen peroxide and ethanedioates [§ 3.15.1].

FIGURE 24.7 Some Reactions of Manganese and its Compounds

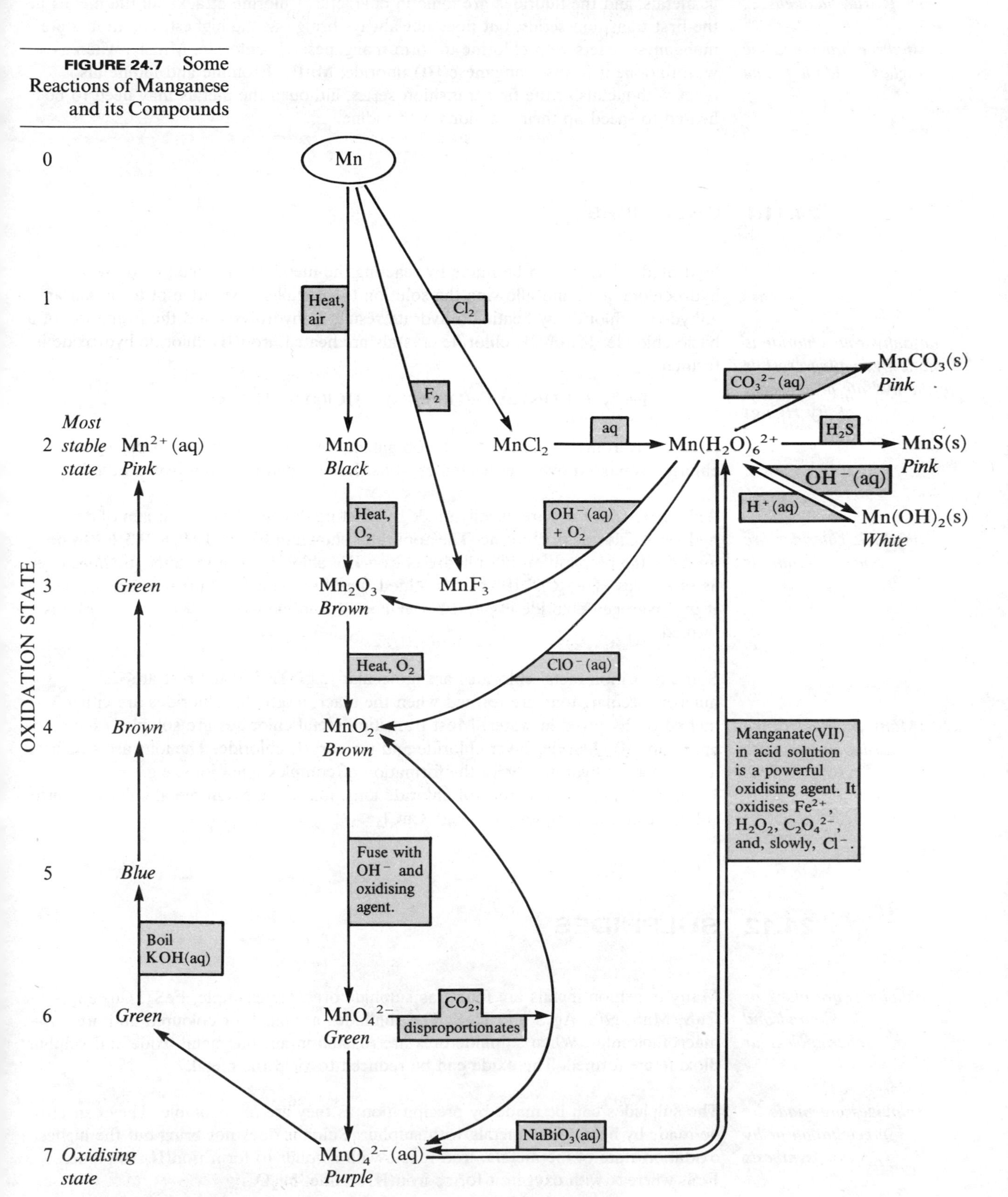

24.11 HALIDES

The reactions of transition metals with halogens...

...fluorine brings out the highest oxidation state

Transition metals react with the halogens. Fluorine brings out high oxidation states in metals, and the fluorides are ionic in character. Chlorine attacks all the metals in the first transition series, but does not always bring out the highest oxidation state: manganese reacts with chlorine to form manganese(II) chloride, $MnCl_2$, whereas with fluorine it forms manganese(III) fluoride, MnF_3. Bromine and iodine also react with metals of the first transition series, although the metals may need to be heated to speed up their reactions with iodine.

24.11.1 CHLORIDES

The anhydrous chloride is made by heating crystals of the hydrate in a stream of dry HCl(g)

Hydrated chlorides can be made by reacting the metal or the metal oxide with hydrochloric acid and allowing the solution to crystallise. An attempt to obtain an anhydrous chloride by heating a hydrate results in hydrolysis and the formation of a basic chloride. If iron(II) chloride crystals are heated, iron(II) chloride hydroxide is formed:

$$FeCl_2 \cdot 6H_2O(s) \rightleftharpoons Fe(OH)Cl(s) + HCl(g) + 5H_2O(g)$$

Anhydrous iron(II) chloride can be obtained, however, if a stream of dry hydrogen chloride is passed over the heated crystals. Equilibrium is driven over to the left.

Anhydrous chlorides are made by synthesis

Anhydrous chlorides are usually made by reacting the metal with a stream of dry hydrogen chloride or chlorine. The apparatus shown in Figure 19.9, § 19.3.1, can be used for the preparation of anhydrous iron(III) chloride. The product sublimes over as molecules of Fe_2Cl_6. If water is added, Fe^{3+}(aq) and Cl^-(aq) ions are formed. If dry hydrogen chloride is passed over heated iron, anhydrous iron(II) chloride is formed.

Many chlorides of the transition metals are soluble

Some transition metal chlorides are molecular (e.g., $TiCl_4$); the rest are macromolecular. Ions are formed when the macromolecular chlorides are either melted or dissolved in water. Most transition metal chlorides are soluble; exceptions are copper(I) chloride, silver chloride and mercury(I) chloride. The ions are stabilised in solution by hydration with the formation of complex aqua ions, e.g., $Fe(H_2O)_6{}^{3+}$. In the presence of chloride ions, many transition metal chlorides form soluble chloride complex ions, e.g., $CuCl_4{}^{2-}$.

24.12 SULPHIDES

Sulphides are black or coloured and macromolecular

Many transition metals are found as sulphide ores, for example, FeS_2, $CuFeS_2$, CuS, MnS, NiS, Ag_2S and HgS. The sulphides are black or coloured and are macromolecular. When sulphide ores are roasted in air, the metal oxide and sulphur dioxide are formed. The oxide can be reduced to yield the metal.

Sulphides are made by precipitation or by synthesis

The sulphides can be made by precipitation as they are all insoluble. They can also be made by heating the metals with sulphur. Sulphur does not bring out the highest oxidation state of the metals: iron reacts with sulphur to form iron(II) sulphide, FeS, whereas with oxygen it forms iron(III) oxide, Fe_2O_3.

24.13 COMPLEX COMPOUNDS

Transition metal ions form complex ions by coordination

Transition metals form complexes or coordination compounds. Such complexes are formed by the coordination of lone pairs of electrons from a donor (called a **ligand**) to an atom or cation (called an **acceptor**) which has empty orbitals to accommodate them. A cation may form a complex with a neutral molecule, e.g.

$$[Cu(NH_3)_4]^{2+}$$

or with an oppositely charged ion, e.g.

$$[CuCl_4]^{2-}$$

An atom may form a complex, e.g., $Ni(CO)_4$. The charge remaining on the central atom or ion when the ligands are removed together with their lone pairs is the **oxidation number** of the metal in the complex. The **coordination number** is the number of atoms forming coordinate bonds with the central atom or ion: 2, 4 and 6 are common.

A ligand shares a lone pair of electrons with a transition metal ion

Ligands must possess one or more unshared pairs of electrons. A ligand which can only form one bond to a central atom or ion is called a **monodentate** (literally, 'one tooth') ligand, e.g., NH_3, CN^-. A **polydentate** ('many teeth') ligand can form more than one bond. Examples are the ethanedioate ion

Polydentate ligands form chelates...

$$\begin{array}{l} CO_2^- \\ | \\ CO_2^- \end{array}$$

and ethane-1,2-diamine $H_2\ddot{N}CH_2CH_2\ddot{N}H_2$

in which the lone pairs on the two nitrogen atoms can form coordinate bonds. The two coordinate bonds formed by each ligand are thought to resemble the claws of a crab (Greek: *chele*), and such compounds are named **chelate compounds** or **chelates** [see Figure 24.14, § 24.13.6]. In the polydentate ligand, 1,2-bis[bis(carboxymethyl)-amino]ethane

...edta is an important ligand

$$\begin{array}{ccc} {}^-O_2C{-}H_2C & & CH_2{-}CO_2^- \\ & N{-}CH_2{-}CH_2{-}N & \\ {}^-O_2C{-}H_2C & & CH_2{-}CO_2^- \end{array}$$

which is called **edta** (short for its old name), there are unshared electron pairs on four oxygen atoms and on the two nitrogen atoms. This ligand forms six coordinate bonds, and its complex ions are very stable. Zinc ions and most other metal ions can be estimated by complexometric titration against a solution of edta [§ 12.7.17].

24.13.1 NAMING

Formula gives central atom followed by ligands

In the formula of a complex ion, the symbol for the central atom appears first and is followed by the anionic ligands and then by neutral ligands, e.g.

$$[CoCl_2(NH_3)_4]^+$$

The formula for the complex ion may be enclosed in square brackets.

The name of the complex gives the name and oxidation state of the central metal cation, e.g., cobalt(III), preceded by the name and number of ligands attached to it, e.g., hexaamminecobalt(III) ion

$$[Co(NH_3)_6]^{3+}$$

The prefixes

di tri tetra penta hexa

are used to show the number of ligands. If several ligands are present, they are listed in alphabetical order, and the prefixes, di, tri, etc., are not allowed to alter this order, e.g.

$[CrCl_2(H_2O)_4]^+$

Tetraaquadichlorochromium(III) ion

The system for naming complex ions

If the complex is an anion, the suffix -ate follows the name of the metal, e.g., zincate and chromate. If the metal has a Latin name, then in the complex anion the Latin name of the metal is used, followed by the suffix -ate, e.g.

$[Fe(CN)_6]^{4-}$

Hexacyanoferrate(II)

Examples are given in Table 24.5.

TABLE 24.5
Complex Ions

Ligand	*Type of complex*	*Example*	*Name of Complex*
Water	*Aqua-*	$[Cr(H_2O)_6]^{3+}$	Hexaaquachromium(III) ion
Ammonia	*Ammine-*	$[Ag(NH_3)_2]^+$	Diamminesilver(I) ion
Hydroxide ion	*Hydroxo-*	$[Zn(OH)_4]^{2-}$	Tetrahydroxozincate(II) ion
Chloride ion	*Chloro-*	$[CuCl_4]^{2-}$	Tetrachlorocuprate(II) ion
Cyanide ion	*Cyano-*	$[Fe(CN)_6]^{3-}$	Hexacyanoferrate(III) ion
Nitrite ion	*Nitro-*	$[Co(NO_2)_6]^{3-}$	Hexanitrocobaltate(III) ion
Carbon monoxide	*Carbonyl-*	$Ni(CO)_4$	Tetracarbonylnickel(0)
Ethane-1,2-diamine	*Ethane-1,2-diamine-*	$[Cr(en)_3]^{3+}$	Tris(ethane-1,2-diamine)-chromium(III) ion
edta	*edta-*	$[Zn(edta)]^{2-}$	edtazincate(II) ion
Cl^-, NH_3	*Mixed*	$[CoCl_2(NH_3)_4]^+$	Tetraamminedichloro-cobalt(III) ion
OH^-, H_2O	*Mixed*	$[Fe(OH)_2(H_2O)_4]^+$	Tetraaquadi-hydroxoiron(III) ion

24.13.2 COLOUR

Transition metal ions are often coloured because electrons move between non-degenerate d orbitals

Transition metal ions are often coloured [see vanadium, Figure 24.8]. In an isolated transition metal atom, the five d orbitals are **degenerate**, that is, they are all at the same energy level. In a complex ion, the d orbitals differ slightly in energy as a result of overlapping differently with the ligands: they are **non-degenerate**. Electrons can jump from one d orbital to another if they absorb energy. For most transition metal complexes, the frequency of light absorbed in these energy transitions is in the visible region of the spectrum, and the ion appears coloured. The colour of the ion is complementary to the colours absorbed. The $Sc^{3+}(aq)$ ion has no d electrons, and is colourless. In the ions Cu^+ and Zn^{2+}, with a d^{10} configuration, no d–d transition is possible, and these ions are colourless. Different ligands affect the energy levels of the d orbitals: $[Cu(H_2O)_4]^{2+}$ is blue, whereas $[Cu(NH_3)_4]^{2+}$ is a very intense deep blue.

FIGURE 24.8 *Some Reactions of Vanadium

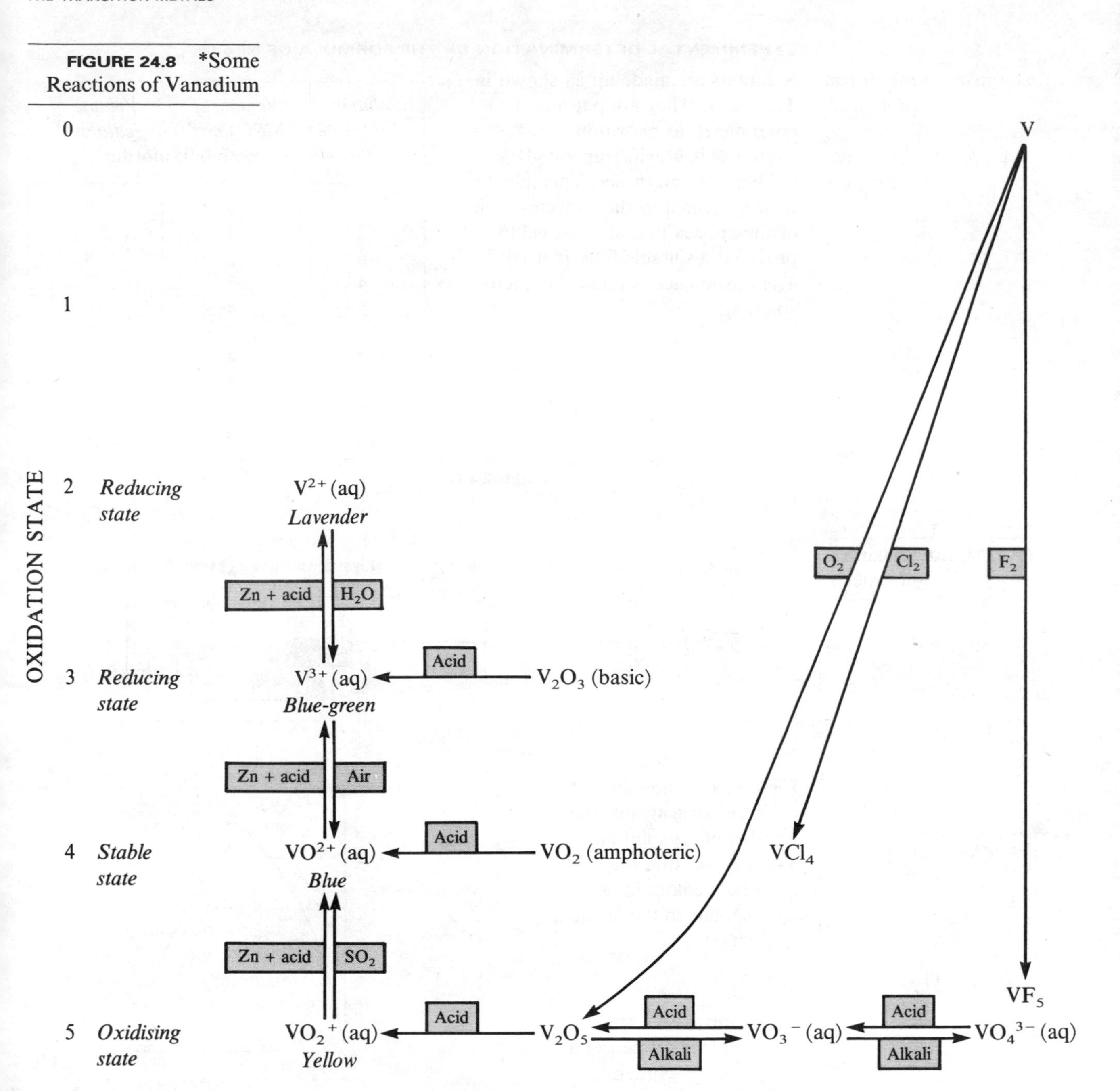

24.13.3 STOICHIOMETRY OF COMPLEX IONS

Colorimetric methods can be used to find the formula of a complex ion . . .

Colorimetric methods are also used for finding the composition of complex ions. When solutions of nickel(II) sulphate and edta are mixed, coloured complex ions are formed. If the formula of the ion is

$$Ni(edta)_n^{x-}$$

then the maximum intensity of colour will be obtained when the two solutions are mixed to give a molar ratio of

$$Ni : edta = 1 : n$$

EXPERIMENTAL DETERMINATION OF THE FORMULA OF $Ni(edta)_n^{x-}$

Colorimetric methods can be used...

...e.g., for the Ni^{2+} edta complex

Solutions are made up as shown in Table 24.6. They are put into a colorimeter, as shown in Figure 24.9. The instrument gives readings of *absorbance*. This quantity is proportional to the concentration of the species that absorbs light, provided a suitable filter is used. The absorbance is found for each solution.

TABLE 24.6

Solution	*Volume of $NiSO_4$*/cm^3 (both 0.05 mol dm^{-3})	*Volume of edta*/cm^3
1	0	10
2	1	9
3	2	8
4	3	7
5	4	6
6	5	5
7	6	4
8	7	3
9	8	2
10	9	1
11	10	0

FIGURE 24.9 Using a Colorimeter

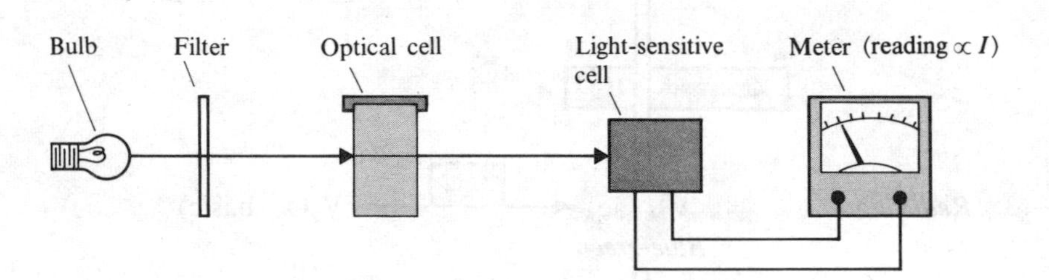

Figure 24.10 shows a plot of absorbance against the number of the solution. You can see that the maximum colour intensity corresponds to the formula $[Ni(edta)]^{2-}$.

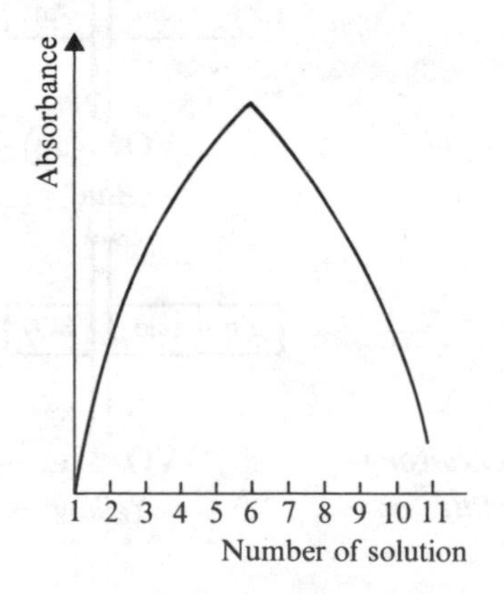

FIGURE 24.10 Results of Colorimetry on $Ni(edta)_n^{x-}$

24.13.4 STABILITY OF COMPLEX IONS

The dissociation constants of complex ions are covered in § 12.7.17.

24.13.5 STEREOCHEMISTRY

The spatial arrangement of the bonds from the ligands to the central atom or ion depends on the identity of the atomic orbitals used by the central atom or ion. The orientation of bond orbitals has been covered in § 5.1, summarised in § 5.1.6 and extended in § 5.2. It is interesting to look at a selection of the many complex ions formed by transition metals.

24.13.6 *COMPLEXES OF CHROMIUM(III)

$Cr^{3+}(aq)$...

...and complex formation

In aqueous solution, chromium(III) ions exist as blue-violet hexaaquachromium(III) ions, $Cr(H_2O)_6^{3+}$. These ions are present in *chrome alum*, $KCr(SO_4)_2 \cdot 12H_2O$. Water molecules can be replaced by other ligands if these form more stable complexes. In ammonia solution, ammonia molecules replace water molecules to form hexaamminechromium(III) ions, $Cr(NH_3)_6^{3+}$. The octahedral arrangement of ligands is shown in Figure 24.11.

FIGURE 24.11 Hexaammine-chromium(III) Ion

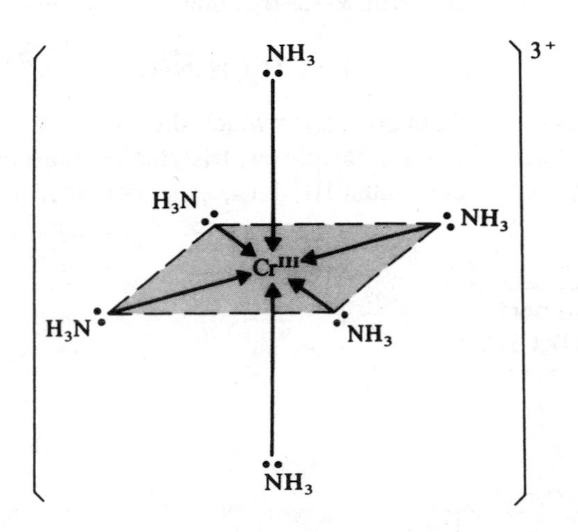

If the six ligands of the octahedron are not identical, isomerism occurs

If the six ligands are not identical, *cis-trans* geometrical isomerism will occur. In the tetraamminedichlorochromium(III) ions, shown in Figure 24.12, the *cis*-form (a) has two chlorine ligands adjacent, whereas in the *trans*-form (b), the chlorine atoms are diagonally opposite.

FIGURE 24.12 *Cis-* and *Trans*-tetraammine-dichlorochromium(III) ions

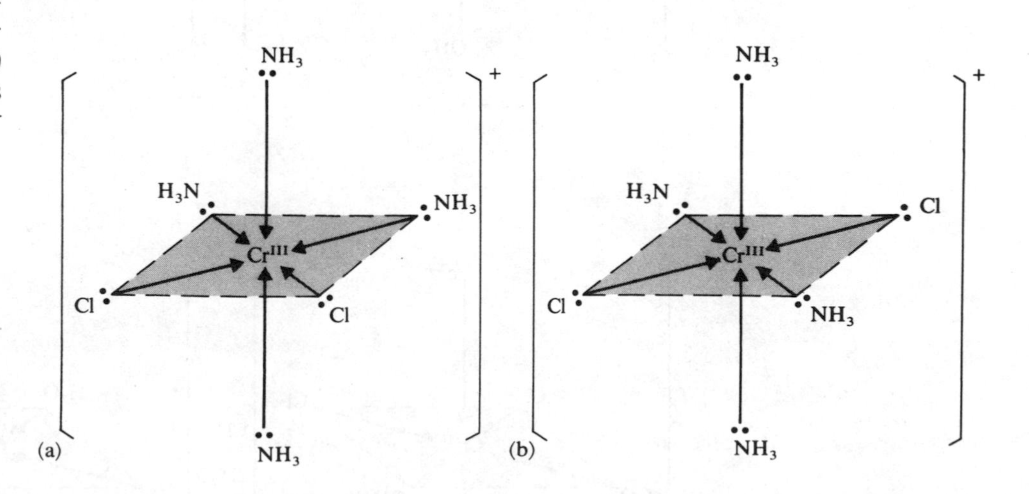

The complex salt $Cr(H_2O)_6Cl_3$ has four isomers. There are three structural isomers, which differ in electrical conductivity and in the fraction of the chloride content that can be precipitated by silver nitrate solution. These three salts are

(*a*) $[Cr(H_2O)_6]^{3+} \cdot 3Cl^-$ *Grey-blue*

(*b*) $[Cr(H_2O)_5Cl]^{2+} \cdot 2Cl^- \cdot H_2O$ *Pale green*

(*c*) $[Cr(H_2O)_4Cl_2]^+ \cdot Cl^- \cdot 2H_2O$ *Green*

Isomer (c) exists as *cis*- and *trans*- geometrical isomers. See Figure 24.13 and Questions 1, 4 and 10, Checkpoint 24A.

Bidentate ligands, such as the ethanedioate ion

$$^-O_2C—CO_2{}^-$$

and ethane-1,2-diamine

$$H_2\ddot{N}CH_2CH_2\ddot{N}H_2$$

Optical isomerism occurs in complexes with bidentate ligands

form complexes which show optical isomerism. Figure 24.14 shows how the two chelate complexes, tris(ethanedioato)chromium(III) and tris(ethane-1,2-diamine)-chromium(III) ions, exist as mirror image forms or **enantiomers**.

FIGURE 24.13 Isomers of $Cr(H_2O)_6Cl_3$

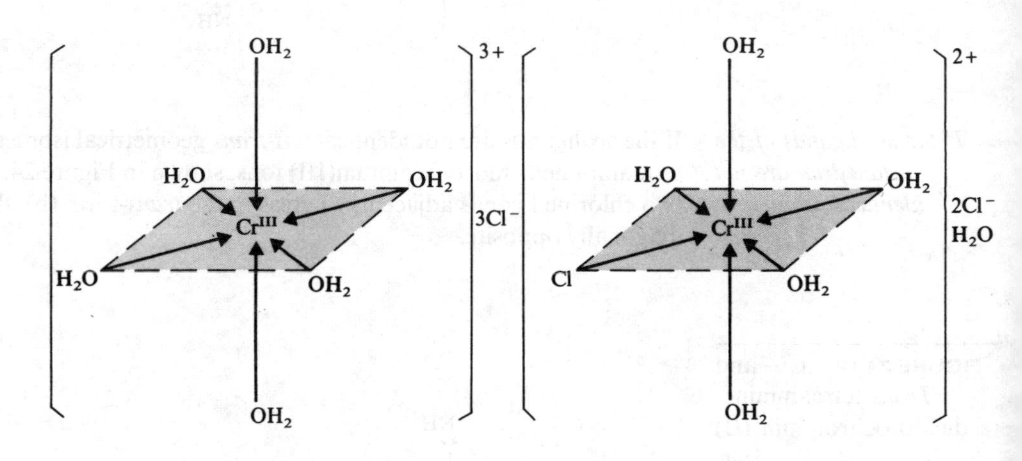

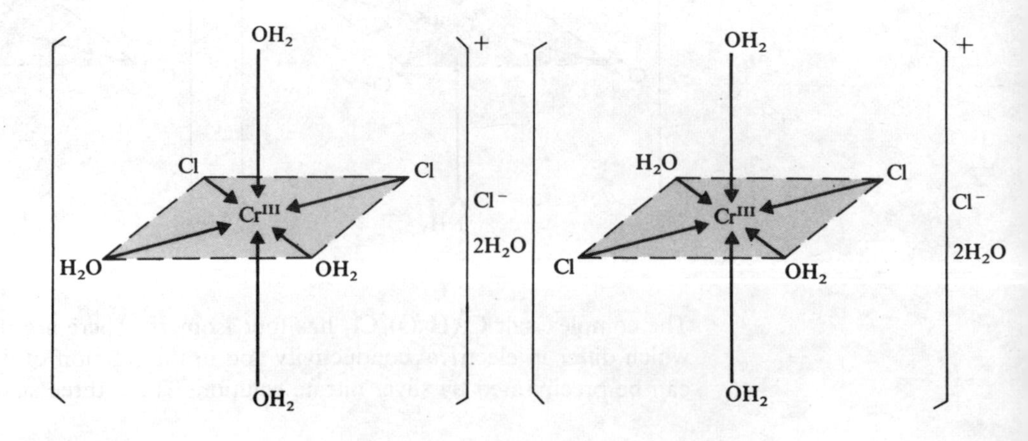

FIGURE 24.14
Enantiomers of $[Cr(C_2O_4)_3]^{3-}$ and $[Cr(C_2N_2H_8)_3]^{3+}$

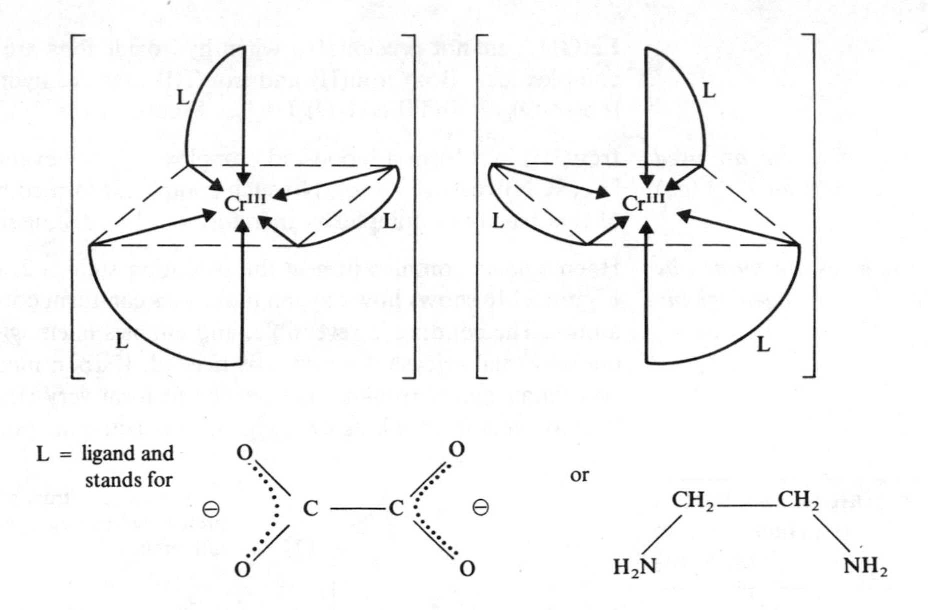

24.13.7 COMPLEXES OF COBALT(II) AND COBALT(III)

Cobalt(III) ions form complexes with various ligands

Cobalt(III) ions form complex ions with the ligands H_2O, NH_3, Cl^-, $C_2O_4^{2-}$, $H_2NCH_2CH_2NH_2$, CN^-, NO_2^- and many others. The complex ions are similar to those of chromium(III). Cobalt(II) ions are more stable than cobalt(III) ions, with the result that Co^{3+} ions are powerful oxidising agents. Some cobalt(II) complex ions are:

$$\underset{brown}{Co(NH_3)_6^{2+}} \underset{H_2O}{\overset{NH_3}{\rightleftharpoons}} \underset{pink}{Co(H_2O)_6^{2+}} \underset{H_2O}{\overset{Cl^-}{\rightleftharpoons}} \underset{blue}{CoCl_4^{2-}}$$

24.13.8 IRON COMPLEXES

Complexes are formed by Fe^{2+} and Fe^{3+} with CN^- and H_2O

Iron(II) and iron(III) ions employ six hybrid bonds to form the complexes hexacyanoferrate(II), $[Fe(CN)_6]^{4-}$, and hexacyanoferrate(III), $[Fe(CN)_6]^{3-}$ [see Figure 24.15]. So stable are these complexes that the hydroxides $Fe(OH)_2$ and

FIGURE 24.15
(a) Hexacyanoferrate(II) Ion
(b) Hexaaquairon(III) Ion

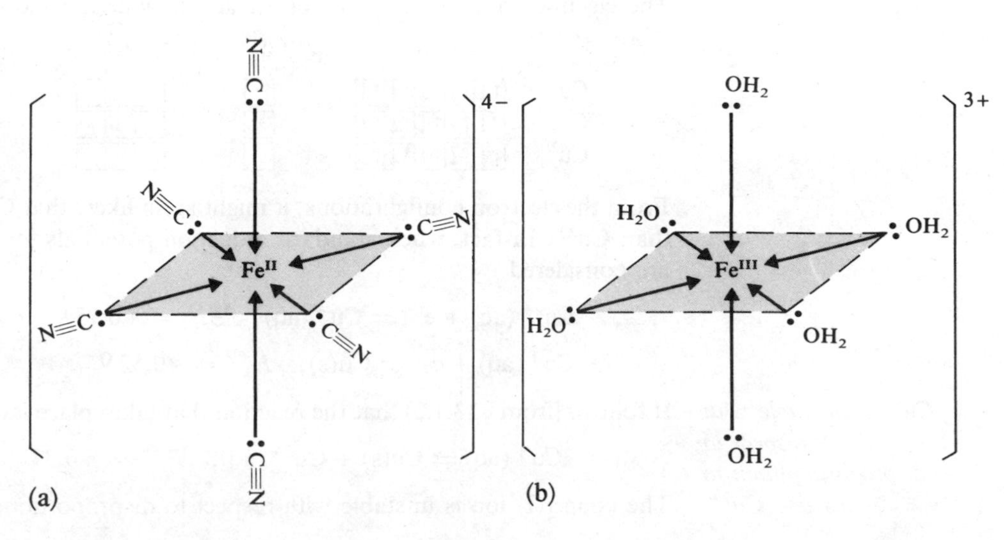

$Fe(OH)_3$ are not precipitated when hydroxide ions are added to solutions of the complex ions. Both iron(II) and iron(III) ions are hydrated in solution as $[Fe(H_2O)_6]^{2+}$ and $[Fe(H_2O)_6]^{3+}$ [see Figure 24.15].

Thiocyanate ions are used in a test for $Fe^{3+}(aq)$

Iron(III) ions form a blood-red complex with thiocyanate ions, $[Fe(SCN)(H_2O)_5]^{2+}(aq)$. No such complex is formed by iron(II) ions. The formation of this blood-red complex is therefore used to distinguish iron(III) from iron(II).

Complexes are formed by haemoglobin

Haemoglobin contains iron in the oxidation state +2, with coordination number 6. Figure 24.16 shows how oxygen molecules can form coordinate bonds to the iron atoms. The bonding is reversible, and enables haemoglobin to carry oxygen around the body and release it where it is needed. Carbon monoxide and cyanide ions coordinate more strongly than oxygen to form very stable complexes. They prevent haemoglobin from taking up oxygen, thus acting as poisons.

FIGURE 24.16 Part of the Haemoglobin Molecule

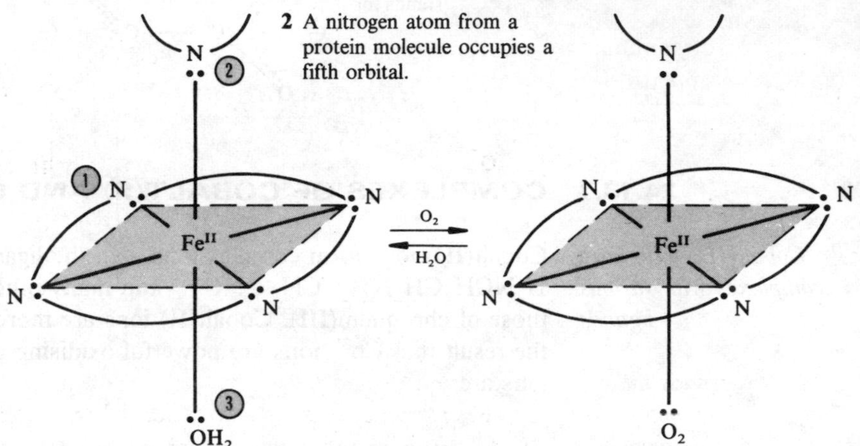

2 A nitrogen atom from a protein molecule occupies a fifth orbital.
1 Four N atoms in a ring structure occupy four of the Fe orbitals.
N
N
N
N
N
Fe^{II}
O_2
H_2O
OH_2
O_2
3 The sixth orbital enables O_2 to be bonded reversibly. CO and CN^- can form stable complexes by coordinating into this orbital. They then prevent the uptake of O_2.

24.13.9 COMPLEXES OF COPPER(I) AND COPPER(II) IONS

The electron configurations of copper and its ions are shown below:

	3d	4s	4p
Cu	↑↓ ↑↓ ↑↓ ↑↓ ↑↓	↑	☐☐☐
Cu^+	↑↓ ↑↓ ↑↓ ↑↓ ↑↓	☐	☐☐☐
Cu^{2+}	↑↓ ↑↓ ↑↓ ↑↓ ↑	☐	☐☐☐

From the electron configurations, it might seem likely that Cu^+ would be more stable than Cu^{2+}. In fact, when standard reduction potentials for the possible reactions are considered

$$Cu^{2+}(aq) + e^- \rightleftharpoons Cu^+(aq); \quad E^\ominus = +0.15\,V$$

$$Cu^+(aq) + e^- \rightleftharpoons Cu(s); \quad E^\ominus = +0.52\,V$$

Cu^+ is unstable with respect to disproportionation to $Cu + Cu^{2+}$

It follows [from § 13.1.2] that the reaction that takes place is

$$2Cu^+(aq) \rightleftharpoons Cu(s) + Cu^{2+}(aq); \quad E^\ominus = +0.37\,V$$

The copper(I) ion is unstable with respect to disproportionation to copper and the copper(II) ion. Copper(I) ions can be stabilised by the formation of complexes such as

$$[CuCl_2]^- \quad [Cu(CN)_4]^{3-} \quad [Cu(NH_3)_2]^+$$

Cu^{2+} uses four orbitals for complex formation

Copper(II) ions use four bonds in complex formation. The spatial arrangement of the bonds in the hydrated $[Cu(H_2O)_4]^{2+}$ ion is square planar. In solution, two water molecules are loosely coordinated at right angles [see Figure 24.17(a)] in a distorted octahedral configuration. Ammonia displaces water molecules from the pale blue tetraaquacopper(II) ions, converting them into the deep blue tetraamminecopper(II) ions, $[Cu(NH_3)_4]^{2+}$ [see Figure 24.17(b)].

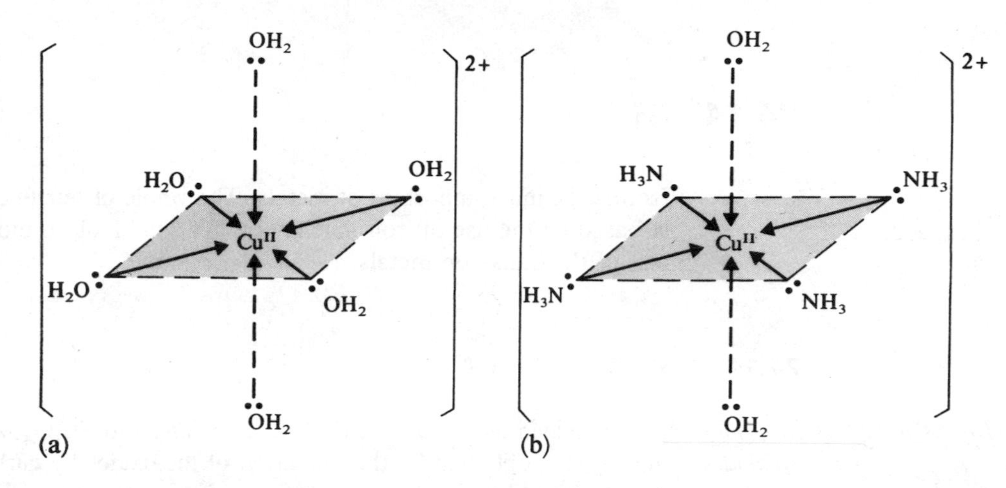

FIGURE 24.17
(a) Tetraaquacopper(II) Ion
(b) Tetraammine-copper(ii) Ion

The six ligands take up an octahedral arrangement

In concentrated hydrochloric acid, water molecules in $[Cu(H_2O)_4]^{2+}$ are replaced by chloride ions, with the formation of yellow tetrachlorocuprate(II) ions, $[CuCl_4]^{2-}$:

$$\underset{blue}{[Cu(H_2O)_4]^{2+}(aq)} + 4Cl^-(aq) \rightleftharpoons \underset{yellow}{[CuCl_4]^{2-}(aq)} + 4H_2O(l)$$

Cl^- and edta complex with Cu^{2+}

In more dilute hydrochloric acid, the solution is green due to the presence of both blue $[Cu(H_2O)_4]^{2+}$ ions and yellow $[CuCl_4]^{2-}$ ions. Copper(II) ions also form complexes with ethane-1,2-diamine and with edta. Copper(II) is the most strongly complexing of all the first long period dipositive cations.

CHECKPOINT 24A: COMPLEX IONS

1. Name the four complex ions shown in Figure 24.13.

2. Name the complex ion $[CrCl_2(H_2NCH_2CH_2NH_2)_2]^+$. Sketch the two geometrical isomers which have this formula. Which of the two is optically active?

3. Sketch the arrangement of bonds in the complex ions
(*a*) hexaaquacobalt(III)
(*b*) hexaamminecobalt(III)
(*c*) hexafluorocobalt(III)
(*d*) tris(1,2-diaminoethane)cobalt(III)
(*e*) tris(ethanedioato)cobalt(III).

***4.** Explain why the ion $[Co(H_2NCH_2CH_2NH_2)_3]^{3+}$ exists in enantiomeric forms.

***5.** How many isomers exist of the complex ion $[Co(H_2NCH_2CH_2NH_2)_2Cl_2]^+$?

6. The addition of ammonia solution to an aqueous solution of copper(II) sulphate resulted in the formation of a pale blue precipitate. This dissolved on the addition of more ammonia solution to give a deep blue solution. The precipitate also dissolved in dilute hydrochloric acid to give a pale blue solution and in concentrated hydrochloric acid to give a green solution. Explain these colour changes, giving the name and formula of each coloured species formed.

7. Two isomers exist of formula $CrCl_2(NO_2)(NH_3)_4$. One gives a precipitate with silver nitrate solution; the other does not. Suggest how this difference in behaviour can arise.

***8.** Name the complex ion $[PtCl_2(NH_3)_4]^{2+}$. Sketch and name the two isomeric forms of this ion.

***9.** There are two isomers of $Pt(NH_3)_2Cl_2$. What does this tell you about the geometry of the compound?

***10.** How many isomers exist of the following compounds?
(*a*) $[Co(NH_3)_6]^{3+}$
(*b*) $[Co(NH_3)_5Cl]^{2+}$
(*c*) $[Co(NH_3)_4Cl_2]^+$
(*d*) $Co(NH_3)_3Cl_3$

***11.** There are five compounds containing platinum(IV) chloride and ammonia. Werner deduced their formulae from the following information. See whether you can do so. The five compounds are:

Compound	*Empirical formula*	*Number of ions*	*Number of Cl^- ions*
1	$PtCl_4 \cdot 6NH_3$	5	4
2	$PtCl_4 \cdot 5NH_3$	4	3
3	$PtCl_4 \cdot 4NH_3$	3	2
4	$PtCl_4 \cdot 3NH_3$	2	1
5	$PtCl_4 \cdot 2NH_3$	0	0

24.14 IRON

Iron is the most important of metals. The whole of our twentieth-century way of life is based on the use of iron machinery. We shall look in more detail at iron than the rest of the transition metals.

24.14.1 EXTRACTION

Iron is mined as its oxides and sulphides

Iron is mined as its oxides, Fe_2O_3, *haematite*, and Fe_3O_4, *magnetite*, and FeS_2, *iron pyrites*. It is obtained by the reduction of the oxides by carbon monoxide [for the theory, see § 10.9.1]:

$$Fe_2O_3(s) + 3CO(g) \rightarrow 2Fe(s) + 3CO_2(g); \quad \Delta H^\ominus = -27\,\text{kJ mol}^{-1}$$

The **blast furnace**, in which this process takes place, is illustrated in Figure 24.18. At the bottom of the furnace, coke is oxidised exothermically to carbon dioxide:

$$C(s) + O_2(g) \rightarrow CO_2(g); \quad \Delta H^\ominus = -392\,\text{kJ mol}^{-1}$$

and the temperature of the furnace is about 1900 °C in this region. Higher up the furnace, carbon dioxide reacts with coke to form carbon monoxide. This reaction is endothermic, and the furnace in this region has a temperature of 1100 °C:

$$CO_2(g) + C(s) \rightarrow 2CO(g); \quad \Delta H^\ominus = 172\,\text{kJ mol}^{-1}$$

Carbon monoxide reduces iron oxides in the blast furnace

Iron oxides are reduced exothermically, and the iron produced falls to the bottom of the furnace, where the temperature is high enough to melt it, and a layer of molten iron lies on the bottom of the furnace. At the same time, the limestone in the charge dissociates to form calcium oxide and carbon dioxide:

$$CaCO_3(s) \rightleftharpoons CaO(s) + CO_2(g); \quad \Delta H^\ominus = 178\,\text{kJ mol}^{-1}$$

Iron oxide + coke + limestone produce iron + 'slag'

Calcium oxide combines with silicon(IV) oxide and aluminium oxide, the impurities in the ore, to form a molten 'slag' of calcium silicate(IV) and calcium aluminate(III), which trickles down the stack:

$$CaO(s) + SiO_2(s) \rightarrow CaSiO_3(l)$$

$$CaO(s) + Al_2O_3(s) \rightarrow CaAl_2O_4(l)$$

At the bottom of the furnace, iron and slag are tapped off every few hours. A modern furnace makes 3000 tonnes of iron in a day, using 3000 tonnes of coke and 4000 tonnes of air. If natural gas is injected with the hot air, the consumption of coke can be halved. The quantity of slag produced is about 1 tonne for every tonne of iron. It is used for road making and in the manufacture of cement. A furnace can operate continuously for several years before it needs relining.

FIGURE 24.18(a) The Blast Furnace

2 The skip discharges its load.

3 The *double-bell* charging system prevents the escape of gases from the furnace. The small bell is lowered to let the charge fall on to the large bell, and then raised. The large bell is lowered to allow the charge to fall into the furnace.

1 A *skip* is loaded with ore, coke and limestone.

4 The *downcomer* takes away exhaust gases to heat the air in 6.

5 The *stack*, a tower of steel plates, lined with heat-resistant bricks, is 30 m high.

700 °C

$CaCO_3 \rightarrow CaO$

$Fe_2O_3 \rightarrow Fe$

1100 °C

$CO_2 \rightarrow CO$

1900 °C

$C \rightarrow CO_2$

6 Hot blasts of air enter through narrow pipes called *tuyeres*, leading from this circular pipe.

7 Molten iron is tapped off into a *ladle*.

8 Slag is run off.

FIGURE 24.18(b) Blast Furnace, Llanwern Works, Newport, Gwent

24.14.2 CAST IRON

Cast iron or pig iron contains carbon, which decreases its ductility

The iron leaving the blast furnace is run into moulds, where it forms solid blocks called 'pigs'. This *pig iron* or *cast iron* contains about 4% carbon. The carbon present lowers the melting temperature of the iron, increases its hardness and decreases its ductility. Strength increases up to 1% carbon and then decreases. If the carbon is removed, a malleable iron is produced, which can be readily worked by ironsmiths. It is called *wrought iron* [see Figure 24.19].

Wrought iron is the purest form of iron

Some of the iron leaving the blast furnace is not cast; it is run into giant crucibles and transported to the steel-making section of the iron and steel works. Steels contain less than 1.5% carbon, and contain added metals. Many thousands of different steels are made, with different properties suited to different uses. Some are shown in Table 24.7.

Steels contain carbon and other metals alloyed with iron

Carbon steels	*Composition/% carbon*	*Uses*
Low carbon Medium carbon High carbon	<0.3 0.3–0.7 0.7–1.5	Boiler plates Motor vehicles Cutting tools, girders
Alloy steels	*Properties*	*Uses*
Titanium steel	Withstands high temperatures	Gas turbines, spacecraft
Tungsten steel	High melting, tough	High speed tools
Chromium steel	Hard	Ball bearings
Cobalt steel	High magnetic permeability	Magnets
Manganese steel	Tough	Earth-moving machinery
Stainless steel (nickel + chromium)	Non-rusting	Cutlery, car accessories, sinks Food, pharmaceutical and chemical industries

TABLE 24.7 Some Different Steels

FIGURE 24.19 Wrought Iron

FIGURE 24.20(a) The LD Process of Steel Making

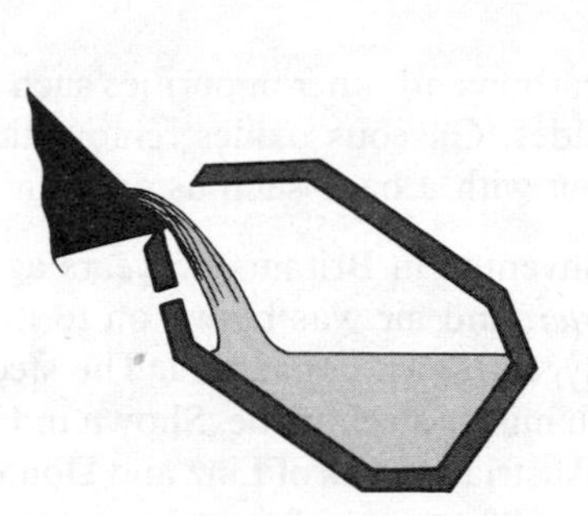

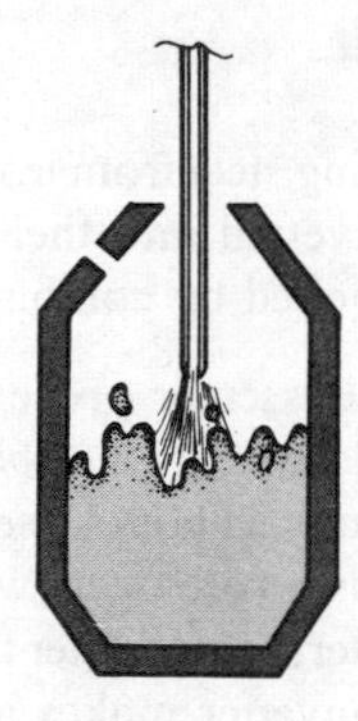

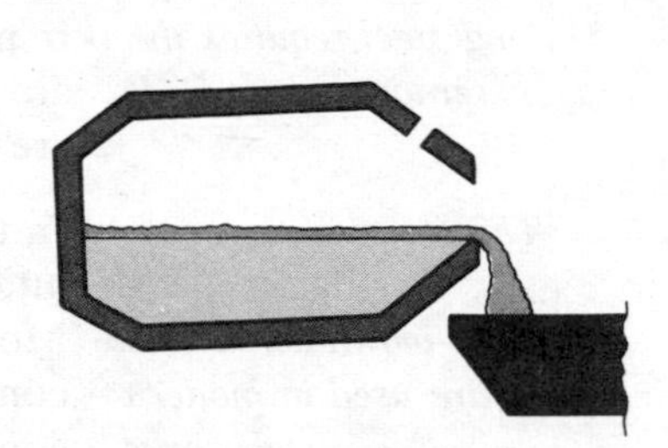

1 Charging.
The *converter* tips to receive a charge of 300 tonnes of molten iron.

2 The first blow.
The water-cooled *lance* directs oxygen and powdered calcium oxide on to the surface at a pressure of 10–15 atm.

3 Slagging.
The converter tips backwards to pour out the primary slag.

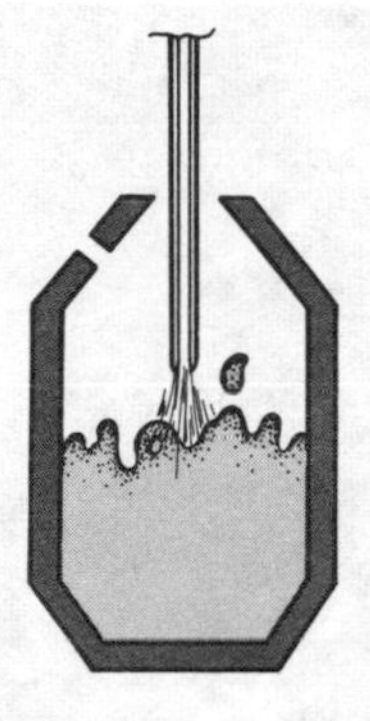

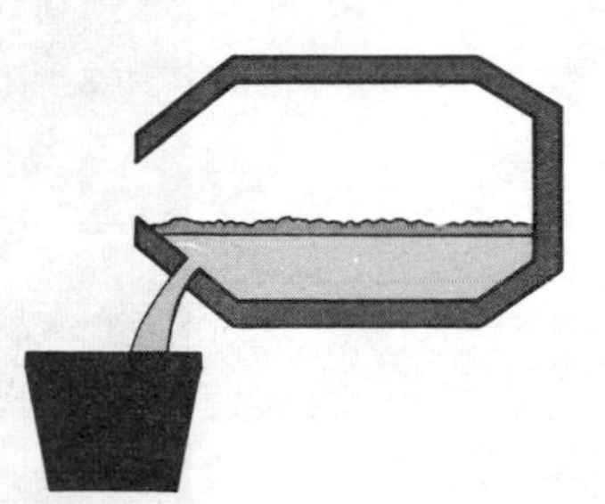

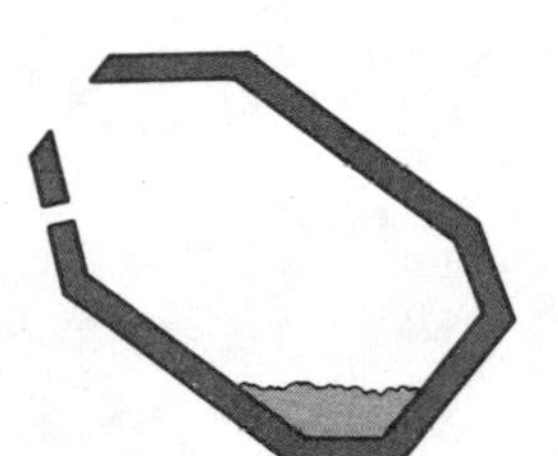

4 The second blow.
The *lance* directs oxygen and lime into the converter.

5 Pouring.
The converter tilts forward to pour the steel into a *ladle*, while the slag remains on top.

6 Slag remains in the converter, which tilts to receive the next charge.

FIGURE 24.20(b) Charging the Converter

24.14.3 STEEL

Making steel requires the removal of C, S, P

In making steel from iron, carbon and other impurities such as sulphur and phosphorus are converted into their oxides. Gaseous oxides remove themselves; other oxides are removed by combination with a base such as calcium oxide to form a slag.

The Bessemer process

In the Bessemer process, invented in Britain 125 years ago, molten iron was poured into a large tub, the *converter*, and air was blown on to it to oxidise the impurities.

Oxygen-blown converters are used in modern steelworks

Modern steelworks use oxygen-blown converters. The steel is better because it contains no nitrogen, which makes steel brittle. Shown in Figure 24.20 is the LD converter, named after the Austrian towns of Linz and Donawitz, where it was invented. One converter makes about 500 tonnes of steel in an hour. The Kaldo process used in Sweden is similar, but employs a converter which can be rotated. Oxygen is blown in at a pressure of only 2 atm; the rotation assists oxidation.

The LD process...

...and the Kaldo process

[See § 24.15 for modern steelmaking.]

FIGURE 24.20(c) Slagging

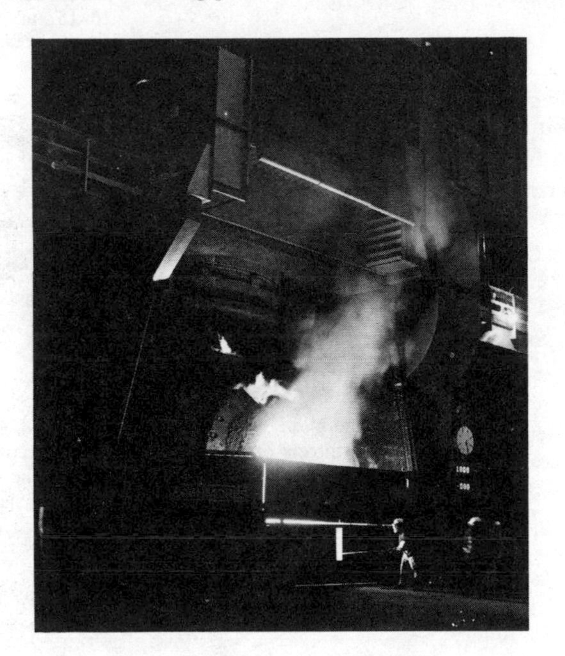

24.14.4 RUSTING

Rusting is a serious problem. A large fraction ($\frac{1}{8}$) of the annual UK production of steel (20 million tonnes) is needed simply to replace iron lost through rusting. Rust is hydrated iron(III) oxide, $Fe_2O_3 \cdot xH_2O$. Both water and air are needed for rusting to occur. It is an electrochemical process, with different parts of an iron structure acting as cathodes and anodes. At an anodic region, the process which occurs is

Rusting is an electrochemical process

The mechanism of rusting

$$Fe(s) \rightarrow Fe^{2+}(aq) + 2e^-$$

At a cathodic region, the process is

$$O_2(aq) + 2H_2O(l) + 4e^- \rightarrow 4OH^-(aq)$$

The presence of dissolved acids and salts in water increases its conductivity and speeds up the process of rusting. If cathodic and anodic areas are close together, precipitation of iron(II) hydroxide, $Fe(OH)_2(s)$, occurs. Air oxidises this to rust, hydrated iron(III) oxide:

$$2Fe(OH)_2(s) + \tfrac{1}{2}O_2(aq) + H_2O(l) \rightarrow Fe_2O_3 \cdot xH_2O(s)$$

24.14.5 PREVENTION OF RUSTING

COATING

Rusting can be prevented by . . .

Various methods are used to provide a protective coat to exclude water and oxygen.

. . . paint . . .

(a) Paint is used for many large objects, e.g., ships and bridges [see Figure 24.21]. A paint containing phosphoric(V) acid is effective as it forms a layer of insoluble iron(III) phosphate(V) with any rust present on the surface.

. . . oil, grease . . .

(b) A coat of grease or oil is used for moving parts of machinery.

. . . zinc coating . . .
. . . tin plating . . .

(c) A coat of another metal may be used; zinc coating, i.e., *galvanising*, and tin plating are used. Since zinc is higher than iron in the Electrochemical Series, even if the coating of zinc is scratched, it will continue to protect the iron underneath from rusting. Tin cans rust if they are scratched because, being higher in the Electrochemical Series, iron is corroded in preference to tin.

. . . chromium plating . . .

(d) Chromium plating is used for many car accessories because it is decorative as well as protective. An electrolytic method is used for plating [see Figure 24.22].

FIGURE 24.21
Galvanised Iron Girders at the Holyhead Ferry Terminal

FIGURE 24.22
Decorative Chromium Plating

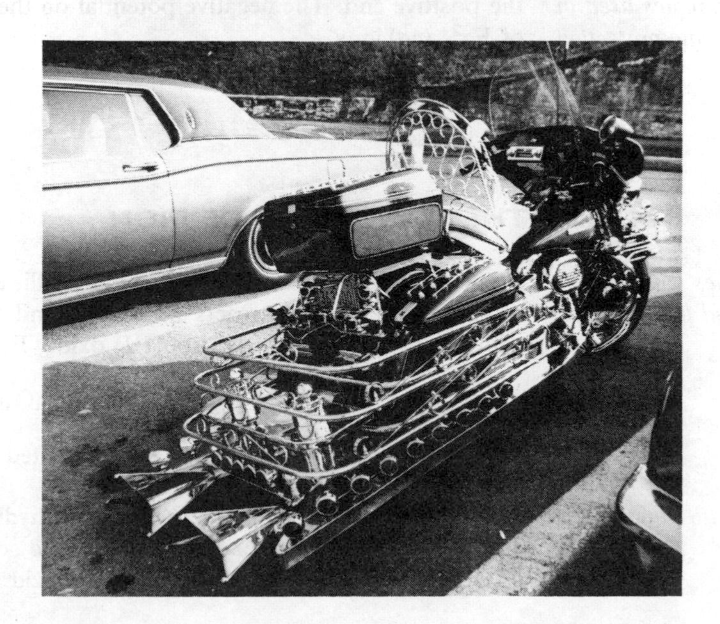

ALLOYING

...forming alloys, e.g., stainless steel...

Stainless steel is an alloy of iron with nickel and chromium (e.g., 18% Ni, 8% Cr). The added metals produce a surface film of metal oxides which is impervious to water.

CATHODIC PROTECTION

...or by cathodic protection

If a block of a metal higher in the Electrochemical Series is connected to iron, then that metal acts as the anode, and it is corroded while iron remains intact. Zinc and magnesium are often used. Underground pipes are protected by attaching bags of magnesium scraps at intervals and replacing these from time to time when they have been corroded. The hulls of ships are protected by attaching blocks of zinc, which are sacrificed to protect the iron. This technique is called **sacrificial protection** [see Figure 24.23].

Sacrificial protection...

FIGURE 24.23 Sacrificial Protection of Iron by Zinc

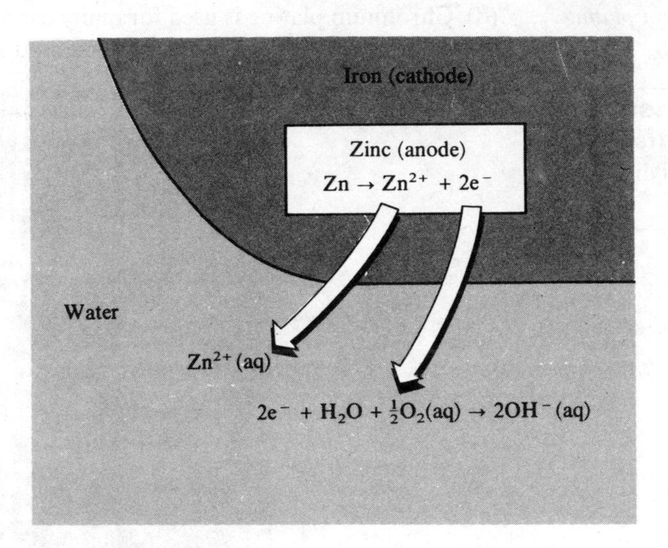

...or the application of a negative potential, are the techniques used in cathodic protection

Another method of cathodic protection is to make iron the cathode by connecting it to the negative side of a battery while a conductor such as graphite is connected to the positive end. The negative potential on the iron structure inhibits the formation of Fe^{2+}(aq) ions.

24.14.6 THE CHEMISTRY OF IRON

Iron combines with O_2, S, N_2, C and the halogens

Iron combines on heating with the non-metallic elements oxygen, nitrogen, the halogens, sulphur and carbon. It reacts with water and air to form rust, $Fe_2O_3 \cdot xH_2O$, and with steam to form iron(II) iron(III) oxide, Fe_3O_4, which is magnetic:

$$3Fe(s) + 4H_2O(g) \rightarrow 4H_2(g) + Fe_3O_4(s)$$

This oxide is also formed when iron is heated in air.

It reacts with dilute acids

Iron reacts with dilute sulphuric acid and hydrochloric acid to form iron(II) salts and hydrogen. It is attacked by dilute nitric acid to form iron(III) nitrate and is rendered 'passive' by concentrated nitric acid.

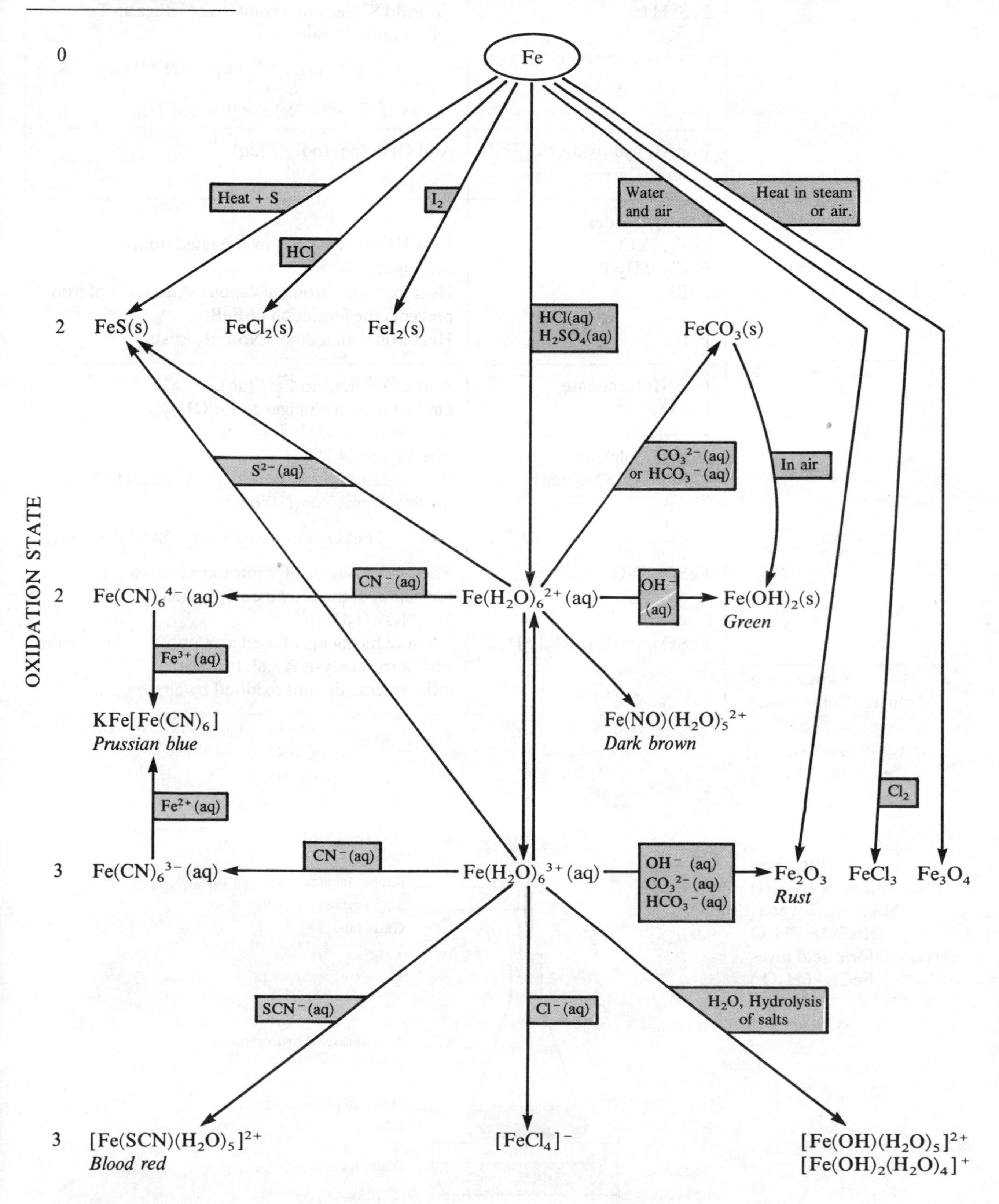

FIGURE 24.24 Some Reactions of Iron

Compound	*Preparation and Comments*
Iron(II) sulphide, FeS, black	(a) Heat iron with sulphur. (b) Add $S^{2-}(aq)$ to a solution of either an Fe^{3+} salt or an Fe^{2+} salt: $2Fe^{3+}(aq) + S^{2-}(aq) \rightarrow 2Fe^{2+}(aq) + S(s)$ $Fe^{2+}(aq) + S^{2-}(aq) \rightarrow FeS(s)$
Iron(II) hydroxide, $Fe(OH)_2$, green	Add $OH^-(aq)$ to $Fe^{2+}(aq)$.
Iron(II) halides	
FeF_2, $FeCl_2$	Pass HF(g) or HCl(g) over heated iron.
$FeCl_2 \cdot 6H_2O$	See Figure 24.25.
$FeBr_2$	Heat iron with bromine vapour. An excess of iron prevents the formation of $FeBr_3$.
FeI_2	Heat iron with iodine. (No FeI_3 exists.)
Iron(II) carbonate, $FeCO_3$	Add $CO_3^{2-}(aq)$ to $Fe^{2+}(aq)$. On standing, it changes to $Fe(OH)_2(s)$.
Iron(II) sulphate $FeSO_4 \cdot 7H_2O$, green	[See Figure 24.25.] When heated, gives the anhydrous salt. On further heating, forms iron(III) oxide: $2FeSO_4(s) \rightarrow Fe_2O_3(s) + SO_2(g) + SO_3(g)$
$FeSO_4 \cdot NO$	The brown ring test for nitrates involves the formation of the brown complex ion, $[Fe(NO)(H_2O)_5]^{2+}$
$FeSO_4 \cdot (NH_4)_2SO_4 \cdot 6H_2O$	It is a reducing agent used as a primary standard in titrimetric analysis [§ 3.8]. It is not efflorescent, and not oxidised by air.

TABLE 24.8
Iron(II) Compounds

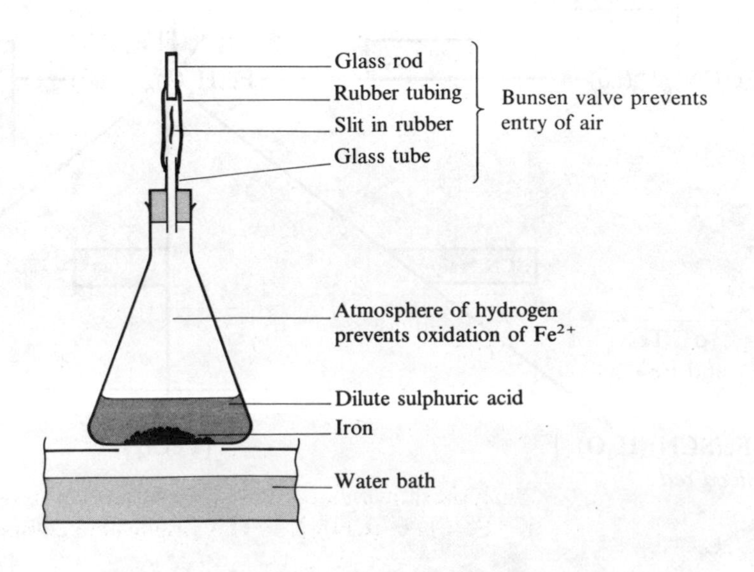

FIGURE 24.25
Preparation of Iron(II) Sulphate-7-water, $FeSO_4 \cdot 7H_2O$ (Hydrochloric acid gives $FeCl_2 \cdot 6H_2O$.)

Compound	Preparation and Comments
Iron(III) oxide, Fe_2O_3, rust	(a) Mined as *haematite*. (b) Heat $FeSO_4 \cdot 7H_2O$. (c) Add $OH^-(aq)$ to $Fe^{3+}(aq)$. No $Fe(OH)_3$ exists; hydrated Fe_2O_3 is precipitated.
Iron(II) iron(III) oxide, Fe_3O_4 blue-black	(a) Mined as *magnetite*. (b) Heat iron in air or steam: $3Fe(s) + 4H_2O(g) \rightarrow Fe_3O_4(s) + 4H_2(g)$ It reacts as a mixture of $FeO \cdot Fe_2O_3$: $Fe_3O_4(s) + 8H^+(aq) \rightarrow Fe^{2+}(aq) + 2Fe^{3+}(aq) + 4H_2O(l)$
Iron(III) halides $FeCl_3$, $FeBr_3$	Pass Cl_2 or $Br_2(g)$ over heated iron. The apparatus in Figure 19.9, § 19.3.1 can be used. Fe_2Cl_6 or Fe_2Br_6 sublimes over into the receiver. (The reaction between Fe and I_2 gives FeI_2 because Fe^{3+} oxidises I^- to I_2.)

TABLE 24.9 Iron(III) Compounds

HYDROLYSIS OF IRON SALTS

Fe^{3+} salts are acidic in solution due to hydrolysis

Iron(II) salts are slightly acidic in solution; iron(III) salts are more acidic as the degree of hydrolysis is greater. The small, highly charged Fe^{3+} ion is hydrated, and, as in the case of aluminium salts [§ 19.2.4], this leads to hydrolysis:

$$[Fe(H_2O)_6]^{3+} + H_2O \rightleftharpoons [Fe(OH)(H_2O)_5]^{2+} + H_3O^+$$
$$[Fe(OH)(H_2O)_5]^{2+} + H_2O \rightleftharpoons [Fe(OH)_2(H_2O)_4]^+ + H_3O^+$$

Reagent	Reaction of $Fe^{2+}(aq)$	Reaction of $Fe^{3+}(aq)$
$OH^-(aq)$	Gelatinous green ppt. of $Fe(OH)_2(s)$	Gelatinous rust ppt. of Fe_2O_3
Potassium hexacyanoferrate(II), $K_4Fe(CN)_6(aq)$	Green or brown colour	Prussian-blue $KFe[Fe(CN)_6](s)$
Potassium hexacyanoferrate(III), $K_3Fe(CN)_6(aq)$	Turnbull's blue, $KFe[Fe(CN)_6](s)$	Green or brown colour
Potassium thiocyanate, KCNS	No reaction	Blood-red colour, $Fe(CNS)^{2+}(aq)$

TABLE 24.10 Tests for Fe^{2+} and Fe^{3+} Ions

Analysis shows that Turnbull's blue and Prussian blue are both $K^+Fe^{3+}[Fe^{II}(CN)_6]^{4-}$. This bright blue compound is used as a pigment.

24.15 STEEL

TOPIC

Metallurgists match the composition of a steel with the job that it has to do [see Table 24.7, § 24.14.2]. They are constantly producing new alloys for particular purposes. For making alloy steels of accurately known composition, the electric arc furnace is used [see Figure 24.26]. It takes only 2 or 3 minutes to observe the emission spectrum of a sample of metal [§ 2.1]. Once the required composition is reached, the alloy can be tapped off.

FIGURE 24.26(a) An Electric Arc Furnace for Making Steel

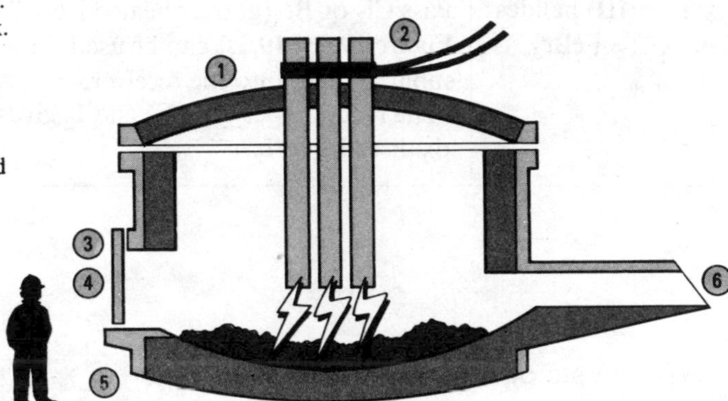

1 The roof is swung to the side. Steel scrap is tipped in. Then the roof is swung back.

2 Power cables and carbon electrodes. The current is switched on. As each electrode is lowered, electric current jumps the gap, forming an arc between the electrode and the metal. The arc makes a crackling noise like gunfire. The intense heat melts the metal.

3 Limestone is added through the furnace door. It combines with impurities to form a liquid slag. Measured quantities of alloy metals are added.

4 Samples of steel are removed and analysed. Eventually, the correct composition is reached.

5 The furnace is tilted to this side so that slag can be poured off.

6 The furnace is tilted to this side so that molten steel can be poured out through the *tapping spout*.

Research workers at the British Iron and Steel Research Association laboratories in Sheffield developed a new way of making steel. Instead of making steel in batches, it works continuously [see Figure 24.27]. The process is fast. The tiny droplets of metal present a vastly increased surface area over which oxygen can reach the impurities, and oxidation occurs rapidly.

FIGURE 24.26(b) Control Panel of an Electric Arc Furnace

FIGURE 24.27 A Spray Steelmaker

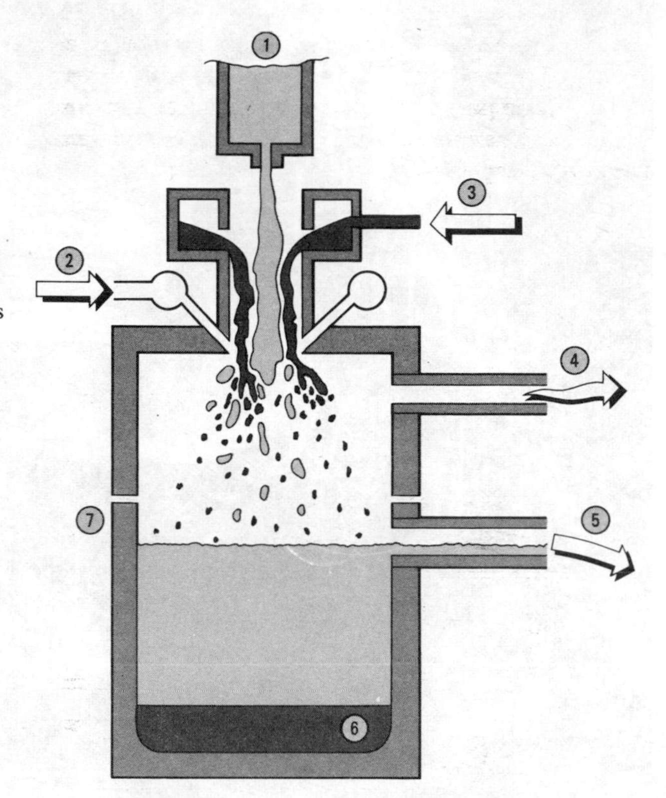

1 Molten iron from blast furnace.

2 Jet of oxygen breaks up the stream of molten iron into droplets of spray. Impurities burn to form oxides.

3 Lime (calcium oxide) is fed in. It forms slag with the oxides.

4 Waste gases leave.

5 Slag is run off.

6 Steel collects.

7 The bottom part of the container is detachable. Scrap iron or steel can be added before or during the process.

A pilot plant in Millom in Cumbria ran successfully from May, 1966 to December, 1966. The Millom Company (a private company) needed to raise capital in order to expand into an industrial scale plant. The Government and the British Steel Corporation had to decide whether to invest in a proven batch process or in the less well tried continuous process. They decided against investing in spray steelmaking, and in 1968 the Millom Steelworks closed.

FIGURE 24.28 Galvanising a Metal Container

FIGURE 24.29
Many Coats of Paint are used to protect a Car Body from Rusting

24.16 COPPER

Copper is low in the Electrochemical Series

Extraction and purification

Copper is low in the Electrochemical Series and is found 'native', i.e., uncombined. The chief ores are *copper pyrites*, $CuFeS_2$, and *copper glance*, CuS. Copper is obtained by roasting copper pyrites with silica and air in a furnace or by roasting copper glance with air. Impure copper is formed. The electrolytic method used to purify copper is illustrated in Figure 24.30.

FIGURE 24.30
Purification of Copper

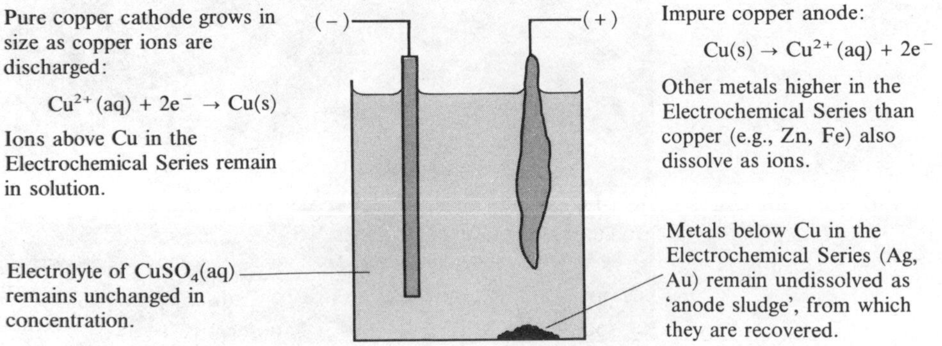

Pure copper cathode grows in size as copper ions are discharged:

$$Cu^{2+}(aq) + 2e^- \rightarrow Cu(s)$$

Ions above Cu in the Electrochemical Series remain in solution.

Electrolyte of $CuSO_4(aq)$ remains unchanged in concentration.

(−) (+)

Impure copper anode:

$$Cu(s) \rightarrow Cu^{2+}(aq) + 2e^-$$

Other metals higher in the Electrochemical Series than copper (e.g., Zn, Fe) also dissolve as ions.

Metals below Cu in the Electrochemical Series (Ag, Au) remain undissolved as 'anode sludge', from which they are recovered.

FIGURE 24.31
Electrolytic Cells at Capper Pass

24.16.1 USES OF COPPER

The uses of copper include cooking ware...

...electrical cables...

...and roofing

The high thermal conductivity of copper leads to its use for cooking ware. The high electrical conductivity makes copper wire admirably suitable for electrical circuits and cables. The resistance to corrosion makes copper useful for water pipes. Copper is used as a roofing material because it weathers to acquire a coating of green basic copper carbonate, $CuCO_3 \cdot Cu(OH)_2 \cdot nH_2O$, which lends a colourful touch to a building. Alloys of copper are *coinage metal* (Cu, Ni), *brass* (Cu, Zn) and *bronze* (Cu, Sn).

24.16.2 REACTIONS OF COPPER

Copper reacts with O_2, S, the halogens, and oxidising acids

Copper is low in the Electrochemical Series. It reacts with the oxidising acids, dilute nitric acid, concentrated nitric acid and concentrated sulphuric acid to form copper(II) salts. It combines directly with oxygen, sulphur and the halogens to form copper(II) compounds, except in the case of iodine, with which it forms copper(I) iodide. At temperatures of 800–1000 °C, it combines with oxygen to form copper(I) oxide.

FIGURE 24.32 Some Reactions of Copper and its Compounds

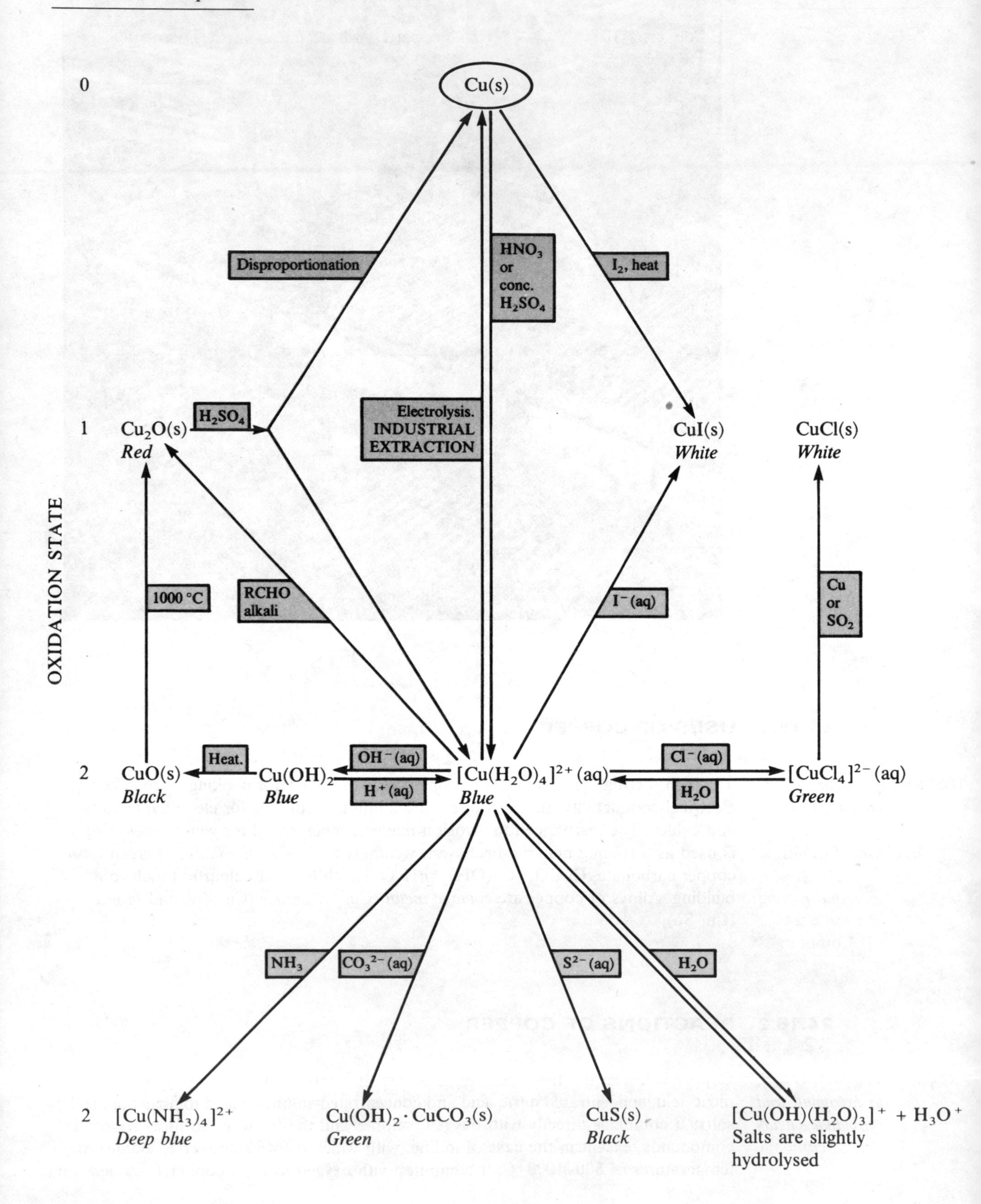

24.16.3 COPPER(II) COMPOUNDS

Compound	*Preparation and Comments*
Copper(II) oxide, CuO, black	Heat copper(II) nitrate, carbonate or hydroxide. Basic
Copper(II) hydroxide, $Cu(OH)_2$, blue	Add OH^-(aq) to Cu^{2+}(aq). A gelatinous blue precipitate forms. It dissolves in NH_3(aq) to form $Cu(NH_3)_4{}^{2+}$(aq). Basic
Copper(II) salts	Warm CuO with a dilute acid.
Anhydrous CuX_2	Heat Cu with halogen (except I_2).
Copper(II) sulphide, CuS, black	Add S^{2-}(aq) to a solution of a Cu^{2+} salt.

TABLE 24.11
Copper(II) Compounds

24.16.4 COPPER(I) COMPOUNDS

The stability of copper(I) compounds and complex ions has been discussed in §24.13.9.

Compound	*Preparation and Comments*
Copper(I) oxide, Cu_2O, reddish solid	Reduce Cu^{2+}(aq) in alkaline solution with an aldehyde or by SO_2: $2Cu^{2+}(aq) + 2OH^-(aq) + 2e^- \rightarrow Cu_2O(s) + H_2O(l)$ The formation of Cu_2O is a test for an aldehyde [§31.6.3]. With dilute acids, Cu(s) and a Cu^{2+} salt are formed.
Copper(I) chloride, CuCl, white solid	(a) Reduce $CuCl_2$(aq) by Cu(s) or by SO_2(g) to $[CuCl_2]^-$(aq). (b) Add oxygen-free water. White CuCl(s) precipitates: $Cu^{2+}(aq) + 4Cl^-(aq) + Cu(s) \rightarrow 2[CuCl_2]^-(aq)$ $[CuCl_2]^-(aq) \rightleftharpoons CuCl(s) + Cl^-(aq)$
Copper(I) bromide	Resembles copper(I) chloride
Copper(I) iodide	(a) Heat copper with iodine. (b) Add I^-(aq) to Cu^{2+}(aq).

TABLE 24.12
Copper(I) Compounds

24.17 ZINC

Zinc is extracted from zinc blende or zinc carbonate

Zinc is mined as *zinc blende*, ZnS, and as zinc carbonate, $ZnCO_3$. The sulphide ore is roasted to form the oxide and sulphur dioxide, which is used in the Contact process. Zinc oxide is reduced by coke, and, being a volatile metal, zinc can be distilled from the furnace, leaving less volatile impurities behind.

24.17.1 USES

Its uses include galvanising steel

Zinc is used in galvanising steel and in the production of *brass* (Cu, Zn).

24.17.2 REACTIONS

Zn and Zn^{2+} both have a full d subshell

Zinc resembles transition metals

Zinc is not a transition metal. It has a full d subshell of electrons, and its ions, Zn^{2+}, which also have a full d subshell, are colourless. The inability of the d electrons to take part in the metallic bonding gives zinc a lower melting temperature and boiling temperature than the transition metals. Zinc resembles transition metals in forming some complex ions and does not resemble the metals of Group 2.

It reacts with acids and alkalis

Zinc reacts with acids to form salts of the hydrated zinc(II) ion, $Zn(H_2O)_6{}^{2+}$. It also reacts with alkalis to form salts of the tetrahydroxozincate ion (or zincate ion), $Zn(OH)_4{}^{2-}$, with the evolution of hydrogen:

$$Zn(s) + 2OH^-(aq) + 2H_2O(l) \rightarrow [Zn(OH)_4]^{2-}(aq) + H_2(g)$$

24.17.3 COMPOUNDS OF ZINC

Compound	*Preparation and Comments*
Zinc oxide, ZnO, *white*	Heat zinc carbonate, nitrate or hydroxide. Amphoteric. It becomes yellow when hot. It loses some of its oxygen content, which recombines on cooling. The absence of atoms from the crystal structure leads, here and in other cases, to colour.
Zinc hydroxide, $Zn(OH)_2$	Add $OH^-(aq)$ to a solution of a Zn^{2+} salt. A gelatinous precipitate appears. Amphoteric.
Zinc chloride. $ZnCl_2 \cdot 2H_2O$, $ZnCl_2$	Allow hydrochloric acid to react with Zn, ZnO or $ZnCO_3$. Attempts to make the anhydrous salt from the hydrate lead to $Zn(OH)Cl$. (a) Heat $ZnCl_2 \cdot 2H_2O$ in dry $Cl_2(g)$ or dry $HCl(g)$. (b) Heat Zn in dry $Cl_2(g)$ or dry $HCl(g)$.
Zinc carbonate, $ZnCO_3$	Add $HCO_3{}^-(aq)$ to a solution of a Zn^{2+} salt (addition of $CO_3{}^{2-}(aq)$ leads to the precipitation of the basic carbonate, $ZnCO_3 \cdot 2Zn(OH)_2$).

TABLE 24.13
Compounds of Zinc

FIGURE 24.33 RTZ Electrolytic Zinc Plant at Budel, Holland

FIGURE 24.34 Some Reactions of Zinc and its Compounds

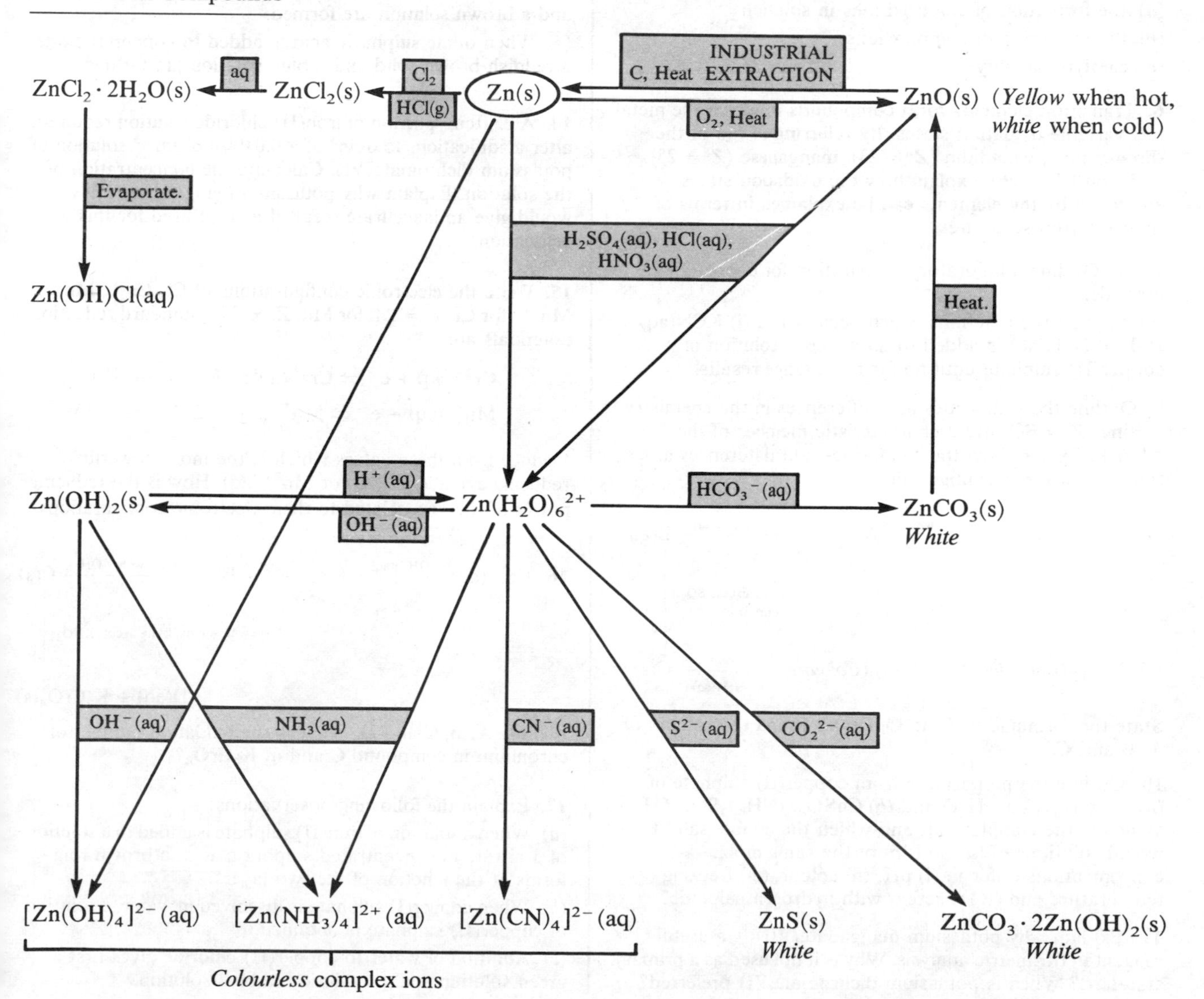

QUESTIONS ON CHAPTER 24

1. Describe the manufacture of steel. What are the advantages and disadvantages of steel, compared with aluminium? Why is the production of steel greater than the total production of all other metals?

2. Compare the chlorides $FeCl_3$ and $AlCl_3$ with respect to (*a*) bond type and (*b*) the behaviour of aqueous solutions. What is the reason for the resemblance?

3. Copper forms a complex

$$[Cu(H_2NCH_2CH_2NH_2)_n]^{2+}$$

which is highly coloured. Explain how the formula of the ion could be found by a colorimetric method.

4. State three characteristics of a transition element, illustrating them by reference to the chemistry of iron. In your answer, refer to the electron configurations of Fe, Fe^{2+} and Fe^{3+}.

5. Give examples of the chemistry of two transition elements other than iron to illustrate

(*a*) the formation of coloured ions in solution

(*b*) the formation of complex ions

(*c*) catalytic activity.

6. Transition elements form compounds in which the metal has variable oxidation states. By referring to *two* of the three metals, vanadium (Z = 23), manganese (Z = 25) and iron (Z = 26), explain how the oxidation states employed by the elements can be explained in terms of their electron structures.

7. (*a*) Outline a laboratory preparation for copper(I) chloride.

(*b*) Explain the reactions which occur when (i) KCN(aq) and (ii) NH_3(aq) is added to an aqueous solution of copper(II) sulphate until no further change results.

8. Outline the similarities and differences in the chemistry of zinc (Z = 30) and a characteristic member of the d block. Explain how the similarities and differences arise from the electron configurations.

9.

$$CuSO_4(s) \xrightarrow{\text{water}} \mathbf{A}(aq) \xrightarrow{\text{excess of NaCl(aq)}} \mathbf{B}(aq)$$

$$\mathbf{B}(aq) \xrightarrow{\text{(i) excess } SO_2 \text{ (ii) water}} \mathbf{C}(s)$$

$$\mathbf{E}(aq)\ (\textit{Blue-violet}) \leftarrow \mathbf{D}(aq)\ (\textit{Colourless}) \xleftarrow{NH_3(aq)} \mathbf{C}(s)$$

State the formulae of **A**, **B**, **C**, **D** and **E** and the colours of **A**, **B** and **C**.

10. Outline the preparation from copper(II) sulphate of (*a*) $Cu(NH_3)_4SO_4 \cdot H_2O$ and (*b*) $CuSO_4 \cdot (NH_4)_2SO_4 \cdot 6H_2O$. Which is the complex salt and which the double salt? How would solutions of (*a*) and (*b*) of the same molar concentrations differ in (i) pH, (ii) colour, (iii) freezing temperature and (iv) reaction with hydrogen sulphide?

11. Explain why potassium manganate(VII) is a useful reagent in titrimetric analysis. Why is it not used as a primary standard? When is potassium dichromate(VI) preferred? Construct equations for (*a*) the oxidation by acidified potassium manganate(VII) of (i) Fe^{2+}(aq) to Fe^{3+}(aq) and (ii) H_2O_2(aq) to O_2(g) + $2H^+$(aq) and (*b*) the reduction by acidified potassium manganate(VII) of BiO_3^-(aq) to Bi^{3+}(aq).

12. Copper ores often contain iron.
Explain why (*a*) iron does not contaminate copper obtained by electrolytic purification and (*b*) why anode sludge contains silver and gold.

13. Explain the following statements:

*(*a*) The compound $CrCl_3 \cdot 6H_2O$ exists in three forms. All produce a precipitate with silver nitrate solution, but in different molar proportions.

(*b*) Addition of alkali to potassium dichromate(VI) solution produces a yellow solution.

(*c*) The addition of barium chloride solution to a solution of potassium dichromate(VI) gives a yellow precipitate.

(*d*) When an excess of potassium iodide solution is added to an aqueous solution of a copper(II) salt, a white precipitate and a brown solution are formed.

(*e*) When dilute sulphuric acid is added to copper(I) oxide, a reddish brown solid and a blue solution are formed.

14. A 25.0 cm^3 portion of iron(II) chloride solution required, after acidification, 15.0 cm^3 of a 0.0100 $mol\,dm^{-3}$ solution of potassium dichromate(VI). Calculate the concentration of the solution. Explain why potassium manganate(VII) would give an inaccurate result if it were used for this estimation.

15. Write the electronic configurations of Cr, Cr^{2+}, Mn, Mn^{2+} (for Cr, Z = 24; for Mn, Z = 25). Standard reduction potentials are:

$$Cr^{3+}(aq) + e^- \rightleftharpoons Cr^{2+}(aq); \quad E^\ominus = -0.41\,V$$

$$Mn^{3+}(aq) + e^- \rightleftharpoons Mn^{2+}(aq); \quad E^\ominus = +1.51\,V$$

Deduce from these values which is the more powerful reducing agent, Cr^{2+}(aq) or Mn^{2+}(aq). How is the reducing power of the ions related to their electronic configurations?

16.

$$Cr^{3+}(aq) \xrightarrow{OH^-(aq)} \mathbf{A}(s) \xrightarrow{\text{Heat}} \mathbf{B}(s) \xrightarrow{\text{Fuse with KOH.}} \mathbf{C}(s)$$

$$\mathbf{C}(s) \xrightarrow{\text{Fuse with the oxidising agent, } KNO_3.} \mathbf{D}(aq) + K_2CrO_4(s)$$

Identify **A**, **B**, **C** and **D**. What is the oxidation number of chromium in compound **C** and in K_2CrO_4?

17. Explain the following observations:

(*a*) When a solution of iron(II) sulphate is added to a solution of a nitrate in concentrated sulphuric acid, a brown ring forms at the junction of the two layers.

(*b*) When copper(I) sulphate is dissolved in water, a solution of copper(II) sulphate is obtained.

(*c*) Addition of water to copper(II) chloride gives first a green solution and, on dilution, a blue solution.

(*d*) When aqueous sodium hydroxide is added to a solution of tetraamminecopper(II) sulphate, only a small quantity of copper(II) hydroxide is precipitated. When hydrogen sulphide is passed through the solution, the copper ions are precipitated as copper(II) sulphide.

18. (a) List the chemical reactions that take place in the extraction of iron in the blast furnace. Discuss the physico-chemical principles involved.

(*b*) Discuss the electrochemical methods used to prevent the rusting of iron and steel.

19. Explain the meaning of each of the following terms. Illustrate your answers with examples from the chemistry of named elements:

(*a*) transition metal

(*b*) oxidation number

(*c*) complex ion

(*d*) coordination number.

20. This question refers to the reactions

$$MnO_2 \xrightarrow{A} K_2MnO_4 \xrightarrow{B} KMnO_4 + MnO_2$$

(*a*) Give the reagents and conditions employed in reactions **A** and **B**.

(*b*) Construct an equation for reaction **B**.

(*c*) State the oxidation number of manganese in each species.

(*d*) What name is given to reactions of type **B**?

(*e*) Why is reaction **B** carried out under acidic conditions?

21. Explain the following statements:

(*a*) $[Fe(CN)_6]^{3-}$ is used to test for iron ions.

(*b*) $[Ag(NH_3)_2]^+$ is used to test for chloride ions.

(*c*) Copper(I) chloride is used to distinguish between alkynes.

(*d*) Copper(I) chloride is used in the preparation of chlorobenzene.

(*e*) Copper(II) sulphate is used in a test for aldehydes.

22. Sketch the arrangement of bonds in the ions

CrO_4^{2-} $Cr_2O_7^{2-}$ $Cu(NH_3)_4^{2+}$
$[Ni(NH_3)_6]^{2+}$ $[Fe(CN)_6]^{3-}$

23. Write an account of the first transition metal series. Consider (*a*) physical properties, (*b*) the oxidation states shown by each element, (*c*) complex formation. Explain how these properties are related to electronic structure.

24. Explain the terms *complex ion* and *ligand*.
Draw the structure of (*a*) NH_3 and (*b*) Cl^-.
Explain why ammonia molecules and chloride ions are able to act as ligands. What type of bonds do they employ in complex formation? Give examples of complex ions containing (*a*) NH_3 and (*b*) Cl^- as ligands. Sketch the arrangement of ligands in the ions you mention.

25. (*a*) State what you would observe and give the formulae of the metal-containing species involved when aqueous solutions of iron(III) nitrate and cobalt(II) nitrate react separately with

(i) concentrated hydrochloric acid

(ii) aqueous sodium carbonate

(iii) concentrated aqueous ammonia.

(*b*) The tests and observations in the reaction scheme shown below were recorded. Use this information to identify the metal ions which can be present in **A**. Give the formulae for *all* the metal-containing species involved in **B** to **I**.

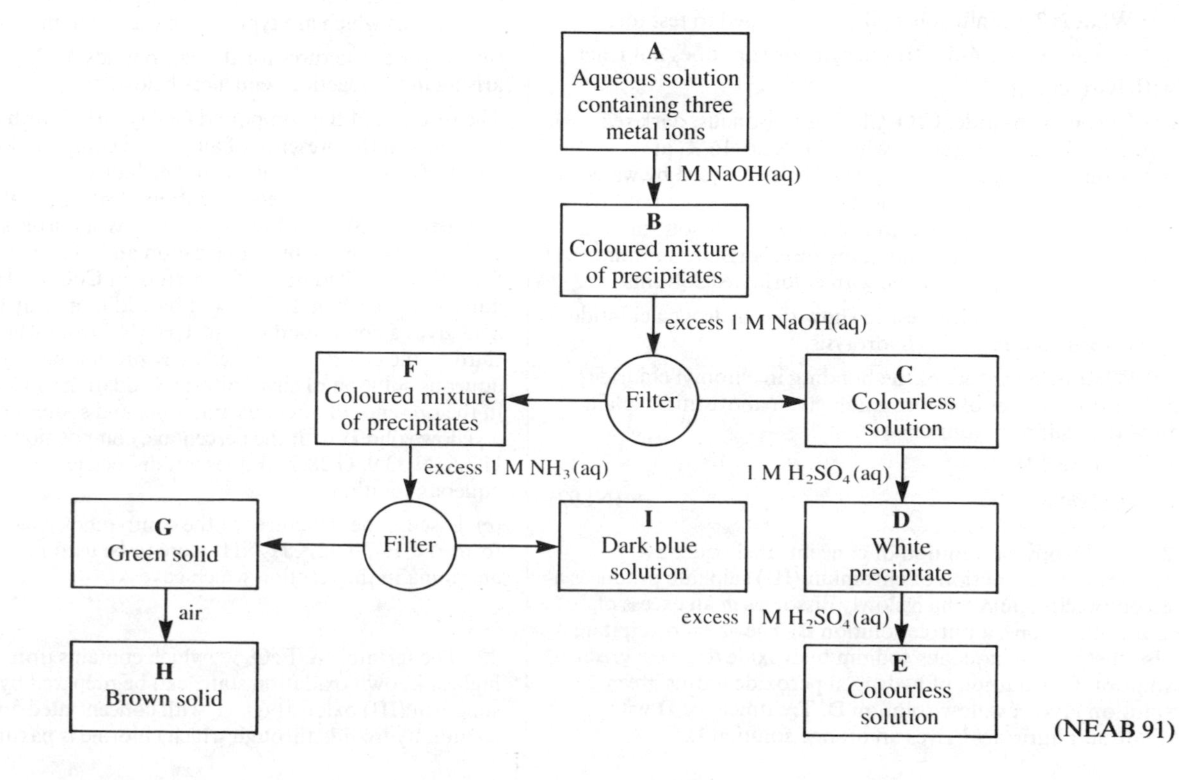

(NEAB 91)

26. Chromium is a transition metal and forms compounds in a variety of oxidation states. Some of the reactions of chromium and its compounds are shown in the scheme below:

$CrCl_2$ ← HCl — Cr

Cr —Cl_2→ $CrCl_3$ —NaOH→ $Cr(OH)_3$ —acac*→ $Cr(acac)_3$

Cr → Cr_2O_3 —Na_2O_2→ $Na_2Cr_2O_4$ —H^+→ $Na_2Cr_2O_7$ —NaCl, H_2SO_4→ CrO_2Cl_2

acac* is acetylacetone

(*a*) State the colour of the following ions in aqueous solution: Cr^{3+}, $Cr_2O_7{}^{2-}$, $CrO_4{}^{2-}$

(*b*) State the oxidation number of chromium in the following compounds: $Cr(OH)_3$, CrO_2Cl_2.

†(*c*) Acetylacetone (acac) has the formula $CH_3COCH_2COCH_3$ and is a bidentate ligand. It reacts with chromium hydroxide to form a red-violet solid $[Cr(acac)_3]$ which is a neutral complex and is unaffected by 2,4-dinitrophenylhydrazine.

(i) Deduce the systematic name (IUPAC) for acetylacetone.

(ii) What is the meaning of the term *bidentate*?

(iii) Draw the three-dimensional structure of $[Cr(acac)_3]$ which is similar to that of $[Ni(en)_3]^{2+}$.

Represent the structure of acac as O O.

(iv) What is 2,4-dinitrophenylhydrazine used to test for?

(v) Suggest why 2,4-dinitrophenylhydrazine does not react with $[Cr(acac)_3]$.

(*d*) Chromyl chloride, CrO_2Cl_2, is a poisonous, dark red liquid of density 2.0 g cm^{-3} which boils at 116 °C at atmospheric pressure. It is vigorously hydrolysed by water to chromic(VI) acid and hydrogen chloride gas. As a powerful oxidising agent it explodes in contact with phosphorus and ignites sulphur, ethanol and many other organic substances. Chromyl chloride is miscible with chlorinated solvents.

(i) Describe a chemical test to show that hydrogen chloride gas is formed during this hydrolysis.

(ii) What is the nature of the bonding in chromyl chloride? Select three pieces of information given above and explain how they support your answer.

[For (*c*) see § 31.2.]

(NI 92)

27. (*a*) Dropwise addition of concentrated aqueous ammonia to a solution of chromium(III) sulphate produces a green precipitate **A** which slowly dissolves in an excess of ammonia to give a purple solution **B**. The green precipitate **A** also dissolves in aqueous sodium hydroxide to give a green solution **C**. Addition of hydrogen peroxide to this green solution gives a yellow solution **D**. Treatment of **D** with dilute sulphuric acid gives an orange solution **E**.

Identify the chromium species present in **A**, **B**, **C**, **D** and **E** and explain the function of hydrogen peroxide in the conversion of **C** into **D**.

(*b*) On dropwise addition of concentrated ammonia solution to a pink aqueous solution of cobalt(II) chloride containing the cobalt species **F**, a green-blue precipitate **G** is first formed. This precipitate dissolves when an excess of ammonia is added forming a pale brown solution containing the cobalt species **H**. When left to stand in air the colour of this solution quickly darkens due to the formation of cobalt species **I**.

Identify the cobalt species **F**, **G**, **H** and **I**. In the case of **F** and **I** state the shape of the complex. Explain why **H** darkens when left to stand in air.

(*c*) A blue aqueous solution of copper(II) sulphate, containing the copper species **J**, turned yellow-green on treatment with an excess of concentrated hydrochloric acid, due to the formation of the copper species **K**. When copper metal was added and the solution boiled for several minutes, a new copper species **L** was produced. When the hot solution containing **L** was poured into an excess of distilled water, a white precipitate **M** was obtained. When **M** was filtered off, and allowed to stand in the air it slowly turned green.

Identify the copper species **J**, **K**, **L** and **M**. Explain the function of the copper metal in the conversion of **K** into **L**, and write equations for the conversions of **J** into **K**, and **K** into **L**. Explain why **M** slowly turns green when allowed to stand in air.

(*d*) Explain what is meant by the term *disproportionation*. Illustrate your answer by referring to the reaction between copper(I) oxide and dilute sulphuric acid. State what you would observe in the reaction, identify the reaction products and write an equation for the reaction occurring.

(NEAB 92)

28. (*a*) Describe the characteristic electronic arrangement in a transition element and state *five* aspects of the properties of copper ions which are typical of a transition metal ion.

(*b*) Propose structures for the compounds **A**, **B**, **C** and **D** arising in the reaction sequences below.

The reaction of the compound $CoCl_2 \cdot 6H_2O$ with aqueous ammonia in the presence of air gives a compound **A**, $CoCl_3H_{17}N_5O$, from which all the chlorine is immediately precipitated by an aqueous solution of silver nitrate. When **A** is heated to 100 °C, it loses water to give a purple solid **B**, $CoCl_3H_{15}N_5$; **B** contains one cation and two anions per formula unit. Oxidation of a mixture of $CoCl_2 \cdot 6H_2O$ and ammonium carbonate followed by addition of hydrochloric acid gives a compound **C**, $CoCl_3H_{12}N_4$, from which only one third of the chlorine is immediately precipitated by an aqueous solution of silver nitrate. Oxidation of $CoCl_2 \cdot 6H_2O$ in the presence of aqueous ammonia and sodium nitrite gives a yellow solid **D** with the percentage composition Co 23.8, H 3.6, N 33.9, O 38.7; **D** does not dissociate into ions in aqueous solution.

(*c*) Discuss the structures of the compounds which might be formed if $H_2NCH_2CH_2NH_2$ were to be used instead of ammonia in the reaction which gave **A**.

(NEAB 91, S)

29. The ferrate ion, $FeO_4{}^{2-}$, which contains iron in its highest known oxidation state, can be prepared by reacting solid iron(III) oxide at 60 °C, with concentrated aqueous sodium hydroxide through which chlorine is passing.

The reaction mixture gradually turns deep purple and is filtered hot through a sintered glass filter to give a solution containing sodium ferrate. This solution is then treated with saturated aqueous barium chloride to precipitate small crystals of deep-red barium ferrate. The precipitate is removed by suction filtration and dried in air. It is stable only in strongly alkaline solution, oxidising water to oxygen if the pH is decreased. The purity of a sample of barium ferrate prepared as above was determined by adding 0.267 g to excess acidified potassium iodide solution. The liberated iodine was titrated with sodium thiosulphate solution of concentration 0.100 mol dm^{-3}, 30.0 cm^3 being required.

(*a*) (i) Draw a labelled diagram of a suitable apparatus in which the reaction might be carried out in the laboratory.

(ii) Why is a filter paper not used for the first filtration?

(iii) Draw a diagram showing the apparatus required for suction filtration and give an advantage of this method.

(*b*) Suggest and justify one safety precaution which you would take.

(*c*) (i) What is the highest oxidation state which iron might reasonably be expected to form? Explain your answer.

(ii) Suggest a reason why compounds containing iron in this oxidation state have not been prepared.

(*d*) (i) Name and give the formula of a chlorine oxo-anion likely to be formed during this preparation.

(ii) Give a balanced ionic equation for the formation of this oxo-anion.

(iii) Write a balanced ionic equation for the production of oxygen by ferrate ions in acidic solution.

(iv) How do you account for the greater stability of ferrate ions at high pH?

(*e*) Using the half-reactions below, calculate the percentage purity of the barium ferrate sample. (Relative atomic masses: Ba = 137, Fe = 56, O = 16)

$$FeO_4^{2-} + 8H^+ + 3e^- \rightarrow Fe^{3+} + 4H_2O$$
$$I_2 + 2e^- \rightarrow 2I^-$$
$$S_4O_6^{2-} + 2e^- \rightarrow 2S_2O_3^{2-}$$

(L 92)

30. Pure iron melts at 1530 °C and the liquid will dissolve carbon causing a lowering of the melting point. A eutectic is formed at 1125 °C.

Ordinary iron (alpha-iron) has a body-centred structure. At 912 °C alpha-iron changes to gamma-iron which has a face-centred structure.

(*a*) (i) What is a eutectic?

(ii) Explain, using a diagram, what is meant by a body-centred cubic structure.

(*b*) Mild steel contains iron and carbon. The percentage of iron in steel may be determined by dissolving the steel in dilute sulphuric acid and titrating the iron(II) produced with potassium manganate(VII).

(i) Write the balanced equation for the reaction of iron(II) ions with manganate(VII) ions using the following 'half equations':

$$Fe^{2+} \rightarrow Fe^{3+} + e^-$$
$$MnO_4^- + 8H^+ + 5e^- \rightarrow Mn^{2+} + 4H_2O$$

(ii) State the colour change expected at the end-point of the titration.

(iii) State how you would show that iron(III) had been produced in the titration. Name the reagent used and the observation expected.

(*c*) Both pure iron and steel rust. The initial step in rusting produces iron(II) ions by an electrochemical process involving air and water.

The iron(II) ions are rapidly oxidised by air to form rust which is hydrated iron(III) oxide.

The electrode reactions and their standard electrode potentials are:

$$Fe^{2+} + 2e^- \rightleftharpoons Fe; \quad E^\ominus = -0.44\,V$$
$$O_2 + 2H_2O + 4e^- \rightleftharpoons 4OH^-; \quad E^\ominus = +0.40\,V$$

(i) Use the information presented to calculate the emf produced in the cell which produces iron(II) ions.

(ii) From the half-cell equations write a (reversible) balanced equation for the overall cell reaction.

(iii) Draw a fully labelled diagram to show how the standard electrode potential of iron can be determined.

(iv) Rusting is slowed down by the presence of alkali. Use the equation obtained in part (ii) to explain this fact.

(NI 92)

31. 'Edta' is a ligand whose structure in alkaline solution is shown below.

$$\begin{array}{ccc} {}^-O_2CCH_2 & & CH_2CO_2{}^- \\ & \diagdown \quad\quad\quad\quad \diagup & \\ & N-CH_2-CH_2-N & \\ & \diagup \quad\quad\quad\quad \diagdown & \\ {}^-O_2CCH_2 & & CH_2CO_2{}^- \end{array}$$

Edta titrations are used in the quantitative analysis of many metal ions, such as the nickel ion. The hydrated nickel ion is $[Ni(H_2O)_6]^{2+}$, and all six water molecule ligands can be replaced by *one* molecule of edta, giving a nickel–edta complex.

The relative molecular mass of nickel salt, containing one mole of nickel ions per mole of salt, was determined as follows:

1.01 g of the salt was dissolved in 100 cm^3 of water. 10.0 cm^3 of this solution was treated with 25.0 cm^3 of 0.05 mol dm^{-3} edta solution, in the presence of an alkaline buffer solution. The unreacted edta was back-titrated with 0.05 mol dm^{-3} $MgSO_4$ solution, in the presence of a little solochrome black. 15.0 cm^3 of $MgSO_4$ were required to react.

(*a*) What is the oxidation state of nickel in the $[Ni(H_2O)_6]^{2+}$ ion?

(*b*) Draw a sketch of the hydrated nickel ion, naming its shape.

(*c*) Make a careful copy of the edta structure. Show clearly the six positions from which the edta is able to form dative bonds.

(*d*) How might the structure of edta change when in solution of pH less than 7?

(*e*) What effect does a *buffer* solution have?

(*f*) Why is it necessary to add *excess* aqueous edta to the nickel salt solution?

(*g*) Suggest the purpose of adding solochrome black before the titration with Mg^{2+}(aq) ions.

(*h*) $MgSO_4$ reacts with edta in a 1 : 1 mole ratio. Use this information to calculate the relative molecular mass of the nickel salt used.

(*i*) Give *one* large-scale use of a nickel alloy.

(O 92, AS)

32. (*a*) Using the Haber and the Kroll processes as examples, discuss the advantages and disadvantages of *continuous* and *batch* manufacturing operations. Start by describing the essential features of these operations, then explain why the particular process is advantaged by the choice of operation. What general features of chemical processes favour the choice of one or other type of operation?

One of the processes above uses a fluidised bed technique. State which of the two processes this is, and indicate *two* advantages of the technique.

(*b*) What is meant by the terms *conversion, yield* and *recycle* as applied to industrial processes?

In a given reaction which is exploited commercially, a molecule of reactant **R** can form either one molecule of wanted product **P**, or one molecule of unwanted by-product **Q**. Under manufacturing conditions, it is found that each 100 mol of **R** produces 70 mol of **P** and 20 mol of **Q**. Calculate the conversion and yield of this reaction. What would be the effect on these quantities if 80% of unreacted **R** could be recovered and recycled? Explain your reasoning.

(*c*) In commercial processes, manufacturing costs can broadly be divided into *capital costs* and *production costs*. Describe and explain the effect of scale of operation on these costs, indicating clearly which one is most immediately affected by an increase from medium to large-scale production.

List *three* factors that tend to increase capital costs and *three* others that can affect production costs. State *two* adverse economic factors to which very large plants are especially vulnerable.

(NEAB 91, S)

33. (*a*) Explain *each* of the following.

(i) Aqueous iron(III) chloride has a pH of less than 7.

(ii) When aqueous ammonium thiocyanate is added to aqueous iron(III) chloride an intense red colour develops which is discharged on the addition of aqueous sodium fluoride.

(*b*) Aqueous iron(III) ions are known to react quantitatively with hydroxylamine, NH_2OH, in acidic solution. This is a redox reaction in which hydroxylamine is oxidised to a simple gaseous oxide of nitrogen and iron(III) ions are reduced to iron(II) ions. There are no other nitrogen-containing products.

In a typical experiment, 50 g of ammonium iron(III) sulphate-12-water were dissolved in dilute sulphuric acid and the solution made up to 1.00 dm^3 with distilled water. 0.8793 g of hydroxylammonium chloride were dissolved in water and the solution made up to 100 cm^3. 10.0 cm^3 of this solution were pipetted into a conical flask, 50.0 cm^3 of the iron(III) solution (an excess) were added and the mixture heated for 15 minutes. After cooling it was found that the solution required 25.0 cm^3 of aqueous potassium manganate(VII) of concentration 0.0200 mol dm^{-3} for complete reaction.

(i) Determine the oxidation state of nitrogen in NH_2OH.

(ii) Calculate the reacting mole ratio of $Fe^{3+} : NH_2OH$ in the reaction.

(iii) Deduce the oxidation state of nitrogen in the final gaseous nitrogen oxide and name the gas.

(iv) Write a balanced equation for the reaction between iron(III) ions and hydroxylamine.

(v) Suggest *two* reasons why the mixture was heated.

(AEB 92, S)

APPENDIX 2: TOPICS WHICH SPAN GROUPS OF THE PERIODIC TABLE

A2.1: SOME DETECTIVE WORK

1. Spot the gas. Identify the following gases, giving your reasons, and explaining all the reactions described.

(*a*) **A** is a colourless, poisonous gas. It burns readily in air and reduces heated copper(II) oxide to copper. **A** reacts with chlorine to form another poisonous gas. When **A** is heated with aqueous sodium hydroxide under pressure, a salt is formed. This salt gives **A** when treated with concentrated sulphuric acid.

(*b*) **B** is a gas with a distinctive smell. It is neutral and turns starch-iodide paper blue. When **B** is heated and then cooled back to room temperature, the volume is found to have increased by 50% (measured at the same pressure).

(*c*) **C** is a colourless gas with an unpleasant smell. It dissolves sparingly in water to give a neutral solution. It burns in air, 1 mole of **C** requiring 2 moles of oxygen for complete combustion. Two products are formed, one of which reacts with water to form an acid. **C** combines with an equal volume of hydrogen iodide to form a crystalline salt.

2. Identify these solids, giving reasons and explaining all the reactions described.

(*a*) **A** is a shiny metal, which is resistant to corrosion. When heated, **A** reacts with chlorine to form a compound which dissolves in water to form a green solution. When treated with sodium peroxide, the green solution turns yellow.

(*b*) **B** is a brown solid. When heated, it gives a colourless, odourless gas and a yellow residue. When heated with concentrated hydrochloric acid, **B** gives a gas which bleaches moist litmus paper and a solution which reacts with potassium iodide solution to give a yellow precipitate.

(*c*) **C** is a white solid which melts at 44 °C. It burns readily in air, producing a white solid that reacts with water to form an acid. When **C** is heated with sodium hydroxide solution, it gives a gas which is spontaneously flammable in air.

(*d*) **D** is a solid which dissolves in water to give a pale green acidic solution. Addition of silver nitrate solution precipitates a white solid, and addition of sodium hydroxide solution precipitates a green solid. When **D** is heated in a stream of chlorine, it forms a dark brown sublimate.

3. Identify the liquid. Giving reasons, and explaining the reactions involved, identify the following liquids.

E is a colourless liquid which boils close to room temperature. It is not oxidised by chlorine or by potassium manganate(VII). **E** etches glas and its vapour reacts with ammonia to form a white solid.

F is a colourless, neutral liquid. It reacts as a base towards hydrogen chloride and as an acid towards ammonia. It has a very low electrolytic conductivity and high boiling and freezing temperatures compared with analogous compounds of other elements in the same group of the Periodic Table. **F** reacts reversibly with red hot iron.

G is a colourless aqueous solution. It reacts with potassium manganate(VII) to give oxygen. **G** reacts with lead(II) sulphide to form a white solid and with iron(II) salts to form iron(III) salts.

4. Some more solids. Identify these solids, with explanations of the chemical reactions described.

J is a solid which colours a flame lilac. Its aqueous solution is yellow, changing to orange when dilute sulphuric acid is added. This orange solution reacts with ethanol when warmed, turning green and evolving a vapour with a distinctive smell.

K is a black solid. When **K** is melted with potassium hydroxide and potassium nitrate, a green solid results. This dissolves in water to form a purple solution and a suspension of **K**.

L is a colourless solid. Its aqueous solution is neutral. When copper(II) sulphate solution is added, a white precipitate and a brown solution are formed. When lead(II) nitrate solution is added, a yellow precipitate appears. When **L** is heated with concentrated sulphuric acid, it evolves a violet vapour.

M is a colourless solid. It combines with chlorine to form a covalent, colourless liquid. **M** dissolves in water to give an acidic solution, which reduces Fe^{3+} salts to Fe^{2+} salts and gives a dark brown precipitate with hydrogen sulphide.

5. Four solids for you to identify. Give your reasons.

P is an orange-red crystalline solid. When heated, it forms three products, including a green solid and a colourless gas. When **P** is heated with sodium hydroxide solution, an alkaline gas is evolved, and a yellow solution is formed.

Q is a white solid which gives a yellow flame test. When boiled with an excess of ammonium chloride, **Q** evolved a gas which was not identified by simple tests. When added to a solution of iron(II) sulphate in dilute sulphuric acid, **Q** gave a brown colour. The brown colour disappeared on heating, with the evolution of a colourless gas which turned brown on meeting the air.

R is a white solid which gives a lilac flame test. The addition of aqueous barium chloride to a solution of **R** gives a white precipitate, which dissolves in dilute hydrochloric acid with the evolution of a gas which decolorises acidified potassium manganate(VII).

S is a crystalline green solid, containing water of crystallisation. With silver nitrate solution, aqueous **S** gives a white precipitate which dissolves in aqueous ammonia. **S** reacts with hydrogen peroxide in alkaline solution to form a yellow solution, which turns orange when acidified. The yellow solution gives a yellow precipitate with lead(II) nitrate solution.

6. Identify **A**, **B**, **C**, **D** and **E**, giving your reasons.

$$\text{A (\textit{Dark brown solid})} \xrightarrow{KNO_3,\ KOH,\ \text{fuse.}} \text{Melt} \xrightarrow[\text{with water}]{\text{extracted}} \text{B(aq) \textit{Green}}$$

$$\text{B(aq)} \xrightarrow{CO_2(g)} \text{C(aq), \textit{Purple}} + \text{A(s)}$$

$$\text{E (s)} \xleftarrow{(NH_4)_2S(aq)} \text{D(aq)} \xleftarrow[\text{agent, acid}]{\text{reducing}} \text{C(aq), \textit{Purple}} + \text{A(s)}$$

E: (*Buff-coloured precipitate*)

7. Identify **P**, **Q**, **R**, **S** and **T**, and give the colours of **Q**, **R**, **S** and **T**.

$$\text{P(aq)} \xleftarrow{\text{conc. HCl(aq)}} Fe(NO_3)_3(aq) \xrightarrow{Na_2CO_3(aq)} \text{Q(s)}$$

P: *Pale yellow solution*; $Fe(NO_3)_3(aq)$: *Orange-yellow solution*

$$Fe(NO_3)_3(aq) \xrightarrow{Zn\ +\ H_2SO_4(aq)} \text{R(aq)}$$

$$\text{S(s)} \xleftarrow{NaOH(aq)} \text{R(aq)} \xrightarrow{Na_2CO_3(aq)} \text{T(s)}$$

8. Describe how you could distinguish between the members of the following pairs:

(*a*) $CO_3^{2-}(aq)$ and $HCO_3^-(aq)$
(*b*) $SO_4^{2-}(aq)$ and $SO_3^{2-}(aq)$
(*c*) $NO_3^-(aq)$ and $NO_2^-(aq)$
(*d*) $S_2O_3^{2-}(aq)$ and $SO_4^{2-}(aq)$
(*e*) $FeCl_3(aq)$ and $K_2Cr_2O_7(aq)$
(*f*) $Ag^+(aq)$ and $Zn^{2+}(aq)$.

9. The following procedures are used to identify some common cations and anions. In each case, describe what you would see in each step of the procedure. Explain the basis of each test in terms of the chemical reactions involved, giving equations.

(*a*) $Cu^{2+}(aq)$: add aqueous ammonia slowly until it is present in excess.

(*b*) $Cr^{3+}(aq)$: add aqueous sodium hydroxide until the solution is strongly alkaline, then add aqueous hydrogen peroxide, boil the mixture and finally acidify the solution with dilute sulphuric acid.

(*c*) $NO_3^-(aq)$: make the solution alkaline with aqueous sodium hydroxide, add zinc (or aluminium or Devarda's alloy) and test for the evolution of an alkaline gas.

†(*d*) $CH_3CO_2^-$ (as in solid sodium ethanoate): add concentrated sulphuric acid to the dry ethanoate salt then add a little ethanol and warm the mixture gently.

[For (*d*) see § 30.7.1.] (O & C 91)

A2.2: YOU HAVE SOME EXPLAINING TO DO

1. Explain the following statements:

(*a*) Aqueous solutions of sodium nitrate, aluminium sulphate and potassium cyanide have different pH values.

(*b*) Although silicon tetrachloride is readily hydrolysed by water, carbon tetrachloride is not.

(*c*) Solid beryllium chloride is soluble in ethoxyethane.

(*d*) Calcium oxide melts at a higher temperature than sodium chloride.

(*e*) Ammonium nitrate dissolves in water, although the process is endothermic.

(*f*) Potassium thiocyanate will detect iron(III) ions in the presence of iron(II) ions in aqueous solution.

(*g*) The standard enthalpy of neutralisation of hydrochloric acid by aqueous potassium hydroxide and sodium hydroxide is the same but is different from that obtained when aqueous ammonia is used.

(*h*) Although methane is unaffected by aqueous alkali, silane is violently hydrolysed.

(*i*) When potassium manganate(VII) is used in redox titrations, acidic conditions are employed. Sulphuric acid is often used.

(*j*) The shapes of the molecules BF_3 and PCl_3 are different.

(*k*) In XeF_4, the arrangement of bonds is square planar.

(*l*) There are two isomers with the formula $[Co(NH_3)_4Cl_2]^+$.

(*m*) The standard enthalpy of solution of alkali metal fluorides changes from endothermic (e.g., LiF, NaF) to exothermic (e.g., KF, RbF, CsF).

(*n*) The decrease in atomic radius from scandium to zinc is very small.

(*o*) Potassium chromate(VI) is used as an indicator in the titration of a solution of chloride ions with silver nitrate solution.

(*p*) The boiling temperature of ammonia (−33 °C) is higher than that of phosphine (−88 °C).

(*q*) Tin(IV) chloride is a fuming liquid; tin(II) chloride is a solid.

(*r*) Silver nitrate solution is acidified with dilute nitric acid before being used to test for the presence of halide ions in solution.

2. Explain the following statements:

(*a*) Although water and ammonia react with anhydrous copper(II) sulphate, methane does not.

(*b*) Dilution of a concentrated solution of copper(II) chloride results in a change of colour from green to blue.

(*c*) The addition of potassium iodide solution to aqueous copper(II) sulphate produces a deep brown solution.

(*d*) Although sulphur forms a hexafluoride, the highest chloride is sulphur tetrachloride.

(*e*) Sulphurous acid is a weaker acid than sulphuric acid.

(*f*) Hydrogen sulphide does not precipitate copper(II) sulphide from a solution of copper(II) sulphate and potassium cyanide.

(*g*) There are many alkanes but few silanes; there are many giant silicate structures, but no analogous carbonates.

(*h*) The properties of beryllium resemble those of aluminium.

(*i*) The dissolution of sodium hydroxide in water is exothermic, whereas the dissolution of sodium nitrite is endothermic.

(*j*) The position of equilibrium in the reaction

$$Mg(s) + Cu^{2+}(aq) \rightleftharpoons Cu(s) + Mg^{2+}(aq)$$

can be predicted from the values of the standard electrode potentials:

$$E^\ominus(Mg^{2+}/Mg) = -1.96\,V;$$
$$E^\ominus(Cu^{2+}/Cu) = +0.34\,V$$

(*k*) A solution of aluminium nitrate boils at a higher temperature than a solution of sodium nitrate of the same molar concentration.

(*l*) A solution of aluminium nitrate is more effective in coagulating a sol than an equimolar solution of sodium nitrate.

(*m*) Aluminium chloride is a catalyst for a number of organic reactions.

A2.3: PATTERNS IN THE PERIODIC TABLE

1. Metallic character increases from right to left and from top to bottom in the Periodic Table. Illustrate this statement by referring to the period from sodium to chlorine and Group 4.

***2.** The first element in a group of the Periodic Table often has properties which differ from those of later members. Illustrate this statement by referring to lithium, beryllium and fluorine.

3. Illustrate one of the trends in the Periodic Table by considering the properties of the oxides: Na_2O, CaO, Al_2O_3, SiO_2 and P_2O_5.

4. Shown below are the first ionisation energies (E/kJ mol^{-1}) of the elements of the second short period (Z = atomic number):

Element	Na	Mg	Al	Si	P	S	Cl	Ar
Z	11	12	13	14	15	16	17	18
E/kJ mol^{-1}	500	740	580	790	1010	1000	1260	1520

(*a*) State the meaning of *first ionisation energy*.

(*b*) Plot the ionisation energies against atomic number.

(*c*) Describe the shape of the plot you obtain, and explain why it has this form.

(*d*) Mention one way in which the behaviour of the elements mirrors the form of the graph you have plotted.

5. Distinguish between electronegativity and electron affinity.
How does electronegativity vary (*a*) from carbon through nitrogen and oxygen to fluorine and (*b*) from fluorine to iodine? Explain the variations.
Which of these elements is the most powerful oxidising agent? Why?

***6.** What is meant by the term *diagonal relationship*? Give two examples of pairs of elements which show a diagonal relationship to one another, and quote two chemical reactions for each pair to illustrate the relationship.

7. Give an account of the hydrides of the elements. Relate the nature of the bonding in the hydrides to the position of the element in the Periodic Table and to the properties of the compounds.

8. Compare the following with respect to (w.r.t.) the features cited:
(*a*) Ammonia and phosphine w.r.t. stability and basicity.
(*b*) Water and hydrogen sulphide w.r.t. boiling temperature and ionisation.
(*c*) The halogens w.r.t. oxidising power.
(*d*) Magnesium and calcium w.r.t. reactivity with water.
(*e*) Carbon dioxide and silicon(IV) oxide w.r.t. structure and acidity.

9. Discuss the elements of the second short period (Na to Ar) with reference to the following features:
(*a*) their action on water
(*b*) their reaction with dilute sulphuric acid
(*c*) methods for the preparation of their chlorides
(*d*) the physical and chemical properties of their chlorides.

10. Explain the following statements:
(*a*) The atomic radii of the first series of transition elements are similar.
(*b*) The ionic radius of sodium is smaller than its atomic radius, but chloride ions have a larger radius than chlorine atoms.
(*c*)• The first ionisation energy of potassium is smaller than that of sodium.
(*d*) The first electron affinity of chlorine is greater (more exothermic) than that of iodine.
(*e*) The ion Pb^{2+} is stable, but C^{2+} does not exist.
(*f*) The salts NaCl and NaF have similar crystal structures, but CsCl and CsF have different crystal structures.
(*g*) Beryllium chloride shows evidence of covalent bonding, but calcium chloride is ionic.

11. Explain the trends in the atomic radii of the elements (*a*) along Period 3 (Na–Cl) and (*b*) down Group 7 (F–I). Show how these trends help to explain the changes in the chemistry of the elements.

12. The bonding in sodium chloride is different from that in aluminium chloride. Discuss this statement, referring to the action of (*a*) heat and (*b*) cold water on the two compounds.

13. Why are elements described as s block, p block or d block?
Which features are characteristic of the elements in each of these three blocks?

14. Discuss the oxides formed by (*a*) the first period (Li–F) and (*b*) Group 4 (C–Pb). Mention (i) the structure and bonding and (ii) the acid–base properties of the oxides.

15. Write an account of the elements in the second short period (Na–Cl). Include a discussion of their structures and bonding and the chemistry of their oxides and hydrides.

16. Give an account of the chlorides of the elements in the second short period from sodium to sulphur. Mention the methods of preparation, the type of bonding and the reaction of the chlorides with water.

17. Plot a graph of ionic radius (vertical axis) against atomic number (horizontal axis) for the elements sodium to chlorine. Explain the variation in ionic radius in terms of electron configurations.

Ion	Na^+	Mg^{2+}	Al^{3+}	Si^{4+}	P^{3-}	S^{2-}	Cl^-
Radius/nm	0.095	0.065	0.050	0.041	0.212	0.184	0.181

***18.** Beryllium and aluminium are positioned diagonally to one another in the Periodic Table. Why would this relationship lead you to expect similar properties for the two elements? Point out ways in which the chemistry of the elements bears out your prediction.

19. (*a*) Give *one* example of each of the following types of oxide: acidic, basic, amphoteric, neutral.
Using your examples, discuss for each type of oxide:
(i) the chemical properties (with balanced equations)
(ii) the structures and bonding.
(*b*) Compare and contrast the reactions of fluorine and chlorine with water.
(*c*) Suggest explanations for the following observations:
(i) There are only two oxidation states of lead, Pb(II) and Pb(IV), and yet Pb_2O_3 and Pb_3O_4 are known.
(ii) When H_2O_2 is added to acidified dichromate(VI) solution, CrO_5 may be obtained and yet Cr has a maximum oxidation state of six.

(O 91)

20. (*a*) Describe the bonding in and the structure of the anhydrous chlorides formed by each of sodium, aluminium and silicon. Explain differences in their volatilities.

(*b*) Distinguish between the terms *hydration* and *hydrolysis*. Illustrate your answer with reference to the changes which occur when sodium chloride and silicon tetrachloride are added separately to water. In *each* case describe what would be observed, identify the products and comment on the pH of any solution formed.

(*c*) When an aqueous solution of sodium carbonate is added to an aqueous solution of aluminium sulphate, a white precipitate is formed and a colourless gas is evolved. The white precipitate is insoluble in an excess of sodium carbonate solution but dissolves in sodium hydroxide solution and in dilute hydrochloric acid.

(i) Identify the gas and the white precipitate. Write ionic equations for the formation of the white precipitate and for its reactions with sodium hydroxide and with hydrochloric acid.

(ii) Name the property exhibited by aluminium in this series of experiments and explain why the white precipitate is soluble in sodium hydroxide but insoluble in sodium carbonate solution.

(NEAB 92)

21. Explain *each* of the following:

(*a*) A blue precipitate is obtained when either aqueous potassium hexacyanoferrate(III) is added to aqueous Fe^{2+} ions, or when aqueous potassium hexacyanoferrate(II) is added to aqueous Fe^{3+} ions.

(*b*) When copper(II) carbonate is dissolved in the minimum quantity of concentrated hydrochloric acid the resulting solution is yellow/brown but addition of water turns the mixture green and on further dilution with water it turns blue.

(*c*) When excess tin is added to a solution of iodine in trichloromethane the colour of the liquid gradually changes from purple to orange/brown.

On filtration and evaporation an orange-brown solid remains with a relative molecular mass of approximately 627.

(*d*) An oxide of barium containing 81.1% barium by mass gives a white precipitate when added to an excess of ice-cold aqueous sulphuric acid. Subsequent filtration gives a colourless liquid which discharges the colour of acidified aqueous potassium manganate(VII) and gives a brown colour with acidified aqueous potassium iodide.

(AEB 92, S)

22. A bright red solid **A** dissolves in water to give a highly acidic solution. When a solution of **A** is made alkaline with aqueous sodium hydroxide, the solution turns yellow. On heating 1.0 g of **A**, 0.76 g of a green powder **B** is formed and 168 cm^3 of oxygen (measured at stp) are given off.

An orange solid **C** may be obtained from the solution made by dissolving 1.0 g of **A** in 5.0 cm^3 of 2.0 mol dm^{-3} ammonia.

When 1.26 g of **C** are warmed, a violent reaction takes place, giving off 112 cm^3 (measured at stp) of an inert gas, as well as steam, and leaving behind the same green powder **B** that was made by heating **A**.

If **A** is dissolved in cold, concentrated hydrochloric acid, and concentrated sulphuric acid is then gradually added, a dark, red-brown oil **D** separates out. **D** has a boiling point of 117 °C and rapidly reacts with water. **D** contains 45.8% of chlorine by mass.

Identify **A, B, C** and **D**, give equations for all the reactions involved and show that these equations are consistent with the quantitative data.

(C 92, S)

23. Identify, as fully as possible, the substances **W, X, Y** and **Z**. Explain each of the reactions described and give equations where possible.

(*a*) Substance **W** is a white powder which dissolves in water to form a blue solution. When concentrated hydrochloric acid is added to aqueous **W** the solution turns green but the blue colour returns when the solution is diluted with water. A white precipitate forms when aqueous barium chloride is added to aqueous **W**. The precipitate is insoluble in dilute hydrochloric acid.

(*b*) Substance **X** is a yellow solid which dissolves in water to give a yellow solution. On the addition of dilute sulphuric acid the solution turns orange. When the acidified aqueous **X** is warmed with ethanol the solution turns green. A flame test carried out on solid **X** produced a lilac flame.

(*c*) Substance **Y** is a pale pink solid which dissolves in water to give an almost colourless solution. A white precipitate forms when aqueous sodium hydroxide is added but this precipitate turns brown on standing in an open test tube. When aqueous silver nitrate is added to aqueous **Y** a white precipitate forms which is insoluble in dilute nitric acid.

(*d*) Substance **Z** is a white solid which dissolves in water to form a colourless solution. When aqueous ammonia is added a white precipitate forms which redissolves in excess of the reagent. When dilute sulphuric acid is added to **Z** and the mixture is warmed, a vinegar-like smell is observed and the vapour turns blue litmus red.

(AEB 90)

24. Suggest explanations for each of the following.

(*a*) Solid sodium chloride has a small, but measurable, electrical conductivity.

(*b*) When $MgCl_2 \cdot 6H_2O$ is heated, water and hydrogen chloride are given off and a solid of composition Mg_2OCl_2 is left.

(*c*) Aluminium metal dissolves in concentrated hydrochloric acid but not in concentrated nitric acid.

(*d*) Phosphorus forms trihalides and pentahalides with fluorine, chlorine and bromine, but it forms only a trihalide with iodine.

(*e*) Sulphur forms compounds with fluorine, chlorine and bromine but none with iodine.

(*f*) The ozone layer is damaged by chlorine atoms, but not fluorine atoms, from chlorofluorocarbons (CFCs).

(Bond energies: C—F 490 kJ mol^{-1}, C—Cl 340 kJ mol^{-1})

(C 92, S)

25. A gas **A**, when burned in an atmosphere of gas **B**, produces gas **C** which dissolves in water to form a strongly acidic solution **D**. **B** and **D** both react with aluminium metal to produce the compound **E** but in two different forms. That made from **B** is in the form of yellow lumps which react vigorously with water to evolve **C** whereas that made from **D** is a white crystalline solid which dissolves quietly in water.

The acidic solution **D** reacts with aqueous sodium hydroxide to produce a solution of compound **F** whereas the gas **B** reacts with aqueous sodium hydroxide to produce a mixture of two compounds **F** and **G**.

B reacts with potassium iodide in the presence of trichloroethane to produce a purple solution whereas, in the same situation, **D** has no effect.

(*a*) Identify **A, B, C, D, E, F** and **G**.

(*b*) Describe the bonding in the yellow form of **E**.

(*c*) Outline why the white form of **E** dissolves in water to give an acidic solution.

(*d*) Why is aluminium often chosen as the material used for the manufacture of the burners for gas cookers?

(C 90, AS)

Part 4

ORGANIC CHEMISTRY

25

ORGANIC CHEMISTRY

25.1 ORGANIC CHEMISTRY

Organic chemistry was originally described as the chemistry of compounds found in living things; in plants and animals. All such naturally occurring compounds contain carbon, and it was thought that some 'vital force' was needed for their formation. Sucrose, $C_{12}H_{22}O_{11}$, was obtained from sugar cane; urea, $CO(NH_2)_2$, was obtained from urine; and glycerol, $CH_2OHCHOHCH_2OH$, was obtained from the saponification of mutton fat. Until 150 years ago, the only carbon compounds that had been made in the laboratory were very simple compounds such as carbon dioxide, calcium carbide, CaC_2, and potassium cyanide, KCN. When F Wöhler, in 1828, made urea from an inorganic salt, ammonium cyanate, NH_4CNO, he changed the definition of organic chemistry. Today, the term organic chemistry refers to the chemistry of millions of carbon compounds. Some of them have been extracted from plant or animal sources, but many more have been made by organic chemists in their laboratories.

Organic compounds were originally obtained from natural sources

... until Wöhler synthesised urea from an inorganic salt

Carbon forms more compounds than any other element. With the ground state electron configuration $1s^22s^22p^2$ [§ 2.7], it has very little tendency to form positive or negative ions. In order to achieve a stable outer octet of electrons, it forms four covalent bonds. The electron pairs in these bonds occupy sp^3 hybrid orbitals. They have axes directed to the corners of a regular tetrahedron with the carbon nucleus at the centre.

Carbon forms four covalent bonds

When a carbon atom combines with four hydrogen atoms, it forms a molecule of methane, CH_4 [see Figure 25.1(a)]. If two carbon atoms join, each can still combine with three hydrogen atoms to form a molecule of ethane, C_2H_6 [see Figure 25.1(b)]. A model of a molecule of propane, C_3H_8, is shown in Figure 25.1(c).

FIGURE 25.1(a) Models of CH_4

FIGURE 25.1(b)
Models of C_2H_6

FIGURE 25.1(c)
Models of C_3H_8

25.2 HYDROCARBONS

Compounds of carbon and hydrogen are called **hydrocarbons**. The compounds shown above, methane, ethane and propane, are **alkanes**. They possess only single bonds and have the general formula

$$C_nH_{2n+2}$$

The alkanes are a homologous series

A series of compounds with similar chemical properties, in which members differ from one another by the possession of an additional CH_2 group, is called a **homologous series**. The first ten members of the unbranched-chain alkane series are:

The names of alkanes and alkyl groups

CH_4	Methane	C_6H_{14}	Hexane
C_2H_6	Ethane	C_7H_{16}	Heptane
C_3H_8	Propane	C_8H_{18}	Octane
C_4H_{10}	Butane	C_9H_{20}	Nonane
C_5H_{12}	Pentane	$C_{10}H_{22}$	Decane

The groups of atoms CH_3—, C_2H_5— and C_nH_{2n+1}— are called *methyl*, *ethyl* and *alkyl* groups.

25.3 ISOMERISM AMONG ALKANES

Figure 25.1 shows models of methane, ethane and propane. The next member of the series, butane, C_4H_{10}, can be modelled in two ways:

```
(a) H        H        H           or (b)  H  H           H  H
     \      /        /                     \ |           | /
 H    C          C—H                         C    H    C
  \  /  \       /  \                        /  \  |   /  \
 H—C      C        H                      H      C        H
  /      /  \                                    |
 H      H    H                                   C
                                               / | \
                                              H  H  H
```

Isomeric compounds have the same molecular formulae but different structural formulae

Molecule (a) is a continuous-chain or unbranched-chain molecule, and (b) is a branched-chain molecule. The two formulae correspond to different compounds. Compound (a) is called butane (or sometimes, *normal*-butane or *n*-butane) and boils at −0.5 °C. Compound (b) is called 2-methylpropane, and boils at −12 °C.

The existence of different compounds with the same molecular formulae but different structural formulae is called **isomerism**. There are five isomeric hexanes of formula C_6H_{14}. They are:

The isomers of hexane

$CH_3CH_2CH_2CH_2CH_2CH_3$ — Hexane (sometimes referred to as *normal*-hexane or *n*-hexane)

$\begin{array}{l} CH_3CH_2CH_2\underset{\displaystyle |\atop \displaystyle CH_3}{C}HCH_3 \end{array}$ — 2-Methylpentane

$CH_3CH_2\underset{\displaystyle |\atop \displaystyle CH_3}{C}HCII_2CH_3$ — 3-Methylpentane

$CH_3CH_2\overset{\displaystyle CH_3\atop \displaystyle |}{\underset{\displaystyle |\atop \displaystyle CH_3}{C}}—CH_3$ — 2,2-Dimethylbutane

$CH_3\underset{\displaystyle |\atop \displaystyle CH_3}{C}H—\underset{\displaystyle |\atop \displaystyle CH_3}{C}HCH_3$ — 2,3-Dimethylbutane

25.4 SYSTEM OF NAMING HYDROCARBONS

The IUPAC system of nomenclature

The names of these isomers are given in accordance with the International Union of Pure and Applied Chemistry (IUPAC) system of nomenclature. The procedure followed is:

1. Name the longest unbranched carbon chain.
2. Name the substituent groups.
3. Give the positions of the substituent groups.

$CH_3—CH_2—CH_2—\underset{\displaystyle |\atop \displaystyle CH_3}{C}H—CH_3$

2-Methylpentane

One could count from the other end and call it 4-methylpentane, but the IUPAC system is to count from the end which will give the lower locant (number) for the position of a substituent group.

$$\begin{array}{l} CH_3{-}CH_2{-}\underset{\displaystyle H_3C}{\underset{|}{CH}}{-}\underset{\displaystyle CH_3}{\underset{|}{CH}}{-}CH_3 \end{array}$$

2,3-Dimethylpentane

$$CH_3{-}CH_2{-}CH_2{-}\underset{\displaystyle H_3C}{\underset{|}{CH}}{-}\underset{\displaystyle C_2H_5}{\underset{|}{CH}}{-}CH_3$$

3,4-Dimethylheptane

In spite of the way it is written, you should be able to see that the longest unbranched chain has seven carbon atoms.

$$CH_3{-}\underset{\displaystyle Cl}{\underset{|}{CH}}{-}CH_2{-}\underset{\displaystyle Br}{\underset{|}{CH}}{-}CH_3$$

2-Bromo-4-chloropentane

The substituent groups are named in alphabetical order.

$$CH_3{-}CH_2{-}CH_2{-}\underset{\displaystyle Br}{\underset{|}{CH}}{-}\underset{\displaystyle Cl}{\underset{|}{CH}}{-}CH_3$$

3-Bromo-2-chlorohexane

The substituents are named in alphabetical order, not in the numerical order of the locants.

To construct the formula of a compound from its name, e.g., 2,3-dichloro-4-methylhexane, first write the carbon atoms of the hexane part of the molecule:

$$C{-}C{-}C{-}C{-}C{-}C$$

Then put in Cl atoms on carbons 2 and 3, and a CH_3 group on carbon 4:

$$C{-}\underset{\displaystyle Cl}{\underset{|}{C}}{-}\underset{\displaystyle Cl}{\underset{|}{C}}{-}\underset{\displaystyle CH_3}{\underset{|}{C}}{-}C{-}C$$

Fill in the hydrogen atoms to give each carbon atom a valency of four:

$$CH_3{-}\underset{\displaystyle Cl}{\underset{|}{CH}}{-}\underset{\displaystyle Cl}{\underset{|}{CH}}{-}\underset{\displaystyle CH_3}{\underset{|}{CH}}{-}CH_2{-}CH_3$$

25.4.1 UNSATURATED HYDROCARBONS

Alkanes are saturated hydrocarbons

Alkenes and alkynes are unsaturated hydrocarbons: they contain multiple bonds

Alkanes, alkenes and alkynes are aliphatic hydrocarbons

The alkanes are not the only hydrocarbons. There are also alkenes and alkynes. Alkanes are said to be **saturated** hydrocarbons as they contain only single bonds between carbon atoms. Alkenes and alkynes are **unsaturated** hydrocarbons: they contain multiple bonds between carbon atoms. The simplest alkene is ethene

$$H_2C{=}CH_2$$

formerly called ethylene. It is the first member of the homologous series of alkenes, which have the general formula C_nH_{2n}. Alkynes contain one or more carbon–carbon triple bonds. Ethyne

$$HC{\equiv}CH$$

is the first member of the homologous series of alkynes, which have the general formula C_nH_{2n-2}.
The names of the other members of the series are given in Table 25.1. All these hydrocarbons are classified as **aliphatic** hydrocarbons. Aliphatic means 'fatty' in Greek, the connection being that fats contain large alkyl groups, e.g. $C_{15}H_{31}{-}$.

TABLE 25.1 Names of Aliphatic Hydrocarbons

No. of C atoms	*Alkane*		*Alkene*		*Alkyne*		*Alkyl group*	
1	CH_4	Methane					$CH_3—$	Methyl
2	C_2H_6	Ethane	C_2H_4	Ethene	C_2H_2	Ethyne	$C_2H_5—$	Ethyl
3	C_3H_8	Propane	C_3H_6	Propene	C_3H_4	Propyne	$C_3H_7—$	Propyl
4	C_4H_{10}	Butane	C_4H_8	Butene	C_4H_6	Butyne	$C_4H_9—$	Butyl
5	C_5H_{12}	Pentane	C_5H_{10}	Pentene	C_5H_8	Pentyne	$C_5H_{11}—$	Pentyl
n	C_nH_{2n+2}		C_nH_{2n}		C_nH_{2n-2}		$C_nH_{2n+1}—$	

In naming alkenes and alkynes, the positions of the multiple bonds must be stated:

How to state the position of the double bond in an alkene

$CH_2{=}CH—CH_2—CH_3$ But-1-ene

The *but*-part of the name shows that there are 4 carbon atoms. The *-ene* suffix shows that there is a C=C double bond. The number 1 indicates that the double bond is between carbon atoms 1 and 2. Count from the end that will give the lowest numbers, not 3 and 4.

$CH_3—CH{=}CH—CH_3$ But-2-ene

$CH_2{=}CH—CH{=}CH_2$ Buta-1,3-diene

$CH_2{=}C(CH_3)—CH_2CH_3$ 2-Methylbut-1-ene

$CH_3—CH{=}CH—CH(CH_3)—CH_3$ 4-Methylpent-2-ene

The double bond is numbered first and then the methyl group.

How to state the position of the triple bond

$CH_3—CH(CH_3)—C{\equiv}C—CH_3$ 4-Methylpent-2-yne

$CH_3—CH{=}CH—C{\equiv}CH$ Pent-2-en-4-yne

The double bond is numbered first, then the triple bond.

25.5 ALICYCLIC HYDROCARBONS

A second set of hydrocarbons is the alicyclic hydrocarbons. They contain rings of carbon atoms. Examples are

Alicyclic hydrocarbons

Cyclopropane: ring of $CH_2—CH_2$ and CH_2

Cyclobutane: ring $CH_2—CH_2$ / $CH_2—CH_2$

Cyclohexene: ring CH_2, CH=CH, CH_2, CH_2, CH_2

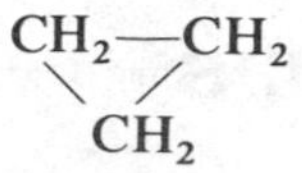

Cyclopropane Cyclobutane 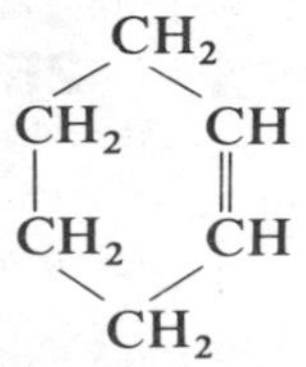 Cyclohexene

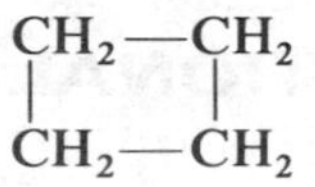

[See Figure 25.2.]

FIGURE 25.2
(a) Cyclopropane
(b) Cyclohexane

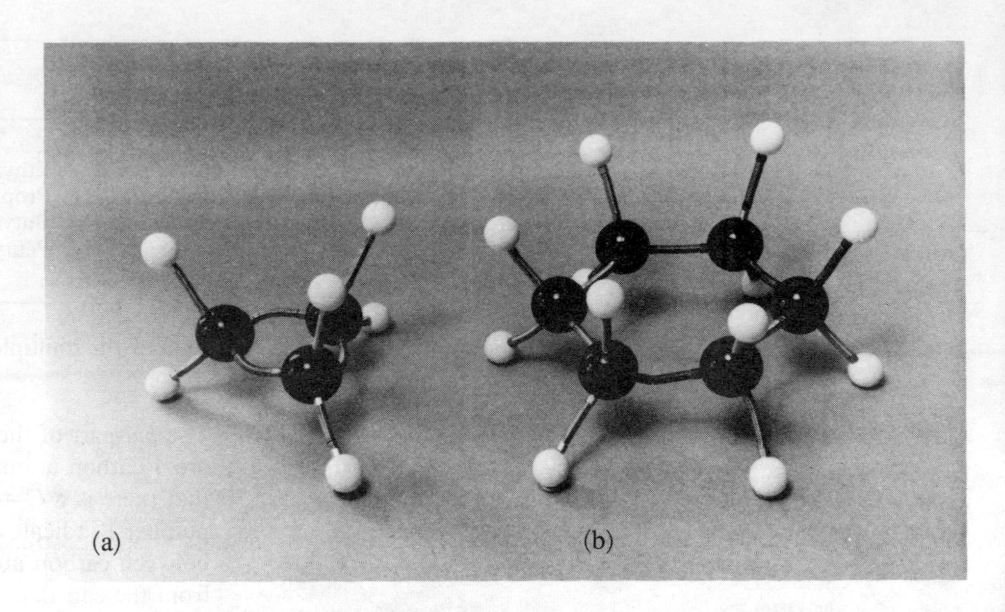

25.6 AROMATIC HYDROCARBONS

Aromatic hydrocarbons are related to benzene; they are called arenes

The phenyl and naphthyl groups are aryl groups

A third group of hydrocarbons is the **aromatic** hydrocarbons. They are related to benzene. The first benzene compounds to be isolated had pleasant aromas, and gave this group of hydrocarbons their name. Members of the group of aromatic hydrocarbons, the **arenes**, are: benzene, C_6H_6; methylbenzene, $C_6H_5CH_3$; and naphthalene, $C_{10}H_8$. The structure of these compounds is discussed § 5.3 and in Chapter 28. The group C_6H_5— is called a **phenyl** group, and the group $C_{10}H_{17}$— is called a **naphthyl** group. Both are **aryl** groups.

The subsets of hydrocarbons are summarised in Figure 25.3.

FIGURE 25.3 Classes of Hydrocarbons

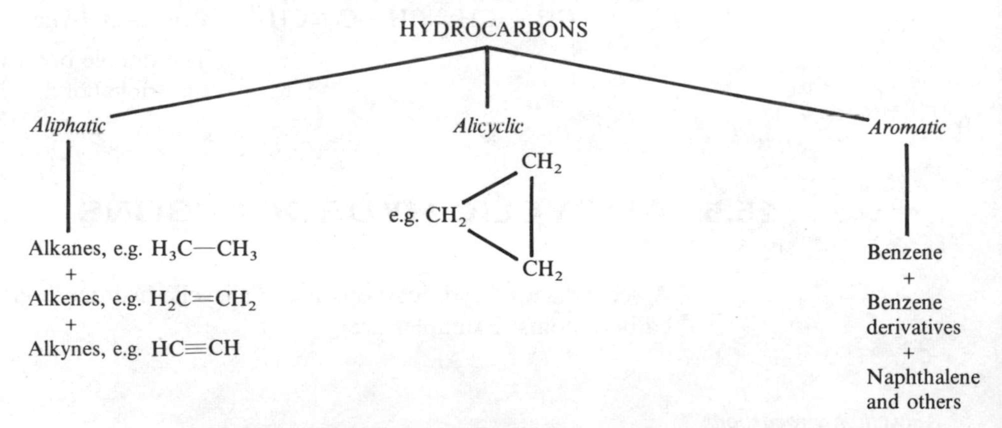

25.7 FUNCTIONAL GROUPS

The double bond in the alkenes is responsible for most of the chemical reactions of these compounds. The group of atoms

$$\gt C=C \lt$$

is called the **functional group** of the alkenes.

Similarly, alkynes possess the functional group

$$—C{\equiv}C—$$

The reactions of a homologous series depend on the functional group

In your study of the reactions of the hydrocarbons, you will come across some other classes of compounds with different functional groups, and it will help to deal with their names in advance.

How to name halogenoalkanes

The halogenoalkanes (or haloalkanes) have the formula RX where R is an alkyl group and X is a halogen:

CH_3Br Bromomethane

C_2H_5Cl Chloroethane

$C_2H_4Cl_2$ Dichloroethane

There are two isomers. Figure 25.4(a) shows 1,2-dichloroethane, $ClCH_2CH_2Cl$, with the chlorine atoms attached to different carbon atoms. Figure 25.4(b) shows 1,1-dichloroethane, CH_3CHCl_2, with the chlorine atoms on the same carbon atom.

FIGURE 25.4 (a) 1,2-Dichloroethane, (b) 1,1-Dichloroethane

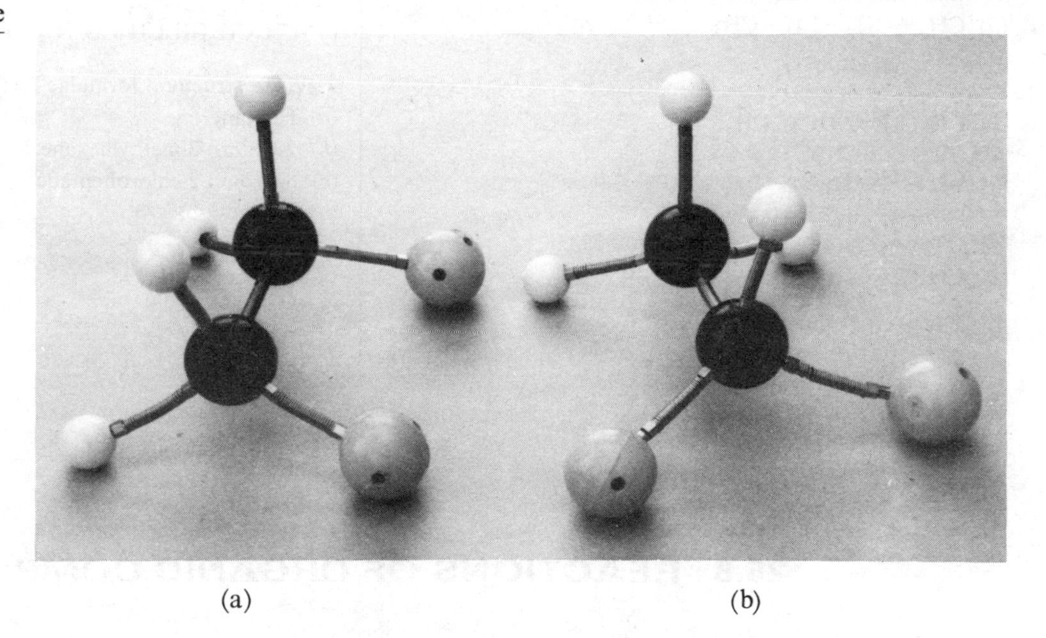

(a) (b)

How to name alcohols...

The **alcohols** or **alkanols** are compounds with the functional group —OH, a hydroxyl group. They are named by taking the name of the alkane with the same number of carbon atoms and changing the ending from *-ane* to *-anol*:

CH_3OH Methanol

C_2H_5OH Ethanol

C_3H_7OH Propanol. There are two isomers

$CH_3CH_2CH_2OH$ Propan-1-ol

$$\begin{array}{l} CH_3CHOH \\ \quad\;\;\; | \\ \quad\;\; CH_3 \end{array}$$ Propan-2-ol

A number must be used to show the position of the hydroxyl group in the carbon chain.

Carboxylic acids or **alkanoic acids** have the functional group

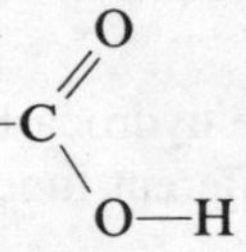

$$-C\begin{matrix} //O \\ \backslash O-H \end{matrix}$$

... and carboxylic acids

They are named by taking the alkane with the same number of carbon atoms and changing the ending from *-ane* to *-anoic acid.*

CH_3CO_2H	Ethanoic acid (There are *two* carbon atoms, that in CH_3 and that in CO_2H.)
$CH_3CH_2CH{=}CHCO_2H$	Pent-2-enoic acid (The C of the CO_2H group is C-1, and the double bond lies between C-2 and C-3.)

CHECKPOINT 25A: NOMENCLATURE

1. Name the following compounds:

(*a*) $CH_3{-}CH_2{-}CH(CH_3){-}CH_3$

(*b*) $CH_3CH{=}CHCH_2CH_3$

(*c*) $CH_3CH(CH_2CH{=}CH_2)CH_3$

(*d*) $CH_3CH_2CH(Cl)CH(Br)CH_3$

(*e*) $CH_3CH_2CH_2OH$

(*f*) $CH_3CH_2CH_2CH_2CO_2H$

2. Write structural formulae for:

(*a*) Heptane

(*b*) 2-Chloro-3-methylhexane

(*c*) 3-Bromo-2-chloroheptane

(*d*) Pentan-2-ol

(*e*) Hex-2-ene

(*f*) Butanoic acid

25.8 REACTIONS OF ORGANIC COMPOUNDS

25.8.1 TYPES OF REACTIONS

The reactions of organic compounds fall into four classes. These are listed below.

SUBSTITUTION

There are four types of organic reactions

An atom or group of atoms replaces another:

$$C_2H_5Cl + OH^- \rightarrow C_2H_5OH + Cl^-$$

Chloroethane + Hydroxide ion → Ethanol + Chloride ion

ADDITION

Two molecules react to form one:

$$Br_2 + CH_2{=}CH_2 \rightarrow BrCH_2CH_2Br$$

Bromine + Ethene → 1,2-Dibromoethane

ELIMINATION

One molecule reacts to form more than one:

$$C_2H_5OH \rightarrow C_2H_4 + H_2O$$

Ethanol → Ethene + Water

REARRANGEMENT

One molecule reacts to give a different molecule:

$$CH_3{-}CH(Cl){-}CH{=}CH_2 \rightleftharpoons CH_3{-}CH{=}CH{-}CH_2{-}Cl$$

3-Chlorobut-1-ene ⇌ 1-Chlorobut-2-ene

25.8.2 TYPES OF BOND FISSION

There are two types of bond fission...

When organic compounds react, their bonds can split in either of two ways, by **homolytic** or **heterolytic** fission.

HOMOLYTIC FISSION

...homolysis...

When the bond breaks, each of the bonded atoms takes one of the pair of electrons. **Free radicals** are formed. These are atoms or groups of atoms with unpaired electrons:

$$CH_3CH_2CH_3 \xrightarrow{\text{heat}} CH_3CH_2\cdot + \cdot CH_3$$

Propane Ethyl radical Methyl radical

$$Cl_2 \xrightarrow{\text{sunlight}} 2Cl\cdot$$

Chlorine molecule → Chlorine atoms or free radicals

Energy must be supplied, either as heat or light, to break the bond. The free radicals formed possess this energy, and are very reactive.

HETEROLYTIC FISSION

...and heterolysis

When the bond breaks, one of the bonded atoms takes both of the bonding electrons to form an anion. The rest of the molecule becomes a cation. An ion with a positively charged carbon atom is called a **carbocation**. An ion with a negatively charged carbon atom is called a **carbanion**:

$$(CH_3)_3C{-}Cl \rightarrow (CH_3)_3C^+ + Cl^-$$

2-Chloro-2-methylpropane → A carbocation + Chloride ion

$$CH_3COCH_2CO_2C_2H_5 + OH^- \rightleftharpoons CH_3COC^-HCO_2C_2H_5 + H_2O$$

Ethyl 3-oxobutanoate A carbanion + Water

25.8.3 TYPES OF REAGENT

In a covalent bond between **A** and **B**, if **A** is more **electronegative** [§ 4.5.3] than **B**, the distribution of bonding electrons can be represented as

$$\overset{\delta-}{\mathbf{A}}{-}\overset{\delta+}{\mathbf{B}}$$

Covalent bonds can be polar

The bond is described as **polar**. The reagents which attack organic compounds seek out either the slightly positive ($\delta+$) end of the bond or the slightly negative ($\delta-$) end of the bond. There are two main classes of reagent:

NUCLEOPHILIC REAGENTS

Nucleophiles attack centres of positive charge

Negative ions, e.g., OH^-, CN^-, and compounds in which an atom has an unshared pair of electrons, e.g., NH_3, are **nucleophilic** (nucleus-seeking). They attack the electron-deficient end of a polar bond.

ELECTROPHILIC REAGENTS

Electrophiles attack centres of negative charge

A reagent which attacks a region where the electron density is high is called an **electrophile**. Examples of electrophilic reagents are the nitryl cation, NO_2^+, and sulphur(VI) oxide, SO_3.

25.8.4 REACTION MECHANISM

The mechanism of a reaction is the sequence of steps from start to finish

The stoichiometric equation for an organic reaction does not tell you how the reaction takes place. There may be a series of reactions in between the mixing of the reactants and the formation of the products. The sequence of steps by which the reaction takes place is called the **reaction mechanism**. The mechanism is worked out from a study of the kinetics of the reaction [see Chapter 14]. Other techniques, such as spectroscopy [see Chapter 34] and the incorporation of radioisotopes into a reactant, are also used.

25.9 ISOMERISM

Isomerism is the existence of different compounds with the same molecular formulae but different structural formulae. There are various types of isomerism.

25.9.1 STRUCTURAL ISOMERISM

A structural formula shows the order in which atoms are bonded together

The structural formula shows the sequence in which the atoms in a molecule are bonded. A structural formula can be written in full, with every bond drawn, or it can be written by joining groups of atoms in sequence, provided the formula for each of the groups is unambiguous. The structural formula for 2-methylpropane is shown in Figure 25.5.

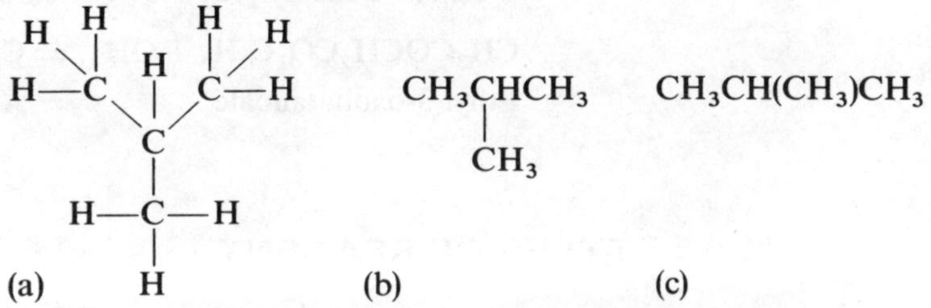

(a) [full structural formula of 2-methylpropane with every bond drawn]

(b) CH_3CHCH_3 with CH_3 attached to the central carbon

(c) $CH_3CH(CH_3)CH_3$

FIGURE 25.5
2-Methylpropane

Formula (b) is unambiguous because there is only one way of writing the bonds in a $-CH_3$ group. Formula (c) is a condensed way of writing formula (b).

CHAIN ISOMERISM

In chain isomers, the carbon 'skeletons' differ

The isomers have different carbon chains. They possess the same functional group, and belong to the same homologous series, e.g.

$CH_3CH_2CH_2CH_3$ — Butane

CH_3CHCH_3 with CH_3 on the central carbon — 2-Methylpropane

POSITIONAL ISOMERISM

The position of a functional group in the carbon skeleton differs between positional isomers

These isomers have a substituent group in different positions in the same carbon 'skeleton'. The isomers are chemically similar because they possess the same functional group, e.g.

(*a*) Propan-1-ol, $CH_3CH_2CH_2OH$ and propan-2-ol, $CH_3CH(OH)CH_3$.

(*b*) Pent-1-ene, $CH_3CH_2CH_2CH{=}CH_2$ and pent-2-ene, $CH_3CH_2CH{=}CHCH_3$.

(*c*) This type of isomerism is found in benzene derivatives. The structural formula of benzene, C_6H_6, was discussed in § 5.3.1. If two chlorine atoms replace two hydrogen atoms to form $C_6H_4Cl_2$, three different compounds can be formed. Their names are given below:

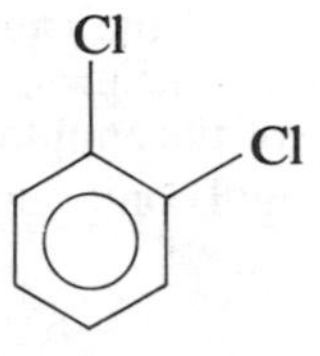

1,2-Dichlorobenzene (*ortho*-dichlorobenzene)

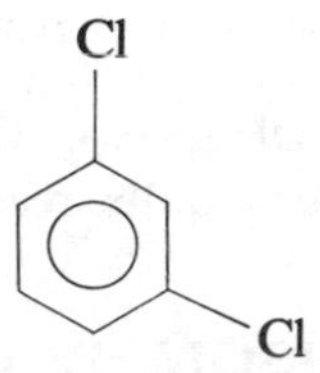

1,3-Dichlorobenzene (*meta*-dichlorobenzene)

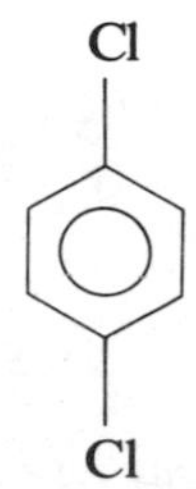

1,4-Dichlorobenzene (*para*-dichlorobenzene)

FUNCTIONAL GROUP ISOMERISM

In another type of isomerism, the functional group is different in the isomers

These isomers have different functional groups and belong to different homologous series. Some examples follow:

(*a*) An alcohol and an ether, e.g.,
ethanol, C_2H_5OH, and methoxymethane (dimethyl ether) CH_3OCH_3.

(*b*) An aldehyde and a ketone, e.g.,
propanal, CH_3CH_2CHO, and propanone, CH_3COCH_3.

(*c*) A carboxylic acid and one or more esters, e.g.,
butanoic acid, $CH_3CH_2CH_2CO_2H$ and methyl propanoate, $C_2H_5CO_2CH_3$, ethyl ethanoate, $CH_3CO_2C_2H_5$, propyl methanoate, $HCO_2CH_2CH_2CH_3$, and methylethyl methanoate, $HCO_2CH(CH_3)_2$.

CHECKPOINT 25B: STRUCTURAL ISOMERISM

1. Draw structural formulae for three compounds of formula C_5H_{12} and for five compounds of formula C_6H_{14}.

2. Draw the formulae of the three structural isomers of formula C_3H_8O. How many can you find for $C_4H_{10}O$?

3. There are ten compounds with the formula $C_4H_8Cl_2$. Draw their structural formulae.

4. The molecular formulae C_4H_8O and $C_5H_{10}O$ each correspond to a number of carbonyl compounds, which contain the group

$$\rangle C{=}O$$

Sketch their structural formulae.

5. Supply structural formulae for the acids and esters which have the molecular formulae (*a*) $C_5H_{10}O_2$ and (*b*) $C_6H_{12}O_2$.

TAUTOMERISM

The position of a hydrogen atom differs between tautomers

Two isomeric forms of a compound may exist in dynamic equilibrium. Aldehydes and ketones exhibit **keto-enol** tautomerism. The equilibrium between the forms is represented

$$CH_3{-}\underset{\underset{O}{\|}}{C}{-}CH_3 \rightleftharpoons CH_2{=}\underset{\underset{O-H}{|}}{C}{-}CH_3$$

the *keto* form of propanone the *enol* form of propanone

The **tautomers** differ in the position of a hydrogen atom. They are called the **keto** form (for the $\rangle C{=}O$ group in ketones) and the **enol** form (*ene* for $-C{=}C-$ and *ol* for the $-OH$ group). The percentage of the enol tautomer is small, but it is invoked to explain some of the reactions of carbonyl compounds [§§ 14.11 and 31.3.3].

25.9.2 STEREOISOMERISM

There are various types of stereoisomerism

Stereoisomers have the same molecular formula and also the same structural formula. The difference between them is the arrangement of the bonds in space. Stereoisomerism can be (*1*) *cis-trans* isomerism, (*2*) conformational isomerism or (*3*) optical isomerism.

CIS-TRANS ISOMERISM

Restriction of rotation about a C=C double bond gives rise to cis-trans isomerism

The planar arrangement of bonds in $R_2C{=}CR_2$ was discussed in § 5.2.7. The CR_2 groups are not free to rotate about the double bond. In a compound

$$R_1R_2C{=}CR_3R_4$$

R_1 and R_3 may be on the same side of the double bond (the *cis*-isomer) or on opposite sides (the *trans*-isomer). *Cis-* and *trans*-butenedioic acid are shown in Figure 25.6.

FIGURE 25.6 Models of Butenedioic Acids

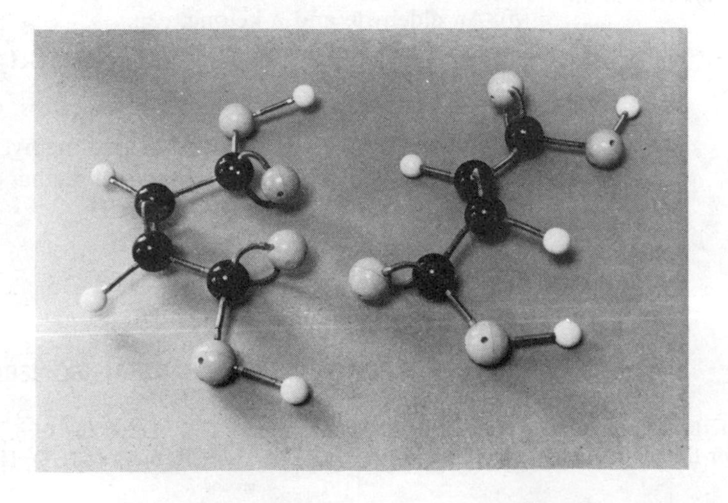

$$\underset{\textit{cis}\text{-Butenedioic acid}}{\begin{matrix} HO_2C & & CO_2H \\ & C{=}C & \\ H & & H \end{matrix}} \qquad \underset{\textit{trans}\text{-Butenedioic acid}}{\begin{matrix} HO_2C & & H \\ & C{=}C & \\ H & & CO_2H \end{matrix}}$$

The physical and chemical properties of cis-trans *isomers differ*

They do not have the same physical and chemical properties. They differ in melting temperature (*cis* 135 °C, *trans* 287 °C), in solubility (*cis* being 100 times more soluble than *trans*) and in dipole moments. The *cis* acid (formerly called maleic acid) forms an anhydride on gentle heating, but the *trans* isomer (fumaric acid) does not:

$$\underset{\textit{cis}\text{-Butenedioic acid}}{\begin{matrix} H & & CO_2H \\ & C & \\ & \| & \\ & C & \\ H & & CO_2H \end{matrix}} \xrightarrow{160\,°C} \underset{\textit{cis}\text{-Butenedioic anhydride}}{\begin{matrix} H & & C{=}O \\ & C & \\ & \| & O \\ & C & \\ H & & C{=}O \end{matrix}} + H_2O$$

Examples are butenedioic acid...

On strong heating, *trans*-butenedioic acid forms the anhydride of *cis*-butenedioic acid, showing that rotation about the double bond is possible at higher temperatures.

...and some inorganic compounds...

For *cis-trans* isomerism in inorganic compounds, see § 24.13.6.

CONFORMATIONAL ISOMERISM

Cyclohexane shows isomerism

Cyclohexane, C_6H_{12}, exists in two forms which differ in conformation and are described as the **boat** and **chair** forms [see Figure 25.7].

FIGURE 25.7(a) The Isomers of Cyclohexane

The boat form

The chair form

FIGURE 25.7(b) Models of Cyclohexane, in its Boat and Chair Forms

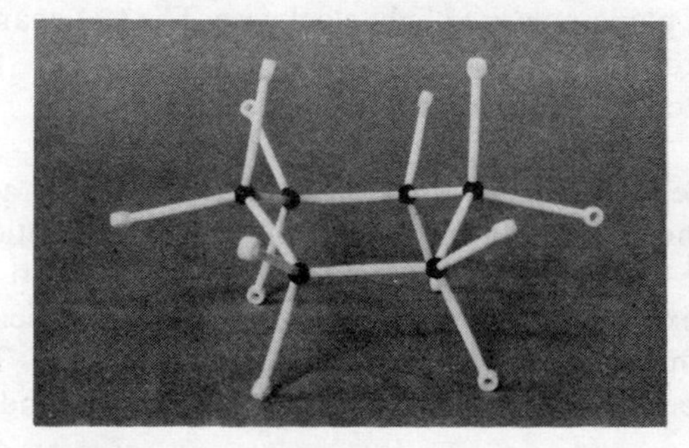

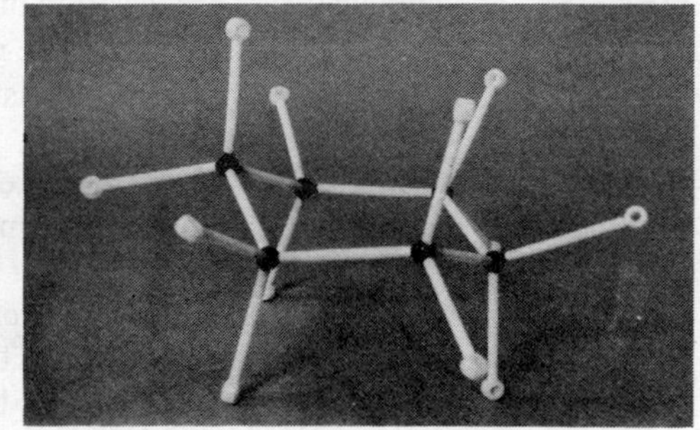

CHECKPOINT 25C: STEREOISOMERISM

1. Write structural formulae for (*a*) $CH_3CH{=}CHCl$, (*b*) $CH_3CH{=}CHC_2H_5$, (*c*) $ClCH{=}C(CH_3)CO_2H$, (*d*) $C_2H_5C(CH_3){=}CHCH_3$.

2. Make a model of cyclopentane, C_5H_{10}. How many ways can you find of replacing 2H by 2Cl to give the structures of the isomers of $C_5H_8Cl_2$?

OPTICAL ISOMERISM

If a beam of light is passed through a Nicol prism (of *calcite*, $CaCO_3$) or a piece of polaroid, the emergent light vibrates in a single plane. It is said to be **plane-polarised** [see Figure 25.8].

FIGURE 25.8
Plane-polarisation of Light

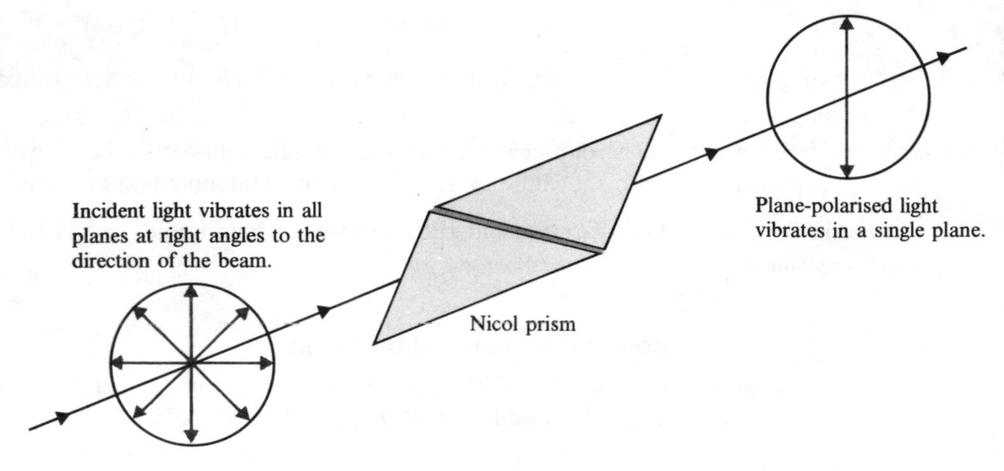

Incident light vibrates in all planes at right angles to the direction of the beam.
Nicol prism
Plane-polarised light vibrates in a single plane.

Optical activity is the ability to rotate the plane of polarisation of plane-polarised light

Certain substances, either in crystalline form or in solution, have the ability to rotate the plane of polarisation of plane-polarised light. They are said to be **optically active**. The effect is measured in an instrument called a **polarimeter**, which sends a beam of plane polarised light through a solution of the substance. If the plane of polarisation rotates in a clockwise direction, viewed from the direction of the emergent beam, the rotation is designated (+); an anticlockwise rotation is designated (−).

Optically active compounds are chiral...

...Their molecules have no plane of symmetry...

...They exist in (+)- and (−)-forms and as optically inactive mixtures called racemates...

Compounds which are optically active are **chiral**. Their molecules have no plane or axis or centre of symmetry [§ 5.1.4]. Chiral compounds have two different types of molecules. They are **enantiomers**, i.e., mirror images of one another [see Figure 5.10, § 5.1.4 and Question 1, Checkpoint 25D]. The (+)-enantiomer rotates the plane of polarisation of plane-polarised light clockwise. The (−)-enantiomer produces an anticlockwise rotation. An equimolar mixture of (+) and (−)-enantiomers is optically inactive and is called a **racemic mixture** or **racemate**.

There are three forms of 2-hydroxypropanoic acid (lactic acid), $CH_3CH(OH)CO_2H$. The racemic mixture is found in sour milk; the (+)-enantiomer is present in muscle. The (−)-enantiomer does not occur naturally and has to be obtained from the racemic mixture. All three forms have the same chemical properties. The physical properties of the (+) and (−)-enantiomers are the same. They differ from those of the racemate because it is a mixture, not a pure compound.

...(+)- and (−)-forms are called enantiomers

Enantiomers react in the same way in chemical reactions with achiral reagents. They may differ in biochemical reactions. Enzymes are catalysts found in plants and animals. An enzyme and its substrate (the substance which requires the enzyme in order to react) fit together 'like a key in a lock'. The geometry of the substrate is important, and enzymes can distinguish between enantiomers. *Penicillium glaucum* (a mould) feeds on (+)-2-hydroxypropanoic acid (lactic acid) but not on its (−)-enantiomer.

INORGANIC COMPOUNDS

Optical isomerism among inorganic compounds is met in § 24.13.6.

CHECKPOINT 25D: OPTICAL ACTIVITY

1. (*a*) Construct models of the enantiomers of 2-hydroxypropanoic acid, $CH_3CH(OH)CO_2H$, and draw them.
(*b*) Find out what happens when you substitute a hydrogen atom for the —OH group.

2. Alanine, an amino acid, has the formula $H_2NCH(CH_3)CO_2H$. Construct a model of the molecule. Do you think that alanine should be optically active? Replace $-CH_3$ by —H. This is the formula of the amino acid, glycine. Do you think that glycine should be optically active?

3. What is *optical activity*? How could you demonstrate that a compound is optically active?
Draw structural formulae for the optical isomers of molecular formula (*a*) $C_3H_6O_3$, (*b*) $C_4H_{10}O$.

4. Discuss the isomerism shown by the following compounds. State the differences (if any) in physical and chemical properties that exist between the isomers of each formula.

(*a*) C_4H_{10}
(*b*) C_4H_8
(*c*) $CH_3CH(OH)CO_2H$
(*d*) $C_6H_5CH{=}CH{-}CO_2H$
(*e*) $C_6H_5{-}CH(OH){-}CN$ (OH on the CH)
(*f*) $CH_2{-}CHCl$ ring with $CHCO_2H$ (cyclopropane: CH_2, CHCl, $CHCO_2H$)
(*g*) $C_3H_6Br_2$
(*h*) C_3H_7Cl
(*i*) $C_6H_5CH_2CH(NH_2)CO_2H$

QUESTIONS ON CHAPTER 25

1. (*a*) (i) Explain what is meant by the term *chiral molecule*.
(ii) The compound $CH_3CH(OH)CO_2H$ exists as two optical isomers.
Draw structures to show *each* isomer. How may the isomers be distinguished from each other?
(*b*) Compounds **A** and **B** both have the same molecular formula, C_4H_8. Compound **A** exists in two forms but compound **B** has only one form. Both **A** and **B** undergo an addition reaction with HBr to give a compound **C** which exists in two forms.
Draw structures for **A**, **B** and **C**.

(AEB 90)

26

THE ALKANES

26.1 PETROLEUM OIL

Alkanes are fuels...

The most important feature of the alkanes is their use as fuels. A huge fraction of the energy we use comes from the combustion of alkanes. The gas in our cookers, the petrol in our cars, aviation fuel and diesel oil for powering ships and electric generators – all these fuels are mixtures of alkanes. The source of these fuels is either crude petroleum oil or natural gas. Deposits of crude oil and natural gas usually occur together as they are formed by the same slow decay of marine animals and plants. Crude oil is found in many parts of the world. Figure 26.1 shows an oil rig in the North Sea.

26.1.1 FRACTIONAL DISTILLATION OF CRUDE OIL

...They are obtained from crude oil by fractional distillation

Crude oil is a mixture of about 150 compounds. It is difficult to ignite. To yield volatile substances which can be used as fuels, crude oil is fractionally distilled. The theory is covered in § 8.4.2. Figure 26.2 represents a fractionating column for use in separating crude oil into fractions. Each fraction is a mixture of hydrocarbons which boil over a limited range of temperature [see Table 26.1].

FIGURE 26.1
An Oil Rig

FIGURE 26.2 An Industrial Fractionating Column

TABLE 26.1 The Fractions obtained from Crude Oil

Fraction	*Boiling temperature/°C*	*Length of carbon chain*	*Use*
Refinery gas	20	C_1–C_4	Fuel: domestic heating, gas cookers
Light petroleum	20–60	C_5–C_6	Solvent
Light naphtha	60–100	C_6–C_7	Solvent
Gasoline (petrol)	40–205	C_5–C_{12}	Fuel for the internal combustion engine (cars etc.)
Kerosene (paraffin)	175–325	C_{12}–C_{18}	Fuel for jet engines
Gas oil	275–400	C_{18}–C_{25}	Does not vaporise easily. Used in diesel engines, where it is injected into compressed air to make it ignite. Used in industrial furnaces, being introduced as a fine mist to help the oil to burn
Lubricating oil	Non-volatile	C_{20}–C_{34}	Lubrication
Paraffin wax	Solidifies from lubricating oil fraction	C_{25}–C_{40}	Polishing waxes, petroleum jelly
Bitumen (asphalt)	Residue	$>C_{30}$	Road surfacing, roofing

26.1.2 PETROCHEMICALS

Besides being used as fuels, all these fractions have another, very important use. They are the foundation of the petrochemicals industry. From them are manufactured thousands of compounds: plastics, paints, solvents, rubbers, detergents and many medicines are petrochemicals.

26.2 PHYSICAL PROPERTIES

The C—H bond has a weak dipole...

Since the electronegativities of carbon and hydrogen are 2.5 and 2.1 respectively

$$\overset{\delta-}{\mathrm{C}}-\overset{\delta+}{\mathrm{H}} \text{ bonds}$$

...and the intermolecular forces are weak

have only very weak dipole moments. Weak attractive forces exist between dipoles in neighbouring molecules [§4.7.1], and van der Waals forces [§4.7.2] also come into play. The attractive forces are so weak that the lower alkanes, from methane to butane, are gases at room temperature and pressure. Linear molecules of higher homologues can align themselves in a parallel arrangement so that dipole–dipole interactions and van der Waals forces can operate along the whole length of the molecule. The alkanes from C_5 to C_{17} are liquids, while those with larger molecules are solids. Since branched-chain molecules are more spherical in shape than unbranched-chain hydrocarbons, the attractive forces between molecules are more restricted. The boiling temperatures of branched alkanes are therefore lower than those of their straight-chain isomers. The boiling temperatures of unbranched-chain alkanes are plotted against molar mass in Figure 26.3.

Branched-chain alkanes are more volatile than the unbranched-chain isomers

FIGURE 26.3 Boiling Temperatures of *n*-Alkanes

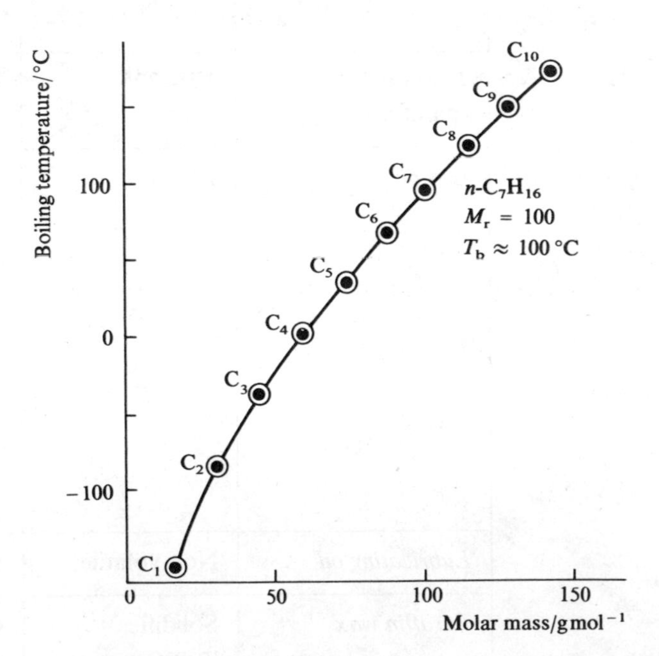

The difference in boiling temperatures between the C_1 and C_2 alkanes is 73 °C, while the C_9 and C_{10} alkanes differ by only 25 °C. It is therefore more difficult to separate the higher members by fractional distillation.

Liquid alkanes float on water...

The liquid alkanes are less dense than water: oil floats on water. The higher members are viscous liquids, the viscosity increasing with increasing molecular mass as the attractive forces between molecules increase.

...Alkanes cannot form hydrogen bonds with water and are therefore insoluble

Alkanes are only slightly soluble in water. Water molecules interact because of the strong dipoles in the

$$\overset{\delta-}{O}-\overset{\delta+}{H} \text{ bonds}$$

[See Figure 4.35, § 4.7.3.] The hydrogen bonds formed are stronger than any interaction which can occur between water molecules and the non-polar alkane molecules. Dissolution is therefore not favoured by energy considerations.

26.3 REACTIONS OF ALKANES

26.3.1 COMBUSTION

Alkanes burn with the release of energy

The most important reaction of the alkanes is combustion. They burn to form the harmless products, carbon dioxide and water, in an exothermic reaction [for $\Delta H^\ominus$, see Chapter 10]:

$$CH_4(g) + 2O_2(g) \rightarrow CO_2(g) + 2H_2O(l); \quad \Delta H^\ominus = -890\,\text{kJ mol}^{-1}$$

Liquid alkanes, such as octane, must be vaporised before they will burn:

$$C_8H_{18}(g) + 12\tfrac{1}{2}O_2(g) \rightarrow 8CO_2(g) + 9H_2O(l); \quad \Delta H^\ominus = -5510\,\text{kJ mol}^{-1}$$

Incomplete combustion gives the poisonous gas, carbon monoxide

This is the reaction which takes place in the internal combustion engine. The hydrocarbons in gasoline have boiling temperatures of around 150 °C and will vaporise in the internal combustion engine. If the supply of oxygen is insufficient, incomplete combustion will take place, and the poisonous gas carbon monoxide will be formed. It is the more dangerous in that, being odourless, it gives no warning of its presence. It combines with haemoglobin, the red pigment in the blood, to form a very stable complex, carboxyhaemoglobin. This cannot combine with oxygen, and haemoglobin is unable to do its job of transporting oxygen around the body [see Figure 24.16, § 24.13.8]. There is always a certain percentage of carbon monoxide, around 5%, in the exhaust gases of motor vehicles.

In the internal combustion engine, compression of petrol vapour and air can lead to auto-ignition...

...which causes 'knocking'

When combustion of petrol vapour occurs inside the cylinders in a car engine, a large volume of hot gases is formed. The gases force the piston down the cylinder, and the power generated is transmitted to the wheels. For smooth running it is essential that the ignition of gasoline vapour and air takes place when the piston is at the right point in the cylinder. To obtain the maximum energy from the fuel, the engine design must ensure that all the petrol vapour is burnt. For this reason the mixture of petrol vapour and air is compressed, and this compression can lead to **auto-ignition.** The ignition of gases takes place before the spark. It results in a sudden rise in pressure, which delivers a blow to the piston. The engine makes a metallic sound called **knocking**. The best fuels for resistance to knocking are branched-chain hydrocarbons, such as 2,2,4-trimethylpentane:

```
                      CH3
                      |
H3C—CH—CH2—C—CH3
        |             |
       CH3           CH3
```

The octane number of a fuel measures its resistance to knocking

This compound, which used to be called *iso*-octane, has been assigned an *octane number* of 100. Heptane, $CH_3CH_2CH_2CH_2CH_2CH_2CH_3$, has very bad knocking properties and has been assigned an *octane number* of 0. The **octane number** of a petrol is found by comparing its performance with a mixture of heptane and 2,2,4-trimethylpentane. If it has the same performance as a mixture of 25% heptane and 75% 2,2,4-trimethylpentane, then its octane number is 75.

TEL is an antiknock

After a great deal of research, it was found that tetraethyllead, $Pb(C_2H_5)_4$, TEL, reduced the knocking properties of unbranched hydrocarbons. It is called an **antiknock**. To prevent lead(II) oxide from coating the cylinders when TEL is added to petrol, 1,2-dibromoethane, $BrCH_2CH_2Br$, must also be added. Waste lead is converted into volatile compounds, chiefly lead bromide chloride, PbBrCl, which pass out with the exhaust gases. Four-star petrol contains 6 g of TEL per gallon, and the number of vehicles on the road is so large that the emission of lead has reached hazardous levels. When ingested in only moderate quantities, it makes people depressed, and increases their reaction times. A growing body of people think that city dwellers are being affected by the lead content of the air. In the UK, since 1990 all new petrol-driven vehicles have been designed to run on lead-free petrol. Another approach to the problem of knocking does exist: this is to convert fuels with poor knocking properties into branched-chain compounds. This is done by *cracking*, which is discussed below. [See also § 26.4.]

Lead compounds in car exhaust gases are becoming a health hazard

Lubricating oil reduces engine wear

Lubricating oil does not vaporise at the engine temperature. It is used to ease the movement of the pistons in the cylinders by reducing friction, thus reducing wear and prolonging the life of the engine.

26.3.2 CRACKING

'Cracking' is used to obtain more of lower molecular mass alkanes, which are more volatile

The petroleum fractions with 1 to 12 carbon atoms in the molecule are in demand in larger quantities than the fractions with bigger molecules. The petroleum industry uses **pyrolysis** (splitting by heat) of high molar mass alkanes to give hydrocarbons with smaller molecules, which are more easily vaporised and are therefore more useful fuels:

$$\text{Alkane with large molecules} \xrightarrow[\text{at 450 °C over catalyst of } Al_2O_3/SiO_2]{\text{Vapour passed}} \text{Alkane with smaller molecules} + \text{Alkene} + \text{Hydrogen}$$

e.g.

$$\underset{\text{Propane}}{2CH_3CH_2CH_3(g)} \rightarrow \underset{\text{Methane}}{CH_4(g)} + \underset{\text{Propene}}{CH_3CH{=}CH_2(g)} + \underset{\text{Ethene}}{CH_2{=}CH_2(g)} + \underset{\text{Hydrogen}}{H_2(g)}$$

The industry calls this type of reaction **cracking**.

26.3.3 ALKYLATION

'Alkylation' is used to make branched-chain alkanes

Since branched-chain compounds have higher octane numbers than straight-chain compounds, a good deal of research has gone into the synthesis of branched-chain compounds. They are made by the **alkylation** of alkenes by alkanes in the presence of a catalyst.

In alkylation reactions

$$\text{Tertiary alkane} + \text{Alkene} \xrightarrow[\text{conc. } H_2SO_4 \text{ as catalyst}]{20\,°C} \text{Branched-chain alkane}$$

e.g.

$$\underset{\text{2-Methylpropane}}{(CH_3)_3CH} + \underset{\text{Pent-2-ene}}{CH_3CH{=}CHCH_2CH_3} \rightarrow \underset{\text{3-Ethyl-2,2-dimethylpentane}}{CH_3CH_2CH(C(CH_3)_3)CH_2CH_3}$$

26.3.4 REFORMING

'Reforming' converts alkanes into aromatic compounds, e.g., benzene

A huge number of important chemicals are derived from benzene. One source of benzene is petroleum. Unbranched-chain alkanes are converted into aromatic compounds by the process of **reforming**. For example

$$\underset{\text{Hexane}}{C_6H_{14}(l)} \xrightarrow[\text{compressed to 40 atm passed over } Al_2O_3 \text{ as catalyst (or Pt, 10 atm)}]{\text{Vaporised at 500 °C}} \underset{\text{Benzene}}{C_6H_6(l)} + \underset{\text{Hydrogen}}{4H_2(g)}$$

'Platforming' uses a platinum catalyst

Much work has gone into the effectiveness of different catalysts. If a platinum catalyst is used, the process is called **platforming**.

26.3.5 THE 'CUMENE' PROCESS FOR MAKING PHENOL

See § 30.11.1.

26.3.6 OTHER REACTIONS OF ALKANES

Alkanes are not reactive

They undergo some substitution reactions

The alkanes used to be called the **paraffins**, a name derived from the Latin for *little liking*, implying that this class of compounds had little liking for the usual chemical reagents. Alkanes do not react with dilute acids or alkalis or with oxidising agents. At high temperatures they react with nitric acid vapour. All the reactions of alkanes, apart from combustion, are **substitution** reactions, of the form

$$RH + XY \rightarrow RX + HY$$

26.3.7 HALOGENATION

Substitution by halogens occurs in sunlight

Under certain conditions, alkanes react with halogens. If a stoppered test-tube containing hexane and a drop of liquid bromine is left to stand at room temperature in the dark, nothing happens. The colour of the bromine is still as intense after three or four days. If the solution is exposed to sunlight, the colour fades in a few minutes, and the acidic, fuming gas hydrogen bromide can be detected. The reaction that has occurred is

$$C_6H_{14}(l) + Br_2(l) \rightarrow C_6H_{13}Br(l) + HBr(g)$$

with the formation of bromohexanes. Since it takes place in the presence of light, it is called a **photochemical reaction**. Alkanes can be chlorinated and brominated photochemically.

Produced by chlorination of CH_4 are CH_3Cl, CH_2Cl_2, $CHCl_3$, CCl_4

When methane reacts with chlorine in sunlight, one or more chlorine atoms may replace hydrogen atoms, depending on the amounts of halogen and alkane present. The formation of chloromethane

$$CH_4(g) + Cl_2(g) \rightarrow CH_3Cl(g) + HCl(g)$$

...They are useful solvents

may be followed by the formation of dichloromethane, CH_2Cl_2, trichloromethane (chloroform), $CHCl_3$, and tetrachloromethane, CCl_4. Chloroalkanes are useful solvents, and the mixture of products formed in the chlorination of an alkane may find use as a solvent, without the need for isolation of individual compounds.

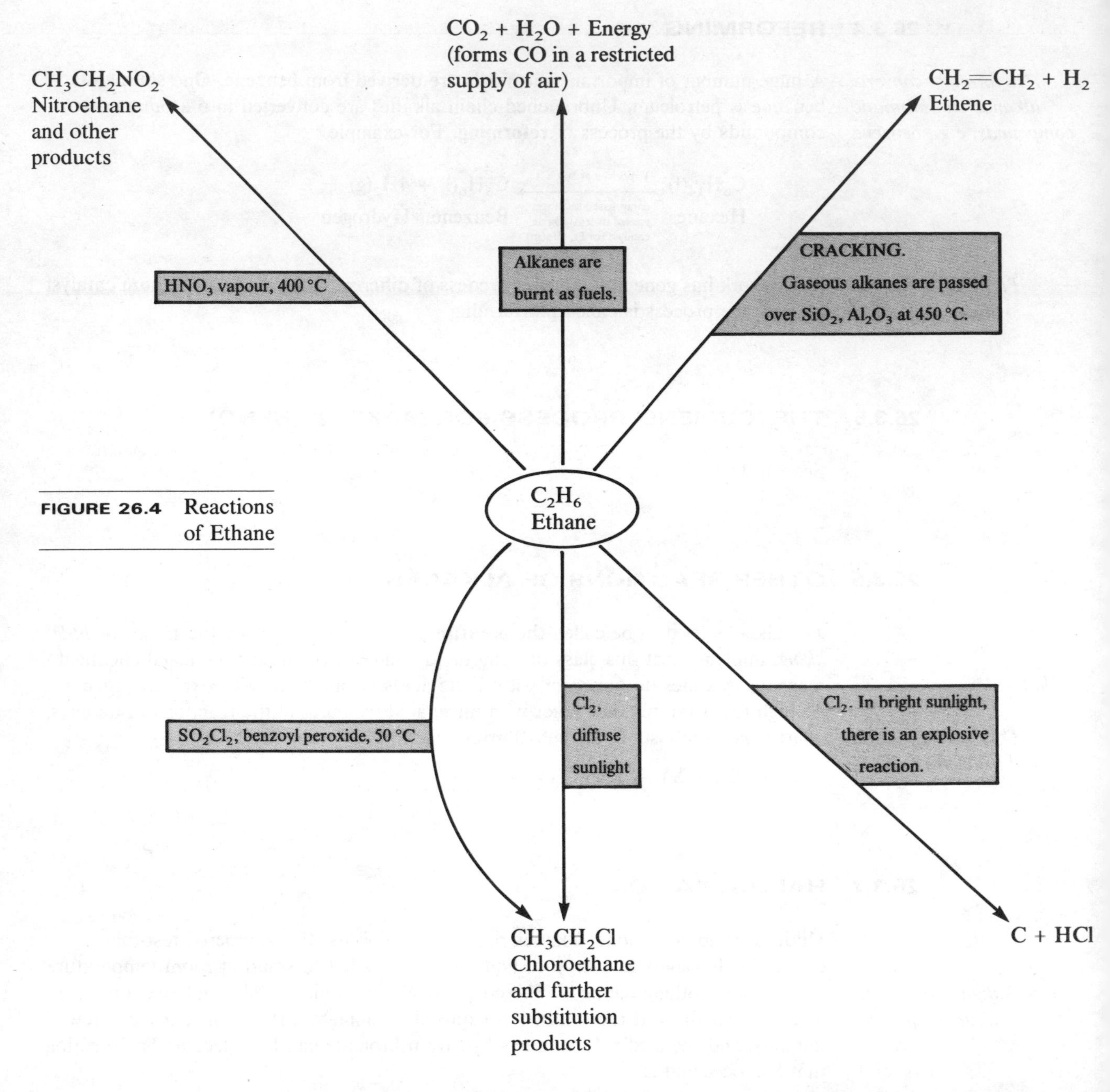

FIGURE 26.4 Reactions of Ethane

Much work has been done on the subject of how this reaction takes place, when methane is so unreactive towards other reagents. The mechanism proposed for the chlorination of methane takes into account the following experimental observations.

Experimental observations on the reaction, $CH_4 + Cl_2$

1. Reaction takes place rapidly in sunlight or above 300 °C but not in the dark at room temperature.

2. Thousands of molecules of chloromethane are formed for each photon of light absorbed.

3. A little ethane is formed.

4. If a trace of tetramethyllead, $Pb(CH_3)_4$, is added, the reaction will take place in the dark or at room temperature. This substance is known to dissociate into methyl radicals, $\cdot CH_3$.

A mechanism which fits these observations is as follows.

26.3.8 THE MECHANISM OF THE CHLORINATION OF METHANE

STEP 1

Photochemical homolysis of Cl_2

Homolysis of the Cl—Cl bond. The necessary energy comes from the light absorbed or the heat supplied. It is easier to split the Cl—Cl bond than the C—H bond. (Bond energy terms are Cl—Cl, 242 kJ mol^{-1}; C—H, 435 kJ mol^{-1} [§ 10.6].)

$$Cl_2 \xrightarrow[\text{or heat}]{\text{light}} 2Cl\cdot$$

STEP 2

There are two possible reactions of $Cl\cdot$ with CH_4

The chlorine atoms formed are very reactive. Since they are surrounded by methane molecules, there are two possible reactions:

$$Cl\cdot + CH_4 \rightarrow CH_3Cl + H\cdot$$

$$Cl\cdot + CH_4 \rightarrow HCl + \cdot CH_3$$

The second possibility is more likely because the formation of an H—Cl bond is more exothermic than the formation of a C—Cl bond. (Bond energy terms are H—Cl, 431 kJ mol^{-1}; C—Cl, 350 kJ mol^{-1}.)

STEP 3

The reaction of $CH_3\cdot$ with Cl_2 results in a chain reaction

The methyl radicals formed collide with methane molecules and chlorine molecules. The reaction

$$\cdot CH_3 + CH_4 \rightarrow CH_4 + \cdot CH_3$$

results in no net change.

The reaction

$$\cdot CH_3 + Cl_2 \rightarrow CH_3Cl + \cdot Cl$$

leads to a chain reaction because the chlorine atoms formed react as in Step 2.

STEP 4

Thousands of molecules of chloromethane are formed for every photon of light absorbed. The high yield is due to the chain reaction — Steps 2 and 3. The reason why the yield is not higher is that radicals can combine with each other and bring the chain to an end. The reactions

There are three chain-terminating reactions

$$2Cl\cdot \rightarrow Cl_2$$

$$2CH_3\cdot \rightarrow C_2H_6$$

$$Cl\cdot + \cdot CH_3 \rightarrow CH_3Cl$$

bring the chain reaction to an end. Some ethane can be detected in the product.

Summary of the mechanism

To summarise, the steps in the chlorination of methane to chloromethane are:

Chain initiation	$Cl_2 \xrightarrow[\text{or heat}]{\text{light}} 2Cl\cdot$
Chain propagation	$Cl\cdot + CH_4 \rightarrow HCl + \cdot CH_3$
	$\cdot CH_3 + Cl_2 \rightarrow CH_3Cl + Cl\cdot$
Chain termination	$2Cl\cdot \rightarrow Cl_2$
	$2\cdot CH_3 \rightarrow C_2H_6$
	$Cl\cdot + \cdot CH_3 \rightarrow CH_3Cl$

FORMATION OF CH_2Cl_2

Further Cl atoms can be introduced

Step 3 can give rise to the chain:

$$CH_3Cl + Cl\cdot \rightarrow HCl + \cdot CH_2Cl$$

$$\cdot CH_2Cl + Cl_2 \rightarrow CH_2Cl_2 + Cl\cdot$$

CH_2Cl_2 can undergo further chlorination to $CHCl_3$ and CCl_4.

26.3.9 BROMINATION

Methane reacts with the other halogens

The yield per photon is less for bromination than for chlorination because the step

$$Br\cdot + CH_4 \rightarrow HBr + \cdot CH_3$$

is more endothermic than the corresponding step in chlorination.

26.3.10 IODINATION

Iodination is slow and reversible:

$$RH + I_2 \rightleftharpoons RI + HI$$

Iodoalkanes can be reduced to alkanes by HI.

26.3.11 FLUORINATION

Fluorination is dangerously exothermic because of the low bond dissociation enthalpy (energy) of fluorine and the high enthalpy (energy) of C—F bonds.

26.3.12 SULPHUR DICHLORIDE DIOXIDE

Chlorination can also be effected by SO_2Cl_2

Sulphur dichloride dioxide, SO_2Cl_2, is another chlorinating agent. In the presence of a source of free radicals (such as benzoyl peroxide) it will chlorinate alkanes at 50 °C.

26.3.13 NITRATION OF ALKANES

Nitration gives a mixture of products

Alkanes can be nitrated by nitric acid vapour at 400 °C. A mixture of products is formed: for example, ethane gives nitromethane as well as nitroethane.

26.4 THE CAR

TOPIC

Two hundred million motor vehicles are travelling the roads of the world. They are covering 2 million million miles a year and burning 70 billion litres of petrol. From their exhausts come 40 million tonnes of carbon monoxide, 4 million tonnes of nitrogen oxide, 4 million tonnes of hydrocarbons and 0.2 million tonnes of lead. These pollutants are discharged at street level where people cannot avoid inhaling them.

FIGURE 26.5 The Four-stroke Spark-ignition Internal Combustion Engine

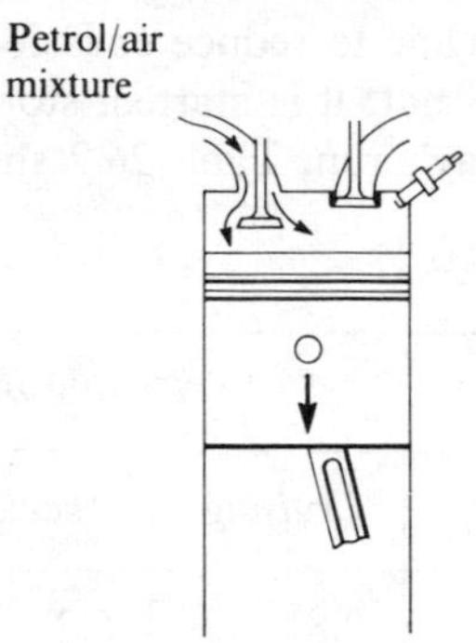

Induction stroke

1. The induction stroke

   The inlet valve is open; the exhaust valve is closed. The piston moves down the cylinder, drawing a mixture of petrol and air into the cylinder. Then the inlet valve closes.

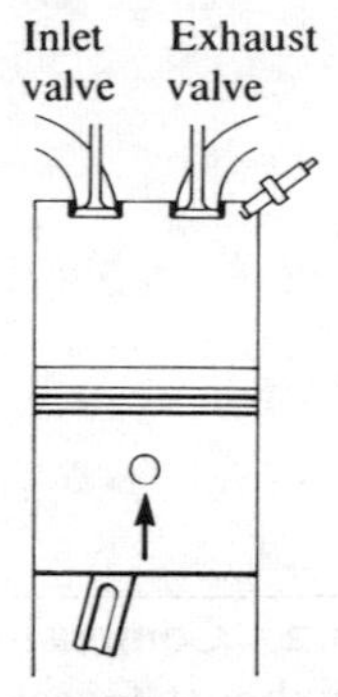

Compression stroke

2. The compression stroke

   Both valves are closed. The rising piston compresses the mixture, and the heat generated by compression vaporises the mixture. (Compression ratio = volume of gas in the cylinder before compression / volume after compression)

Vehicle engines discharge pollutants...

...carbon monoxide, nitrogen oxides, hydrocarbons and lead compounds

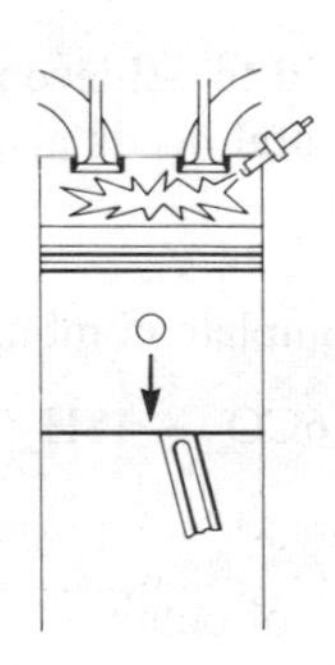

Power stroke

3. The power stroke

   Both valves remain closed. A spark from the spark plug ignites the mixture of gases. Expansion of burning gas drives the piston down the cylinder, turning the crank shaft and providing the power to move the engine. The exhaust valve opens.

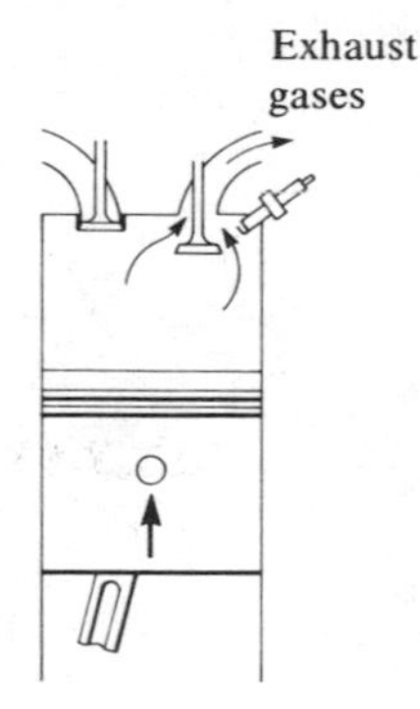

Exhaust stroke

4. Exhaust stroke

   The inlet valve is closed, and the exhaust valve is open. The piston rises to expel the products of combustion. The inlet valve opens and the exhaust valve closes. Then the cycle starts again.

High engine temperatures reduce the emission of carbon monoxide and unburnt hydrocarbons...

The petrol engine powers most of our motor vehicles. § 26.3.1 gave a brief description of the combustion of petrol in a vehicle engine. The burning of the fuel in the cylinder after ignition by the spark is sudden and intense. The temperature soars to about 2800 °C, and some of the nitrogen and oxygen in the cylinder combine to form nitrogen oxide. As the piston is pushed out of the cylinder, the combustion gases expand and cool in less than a hundredth of a second. The heating–cooling cycle occurs so rapidly that much of the fuel is not completely oxidised to carbon dioxide and water. Some carbon monoxide is formed, and some hydrocarbons remain uncombusted. Features of the petrol engine are that the high temperature promotes oxidation of nitrogen to nitrogen oxide and the rapid cycling leads to incomplete oxidation of the fuel. The simultaneous production of overoxidised contaminants, e.g. nitrogen oxide, and underoxidised contaminants, e.g. hydrocarbons and carbon monoxide, make it difficult to clean up the exhaust with a single chemical treatment.

...but increase the formation of nitrogen oxides

Adjustments can be made to the fuel supply, fuel/air ratio and spark timing of an engine. These adjustments are called 'tuning' the engine. It would be possible to work out how to tune the engine to reduce emissions *if* the engine were running at a steady pace. In practice, it is not: it is started, stopped, accelerated, decelerated and idled many times during a day's run. Table 26.2 shows how exhaust levels change with the mode of driving.

TABLE 26.2 Composition of Exhaust Gases with Different Driving Modes

Exhaust component/%	*Driving mode*			
	Idling	*Cruising*	*Accelerating*	*Decelerating*
Carbon monoxide	5.2	0.8	5.2	4.2
Hydrocarbons	0.075	0.030	0.040	0.40
Nitrogen oxide	0.0030	0.15	0.30	0.0060

The combustion of fuel in the petrol engine contrasts with that in power stations [see § 21.14]. The burning of coal or oil in a power station is slow, steady and nearly complete. The pollutants arise from impurities in the fuel itself. Combustion in the petrol engine is rapid and erratic, and these characteristics favour the formation of pollutants.

In the past, engines were tuned to achieve maximum performance. Now that public opinion weighs up environmental needs also, manufacturers have to balance performance against pollution.

From the equation for the complete combustion of octane,

$$2C_8H_{18} + 25O_2 \rightarrow 16CO_2 + 18H_2O$$

you can work out the ratio

mass of oxygen/mass of octane.

Knowing that air is 21% by volume of oxygen, you can work out the ratio

mass of air/mass of octane.

This is called the air/fuel ratio; the value is 14.7. According to the precise composition of the petrol, an air/fuel ratio of 14 or 15 will give complete combustion. Figure 26.6 shows the way in which the production of each major pollutant depends on the air/fuel ratio.

The stoichiometric air/fuel ratio (by mass) is 14.7. A rich mixture (with more fuel) gives more carbon monoxide and hydrocarbons. A lean mixture (with more air) produces more nitrogen oxides

You can see that a rich mixture (with less than the stoichiometric amount of air) leads to the production of carbon monoxide and hydrocarbons. A lean mixture (with an excess of air) leads to the formation of nitrogen oxide. A fuel-rich mixture is better for smooth engine performance. You can see from the figure that altering the fuel/air ratio will not solve the problem: it merely trades one set of pollutants for another.

Another variable is the engine's **compression ratio**. It is the factor by which the fuel–air mixture is compressed before spark ignition. Since any gas gets hotter when

FIGURE 26.6 Exhaust Gases from Vehicle Engines

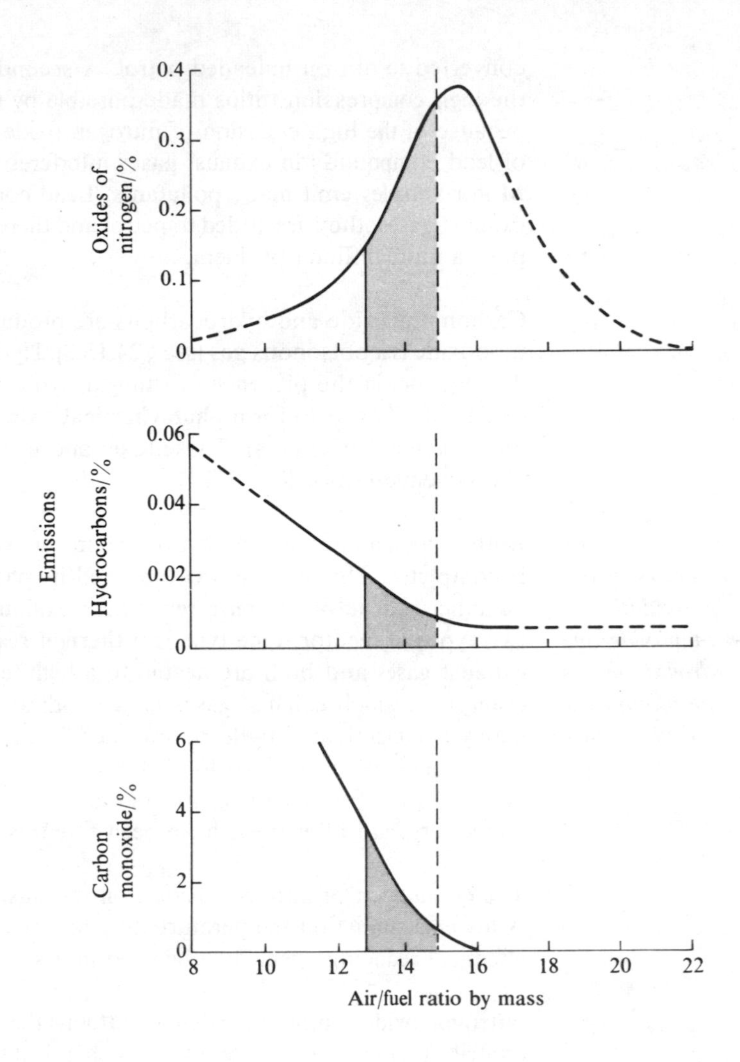

A high compression ratio improves engine efficiency but produces more nitrogen oxides

it is compressed, high compression ratios heat the fuel–air mixture before combustion, and therefore produce more nitrogen oxide. The higher temperature improves engine efficiency, giving more kilometres to the litre. Whereas in the 1960s a compression ratio of 11 was common, now a value of 8.5 is favoured to reduce the emission of nitrogen oxide.

Tetraethyllead, TEL, is added to much of the petrol sold in the UK. After 1990, new cars will be designed to run on unleaded petrol

The octane number of a fuel [see § 26.3.1] can be increased by the addition of 0.5–1.0 g of lead per litre. This is added as tetraethyllead, $Pb(C_2H_5)_4$ (TEL). If the fuel gains 10 points on the octane scale, the compression ratio can be increased from 8 to 11. High-octane hydrocarbons are expensive, but TEL is cheap (about 0.1p per litre of petrol). TEL has been an important ingredient of petrol for 40 years. During combustion, TEL reacts with other additives to form volatile lead compounds which escape through the exhaust pipe. Lead and its compounds have long been known to be poisonous. When lead is ingested in only moderate quantities it makes people depressed, and increases their reaction times. A vigorous compaign over a period of 20 years by a number of organisations has convinced people that city dwellers are being affected by the lead content of the air and that vegetation is being contaminated by lead. Concern over the release into the air of toxic lead compounds has led to the banning of TEL in a number of countries. The UK has followed suit, and from 1990 all new cars have been designed to run on lead-free petrol. Early in 1989 the Queen decided to have all the vehicles in the royal household converted to run on unleaded petrol. A second reason for eliminating TEL is that the high compression ratios made possible by the use of TEL are now less popular because of the high emission of nitrogen oxide which results. Thirdly, the presence of lead compounds in exhaust gases interferes with catalytic converters [see below]. Motor vehicles emit many pollutants. Lead compounds need not be present in exhaust gases; they are added to petrol and therefore easy to eliminate. Other pollutants pose a more difficult problem.

Carbon monoxide and hydrocarbons are products of incomplete combustion. Carbon monoxide is a poisonous gas [see § 24.13.8]. Hydrocarbons by themselves cause little damage, but in the presence of sunlight hydrocarbons react with oxygen, ozone and oxides of nitrogen to form **photochemical oxidants** (oxidising agents formed by photochemical reactions). These toxic and irritating compounds are ingredients of photochemical smog [§ 27.11].

Vehicles can be fitted with catalytic converters, which catalyse the oxidation of carbon monoxide and unburnt hydrocarbons. A different type of catalytic converter can reduce oxides of nitrogen

Both carbon monoxide and hydrocarbons are rendered harmless when their oxidation is completed. This can be accomplished by passing exhaust gases through a special chamber (a **reactor** or **converter**) where oxidation reactions are promoted. There are two types of reactor. One type is a **thermal reactor** in which air is added to the exhaust gases and both are heated to a high temperature. The other is a **catalytic reactor**, in which exhaust gases mixed with air are passed over a solid catalyst. The catalyst is inactivated by lead, and therefore requires unleaded petrol. **Catalytic converters** have been fitted to all US cars since 1975 and are also fitted to West German cars. Some British drivers decide to have their cars fitted with catalytic converters, and after these have been fitted use only unleaded petrol.

The elimination of nitrogen oxide from exhaust gases poses a more difficult problem. A lower combustion temperature may be employed to suppress its formation, or a chemical reducing agent may be used in a separate reactor to remove it.

Nitrogen oxide is a pollutant which attacks the ozone layer [see § 21.5]. It also contributes to the formation of photochemical smog [see § 27.11], and is oxidised to nitrogen dioxide, which reacts with oxygen and water vapour to form nitric acid, one of the ingredients of acid rain [see § 21.14].

Table 26.3 shows the effect that catalytic converters have had on vehicle exhaust gases in the USA.

TABLE 26.3 Emissions from US Vehicles

Time	*Pollutant/g mile^{-1}*		
	Carbon monoxide	*Nitrogen oxides*	*Hydrocarbons*
Before controls	80	4.6	11
1970	23	4.0	2.2
1973	39	3.0	3.4
1977	3.4	0.4	0.4

There are alternatives to the petrol spark-ignition engine which create less pollution. These include the Rankine engine, the gas turbine, rotary engines of the Wankel type and battery-driven engines. A lot of development work on these engines is in progress.

CHECKPOINT 26A: THE CAR

1. It has been suggested that battery-driven cars, which do not emit pollutants, are a solution to the problems outlined above. What effect would the increased use of battery-powered cars have on the demand for electricity? What would be the consequences for the environment of this demand?

2. (*a*) Explain why the concentration of carbon monoxide in exhaust gases is (i) higher during acceleration than during cruising and (ii) higher during idling than in cruising.
(*b*) Explain why the concentration of hydrocarbons in exhaust gases is greater during deceleration than in acceleration.
(*c*) Explain why the concentration of nitrogen oxide is greater when a car is accelerating than when it is idling.

3. Give three advantages of running cars on lead-free petrol.

4. The US Clean Air Act of 1970 stated that automobile emissions of oxides of nitrogen should be reduced by 90% by 1976.

(*a*) Why was the US keen to reduce the emission of oxides of nitrogen?
(*b*) How was this accomplished in US vehicles?

5. Why is carbon monoxide much better at binding to haemoglobin than oxygen is? [See § 24.13.8.]

6. The mixture of gases in the cylinder expands suddenly after ignition and pushes the piston down the cylinder. Why does this sudden expansion cool the gases?

QUESTIONS ON CHAPTER 26

1. Give the names of the following compounds:
(*a*) $CH_3CH_2CH_2CH_3$
(*b*) $CH_3CH_2CH_2CH_2CH_2CH_3$
(*c*) $CH_3CH_2CH(CH_3)CH_3$
(*d*) $CH_3C(CH_3)_2CH_2CH_3$
(*e*) $CH_3CH_2CHCH_3$ with CH_2CH_3 on the third carbon
(*f*) CH_3CHCH_2Cl with $CH_2CH_2CH_3$ on the second carbon
(*g*) $CH_3CH_2CCH_2CH_2CH_3$ with Cl and CH_3 on the third carbon
(*h*) $CH_3CH—CH—CHCH_3$ with CH_2CH_3 on the second carbon, Cl on the third carbon and OH on the fourth carbon
(*i*) $(CH_2)_6$
(*j*) $C_2H_5CH_2CHCH_2CHCH_3$ with Br on the first CH and I on the second CH

2. Write structural formulae for the following:
(*a*) 2,3-dimethylpentane,
(*b*) 2,4,5-trimethylheptane,
(*c*) 3-ethyl-2,4-dimethylheptane,
(*d*) 2,2,4-trimethylhexane,
(*e*) 3-bromo-2-chloropentane.

3. Which member of the following pairs of alkanes has the higher boiling temperature?
(*a*) (i) butane and (ii) heptane,
(*b*) (i) 2-methylbutane and (ii) pentane,
(*c*) (i) hexane and (ii) 2,3-dimethylbutane,
(*d*) (i) hexane and (ii) cyclohexane.

4. Name all the products formed in the following reactions:
(*a*) the combustion of octane,
(*b*) the bromination of methane,
(*c*) the reaction between sulphur dichloride dioxide and methane,
(*d*) the reactions between $CH_3\cdot$ and $H\cdot$.

5. How can butane be prepared from (*a*) pentanoic acid, (*b*) propanoic acid and (*c*) bromobutane?

6. Write structural formulae for the compounds with the formulae
(*a*) C_4H_{10} (*b*) $C_2H_4Cl_2$ (*c*) C_5H_{12} (*d*) $C_4H_8Cl_2$

7. Write equations for all the steps that occur in the reaction

$$CH_3Cl + Cl_2 \rightarrow CH_2Cl_2 + HCl$$

What names are given to these steps?

8. Explain briefly what is meant by the following terms:
(*a*) fractional distillation
(*b*) pyrolysis
(*c*) catalytic cracking
(*d*) reforming
(*e*) platforming
(*f*) knocking
(*g*) octane number

9. Name the substitution products formed when propane reacts with (*a*) concentrated nitric acid, (*b*) bromine, (*c*) sulphur dichloride dioxide. State the conditions under which these reactions will occur.

10. A car drives 100 000 miles in its lifetime. It does 35 miles to the gallon, and runs on three-star petrol, which contains 3.5 g of TEL per gallon. How much lead does this car use in its lifetime? What happens to this lead?

11. (*a*) In considering the economics of the *refining* of crude petroleum (*not* its production), explain what is meant by the following terms:
(i) fixed costs

(ii) variable costs?

(*b*) What are the *major* factors which influence the choice of a site for an oil refinery?

(*c*) The two major processes used in the initial refining of crude petroleum are (i) fractional distillation, (ii) cracking. *Briefly* describe the principles of each process, and explain why each process is necessary.

(*d*) Some of the compounds found in crude petroleum contain sulphur. Petrol is obtained by refining crude petroleum. State and explain *two* reasons why sulphur compounds are removed from petrol during its manufacture.

(*e*) Outline the major disadvantages of the world's dependence on refined crude petroleum as a source of energy.

(AEB 92)

12. (*a*) Briefly describe the processes by which mixtures of hydrocarbons containing 5 to 15 carbon atoms are manufactured. Give common names for *two* of these mixtures.

(*b*) What is meant by the 'cracking' of hydrocarbons? Give a balanced equation to illustrate your answer, and state and explain the type of reactive intermediaries involved in these reactions.

(*c*) Compound **B** contains carbon, 85.7%, and hydrogen, 14.3%. **B** can exist as two geometrical isomers but there are no optical isomers. **B** reacts with hydrogen in the presence of a palladium catalyst to give **C**. **C** gives a molecular ion in the mass spectrometer at a mass : charge ratio of 100. **C** can exist as two optical isomers but there are no geometrical isomers. Suggest, with reasons

(i) a possible formula for **B**

(ii) the full structures of the isomers of **B** and **C**

(iii) the equation for the conversion of **B** into **C**.

(O 92)

13. Petrol is a mixture of hydrocarbons together with additives. In a car engine, petrol is mixed with air and burned. The combustion of hydrocarbons provides the energy for the car to move. Waste gases leave the car through the exhaust.

The octane number of the petrol measures the efficiency of combustion in the engine of the car. Iso-octane is given an octane rating of 100. Heptane has a rating of zero. Small quantities of tetraethyllead, (TEL), $Pb(C_2H_5)_4$, may be added to petrol to increase the octane number.

(*a*) (i) Name the process by which the alkanes are first obtained from petroleum (crude oil).

(ii) The higher alkanes are not suitable for use in petrol. Name the process by which they are converted into lower alkanes.

(*b*) Iso-octane is 2,2,4-trimethylpentane.

(i) Draw the structure of iso-octane.

(ii) Write a balanced equation for the complete combustion of iso-octane, C_8H_{18}.

(*c*) TEL speeds up the reaction between methane and chlorine in the dark because it produces free radicals.

$$Pb(C_2H_5)_4 \rightarrow Pb\cdot + 4C_2H_5\cdot$$

(i) What is a free radical?

(ii) Write down the mechanism for the reaction between chlorine and methane in ultraviolet light.

(iii) Suggest, using a balanced equation, how heated TEL speeds up the reaction of methane with chlorine.

(iv) Explain why TEL can be considered to be behaving as a catalyst.

(*d*) The formula of any single hydrocarbon in petrol can be determined by reacting a mixture of the hydrocarbon with oxygen.

It was found that 5 cm^3 of a hydrocarbon reacted exactly with 45 cm^3 of oxygen to give 30 cm^3 of carbon dioxide which were removed using aqueous sodium hydroxide. These volumes were measured under the same conditions of temperature and pressure. Calculate the formula of the hydrocarbon from the results of the experiment.

(*e*) The catalytic converters which are fitted to the exhaust systems of cars remove nitrogen oxides and carbon monoxide but are 'poisoned' by lead compounds in petrol.

(i) Suggest how the catalyst is affected by 'poisoning'.

(ii) State and explain whether the catalysis, in the exhaust system, is homogeneous or heterogeneous.

(NI 91)

27

ALKENES AND ALKYNES

27.1 ALKENES

Alkenes are a homologous series of aliphatic hydrocarbons with the general formula C_nH_{2n}. Typical members of the series are listed below:

The names of alkenes

$CH_2{=}CH_2$	Ethene
$CH_3CH{=}CH_2$	Propene
$CH_3CH_2CH{=}CH_2$	But-1-ene
$CH_3CH{=}CHCH_3$	But-2-ene

They are called **unsaturated** hydrocarbons because the double bond can open to allow them to take up more hydrogen atoms or other species.

27.2 SOURCE OF ALKENES

Alkenes are obtained in the petroleum industry by the process of cracking [§ 26.3.2]. The laboratory preparation is usually the dehydration of an alcohol. Figure 27.1 summarises the methods employed.

27.3 PHYSICAL PROPERTIES OF ALKENES

The volatility and solubility of alkenes are similar to those of the corresponding alkanes.

27.4 STRUCTURAL ISOMERISM

27.4.1 CARBON CHAIN

Both branched-chain and unbranched-chain isomers of alkenes exist, as with alkanes.

27.4.2 POSITIONAL ISOMERISM

Isomerism can involve the carbon chain or the position of the double bond

The position of the double bond must be given in the name of an alkene. Butene has two isomers, but-1-ene, with the double bond between carbon atoms 1 and 2; and but-2-ene, with the double bond between carbon atoms 2 and 3.

27.4.3 *CIS–TRANS* ISOMERISM OF ALKENES

Inhibition of rotation about the C=C bond gives rise to cis-trans *isomerism*

In ethene [see Figure 27.2(a)], the four hydrogen atoms all lie in the same plane. Rotation of the CH_2 groups about the C=C bond is inhibited by the requirements for the formation of a strong π bond [§ 5.2.7]. The restriction of rotation is the reason why but-2-ene has two isomers, which are shown in Figure 27.2(b) and (c). The geometry of the molecules is different. The isomer with both CH_3— groups on the same side of the double bond is called the *cis*-isomer, and the isomer with the CH_3— groups on opposite sides of the double bond is called the *trans*-isomer.

The cis-trans *isomers of but-2-ene*

```
H3C     CH3          H3C     H
   \   /                \   /
    C=C                  C=C
   /   \                /   \
  H     H              H     CH3
 cis-But-2-ene       trans-But-2-ene
```

FIGURE 27.1
Preparations of Alkenes

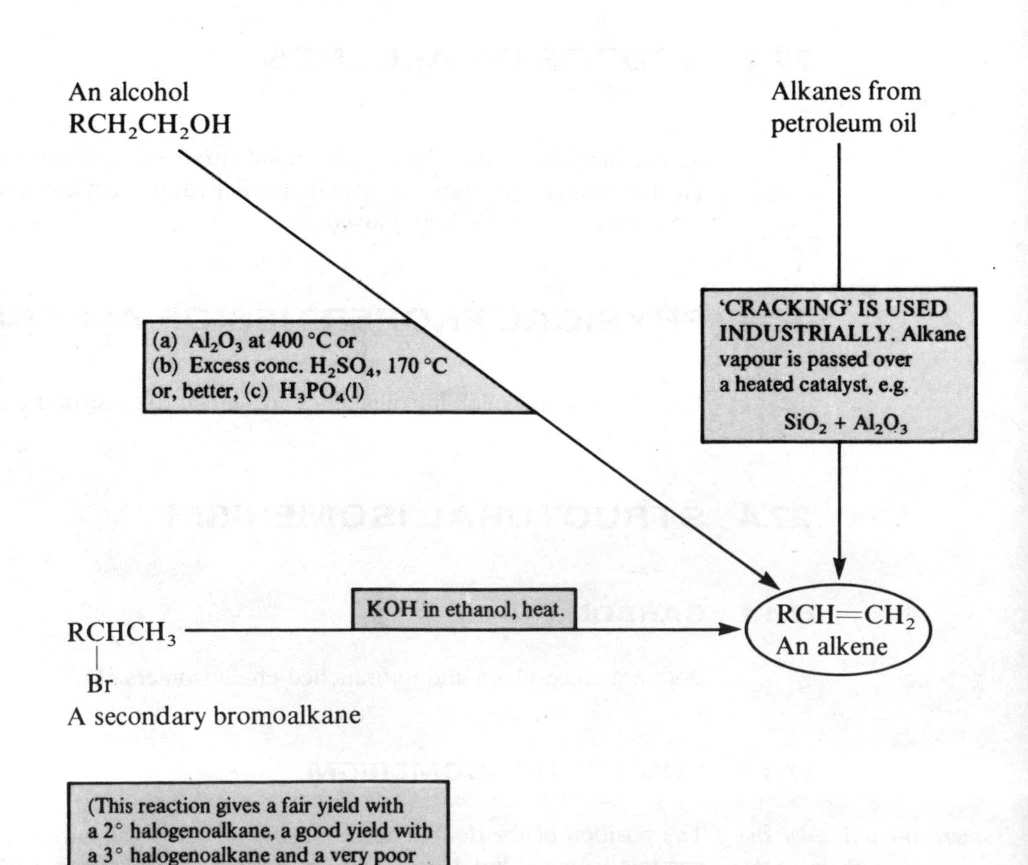

An alcohol
RCH_2CH_2OH
Alkanes from petroleum oil
(a) Al_2O_3 at 400 °C or
(b) Excess conc. H_2SO_4, 170 °C
or, better, (c) H_3PO_4(l)
'CRACKING' IS USED INDUSTRIALLY. Alkane vapour is passed over a heated catalyst, e.g.
$SiO_2 + Al_2O_3$
KOH in ethanol, heat.
$RCHCH_3$
Br
A secondary bromoalkane
$RCH{=}CH_2$
An alkene
(This reaction gives a fair yield with a 2° halogenoalkane, a good yield with a 3° halogenoalkane and a very poor yield with a 1° halogenoalkane.)

FIGURE 27.2
(a) Ethene,
(b) *cis*-But-2-ene,
(c) *trans*-But-2-ene

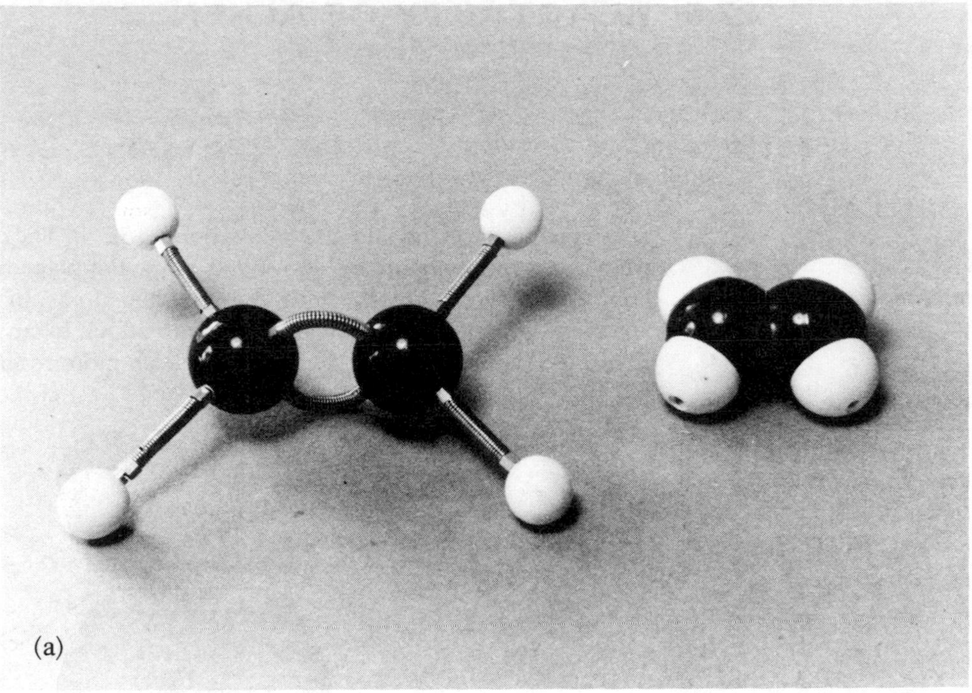

(a)

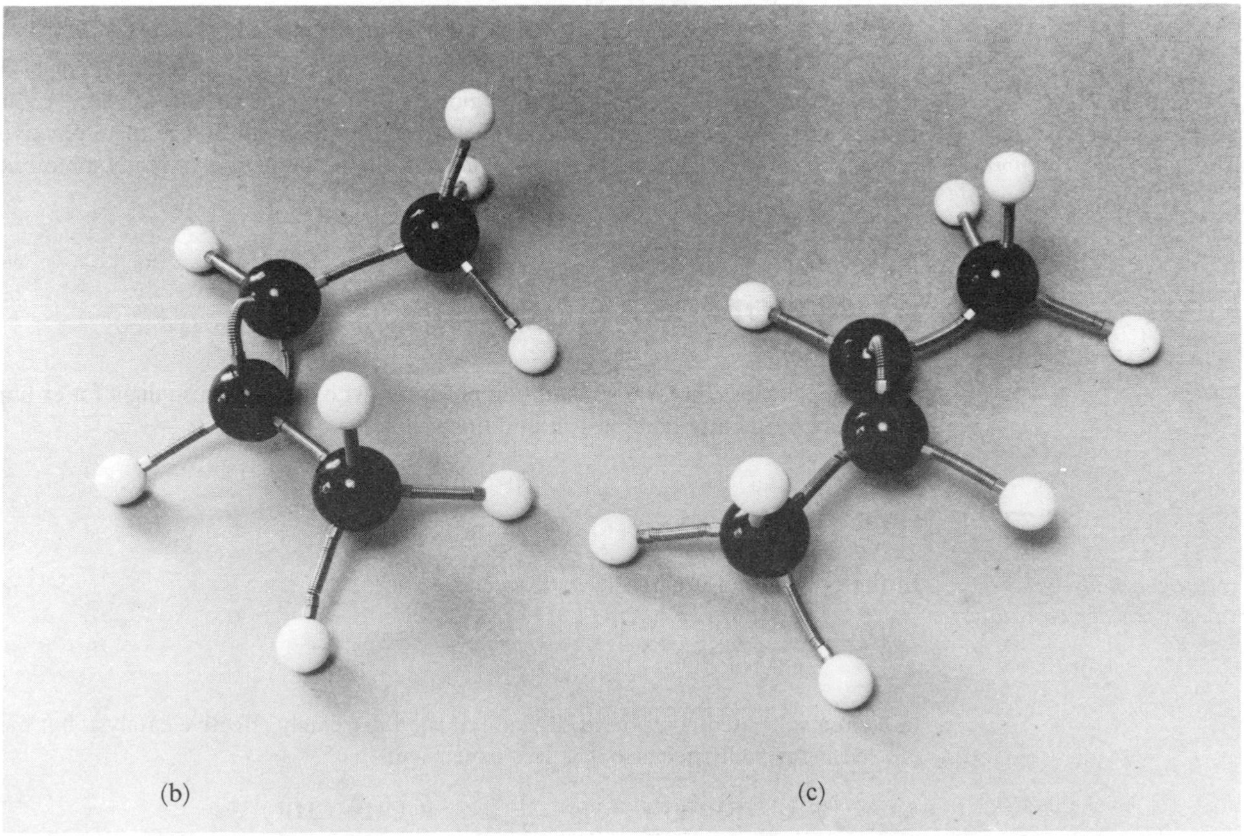

(b) (c)

27.5 REACTIVITY OF ALKENES

The two unsaturated carbon atoms in the

$$\rangle C{=}C\langle \text{ bond}$$

Alkenes are more reactive than alkanes...

are joined by a σ bond and also by a π bond [§ 5.2.7]. The cloud of electrons which forms the π bond lies above and below the plane of the three sp^2 hybrid bonds formed by each of the unsaturated carbon atoms. In this position, the π electrons are more susceptible than the σ electrons to attack by an electrophilic reagent [see Figure 27.3]. This is why alkenes are so much more reactive than alkanes.

FIGURE 27.3
The Reactivity of Ethene

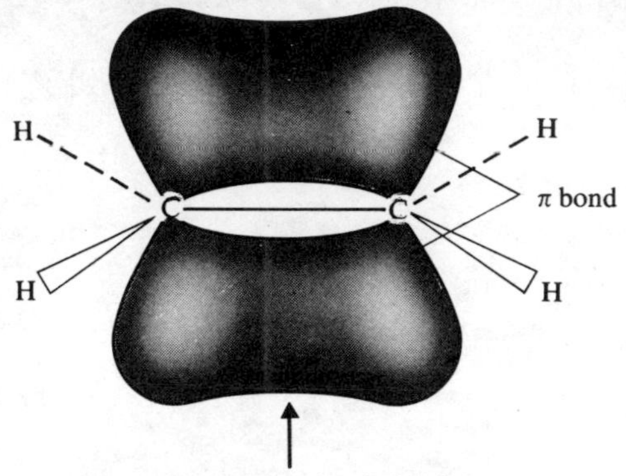

27.6 COMBUSTION

...and are not used as fuels, because they can be used instead for the manufacture of important chemicals

Alkenes burn to form carbon dioxide and water. They show a greater tendency than alkanes to undergo incomplete combustion to form carbon monoxide and carbon. In any case, alkenes are not used as fuels. The double bond makes them reactive compounds, and they are the starting point in the manufacture of many important chemicals.

27.7 ADDITION REACTIONS

Although alkenes can form substitution products, with chlorine in sunlight for example, they more frequently take part in addition reactions.

27.7.1 HYDROGEN

Hydrogen adds to alkenes with Ni powder as catalyst

In the presence of a catalyst, hydrogen adds across a

$$\rangle C{=}C\langle \text{ bond}$$

to form a saturated compound. Palladium is an extremely effective catalyst, but nickel is used industrially because it is less expensive:

$$R_2C{=}CR'_2(g) + H_2(g) \xrightarrow[100\,°C,\ 4\ atm]{Ni\ powder} R_2CH{-}CHR'_2$$

Catalytic hydrogenation, using nickel powder, is an important process in the food industry. Plant oils, such as sunflower seed oil and peanut oil, are 'polyunsaturates': they are esters of carboxylic acids which contain more than one carbon–carbon double bond. Although they are useful for cooking, these oils are less valuable commercially than expensive animal fats such as butter. Animal fats are esters of saturated carboxylic acids. Hydrogenation is used to convert unsaturated edible oils into edible fats. In soft margarine some of the double bonds remain: the degree of softness can be controlled by regulating the amount of hydrogenation.

Hydrogenation converts unsaturated edible oils into edible fats

...e.g., margarine

27.7.2 ALKANES

Tertiary alkanes add to alkenes

Tertiary alkanes will form adducts with alkenes [§ 26.3.3].

27.7.3 HALOGENS

Chlorine adds across a double bond...

Chlorine gas will react with ethene at room temperature, without catalysis by light or peroxides, to form liquid 1,2-dichloroethane:

$$H_2C{=}CH_2(g) + Cl_2(g) \rightarrow ClCH_2CH_2Cl(l)$$

...as do bromine and iodine

Bromine and iodine will also add across the double bond. When an alkene is bubbled through bromine in a solvent, such as tetrachloromethane, the bromine colour disappears as bromine adds across the double bond:

A test for a multiple bond is the decolorisation of a bromine solution without gas evolution

$$CH_2{=}CH_2(g) + Br_2(CCl_4) \rightarrow \underset{\text{1,2-Dibromoethane}}{BrCH_2CH_2Br(CCl_4)}$$

The decolorisation of bromine solutions without the evolution of hydrogen bromide (a fuming, pungent gas) is a test for a multiple bond. Bromine water is also decolorised by alkenes to give a different product, a **bromoalcohol**:

$$CH_2{=}CH_2(g) + Br_2(aq) + H_2O(l) \rightarrow \underset{\text{2-Bromoethanol}}{BrCH_2CH_2OH(aq)} + HBr(aq)$$

27.7.4 HYDROGEN HALIDES

Hydrogen halides add across the double bond

The hydrogen halides, HCl, HBr and HI, all add across the double bond. When hydrogen bromide adds to ethene, bromoethane is formed:

$$CH_2{=}CH_2(g) + HBr(g) \rightarrow CH_3CH_2Br(g)$$

Addition of hydrogen halide

When hydrogen halides add to an unsymmetrical alkene, two products are possible. The addition of hydrogen bromide to propene could give either

$$CH_3CH{=}CH_2(g) + HBr(g) \rightarrow \underset{\text{2-Bromopropane}}{CH_3\underset{\substack{|\\ Br}}{C}HCH_3(l)}$$

$$\textit{or}\ \underset{\text{1-Bromopropane}}{CH_3CH_2CH_2Br(l)}$$

Markovnikov's Rule predicts the product of the addition of HX to an alkene

In fact, the product is almost entirely 2-bromopropane. The Russian chemist, Markovnikov, formulated a rule for predicting which addition product would be formed. **Markovnikov's Rule** can be stated in this form: in addition of a compound HX to an unsaturated compound, hydrogen becomes attached to the unsaturated carbon atom which carries the larger number of hydrogen atoms. Thus, in addition of hydrogen iodide to 2-methylbut-2-ene, the Markovnikov Rule predicts the formation of 2-iodo-2-methylbutane:

$$(CH_3)_2C{=}CH(CH_3)(g) + HI(g) \rightarrow CH_3{-}C(CH_3)(I){-}C(CH_3)(H){-}H(l)$$

with hydrogen adding to the carbon atom which already has one hydrogen atom bonded to it. This is in fact the major product of the reaction, with a trace of 2-iodo-3-methylbutane also being formed:

$$CH_3{-}C(CH_3)(H){-}C(CH_3)(I){-}H(l)$$

27.7.5 MECHANISM OF ADDITION OF BROMINE TO ALKENES

At first sight it might seem that the reaction could take place in one step if the bromine molecule approaches the alkene from the right direction:

A simple one step mechanism for the addition of bromine to alkenes does not fit the facts...

$$H_2C{=}CH_2 + Br{-}Br \rightarrow H{-}CH(Br){-}CH(Br){-}H$$

A number of workers have found evidence that this is not the course of the reaction. Only two of these pieces of evidence are given here.

EVIDENCE

...For example, these two experimental observations on the addition of bromine to alkenes

(a) Ethene reacts with bromine in a solvent such as tetrachloromethane to give 1,2-dibromoethane. With bromine water, it gives a mixture of 2-bromoethanol, 1,2-dibromoethane and hydrogen bromide:

$$CH_2{=}CH_2 + Br_2 \xrightarrow{\text{in } CCl_4} BrCH_2CH_2Br$$
$$CH_2{=}CH_2 + Br_2 \xrightarrow{\text{in } H_2O} BrCH_2CH_2OH + BrCH_2CH_2Br + HBr$$

(The yield of $BrCH_2CH_2OH$ is too great to be explained by hydrolysis of $BrCH_2CH_2Br$).

(b) Ethene and bromine water containing sodium chloride give three products, 2-bromoethanol, 1-bromo-2-chloroethane and 1,2-dibromoethane:

$$CH_2{=}CH_2 + Br_2 + H_2O + NaCl \rightarrow HOCH_2CH_2Br \text{ (major product)} + ClCH_2CH_2Br \text{ (smaller amount)} + BrCH_2CH_2Br \text{ (a trace)}$$

THE PROPOSED MECHANISM

The formation of the chloro-compound in (b) needs some explaining. Since sodium chloride does not react with ethene, there must be formed from bromine and ethene a positively charged species which will react with chloride ions:

$$CH_2{=}CH_2 \xrightarrow{Br_2} \underset{\overset{+}{Br}}{CH_2{-}CH_2} \xrightarrow{Cl^-} ClCH_2CH_2Br$$

The following mechanism is proposed for the formation of the positive ion.

1. Ethene has both a σ bond and a π bond between the two carbon atoms. Bromine is an electrophile. When a bromine molecule approaches an ethene molecule, the π electron cloud interacts with the approaching bromine molecule, causing a polarisation of the Br—Br bond:

When the π electron cloud of ethene reacts with bromine...

$$H_2C{=}CH_2 + Br{-}Br \rightarrow H_2C{=}CH_2 \cdots \overset{\delta+}{Br}{-}\overset{\delta-}{Br}$$

2. The π electrons become gradually more attached to the $\delta+$ Br atom, and the electrons of the Br—Br bond become gradually more polarised until the association of ethene and bromine is transformed into a positive ion, called a *bromonium ion*, and a bromide ion:

...the bromonium ion intermediate which is formed...

$$H_2C{=}CH_2 \cdots \overset{\delta+}{Br}{-}\overset{\delta-}{Br} \rightarrow \underset{\overset{+}{Br}}{H_2C{-}CH_2} + Br^-$$

The positive charge is stabilised by delocalisation. The negative charge resides on Br^-, which is a good leaving group. These factors stabilise the transition state that precedes the bromonium ion intermediate [§ 14.9.2].

3. The positive bromonium ion is immediately attacked by a bromide ion to form the product, 1,2-dibromoethane:

...quickly reacts to form the product

$$Br^- + \underset{\overset{+}{Br}}{H_2C{-}CH_2} \rightarrow BrCH_2{-}CH_2Br$$

This reaction step is very fast compared with the other steps.

4. In reaction (b), i.e., with $Br_2(aq) + NaCl(aq)$, the cation reacts in three ways:

$$\underset{\overset{+}{Br}}{CH_2{-}CH_2} + Br^- \longrightarrow BrCH_2CH_2Br \quad [1]$$

$$\underset{\overset{+}{Br}}{CH_2{-}CH_2} + H_2O \longrightarrow H_2\overset{+}{O}{-}CH_2CH_2Br \xrightarrow{H_2O} HOCH_2CH_2Br + H_3O^+ \quad [2]$$

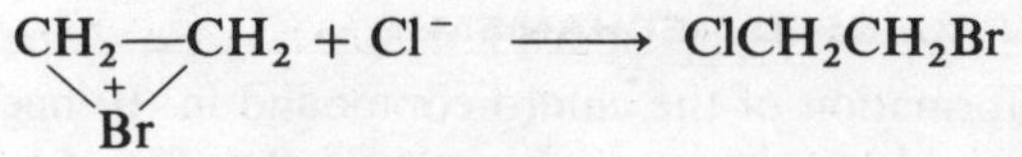

$$CH_2\text{—}CH_2 (\overset{+}{Br} \text{ bridged}) + Cl^- \longrightarrow ClCH_2CH_2Br \quad [3]$$

A summary

To summarise, the addition of halogens to alkenes can be formulated as

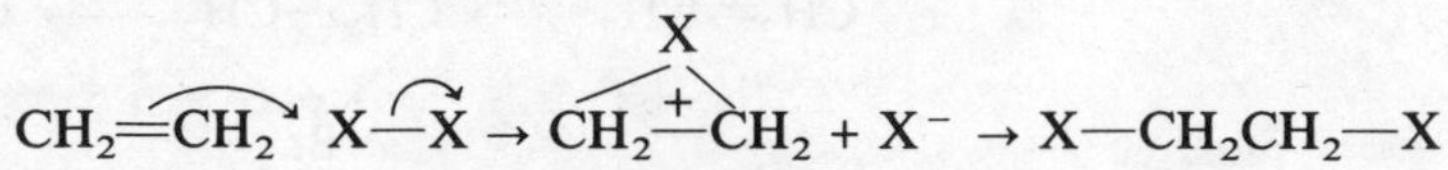

$$CH_2{=}CH_2 \; X\text{—}X \rightarrow CH_2\overset{+}{\text{—}}CH_2 (\text{X bridged}) + X^- \rightarrow X\text{—}CH_2CH_2\text{—}X$$

A curly arrow represents the movement of a pair of electrons from the tail of the arrow to its tip. The addition of an electrophile to an alkene results in the formation of a cation which can accept electrons from a nucleophile, such as Br^-, Cl^-, NO_3^-, HSO_3^- or H_2O.

CONFIRMATION OF THE MECHANISM

Methyl groups stabilise the carbocation intermediate and the transition state that precedes it

Confirmation of this mechanism is seen in the effect of alkyl groups on the rate. Propene, $CH_3CH{=}CH_2$, reacts twice as fast as ethene does, and 2,3-dimethylbut-2-ene, $(CH_3)_2C{=}C(CH_3)_2$, reacts 14 times faster than ethene does. The reason is that methyl groups are electron-releasing [§ 12.7.7]. They tend to decrease the positive charge on a carbocation and thus stabilise it:

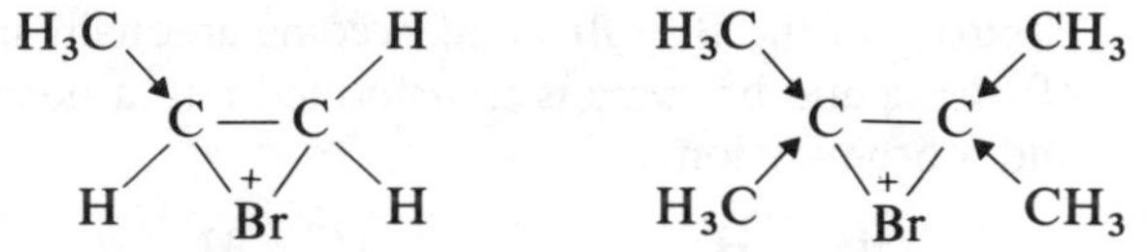

$$\text{H}_3\text{C}(\text{H})\text{C—C}(\text{H})(\text{H}) \text{ with } \overset{+}{Br} \text{ bridge} \qquad (\text{H}_3\text{C})_2\text{C—C}(\text{CH}_3)_2 \text{ with } \overset{+}{Br} \text{ bridge}$$

The more stable a carbocation is, the lower is the energy of the transition state that precedes it [§ 14.9.2], assuming that the transition state is similar to the carbocation intermediate. The rate of formation of the transition state is therefore increased by methyl groups. Propenoic acid, $CH_2{=}CHCO_2H$, adds bromine slowly. The reason is that the carboxyl group, by withdrawing electrons, decreases the stability of the carbocation.

27.7.6 MECHANISM OF ADDITION OF HYDROGEN HALIDE TO ALKENES

Addition of HX to an alkene involves interaction of the electrophile with the double bond

A mechanism similar to that for the addition of bromine is proposed. A molecule of hydrogen halide, H—X is permanently polarised as $\overset{\delta+}{H}\text{—}\overset{\delta-}{X}$. The π electrons of the alkene bond to the electrophilic H atom:

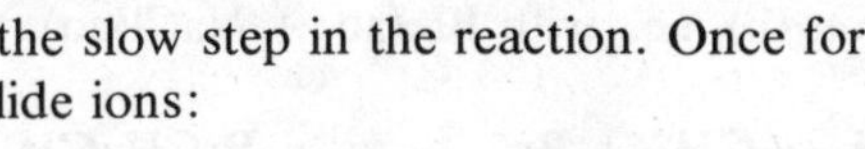

$$CH_2{=}CH_2 \; H \rightarrow X \rightarrow \overset{+}{C}H_2\text{—}CH_3 + X^-$$

This is the slow step in the reaction. Once formed, the carbocation reacts rapidly with halide ions:

$$\overset{+}{C}H_2CH_3 + X^- \rightarrow XCH_2CH_3$$

EVIDENCE

The rate of addition increases with increasing acid strength of HX

(a) The rate of addition increases in the order

$$HF < HCl < HBr < HI$$

Markovnikov's Rule is explained by the carbocation theory

This is the order of increasing acid strengths, i.e., the order of readiness to release a proton. The reaction of the carbocation with X^- is rapid, and its speed is much the same for all X^- ions.

(b) Addition follows Markovnikov's Rule:

$$CH_3CH{=}CH_2 + HCl \rightarrow CH_3CHClCH_3(90\%) + CH_3CH_2CH_2Cl(10\%)$$

Carbocations are produced during the formation of the two products:

$$CH_3{-}CH{=}CH_2 + HCl \rightarrow CH_3{-}\overset{+}{C}H{-}CH_3 + Cl^- \quad [1]$$

$$CH_3{-}CH{=}CH_2 + HCl \rightarrow CH_3{-}CH_2{-}\overset{+}{C}H_2 + Cl^- \quad [2]$$

Since alkyl groups are electron-releasing [§ 12.7.7], they tend to decrease the charge on a cation, and thus stabilise it. The carbocation

$$CH_3 \longrightarrow \overset{+}{C}H \longleftarrow CH_3 \ [1]$$

is therefore more stable than

$$CH_3 \longrightarrow CH_2 \longrightarrow \overset{+}{C}H_2 \ [2]$$

The more stable the carbocation, the lower is the energy of the transition state that precedes it. The rate of formation of carbocation [1] is therefore the greater, and reaction [1] predominates.

Alkyl groups increase the rate of addition

(c) The rate of reaction of alkenes increases in the order

$$CH_2{=}CH_2 < RCH{=}CH_2 < R_2C{=}CH_2$$

(where R is an alkyl group).
This is in accordance with the carbocation theory since the alkyl groups confer stability on the carbocations in the order

$$R_2\overset{+}{C}CH_3 > R\overset{+}{C}HCH_3 > \overset{+}{C}H_2CH_3$$

This leads to reaction rates in the same order (as discussed in (b) above).

27.7.7 CONCENTRATED SULPHURIC ACID

Addition of conc. H_2SO_4 yields the hydrogensulphate...

The addition of sulphuric acid across a double bond is similar to that of a hydrogen halide. Markovnikov's Rule is followed. When ethene is bubbled into concentrated sulphuric acid at room temperature, ethyl hydrogensulphate is formed:

$$CH_2{=}CH_2(g) + H_2SO_4(l) \xrightarrow{\text{cold}} \underset{\underset{H}{|}}{H_2C}{-}\underset{\underset{OSO_2OH}{|}}{CH_2}(l)$$

Ethyl hydrogensulphate

... which can be hydrolysed to the alcohol

The product, ethyl hydrogensulphate, when added to water and warmed, is hydrolysed to ethanol:

$$CH_2{=}CH_2 + H_2SO_4 \rightarrow C_2H_5SO_4H \xrightarrow[\text{warm}]{H_2O} C_2H_5OH + H_2SO_4$$

The net result is the addition of H·OH across the double bond. The industrial method of accomplishing this is the catalytic hydration of ethene. Ethene and steam are passed at 300 °C and 60 atm over phosphoric acid absorbed on silica pellets:

$$\underset{\text{Ethene}}{CH_2{=}CH_2(g)} + H_2O(g) \xrightarrow[H_3PO_4\text{ catalyst}]{300\,°C,\ 60\,atm} \underset{\text{Ethanol}}{C_2H_5OH(g)}$$

The reaction makes possible the manufacture of the important solvent ethanol from ethene, which is a product of the petroleum industry.

27.7.8 ALKALINE POTASSIUM MANGANATE(VII)

Oxidation of C=C by alkaline $KMnO_4$...

Potassium manganate(VII) in alkaline solution is a weak oxidising agent. When ethene is bubbled into alkaline potassium manganate(VII) solution, the purple colour fades as ethene is oxidised to ethane-1,2-diol:

...forming a diol...

$$\underset{\text{Ethene}}{H_2C{=}CH_2} \xrightarrow[KMnO_4,\ OH^-]{} \underset{\text{Ethane-1,2-diol}}{HOCH_2{-}CH_2OH}$$

...with decolorisation is a test for C=C

A diol is a compound containing two hydroxyl groups [§ 30.1.3]. A brown suspension of manganese(IV) oxide, MnO_2, appears. The disappearance of the purple colour of ice-cold, dilute alkaline manganate(VII) solution is a test for a carbon–carbon multiple bond as few organic compounds are oxidised by this weak oxidising agent.

27.7.9 OZONE

Ozone adds to alkenes to form ozonides

Ozone, or trioxygen, O_3, adds across the double bond to form an **ozonide**. Since the ozonide formed is an explosive compound, it is prepared below 20 °C in a non-aqueous solvent:

$$(R)(H)C{=}C(R')(R'') + O_3 \xrightarrow[\text{in } CCl_4]{<20\ ^\circ C} (R)(H)C\!\begin{smallmatrix}O\\ \\ O{-}O\end{smallmatrix}\!C(R')(R'')$$

On hydrolysis, it splits up into two carbonyl compounds [see Chapter 31], and hydrogen peroxide, which may oxidise the other products:

$$(R)(H)C\!\begin{smallmatrix}O\\ \\ O{-}O\end{smallmatrix}\!C(R')(R'') + H_2O \longrightarrow (R)(H)C{=}O + O{=}C(R')(R'') + H_2O_2$$

Ozonolysis can be used to locate the position of the C=C bond...

When zinc and ethanoic acid are used as a combined hydrolysing and reducing agent, the possibility of oxidation is avoided. The carbonyl compounds formed in the hydrolysis of the ozonide can be identified. The usefulness of **reductive ozonolysis**, i.e., the formation of an ozonide followed by its hydrolysis under mildly reducing conditions, is that it can be used to locate the position of the double bond in an alkene. If the ozonolysis of hexene gives the carbonyl compounds

$$CH_3{-}C(H){=}O \quad \text{and} \quad CH_3{-}CH_2{-}C(CH_3){=}O$$

then putting these compounds together gives the formula of the original hexene:

$$CH_3{-}\overset{H}{\overset{|}{C}}{=}O \quad \text{and} \quad O{=}\overset{CH_3}{\overset{|}{C}}{-}CH_2CH_3 \quad \text{come from} \quad CH_3{-}\overset{H}{\overset{|}{C}}{=}\overset{CH_3}{\overset{|}{C}}{-}CH_2CH_3$$

...The amount of ozone absorbed gives the number of C=C bonds per molecule

Ozone reacts quantitatively with alkenes: 1 mole of ozone is absorbed by 1 mole of an alkene which possesses one double bond per molecule. This reaction can be used to find the number of double bonds in a compound.

FIGURE 27.4 Reactions of Alkenes

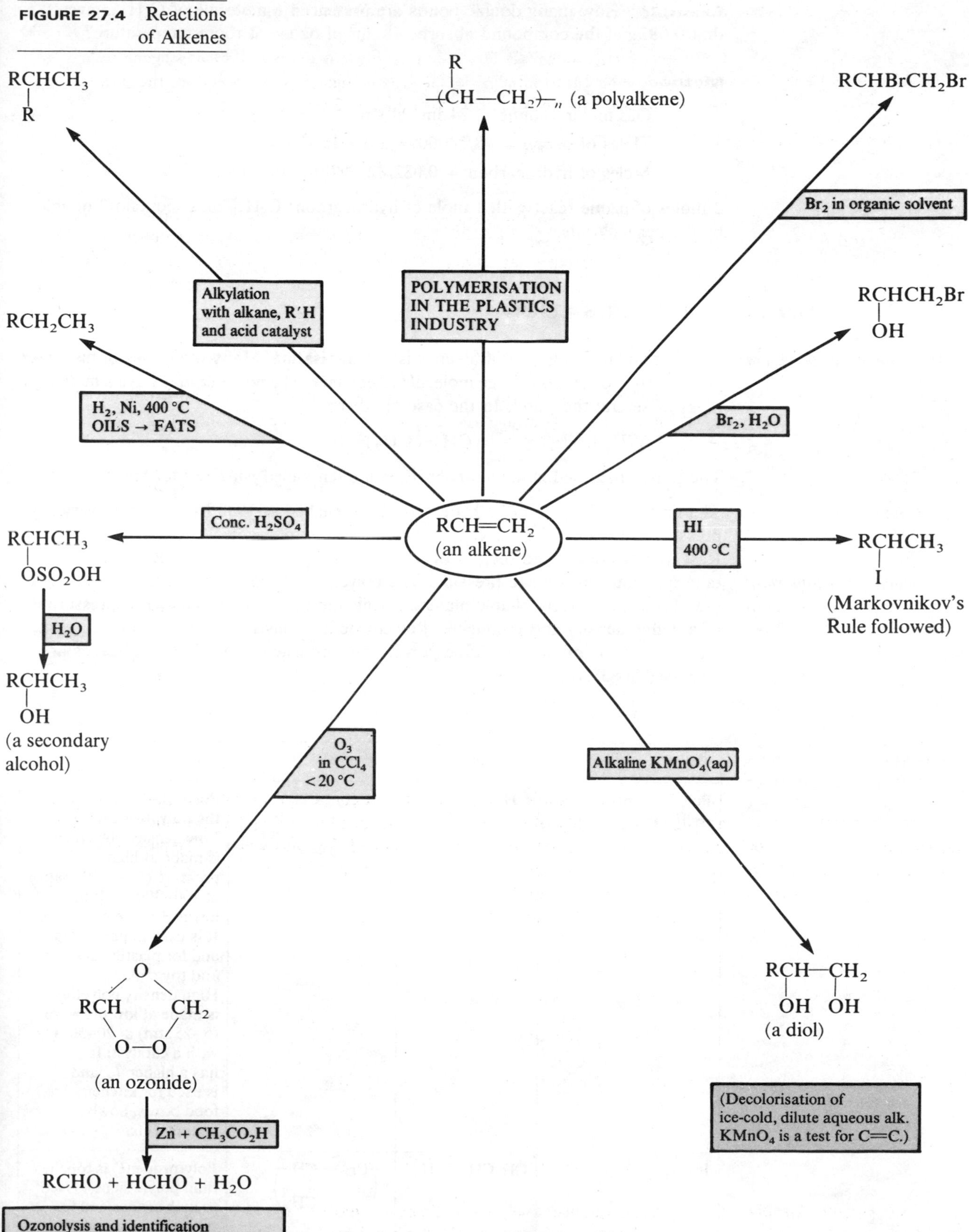

Example How many double bonds are present in a molecule of C_6H_{10}, given that 0.082 g of the compound absorbs 48 cm^3 of ozone at room temperature?

Method

Gas molar volume = 24 dm^3 at rtp

Moles of ozone = 48/24000 = 2.0×10^{-3} mol

Moles of hydrocarbon = 0.082/82 = 1.0×10^{-3} mol

2 moles of ozone react with 1 mole of hydrocarbon: C_6H_{10} must contain 2 double bonds per molecule.

27.7.10 POLYMERISATION

Monomers and polymers

Another addition reaction of alkenes is **polymerisation**. Many molecules of **monomer** add together to form one large molecule of **polymer**. The polymer may have a molecular mass of several thousand. In the case of ethene

$$nCH_2{=}CH_2(g) \rightarrow {+}CH_2{-}CH_2{+}_n(s)$$

The polymer formed is named *poly(ethene)*, often called *polythene* for short.

Ethene polymerises to form polyethene . . .

. . . which is an important plastics material

The first polymerisation of ethene was accomplished in 1933 by the use of very high pressure (1000 atm) and oxygen as the catalyst. A free radical mechanism operates. Research has developed the use of powerful catalysts, which enable the addition to take place at atmospheric pressure. The polyethene formed under high pressure is a low density, extremely pliable material, while the polymer formed at low pressure is of a higher density and is tougher. Polyethene is a **plastic**: it can be moulded easily into a multitude of shapes. Some polyalkenes are listed in Table 27.1. Plastics are discussed in § 35.1.

TABLE 27.1 Polymers of Ethene and other Alkenes

Name	*Monomer*	*Polymer*	*Uses*
Poly(ethene) (polythene)	$CH_2{=}CH_2$	${+}CH_2{-}CH_2{+}_n$	Polyethene has 40% of the polyalkene market. Low density polyethene is made at high pressure (1000–2000 atm) at 100–300 °C. It has a low melting temperature. It is used in packaging and for plastic bags and toy making. High density polyethene is made at low pressure (5–25 atm) at 20–50 °C, with a catalyst. It has a higher T_m and is used for kitchenware, food boxes, bowls, buckets, etc.
Poly(propene)	$CH_3CH{=}CH_2$	${+}CH_2{-}CH(CH_3){+}_n$	Poly(propene) is tougher than poly(ethene). Used to make ropes and for packaging. Its high T_m enables it to resist boiling water.

Name	*Monomer*	*Polymer*	*Uses*
Poly(chloroethene) (polyvinylchloride, PVC)	$CH_2{=}CHCl$	$-(CH_2-CHCl)-_n$	PVC is more rigid than polyethene. Used as a building material, e.g., guttering and electrical insulation. With added plasticisers, PVC is used for macintoshes, wellingtons, etc.
Poly(tetrafluoro-ethene) (PTFE, Teflon)	$CF_2{=}CF_2$	$-(CF_2-CF_2)-_n$	Used for coating surfaces to reduce friction, e.g., non-stick frying pans
Poly(phenylethene) (polystyrene)	$CH_2{=}CH$–C_6H_5 (benzene ring on CH)	$-(CH_2-CH)-_n$ with C_6H_5 (benzene ring) on CH	Made into polystyrene foam by dissolving the polymer in a solvent and then vaporising the solvent. Used as insulation and for packaging. Polystyrene is flammable.
Poly(2-chloro-butadiene) (Neoprene)	$CH_2{=}CH$–$ClC{=}CH_2$	$\left(-CH_2-CH(ClC{=}CH_2)-\right)_n$	Used in the manufacture of synthetic rubber
Poly(ethenyl-ethanoate) (polyvinylacetate, PVA)	CH_3CO_2–$CH{=}CH_2$	$-(CH(CH_3CO_2)-CH_2)-_n$	Records. More flexible than PVC. Used in emulsion paints
Poly(methyl 2-methyl-propenoate) (Perspex®, Plexiglass®)	$CH_2{=}C(CH_3)CO_2CH_3$	$\left(-C(CH_3)(CO_2CH_3)-CH_2-\right)_n$	Transparent. Used as a substitute for glass
Poly(propenonitrile) (polyacrylonitrile) (Acrilan®, Orlon®, Courtelle®)	$CH_2{=}CHCN$	$\left(-CH_2-CH(CN)-\right)_n$	Used as *acrylic* fibres for making clothing, blankets and carpets

CHECKPOINT 27A: ALKENES

1. Name the following compounds:

(*a*) $CH_3CH{=}CHCH_2CH_2CH_3$

(*b*) $CH_3C(CH_3){=}CHCH_2CH(CH_3)CH_3$

(*c*) $CH_3CH_2(H)C{=}C(H)CH_2CH_2Cl$ (CH₃CH₂ and H on one carbon; H and CH₂CH₂Cl on the other; CH₃CH₂ and CH₂CH₂Cl trans)

(*d*) $CH_3CHClCH{=}CH_2$

(*e*) $CH_3(H)C{=}C(H)CH_2CH_2CH_3$ (CH₃ and CH₂CH₂CH₃ cis)

2. Write structural formulae for the following:

(*a*) Pent-1-ene

(*b*) 3-Chlorohexa-2,4-diene

(*c*) Buta-1,3-diene

(*d*) 4,4-Dimethylpent-2-ene

3. A compound C_6H_{12}, after ozonolysis, gives two products, one of which is propanone, $(CH_3)_2CO$. Which of the following is its formula?

(*a*) $CH_3CH{=}CHCH(CH_3)_2$

(*b*) $CH_3CH_2CH{=}C(CH_3)_2$

(*c*) $CH_3C(C_2H_5){=}CHCH_3$

(*d*) $(CH_3)_2C{=}C(CH_3)_2$ or

(*e*) $CH_3CH{=}C(CH_3)C_2H_5$

4. Give the name and formula of the product from each of the reactions

(*a*) $CH_3CH{=}CH_2 + HBr \rightarrow$

(*b*) $(CH_3)_2C{=}CH_2 + Br_2 + NaOH \rightarrow$

(*c*) $(CH_3)_2C{=}CH_2 + H_2SO_4 \rightarrow$

5. Give the names and formulae of the products formed when the following reagents add to propene:
(*a*) chlorine in tetrachloromethane
(*b*) chlorine water.
Write the mechanism for each reaction.

6. Complete the following equations. Indicate the conditions needed for reaction.

(*a*) $CH_2{=}CH_2 + \quad \rightarrow CH_3CH_2OH$

(*b*) $CH_3CH{=}CH_2 + HBr \rightarrow$

(*c*) $(CH_3)_2C{=}CH_2 + Br_2 + H_2O + HNO_3 \rightarrow$

(*d*) $CH_3CH{=}CH_2 + H_3PO_4 \rightarrow$

(*e*) $CH_3CH_2CH{=}CH_2 + \rightarrow CH_3CH_2CHOHCH_2OH$

(*f*) $(CH_3)_2CHCH{=}CHCH_3 + \quad \rightarrow (CH_3)_2CHCHO + CH_3CHO$

7. When propene is bubbled through chlorine water containing nitrate ions, three products are formed. Give the names and formulae of the three products, and explain how they come to be formed. What product do you think would be formed in the reaction between propene and nitrogen chloride oxide, NOCl, which reacts as NO^+Cl^-?

8. Why do alkenes show geometrical isomerism, whereas alkanes do not? Draw and name the isomers of $CH_3CH{=}CHCl$, $CH_3CH{=}CHC_2H_5$, $ClCH{=}CHBr$ and $ClCH{=}CBrCH_3$.

27.8 ALKYNES

Alkynes are aliphatic hydrocarbons with a C≡C triple bond.
The general formula of this homologous series is C_nH_{2n-2}. They are named by changing the name of the corresponding alk*ane* to end in *-yne*. Some members are:

The names of alkynes

$HC{\equiv}CH$	Ethyne
$CH_3{-}C{\equiv}C{-}CH_3$	But-2-yne
$CH_3{-}CH{=}CH{-}C{\equiv}CH$	Pent-2-en-4-yne

The model of ethyne in Figure 27.5 shows that the four atoms lie in a straight line.

FIGURE 27.5 Ethyne

27.9 SOURCE OF ETHYNE

The combustion of ethyne . . .

Ethyne (formerly called acetylene) is a valuable compound. It burns in oxygen in an extremely exothermic reaction.

$$CH{\equiv}CH(g) + 2\tfrac{1}{2}O_2(g) \rightarrow 2CO_2(g) + H_2O(l); \quad \Delta H^{\ominus} = -1300\,kJ\,mol^{-1}$$

. . . is used in welding

Ethene is used in some syntheses

Oxy-acetylene torches are used for cutting and welding metals. Ethyne is also valued because its unsaturated nature makes it a starting-point for the manufacture of a variety of organic compounds. It is, however, being superseded by ethene as a starting-point in organic syntheses as ethene is a cheaper starting-material.

It is made from natural gas

Ethyne is manufactured industrially from natural gas. The partial combustion at high temperature of methane [see Figure 27.6], the main component of natural gas, gives ethyne:

$$2CH_4(g) + 1\tfrac{1}{2}O_2(g) \rightarrow C_2H_2(g) + 3H_2O(l)$$

Ethyne may undergo an explosive reaction

Work with ethyne is difficult as it tends to undergo an explosive reaction to form carbon and hydrogen. It is stored in solution in propanone under pressure.

27.10 *REACTIONS OF ALKYNES

Accessibility of π electrons explains the reactivity of alkynes

The unsaturated carbon atoms in the $-C{\equiv}C-$ bond are joined by a σ bond and two π bonds [§ 5.2.7]. The two π bonds lie in mutually perpendicular planes. The exposed position of the π electron clouds makes them accessible to attack by electrophilic reagents [see Figure 27.6]. Alkynes react with electrophilic reagents such as hydrogen, halogens and hydrogen halides, to form adducts. The reactions are similar to those of alkenes (see Figure 27.7).

FIGURE 27.6
Bonding in Ethyne

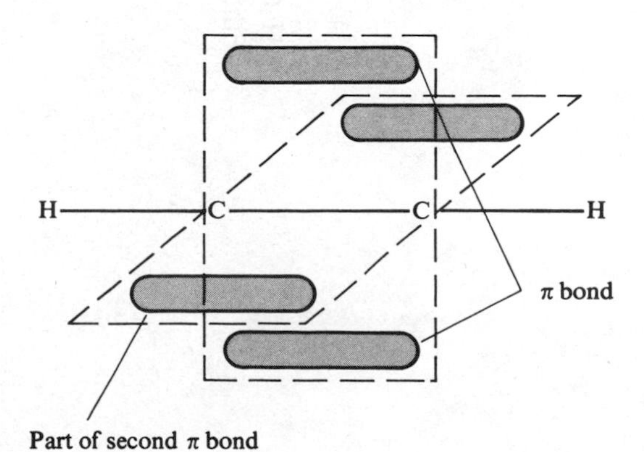

FIGURE 27.7
Reactions of Ethyne

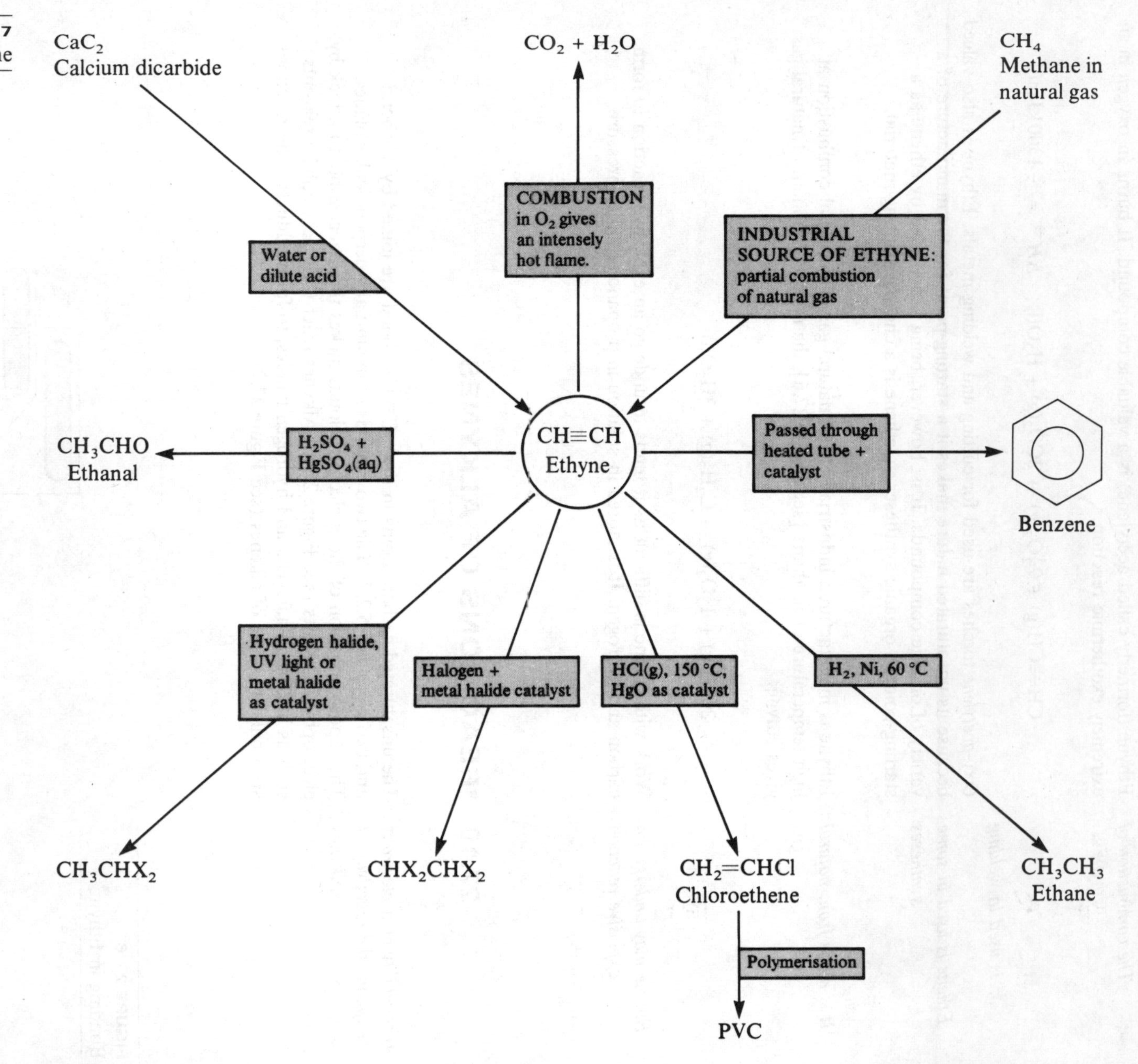

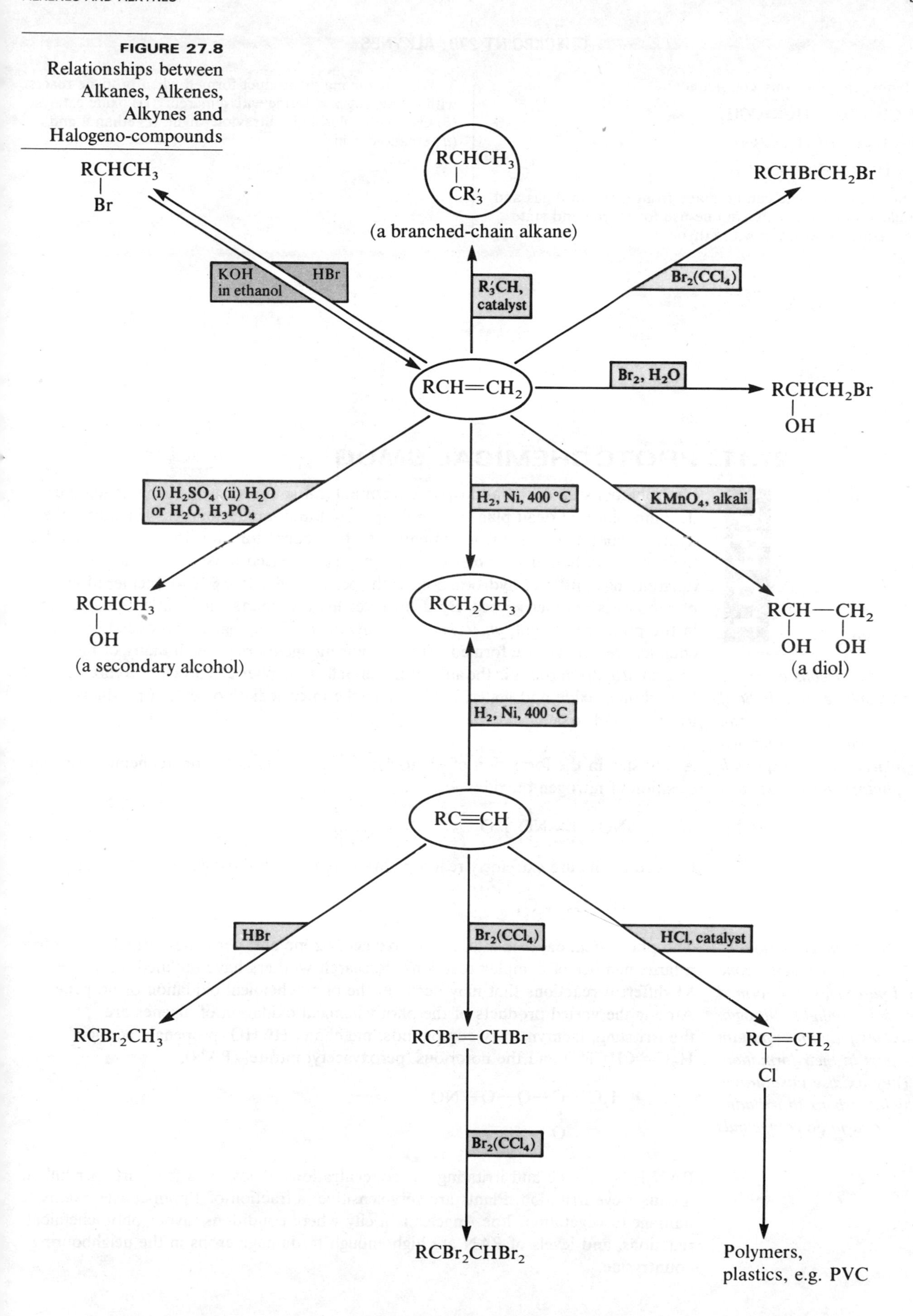

FIGURE 27.8 Relationships between Alkanes, Alkenes, Alkynes and Halogeno-compounds

CHECKPOINT 27B: ALKYNES

1. Name the following compounds:

(*a*) $CH_3CH{=}CHC{\equiv}CCH_3$

(*b*) $CH_2{=}CHCH_2C{\equiv}CH$

(*c*) $CH{\equiv}CCHBrCH{=}CH_2$

2. State how ethyne can be made from (a) natural gas and (b) calcium dicarbide. Give one use for ethyne and state one hazard of working with ethyne.

3. What is the major product formed when propyne reacts with (*a*) hydrogen chloride with a mercury(II) oxide catalyst, (*b*) hydrogen chloride in ultraviolet light, (*c*) ethanol and (*d*) ethanoic acid?

27.11 PHOTOCHEMICAL SMOG

TOPIC

Hydrocarbons enter the atmosphere from natural sources and from petrol engines. Unsaturated hydrocarbons react with photochemical oxidants

Hydrocarbons enter the atmosphere from natural sources, such as the anaerobic decomposition of dead plant material, and from man-made sources. The atmospheric hydrocarbons that come from human activity account for only 15% of the total, but their effect on humans is great because they are released mostly in urban air. The outstanding emitter of hydrocarbons is the petrol engine [see § 26.4]. Incineration of rubbish is another source. By themselves hydrocarbons cause little damage, but in the presence of light, *photochemical oxidants* (oxidising agents produced by photochemical reactions) are formed. These oxidising agents react with many of the unsaturated hydrocarbons in the air. Ultimately, all atmospheric hydrocarbons are oxidised to carbon dioxide and water, but some of the intermediate oxidation products are irritating and toxic.

A key step in the formation of photochemical oxidants is the photochemical decomposition of nitrogen dioxide:

$$NO_2 \xrightarrow{h\nu} NO + O$$

Oxygen atoms are extremely reactive. Many react with dioxygen to form ozone:

$$O + O_2 \rightarrow O_3$$

Photochemical oxidants, e.g. atomic oxygen, ozone and peroxides, are formed in sunlight. Nitrogen oxides play an important part in their formation. They oxidise unsaturated hydrocarbons to irritating and toxic compounds

The photochemical oxidants, atomic oxygen, ozone and peroxides, attack alkenes in a large number of complex reactions. Research workers have outlined a scheme of 81 different reactions that may occur in the photochemical oxidation of propene! Among the varied products of the photochemical oxidation of alkenes are the irritating, lachrymatory compounds, methanal, HCHO, propenal, 'acrolein', $H_2C{=}CHCHO$ and the notorious 'peroxyacetyl nitrate' (PAN),

$$H_3C{-}\underset{\underset{O}{\|}}{C}{-}O{-}O{-}NO_2.$$

PAN is both toxic and irritating. At concentrations as low as a few parts per billion it causes eye irritation. Plants are very sensitive: a fraction of 1 ppm causes extensive damage to vegetation. Los Angeles is a city where conditions favour photochemical reactions, and levels of PAN are high enough to damage crops in the neighbouring countryside.

FIGURE 27.9
The Formation of Photochemical Smog

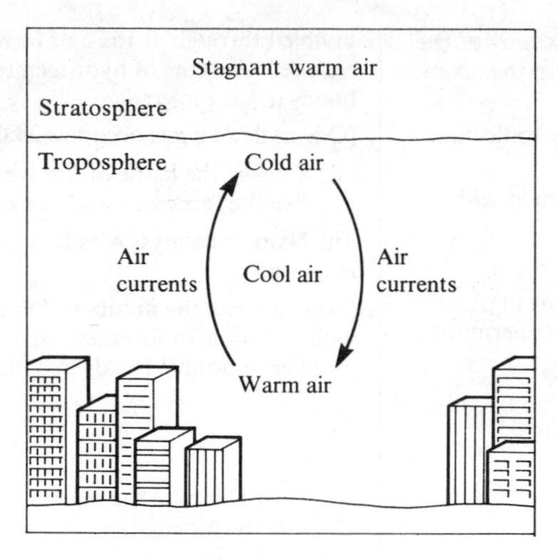

Air near the ground is warmed by the earth. As warm air rises and cold air descends, vertical air currents carry pollutants into the upper layers of the atmosphere.

Sometimes a temperature inversion occurs. Brilliant sunshine makes a layer of air above the ground warmer than air at ground level. The stagnant air traps pollutants near the ground. These conditions favour the formation of photochemical smog.

A photochemical smog of oxidants and oxidation products is formed. It includes irritating and toxic compounds

As chemical reactions continue in the atmosphere, more ozone is produced and ozone levels soar [see § 21.5]. Small molecules combine to form larger molecules and eventually tiny particles. These particles and the brown colour of nitrogen dioxide give the air a dirty, 'smoggy' appearance. Eventually, the intermediate oxidation products are converted into carbon dioxide and water. Until that happens, they are irritating and toxic to the population. This type of pollution is called *photochemical smog*.

The control of photochemical smog can be achieved by reducing emission of hydrocarbons and oxides of nitrogen. This is discussed in § 26.4.

CHECKPOINT 27C: PHOTOCHEMICAL SMOG

1. What is a 'lachrymatory substance'? One of the lachrymators mentioned is methanal. It used to be called 'formaldehyde', and solutions of methanal are still often called 'formalin'. State one use of formalin in school laboratories. Describe the smell.

2. (*a*) What kind of weather favours the formation of photochemical smog?

(*b*) What kind of geographical locations are prone to photochemical smog?

(*c*) Explain why the conditions you mention in (*a*) and (*b*) favour the formation of photochemical smog.

QUESTIONS ON CHAPTER 27

1. When ethene is bubbled through a solution containing bromine and potassium chloride, the products are CH_2BrCH_2Br and CH_2BrCH_2Cl, but no CH_2ClCH_2Cl is formed. Explain this behaviour.

2. What is the major product of the reaction between hydrogen bromide and but-1-ene in solution in tetrachloromethane at room temperature? What other product is formed in small amount? Explain, by referring to the mechanism of the reaction, why the first product is the major one.

3. What is formed when propene reacts with (*a*) bromine water and (*b*) concentrated sulphuric acid? What is the product of hydrolysis of the compound formed in reaction (*b*)? Discuss the mechanisms of these two reactions, pointing out the similarity between them.

4. Some hydrocarbons are described as 'saturated'; others as 'unsaturated'. What are the differences between these classes of hydrocarbons, in structure and in reactivity? Illustrate your answer by referring to the reactions of two named aliphatic hydrocarbons.

5. Measurements on an alkene showed that $100\,cm^3$ of the gas weighed 0.231 g at 25 °C and 1 atm. $25.0\,cm^3$ of the alkene reacted with $25.0\,cm^3$ of hydrogen.

Find the molar mass of the alkene, and give its molecular formula.

Give the names and structural formulae of alkenes with this formula.

6. Terpenes are unsaturated hydrocarbons present in plants. Limonene is a terpene found in lemon, orange, peppermint and turpentine oils.

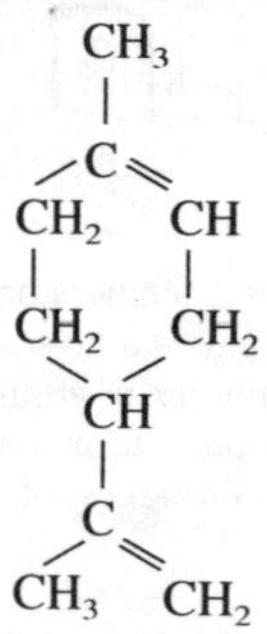

```
        CH3
        |
       /C\\
    CH2    CH
    |      |
    CH2    CH2
      \   /
       CH
        |
        C
      /  \\
   CH3    CH2
```

Limonene

(*a*) The following is an account of the extraction of limonene from oranges.

'The outer rind of two oranges was finely ground and placed in $100\,cm^3$ of water contained in a round-bottomed flask. The flask was arranged for distillation and heated on a wire gauze. About $50\,cm^3$ of distillate were collected in a measuring cylinder. The distillate was transferred to a separating funnel and $20\,cm^3$ of dichloromethane were added. After shaking the funnel for a minute and allowing the layers to separate, the dichloromethane layer was run into a small conical flask, and a few spatula measures of anhydrous sodium sulphate were added. When the solution was clear it was filtered into a distillation flask. The dichloromethane was distilled off leaving $1\,cm^3$ of limonene.'

(i) Draw a diagram to represent the first distillation stage of the experiment.

(ii) What is the purpose of adding anhydrous sodium sulphate to the dichloromethane solution?

(iii) Suggest and explain *two* reasons why the amount of limonene collected is not the amount actually present in the peel.

(*b*) Limonene reacts with excess bromine to produce a crystalline solid.

(i) Draw the structure of this product.

(ii) Suggest why the bromo-derivative is a solid and not a liquid like limonene.

(iii) After purification the crystalline solid may be used to prove the terpene was limonene. Suggest how this may be done.

(*c*) The number of double bonds in a terpene can be determined by the amount of hydrogen absorbed when the terpene is fully hydrogenated.

In a hydrogenation experiment the terpene farnesene was dissolved in a solvent, a catalyst added, and hydrogen bubbled through. 0.162 g of farnesene, molar mass 324, required $48.0\,cm^3$ of hydrogen to react fully with the double bonds in the molecule.

(One mole of a gas occupies $24.0\,dm^3$ at room temperature.)

(i) Suggest the name of a solvent which may be used to dissolve the farnesene and not react with hydrogen.

(ii) Name a catalyst which may be used in a hydrogenation reaction.

(iii) Calculate the number of moles of hydrogen which react with one mole of farnesene and from this work out the number of double bonds in a molecule of farnesene.

(NI 91)

7. (*a*) (i) Describe the nature of the bonding in simple alkenes, explaining how this governs their shape and chemical behaviour.

(ii) State the most common type of reaction undergone by alkenes, the type of reagents with which they commonly react, and give *two* distinct examples of reactions of alkenes.

(*b*) Describe the mechanism and direction (orientation) of the addition of hydrogen bromide to propene.

(*c*) State the types of isomerism which exist in simple alkenes, sketch the structures of the four different isomeric forms of the alkene C_4H_8, and write the systematic names of *three* of them beneath the appropriate sketch.

(*d*) The molar enthalpy change for the hydrogenation of ethene, $\Delta H^{\ominus}$ (298 K) for

$$C_2H_4(g) + H_2(g) \rightarrow C_2H_6(g)$$

is $-137\,kJ\,mol^{-1}$.

Use this in conjunction with the mean molar bond dissociation enthalpies given below, to find

(i) the individual strengths of the two bonds making up the carbon–carbon double bond

(ii) the *total* strength of the carbon–carbon bond in ethene.

Bond	C—C	C—H	H—H
Bond dissociation enthalpy/ $kJ\,mol^{-1}$	348	413	436

(*e*) Relate your answers in (*d*) to what you have said about the nature of the bonding and reactivity of alkenes in section (*a*) above.

(WJEC 90)

8. (*a*) 10 g of but-1-ene, C_4H_8, were treated with an excess of liquid bromine. From the reaction mixture 25 g of a pure organic product were eventually isolated.

Give the name of the product, write an equation showing the structures of reactant and product, and calculate the percentage yield of product in this reaction.

(*b*) Discuss the mechanism of the reaction in (*a*).

(*c*) Including but-1-ene, there are four non-cyclic isomers of molecular formula C_4H_8. Draw the full graphical structures of the four isomers, clearly name each isomer, and discuss the type(s) of isomerism which they illustrate.

(*d*) Assuming that but-1-ene may be polymerised by a method similar to that used to polymerise ethene, suggest conditions which may be used to achieve polymerisation of but-1-ene.

(AEB 92)

9. (*a*) (i) What is the chemical nature of crude oil?

(ii) Why is it classed as a 'fossil fuel'?

(iii) What are the other main 'fossil fuels'?

(iv) How is crude oil made into fuels suitable for domestic heating and jet aircraft and liquid petroleum gas for motor vehicles?

(*b*) Propene is a much more useful source of low molecular mass organic chemicals than propane. Why should this be?

(*c*) What is the industrial source of propene?

(*d*) Under what conditions can propene be converted into poly(propene) and what is the structure of poly(propene)?

(*e*) Poly(propene) can be made into fibres, which can be made into string and ropes, but which have properties that make them unsuitable for the manufacture of clothing. Suggest what these undesirable properties are.

(O 91, S)

10. (*a*) A gaseous, unsaturated hydrocarbon **X** was found to contain 88.9% carbon. Show that the empirical formula of the compound is C_2H_3.

(*b*) Given that the relative molecular mass of the hydrocarbon is 54, show that its molecular formula is C_4H_6.

(*c*) 5.4 g of **X** was found to react with 32 g of Br_2. Calculate the number of moles of **X** and Br_2 used during the reaction and hence calculate the ratio by moles in which they react.

(*d*) (i) Assuming the unsaturation arises only from double bonds, deduce from (*c*) the number of double bonds present in **X**.

(ii) Write an equation for the reaction with Br_2.

(iii) What would you *see* when **X** reacts with bromine?

(*e*) Draw the structural formulas of two possible isomers of **X**.

(O 92, AS)

28

BENZENE AND OTHER ARENES

28.1 BENZENE

Benzene is an arene

Benzene, C_6H_6, is the simplest member of the class of hydrocarbons called **aromatic hydrocarbons** or **arenes**. It is a colourless liquid with a characteristic smell.

For many years there was speculation over the structure of benzene. After 1834, when the molecular formula was established as C_6H_6, people put forward unsaturated formulae, such as

From the formula, C_6H_6, one might expect it to be an unsaturated compound...

$$CH_2{=}C{=}CH{-}CH{=}C{=}CH_2$$

In fact, benzene appeared to be strangely unreactive in comparison with the alkenes. In 1865, A Kekulé suggested the cyclic hexatriene structure, written in full in (a) and schematically in (b).

...but in fact it does not show much resemblance to the alkenes

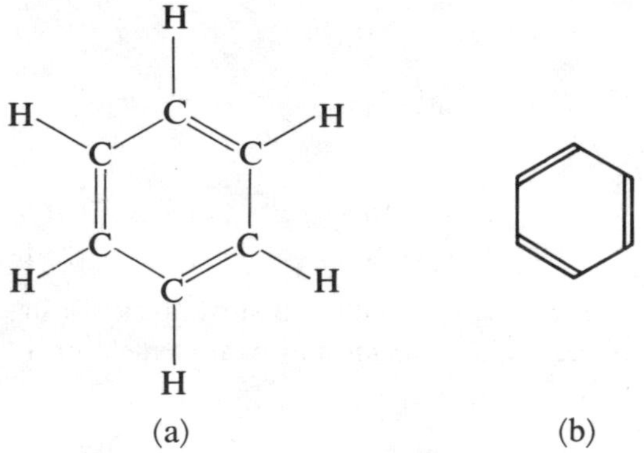

(a) (b)

The Kekulé formula explains the number of isomeric substitution products formed by benzene...

Kekulé wanted to explain why benzene gave only one monosubstitution product, C_6H_5X, and three isomeric disubstitution products, $C_6H_4X_2$. No linear formula would correctly account for the number of substitution products, but the cyclic structure solved the problem [see Question 9, Checkpoint 28A].

...but does not explain its lack of reactivity

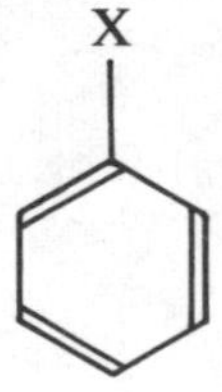

C_6H_5X: no isomerism is possible

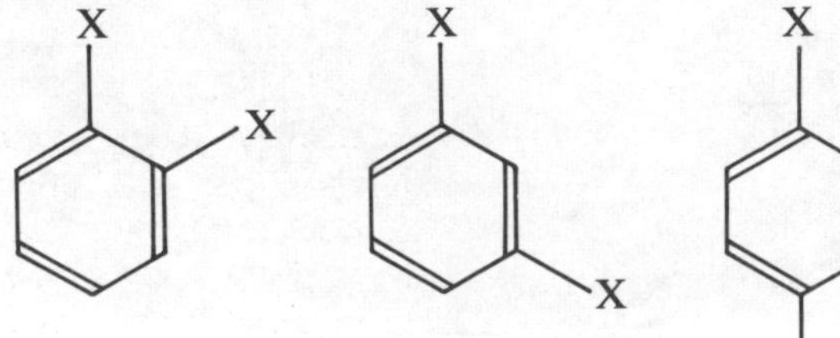

$C_6H_4X_2$: the three isomers

The Kekulé formula was not accepted by all chemists because it implies that benzene should show the same addition reactions as alkenes.

The Kekulé formula has now been superseded by the delocalised formula

Years later, X ray work showed that all the C—C bonds in benzene have the same length, 0.139 nm. This is intermediate between the C—C single bond length of 0.154 nm and the C=C double bond length of 0.133 nm. The Kekulé structure was superseded by the delocalised structure, which has been described and shown in Figure 5.27 in § 5.3.1. The delocalised formulae for benzene and other arenes are shown below:

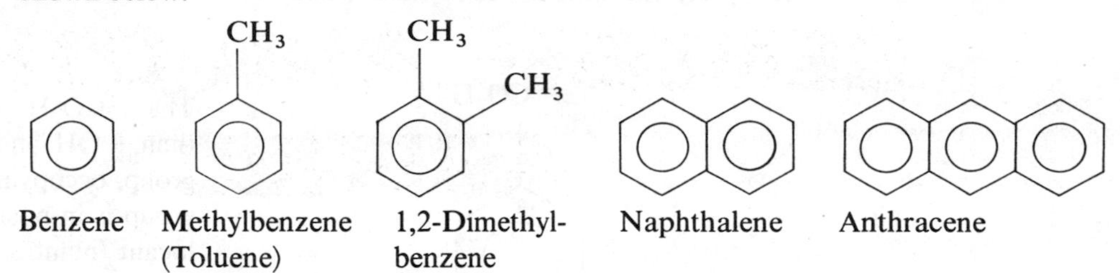

Other arenes

The group C_6H_5- is called a *phenyl* group. Phenyl and substituted phenyl groups are called *aryl* groups. Some derivatives of benzene are:

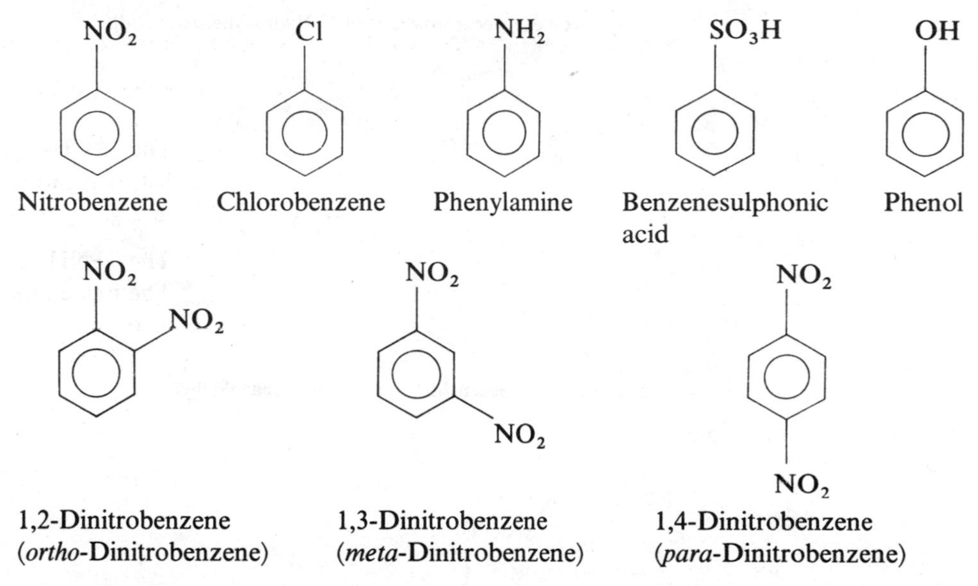

Naming substituted benzenes

Isomers with two substituents in the 1,2-, 1,3-, and 1,4-positions are often called *ortho-*, *meta-* and *para-*isomers or *o-*, *m-* and *p-* for short.

28.2 SOME MORE NAMES OF AROMATIC COMPOUNDS

Derivatives of benzene are:

Formula	Name	Other name
$C_6H_5NH_3^+Cl^-$	Phenylammonium chloride	(Anilinium chloride)
$C_6H_5CO_2H$	Benzenecarboxylic acid	(Benzoic acid)
$C_6H_5CO_2C_2H_5$	Ethyl benzenecarboxylate	(Ethyl benzoate)
C_6H_5COCl	Benzenecarbonyl chloride	(Benzoyl chloride)
$C_6H_5CONH_2$	Benzenecarboxamide	(Benzamide)
C_6H_5CN	Benzenecarbonitrile	(Benzonitrile)
C_6H_5CHO	Benzenecarbaldehyde	(Benzaldehyde)
$C_6H_5COCH_3$	Phenylethanone	
C_6H_5OH	Phenol	
$C_6H_5NH_2$	Phenylamine	(Aniline)
$C_6H_5OCH_3$	Methoxybenzene	

The names given are the IUPAC names. The names in parentheses are traditional names which are still in widespread use, and are acceptable. The naming of benzene derivatives containing two or more substituent groups is done by giving the group which is nearest to the top of the list the number 1 position in the benzene ring. Other groups are numbered by counting from position 1 in the manner which gives them the lowest **locants** (numbers). Some examples are given below:

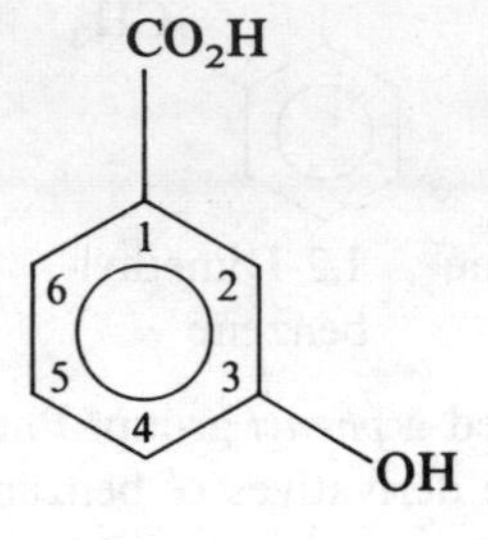

The —CO_2H group is higher up the list than —OH, and is regarded as the principal group, occupying position 1. The —OH group is in position 3, not 5, as the lower locant (number) is used.

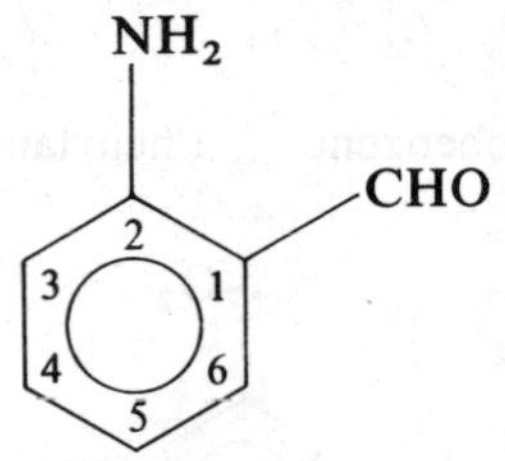

3-Hydroxybenzenecarboxylic acid (3-Hydroxybenzoic acid)

The —CHO group, being higher up the list, is regarded as the principal group and given position 1.

The —NH_2 group is in position 2, not 6. The name *ortho*-aminobenzaldehyde could also be used.

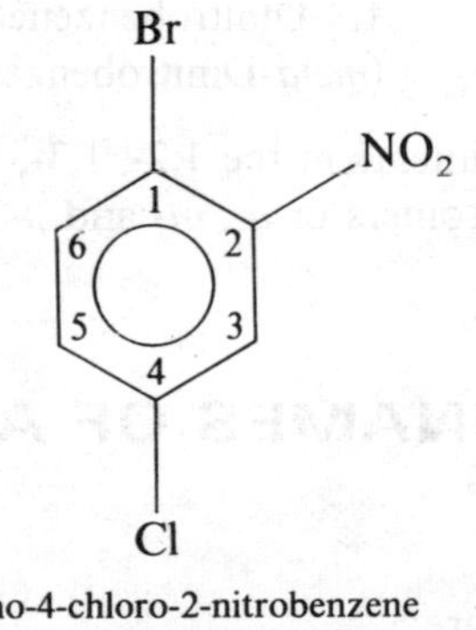

2-Aminobenzenecarbaldehyde (2-Aminobenzaldehyde)

These groups are not on the list. They are listed alphabetically.

1-Bromo-4-chloro-2-nitrobenzene

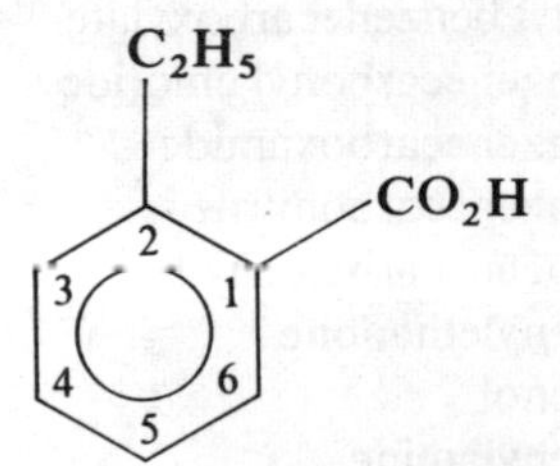

The functional group —CO_2H, is given the locant 1.

2-Ethylbenzenecarboxylic acid (2-Ethylbenzoic acid)

FIGURE 28.1 Sources of Benzene

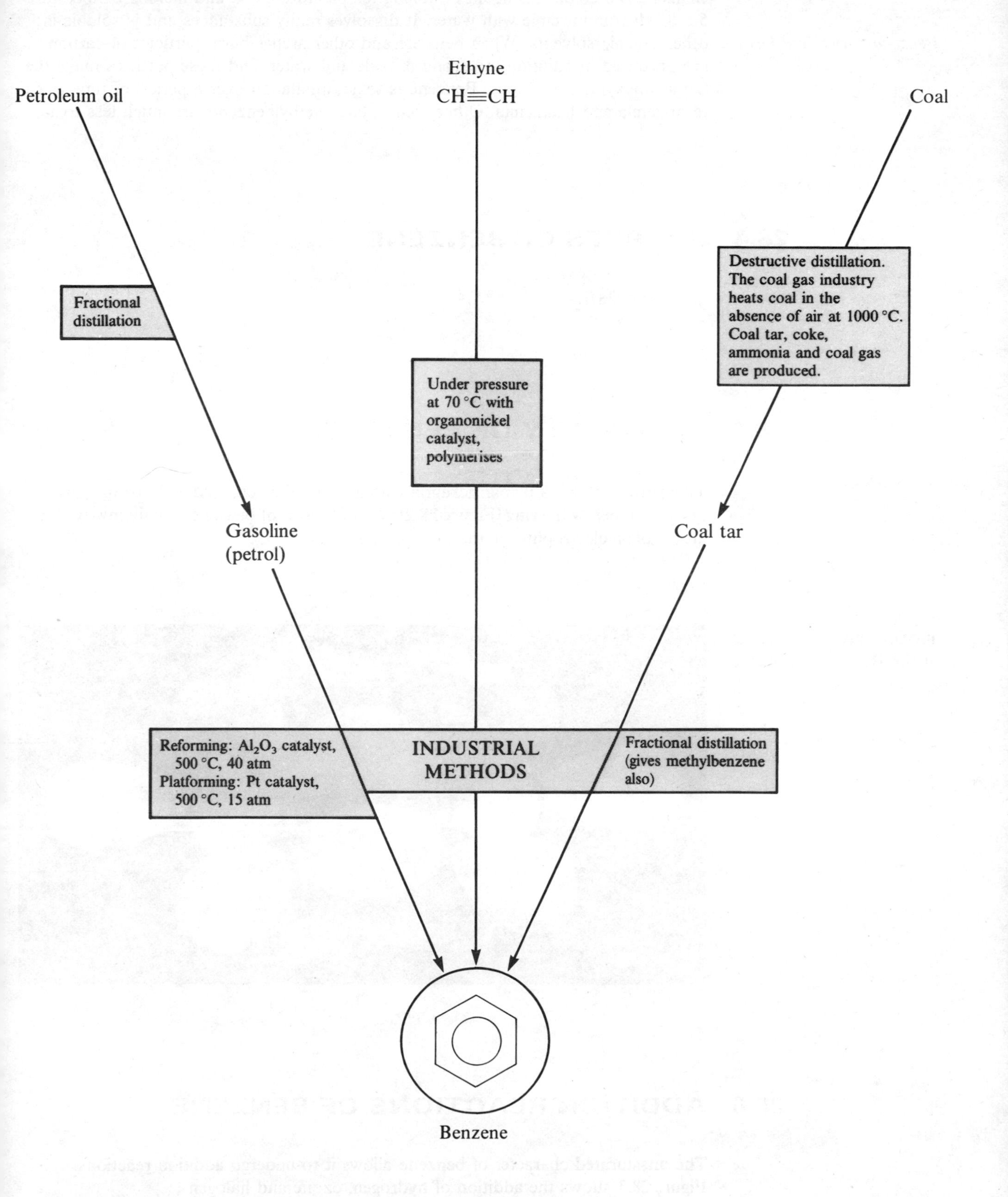
Petroleum oil
Ethyne
CH≡CH
Coal
Fractional distillation
Under pressure at 70 °C with organonickel catalyst, polymerises
Destructive distillation. The coal gas industry heats coal in the absence of air at 1000 °C. Coal tar, coke, ammonia and coal gas are produced.
Gasoline (petrol)
Coal tar
Reforming: Al_2O_3 catalyst, 500 °C, 40 atm
Platforming: Pt catalyst, 500 °C, 15 atm
INDUSTRIAL METHODS
Fractional distillation (gives methylbenzene also)
Benzene

28.3 PHYSICAL PROPERTIES OF BENZENE

Benzene burns, and has a toxic vapour

Benzene is a colourless liquid of boiling temperature 80 °C and melting temperature 5.5 °C. It is immiscible with water. It dissolves many substances and is soluble in other organic solvents. When benzene and other arenes burn, particles of carbon are produced in addition to carbon dioxide and water, and these particles make the flame smoky and luminous. Benzene is toxic; inhalation over a period of time leads to anaemia and leukaemia. Other arenes, e.g., methylbenzene, are much less toxic.

28.4 SOURCES OF BENZENE

See Figure 28.1.

28.5 REACTIVITY OF BENZENE

The benzene ring is a planar hexagon with a cloud of delocalised π electrons lying above and below the ring [Figure 28.2]. The reactions of benzene usually involve the attack of an electrophile on the cloud of π electrons.

FIGURE 28.2 A Model of the Benzene Molecule

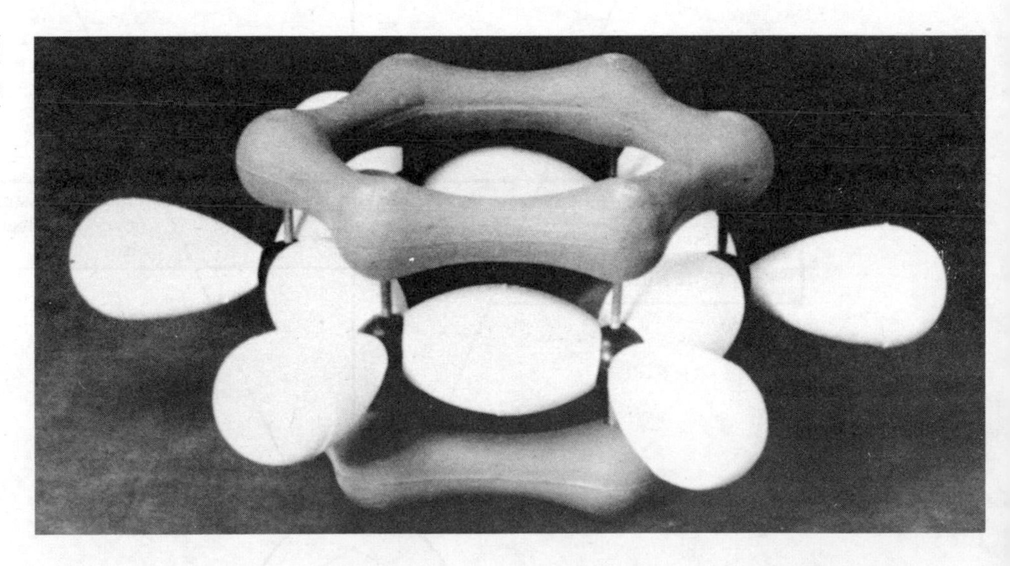

28.6 ADDITION REACTIONS OF BENZENE

The unsaturated character of benzene allows it to undergo addition reactions. Figure 28.3 shows the addition of hydrogen, ozone and halogen.

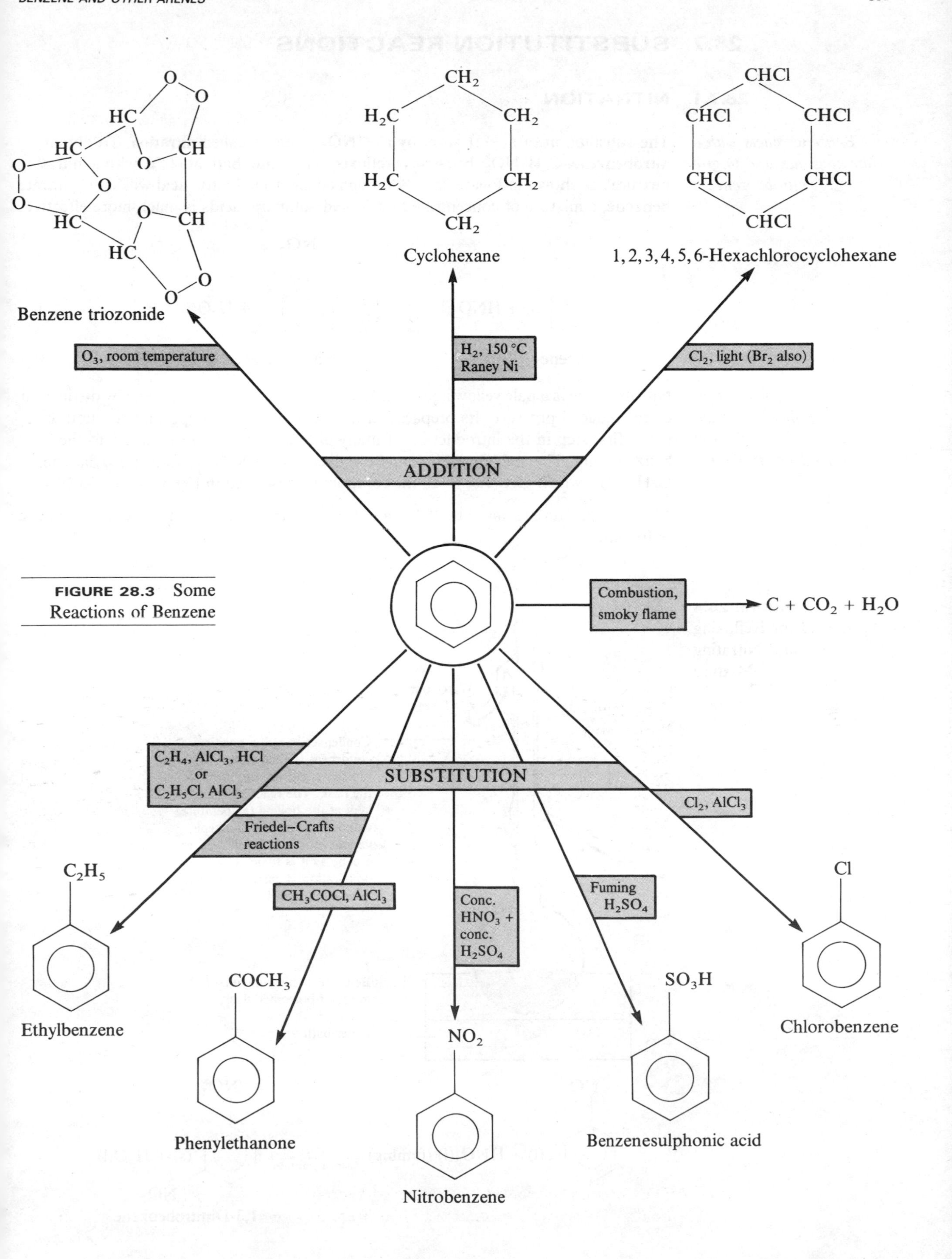

FIGURE 28.3 Some Reactions of Benzene

28.7 SUBSTITUTION REACTIONS

28.7.1 NITRATION

Benzene reacts with a 'nitrating mixture' to give nitrobenzene...

The substitution of a —H atom by a $—NO_2$ group is called **nitration**. To obtain nitrobenzene, $C_6H_5NO_2$, benzene is refluxed on a water bath at 60 °C with a 'nitrating mixture', as shown in Figure 28.4. Although concentrated nitric acid will slowly nitrate benzene, a mixture of concentrated nitric and sulphuric acids is much more effective:

$$C_6H_6(l) + HNO_3(l) \xrightarrow[\text{conc. } HNO_3 + \text{ conc. } H_2SO_4]{\text{Reflux, 60 °C}} C_6H_5NO_2(l) + H_2O(l)$$

Benzene — Nitrobenzene

...which is an intermediate in the preparation of many derivatives of benzene

Nitrobenzene is a pale yellow liquid which can be separated from benzene by distillation under reduced pressure. Its preparation from benzene is an important reaction as it is the first step in the introduction of many different substituent groups into the benzene ring. The $—NO_2$ group can be reduced to $—NH_2$, to form phenylamine, $C_6H_5NH_2$, which gives rise to all the compounds described in Figure 32.6, § 32.10.4.

If the temperature is raised to 95 °C, and fuming nitric acid is used, 1,3-dinitrobenzene is formed.

FIGURE 28.4
Apparatus for Refluxing Benzene and Nitrating Mixture

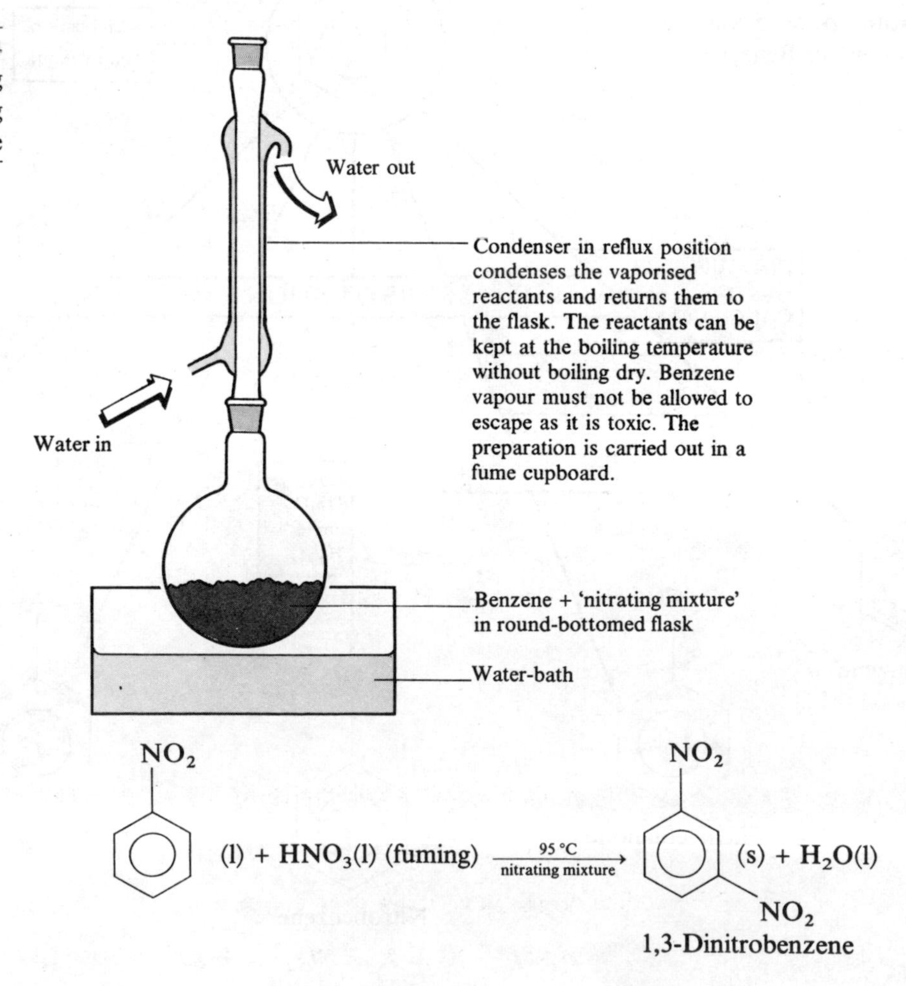

Water out
Condenser in reflux position condenses the vaporised reactants and returns them to the flask. The reactants can be kept at the boiling temperature without boiling dry. Benzene vapour must not be allowed to escape as it is toxic. The preparation is carried out in a fume cupboard.
Water in
Benzene + 'nitrating mixture' in round-bottomed flask
Water-bath

$$C_6H_5NO_2(l) + HNO_3(l)\ (\text{fuming}) \xrightarrow[\text{nitrating mixture}]{95\ °C} C_6H_4(NO_2)_2(s) + H_2O(l)$$

1,3-Dinitrobenzene

Dinitrobenzene can also be formed by nitrating benzene

Nitrobenzene is less reactive than benzene, and 1,3-dinitrobenzene is obtained together with nitrobenzene and benzene. Since it is a solid, 1,3-dinitrobenzene crystallises out when the mixture is poured into cold water. No trinitrobenzene is formed.

Nitration of methylbenzene produces TNT

When methylbenzene (toluene), $C_6H_5CH_3$, is nitrated, 2,4,6-trinitromethylbenzene (2,4,6-trinitrotoluene) is formed.

CH_3, O_2N, NO_2, NO_2

This shows that the methyl group makes methylbenzene more reactive than benzene. The product, which is called TNT for short, is a powerful explosive.

Nitration of aromatic compounds is important for the production of this and other explosives.

28.7.2 *MECHANISM OF NITRATION

What is the nitrating agent?

Chemists were intrigued by the puzzle that, although concentrated nitric acid alone is a poor nitrating agent, a mixture of concentrated nitric and sulphuric acids is a powerful reagent. It seemed that the two acids must react to produce a species which can attack benzene. Much research work has gone into the unravelling of this mystery, and only a glimpse of this exciting area of study can be given here.

It is suggested that sulphuric acid, being a stronger acid than nitric acid, can to some extent donate a proton to nitric acid:

$$H_2SO_4 + HNO_3 \rightleftharpoons HSO_4^- + H_2\overset{+}{O}{-}NO_2$$

Nitric acid, in accepting a proton, is acting as a base. The protonated nitric acid splits up to form a molecule of water and the cation, NO_2^+, which is called a *nitryl cation*.

Formation of the nitryl cation...

$$H_2\overset{+}{O}{-}NO_2 \rightleftharpoons H_2O + NO_2^+$$

The molecule of water is protonated by a second molecule of sulphuric acid:

$$H_2O + H_2SO_4 \rightleftharpoons H_3O^+ + HSO_4^-$$

Adding up the three steps gives

$$HNO_3 + 2H_2SO_4 \rightleftharpoons NO_2^+ + H_3O^+ + 2HSO_4^-$$

... The nitryl cation as a nitrating agent

The salt nitryl chlorate(VII), $NO_2^+ ClO_4^-$, has been prepared and shown to nitrate benzene. The nitryl cation must be a nitrating agent.

NO_2^+ is an electrophile ...

... which adds to benzene to form a reactive intermediate ...

... this intermediate rapidly loses a proton to form the product

Once formed, the nitryl cation is attracted by the cloud of delocalised π electrons which lies above and below the plane of the benzene ring [see Figure 28.2, § 28.5]. The NO_2^+ cation adds to the ring to form a short-lived intermediate, in which both the entering $—NO_2$ group and the leaving —H atom are bonded to the ring. The symmetry of the annular cloud of delocalised π electrons is disrupted. The positive charge contributed by the NO_2^+ ion is distributed over the ring. The intermediate rapidly loses a proton, restoring the symmetry and stability of the benzene ring. The proton is immediately picked up by a hydrogensulphate ion:

$$C_6H_6 + NO_2^+ \xrightarrow{\text{slow}} [C_6H_6NO_2]^+ \xrightarrow{\text{fast}} C_6H_5NO_2 + H^+$$

Reactive intermediate

$$H^+ + HSO_4^- \xrightarrow{\text{fast}} H_2SO_4$$

The overall rate of the reaction is determined by the slowest step in the sequence. This is the rate at which the $C—NO_2$ bond is formed in the intermediate.

28.7.3 SULPHONATION

Benzene can be sulphonated by fuming H_2SO_4

Sulphonation is the substitution of a —H atom by an $—SO_3H$ group. Benzenesulphonic acid, $C_6H_5SO_3H$, is obtained by refluxing benzene with concentrated sulphuric acid for many hours, or warming with fuming sulphuric acid (which contains SO_3) at 40 °C for 20–30 minutes:

$$C_6H_6(l) + H_2SO_4(l) \xrightarrow[\text{warm with fuming acid}]{\text{reflux with conc. acid}} C_6H_5SO_3H(l) + H_2O(l)$$

Benzenesulphonic acid

Sulphonic acids are important in the preparation of phenols and in the manufacture of detergents

The importance of this reaction is that the $—SO_3H$ group can be replaced by hydrolysis to give an —OH group. Benzenesulphonic acid is converted into phenol, C_6H_5OH. This is the easiest way of preparing phenol in the laboratory. Sulphonation is also important in the manufacture of detergents. It is because detergents contain $—SO_3H$ groups that they lather even in hard water, as the calcium and magnesium salts of sulphonic acids are soluble.

28.7.4 *ALKYLATION

An alkyl group can be introduced into the benzene ring by the reaction of a halogenoalkane with benzene:

$$C_6H_6(l) + C_2H_5Br(l) \xrightarrow[\text{warm}]{FeBr_3} C_6H_5C_2H_5(l) + HBr(g) \rightarrow \text{Further substitution}$$

Benzene Bromoethane Ethylbenzene

Alkylation needs a catalyst + RX

The reaction takes place under the influence of a catalyst, such as aluminium chloride or iron(III) bromide. More than one alkyl group enters the ring as the alkylbenzene formed is more reactive than benzene. The effectiveness of the catalysts was discovered by C Friedel and J M Crafts, and reactions of this type are called **Friedel–Crafts reactions**. The mechanism is discussed in § 28.7.7.

Alkenes are alkylating agents...

Alkylation can also be effected by the use of an alkene and a Friedel–Crafts catalyst together with an acid such as HCl or H_3PO_4:

$$C_6H_6(l) + CH_2{=}CH_2(g) \xrightarrow[\text{HCl or } H_3PO_4]{AlCl_3 \text{ catalyst}} C_6H_5C_2H_5(l)$$

Benzene Ethene Ethylbenzene

...used to make styrene and cumene

This is the method used industrially to make ethylbenzene. The product is dehydrogenated to give phenylethene (styrene), from which polystyrene is made [see Table 27.1, § 27.7.10]. Another Friedel–Crafts reaction used in industry is the reaction between benzene and propene to make cumene and thence phenol [§ 30.11.1].

Dimethyl sulphate, $(CH_3O)_2SO_2$, is a popular reagent for the introduction of methyl groups. It is toxic, and must be used in a fume cupboard and prevented from touching the skin.

$$2\,C_6H_6(l) + (CH_3O)_2SO_2(l) \rightarrow 2\,C_6H_5CH_3(l) + H_2SO_4(l)$$

Benzene Dimethyl sulphate Methylbenzene

28.7.5 *ACYLATION

Acylation, the introduction of an acyl group

$$-C(=O)R$$

is another Friedel–Crafts reaction:

Acylation needs RCOCl + catalyst

$$C_6H_6(l) + CH_3COCl(l) \xrightarrow[40\,°C]{AlCl_3} C_6H_5COCH_3(l) + HCl(g)$$

Ethanoyl chloride Phenylethanone

An acid chloride or acid anhydride, in the presence of a Friedel–Crafts catalyst, will acylate benzene and its derivatives [see Figure 28.5]. Only one acyl group enters the ring as the acyl compound formed is less reactive than benzene. This gives acylation an advantage over alkylation, which yields polyalkyl derivatives. A method of preparing pure ethylbenzene is to introduce the group $—COCH_3$ by acylation and then reduce it to $—CH_2CH_3$:

The reduction of an acyl group gives an alkyl group

$$C_6H_6 + CH_3COCl \xrightarrow{AlCl_3} C_6H_5COCH_3 \xrightarrow{reduce} C_6H_5CH_2CH_3$$

$$C_6H_6 \xrightarrow{C_2H_5Cl,\ AlCl_3} \text{Ethylbenzene}$$

Gives some polyethylbenzenes in addition.

FIGURE 28.5
Friedel–Crafts Acylation

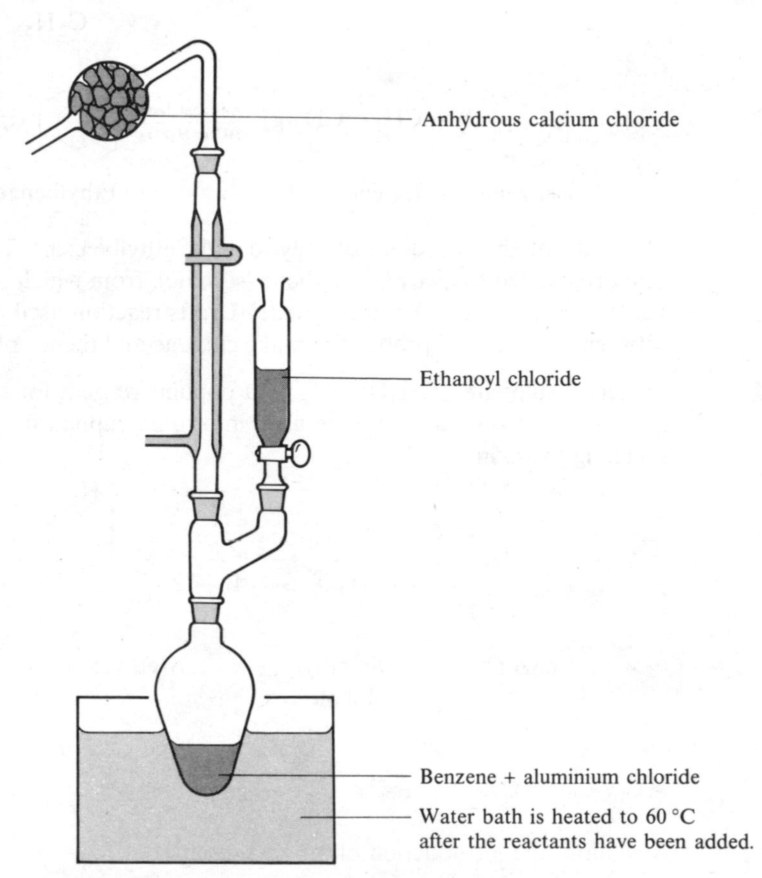
Anhydrous calcium chloride
Ethanoyl chloride
Benzene + aluminium chloride
Water bath is heated to 60 °C
after the reactants have been added.

28.7.6 *HALOGENATION

The addition of halogens to the benzene ring [§ 28.6] takes place under conditions which favour the formation of free radicals, i.e., sunlight or high temperature. The first step in the addition is the homolysis of the halogen molecule, e.g.

$$Cl_2 \xrightarrow{h\nu} 2Cl$$

Chlorine and bromine can be substituted in the ring in the presence of a Friedel–Crafts catalyst . . .

A different type of reaction takes place at room temperature in the presence of a Friedel–Crafts catalyst. Substitution then occurs. In recognition of the way in which they ease the path of halogens into the benzene ring the Friedel–Crafts catalysts are referred to as **halogen-carriers**. Aluminium and iron act as catalysts because during the reaction they are converted into their chlorides or bromides, which act as halogen-carriers:

$$C_6H_6(l) + Cl_2(g) \xrightarrow[\text{room temperature}]{\text{Al or } AlCl_3} C_6H_5Cl(l) + HCl(g)$$

Chlorobenzene

$$C_6H_6(l) + Br_2(l) \xrightarrow[\text{room temperature}]{\text{Fe or } FeBr_3} C_6H_5Br(l) + HBr(g)$$

Bromobenzene

Iodine cannot be introduced in this way.

...which is called a halogen-carrier

Why does the presence of a halogen-carrier result in a completely different reaction—substitution instead of addition? Much work has been done on the mode of action of Friedel–Crafts catalysts, and only an outline can be given here.

28.7.7 MECHANISM OF ALKYLATION, ACYLATION AND HALOGENATION

The mechanism: Friedel–Crafts catalysts act as Lewis acids...

Friedel–Crafts catalysts are halides like $AlCl_3$, $AlBr_3$, BF_3 and $FeBr_3$. They function as Lewis acids [§ 12.7.1] since the central atom can accept a pair of electrons. Aluminium chlorice can accept a pair of electrons from a chloride ion. In the complex ion formed, $AlCl_4^-$, aluminium has eight electrons in its outer shell:

$$AlCl_3 + Cl^- \rightarrow [AlCl_4]^-$$

Let us see how the ability to form complex ions by acting as Lewis acids can explain the catalytic effect of these halides.

ALKYLATION

...They form complexes with halogenoalkanes

When a Friedel–Crafts catalyst is dissolved in a halogenoalkane, the solution formed is a weak electrical conductor. This could be explained by the formation of an ionic complex. In the case of bromoethane and aluminium bromide

$$C_2H_5Br + AlBr_3 \rightleftharpoons C_2H_5^+\ AlBr_4^-$$

A carbocation complex attacks benzene to form an intermediate...

The attack on the π electron cloud of the benzene ring could be by the carbocation, $C_2H_5^+$ or by the complex, $C_2H_5^+\ AlBr_4^-$. In the case of primary and secondary carbocations, it is thought that the complex is the attacking electrophile. Tertiary carbocations are more stable, and in the case of $(CH_3)_3C^+$, for example, it may be the carbocation that attacks benzene:

$$\text{C}_6\text{H}_6 + C_2H_5^+ \, AlBr_4^- \xrightarrow{\text{slow}} [C_6H_6(C_2H_5)]^+ \, AlBr_4^-$$

... which loses a proton to form the product

The reactive intermediate that is formed quickly loses a proton to form the product, ethylbenzene. The catalyst, aluminium bromide, is regenerated, and hydrogen bromide is evolved:

$$[C_6H_6(C_2H_5)]^+ \, AlBr_4^- \xrightarrow{\text{fast}} C_6H_5C_2H_5 + HBr + AlBr_3$$

Reactive intermediate Ethylbenzene

When alkenes are used for alkylation a proton acid must be present, in addition to a Lewis acid. The first step in the reaction is protonation of the alkene to form a carbocation which attacks the benzene ring.

ACYLATION

The mechanism of acylation: the F-C catalyst + RCOCl form a complex...

Again, the Friedel–Crafts catalyst acts as a Lewis acid. In the reaction with an acyl chloride, a complex containing an acyl cation, $R\overset{+}{C}O$, is formed:

$$RCOCl + AlCl_3 \rightleftharpoons R\overset{+}{C}O \, AlCl_4^-$$

The electrophile which attacks the benzene ring could be either the acyl cation, $R\overset{+}{C}O$, or the complex, $R\overset{+}{C}O \, AlCl_4^-$. There is evidence that the attacking reagent is the whole complex:

... the electrophilic complex attacks benzene...

$$C_6H_6 + R\overset{+}{C}O \, AlCl_4^- \xrightarrow{\text{slow}} [C_6H_6(COR)]^+ \, AlCl_4^- \xrightarrow{\text{fast}} C_6H_5COR + HCl + AlCl_3$$

Reactive intermediate Phenylketone

... to form a reactive intermediate which is rapidly converted into the product

The rate-determining step is the formation of the RCO—ring bond. The intermediate that is formed rapidly loses a proton to yield the product.

HALOGENATION WITH THE AID OF A HALOGEN-CARRIER

In the bromination of benzene a bromine molecule approaches a benzene molecule and encounters the annular cloud of delocalised π electrons above and below the plane of the ring [see Figure 28.2, § 28.5]. As the π electron cloud interacts with the bromine molecule, the Br—Br bond becomes polarised:

The role of a halogen-carrier is explained in terms of polarising the X—X bond

$$C_6H_6 + Br{-}Br \rightarrow C_6H_6 \rightarrow \overset{\delta+}{Br}{-}\overset{\delta-}{Br}$$

This will remind you of the interaction between Br_2 and the π electrons of the C=C bond in alkenes [§ 27.7.5]. Benzene is less reactive than an alkene, and the reaction proceeds only if there is a Lewis acid (e.g., $FeBr_3$) present. This accepts a pair of electrons from the $\delta-$ Br atom in the polarised Br_2 molecule, and enables the Br—Br bond to split. A bromonium ion and a $FeBr_4^-$ complex ion are formed.

The bromonium ion rapidly loses a proton to form bromobenzene, with the regeneration of the catalyst, $FeBr_3$:

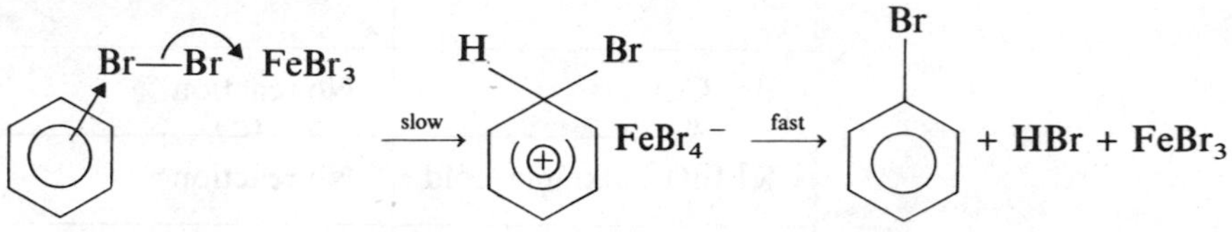

$$C_6H_6 + Br{-}Br \cdot FeBr_3 \xrightarrow{\text{slow}} [C_6H_6Br]^+ FeBr_4^- \xrightarrow{\text{fast}} C_6H_5Br + HBr + FeBr_3$$

Iodination of benzene does not occur, but some benzene derivatives, containing activating substituents, are iodinated.

A SUMMARY

Alkylation, first step

$$RX + AlX_3 \rightleftharpoons R^+ AlX_4^-$$

followed by

Summary of the role of Friedel–Crafts catalysts

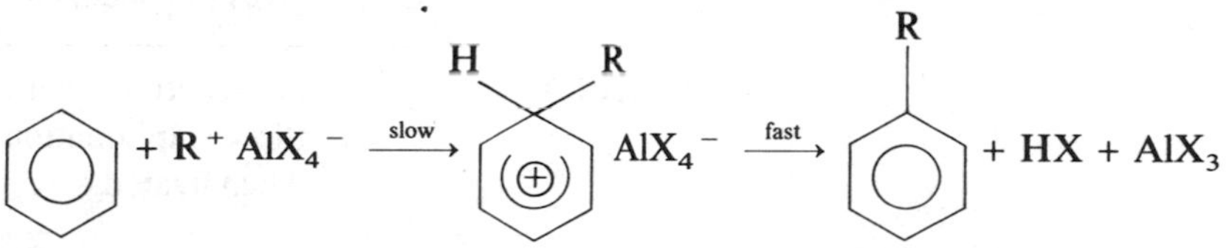

$$C_6H_6 + R^+ AlX_4^- \xrightarrow{\text{slow}} [C_6H_6R]^+ AlX_4^- \xrightarrow{\text{fast}} C_6H_5R + HX + AlX_3$$

Acylation, first step

$$RCOX + AlX_3 \rightleftharpoons R\overset{+}{C}O\ AlX_4^-$$

followed by

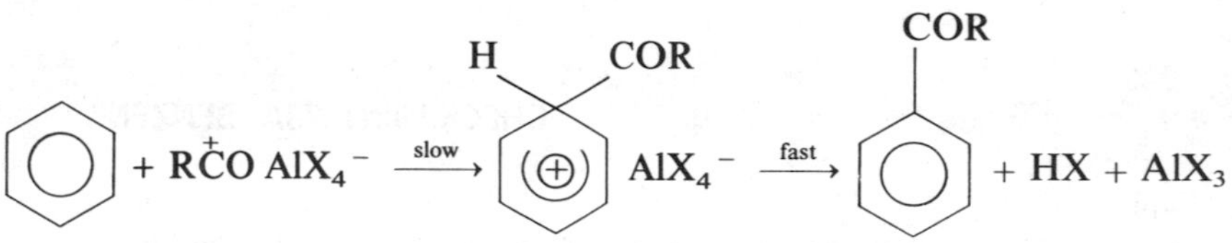

$$C_6H_6 + R\overset{+}{C}O\ AlX_4^- \xrightarrow{\text{slow}} [C_6H_6COR]^+ AlX_4^- \xrightarrow{\text{fast}} C_6H_5COR + HX + AlX_3$$

Halogenation, first step

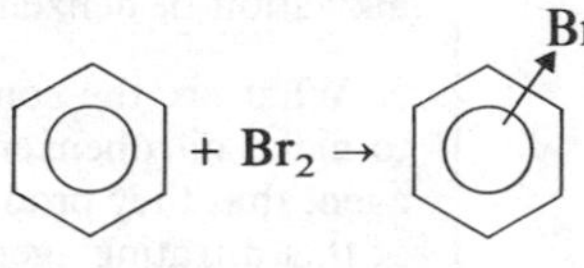

$$C_6H_6 + Br_2 \rightarrow C_6H_6 \cdot Br{-}Br$$

followed by

$$C_6H_6 \cdot Br{-}Br \cdot FeBr_3 \xrightarrow{\text{slow}} [C_6H_6Br]^+ FeBr_4^- \xrightarrow{\text{fast}} C_6H_5Br + HBr + FeBr_3$$

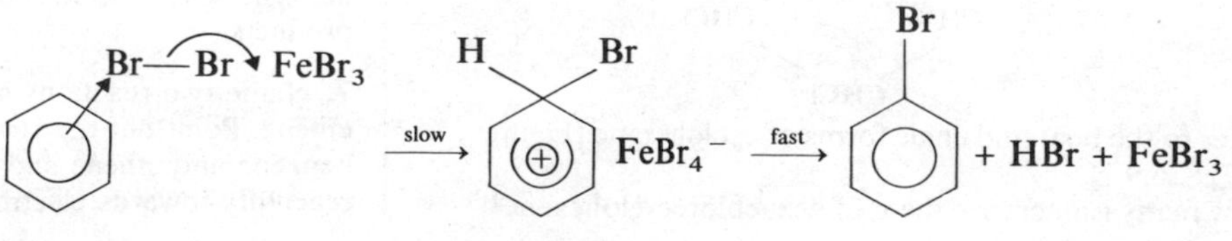

Reagent	Benzene	Cyclohexene
$Br_2(CCl_4)$	No reaction	Rapid decolorisation
$KMnO_4$, alkaline, cold	No reaction	Rapid decolorisation
HBr(g)	No reaction	Rapid addition
Air, catalyst, 150 °C	No reaction	Oxidised to hexane-1,6-dioic acid
H_2, catalyst, 200 °C	Slowly adds to give cyclohexane	Adds rapidly to give cyclohexane
Halogen, sunlight, boil	Slow addition gives $C_6H_6X_6$	Rapid addition gives $C_6H_{10}X_2$
Conc. HNO_3, H_2SO_4	Substitution gives nitrobenzene	Oxidation gives hexane-1,6-dioic acid
Conc. H_2SO_4	No reaction when cold. Slow sulphonation when heated	Absorbed by H_2SO_4
Friedel–Crafts reagents	Alkylation or acylation	No reaction

TABLE 28.1
A Comparison of an Arene and a Cycloalkene

CHECKPOINT 28A: BENZENE

1. Explain why:
(*a*) 1 mole of benzene reacts with 3 moles of chlorine in the presence of ultraviolet light, without the formation of hydrogen chloride, and
(*b*) 1 mole of benzene reacts with 1 mole of chlorine in the presence of iron(III) chloride with the formation of hydrogen chloride.

2. Gammexane is an insecticide, with the formula

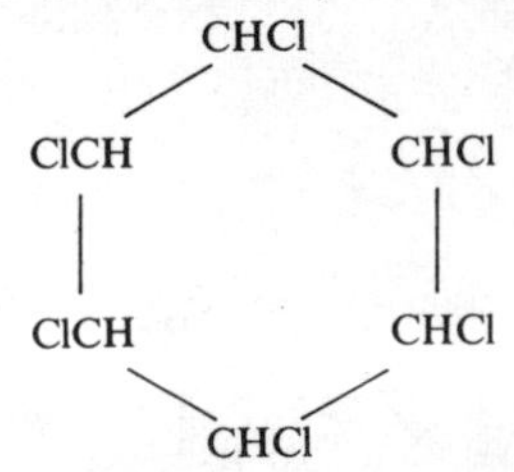

Refer to the boat and chair forms of cyclohexane [Figure 25.7, § 25.9.2].
How many isomers are there of hexachlorocyclohexane?

3. Compare the reactions of benzene with those of cyclohexene.

4. What type of reagent attacks the benzene ring to form substitution compounds? State the attacking reagents which are involved in (*a*) the nitration of benzene, (*b*) the alkylation of benzene, and (*c*) the acylation of benzene.

5. What are the components of the 'nitrating mixture' used to make nitrobenzene from benzene? What is the nitrating agent that they produce? Give an equation for the formation of this nitrating agent from the nitrating mixture, and explain how benzene reacts with it.

6. Name two reagents that react with both ethene and benzene. Give the structural formulae and names of the products.

7. Name two reactions in which benzene differs from ethene. Point out the structural resemblance between benzene and ethene and the reason for the difference in reactivity towards electrophiles.

8. Before Kekulé proposed his formula for benzene, people tried out formulae such as

$$CH{\equiv}C{-}CH{=}CH{-}CH{=}CH_2$$

State two reactions which you would expect of a compound with this formula and which are not typical of benzene.

9. Long ago, the formula

$$CH_2{=}C{=}CH{-}CH{=}C{=}CH_2$$

was suggested for benzene. On the basis of this formula, how many isomers would you expect of the substitution products (*a*) C_6H_5X, (*b*) $C_6H_4X_2$, (*c*) $C_6H_3X_3$? How many isomeric substitution products with each of these formulae are formed by benzene?

28.8 METHYLBENZENE (TOLUENE): PHYSICAL PROPERTIES

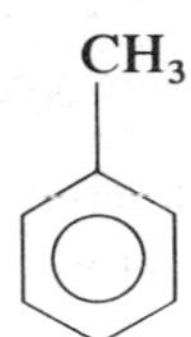

Methylbenzene, or toluene, $C_6H_5CH_3$, resembles benzene. Its physical properties are similar: the boiling temperature is higher (111 °C) and the melting temperature lower (− 95 °C).

28.9 INDUSTRIAL SOURCE AND USES

Methylbenzene is used as a solvent and a petrol additive

Like benzene, methylbenzene is obtained from petroleum oil and from coal tar [see Figure 28.1, § 28.4].

Methylbenzene is used as a solvent, as a source of the explosive trinitrotoluene (TNT), and as an additive to petrol, in which it improves the antiknock quality.

28.10 REACTIONS OF THE RING

There are two sets of reactions, one set involving the aromatic ring in substitution or addition and the other involving reactions of the methyl group or 'side chain', as it is called.

28.10.1 SUBSTITUTION

Methylbenzene is more reactive than benzene towards electrophiles...

Methylbenzene is more reactive than benzene towards the electrophilic reagents which substitute in the benzene ring. This is because the methyl group pushes electrons into the ring [§ 28.12.1]. Milder conditions are employed than in the reactions of benzene. They are:

... Conditions for substitution are milder...

Chlorination/bromination: Cl_2/Br_2 with halogen carrier at 20 °C

Nitration: conc. HNO_3 + conc. H_2SO_4 at 30 °C

Sulphonation: fuming H_2SO_4 (containing SO_3) at 0 °C

Alkylation: RCl, $AlCl_3$ at 20 °C

Acylation: RCOCl or $(RCO)_2O$, $AlCl_3$, warm

A mixture of 1,2- and 1,4-substituted methylbenzenes is obtained in each case, e.g.

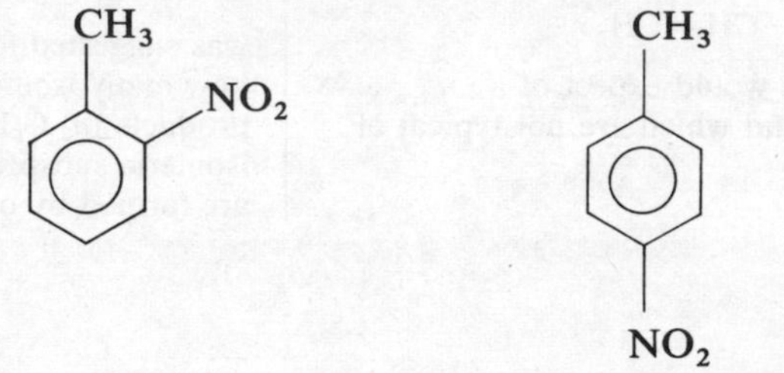

1-Methyl-2-nitrobenzene and 1-Methyl-4-nitrobenzene

If the temperature is raised, two groups or three groups are introduced, e.g.

...More than one group is substituted

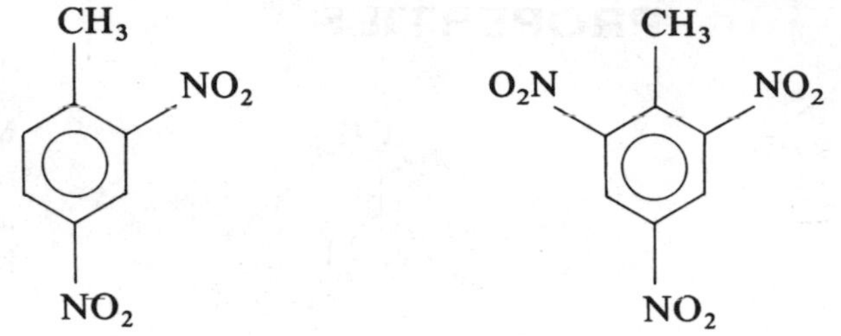

1-Methyl-2,4-dinitrobenzene and 1-Methyl-2,4,6-trinitrobenzene

28.10.2 ADDITION

Addition reactions are the same as those of benzene [see Figure 28.6].

28.11 REACTIONS OF THE SIDE CHAIN

28.11.1 HALOGENATION

In sunlight, Cl_2 and Br_2 substitute the CH_3— group, not the ring

When chlorine is bubbled through boiling methylbenzene in strong sunlight or ultraviolet light, substitution occurs in the side chain:

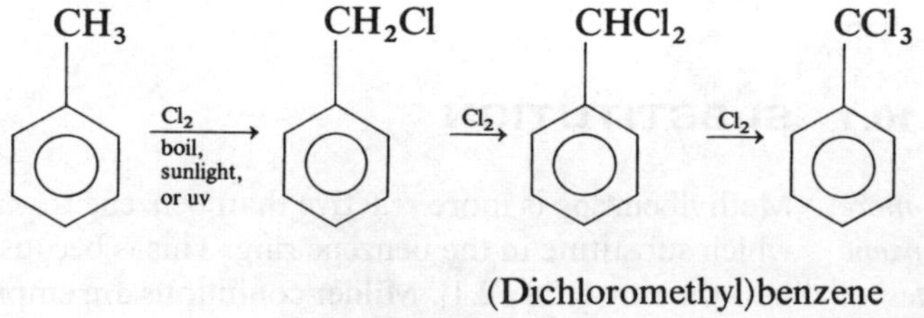

CH_3 —(Cl_2, boil, sunlight, or uv)→ CH_2Cl —Cl_2→ $CHCl_2$ —Cl_2→ CCl_3

(Chloromethyl)benzene (Dichloromethyl)benzene (Trichloromethyl)benzene

Reaction involves free radicals, as does the halogenation of methane

Bromination occurs under the same conditions to give similar products. The conditions employed here favour the formation of free radicals, and the reaction proceeds in a similar way to the halogenation of alkanes [§26.3.8]. Note the difference between the product of free radical halogenation in the side chain and halogenation in the presence of a halogen carrier, when the halogen substitutes in the aromatic ring.

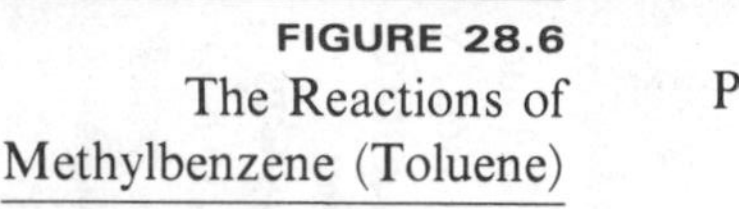

FIGURE 28.6 The Reactions of Methylbenzene (Toluene)

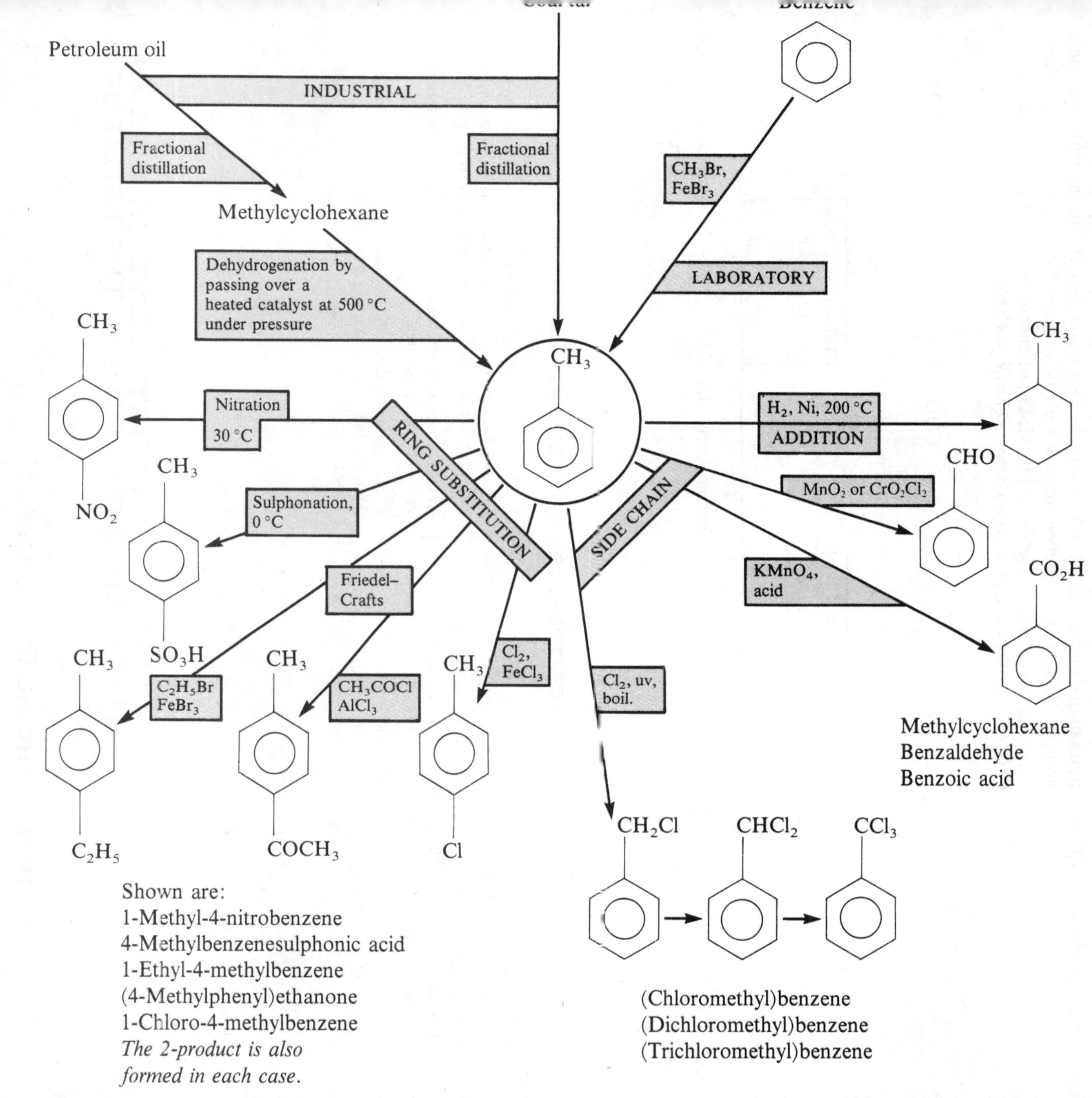

28.11.2 OXIDATION

Oxidation of —CH_3 gives —CHO or —CO_2H

The powerful oxidising agents acidified potassium manganate(VII) and acidified potassium dichromate(VI) will oxidise the side chain, —CH_3, to the carboxylic acid group, —CO_2H. The reaction mixture must be refluxed for several hours. A milder oxidising agent, manganese(IV) oxide, MnO_2, or chromium dichloride dioxide, CrO_2Cl_2, oxidises —CH_3 to the aldehyde group, —CHO:

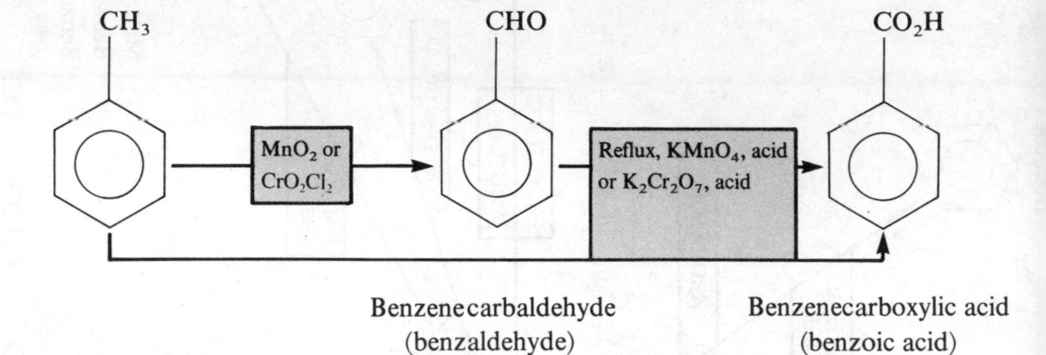

CHECKPOINT 28B: METHYLBENZENE

1. Explain why alkaline potassium manganate(VII) attacks methylbenzene but not benzene.

2. What are the products of the reactions between (i) benzene and (ii) methylbenzene with
(*a*) chloromethane and aluminium chloride
(*b*) ethanoyl chloride and iron(III) chloride
(*c*) concentrated nitric and concentrated sulphuric acid.

3. Write structural formulae for 1-chloro-2-methylbenzene, 1-chloro-4-methylbenzene and (chloromethyl)benzene. How can these compounds be made from methylbenzene?

4. How do (*a*) chlorine and (*b*) nitric acid react with benzene and with methylbenzene?

5. How is methylbenzene obtained (*a*) industrially and (*b*) in the laboratory, starting from benzene?

28.12 *THE EFFECT OF SUBSTITUENT GROUPS ON FURTHER SUBSTITUTION IN THE BENZENE RING

28.12.1 THE INDUCTIVE EFFECT

Polarisation of a σ bond gives rise to an inductive effect

In a bond between carbon and a more electronegative element X the electron cloud will be denser at the X end of the bond than at the C end. The bond is polarised. This polarising effect of X is called an **inductive effect**. If X is more electronegative than carbon (e.g., F, Cl, Br), X is said to have a negative inductive (−I) effect. Alkyl groups are electron-donating, and are said to have a positive inductive (+I) effect [§ 12.7.7]:

$$C \rightarrow X \qquad C \leftarrow R$$

−I effect   +I effect

28.12.2 THE MESOMERIC EFFECT

Polarisation of a π bond gives rise to a mesomeric effect

In a multiple bond, the π electrons shift towards the more electronegative of the bonded atoms. In a carbonyl group, the arrangement of electrons is in between the two structures

$$\rangle C{=}O \quad \text{and} \quad \rangle \overset{+}{C}{-}O^-$$

The actual structure is represented as

$$\rangle C{=}O \quad \text{or} \quad \rangle \overset{\delta+}{C}{=}\overset{\delta-}{O}$$

The shift in π electrons is called a **mesomeric effect**.

28.12.3 SUBSTITUENTS IN THE BENZENE RING

The reagents which attack the benzene ring are mainly electrophilic reagents (such as NO_2^+). If one of the hydrogen atoms in the benzene ring is replaced by a substituent group, the compound formed will differ in reactivity from benzene. If the substituent withdraws electrons, the compound will be less reactive than benzene.

A substituent with a +I effect, such as CH_3—, donates electrons to the benzene ring. It is said to **activate** the ring. A substituent with a −I effect, such as $-NO_2$, withdraws electrons and **deactivates** the ring:

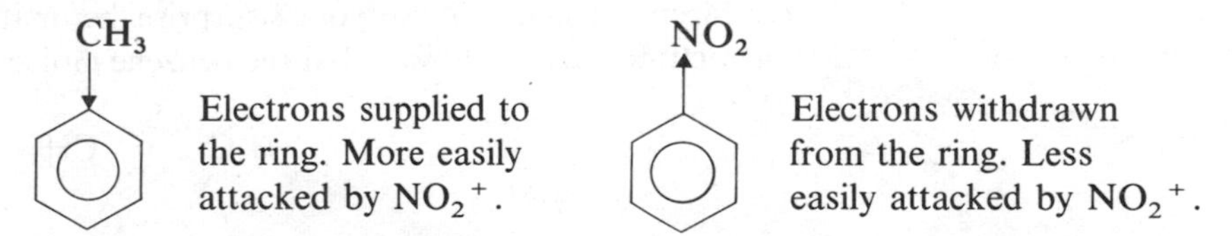

CH3
Electrons supplied to the ring. More easily attacked by NO2+.
NO2
Electrons withdrawn from the ring. Less easily attacked by NO2+.

The substituent —OH is more complicated. In addition to its −I effect, withdrawing electrons from the ring, another effect operates. The oxygen atom has two lone pairs of electrons. It can feed these electrons into the π orbitals of the benzene ring by means of a mesomeric effect, which is shown by a curved arrow representing the movement of a pair of electrons:

OH

−I effect withdraws electrons.
+M effect contributes electrons.
+M > −I, and the ring is activated.

The +M effect of the hydroxyl group is greater than the −I effect, and a hydroxyl group therefore activates the benzene ring.

Activating substituents, e.g. $-CH_3$, direct groups entering the benzene ring into the 2- and 4- positions. Thus methylbenzene yields on nitration a mixture of 1-methyl-2-nitrobenzene and 1-methyl-4-nitrobenzene. Deactivating substituents, e.g. $-CO_2H$, direct entering groups into the 3-position in the ring. Thus benzoic acid is nitrated to 3-nitrobenzoic acid.

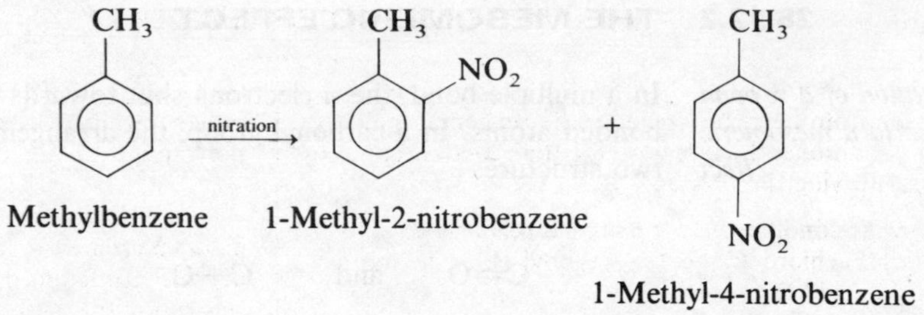

CH_3 $\xrightarrow{\text{nitration}}$ CH_3 NO_2 + CH_3 NO_2

Methylbenzene 1-Methyl-2-nitrobenzene 1-Methyl-4-nitrobenzene

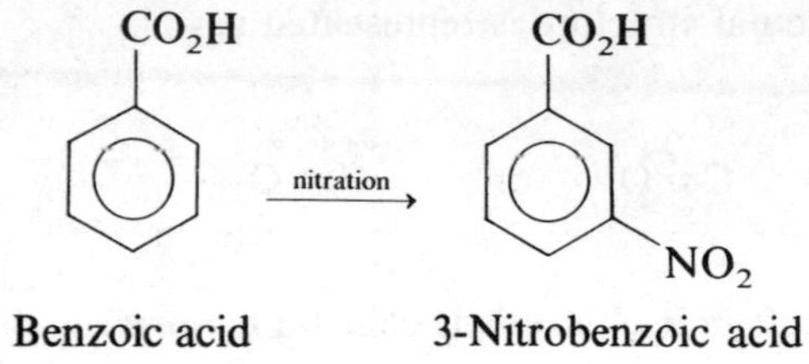

CO_2H $\xrightarrow{\text{nitration}}$ CO_2H NO_2

Benzoic acid 3-Nitrobenzoic acid

Profile: Kathleen Lonsdale (1903–1971)

Benzene is the characteristic part of all aromatic compounds. Identifying its structure was therefore a matter of some importance. The German chemist Kekulé had suggested a ring structure for benzene, a hexagonal ring with alternate single and double bonds. Benzene, being a liquid, was not suitable for crystallographic work. Professor Ingold of London University prepared some crystals of hexamethylbenzene and sent them to the crystallographers in the University of Leeds for analysis. The scientist who was entrusted with the job was a newly married young woman called Kathleen Lonsdale. To everyone's surprise, her mathematical analysis of her X ray diffraction patterns showed that the benzene molecule was planar.

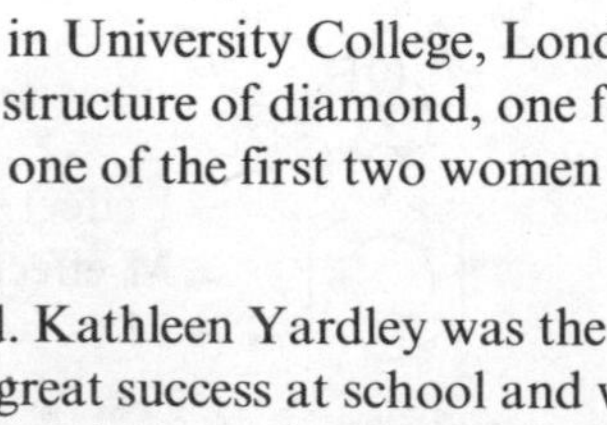

H_3C CH_3 H_3C CH_3 H_3C CH_3

Hexamethylbenzene

Kathleen became a professor in University College, London. In recognition of her crystallographic work on the structure of diamond, one form of diamond is named lonsdaleite after her. She was one of the first two women to be elected Fellow of the Royal Society.

Kathleen's early life was hard. Kathleen Yardley was the youngest of ten children in a poor family. Kathleen was a great success at school and went on to read physics in the University of London. While working for her PhD, Kathleen married an engineering student called Thomas Lonsdale. Thomas never failed to support his wife in her work, recognising the problems which she had to cope with as wife, mother and scientist and sharing them with her. The Lonsdales were active members of the Society of Friends. During the Second World War, they were pacifists. Kathleen was sent to prison for a month for refusing to register for Civil Defence. After the war, she was made a Dame Commander of the Order of the British Empire. She travelled widely in support of peace and better East–West relations.

QUESTIONS ON CHAPTER 28

1. Describe a laboratory method for the preparation of nitrobenzene from benzene. Include: the reagents, the necessary conditions, the equation for the reaction and the method of purifying the product.

2. Under what conditions does benzene react with (*a*) chlorine, (*b*) chloroethane, (*c*) ethanoyl chloride and (*d*) sulphuric acid?

3. Discuss the mechanisms of the Friedel–Crafts alkylation of benzene.

4. Describe the reactions of (*a*) propene and (*b*) benzene with (i) H_2, (ii) Cl_2 and (iii) H_2SO_4. State the conditions required for reaction, and give equations.
Discuss the mechanisms of the reactions of chlorine with propene and with benzene.

5. List the reactions of methylbenzene which are (*a*) typical of benzene and (*b*) not shared by benzene.

***6.** Explain the reason for the difference in the reactivity towards electrophiles of benzene and methylbenzene.

7. (*a*) Explain the following observations, giving equations for the reactions and structural formulae for the products.

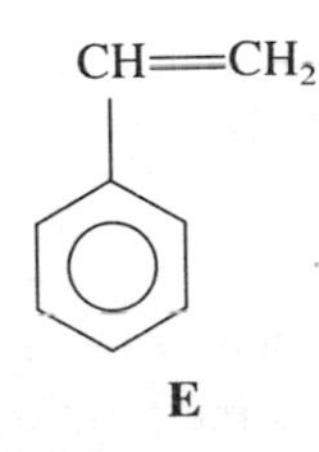

E

(i) Treatment of **E** with hydrogen at room temperature and pressure in the presence of a metal catalyst gives **F**. A similar reaction at high temperature and pressure gives a different compound **G**.
(ii) Ethene decolorises bromine rapidly but reaction of benzene with bromine is very slow unless $AlBr_3$ is also present.
(iii) Reaction of benzene with DBr in the presence of $AlBr_3$ gives C_6H_5D.

(*b*) Compound **E** above is used in industry in an addition polymerisation. State the *type* of catalyst used in this polymerisation and draw the structure of the product.

(O 90)

8. (*a*) Benzene can be nitrated by treating it with a mixture of concentrated nitric acid and sulphuric acid. After separation from the reaction mixture, the nitrobenzene formed can be purified by steam distillation.

(i) What type of mechanism is involved in nitrations of this type? Give the name of the nitrating species and write equations to show how it is formed in the nitrating mixture.

(ii) What reaction conditions are necessary to ensure a good yield of nitrobenzene? Explain your answer.

(iii) Outline the technique and theory of steam distillation.

(*b*) The nitration of phenol results in the formation of 2-nitrophenol and 4-nitrophenol.

(i) Draw the structural formula of each of the two isomers.

(ii) Describe how the technique of thin layer chromatography could be used to show the presence of both isomers in the reaction mixture.

(AEB 91)

9. (*a*) Compare and contrast the chemistry of benzene and alkenes by commenting on the following observations:

(i) The carbon–carbon bond length in benzene is 139 pm and that of ethene is 133 pm.

(ii) Both benzene and ethene react with bromine, but the conditions and type of reaction are different.

(iii) Benzene can be used as the solvent in some reactions involving $KMnO_4$ but the liquid alkene, cyclohexene, cannot.

(*b*) Define the term *electrophile*.

Predict the products, if any, obtained by treating ethene with each of the following reagents. Give your reasoning.

(i) HCN

(ii) HBr

(iii) NaBr

(O 91)

29

HALOGENOALKANES AND HALOGENOARENES

29.1 HALOGENOALKANES

The homologous series called **halogenoalkanes** (or haloalkanes) have the functional group

$$—C—X$$

where X = F, Cl, Br or I. Some members of the series are listed:

Names of some halogenoalkanes...

C_2H_5Cl — Chloroethane.
The name is derived from ethane, as the compound can be thought of as ethane, C_2H_6, with a Cl atom substituted for an H atom.

$CH_3CH_2CHCH_3$ (Cl on C-2) — 2-Chlorobutane.
The name is taken from that of the longest unbranched chain. The **locant**, 2, gives the position of the chlorine atom, counting from the end which gives the lower number for the locant.

...and halogenoalkenes

$CH_3CH{=}CHCHCH_3$ (Cl on C-4) — 4-Chloropent-2-ene.
The double bond is numbered first; then the chlorine atom.

$CH_3CH_2CHCH_2CHCH_3$ (CH_3 on C-4, Br on C-2) — 2-Bromo-4-methylhexane.
The longest chain is 6C: the compound is a derivative of hexane. The substituents are named in alphabetical order and then numbered.

CH_3CHCl_2 — 1,1-Dichloroethane

CH_2ClCH_2Cl — 1,2-Dichloroethane

$CHCl_3$ — Trichloromethane (chloroform)

CCl_4 — Tetrachloromethane (carbon tetrachloride)

$(CH_3)_2CHI$ (C bonded to CH_3, CH_3, H, I) — 2-Iodopropane

$(CH_3)_3CI$ (C bonded to three CH_3 and I) — 2-Iodo-2-methylpropane

29.1.1 PRIMARY, SECONDARY AND TERTIARY HALOGENOALKANES

Primary, secondary and tertiary

There are three types of halogenoalkanes:

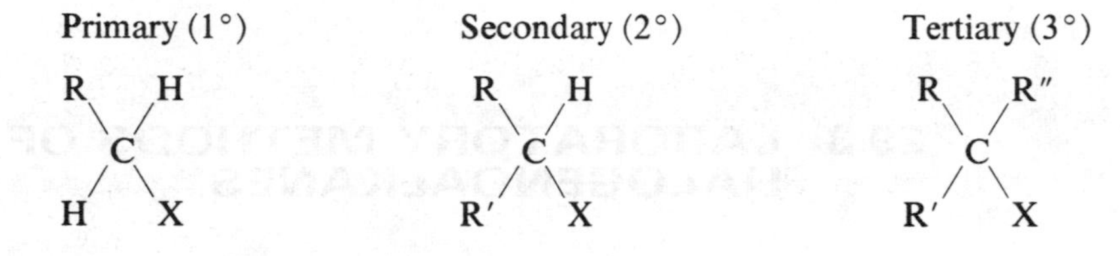

29.2 PHYSICAL PROPERTIES

29.2.1 VOLATILITY

Fluoroalkanes and chloroalkanes have polar molecules . . .

Electronegativities are F = 4.0, Cl = 3.0, Br = 2.8, I = 2.5, C = 2.5. The strong polarity of C—F and C—Cl bonds gives rise to attraction between the dipoles in neighbouring molecules:

$$\overset{\delta_+}{CH_3}—\overset{\delta_-}{Cl} \cdots \overset{\delta_+}{CH_3}—\overset{\delta_-}{Cl} \cdots \overset{\delta_+}{CH_3}—\overset{\delta_-}{Cl}$$

. . . As a result, their boiling temperatures are higher than those of the corresponding alkanes . . .

Energy must be supplied to separate the molecules, and the boiling temperatures of fluoroalkanes and chloroalkanes are therefore higher than those of alkanes of similar molecular mass.

. . . This is not the case for bromoalkanes and iodoalkanes

One bromine atom has the same mass as six CH_2 groups. A bromoalkane therefore has a smaller volume than an alkane of the same molecular mass. A smaller volume means less interaction between molecules and a lower boiling temperature. The boiling temperatures of bromoalkanes and iodoalkanes are lower than those of alkanes of the same molecular mass.

29.2.2 SOLUBILITY

Halogenoalkanes have a low solubility in water

The polar molecules can interact with water molecules, but the attractive forces set up are not as strong as the hydrogen bonds present in water. Halogenoalkanes therefore, although they dissolve more than alkanes, are only slightly soluble in water.

29.2.3 SMELL

These compounds have a sweet, slightly sickly smell.

29.2.4 DENSITY

Chloroalkanes are less dense than water; bromoalkanes and iodoalkanes are denser than water.

29.3 LABORATORY METHODS OF PREPARING HALOGENOALKANES

Preparative methods are summarised in Figures 29.1 and 29.2.

FIGURE 29.1 Laboratory Preparations of Halogenoalkanes

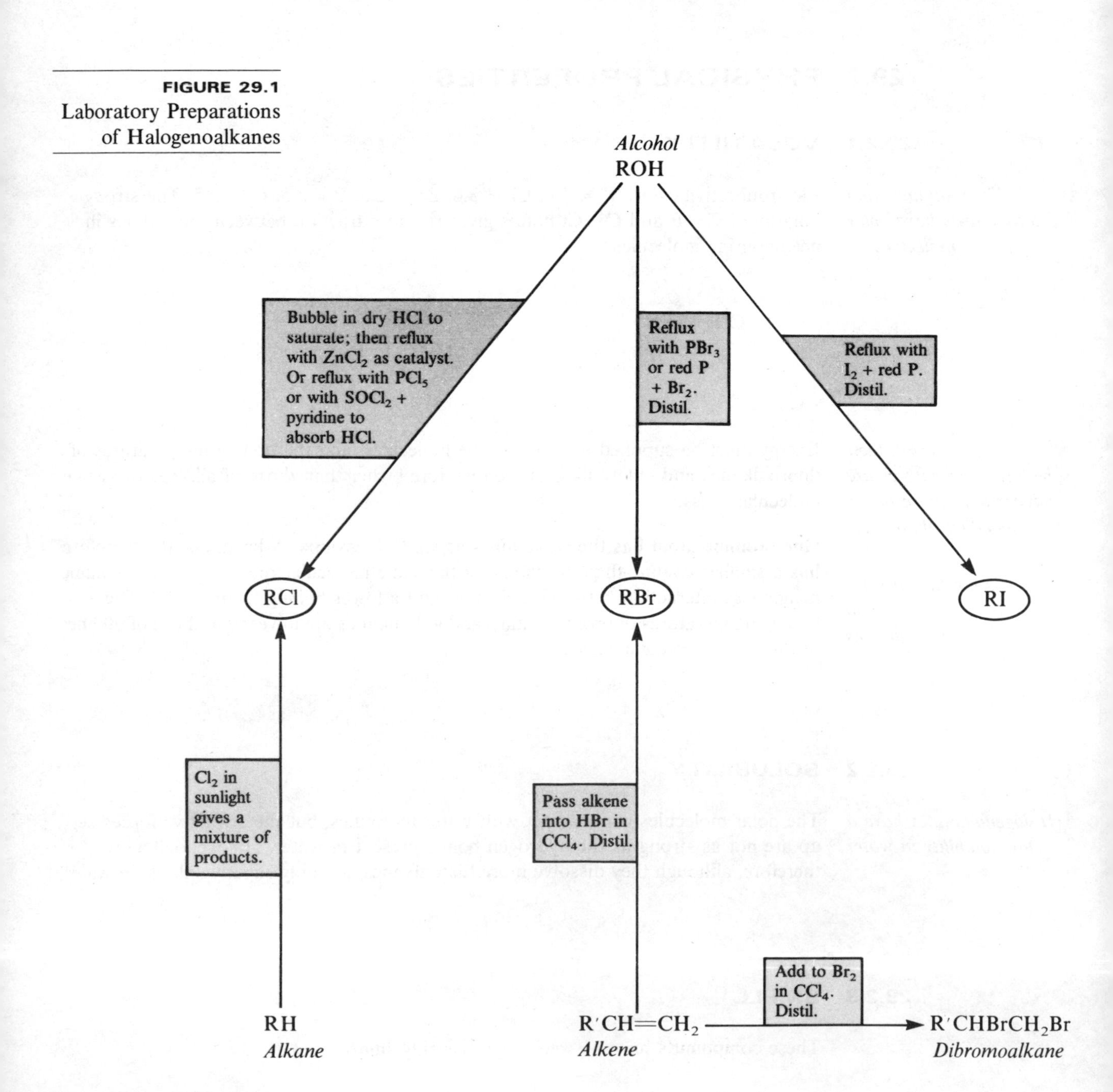

FIGURE 29.2 Preparation of Bromoethane from Ethanol

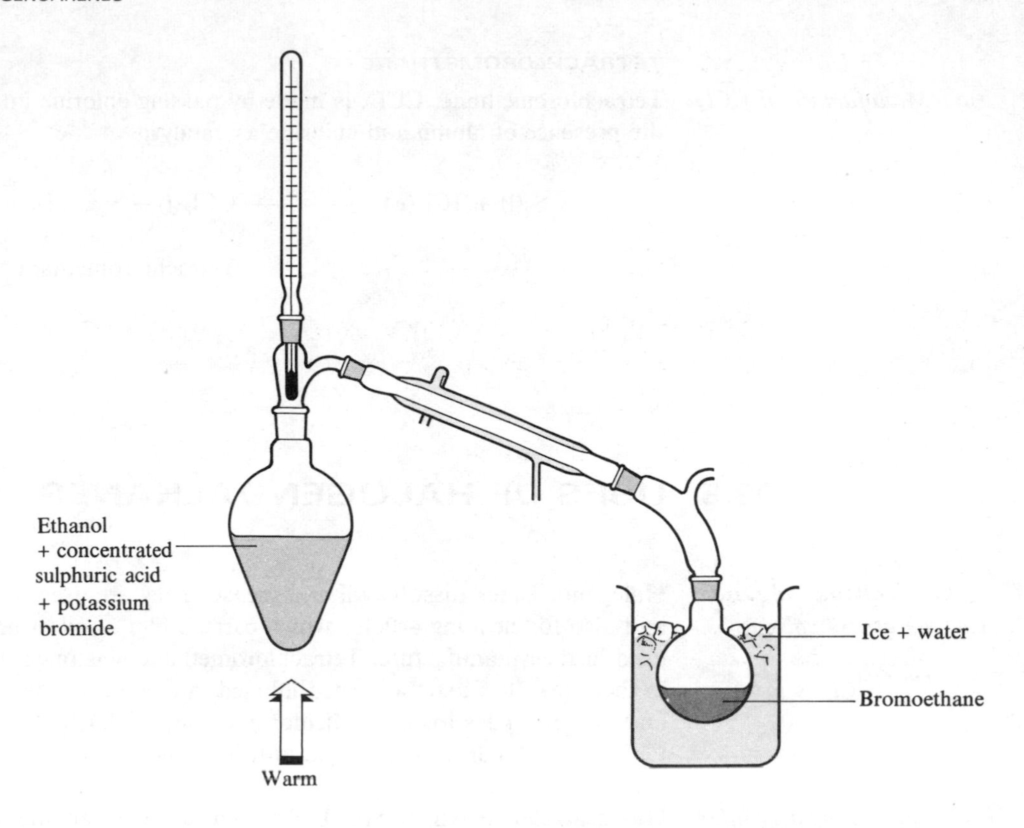

29.4 INDUSTRIAL MANUFACTURE

Industrial manufacture from Cl_2 + alkane...

Industrially, chloroalkanes are the most readily manufactured as chlorine is available from the electrolysis of brine. The chlorination of alkanes yields a mixture of monochloroalkanes and polychloroalkanes, from which the components can be obtained by distillation. If the product is to be used as a solvent, the fact that it consists of a mixture of chloroalkanes does not matter:

$$C_2H_6(g) + Cl_2(g) \xrightarrow{\text{light}} C_2H_5Cl(g) + HCl(g) \xrightarrow{Cl_2} C_2H_4Cl_2(g) \xrightarrow{Cl_2}$$

Ethane   Chloroethane   Dichloroethane

The addition of hydrogen chloride to an alkene is the method used when a monochloroalkane is required:

...and from HCl + alkene...

$$CH_3CH{=}CH_2(g) + HCl(g) \rightarrow CH_3CHClCH_3(l)$$

Propene   2-Chloropropane

Alkenes are obtained from the petrochemicals industry.

Alternatively, an alcohol can be vaporised and passed with hydrogen chloride over a heated catalyst:

...and from alcohol + HCl

$$CH_3OH(g) + HCl(g) \xrightarrow[\text{catalyst}]{\text{heat}} CH_3Cl(g) + H_2O(g)$$

Methanol   Chloromethane

TETRACHLOROMETHANE

Manufacture of CCl_4

Tetrachloromethane, CCl_4, is made by passing chlorine into carbon disulphide in the presence of aluminium chloride as catalyst:

$$CS_2(l) + 3Cl_2(g) \xrightarrow{AlCl_3\text{ catalyst}} \underset{\text{Tetrachloromethane}}{CCl_4(l)} + S_2Cl_2(l)$$

29.5 USES OF HALOGENOALKANES

Halogenoalkanes are used as grease solvents... ...CCl_4 is toxic...

Halogenoalkanes dissolve oil and grease. They are used in the dry-cleaning industry and also for cleaning articles which carry a film of oil or grease from the machinery used in their manufacture. Tetrachloromethane was once the prime halogenoalkane in this area. It is toxic as, when inhaled in quantity, it dissolves fat from the liver and kidneys. Less toxic are dichloromethane, CH_2Cl_2, trichloroethene, $CCl_2{=}CHCl$, called 'trichlor' in industry, and tetrachloroethene, $CCl_2{=}CCl_2$.

...Trichlor is safer

They are also used as refrigerant liquids

Halogenoalkanes which have boiling temperatures just below room temperature can easily be liquefied by a slight increase in pressure. This property makes them suitable for use as the liquid in refrigerators. The most suitable refrigerator liquids are fluoro-compounds, called fluorocarbons for short, or 'freons'. Dichlorodifluoromethane, CCl_2F_2, and 1,2-dichloro-1,1,2,2,-tetrafluoroethane, $CClF_2CClF_2$, are used.

Freons are stable and used in aerosol cans... ...they react with ozone in the upper atmosphere

Being easily liquefied by pressure, chlorofluorohydrocarbons vaporise when the pressure is reduced and are suitable for use as propellant liquids in aerosol sprays. They are very unreactive towards most reagents. They reach the upper atmosphere before they find anything with which they will react. Here, they undergo a photochemical reaction with ozone, converting it to oxygen. The ozone layer performs a useful function in absorbing ultraviolet rays from the Sun [see § 21.5]. Without this absorption, we should receive too much ultraviolet radiation, and this would cause skin cancer. Chlorofluorohydrocarbons in aerosol sprays are using up too much of the ozone layer, and endangering us from ultraviolet radiation [see § 29.9].

Halogenoalkanes are also used as fire-extinguishers

Fully halogenated alkanes, being non-flammable, volatile and dense, are used in fire extinguishers. Tetrachloromethane, CCl_4, was used as a fire extinguisher, but, at high temperatures, there is a danger of its being oxidised to the poisonous gas phosgene, $COCl_2$. It also has a toxic effect on the liver and kidneys and has been replaced by the safer gas dibromochlorofluoromethane, CBr_2ClF, called 'BCF'.

29.6 REACTIONS

The reactions of halogenoalkanes are shown in Figure 29.3.

FIGURE 29.3 Reactions of Halogenoalkanes

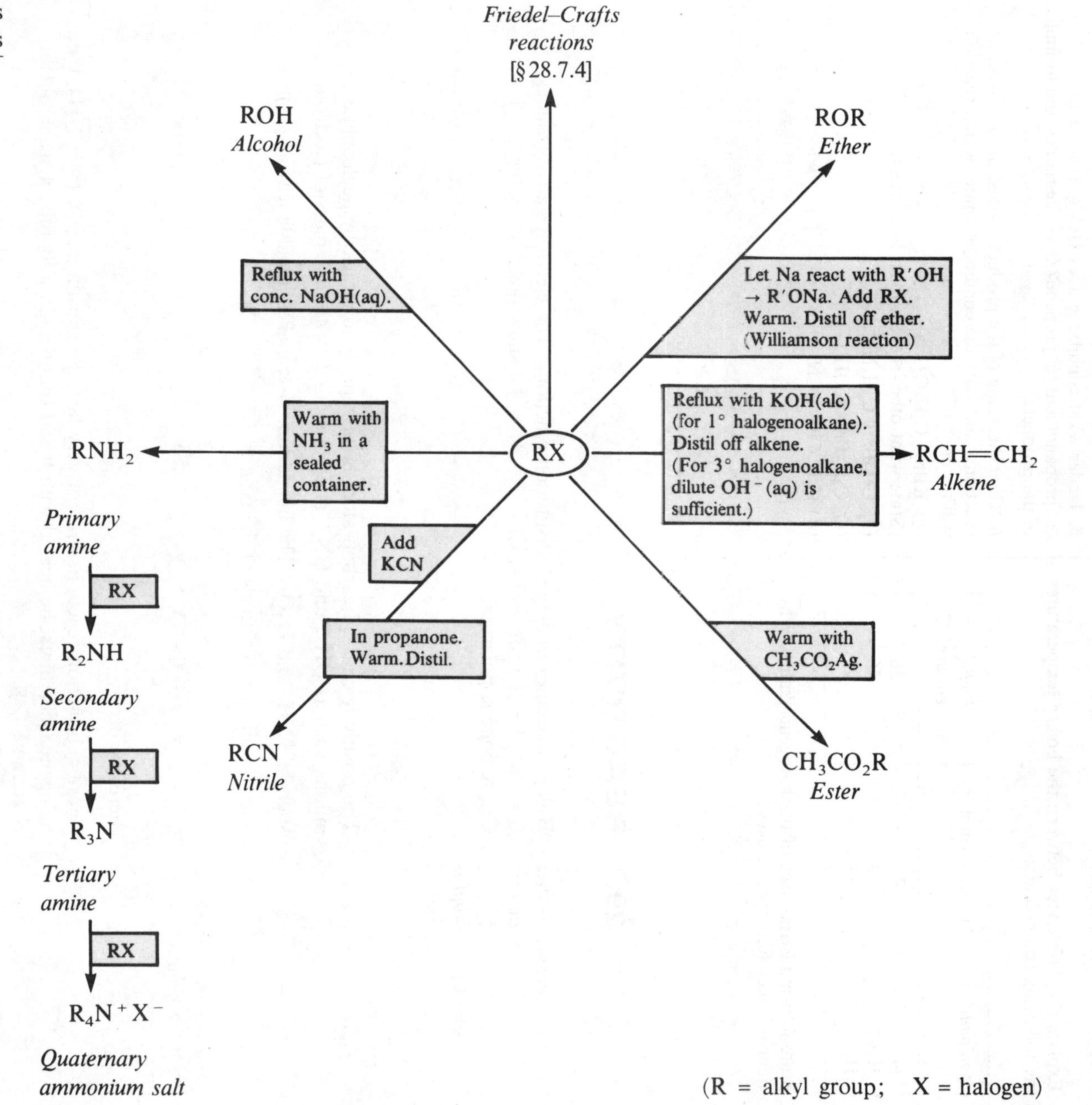

CHECKPOINT 29A: HALOGENOALKANES I

1. Name the following:

(*a*) $CH_3CH_2CHC(CH_3)_3$ (with I on the third carbon) (*b*) $(CH_3)_3CBr$

(*c*) $CH_3CH{=}CHC(Cl)(CH_3)_2$

2. Explain the differences between the boiling temperatures of the following compounds:

Compound	*Molar mass*/g mol^{-1}	*Boiling temperature*/°C
C_5H_{12}	72	36
C_3H_7Cl	79	46
C_3H_8	44	−42

3. State four industrial uses of the halogenoalkanes. Why do fluoroalkanes find special uses?

4. What reagents and what conditions are needed to bring about these reactions?

(*a*) $C_2H_5OH \rightarrow C_2H_5Cl$

(*b*) $C_3H_7OH \rightarrow C_3H_7I$

(*c*) $(CH_3)_2CHOH \rightarrow (CH_3)_2CHBr$

5. Under what conditions does HCl(g) react with (*a*) propanol and (*b*) propene? Give the names and formulae of the products.

6. Give the name of the product formed in the following reactions, and state the conditions under which reaction will occur:

(*a*) $KBr + C_3H_7OH \rightarrow$
(State what other reagent must be present.)

(*b*) $CH_3Br + C_2H_5NH_2 \rightarrow$

(*c*) $C_3H_7Br + KOH \rightarrow$

(*d*) $C_2H_5Cl + KCN \rightarrow$

29.7 REACTIVITY

Why are halogenoalkanes reactive?

Halogenoalkanes undergo substitution reactions. The electronegativity values C = 2.5, I = 2.5, Br = 2.8, Cl = 3.0, F = 4.0 show that, except for C—I, the C—X bond is polarised:

The C—X bond is polarised...

$$\geq\overset{\delta+}{C}-\overset{\delta-}{X}$$

The positively charged carbon atom is susceptible to attack by nucleophiles, anions such as OH^- and CN^- and compounds with lone pairs of electrons such as $:NH_3$ and $H_2\ddot{O}:$. The halide ion, X^-, makes a stable leaving group, and substitution reactions of the type

...Nucleophiles attack the carbon in C—X

$$Y: + \geq\overset{\delta+}{C}-\overset{\delta-}{X} \rightarrow Y-C\leq + X^-$$

therefore occur.

These substitution reactions are not very fast. To substitute —X by —NH_2, the halogenoalkane must be heated with 0.880 ammonia in a 'bomb', a sealed metal container:

Conditions needed for reaction

$$C_2H_5Cl(g) + NH_3(aq) \xrightarrow[\text{pressure develops}]{\text{heat in a 'bomb'}} C_2H_5NH_2(g) + HCl(g)$$

Chloroethane Ethylamine or Aminoethane

To substitute —X by —OH, a primary halogenoalkane must be refluxed with aqueous alkali for about an hour:

$$C_2H_5Cl(g) + OH^-(aq) \xrightarrow{\text{reflux}} C_2H_5OH(g) + Cl^-(aq)$$

Chloroethane Ethanol

A suitable apparatus is shown in Figure 28.4.

The ease of reaction depends on the ease of breaking the C—X bond. Average standard bond enthalpies in $kJ\,mol^{-1}$ are C—F = 484, C—Cl = 338, C—Br = 276, and C—I = 238. Thus, the ease of bond-breaking is

C—I > C—Br > C—Cl > C—F

The ease of reaction also depends on the stability of the *leaving group*. The ease of formation of X^-(aq) is

$$F^-(aq) > Cl^-(aq) > Br^-(aq) > I^-(aq)$$

The two factors are in opposition. The C—F bond is so strong that fluoroalkanes are extremely unreactive. The order of reactivity of the other halogenoalkanes towards nucleophiles is usually

RI > RBr > RCl

Elimination of HX with the formation of an alkene takes place in reactions with strong bases

Halogenoalkanes also undergo reactions in which hydrogen halide is eliminated and an alkene is formed. The nucleophiles which take part in substitution reactions (e.g., OH^-, CN^-) have basic properties. They are able to abstract a hydrogen atom from a halogenoalkane, while the halogen is expelled as halide ion. Thus, two reactions, substitution and elimination, are in competition:

$$RCH_2CH_2X + OH^-(aq) \rightarrow RCH_2CH_2OH + X^-(aq);\ \textit{Substitution}$$

$$RCH_2CH_2X + OH^-(aq) \rightarrow RCH{=}CH_2 + H_2O + X^-(aq);\ \textit{Elimination}$$

Substitution reactions are favoured by weakly basic nucleophiles, e.g., CN^-; elimination by strongly basic reagents, e.g., OH^-. A high concentration of base in a non-aqueous solvent (e.g., potassium hydroxide in ethanol) at a reflux temperature favours elimination.

The relative extents to which substitution and elimination take place depend on the structure of the molecule, as well as on the basic strength of the nucleophile. Elimination becomes progressively more important in the order

Primary < Secondary < Tertiary halogenoalkane

When a tertiary halogenoalkane reacts with a base, the main product is an alkene:

$$(CH_3)_3CCl(l) + OH^-(aq) \rightarrow (CH_3)_2C{=}CH_2(l) + H_2O(l) + Cl^-(aq)$$

Tertiary halogenoalkanes tend to undergo elimination reactions

To replace —X by —OH in a tertiary halogenoalkane, no base is needed:

$$(CH_3)_3CCl(l) + H_2O(l) \xrightarrow[\text{at 25 °C}]{\text{80\% aqueous ethanol}} (CH_3)_3COH(l) + HCl(aq)$$

For a given aqueous base, the reactions which take place are
Primary halogenoalkane: Substitution
Secondary halogenoalkane: Substitution/Elimination
Tertiary halogenoalkane: Elimination.

CHECKPOINT 29B: HALOGENOALKANES II

1. How could the following reactions be carried out? State the necessary reagents and conditions:

(*a*) $C_4H_9OH \rightarrow C_4H_9Cl$
(*b*) $(C_2H_5)_3COH \rightarrow (C_2H_5)_3CCl$
(*c*) $C_3H_7I \rightarrow C_3H_7OH$
(*d*) $C_3H_7Br \rightarrow CH_3CO_2C_3H_7$

2. Explain why halogenoalkanes undergo substitution reactions (*a*) more readily than alkanes and (*b*) fairly slowly.

3. How can the following compounds be made from bromopropane?

(*a*) propylamine (*c*) propanol
(*b*) propene (*d*) butanonitrile

4. Give named examples of primary, secondary and tertiary halogenoalkanes.

29.8 THE MECHANISM OF HYDROLYSIS OF HALOGENOALKANES

Hydrolyses of halogenoalkanes are S_N reactions

The hydrolysis of a halogenoalkane is a **substitution** reaction. The attacking species is a **nucleophile**. Hydrolyses are therefore described as **S_N reactions**. A great deal of work has been done on the mechanisms of hydrolysis reactions.

29.8.1 HYDROLYSIS OF PRIMARY HALOGENOALKANES

EXPERIMENTAL EVIDENCE

The alkaline hydrolysis of a primary halogenoalkane

$$RX(l) + OH^-(aq) \rightarrow ROH(aq) + X^-(aq)$$

is second-order, following the rate expression

$$\text{Rate} = k\,[RX][OH^-]$$

where [RX] = concentration of RX and k is a constant.

Experimental evidence on the hydrolysis of primary halogenoalkanes

One can deduce from this evidence that a molecule of RX and an OH^- ion must collide before reaction will occur. Calculation shows that the rate of reaction is much less than the rate at which RX and OH^- collide. Only a small fraction of the collisions result in reaction. It must be necessary for the two species not only to collide but also to collide with enough energy to overcome the repulsion between the hydroxide ion and the halogenoalkane molecule.

MECHANISM

The mechanism . . .

. . . The making of a new bond eases the breaking of the old bond . . .

. . . A transition state is formed

It is known that, in order to minimise repulsion between RX and OH^-, OH^- approaches the carbon atom attached to X on the opposite side of the molecule from X:

$$HO^- \text{ approaches } H{-}\underset{CH_3}{\overset{H}{\underset{|}{\overset{|}{C}}}}{-}X \text{ to form } \left[H{-}O\cdots\cdots\underset{CH_3}{\overset{H\quad H}{\underset{|}{\overset{\diagdown\ \diagup}{C}}}}\cdots\cdots X \right]^-$$

As a bond forms between O and C, the C—X bond weakens, and a transition state is reached in which C is partially bonded both to O and X. Once the transition state has been formed, it is rapidly converted into the products. A transition state has a momentary existence [§ 14.9.2]. The reactive intermediates that have been postulated in other reactions are longer-lived.

A summary of the mechanism

The proposed mechanism for the alkaline hydrolysis of primary halogenoalkanes is

$$HO^- + H{-}\underset{CH_3}{\overset{H}{C}}{-}X \xrightarrow[\text{step}]{\text{rate-determining}} \left[H{-}O\cdots\underset{CH_3}{\overset{H\ \ H}{C}}\cdots X \right]^- \xrightarrow{\text{fast}} H{-}O{-}\underset{CH_3}{\overset{H}{C}}{-}H + X^-$$

Transition state

The movement of the bonding electrons is often shown by a curved arrow

$$HO^- \ \underset{CH_3}{CH_2}{-}I \rightarrow HO{-}\underset{CH_3}{CH_2} + I^-$$

29.8.2 HYDROLYSIS OF TERTIARY HALOGENOALKANES

The hydrolysis of tertiary halogenoalkanes is first-order

An example of the hydrolysis of a tertiary halogenoalkane is the hydrolysis of 1,1-dimethylchloroethane in aqueous ethanol:

$$(CH_3)_3CCl(l) + H_2O(l) \rightarrow (CH_3)_3COH(l) + HCl(aq)$$

Experimental results show that the reaction is first order with respect to the halogenoalkane:

$$\text{Rate} = k[(CH_3)_3CCl]$$

It is inferred that the slow, rate-determining step must involve the halogenoalkane alone. It is suggested that this step is the dissociation of the halogenoalkane into a carbocation and a halide ion:

A carbocation intermediate is postulated

$$(CH_3)_3CCl \xrightarrow{\text{slow, rate-determining step}} (CH_3)_3C^+ + Cl^-$$

The carbocation formed reacts rapidly with water molecules:

$$(CH_3)_3C^+ + H_2O \xrightarrow{\text{fast step}} (CH_3)_3COH + H^+(aq)$$

This step is fast, and the rate of the overall reaction is determined by the rate at which the carbocations are formed.

A summary of the mechanism

The proposed mechanism for the hydrolysis of tertiary halogenoalkanes is

$$R_3CX \xrightarrow{\text{rate-determining step}} R_3C^+ + X^-$$

followed by $R_3C^+ + H_2O \xrightarrow{\text{fast}} R_3COH + H^+(aq)$

29.8.3 HYDROLYSIS OF SECONDARY HALOGENOALKANES

Secondary halogenoalkanes

The behaviour of secondary halogenoalkanes is intermediate between primary and tertiary halogenoalkanes, depending on the nature of the halogenoalkane and the solvent.

29.8.4 *DESCRIPTION OF MECHANISM

S_N1 and S_N2 reactions

The hydrolyses of halogenoalkanes are S_N reactions, nucleophilic substitution reactions. In the hydrolysis of primary (and some secondary) halogenoalkanes, two species, RX and OH^-, are involved in the formation of the transition state: the rate-determining step is bimolecular. The reaction is described as an **S_N2 reaction**. In the hydrolysis of tertiary (and some secondary) halogenoalkanes, only one species is required for the formation of the transition state: the rate-determining step is unimolecular. Such reactions are designated as **S_N1 reactions**.

***CHECKPOINT 29C: HALOGENOALKANES III**

***1.** Sketch the transition state which is thought to be formed in the alkaline hydrolysis of 1-bromopropane. What leads a molecule of bromopropane to form this transition state? Why do hydroxide ions not attack alkanes? What is the difference between an S_N1 reaction and an S_N2 reaction? Explain how the hydrolysis of 1-bromo-1,1-dimethylethane differs from that of 1-bromopropane in the manner in which the rate of the reaction depends on the concentrations of the reactants. What reason has been suggested to account for the difference?

29.9 CHLOROFLUOROCARBONS

TOPIC

During the Second World War, 'bug bombs' were issued to US soldiers. They were portable aerosol insecticides made from two shell cases welded together. The first aerosols for consumer use came out in 1947.

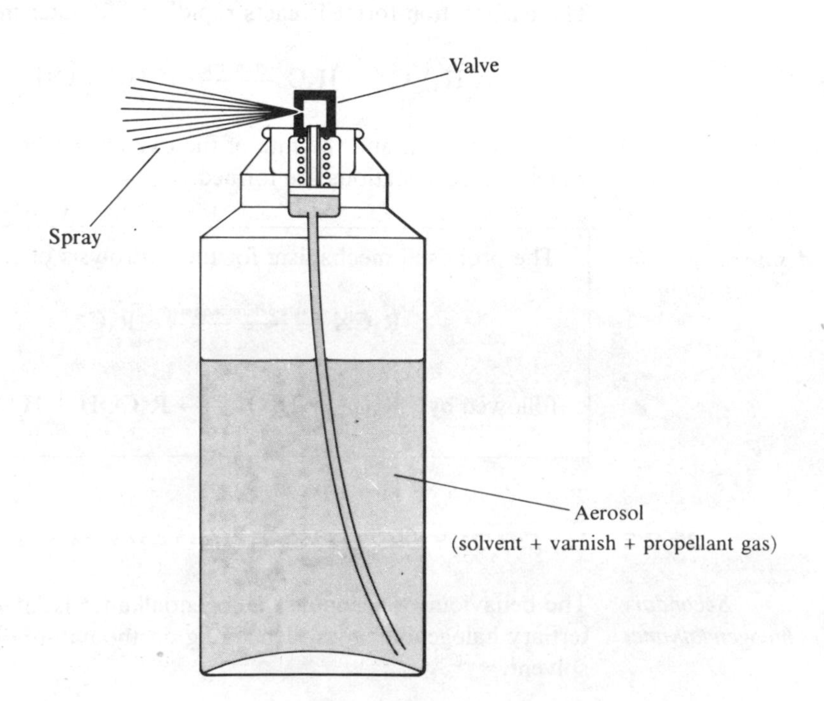

FIGURE 29.4
An Aerosol Can

Aerosol cans employ 'chlorofluorocarbons' (CFCs), which are properly called chlorofluoroalkanes, as propellants. These fully halogenated alkanes are very stable

A low pressure propellant was developed so that lighter, low pressure containers and valves could be used without danger of explosion. The most suitable propellants were 'chlorofluorocarbons' (CFCs), These compounds are properly called chlorofluoroalkanes. Examples are trichlorofluoromethane, $CFCl_3$, dichlorodifluoromethane, CF_2Cl_2 and 1,2-dichloro-1,1,2,2-tetrafluoroethane, $CClF_2CClF_2$. They are manufactured under the trade name Freon®. Chlorofluoroalkanes are suitable for use as aerosols because they are good solvents with low boiling temperatures and are odourless, non-flammable, stable and non-toxic.

The first aerosols dispensed insecticides. These were followed by personal products, e.g. hairsprays, perfumes and deodorants, and household products, e.g. window-cleaning liquids, polishes, waxes and laundry products.

CFCs are also used in refrigerators, air conditioners and the plastic foams used as thermal insulators in buildings and in food packaging

Chlorofluoroalkanes were originally developed in the 1920s for use as refrigerant fluids. They have the same thermal properties as the original refrigerants, sulphur dioxide and ammonia, but their non-toxic and non-corrosive character made them much safer. They are used in refrigerators and air conditioners and as blowing agents in the manufacture of thermal foam insulation for buildings and foamed plastic cups and food packaging. As more and more uses were found for these remarkable compounds, CFCs became big business, with hundreds of thousands of tonnes being produced yearly. Now they are being phased out. These stable, non-toxic compounds are dangerous!

CFCs are so stable that they reach the stratosphere unchanged. There the intense UV light photolyses CFCs, with the production of chlorine radicals and chlorine oxide. These species convert ozone into dioxygen

During all the time that the use of chlorofluoroalkanes was increasing, no one thought about what would happen to the gases in the atmosphere. Because of their lack of reactivity and insolubility in water, there is no natural process for removing chlorofluoroalkanes. In fact they drift up into the stratosphere (the upper atmosphere; see Figure 21.5, § 21.5), where they receive sufficient ultraviolet light to cause photolysis. The chlorine radicals formed in photolysis take part in reactions which convert ozone into dioxygen.

(*a*) $Cl\cdot + O_3 \rightarrow ClO + O_2$

(*b*) $ClO + O\cdot \rightarrow Cl\cdot + O_2$

(*c*) $ClO + O_3 \rightarrow ClO_2 + O_2$

Notice that reactions (*a*) and (*b*) form a chain. This is why one chlorine radical can destroy thousands of ozone molecules.

The rise in CFC concentration in the stratosphere will deplete the ozone layer, decreasing its ability to absorb UV light and increasing the incidence of skin cancer. It will also increase the ability of the stratosphere to absorb IR radiation and thus contribute to the greenhouse effect

The fall in ozone concentration is serious because it decreases the absorption of UV light by the stratosphere. One result could be an increase in skin cancer among humans [see § 21.5].

There are also two ways in which CFCs may affect the climate. One is by depleting the ozone layer. Because the ozone layer warms the stratosphere, a decrease in its thickness should cause a drop in the stratosphere temperature. What effect this would have on the troposphere (lower atmosphere; see Figure 21.5, § 21.5) is not known: the relationship between the troposphere and the stratosphere is poorly understood. A second effect of CFCs is due to their strong absorption of infrared light. By absorbing infrared radiation given off by the earth, CFCs contribute to the 'greenhouse effect' which warms the earth [§ 23.9]. CFCs are 10 000 times more effective than carbon dioxide in absorbing infrared radiation.

The search is on for alternatives to CFCs. Some personal products and household products can be packaged in a container fitted with a pump.

Alternatives to CFCs are nitrogen, carbon dioxide, hydrocarbons and alkanes which are not fully halogenated (HCFCs and HFCs)

The use of different gases in aerosols is being investigated. Of the compressed gases, nitrogen and carbon dioxide are used. They are inexpensive, non-flammable and non-toxic. They are not ideal for personal products because they do not produce as fine a mist as CFCs. Liquefied petroleum gases, e.g. propane and methylpropane, are also used, but they have the disadvantage of being flammable. Chemical firms are looking at halogenoalkanes which are not fully halogenated, which industry calls 'hydrochlorofluorocarbons' (HCFCs) and 'hydrofluorocarbons' (HFCs). The presence of a hydrogen atom or atoms in the molecule makes these compounds less stable than CFCs. and they do not reach the stratosphere in such large concentrations. A programme of tests, including toxicity tests, is being carried out on these compounds.

In 1987 an international agreement to reduce CFC emission by 50% was signed in Montreal. The UK is now aiming for a reduction of 85%

The CFC problem cannot be tackled on a national scale. Only if their release is controlled worldwide can a solution be achieved. In September 1987, 38 nations signed the Montreal Protocol, an agreement to reduce their consumption of CFCs by 50% by the year 2000. The Commonwealth Secretary General, Sir Shridath Ramphal, described the agreement as 'the first time governments have acted together not in response to demonstrated calamity but to the predictions and warnings of scientists'. Early in 1988 Prince Charles, in a speech to the Royal Society of Arts, criticised Britain's record on environmental legislation and said that he had banned aerosols in his household. Also early in 1988, ICI announced that it was phasing out CFCs in aerosols and increasing its research programme on alternatives to CFCs. The chemical firm of Dupont in the USA is also well advanced in the search for alternatives. Later in 1988 the chairman of the Consumers' Association said that a survey showed that many people had stopped using aerosols. Until 1986 the UK and other members of the European Community (EC) resisted efforts to ban CFCs, saying that their chemical industries were not able to offer substitutes. By 1989, however, the UK Government had become convinced that a reduction in CFC emission was urgent, and promised an 85% cut in CFC emissions by the year 2000, an even bigger reduction than in the Montreal agreement. At the second meeting of the Montreal Protocol parties in 1990 the 50% cut was converted into a total phase-out of CFCs by the year 2000.

CHECKPOINT 29D: CHLOROFLUOROCARBONS

1. Why does a liquid need to have a low boiling temperature for use in an aerosol? Why does it have to be a good solvent? Why must it be chemically unreactive? What other properties should an aerosol liquid have?

2. Make a list of (*a*) personal products and (*b*) household products in aerosol cans. Suggest alternative ways of using the different products, and say whether or not the alternatives would be inferior to aerosols.

3. What thermal properties must a refrigerant liquid have? Why are CFCs better than (*a*) sulphur dioxide and (*b*) ammonia?

4. What effect could the increasing concentration of CFCs in the stratosphere have on (*a*) human health and (*b*) the climate? [See also §§21.5 and 21.14.]

5. Why has the move to phase out CFCs gained momentum in recent years? [See also §21.5.]

29.10 HALOGENOALKENES

Halogenoalkenes are unreactive

The C—Cl bond is strong as it interacts with π electrons of C=C

Halogenoalkenes, e.g., chloroethene, $CH_2{=}CHCl$, are very unreactive. They are not hydrolysed by aqueous alkali or attacked by other nucleophiles. The reason is that the p orbitals of the doubly bonded carbon atoms overlap with the p orbitals of the chlorine atom to form a π bond [see Figure 29.5]. This makes the C—Cl bond much stronger than that in chloroalkanes, where there is only a σ bond between carbon and chlorine.

FIGURE 29.5 Bonding in Chloroethene

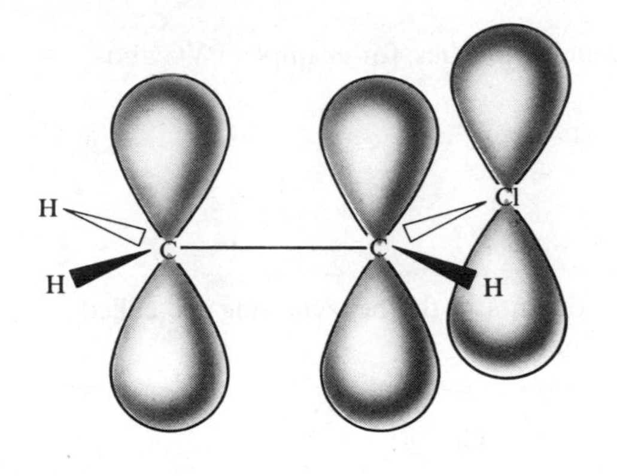

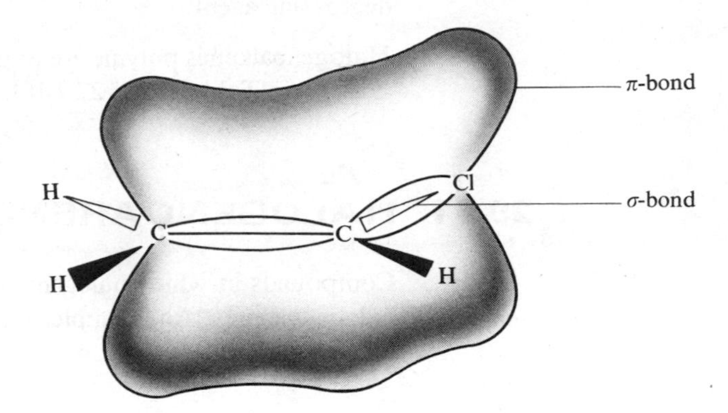

FIGURE 29.6 Relating Halogenoalkanes to Other Series of Compounds

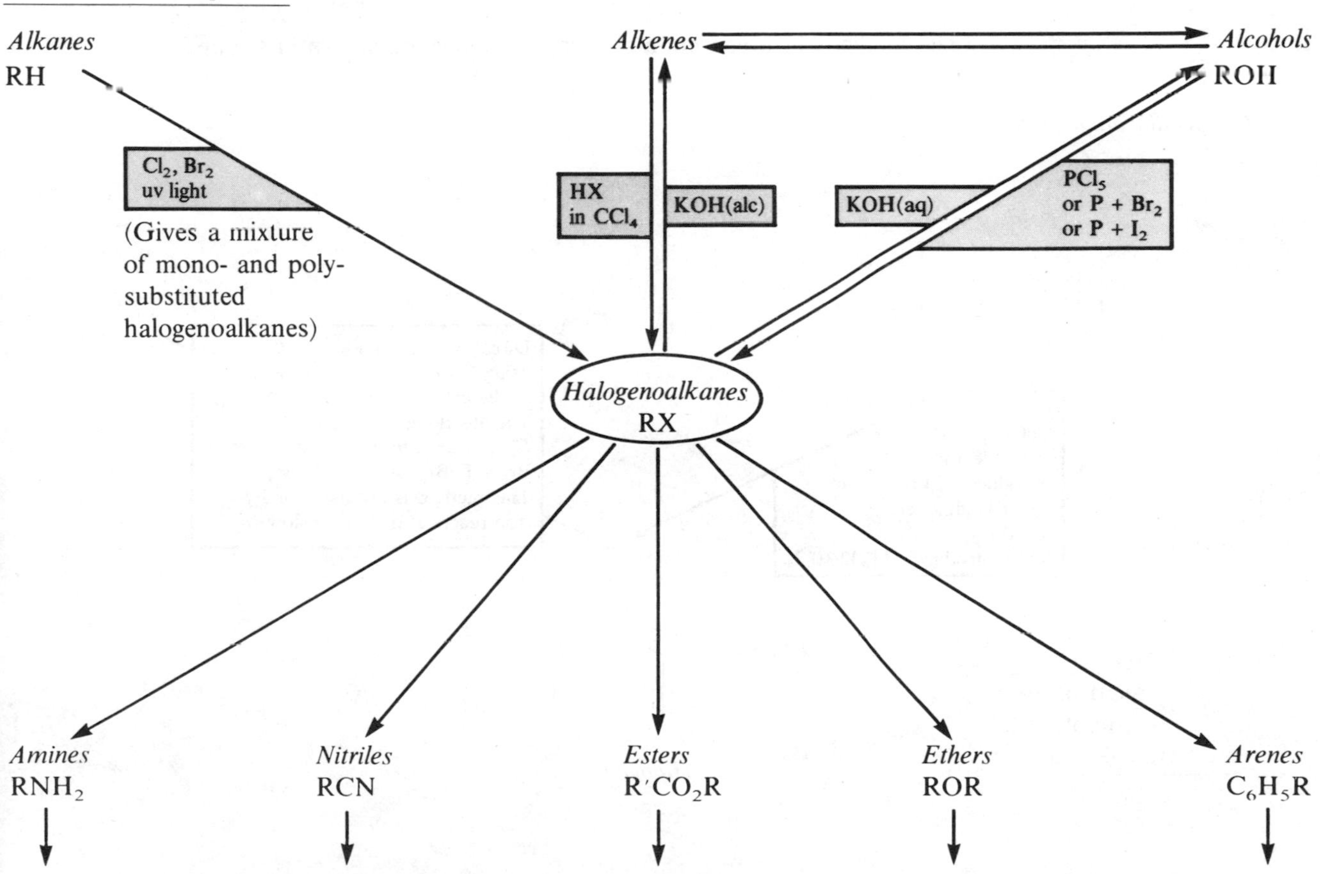

All these compounds contain functional groups which enable them to react to form a variety of other compounds.

Note Alkanes and alkenes are obtained from the petroleum industry. They can be converted to halogenoalkanes, from which a variety of different compounds can be made. The halogenation of alkanes gives a mixture of halogenoalkanes. If the appropriate alcohol is available, it is a more convenient source of a monohalogenoalkane.

The chemical stability of chloroalkenes and their solvent properties explain their use as degreasing agents. 'Trichlor', trichloroethene, $CHCl{=}CCl_2$, is much used as a degreasing agent.

Halogenoalkenes polymerise to give valuable plastics, for example, PVC and Neoprene [Table 27.1, § 27.7.10].

29.11 HALOGENOARENES

Halogenoarenes

Compounds in which halogens are substituents in the benzene ring are called halogenoarenes. An example is chlorobenzene:

29.12 PREPARATION

The methods of preparation of halogenoarenes are shown in Figure 29.7.

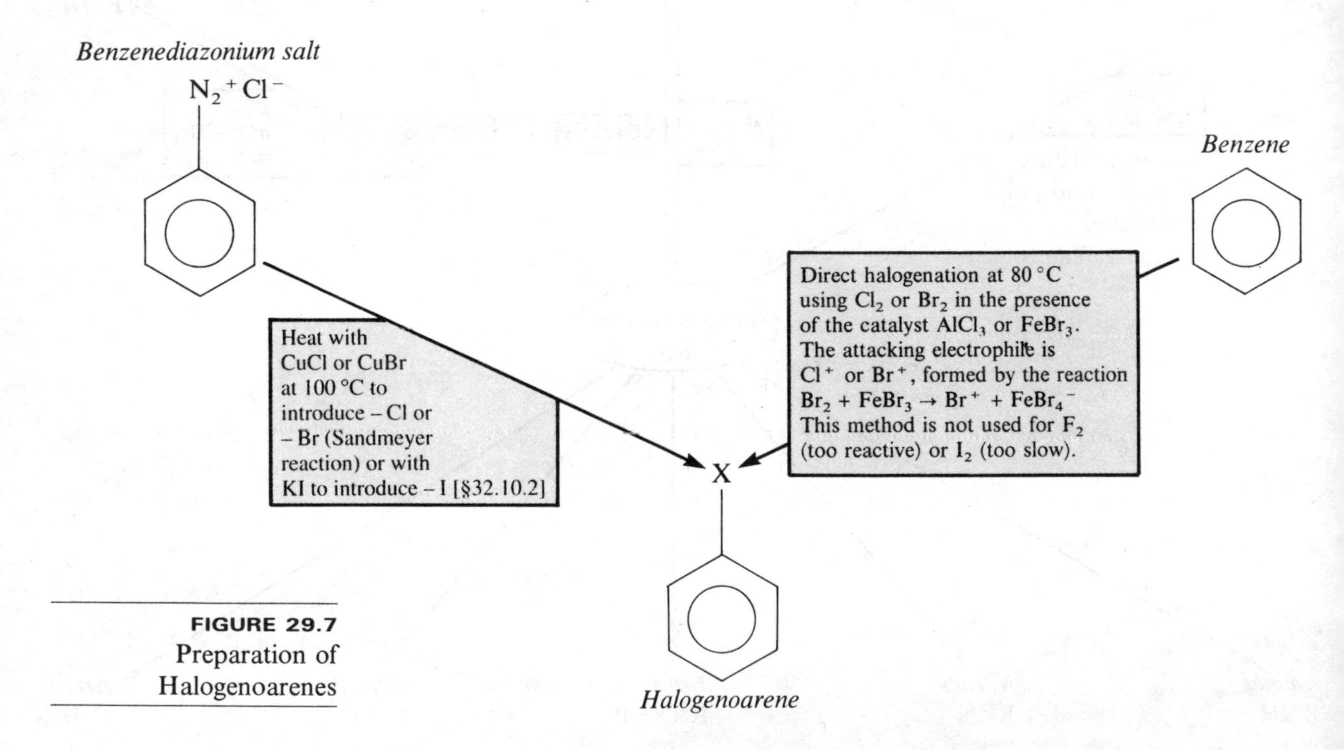
Benzenediazonium salt
$N_2^+Cl^-$
Benzene
Heat with CuCl or CuBr at 100 °C to introduce – Cl or – Br (Sandmeyer reaction) or with KI to introduce – I [§32.10.2]
Direct halogenation at 80 °C using Cl_2 or Br_2 in the presence of the catalyst $AlCl_3$ or $FeBr_3$. The attacking electrophile is Cl^+ or Br^+, formed by the reaction $Br_2 + FeBr_3 \rightarrow Br^+ + FeBr_4^-$ This method is not used for F_2 (too reactive) or I_2 (too slow).
X
Halogenoarene

FIGURE 29.7 Preparation of Halogenoarenes

29.13 REACTIVITY OF HALOGENOARENES

The halogen atom in a halogenoarene is very much less reactive than that in a halogenoalkane. The reason is similar to the reason for the lack of reactivity of the halogenoalkenes. The p orbitals of the halogen atom interact with the p orbitals of the six carbon atoms in the benzene ring to form a delocalised cloud of π electrons [see Figure 29.8]. This π bond adds to the strength of the σ bond between the halogen atom and the ring, and makes the halogen atom very difficult to displace.

FIGURE 29.8 Electron Distribution in Halogenoarenes

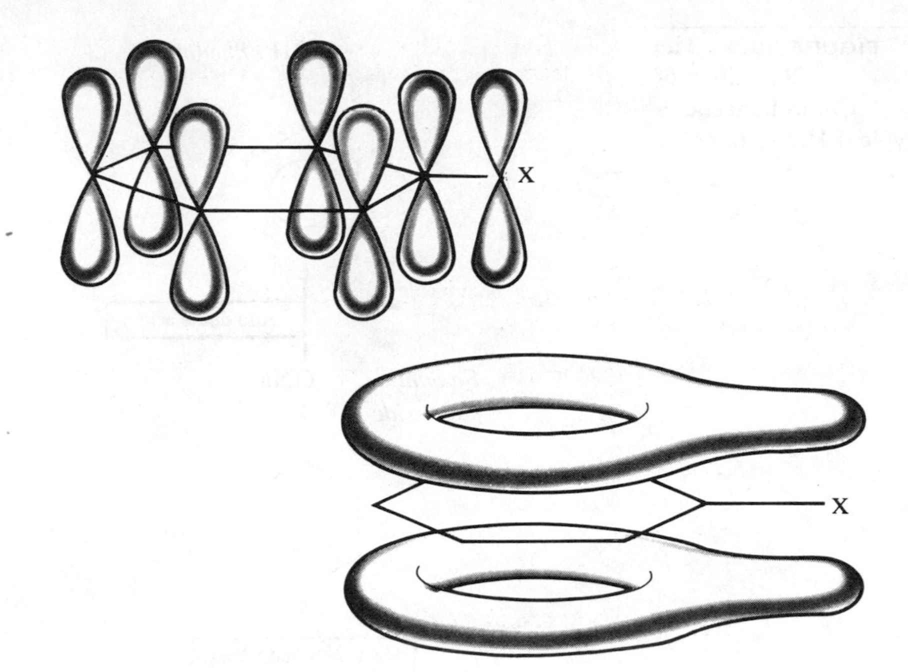

The C—X bond interacts with the ring...

...π bonding adds to the strength of the σ bond...

29.14 REACTIONS OF HALOGENOARENES

...The halogen is difficult to displace

There are two kinds of reactions: replacement of the halogen atom and substitution in the benzene ring.

29.14.1 REPLACEMENT OF THE HALOGEN ATOM

Stringent conditions are needed for hydrolysis of halogenoarenes

The conditions needed for the replacement of a halogen atom are exemplified in the manufacture of phenol, C_6H_5OH. One industrial method is to heat chlorobenzene and aqueous sodium hydroxide together at 150 atm and 350 °C. The sodium phenoxide formed is converted into phenol by the action of dilute acid:

$$C_6H_5Cl(l) + 2NaOH(aq) \xrightarrow[350\,°C]{150\,atm} C_6H_5ONa(aq) + NaCl(aq) + H_2O(l)$$

Chlorobenzene Sodium phenoxide

$$C_6H_5OHa(aq) + HCl(aq) \rightarrow C_6H_5OH(l) + NaCl(aq)$$

Sodium phenoxide Phenol

29.14.2 SUBSTITUTION IN THE BENZENE RING

Halogenoarenes are less reactive than benzene

The benzene ring undergoes the usual reactions [see Figure 29.9, § 28.12.3]. Halogen substituents withdraw electrons, thus deactivating the benzene ring [§ 28.12.3] and are 2/4-directing [§ 28.12.3]. A comparison of a halogenoalkane and a halogenoarene is made in Table 29.1, § 29.15.

FIGURE 29.9 The Reactions of Chlorobenzene, a Typical Halogenoarene

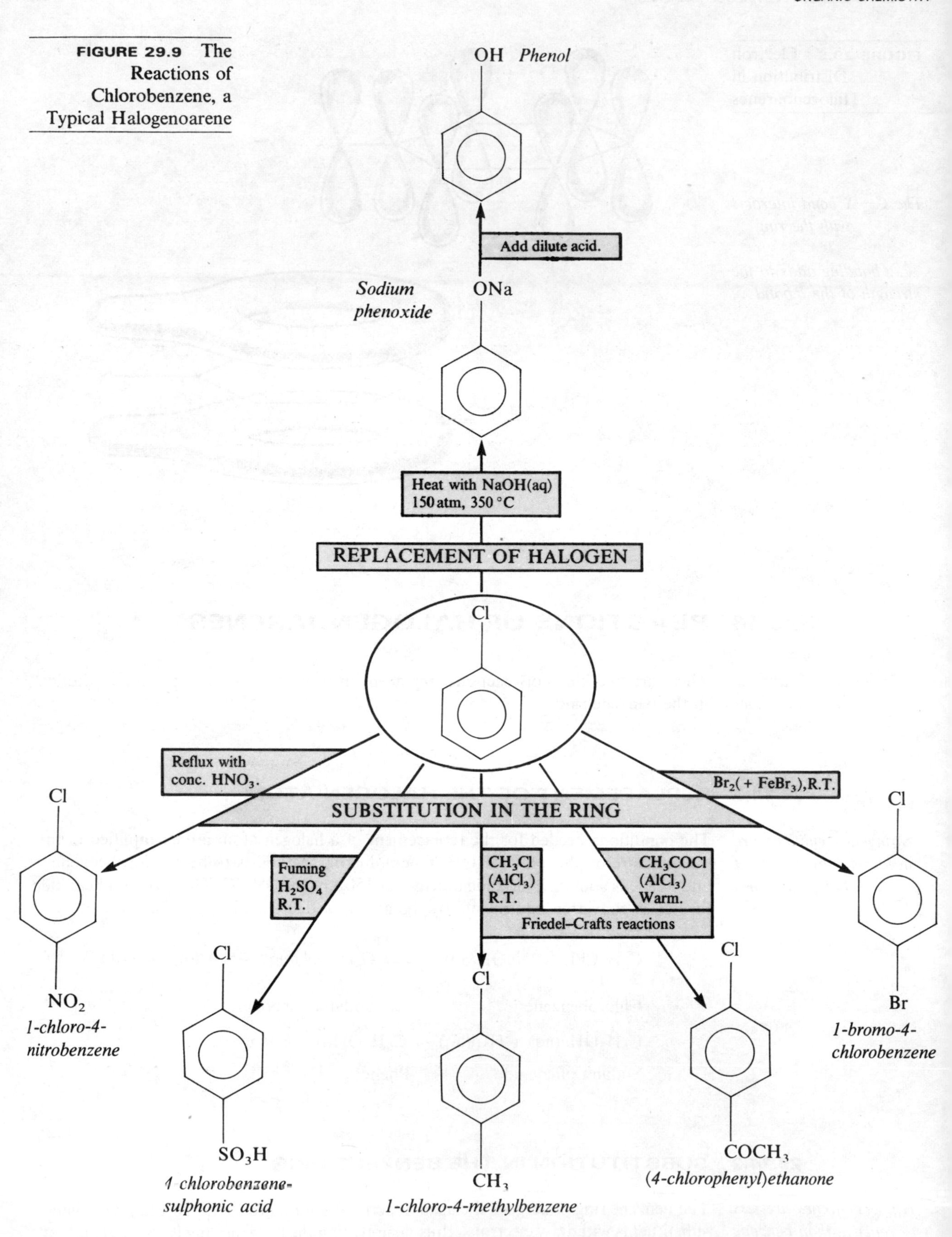

In each substitution reaction, a mixture of the 2- and 4-isomers is formed.

29.15 COMPARISON OF HALOGENOALKANES AND HALOGENOARENES

TABLE 29.1 Comparison of Halogenoalkanes and Halogenoarenes

	Halogenoalkanes, RX	*Halogenoarenes*, ArX
(1)	—OH replaces —X → Alcohol $RX + NaOH(aq) \rightarrow ROH + NaX$	—ONa replaces —X : NaOH(conc.) 150 atm, 400 °C → ArONa
(2)	—OR′ replaces X → Ether (Williamson synthesis) $RX + R'ONa \rightarrow ROR' + NaX$	No such reaction
(3)	R′ replaces X → Alkane (Wurtz reaction) $R'X + RX + 2Na \xrightarrow{\text{dry ether}} RR' + 2NaX$	R replaces X → Arene (Fittig reaction) $ArX + RI + 2Na \rightarrow ArR + NaX + NaI$
(4)	—CN replaces X → Nitrile $RX + NaCN \rightarrow RCN + NaX$	No comparable reactions
(5)	$—CO_2R'$ replaces X → Ester $RX + R'CO_2Ag \rightarrow R'CO_2R + AgX$	
(6)	$—NH_2$ replaces X → Amine $RX + NH_3(alc) \rightarrow RNH_3^+X^-$	
(7)	Conversion to alkenes $RCH_2CH_2X + KOH(alc) \rightarrow RCH{=}CH_2$	
		Substitution in the aromatic ring
(8)	No comparable reactions	*Nitration*: conc. HNO_3/H_2SO_4, R.T.
(9)		*Sulphonation*: fuming H_2SO_4, R.T.
(10)		*Halogenation*: X_2/FeX_3, R.T.
(11)		*Alkylation*: $RCl/AlCl_3$, R.T.
(12)		*Acylation*: $RCOCl/AlCl_3$, R.T.

CHECKPOINT 29E: HALOGENOARENES

1. How does chlorobenzene differ in reactivity from benzene? Give two examples of reactions of the two compounds which illustrate the difference in reactivity. What explanation can you give for the difference?

2. What examples can you find of compounds C_6H_5X which can be made by the route

$$C_6H_6 \rightarrow C_6H_5Cl \rightarrow C_6H_5X$$

and which cannot be made by the direct route

$$C_6H_6 \rightarrow C_6H_5X?$$

3. Compare the rates of reaction of aqueous alkali with C_6H_5Br, with C_3H_7Br and with $CH_3CH{=}CHBr$. Give a reason for the differences.

29.16 ANAESTHETICS

TOPIC

Trichloromethane (chloroform), $CHCl_3$, is a clear dense liquid with a pleasant smell and a boiling temperature of 61 °C. Its first use in medicine was as an inhalant for people suffering from asthma. In 1847, chloroform was used as a total anaesthetic by Dr Simpson in Edinburgh. Ethoxyethane (ether), $C_2H_5OC_2H_5$, was first used an as anaesthetic in dentistry in the same year. Dinitrogen oxide (laughing gas), N_2O, was another pioneering anaesthetic. These anaesthetics revolutionised surgery. Before total anaesthesia, a surgeon adopted the fastest possible procedure so as to minimise pain and shock to the patient. With the patient unconscious, a surgeon could undertake more complicated operations and explore new methods of surgery. The risk of the patients dying of shock was much reduced.

Trichloromethane (chloroform), dinitrogen oxide (laughing gas) and ethoxyethane (ether) were the first anaesthetics to be used. They revolutionised surgery

The night before surgery, the patient is usually given a tranquilliser to relax him and ensure a good night's sleep. During surgery, a combination of drugs is given to achieve anaesthesia, analgesia (relief of pain) and muscle relaxation. Following surgery, a drug may be given for the relief of severe pain. All these drugs are part of the chemical industry's contribution to medicine. Life is now longer, safer and freer from pain than in any previous century.

Ethoxyethane (ether), $C_2H_5OC_2H_5$, is a good anaesthetic and there is a big difference between the amount that causes unconsciousness and the lethal dose. It is given in a stream of oxygen containing 10–30% ether. The disadvantage of ether is that it is extremely flammable and precautions have to be taken over its use and storage, and surgeons cannot use electrical equipment which may cause a spark and ignite it.

Ethoxyethane is flammable, trichloromethane causes liver damage, and dinitrogen oxide is still regarded as an excellent anaesthetic. Fluothane® or Halothane® is now the most popular anaesthetic

Trichloromethane (chloroform), $CHCl_3$, is non-flammable, but it causes liver damage and there is a small difference between the dose which produces anaesthesia and the lethal dose. It is seldom used now.

Dinitrogen oxide (laughing gas), N_2O, is an excellent anaesthetic.

The majority of operations use 2-bromo-2-chloro-1,1,1-trifluoroethane, $CF_3CHBrCl$, which has the trade names Fluothane® and Halothane®. It was introduced by ICI in 1957. ICI was looking for a new anaesthetic which would satisfy these requirements:

1. It should work rapidly and smoothly.
2. It should not be unpleasant or irritating to breathe.
3. There should be a good safety margin between the anaesthetic dose and the lethal dose.
4. It should not affect the heart or other organs.
5. Recovery should not have unpleasant effects such as nausea.
6. It should be non-flammable and non-explosive.

An anaesthetic must satisfy medical requirements. Chemists have to find a substance with the right chemical properties to match these medical criteria

In order to meet these medical requirements, the research chemists at ICI drew up a list of chemical properties which a new anaesthetic must have:

1. It must be a volatile liquid, boiling at about 60 °C.
2. It must be chemically inert in the body and must not react with oxygen.
3. It must be stable to soda lime because during surgery the anaesthetic is conserved by passing the patient's exhaled breath over soda lime to remove carbon dioxide and water and then recycling it.
4. It must be non-flammable and non-explosive.
5. It must be potent.

The chemists realised that fluoroalkanes satisfy these requirements. They chose fluoroalkanes with:

1. the required boiling temperature
2. a CF_3 or CF_2 group to confer stability
3. no capability of eliminating HCl or HBr (stable to soda lime)
4. a low percentage of hydrogen (to reduce flammability)
5. a high saturation vapour pressure (for potency)

Fluoroalkanes satisfy the criteria for a good anaesthetic. The best one is $CF_3CHBrCl$, Fluothane® (or Halothane®)

Selected fluoroalkanes were tested on animals and the one which emerged as the best was 2-bromo-2-chloro-1,1,1-trifluoroethane. Clinical trials on humans showed that the new anaesthetic was excellent and it was marketed as Fluothane®.

For minor surgery, a local anaesthetic is sometimes used. Novocaine® and Xylocaine® may be applied to the surface before surgery on the eye, nose or throat.

Local aenaesthetics are Novocaine®, Xylocaine® and chloroethane

$$NH_2-C_6H_4-C(=O)-OCH_2CH_2N(CH_2CH_3)_2$$

Novocaine®

$$2,6-(CH_3)_2C_6H_3-NH-C(=O)-CH_2N(CH_2CH_3)_2$$

Xylocaine®

Chloroethane, C_2H_5Cl, is a gas at room temperature. With boiling temperature 12 °C, it can be liquefied under pressure in a sealed container. When the liquid is sprayed on to the surface of the skin, it evaporates so quickly and cools the surface so rapidly that it freezes tissues near the surface, making them insensitive to pain. It is used in minor surgery and in sporting injuries.

CHECKPOINT 29F: ANAESTHETICS

1. What are the medical requirements which a general anaesthetic must satisfy?

2. How do the chemical properties of Fluothane® enable it to meet these requirements?

3. What else had to be found out about Fluothane® before it could be used as an anaesthetic?

4. In what ways is Fluothane® an improvement on (*a*) chloroform and (*b*) ether?

5. What medical requirements must a local anaesthetic satisfy?

QUESTIONS ON CHAPTER 29

1. State the conditions needed for bromine to react with
(*a*) ethane to form bromoethane
(*b*) benzene to form bromobenzene
(*c*) benzene to form hexabromocyclohexane
(*d*) methylbenzene to form (bromomethyl)benzene
(*e*) methylbenzene to form 2-bromomethylbenzene

Compare the reactions of (i) silver nitrate solution and (ii) aqueous alkali with the products in (*a*), (*b*) and (*d*).

2. Describe a laboratory preparation for bromoethane. Write an equation for the reaction. State what impurities may be present, and how you would purify a sample of bromoethane made in this way.

3. (*a*) Describe how you could make 2-bromobutane from a named alcohol.

(*b*) Give the structures and names of the isomers of 2-bromobutane. Compare the rates at which they react with aqueous sodium hydroxide.

(*c*) State the conditions under which 2-bromobutane reacts with the following reagents. Name the products, and write equations for the reactions.

(i) OH^-(aq) (iii) KCN
(ii) KOH(ethanol) (iv) NH_3

4. What chemical tests could you do to distinguish between the members of the following pairs of compounds?

(*a*) $CH_3CH_2CH_2CH_2Cl$ and $CH_3CH_2CH{=}CHCl$

(*b*) (benzene ring)—Br and (cyclohexane ring with H and Br on the same carbon)

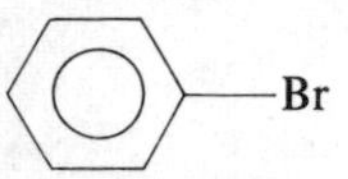

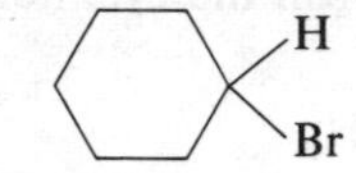

5. (*a*) Distinguish between *homolytic fission* and *heterolytic fission*, illustrating your answer by means of one example in each case.

(*b*) Describe the mechanism by which methane reacts with chlorine in the presence of light and show how a variety of chlorinated products can be formed.

(*c*) Discuss the mechanism by which 1-bromopropane reacts with warm aqueous sodium hydroxide.

(*d*) Show how the following compounds can be synthesised from 1-bromopropane. For each step in each synthesis give the reagent(s) and an equation.

(i) $CH_3CH_2CO_2H$
(ii) $CH_3CH_2CH_2CO_2H$

(NEAB 91, AS)

6. (*a*) Write overall balanced equations for the following reactions, and for each draw the structure of the organic intermediate:

(i) chlorine and methane (1 : 1 molar ratio), in the presence of ultraviolet light
(ii) bromine and ethene, in the dark
(iii) bromine and benzene, in the presence of $FeBr_3$

(*b*) (i) Give a systematic name for compound **E**.

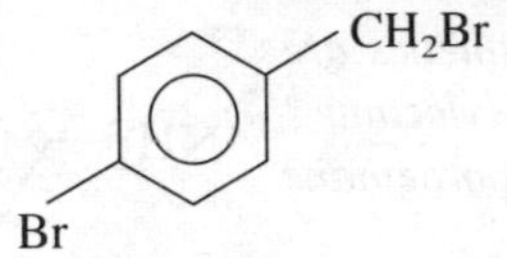

(benzene ring with CH_2Br and Br substituents in para positions)

E

(ii) Compound **E** (2.5 g) is heated under reflux with aqueous sodium hydroxide solution (1.0 mol dm^{-3}, 25 cm^3).

Calculate the volume of aqueous sulphuric acid (1.0 mol dm^{-3}) needed to neutralise the mixture after reaction, explaining your reasoning.

(O 92)

30

ALCOHOLS, PHENOLS AND ETHERS

30.1 ALCOHOLS

30.1.1 NOMENCLATURE

Aliphatic alcohols have the general formula $C_nH_{2n+1}OH$

Aliphatic alcohols are a homologous series. They all possess the same functional group, a hydroxyl group attached to a saturated carbon atom

$$\begin{array}{c}\diagdown \\ -C-OH \\ \diagup\end{array}$$

They all have the formula ROH, where R is an alkyl group

$$C_nH_{2n+1}-$$

They can be regarded as alkanes in which one H atom is replaced by a —OH group, and are often called the **alkanols**. The names are arrived at by adapting the name of the parent alkane by changing the terminal *-ane* to *-anol*. The first members of the series are:

Names

CH_3OH	Methanol	C_4H_9OH	Butanol
C_2H_5OH	Ethanol	$C_5H_{11}OH$	Pentanol
C_3H_7OH	Propanol	$C_6H_{13}OH$	Hexanol

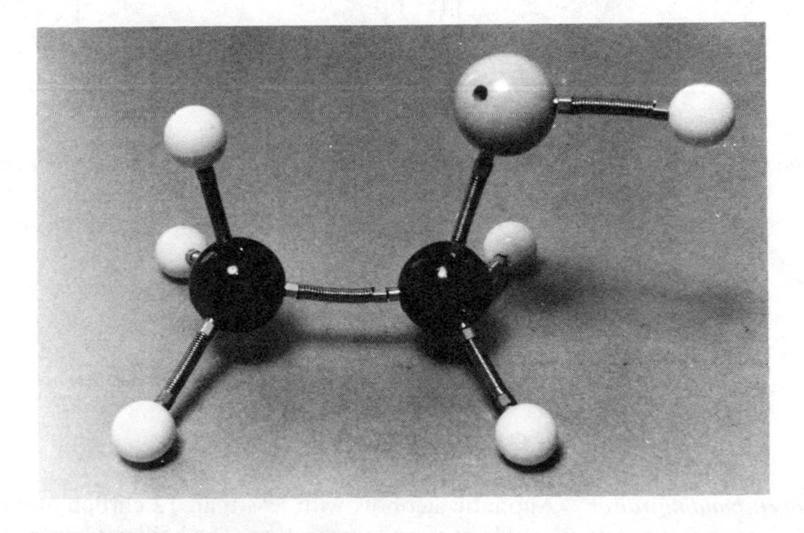

FIGURE 30.1 Ethanol

'Alcohol'

Ethanol is by far the most important member of the series [see Figure 30.1], and is often referred to simply as 'alcohol'.

30.1.2 ISOMERISM

Primary, secondary and tertiary alcohols

Isomerism occurs. Primary (1°), secondary (2°) and tertiary (3°) alcohols exist. Their formulae are:

$$\text{R(H)C(H)OH} \qquad \text{R(R')C(H)OH} \qquad \text{R(R')C(R'')OH}$$

Primary (1°) Secondary (2°) Tertiary (3°) alcohols

The three isomeric butanols, of formula C_4H_9OH, are:

$$CH_3CH_2CH_2CH_2OH \qquad CH_3CH_2CH(OH)CH_3 \qquad (CH_3)_3COH$$

Butan-1-ol (1°) Butan-2-ol (2°) 2-Methylpropan-2-ol (3°)

30.1.3 POLYHYDRIC ALCOHOLS

Polyhydric alcohols contain more than one —OH group. Diols contain two, and triols contain three hydroxyl groups:

$$CH_2(OH){-}CH_2(OH) \qquad CH_2(OH){-}CH(OH){-}CH_2(OH)$$

Ethane-1,2-diol Propane-1,2,3-triol (glycerol)

30.1.4 ARYL ALCOHOLS

Aromatic alcohols or aryl alcohols contain a benzene ring. The simplest are:

$$C_6H_5CH_2OH \qquad C_6H_5CH_2CH_2OH$$

Phenylmethanol (benzyl alcohol) 2-Phenylethanol

Aryl alcohols differ from phenols

The hydroxyl group is separated from the benzene ring by a saturated carbon atom and behaves as it does in aliphatic alcohols. If the hydroxyl group is in the ring, as in C_6H_5OH, the compound is not an alcohol: it is a **phenol**, and its reactions are quite different.

30.2 PHYSICAL PROPERTIES

30.2.1 VOLATILITY

Hydrogen bonding raises the boiling temperatures of alcohols...

Aliphatic alcohols with less than 12 carbon atoms and the lower aryl alcohols are liquids at room temperature. The boiling temperatures are higher than those of alkanes of comparable molecular mass (e.g., values of T_b/°C are: C_2H_5OH = 78° and C_3H_8 = −42°; $C_7H_{15}OH$ = 180° and C_8H_{18} = 126°). The reason is that the highly

FIGURE 30.2 Hydrogen Bonding between Alcohol Molecules

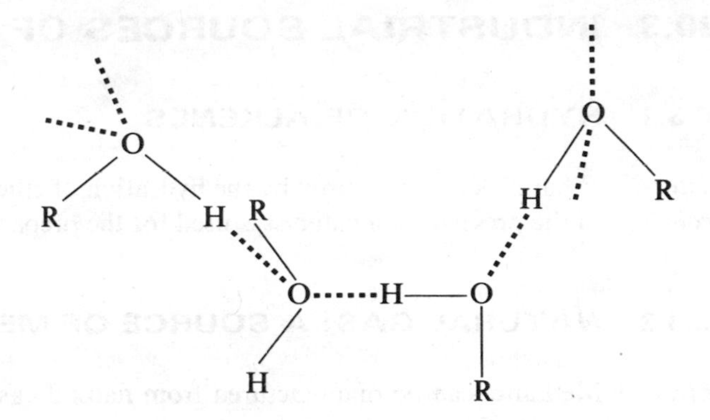

polar nature of the $-\overset{\delta-}{O}-\overset{\delta+}{H}$ bond leads to hydrogen bonding between molecules of alcohols [see Figure 30.2].

When the liquid is vaporised, energy must be supplied to break the hydrogen bonds and to convert the association of molecules into monomers in the vapour phase.

. . . as do an increasing molecular mass and reduced branching

The boiling temperatures increase with increasing molecular mass, and branched-chain isomers have lower boiling temperatures than unbranched-chain isomers.

30.2.2 SOLUBILITY

Lower alcohols are soluble in water. Molecules of alcohol can displace molecules of water in the hydrogen-bonded association of water molecules.

FIGURE 30.3 Hydrogen Bonding in Aqueous Ethanol

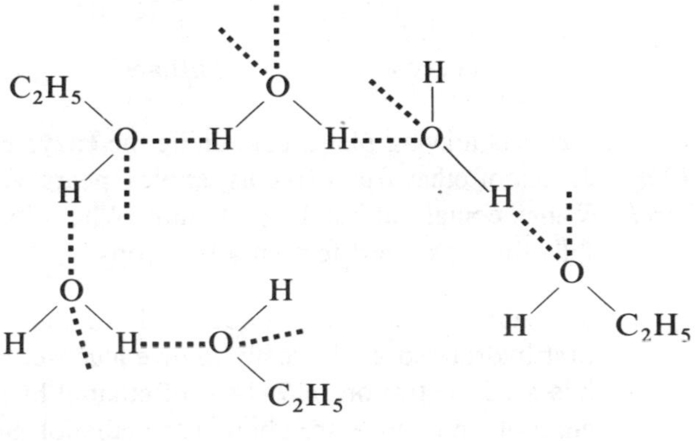

Alcohols of high molecular mass form fewer hydrogen bonds than the numerous water molecules they displace. Energy considerations therefore make the higher alcohols less soluble than the lower members.

An azeotrope is formed by ethanol and water

Ethanol forms a constant-boiling azeotropic mixture with water [§ 8.4.4]. It contains 95.6% ethanol and boils at 78.1 °C. Pure ethanol can be obtained by distilling the azeotrope from a drying agent such as calcium oxide. Other alcohols also form azeotropes with water.

Alcohols dissolve polar solutes and non-polar solutes

Alcohols are good solvents. The polar —OH group enables them to dissolve sodium hydroxide and potassium hydroxide. The non-polar hydrocarbon part of the molecule enables alcohols to dissolve substances like hexane.

30.3 INDUSTRIAL SOURCES OF ALCOHOLS

30.3.1 HYDRATION OF ALKENES

Catalytic hydration of alkenes gives alcohols

Ethanol is manufactured by the hydration of ethene [§ 27.7.7]. Hydration of alkenes in the presence of a catalyst is used for the preparation of other alcohols too.

30.3.2 NATURAL GAS: A SOURCE OF METHANOL

Methanol is obtained from synthesis gas

Methanol can be manufactured from natural gas. First, methane from natural gas is passed with steam over a catalyst to convert it into a mixture of carbon monoxide and hydrogen known as ***synthesis gas***:

$$CH_4(g) + H_2O(g) \xrightarrow[\text{Ni as catalyst}]{900\,°C,\ \text{high pressure}} CO(g) + 3H_2(g)$$

Then synthesis gas is passed over another catalyst at high pressure and moderate temperature to give methanol:

$$2H_2(g) + CO(g) \xrightarrow[Cr_2O_3\ +\ ZnO\ \text{as catalyst}]{400\,°C,\ \text{high pressure}} CH_3OH(l)$$

30.3.3 FERMENTATION: A METHOD USED FOR ETHANOL

Ethanol is made by the fermentation of sugars

The fermentation method is used to make alcoholic drinks. Fruit juices such as grape juice contain the sugar glucose, $C_6H_{12}O_6$. When yeast is added, the sugar 'ferments' to form wine (a solution of ethanol) and carbon dioxide:

$$\underset{\text{Glucose}}{C_6H_{12}O_6(aq)} \xrightarrow{\text{yeast}} \underset{\text{Ethanol}}{2C_2H_5OH(aq)} + 2CO_2(g)$$

Wines contain about 12% ethanol

Yeast is a living plant, containing the enzyme zymase, which catalyses the reaction. Juices of other fruits (plums, apples, pears, etc.) can be fermented to give wine. Wines contain about 12% ethanol. When the ethanol content reaches this level it kills the yeast, and fermentation stops.

Fractional distillation of fermented liquors gives 'spirits'

The starch present in potatoes and grain is also used as a source of ethanol. It is first hydrolysed enzymically to give glucose. Ethanol produced from fermentation has a concentration of 7–14% of ethanol by volume. Beer and cider are 7–8% ethanol, and wines are about 12% ethanol. Some people like drinks with a higher ethanol content. Whisky, gin and brandy have 40% ethanol. These 'spirits' are made by fractional distillation of fermented liquors [see Figure 8.10, § 8.4.2 for fractional distillation].

'Meths' is a mixture of methanol and ethanol

Ethanol is the only alcohol which people drink, and it is referred to as 'alcohol'. It is a toxic substance if taken in large quantities. Methanol is much more toxic than ethanol: drinking methanol leads to blindness and then to death. Methanol is added to ethanol to give 'industrial methylated spirit', which is unfit to drink and therefore carries tax at a lower rate than ethanol. For domestic sale, industrial methylated spirit has some unpleasant-tasting substance added to it and also a purple dye as a warning. This liquid is referred to as 'meths'. Desperate alcoholics can be reduced to drinking meths; the effects prove fatal. The higher alcohols are unpleasant-tasting liquids of moderate toxicity. See § 30.5 for more about 'alcohol'.

'Degrees proof' and the gunpowder test

An old-fashioned way of measuring the ethanol concentration of spirits is to quote the 'degrees proof' (° proof). The measurement was made by pouring the spirit over gunpowder. If the gunpowder would not burn, the spirit was 'under-proof': it contained too much water. The spirit that would just allow gunpowder to burn was called 100° proof spirit. It is 60% ethanol. Whisky is 70° proof, i.e., 42% ethanol.

30.4 LABORATORY METHODS OF PREPARATION

30.4.1 FROM HALOGENOALKANES

Alcohols can be prepared by hydrolysis of RX...

Hydrolysis of a halogenoalkane, RX, gives the alcohol ROH [§ 29.6].

30.4.2 FROM ESTERS

... and of esters, RCO_2R' ...

When an ester, RCO_2R', is refluxed with dilute acid or alkali [§ 33.13.1], a carboxylic acid or its salt and an alcohol are formed:

$$\underset{\text{Ethyl ethanoate}}{CH_3CO_2C_2H_5(l)} + NaOH(aq) \rightarrow \underset{\text{Sodium ethanoate}}{CH_3CO_2Na(aq)} + \underset{\text{Ethanol}}{C_2H_5OH(l)}$$

30.4.3 HYDRATION OF ALKENES

... by the hydration of alkenes...

The alkene is bubbled into concentrated sulphuric acid to form an alkyl hydrogensulphate. When this is diluted with water and distilled, an alcohol is obtained:

$$CH_2{=}CH_2(g) + H_2SO_4(l) \rightarrow CH_3CH_2HSO_4(l)$$

$$\underset{\text{Ethyl hydrogensulphate}}{CH_3CH_2HSO_4(l)} + H_2O(l) \rightarrow \underset{\text{Ethanol}}{CH_3CH_2OH(l)} + H_2SO_4(l)$$

30.4.4 REDUCTION OF ALDEHYDES, KETONES AND CARBOXYLIC ACIDS

... by reduction of carbonyl compounds

The reduction of aldehydes and ketones is covered in § 31.6.1. Carboxylic acids, acid chlorides and anhydrides and esters can be reduced, as described in §§ 33.8.6, 33.11 and 33.13.

30.4.5 ARYL ALCOHOLS

The Cannizzaro reaction gives aryl alcohols

Aryl alcohols can be made from aryl aldehydes by the Cannizzaro reaction [§ 31.6.9].

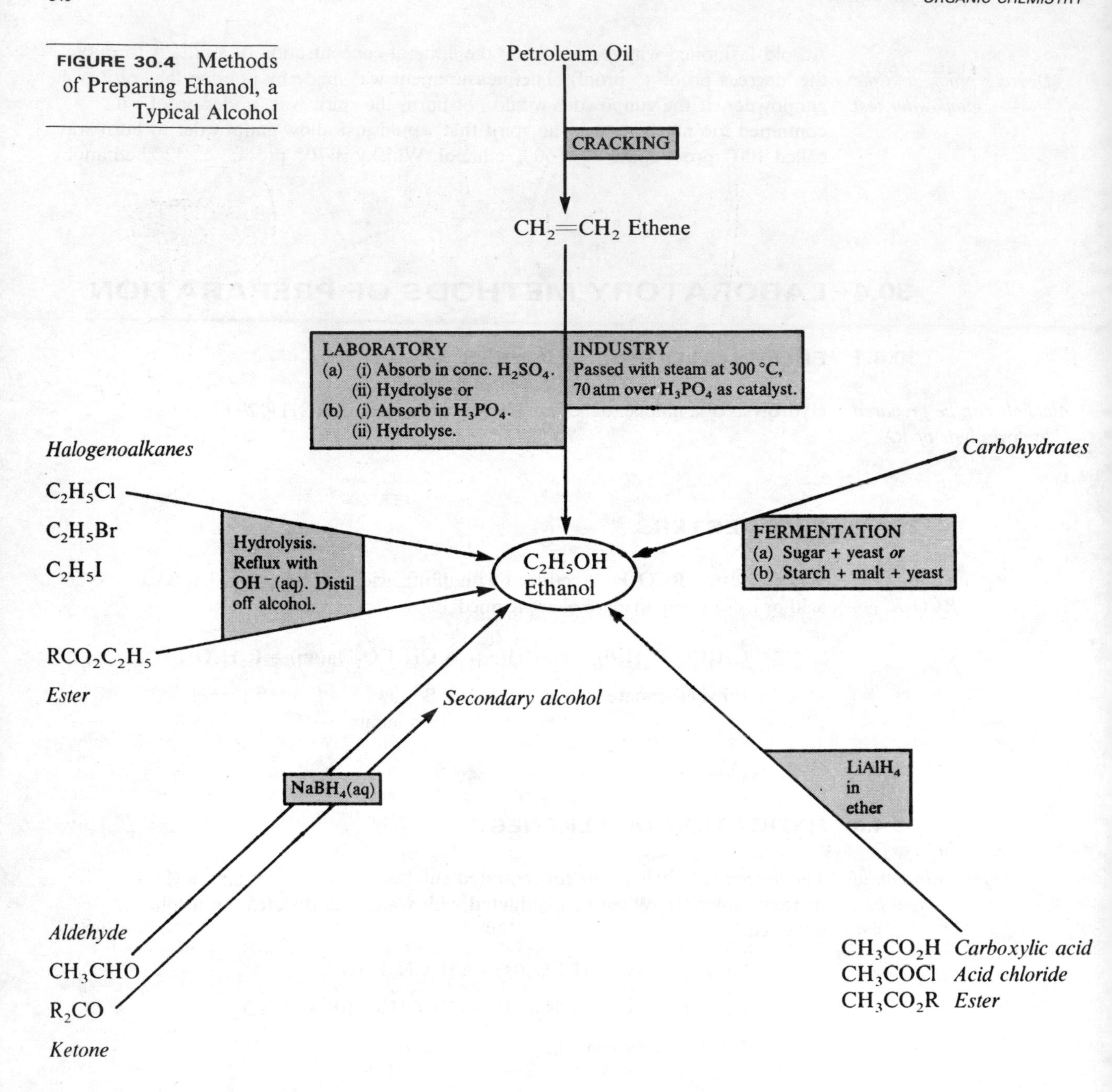

FIGURE 30.4 Methods of Preparing Ethanol, a Typical Alcohol

30.5 ALCOHOL

TOPIC

Ethanol, C_2H_5OH, commonly referred to as 'alcohol', is the most commonly used drug used legally by adults and illegally by young people. It is a depressant of the central nervour system. Addiction to ethanol is the greatest medical problem resulting from the use of drugs: there are 30 000 alcoholics in the UK alone. Alcohol addiction develops slowly and is a disease of middle age. Addiction to heroin and other drugs develops quickly and is a disease of young people.

Ethanol is a depressant of the CNS. It is an addictive drug

Ethanol does not have to be digested: it can be absorbed into the blood stream through the walls of the stomach and intestine. Carbon dioxide relaxes the pyloric valve, between the stomach and the small intestine, and so sparkling wines have a faster effect than still wines.

FIGURE 30.5
One 'Unit' of Alcohol

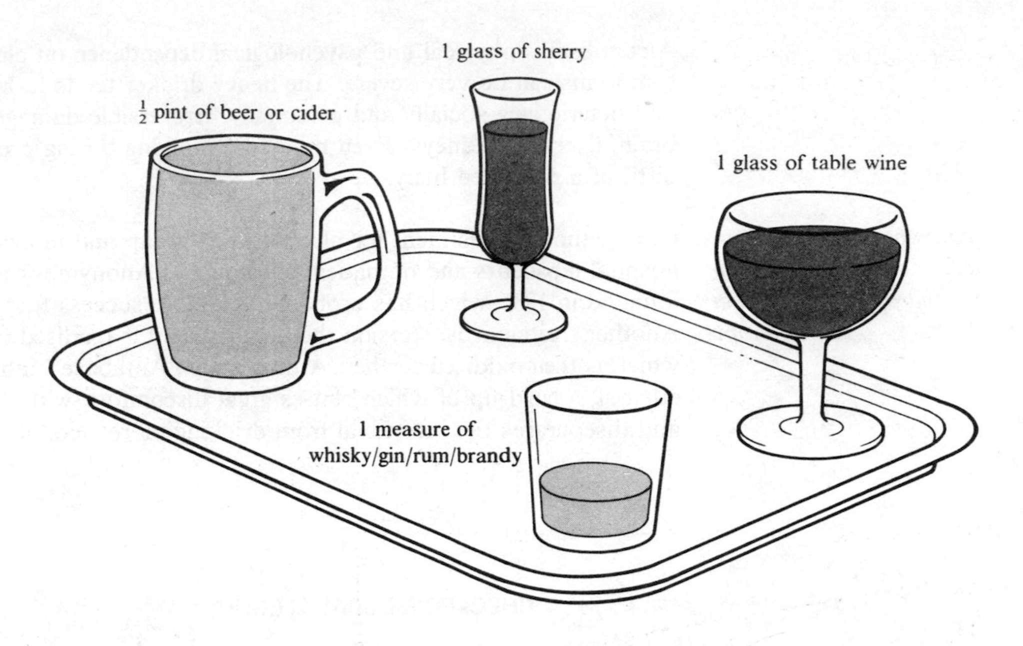

Heavy drinking damages many parts of the body. Drinking during pregnancy is a hazard to the unborn baby

Five 'units' of alcohol will take a man of average weight up to the legal limit for driving. Women and lighter men can tolerate less alcohol. It takes many hours to eliminate alcohol from the blood

For convenience, the amount of ethanol in drinks is measured in 'units'. The number of units in some common drinks is shown in Figure 30.5. The effects of a given amount of ethanol depend on the rate of drinking, the size of the drinker and other factors. Some people are visibly affected after 2 units of alcohol; others can drink 5 units over a period of 2–3 hours before they appear drunk. In an average man, at blood alcohol levels of up to 0.05% by volume, ethanol produces a sense of well being and some impairment of coordination. A level of 0.08% is the legal limit for driving. A man of average weight can drink 5 units of alcohol before he reaches this level. A lighter man could drink less, and a woman of the same weight could drink less before reaching this level. Above 0.08% accidents can result from faulty judgement and coordination and increased aggressiveness. At 0.15% drowsiness and vomiting occur, and at 0.30% breathing difficulties may cause death. People usually become unconscious before they can take a lethal dose. Figure 30.6 shows how long it takes for the blood alcohol level to drop from the legal limit for driving to zero.

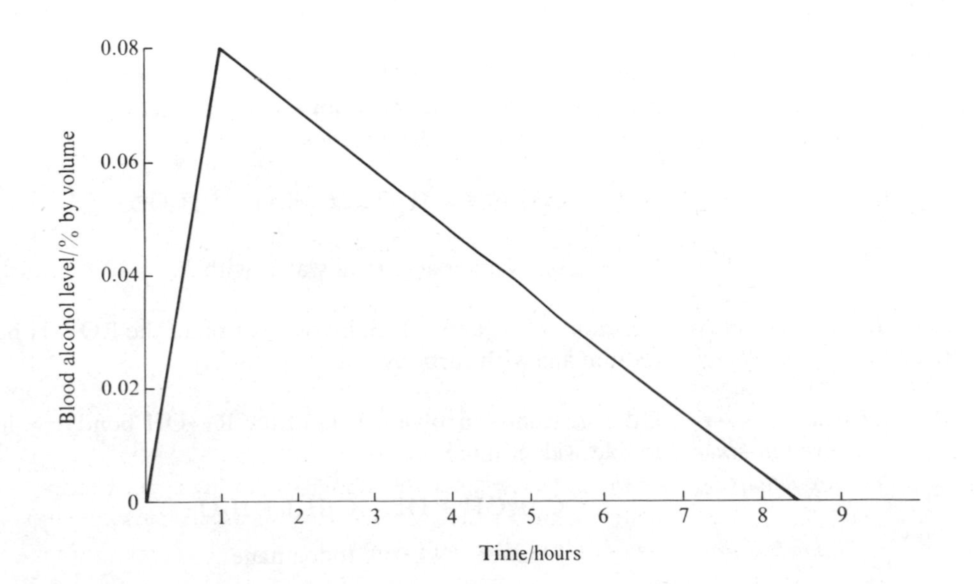

FIGURE 30.6
How Long it Takes for Alcohol to be Eliminated from the Blood

Alcoholism is physical and psychological dependence on alcohol. Withdrawal symptoms can be very severe. The heavy drinker tends to behave in antisocial ways and deteriorates socially and personally. Irreversible damage is caused to the heart, brain, liver and kidneys. Even moderate drinking during pregnancy may result in the birth of a deformed baby.

Treatments for alcoholism include counselling, group therapy and aversion therapy

One method of treatment for alcoholism is group and individual psychotherapy for hospital inpatients and outpatients. Alcoholics Anonymous is a self-help organisation founded in 1935 which has probably had more success than any other treatment. Another treatment is aversion therapy. Ethanol is oxidised in the liver to ethanal, which is then oxidised further. A drug called Antabuse® inhibits the oxidation of ethanal, a build-up of which causes great discomfort, with sweating and nausea, and discourages the individual from drinking more alcohol.

CHECKPOINT 30A: ALCOHOL

1. It has been suggested that people drink alcohol to relax, promote sleep, relieve social and physical discomforts and tension, quench thirst, increase appetite, satisfy curiosity, relieve boredom, gain courage, escape personal responsibility and imitate parents or peers.

(*a*) What alternatives to alcohol can you suggest for bringing each of the results listed, i.e. relaxation, sleep, etc.

(*b*) Write a questionnaire and conduct a survey to find out as much as you can about people's reasons for drinking alcohol. From your survey, see whether you can reach any conclusions about the attitudes to drinking of (i) your age group and (ii) your parents' age group.

30.6 REACTIVITY OF ALCOHOLS

Alcohols are weak acids with very low values of K_a

In alcohols, the —O—H bond is polarised as

$$\overset{\delta-}{—\text{O}}—\overset{\delta+}{\text{H}}$$

There is a tendency for H^+ to dissociate in the presence of a base [§ 12.7.1]. An example of ethanol acting as an acid is its slight dissociation in water to form ethoxide ions and hydrogen ions:

$$C_2H_5OH + H_2O \rightleftharpoons C_2H_5O^- + H_3O^+$$

It is an even weaker acid than water, with $K_a = 10^{-16}\ \text{mol dm}^{-3}$ at 25 °C [§ 12.7.4].

RO—H bond fission is involved in some reactions

Reactions of ethanol which involve fission of the RO—H bond are the reactions with sodium and with carboxylic acids [§ 30.7.1].

R—OH bond fission is involved in other reactions...

Other reactions involve fission of the R—OH bond, e.g. in the reaction of hydrogen iodide with ethanol:

$$C_2H_5OH + HI \rightarrow C_2H_5I + H_2O$$

Ethanol Iodoethane

30.7 REACTIONS OF ALCOHOLS

30.7.1 FISSION OF RO—H

REACTION WITH SODIUM

Alcohols react with sodium

Alcohols react with sodium to give hydrogen and a sodium alkoxide. The reaction is much slower than the reaction of water and sodium:

$$2C_2H_5OH(l) + 2Na(s) \rightarrow 2C_2H_5O^-Na^+(s) + H_2(g)$$

Ethanol Sodium ethoxide

ESTERIFICATION

They react with carboxylic acids to give esters

Alcohols and carboxylic acids react to give esters. The functional groups of acids and esters are (where R is an alkyl group)

Carboxylic acid Carboxylic ester

Esterification takes place much faster in the presence of a catalyst, such as hydrogen chloride or concentrated sulphuric acid. Acid anhydrides and acid chlorides also react with alcohols to give esters:

$$CH_3CO_2H(l) + C_2H_5OH(l) \xrightleftharpoons[\text{conc. } H_2SO_4]{HCl(g) \text{ or}} CH_3CO_2C_2H_5(l) + H_2O(l)$$

Ethanoic acid Ethanol Ethyl ethanoate

30.7.2 FISSION OF R—OH BOND

HALOGENATION

(a) Chlorination

Dry HCl(g) or conc. HCl replaces OH by Cl

Dry hydrogen chloride is bubbled through the anhydrous alcohol in the presence of anhydrous zinc chloride as catalyst. When the solution is saturated, it is refluxed on a water bath. For secondary and tertiary alcohols concentrated hydrochloric acid can be used. For tertiary alcohols no zinc chloride catalyst is needed:

$$C_2H_5OH(l) + HCl(g) \xrightarrow[ZnCl_2 \text{ catalyst}]{\text{reflux}} C_2H_5Cl(l) + H_2O(l)$$

The order of rates of reaction is

tertiary > secondary > primary alcohol

(b) Iodination

Red P + I_2 replace OH by I

Red phosphorus and iodine are used as a source of hydrogen iodide (concentrated sulphuric acid cannot be used because it oxidises HI to I_2):

$$C_2H_5OH(l) + HI(g) \xrightarrow{\text{distil}} C_2H_5I(l) + H_2O(l)$$

Ethanol — Iodoethane

(c) Halogenation using Phosphorus Halides

RBr and RI are formed by reaction with PBr_3 and PI_3

If it is available, phosphorus(III) bromide is used. Alternatively, red phosphorus and bromine are used to generate PBr_3 in the reaction mixture. Phosphorus(III) iodide is generated by the addition of iodine to red phosphorus and the alcohol. The mixture is refluxed on a water bath, and then the halogenoalkane is distilled over:

$$3C_2H_5OH(l) + PBr_3(l) \xrightarrow[\text{distil}]{\text{reflux}} 3C_2H_5Br(l) + H_3PO_3(l)$$

Ethanol — Bromoethane — Phosphonic acid

PCl_5 replaces OH by Cl

Phosphorus(V) chloride reacts in the cold with alcohols. Phosphorus(III) chloride cannot be used:

$$C_2H_5OH(l) + PCl_5(l) \xrightarrow{\text{room temperature}} C_2H_5Cl(l) + POCl_3(l) + HCl(g)$$

Ethanol — Chloroethane — Phosphorus trichloride oxide

A test for the —OH group

The formation of hydrogen chloride on reaction with phosphorus(V) chloride is a test for the presence of a hydroxyl group in a compound.

(d) Chlorination with Sulphur Dichloride Oxide

$SOCl_2$ is a convenient chlorinating agent

A halogenoalkane is formed when an alcohol is refluxed with sulphur dichloride oxide, $SOCl_2$, in the presence of pyridine. This organic base assists the reaction by absorbing the hydrogen chloride as it is formed:

$$C_6H_5CH_2OH(l) + SOCl_2(l) \xrightarrow[\text{pyridine}]{\text{reflux with}} C_6H_5CH_2Cl(l) + HCl(g) + SO_2(g)$$

The convenient feature of this method is that the by-products, being gases, do not contaminate the product.

DEHYDRATION

Reaction with conc. H_2SO_4...

A primary alcohol reacts with cold concentrated sulphuric acid to form an alkyl hydrogensulphate:

$$C_2H_5OH(l) + (HO)(OH)SO_2(l) \rightarrow (C_2H_5O)(OH)SO_2(l) + H_2O(l)$$

Ethanol — Sulphuric acid — Ethyl hydrogensulphate

...results in dehydration to give an ether...

If an excess of alcohol is used and the reaction mixture is warmed to 140 °C, an ether is formed:

(excess alcohol, <170 °C)

$$C_2H_5OH(l) + C_2H_5HSO_4 \xrightarrow[C_2H_5OH \text{ (in excess)}]{140\,^{\circ}C} C_2H_5OC_2H_5(g) + H_2SO_4(l)$$

Ethoxyethane (or diethyl ether)

FIGURE 30.7
Dehydration of Ethanol to Ethene

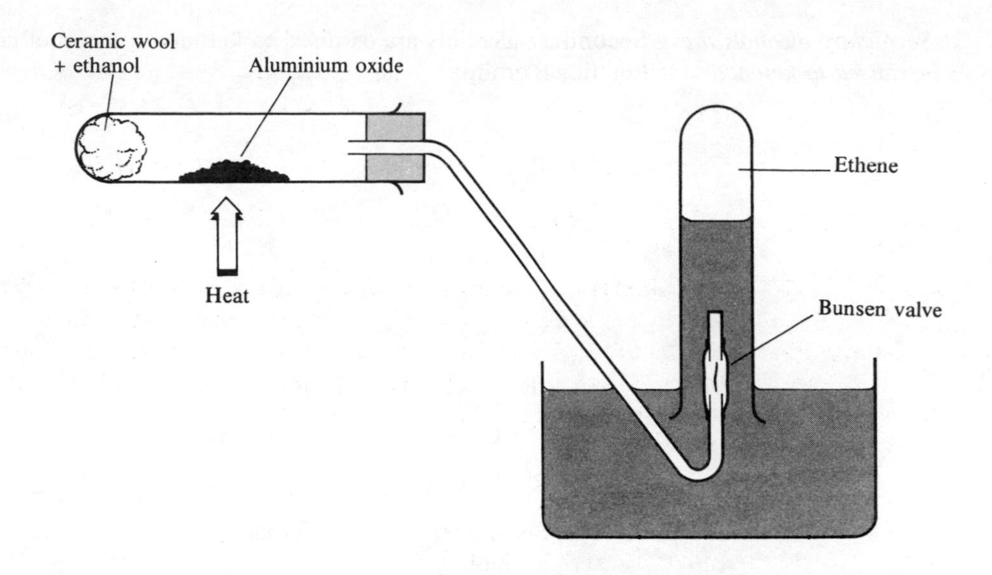

Ethers possess the group

$$\geq C—O—C \leq$$

... or to give an alkene (excess conc H_2SO_4, >170 °C)

If an excess of concentrated sulphuric acid is used, and the temperature is raised to 170 °C, water is eliminated, with the formation of an alkene:

$$C_2H_5HSO_4(l) \xrightarrow[\text{conc. } H_2SO_4 \text{ (in excess)}]{170\ °C} C_2H_4(g) + H_2SO_4(l)$$

Ethyl hydrogensulphate — Ethene

Other dehydrating agents are Al_2O_3 and H_3PO_4

Phosphoric(V) acid and aluminium oxide at 300 °C [see Figure 30.7] also act as dehydrating agents.

OXIDATION

Primary alcohols are oxidised to aldehydes and to acids...

Primary alcohols are oxidised to **aldehydes**, a series of compounds which have the functional group

$$R—C(H)=O$$

They can be further oxidised to carboxylic acids:

$$R—CH_2OH \xrightarrow{[-2H]} R—C(H)=O \xrightarrow{[+O]} R—C(O—H)=O$$

Primary alcohol — Aldehyde — Carboxylic acid

... Secondary alcohols are oxidised to ketones...

Secondary alcohols are oxidised to ketones, a series of compounds possessing the functional group

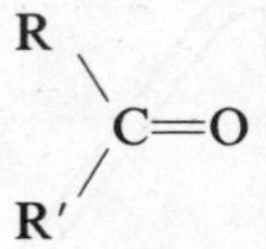

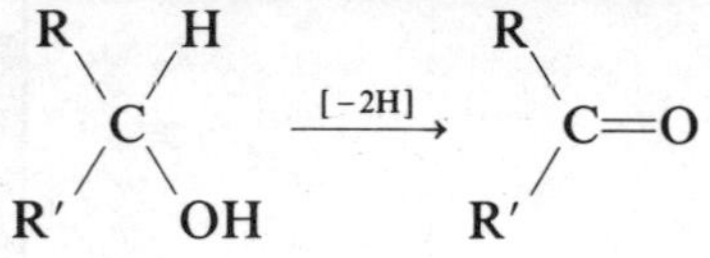

$$\begin{matrix} R \\ & \diagdown \\ & & C{=}O \\ & \diagup \\ R' \end{matrix}$$

$$\underset{\text{Secondary alcohol}}{R(R')C(H)OH} \xrightarrow{[-2H]} \underset{\text{Ketone}}{R(R')C{=}O}$$

... Tertiary alcohols resist oxidation

Tertiary alcohols are resistant to oxidation. A powerful acidic oxidising agent converts them into a mixture of carboxylic acids.

A number of oxidising agents can be used. Acidified sodium dichromate solution at room temperature will oxidise primary alcohols to aldehydes and secondary alcohols to ketones. At higher temperatures primary alcohols are oxidised further to acids:

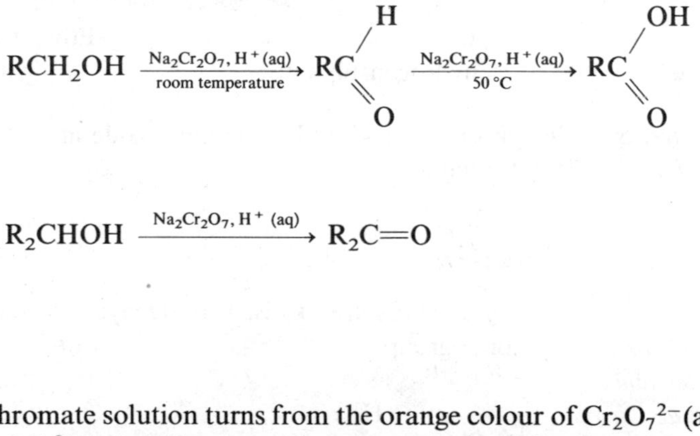

$$RCH_2OH \xrightarrow[\text{room temperature}]{Na_2Cr_2O_7,\ H^+(aq)} RC(H){=}O \xrightarrow[50\ °C]{Na_2Cr_2O_7,\ H^+(aq)} RC(OH){=}O$$

$$R_2CHOH \xrightarrow{Na_2Cr_2O_7,\ H^+(aq)} R_2C{=}O$$

Acid + dichromate were used in a breathalyser test

The dichromate solution turns from the orange colour of $Cr_2O_7^{2-}(aq)$ to the blue colour of $Cr^{3+}(aq)$. This colour change was the basis for the original 'breathalyser test'. The police could ask a motorist to exhale through a tube containing some orange crystals. If the crystals turned blue, it showed that the breath contained a considerable amount of ethanol vapour.

Acid + $KMnO_4$...

Acidified potassium manganate(VII) solution. This is too powerful an oxidising agent to stop at the aldehyde: it oxidises primary alcohols to acids. It oxidises secondary alcohols to ketones.

All secondary alcohols and one primary alcohol, ethanol, give a positive result in the haloform reaction [§ 31.6.7].

The reactions of alcohols are summarised in Figure 30.8 and Table 30.1.

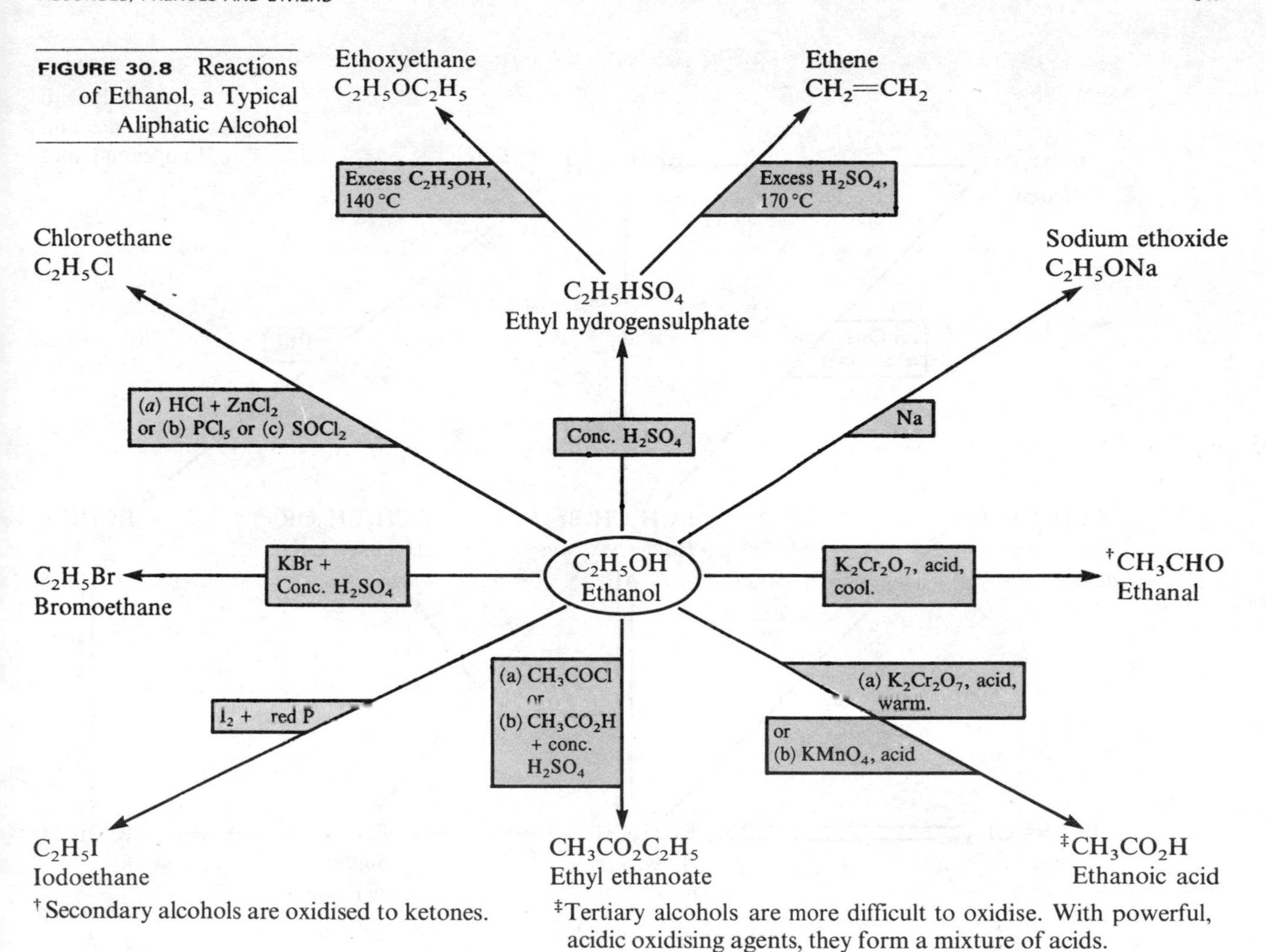

FIGURE 30.8 Reactions of Ethanol, a Typical Aliphatic Alcohol

† Secondary alcohols are oxidised to ketones.

‡ Tertiary alcohols are more difficult to oxidise. With powerful, acidic oxidising agents, they form a mixture of acids.

TABLE 30.1 Methods of Distinguishing between Primary, Secondary and Tertiary Alcohols

Reagent	*Primary alcohol*	*Secondary alcohol*	*Tertiary alcohol*
Acidified $K_2Cr_2O_7$ (orange)	Aldehyde, RCHO formed (blue Cr^{3+}(aq) formed)	Ketone, R_2CO formed (blue Cr^{3+}(aq) formed)	Resists oxidation
Conc. H_2SO_4	Alkene formed slowly	Intermediate in speed	Alkene formed fast
Conc. HCl + $ZnCl_2$. Add to alcohol and place in boiling water bath.	Cloudiness due to formation of RCl is slow to appear. (Anhydrous conditions are needed for primary alcohols.)	Cloudiness appears in 5 minutes	Cloudiness appears in 1 minute owing to the formation of RCl, which is insoluble in water.

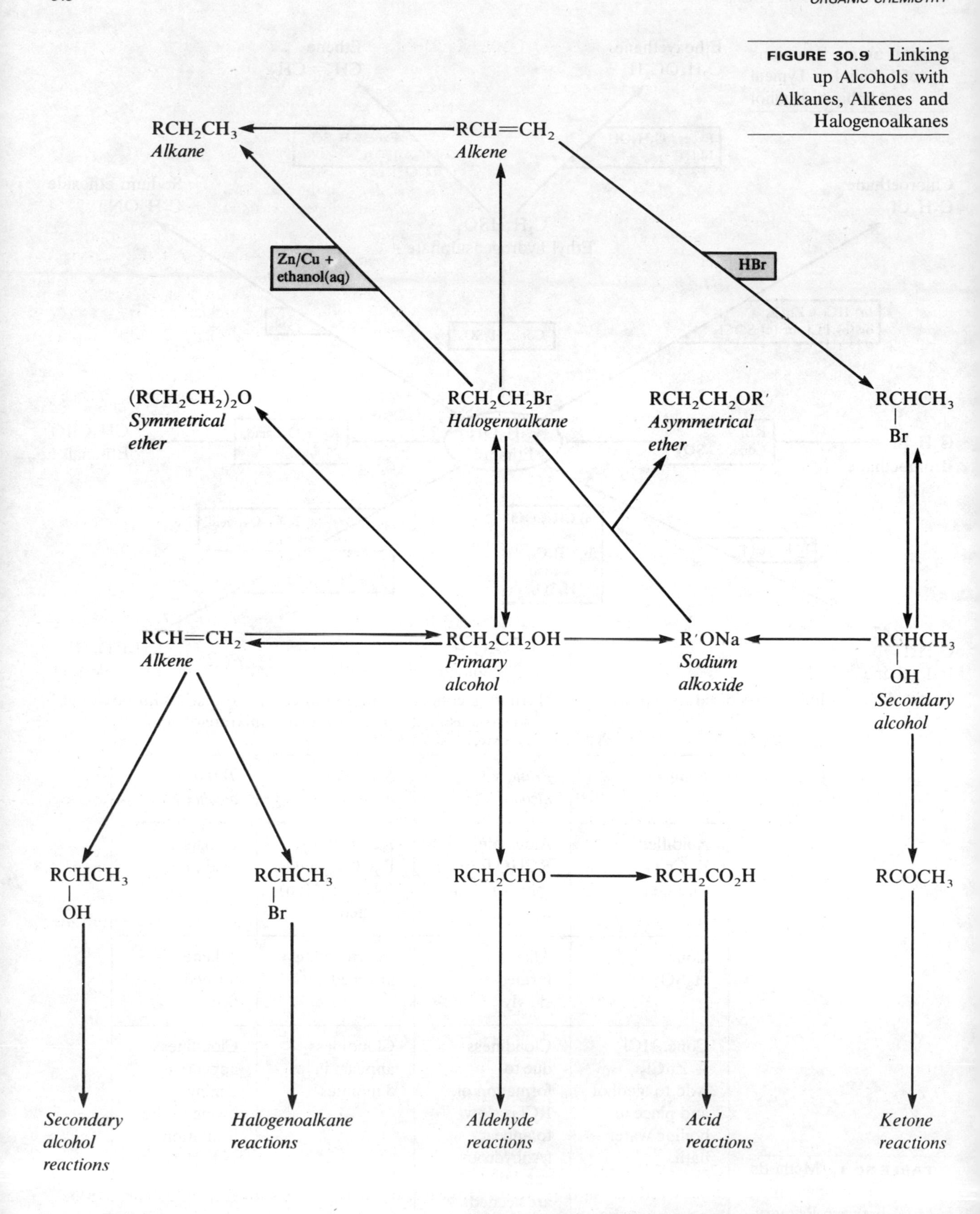

FIGURE 30.9 Linking up Alcohols with Alkanes, Alkenes and Halogenoalkanes

Note that the group —CH_2CH_2OH can be converted into a number of different functional groups. All these groups have characteristic reactions, to be studied in later chapters. By means of these reactions it is possible to make a large number of compounds in two stages starting from alcohols.

CHECKPOINT 30B: ALCOHOLS

1. Name the reagents and describe the conditions needed to convert C_2H_5OH into

(*a*) C_2H_4
(*b*) C_2H_5I
(*c*) C_2H_5Br
(*d*) C_2H_5Cl
(*e*) $C_2H_5OCOCH_3$
(*f*) $C_2H_5OC_2H_5$

2. Name and give the structural formulae of the products formed when propan-1-ol reacts with

(*a*) Na
(*b*) PCl_5
(*c*) conc. $H_2SO_4 < 170\,°C$
(*d*) conc. $H_2SO_4 > 170\,°C$
(*e*) CH_3CO_2H

3. **A** $CH_3CH_2CH_2CH_2OH$
B $CH_3CH_2CH(OH)CH_3$
C $CH_3CH(CH_3)CH_2OH$
D $(CH_3)_3COH$

Name these four alcohols, and explain how you could distinguish between

(*a*) **A** and **B**
(*b*) **B** and **D**
(*c*) **C** and **D**

4. Explain how you could synthesise from propan-1-ol

(*a*) propan-2-ol
(*b*) 1,2-dichloropropane
(*c*) propane-1,2-diol

5. Outline a way in which you could obtain propanonitrile, C_2H_5CN, starting from ethanol.

6. Give the names and structural formulae of a primary, a secondary and a tertiary alcohol, and explain how you could distinguish between them.

7. Compare the reactivity of (*a*) C_2H_5OH and H_2O, (*b*) C_2H_5OH and C_2H_5Cl, and (*c*) C_2H_5OH and $C_2H_5OC_2H_5$. Explain the reasons why the members of each pair of compounds react differently.

8. Compounds **A** and **B** are colourless liquids with the formula $C_4H_{10}O$. **A** reacts with sodium, with the formation of hydrogen. It reacts with conc. HI to form **C**, C_4H_9I, and with conc. H_2SO_4 to form **D**, C_4H_8. **B** does not react with sodium, but reacts with conc. HI to form **E**, C_2H_5I. Identify **A**, **B**, **C**, **D** and **E**, and write equations for all the reactions.

9. Identify the intermediate compound in each of the following syntheses. State the reagents and the conditions which are needed for each step in the synthesis:

$$C_2H_5OH \rightarrow \textbf{A} \rightarrow C_2H_5CN$$
$$C_2H_5OH \rightarrow \textbf{B} \rightarrow C_2H_5NH_2$$
$$CH_3CH(Br)CH_3 \rightarrow \textbf{C} \rightarrow CH_3COCH_3$$
$$CH_3CH{=}CH_2 \rightarrow \textbf{D} \rightarrow CH_3CH(OH)CH_3$$

30.8 POLYHYDRIC ALCOHOLS

Ethane-1,2-diol is used as antifreeze

Ethane-1,2-diol, $HOCH_2CH_2OH$, is used in car radiators as antifreeze and as a de-icing fluid on aeroplane wings. It is often called by its old name of ethylene glycol. Its reactions are similar to those of monohydric alcohols.

Glycerol is a triol

Propane-1,2,3-triol is usually known by its old name of glycerol. It is a by-product in the manufacture of soap [§ 33.13.3]. Glycerol is used for the manufacture of the ester which it forms with nitric acid, propane-1,2,3-triyl trinitrate or glyceryl trinitrate:

$$\begin{array}{l} CH_2OH \\ | \\ CHOH \\ | \\ CH_2OH \end{array} + 3HNO_3 \rightarrow \begin{array}{l} CH_2ONO_2 \\ | \\ CHONO_2 \\ | \\ CH_2ONO_2 \end{array}$$

Propane-1,2,3-triol (*Glycerol*) — Propane-1,2,3-triyl trinitrate (*Glyceryl trinitrate*)

Nitroglycerine is the major component of dynamite

This ester is also called *nitroglycerine*. It is an explosive which is detonated by shock. When nitroglycerine is absorbed on kieselguhr (a type of clay), it is called *dynamite*. It is still a powerful explosive, but it is less sensitive to shock and can be handled with safety. Dynamite was invented by the Swedish chemist, A B Nobel, who invested some of the rewards for his invention to endow the Nobel prizes for great achievements in science and the arts.

30.9 ARYL ALCOHOLS

Benzyl alcohol is an aryl alcohol

The reactions of phenylmethanol (benzyl alcohol) are very similar to those of primary aliphatic alcohols. (See Figure 30.10.)

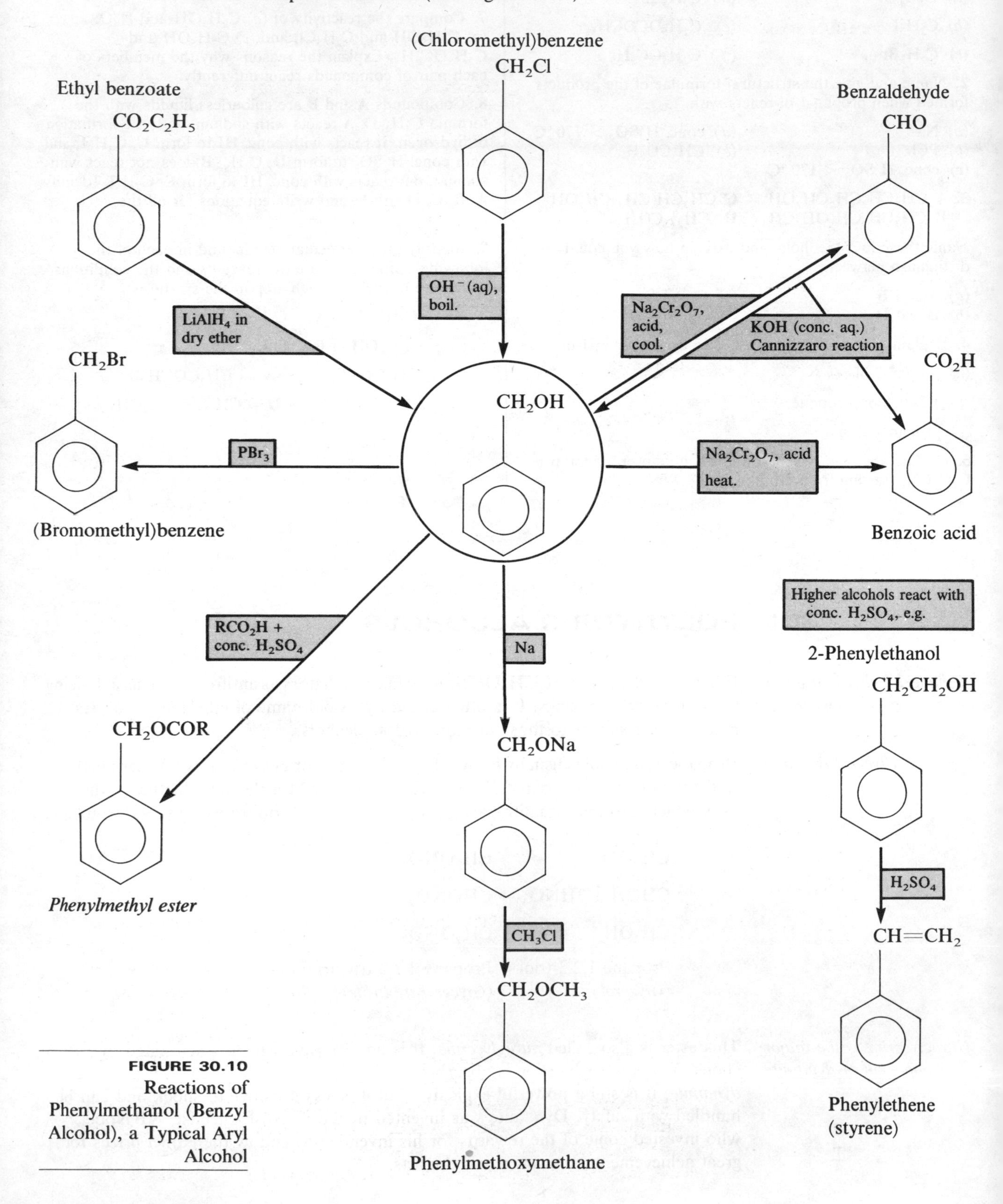

FIGURE 30.10 Reactions of Phenylmethanol (Benzyl Alcohol), a Typical Aryl Alcohol

CHECKPOINT 30C: ARYL ALCOHOLS

1. How would you distinguish between the following?

C_6H_5–CH_2Cl and CH_3–C_6H_4–Cl

2. A liquid **P**, of formula $C_8H_{10}O$, reacted with sodium to give hydrogen. With concentrated sulphuric acid, **P** gave a liquid, **Q**, which was separated by distillation. When a trace of benzoyl peroxide (a source of free radicals) was added, **Q**, with a molar mass of 104 g mol^{-1}, was converted into a solid foam, **R**, with a very high molar mass. Identify **P**, **Q** and **R**, and explain the reactions which have taken place, giving equations.

3. In the synthetic route

Benzene → **A** → **B** → **C** → **D** → $C_6H_5CH_2OCH_3$

identify **A**, **B**, **C** and **D**. State the reagents and conditions needed for each step in the route.

4. Identify **E** and **F** in the synthetic pathway shown, and state how each step may be carried out:

$C_6H_5CH_2Cl$ → **E** → **F** → $C_6H_5CH_2O_2CC_6H_5$

30.10 PHENOLS

30.10.1 FUNCTIONAL GROUP

Phenols

Phenols are compounds containing a hydroxyl group attached to an aromatic ring. Some members of the group are listed below:

OH (phenol ring) — The systematic name is benzenol; always called phenol.

OH, CH_3 (ring) — 2-Hydroxymethylbenzene or 2-Methylphenol

OH, Cl (ring) — 4-Chlorophenol

OH, NO_2 (ring) — 4-Nitrophenol

OH — Naphthalen-1-ol

OH — Naphthalen-2-ol

30.10.2 PHYSICAL PROPERTIES

Phenol + water = the antiseptic, 'carbolic acid'

The vapour is toxic. Solid or liquid phenol burns the skin

Most phenols are colourless solids. Phenol itself melts at 42 °C. The presence of water lowers its melting temperature, and a mixture of phenol and water is a liquid at room temperature. It is called 'carbolic acid', and is a powerful antiseptic. A century ago, a surgeon called J Lister began to wonder why patients were dying some time after operations when the initial surgery appeared to have been satisfactory. He realised that their wounds were becoming infected both during the operation and afterwards in the ward. By spraying phenol mist around the operating theatre and the wards, he was able to reduce the death rate among his patients. Phenol is not a convenient antiseptic as the vapour is toxic, and if the solid or liquid comes into contact with the skin, it causes burns. Research workers have come up with a number of substituted phenols which are better at killing bacteria and less caustic to the skin. One of these is 2,4,6-trichlorophenol, which is present in TCP® and Dettol®.

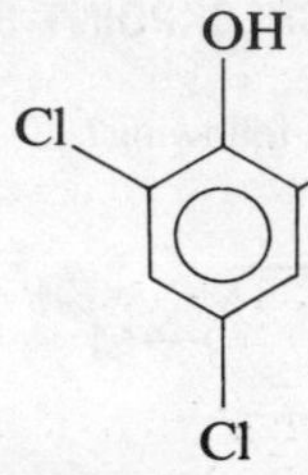

OH, Cl, Cl, Cl (2,4,6-trichlorophenol structure)

TCP® Hydrogen bonding raises the melting temperatures

The presence of hydrogen bonds makes the melting temperatures of phenols higher than those of hydrocarbons of comparable molecular mass.

30.11 SOURCES OF PHENOL

30.11.1 PETROLEUM OIL

The major source of phenol is the petrochemical industry. The *cumene* process is used to make phenol. Benzene and propene are obtained from petroleum oil by cracking. They take part in the reactions shown below:

Phenol is manufactured by the 'cumene' process...

$$C_6H_6 + CH_3CH{=}CH_2 \xrightarrow[\text{over } H_3PO_4 \text{ as catalyst}]{\text{vapours passed at 250 °C, 30 atm,}} C_6H_5CH(CH_3)_2$$

Benzene Propene (1-Methylethyl)benzene 'Cumene'

Oxidised by air (+ catalyst)

$$C_6H_5OH + (CH_3)_2CO \xleftarrow{H_2SO_4(aq)} C_6H_5{-}C(CH_3)_2{-}O{-}OH$$

Phenol Propanone 'Cumene' hydroperoxide

30.11.2 SODIUM BENZENESULPHONATE

...from sodium benzenesulphonate...

Both in industry and in the laboratory, fusion of sodium benzenesulphonate with sodium hydroxide is used to make sodium phenoxide. This is dissolved in water and acidified to give phenol:

$$C_6H_5SO_3Na + 2NaOH(s) \xrightarrow{\text{fuse}} C_6H_5ONa(s) + Na_2SO_3(s) + H_2O(g)$$

Sodium benzenesulphonate Sodium phenoxide

$$2C_6H_5ONa(aq) + CO_2(g) + H_2O(l) \rightarrow 2C_6H_5OH(l) + Na_2CO_3(aq)$$

Sodium phenoxide Phenol

30.11.3 CHLOROBENZENE

...or from chlorobenzene

Under drastic conditions, 400 °C and 150 atm, aqueous sodium hydroxide will hydrolyse chlorobenzene. The product, sodium phenoxide, is converted into phenol by acidification:

$$C_6H_5Cl(l) + 2NaOH(aq) \xrightarrow{400\text{ °C, }150\text{ atm}} C_6H_5ONa(aq) + NaCl(aq) + H_2O$$

Chlorobenzene Sodium phenoxide

This is a method of preparation which is used industrially. The conditions are difficult to obtain in the laboratory. (You will remember from § 29.13 why the hydrolysis of halogenoarenes is so difficult.)

30.11.4 DIAZONIUM COMPOUNDS

The laboratory preparation of phenol uses diazonium compounds

The easiest method to use in the laboratory is the hydrolysis of diazonium compounds [§ 32.10.1].

30.12 REACTIONS OF PHENOL

The reactions of phenol are of two kinds: (a) the reactions of the hydroxyl group and (b) substitution in the aromatic ring.

30.12.1 REACTIONS OF THE —OH GROUP

DISSOCIATION

Phenol dissociates in water:

$$C_6H_5OH(aq) + H_2O(l) \rightleftharpoons C_6H_5O^-(aq) + H_3O^+(aq)$$

Phenol is a weak acid...

Dissociation occurs to only a slight extent: phenol is a very weak acid, with $pK_a = 10.0$. The fact that it is an acid makes it dissolve more readily in sodium hydroxide solution than in water. It forms a solution of sodium phenoxide, $C_6H_5O^-\ Na^+$:

$$C_6H_5OH(l) + NaOH(aq) \rightarrow C_6H_5ONa(aq) + H_2O(l)$$

...weaker than H_2CO_3...

...but stronger than C_2H_5OH

Since phenol is a weaker acid than carbonic acid, it will not displace carbon dioxide from carbonates as carboxylic acids do. Phenols are stronger acids than alcohols. The ethoxide ion, $C_2H_5O^-$, readily accepts a proton to form C_2H_5OH: ethanol is a very weak acid, with $pK_a = 16.0$. In the phenoxide ion, a p orbital of the oxygen atom overlaps with the π orbital of the ring carbon atoms [see Figure 30.11].

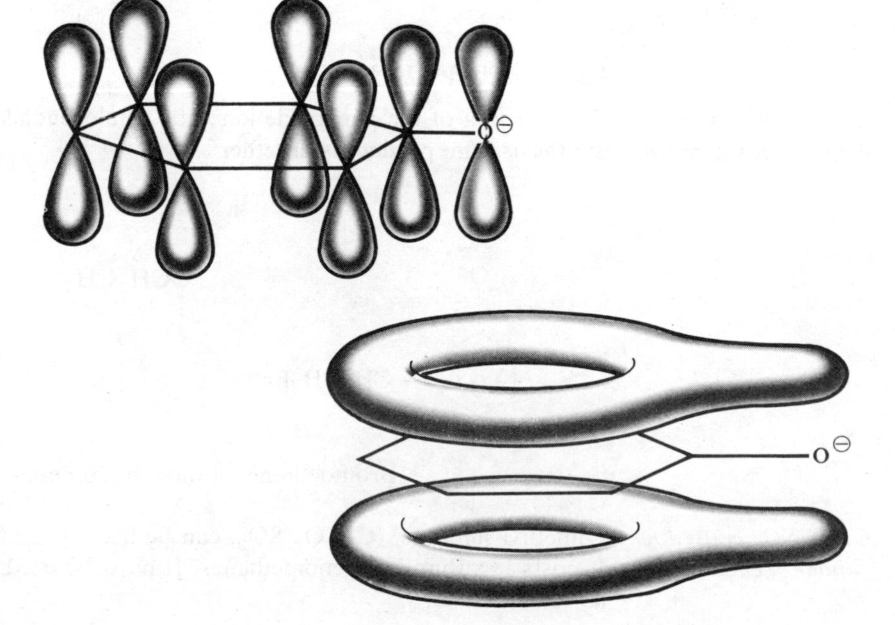

FIGURE 30.11 Interaction between the Charge on the Oxygen Atom and the Ring in Phenol

The phenoxide ion is stabilised by charge delocalisation

The negative charge is to some extent spread around the ring, and the —O⁻ group is less likely to accept a proton to form C_6H_5OH. The presence of electron-withdrawing groups in the ring attracts electrons away from the oxygen atom and so stabilises the phenoxide ion. This is why 4-chlorophenol is a stronger acid than phenol and why 2,4,6-trinitrophenol is a very much stronger acid than phenol, with $pK_a = 0.42$. Electron-donating groups, such as CH_3, in the ring decrease the acid strength of phenol:

Substituents affect the acid strength of phenol

O^- / Cl O^- / O_2N, NO_2 / NO_2 O^- / CH_3

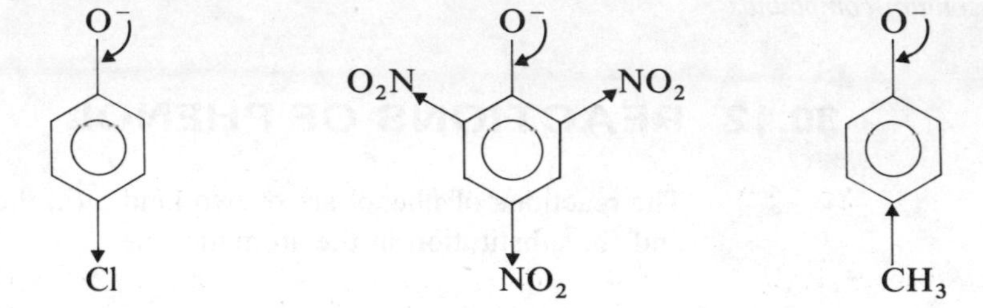

—Cl and —NO_2 assist the interaction of —O^- with the ring.

—CH_3 opposes the interaction of —O^- with the ring.

ESTERIFICATION

Before esterification, a phenol, ArOH, must be converted into the ion ArO^-

Esterification occurs less readily than with alcohols. First, the phenol must be converted into the phenoxide ion; then it will react with acid chlorides and anhydrides:

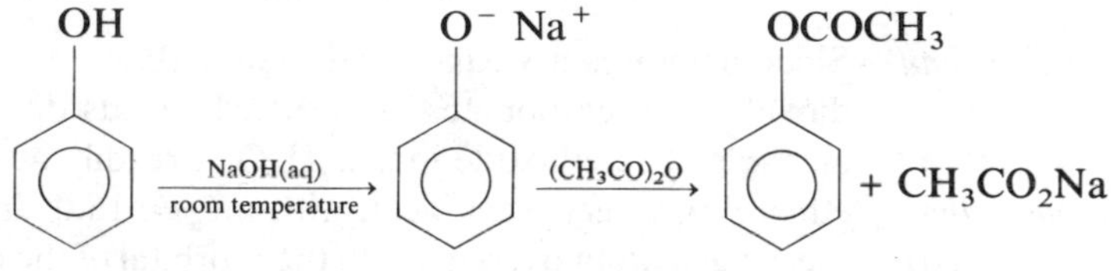

$$C_6H_5OH \xrightarrow[\text{room temperature}]{NaOH(aq)} C_6H_5O^-Na^+ \xrightarrow{(CH_3CO)_2O} C_6H_5OCOCH_3 + CH_3CO_2Na$$

Phenol   Sodium phenoxide   Phenyl ethanoate

ETHER FORMATION

The reaction between ArO^- + RX gives an ether

The reaction of the phenoxide ion with a halogenoalkane is known as Williamson's synthesis. The product is an ether:

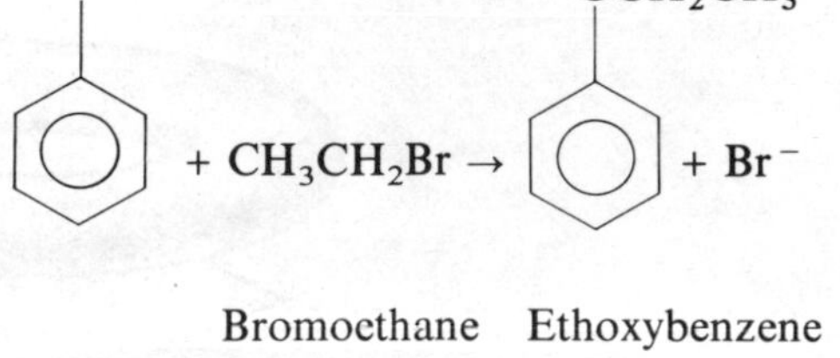

$$C_6H_5O^- + CH_3CH_2Br \rightarrow C_6H_5OCH_2CH_3 + Br^-$$

Bromoethane   Ethoxybenzene

$(CH_3)_2SO_4$ is used to make methyl ethers

Dimethyl sulphate, $(CH_3O)_2SO_2$, can be used as a methylating agent. It costs less than halogenomethanes. It must be used with extreme care because of its toxic nature.

REPLACEMENT BY HALOGENS

Reaction with PCl_5 is slow

The replacement of —OH by a halogen takes place much less readily than with alcohols. Hydrogen halides do not react, and phosphorus(V) halides react slowly to give a poor yield of the halogenoarene:

$$C_6H_5OH(s) + PCl_5(s) \rightarrow C_6H_5Cl(l) + POCl_3(s) + HCl(g)$$

Phenol	Phosphorus(V) chloride	Chloro-benzene	Phosphorus trichloride oxide

30.12.2 SUBSTITUTION IN THE RING

Phenol is more reactive than benzene towards electrophiles...

Phenol is more reactive than benzene towards electrophilic reagents. The reason is that the interaction between the lone pairs on the oxygen atom in —OH or —O^- and the ring increases the availability of electrons in the aromatic ring [§ 28.12.3]. Milder conditions are employed in substitution reactions than are needed for benzene. These are:

NITRATION

...Dilute HNO_3 will nitrate phenol...

Dilute nitric acid at room temperature.
When concentrated nitric acid is used, 2,4,6-trinitrophenol is formed. This is a powerful explosive known as 'picric acid'.

SULPHONATION

Concentrated sulphuric acid at room temperature.

HALOGENATION

...No halogen-carrier is needed for halogenation with Cl_2, Br_2...

Cl_2/Br_2, no halogen-carrier, at room temperature.
Chlorine, in the absence of solvent, gives 2- and 4-chlorophenol. Bromine, in a non-polar solvent such as carbon disulphide or tetrachloromethane, gives 2- and 4-bromophenol. Bromine water gives a precipitate of 2,4,6-tribromophenol. The faster reaction in water is due to the presence of phenoxide ions. In non-polar solvents, phenol molecules are the reacting species.

FRIEDEL–CRAFTS REACTIONS

...Alkylation uses an alcohol or an alkene...

Alkylation: Alcohol or alkene in the presence of sulphuric acid as catalyst, on gentle warming:
A low yield is obtained. Halogenoalkanes give poorer results.

OH (phenol) $\xrightarrow[H_2SO_4]{ROH}$ OH, 2-R (2-alkylphenol) + OH, 4-R (4-alkylphenol)

Acylation: Acid chloride or anhydride with Friedel–Crafts catalyst.

REACTION WITH DIAZONIUM COMPOUNDS

See Figure 32.6, § 32.10.4.

REACTION WITH METHANAL

See § 31.6.5.

30.12.3 TEST FOR PHENOL

A test for phenol

In solution, phenol reacts with iron(III) chloride to form a violet complex. Other phenols give different colours, and this reaction is used as a test for phenols.

TABLE 30.2 Comparison of Ethanol and Phenol

	Reagent	*Ethanol*, C_2H_5OH	*Phenol*, C_6H_5OH
		Liquid at room temperature. Characteristic smell.	Solid at room temperature. Characteristic smell.
(1)	Water	Miscible. Neutral solution	Partially miscible. Weakly acidic solution
(2)	Sodium	$C_2H_5ONa + H_2(g)$ formed	$C_6H_5ONa + H_2(g)$ formed
(3)	PCl_5	Vigorous reaction $\rightarrow$ HCl(g)	Slow reaction. Poor yield
(4)	CH_3COCl, $(CH_3CO)_2O$	Readily $\rightarrow$ ester	Readily $\rightarrow$ ester
(5)	$C_6H_5COCl + NaOH$	Readily $\rightarrow$ ester	Readily $\rightarrow$ ester
(6)	Organic acid	Ethyl ester formed	No reaction
(7)	HCl, HBr, HI	Halogenoethanes formed	No reaction
(8)	H_2, Ni catalyst	No addition	Addition $\rightarrow C_6H_{11}OH$
(9)	$HNO_3(aq)$	No reaction	$O_2NC_6H_4OH$ formed
(10)	Conc. H_2SO_4	Dehydration $\rightarrow$ ethene or ether	Sulphonation $\rightarrow HOC_6H_4SO_3H$
(11)	Br_2, H_2O	No reaction	White ppt. of $HOC_6H_2Br_3$
(12)	I_2, $OH^-(aq)$	CHI_3, yellow ppt. formed	No reaction
(13)	Distil with Zn.	No reaction	Benzene formed
(14)	Acid $KMnO_4$	Aldehyde, CH_3CHO formed	Mixture of oxidation products
(15)	Diazonium salt	No reaction	Yellow dye produced
(16)	$FeCl_3$	No colour produced	Violet colour produced

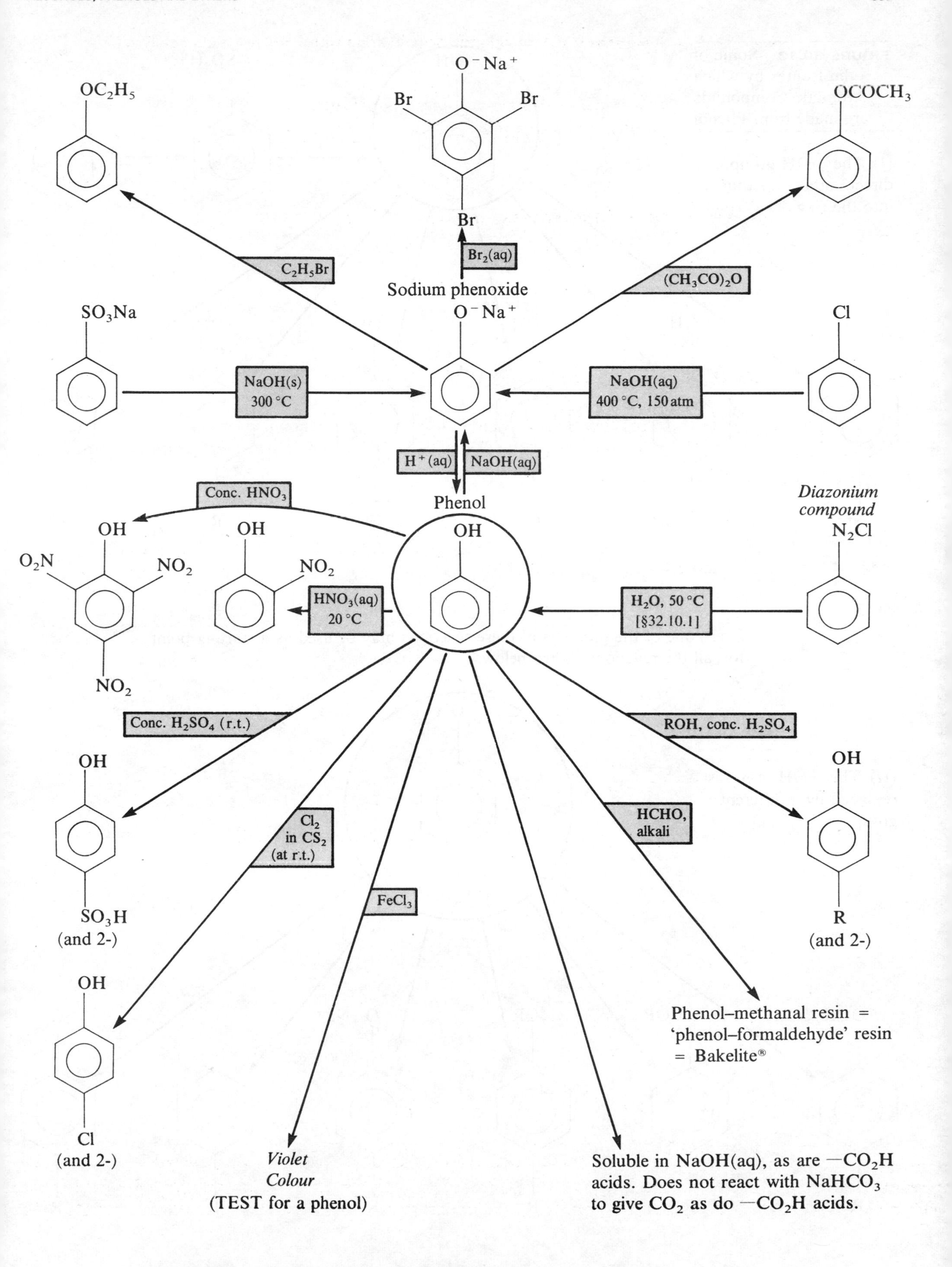

O⁻Na⁺
Br
Br
Br
OC_2H_5
$OCOCH_3$
C_2H_5Br
$Br_2(aq)$
$(CH_3CO)_2O$
Sodium phenoxide
O⁻Na⁺
SO_3Na
Cl
NaOH(s)
300 °C
NaOH(aq)
400 °C, 150 atm
H⁺(aq)
NaOH(aq)
Conc. HNO_3
Phenol
Diazonium compound
N_2Cl
OH
O_2N
NO_2
OH
NO_2
OH
$HNO_3(aq)$
20 °C
H_2O, 50 °C
[§32.10.1]
NO_2
Conc. H_2SO_4 (r.t.)
ROH, conc. H_2SO_4
OH
OH
Cl_2
in CS_2
(at r.t.)
HCHO,
alkali
$FeCl_3$
SO_3H
(and 2-)
R
(and 2-)
OH
Phenol–methanal resin = 'phenol–formaldehyde' resin = Bakelite®
Cl
(and 2-)
Violet Colour
(TEST for a phenol)
Soluble in NaOH(aq), as are —CO_2H acids. Does not react with $NaHCO_3$ to give CO_2 as do —CO_2H acids.

FIGURE 30.13 Some of the Routes by which Aromatic Compounds are made from Phenol

(I) The —OH group directs other substituents into the ring.

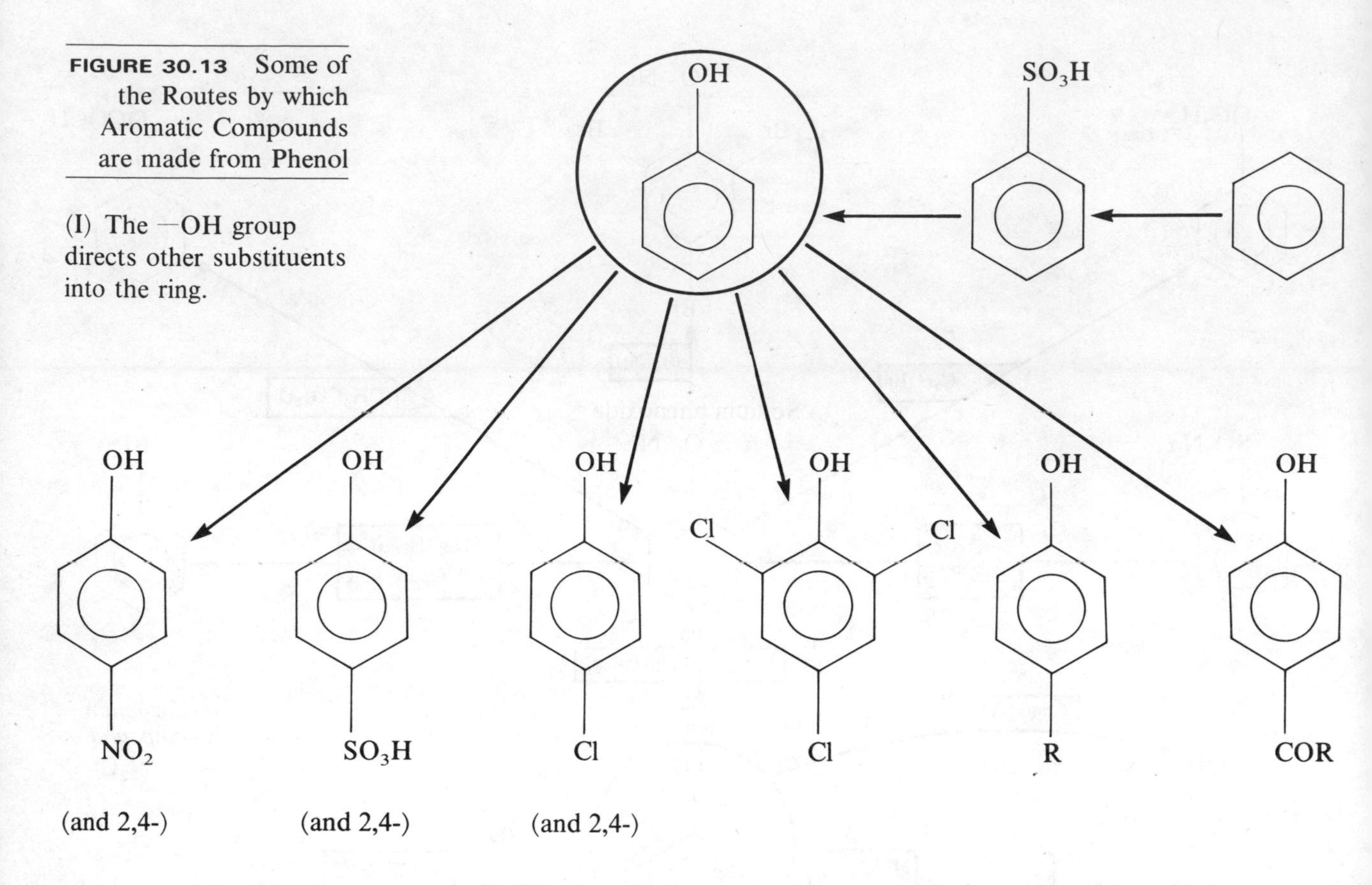

Every one of the products of these reactions may be used as a starting point for all the reactions shown below.

(II) The —OH group is replaced by a different group.

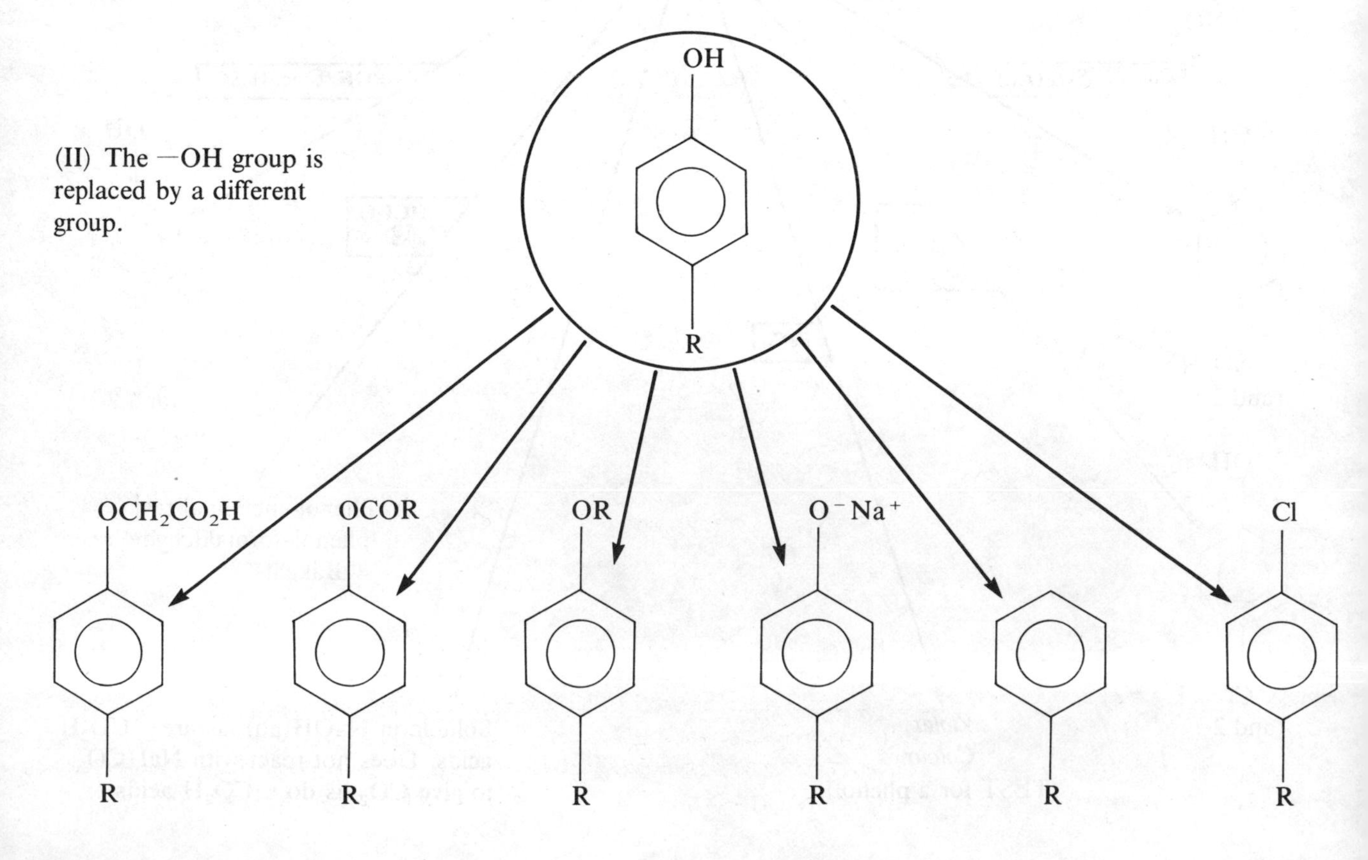

CHECKPOINT 30D: PHENOLS

†**1.** Copy the scheme shown below and fill in the necessary reagents and conditions. After reading Chapter 32, come back and fill in **X** and **Y**.

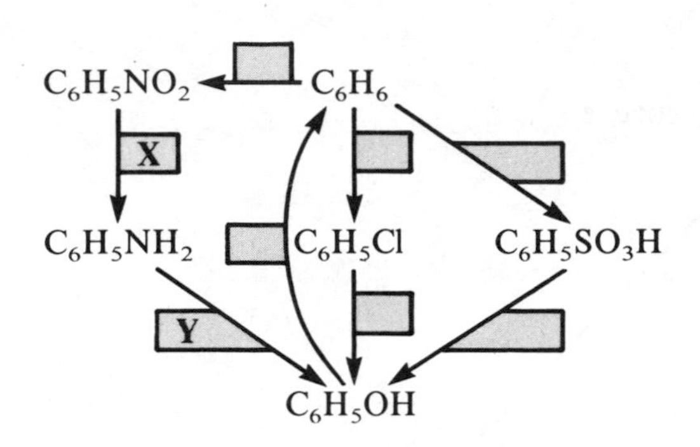

2. Compare the reactions of phenol and benzyl alcohol.

3. Starting from phenol, how could you make the following compounds?

(*a*) OH, SO_3H

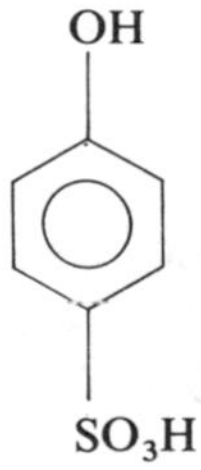

(*b*) OH, $COCH_3$

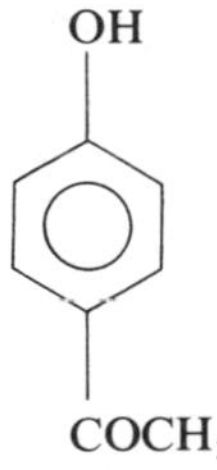

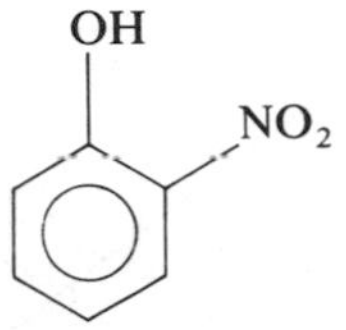

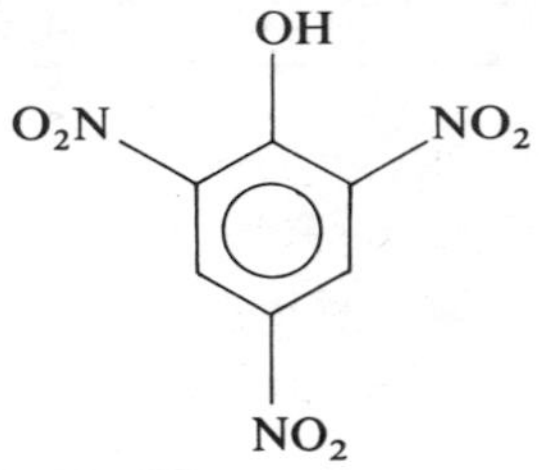

(*c*) OH, C_2H_5

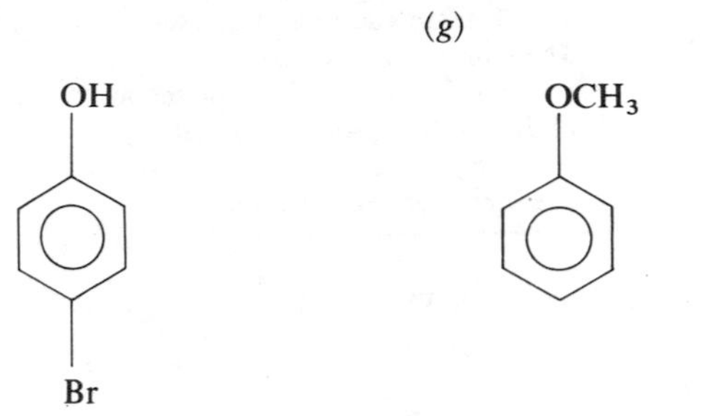

(*e*) OH, NO_2

(*f*) OH, O_2N, NO_2, NO_2

(*h*) $OCOCH_3$

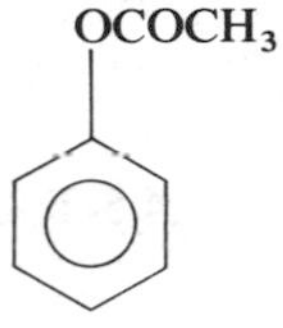

4. Explain why phenol is a stronger acid than ethanol and why 'picric acid', 2,4,6-trinitrophenol, is stronger still.

30.13 THE EXPLOSION AT SEVESO

TOPIC

In 1976 an explosion at a chemical plant in Seveso, Italy, released into the air a mixture of pollutants, including dioxin

The Seveso incident focused world attention on the dangers connected with the chemical industry. A chemical plant in Seveso, Italy, employed 200 people in the production of 2,4,5-trichlorophenol for the Swiss Givaudan Corporation, which is a subsidiary of Hoffman-LaRoche. On 10 July 1976 a build-up of pressure in a reaction vessel caused a safety valve to rupture, and a cloud of 2,4,5-trichlorophenol and other chemicals was released into the atmosphere.

One of the chemicals released was 2,3,6,7-tetrachlorodibenzo-4-dioxin, known as TCDP and as dioxin. It is an accidental by-product of the manufacture of 2,4,5-trichlorophenol [see Figure 30.14]. 2,4,5-Trichlorophenol is an important intermediate in the manufacture of 2,4,5-trichlorophenoxyethanoic acid (2,4,5-T), an agricultural herbicide used to control brushwood, and 'hexachlorophene', a bactericide which is used for the treatment of acne, for sterilising wounds and in skin cleansers.

FIGURE 30.14
Manufacture of Hexachlorophene and 2,4,5-T

1,3,4,5-Tetrachlorobenzene
Hydrolysis by NaOH (in methanol) at high temperature and pressure. The temperature is carefully controlled to prevent the formation of dioxin, but, more often than not, the product is contaminated with dioxin.
2,4,5-Trichlorophenol
Dioxin
Bis(2,3,5-trichloro-6-hydroxyphenyl)methane (Hexachlorophene)
2,4,5-Trichlorophenoxyethanoic acid (2,4,5-T)

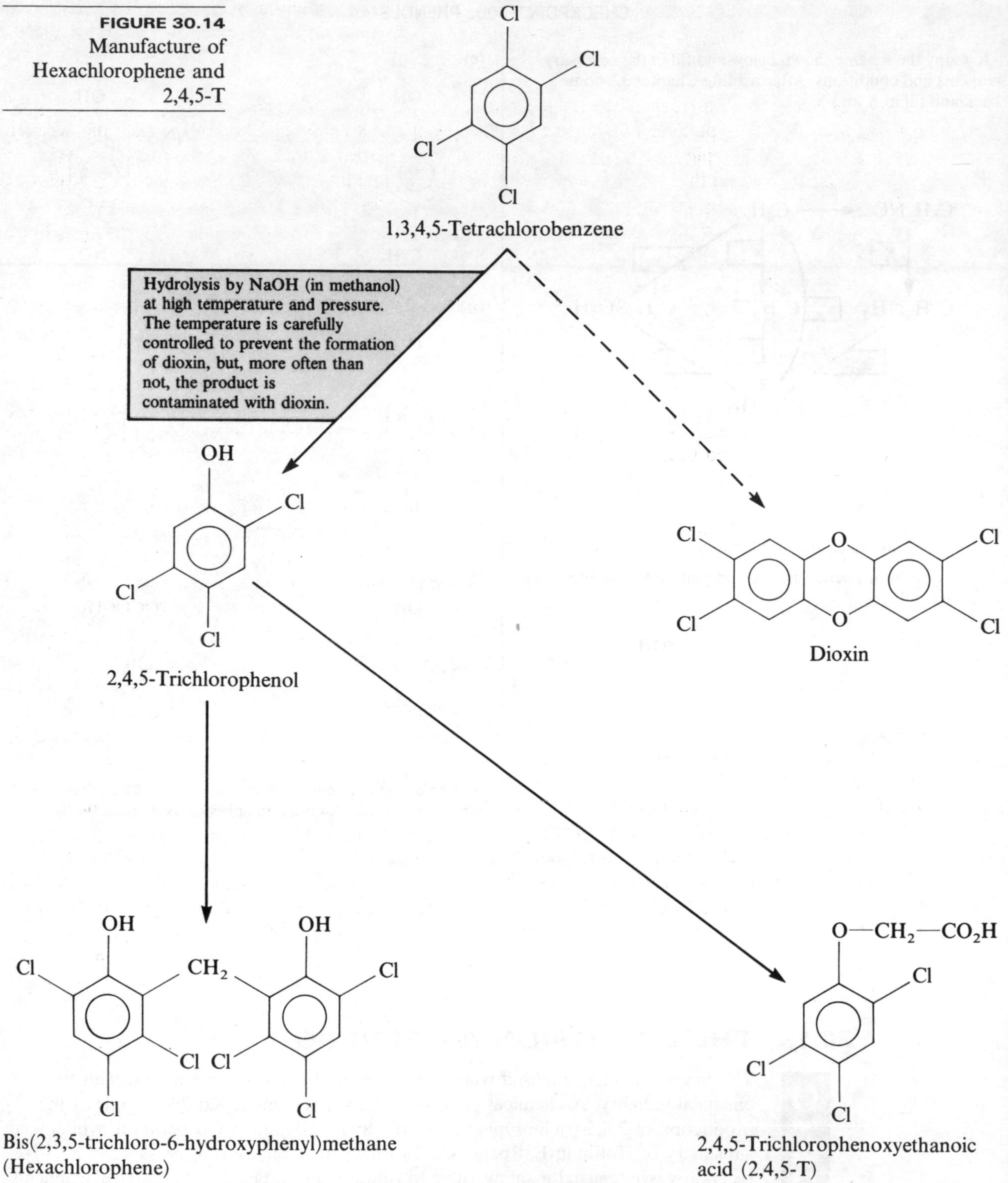

Dioxin is extremely toxic. Chloracne is a symptom of exposure to dioxin. Dioxin also has a teratogenic effect

Dioxin is stable to heat, acids and alkalis, is almost insoluble in water but soluble in some organic solvents. It is 500 times more poisonous than strychnine and 10 000 times more poisonous than cyanide ion. When a person is exposed to dioxin over a long time, dioxin residues accumulate in the liver and fat cells. Symptoms are cirrhosis of the liver, damage to the heart, kidney, spleen, central nervous system, lungs and pancreas, memory and concentration disturbances and depression. The skin disease, *chloracne*, is caused by the body's attempt to get rid of the poison through the skin. Dioxin also has a *teratogenic* effect, an effect on the genes, which results in birth defects [see below].

The population of Seveso suffered from their exposure to dioxin. Thousands of animals died. The area was so badly contaminated with dioxin that it had to be evacuated

As a result of the accident, thousands of sheep and cows died, and people became ill, particularly with the terrible skin sores of chloracne. Nine days after the accident, people were told that the dust that had settled all over the town contained dioxin, and the town was evacuated. Among the children of Seveso, there were 134 confirmed cases of chloracne and 600 suspected cases. Of the 730 pregnant women in the town, 250 applied for abortions, and the Italian Government changed the law to allow the women to end their pregnancies.

The contaminated area of Seveso was sealed off, and experts debated how to tackle the pollution. Incineration of the contaminated soil and bacterial degradation of the dioxin in the soil were suggested. One expert recommended dismantling buildings and planting forests, rather than moving earth and washing buildings. Over a period of many years, layers of topsoil were removed and buried 10 m down beneath plastic and cement. Hoffman-LaRoche agreed to pay for all material damage and set up a fund to pay compensation to individuals.

The toxic effects of dioxin were already known because the defoliant, Agent Orange, which was used in Vietnam, contained dioxin as impurity

The population of Seveso were not the first people to suffer from the effects of dioxin. It was already notorious as a result of its effects on the population of Vietnam. 2,4,5-Trichlorophenoxyethanoic acid (2,4,5-T) was the first herbicide to be used for military purposes when Britain used it in Malaya in the early 1950s. 2,4,5-T was sprayed along the sides of roads to destroy bushes which provided cover for guerilla fighters lying in wait to ambush troops. 2,4,5-T achieved notoriety through its use by the USA as a defoliant in Vietnam in the 1960s. Mixed with 2,4-dichlorophenoxyethanoic acid (2,4,-D), it was known as *Agent Orange*. American planes sprayed Agent Orange over the jungle. By destroying the leaves on the trees, it removed the cover of the Viet Cong. It is estimated that 40 million litres of Agent Orange were sprayed on Vietnam.

Many countries now ban the use of the herbicide 2,4,5-T because it contains dioxin

The defoliant was contaminated with dioxin at levels of 0.07–50 ppm. It had the dreadful effects already described both on the Vietnamese and also on American servicemen who came into contact with it. As a result of the *teratogenic* effect, grossly deformed babies were born to Vietnamese mothers and to the wives of American servicemen. In 1977 the US Army decided to destroy stocks of Agent Orange left over from the war in Vietnam because the stocks contained about 40 ppm of dioxin. They were unable to find any treatment which would remove dioxin from the defoliant. Since 1979 the UK, the USA and other countries have banned the use of 2.4.5-T because it always contains some dioxin as impurity. Alternatives are Amcide®, Glyphosphate® and Krenite®, which are more expensive than 2,4,5-T but less hazardous.

Hexachlorophene has been largely replaced by other bactericides

The trichlorophenol made at Seveso was used in the manufacture of hexachlorophene, which is used for the treatment of acne. Ironically, the most common symptom of dioxin poisoning in humans is a form of acne. In 1972, in France, talcum powder accidentally contaminated with 6% hexachlorophene caused the death of 35 babies. Since then, hexachlorophene has been used only in prescription drugs. A replacement bactericide is chlorhexidine.

The Seveso Directive is the EC code for controlling hazardous chemical processes

The Seveso accident moved the European Community (EC) to bring out a set of guidelines aimed at preventing similar accidents. This legislation, known as the *Seveso Directive*, laid down regulations for the control of hazardous industrial activities in member countries of the EC.

CHECKPOINT 30E: SEVESO

1. Refer to the reaction scheme for the manufacture of 2,4,5-T. Explain why dioxin is present as an impurity.

2. When so much was known about the danger of the dioxin impurity in 2,4,5-T, following its use in Vietnam, why do you think it took the authorities nine days to decide to evacuate Seveso?

3. The accident at Seveso started with a safety valve giving way under pressure.

(*a*) What improvements do you think chemical engineers would make to the plant after such an accident?

(*b*) What chemical could be used to absorb escaping trichlorophenol vapour?

(*c*) Which chemical properties of dioxin make it difficult to remove?

4. Why did the Italian Government change the abortion law after the Seveso accident?

30.14 ETHERS

Saturated aliphatic ethers form a homologous series of formula

$$R—O—R'$$

where R and R′ are alkyl groups. Examples are:

Names of aliphatic ethers

CH_3OCH_3	Methoxymethane (or dimethyl ether)
$CH_3OC_2H_5$	Methoxyethane (or ethyl methyl ether)
$C_2H_5OC_2H_5$	Ethoxyethane (or diethyl ether)

If the two alkyl groups are the same, the ether is described as **symmetrical**; if R and R′ are different, the ether is said to be **unsymmetrical**. The examples show the method of naming: the ethers are named as alkoxy derivatives of alkanes. Ethoxyethane is well known as a solvent and as an anaesthetic, and it is often referred to as simply 'ether'. Aliphatic ethers have the general formula

'Ether'

$$C_nH_{2n+2}O$$

They are isomeric with alcohols.

Aromatic ethers have the formula

$$ArOR$$

where Ar is an aryl group, such as phenyl or substituted phenyl or naphthyl, and R is either an alkyl group or an aryl group. Examples are:

Aromatic ethers

—OCH_3 Methoxybenzene (or methyl phenyl ether)

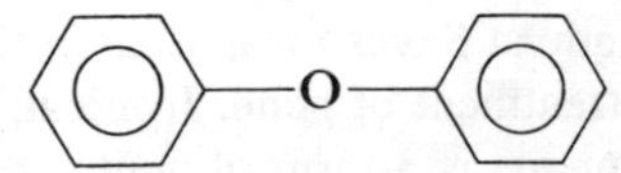

Phenoxybenzene (or diphenyl ether)

30.14.1 PHYSICAL PROPERTIES

VOLATILITY

Ethers are volatile, highly flammable...

The lower aliphatic ethers are gases or highly volatile liquids. Since ethers contain polar

$$\overset{\delta+}{C}—\overset{\delta-}{O} \text{ bonds}$$

the dipoles in neighbouring molecules interact, and the resulting attractive forces make the boiling temperatures of ethers a little higher than those of alkanes. Phenoxybenzene is a solid which melts at 28 °C. The vapours of ethers are highly flammable. As there is no possibility of hydrogen bonding, ether molecules are not associated, and the boiling temperatures are much lower than those of corresponding alcohols.

SOLUBILITY

... and only slightly soluble in water

The lower ethers are slightly soluble in water. The hydrogen bonds formed between water molecules and ether molecules are weak, and, as the hydrocarbon part of the molecule increases, solubility decreases. Ethers are important solvents for covalent substances.

DENSITY

There is danger of fire from the dense, flammable vapour

Ether vapours are denser than air. If ether is spilt in a laboratory the vapour, being denser than air, lingers at bench level and is therefore easily ignited. This is why chemists never handle ether anywhere near a flame.

ETHER EXTRACTION

Ether extraction is used to separate a covalent substance from water

The technique of **ether extraction** is used to obtain a covalent substance from solution in water. It depends on two factors:

(*a*) Covalent substances are more soluble in ethoxyethane than in water.
(*b*) Ethoxyethane and water are immiscible.

When an aqueous solution of a covalent substance is shaken with ethoxyethane in a separating funnel [see Figure 30.15], the covalent substance passes into the ether layer. This is separated and dried with, e.g., anhydrous sodium sulphate. The ethoxyethane is removed by distillation over a water-bath. Since it boils at 38 °C, this is easy to accomplish [for the theory, see § 8.6].

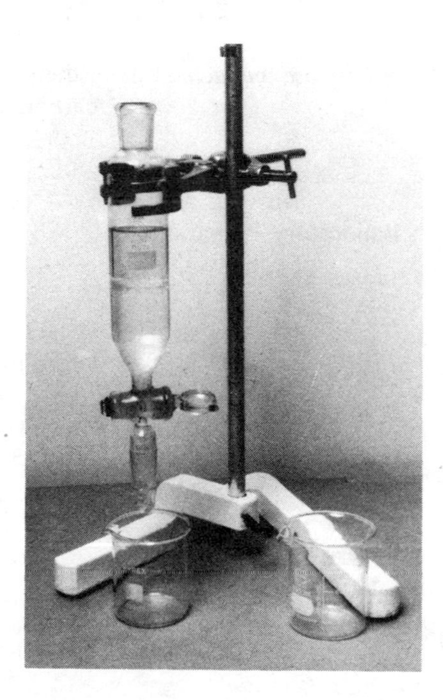

FIGURE 30.15
A Separating Funnel

30.14.2 INDUSTRIAL SOURCE OF ETHOXYETHANE

Ethoxyethane is manufactured from ethanol

The industrial method of manufacture of ethoxyethane is to pass ethanol vapour into a mixture of ethanol and concentrated sulphuric acid at 140 °C:

$$2C_2H_5OH(g) \xrightarrow[140\,°C]{\text{conc. } H_2SO_4} C_2H_5OC_2H_5(g) + H_2O(g)$$

Ethanol must be in excess, and the temperature must be kept below 170 °C; otherwise further dehydration to ethene occurs.

30.14.3 LABORATORY METHODS OF PREPARATION

The industrial method can be adopted in the laboratory [see Figure 30.16].

30.14.4 WILLIAMSON'S SYNTHESIS

RONa + R′X give unsymmetrical ethers

This method will give unsymmetrical ethers, ROR′, as well as symmetrical ethers, ROR. The sodium derivative of an alcohol or phenol is boiled with a halogenoalkane until sodium halide crystals appear. Then the mixture is distilled to give an ether:

$$C_2H_5ONa(alc) + C_3H_7Br(alc) \rightarrow C_2H_5OC_3H_7(g) + NaBr(s)$$

Sodium ethoxide — 1-Bromopropane — Ethoxypropane

30.14.5 REACTIONS OF ETHERS

Ethers are unreactive...

Ethers are unreactive. They burn readily, on account of their high volatility, to form carbon dioxide and water.

FIGURE 30.16
Laboratory Preparation of Ethoxyethane

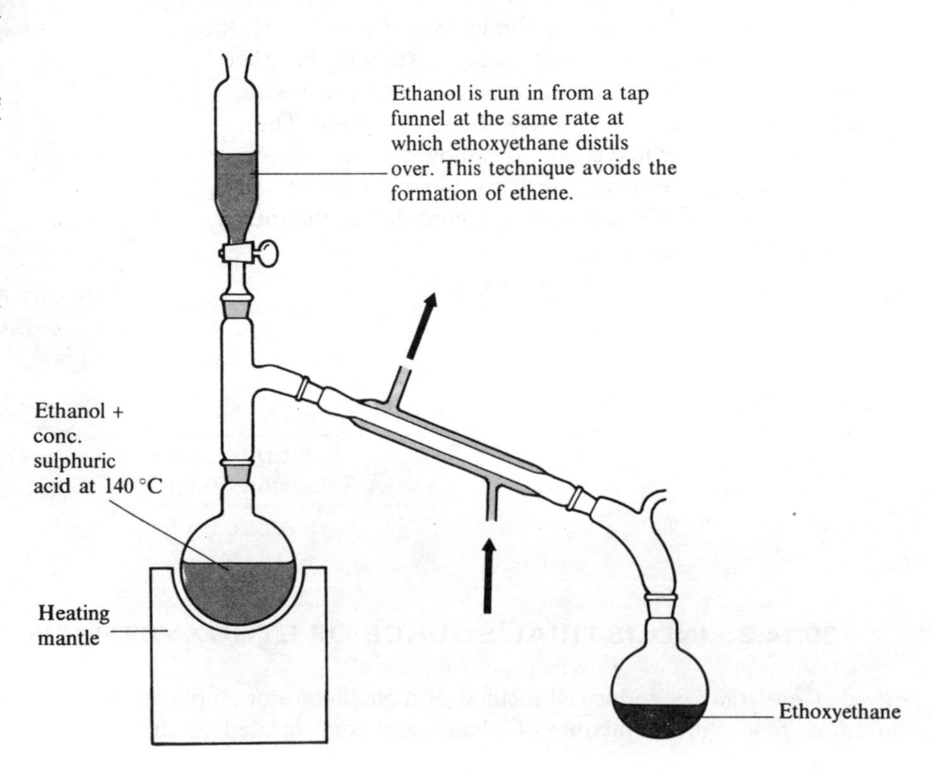

Ethanol is run in from a tap funnel at the same rate at which ethoxyethane distils over. This technique avoids the formation of ethene.
Ethanol + conc. sulphuric acid at 140 °C
Heating mantle
Ethoxyethane

In the presence of air and sunlight, ethers form peroxides. These compounds are unstable and explode if heated. To avoid their formation, ethers are stored in brown, airtight bottles.

FIGURE 30.17
Preparations of Ethers

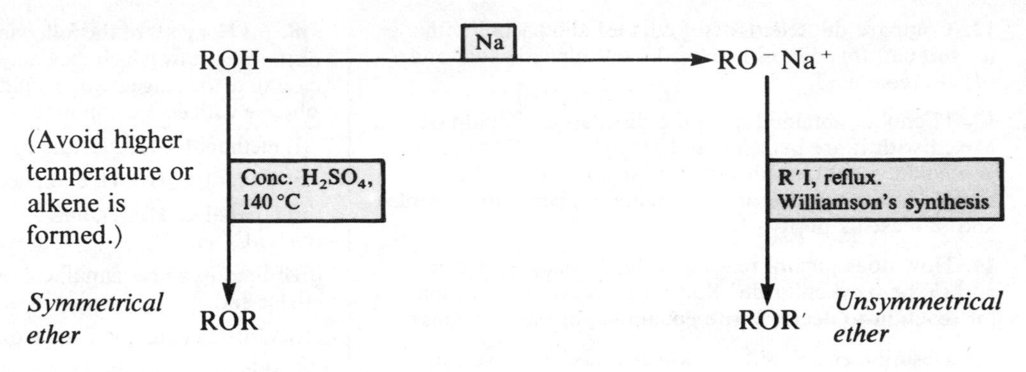

CHECKPOINT 30F: ETHERS

1. Explain why ethoxyethane is (*a*) more volatile than ethanol, (*b*) less dense than water and (*c*) a better solvent than water for organic compounds.

2. Say how you could prepare the following from propan-1-ol:
(*a*) $CH_3CH_2CH_2OCH_2CH_2CH_3$
(*b*) $CH_3CH_2CH_2OCH_2CH_3$

3. Give the name and formula of the ether formed as a result of the reaction between concentrated sulphuric acid and butan-2-ol. What other substance is formed? What precautions are taken to ensure a good yield of the ether?

4. What precautions should be taken over (*a*) storing ethers and (*b*) using ethers as solvents?

QUESTIONS ON CHAPTER 30

1. Explain why ethanol dissolves in water, whereas ethane does not.

2. Outline an industrial method for the manufacture of ethanol. What products are formed when ethanol reacts with (*a*) concentrated sulphuric acid and (*b*) ethanoic acid? State the conditions under which the reactions occur.

3. Compare and contrast the reactions of phenol and ethanol with (*a*) concentrated sulphuric acid, (*b*) ethanoic acid, (*c*) ethanoyl chloride, (*d*) iron(III) chloride solution.

4. There are four isomeric butenes. On reaction with concentrated sulphuric acid, followed by hydrolysis, three of the four give the same alcohol, while the fourth butene gives a different alcohol. Deduce the structures of the two alcohols, and show how they are formed from the butenes.

5. Write structural formulae for **A**, **B** and **C**:

```
   H  H  H
   |  |  |
H—C—C—C—H  --Na·Hg + ethanol-->  A(C3H6) → B(C3H8O)
   |  |  |
   H  Br Br
             ↘ C(C3H8O2)
```

State how **A** is converted to **B** and how **C** is made from the starting material.

6. Outline two methods of preparing phenol, starting from benzene. What products are formed when phenol reacts with (*a*) benzoyl chloride, (*b*) ethanoic anhydride, (*c*) concentrated sulphuric acid, (*d*) bromine water?

7. Complete the following reactions. State the necessary conditions for reaction to occur:

(*a*) $CH_3CH(OH)CH_3 + H_2SO_4 \rightarrow$

(*b*) $C_6H_5SO_3Na + NaOH \rightarrow$

(*c*) $CH_3CH_2OH + Na \rightarrow \xrightarrow{CH_3I}$

(*d*) $CH_3CH_2C(CH_3){=}CH_2 + H_3PO_4 \rightarrow$

(*e*) $(CH_3)_3COH + HCl \rightarrow$

8. Give the names and structural formulae of the alcohols with the molecular formula C_3H_8O. Describe how you could distinguish between them by means of chemical tests.

9. How can ethanol be made from (*a*) ethene and (*b*) bromoethane? How and under what conditions does ethanol react with (i) ethanoic acid and (ii) sulphuric acid?

10. How would you prepare the following compounds from phenol?
(*a*) sodium phenoxide, (*b*) 2-nitrophenol, (*c*) 2,4,6-tribromophenol, (*d*) ethoxybenzene.

11. Name the compounds **A** and **B**. Both are soluble in ethoxyethane. Neither is soluble in water. Describe how you could separate **A** and **B** from a mixture of the two compounds, using only sodium hydroxide, hydrochloric acid and ethoxyethane.

A: benzene ring with OH and, para, $CH(CH_3)_2$

B: benzene ring with OCH_3 and, para, $CH(CH_3)_2$

12. Compare the reactions of ethanol and phenol with (*a*) sodium, (*b*) sodium hydroxide, (*c*) sulphuric acid and (*d*) ethanoic acid.

13. Phenol is obtained from the distillation of coal tar. Mixed with it are benzene and phenylamine. Describe how you could (*a*) obtain phenol from this mixture, (*b*) test the sample to see whether it is a phenol, (*c*) purify the sample, and (*d*) test its purity.

14. How does phenol react with (*a*) ethanoic anhydride and (*b*) benzoyl chloride? State the necessary conditions for reaction to occur. Write equations for the reactions.

15. Describe chemical tests which would allow you to distinguish between the following pairs:

(*a*) Cyclohexanol and methoxybenzene.

(*b*) Ethane-1,2-diol and propan-2-ol.

(*c*) Cyclohexanol and cyclohexene.

(*d*) Cyclohexanol and phenol.

16. Outline the *cumene* process for the manufacture of phenol. State the sources of the raw materials, and the conditions under which the steps in the process are carried out.

17. A secondary alcohol, **P**, ($C_4H_{10}O$) can be dehydrated to give two isomeric alkenes, **Q** and **R**, one of which (**Q**) can exhibit geometrical isomerism. **Q** reacts with hydrogen bromide to give **S**, which can undergo the following sequence of reactions.

$$\mathbf{Q} \xrightarrow{\text{Step 1}} \underset{\mathbf{S}}{CH_3-\underset{H}{\overset{H}{C}}-\underset{Br}{\overset{H}{C}}-CH_3} \xrightarrow{\text{Step 2}} \underset{\mathbf{T}}{CH_3CH_2-\underset{CN}{\overset{H}{C}}-CH_3}$$

$$\mathbf{T} \xrightarrow{\text{Step 3}} \mathbf{U}$$

$$\underset{\mathbf{V}}{CH_3CH_2-\underset{CO_2C_2H_5}{\overset{H}{C}}-CH_3} \xleftarrow{\text{Step 4}} \underset{\mathbf{U}}{CH_3CH_2-\underset{CO_2H}{\overset{H}{C}}-CH_3}$$

(*a*) Name a reagent which could bring about the dehydration of **P**.

(*b*) (i) Draw the structure of **P**.

(ii) Draw the structures of the geometrical isomers of **Q**.

(iii) Explain why **Q** can exist as geometrical isomers.

(*c*) Give the reagents and conditions required for Steps 2, 3 and 4.

(*d*) (i) Show the mechanism of the reaction of **Q** with hydrogen bromide.

(ii) Step 2 proceeds by an S_N2 mechanism. Show this mechanism in full.

(*e*) State, giving an explanation, whether or not **S** would be chiral.

(L 92)

†**18.** (*a*) For *each* of the following pairs of compounds give a chemical test by which they may be distinguished. In *each* case state the reagent you would use and what you would observe with *each* compound.

(i) methanol, CH_3OH, and propan-2-ol, $CH_3CH(OH)CH_3$

(ii) ethanol, C_2H_5OH, and phenol, C_6H_5OH

(iii) ethanol, C_2H_5OH, and 2-methylpropan-2-ol, $(CH_3)_3COH$

(*b*) Ethanol can be manufactured by the direct hydration of ethene and also by fermentation.

(i) Write an equation for the direct hydration of ethene.

(ii) What is the source of the ethene in this process?

(iii) Write a reaction scheme for preparing ethanol from ethene other than by direct hydration.

(*c*) Ethanol can be used as a motor fuel blending agent. In Brazil ethanol, produced by fermentation of vegetation such as sugar cane, is blended with lead-free petrol.

(i) Give *three* advantages of using this fuel rather than normal petrol.

(ii) Write an equation for the complete combustion of ethanol.

(AEB 90)

19. (*a*) (i) State and give equations for *two* typical reactions of alkenes.

(ii) State and give equations for *two* typical reactions of benzene.

(iii) Explain why the typical reactions are different in the two cases.

(iv) State how alkenes are produced industrially.

(*b*) (i) Give equations for the reaction of a halogenoalkane (alkyl halide) with

(*1*) hydroxide ion, OH^-

(*2*) cyanide ion, CN^-.

(ii) State the type of reaction occurring with the cyanide ion.

(iii) Give the mechanism of the reaction with OH^- if 1-bromobutane is the halogenoalkane.

(iv) Explain why the reaction with CN^- is useful.

(*c*) (i) Write down the structural formulae of the *four* alcohols of molecular formula $C_4H_{10}O$.

Label any primary alcohol with the letters 'PR'.

Label any secondary alcohol with the letters 'SEC'.

Label the chiral centre of any chiral alcohol with a star '*'.

(ii) A chemical 'breathalyser' test for ethanol involves blowing air from the lungs through acidified potassium dichromate(VI) crystals.

State what is observed when a positive test results, i.e. excess alcohol is present, and write an equation (which need *not* be balanced) to show the reactants and products.

(WJEC 91)

31

ALDEHYDES AND KETONES

31.1 THE FUNCTIONAL GROUP

In aldehydes and ketones the functional group is the carbonyl group

$$\gt C{=}O$$

The formulae of aldehydes and ketones

Aldehydes have a hydrogen atom attached to the carbonyl carbon atom; ketones have two alkyl or aryl groups:

R—C(=O)—H	R′—C(=O)—R
Aldehyde	Ketone
(In the first member of the series, methanal, R = H.)	(R and R′ may be aliphatic or aromatic.)

The general formula of saturated aliphatic aldehydes and ketones is $C_nH_{2n}O$. In addition, there are cyclic ketones of formula $C_nH_{2n-2}O$, for example, cyclohexanone. The simplest aromatic aldehyde and ketone are benzenecarbaldehyde and phenylethanone:

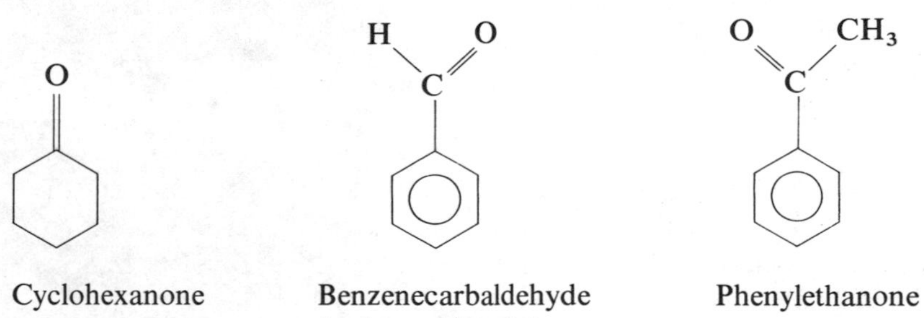

Cyclohexanone — Benzenecarbaldehyde (or benzaldehyde) — Phenylethanone

31.2 NOMENCLATURE FOR ALDEHYDES AND KETONES

The system of naming carbonyl compounds

Aldehydes and ketones are named after the hydrocarbon with the same number of carbon atoms. The name of the parent alkan*e* is changed to end in *-al* for aldehydes and *-one* for ketones. The position of the carbonyl group in ketones is indicated by a locant preceding the suffix, *-one* [see Table 31.1(b)].

(a) *Aldehydes*

Formula	*Name*
$HCHO$	Methanal (formerly formaldehyde)
CH_3CHO	Ethanal (formerly acetaldehyde)
$CH_3(CH_2)_3CHO$	Pentanal
$C_6H_5CH_2CHO$	Phenylethanal
C_6H_5CHO	Benzenecarbaldehyde (or benzaldehyde)

(b) *Ketones*

Formula	*Name*
CH_3COCH_3	Propanone (formerly acetone)
$CH_3COCH_2CH_3$	Butanone (formerly ethyl methyl ketone)
$CH_3CO(CH_2)_2CH_3$	Pentan-2-one
$CH_3CH_2COCH_2CH_3$	Pentan-3-one
$C_6H_5COCH_3$	Phenylethanone

TABLE 31.1 The Names of some Aldehydes and Ketones

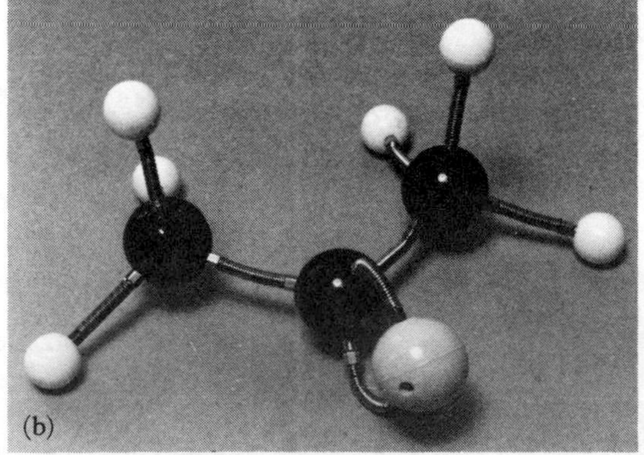

FIGURE 31.1 Models of (a) Ethanal, CH_3CHO and (b) Propanone, $(CH_3)_2CO$

31.3 THE CARBONYL GROUP

In the group $R_2C{=}O$, the carbon atom forms three single σ bonds, which are coplanar. The unused p orbital of the carbon atom overlaps with one of the unused p orbitals of the oxygen atom to form a π orbital [see Figure 31.2].

FIGURE 31.2 The Carbonyl Group, $\diagdown C=O$

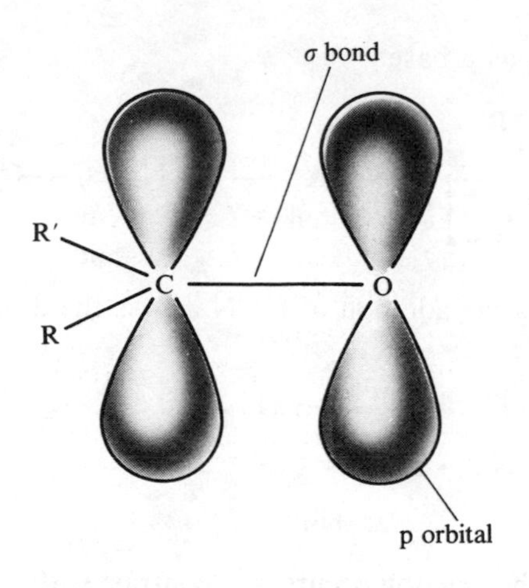

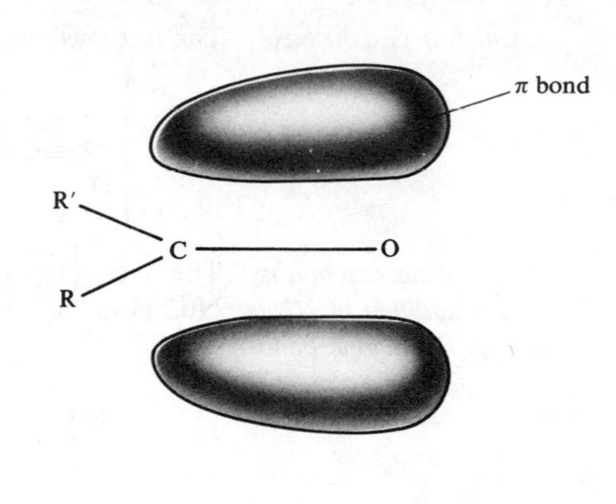

The π electrons in the C=O bond are extensively polarised

The electrons in the σ and π bonds are drawn towards the more electronegative oxygen atom. To represent this polarisation the carbonyl group is written as

$$\diagdown C \overset{\frown}{=} O \text{ or } \diagdown \overset{\delta+}{C} = \overset{\delta-}{O}$$

Since π electrons are polarised much more extensively than σ electrons, the dipole in a

$$\diagdown C \overset{\frown}{=} O \text{ bond}$$

is greater than that in a

$$-C \rightarrow Cl \text{ bond}$$

The electronic structure of the carbonyl group is responsible for the reactions of aldehydes and ketones.

31.3.1 ATTACK BY NUCLEOPHILES

Nucleophiles attack $\overset{\delta+}{C} = \overset{\delta-}{O}$ to form an intermediate...

Nucleophilic reagents are attracted by the partial positive charge on the carbon atom of the

$$\diagdown C \overset{\frown}{=} O \text{ group}$$

At the approach of a nucleophile (e.g., CN^-), the π electrons are repelled away from the carbon atom towards the oxygen atom. In the reactive intermediate which is formed, oxygen bears a negative charge, and the carbon atom is surrounded by four electron pairs.

$$N{\equiv}\overset{\ominus}{C}\!:\ \curvearrowright \ (R)(R')C \overset{\frown}{=} \ddot{O}: \longrightarrow \left[N{\equiv}C{-}C(R)(R'){-}\ddot{\underset{\cdot\cdot}{O}}:^{\ominus} \right]$$

Nucleophile attacks $\overset{\delta+}{C}$ C is 4-valent; O is negative

... which acts as a base

The intermediate acts as a base:

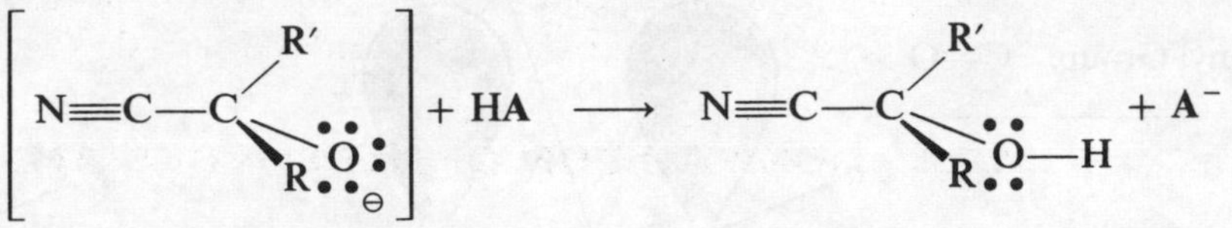

$$\left[N{\equiv}C{-}C(R')(R){-}\ddot{\underset{\cdot\cdot}{O}}{:}^{\ominus} \right] + HA \longrightarrow N{\equiv}C{-}C(R')(R){-}\ddot{O}{-}H + A^-$$

The complete reaction is the addition of HNu across the double bond

The overall reaction is the addition of HCN across the double bond. In general, if Nu**:** is the nucleophile

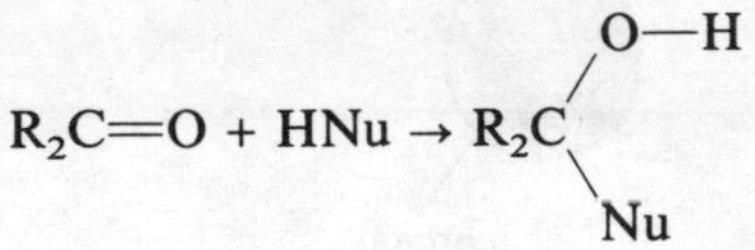

$$R_2C{=}O + HNu \rightarrow R_2C(O{-}H)(Nu)$$

The ease with which the reaction occurs is in contrast with the attack by a nucleophile on a saturated carbon atom [see the hydrolysis of halogenoalkanes, § 29.8].

31.3.2 COMPARISON WITH THE ALKENE GROUP

>C=O does not react with electrophiles as does >C=C<

The carbonyl group resembles the alkene group in that both groups contain a σ bond and a π bond between the bonded atoms. One might expect that the π electrons in the

>C=O bond like those in

>C=C<

would react with electrophiles, such as Br_2 and HBr. This does not happen. The reason is that the oxygen atom in the carbonyl group, being electronegative, is able to keep control of the π electrons and does not make them available for bonding to an electrophile. To summarise:

$>\overset{\delta+}{C}=\overset{\delta+}{O}$	>C=C<
polar bond	non-polar bond
The electron-deficient carbon atom is attacked by nucleophiles.	The double bond is attacked by electrophiles.

31.3.3 KETO–ENOL TAUTOMERISM

The equilibrium mixture of tautomers gives the reactions of >C=C<, >C=O and —OH

There is another set of reactions. These originate in the tautomerism shown by aldehydes and ketones [§ 25.9.1]. In propanone, the tautomers are

$$CH_3{-}\underset{\underset{O}{\|}}{C}{-}CH_3 \rightleftharpoons CH_3{-}\underset{\underset{OH}{|}}{C}{=}CH_2$$

keto-Propanone enol-Propanone

There is only a fraction of a per cent of the enol present. The enol has a different set of reactions: it has the reactions of the hydroxyl group and the reactions of the alkene group. If the small fraction of enol present is used up in a reaction, more will be formed from the keto-tautomer to maintain equilibrium. The whole of the equilibrium mixture can therefore undergo the reactions of the enol-tautomer. The iodination of propanone proceeds through the formation of the enol [§ 14.11].

31.3.4 COMPARISON OF ALDEHYDES AND KETONES

Aldehydes are more reactive than ketones

Aldehydes are more reactive than ketones. The reason is chiefly that the presence of two alkyl groups in ketones hinders the approach of attacking reagents to the carbonyl group. Another factor is that alkyl groups are electron-donating and reduce the partial positive charge on the carbonyl atom [see Figure 31.3].

FIGURE 31.3 Hindrance by Alkyl Groups to Nucleophilic Attack

Approach of nucleophile is hindered.

CHECKPOINT 31A: CARBONYL COMPOUNDS

1. Write structural formulae for

(*a*) butanal
(*b*) pentan-2-one
(*c*) pentan-3-one
(*d*) cyclohexanone
(*e*) 3-methylhexanal
(*f*) pentane-2,4-dione
(*g*) 4-chloro-2-methylhexan-3-one
(*h*) *trans*-but-2-enal
(*i*) 4-hydroxybenzaldehyde
(*j*) 2-chlorobenzaldehyde

2. Name the following:

(*a*) $CH_3CH_2CH_2CHO$
(*b*) $CH_3CH_2COCH_3$
(*c*) $C_2H_5COC_2H_5$
(*d*) $(CH_3)_2CHCHO$
(*e*) $(CH_3)_2CHCOCH_3$
(*f*) $CH_3CH{=}CHCHO$
(*g*) $CH_3CH{=}CHCOCH_3$
(*h*) $Cl-C_6H_4-COCH_3$ (para-substituted benzene ring)
(*i*) $O_2N-C_6H_4-CHO$ (para-substituted benzene ring)

3. What feature of the carbonyl group is chiefly responsible for the reactive nature of aldehydes and ketones?

4. Why does bromine attack

$$>C{=}C<$$

but not

$$>C{=}O\ ?$$

***5.** Propanone contains only 0.01% of the enol tautomer, yet pentane-2,4-dione contains 80% enol. Draw the *keto* and *enol* forms of this dione. Can you explain how the stability of the enol tautomer arises?

6. Explain why there is intermolecular hydrogen bonding in $C_2H_5CH_2OH$ but not in C_2H_5CHO.

31.4 INDUSTRIAL MANUFACTURE AND USES

31.4.1 METHANAL

Methanal is made from methanol by aerial oxidation...

Methanal (*formaldehyde*) is obtained by the aerial oxidation of methanol in the presence of a silver catalyst.

...It is used as formalin and in the manufacture of plastics, antiseptics and explosives

Methanal is a gas (T_b = −21 °C). It is used as a 40% aqueous solution, called *formalin*, for the preservation of biological specimens. The solid trimethanal (*paraformaldehyde*), $(CH_2O)_3$, is a convenient source of methanal, which it yields on heating. Methanal is used in the manufacture of the tough plastics, Delrin® [§ 31.6.4] and Bakelite® [§ 31.6.5], the antiseptic hexamethylene tetramine and the explosive *cyclonite*.

31.4.2 ETHANAL

Ethanal is made from ethanol...

Ethanal is made from ethanol (*a*) by aerial oxidation in the presence of silver as catalyst or (*b*) by catalytic dehydrogenation over copper.

...It is used to make chloral and DDT

Ethanal is used as a starting point in the manufacture of certain organic compounds. Examples are ethanoic acid and chloral ('knockout drops').

31.4.3 PROPANONE

Propanone is made from propan-2-ol by dehydrogenation and is used as a solvent

Propanone is made by the dehydrogenation of propan-2-ol using a copper catalyst. It is a widely used, inexpensive industrial solvent, with T_b = 56 °C. If a solvent with a higher boiling temperature is needed, butanone is used (T_b = 80 °C). Propanone is formed in the fermentation of some sugars and starches and is found in the breath and urine of diabetics.

31.4.4 OCCURRENCE

Some carbonyl compounds are found in natural products

Many aldehydes and ketones occur naturally. A few examples are:

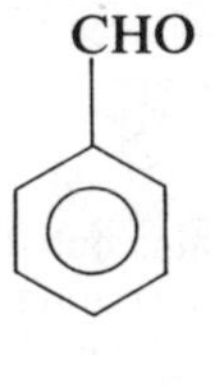

CHO | CH=CHCHO | CHO, OCH$_3$, OH | OH, CH$_3$, CH$_3$, O

Benzaldehyde (A derivative occurs in almonds.)

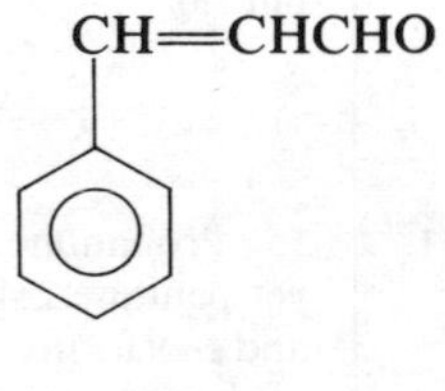

3-Phenylpropenal (In cinnamon)

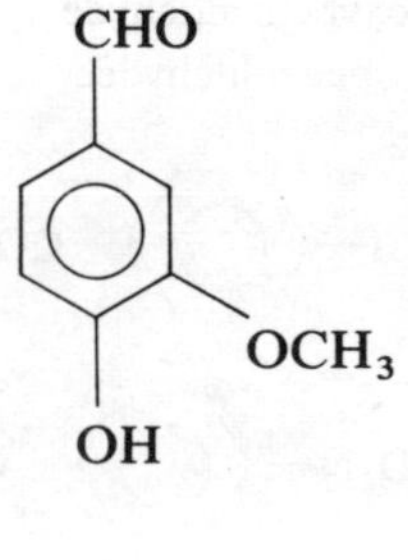

Vanillin (Vanilla flavouring)

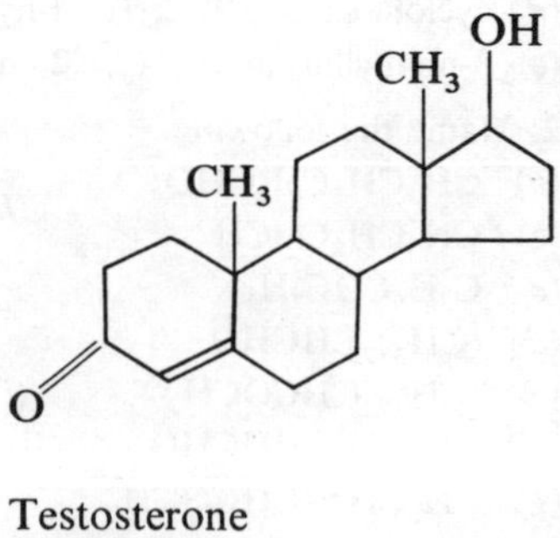

Testosterone (A male sex hormone)

31.5 LABORATORY PREPARATIONS

31.5.1 FROM ALCOHOLS

OXIDATION OF ALCOHOLS

A laboratory method of preparation is the oxidation of alcohols to aldehydes and ketones...

Oxidation of primary alcohols gives aldehydes, provided that conditions are controlled to avoid further oxidation to a carboxylic acid [§ 30.7.2]. Oxidation of secondary alcohols gives ketones. Acidified sodium dichromate(VI) is often employed for oxidation in the laboratory. The oxidising agent is added slowly to the alcohol [see Figure 31.4]. The temperature is kept below the boiling temperature of the alcohol and above that of the carbonyl compound. (Carbonyl compounds are more volatile than the corresponding alcohols.) Arranging for an excess of alcohol over oxidant and distilling off the aldehyde as it is formed avoid further oxidation. Ketones are in little danger of further oxidation:

$$C_2H_5OH(l) + [O] \xrightarrow[\text{acid, warm}]{Na_2Cr_2O_7} CH_3CHO(l) + H_2O(l)$$

Ethanol Ethanal

$$(CH_3)_2CHOH(l) + [O] \xrightarrow[\text{acid, warm}]{Na_2Cr_2O_7} (CH_3)_2CO(l) + H_2O(l)$$

Propan-2-ol Propanone

FIGURE 31.4 Preparation of Aldehydes and Ketones

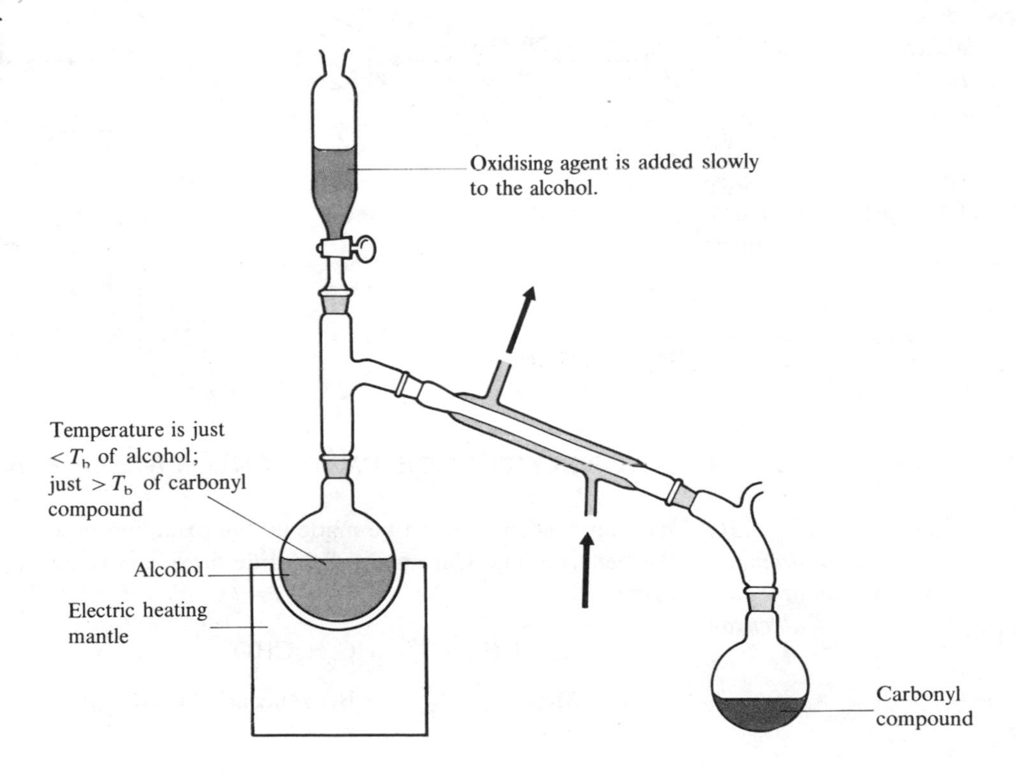
Oxidising agent is added slowly to the alcohol.
Temperature is just $< T_b$ of alcohol; just $> T_b$ of carbonyl compound
Alcohol
Electric heating mantle
Carbonyl compound

DEHYDROGENATION OF ALCOHOLS

Carbonyl compounds are made from alcohols...

Catalytic dehydrogenation of alcohols can be used in the laboratory as well as in industry [§ 31.4].

31.5.2 AROMATIC KETONES BY FRIEDEL–CRAFTS ACYLATION

Friedel–Crafts acylation gives aromatic ketones

Aromatic ketones are obtained by the Friedel–Crafts acylation reactions [§ 28.7.5].

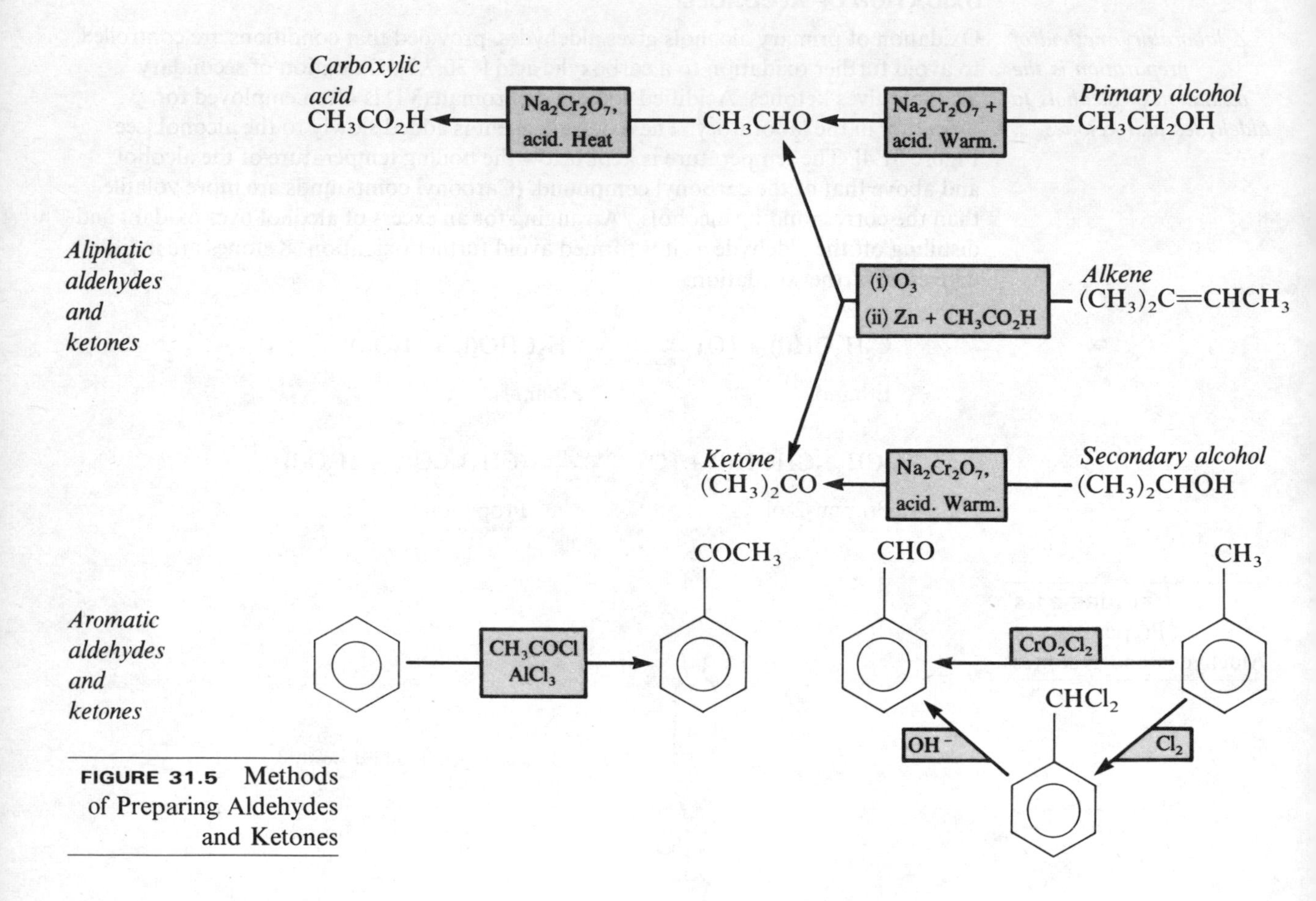

FIGURE 31.5 Methods of Preparing Aldehydes and Ketones

31.5.3 AROMATIC ALDEHYDES FROM METHYLARENES

Oxidation of —CH_3 attached to an aromatic ring gives an aromatic aldehyde

Aromatic aldehydes can be made by the oxidation of a methyl group attached to the benzene ring. Chromium dichloride dioxide is often employed as the oxidising agent:

$$C_6H_5CH_3 \xrightarrow{CrO_2Cl_2} C_6H_5CHO$$

Methylbenzene → Benzenecarbaldehyde (benzaldehyde)

Another route is the alkaline hydrolysis of 1,1-dihalogenocompounds:

$$C_6H_5CH_3 \xrightarrow{Cl_2} C_6H_5CHCl_2 \xrightarrow{OH^-(aq)} C_6H_5CHO$$

Aliphatic aldehydes can be made in this way, but the dihalogenocompounds are usually made from the aldehydes.

CHECKPOINT 31B: PREPARATIVE METHODS

1. Aldehydes are often prepared from alcohols. What difficulty arises in this method of preparation? How is it overcome?

2. Draw the structural formulae of the alkenes that yield the following products on ozonolysis:

(*a*) ethanal and propanone
(*b*) propanal, propanone and butanedial
(*c*) methanal and pentane-2,4-dione
(*d*) propanedial and methanal.

31.6 REACTIONS OF CARBONYL COMPOUNDS

31.6.1 REDUCTION

Aldehydes are reduced to primary alcohols by H_2 or by $LiAlH_4$ in ether or by $NaBH_4(aq)$

Hydrogen adds across the carbonyl group, reducing aldehydes to primary alcohols and ketones to secondary alcohols. The reaction is more difficult than addition to a carbon–carbon double bond, and usually requires heat, pressure and a metal catalyst, (Pt or Ni):

$$\mathrm{R'(R)C{=}O} + \mathrm{H_2} \xrightarrow[\text{Pt or Ni}]{\mathrm{H_2},\ \text{pressure}} \mathrm{R'(R)C(H)OH}$$

(R′ can be H)

In the laboratory, reduction is usually effected by the use of a metal hydride. Lithium tetrahydridoaluminate, $LiAlH_4$, is used in solution in ethoxyethane, which must be dry because the hydride reacts with water. It releases hydride ions which attack the carbonyl carbon atom. The reaction with aldehydes is very vigorous, and the less powerful reducing agent, sodium tetrahydridoborate, $NaBH_4$, can be used. This is a more convenient reagent as it can be used in aqueous or methanolic solution:

Ketones are reduced by H_2 or by $LiAlH_4$ in ether to give secondary alcohols

$$\mathrm{R(H)C{=}O} \xrightarrow[\text{or } \mathrm{NaBH_4}\ \text{(methanol/water)}]{\mathrm{LiAlH_4}\ \text{(ethoxyethane)}} \mathrm{R(H)C(H)OH}$$

$$\mathrm{R'(R)C{=}O} \xrightarrow{\mathrm{LiAlH_4}\ \text{(ethoxyethane)}} \mathrm{R'(R)C(H)OH}$$

31.6.2 OXIDATION

Aldehydes are oxidised to carboxylic acids...

Aldehydes are readily oxidised to carboxylic acids by acidified potassium dichromate(VI) or acidified potassium manganate(VII). Alternatively, the aldehyde vapour can be passed with air over a heated catalyst or over silver oxide:

$$\mathrm{R{-}C(H){=}O(g)} + \mathrm{O_2(g)} \xrightarrow[\text{Pt or Cu as catalyst}]{\text{heat}} \mathrm{R{-}C({=}O)OH}\ \text{(g)}$$

Aromatic aldehydes are less easily oxidised than aliphatic aldehydes.

. . . Ketones are more difficult to oxidise

Ketones are more difficult to oxidise than aldehydes. With a powerful oxidising agent, in acid conditions, they are oxidised to a mixture of carboxylic acids.

31.6.3 DISTINGUISHING BETWEEN ALDEHYDES AND KETONES

The difference in the ease of oxidation provides a means of distinguishing between aldehydes and ketones

The relative ease with which aldehydes are oxidised gives a means of distinguishing them from ketones. The gentle oxidising agent, Fehling's solution, contains complex copper(II) ions. Warm Fehling's solution is reduced to a reddish precipitate of copper(I) oxide, Cu_2O, by aldehydes but not by ketones. Tollens' reagent, a solution of complex silver ions, $[Ag(NH_3)_2]^+$, is reduced to a silver mirror when warmed with aldehydes but not by ketones.

31.6.4 ADDITION REACTIONS

HYDROGEN CYANIDE

The carbonyl group,

$$>C{=}O$$

is attacked by nucleophiles. As mentioned in § 31.3.1, the cyanide ion, CN^-, is such a nucleophile. When hydrogen cyanide is added to a cold (10–20 °C) solution of a carbonyl compound, it adds across the double bond. The reaction occurs rapidly in the presence of a base and is inhibited by acids:

HCN adds across $>C{=}O$ to give a hydroxynitrile (a cyanohydrin) . . .

$$(CH_3)(C_2H_5)C{=}O(l) + HCN(g) \xrightarrow{\text{base}} (CH_3)(C_2H_5)C(OH)(CN)(s)$$

Butanone → 2-Hydroxy-2-methylbutanonitrile (the cyanohydrin of butanone)

The product that is formed is described as the **cyanohydrin** of the carbonyl compound. It is a hydroxynitrile. The **nitrile** group

$$-C{\equiv}N$$

can be hydrolysed to the carboxyl group

$$-CO_2H$$

The hydrolysis of hydroxynitriles is a useful way of making hydroxycarboxylic acids:

. . . which can be hydrolysed to a hydroxyacid

$$(CH_3)(C_2H_5)C(OH)(CN) \xrightarrow[\text{reflux}]{\text{conc. HCl(aq)}} (CH_3)(C_2H_5)C(OH)(CO_2H)$$

2-Hydroxy-2-methylbutanonitrile → 2-Hydroxy-2-methylbutanoic acid

To avoid the danger of working with the poisonous gas hydrogen cyanide, it is generated in the reaction mixture by the action of dilute sulphuric acid on potassium cyanide.

NaHSO₃ adds across the C=O bond to give a crystalline hydrogensulphite compound, from which the carbonyl compound can be regenerated

*SODIUM HYDROGENSULPHITE

Another reagent which adds across the carbonyl group is sodium hydrogensulphite:

$$(C_2H_5)_2C{=}O + NaHSO_3 \underset{\text{acid or alkali}}{\overset{\text{excess of NaHSO}_3}{\rightleftharpoons}} (C_2H_5)_2C(OH)SO_3Na$$

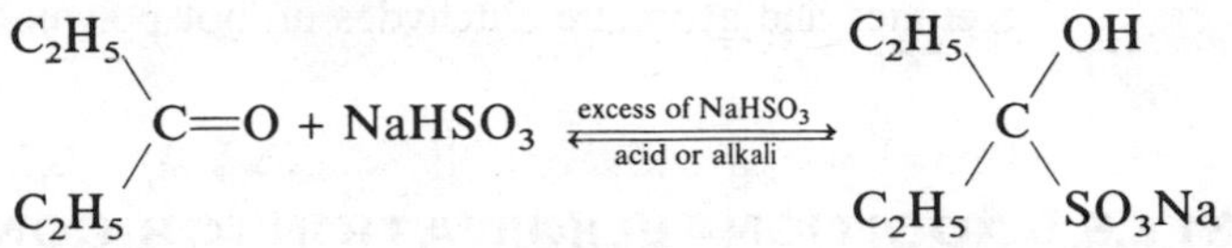

Pentan-3-one → Sodium hydrogensulphite compound of pentan-3-one

These derivatives are used to purify carbonyl compounds . . .

Aldehydes react more readily than ketones, possibly on account of the steric hindrance to the approach of the attacking reagent offered by two alkyl groups. The reaction is carried out by shaking together the carbonyl compound and an excess of a saturated sodium hydrogensulphite solution at room temperature. The hydrogensulphite derivative crystallises out of the solution. On treatment with acid or alkali, it liberates the free aldehyde or ketone. The pair of reactions are used to purify carbonyl compounds. The hydrogensulphite compound is precipitated, recrystallised from water and then decomposed.

. . . and for conversion into hydroxynitriles

With sodium cyanide, hydrogensulphite compounds react to give hydroxynitriles (cyanohydrins):

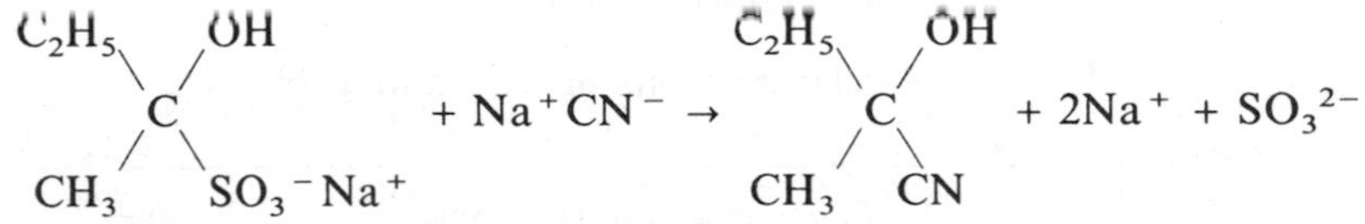

$$C_2H_5C(CH_3)(OH)SO_3^-Na^+ + Na^+CN^- \rightarrow C_2H_5C(CH_3)(OH)CN + 2Na^+ + SO_3^{2-}$$

Sodium hydrogensulphite compound of butanone → 2-Hydroxy-2-methylbutanonitrile

Hydroxynitriles are often made from carbonyl compounds by these two steps in order to avoid the use of hydrogen cyanide.

*WATER

H_2O adds across >C=O

In aqueous solution, carbonyl compounds exist in equilibrium with diols:

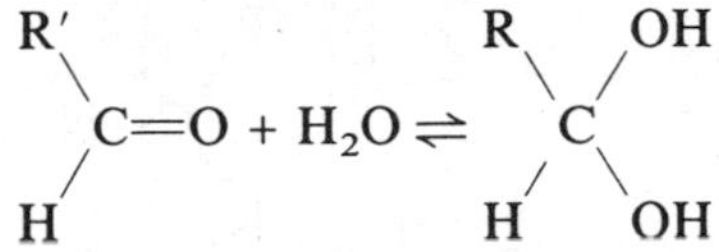

$$R'(H)C{=}O + H_2O \rightleftharpoons R(H)C(OH)_2$$

POLYMERISATION

Methanal polymerises to form poly(methanal) . . .

. . . a useful plastics material

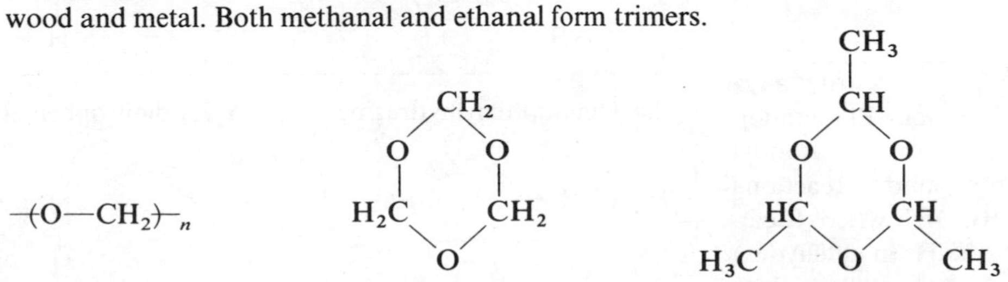

If a solution of methanal is evaporated, poly(methanal), $(CH_2O)_n$, is formed. This polymer (Delrin®) is a tough plastic material which is used as a substitute for wood and metal. Both methanal and ethanal form trimers.

$$-\!\!(O{-}CH_2)\!\!-_n$$

Poly(methanal)

Methanal trimer (*trioxane*): six-membered ring of alternating O and CH_2 groups, $(CH_2O)_3$

Ethanal trimer (*paraldehyde*): six-membered ring of alternating O and CH groups, each CH bearing a CH_3, $(CH_3CHO)_3$

Ethanal trimerises

Ethanal trimer (*paraldehyde*) is a liquid which is made by adding a few drops of concentrated sulphuric acid to ethanal. It has been used as a hypnotic (for inducing sleep).

Ketones and aromatic aldehydes do not polymerise.

31.6.5 ADDITION–ELIMINATION (OR CONDENSATION) REACTIONS

Compounds XNH_2 give products as the result of addition followed by elimination

Carbonyl compounds form adducts with many derivatives of ammonia, $\mathbf{XNH_2}$, adducts which immediately eliminate a molecule of water to form a product $R_2C{=}NX$. In the reactions with hydrazine, the products are **hydrazones**:

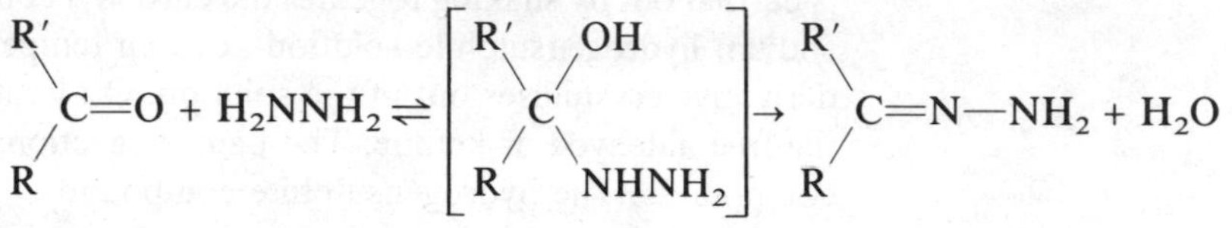

$$\underset{R}{\overset{R'}{\diagdown}}C{=}O + \underset{\text{Hydrazine}}{H_2NNH_2} \rightleftharpoons \underset{\text{Intermediate}}{\left[\begin{matrix} R' & & OH \\ & C & \\ R & & NHNH_2 \end{matrix}\right]} \rightarrow \underset{\text{A hydrazone}}{\underset{R}{\overset{R'}{\diagdown}}C{=}N{-}NH_2 + H_2O}$$

Since H_2O is eliminated, these **addition–elimination** reactions are described as **condensation** reactions.

Similar reactions are listed in Table 31.2.

Reaction with a weak base	*Product*
*Hydroxylamine H_2NOH	An oxime $(R)(R')C{=}N{-}OH + H_2O$
Hydrazine H_2NNH_2	A hydrazone $(R)(R')C{=}N{-}NH_2 + H_2O$
Phenylhydrazine 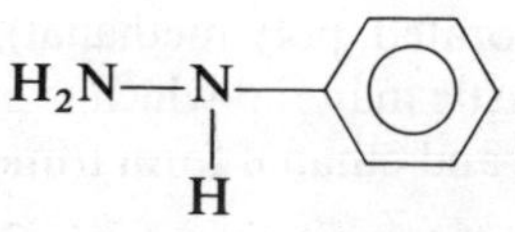 $H_2N{-}N(H){-}C_6H_5$	A phenylhydrazone $(R)(R')C{=}N{-}N(H){-}C_6H_5 + H_2O$ 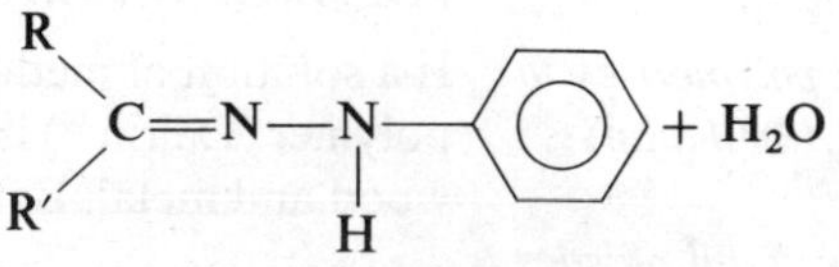
2,4-Dinitrophenylhydrazine 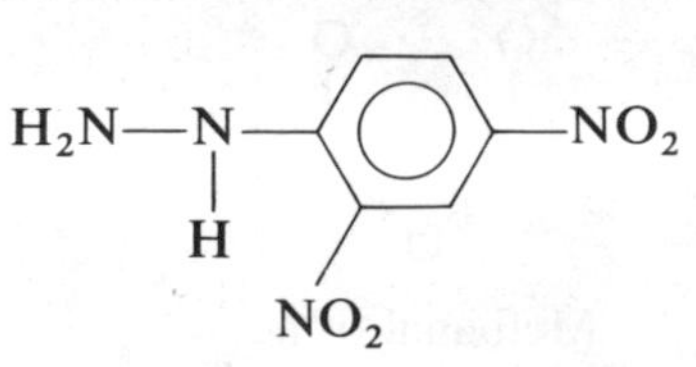 $H_2N{-}N(H){-}C_6H_3(NO_2)_2$	A 2,4-dinitrophenylhydrazone 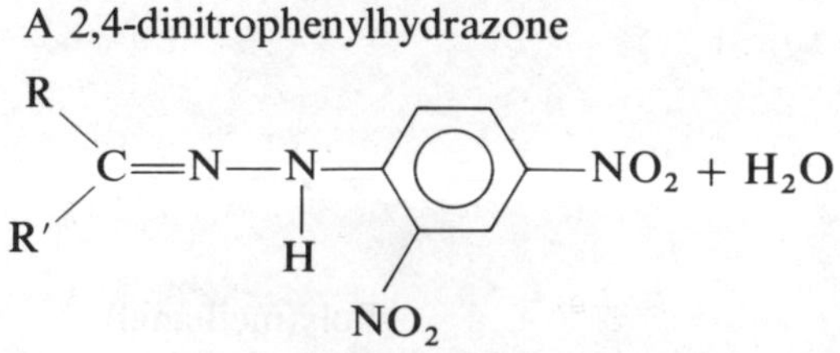 $(R)(R')C{=}N{-}N(H){-}C_6H_3(NO_2)_2 + H_2O$

TABLE 31.2 Addition–Elimination Reactions of Carbonyl Compounds (Reactions of RCOR′, where R can be H (in aldehydes) or an alkyl group (in ketones).)

...Brady's reagent is used to detect carbonyl compounds

A solution of 2,4-dinitrophenylhydrazine in methanol and sulphuric acid is called Brady's reagent. When it is added to a solution of a carbonyl compound, the 2,4-dinitrophenylhydrazone separates rapidly. Aliphatic carbonyl compounds give orange or yellow derivatives, while those of aromatic carbonyl compounds are darker in colour. Brady's reagent can therefore be used to test for the presence of carbonyl compounds. The reaction is quantitative, and the precipitate can be dried and weighed to reveal the amount of carbonyl compound present.

PLASTICS

Urea-methanal polymers are useful plastics

A number of useful plastics are formed by addition–elimination reactions. The *urea-formaldehyde* resins (such as melamine and Formica®) are polyamides made from carbamide (urea) and methanal (formaldehyde):

$$nHCHO + nH_2N{-}\underset{\underset{O}{\|}}{C}{-}NH_2 \rightarrow {-}(CH_2{-}NHCONH){-}_n + nH_2O$$

Methanal Carbamide 'Urea-formaldehyde' resin

Bakelite® is a phenol-methanal polymer

Bakelite® is formed by **condensation polymerisation** from phenol and methanal. It is a thermosetting resin [§ 35.1.1].

The structural formula of Bakelite®

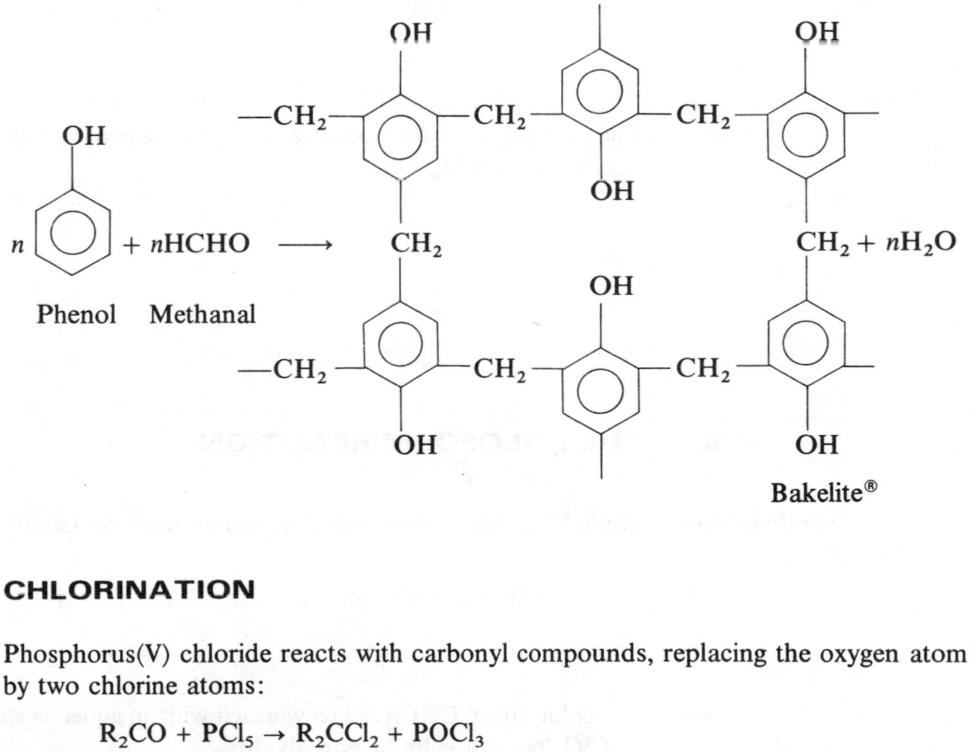

n (phenol, C_6H_5OH) + nHCHO ⟶ [cross-linked network of phenol rings joined by CH_2 bridges, each ring bearing OH] + nH_2O

Phenol Methanal

Bakelite®

31.6.6 CHLORINATION

PCl_5 converts $\backslash C{=}O$ to $\backslash CCl_2$

Phosphorus(V) chloride reacts with carbonyl compounds, replacing the oxygen atom by two chlorine atoms:

$$R_2CO + PCl_5 \rightarrow R_2CCl_2 + POCl_3$$

Chlorine replaces the hydrogens on the α carbon atoms...

Chlorine reacts with carbonyl compounds. The hydrogen atoms on the **α carbon atoms** (those adjacent to the carbonyl atoms) are easily substituted by chlorine atoms. This is because the electron-withdrawing effect of the carbonyl group weakens the C—H bonds. When chlorine is bubbled through ethanal, trichloroethanal, CCl_3CHO, is formed. Propanone forms a mixture of products, and benzaldehyde forms benzoyl chloride:

...e.g., in the formation of CCl_3CHO, chloral

$$CH_3\overset{H}{C}{=}O + 3Cl_2 \rightarrow CCl_3\overset{H}{C}{=}O + 3HCl$$

Ethanal 2,2,2-Trichloroethanal

$$CH_3COCH_3 \xrightarrow{Cl_2} CH_3COCH_2Cl \xrightarrow{Cl_2} CH_3COCHCl_2 \rightarrow \text{etc.}$$

Propanone Chloropropanone 1,1-Dichloropropanone

$$C_6H_5C(H){=}O + Cl_2 \rightarrow C_6H_5C(Cl){=}O + HCl$$

Benzaldehyde Benzoyl chloride

Trichloroethanal is a colourless liquid called *chloral.* It combines with water to form a colourless crystalline solid, *chloral hydrate*, $CCl_3CH(OH)_2$. Chloral hydrate was used to induce sleep.

...Chloral is used in medicine...

Trichloroethanal is used industrially. It condenses with chlorobenzene in the presence of concentrated sulphuric acid:

...and in the manufacture of DDT

$$Cl{-}C_6H_4 + O{=}CH{-}CCl_3 + C_6H_4{-}Cl \xrightarrow[H_2SO_4]{\text{conc.}} Cl{-}C_6H_4{-}CH(CCl_3){-}C_6H_4{-}Cl + H_2O$$

The product, 1,1,1-trichloro-2,2-bis(4-chlorophenyl)ethane (DDT) is a powerful insecticide [§ 20.15].

31.6.7 THE HALOFORM REACTION

Chloroform from CH_3COR, chlorine and alkali

Methyl carbonyl compounds, i.e., compounds containing a

$$CH_3{-}\underset{\underset{O}{\|}}{C}{-}\overset{/}{\underset{\backslash}{C}}{-} \text{ group} \quad \text{and} \quad CH_3{-}\underset{\underset{O}{\|}}{C}{-}H$$

give chloroform, $CHCl_3$, when warmed with an aqueous solution of sodium chlorate(I), NaClO (or with chlorine and alkali):

$$(CH_3)(R)C{=}O + 3Cl_2 \rightarrow (CCl_3)(R)C{=}O + 3HCl$$

$$(CCl_3)(R)C{=}O + OH^- \rightarrow CHCl_3 + RCO_2^-$$

Iodoform from CH_3COR and I_2 and alkali or NaIO...

If potassium bromide or potassium iodide is dissolved in the reacting mixture, bromoform, $CHBr_3$, or iodoform, CHI_3, is formed. Iodoform, tri-iodomethane, is precipitated as fine yellow crystals with a characteristic smell. The reaction is used as a test for the group

$$CH_3-\underset{\underset{O}{\|}}{C}-\overset{|}{\underset{|}{C}}-$$

...and from $CH_3CH(OH)R$

Alcohols of formula $CH_3CH(OH)R$ are oxidised by sodium iodate(I) to CH_3COR and therefore give a positive iodoform test:

$$CH_3CH_2OH \xrightarrow{NaIO} CH_3CHO \xrightarrow{NaIO} CI_3CHO \xrightarrow{OH^-} CHI_3$$

Ethanol | Ethanal | Tri-iodo-ethanal | Tri-iodomethane

31.6.8 *ALDOL REACTION

Aldehydes and ketones with α H atoms dimerise in the aldol reaction...

In the presence of a dilute alkali, some aldehydes and ketones dimerise to form a hydroxycarbonyl compound called an **aldol**. The carbonyl compound must possess an α hydrogen atom (on the carbon atom adjacent to the carbonyl group):

$$2CH_3CHO \xrightarrow{OH^-(aq)} CH_3\underset{\underset{OH}{|}}{C}HCH_2CHO$$

Ethanal | 3-Hydroxybutanal

Methanal and benzaldehyde do not have α hydrogen atoms and do not dimerise.

31.6.9 *CANNIZZARO REACTION

...Those without α H atoms undergo the Cannizzaro reaction

Carbonyl compounds which contain no 2-hydrogen atoms react differently with alkali. They undergo the Cannizzaro reaction, in which both oxidation to an acid and reduction to an alcohol occur:

$$2C_6H_5CHO + NaOH \rightarrow C_6H_5CO_2Na + C_6H_5CH_2OH$$

Benzaldehyde | Sodium benzoate | Benzyl alcohol

$$2HCHO + NaOH \rightarrow HCO_2Na + CH_3OH$$

Methanal | Sodium methanoate | Methanol

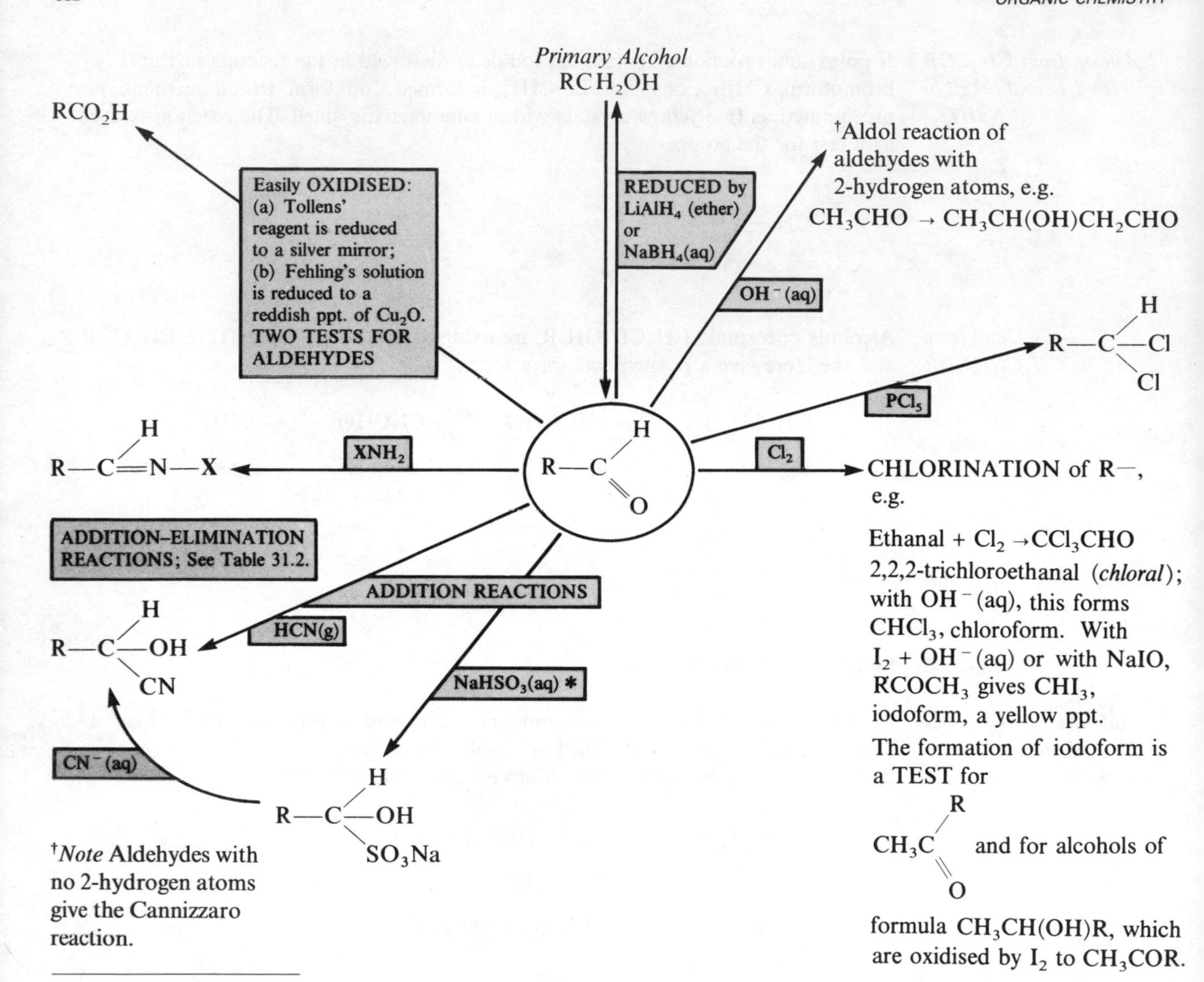

FIGURE 31.6
A Summary of the Reactions of Aldehydes

TABLE 31.3
A Comparison of Aliphatic and Aromatic Aldehydes

	Ethanal, CH_3CHO	*Benzaldehyde, C_6H_5CHO*
Reagent	*Product*	*Product*
$KMnO_4$, acid	CH_3CO_2H	$C_6H_5CO_2H$
$LiAlH_4$ (ethoxyethane)	CH_3CH_2OH	$C_6H_5CH_2OH$
PCl_5 or $SOCl_2$	CH_3CHCl_2	$C_6H_5CHCl_2$
HCN	$CH_3CH(OH)CN$	$C_6H_5CH(OH)CN$
*$NaHSO_3$	$CH_3CH(OH)SO_3Na$	$C_6H_5CH(OH)SO_3Na$
$C_6H_5NHNH_2$	$CH_3CH{=}NNHC_6H_5$	$C_6H_5CH{=}NNHC_6H_5$
*NH_2OH	$CH_3CH{=}NOH$	$C_6H_5CH{=}NOH$
*NaOH(aq)	$CH_3CH(OH)CH_2CHO$	$C_6H_5CH_2OH + C_6H_5CO_2Na$
Cl_2	CCl_3CHO	C_6H_5COCl
Cl_2 + Halogen-carrier	—	$3\text{-}ClC_6H_4CHO$
Conc. HNO_3 + Conc. H_2SO_4	—	$3\text{-}O_2NC_6H_4CHO$
Hot conc. H_2SO_4	—	$3\text{-}HO_3SC_6H_4CHO$
Fehling's solution	Red Cu_2O	—
Tollens' reagent	Ag mirror	—
KI, NaClO	CHI_3, iodoform	—
*Conc. H_2SO_4	$(C_2H_4O)_3$, paraldehyde	—

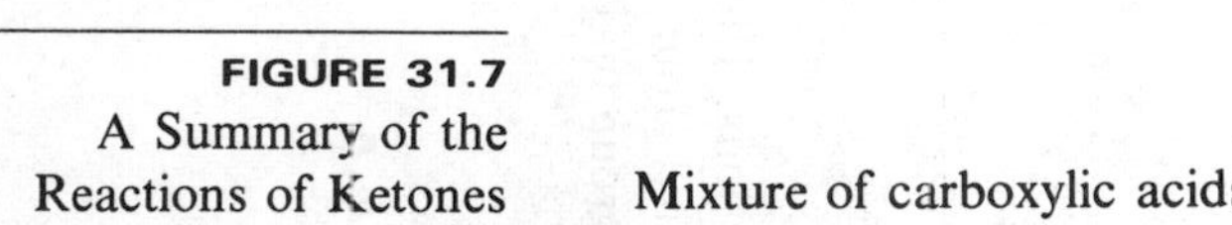
FIGURE 31.7 A Summary of the Reactions of Ketones

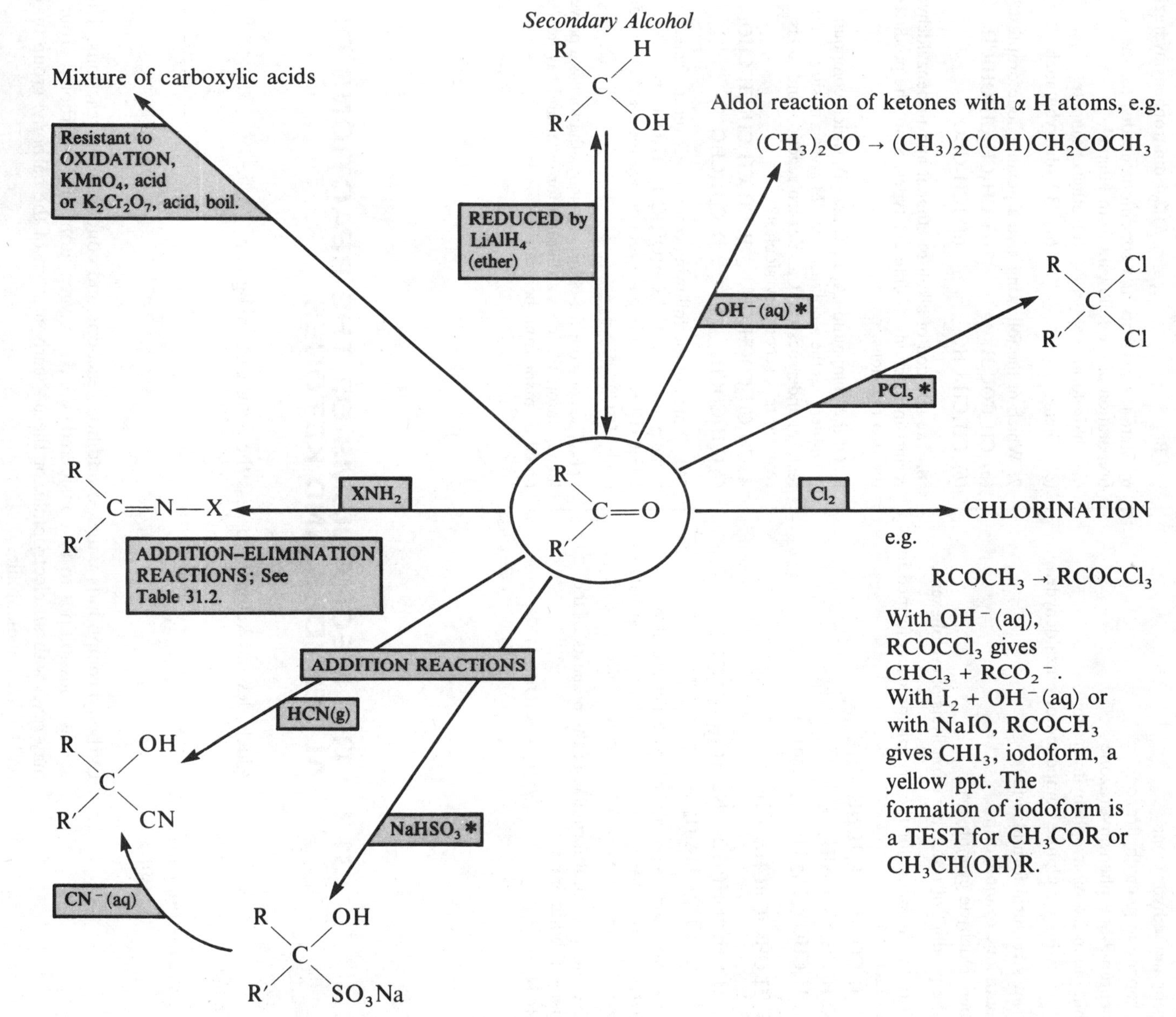

CHECKPOINT 31C: REACTIONS OF CARBONYL COMPOUNDS

1. Draw structural formulae for
(*a*) propanone hydrazone
(*b*) propanone phenylhydrazone
(*c*) pentan-2-one phenylhydrazone
(*d*) butanone 2,4-dinitrophenylhydrazone
(*e*) *the hydrogensulphite compound of benzaldehyde.

2. Draw structural formulae for
(*a*) butanone cyanohydrin
(*b*) benzaldehyde cyanohydrin
(*c*) the product of reductive ozonolysis of cyclohexene.

3. Write the formulae for the organic products formed in these reactions:
(*a*) $C_2H_5COCH_3 + C_6H_5NHNH_2$
(*b*) $C_6H_5CHO + LiAlH_4 \rightarrow$
(*c*) $C_2H_5CHO + [Ag(NH_3)_2]^+(aq) \rightarrow$
(*d*) $C_6H_5CH{=}CHCHO + H_2 \xrightarrow{Pd}$
(*e*) $C_6H_5CH{=}CHCHO + NaBH_4 \rightarrow$
(*f*) $C_2H_5COCH_3 + LiAlH_4 \rightarrow$

4. State two reactions of aldehydes which are also characteristic of ketones and two which are not shared by ketones. What is the reason for the difference?

5. Give the structural formulae of the products of the reactions of butanone with
(*a*) HCN
(*b*) $C_6H_5NHNH_2$
(*c*) Br_2
(*d*) $I_2 + NaOH(aq)$
(*e*) $LiAlH_4$
(*f*) 2,4-dinitrophenylhydrazine

6. State the reagents and conditions needed for the preparation of the following from ethanal:
(*a*) chloroform
(*b*) ethene
(*c*) ethyl ethanoate
(*d*) *3-hydroxybutanal.

7. Which of the following give a positive iodoform test?
(*a*) $CH_3COC_2H_5$
(*b*) CH_3CH_2CHO
(*c*) $CH_3CH_2CH(OH)CH_3$
(*d*) ICH_2CHO

***8.** The addition of alkali to ethanal and to benzaldehyde results in reactions of different types. State the products of the two reactions.

9. Of the compounds **A**, **B**, **C** and **D**, which compound (*a*) undergoes the aldol reaction, (*b*) gives a positive iodoform test, (*c*) undergoes the Cannizzaro reaction and (*d*) is reduced to a secondary alcohol?
A, $CH_3CH_2COCH_2CH_3$ **B**, $CH_3CH_2CH_2CH_2CHO$
C, $C_6H_5COCH_3$ **D**, C_6H_5CHO

***10.** Which of the following compounds reacts faster with sodium iodate(I): $CH_3CH(OH)CH_2CH_2OH$ or $CH_3CH_2COCH_3$? Explain your choice and state the products of the reactions.

***11.** The compound 2-chloro-1-phenylethanone is a powerful lachrymator. Write its structural formula, and say how it could be made from benzene.

31.7 *THE MECHANISMS OF THE REACTIONS OF ALDEHYDES AND KETONES

In the $\geq\overset{\delta+}{C}=\overset{\delta-}{O}$ *group, there are two reactive sites...*

Aldehydes and ketones possess the polar carbonyl group

$$\geq\overset{\delta+}{C}=\overset{\delta-}{O}$$

Carbonyl compounds undergo addition reactions and condensation reactions. There are two possibilities to be considered: the δ− oxygen atom of the carbonyl group may react with an electrophile, or the δ+ carbon atom of the carbonyl group may react with a nucleophile.

31.7.1 THE ADDITION OF HYDROGEN CYANIDE TO FORM A CYANOHYDRIN

...In the addition of HCN, it is the $\overset{\delta+}{C}$ *that is attacked by* CN^-*...*

$$RR'C{=}O + HCN \rightarrow RR'C(OH)CN$$

The rate of reaction is increased by the presence of a base and decreased by hydrogen ions. This is what would happen if the concentration of cyanide ions were a crucial factor. Hydrogen cyanide dissociates:

$$HCN(aq) + H_2O(l) \rightleftharpoons H_3O^+(aq) + CN^-(aq)$$

... The evidence comes from kinetic studies

Addition of hydrogen ions suppresses dissociation, while addition of hydroxide ions removes hydrogen ions and increases the degree of dissociation and the concentration of CN^- ions. It is likely that the rate-determining step (R.D.S.) involves CN^- ions. There are two possibilities:

$$\rangle C{=}O + H^+(aq) \underset{\text{fast}}{\rightleftharpoons} \rangle \overset{+}{C}{-}O{-}H \xrightarrow[\text{R.D.S.}]{CN^-} \rangle C(OH)(CN) \quad [1]$$

$$\rangle C{=}O + CN^-(aq) \xrightarrow[\text{R.D.S.}]{} \rangle C(CN)(O^-) \xrightarrow[\text{fast}]{H^+(aq)} \rangle C(CN)(OH) \quad [2]$$

In [1], the step involving CN^- is the attack on a positive ion. This type of reaction is fast, and is unlikely to be the rate-determining step. In [2], CN^- is involved in an attack on the polar

$$\overset{\delta+}{C}{=}\overset{\delta-}{O} \text{ group}$$

The second step is a fast reaction between $H^+(aq)$ and a negatively charged intermediate. Mechanism [2] involves CN^- in the rate-determining step, and is more likely to be correct.

A summary of the mechanism of cyanohydrin formation

The proposed mechanism for the addition reactions of aldehydes and ketones is illustrated by the formation of a cyanohydrin:

$$HCN(aq) \rightleftharpoons H^+(aq) + CN^-(aq)$$

$$RR'C{=}O + CN^- \xrightarrow[\text{R.D.S.}]{} RR'C(CN)(O^-) \xrightarrow[\text{fast}]{H^+(aq)} RR'C(CN)(OH)$$

31.7.2 A COMPARISON OF ALKENES AND CARBONYL COMPOUNDS

Addition to C=O starts with attack by a nucleophile; addition to C=C starts with attack by an electrophile

The addition reactions of carbonyl compounds contrast with those of alkenes. The first step in addition to

$$\rangle C{=}O$$

is the addition of a nucleophile, Nu: (e.g. CN^-) to form

```
\
 C—O⁻
/|
 Nu
```

In the addition to

```
\   /
 C=C
/   \
```

the first step is the addition of an electrophile (e.g. $Br^{\delta+}$—$Br^{\delta-}$). A nucleophile does not add to a carbon–carbon double bond because it is repelled by the unpolarised π electrons of the C=C bond.

The hypothetical adduct

```
\   /
 C—C
/|  -\
 Nu
```

is not formed.

Carbon is less electronegative than oxygen, and this type of species is much less stable than

```
\
 C—O
/|   -
 Nu
```

where the negative charge is carried by an oxygen atom.

CHECKPOINT 31D: REACTIVITY

1. The alkene group reacts with compounds of formula **HX**. What can **X** be?
The carbonyl group reacts with compounds of formula HY. What can **Y** be? Compare the reactivity of the alkene group and the carbonyl group in reactions of this type.

2. The formation of butanone cyanohydrin is speeded up by the addition of cyanide ions and retarded by the addition of hydrogen ions. Explain these observations.

3. (*a*) Sketch the activated complex that is formed in the reaction

$$OH^- + CH_3CH_2Br \rightarrow CH_3CH_2OH + Br^-$$

How many pairs of bonding electrons surround the carbon atom attached to the halogen? Which atoms carry negative charge?

(*b*) Sketch the intermediate in the reaction

$$HCN + CH_3CHO \rightarrow CH_3CH(OH)CN$$

How many pairs of bonding electrons surround the central carbon atom? Which atoms carry negative charge?

(*c*) Which of the two species, the activated complex in (*a*) or the intermediate in (*b*), appears to you to be the more stable? Explain your answer.

***4.** Why does HBr attack >C=C< but not >C=O?

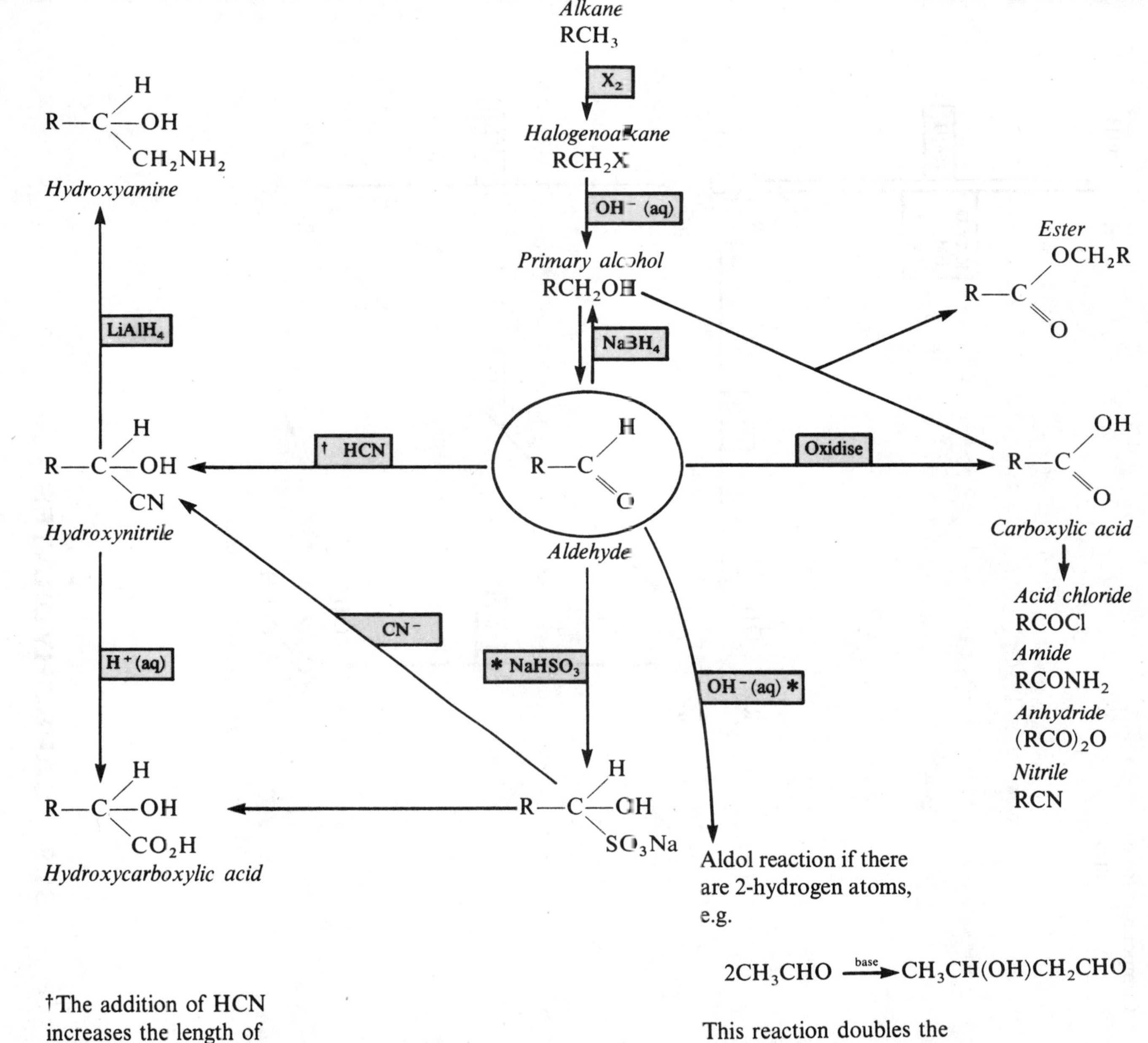

FIGURE 31.8 Aldehydes in the Preparation of Other Classes of Compounds

FIGURE 31.9 Some of the Ways of Making other Classes of Compounds from Ketones

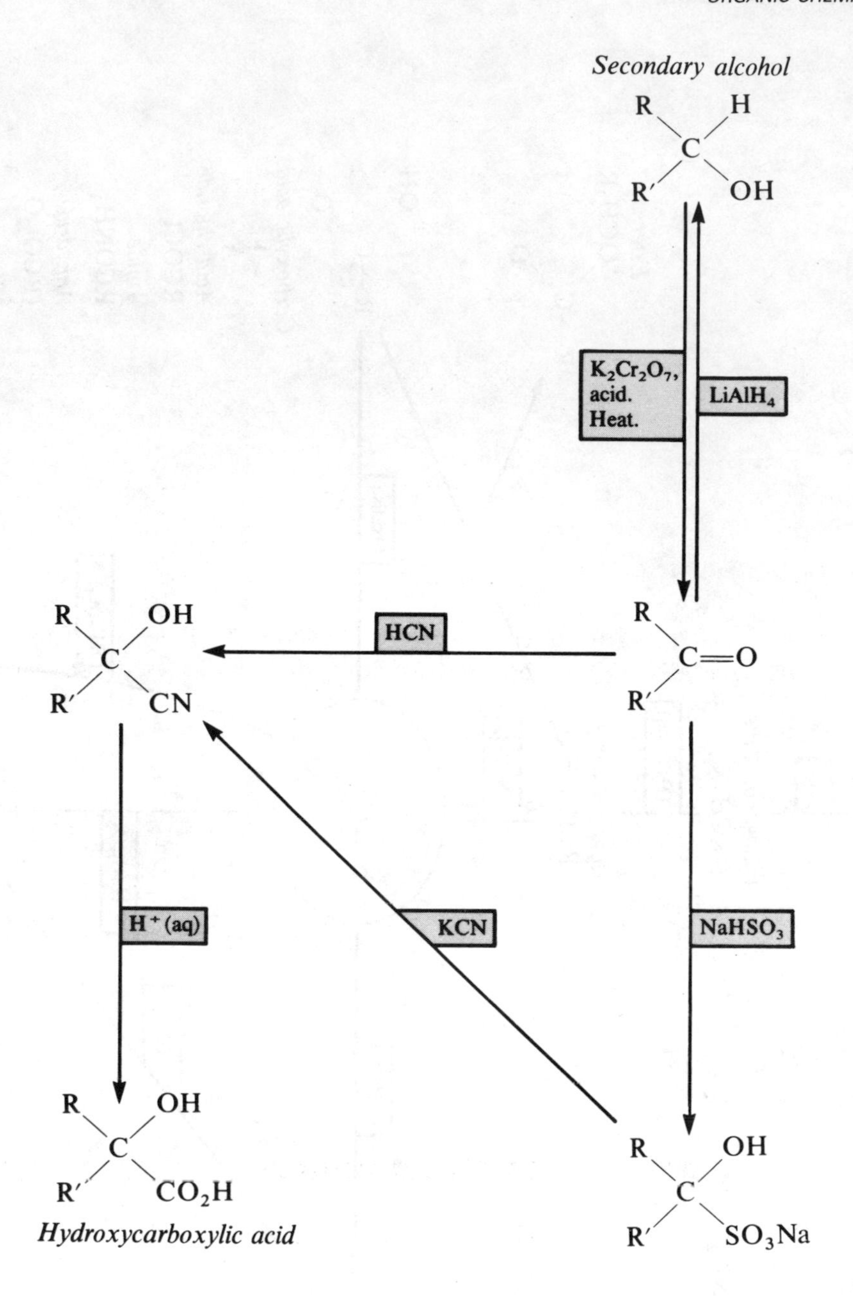

31.8 *CARBOHYDRATES

The general formula of carbohydrates is $C_m(H_2O)_n$

Carbohydrates are an important, naturally occurring group of compounds which contain carbonyl groups. This group includes sugars and starch, cellulose and a number of antibiotics. The name carbohydrate is derived from the molecular formulae of these compounds, which in many cases can be written $C_m(H_2O)_n$, e.g., glucose, $C_6H_{12}O_6$.

Sugars contain one or more carbonyl groups and other functional groups

Carbohydrates are polyfunctional. Glucose contains five hydroxyl groups as well as an aldehyde group. Fructose contains a ketone group. Both glucose and fructose have the formula $C_6H_{12}O_6$ and are monosaccharides:

```
        CHO                      CH2OH
         |                         |
     H—C—OH                      C=O
         |                         |
    HO—C—H                   HO—C—H
         |                         |
     H—C—OH                   H—C—OH
         |                         |
     H—C—OH                   H—C—OH
         |                         |
       CH2OH                     CH2OH
```

(+)-Glucose (an **aldose**) (−)-Fructose (a **ketose**)

31.8.1 DISACCHARIDES

Sucrose, maltose and lactose are disaccharides

Disaccharides have the formula $C_{12}H_{22}O_{11}$. They are hydrolysed by dilute acids and by enzymes to monosaccharides. Sucrose (cane or beet sugar), maltose (in malt) and lactose (in milk) are the commonest disaccharides.

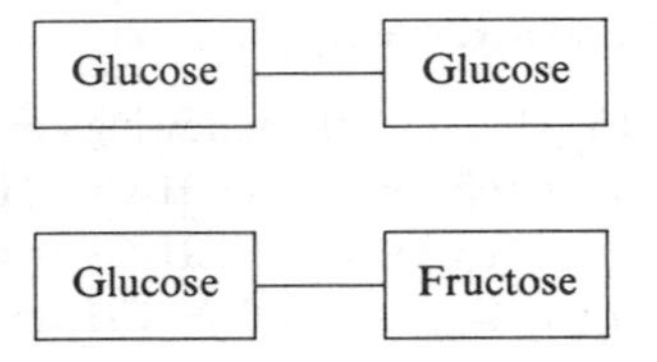

Maltose: 2 glucose ring molecules linked by the elimination of H_2O between 2OH groups

Sucrose: a glucose molecule and a fructose molecule linked as in maltose

FIGURE 31.10 The Disaccharides Maltose and Fructose

31.8.2 POLYSACCHARIDES

Starch is made by photosynthesis...

Starch and cellulose are polysaccharides. They are polymers of glucose. Starch is formed as a result of photosynthesis in green plants:

$$6CO_2(g) + 6H_2O(l) \xrightarrow[\text{sunlight and chlorophyll in green plants}]{\text{photosynthesis}} \underset{\text{Glucose}}{C_6H_{12}O_6(aq)} + 6O_2(g)$$

$$\underset{\text{Glucose}}{nC_6H_{12}O_6(aq)} \rightarrow \underset{\text{Starch}}{(C_6H_{10}O_5)_n(s)} + nH_2O(l)$$

...A test for starch is the formation of a blue colour with iodine

Starch is found in potatoes, cereals and rice. With iodine, it gives an intense blue colour, which is used as a test for iodine or for starch. Starch does not reduce Fehling's solution or Tollens' reagent.

Cellulose is another polymer of glucose

Cellulose is the main structural component of cell walls, and wood contains large quantities of cellulose. It cannot be digested by animals: they do not produce the enzymes which will hydrolyse it. Cows have living in their alimentary canals bacteria which hydrolyse cellulose.

QUESTIONS ON CHAPTER 31

1. Pentan-2-ol yields a ketone when heated with potassium dichromate solution under acid conditions.

(*a*) Give the structural formula and name of the ketone.

(*b*) What is the change in the oxidation number of chromium during the reaction?

(*c*) What other reagents will oxidise the alcohol?

(*d*) How can the product be identified as a ketone?

2. Compare the reactions of ethanal and propanone with (*a*) a few drops of concentrated hydrochloric acid, (*b*) aqueous ammoniacal silver nitrate, (*c*) 2,4-dinitrophenylhydrazine, (*d*) sodium hydrogensulphite.

3. Compare and contrast the addition reactions of ethene and ethanal. Include a discussion of the mechanisms of the reactions.

4. Suggest a method of synthesis for phenylethanone, $C_6H_5COCH_3$. Summarise the addition and addition–elimination reactions of the compound.

5. State the reactions of propanone with (*a*) acidified potassium manganate(VII), (*b*) lithium tetrahydridoaluminate, (*c*) iodine and alkali, (*d*) hydrogen cyanide. Why is reaction (*d*) so slow in the absence of a catalyst? What catalyst is used to speed up the reaction? How does it function?

6. Outline a method for the preparation of propanone from propene. What structural isomer of propanone exists? How would you distinguish between the two isomers?

7. Three different compounds of formula C_8H_8O give precipitates with 2,4-dinitrophenylhydrazine and are reduced by $LiAlH_4$ to compounds of formula $C_8H_{10}O$. Suggest structures for the three compounds, and say how you could distinguish between them. The isomerism does not arise from a difference in the orientation of substituents in the benzene ring.

8. A compound, **P**, of formula C_6H_{10}, was subjected to reductive ozonolysis. **Q**, $C_6H_{10}O_2$, was formed. **Q** gave a silver mirror with ammoniacal silver nitrate solution and reacted with 2,4-dinitrophenylhydrazine. One mole of **Q** reacted with 2 moles of DNP to form a yellow, crystalline solid. Deduce the identity of **P** and **Q**.

***9.** A liquid, **A**, of formula C_7H_6O, was refluxed with concentrated aqueous sodium hydroxide. Ethoxyethane extraction of the resulting solution, followed by drying and distillation, gave as the main product a liquid, **B**, of boiling temperature 205 °C. **B** reacted with phosphorus(V) chloride to give a pungent gas. When the solution from which **B** had been extracted was acidified, a white crystalline solid, **C**, precipitated. On recrystallisation from hot water it appeared as flaky white crystals. A solution of **C** effervesced when sodium hydrogencarbonate was added. Identify **A**, **B** and **C**, and explain the reactions described.

10. (*a*) What type of reaction takes place between benzenecarbaldehyde (benzaldehyde) and 2,4-dinitrophenylhydrazine? Under what conditions does reaction occur?

(*b*) Write equations for the reactions of benzenecarbaldehyde with (i) 2,4-dinitrophenylhydrazine, (ii) hydroxylamine, (iii) hydrazine. In reaction (iii), two products may be formed, with molecular formulae $C_7H_8N_2$ and $C_{14}H_{12}N_2$. Write structural formulae for these compounds.

11. **A** has the formula $C_5H_{12}O$. On oxidation it gives **B**, of formula $C_5H_{10}O$. **B** reacts with phenylhydrazine and gives a positive result in an iodoform test. **A** is dehydrated by concentrated sulphuric acid to **C**, C_5H_{10}. Reductive ozonolysis of **C** gives butanal. What is **A**?

12. **P** has the formula $C_5H_8O_2$. It forms a compound by reaction with hydrogen cyanide which has the formula $C_7H_{10}O_2N_2$. **P** gives a positive iodoform test, a silver mirror with Tollens' reagent and can be reduced to pentane. What is **P**?

13. **Q**, of formula $C_5H_{12}O_2$, reacts with ethanoic anhydride to form **R**, formula $C_9H_{16}O_4$. **Q** does not react with phenylhydrazine, but on oxidation it forms **S**, $C_5H_8O_2$, which reacts with phenylhydrazine to form $C_5H_8({=}NNHC_6H_5)_2$, and also reduces Fehling's solution. **S** reacts with sodium iodate(I) to give iodoform and butane-1,4-dioic acid. Deduce the identity of **Q**, and explain the reactions described.

14. Describe chemical tests which you could do to distinguish between the following compounds:

(*a*) C_2H_5CHO, CH_3COCH_3, $C_2H_5CH_2OH$ and $(CH_3)_2CHOH$

(*b*) C_6H_5CHO, $C_6H_5CH_2OH$, $C_6H_5COCH_3$ and $C_6H_5CH{=}CH_2$

15. How can the following conversions be brought about?

(*a*) $CH_3COCH_3 \rightarrow CH_3COCH_2OH$

(*b*) $CH_3COCH_3 \rightarrow CH_3CH{=}CH_2$

(*c*) $C_6H_5CHO \rightarrow C_6H_5CH_2Br$

16. (*a*) Compare the reactions of ethanal and propanone with (i) aqueous ammoniacal silver nitrate, (ii) acidified potassium dichromate((VI), (iii) hydrogen cyanide, (iv) 2,4-dinitrophenylhydrazine.

(*b*) How does the presence of a catalyst assist reaction (iii)?

(*c*) Explain why there is only one oxime of propanone, while there are two isomeric oximes of ethanal.

***17.** A compound **X** has the formula $C_5H_{10}O$. It gives a 2,4-dinitrophenylhydrazone. Ammoniacal silver nitrate oxidises **X** to **Y**. Sodium tetrahydridoborate reduces **X** to **Z**. With aqueous potassium hydroxide, **X** gives a mixture of **Z** and the potassium salt of **Y**. Deduce the identity of **X**, **Y** and **Z**, and explain the reactions described.

***18.** The formula of glucose is

$$\begin{array}{c} CHO \\ | \\ (CHOH)_4 \\ | \\ CH_2OH \end{array}$$

From your knowledge of the reactions of —OH groups and —CHO groups, say how you would prepare the following substances from glucose:

(*a*) hexane-1,2,3,4,5,6-hexol

(*b*) 2,3,4,5,6-pentahydroxyhexanoic acid

(*c*) sugar charcoal

(*d*) glucose hexaethanoate

19. Compound **A** is an aldehyde of general formula RCHO, where R is an alkyl group.

(*a*) (i) Write the structural formula of the compound **B**, produced by the treatment of **A** with warm acidified potassium dichromate solution.

(ii) State, with an explanation, whether you would expect **B** to have a higher or a lower boiling point than **A**.

(*b*) (i) State what reaction you would expect to occur on reacting **A** with 2,4-dinitrophenylhydrazine.

(ii) Name the type of reaction that is taking place in (*b*)(i) above.

(iii) Explain how the product of this reaction may be used to identify carbonyl compounds such as **A**.

(*c*) (i) Write the structural formula of the compound **C**, which you would expect to be formed by reacting **A** with lithium tetrahydridoaluminate(III) in ether.

(ii) State what type of reaction is taking place in (*c*)(i) above.

(iii) Suggest, with the appropriate conditions, another reagent which would bring about the same transformation, **A** to **C**.

(iv) State, with an explanation, whether you would expect **C** to have a higher or a lower boiling point than **A**.

(*d*) (i) Describe the reaction which you would expect to take place on *gently* warming **A** with diamminesilver(I) solution State what you would expect to *observe*.

(ii) State what type of reaction is taking place in (*d*)(i) above.

(*e*) (i) Write the structural formula of the compound **D**, which you would expect to be formed on reacting compound **B** with compound **C**.

(ii) State the appropriate conditions for the reaction in (*e*)(i).

(iii) State, with an explanation, whether you would expect **D** to have a higher or a lower boiling point than **B** and **C**.

(WJEC 90)

20. (*a*) Describe the manufacture of chloroethene (vinyl chloride monomer) from ethene.

(*b*) Describe *one* method of manufacturing poly(chloroethene) (PVC) from the monomer.

(*c*) State *two* problems associated with the handling and use of chloroethene.

(*d*) The world production of PVC is some 17 million tons annually.

(i) State the general properties of the pure polymer and discuss the way in which these may be modified.

(ii) Give *two* uses of PVC and relate the properties in (*d*)(i) to these uses.

(*e*) An industrially important monomer **A**, which can undergo addition polymerisation, has the empirical formula C_2H_3O. On heating with aqueous sodium hydroxide it forms two products **B** and **C**. **B** yields methane on heating with soda lime. **C** rapidly undergoes structural isomerisation into a liquid **D**. **D** reacts with a diamminesilver(I) solution to give a silver mirror and a product **E**. Addition of cold dilute sodium hydroxide to **E** converts it into **B**.

Identify the compounds **A** to **E** by name or (shortened) structural formula, giving your reasoning at *each* stage.

(WJEC 90)

21. (*a*) Ethanal, b.p. 21 °C, can be made from ethanol by slow addition of acidified $K_2Cr_2O_7$ solution while heating the mixture to distil off the product.

(i) Draw a labelled diagram to show the apparatus used for this experiment.

(ii) Write a balanced half-equation to show

(*1*) the oxidation of ethanol to ethanal

(*2*) the reduction of $K_2Cr_2O_7$.

(iii) Why is it necessary continuously to distil off the product?

(*b*) (i) Write down the equation and draw the mechanism for the addition of HCN (in the presence of a small amount of KCN) to ethanal, and explain why the presence of KCN is necessary.

(ii) The addition product from (*b*)(i) can be used to make other compounds:

With hydrogen and a catalyst it gives a base, **P**, C_3H_9NO.

With hot aqueous acid it gives an acid, **Q**, $C_3H_6O_3$.

Draw structural formulae for **P** and **Q** and *classify* the type of reaction

(O 91)

22. A certain industrial cleaner and paint solvent was distilled to produce a single compound **D**. When **D** reacted with 2,4-dinitrophenylhydrazine, an orange precipitate was produced. With alkaline aqueous iodine, **D** gave a pale yellow precipitate. **D** did *not* react either with warm acidified potassium dichromate(VI) or with aqueous bromine. Reduction of **D** with hydrogen over a catalyst produced an equimolecular mixture of two isomers, **E** and **F**, with the molecular formula $C_4H_{10}O$.

Suggest structural formulae for **D**, **E** and **F** and explain the reactions involved.

(C 91)

23. (*a*) (i) Both propene and propanal undergo addition reactions. By describing the mechanisms of the reactions, compare the addition of hydrogen bromide to propene with the addition of hydrogen cyanide to propanal.

(ii) Both propanal and propanone show similar addition reactions but differ in their behaviour towards oxidising reagents. Choose a specific oxidising reagent which may be used in a test tube reaction to distinguish propanal from propanone. Give the changes which would be observed with *each* compound and the change in the oxidation number of the oxidising reagent.

(*b*) (i) Outline a mechanism by which ethanal reacts with hydroxylamine, NH_2OH.

(ii) Suggest a method by which the products formed when a mixture of carbonyl compounds reacts with 2,4-dinitrophenylhydrazine may be separated and the individual aldehydes or ketones characterised.

(iii) There are two possible organic products of the reaction between ethanal and hydroxylamine but only one possible product between propanone and hydroxylamine. Suggest an explanation for this.

(AEB 92, S)

32

AMINES

32.1 NOMENCLATURE FOR AMINES

Primary, secondary and tertiary (1°, 2°, 3°) amines...

Amines are organic derivatives of ammonia. They are classified as primary, secondary or tertiary amines according to the number of alkyl or aryl groups attached to the nitrogen atom. Examples are:

... some examples...

CH_3NH_2	$(CH_3)_2NH$	$(CH_3)_3N$	$C_6H_5CH_2NH_2$
Methylamine a primary (1°) amine	Dimethylamine a secondary (2°) amine	Trimethylamine a tertiary (3°) amine	(Phenylmethyl)amine (or benzylamine) 1° amine

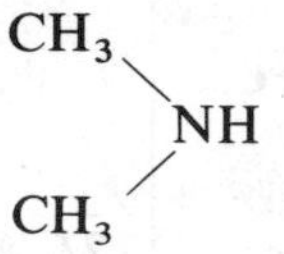

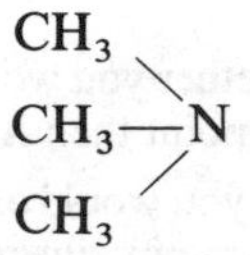

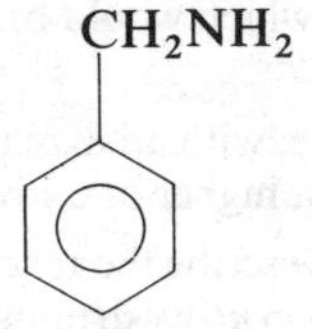

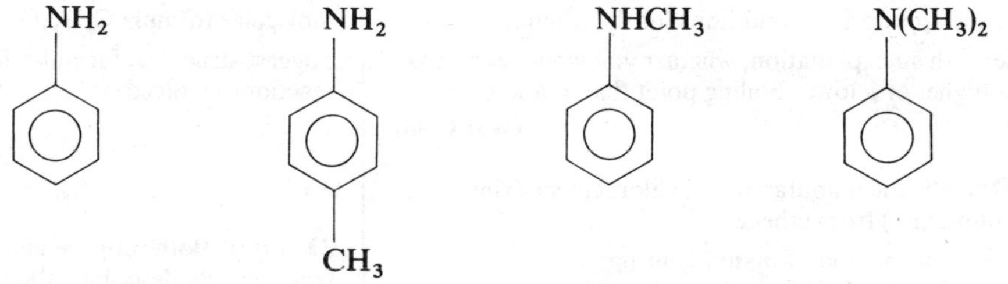

$C_6H_5NH_2$	$4\text{-}CH_3C_6H_4NH_2$	$C_6H_5NHCH_3$	$C_6H_5N(CH_3)_2$
Phenylamine (or aniline) 1° aromatic amine	4-Methyl-phenylamine	*N*-Methyl-phenylamine 2° aromatic amine	*N*,*N*-Dimethyl-phenylamine 3° aromatic amine

An aromatic amine has the nitrogen atom directly attached to the aromatic ring. In phenylmethylamine the nitrogen atom is not directly attached to the ring: this is not an aromatic amine; it is a phenyl-substituted alkylamine.

... amino-compounds

The system of naming is illustrated by the examples given above. The alkyl groups or aryl groups attached to the nitrogen atom are named, and the ending *-amine* is added. Some amines are named as amino-substituted compounds:

$H_2NCH_2CO_2H$	$HO\text{—}C_6H_4\text{—}NH_2$
2-Aminoethanoic acid	4-Aminophenol

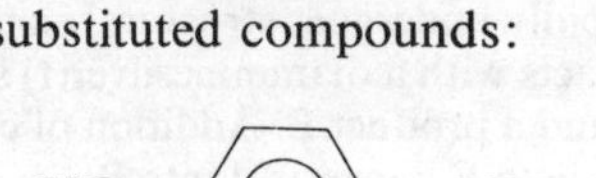

Related to the amines are the **quaternary** (4°) ammonium compounds, in which nitrogen is tetravalent. They are named as alkyl-substituted ammonium salts:

Quaternary ammonium compounds

$$(CH_3)_4N^+I^-$$

Tetramethylammonium iodide

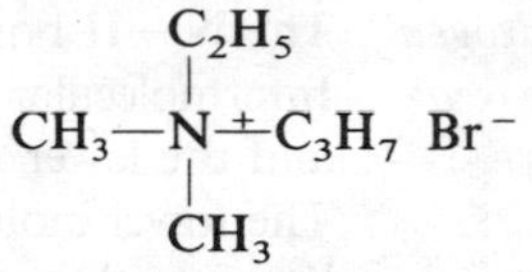

$$CH_3-\overset{C_2H_5}{\underset{CH_3}{N^+}}-C_3H_7\ Br^-$$

Ethyldimethylpropylammonium bromide

Other nitrogen compounds will be covered in Chapter 33. They are the amides and nitriles. They are named as derivatives of acids:

Amides, nitriles

Carboxylic acid $R-C(=O)-OH$ e.g. $CH_3CH_2C(=O)-OH$
Propanoic acid

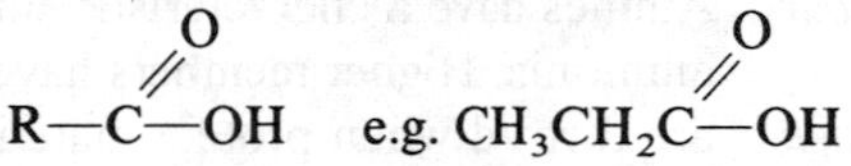

Amide $R-C(=O)-NH_2$ e.g. $CH_3CH_2C(=O)-NH_2$
Propanamide

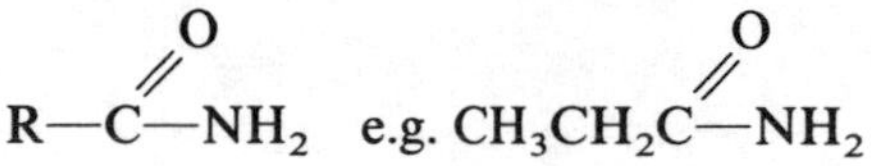

Nitrile $R-C\equiv N$ e.g. $CH_3CH_2C\equiv N$
Propanonitrile

Other derivatives of acids which will be met in this chapter are:

Acid chlorides and anhydrides

Acid chloride $R-C(=O)-Cl$ e.g. $CH_3CH_2C(=O)-Cl$
Propanoyl chloride

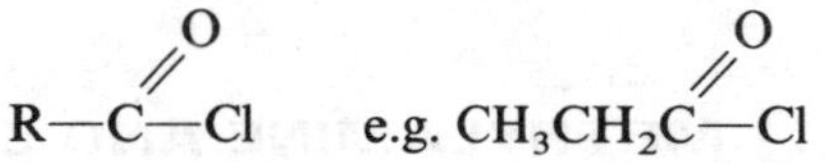

Acid anhydride $R-C(=O)-O-C(=O)-R$ e.g. $CH_3CH_2C(=O)-O-C(=O)CH_2CH_3$
Propanoic anhydride

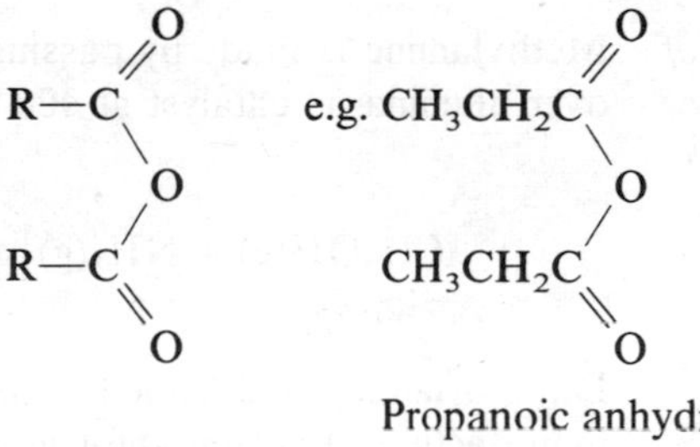

32.2 NATURAL OCCURRENCE

Proteins . . . *. . . amino acids . . .* *. . . DNA . . .*

Compounds with amino groups are widely distributed in nature. Amino acids, of formula $H_2NCH(R)CO_2H$, are the building blocks from which proteins are made. Their chemistry is described in § 33.17. The way in which hydrogen bonding between amino groups and other groups maintains the three-dimensional configuration of proteins is illustrated in Figure 4.39, § 4.7.3. The bases in DNA (deoxyribonucleic acid) are amines. Hydrogen bonding between the bases maintains the double helical structure of DNA, as shown in Figures 4.40 to 4.43, § 4.7.3.

. . . drugs

Many powerful drugs and medicines are amines, e.g., morphine (a powerful painkiller) [§ 35.2], strychnine (a powerful poison) and LSD (a hallucinogen) [§ 35.2].

32.3 PHYSICAL PROPERTIES

Intermolecular hydrogen bonding is weak

The N—H bond is polar, more polar than C—H but less polar than O—H. Intermolecular hydrogen bonding in amines is therefore weaker than in alcohols, and the lower molecular mass amines (up to C_3) are gases at room temperature. The lower molecular mass amines dissolve in water as they can form hydrogen bonds with water molecules. Phenylamine and other amines with a large hydrocarbon part of the molecule are only sparingly soluble in water but are soluble in organic solvents. Since phenylamine is soluble in fatty tissues, it can be absorbed through the skin. The ease of absorption, combined with its toxicity, makes phenylamine a somewhat dangerous substance.

Solubility

Toxicity

Odour

Amines are formed when proteins decay

Amines have a characteristic smell. That of the lower members of the series resembles ammonia. Higher members have an odour described as 'fishy'. This is because amines are formed when protein material decomposes; dimethylamine and trimethylamine are found in rotting fish. Higher amines are found in decaying animal flesh.

32.4 INDUSTRIAL MANUFACTURE AND USES

32.4.1 METHYLAMINE AND ETHYLAMINE

Alcohols are a source of amines

Methylamine is made by passing methanol vapour with ammonia under pressure over alumina as catalyst at 400 °C:

$$CH_3OH(g) + NH_3(g) \xrightarrow[400\,°C]{Al_2O_3} CH_3NH_2(g) + H_2O(g)$$

Dimethylamine and trimethylamine are also formed. Dimethylamine is used for the manufacture of solvents, for jet fuel and rocket fuel.

Ethylamine is made in a similar way from ethanol and ammonia.

32.4.2 PHENYLAMINE

Phenylamine is used to make dyes and rubber...

Phenylamine is used for the manufacture of dyes [§32.10.3]. Its derivatives find use in the rubber industry. There are two chief industrial methods of manufacture:

...It is manufactured industrially from nitrobenzene or chlorobenzene

(*a*) Reduction of nitrobenzene, $C_6H_5NO_2$, by
(i) catalytic hydrogenation, or
(ii) iron and hydrochloric acid (similar to the laboratory method [§ 32.7.2]).

(b) Reaction between ammonia and chlorobenzene at 200 °C, under high pressure, in the presence of copper(I) oxide as catalyst.

32.5 DISASTER AT BHOPAL

TOPIC

In December 1984 a Union Carbide plant in Bhopal, India, released a cloud of poisonous gas which killed 2500 people and injured 200 000 more

In December 1984 a cloud of poisonous gas swept through the Indian city of Bhopal, killing 2500 people and injuring 200 000 more. The gas came from a Union Carbide factory making a pesticide called Carbaryl® or Sevin®. On the night of the accident, Union Carbide did not inform the authorities or tell the hospitals what to do. As victims crowded into the hospital, the company medical officer told doctors that the gas was methyl isocyanate, that it was non-poisonous, that it was like tear gas, making the eyes water, and that applying water would bring relief. Even 15 days later, when thousands had died, the works manager was still defending this statement and saying that he knew of no fatalities from methyl isocyanate. The doctors trying to treat the victims did not know what had poisoned them.

In the manufacture of the pesticide the following reactions take place:

$$COCl_2(g) + CH_3NH_2(g) \rightarrow CH_3NCO(l) + 2HCl(g)$$

Carbon dichloride oxide (phosgene) + Methylamine → Methyl isocyanate

$$CH_3NCO + \text{1-Naphthol (}C_{10}H_7OH\text{)} \rightarrow \text{Carbaryl (}C_{10}H_7\text{–O–C(=O)–N(H)CH}_3\text{)}$$

Methyl isocyanate + 1-Naphthol → Carbaryl

The escaping gas was said by Union Carbide to be methyl isocyanate and to be non-poisonous. Doctors thought that it might have contained phosgene because it killed foliage. Some of the symptoms resembled cyanide poisoning. At 400 °C, methyl isocyanate decomposes to give hydrogen cyanide. When victims were treated with sodium thiosulphate, their symptoms were alleviated and they excreted thiocyanate faster in their urine. Union Carbide gave no help or advice to the doctors treating the injured

Phosgene is a poisonous gas that has been used in chemical warfare, methylamine is a non-toxic gas with a fishy smell, and the effects of methyl isocyanate were unknown. At first doctors in Bhopal thought the gas was phosgene (T_b 8 °C) because phosgene would be more likely to vaporise on a cool evening than methyl isocyanate (T_b 39 °C) and also because the gas had damaged plants for miles around the factory, just as phosgene does. The effects of phosgene are known because it was used in the First World War. It causes mild irritation at first, until it is hydrolysed in the body to carbon monoxide and hydrogen chloride which cause pulmonary oedema: lungs swollen up with water. Phosgene kills the victim within two days or more, but in Bhopal many people were killed immediately. As days went by, it seemed that a mixture of gases was at work, one acting quickly and one producing symptoms after two to three days.

Local investigators began to think that methyl isocyanate had broken down at the raised temperature of the storage tank to form hydrogen cyanide; this would explain the speed with which some of the victims died. There are research papers which describe decomposition of methyl isocyanate at 400°C. The Union Carbide report on the accident did not identify the mixture of gases that escaped. Union Carbide maintained that methyl isocyanate cannot lead to permanent damage or long-term effects. The victims suffered in the uncertainty about what gases they had inhaled. Sodium thiosulphate is used as a treatment for cyanide poisoning; it causes excretion of cyanide as thiocyanate, CNS^-. Union Carbide said that it was not necessary or advisable to use sodium thiosulphate to treat the survivors. The blood of victims was dark cherry-red, showing that some poison was blocking the use of oxygen; this would fit in with cyanide poisoning. The possibility that methyl isocyanate breaks down in the body to cyanide was suggested. When survivors were treated with thiosulphate, their use of oxygen improved. Two months after the disaster, the Indian Council for Medical Research started treating survivors with thiosulphate, and found that it alleviated their symptoms and increased the excretion of thiocyanate in their urine. Even after this, very few people received thiosulphate treatment.

The death toll continues to rise as more people die of lung diseases. Thousands of people suffered impairment of their eyesight

The number of casualties is still increasing as people continue to die from lung diseases. Thousands of survivors have been unable to work because of their failing lungs. For some months after the accident there was an increase in women's diseases and in still births and in the number of babies born with defects. The eyes of over 70% of the population were affected, and at first it was thought that thousands would go blind. Although thousands of people now have seriously impaired eyesight and eye irritation, fortunately blindness has not been one of the permanent injuries inflicted by the gas.

How did the accident happen? At 11 p.m. on 2 December 1986 the temperature of a tank containing 3840 gallons of methyl isocyanate rose to 38 °C. As the temperature rose, the liquid vaporised and the pressure in the tank increased. Valves are fitted to the storage tanks to open automatically if the pressure rises and allow gas to escape into 'scrubbers', filled with sodium hydroxide which converts the gas into harmless products:

$$\underset{\text{Methyl isocyanate}}{CH_3NCO(g)} + 2OH^-(aq) \rightarrow \underset{\text{Methylamine}}{CH_3NH_2(g)} + \underset{\text{Carbonate ion}}{CO_3^{2-}(aq)}$$

A rise in temperature of a storage tank holding methyl isocyanate started the chain of events which led to the accident. Valves failed to open to allow gas to escape into 'scrubbers' and release the pressure. The refrigeration unit did not work, one scrubber was out of action, and the flare tower failed to ignite and burn the escaping gas

The automatic relief valves were not working, and the emergency valves did not work either. The refrigeration plant which might have cooled the storage tank was not working. Two men were sent with hosepipes to cool the tank, but as the pressure inside the tank continued to rise, they ran away. At 1 a.m. a valve ruptured and methyl isocyanate gas surged towards the two scrubber tanks. Only one scrubber was working: the other had been shut down for repairs. The one scrubber could not cope, and gas poured out of it. There was a last line of defence, a flare tower which was designed to burn off escaping gas; this flare failed to ignite. The nightshift fled, while a lone supervisor struggled for 45 minutes to stem the flow of gas. All 3840 gallons of methyl isocyanate escaped. Its dense vapour rolled across the ground to the shanty towns just across the road. It left a trail of dead, blinded and choking people.

There must have been a reason for the liquid methyl isocyanate to have reached 38 °C on a cool evening. A chemical reaction must have been taking place in the storage tank. Methyl isocyanate may have reacted with impurities present in it or in the nitrogen that was fed into the tank. There must have been serious failures in the purification and detection systems for this to happen — unless warnings from the detection systems were ignored.

Union Carbide gave little assistance to the injured. The local emergency services were unable to cope with a disaster of this magnitude

Union Carbide had never instructed the population of Bhopal what to do if a leak occurred. If people had known that the best thing to do was to stay indoors and cover their faces with wet towels, hundreds of lives could have been saved. The plant's alarm system did not go off until three hours after the accident. Local government proved to be incapable of tackling a major emergency. India does not have the kind of emergency services that exist in advanced countries. It is interesting that in the Union Carbide plant in France there is a computerised system which is sensitive enough to detect 0.3 ppm of methyl isocyanate and to switch on sprinklers if the temperature of the liquid rises. At the Union Carbide plant in West Germany there is a mobile chemical emergency unit which can smother a leaking tank with foam and suck up gallons of liquid or gas, yet no one is allowed to live within 2 km of the plant.

CHECKPOINT 32A: BHOPAL

1. (*a*) Why do people in developing countries live near the factory?
(*b*) What can be done to improve the chances of the local people if there is an accident?
(*c*) Why is Union Carbide able to get away with less stringent safety precautions in India than in France and Germany?

2. (*a*) What precautions could Union Carbide have taken to avoid the accident that happened at Bhopal?
(*b*) After the accident had happened, what assistance could Union Carbide have given to the injured?

3. How does the accident in Bhopal in 1984 compare with the accident at Seveso in 1976 [§ 30.13]?

32.6 BASICITY OF AMINES

Amines are weak bases, resembling ammonia

The nitrogen atom in ammonia has a lone pair of electrons, which enable it to act as a base ($pK_b = 4.74$) [§ 12.7.1]. In the same way, amines are weak bases. Their solutions are alkaline, and they react with acids to form salts:

$$CH_3NH_2{:} + H_2O \underset{pK_b\,=\,3.36}{\rightleftharpoons} CH_3\overset{+}{N}H_3 + OH^-$$

$$(CH_3)_3N{:} + H_2O \underset{pK_b\,=\,4.20}{\rightleftharpoons} (CH_3)_3\overset{+}{N}{-}H + OH^-$$

$$C_6H_5NH_2{:} + H_2O \underset{pK_b\,=\,9.4}{\rightleftharpoons} C_6H_5\overset{+}{N}H_3 + OH^-$$

The gas methylamine reacts with hydrogen chloride to form a white crystalline solid, methylammonium chloride, $CH_3NH_3^+Cl^-$. The salt is involatile, odourless and soluble in water. The addition of a strong base, such as sodium hydroxide, liberates the weak base methylamine from its salt:

$$CH_3NH_3^+Cl^-(s) + OH^-(aq) \rightarrow CH_3NH_2(g) + H_2O(l) + Cl^-(aq)$$

Trimethylamine forms trimethylammonium salts, e.g., $(CH_3)_3\overset{+}{N}H\ Br^-$, and the salts of phenylamine are called phenylammonium salts, e.g., $C_6H_5NH_3^+I^-$.

Aliphatic amines are more basic than ammonia. Aromatic amines are weaker bases because of the delocalisation of the N lone pair

Aliphatic amines are stronger bases than ammonia in aqueous solution because the positive charge in the alkylammonium ion can be shared between the nitrogen atom and a carbon atom. This distribution of charge stabilises the cation. Aromatic amines are weaker bases. The electron pair on the nitrogen atom is partially delocalised by interaction with the π electron cloud of the benzene ring [see Figure 32.1]. As a result of delocalisation, the lone pair is less available for coordination to a proton.

FIGURE 32.1
Phenylamine

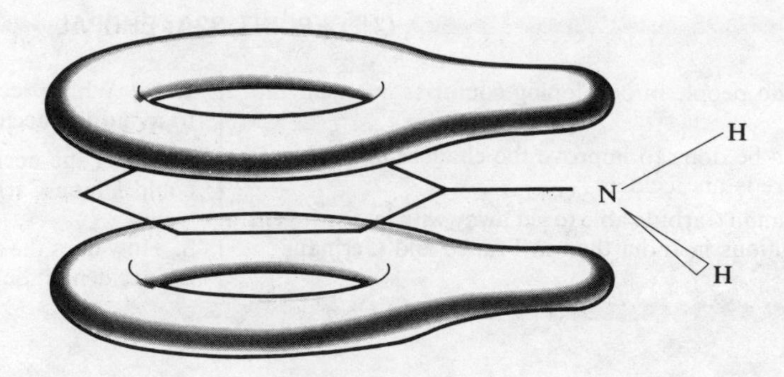

Substituents affect basicity

Substituents in the benzene ring affect the basicity of phenylamine. Electron-donating groups, e.g. CH_3O— and HO—, especially in the 4-position, increase the basicity; electron-withdrawing groups, e.g. —NO_2 and —Cl, decrease the basicity.

32.6.1 NOTE ON AMIDES

Amides, $RCONH_2$, are not bases

Amides

$$R{-}\overset{\overset{\displaystyle O}{\|}}{C}{-}NH_2$$

are not basic. The electron-withdrawing character of the carbonyl group reduces the electron density on the nitrogen atom:

$$-C(=O)-\ddot{N}H_2$$

CHECKPOINT 32B: FORMULAE

1. Draw structural formulae for (*a*) ethylmethylpropylamine, (*b*) (1-methylhexyl)amine, (*c*) (2-chloro-1-ethylhexyl)amine, (*d*) diethyldimethylammonium chloride, (*e*) *N,N*-dimethylphenylamine and (*f*) 4-nitrophenylamine.

2. Name the following compounds, and say whether they are 1°, 2°, 3° or 4° amines:

(*a*) $CH_3{-}N(CH_3){-}H$

(*b*) $C_2H_5{-}N(CH_3){-}C_2H_5$

(*c*) $C_6H_5{-}NH{-}C_6H_5$

(*d*) $I{-}C_6H_4{-}NHC_2H_5$

(*e*) $CH_3{-}\overset{+}{N}(CH_3)_2{-}C_2H_5\ Br^-$

(*f*) $H_3C{-}C_6H_4{-}NHCH_3$

3. Which of the following has the highest boiling temperature?

$(C_2H_5)_3N$ $(CH_3CH_2CH_2)_2NH$ $CH_3(CH_2)_5NH_2$

32.7 LABORATORY PREPARATIONS

32.7.1 THE REACTION OF AMMONIA AND A HALOGENOALKANE

RX + NH_3 form a mixture of 1°, 2°, 3° and 4° ammonium salts

Ammonia and a halogenoalkane (in solution in ethanol) are heated together in a **bomb** (a metal container which will withstand high pressure):

$$RX(alc) + NH_3(alc) \xrightarrow{\text{heat in a bomb}} R\overset{+}{N}H_3\,X^-(s)$$

A mixture of primary, secondary, tertiary and quaternary ammonium salts is formed:

$$NH_3 \xrightarrow{CH_3I} CH_3\overset{+}{N}H_3I^- \xrightarrow{CH_3I} (CH_3)_2\overset{+}{N}H_2I^- + HI \xrightarrow{CH_3I} (CH_3)_3\overset{+}{N}H\,I^- + HI$$

$$\xrightarrow{CH_3I} (CH_3)_4\overset{+}{N}\,I^- + HI$$

The products are separated by the addition of alkali to liberate the free amines, followed by fractional distillation.

32.7.2 REDUCTION OF NITROGEN COMPOUNDS

(a) Reduction of a Nitrile, R—C≡N

(b) Reduction of an Amide, $RCONH_2$

Lithium tetrahydridoaluminate in ethoxyethane solution is used as the reducing agent for (**a**) and (**b**). The use of an acidic reagent which would hydrolyse the starting material must be avoided.

(c) Reduction of a Nitro-compound

$ArNH_2$ is made by the reduction of $ArNO_2$

Practical details for the preparation of phenylamine

Phenylamine, $C_6H_5NH_2$, is made by the reduction of nitrobenzene, $C_6H_5NO_2$. Tin(II) chloride is made in the reaction mixture from tin and hydrochloric acid [see Figure 32.2]. In reducing the nitro-compound, tin(II) chloride is oxidised to tin(IV) chloride, which reacts with hydrochloric acid to form the complex ion, $[SnCl_6]^{2-}$. The reaction product is therefore $(C_6H_5\overset{+}{N}H_3)_2\,[SnCl_6]^{2-}$. The addition of concentrated alkali liberates phenylamine, and the addition of sodium chloride to the mixture reduces the solubility of phenylamine in water. Steam distillation is employed to separate phenylamine. Extraction with ethoxyethane separates phenylamine from the water in the distillate. Distillation removes ethoxyethane from the dried extract. The product can be purified by distillation under reduced pressure. Phenylamine decomposes when heated to its normal boiling temperature.

FIGURE 32.2 Reduction of Nitrobenzene to Phenylamine

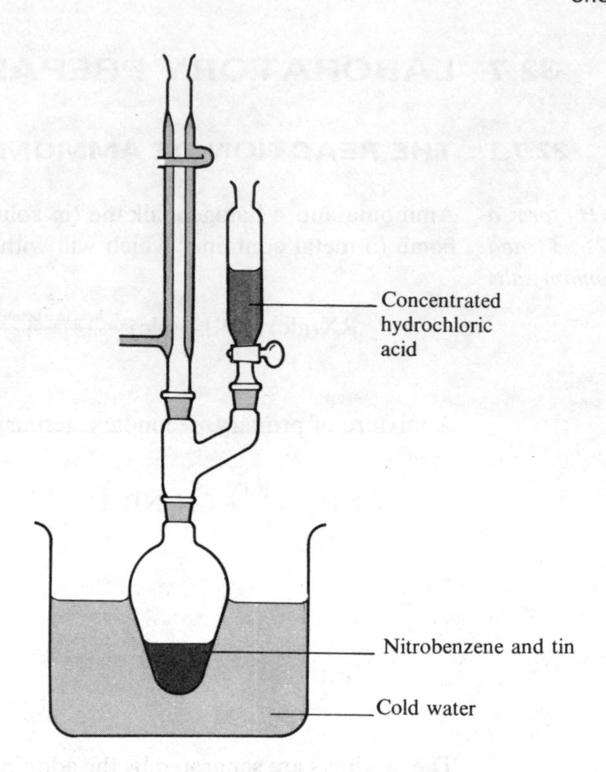

32.7.3 FROM AN AMIDE BY THE HOFMANN DEGRADATION

$RCONH_2$ can be converted into RNH_2 by the Hofmann degradation

The action of bromine and concentrated alkali is to convert an amide of formula $RCONH_2$ into an amine of formula RNH_2:

$$\underset{\text{Ethanamide}}{CH_3CONH_2} + Br_2 + 4NaOH \rightarrow \underset{\text{Methylamine}}{CH_3NH_2} + 2NaBr + Na_2CO_3 + 2H_2O$$

This reaction, the Hofmann degradation, reduces the number of atoms in the carbon chain [§ 35.3].

CHECKPOINT 32C: PREPARATIONS OF AMINES

1. Explain how you would carry out the following conversions. Some of them involve more than one step:

(*a*) $CH_3CH_2Cl \rightarrow CH_3CH_2NH_2$

(*b*) $CH_3CH_2Cl \rightarrow CH_3CH_2CH_2NH_2$

(*c*) C_6H_5–CHO $\longrightarrow$ C_6H_5–CH_2NH_2

(*d*) C_6H_5–CH_3 $\longrightarrow$ $C_6H_4(CH_3)(NH_2)$ (2-methylphenylamine: NH_2 ortho to CH_3)

(*e*) $CH_3CH{=}CH_2 \rightarrow (CH_3)_2CHNH_2$

(*f*) C_6H_5–CH_3 $\longrightarrow$ H_2N–C_6H_4–CH_3 (4-methylphenylamine)

2. Phenylamine can be made from nitrobenzene. Describe how the reduction is carried out. Explain why the product of the reaction is made alkaline before being steam-distilled. Describe how the product is extracted from the steam-distillate, and how it is finally purified.

3. Complete the following equations, indicating reagents and conditions:

(*a*) $(CH_3)_3N + C_2H_5I \xrightarrow{\text{conditions}}$

(*b*) C_6H_5–$NO_2 + Zn + HCl(aq) \rightarrow$

(*c*) $C_3H_7OH + NH_3 \xrightarrow[\text{conditions}]{\text{catalyst}}$

(*d*) $C_2H_5CONH_2 \xrightarrow[\text{solvent}]{\text{reducing agent}} C_2H_5CH_2NH_2$

(*e*) $C_2H_5Br \rightarrow \mathbf{X} \rightarrow (C_2H_5)_4NBr$

4. A is an amine which is insoluble in water. **P** is a phenol which is insoluble in water. **E** is a solution of **A** and **P** in ethoxyethane. How could you separate **A** and **P** from the solution **E**, using only acid and alkali and ethoxyethane?

FIGURE 32.3 Methods of Preparing Amines

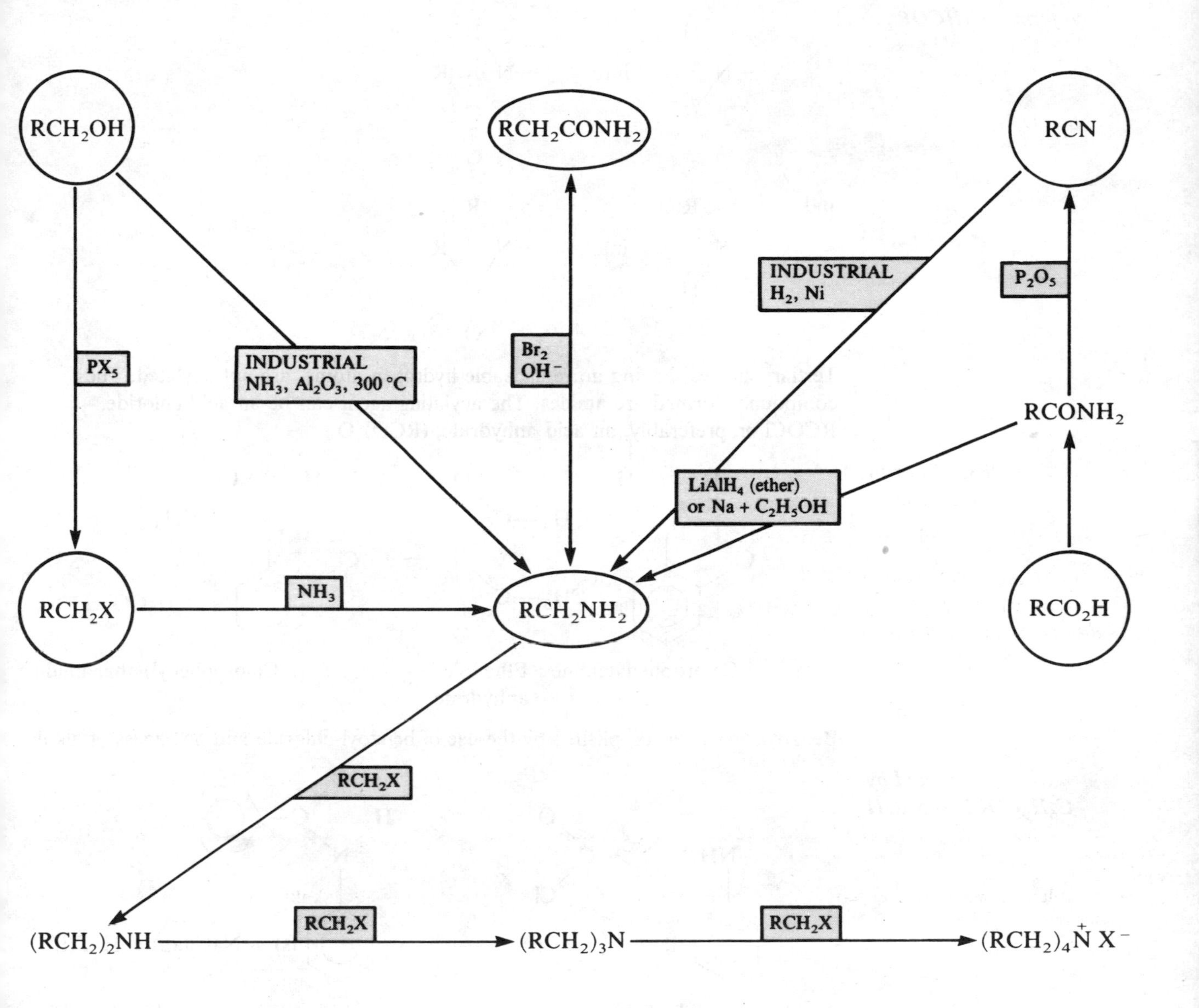

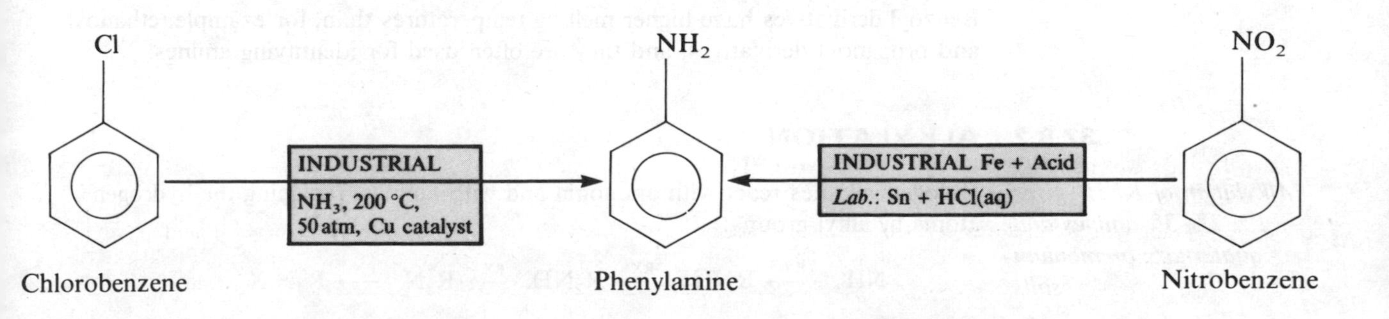

32.8 THE REACTIONS OF AMINES

32.8.1 ACYLATION

—NH_2 can be acylated to form —NHCOR...

Acylation is the conversion of the groups

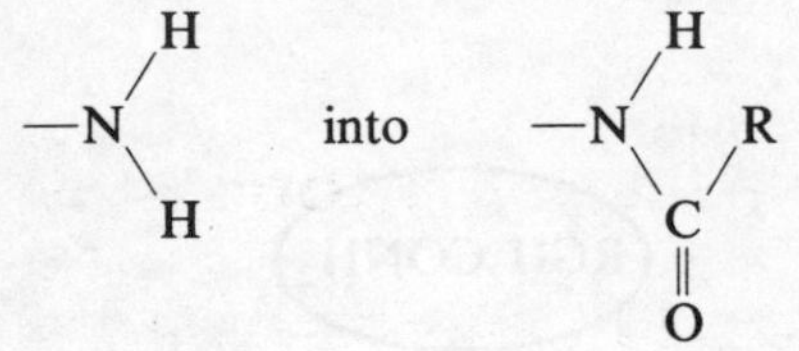

$$-N(H)H \quad \text{into} \quad -N(H)C(=O)R$$

and

$$-N(R')H \quad \text{into} \quad -N(R')C(=O)R$$

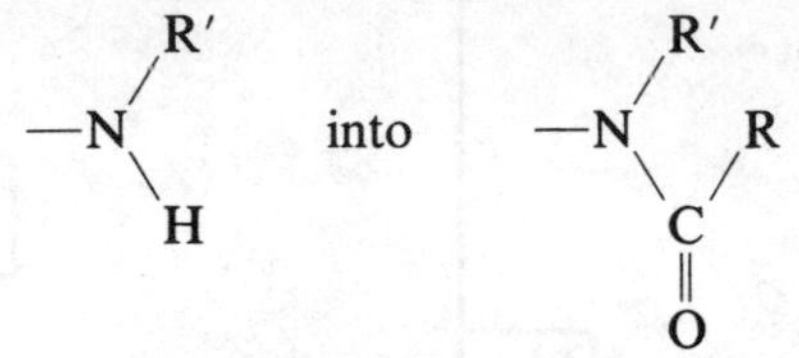

Tertiary amines, having no replaceable hydrogen atoms, are not acylated. The compounds formed are **amides**. The acylating agent can be an acid chloride, RCOCl or, preferably, an acid anhydride, $(RCO)_2O$:

...by RCOCl or by $(RCO)_2O$...

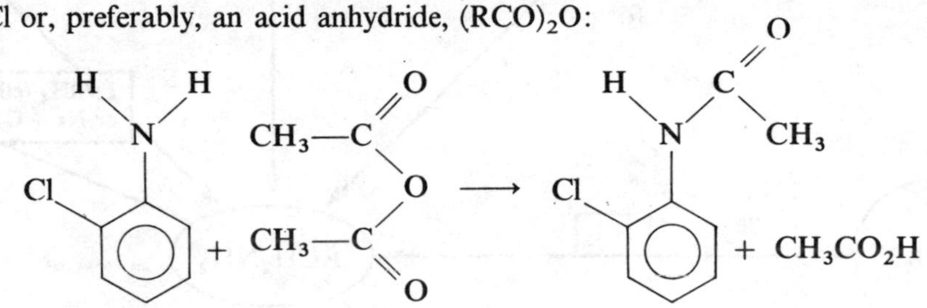

$$2\text{-}ClC_6H_4NH_2 + (CH_3CO)_2O \longrightarrow 2\text{-}ClC_6H_4NHCOCH_3 + CH_3CO_2H$$

2-Chlorophenylamine; Ethanoic anhydride; *N*-(2-Chlorophenyl)ethanamide

Benzoylation is accomplished by the use of benzoyl chloride and an excess of alkali:

and by C_6H_5COCl + NaOH

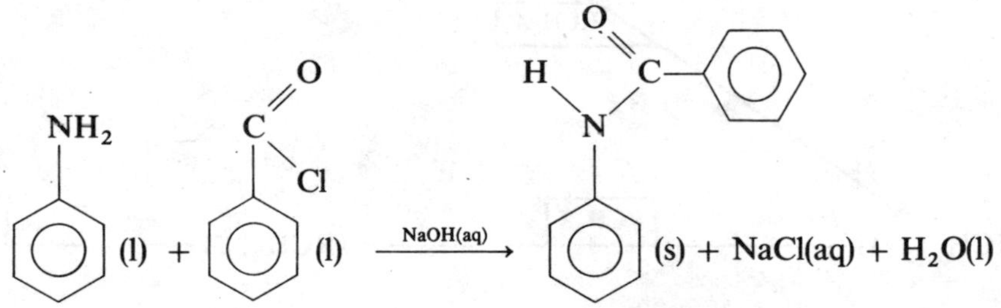

$$C_6H_5NH_2(l) + C_6H_5COCl(l) \xrightarrow{NaOH(aq)} C_6H_5NHCOC_6H_5(s) + NaCl(aq) + H_2O(l)$$

Amides are used for identification of amines

Amides are solids. When pure, they have sharp melting temperatures. Chemists often prepare an amide when they want to identify an amine. After purifying the derivative they find its melting temperature. Then they look at tables which list the melting temperatures of many amides. By identifying the derivative, they have succeeded in identifying the parent amine.

Benzoyl derivatives have higher melting temperatures than, for example, ethanoyl and propanoyl derivatives, and they are often used for identifying amines.

32.8.2 ALKYLATION

Alkylation of RNH_2 gives 1°, 2°, 3° amines and quaternary ammonium salts

Halogenoalkanes react with ammonia and with amines, replacing the hydrogen atoms by alkyl groups:

$$NH_3 \xrightarrow{RX} RNH_2 \xrightarrow{RX} R_2NH \xrightarrow{RX} R_3N \xrightarrow{RX} R_4\overset{+}{N}X^-$$

Halogenoarenes do not react in this way.

Primary amines are converted into secondary amines and secondary amines into tertiary amines. The final product is a quaternary ammonium salt, e.g., $(CH_3)_4\overset{+}{N}\,I^-$, tetramethylammonium iodide.

32.8.3 REACTIONS WITH NITROUS ACID

Nitrous acid HNO_2 reacts with aliphatic 1° amines to form unstable $R—\overset{+}{N}═N$ ions which decompose to give N_2...

Amines undergo a number of different reactions with nitrous acid. Since it is an unstable compound, nitrous acid is generated in the reaction mixture by the action of a mineral acid on sodium nitrite at 5 °C [Figure 32.4].

Aliphatic primary amines and aromatic primary amines react to form a cation

$$R—\overset{+}{N}≡N$$

called a **diazonium ion**. The diazonium compounds from primary aromatic amines are stable in solution at 5 °C. Their reactions are important, and are described in § 32.10. The diazonium compounds from primary aliphatic amines decompose to form carbocations, with the elimination of nitrogen:

$$R\overset{+}{N}H_3(aq) + HNO_2(aq) \rightarrow R—\overset{+}{N}≡N(aq) + 2H_2O(l)$$

$$R—\overset{+}{N}≡N(aq) \rightarrow R^+(aq) + N_2(g)$$

FIGURE 32.4 Reactions of Nitrous Acid with Amines

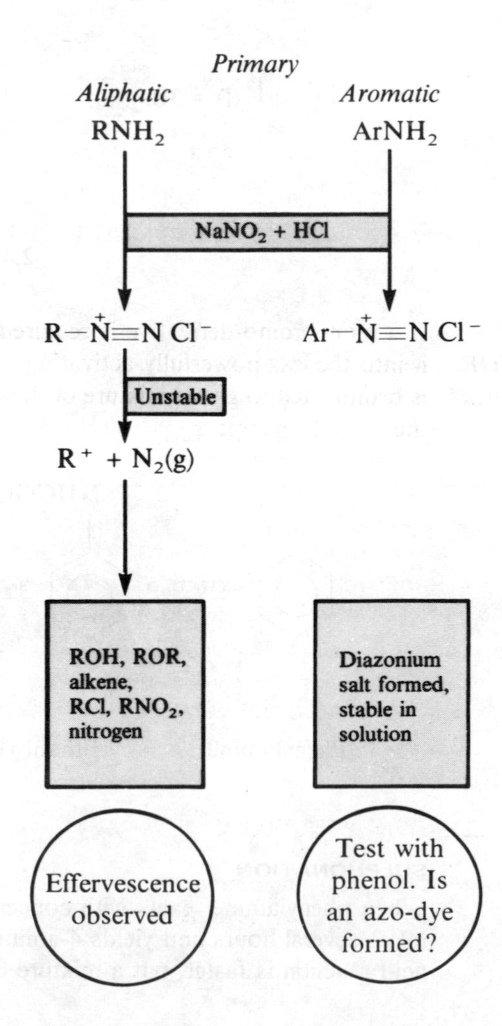

. . . with aromatic 1° amines to form more stable diazonium compounds . . .

The reaction is quantitative: the volume of nitrogen evolved is a measure of the amount of primary aliphatic amine present. The carbocation, R^+, can react in a number of ways, to form an alkene, an alcohol, ROH, an ether, ROR, and also, by reaction with the $NaNO_2$ + HCl(aq) present, RCl and RNO_2.

Secondary amines, both aliphatic and aromatic, are nitrosated (i.e. a —NO group is introduced) by nitrous acid to form *N*-nitrosoamines, which are yellow oils of a highly carcinogenic nature.

Aromatic tertiary amines are nitrosated in the ring.

32.8.4 SUBSTITUTION IN THE AROMATIC RING

The —NH_2 group activates the ring

Aromatic amines undergo substitution in the ring. The —NH_2 group activates the ring, and is 2/4-directing (like —OH [§ 28.12.3]). In many cases, it is difficult to obtain a monosubstituted compound.

HALOGENATION

Phenylamine reacts quantitatively with bromine water to give a white precipitate of 2,4,6-tribromophenylamine:

Halogenation gives a tri-halogeno-derivative . . .

$$C_6H_5NH_2(l) + 3Br_2(aq) \rightarrow C_6H_2Br_3NH_2(s) + 3HBr(aq)$$

2,4,6-Tribromophenylamine

. . . unless the —NHCOR derivative is halogenated

If a monobromo-derivative is required, the —NH_2 group is ethanoylated to convert it into the less powerfully activating —$NHCOCH_3$ group. Then the acyl derivative is brominated to give a mixture of the 2- and 4-bromo-derivatives. Hydrolysis restores the —NH_2 group:

$$C_6H_5NH_2 \xrightarrow{(CH_3CO)_2O} C_6H_5NHCOCH_3 \xrightarrow{Br_2,\ H_2O} BrC_6H_4NHCOCH_3 \xrightarrow[\text{boil}]{H_2SO_4(aq)} BrC_6H_4NH_2$$

Phenylamine — *N*-Ethanoylphenylamine — *N*-Ethanoyl-4-bromophenylamine — 4-Bromophenylamine (and 2-

SULPHONATION

When phenylamine reacts with concentrated sulphuric acid at 180 °C, the reaction takes several hours and yields 4-aminobenzenesulphonic acid. With fuming sulphuric acid reaction is faster, but a mixture of products is obtained

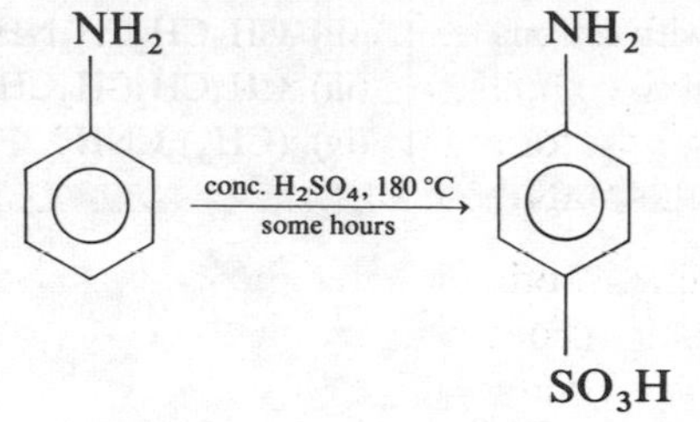

$$C_6H_5NH_2 \xrightarrow[\text{some hours}]{\text{conc. } H_2SO_4,\ 180\ ^\circ C} 4\text{-}H_2NC_6H_4SO_3H$$

4-Aminobenzenesulphonic acid

NITRATION

The —NH_2 group must be converted to —NHCOR before nitration

A nitrating mixture of concentrated nitric and sulphuric acids may oxidise the —NH_2 group as well as nitrating the ring. If the —NH_2 group is acylated, the acyl-derivative can be nitrated at room temperature, to give a mixture of 2- and 4-nitro-compounds. Hydrolysis with sulphuric acid yields the nitrophenylamines:

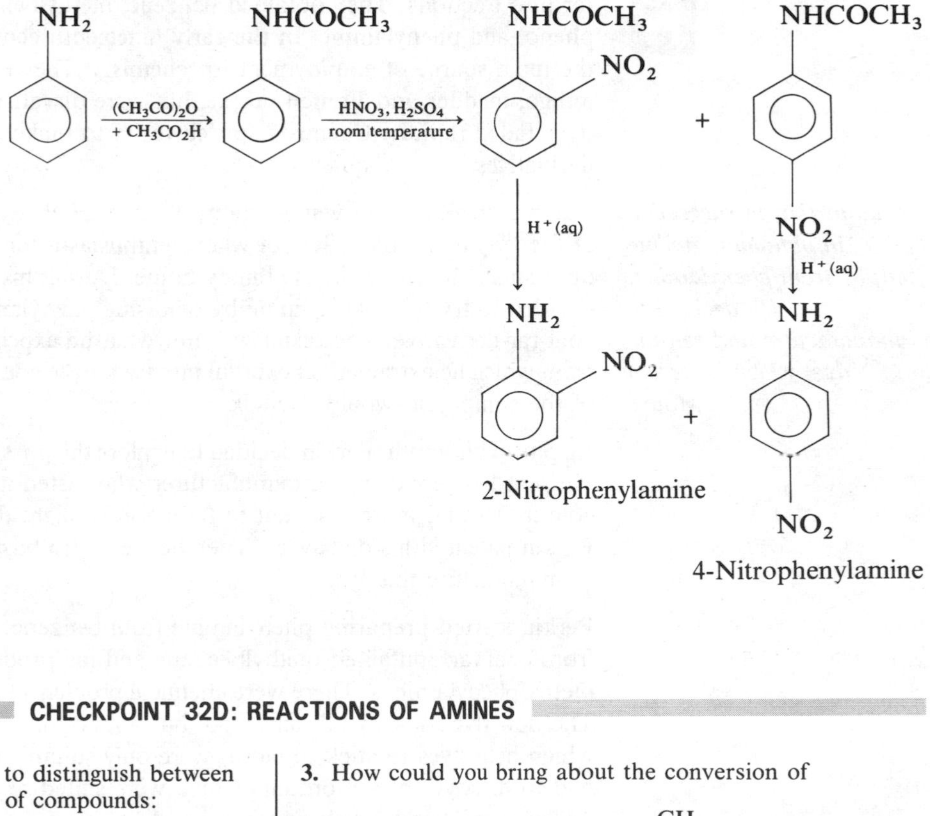

$$C_6H_5NH_2 \xrightarrow[+\ CH_3CO_2H]{(CH_3CO)_2O} C_6H_5NHCOCH_3 \xrightarrow[\text{room temperature}]{HNO_3,\ H_2SO_4} 2\text{-}O_2NC_6H_4NHCOCH_3 + 4\text{-}O_2NC_6H_4NHCOCH_3$$

$$2\text{-}O_2NC_6H_4NHCOCH_3 \xrightarrow{H^+(aq)} 2\text{-}O_2NC_6H_4NH_2 \qquad 4\text{-}O_2NC_6H_4NHCOCH_3 \xrightarrow{H^+(aq)} 4\text{-}O_2NC_6H_4NH_2$$

2-Nitrophenylamine + 4-Nitrophenylamine

CHECKPOINT 32D: REACTIONS OF AMINES

1. Describe the tests you could do to distinguish between the members of the following pairs of compounds:

(*a*) $CH_3CH_2CH_2NH_2$ and $CH_3CH_2NHCH_3$

(*b*) $C_4H_9\overset{+}{N}H_3\ Cl^-$ and $(CH_3)_4\overset{+}{N}\ Cl^-$

(*c*) H_3C—C_6H_4—NH_2 (benzene ring) and H_3C—C_6H_{10}—NH_2 (cyclohexane ring)

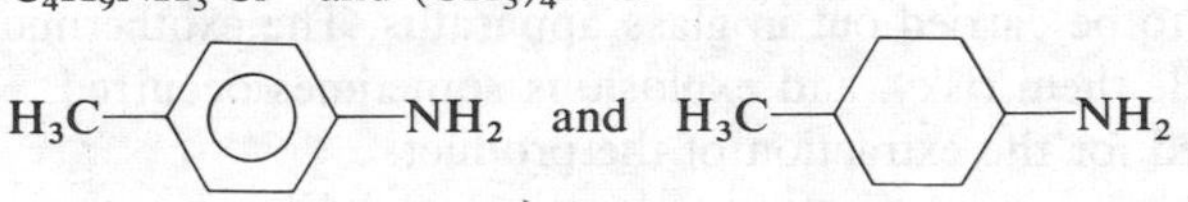

(*d*) C_6H_5—NO_2 and C_6H_5—NH_2

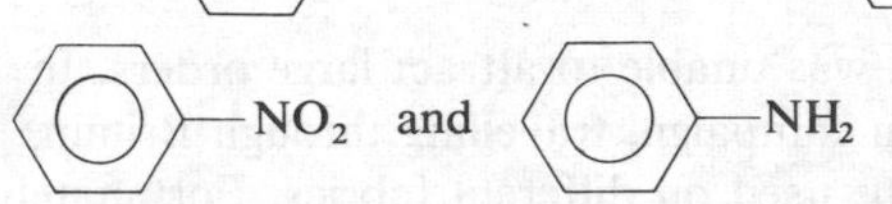

2. (*a*) How can phenylamine be made from benzene?

(*b*) How does phenylamine react with the following?

(i) hydrochloric acid

(ii) ethanoyl chloride

(iii) cold sodium nitrite and hydrochloric acid.

3. How could you bring about the conversion of

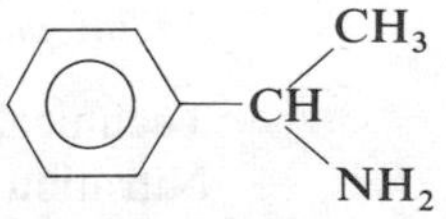

C_6H_5—$CH(CH_3)$—NH_2

into the following?

(*a*) C_6H_5—$CH(CH_3)$—$NHCOCH_3$

(*b*) C_6H_5—$CH(CH_3)$—$\overset{+}{N}\ Cl^-$

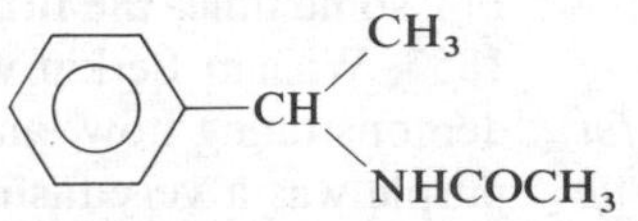

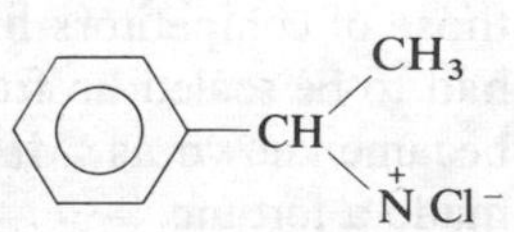

4. Which of the following compounds reacts with nitrous acid to give (*a*) nitrogen, (*b*) no visible change?
(i) $CH_3CH_2N(CH_3)_2$
(ii) $CH_3CH_2CH_2NHCH_3$
(iii) $CH_3CH_2CH_2CH_2NH_2$
(iv) $(CH_3)_3CNH_2$

32.9 WILLIAM PERKIN AND MAUVE

TOPIC

In the early nineteenth century chemists did not know what to do with coal tar. The destructive distillation of coal (in the absence of air) gave coke, which they needed for smelting iron ores, coal gas, ammonia and coal tar. Some of the coal tar was used to make creosote, which is a wood preservative, but most of it was dumped in rivers and on waste ground. From 1830 onwards, chemists began to separate coal tar into fractions. They obtained benzene, methylbenzene, naphthalene, anthracene, phenol and phenylamine. In the early nineteenth century, the textile industry was the main source of employment for chemists. They extracted dyes from plants, e.g. indigo, madder and French purple, but were dissatisfied with the way many of the dyes faded rapidly. Chemists started to try to make dyes from their new coal tar derivatives.

William Perkin succeeded in obtaining 'aniline purple' from phenylamine. He decided to manufacture and sell the dye, which he called Mauve

The big breakthrough was made by William H Perkin. He was an 18 year old pupil at the City of London School whose enthusiasm for chemistry had led him to set up a small laboratory in the family home. During his Easter holidays in 1856, Perkin decided to try to make quinine by oxidising phenylamine (then called aniline), a coal tar derivative. The result was not what he expected: he obtained a black solid from which he extracted a beautiful intense purple compound. He found that a solution of the compound would dye silk.

Inspite of his youth, Perkin decided to explore the possibility of marketing his discovery. He sent his new dye to a manufacturer who tested it and found that it was more colour-fast and more resistant to fading in sunlight than any similar dye available. Perkin patented his discovery. Then he went into business with his father and brother to manufacture the dye.

Perkin started preparing phenylamine from benzene. He was unaware that benzene from coal tar contained methylbenzene and his product therefore contained methylphenylamines. There were chemical problems to be solved in the dyeing process. The new dye took to silk and wool but not cotton. Existing mordants (substances which help dyes to stick to cloth) were only suitable for the acidic plant dyes. Perkin had to discover new mordants which were suited to his new dye, which he called *aniline purple*, and which was basic.

Technical problems had to be overcome. Benzene had to be redistilled in iron stills. Nitration and reduction had to be carried out in glass apparatus. The exothermic nature of these reactions made them risky, and explosions sometimes occurred. Airtight apparatus was needed for the extraction of the product.

Perkin overcame the technical problems of finding mordants for Mauve and working up the preparation to a manufacturing scale. He went on a promotional tour to secure orders

For some time, the firm of Perkin & Sons was unable to attract large orders. In 1857, William Perkin went on a promotion campaign, travelling through Britain, demonstrating how 'aniline purple' could be used on different fabrics. Fortunately, purple was a very fashionable colour. By 1858, the superiority of Perkin's dye over those of competitors had been demonstrated, and orders poured in. The apparatus had to be scaled up from bench scale to commercial scale. Perkin's 'aniline purple' became known as 'Mauve'. The fashion for Mauve lasted until 1863, and Perkin made a fortune.

All this work was done before there was any theory of organic chemistry. It was not until 1865 that Kekulé put forward his theory of the benzene ring. Eventually, Perkin showed that crude Mauve was a mixture. The most important component was *mauveine*, the structure of which was worked out in 1888.

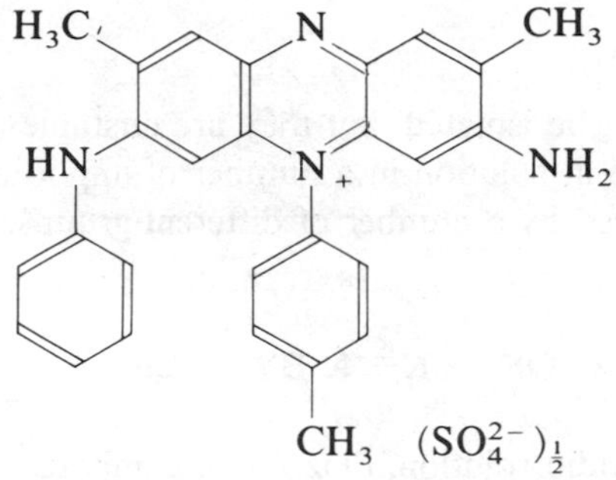

H_3C N CH_3 HN N_+ NH_2 CH_3 $(SO_4^{2-})_{\frac{1}{2}}$

Mauveine

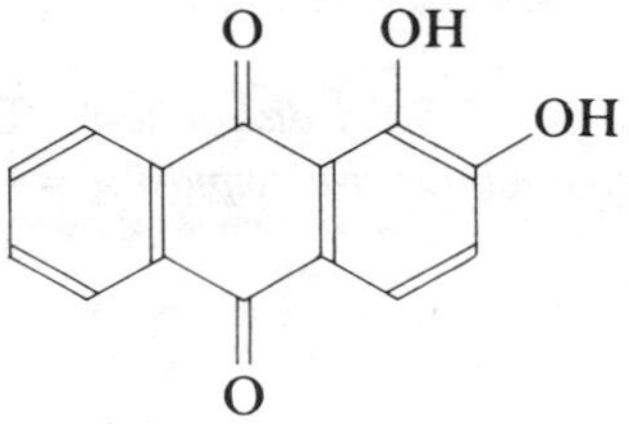

O OH OH O

Alizarin

Perkin's work stimulated interest in organic chemistry, and many new compounds were synthesised. The firm of Perkin & Sons became part of ICI

Perkin's discovery of Mauve stimulated interest in organic chemistry and the commercial possibilities which it offered. The oxidation of phenylamine yielded a second dye, magenta. Peter Greiss discovered the diazotisation reaction in 1858 and opened the route to the azo dyes. In 1869, Perkin filed a patent for the manufacture of the red dye alizarin. He was disappointed to find that Heinrich Caro of Germany had beaten him to it by one day. After further work, Perkin was able to devise and patent a different route to alizarin.

Perkin and his brother sold their business in 1873. Eventually it became part of ICI. William Perkin spent the rest of his life in scientific research. He was knighted in 1906 and died in 1907

CHECKPOINT 32E: WILLIAM PERKIN AND MAUVE

1. Why did the need for coke increase in the late eighteenth and early nineteenth centuries? What use was made of (*a*) coal gas and (*b*) ammonia?

2. Outline a method which Perkin could have used for making phenylamine from benzene. Why did it contain methylphenylamines? Write the formulae of the methylphenylamine impurities.

3. Why was Perkin's dye called 'aniline purple'?

4. Explain why, unlike distillation, nitration and reduction were not carried out in iron apparatus.

5. State one chemical problem, one technical problem and one commercial problem that Perkin had to tackle. What do you think were the ingredients of his success?

32.10 DIAZONIUM COMPOUNDS

$ArNH_2 + HNO_2$ form a diazonium compound

Diazonium compounds result from the reaction between primary aromatic amines and nitrous acid. The reaction is carried out below 10 °C to avoid the decomposition of both nitrous acid and the product:

$$C_6H_5NH_2(l) + NaNO_2(aq) + 2HCl(aq) \xrightarrow{<10\,°C} C_6H_5\overset{+}{N}{\equiv}N\ Cl^- + NaCl(aq) + 2H_2O$$

Phenylamine Benzenediazonium chloride

Solid diazonium compounds are explosive

Diazonium salts can be isolated, but they are unstable and explosive in the solid state. They are used in solution in a number of important preparations. The $-\overset{+}{N}_2$ group can be replaced by a number of different groups.

32.10.1 REPLACEMENT OF $-\overset{+}{N}{\equiv}N$ BY $-OH$

They are used in solution

When warmed in acidic solution, diazonium compounds form phenols, with the evolution of nitrogen:

$$C_6H_5-\overset{+}{N}{\equiv}N + H_2O \xrightarrow[H_2SO_4\text{ (aq, dilute)}]{\text{warm } >10\,°C \text{ with}} C_6H_5-OH + N_2(g) + H^+(aq)$$

On warming, they give phenols

To avoid a reaction between the phenol formed and the diazonium compound, [§ 32.10.3] the diazonium compound is added slowly to a large excess of boiling dilute sulphuric acid. The volatile phenol distils over.

32.10.2 REPLACEMENT OF $-\overset{+}{N}{\equiv}N$ BY —HALOGEN OR —CN

$-\overset{+}{N}{=}N$ can be replaced by —Cl, —Br, —I or —CN . . .

In a set of reactions named after T Sandmeyer, $-\overset{+}{N}{\equiv}N$ is replaced by —Cl, —Br or —CN. The diazonium compound is warmed to 100 °C with the appropriate reagent and catalyst. These are shown in Figure 32.5. If potassium iodide solution is added, $-\overset{+}{N}{\equiv}N$ is replaced by —I.

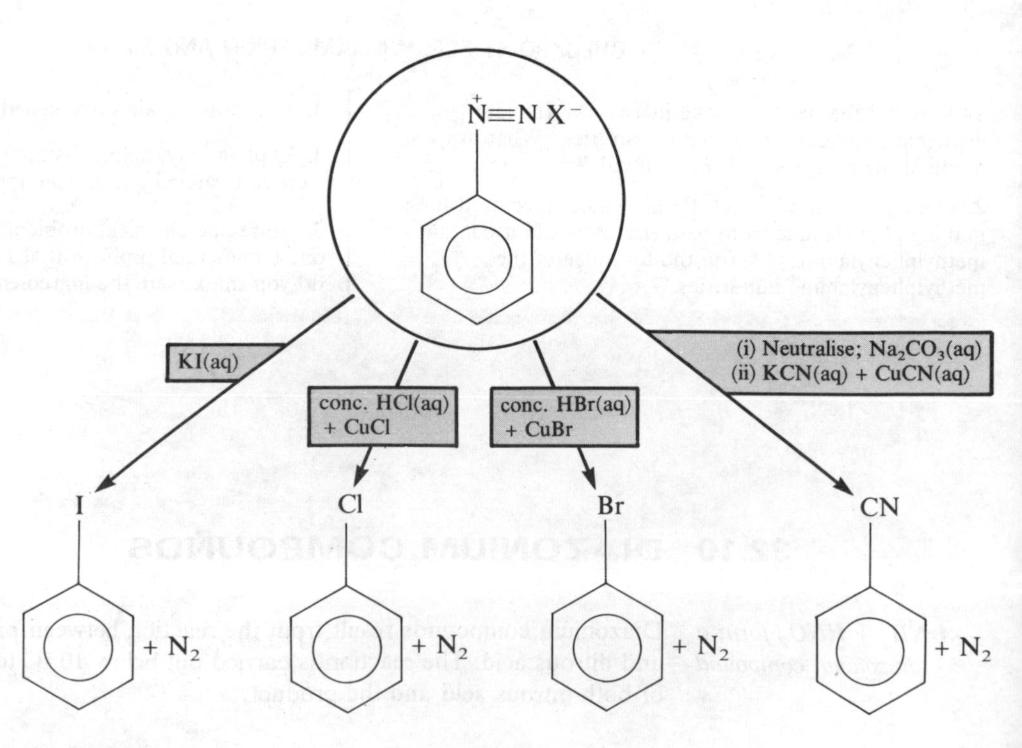

FIGURE 32.5 Replacement of $-\overset{+}{N}_2$ by —Cl, —Br, —I, —CN

32.10.3 COUPLING REACTIONS

Diazonium compounds will react with phenols and aromatic amines to form azo-dyes

Diazonium ions are electrophiles. The diazonium cation will attack reactive nucleophilic sites, such as the 4-positions in phenols and aromatic amines. The reaction results in the formation of a —N=N— bond between two aromatic rings, and is called **azo-coupling**. The compounds formed are highly coloured, and many are used as dyes.

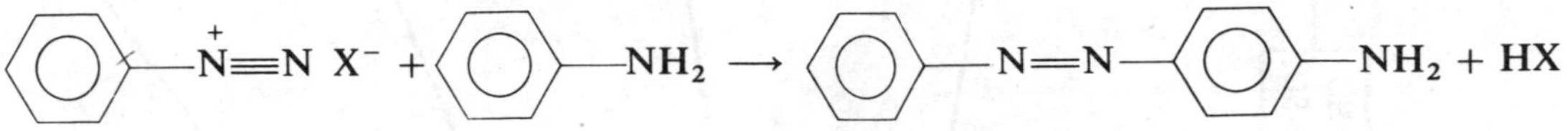

$$C_6H_5-\overset{+}{N}\equiv N\ X^- + C_6H_5-NH_2 \longrightarrow C_6H_5-N{=}N-C_6H_4-NH_2 + HX$$

Phenylamine (aniline)

4-(Phenylazo)phenylamine (aniline yellow, the first azo-dye to be made)

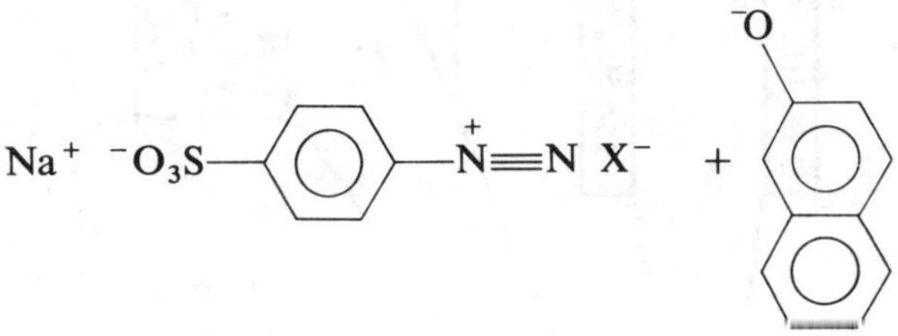

$$Na^+\ {}^-O_3S-C_6H_4-\overset{+}{N}\equiv N\ X^- \quad + \quad C_{10}H_7O^-$$

Sodium 4-sulphonatobenzenediazonium halide   Naphthalen-2-ol

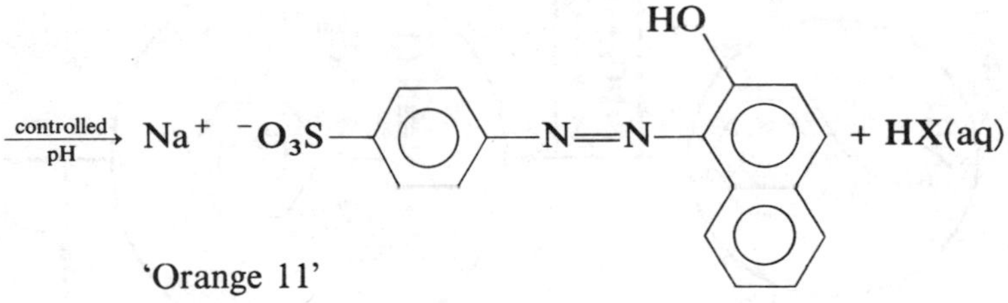

$$\xrightarrow[\text{pH}]{\text{controlled}} Na^+\ {}^-O_3S-C_6H_4-N{=}N-C_{10}H_6(OH) \quad + HX(aq)$$

'Orange 11'

(The $—SO_3^-\ Na^+$ substituent makes this dye soluble in water.)

32.10.4 THE IMPORTANCE OF DIAZONIUM COMPOUNDS

Diazonium compounds are used in synthesis

Diazonium compounds open up the possibility of making a number of benzene derivatives which cannot be made directly from benzene [See Figure 32.6.] You saw how halogenoalkanes opened up the route

Alkane, RH → *Halogenoalkane*, RX → RY

where Y can be any of a number of groups. Halogenoarenes are not very reactive, and it is often diazonium compounds that open up routes to aromatic compounds [see Figure 32.7]:

Arene → *Nitroarene* → *Arylamine* → *Diazonium compound* → *Substituted arene*

$$ArH \rightarrow ArNO_2 \rightarrow ArNH_2 \rightarrow Ar\overset{+}{N}_2\,X^- \rightarrow ArY$$

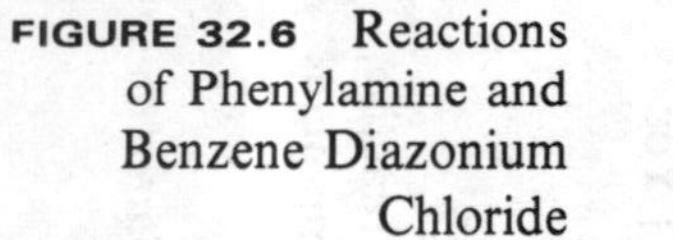
FIGURE 32.6 Reactions of Phenylamine and Benzene Diazonium Chloride

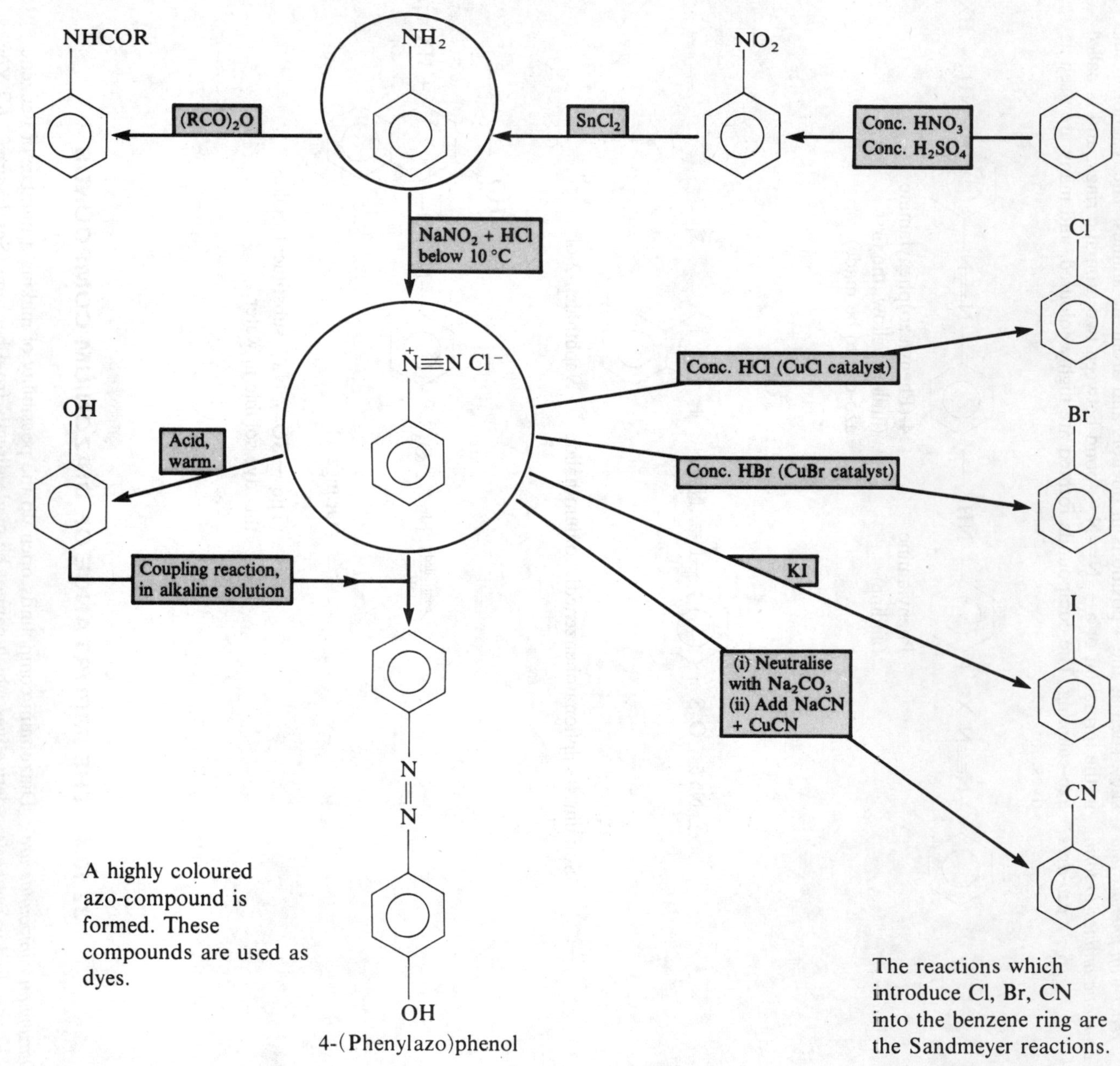

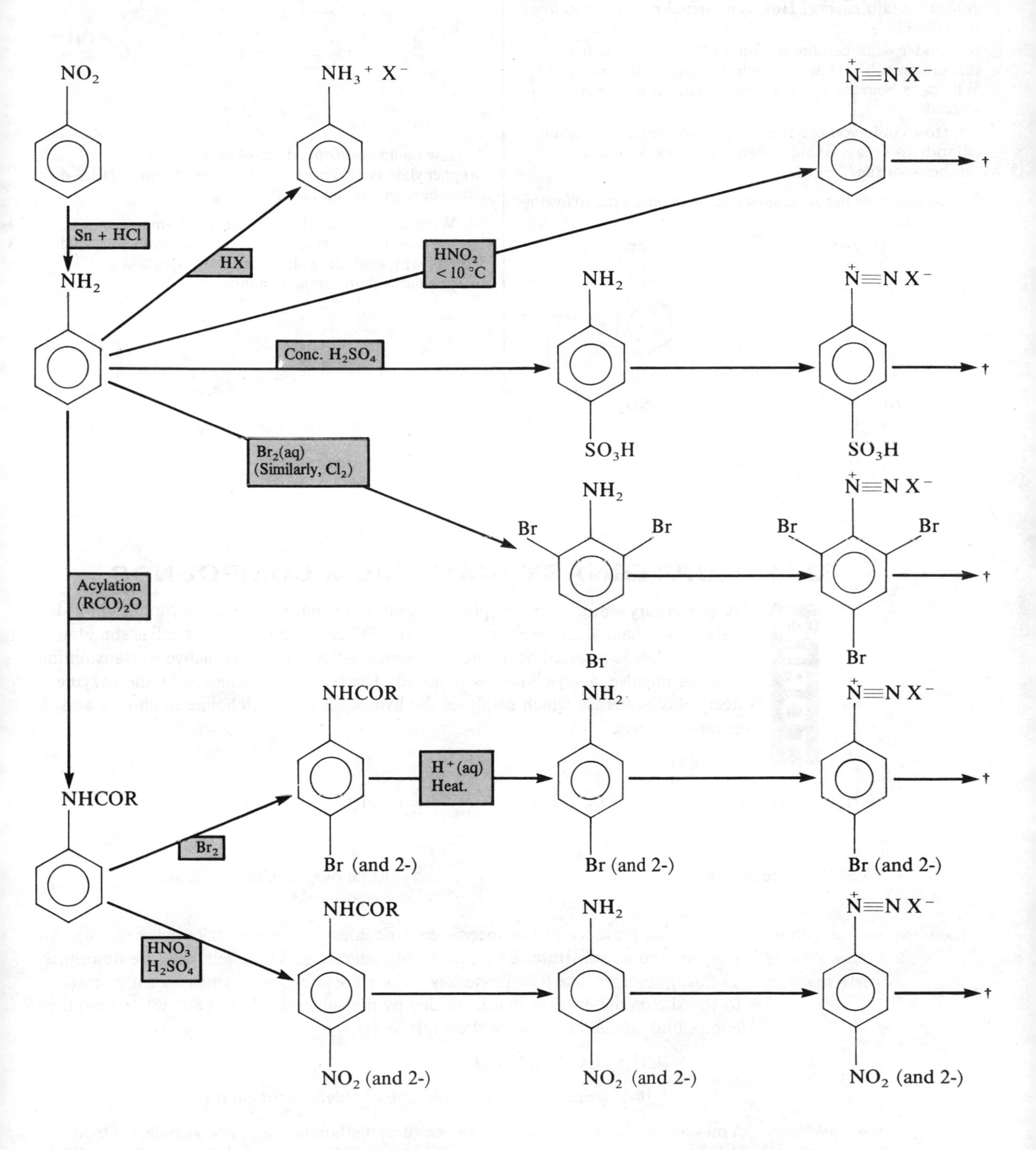

FIGURE 32.7 Routes from Aromatic Amines to Other Compounds

† See reactions of diazonium compounds. Figure 32.6.

CHECKPOINT 32F: DIAZONIUM COMPOUNDS

1. (*a*) Describe how you could prepare a solution of benzenediazonium chloride. How can the compound be converted into phenol? How can phenol be separated from the solution?

(*b*) Under what conditions will a phenol react with a diazonium salt? What is the electrophile in this reaction? Why does benzene not react with benzenediazonium chloride?

(*c*) How could you use the solution of benzenediazonium chloride to obtain (i) chlorobenzene, (ii) iodobenzene, (iii) benzonitrile?

2. Starting from benzene, how could you make the following compounds?

(*a*) $NHCOCH_3$ (para) NO_2 benzene ring

(*b*) NH_2 (para) NO_2 benzene ring

(*c*) (*d*)

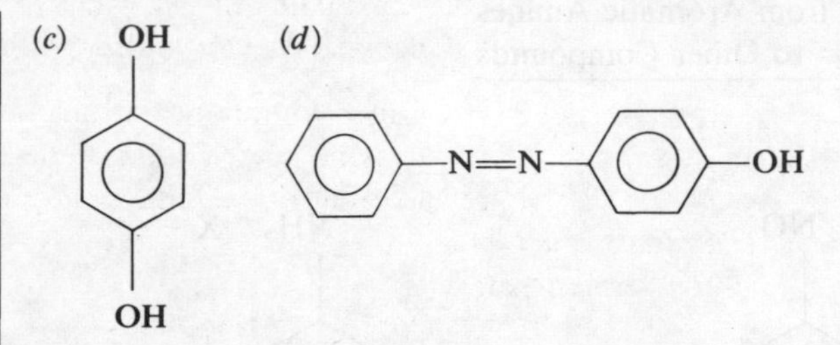

3. How could you convert nitrobenzene into (*a*) phenylamine, (*b*) phenol, (*c*) chlorobenzene, (*d*) 1,3-dichlorobenzene, (*e*) an azo-dye?

***4.** Write an essay on the importance of amines as intermediates in the synthesis of other organic compounds. Explain why aliphatic amines are less important as intermediates than aromatic amines.

32.11 QUATERNARY AMMONIUM COMPOUNDS

TOPIC

A quaternary ammonium compound which occurs naturally is acetylcholine chloride. (*Acetyl* = ethanoyl; *choline* is shown below.) When a voluntary nerve cell is stimulated, acetylcholine is released at the nerve endings. After it has been active in transmitting a nerve impulse, acetylcholine is removed. The removal is achieved by the enzyme acetylcholinesterase which catalyses the hydrolysis of acetylcholine to choline and ethanoic (acetic) acid:

$$CH_3CO_2CH_2CH_2\text{—}\overset{+}{N}(CH_3)_3\ Cl^- + H_2O \xrightarrow[\text{acetylcholinesterase}]{\text{catalysed by}} CH_3CO_2H + HOCH_2CH_2\text{—}\overset{+}{N}(CH_3)_3\ Cl^-$$

Acetylcholine chloride — Ethanoic (acetic) acid — Choline chloride

Acetylcholine is important in the transmission of nerve impulses

Only in the presence of the specific enzyme does the reaction take place rapidly. An enzyme and its **substrate** (e.g., acetylcholinesterase and acetylcholine) are described as fitting together 'like a lock and key'. There are substances which attach themselves to the site on the enzyme that is needed by the substrate. The quaternary ammonium compound, '*decamethonium*' will do this:

$$(CH_3)_3\overset{+}{N}(CH_2)_{10}\overset{+}{N}(CH_3)_3$$

1,10-Di(trimethylammonio)decane (or 'decamethonium')

Enzyme inhibitors interfere with the transmission of nerve impulses

You can see the resemblance between 'decamethonium' and acetylcholine. Once decamethonium is attached to the enzyme, acetylcholinesterase is prevented or **inhibited** from doing its job, and nerve impulses are not transmitted. Decamethonium was used as a muscle relaxant. The naturally occurring base, *curare*, which acts in a similar way, was also used medicinally. Curare has other uses too: some South American Indians use it to tip their poison arrows. The quantities which they use cause paralysis and death in their victims.

Nerve gases, e.g. DFP, and insecticides, e.g., Parathion®, act on acetylcholinesterase

Other compounds which do not resemble acetylcholine in structure also inhibit the enzyme acetylcholinesterase. Nerve gases (e.g., DFP) work in this way. Insecticides like Parathion® have the same effect on insects:

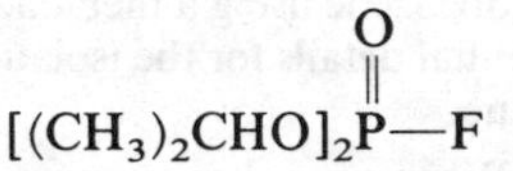

$$[(CH_3)_2CHO]_2P(=O)—F$$

Di(1-methylethoxy)fluorophosphate (DFP)

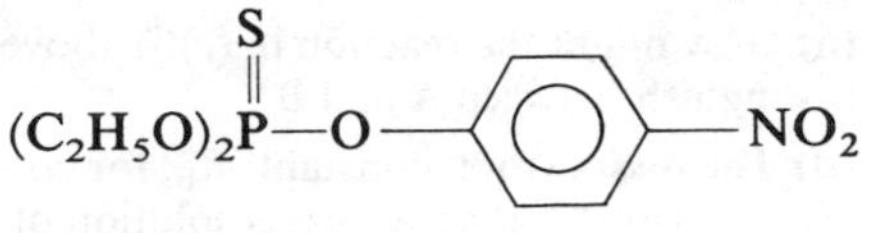

$$(C_2H_5O)_2P(=S)—O—C_6H_4—NO_2$$

Parathion®

Parathion® and similar insecticides can affect human beings. Agricultural workers who use the compounds regularly and who do not wear protective clothing can absorb sufficient insecticide to damage their health.

QUESTIONS ON CHAPTER 32

1. Describe the preparation of (*a*) 2-aminopropane from propane and (*b*) phenylamine from benzene. Give any details you can about the mechanisms of the reactions involved.

2. Outline the preparation, starting from benzene, of (*a*) nitrobenzene, (*b*) phenylamine, (*c*) bromobenzene, (*d*) iodobenzene and (*e*) a named azo-compound.

3. Suggest a synthesis for choline chloride, $HOCH_2CH_2\overset{+}{N}(CH_3)_3\ Cl^-$, starting from ethene.

***4.**

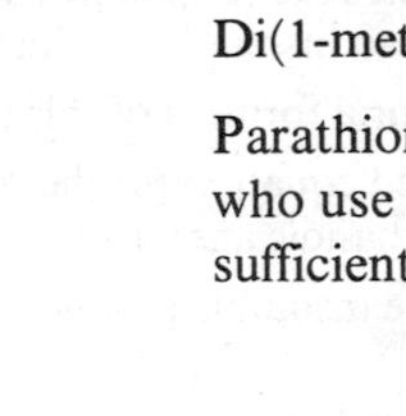

$$C_6H_5—CH_2—CH(NH_2)—CH_3$$

The formula given above is that of the stimulant, *Benzedrine*®. Give it a systematic name. Can you suggest how it could be prepared?

5. How can iodo-4-nitrobenzene be made, using benzene as the starting point?

6. '*Sulphanilic acid*', 4-aminobenzenesulphonic acid, is used in the manufacture of sulphanilamide drugs. How can it be made from benzene?

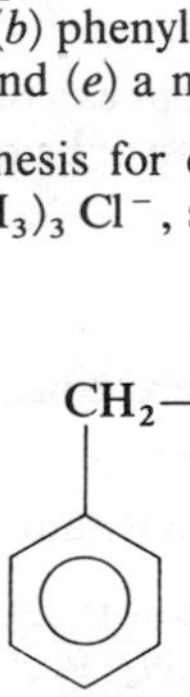

$$NH_2—C_6H_4—SO_3H$$

***7.** Compare and contrast the reactions of phenol and phenylamine.

***8.** A compound, **A** (C_4H_7N) is reduced by $LiAlH_4$ to **B** ($C_4H_{11}N$). Ethanoylation of **B** gives **C** ($C_6H_{13}NO$), and treatment of **B** with iodomethane followed by aqueous NaOH gives **D** ($C_5H_{13}N$). Further treatment of **D** with iodomethane gives a white solid **E** containing I^- ions. When **A** is treated with hot aqueous sulphuric acid, **F** ($C_4H_8O_2$) is formed. Reduction of **F** with $LiAlH_4$ gives **G** ($C_4H_{10}O$), and concentrated H_2SO_4 converts **G** into **H** (C_4H_8). Treatment of **H** with HBr gives $(CH_3)_3CBr$. Identify the compounds **A** to **H**, giving your reasons, and explaining the reactions mentioned.

***9.** 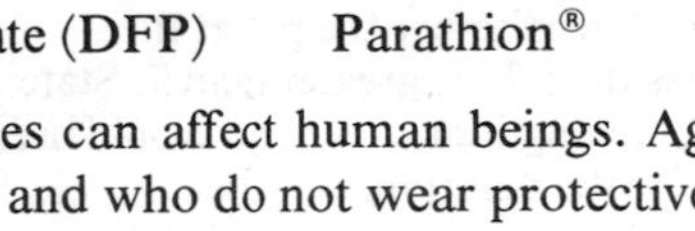 $C_2H_5O—C_6H_4—NHCOCH_3$

Shown above is the formula of Phenacetin®. Give its IUPAC name.
Suggest a synthesis of phenacetin from 4-nitrophenol.

10. 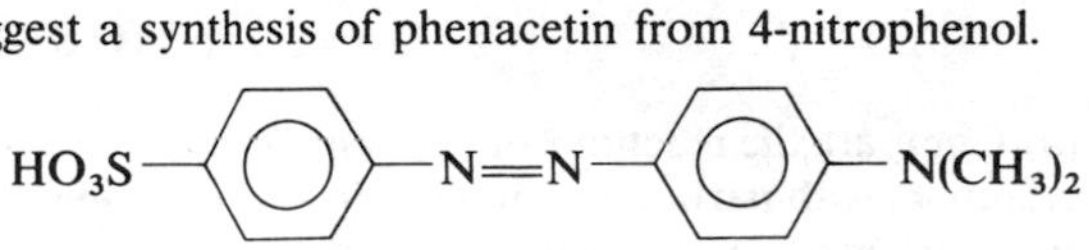 $HO_3S—C_6H_4—N{=}N—C_6H_4—N(CH_3)_2$

Shown above is the formula of methyl orange. Suggest a method of making the compound from 4-aminobenzenesulphonic acid, $4\text{-}HO_3SC_6H_4NH_2$, and phenylamine.

11. A is phenylamine (aniline) $C_6H_5NH_2$, a typical aromatic amine with the $—NH_2$ group directly attached to the ring. **B** is (phenylmethyl)amine, $C_6H_5CH_2NH_2$, which may be considered to have properties of an aliphatic amine with the phenyl group as a substituent.

(*a*) Calculate the percentage by mass of C, H and N in both **A** and **B**.

$$(A_r(C) = 12.01,\ A_r(H) = 1.01,\ A_r(N) = 14.01)$$

Comment on the usefulness of an elemental analysis giving C = 78.10%, H = 8.00% and N = 13.9% for distinguishing between **A** and **B**.

(*b*) (i) Outline the preparation of

(*1*) **A** from nitrobenzene, $C_6H_5NO_2$

(2) **B** from (phenylmethyl) bromide, $C_6H_5CH_2Br$

(ii) Give the conditions under which both **A** and **B** may be acylated, together with formulae for the acylating agent and for the products formed.

(*c*) (i) State the reagents and conditions required for the conversion of **A** into a diazonium salt. Give *one* example of an azo coupling reaction for the diazonium salt and explain the significance of this reaction for the dyestuff industry.

(ii) How might the reaction in (*c*)(i) above be used to distinguish between **A** and **B**?

(*d*) The dissociation constant, K_b, for compound **B** is 10^{-5} mol dm^{-3}. If an aqueous solution of **B** is titrated against dilute hydrochloric acid, state whether the pH at the end-point would be 7, less than 7, or greater than 7. State therefore which of the following indicators you would use to determine the end-point of the titration:

Bromophenol blue (pH range 3.0–4.6)

Bromothymol blue (pH range 6.2–7.6)

Phenolphthalein (pH range 8.0–10.0)

(WJEC 90)

12. (*a*) What is meant by the terms *structural isomer* and *stereoisomer*?

(*b*) (i) Using standard conventions to represent the three-dimensional arrangement of groups, give diagrams of the pair(s) of enantiomers of any chiral primary amines which have the molecular formula $C_4H_9NH_2$.

(ii) Similarly, give diagrams of all the structural and stereoisomers of molecular formula C_3H_5Br.

(*c*) A student, asked to devise a synthesis of the primary amine $CH_3CH(NH_2)CH_3$, decided to use a method involving the reduction of a nitrile as the final step.

(i) What reagent will reduce a nitrile to a primary amine?

(ii) What would make such a method impossible in this case?

(iii) Give a structural isomer of this primary amine which *could* be synthesised by reduction of a nitrile.

(*d*) Give the products obtained by treating a primary amine, such as that referred to in (*c*), with ethanoyl chloride.

(O & C 92)

13. (*a*) Compare the reactions of ethylamine and phenylamine, with particular reference to the following:

(i) their relative basicities

(ii) their reaction (if any) with bromine

(iii) their reaction with ethanoyl chloride.

(*b*) 4-Aminophenol, **B**, is used as a photographic developer. In alkaline solution, it acts as a mild reducing agent, being oxidised to quinone, **C**.

This allows it to reduce silver ions in the photographic emulsion to atoms of silver.

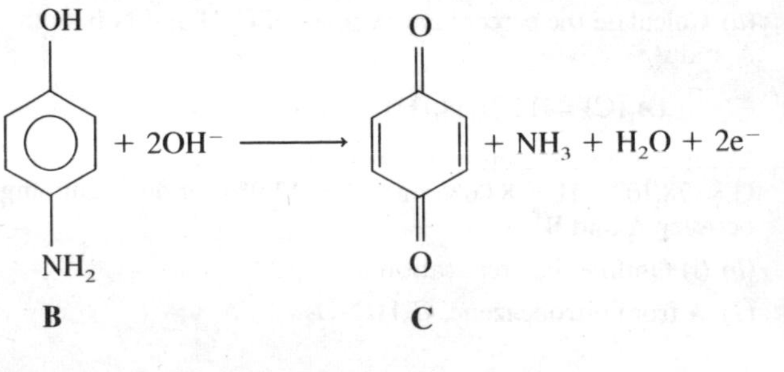

OH / NH$_2$ (**B**) + 2OH$^-$ ⟶ O / O (**C**) + NH$_3$ + H$_2$O + 2e$^-$

(i) Suggest reasons why **B** is much more soluble in alkaline or acidic solutions than in neutral water.

(ii) Write a balanced equation for the reaction between **B** and silver ions in alkaline solution.

(C 91)

14. (*a*) Outline the preparation of phenylamine from nitrobenzene using a metal/acid reducing system. Give essential details for the isolation and purification of the amine.

(*b*) When phenylamine reacts with ethanoic anhydride in ethanoic acid and the reaction mixture is then poured into water with stirring, an insoluble product is formed.

(i) Give the structural formula of this product.

(ii) Write a balanced equation for the reaction between phenylamine and ethanoic anhydride.

(iii) Explain why the insoluble product is less basic than phenylamine.

(iv) The compound **A** is a second, but very minor product of the reaction between phenylamine and ethanoic anhydride.

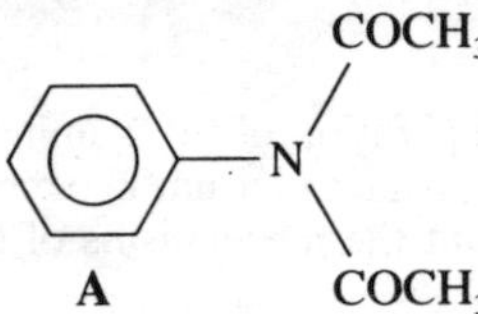

$C_6H_5N(COCH_3)_2$ (**A**)

Explain how this product could arise, and suggest why it is not the major product of the reaction.

(*c*) How and under what conditions can phenylamine be converted into an aqueous solution of benzenediazonium chloride?

(O 92, S)

15. The decomposition of the benzenediazonium ion at 35 °C to phenol obeys first-order kinetics.

$$C_6H_5N_2^+(aq) + H_2O(l) \rightarrow C_6H_5OH(aq) + H^+(aq) + N_2(g)$$

(*a*) (i) Write a rate equation to show that the above reaction is first-order with respect to $C_6H_5N_2^+(aq)$.

(ii) Sketch the shape of the graph obtained by plotting the volume of nitrogen gas collected as the decomposition of the diazonium ion continues to completion.

(iii) Describe how you would use the results from a series of experiments to show that the decomposition of the benzenediazonium ion is first order.

(*b*) Benzenediazonium chloride may be converted into iodobenzene. Name the reagent(s) and the conditions needed.

(*c*) Benzenediazonium chloride may be coupled with phenol. Draw the structure of the product and state its colour.

(*d*) Diazonium salts are prepared by dissolving or suspending an aromatic amine in excess dilute inorganic acid cooled in ice and adding a cooled aqueous solution of sodium nitrite, e.g. benzenediazonium chloride is formed by the reaction of phenylamine (aniline) with nitrous acid in the presence of hydrochloric acid.

(i) Suggest why ethyldiazonium chloride is not obtained when ethylamine is used in the above experimental procedure.

(ii) Name the organic product of the reaction of nitrous acid with ethylamine.

(*e*) Phenylhydrazine may be prepared by the reduction of benzenediazonium chloride. It is a base and forms salts with acids.

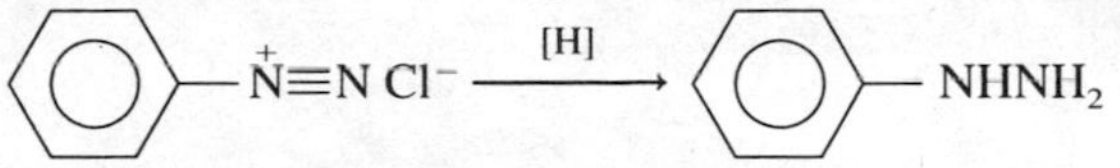

$$C_6H_5-\overset{+}{N}\equiv N\ Cl^- \xrightarrow{[H]} C_6H_5-NHNH_2$$

Both 2,4-dinitrophenylhydrazine and phenylhydrazine react with carbonyl compounds and may be used to identify aldehydes and ketones.

(i) Write a balanced equation for the reaction of one mole of phenylhydrazine with one mole of hydrogen chloride.

(ii) Write a balanced equation for the reaction of one mole of phenylhydrazine with one mole of methanal.

(iii) Suggest why phenylhydrazine has been replaced by 2,4-dinitrophenylhydrazine in identifying aldehydes and ketones.

(NI 90)

16. A dye used in instant colour films has the formula

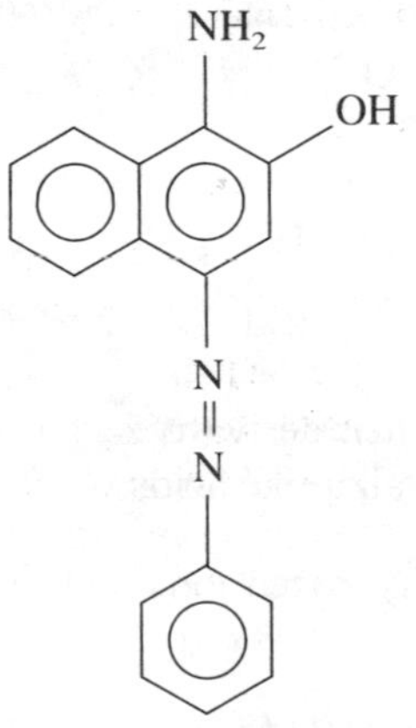

[structure: naphthalene ring bearing NH_2 at C-1, OH at C-2, and $-N=N-C_6H_5$ at C-4]

Would you expect the dye to be water soluble? Explain your answer.

Suggest the formula of the product formed when the dye is treated with:

(*a*) dilute hydrochloric acid

(*b*) aqueous sodium hydroxide

Suggest how you could prepare 5 g of this dye starting from 1-aminonaphthalen-2-ol, [structure: naphthalene with NH_2 at C-1 and OH at C-2], and nitrobenzene, [structure: $C_6H_5NO_2$], assuming a 75% yield at each of the two stages. Give sufficient details, including quantities of chemicals, suitable apparatus and reaction conditions, to enable an average A-level student to follow your instructions.

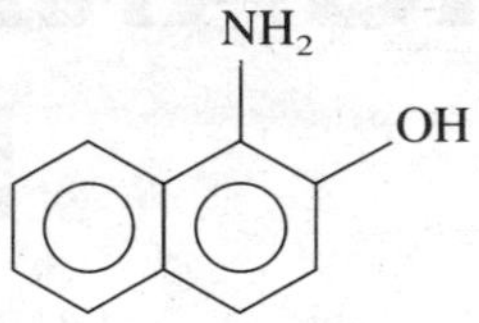

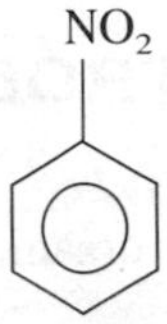

[Nitrobenzene $C_6H_5-NO_2$ can be reduced to phenylamine $C_6H_5-NH_2$ by heating with zinc and dilute acid.]

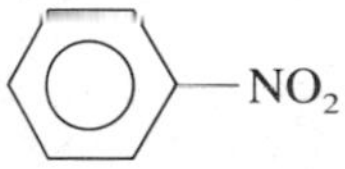

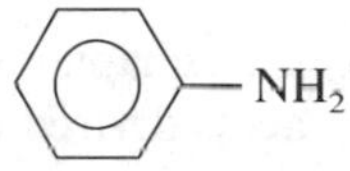

(L(N) 91, S)

33

ORGANIC ACIDS AND THEIR DERIVATIVES

33.1 INTRODUCTION

Most organic acids contain a carboxyl group or a sulphonic acid group:

```
      O               R   O
     //                \ //
R—C                     S
     \                 // \
      O—H             O    O—H
Carboxylic acid     Sulphonic acid
```

This chapter is devoted to carboxylic acids and their derivatives. The carboxylic acids derived from alkanes are sometimes called alkanoic acids.

Carboxylic acids are weakly ionised...

Carboxylic acids ionise to some extent to give hydrogen ions, and are neutralised by bases to form salts:

$$RCO_2H(aq) + H_2O(l) \rightleftharpoons RCO_2^-(aq) + H_3O^+(aq)$$

$$RCO_2H(aq) + OH^-(aq) \rightarrow RCO_2^-(aq) + H_2O(l)$$

...They form salts, many of which are soluble

The salts are usually soluble in water. This property makes carboxylic acids easy to extract from natural sources, and is the reason why they were among the first organic compounds to be isolated. Ethanoic (acetic) acid was obtained from sour wine; butanoic (butyric) acid from butter; 2-hydroxypropanoic (lactic) acid from sour milk and benzoic acid from gum benzoin. Aliphatic carboxylic acids were called 'fatty acids' because esters of several of the higher members are fats.

33.2 NOMENCLATURE FOR ORGANIC ACIDS AND THEIR DERIVATIVES

Aliphatic carboxylic acids are named after the alkane with the same number of carbon atoms. The ending *-ane* is changed to *-anoic acid*. Examples are:

The IUPAC names of some acids

HCO_2H Methanoic acid (formerly called formic acid)

CH_3CO_2H Ethanoic acid (formerly acetic acid)

$CH_3CH_2CO_2H$ Propanoic acid (formerly propionic acid)

The C in the $-CO_2H$ group is always given the number 1, and substituents are given locants:

$CH_3-CH(OH)-CO_2H$ 2-Hydroxypropanoic acid

$CH_3CH{=}CHCH_2CO_2H$ Pent-3-enoic acid

HO_2C-CO_2H Ethanedioic acid (formerly oxalic acid)

$CH_2CO_2H-CH_2CO_2H$ Butanedioic acid (formerly succinic acid)

Aromatic carboxylic acids are named by adding the suffix *-carboxylic acid* to the name of the parent hydrocarbon. Alternatively, the suffix *-oic acid* can be used:

$C_6H_5-CO_2H$ Benzenecarboxylic acid (or benzoic acid)

$O_2N-C_6H_4-CO_2H$ 4-Nitrobenzenecarboxylic acid (or 4-nitrobenzoic acid)

The derivatives of carboxylic acids covered in this chapter contain the **acyl** group

$R-C(=O)-$

and are listed in Table 33.1. The nitriles are included because their reactions link up with those of acids.

33.3 PHYSICAL PROPERTIES OF ACIDS AND THEIR DERIVATIVES

33.3.1 ACIDS

The lower aliphatic acids are liquids...

The aliphatic acids C_1–C_{10} are liquids. Anhydrous ethanoic acid freezes at 17 °C, and is often called *glacial* ethanoic acid. The boiling temperatures increase with increasing molecular mass. Aromatic acids are crystalline solids with melting temperatures above those of aliphatic acids of comparable molecular mass.

The lower members of the aliphatic carboxylic acids have penetrating odours. Vinegar is a 3% solution of ethanoic acid. Butanoic acid is the substance you smell in rancid butter.

Hydrogen bonding takes place between molecules of carboxylic acids. In the vapour phase and in solution in organic solvents, dimerisation occurs:

... associated by hydrogen bonds...

$$R-C(=O\cdots H-O)(-O-H\cdots O=)C-R$$

The facility which acids have for forming two hydrogen bonds per molecule makes their boiling temperatures higher than those of corresponding alcohols.

TABLE 33.1 The Names and Formulae of some Derivatives of Acids

Formula	Name	Example	
R—C(=O)—	Acyl group	CH_3CO—	Ethanoyl group
$[R—C(\cdots O)_2]^-$	Carboxylate ion	$CH_3CO_2^-$	Ethanoate ion
R—C(=O)Cl	Acid chloride	CH_3COCl	Ethanoyl chloride
R—C(=O)—O—C(=O)—R	Acid anhydride	$(CH_3CO)_2O$	Ethanoic anhydride
R—C(=O)NH_2	Amide	CH_3CONH_2	Ethanamide
R—C(=O)OR′	Carboxylic ester	$CH_3CO_2C_2H_5$	Ethyl ethanoate
R—C≡N	Nitrile	CH_3CN	Ethanonitrile

...and dissolving in water as they form hydrogen bonds to water molecules

Carboxylic acids of fairly low molecular mass dissolve in water. The dimers dissociate to form monomers in order to form hydrogen bonds to water molecules:

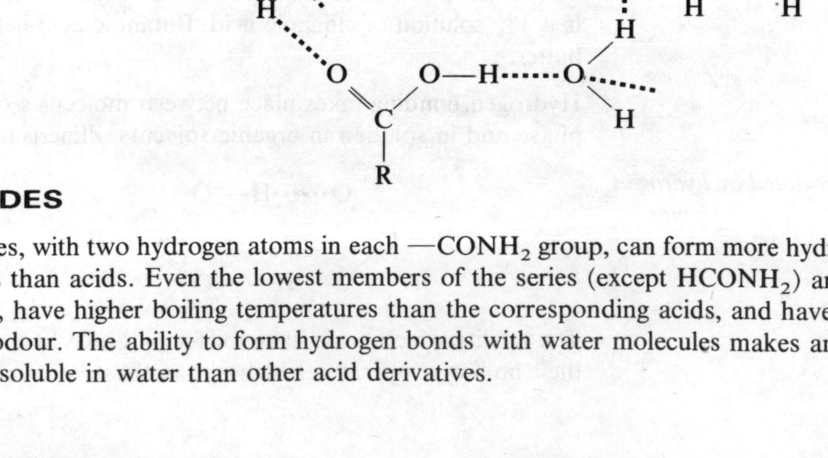

33.3.2 AMIDES

Amides are solids...

...which dissolve in water

Amides, with two hydrogen atoms in each —$CONH_2$ group, can form more hydrogen bonds than acids. Even the lowest members of the series (except $HCONH_2$) are solids, have higher boiling temperatures than the corresponding acids, and have little odour. The ability to form hydrogen bonds with water molecules makes amides more soluble in water than other acid derivatives.

33.3.3 ESTERS, CHLORIDES, ANHYDRIDES AND NITRILES

Other derivatives are volatile liquids...

Esters, chlorides, anhydrides and nitriles form no hydrogen bonds. Except for anhydrides, their boiling temperatures are lower than those of the corresponding acids. These derivatives are volatile and odorous. There are, however, strong dipole–dipole interactions between the molecules, arising from the polarity of the

$$\rangle C{=}O \text{ group}$$

Even the lowest members of the series are liquids at room temperature.

...Chlorides are fuming liquids...

Acid chlorides are colourless liquids with a pungent smell and a lachrymatory action. Anhydrides also have a pungent smell. Aromatic anhydrides are solids.

...Esters have a fruity smell

Esters are colourless liquids with pleasant fruity odours. Examples are:

$CH_3CH_2CH_2CO_2C_2H_5$	Ethyl butanoate	*apple odour*
$CH_3CO_2(CH_2)_7CH_3$	Octyl ethanoate	*orange odour*

33.3.4 SALTS

The salts of carboxylic acids are ionic solids

The salts of carboxylic acids are electrovalent compounds and are therefore involatile crystalline solids. They are often soluble as, on dissolution, the ions are hydrated as the $—CO_2^-$ group forms hydrogen bonds with water molecules. The lower members of the series are soluble, but in higher members the large hydrocarbon part of the molecule makes the compounds insoluble.

33.4 REACTIVITY OF CARBOXYLIC ACIDS

The $\rangle C{=}O$ and —OH groups in —CO_2H do not react as they do in carbonyl compounds and alcohols

The carboxyl group is so named because it contains a *carb*onyl group and a hydr*oxyl* group. The two groups influence each other to such an extent that the reactions of carboxylic acids bear little resemblance to those of either carbonyl compounds or alcohols

The carboxylic acid group is

$$—C(=O)—O—H$$

The polar

$$\rangle C{=}O \text{ group}$$

attracts electrons away from the —O—H bond, and makes it easier for the hydrogen atom to ionise than is the case in the —O—H bond in an alcohol. The flow of electrons from the —OH group towards the carbonyl carbon atom reduces the $\delta+$ charge on the carbonyl carbon atom, with the result that it is not attacked by the nucleophiles that attack carbonyl compounds.

The charge in RCO_2^- is delocalised. RCO_2^- is a weaker base than RO^-

When a carboxylate ion, RCO_2^-, is formed, the negative charge on the ion is shared equally between two oxygen atoms. The structure of RCO_2^- can be represented by a molecular orbital picture:

$$\left[R—C \begin{matrix} \overset{\cdot\cdot}{\diagup} O \\ \underset{\cdot\cdot}{\diagdown} O \end{matrix} \right]^-$$

The delocalisation of the charge makes the carboxylate ion less ready to accept a proton than is an alkoxide ion, $R—O^-$. The carboxylate ion, RCO_2^- is therefore a weaker base than the alkoxide ion, RO^-, and RCO_2H is a stronger acid than ROH.

Carboxylic acids are weak acids...

...Substituents affect the strength of acids...

Carboxylic acids are much weaker than the common mineral acids. The dissociation constant, K_a, for ethanoic acid is 1.8×10^{-5} mol dm^{-3} [§ 12.7.4]. This means that in a 1 mol dm^{-3} solution of the acid, 3 molecules in a thousand are ionised. Substituents affect the strength of acids [see § 12.7.7].

...Benzoic acid is stronger than aliphatic acids...

Benzoic acid is stronger than ethanoic acid because the negative charge on the $—CO_2^-$ group, being delocalised by interaction with the π electron cloud of the benzene ring, is less available for attaching a proton to form $—CO_2H$ [see Figure 33.1].

FIGURE 33.1 The Benzoate Ion

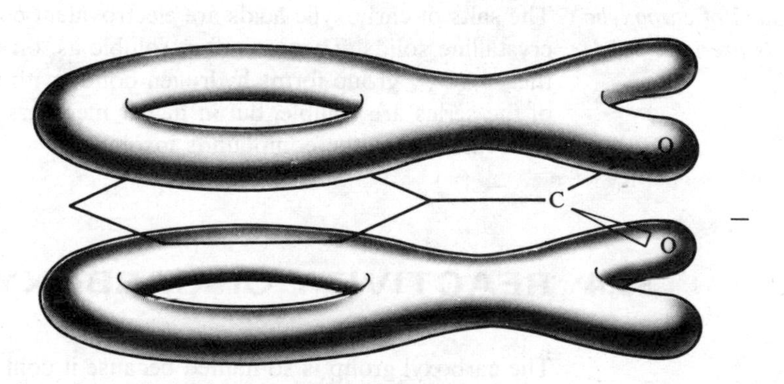

...Carboxylic acids are stronger than carbonic acid

Carboxylic acids are stronger than carbonic acid. The evolution of carbon dioxide that occurs when a carboxylic acid reacts with sodium hydrogencarbonate is used to distinguish carboxylic acids from weaker acids, such as phenols. Some substituted phenols also give a positive result.

33.5 INDUSTRIAL SOURCE AND USES

33.5.1 METHANOIC ACID

HCO_2H is obtained from CO

Carbon monoxide and aqueous sodium hydroxide react at 200 °C under pressure to form sodium methanoate. On acidification with mineral acid, this salt gives methanoic acid:

$$CO(g) + NaOH(aq) \xrightarrow[200\,°C]{pressure} HCO_2Na(aq) \xrightarrow{acid} HCO_2H(aq)$$

33.5.2 ETHANOIC ACID

1. The aerial oxidation of alkanes (C_5–C_7) from petroleum can be carried out at high temperature and pressure to yield ethanoic acid.

CH_3CO_2H is made from petroleum and from ethanal

2. Ethanal can be oxidised by air in the presence of a catalyst to give ethanoic acid:

$$CH_3CHO \xrightarrow[\text{catalyst}]{\text{air}} CH_3CO_2H$$

USE

Much of the ethanoic acid produced is converted into ethanoic anhydride, which is used in the manufacture of the fabric acetate rayon and the drug aspirin [§ 33.12]. Chloroethanoic acid is used in the manufacture of the weedkiller 2,4-D [§ 20.15].

33.5.3 BENZOIC ACID

See Figure 33.2.

33.6 LABORATORY PREPARATIONS OF CARBOXYLIC ACIDS

33.6.1 OXIDATION

PRIMARY ALCOHOLS AND ALDEHYDES

Oxidation of alcohols, aldehydes...

Potassium dichromate(VI) and potassium manganate(VII), both in acid solution, are often used as oxidising agents. Primary alcohols are oxidised via aldehydes to carboxylic acids:

$$RCH_2OH \rightarrow RCHO \rightarrow RCO_2H$$

ALKENES

...and alkenes gives acids

Alkenes are oxidised by acidified potassium manganate(VII):

$$\text{Cyclohexene} \xrightarrow[\text{acid}]{KMnO_4} \text{Hexane-1,6-dioic acid } (CO_2H, CO_2H)$$

Cyclohexene Hexane-1,6-dioic acid

METHYLBENZENE

Aromatic side chains are oxidised to CO_2H

Benzoic acid is made by oxidising methylbenzene. Industrially, air is used as the oxidising agent. Even if the group attached to the aromatic ring is larger than —CH_3, benzoic acid is the oxidation product:

$$C_6H_5{-}CH_3 \xrightarrow[\text{acid, heat}]{KMnO_4} C_6H_5{-}CO_2H \xleftarrow[\text{acid, heat}]{KMnO_4} C_6H_5{-}CH_2CH_2OH$$

FIGURE 33.2
Preparations of Benzoic Acid

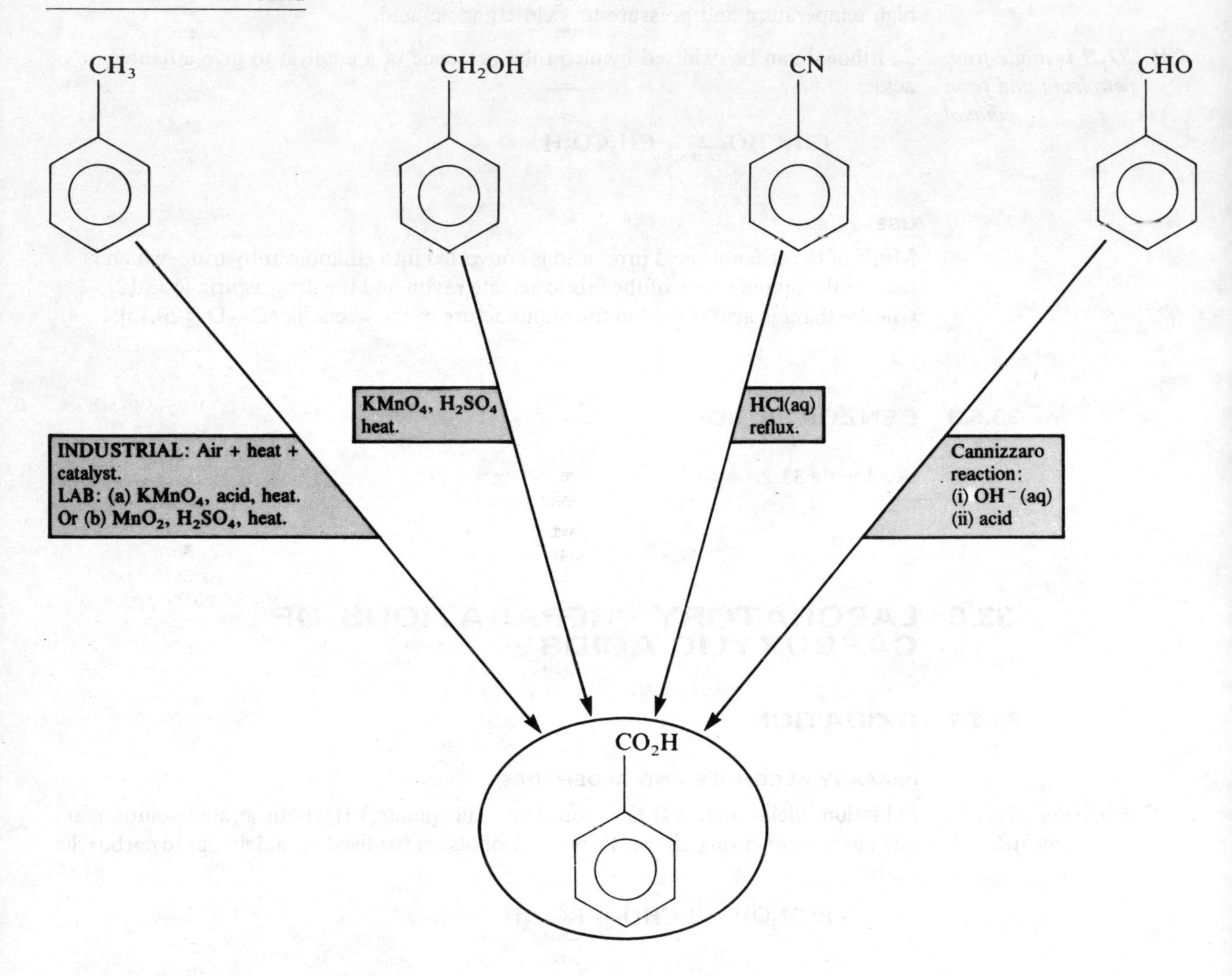

33.6.2 HYDROLYSIS

NITRILES

Hydrolysis of RCN gives RCO_2H

Nitriles are hydrolysed to carboxylic acids by being boiled under reflux with aqueous alkali or mineral acid. The amide, $RCONH_2$, is an intermediate:

$$RCN(l) \xrightarrow[\text{or } H^+(aq)]{NaOH(aq)} RCONH_2(aq) \xrightarrow{H_2O} RCO_2Na(aq) + NH_3(g) \text{ or } RCO_2NH_4(aq)$$

ESTERS

Esters are hydrolysed to acids

Esters

$$R{-}\overset{\overset{\displaystyle O}{/\!/}}{C}{-}OR'$$

are hydrolysed to alcohols and carboxylic acid salts when they are boiled under reflux with aqueous alkali:

$$RCO_2R'(l) + NaOH(aq) \rightarrow RCO_2Na(aq) + R'OH(aq)$$

The carboxylic acid is obtained by acidifying the salt with a mineral acid:

$$RCO_2Na(aq) + HCl(aq) \rightarrow RCO_2H(aq) + NaCl(aq)$$

33.6.3 BENZOIC ACID

See Figure 33.2.

CHECKPOINT 33A: ACIDS

1. Name the following acids:
(*a*) $CH_3(CH_2)_5CO_2H$ (*b*) $(CH_3)_2CHCO_2H$ (*c*) CF_3CO_2H

(*d*) CO_2H | CH_2 | CO_2H (*e*) CO_2H on a benzene ring with CH_3 para (*f*) benzene ring with two adjacent CO_2H groups

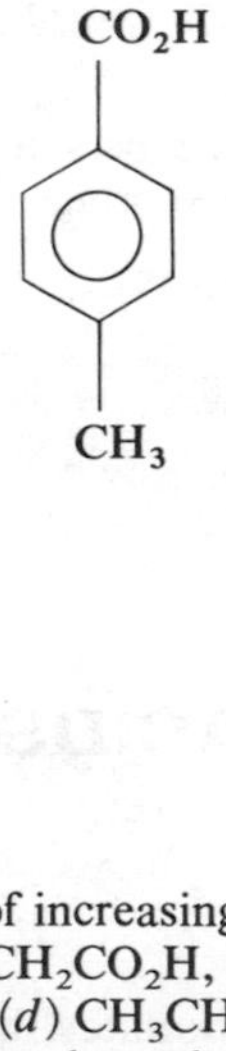

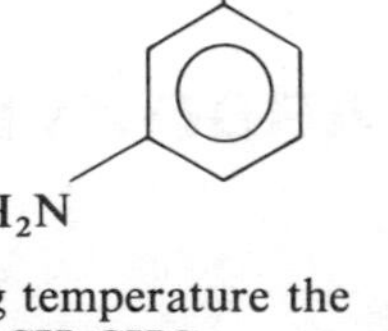

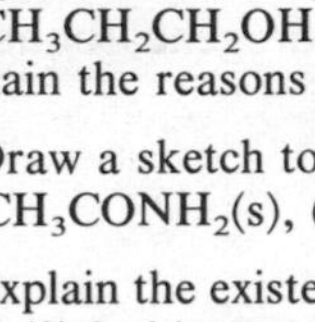

(*g*) CH_2CO_2H on a benzene ring (*h*) CO_2H on a benzene ring with H_2N meta

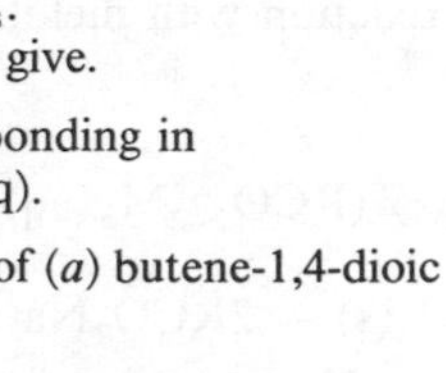

2. Arrange in order of increasing boiling temperature the compounds: (*a*) $CH_3CH_2CO_2H$, (*b*) CH_3CH_2CHO, (*c*) $CH_3CH_2CH_2OH$, (*d*) $CH_3CH_2CH_3$. Explain the reasons for the order you give.

3. Draw a sketch to show hydrogen bonding in (*a*) $CH_3CONH_2(s)$, (*b*) $CH_3CONH_2(aq)$.

4. Explain the existence of two forms of (*a*) butene-1,4-dioic acid, (*b*) 2-chloropropanoic acid.

***5.** Arrange in order of increasing acid strength: (*a*) phenol, (*b*) benzoic acid, (*c*) carbonic acid, (*d*) trichloroethanoic acid.

6. What products are formed by the reaction of pent-2-ene with (*a*) cold, alkaline potassium manganate(VII), (*b*) hot acidic potassium manganate(VII)?

†**7.** Complete the following equations:

(*a*) $RCH_2Cl + \quad \rightarrow RCH_2CN$

(*b*) $HC{\equiv}CH + \quad \rightarrow H_2C{=}CHCN$

(*c*) $H_2C{=}CHCN + \quad \rightarrow H_2C{=}CHCO_2H$

Indicate the conditions needed for each reaction. What is the IUPAC name for $H_2C{=}CHCO_2H$? The trivial name for this compound is acrylic acid. What is its industrial importance? Why cannot it be made from $H_2C{=}CHCl$? [For help, see § 29.10].

8. A liquid **A** of formula C_7H_6O, is boiled with concentrated aqueous sodium hydroxide. Ethoxyethane extraction of the products, followed by drying and distillation, gave as the main product a liquid **B** of boiling temperature 205 °C. **B** reacted with phosphorus(V) chloride to give a pungent gas. When the solution from which **B** had been extracted was acidified, a white crystalline solid **C** was precipitated. On recrystallisation from hot water, it appeared as flaky white crystals. A solution of **C** effervesced when sodium carbonate was added. Identify **A**, **B** and **C**, and explain the reactions described.

9. (*a*) Outline a method for the conversion

$$CH_3CHO \rightarrow CH_3CH(OH)CO_2H$$

(*b*) Name the product.

(*c*) How could you show that the product is (i) a secondary alcohol and (ii) an acid?

(*d*) What effect would you expect the acid to have on plane-polarised light? Explain your answer.

33.7 VITAMIN C: ASCORBIC ACID

TOPIC

Vitamin C is the most unstable of the vitamins: it is easily destroyed by oxidation, which is speeded up by heat or by a base.

Vitamin C is an acid. It is easily oxidised

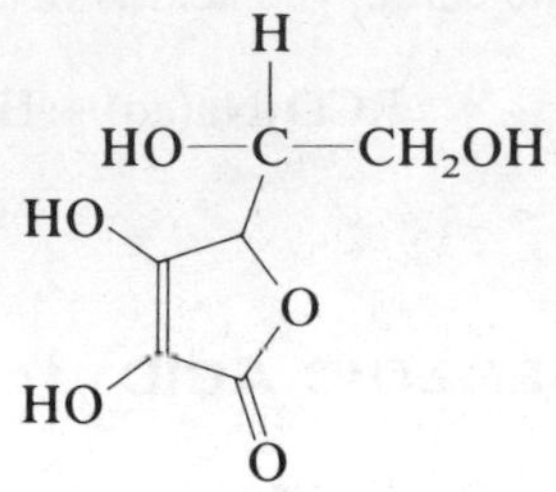

H
HO—C—CH_2OH
HO
O
HO
O

Ascorbic acid

Vitamin C is needed for the formation of tissues, teeth and bone. It plays a role in the metabolism of carbohydrates and proteins and in the formation of adrenalin, serotonin, haemoglobin and collagen. Lack of collagen results in the symptoms of scurvy.

Many people, including Linus Pauling [see § 4.2.8], believe that large doses of Vitamin C prevent colds. It has been proved that it helps wounds to heal.

CHECKPOINT 33B: VITAMIN C

1. What is the first oxidation product of Vitamin C?

2. Why is it more likely that a person will suffer from a Vitamin C deficiency than a shortage of other vitamins?

33.8 REACTIONS OF CARBOXYLIC ACIDS

33.8.1 SALT FORMATION

Carboxylic acids form salts by reaction with metals, carbonates, hydrogencarbonates and alkalis:

A test for —CO_2H is the evolution of CO_2 from $NaHCO_3$

$$2RCO_2H(aq) + Mg(s) \rightarrow (RCO_2)_2Mg(aq) + H_2(g)$$

$$2RCO_2H(aq) + Na_2CO_3(s) \rightarrow 2RCO_2Na(aq) + CO_2(g) + H_2O(l)$$

$$RCO_2H(aq) + NaOH(aq) \rightarrow RCO_2Na(aq) + H_2O(l)$$

The evolution of carbon dioxide from sodium hydrogencarbonate is used as a test to distinguish carboxylic acids from weaker acids, such as phenols. The reaction with alkali can be carried out by titration against a standard alkali to find the concentration of a solution of the acid. A suitable indicator must be used [§ 12.7.8].

Stronger acids displace carboxylic acids from their salts

The salts are crystalline solids, most of which are soluble in water. Organic acids can be extracted from a mixture of products by dissolving them in aqueous sodium hydroxide. Acidification with a mineral acid may precipitate the free acid from its salt:

$$C_6H_5CO_2H(s) + NaOH(aq) \rightarrow C_6H_5CO_2^- Na^+(aq) + H_2O(l) \xrightarrow{HCl(aq)} C_6H_5CO_2H(s) + NaCl(aq)$$

33.8.2 ESTERIFICATION

Reaction with an alcohol gives an ester

Carboxylic acids react with alcohols in the presence of concentrated sulphuric acid to form esters. The acid and alcohol are boiled under reflux with the concentrated acid or, alternatively, hydrogen chloride can be bubbled through the mixture. If the alcohol is labelled with ^{18}O, analysis of the products in a mass spectrometer shows that all the ^{18}O is present in the ester and none is in the water. This proves that the bonds that are broken in the reaction are:

$$R{-}C(=O){\vdots}OH + R'{-}^{18}O{\vdots}H \rightarrow R{-}C(=O){-}^{18}O{-}R' + H_2O$$

In the esterification of tertiary alcohols and some secondary alcohols, there is fission of the

$$R{\vdots}O{-}H$$

bond, and the yield of ester is low.

33.8.3 CONVERSION INTO ACID CHLORIDES

RCOCl is formed by reaction with PCl_5 or $SOCl_2$

The conversion

$$R{-}C(=O){-}OH \rightarrow R{-}C(=O){-}Cl$$

is accomplished by the action of phosphorus(V) chloride, PCl_5, or by sulphur dichloride oxide, $SOCl_2$. The latter is more convenient as the by-products are gaseous:

$$C_2H_5CO_2H(l) + PCl_5(l) \rightarrow C_2H_5COCl(l) + POCl_3(l) + HCl(g)$$

Propanoic acid → Propanoyl chloride

$$C_6H_5CO_2H(s) + SOCl_2(l) \rightarrow C_6H_5COCl(l) + SO_2(g) + HCl(g)$$

Benzoic acid → Benzoyl chloride

33.8.4 CONVERSION INTO AMIDES

The conversion

$$R-C(=O)-O-H \rightarrow R-C(=O)-NH_2$$

Amides are made from ammonium salts by distillation with acid

is accomplished via the ammonium salt of the acid. The ammonium salt is made by the action of the acid on ammonium carbonate (not ammonia solution because of salt hydrolysis). When heated to 100–200 °C, with an excess of the parent acid, the ammonium salt is dehydrated to give the amide:

$$R-C(=O)-O^- \ NH_4^+ \xrightarrow[100-200\,°C]{RCO_2H} R-C(=O)-NH_2 + H_2O$$

The reason for the presence of an excess of the free acid is that ammonium salts dissociate on being heated:

$$RCO_2^- \, NH_4^+(s) \rightleftharpoons RCO_2H(g) + NH_3(g)$$

The addition of acid displaces the equilibrium to the left.

33.8.5 HALOGENATION

Hydrogen atoms in the α position to $-CO_2H$ can be replaced by halogens

When chlorine is bubbled through boiling ethanoic acid, in sunlight and in the presence of iodine or red phosphorus as catalyst, chloroethanoic acid is formed. It is a crystalline solid which melts at 61 °C:

$$CH_3CO_2H(l) + Cl_2(g) \xrightarrow[I_2 \text{ or red P}]{\text{boil}} CH_2ClCO_2H(s) + HCl(g)$$

If the flow of chlorine is continued and the temperature is raised to the boiling temperature of the product, further substitution occurs to give dichloro- and trichloroethanoic acid:

$$CH_2ClCO_2H(l) \xrightarrow{Cl_2} CHCl_2CO_2H(l) + HCl(g) \xrightarrow{Cl_2} CCl_3CO_2H(s) + HCl(g)$$

Homologues of ethanoic acid are halogenated only in the α position. A phosphorus(III) halide is used as a catalyst:

$$RCH_2CO_2H(l) + Br_2(l) \xrightarrow{PBr_3} RCHBrCO_2H \xrightarrow[PBr_3]{Br_2} RCBr_2CO_2H + HBr(g)$$

As mentioned previously, halogenoacids are stronger acids than the parent acids.

2,4-D is made from chloroethanoic acid

Chloroethanoic acid is used in the manufacture of the weedkiller, 2,4-dichlorophenoxyethanoic acid, which is marketed as 2,4-D [§ 20.15]:

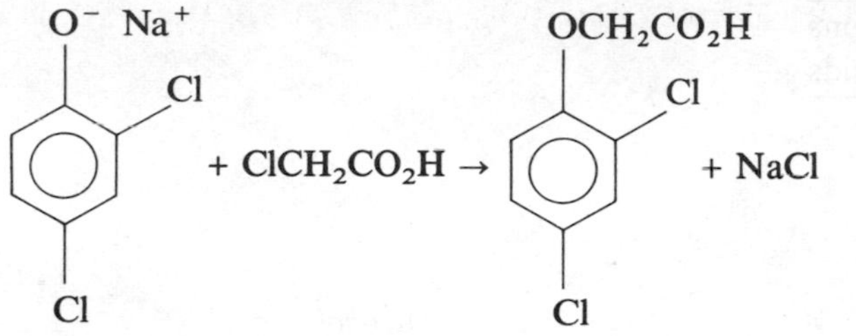

$$C_6H_3Cl_2O^-Na^+ + ClCH_2CO_2H \rightarrow C_6H_3Cl_2OCH_2CO_2H + NaCl$$

Sodium 2,4-dichlorophenoxide → 2,4-Dichlorophenoxyethanoic acid

33.8.6 REDUCTION

Acids can be reduced to alcohols...

To convert a carboxylic acid into an alcohol, a powerful reducing agent is needed. Lithium tetrahydridoaluminate dissolved in ethoxyethane will effect reduction:

$$RCO_2H \xrightarrow[\text{in ethoxyethane}]{LiAlH_4} RCH_2OH$$

33.8.7 DECARBOXYLATION OF A SODIUM SALT

...decarboxylated to give a hydrocarbon

When the sodium salt of a carboxylic acid is heated with soda lime, the carboxyl group is removed as a carbonate and a hydrocarbon is formed:

$$C_6H_5CO_2Na(s) + NaOH(s) \rightarrow C_6H_6 + Na_2CO_3$$

CHECKPOINT 33C: REACTIONS OF ACIDS

1. A student has read that rancid butter contains butanoic acid. She has supplies of ethoxyethane and laboratory acids and alkalis. How can she obtain a sample of butanoic acid? How can she (*a*) purify her sample and (*b*) test its purity? Butanoic acid is a liquid at room temperature.

2. How could propanoic acid be converted into (*a*) a stronger acid, (*b*) a buffer solution? [§ 12.7.12]

3. Suggest a method for the conversion

$$C_2H_5OH \rightarrow CH_3CH(OH)CO_2H$$

4. Explain why the 2-hydrogen atoms in $CH_3CH_2CO_2H$ can be substituted by halogens but not the 3-hydrogen atoms.

5. Butene-1,4-dioic acid exists in two isomeric forms. Draw their structural formulae. What type of isomerism is involved? Which of these isomers will not readily form an anhydride? [§ 25.9.2]

On the addition of hydrogen bromide, both acids give two isomeric compounds. The same two products are obtained from both acids. Draw the structural formulae of the two adducts. Explain the type of isomerism involved.

6. Predict the results of the reactions of the compound shown below with (*a*) bromine water, (*b*) ethanoic anhydride, (*c*) hydrogen and nickel, (*d*) lithium tetrahydridoaluminate, (*e*) ozone.

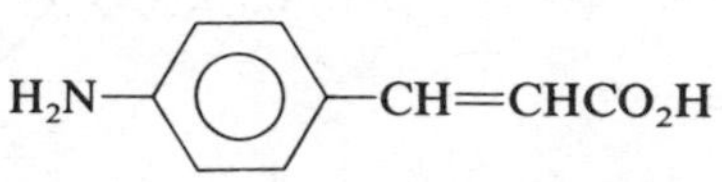

$$H_2N{-}C_6H_4{-}CH{=}CHCO_2H$$

7. Explain why the carbonyl group in a carboxylic acid is less reactive than that in an aldehyde or a ketone.

***8.** Compare the —OH group in ethanol, ethanoic acid and phenol.

FIGURE 33.3 Reactions of Carboxylic Acids

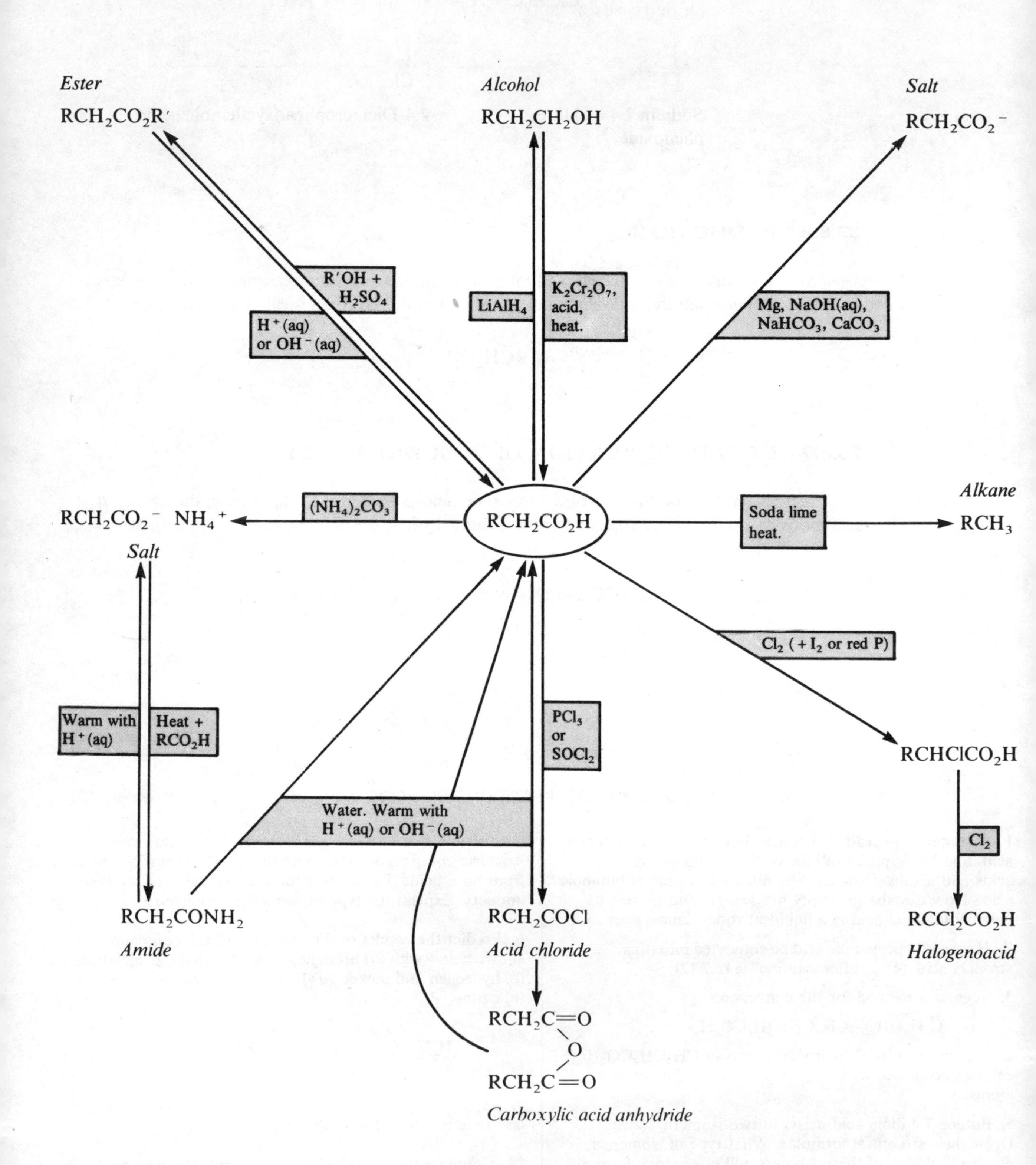

33.9 DERIVATIVES OF CARBOXYLIC ACIDS

The derivatives of carboxylic acids that contain an acyl group are shown in Figure 33.4 in order of decreasing reactivity.

FIGURE 33.4 Acid Derivatives in Order of Decreasing Reactivity

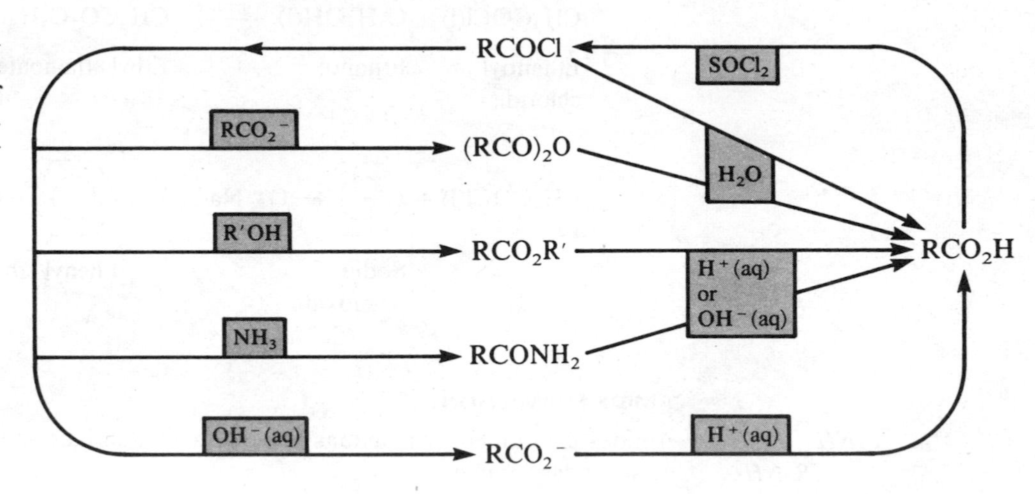

33.10 ACID CHLORIDES

33.10.1 REACTIVITY

With the functional group

$$-\overset{\delta+}{C}(=O^{\delta-})Cl$$

Nucleophiles attack the carbonyl C atom in RCOCl...

the reactions of acid chlorides depend upon the attack by nucleophiles on the $\delta+$ carbon atoms. In benzoyl chloride and other aromatic acid chlorides, the $\delta+$ charge is spread over the ring, and the carbon atom is less susceptible to nucleophilic attack than it is in aliphatic acid chlorides.

33.10.2 REACTIONS OF ACID CHLORIDES

HYDROLYSIS

...H_2O is an example

Acid chlorides are insoluble in water but are hydrolysed quickly to a carboxylic acid and hydrochloric acid. Aliphatic acid chlorides fume in moist air to give hydrogen chloride:

$$CH_3COCl(l) + H_2O(l) \rightarrow CH_3CO_2H(l) + HCl(g)$$

ESTER FORMATION

...ROH and ArO^-...

Acid chlorides, RCOCl, and aroyl chlorides, ArCOCl, react readily with phenols in alkaline solution and with alcohols in the presence of pyridine (a base which removes the HCl produced) to form esters:

$$CH_3COCl(l) + C_2H_5OH(l) \xrightarrow{\text{pyridine}} CH_3CO_2C_2H_5(l) + HCl$$

Ethanoyl chloride — Ethanol — Ethyl ethanoate

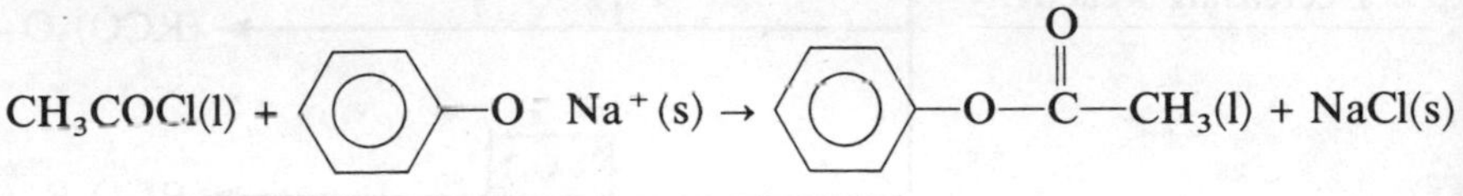

$$CH_3COCl(l) + C_6H_5O^- Na^+(s) \rightarrow C_6H_5\text{—O—}\overset{O}{\overset{\|}{C}}\text{—}CH_3(l) + NaCl(s)$$

Sodium phenoxide — Phenyl ethanoate

AMIDE FORMATION

...NH_3, RNH_2, R_2NH...

Amides are formed by reactions between acid chlorides and ammonia or a primary or secondary amine:

$$RCOCl + R'_2NH \rightarrow RCONR'_2 + HCl$$

The reaction takes place readily at room temperature.

ANHYDRIDE FORMATION

...and RCO_2^-

When an acid chloride is heated with the sodium salt of the carboxylic acid, the acid anhydride distils over:

$$C_2H_5CO_2Na(s) + C_2H_5COCl(l) \xrightarrow{\text{distil}} (C_2H_5CO)_2O(l) + NaCl(s)$$

Sodium propanoate — Propanoyl chloride — Propanoic anhydride

KETONE FORMATION

See Friedel–Crafts acylation, §28.7.5 and Figure 33.5.

33.11 ACID ANHYDRIDES

The reactions of acid anhydrides are similar to those of acid chlorides

Anhydrides are insoluble in water but react with it to form the parent acid:

$$(RCO)_2O(l) + H_2O(l) \rightarrow 2RCO_2H(aq)$$

Anhydrides react in a similar way to acid chlorides. They form esters with alcohols and phenols [see Figure 33.5]; they form amides with ammonia and primary and secondary amines and are Friedel–Crafts acylating agents:

$$(CH_3CO)_2O + C_6H_5O^- Na^+ \rightarrow C_6H_5O_2CCH_3 + CH_3CO_2^- Na^+$$

Ethanoic anhydride, Sodium phenoxide, Phenyl ethanoate

$$(CH_3CO)_2O + C_2H_5NHCH_3 \rightarrow C_2H_5N(COCH_3)CH_3 + CH_3CO_2H$$

N-Methylethylamine, *N*-Ethyl-*N*-methylethanamide

The industrial importance of ethanoic anhydride is featured in § 33.12.

Methanoic acid does not form an anhydride. Dehydration of this acid yields carbon monoxide [§ 23.7.3].

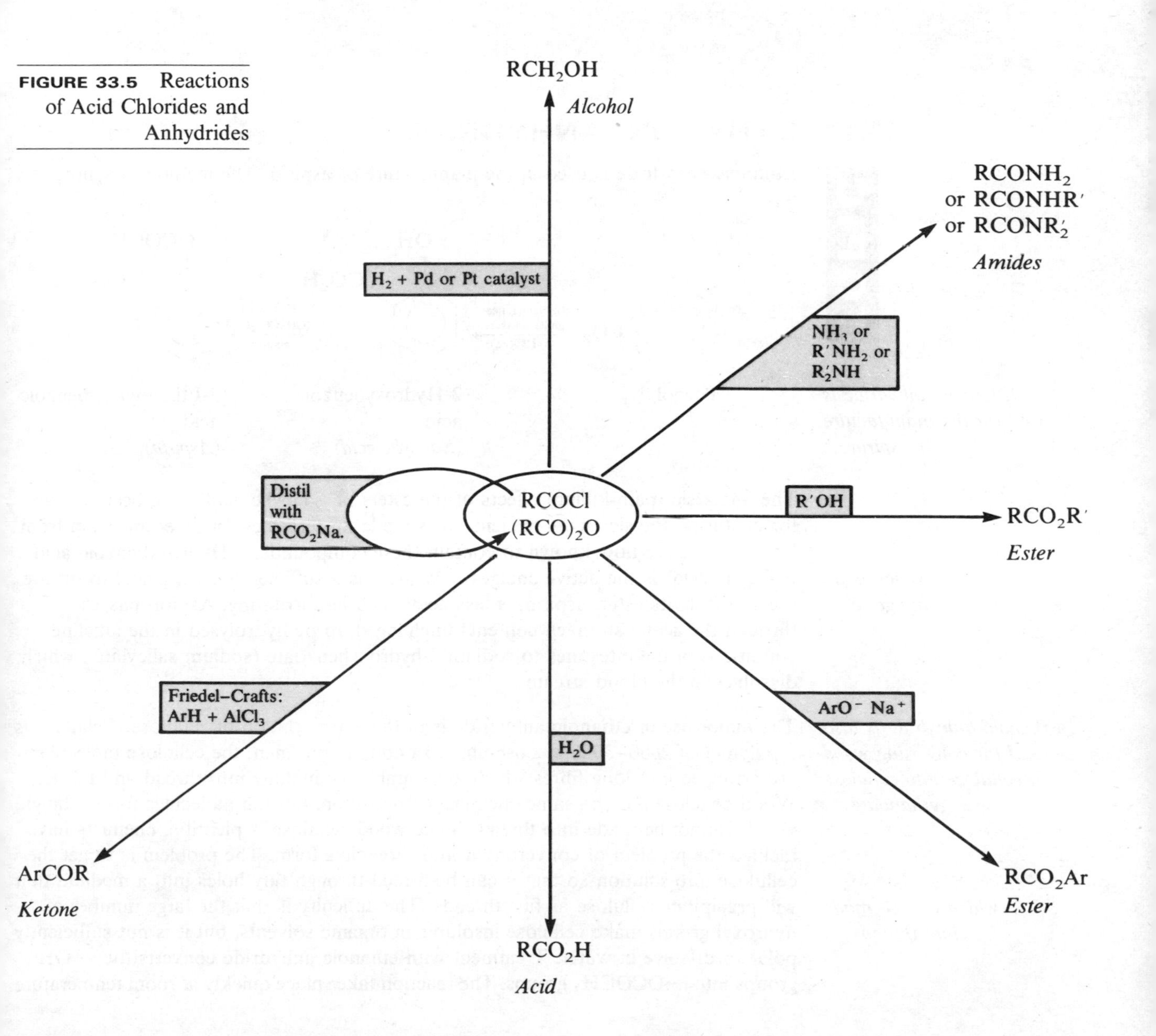

FIGURE 33.5 Reactions of Acid Chlorides and Anhydrides

CHECKPOINT 33D: ACID CHLORIDES AND ANHYDRIDES

1. Complete the following equations:

(*a*) $C_2H_5CO_2H + PCl_5 \rightarrow$

(*b*) $C_6H_5CO_2H + SOCl_2 \rightarrow$

(*c*) $(C_6H_5CO)_2O + NH_3 \rightarrow$

(*d*) $(CH_3)_2CHCOCl + C_2H_5OH \rightarrow$

2. State how you could prepare ethanoyl chloride and ethanoic anhydride. Compare the reactions of the two compounds with (*a*) water, (*b*) ethanol, (*c*) phenol, (*d*) phenylamine.

3. Compare the reactions of chloroethane and ethanoyl chloride. Explain the reason for the differences in behaviour.

4. Into a large excess of water were put *m* grams of ethanoyl chloride. The products remained in solution and were neutralised by $50.0\,cm^3$ of a solution of sodium hydroxide of concentration $0.100\,mol\,dm^{-3}$. Find *m*.

5. Describe how you could use benzoyl chloride, C_6H_5COCl, to distinguish between phenylamine and phenylmethylamine.

†6. Describe how you could distinguish ethanoyl chloride from benzoyl chloride. [Remember Chapter 32.]

33.12 ETHANOIC ANHYDRIDE

TOPIC

Ethanoic anhydride is used in the manufacture of aspirin. The method of synthesis is

$$C_6H_5OH + CO_2 \xrightarrow[\text{(ii) HCl(aq)}]{\text{(i) NaOH, heat under pressure}} C_6H_4(OH)CO_2H \xrightarrow[\text{alkali}]{(CH_3CO)_2O} C_6H_4(OCOCH_3)CO_2H$$

Phenol — 2-Hydroxybenzoic acid (*Salicylic acid*) — 2-Ethanoyloxybenzoic acid (*Aspirin*)

Ethanoic anhydride is used for the manufacture of aspirin...

... which is an analgesic...

The analgesic (pain-killing) effects of the esters of 'salicylic acid' have been known for centuries. Physicians in ancient times made up remedies for fever and pain from leaves which are now known to contain these compounds. 2-Hydroxybenzoic acid (salicylic acid) is the active analgesic. While it is a sufficiently strong acid to irritate the stomach, its ester, aspirin, is less acidic and less irritating. Aspirin passes through the acidic stomach contents unchanged, to be hydrolysed in the alkaline conditions of the intestines to sodium 2-hydroxybenzoate (sodium salicylate), which dissolves in the blood stream.

Ethanoic anhydride is also used for converting wood cellulose into cellulose ethanoate...

... which can be made into threads...

The major use of ethanoic anhydride is in the ethanoylation of cellulose. Cellulose is a polymer of 2000–3000 glucose units. In cotton and linen, the cellulose molecules are arranged into long fibres which are suitable for making into thread and fabric. Wood cellulose has the same chemical composition, but the molecules form a tangle which cannot be made into thread. Since wood cellulose is plentiful, chemists have tackled the problem of converting it into threadlike form. The problem is to get the cellulose into solution so that it can be forced through tiny holes into a medium that will precipitate cellulose as fine threads. The difficulty is that the large number of hydroxyl groups make cellulose insoluble in organic solvents, but it is not sufficiently polar to dissolve in water. Treatment with ethanoic anhydride converts the —OH groups into $—OCOCH_3$ groups. The reaction takes place quickly at room temperature

... to be used in the manufacture of acetate rayon

to give a high yield of cellulose ethanoate. If ethanoic acid is used, prolonged heating and the presence of an acid catalyst are necessary, and degradation of the cellulose molecules results. With the hydroxyl groups esterified, cellulose ethanoate is soluble in organic solvents. From solution, it can be precipitated as fine threads of *acetate rayon*, which are made into fabric [see Figure 33.6].

FIGURE 33.6 Cellulose Ethanoate (Acetate Rayon)
Ac is $CH_3C(=O)—$
The monomer is shown

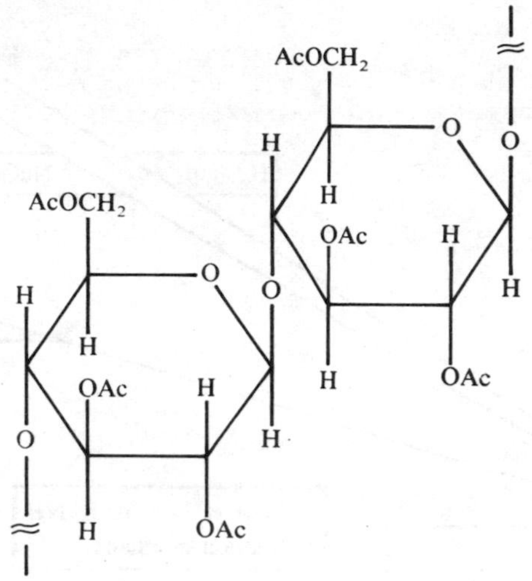

Problem (a)
Methylcellulose is used in the paper industry for imparting a resistance to oil and grease. It is made by converting the —OH groups in cellulose to $—OCH_3$ groups. Suggest how this could be done, and draw a structural formula for methyl cellulose.

Problem (b)
'Oil of wintergreen' has been used for centuries as a soothing liniment. Its IUPAC name is methyl 2-hydroxybenzoate. Devise a synthesis for this compound from phenol.

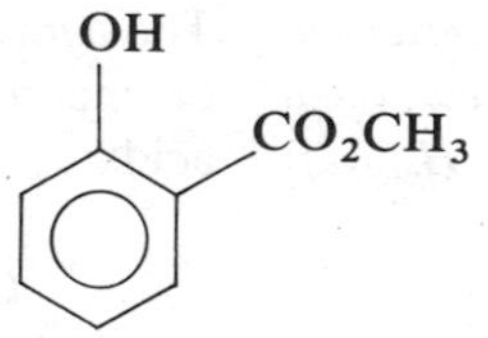

Methyl 2-hydroxybenzoate

33.13 ESTERS

Esters contain the functional group

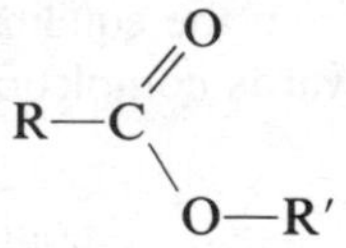

$$R—C(=O)—O—R'$$

where R and R′ may be alkyl or aryl groups and may be the same or different. The reactions of esters are summarised in Figure 33.7.

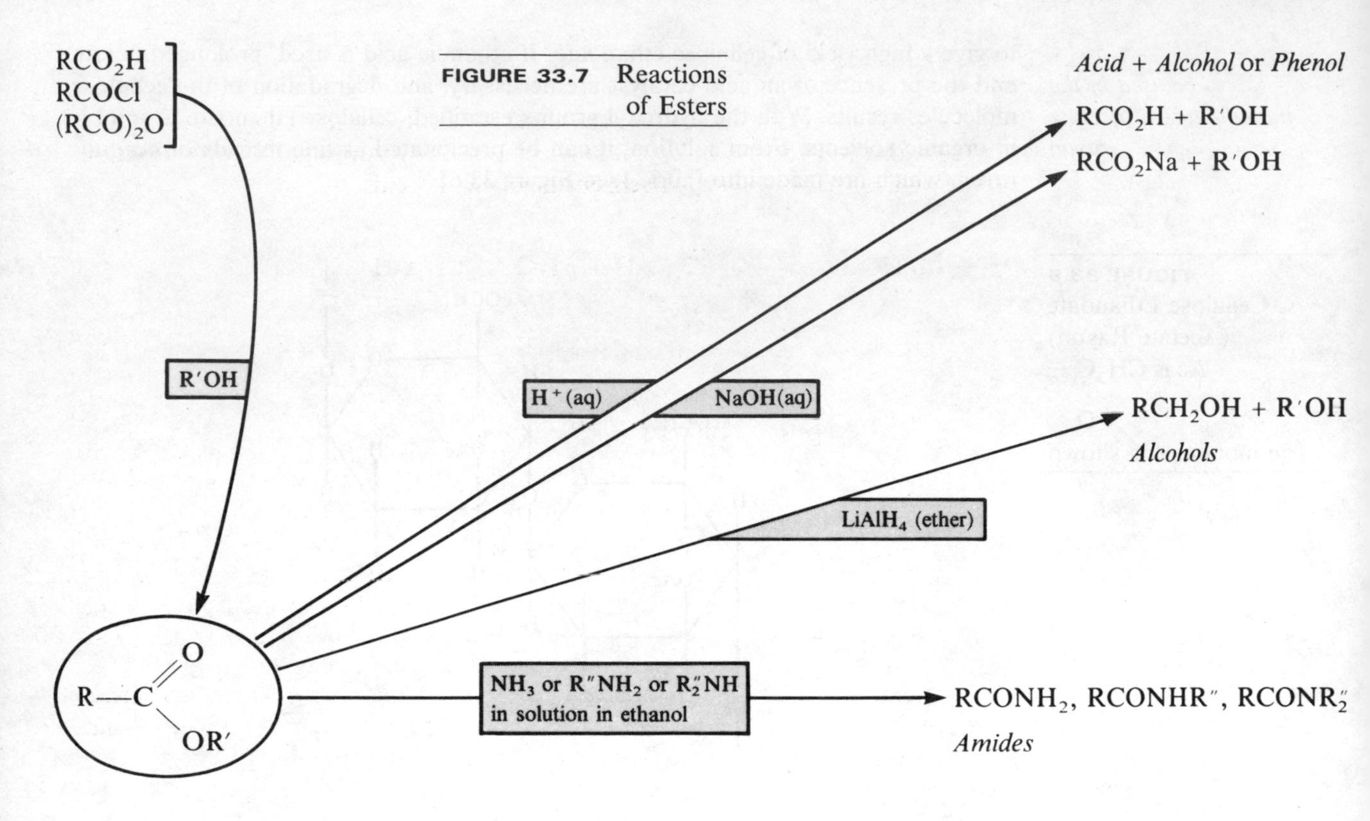

FIGURE 33.7 Reactions of Esters

33.13.1 ESTER HYDROLYSIS

The hydrolysis of esters to acid + alcohol is catalysed by $H^+(aq)$

The hydrolysis of an ester gives an equilibrium mixture of carboxylic acid and alcohol or phenol. The attainment of equilibrium is catalysed by the presence of mineral acids:

$$CH_3CO_2C_2H_5(l) + H_2O(l) \xrightleftharpoons{H^+(aq)} CH_3CO_2H(aq) + C_2H_5OH(aq)$$

OH^- ions catalyse the R.D.S. and also move the position of equilibrium in favour of the products

The reaction also proceeds faster in alkaline conditions. Hydroxide ions catalyse the rate-determining step in the hydrolysis:

$$RCO_2R' + H_2O \xrightleftharpoons{OH^-} RCO_2H + R'OH$$

When some RCO_2H molecules have been formed, OH^- ions react with them to form a salt:

$$RCO_2H + OH^- \rightarrow RCO_2{}^- + H_2O$$

Salt formation removes RCO_2H from the equilibrium mixture. The hydrolysis reaction therefore goes further towards completion in the presence of a base. The base is used up in the reaction:

$$RCO_2R' + NaOH \rightarrow RCO_2Na + R'OH$$

For the alkaline hydrolysis of oils see § 33.13.3.

33.13.2 POLYESTERS

Dicarboxylic acids and diols react to form polyesters...

When esterification takes place between an acid with two carboxyl groups and an alcohol with two hydroxyl groups, a polymer is formed. An example is the reaction between ethane-1,2-diol and 1,4-benzenedicarboxylic acid:

$$n\,HO_2C{-}C_6H_4{-}CO_2H + n\,HOCH_2CH_2OH \xrightarrow{\text{heat with acid catalyst}} {-}\!(O_2C{-}C_6H_4{-}CO_2CH_2CH_2)\!{-}_n + 2n\,H_2O$$

...e.g., Terylene®

The method of manufacture of Terylene®

The polyester formed has the trade names of Terylene® and Dacron®, and is used for making fabrics. Terylene® has about 80 units per chain. It can be melted without decomposition. The molten polymer is forced through tiny holes and cooled to form thin fibres. When the fibres are stretched, polymer molecules become aligned parallel to the axis of the fibre, packing closely together to make a strong fibre. Terylene® clothing is both soft and hard-wearing.

Nucleic acids are polyesters of vital importance

Nucleic acids are polyesters of phosphoric acid. Carbohydrates, proteins and nucleic acids are the three enormous, naturally occurring groups of polymers. Nucleic acids are not covered in this book, except for a reference to the structure of DNA [§4.7.3].

33.13.3 FATS AND OILS; SOAPS AND DETERGENTS

Fats and oils are esters of propane-1,2,3-triol and acids

Fats and oils are together classified as **lipids**. They are esters of propane-1,2,3-triol (*glycerol*) and a long chain carboxylic acid. Two examples are:

$$\begin{array}{l} CH_2OCOC_{15}H_{31} \\ | \\ CHOCOC_{15}H_{31} \\ | \\ CH_2OCOC_{15}H_{31} \end{array}$$

Propane-1,2,3-triyl trihexadecanoate (*Glyceryl palmitate*) is a component of animal fats.

$$\begin{array}{l} CH_2OCO(CH_2)_7CH{=}CH(CH_2)_7CH_3 \\ | \\ CHOCO(CH_2)_7CH{=}CH(CH_2)_7CH_3 \\ | \\ CH_2OCO(CH_2)_7CH{=}CH(CH_2)_7CH_3 \end{array}$$

Propane-1,2,3-triyl trioctadec-9-enoate (*Glyceryl oleate*) occurs in olive oil.

Fats are saturated; oils are unsaturated...

...Hydrogenation converts oils into fats

Fats occur both in plants and animals, and are a source of energy in our diet. If the carboxylic acid groups are largely saturated, the ester is a solid fat; if the hydrocarbon chains are highly unsaturated, the ester is an oil. The difference in melting temperatures arises from the fact that saturated hydrocarbon chains can pack more closely together than unsaturated chains. In these esters, the configuration at the double bonds is always *cis*, so that the molecules are bent, and cannot pack closely. There is a better sale for solid fats than for oils, and the conversion of oils to fats is a profitable business. It is accomplished by hydrogenation in the presence of nickel as a catalyst. Margarine is made in this way from corn oil and soya bean oil. Recently, it has been suggested that unsaturated esters are metabolised more easily than saturated fats, and lead to less cholesterol in the blood. Sales of unsaturated fats have increased, but the matter is far from settled, and debate continues.

Oils are used in varnishes and paints

Saturated fats are very stable compounds. Unsaturated oils undergo oxidation by air. This is what happens when fats and oils turn rancid. The oxidation of oils is put to good use in paints and varnishes. When the oil is spread in a thin film, the surface is exposed to oxygen, and a free radical chain reaction is initiated by oxygen. The surface layer of oil is converted into giant polymer molecules to give a hard, tough surface. Catalysts are added to assist polymerisation.

Saponification is the alkaline hydrolysis of fats to give soaps

An important use of fats is soap-making. The alkaline hydrolysis or **saponification** (Latin: soap-making) of fats gives glycerol and the sodium or potassium salt of the carboxylic acid. This is a soap. On addition of salt, the soap separates from solution, and can be skimmed off the surface:

$$\begin{array}{l} CH_2OCOC_{17}H_{35} \\ | \\ CHOCOC_{17}H_{35}(s) + 3NaOH(aq) \\ | \\ CH_2OCOC_{17}H_{35} \end{array} \rightarrow \begin{array}{l} CH_2OH \\ | \\ CHOH(l) + 3C_{17}H_{35}CO_2Na(s) \\ | \\ CH_2OH \end{array}$$

Propane-1,2,3-triyl trioctadecanoate (*Glyceryl tristearate*) — Sodium octadecanoate (*stearate*), Propane-1,2,3-triol (*glycerol*)

This process has been carried out since the Iron Age. Animal fats were boiled with water and the ashes of a wood fire, which contained potassium carbonate.

Soaps cannot work in hard water

Sodium soaps have limited solubility in water and can be obtained as solid cakes. Potassium soaps are more soluble and are used as gels in shampoos and shaving creams. The calcium and magnesium salts of soaps are insoluble. When a soap, such as sodium hexadecanoate, is added to hard water, an insoluble 'scum' of calcium and magnesium hexadecanoates is formed. To allow soap to lather readily, hard water must be softened by the removal of Ca^{2+} and Mg^{2+} ions [§ 17.10].

Soaps can emulsify fats and oils because they contain both hydrophilic and lipophilic groups

Soap cleans because it can **emulsify** fats and oils, i.e., convert them into a suspension of tiny droplets in water. Dirt is held to fabrics by a thin film of oil or grease, and this film must be removed before the dirt can be rinsed away. Soap owes its emulsifying action to the combination of polar and non-polar groups in its structure. At one end is a highly polar carboxylate ion, which is **hydrophilic** (attracted to water) and **lipophobic** (repelled by oils and fats). At the other end is a long hydrocarbon chain which is **hydrophobic** and **lipophilic**. When soap is added to water, the hydrophilic carboxylate ions dissolve in the water, while the hydrophobic hydrocarbon ends do not. The result is a surface layer one molecule thick [see Figure 33.8] which reduces the surface tension of water.

FIGURE 33.8
A Monomolecular Layer of Soap on the Surface of Water

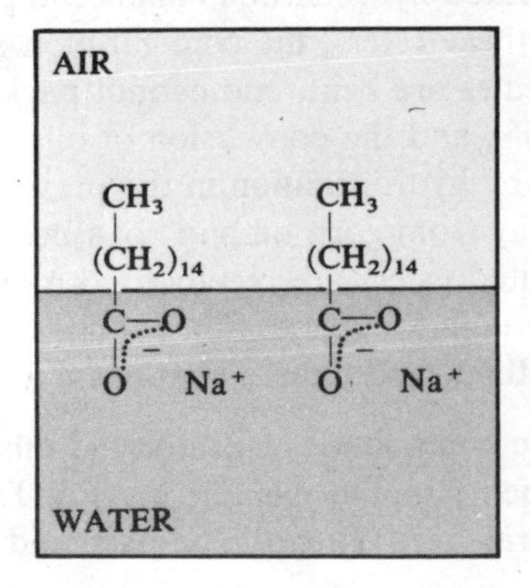

FIGURE 33.9
A Drop of Oil surrounded by Soap Ions

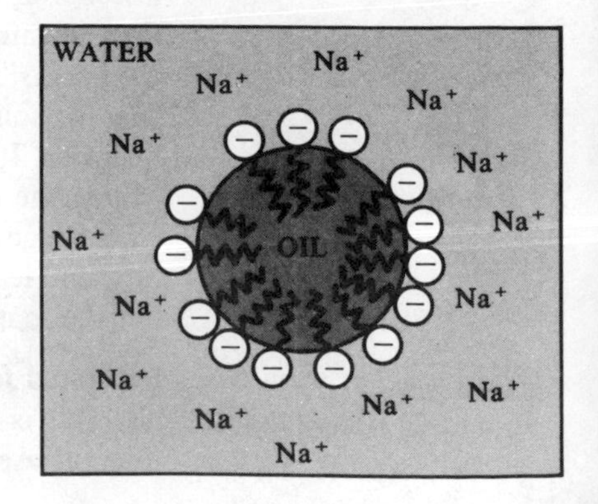

Soap interacts with an interface between oil and water when clothes are washed. The carboxylate end dissolves in the water, and the hydrocarbon end dissolves in the oil. Soap ions arrange themselves round each droplet of oil [see Figure 33.9]. If the water is agitated, the emulsified oil droplet can float free of the fabric. As the surface of each droplet is negatively charged, the drops repel one another and do not coalesce.

Synthetic detergents are the sodium salts of sulphonic acids...

The sodium salts of certain alkyl hydrogensulphates and benzenesulphonic acids also have detergent (cleaning) properties:

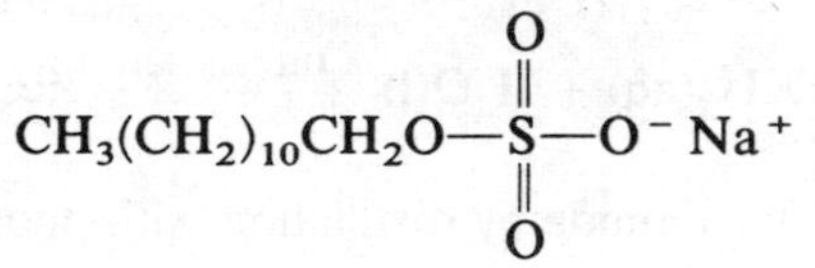

$$CH_3(CH_2)_{10}CH_2O{-}\overset{O}{\underset{O}{\overset{\|}{\underset{\|}{S}}}}{-}O^-\,Na^+$$

Sodium dodecanyl sulphate

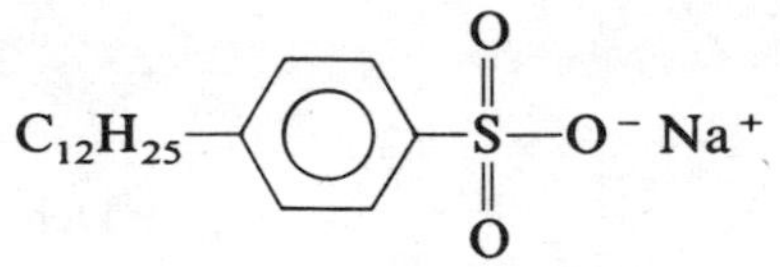

$$C_{12}H_{25}{-}C_6H_4{-}\overset{O}{\underset{O}{\overset{\|}{\underset{\|}{S}}}}{-}O^-\,Na^+$$

Sodium 4-dodecanylbenzenesulphonate

...They can work in hard water

They have an advantage over soaps in that their calcium and magnesium salts are soluble, allowing them to work in hard water [§ 17.10]. A disadvantage of the first soapless detergents is that they were not biodegradable. When these detergents were discharged into rivers, the microorganisms present in the water could not destroy them. This problem has been solved by modifying the structure of the detergents.

CHECKPOINT 33E: ESTERS

1. Write the names and structural formulae of nine isomeric esters with the formula $C_5H_{10}O_2$.

2. What products are formed in the alkaline hydrolysis of this ester?

$$C_3H_7\underset{\underset{O}{\|}}{C}{-}^{18}OC_2H_5$$

3. From which alcohol and acid are the following esters made?

(*a*) propyl propanoate
(*b*) methyl benzoate
(*c*) $HCO_2C_2H_5$
(*d*) $CH_3(CH_2)_3CO_2CH_3$
(*e*) $(CH_3)_2CHO_2CCH_3$
(*f*) $C_6H_5CO_2CH(CH_3)_2$

4. Outline how you would carry out the following conversions:

(*a*) $C_6H_5NH_2 \rightarrow C_6H_5OH$

(*b*) $C_6H_5NH_2 \rightarrow C_6H_5COCl$

More than one step is needed in each case.

(i) How could you obtain from the products of (*a*) and (*b*) a sample of phenyl benzoate, $C_6H_5CO_2C_6H_5$?

(ii) Starting from 10.0 g of phenylamine, and using no other organic compound, what is the maximum yield of the ester?

5. The complete hydrolysis of 1.76 g of an ester of a monocarboxylic acid and a monohydric alcohol required 2.0×10^{-2} mol of sodium hydroxide. Find the molar mass of the ester. Deduce its molecular formula, and write the names and structural formulae of all the esters with this molecular formula.

6. Compare the reactions of the —C—O—C— group in ethers and esters, explaining the reason for the differences.

***7.** Mutton fat is largely an ester of *glycerol* and *stearic acid*

$$CH_3(CH_2)_{16}CO_2H$$

Linseed oil is the *glyceryl* ester of *linolenic* acid

$$CH_3CH_2CH{=}CHCH_2CH{=}CHCH_2CH{=}CH(CH_2)_7CO_2H$$

(*a*) Give the IUPAC names of the acids.

(*b*) The configuration at each double bond is *cis*. Write the formula of *linolenic* acid to show this. Can you explain why linseed oil is a liquid at room temperature while mutton fat is solid?

***8.** Write the structural formula for the ester of propane-1,2,3-triol with *linolenic* acid [see Question 7]. What are the products of the reactions of this ester with the following reagents?

(*a*) NaOH, heat
(*b*) H_2, Ni
(*c*) Br_2
(*d*) O_3 followed by Zn + CH_3CO_2H
(*e*) $LiAlH_4$

9. Naturally occurring fats and oils are the esters of acids with an even number of carbon atoms. Acids with an odd number of carbon atoms are rare. Suggest a method of increasing the length of an aliphatic acid chain by one carbon atom:

$$RCH_2CO_2H \rightarrow RCH_2CH_2CO_2H$$

More than one step will be needed.

33.14 AMIDES

The methods of preparation have been described, and are summarised in Figure 33.12, § 33.16.

Amides are hydrolysed to ammonium salts or acids...

Amides are hydrolysed in a similar way to esters:

$$RCONH_2(aq) + H_2O(l) \xrightarrow{H^+(aq),\ heat} RCO_2NH_4(aq)$$

$$RCONH_2(aq) + H_2O(l) \xrightarrow{OH^-(aq),\ heat} RCO_2^-(aq) + NH_3(aq)$$

Dehydration of an amide by distillation with phosphorus(V) oxide gives a nitrile:

$$RCONH_2(s) \xrightarrow{P_2O_5\ distil} RCN(l) + H_2O(l)$$

...and converted to nitriles, acids, amines...

The action of nitrous acid ($NaNO_2$ + HCl) is to replace —NH_2 by —OH, giving a carboxylic acid and nitrogen:

$$RCONH_2(aq) + HNO_2(aq) \rightarrow RCO_2H(aq) + N_2(g) + H_2O(l)$$

Reduction of an amide to an amine can be accomplished by using hydrogen with nickel as a catalyst or an ethoxyethane solution of lithium tetrahydridoaluminate:

$$RCONH_2(s) \xrightarrow[or\ LiAlH_4(ether)]{H_2,\ Ni,\ heat} RCH_2NH_2(l)$$

...and into the amines which have one carbon atom less than the amides

In the Hofmann degradation [see Figure 33.12], reaction with bromine and concentrated alkali gives an amine with one carbon atom less than the parent amide:

$$RCONH_2(aq) + Br_2(aq) + 4OH^-(aq) \rightarrow RNH_2(l) + 2Br^-(aq) + CO_3^{2-}(aq) + 2H_2O(l)$$

33.14.1 POLYAMIDES

Polyamides are made by condensation, e.g. nylon 66

Nylon is a polyamide, made by **condensation polymerisation** from a dicarboxylic acid and a diamine. Nylon 66 is the type that is manufactured in the largest quantities. It is made from a six-carbon dicarboxylic acid and a six-carbon diamine:

$$n\,H_2N(CH_2)_6NH_2 + n\,HO\overset{\overset{O}{\|}}{C}(CH_2)_4\overset{\overset{O}{\|}}{C}OH \rightarrow \!\left(\!HN(CH_2)_6NH\overset{\overset{O}{\|}}{C}(CH_2)_4\overset{\overset{O}{\|}}{C}\!\right)_n + 2n\,H_2O$$

Hexane-1,6-diamine (in water) Hexane-1,6-dioic acid (in trichloroethane)

Nylon forms at the interface between the aqueous solution of the diamine and the 1,1,1-trichloroethane solution of the dicarboxylic acid. It is removed as it is formed, to allow further reaction to take place [see Figure 33.10].

33.14.2 UREA

Urea

$$H_2N-\overset{\overset{O}{\|}}{C}-NH_2$$

is the diamide of carbonic acid, carbamide. It was the first organic compound to be

Urea is a product of protein metabolism...

made from inorganic material. F Wöhler, in 1828, made it by evaporating an aqueous solution of ammonium cyanate:

$$NH_4CNO(aq) \xrightarrow{\text{evaporate}} H_2NCONH_2(s)$$

...and is also made synthetically for use as a fertiliser...

Urea is formed as an end-product of protein metabolism. It is widely used as a fertiliser. It is made commercially by heating carbon dioxide and ammonia under pressure:

$$CO_2(g) + 2NH_3(g) \xrightarrow[\text{pressure}]{\text{heat under}} H_2NCONH_2(s)$$

As the reaction proceeds slowly in the reverse direction, urea slowly releases ammonia into the soil.

...for the manufacture of plastics, and sleeping pills

Urea is also required in quantity for the manufacture of urea-methanal polymers [§ 31.6.5].

The barbiturate drugs which are used for inducing sleep are amides made from urea [see § 35.2].

FIGURE 33.10
Making Nylon

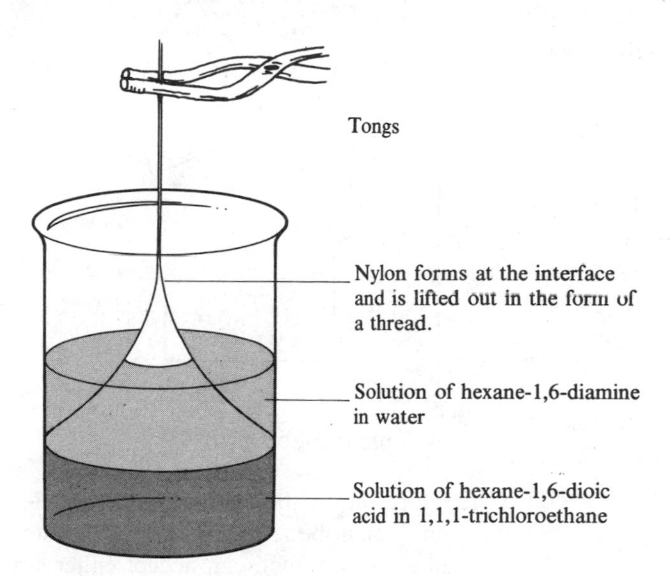

33.15 THE SULPHONAMIDE ANTIBIOTICS

TOPIC

In 1932, a new drug, called Prontosil®, was patented. It had been developed for use as a red dye by a German manufacturer and then found to have an antibacterial action. Gerhard Domagh showed that Prontosil® was good at fighting streptococcal infections. Soon afterwards, it was found that Prontosil is converted into 4-aminobenzenesulphonamide (sulphanilamide) in the body and that sulphanilamide has the same antibacterial activity as Prontosil®. The first person to receive the new antibiotic was Domagh's daughter. As she lay close to death with 'child bed fever', Domagh decided to risk treating her with his discovery. Regulations for testing new drugs were not as strict then as they are now. Fortunately, the antibiotic cured her infection, and she recovered. In 1939, Domagh received the Nobel Prize for his work on antibiotics. A number of derivatives of sulphanilamide, called the *sulphonamides* or *sulpha-drugs* were synthesised.

H_2N—$C_6H_3(NH_2)$—N=N—C_6H_4—SO_2NH_2

Prontosil

H_2N—C_6H_4—SO_2NH_2

4-Aminobenzenesulphonamide (sulphanilamide)

Sulphonamides (sulpha-drugs) inhibit the enzyme which catalyses the synthesis of folic acid

Antibiotics are chemicals extracted from moulds, bacteria and yeasts that destroy or inhibit the growth of other micro-organisms. They act by inhibiting the action of enzymes. Strictly speaking, sulphonamides are not antibiotics because they are not produced in any living organism. The sulphonamides (sulpha-drugs) interfere with the metabolism of bacteria. Bacteria use 4-aminobenzoic acid in the synthesis of folic acid, which is a vital substance for them and a vitamin for animals. 4-Aminobenzenesulphonamide, 'sulphanilamide', bears a close structural resemblance to 4-aminobenzoic acid. Molecules of this and other similar sulpha-drugs compete with 4-aminobenzoic acid for the active site on the bacterial enzyme and inhibit the synthesis of folic acid.

FIGURE 33.11 The Action of Sulphonamide Drugs

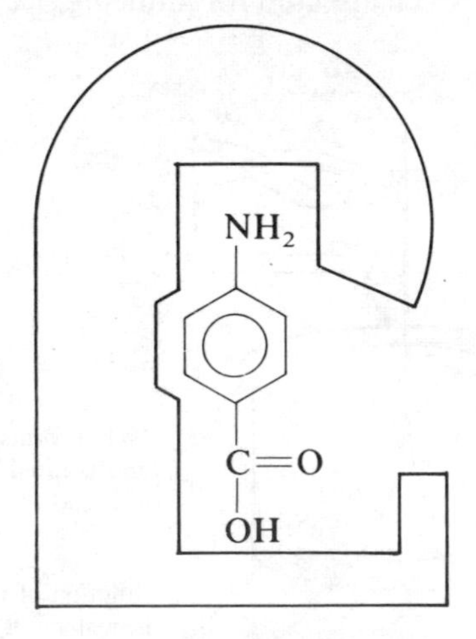

4-Aminobenzoic acid

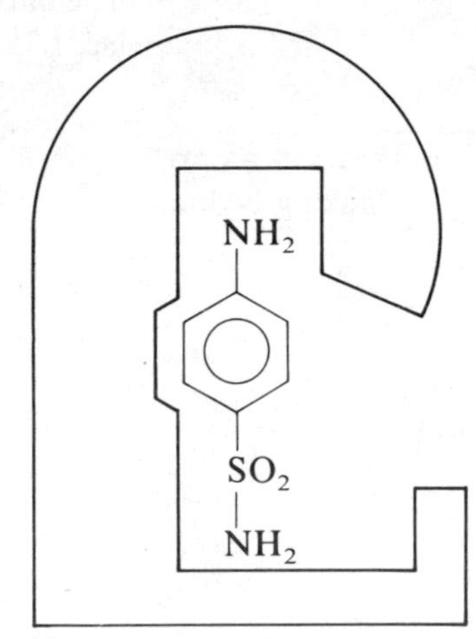

4-Aminobenzenesulphonamide

The figure shows the lock and key model for the action of sulpha-drugs. An enzyme acts on 4-aminobenzoic acid during one step in the synthesis of folic acid. The enzyme acts like a lock which can accept either 4-aminobenzoic acid or a sulpha-drug that fits. When a sulpha-drug enters, it blocks the enzyme and prevents it from acting on 4-aminobenzoic acid.

The sulphonamides have been superseded by penicillin and other antibiotics

Sulpha-drugs were widely used in the Second World War. They were sprinkled on open wounds to prevent infection. They produce a number of side effects with prolonged use, including kidney damage, and have now given way to penicillin and other antibiotics.

33.16 NITRILES

Aliphatic nitriles, RCN are made from halogenoalkanes, RX . . .

. . . Aromatic nitriles, ArCN, are made from diazonium compounds

Nitriles

$$R—C{\equiv}N$$

differ from the other acid derivatives in that they do not possess a carbonyl group. They are readily prepared from acids and readily yield acids on hydrolysis. Aliphatic nitriles are prepared from halogenoalkanes; aromatic nitriles from diazonium compounds [see Figure 32.5, § 32.10.2 and Figure 33.12].

FIGURE 33.12
Reactions of Amides and Nitriles

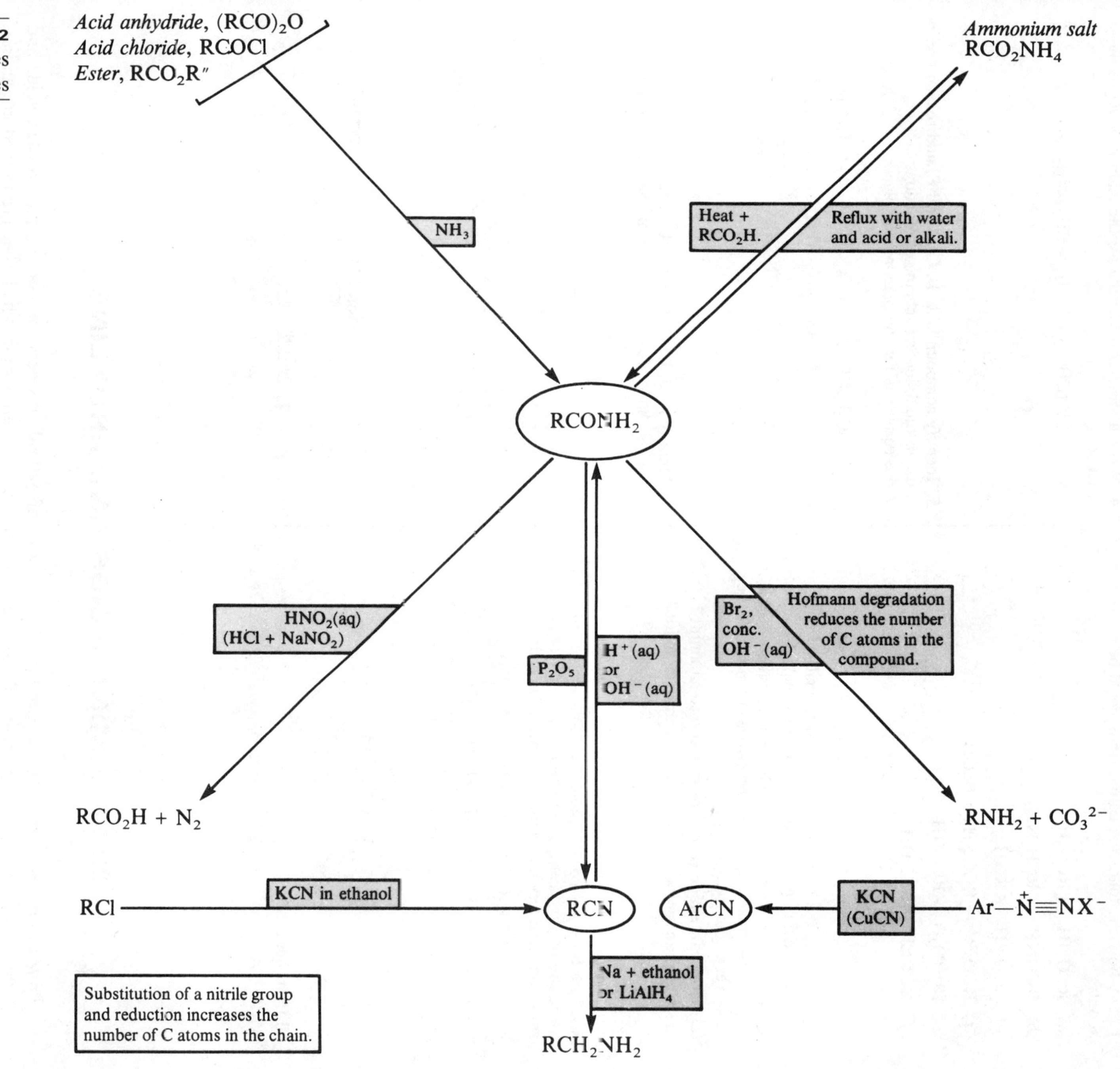

CHECKPOINT 33F: AMIDES AND NITRILES

1. Say how the following conversions could be brought about:

(*a*) $RCONH_2 \rightarrow RCO_2H$

(*b*) $RCONH_2 \rightarrow RCH_2NH_2$

(*c*) $RCONH_2 \rightarrow RNH_2$

(*d*) $RCO_2R' \rightarrow RCONH_2 + R'OH$

(*e*) $(RCO)_2O \rightarrow RCONH_2$

(*f*) $RCOCl \rightarrow RCH_2OH$

2. (*a*) Write the name and structural formula of (i) a carboxylic acid amide and (ii) a primary aliphatic amine.

(*b*) Describe two ways in which the properties of amides differ from those of amines and two ways in which they show resemblance.

(*c*) How can amide (i) be converted into an amine?

(*d*) How can amine (ii) be converted into an amide?

3. State how you could prepare the following compounds, using the organic compound cited as the only organic substance and any inorganic reagents you need:

(*a*) $CH_3CH_2CH_2CO_2H$ from $CH_3CH_2CH_2OH$

(*b*) $CH_3CO_2C_2H_5$ from CH_3CH_2OH

(*c*)

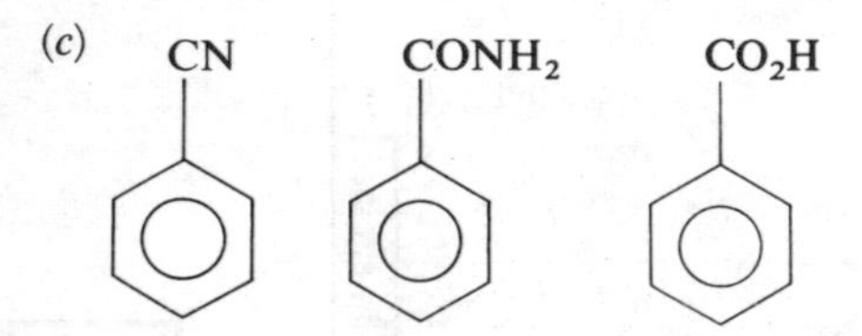

all from benzene.

4. Write a structural formula for the main product of the reaction

$$ClC(=O)(CH_2)_4C(=O)Cl + H_2N(CH_2)_6NH_2 \rightarrow$$

5. Identify compounds **A**, **B**, **C**, **D** and **E**, and reagents *x*, *y* and *z* in the following scheme of reactions.
Write equations for the reactions involved:

$$C_2H_5CO_2H \xrightarrow{PCl_5} \mathbf{A} \xrightarrow{NH_3} \mathbf{B} \xrightarrow{x} C_2H_5NH_2$$

$$C_2H_5CO_2H \xrightarrow{z} \mathbf{E}(C_4H_8O_2)$$

$$C_2H_5NH_2 \xrightarrow{y} \mathbf{C}$$

$$\mathbf{D}(C_4H_8O_2) \xleftarrow[(+\text{ acid})]{CH_3CO_2H} \mathbf{C}$$

6. Identify **A**, **B**, **C**, **D**, **E**, **F** and **G** in the scheme of reactions shown below. Write equations for the reactions portrayed:

$$\mathbf{A} \xrightarrow{LiAlH_4} \mathbf{B}$$

$$\mathbf{A} \xrightarrow{air} C_6H_5CO_2H$$

$$C_6H_5CO_2H \xrightarrow{SOCl_2} \mathbf{C}$$

$$\mathbf{B} + \mathbf{C} \xrightarrow{NaOH} \mathbf{D}$$

$$\mathbf{C} \xrightarrow{NH_3} \mathbf{E}$$

$$\mathbf{E} \xrightarrow{Br_2,\ OH^-(aq)} \mathbf{F}$$

$$\mathbf{C} + \mathbf{F} \rightarrow \mathbf{G}$$

33.17 AMINO ACIDS AND PROTEINS

Proteins are one of the three large classes of natural polymers

There are three large classes of natural polymers: polysaccharides, nucleic acids and proteins. Proteins are polyamides. The units from which they are formed are amino acids, of formula

$$H_2N\underset{\displaystyle R}{\underset{|}{C}}HCO_2H$$

Since the $—NH_2$ group is on the α carbon atom (that adjacent to the $—CO_2H$ group), they are correctly known as α amino acids. Table 33.2 lists some of the 20 naturally occurring α amino acids.

FIGURE 33.13
Relationships between Carboxylic Acids and other Series

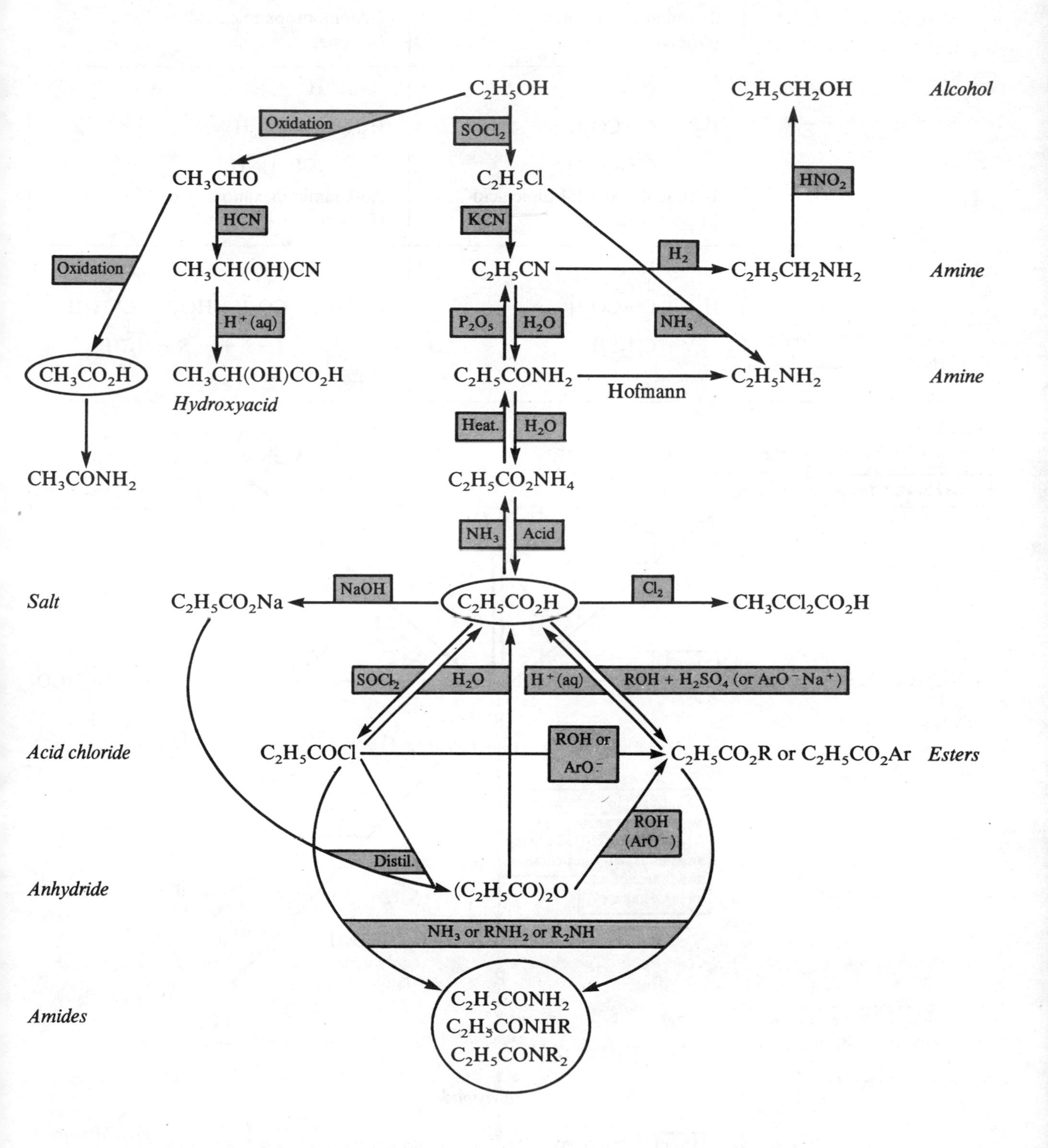

TABLE 33.2 Some Naturally occurring α Amino Acids

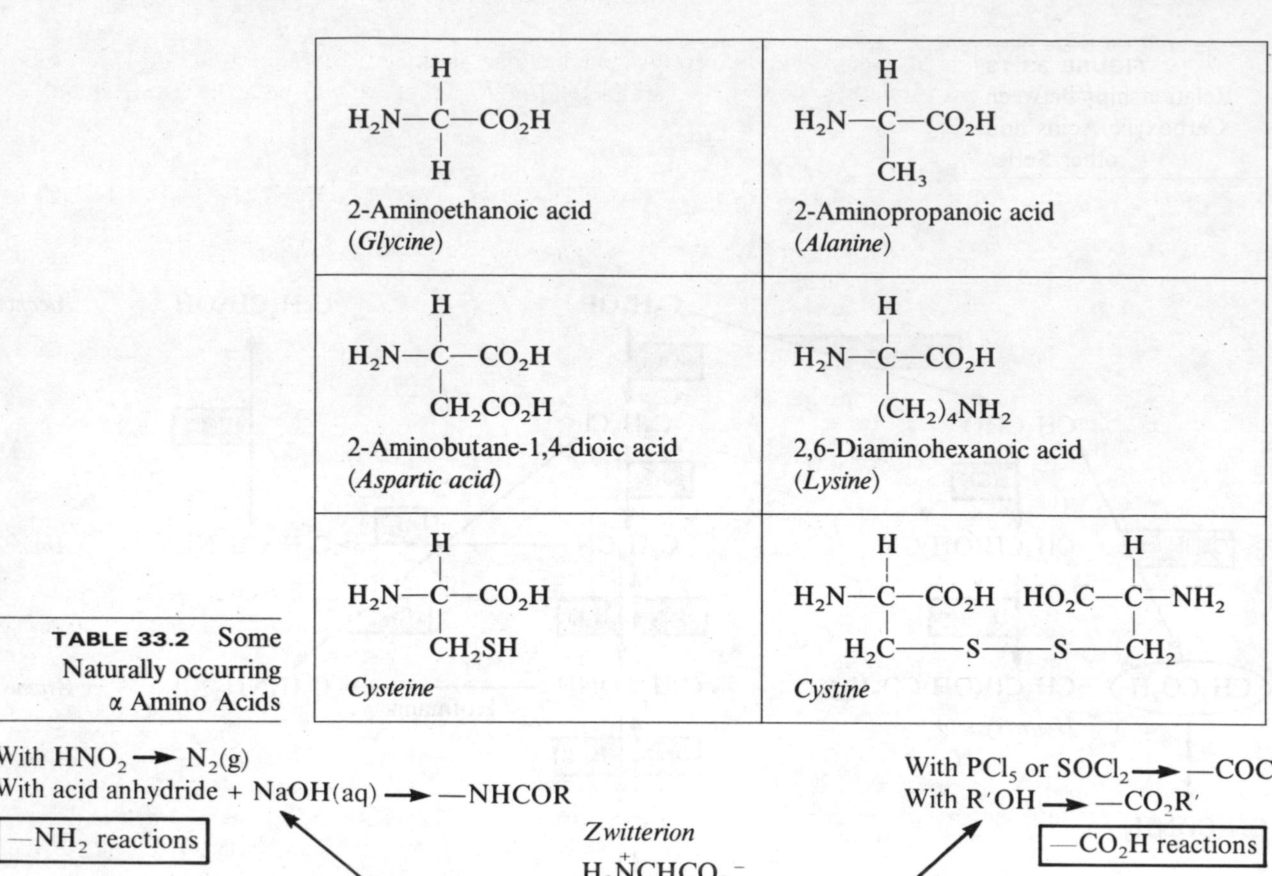

$H_2N-CH(H)-CO_2H$ 2-Aminoethanoic acid (*Glycine*)	$H_2N-CH(CH_3)-CO_2H$ 2-Aminopropanoic acid (*Alanine*)
$H_2N-CH(CH_2CO_2H)-CO_2H$ 2-Aminobutane-1,4-dioic acid (*Aspartic acid*)	$H_2N-CH((CH_2)_4NH_2)-CO_2H$ 2,6-Diaminohexanoic acid (*Lysine*)
$H_2N-CH(CH_2SH)-CO_2H$ *Cysteine*	$H_2N-CH(CO_2H)-H_2C-S-S-CH_2-CH(NH_2)-CO_2H$ *Cystine*

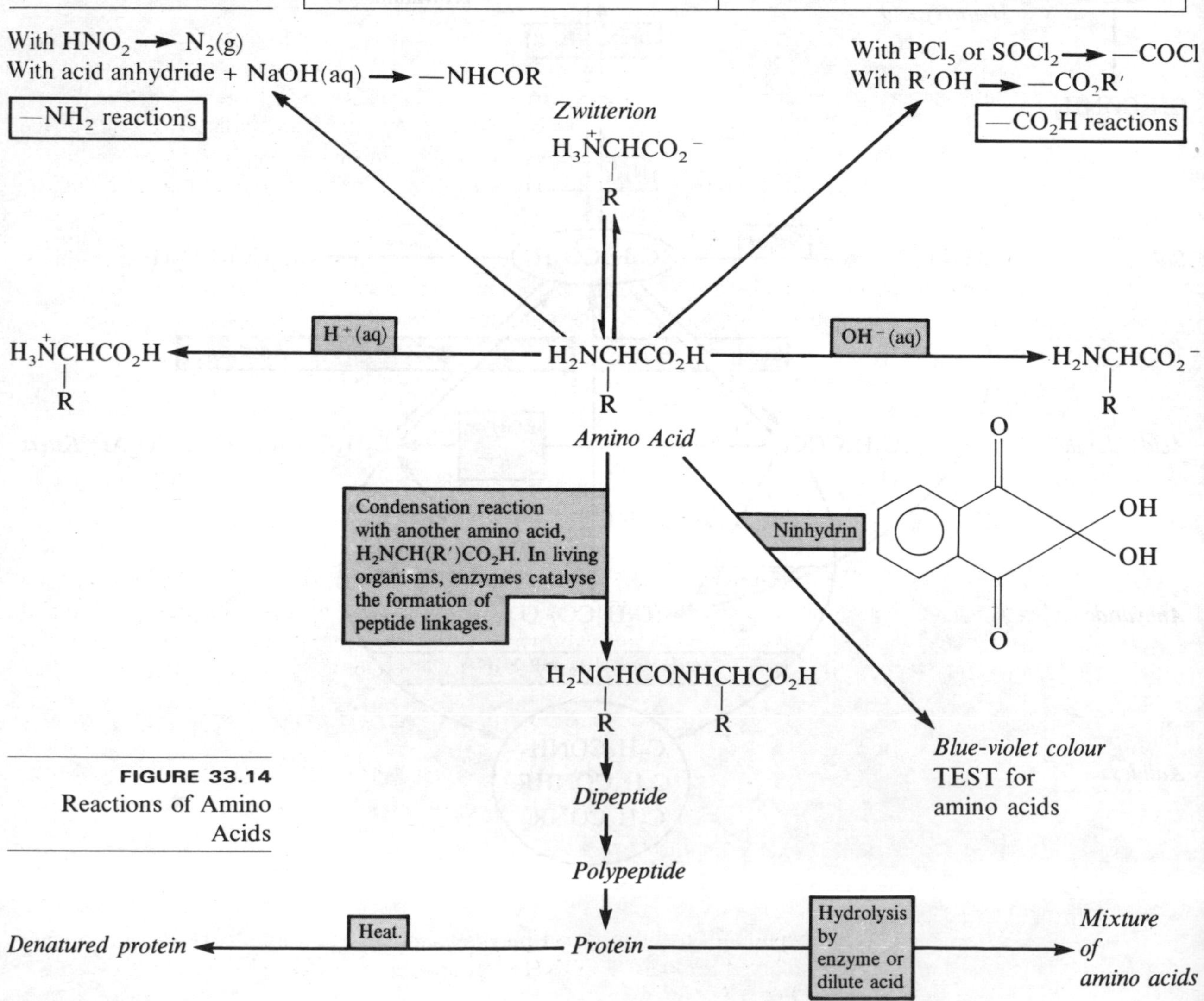

FIGURE 33.14 Reactions of Amino Acids

Glycine can be made from ethanoic acid

All amino acids, except glycine, have an asymmetric carbon atom and show optical isomerism. Glycine is a white solid of T_m 232 °C. It can be made from ethanoic acid:

$$\underset{\text{Ethanoic acid}}{CH_3CO_2H} \xrightarrow[\text{pass } Cl_2]{\text{boil}} \underset{\text{Chloroethanoic acid}}{ClCH_2CO_2H} \xrightarrow[\text{large excess}]{\text{conc. } NH_3(aq)} \underset{\textit{Glycine}}{H_2NCH_2CO_2H} + NH_4Cl$$

33.17.1 REACTIONS OF AMINO ACIDS

There are two functional groups, $—NH_2$ and $—CO_2H$, each of which is responsible for a typical set of reactions [see Figure 33.14].

Amino acids form zwitterions, containing $—NH_3^+$ and $—CO_2^-$ exist

In acidic solution, the $—NH_2$ group ionises as $—\overset{+}{N}H_3$; in alkaline solution, the $—CO_2H$ group ionises as $—CO_2^-$. At intermediate values of pH, the **zwitterion**

$$H_3\overset{+}{N}CH(R)CO_2^-$$

exists (German: *Zwitter*, mongrel). The reason why amino acids are high-melting solids is that they crystallise as zwitterions. In solution, the pH at which the concentration of the zwitterion is maximal is called the **isoelectric point**. It differs for different amino acids. In **electrophoresis**, an amino acid will move towards the cathode or the anode, depending on the pH. This behaviour is the basis of the electrophoretic method (Greek: *phoresis*, being carried) of separating amino acids according to their isoelectric points.

Amino acids can be separated by electrophoresis or by chromatography

Another method of separating amino acids is chromatography [§ 8.7]. The positions of the colourless amino acids on the chromatogram are revealed by spraying the paper with ninhydrin. This reagent gives a blue-violet colour with amino acids.

33.17.2 PEPTIDES AND PROTEINS

The polyamides formed by amino acids are peptides and proteins

The most important reaction of amino acids is polymerisation. In living organisms, enzymes catalyse the polymerisation of amino acids to form peptides and proteins. Polymerisation occurs through the formation of amide groups by the reaction of an $—NH_2$ group in one molecule with a $—CO_2H$ group in another molecule:

$$H_2N—\overset{R}{\underset{H}{C}}—\overset{O}{\overset{\|}{C}}—OH + H—\overset{H}{N}—\overset{R}{\underset{H}{C}}—\overset{O}{\overset{\|}{C}}—OH \rightarrow \underset{\text{A dipeptide}}{H_2N—\overset{R}{\underset{H}{C}}—\overset{O}{\overset{\|}{C}}—\overset{H}{N}—\overset{R}{\underset{H}{C}}—\overset{O}{\overset{\|}{C}}—OH} + H_2O$$

The dipeptide formed can go on polymerising to form a long chain of amino acid residues linked through —CONH— groups. The —CONH— group is called a **peptide linkage**.

Fibrous proteins have linear molecules...

When the number of amino acid residues is 100 or more, the polymer is called a **protein**. Proteins are a diverse group of polymers. Fibrous proteins have linear molecules, and are insoluble in water and resistant to acids and alkalis. Examples are *keratin* (in hair, nails, horn and feathers), *collagen* (in muscles and tendons), *elastin* (in arteries and tendons) and *fibroin* (in silk).

... Globular proteins have a complicated 3D structure...

The molecules of globular proteins, e.g., *albumin* (in egg white) and *casein* (in milk), adopt a complicated three-dimensional structure. The polypeptide chain is held in position by attractions between polar groups, e.g., $—CO_2^-$ in aspartic acid and $—\overset{+}{N}H_3$ in lysine and also by cystine bridges [see Table 33.2]. The —SH groups in two cysteine amino acid residues can be oxidised to form an —S—S— bridge between two strands of polypeptide. Anything which interferes with the three-dimensional configuration, such as acids, alkalis and a rise in temperature, is said to **denature** the protein. Globular proteins are soluble in water and are easily denatured.

...e.g., enzymes

Enzymes are globular proteins, and their catalytic activity depends on their three-dimensional structure. If the enzyme is denatured by a rise in temperature or an extreme of pH, or even by violent agitation of the solution, it is no longer able to function as a catalyst.

Hair protein can be permanently waved

Hair protein owes some of its texture to —S—S— bonds in cystine. Permanent waving techniques utilise this fact. When hair is soaked in a gentle reducing agent, the —S—S— bridges are broken by reduction to —SH groups. While the hair proteins are no longer cross-linked, they are arranged into curls. Re-oxidation by soaking the hair in a mild oxidising agent reforms the disulphide bonds and holds the hair in its new configuration [see Figure 33.15].

FIGURE 33.15
Permanent Waving

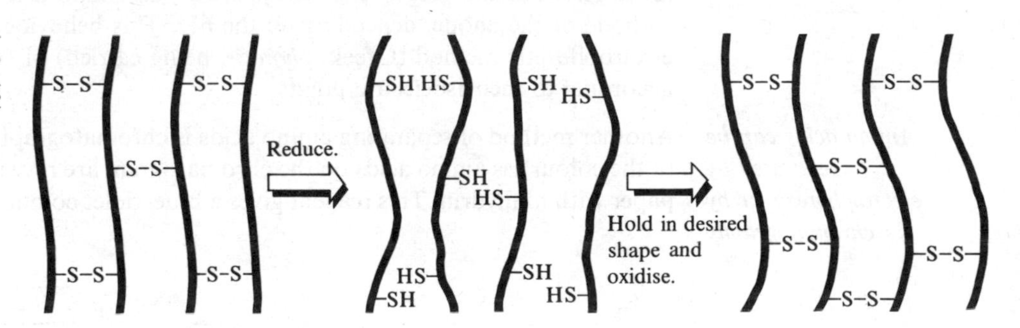

CHECKPOINT 33G: AMINO ACIDS AND PROTEINS

1. Outline how you could make *alanine*, $H_2NCH(CH_3)CO_2H$, from propanoic acid. Give the IUPAC name for alanine. Predict two reactions which alanine would undergo.

2. (*a*) Outline the preparation of *glycine*, aminoethanoic acid, starting from ethanoic acid.

(*b*) Which of these formulae is the better representation of the structure of glycine?

$$H_2NCH_2CO_2H \quad \text{or} \quad H_3\overset{+}{N}CH_2CO_2^-$$

Explain your choice.

(*c*) Write equations for two reactions of glycine.

3. What class of polymers are made from monomers such as glycine and alanine? Give the names of three of these polymers. What is the name and the structure of the functional group in these polymers? State one reaction which is typical of this functional group.

4. How could you distinguish between aminoethanoic acid and ethanamide?

5. A 0.1110 g specimen of an amino acid was dissolved in water and treated with nitrous acid. The volume of nitrogen produced was 16.0 cm^3 at 1 atm and 293 K. Calculate the molar mass of the amino acid. (GMV = 24.0 dm^3 at rtp.)

6. A biochemist is analysing a mixture of amino acids. She adds 1.00×10^{-3} g of pure deuterioalanine, $H_2NCH(CD_3)CO_2H$, to the mixture. After separating the alanine present by chromatography, she finds that it contains 9.5×10^{-3}% deuterium by mass. What mass of alanine was present in the original mixture?

QUESTIONS ON CHAPTER 33

1. How can you distinguish between the members of the following pairs of compounds? Suggest two tests, one of which will give a positive result with the first compound, and the other of which will give a positive result with the second compound.

(*a*) CH_3CH_2CHO and $CH_3CH_2CO_2H$

(*b*) $CH_3COC_2H_5$ and $CH_3CO_2C_2H_5$

(*c*) CO_2H and OH

(*d*) NH_2 and CO_2CH_3

(*e*) CO_2H and OH

(*f*) HCO_2H and CH_3CO_2H

(*g*) CH_3CO_2H and Cl_2CHCO_2H

(*h*) $C_6H_5NH_2$ and $C_6H_5CONH_2$

2.

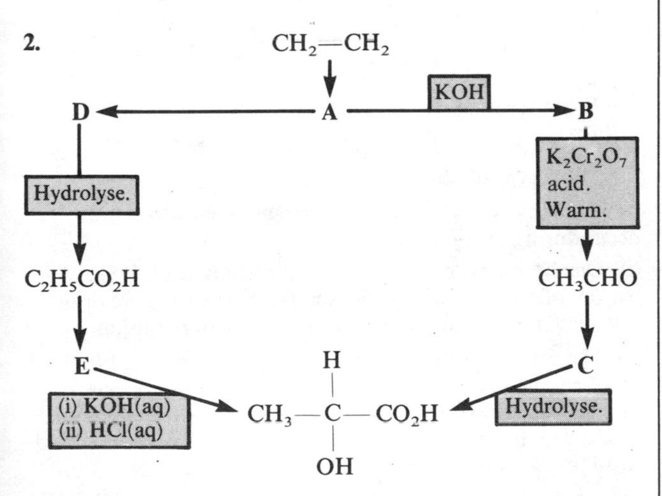

FIGURE 33.16

Identify the compounds **A**, **B**, **C**, **D** and **E** in Figure 33.16. Name the product. How would you expect this acid to react with (*a*) Cl_2, (*b*) $SOCl_2$, (*c*) Na_2CO_3, (*d*) plane-polarised light?

3. What is formed when bromine reacts with (*a*) methane, (*b*) benzene, (*c*) propene, (*d*) phenol, (*e*) ethanoic acid and (*f*) ethanamide? State the conditions under which the reactions occur.

4. How could you carry out the following conversions? Use no organic compounds other than the one cited. Use any inorganic substances you need.

(*a*) $CH_3CH_2CH_2OH \rightarrow CH_3CH_2CO_2CH_2CH_2CH_3$

(*b*) $CH_3CH_2CH_2OH \rightarrow CH_3CH_2CH_2CO_2CH_2CH_2CH_3$

(*c*) $(CH_3)_2CHOH \rightarrow (CH_3)_2C(OH)CO_2H$

(*d*) $CH_3CHO \rightarrow CH_3CO_2CH_2CH_3$

More than one step will be needed.

5. C is an organic compound which dissolves in warm water to form an acidic solution. **C** reacts with phenol in the presence of a base to form an ester, and reacts with diethylamine to form an *N*,*N*-disubstituted amide. Benzene reacts with **C** in the presence of aluminium chloride to form diphenylmethanone.

Deduce the identity of **C**. Explain the reactions described.

6. Consider the three compounds $C_2H_5NH_2$, $C_6H_5NH_2$ and CH_3CONH_2.

(*a*) Name the compounds, and give the names of the series to which they belong.

(*b*) Compare the basicity of the $—NH_2$ group in the compounds. Explain how the differences arise from differences in structure.

(*c*) Describe the behaviour of the compounds with a solution of sodium nitrite in dilute hydrochloric acid.

7. Explain why the melting temperature of aminoethanoic acid (262 °C) is so much higher than that of chloroethanoic acid (62 °C).

8. Starting from an aliphatic carboxylic acid, how could you make (*a*) an acid chloride, (*b*) an acid anhydride, (*c*) an amide, (*d*) an ester? Give the necessary conditions for reaction, and give equations for the reactions. State the conditions under which each of the products will react with water. Write equations, and name the products of the hydrolyses.

9.

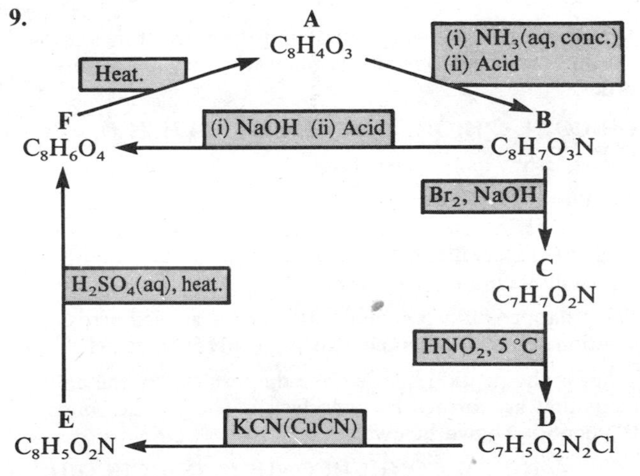

FIGURE 33.17

Write structural formulae for compounds **A–F**. Explain the reactions shown in Figure 33.17.

10. Explain why (*a*) amines are basic and (*b*) amides are not basic.

11. Explain why a carboxylic acid, although it possesses a

$>C=O$ group

does not react with the nucleophiles which attack aldehydes and ketones.

12. When a solution of chlorine in hot ethanoic acid is irradiated with ultraviolet light, 2-chloroethanoic acid is formed. Write equations for all the steps that occur in this photochemical reaction.

13. Two esters have the molecular formula $C_6H_{12}O_2$. Both show optical isomerism. When heated with aqueous sodium hydroxide, **A** gives sodium ethanoate and another product, and **B** gives methanol and another product. Write structural formulae for **A** and **B**.

14. **P** = CHCO₂H=CHCO₂H (i.e. $CHCO_2H \| CHCO_2H$)   **Q** = CH_2CO_2H—CH_2CO_2H

R = CH_3—C(H)(OH)—CO_2H **S** = CH_3—C(H)(CO_2H)—CO_2H

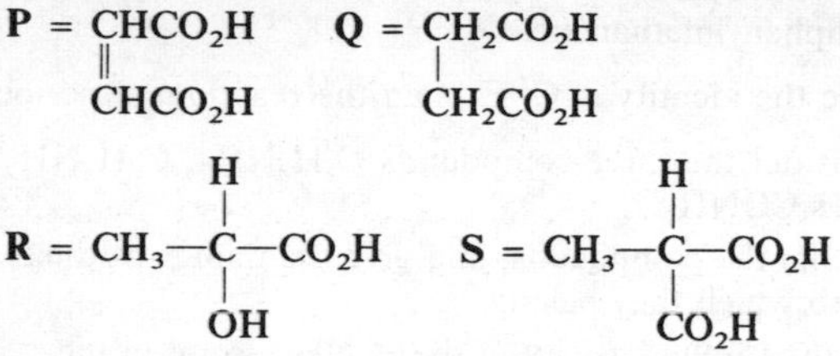

(*a*) Why does **P** exist in isomeric forms, whereas **Q** does not? What name is given to this type of isomerism?

(*b*) Why does **R** display optical isomerism, whereas **S** does not?

15. An optically active ester of molecular formula $C_8H_{16}O_2$ was hydrolysed by aqueous alkali to give as one product a liquid of molecular formula C_3H_8O, which gave a positive iodoform test. Write a structural formula for the ester.

16. Explain these statements:

(*a*) Ethanoyl chloride is more reactive towards water than is chloroethane.

(*b*) Laboratory preparations of 2-hydroxypropanoic acid yield a racemic mixture.

***17.** Place the following compounds in order of increasing acid dissociation constant. Explain why they follow this order.

CH_3CO_2H, C_2H_5OH, C_6H_5OH, Cl_2CHCO_2H, H_2O, $C_6H_5CO_2H$

18. Write a structural formula for aminoethanoic acid (glycine). How does this structure explain why the amino acid is (*a*) a crystalline solid, (*b*) soluble in water and (*c*) insoluble in ethoxyethane?

What happens if a potential difference is applied across a solution of aminoethanoic acid (*a*) at pH 5, (*b*) at pH 9?

What is the peptide bond? Give the formulae of the amino acids that are formed by hydrolysis of the peptide bonds in the peptide shown below:

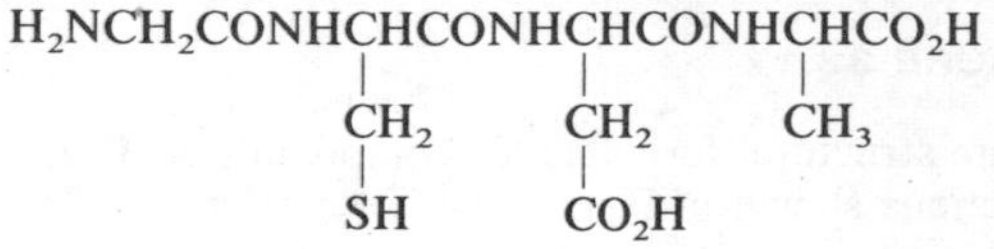

$H_2NCH_2CONHCH(CH_2SH)CONHCH(CH_2CO_2H)CONHCH(CH_3)CO_2H$

19. Describe how, given ethanol as your only organic compound, you would prepare ethyl ethanoate.

Explain why the hydrolysis of esters is usually carried out in the presence of a base, rather than an acid.

Explain why ketones are not hydrolysed, although, like esters, they possess a carbonyl group.

20. Paracetamol is a widely used analgesic drug with a formula

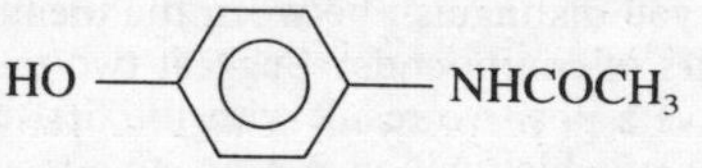

HO—C_6H_4—$NHCOCH_3$

Its 'systematic' chemical name is *N*-(4-hydroxyphenyl)-ethanamide.

(*a*) Draw the full structural formula of paracetamol, showing all the bonds in the side chains attached to the benzene ring.

(*b*) On your diagram, label clearly

(i) *one* bond angle which is 120°

(ii) *two* different examples of bond angles of approximately 109°

(*c*) Explain the reason for any *one* of the bond angles you have labelled.

(*d*) What is the meaning of the '4' in the chemical name of paracetamol?

(*e*) What effect do paracetamol tablets have when taken?

(*f*) The last stage of the synthesis of paracetamol involves the reaction of 4-hydroxyphenylamine with ethanoic acid. (The by-product is water.) Construct the balanced equation for this reaction.

(*g*) This type of reaction is called *condensation*. Why?

(*h*) Paracetamol undergoes hydrolysis in the presence of alkali. The first stage of the mechanism of this hydrolysis includes the following electron pair movement:

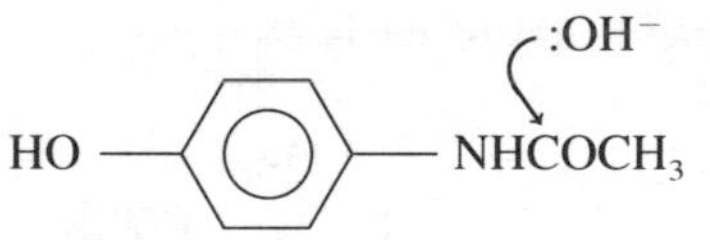

HO—C_6H_4—$NHCOCH_3$ (with :OH^- attacking the carbonyl carbon)

(i) What is meant by *hydrolysis*?

(ii) Why does the OH^- ion attack the carbon shown?

(iii) What type of attack is this?

(iv) Show clearly what other movement of electrons must occur during the first stage.

(*i*) One brand of paracetamol tablets retails at £1.39 per packet of 16. Each tablet contains 0.500 g of the pure drug. Calculate the cost of purchase of one mole of paracetamol.

(*j*) The cost of synthesis of paracetamol by the drug firm is £1.72 per mole. (This sum includes the cost of chemicals, salaries and equipment used.) Give *two* reasons, apart from profit, why there is such a big difference between the cost and retail prices.

(O 91, AS)

21. Aspirin (acetylsalicylic acid), an ester, is made commercially by the reaction of ethanoic anhydride with salicyclic acid (2-hydroxybenzoic acid).

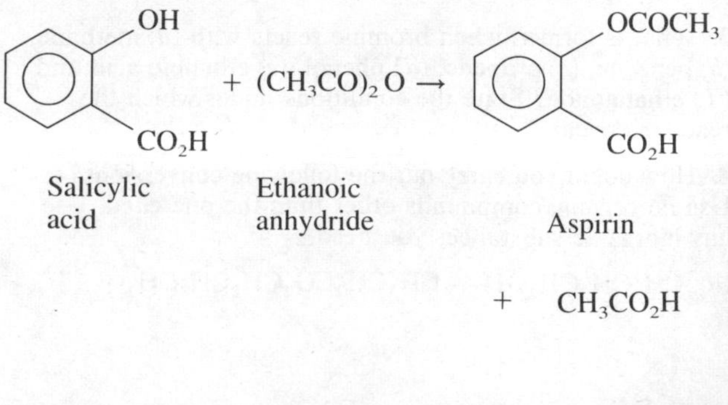

$C_6H_4(OH)CO_2H$ + $(CH_3CO)_2O$ → $C_6H_4(OCOCH_3)CO_2H$ + CH_3CO_2H

Salicylic acid   Ethanoic anhydride   Aspirin

Ethanoic anhydride reacts with the —OH group of the salicylic acid.

(*a*) Name a reagent which could be used instead of ethanoic anhydride to react with the —OH group.

(*b*) In a small batch preparation of aspirin, 185 kg of salicylic acid were reacted with 200 kg of ethanoic anhydride to yield 217 kg of pure aspirin.

(Relative molecular masses: salicylic acid = 138, ethanoic anhydride = 102; relative atomic masses: H = 1, C = 12, O = 16)

(i) Calculate the relative molecular mass of aspirin.

(ii) Calculate the percentage yield of aspirin.

(*c*) Aspirin that has been kept too long in damp conditions undergoes hydrolysis to salicylic acid.

(i) Explain what is meant by *hydrolysis.*

(ii) Giving experimental details suggest how you could quickly hydrolyse an aspirin tablet in the laboratory.

(*d*) *Soluble aspirin* is either the sodium or calcium salt of acetylsalicylic acid.

$OCOCH_3$

$CO_2^-Na^+$

Sodium salt of aspirin

Suggest why soluble aspirin is more soluble in water than normal aspirin.

(*e*) Aspirin may be made in the laboratory from methylbenzene according to the following flow scheme:

Methylbenzene (CH_3) —Step 1→ (NO_2, CH_3) —Step 2→ (NO_2, CO_2H) —Step 3→ ($NH_3^+Cl^-$, CO_2H) —Step 4→ (NH_2, CO_2H) —Step 5→ (OH, CO_2H) —Step 6→ ($OCOCH_3$, CO_2H)

Give the names, or formulae, of the reagents and the conditions needed for Steps I, III and V.

(NI 91)

22. Propenenitrile (acrylonitrile) is used in the manufacture of acrylic fibres.

$H_2C{=}C(H)CN$

Propenenitrile

(*a*) Industrially, propene is converted to propenenitrile as shown by the following equation:

$$2CH_2{=}CHCH_3(g) + 3O_2(g) + 2NH_3(g) \rightleftharpoons 2CH_2{=}CHCN(g) + 6H_2O(g)$$

The process is carried out at a constant operating pressure.

(i) By what process is propene obtained from high molecular weight alkanes?

(ii) Deduce whether there is a change in volume when propenenitrile is formed in the reaction above.

(iii) Explain the effect of an increase in operating pressure on the yield of propenenitrile.

(iv) The conversion of propene to propenenitrile is highly exothermic. Suggest a danger that may be caused by the exothermic reaction and how it may be resolved or controlled.

(*b*) Propenenitrile may be prepared in the laboratory using the following scheme:

$$CH_2{=}CHCHO \xrightarrow{A} CH_2{=}CHCOOH \xrightarrow{B} CH_2{=}CHCOCl \xrightarrow{C} CH_2{=}CHCONH_2 \xrightarrow{D} CH_2{=}CHCN$$

(i) Name, or give the formulae of, the reagents **A**, **B**, **C** and **D**.

(ii) Propenenitrile is toxic. State the major precaution you would take during its preparation in the laboratory.

(*c*) Propenenitrile contains a nitrile group and a double bond. Give the structures of the organic products from the following reactions of propenenitrile:

$CH_2 = CHCN$ — H^+/H_2O → ?; Br_2 → ?; H_2/Ni → ?; $LiAlH_4$ → ?

(*d*) Propenenitrile may be polymerised in a similar way to ethene.

Draw the structure of the polymer produced from propenenitrile showing at least three repeating units.

(NI 92)

23. In 1985, at Bhopal in India, there was a serious accident at a chemical plant which manufactured insecticides.

The first stage of the process involves the reaction of phosgene (carbon dichloride oxide, $COCl_2$) with aminomethane (CH_3NH_2). The reaction produces methyl isocyanate (CH_3NCO) with hydrogen chloride as a by-product. The methyl isocyanate is subsequently converted into the insecticides.

The accident is believed to have been caused during storage of the methyl isocyanate in a cooled storate tank. Some water entered the tank, causing an exothermic reaction with the methyl isocyanate. As a result, volatile unreacted methyl isocyanate evaporated through an escape valve into the atmosphere, causing disaster for local people.

(*a*) Draw the structural formula of phosgene ($COCl_2$), and predict and explain the bond angles in this molecule.

(*b*) Write down the balanced equation for the production of methyl isocyanate.

(*c*) State *one* piece of evidence in the passage which confirms that methyl isocyanate is a covalent compound.

(*d*) Methyl isocyanate reacts readily with nucleophiles such as water. Why is a water molecule able to act as a nucleophile?

(*e*) The structure of methyl isocyanate is

```
     H
     |
H—C—N=C=O
     |
     H
```

Predict and explain which atom in this molecule would be attacked by water.

(*f*) The hydrolysis of methyl isocyanate produces the gas aminomethane plus one other product. Write a balanced equation for this hydrolysis and use it to suggest the identity of the other product.

(*g*) Calculate the *total* volume of gas produced at rtp by the reaction of 10.0 kg of water with excess methyl isocyanate. (The volume of one mole of gas is 24.0 dm^3 at rtp.)

(*h*) State *two* reasons why the hydrolysis of methyl isocyanate led to a build-up of pressure inside the storage tank.

(O 92, AS)

24. The carboxylic acids of general formula $C_4H_7CO_2H$ all contain a carbon–carbon double bond in the C_4H_7 fragment; they may all be catalytically hydrogenated without affecting the carboxylic acid group.

(*a*) State for *each* functional group, *one* experimental test which can be applied to detect the presence of the following:

(i) a carbon–carbon double bond, $>C=C<$

(ii) a carboxylic acid group, $—CO_2H$

In each case state the expected *observations*.

(*b*) (i) Write down shortened structural formulae for the *eight* carboxylic acid isomers of formula $C_4H_7CO_2H$. It may help you to give first all the unbranched structures, then branched structures with methyl substituents and finally any other structures. (Do *not* write down both isomers of a geometric pair, for which one formula only is required.)

(ii) Mark with an asterisk (*) any chiral centres in the formulae that you have given in (*b*)(i) above.

(iii) Write down the formulae of those isomers which exhibit geometric isomerism.

(iv) Write down the shortened structural formulae for the products of hydrogenation of the isomers you have listed in (*b*)(i) above: list these in the same order.

(v) Mark with an asterisk (*) any chiral centres in the formulae you have given in (*b*)(iv) above.

(*c*) State which *one* of the eight original $C_4H_7CO_2H$ isomers displays geometric isomerism *and* gives a hydrogenation product containing a chiral centre.

(WJEC 92)

25. Compound **B**, a diacid that occurs in apples and other fruit, has the following composition by mass: C 35.8%, H 4.5%, O 59.7%.

B reacts with ethanol in the presence of concentrated sulphuric acid under reflux to give **C**, $C_8H_{14}O_5$. Compound **C** evolves hydrogen gas when treated with sodium metal and reacts with acidified potassium dichromate(VI) to give compound **D**. Compound **D** produces an orange precipitate with 2,4-dinitrophenylhydrazine but has no reaction with Fehling's or Tollens' reagent.

(*a*) Calculate the empirical formula of **B**.

(*b*) Suggest structures for compounds **B**, **C** and **D** and explain the reactions described.

(C 92)

26. (*a*) Explain the meanings of the following terms as applied to proteins:

(i) the peptide bond

(ii) an amino acid residue

(iii) primary structure

(iv) the disulphide link

(*b*) The technique of electrophoresis involves applying a potential difference across a starch gel. If the amino acid glycine is dissolved in the gel and a potential applied, it is observed that

(i) in strongly acidic solution the amino acid moves towards the cathode

(ii) in strongly alkaline solution the amino acid moves towards the anode

(iii) at pH = 6 the glycine will not move in either direction.

Explain these observations.

(*c*) The determination of the protein content of food is an important step in judging the quality of the food. In such an experiment 25 g of a food sample was decomposed by heating with hot concentrated sulphuric acid to convert all the nitrogen from proteins into ammonium sulphate. The resulting solution was treated with sodium hydroxide solution which liberated sufficient ammonia gas to neutralise 10 cm^3 of 0.1 M hydrochloric acid.

(i) Calculate the percentage of nitrogen in the food.

(ii) What other factors would you consider to be important in deciding on the quality of food protein?

(O & C 90, AS)

27. For *both* part (*a*) and part (*b*), draw structural formulae for the lettered compounds **C, D, E, F, G, H, J, K, L**. Give your reasoning.

(*a*) A liquid **C**, $C_5H_{10}O_2$, reacts with $LiAlH_4$ to give a mixture of two alcohols, **D** and **E**. Both **D** and **C** give a pale yellow crystalline product **F** when treated with iodine in alkaline solution. The liquid **C** is insoluble in cold, dilute aqueous NaOH; but on boiling, the mixture gradually becomes homogeneous.

(*b*) A hydrocarbon, **G**, does not react with chlorine in the dark. When a mixture of chlorine and **G** is irradiated with ultraviolet light, only *two* monochlorinated products, **H** and **J**, are formed. When **H** and **J** are each treated with warm aqueous NaOH solution, and then with $KMnO_4$ solution, **H** gives an acid, **K**, whereas **J** gives a ketone **L**.

(O 91)

28. When 0.440 g of a solid, **A**, is completely burned in oxygen, 0.590 g of carbon dioxide, 0.240 g of water and 0.094 g of nitrogen are produced. Calculate the empirical formula of **A**.

When **A** is refluxed with moderately concentrated hydrochloric acid just one substance, **B**, is formed, with the molecular formula $C_2H_5O_2N$.

B, on treatment with lithium tetrahydridoaluminate, $LiAlH_4$, forms **C**, C_2H_7ON.

C, on refluxing with sodium bromide and sulphuric acid forms **D**, C_2H_6BrN.

D, on heating under pressure with alcoholic ammonia forms **E**, $C_2H_8N_2$.

Deduce the structural formulae and names of the compounds **B** to **E** and suggest a displayed formula for **A**.

Suggest the chemical and physical properties of **E**. In particular, suggest how **E** will react with

(i) Cu^{2+} ions

(ii) decanedioyl dichloride

Give practical observations and chemical formulae where appropriate.

(L(N) 91, S)

34

THE IDENTIFICATION OF ORGANIC COMPOUNDS

Before a compound can be identified, it must be obtained in a pure state.

34.1 METHODS OF PURIFYING ORGANIC COMPOUNDS

Recrystallisation...

1. Recrystallisation and fractional recrystallisation are used for solids [§ 9.2].

...sublimation...

2. Sublimation can be used for a few substances, e.g., benzoic acid [see Figure 34.1].

...filtration...

3. Filtration will remove suspended matter from liquids. Figure 9.2, § 9.2 shows an apparatus which can be used for filtration under reduced pressure, in order to speed up the process.

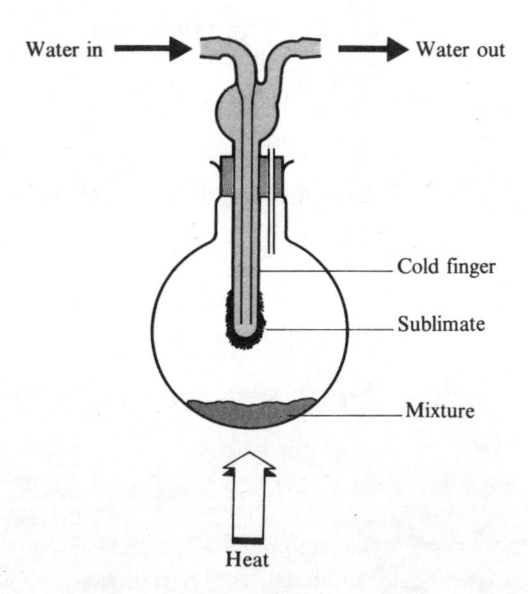

FIGURE 34.1 Sublimation

...and centrifugation...

...are all methods used for purifying organic compounds

4. Centrifuging is used to remove solid particles from liquids. When a liquid is spun round in a centrifuge, the force of gravity acting on suspended particles is increased by a centrifugal force. The particles settle to the bottom of the liquid, and the liquid can be poured off, i.e., **decanted**.

FIGURE 34.2
An Analytical Chemist at Work

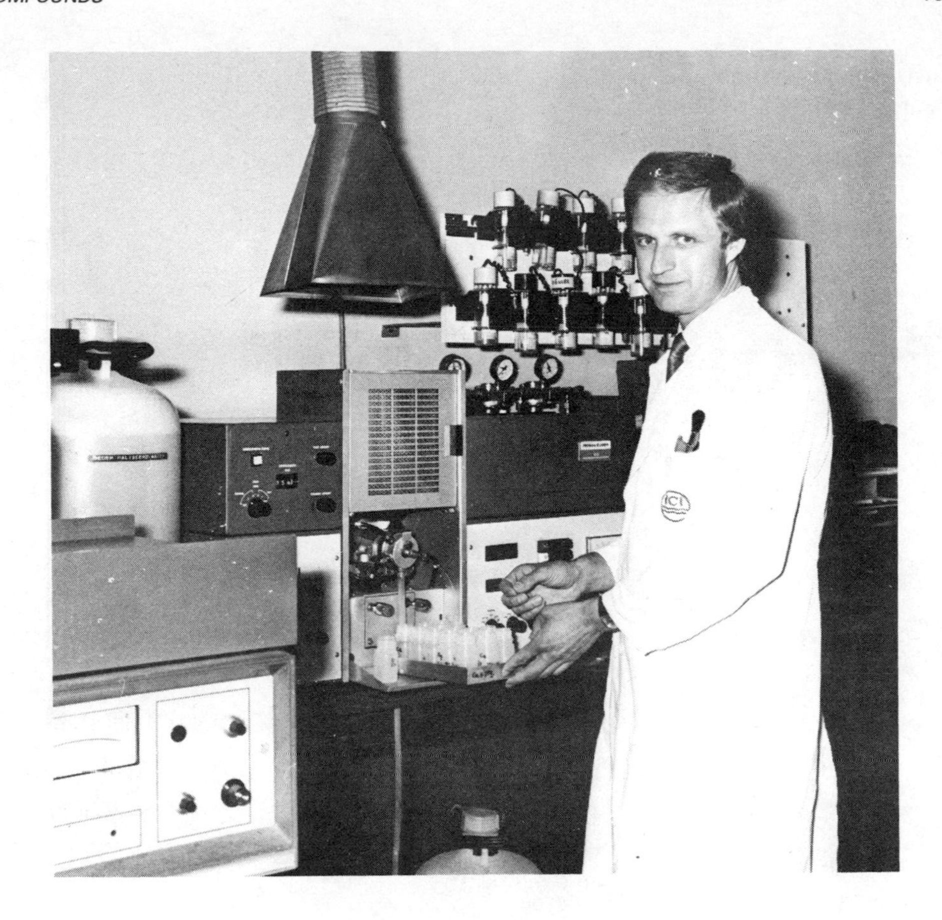

Other methods are distillation: simple, fractional and in steam . . .

. . . solvent extraction and chromatography

5. Distillation is used for liquids [§ 8.2].

6. Fractional distillation is used for mixtures of liquids [§ 8.4.2].

7. Steam distillation is used for mixtures of liquids [§ 8.5.1].

8. Solvent extraction is widely used for liquids and solids [§ 8.6].

9. Various types of chromatography are widely used [§ 8.7].

34.2 CRITERIA OF PURITY

34.2.1 SOLIDS: MELTING TEMPERATURE

A pure solid has a sharp melting temperature

A pure solid melts at a certain temperature, at which it changes sharply from solid to liquid. The presence of impurity lowers the melting temperature, and also results in a more gradual melting over a range of temperature. Figure 34.3 shows an apparatus which can be used for finding a melting temperature.

A mixed melting temperature with an authentic specimen can be used to identify a compound

A still better test of purity is a mixed melting temperature determination. Imagine you have a compound **C** which melts at 93 °C. It could be impure **A**, of T_m 94 °C or impure **B** of T_m 97 °C. A mixed melting temperature determination on a mixture of **A** and **C** gives 85 °C, while a mixture of **B** and **C** melts at 95 °C. The compound you have made is a slightly impure specimen of **B**.

FIGURE 34.3 Apparatus for Finding a Melting Temperature

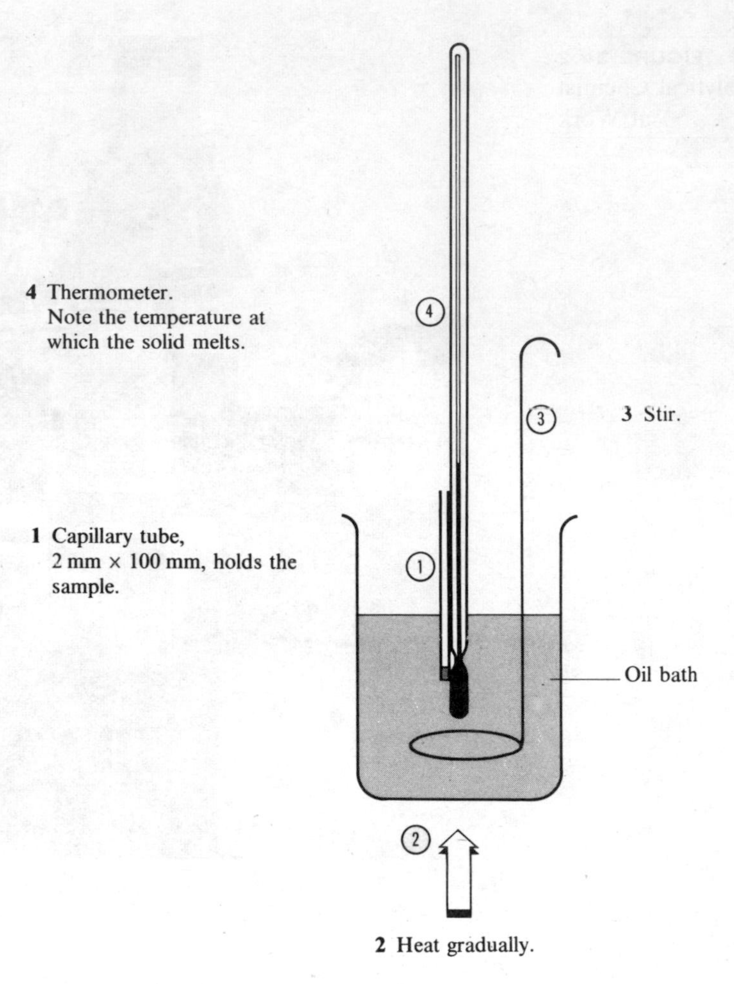
4 Thermometer.
Note the temperature at which the solid melts.
3 Stir.
1 Capillary tube,
2 mm × 100 mm, holds the sample.
Oil bath
2 Heat gradually.

34.2.2 LIQUIDS: BOILING TEMPERATURE

Boiling temperatures are less reliable indicators of purity than melting temperatures

A pure liquid can be identified by its boiling temperature. An apparatus such as that in Figure 8.5, § 8.2 can be used. The thermometer measures the temperature of the vapour in equilibrium with the boiling liquid. It should not be immersed in the liquid because **superheating**, i.e., heating above the boiling temperature of the liquid, can occur if the liquid is heated rapidly. Boiling temperatures tend to be less reliable as a test of purity than melting temperatures because of the danger of superheating.

34.3 ANALYSIS TO FIND OUT WHICH ELEMENTS ARE PRESENT

The Lassaigne test for sulphur, nitrogen and the halogens

Once a pure sample of the compound has been obtained, the next step is to find out which elements it contains. It is assumed that carbon and hydrogen are present in an organic compound. The **Lassaigne test** is done to test for the presence of nitrogen, sulphur and the halogens. Fusion with sodium converts nitrogen into cyanide, sulphur into sulphide and halogens into halides. These ions are then identified by chemical tests.

Alternatively, mass spectrometry [§§ 1.8 and 34.9.5] will detect the elements present.

34.4 QUANTITATIVE ANALYSIS

The percentage by mass of each element present can be found by quantitative analysis.

34.4.1 CARBON AND HYDROGEN

The masses of C and H in a compound can be found by weighing CO_2 and H_2O formed on combustion...

A known mass of the compound is heated in a stream of pure, dry oxygen in the presence of copper(II) oxide. Hydrogen is oxidised to steam, which is absorbed in weighed calcium chloride tubes. Carbon is oxidised to carbon dioxide, which is absorbed in weighed bulbs of concentrated potassium hydroxide solution. From the increases in mass, the masses of carbon and hydrogen in the sample can be found [see Question 1, Checkpoint 34A].

34.4.2 NITROGEN

...N can be estimated after conversion to NH_4^+...

Nitrogen can be converted into ammonium sulphate by boiling a known mass of the compound with concentrated sulphuric acid and sodium sulphate. The ammonium salt is estimated by a titrimetric method [see Question 3, Checkpoint 34A].

34.4.3 HALOGENS

...Halogens can be determined as AgX...

Heating a known mass of the compound in a sealed tube with fuming nitric acid and solid silver nitrate converts the halogens into a precipitate of silver halide. This can be filtered off, dried and weighed [see Question 2, Checkpoint 34A].

34.4.4 SULPHUR

...S can be determined by precipitation as $BaSO_4(s)$...

The sulphur content of a known mass of the compound can be converted into sulphate ions by heating the sample with fuming nitric acid. The sulphate can be precipitated as barium sulphate, filtered off, dried and weighed [see Question 2, Checkpoint 34A].

34.4.5 OXYGEN

...Oxygen is found by difference

The percentage by mass of oxygen is difficult to find by experiment, and is obtained by difference, once the other elements have been assayed.

34.5 EMPIRICAL FORMULA

The empirical formula is calculated from the percentage composition by mass, as outlined in § 3.7.

34.6 MOLECULAR FORMULA

To convert the empirical formula into a molecular formula, the molar mass must be known. Molar mass determinations have been described for gases [§ 7.2]; for volatile liquids [§ 8.3]; for solutes [§ 9.4.1] and the mass spectrometric method is covered in § 1.8 and § 34.9.5.

34.7 QUALITATIVE ANALYSIS

Chemical tests reveal the presence of hydroxyl groups, carbonyl groups, etc. You will be familiar with one of the schemes of qualitative analysis which arrange the tests for functional groups in a logical order.

34.8 STRUCTURAL FORMULA

The molecular formula and the results of chemical tests often enable a compound to be identified

The structural formula can often be worked out from the results of chemical tests and the molecular formula. There are, in addition, a number of physical methods available for identifying the groups present.

Once an opinion on the identity of the compound has been reached, it can be confirmed by matching the melting or boiling temperature with that of a known compound. Boiling temperatures are less reliable than melting temperatures. It is better to convert a liquid into a solid derivative, and find the melting temperature of the derivative. For example, ketones can be converted into solid 2,4-dinitrophenylhydrazones, and amines into solid benzoyl derivatives.

CHECKPOINT 34A: ANALYSIS

1. (*a*) When 0.200 g of a compound **A**, which contains only carbon and hydrogen, was burned completely in a stream of dry oxygen, 0.629 g of carbon dioxide and 0.257 g of water were formed. Find the empirical formula of the compound.

(*b*) When 0.200 g of **A** is vaporised, the volume which it occupies (corrected to stp) is 53.3 cm^3.
The GMV = 22.4 dm^3 at stp. Find the molar mass of **A**.

(*c*) Use the molar mass of **A** to convert the empirical formula, from (a), into a molecular formula.

2. A Lassaigne test showed that a compound **X** contained sulphur and chlorine. From 0.1000 g of **X** was obtained 0.1322 g of $BaSO_4$. Another 0.1000 g sample of **X** gave 0.0813 g of AgCl. The combustion of 0.2000 g of **X** gave 0.2992 g of CO_2 and 0.0510 g of H_2O. Find the empirical formula of **X**. (Remember that, if the percentages of C, H, S and Cl do not total 100%, the difference is due to oxygen.)

3. The compound **Y** contains nitrogen. After conversion of 0.1850 g of **Y** into an ammonium salt, ammonia was expelled by warming the salt with 75.0 cm^3 of sodium hydroxide solution of concentration 0.100 mol dm^{-3}. Titration showed that 50.0 cm^3 of sodium hydroxide remained at the end of the reaction. Find the percentage by mass of nitrogen in **Y**.

When 0.146 g of **Y** was burned completely in dry oxygen, 0.264 g of CO_2 and 0.126 g of H_2O were formed. Find the percentage by mass of C and H in the compound.

Find the empirical formula of the compound.

4. 10 cm^3 of a hydrocarbon C_aH_b was exploded with an excess of oxygen. When the mixture of gaseous products was cooled, 50 cm^3 of steam condensed. When the mixture was treated with sodium hydroxide solution, a contraction of 40 cm^3 occurred. Obtain the formula of the compound.

34.9 PHYSICAL METHODS OF STRUCTURE DETERMINATION

34.9.1 MOLECULAR SPECTROSCOPY

The electromagnetic spectrum is shown in Figure 34.4.

FIGURE 34.4
The Electromagnetic Spectrum

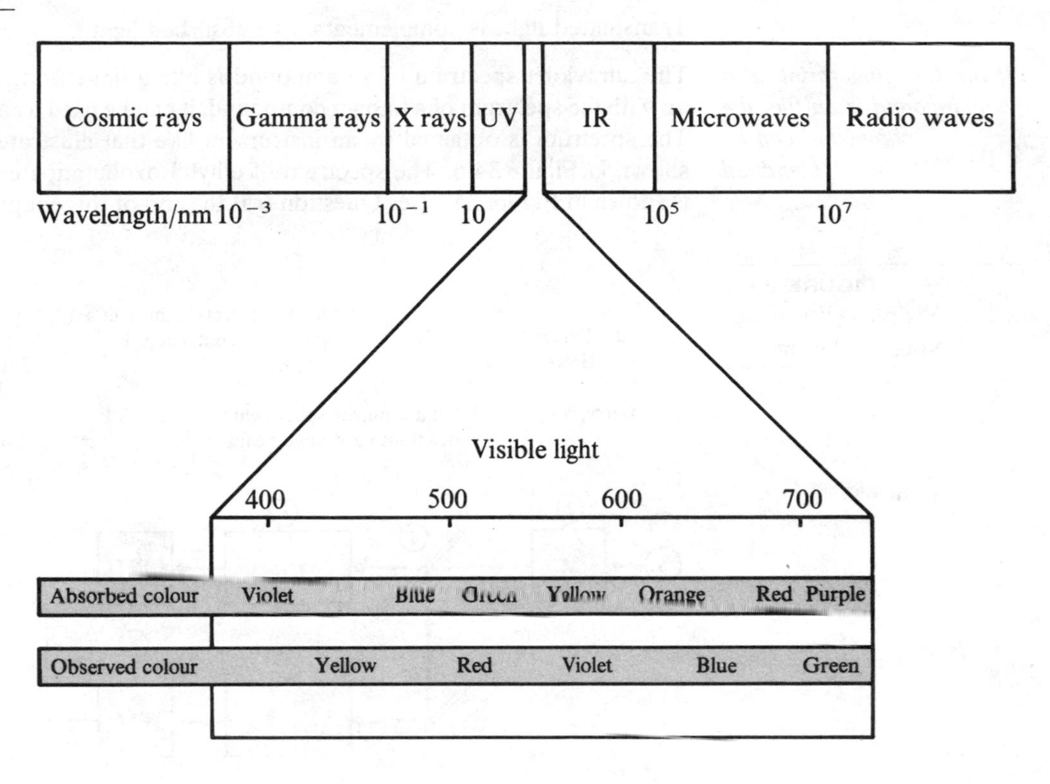

When electromagnetic radiation falls on to a compound, radiation of certain wavelengths is absorbed. In absorbing radiation, the molecules are excited to a higher energy level. The relationship expressed by Bohr is

$$h\nu = E_{\text{upper level}} - E_{\text{lower level}}$$

Molecular spectra arise from electron transitions...

...from changes in the extent of bond vibration and in the frequency of rotation

where ν = frequency of radiation absorbed and h = Planck's constant [§ 2.2.1]. In atomic spectra, the transitions arise from changes in the distribution of electrons. In molecules, in addition to the excitation of electrons, the absorption of energy can increase the extent of vibration in the bonds and the speed of rotation of the molecule. The energy changes associated with changes in vibration and rotation are smaller than those that result in the excitation of electrons, and the frequency of radiation needed is less [see Figure 34.8, § 34.9.3, for frequencies associated with vibration]. Electron transitions require radiation in the visible or ultraviolet region, vibrational changes require infrared light, and rotational changes require microwaves.

34.9.2 VISIBLE–ULTRAVIOLET SPECTROPHOTOMETRY

The promotion of electrons gives rise to UV spectra

Since all matter contains electrons, all substances absorb light at some wavelength in the visible–ultraviolet spectrum. Compounds with π electrons, since these are less strongly held than σ electrons, absorb light intensely. Benzene absorbs in the UV region, and appears colourless, naphthacene absorbs in the blue-violet region of the spectrum, and appears yellow.

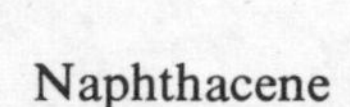

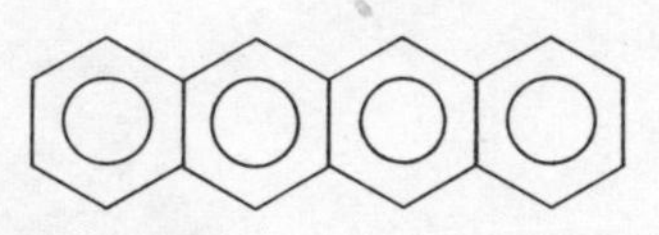

Transmitted light is complementary to absorbed light.*

If the UV 'fingerprint' of a compound is on file, the compound can be identified

The ultraviolet spectrum of a compound is like a fingerprint. If it can be matched up with the spectrum of a known compound, it can be used as a means of identification. The spectrum is obtained by an instrument like that illustrated in Figure 34.5 and shown in Figure 34.6. The spectrum of ethyl 3-oxobutanoate in a number of solvents is shown in Figure 34.7 [see Question 6 at the end of this chapter].

FIGURE 34.5 A Visible–Ultraviolet Spectrophotometer

1 Light source, S

2 Monochromator, M, selects wavelength.

3 Quartz mirror splits light beam into a double beam.

4 One beam travels through a quartz cell containing the sample.

7 Electric circuitry compares the two currents. The difference depends on the absorption of light by the sample.

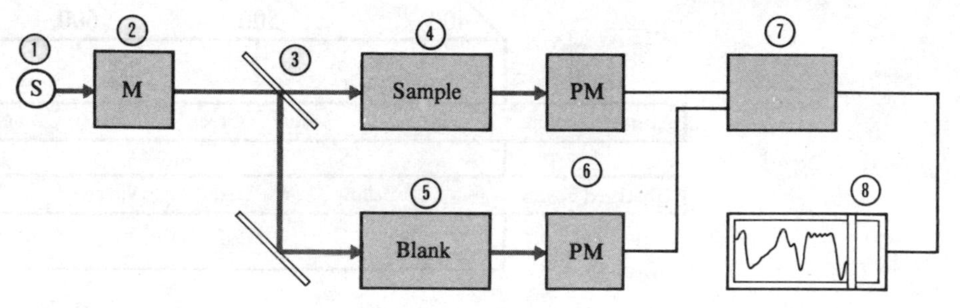

5 Second beam passes through a quartz cell containing the solvent.

8 Recorder. A pen traces the absorption spectrum.

6 Photomultiplier, PM, converts light into electric current.

FIGURE 34.6 A UV Spectrophotometer

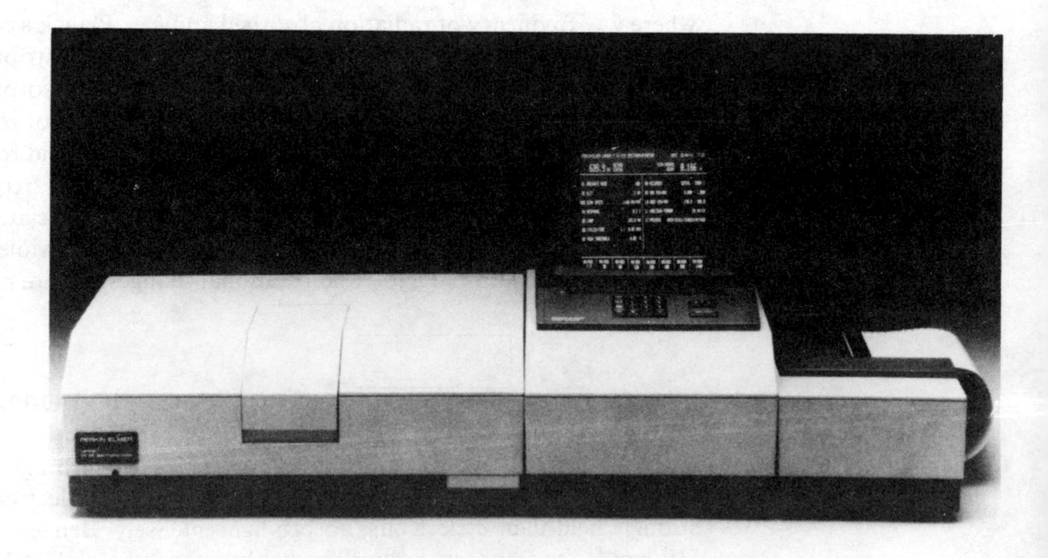

*See R Muncaster, *A-Level Physics* (Stanley Thornes)

FIGURE 34.7 Ultraviolet Spectrum of Ethyl 3-oxobutanoate in A, Hexane; B, Ethoxyethane; C, Ethanol; D, water

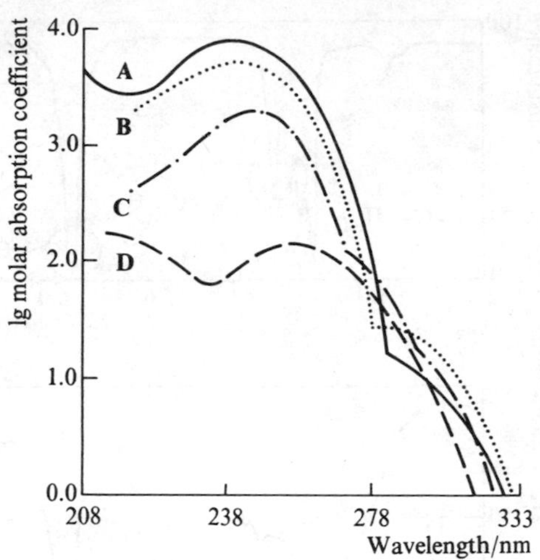

Molar absorption coefficient = $\lg(I_0/I)/cl$
I_0 and I are the intensities of the incident and transmitted light respectively
c = concentration of solution/mol dm^{-3}
l = length of cell/cm

34.9.3 INFRARED SPECTROSCOPY

All organic compounds absorb in the infrared

All organic molecules absorb strongly in the infrared [see Figure 34.8]. The operation of an infrared spectrophotometer is similar to that of a visible–UV spectrophotometer. The source of radiation is a hot wire. The detector is a **thermocouple** (instead of a photomultiplier), which converts heat into an electric current. Since glass and quartz absorb IR radiation, the cells, mirrors and prisms are cut from large sodium chloride crystals.

There are many peaks in an IR spectrum, and it is a good 'fingerprint'

Different bonds give rise to different absorption lines

There are many more peaks in an IR spectrum: it serves better than the UV spectrum to identify an organic compound. Books of recorded spectra are available for comparison. If the unknown compound is a new substance, its spectrum will not match up with any recorded IR spectrum; yet it is still possible to infer a great deal about the structure of the compound. The C═O bond and the C—OH bond and others have characteristic absorption frequencies, and can be identified. Figures 34.9 and 34.10 show the IR spectra of butanal and propanoic acid. It is customary to record IR spectra as percentage transmittance, so that the peaks extend from the top of the trace downwards.

FIGURE 34.8 Some Vibrations and Their Wavelengths

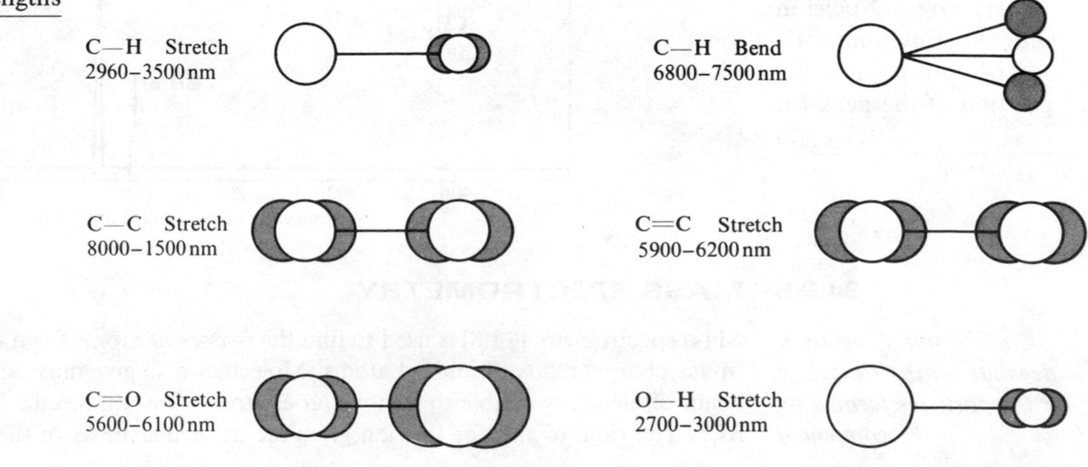

FIGURE 34.9 IR Spectrum of $CH_3CH_2CH_2CHO$

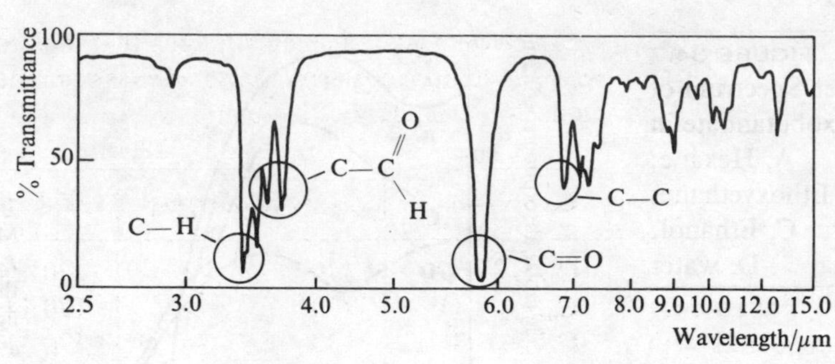

FIGURE 34.10 IR Spectrum of $CH_3CH_2CO_2H$

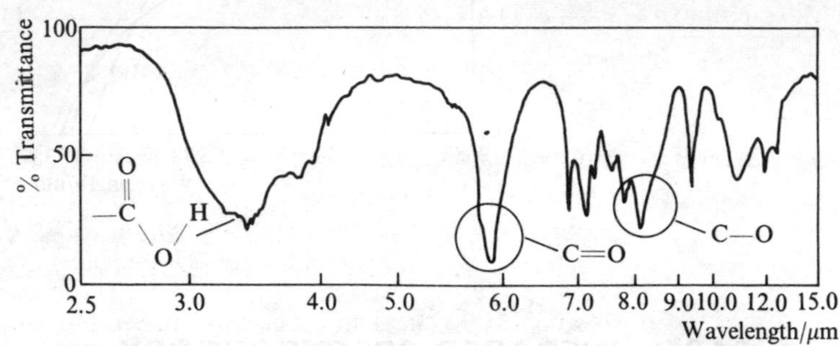

34.9.4 NUCLEAR MAGNETIC RESONANCE SPECTROSCOPY

In a magnetic field, H atoms absorb radio waves...

...The frequency of radio waves absorbed gives information about the bonding of H atoms

Radio frequency spectroscopy becomes possible when substances are positioned in a strong magnetic field. In order to be detected by nuclear magnetic resonance spectroscopy, NMR, atoms must have a magnetic moment. Then the absorption of radio waves changes the orientation of the nuclei in the magnetic field. Any compound which has atoms of ^1_1H, which are magnetic because they have a nuclear spin, will absorb radio waves. The frequency of the radio waves absorbed depends on the magnetic moment of the nucleus and the size of the applied magnetic field.

The value of NMR as a tool in organic chemistry is that ^1_1H atoms in different chemical environments, e.g., in OH and in CH_3, absorb radio waves of different frequencies. In the NMR spectrum, the wavelength of the absorption reveals the environment of the H atoms, and the area of the peak reveals the number of H atoms in this environment. In the NMR spectrum for $C_6H_{12}O_2$ [see Figure 34.11], there are (a) 9H in CH_3—C groups and (b) 3H in —$COCH_3$ groups. The structural formula must be $(CH_3)_3COCOCH_3$.

FIGURE 34.11 The Shifts in Frequency for Hydrogen Nuclei in Different Environments. The zero point is the position of the peak for $(CH_3)_4Si$

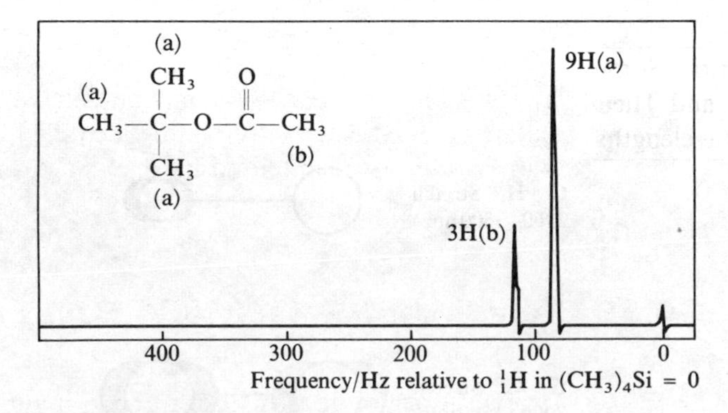

34.9.5 MASS SPECTROMETRY

The mass spectrum depends on the ratio m/e for the particles formed by the compound

Mass spectrometry [§ 1.8] is used to find the masses of atoms from the *m/e* (mass/charge) ratios of ionised atoms. Molecules also give mass spectra. The ionising beam of electrons is able to remove an electron from a molecule **M** to form the ion M^+. The ratio of *m/e* for this ion gives the molecular mass of the compound. The accuracy of the determination is such that molecular masses can be used to distinguish between compounds with the same mass number (nucleon number). The example in Table 34.1 will clarify the point.

TABLE 34.1 Compounds with the Same Mass Number

Compound	*Mass number*/m_u	*Molecular mass*/m_u
$C_{10}H_{20}NO$	170	170.1540
$C_9H_{18}N_2O$	170	170.1415
$C_9H_{16}NO_2$	170	170.1177

The accurate value of the molecular mass will enable a distinction to be made between the compounds.

Molecular masses are found

The ionising beam of electrons has enough energy to split molecules into fragments. One of the fragments formed in each dissociation is ionised. The mass spectrum of methanol shows peaks at 32, 31, 29 and 28. These are produced by the ions CH_3OH^+ (the molecular ion), and CH_3O^+, CHO^+ and CO^+ [see Figure 34.12].

Fragmentation patterns can be used for fingerprinting

Isomers have the same molecular mass, but since the bonds are different, they often break up differently to give different ions. The **fragmentation patterns** of the isomeric pentanols shown in Figure 34.12 differ.

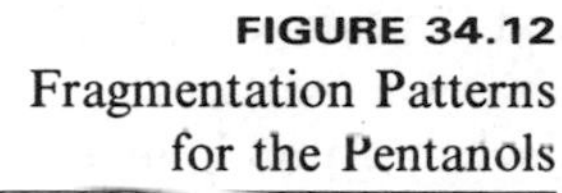

FIGURE 34.12 Fragmentation Patterns for the Pentanols

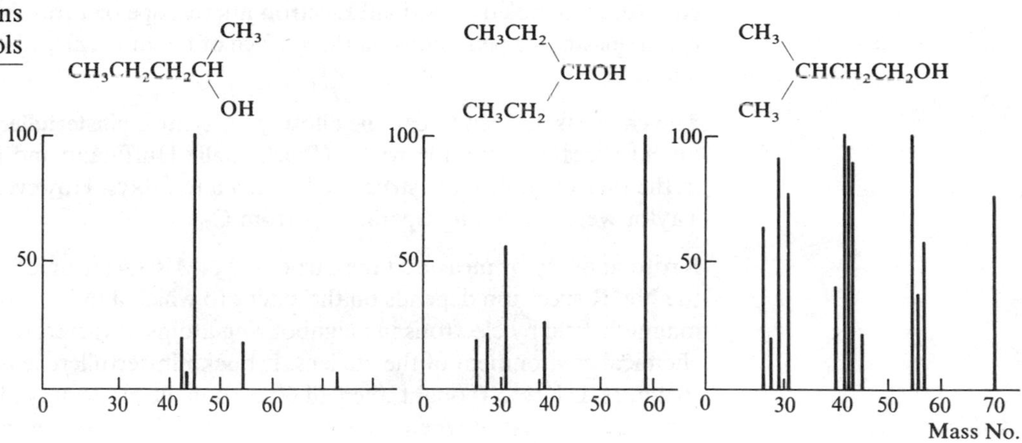

Fragmentation patterns of compounds have been recorded and catalogued. An unknown compound can often be identified by a comparison of its fragmentation pattern with a computerised bank of fragmentation patterns.

34.10 NMR AND 'BUCKY BALLS'

TOPIC

The carbon allotropes, diamond and graphite, have been known for centuries. In 1985 scientists were excited by the discovery of a new family of carbon allotropes. They consist of molecules containing 30–70 carbon atoms and have been named **fullerenes**.

The discovery had its origins in space research. In 1980, Donald Huffmann at Arizona University, USA and Wolfgang Kratschmer at Heidelberg University, Germany, wanted to find out whether soot could form in outer space from heated graphite. They struck an electric arc between a graphite rod and a graphite disc in an inert atmosphere. Graphite vaporised, and soot was deposited on the container. When they studied the ultraviolet absorption spectrum of the soot, they were surprised to find some unexpected peaks, which corresponded to large molecules. They took the matter no further at that time.

In 1984, Harold Kroto and David Walton of Sussex University, UK were investigating their theory that molecules containing up to 33 carbon atoms might exist in the space surrounding red giant stars. In collaboration with Richard Smalley of Texas University, USA, they directed a laser at a graphite target and obtained the molecules they were looking for. They also detected by mass spectrometry some molecules of mass 720 u. They wondered whether these were molecules of formula C_{60}, but they did not have enough material to do a spectroscopic study. When Huffmann and Kratschmer learned about this discovery, they had another look at the soot which they had made in 1980. They measured the mass spectrum and the ultraviolet and infrared spectra and obtained results which agreed with the formula C_{60}. From the soot they separated by sublimation and dissolving in benzene a mixture of 90% C_{60} and 10% C_{70}. Using X ray and electron diffraction measurements, they showed that the C_{60} molecules are spherical and packed 1.04 nm apart.

Kroto discussed the possible structures of C_{60} with his team. Was a layer structure, like graphite, or a spherical structure more likely? Kroto recalled the geodesic domes created by the architect Buckminster Fuller. He suggested that C_{60} might be a perfect sphere of carbon atoms. The team tried to make a model of C_{60} from hexagons of carbon atoms, but they could not make a closed structure in this way. However, success came when they used a combination of 20 hexagons and 12 pentagons. The structure resembled a football! Electron microscope pictures show a mosaic of 20 hexagons and 12 pentagons on the surface of the molecule, which measures 1 nm in diameter.

The chemists decided to call the allotrope C_{60} **buckminsterfullerene**. The new molecule quickly became better known as a **'bucky ball'**. Huffmann and Kratschmer beat Kroto in the race to publish the structure by only a few days. However, Kroto and Roger Taylor were the first to separate C_{60} from C_{70}.

Kroto and Taylor measured the carbon-13 NMR spectrum of a solution of C_{60}. Since the NMR spectrum depends on the extent to which a nucleus is screened from the magnetic field by electrons in neighbouring atoms, it therefore depends on the precise chemical environment of the nucleus. If buckminsterfullerene is a highly symmetrical structure as Kroto thought, then all 60 carbon atoms occupy identical positions in the structure. No matter where a carbon-13 nucleus is located in the C_{60} molecule, it will be in the same environment and will produce the same NMR frequency, and the spectrum should consist of a single line. What a triumph it was when the team discovered that the carbon-13 NMR spectrum consisted of a single line! Their structure was correct!

There are other fullerenes [see Figure 34.13]. In C_{70}, the extra ten atoms run round the outside of the cage. There are five different environments for carbon atoms in this structure. The differently placed carbon atoms are present in the ratio 10 : 10 : 10 : 20 : 20. Kroto found five peaks in the carbon-13 NMR spectrum of C_{70} with intensities in the calculated ratio. Again his structure was proved correct!

FIGURE 34.13
Fullerenes C_{28}, C_{32}, C_{50}, C_{60} and C_{70}

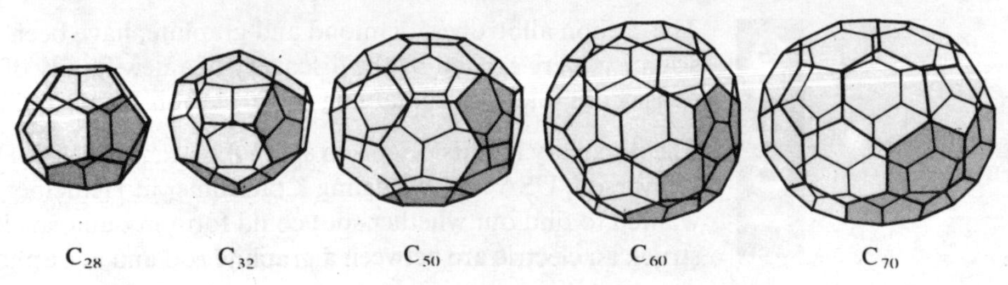

Buckminsterfullerene is now manufactured and sold in the USA for $1250 a gram. To obtain impure C_{60} is less expensive and it is expected that the price will drop to perhaps 10c a gram for 90% pure C_{60}. Scientists believe that fullerenes will find important applications as semiconductors, superconductors, lubricants, catalysts and in batteries. Derivatives of fullerenes, such as $C_{60}F_{60}$, have already been made.

QUESTIONS ON CHAPTER 34

1. An organic dibasic acid has the composition by mass: C 41.4%, H 3.4%, O 55.2%. A sample of the acid of mass 0.189 g neutralised 32.5 cm^3 of 0.100 mol dm^{-3} potassium hydroxide solution.

(*a*) Calculate the empirical formula and molecular formula of the acid.

(*b*) Write the names and structural formulae for the isomers with this molecular formula. State how you could distinguish between them.

2. What precautions must be taken to ensure that a melting temperature determination is accurate?

3. At room temperature, 50 cm^3 of an unsaturated hydrocarbon weigh 0.1167 g and react with 50 cm^3 of hydrogen. Given the GMV at room temperature = 24.0 dm^3, find the molar mass of the hydrocarbon. Suggest a formula for this compound.

4. A chemist is oxidising cyclohexanol. How can he find out when the conversion into cyclohexanone is complete?

5. Explain the reasons for the use of the following techniques in organic chemistry:
(*a*) distillation under reduced pressure,
(*b*) paper chromatography,
(*c*) steam distillation,
(*d*) recrystallisation,
(*e*) mixed melting temperature determinations,
(*f*) ethoxyethane extraction.

6. In Figure 34.7, § 34.9.2, the peak at 238 nm is due to absorption by the $\rangle C{=}O$ group. Can you explain why the absorption of light by ethyl 3-oxobutanoate, $CH_3COCH_2CO_2C_2H_5$, differs in different solvents, and is less in non-polar solvents than in ethanol and water? [See p. 618 if you need help.]

7. Synthesis is only the initial stage in the preparation of a pure organic compound and must be followed by separation, purification and characterisation. Discuss general methods for carrying out these steps subsequent to synthesis, illustrating your answer by reference to appropriate examples.

(NEAB 81, S)

8. Ripening tomatoes produce a gaseous hydrocarbon, **A**, which itself assists the ripening process. The gas **A** reacts in a 1 : 1 mole ratio with hydrogen bromide to give a liquid, **B**. Treatment of 1.0000 g of **B** with hot aqueous sodium hydroxide yields a volatile product, **C**. Acidification of the residual alkaline solution with dilute nitric acid and the addition of excess silver nitrate solution affords 1.7230 g of silver bromide. Mild oxidation of **C** yields a volatile product, **D**, which on treatment with ammoniacal silver nitrate produces a silver mirror.

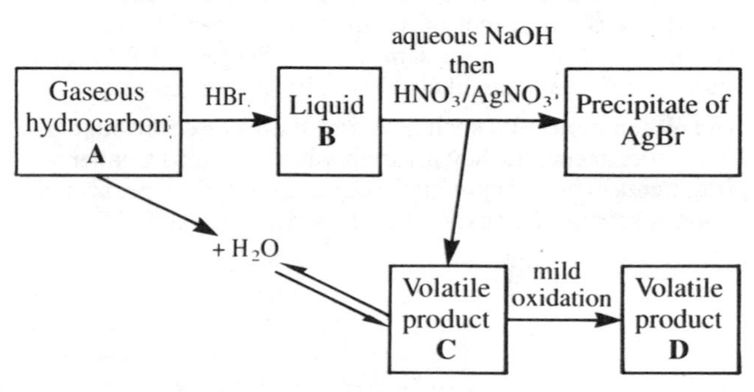

Compound **A** may be converted industrially into **C** by the reversible gas phase addition of one molecule of water. For the equilibrium

$$A(g) + H_2O(g) \rightleftharpoons C(g); \quad \Delta H^{\ominus} = -46.0\,kJ\,mol^{-1}.$$

$$(A_r(H) = 1.01,\ A_r(C) = 12.01,\ A_r(Br) = 79.91,\ A_r(Ag) = 107.87)$$

(*a*) (i) State the functional group present in each of **A**, **B**, **C** and **D**.

(ii) Give your reasoning for these conclusions.

(iii) Calculate the number of moles of silver bromide produced.

(iv) Hence calculate the relative molecular mass of **B** and deduce the identity of the gas **A**.

(*b*) (i) *Briefly* describe a process for the direct conversion of **A** into **C**, giving the appropriate conditions.

(ii) Describe and explain the effects of variation of temperature and pressure on the equilibrium yield of **C** obtained.

(iii) Indicate how the use of excess water (as steam) might influence the equilibrium yield of **C**.

(WJEC 92)

9. (*a*) (i) Describe the basic assumptions of the kinetic theory of ideal gases.

(ii) Suggest the experimental conditions under which these assumptions are no longer likely to be valid. Give your reasons.

(*b*) (i) A known mass of an organic liquid **A** was allowed to evaporate completely and the volume of vapour produced was recorded. The experiment was repeated with different masses of **A**. The following data were produced at 100 °C and 1 atmosphere pressure (1.01×10^5 N m^{-2}).

Mass of liquid/g	*Volume of vapour*/cm³
0.012	5
0.052	21
0.080	33
0.120	50

Using a graphical method, determine the molar mass of the unknown liquid, stating two assumptions that you make.

(ii) The quantitative analysis of **A** gives C 64.9%, H 13.5%, O 21.6%. Determine the empirical and hence the molecular formula of **A**.

(iii) When **A** is passed over heated aluminium oxide a gas is formed which on hydrogenation produces an alkane. Substance **B**, an isomer of **A**, when treated overall in a similar way produces the same alkane. Suggest what types of molecule **A** and **B** are likely to be, giving your reasons.

(iv) When molecules such as **A** and **B** are passed through a mass spectrometer, they invariably break up into smaller fragments. These fragments produce appropriate peaks in a mass spectrum. For example, a fragment such as $C_2H_5^+$ produces a peak at 29 *m*/*e* ratio, and $C_3H_7^+$ at 43 *m*/*e* ratio. Substance **A** produces a mass spectrum which includes peaks at 31 and 43 *m*/*e* ratio. However, the isomeric compound **B** has peaks at 29, 45 and 59 *m*/*e* ratio.

Deduce the species responsible for the peaks in the mass spectra and hence suggest suitable molecular structures for **A** and **B**.

(O & C 91, S)

10. (*a*) Explain the origin of the absorption of energy in the infrared region of the spectrum, using chlorine and hydrogen chloride gas as examples.

(*b*) How many absorptions would you expect to find in the infrared spectrum of water in the gas phase? Draw and label the vibrations which would lead to these absorptions.

(*c*) The two spectra P and Q below were obtained from compounds with the formula $C_4H_{10}O$.

Deduce a structure for the compound producing each spectrum, indicating the evidence you have used in your deduction. What further spectroscopic evidence would enable you positively to identify the compounds concerned?

(C 91)

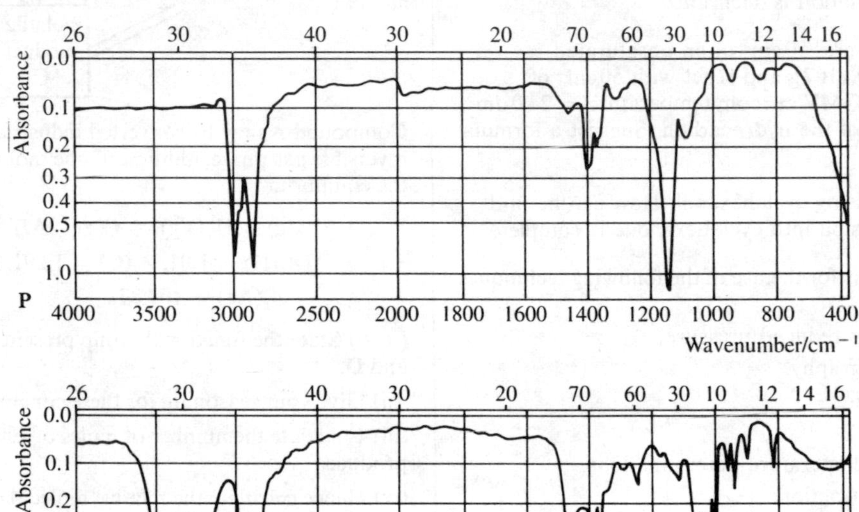

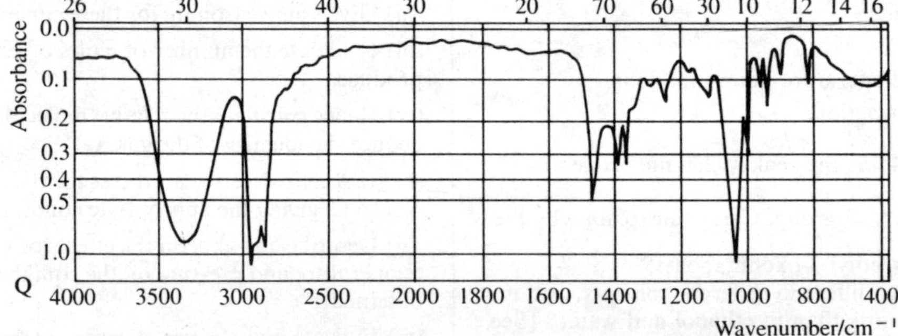

35

SOME GENERAL TOPICS

This chapter deals with a number of topics which have cropped up in different chapters. Polymerisation has been encountered in a number of chapters, and the pieces are here fitted together. Students complain of difficulty in getting from compound **A** to compound **D** via **B** and **C**; the pages on synthetic routes are an attempt to tackle this difficulty. The section 'What are these reagents used for?' is a summary of the plethora of different reagents and conditions which appear over the top of arrows, and which students find onerous to remember if tackled separately. There is a selection of questions on topics which span chapters, such as methods of distinguishing between compounds, reaction mechanisms, comparisons between functional groups and the identification of 'mystery' compounds from descriptions of their reactions.

35.1 PLASTICS

35.1.1 USES OF PLASTICS

The properties of plastics find them many uses

Plastics are polymers of carbon compounds. They are macromolecular substances with molar masses of some thousands. Plastics are low in density and strong, and soften when heated so that they can be moulded into a variety of shapes. They are resistant to chemical attack and therefore to corrosion. They are electrical insulators. A multitude of uses have been found for plastics, and the list grows daily.

Thermosoftening plastics are used when ease of moulding is important...

...Thermosetting plastics are used for articles which must not soften when heated

When plastics are heated, they soften, and in some cases melt. When they are cooled again, they behave in one of two ways. Some plastic materials harden when they are cooled, but soften again if reheated. Plastics which do this are called **thermoplastics**. Examples are polyethene, polystyrene and poly(chloroethene). Other plastic materials can be heated only once: on cooling, they harden and will not soften again when reheated. These are **thermosetting plastics** or **thermosets**. Examples are: polyurethane; phenolic resins, including Bakelite®; and melamine. Both types of plastics have their advantages. A manufacturer can buy a thermosoftening plastic in the form of granules, soften it, and mould it into articles by a mass production technique. Thermosetting plastics are used for articles which must not soften when heated, for example, electric light fittings, saucepan handles and bench tops.

35.1.2 THE STRUCTURE OF PLASTICS

There are differences in structure between thermoplastics and thermosetting plastics. Various types of structure are illustrated below:

There are linear and cross-linked polymers

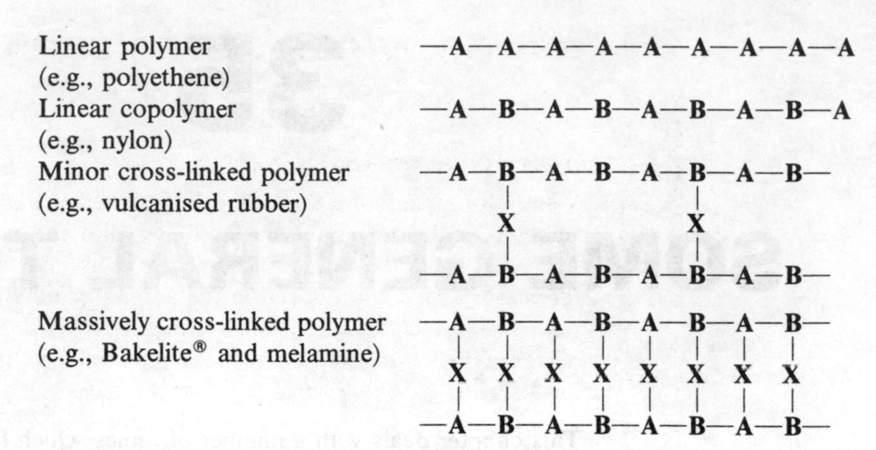

Linear polymers have no cross-links between chains. On heating, the distance between chains increases and the polymer softens, becoming more flexible. On cooling, the process is reversed. Polymers of this type are thermoplastics. Cross-linked polymers are not softened easily. When sufficient heat has been supplied to break the cross-linkages, the whole polymer has decomposed and cannot be reformed on cooling. Polymers of this type are thermosetting, e.g., urea-methanal and phenol-methanal plastics.

35.1.3 METHODS OF POLYMERISATION

Addition polymerisation of alkenes, such as ethene, chloroethene and many others, has been described in § 27.7.10.

References to material in other chapters

Condensation polymerisation to give polyamides, such as nylon, has been described in § 33.14.1. The formation of polyesters, such as Terylene® or Dacron®, by condensation polymerisation has been covered in § 33.13.2. The important thermosetting plastics derived from methanal by polymerisation (Delrin®) and by condensation with phenol (Bakelite®) and with urea (melamine and Formica®) have been covered in §§ 31.6.4 and 31.6.5.

CHECKPOINT 35A: PLASTICS

Some of the material covered is in the chapters referred to above.

1. Polymerisation reactions may be classified as **addition** or **condensation** reactions. Explain the meanings of these terms, and give two examples of each.

2. Give an example of, and explain the structure of (*a*) a polyalkene, (*b*) a polyester and (*c*) a polypeptide. Give examples of the everyday uses of all three.

3. Explain the terms *crosslinking* and *thermosetting* with reference to condensation polymers. For what purposes are thermosetting polymers suitable?

4. How does the chemical inertness of polyethene arise? How does it increase the usefulness of the material? How does it affect the disposal of waste polyethene?

5. (*a*) What type of functional group joins the repeating units in nylon?

(*b*) In what way does the structure of nylon resemble that of a polypeptide?

(*c*) What type of interaction takes place between polymer molecules which contain the functional group present in polypeptides?

6.

A = $\text{-(CH-CH}_2\text{)-}_n$ (with C_6H_5 ring on CH)

B = $\text{-(O}_2\text{C-C}_6\text{H}_4\text{-CO}_2\text{CH}_2\text{CH}_2\text{)-}_n$

Outline the preparation of the polymers **A** and **B** from $C_6H_5COCH_3$ (which must first be converted into phenylethene) and $H_3CC_6H_4COCH_3$ (which must first be converted into benzene-1,4-dicarboxylic acid). If you wanted to manufacture plastics bottles, which would you choose, **A** or **B**, for making bottles to contain (*a*) concentrated sodium hydroxide, (*b*) concentrated hydrochloric acid and (*c*) concentrated sulphuric acid?

35.2 DRUGS WHICH ALTER BEHAVIOUR

TOPIC

STIMULANTS

EPINEPHRINE OR ADRENALIN

Adrenalin is a hormone, a chemical messenger. It stiumates the CNS

A stressful situation, a loud squeal of brakes, a frightening film, the start of a race, makes your heart pound, your body perspire and your stomach feel as if it is rising into your throat. All these reactions are caused by the **epinephrine**, also called **adrenalin**, which is produced in the adrenal glands. It is a hormone, a chemical messenger which travels through the blood stream and causes changes in the body cells. Small amounts of adrenalin are released into the blood to control blood pressure and to maintain the level of sugar in the blood. The compound also enables the body to deal with stress, such as emotional excitement, exercise, extreme temperature changes and severe bleeding. Adrenalin increases the rate and strength of the heart beat, raises blood pressure by constricting blood vessels (except those in the heart, brain and liver), relaxes the smooth muscle of the lungs, increases the rate and depth of breathing, causes an increase in blood sugar and delays the fatigue of skeletal muscles, thus allowing people to show amazing strength under great stress — for example after a car accident a man lifted a car to free his son who was trapped underneath. Each of these changes lasts a short time: adrenalin released into the body is inactivated by the liver in about three minutes.

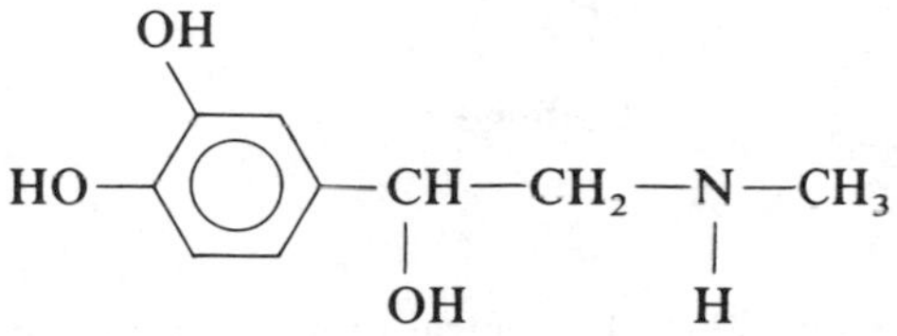

Epinephrine (adrenalin)

AMPHETAMINES

Amphetamines ('uppers') stimulate the CNS. The once widely used drug benzedrine is addictive

Amphetamines (often called 'uppers') stimulate the central nervous system in a similar manner to adrenalin. They make the user feel less tired and more alert. In the past an amphetamine called benzedrine was prescribed medicinally for suppressing appetite and for reducing bronchial congestion. Amphetamines were taken by some students when they wanted to stay awake all night to study for an exam and by some long distance lorry drivers. The drug was found to have another effect: people became **addicted** to benzedrine. The results of using the drug over a period of time are **dependence** and **tolerance** (see next page). The drug is now strictly controlled.

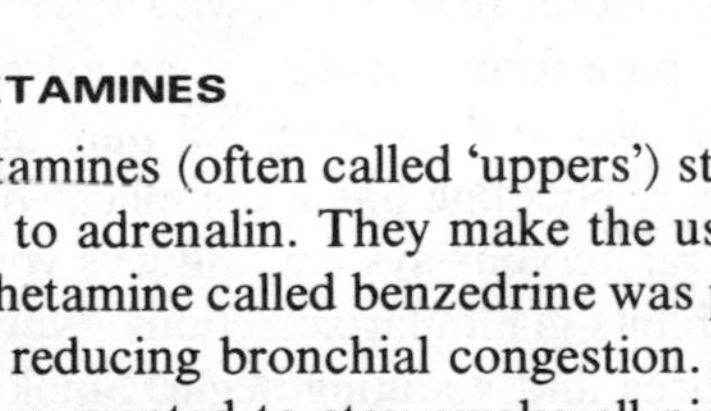

Benzedrine

Note the meanings of these terms:

Dependence on a drug may be physical, psychological or both.

Dependence The user experiences withdrawal symptoms if he or she stops using the drug to which he or she has become addicted. Dependence may be physical or psychological or both. If the dependence is physical, withdrawal may lead to very severe symptoms. If the dependence is psychological, i.e. **habituation**, the user finds life unbearable unless he or she is experiencing the effects of the drug.

Tolerance is the user's need for increasing doses to produce the same 'high'

Tolerance The body becomes increasingly able to tolerate a drug without ill effects. The user requires increasing amounts of the drug in order to experience the effects.

The most serious drug problems arise with depressants of the central nervous system: the narcotics (heroin, morphine, etc.), the barbiturates, tranquillisers and alcohol.

CAFFEINE

Caffeine is found in coffee beans, tea leaves and chocolate.

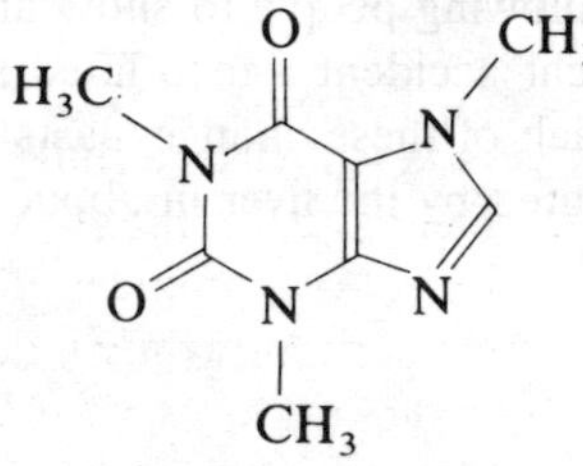

O CH$_3$ H$_3$C N N O N N CH$_3$

Caffeine

Caffeine is a non-addictive stimulant

Caffeine stimulates the central nervous system, causing mental alertness and restlessness. Taking caffeine can become a habit: some people insist that they cannot function without their morning coffee, but it is not addictive.

NICOTINE

Nicotine is found in the leaves of tobacco. Nicotine was isolated in 1828 and named after a French ambassador, Jean Nicot, who was convinced that tobacco had medicinal uses. In fact, in its pure form, nicotine is an extremely toxic drug which acts as quickly as cyanide. Two to three drops of pure nicotine (about 60 mg) will kill if placed on the tongue. Nicotine has a complicated action on the human body. It stimulates the central nervous system, induces vomiting and diarrhoea, and first stimulates, then inhibits glandular secretions. Non-smokers can absorb only 4 mg of nicotine before nausea and vomiting begin. Smokers build up a tolerance: they can absorb twice as much without noticeable effects. The smoke from one cigarette may contain about 6 mg of nicotine, of which 0.2 mg is absorbed into the body. (How many cigarettes would it take to make a person ill if he or she had never smoked before?)

Nicotine is a toxic substance in tobacco. Smokers build up a tolerance to the drug. The tar which condenses from cigarette smoke is carcinogenic

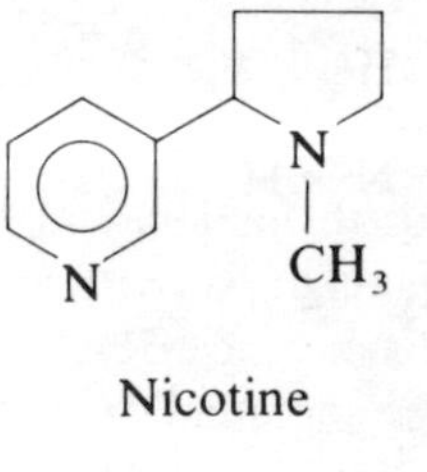

N N CH$_3$

Nicotine

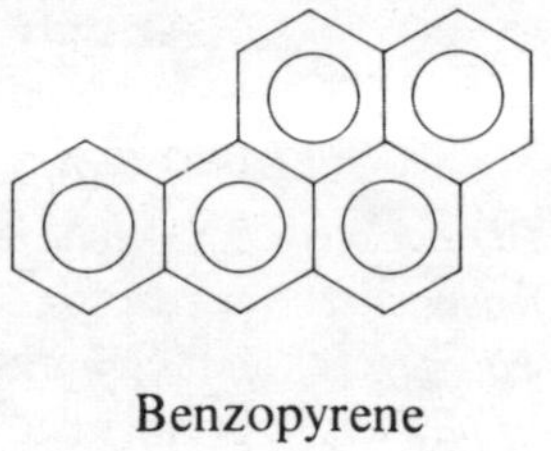

Benzopyrene

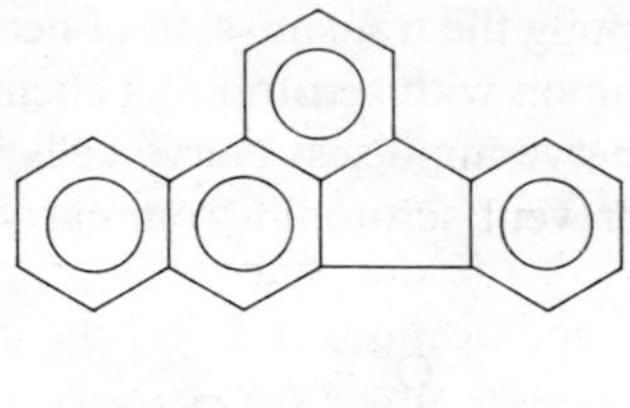

Benzofluoranthene 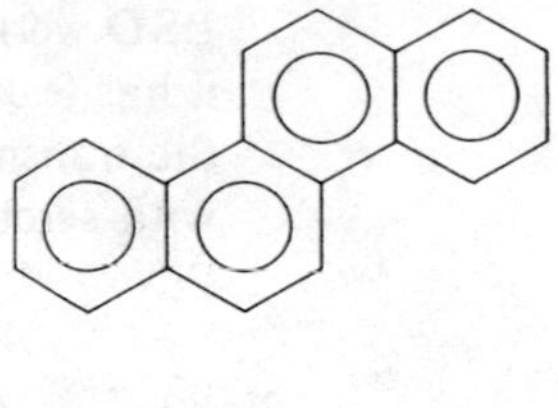 Chrysene

Cigarette smoke condenses to form tar. The lungs of a smoker receive about 100 g of tar a year. This tar has been shown to be carcinogenic in all kinds of tissue. In laboratory tests on animals, it has induced cancer in epithelial tissues and in connective tissues, e.g. skin and bone. Known carcinogens in tobacco tar are **benzopyrenes**, **benzofluoranthenes** and **alkylated chrysenes**.

COCAINE

Cocaine stimulates the CNS. It has some use in medicine. People use it to reduce fatigue, but it gives rise to physical symptoms and psychological dependence

Cocaine, also called 'coke' and 'snow', is a fluffy white powder obtained from the leaves of the coca plant, which grows in South America, especially in Peru and Bolivia. It is used as a local anaesthetic, for example in the nose and throat. Cocaine stimulates the central nervous system even more strongly than the amphetamines do. It quickens reflexes, reduces fatigue and makes the user feel exhilarated without appearing intoxicated. It has been used by some show business people to improve the sparkle of their performance. Although it is even more expensive than heroin, cocaine is popular among drug users because it does not result in tolerance. However, it does cause nausea, weight loss, insomnia and psychological dependence. Chemists have tried to synthesise compounds with the local anaesthetic qualities of cocaine but without its other effects. Xylocaine® and Novocaine® have resulted from this research. They are used in dentistry [see § 29.16].

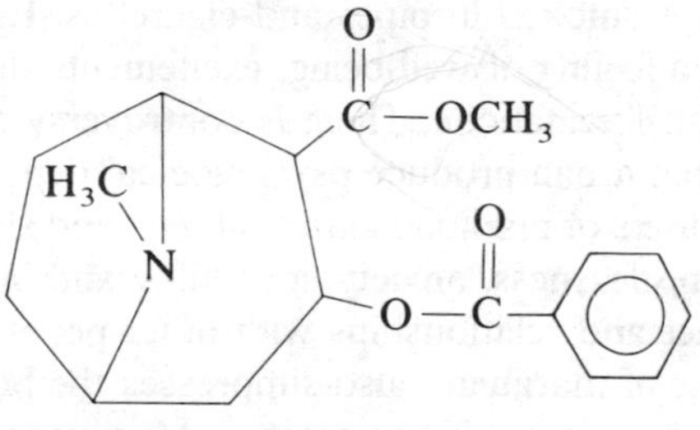

Cocaine

HALLUCINOGENS

LSD

LSD causes hallucinations. Dependence is psychological, rather than physical

LSD, short for lysergic acid diethylamide, is a **hallucinogen**. It causes changes in perception, especially visual perception, with vivid, often brightly coloured hallucinations. No physical dependence develops, but bizarre mental effects and permanent personality changes sometimes occur.

LSD works by disrupting the transmission of nerve impulses to the brain. In structure, it has features in common with serotonin, a chemical produced by the body to control the transmission of nerve impulses. Nerve cells in the brain probably confuse LSD with serotonin and prevent serotonin from carrying out its control function.

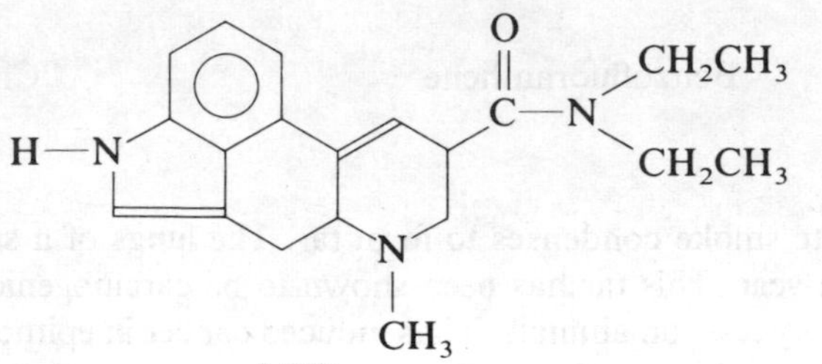

O, CH_2CH_3, C—N, H—N, CH_2CH_3, N, CH_3

LSD

OH, H—N, $CH_2CH_2NH_2$

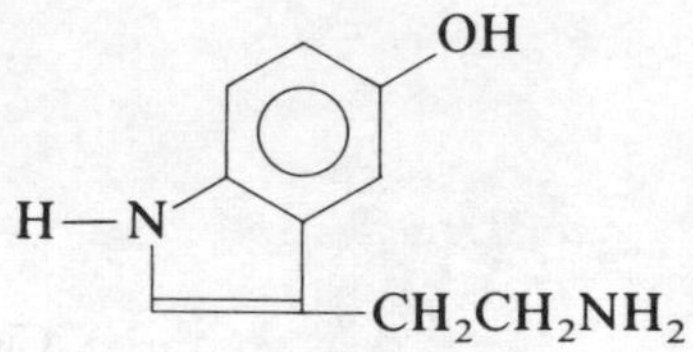

Serotonin

MARIJUANA

Marijuana is the dried leaves, flowering parts, stems and seeds of the Indian hemp plant, *Cannabis*. The stems yield tough fibres that are used for making rope. Marijuana is often called 'cannabis', 'weed', 'grass', 'pot' or 'Mary Jane'. The dried resin from the flowering part of the plant is called **hashish**. It is more potent than marijuana.

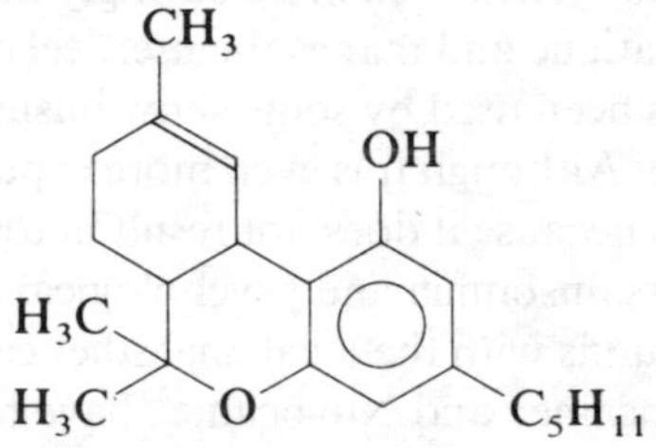

CH_3, OH, H_3C, H_3C, O, C_5H_{11}

Marijuana

Marijuana is a hallucinogen. Users develop psychological dependence. Tar from marijuana cigarettes is carcinogenic. Many marijuana users switch to heroin for a bigger 'high'

Marijuana is smoked in pipes and cigarettes. Its effects last two to three hours, producing a feeling of well-being, excitement, alterations in appreciation of time and space and hallucinations. There is controversy over its safety. It is not regarded as addictive but it can produce psychological dependence. There is some evidence that long-term users of marijuana are apathetic and sluggish. The psychological dependence shows as restlessness, anxiety, irritability and insomnia. Driving performance, work performance and relationships with other people are all affected by marijuana use. Regular use of marijuana also suppresses the body's immune response and makes the user more susceptible to disease. Marijuana cigarettes contain 50% more tar than ordinary cigarettes, and smoking them affects the bronchial tract and lungs. After a while, many marijuana users crave a better or a different sort of 'high', and turn to other drugs. Suppliers are keen to get their customers to switch from marijuana to heroin so that they will quickly become addicted to heroin.

CHECKPOINT 35B: DRUGS I

1. Explain the difference between physical addiction and psychological addiction.

2. Marijuana does not result in physical addiction. List six dangers of using marijuana. Why do you think it is a popular drug in spite of these dangers?

DEPRESSANTS

THE OPIATES

Heroin, morphine and codeine are analgesics (painkillers). Heroin and morphine are narcotics: they produce mental fogginess. Both are addictive drugs

Analgesics are substances that relieve pain; we call them 'painkillers'. **Narcotics** are analgesics which also produce euphoria, a feeling of peace and tranquillity. They are addictive. Heroin and morphine are narcotics. Heroin is the more dangerous of the two because it causes rapid addiction. Heroin, morphine and codeine (which is not addictive) are called **opiates** because they are obtained from opium. Opium is the residue obtained by evaporating the juice of the opium poppy, which is grown in the Orient and the Middle East. It contains 10% morphine and 5% codeine. Heroin is made from morphine.

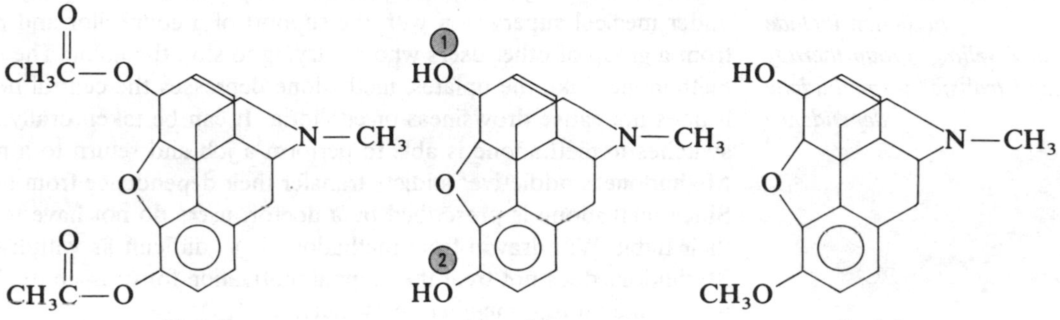

Heroin Morphine Codeine

Morphine has some use in medicine as an analgesic, but it is addictive. Codeine is widely used

Morphine is a wonderful pain reliever; it is used in cases of radical surgery and on the battlefield. Morphine is 50 times as potent as aspirin. It was widely used in the American Civil War. Sadly, it had another effect: 100 000 soldiers became addicted to it. People who take morphine suffer from changes in mood and mental fogginess.

Codeine is only one-sixth as effective as morphine, but it does not cause addiction. While morphine must be injected, codeine can be taken by mouth. Codeine is the least potent of the opiates; it is used as a cough suppressant. A synthetic compound Pethidine® or Demerol®, which is used to treat severe pain, is between morphine and codeine in potency. It is addictive.

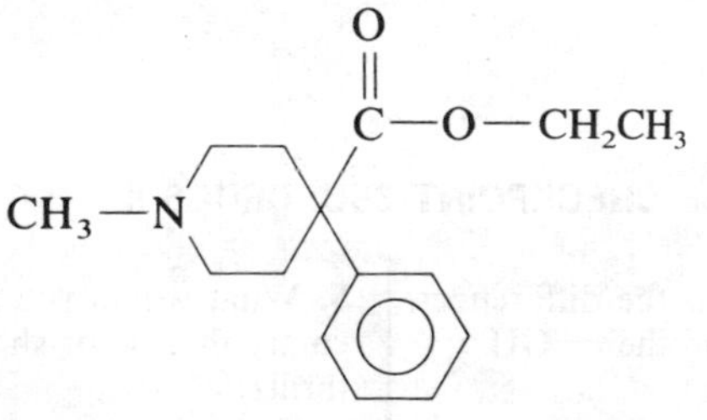

Pethidine® or Demerol®

Heroin is never used in medicine. It rapidly causes addiction. Users inject the drug, and infectious diseases spread through shared needles

Heroin is far more potent than morphine and much more addictive. As heroin depresses the central nervous system, it causes drowsiness, respiratory depression (which is why an overdose causes death) and decreased gastrointestinal movement which leads to constipation, nausea and vomiting. Heroin is rejected for use by the medical profession. It is very popular with 'drug pushers' because users very soon become addicted. The cost of supporting the heroin habit is about £50 a day. The high price encourages the drug user to administer the drug in the most effective manner: by intravenous injection 'mainlining'. The practice of giving injections under insanitary conditions and sharing needles leads to the spread of infectious hepatitis and

AIDS. It is estimated that 60 000 people in Britain are hooked on addictive drugs such as heroin.

The high cost of the heroin habit leads to crime

Taking heroin is illegal, and the price of heroin from drug pushers is high. Many users resort to crime to find the funds to keep up their habit. Normal employment which might provide the funds to support the drug habit is impossible because of the narcosis (drowsiness) induced by the drug. It is, however, possible for heroin addicts to register with a doctor or clinic and obtain free injections. Then the addicts are not reduced to thinking only of how to get their next fix, and they have a chance to lead more normal lives. The hope is that they may eventually choose to live without the drug.

Treatments for heroin addiction include counselling, group therapy and transfer to methadone dependence

There are two ways of treating heroin addiction. One is a programme of withdrawal under medical supervision with the support of a counsellor and possibly support from a group of other users who are trying to stop the habit. The other is to substitute methadone. Like the opiates, methadone depresses the central nervous system, but it does not cause drowsiness or euphoria. It can be taken orally. An addict who switches to methadone is able to perform a job and return to a normal way of life. Methadone is addictive: addicts transfer their dependence from heroin to methadone. Since methadone is prescribed by a doctor, users do not have to steal to support their habit. Withdrawal from methadone is as difficult as withdrawal from heroin. Methadone does not treat the original motivation for drug abuse, but it allows addicts to live like normal members of society.

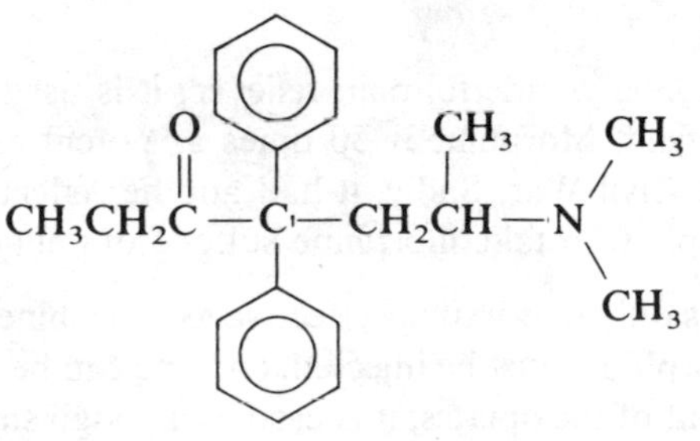

$$CH_3CH_2C(=O)-C(C_6H_5)_2-CH_2CH(CH_3)-N(CH_3)_2$$

Methadone

CHECKPOINT 35C: DRUGS II

1. Refer to the formula of morphine. What is the difference between the —OH group in position (1) and the —OH group in position (2)?
2. How can heroin be made from morphine?
3. How can codeine be made from morphine?
4. How can morphine be made from codeine?
5. What would you say to someone whose life was so empty that he or she was thinking of trying out drugs for a thrill?
6. What would you say to someone who said that marijuana was a safe drug as long as you stick to it and don't switch to more addictive drugs?

BARBITURATES

Barbiturates depress the CNS. They are called 'downers' and are used as sleeping pills

Barbiturates are sometimes called 'downers' because they act in the opposite sense to 'uppers' [amphetamines, see earlier]. They depress the central nervous system. They are prescribed as sleep-inducing drugs, but they can produce dependence. Seconal® (called 'red devils') and Amytal® (called 'blue heavens') are popular among users because they take effect within a few minutes and the effects last for only a few hours.

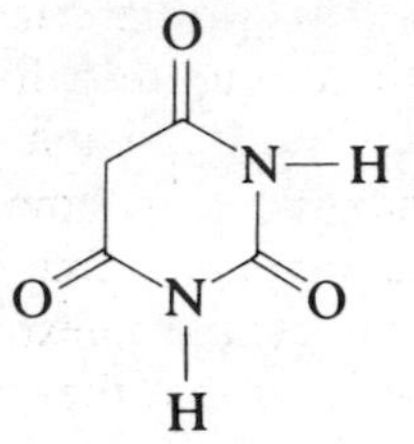

Barbituric acid

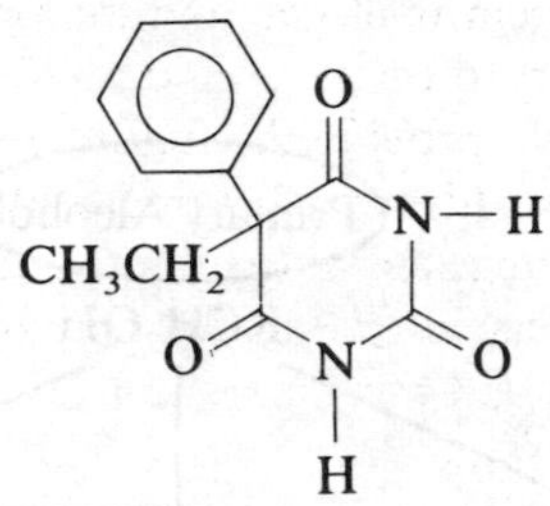

Phenobarbital (a tranquilliser)

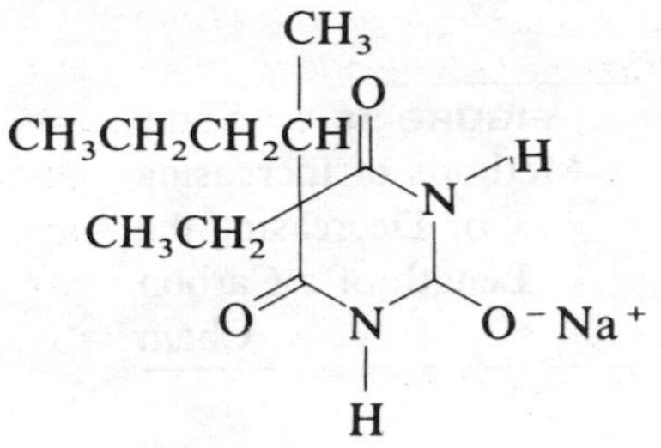

Sodium pentobarbital (Nembutal – sleeping pills)

VALIUM

Tranquillisers such as Valium® are widely prescribed

Valium® (diazepam) is a depressant of the central nervous system, which is prescribed medicinally as a tranquilliser. It is prescribed for many complaints, such as muscular disorders and spasms and to promote sleep. A similar drug called Mogadon® (nitrazepam) is used in sleeping pills. Sometimes a physical dependence develops, but it is usually a psychological need that makes a person start using Valium®.

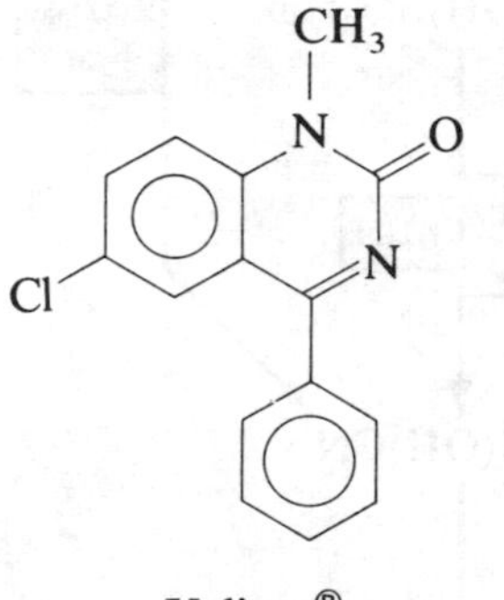

Valium®

ALCOHOL

Alcohol (ethanol) is the most widely used drug. Its effects as a depressant of the central nervous system were described in § 30.5.

DRUG ABUSE

The problem of drug abuse is not confined to a small group of drug addicts. A large percentage of the population takes pills of all kinds to deal with all sorts of problems, both real and imaginary.

Drugs never solve problems

A person can become addicted to a drug if he or she gets into the habit of turning to the drug for relief from every problem, mental, physical and social. If a person is nervous, worried, upset or depressed, drugs are not the cure. It is better to try to cope with realities of life without the assistance of mind-altering and mood-altering chemicals. Once a person tries to solve a problem by escaping through a drug experience, he or she may want to repeat that drug experience every time a similar problem arises. Mood-altering drugs give us a holiday from problems; they do not solve our problems.

35.3 SYNTHETIC ROUTES

FIGURE 35.1 Some Methods of Increasing or Decreasing the Length of a Carbon Chain

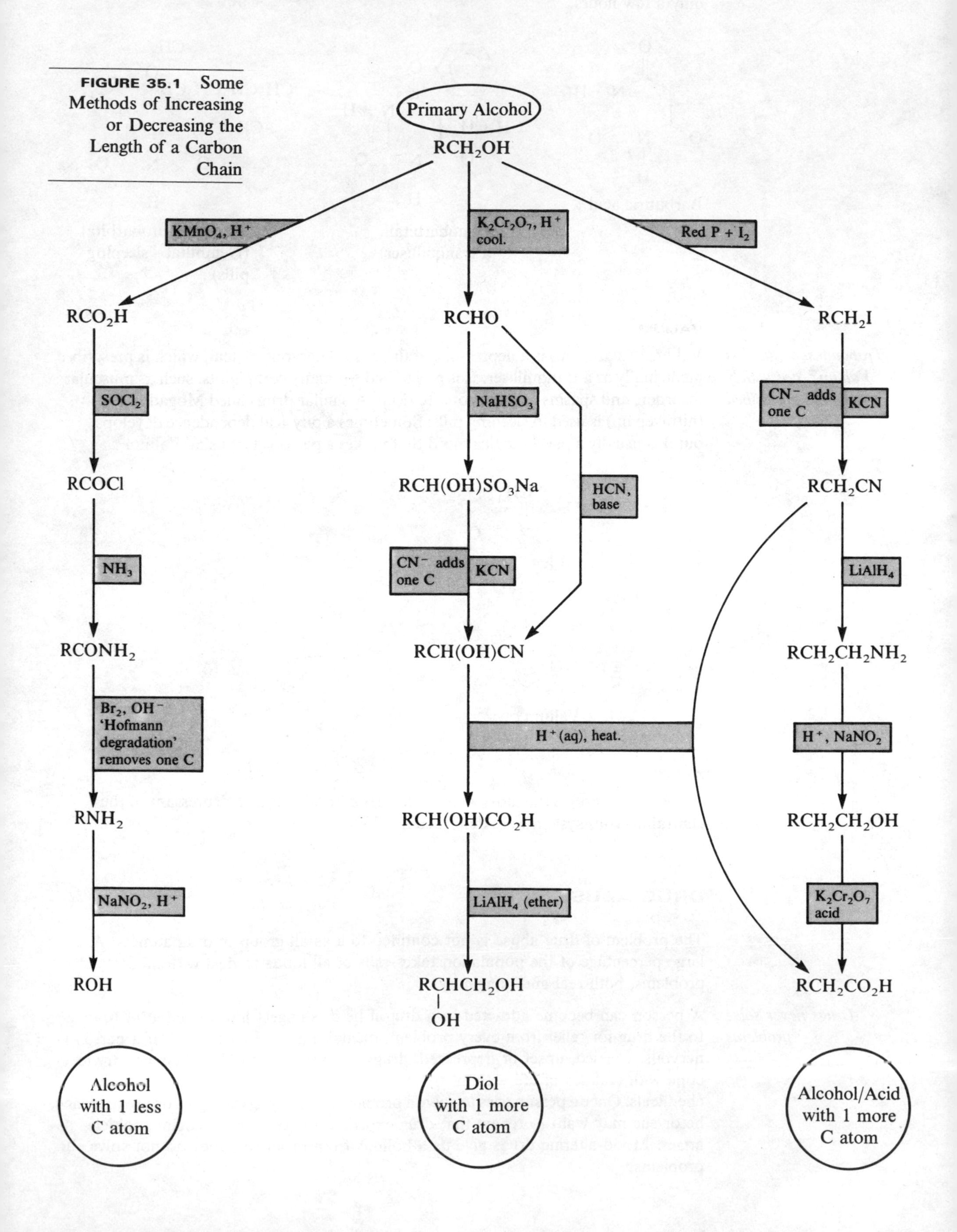

FIGURE 35.2
Synthetic Routes from a Primary Alcohol in Two Steps or More

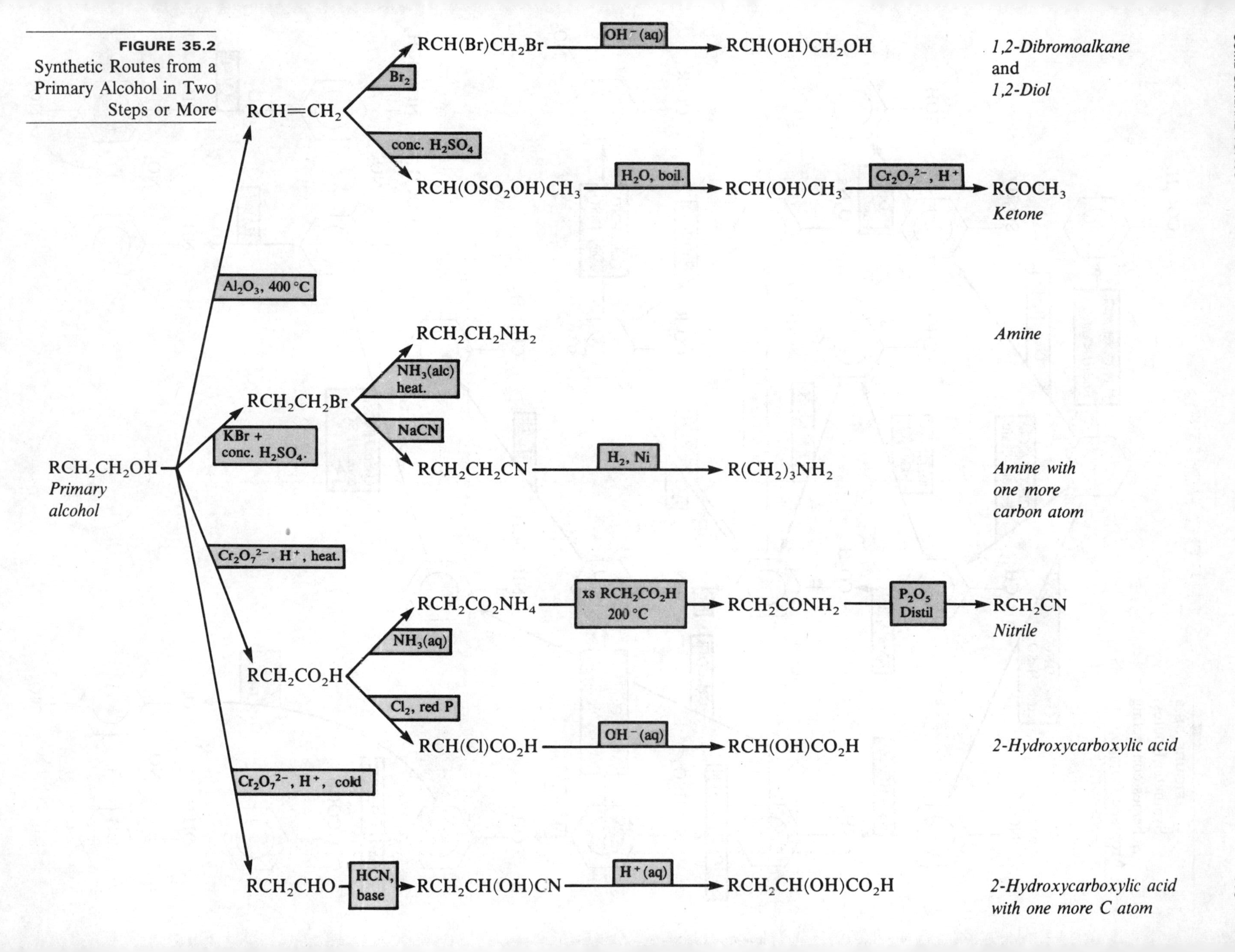

FIGURE 35.3 Synthetic Routes: Schemes connecting Aromatic Compounds

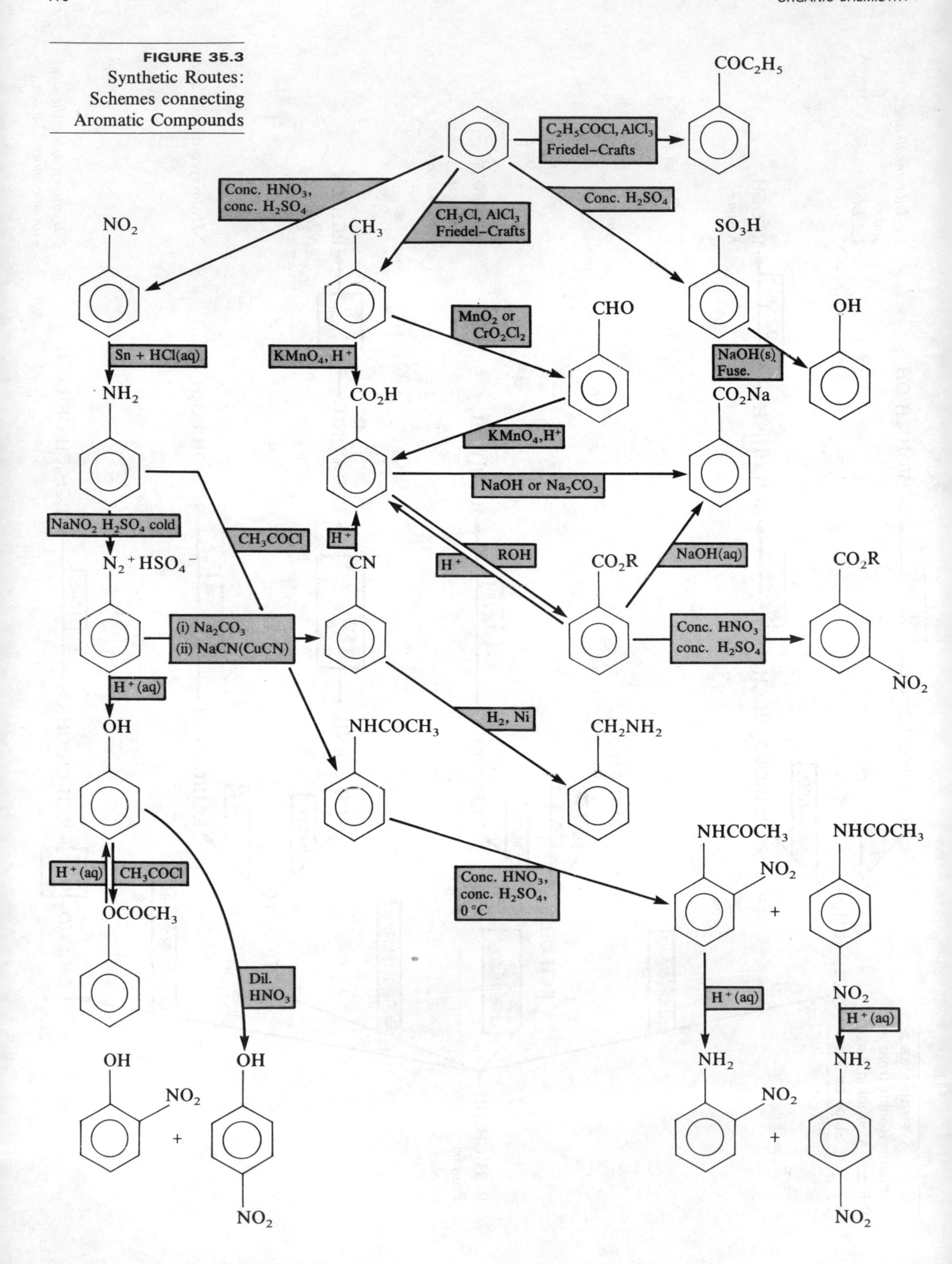

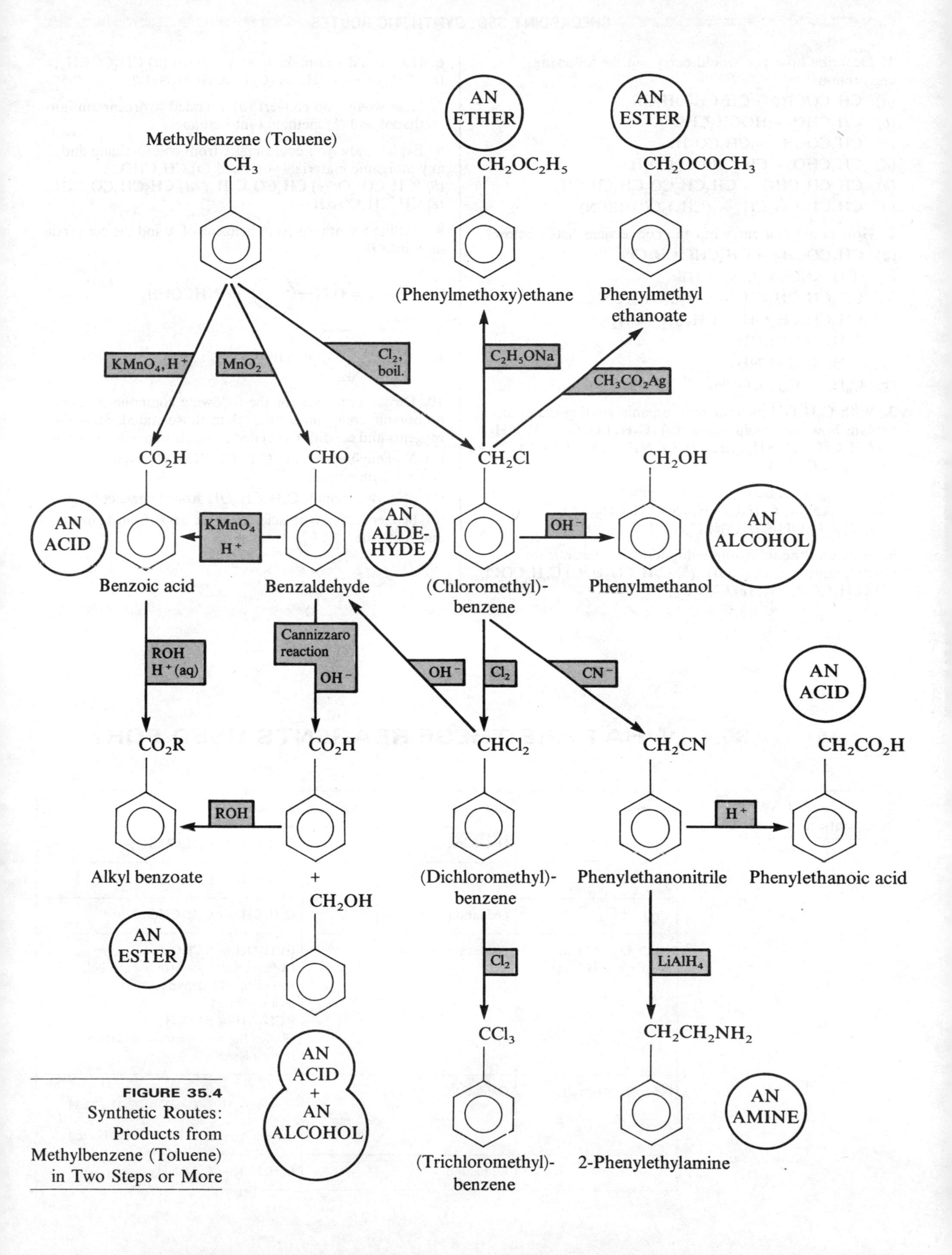

FIGURE 35.4 Synthetic Routes: Products from Methylbenzene (Toluene) in Two Steps or More

CHECKPOINT 35D: SYNTHETIC ROUTES

1. Describe how you would carry out the following conversions:
(*a*) $CH_3COCH_3 \rightarrow CH_3CH(OH)CH_3$
(*b*) $CH_3CHO \rightarrow HOCH_2CHO$
(*c*) $CH_3COCH_3 \rightarrow CH_3CO_2H$
(*d*) $CH_3CHO \rightarrow CH_3CH(OH)CO_2H$
(*e*) $CH_3CH_2CHO \rightarrow CH_3CH_2CO_2CH_2CH_2CH_3$
(*f*) $CH_3CH(OH)CH_3 \rightarrow (CH_3)_2C(OH)(CN)$

2. How could you carry out the conversions listed below?
(*a*) $CH_3COCH_3 \rightarrow CH_3CHClCH_2Cl$
(*b*) $(CH_3)_2CO \rightarrow (CH_3)_2CHBr$
(*c*) $CH_3CH(OH)CH_3 \rightarrow CH_3C(OH)(CN)CH_3$
(*d*) $CH_3CH_2CH_2OH \rightarrow CH_3CH{=}CH_2$
(*e*) $C_2H_6 \rightarrow C_2H_5OH$
(*f*) $C_2H_6 \rightarrow C_2H_5NH_2$
(*g*) $C_6H_6 \rightarrow C_6H_5COCH_3$

3. With C_2H_5OH as your only organic starting-material, explain how you could make: (*a*) $(C_2H_5)_2O$, (*b*) $C_2H_5NH_2$, (*c*) $CH_3CH_2CH_2NH_2$, (*d*) CH_3CONH_2, (*e*) CH_3COCl, (*f*) $C_2H_5CO_2C_2H_5$.

4. Starting from benzene or methylbenzene, say how you could make: (*a*) $C_6H_5CO_2H$, (*b*) $C_6H_5CH_2CO_2H$, (*c*) $C_6H_5CH(OH)CH_3$, (*d*) $C_6H_5CH_2CH_2NH_2$.

5. How can the following compounds be made from phenylamine? (*a*) C_6H_5OH, (*b*) $C_6H_5CO_2H$, (*c*) C_6H_5COCl, (*d*) $C_6H_5CO_2C_6H_5$, (*e*) C_6H_5I.

6. How could you make from propene: (*a*) CH_3COCH_3, (*b*) $(CH_3)_2CHCO_2H$, (*c*) $(CH_3)_2CHCH_2NH_2$?

7. How would you convert (*a*) a named hydrocarbon into methanol and (*b*) methanol into ethanol?

8. Explain how you could make from chloroethane and any inorganic materials you need (*a*) CH_3CHO, (*b*) $(CH_3CO)_2O$, (*c*) $CH_3CO_2C_2H_5$, (*d*) $CH_3CH_2CO_2C_2H_5$, (*e*) $CH_3CH_2CONH_2$.

9. Outline a synthesis from benzene of **A** and the conversion of **A** into **B**:

A = O_2N—⟨benzene ring⟩—$NHCOCH_3$

B = HO—⟨benzene ring⟩—OH

10. Devise syntheses for the following compounds, using no organic compounds other than those stated. State the reagents and conditions needed for each step in the synthesis:
(*a*) *N*-phenylbenzamide, $C_6H_5CONHC_6H_5$ from benzene and methylbenzene,
(*b*) phenylmethanol, $C_6H_5CH_2OH$ from benzene,
(*c*) benzenecarboxylic acid (benzoic acid) from benzene and methane.

35.4 WHAT ARE THESE REAGENTS USED FOR?

TABLE 35.1
Reagents and their Uses

Reagent	*Use*	*Example*
$KMnO_4$, $H^+(aq)$	Oxidation	$C_6H_5CH_3 \rightarrow C_6H_5CO_2H$
MnO_2	Oxidation	$C_6H_5CH_3 \rightarrow C_6H_5CHO$
CrO_3	Oxidation	$C_6H_5CH_3 \rightarrow C_6H_5CHO$
$K_2Cr_2O_7$, $H^+(aq)$ $Na_2Cr_2O_7$, $H^+(aq)$	Oxidation	$RCH_2OH \rightarrow RCHO$ (Add oxidant slowly to hot alcohol, and distil off aldehyde as fast as it is formed.) $RCH_2OH \rightarrow RCO_2H$ (Reflux the mixture until oxidation is complete.)
$KMnO_4$ (neutral)	Oxidation	Reflux: $C_6H_5\text{—}\overset{\vert}{\underset{\vert}{C}}\text{—}X \rightarrow C_6H_5CO_2H$ (where X = $\text{—}CH_3$, $\text{—}CH_2OH$ etc.)
H_2, Pt	Reduction	$C_6H_5NO_2 \rightarrow C_6H_5NH_2$

Reagent	*Use*	*Example*
$NaBH_4(aq)$	Reduction of C=O, not C=C	$RCHO \rightarrow RCH_2OH$ $R_2C{=}O \rightarrow R_2CHOH$
Fe + HCl(aq)	Reduction	$C_6H_5NO_2 \rightarrow C_6H_5NH_2$
Sn + HCl(aq)	Reduction	$C_6H_5NO_2 \rightarrow C_6H_5NH_2$
H_2, Ni	Reduction of C=C, not C=O	$>C{=}CH{-}C(R){=}O \rightarrow >CH{-}CH_2{-}C(R){=}O$ $-C{\equiv}C- \rightarrow -CH_2CH_2-$ $RCN \rightarrow RCH_2NH_2$ Hydrogenation of oils to form fats
$LiAlH_4$ (ether)	Reduction of C=O, not C=C	$RCO_2H \rightarrow RCH_2OH$ $RCO_2R' \rightarrow RCH_2OH$ $RCl \rightarrow RH$ $RCONH_2 \rightarrow RCH_2NH_2$
O_3 followed by $Zn + CH_3CO_2H(aq)$	Ozonolysis to determine position of C=C bond	$R{-}CH{=}CH{-}R' \rightarrow RCHO + R'CHO$ $RCH{=}CH_2 \rightarrow RCHO + HCHO$
$Br_2(CCl_4)$	Test for unsaturation	$RCH{=}CHR' \rightarrow RCH(Br){-}CHR'(Br)$ (No HBr evolved)
$KMnO_4$, $OH^-(aq)$ dilute, 0 °C	Test for unsaturation	$RCH{=}CHR' \rightarrow RCH(OH){-}CHR'(OH)$
NaOH, I_2	Iodoform test for CH_3CO- or $CH_3CH(OH)-$	$CH_3CHO + I_2 + NaOH \rightarrow CHI_3$
$NaNO_2$, $H^+(aq)$, $<10\,°C$	Diazotisation	$C_6H_5NH_2 \rightarrow C_6H_5\overset{+}{N}{\equiv}N$
Al_2O_3	Dehydration	$C_2H_5OH \rightarrow CH_2{=}CH_2$
P_2O_5	Dehydration	$RCONH_2 \rightarrow RCN$
Hot conc. H_2SO_4	Dehydration	$RCH(OH)CH_3 \rightarrow RCH{=}CH_2$
Conc. H_2SO_4	Esterification	$ROH + R'CO_2H \rightarrow R'CO_2R + H_2O$
Cold conc. H_2SO_4 then H_2O	Hydration	$RCH{=}CHR' \rightarrow RCH(H){-}CHR'(OH)$
H_3PO_4, then H_2O	Hydration	$RCH{=}CHR' \rightarrow RCH(H){-}CHR'(OH)$
$H_2SO_4(aq)$, $HgSO_4$	Hydration of an alkyne → ketone	$R{-}C{\equiv}CH \rightarrow RCOCH_3$
$FeCl_3$, $FeBr_3$	Halogen carriers	$C_6H_6 + Cl_2 \xrightarrow{FeCl_3} C_6H_5Cl + HCl$
$AlCl_3$	Friedel–Crafts catalyst	$C_6H_6 + RCl \xrightarrow{AlCl_3} C_6H_5R + HCl$ $C_6H_6 + RCOCl \xrightarrow{AlCl_3} C_6H_5COR + HCl$
Conc. HNO_3 + conc. H_2SO_4	Nitration	$C_6H_6 + NO_2^+ \rightarrow C_6H_5NO_2$

Reagent	*Use*	*Example*
SO_3 in conc. H_2SO_4	Sulphonation	$C_6H_6 \rightarrow C_6H_5SO_3H$
Conc. H_2SO_4	Sulphonation of phenols and aromatic amines	$C_6H_5OH \rightarrow HOC_6H_4SO_3H$ $C_6H_5NH_2 \rightarrow H_2NC_6H_4SO_3H$
'Soda lime'	Decarboxylation	$RCO_2H \rightarrow RH$
OH^- (ethanol)	Elimination of hydrogen halide	$RCH_2CH_2X \rightarrow RCH{=}CH_2$
OH^-(aq)	Hydrolysis	$RCO_2R' \rightarrow RCO_2^-(aq) + R'OH$ $RCN \rightarrow RCO_2^-(aq)$ $RCONH_2 \rightarrow RCO_2^-(aq) + NH_3(g)$
C_6H_5COCl, NaOH	Benzoylation	$C_6H_5OH \rightarrow C_6H_5OCOC_6H_5$
$C_6H_5COOOCOC_6H_5$ (Dibenzoyl peroxide)	Initiating chain reactions	Chlorination of alkanes
$SOCl_2$ (Sulphur dichloride oxide)	Replacement of —OH by —Cl	$RCH_2OH \rightarrow RCH_2Cl$
H^+(aq)	Hydrolysis	RCO_2R', $RCONH_2$, $RCN \rightarrow RCO_2H$
2,4-Dinitrophenylhydrazine (DNP)	Test for a C=O group	Gives an orange ppt. with an aldehyde or ketone
Fehling's solution (Cu^{2+} complex ions)	Test for an aliphatic aldehyde	Gives a reddish ppt. of Cu_2O with aliphatic aldehydes, but not with aromatic aldehydes or with ketones
Tollens' reagent ($Ag(NH_3)_2^+$(aq))	Test for an aldehyde	Reduced to a silver mirror by aliphatic aldehydes and, when warmed, by aromatic aldehydes, but not by ketones

CHECKPOINT 35E: REAGENTS

1. Compare the reactions of bromine with the following compounds. State the products of the reactions and the conditions necessary for reaction.
(*a*) methane, (*b*) benzene, (*c*) phenol, (*d*) ethene, (*e*) ethanamide.

2. Explain the following terms. Give an example of each type of reaction, indicating the reactants and the conditions needed for reaction.
(*a*) reduction, (*b*) ozonolysis, (*c*) alkylation, (*d*) acylation, (*e*) decarboxylation.

3. Give an example of each of the following types of reaction. Indicate the necessary conditions.
(*a*) nitration, (*b*) sulphonation, (*c*) oxidation, (*d*) cracking, (*e*) the halogenation of an aromatic ring.

4. Give examples of the reduction of organic compounds by (*a*) hydrogen and a catalyst, (*b*) zinc and hydrochloric acid, (*c*) lithium tetrahydridoaluminate, (*d*) sodium tetrahydridoborate.

5. Illustrate the use of the following reagents in organic chemistry. State the conditions necessary for reaction, and give equations.
(*a*) bromine, (*b*) aluminium chloride, (*c*) sodium nitrite, (*d*) hydrogen cyanide, (*e*) nickel, (*f*) alkaline potassium manganate(VII), (*g*) ozone.

6. State the products of the reaction between sodium hydroxide and each of the following compounds. For each reaction, state the necessary conditions, and write the equation.
(*a*) ethanal, (*b*) benzenecarbaldehyde (benzaldehyde), (*c*) ethyl ethanoate, (*d*) 1-bromobutane, (*e*) 1,1,1-trichloropropanone, CCl_3COCH_3.

7. The following are well-known reagents. State what use each finds in organic chemistry. Describe the bonding in each compound. Explain how the type of bonding enables the reagent to function as it does.
(*a*) $AlCl_3$, (*b*) HCN, (*c*) KOH(aq), (*d*) $LiAlH_4$.

8. You are supplied with unlabelled samples of six organic compounds. You are also provided with the following reagents:

1. 2,4-dinitrophenylhydrazine solution
2. ammoniacal silver nitrate solution
3. red litmus paper
4. dilute sodium hydroxide solution
5. dilute hydrochloric acid solution

Complete a copy of the table, indicating initially the results you would expect to *observe* on interacting each of the organic compounds separately with reagents 1 and 2. Then devise a scheme of three further tests on each compound using *only* the other reagents listed above, so that you would be able to identify all six organic compounds. Any test which you suggest *must* lead to some *observable* result. In the spaces marked *, indicate the reagent(s) and conditions used.

Unlabelled compound		*Reagent 1*	*Reagent 2*	*	*	*
Name	*Formula*					
Propanal	C_2H_5CHO					
Propanone	CH_3COCH_3					
Methyl benzoate	$C_6H_5CO_2CH_3$					
Ethanamide	CH_3CONH_2					
Ethanoic acid	CH_3CO_2H					
1-Aminopropane	$CH_3CH_2CH_2NH_2$					

(WJEC 91)

35.5 HOW WOULD YOU DISTINGUISH BETWEEN THE MEMBERS OF THE FOLLOWING PAIRS OF COMPOUNDS?

CHECKPOINT 35F: PAIRS OF COMPOUNDS

1.
A CH_3COCH_3
B C_2H_5OH

2.
A CH_3COCH_3
B C_2H_5CHO

3.
A $C_2H_5COCH_3$
B $C_2H_5CO_2H$

4.
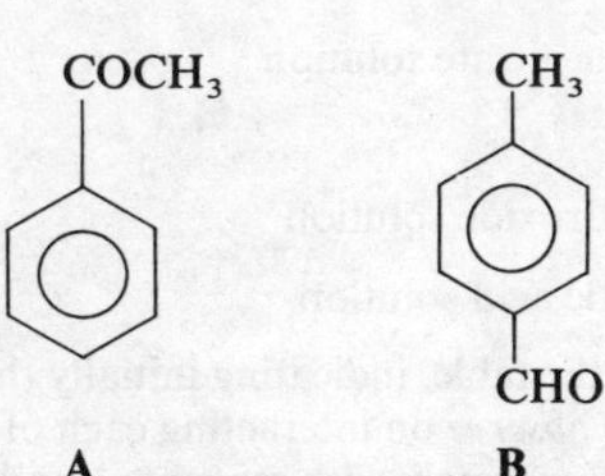

5.
A CH_3CHO
B $CH_3CO_2CH_3$

6.
A $CH_3CH_2CH_2OH$
B $CH_3CH(OH)CH_3$

7.
A $(CH_3)_3COH$
B $(CH_3)_2CHCH_2OH$

8. 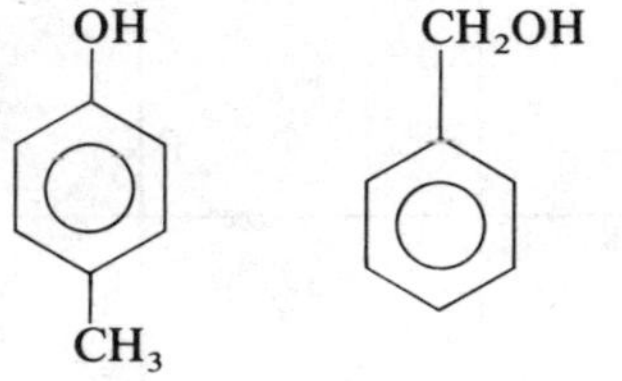
OH CH$_2$OH
CH$_3$
A **B**

9. 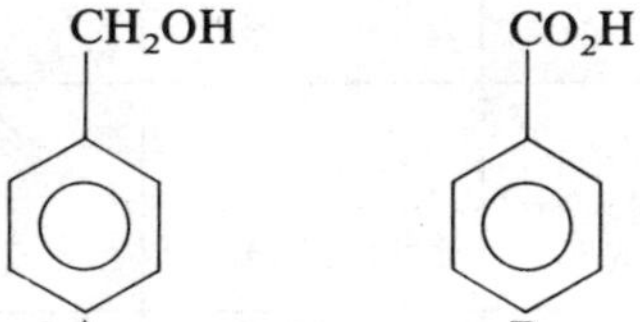
CH$_2$OH CO$_2$H
A **B**

10. 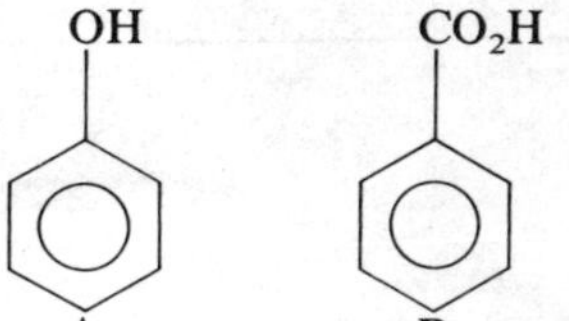
OH CO$_2$H
A **B**

11.
A CH_3CHO
B CH_3CO_2H

12.
A CH_3COCH_3
B $C_2H_5CO_2H$

13.
A CH_3CO_2H
B HCO_2H

14.
A CO_2Na–CO_2Na
B CH_3CO_2Na

15.
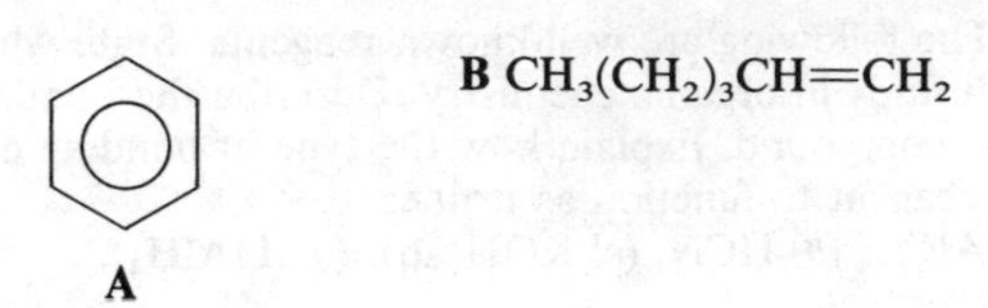

16.
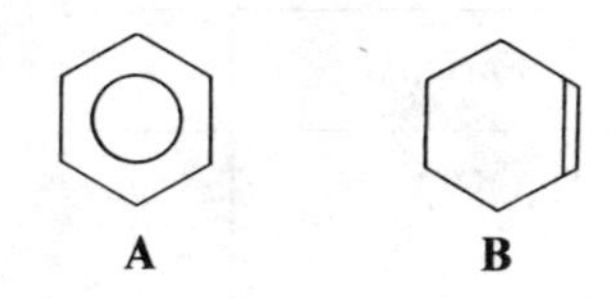

17.
A $C_6H_5CH{=}CH_2$
B C_6H_5OH

18.
A CH_4
B CH_3Cl

19.
A CH_3COCl
B C_6H_5COCl

20.
A $ClCH_2CO_2H$
B CH_3COCl

21.
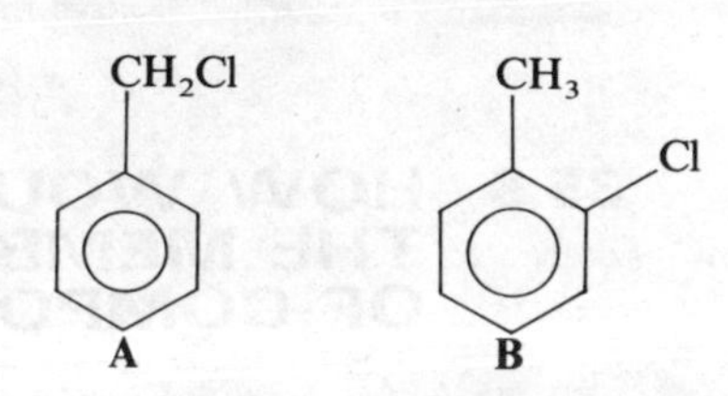

22.
A CH_3CH_2I
B CH_3CH_2Br

23.

A $HCO_2C_2H_5$

B $CH_3CO_2CH_3$

24.

A $CH_3CO_2C_6H_5$

B $C_6H_5CO_2CH_3$

25.

A $ClCH_2CO_2H$

B CH_3CO_2H

26.

A $CH_3CH_2NH_2$

B CH_3CN

27.

A $CH_3CH_2NH_2$

B $(CH_3)_2NH$

28.

A CH_3CONH_2

B $CH_3CH_2NH_2$

29.

A CH_3CONH_2

B $CO(NH_2)_2$

30.

A $C_2H_5CONH_2$

B $C_6H_5NH_2$

31.

A $C_3H_7NH_3{}^+Cl^-$

B $C_3H_7CO_2{}^-Na^+$

32.

A $CH_3CO_2NH_4$

B CH_3CONH_2

33.

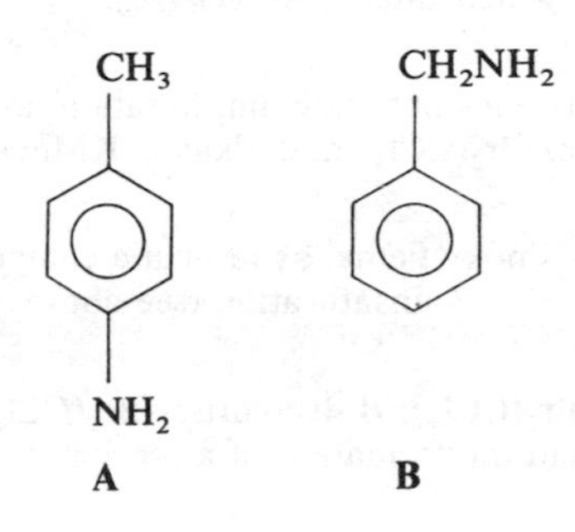

34.

A C_2H_5OH

B $(CH_3)_3COH$

35.

A CH_3CH_2COCl

B $CH_3CH_2CH_2Cl$

ANSWERS TO CHECKPOINT 35F: PAIRS OF COMPOUNDS

1. A, a ketone, reacts with 2,4-dinitrophenylhydrazine, DNP, to give an orange-coloured precipitate of the 2,4-dinitrophenylhydrazone.
B, an alcohol, reacts with $CH_3CO_2H(l)$ + conc. H_2SO_4 to give a fruity, ester smell of $CH_3CO_2C_2H_5$.

2. Both react with DNP. **A**, a ketone, is not easily oxidised. **B**, an aldehyde, reduces Tollens' reagent to silver and Fehling's solution to red Cu_2O.

3. A is a ketone and reacts with DNP. **B** is an acid, and liberates CO_2 from $NaHCO_3$.

4. A has no reducing action; **B** reduces Tollens' reagent but not Fehling's solution.

5. A + NaOH(aq, conc.) forms a brown resin, $(CH_3CHO)_n$, with a characteristic smell.
B + NaOH(aq) when warmed give $CH_3CO_2Na + CH_3OH$. On addition of HCl(aq), CH_3CO_2H, with a characteristic smell, is formed.

6. A, a primary alcohol, is oxidised by acidified $Na_2Cr_2O_7$ in the cold to an aldehyde. (Test as in 2.)
B, a secondary alcohol, is oxidised to a ketone. (Test as in 2.)

7. A, a tertiary alcohol, is oxidised by acid $K_2Cr_2O_7$ to a mixture of products, including an acid.
B, a primary alcohol, is oxidised to an aldehyde. (Test with DNP.)

8. A is a phenol. Its solution is acidic, reacting with NaOH(aq) to give the phenoxide, $H_3CC_6H_4O^-Na^+$. **A** gives a purple colour with $FeCl_3$(aq) and a white precipitate with Br_2(aq).
B is an alcohol, neutral in solution. It does not react with NaOH(aq), but does react with Na(s) to give H_2(g) and the alkoxide, $C_6H_5CH_2O^-Na^+$.

9. A is an alcohol, neutral in solution.
B is an acid: its solution is acidic, and it reacts with Mg to give H_2(g) and with $NaHCO_3$ to give CO_2(g).

10. A is weakly acidic: it dissolves in NaOH(aq), but does not give CO_2(g) with $NaHCO_3$. **B** is an acid, and liberates CO_2(g) from $NaHCO_3$.
A gives a purple colour with $FeCl_3$(aq) and a white ppt. with Br_2(aq).

11. A + $AgNO_3$(aq) give a silver mirror when warmed.
B smells of vinegar and liberates CO_2 from $NaHCO_3$.

12. A with DNP gives an orange crystalline ppt.
B + C_2H_5OH + conc. H_2SO_4 gives, when warmed, an ester with a characteristic, fruity smell. **B** liberates $CO_2(g)$ from $NaHCO_3$.

13. A smells of vinegar, **B** of formalin.
When warmed with $AgNO_3$, $NH_3(aq)$, **B** gives a silver mirror. Conc. H_2SO_4 + **B** give CO(g), which burns with a blue flame.
B + Fehling's solution give a red ppt. of Cu_2O.

14. Conc. H_2SO_4 reacts with **A** to give CO_2 (limewater test) and CO (which burns with a blue flame), and with **B** to give CH_3CO_2H (which smells like vinegar).

15. A is benzene. It does not show unsaturation, as does **B**, which decolorises $Br_2(CCl_4)$ and alkaline $KMnO_4(aq)$.

16. A burns with a smoky flame, as aromatic compounds do. **B**, cyclohexene, shows unsaturation (see above).

17. A decolorises $Br_2(CCl_4)$; **B** decolorises $Br_2(CCl_4)$ with evolution of HBr and the formation of a precipitate.

18. A is unreactive. **B** is hydrolysed by alkali to form methanol (Test for alcohol) and Cl^- ions (Test with $AgNO_3$, $HNO_3(aq)$).

19. In water, **A** immediately forms CH_3CO_2H and HCl. A strongly acidic solution is formed. (Test with $NaHCO_3$).
B is hydrolysed when warmed with NaOH(aq) to sodium benzoate. On acidification with HCl(aq), solid benzoic acid comes out of solution.

20. A, chloroethanoic acid, is a stronger acid than ethanoic acid. (Test: Mg; $NaHCO_3$).
B is not acidic, but it reacts with water to form a strongly acidic solution of CH_3CO_2H + HCl. Addition of $AgNO_3$, $HNO_3(aq)$ will detect Cl^- ions, which are not present in a solution of **A**.

21. A —Cl in the side chain, as in **A**, is hydrolysed off by warming **A** with alkali, to give an alcohol (Test by oxidising it to an aldehyde), and Cl^- ions (Test with $AgNO_3$, $HNO_3(aq)$).
The —Cl in the ring in **B** cannot be removed in this way.

22. On hydrolysis with NaOH(aq), **A** gives I^- ions, and **B** gives Br^- ions. $Pb(NO_3)_2(aq)$ gives a yellow ppt. of PbI_2 with the first, and a white ppt. of $PbBr_2$ with the second.
With **A**, $CH_3CO_2Ag(aq)$ gives a yellow ppt. of AgI; with **B** it gives a pale yellow ppt. of AgBr.

23. A is hydrolysed when warmed with $OH^-(aq)$ to HCO_2H + C_2H_5OH. Ethanol gives the iodoform test: with I_2 + $OH^-(aq)$, a ppt. of CHI_3 is formed.
B is hydrolysed to CH_3CO_2H (which smells of vinegar) and CH_3OH. On oxidation, CH_3OH gives methanal, HCHO, with the smell characteristic of formalin.

24. A is an ester which is hydrolysed to CH_3CO_2H and C_6H_5OH. Phenol in alkaline solution couples with a diazonium salt to give a dye.
B is hydrolysed to $C_6H_5CO_2H$ + CH_3OH. Benzoic acid liberates CO_2 from $NaHCO_3$.

25. A is a stronger acid than **B** (e.g. in reaction with $NaHCO_3$). **A** is hydrolysed by $OH^-(aq)$ to give $Cl^-(aq)$; tested with $AgNO_3$ (aq).

26. A is an amine, with a characteristic smell. Its aqueous solution is alkaline. With acids, it forms salts, which are crystalline solids. With nitrous acid, **A** gives C_2H_5OH + $N_2(g)$.
B is a nitrile, hydrolysed by $H^+(aq)$ to CH_3CO_2H (Test for carboxylic acid).

27. A is a primary amine. With nitrous acid, it gives C_2H_5OH + $N_2(g)$.
B is a secondary amine. With nitrous acid, it gives an oily nitrosoamine.

28. A is an amide. It is hydrolysed by NaOH(aq) to give $NH_3(g)$ (which is basic) and $CH_3CO_2Na(aq)$.
B is an amine, and reacts with nitrous acid to give $N_2(g)$.

29. A, an amide, is hydrolysed by $OH^-(aq)$ to $NH_3(g)$.
B, urea, is hydrolysed by $OH^-(aq)$ to $NH_3(g)$ and by $H^+(aq)$ to $CO_2(g)$.

30. A + $Br_2(aq)$ + $OH^-(aq)$ give an amine, which is basic and has a 'fishy' smell.
B + $Br_2(aq)$ gives a white ppt. of $C_6H_2Br_3NH_2$.

31. A solution of **A** in water gives a white ppt. of AgCl on treatment with $AgNO_3$, $HNO_3(aq)$. Addition of alkali to **A** liberates the free amine, $C_3H_7NH_2$, which has the characteristic smell.
B gives a yellow flame test. Addition of HCl(aq) to **B** liberates butanoic acid, $C_3H_7CO_2H$, with a rancid smell.

32. With cold NaOH(aq), **A** gives $NH_3(g)$.
When strongly heated with NaOH(aq), **B** gives $NH_3(g)$.

33. A is an aryl amine, which can be diazotised with cold $NaNO_2$ + HCl(aq). The diazonium salt formed gives, when warmed, $N_2(g)$ and a smell of phenol. It gives a dye when coupled with phenol in alkaline solution. **B** reacts with $NaNO_2$ + HCl(aq) to give $N_2(g)$.

34. A gives a yellow precipitate of iodoform with I_2 + OH (aq). **B** is not easily oxidised.

35. A is a fuming lachrimatory liquid easily hydrolysed to propanoic acid + HCl. **B** is a volatile liquid which is hydrolysed only by refluxing with alkali.

35.6 QUESTIONS ON SOME TOPICS WHICH SPAN CHAPTERS

CHECKPOINT 35G: QUESTIONS ON FUNCTIONAL GROUPS AND REACTION MECHANISMS

1. Explain these statements:
(*a*) Water and petrol do not mix.
(*b*) Halogenoalkanes are more reactive than alkanes towards nucleophilic reagents.
(*c*) Alkenes, unlike alkanes, react readily with bromine water.
(*d*) Phenol, unlike benzene, reacts readily with bromine water.
(*e*) Ethanoyl chloride is more reactive towards water than is chloroethane.

2. Compare and contrast the reactions of:
(*a*) sulphuric acid with ethanol and phenol,
(*b*) phosphorus(V) chloride with ethanol and phenol,
(*c*) sodium hydroxide with 1-chlorohexane and chlorobenzene,
(*d*) nitrous acid with ethylamine and phenylamine.

3. Describe the mechanisms of the reactions between the following substances:
(*a*) ethene and bromine,
(*b*) methane and chlorine in sunlight,
(*c*) benzene and chlorine in the presence of iron filings,
(*d*) benzene and ethanoyl chloride in the presence of aluminium chloride,
(*e*) 1-iodopropane and aqueous sodium hydroxide,
(*f*) propanal and hydrogen cyanide.

4. Explain, by means of examples, what is meant by the following terms:
(*a*) homolytic fission,
(*b*) heterolytic fission,
(*c*) nucleophilic substitution,
(*d*) an addition–elimination reaction,
(*e*) reductive ozonolysis,
(*f*) catalytic cracking.

5. Explain the following observations:
(*a*) Ethene does not react with water, but, when ethene is passed into concentrated sulphuric acid and water is added to the solution, ethanol is formed.
(*b*) Hydrogen cyanide adds to propanone in the presence of bases but not in the presence of acids.
(*c*) Both propanone and ethanol give yellow precipitates when treated with iodine and alkali. Neither pentan-3-one nor pentan-3-ol gives a positive result in this test.

6. Carboxylic acids, acid halides, amides, esters, aldehydes and ketones all contain the group

$$\gt C{=}O$$

Point out the similarities in the behaviour of these groups of compounds which arise from the possession of a carbonyl group. Why do carboxylic acids and their derivatives not show the addition reactions of aldehydes and ketones with nucleophiles?

7. Adrenalin has the formula shown below. It is a water-soluble hormone:

$$(HO)_2C_6H_3{-}CH(OH){-}CH_2{-}NH{-}CH_3$$

(*a*) Name the functional groups in the molecule.
(*b*) Describe how you could test for the presence of each of these groups.
(*c*) Where does the optical activity of adrenalin arise?

8. Discuss the way in which the properties of the groups (*a*) —Cl and (*b*) $-NH_2$ are influenced by the remainder of the molecule in which they are present.

9.

$$H_2N{-}C_6H_4{-}CH{=}CH{-}CHBr{-}CO_2H$$

Imagine that you have synthesised the compound which has the formula shown above. Describe how you could test for each of the functional groups, and say what the results of these tests would be if the compound you have made is the correct substance.

Say what further tests you could do to confirm that your product is indeed the substance you intended to make.

10. (*a*) For each of the following pairs of compounds, describe a single chemical test which would enable you to distinguish one member of the pair from the other:
(i) propanone and propanal
(ii) propanoyl chloride and 1-chloropropane
(iii) phenol and benzenecarboxylic (benzoic) acid
(iv) ethanamide and ethylamine

(*b*) Giving necessary reagents and conditions, outline how the following conversions might be accomplished in the laboratory:
(i) ethanal, CH_3CHO, to 2-hydroxypropanoic acid, $CH_3CH(OH)CO_2H$
(ii) ethanoic acid, CH_3CO_2H, to *N*-phenylethanamide, $CH_3CONHC_6H_5$
(iii) methylbenzene, $C_6H_5CH_3$, to phenylmethanol, $C_6H_5CH_2OH$
(iv) phenylamine, $C_6H_5NH_2$, to iodobenzene, C_6H_5I

(AEB 90)

11. For *five* of the following reactions, discuss whether you think the yield would be good, fair or nil. Explain, using mechanistic arguments, the reasons for your choice.
(i) $C_2H_5F + NH_3 \rightarrow C_2H_5NH_2 + HF$

(ii) $(CH_3)_3CBr + KCN \rightarrow (CH_3)_3CCN + KBr$

(iii) $CH_2{=}CH_2 + H_2S \rightarrow CH_3CH_2SH$

(iv) $2C_2H_5Br + 2Na \rightarrow C_4H_{10} + 2NaBr$

(v) $C_2H_5Br + CH_3CO_2Ag \rightarrow CH_3CO_2C_2H_5 + AgBr$

(vi) $C_6H_6 + I{-}Cl \rightarrow C_6H_5Cl + HI$

(C 91, S)

12. (*a*) Suggest reagents and conditions by which you could carry out the following conversions:

(i) $CH_3CH_2CH_2OH$ into $CH_3CHOHCH_3$

(ii) C_6H_6 into C_6H_5CN

(iii) $C_6H_5COCH_3$ into $C_6H_5CO_2CH_3$

(*b*) Describe how to obtain a pure sample of each component from a mixture of benzenecarboxylic acid and phenol.

(AEB 92, S)

CHECKPOINT 35H: DRAW YOUR OWN CONCLUSIONS

1.

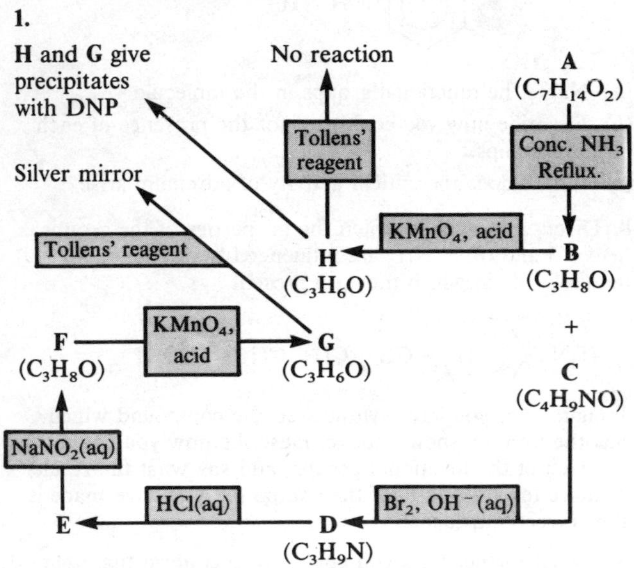

Identify the coumpounds **A** to **H**. Explain the reactions depicted.

2.

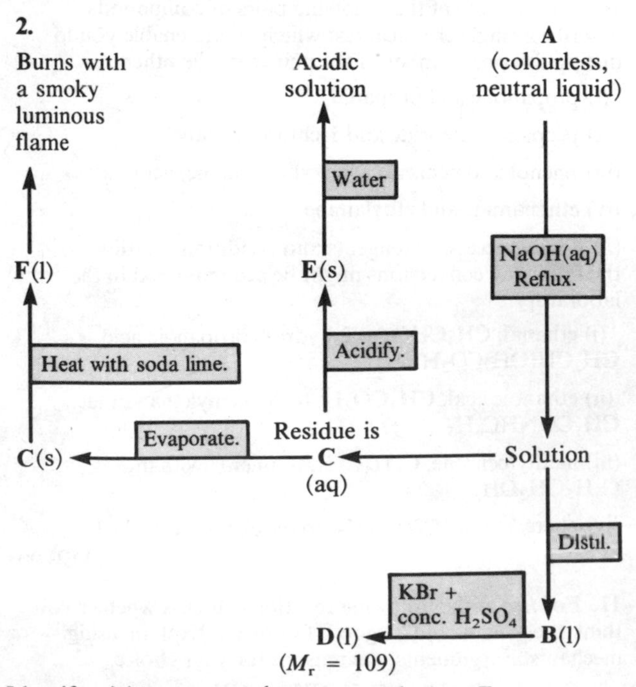

Identify, giving reasons, the compounds **A** to **F**.

3. The isomers **A** and **B** have the molecular formula $C_9H_8O_2$. They are insoluble in water but soluble in aqueous sodium carbonate. Both **A** and **B** are reduced by hydrogen in the presence of a platinum catalyst to **C**, $C_9H_{10}O_2$. Benzoic acid is formed when **A** or **B** or **C** is oxidised by alkaline potassium manganate(VII), and the solution is later acidified. **C** reacts with phosphorus(V) chloride to give **D**, and reacts with sodium carbonate solution to give carbon dioxide.

Identify the compounds **A** to **D**, giving reasons, and explaining the reactions mentioned.

4. P is a crystalline solid which melts at 200 °C. It dissolves in water to give a neutral solution. **P** forms salts with both acids and bases. When heated with soda lime, it gives **Q**, a pungent-smelling gas. **Q** burns in air and dissolves in water to give an alkaline solution. When treated with nitrous acid, **P** gives nitrogen and an acid, **R**, with a molar mass of $76\,g\,mol^{-1}$. Identify, with reasons, **P**, **Q** and **R**.

5. Identify compounds **A** to **G**. Explain the reactions mentioned.

A is a crystalline solid which is very soluble in water and gives a yellow flame test. When **A** is heated with ethanoyl chloride, **B** distils. **B** is a neutral liquid with a relative molar mass of 102. **B** reacts with water to give an acidic solution.

C is a colourless liquid which is sparingly soluble in water but dissolves readily in hydrochloric acid. Evaporation of the solution yields **D**, a crystalline solid. **D** reacts with alkali to give **C**. A solution of **C** in hydrochloric acid reacts with a cold solution of sodium nitrite, followed by an alkaline solution of phenol to give an orange dye. When **C** is treated with bromine water, it gives a white precipitate of **E**, which has a relative molar mass of 330.

F is a colourless solid with a distinctive smell. It dissolves in water to give a weakly acidic solution, which does not liberate carbon dioxide from a solution of sodium hydrogencarbonate. When an alkaline solution of **F** is shaken with benzoyl chloride, a solid, **G**, of molar mass $198\,g\,mol^{-1}$, is formed.

6. W is a colourless aliphatic liquid with a relative molar mass of 123. **W** is insoluble in water. When refluxed with aqueous sodium hydroxide, it gives a solution of **X** and **Y**.

A solution of **X** gives a creamy yellow precipitate with silver nitrate solution.

Y can be distilled from the solution. It gives a positive iodoform test and is oxidised by chromic acid to **Z**. **Z** gives a positive iodoform test, and reacts with 2,4-dinitrophenylhydrazine but not with ammoniacal silver nitrate.

Identify **W**, **X**, **Y** and **Z**. Write equations for the reactions involved.

7. **A** ($C_7H_7NO_2$) is reduced by tin and concentrated hydrochloric acid, followed by alkali, to **B** (C_7H_9N).
B is converted by sodium nitrite and dilute hydrochloric acid into **C** (C_7H_8O).
C is converted by (i) sodium, (ii) iodomethane into **D** ($C_8H_{10}O$).
D is oxidised by acidified dichromate to **E** ($C_8H_8O_3$).
On treatment with concentrated hydriodic acid, **E** gives **F** ($C_7H_6O_3$).
When heated with soda lime and acidified, **F** gives **G** (C_6H_6O).
B dissolves in acid; **E**, **F** and **G** dissolve in alkali; **C** and **G** give a violet colour with iron(III) chloride.

Give the names and structural formulae for **A** to **G**. Explain what happens in each of the reactions mentioned.

8.

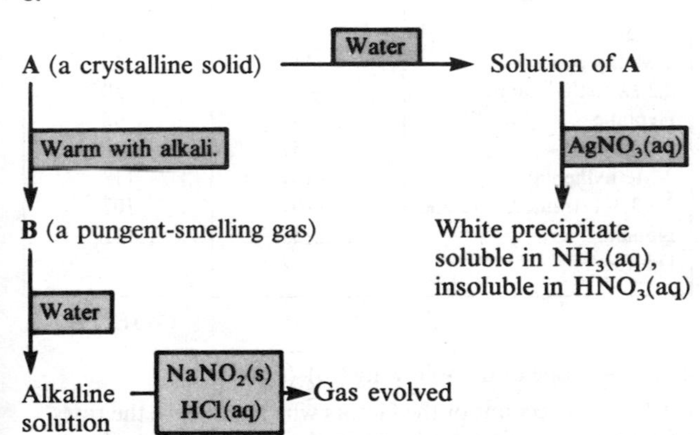

Name two compounds which could be **A** and **B**. Explain the reactions depicted above.

9. Refer to the formula of thyroxine in § 20.11.
(*a*) How could you demonstrate the presence of an —OH group in thyroxine?
(*b*) How does thyroxine fit the formula $H_2NCHRCO_2H$? What name is given to compounds of this type? What properties would you expect a member of this group of compounds to have?
(*c*) What kind of reaction would you expect of the iodine atoms in thyroxine?
Explain your answer.

10. **X** is a volatile liquid. Its aqueous solution is neutral, but becomes acidic on exposure to the air. **X** reduces ammoniacal silver nitrate, and reacts with hydrogen cyanide to form **Y**. When a drop of concentrated hydrochloric acid is added to **X**, an oily product, **Z**, is formed. **Z** is almost insoluble in water, and has a molar mass of 132 g mol^{-1}.

Identify **X**, **Y** and **Z**, and explain the reactions described.

11. An organic compound **H** ($C_2H_4O_3$) is oxidised to **I** ($C_2H_2O_3$), which can be oxidised to **J** ($C_2H_2O_4$). **H**, **I** and **J** dissolve in water to give acidic solutions which decolorise potassium manganate(VII) on warming. On treatment with phosphorus(V) chloride, all three react, **H** and **J** giving 2 moles of hydrogen chloride per mole, and **I** giving 1 mole of hydrogen chloride per mole. **I** gives a precipitate with 2,4-dinitrophenylhydrazine.

Deduce the identity of **H**, **I** and **J**. Explain the reactions described.

12. Identify the compounds **A** to **J** in the reaction scheme below. Draw their structures and write appropriate equations. Give a use for **G**.

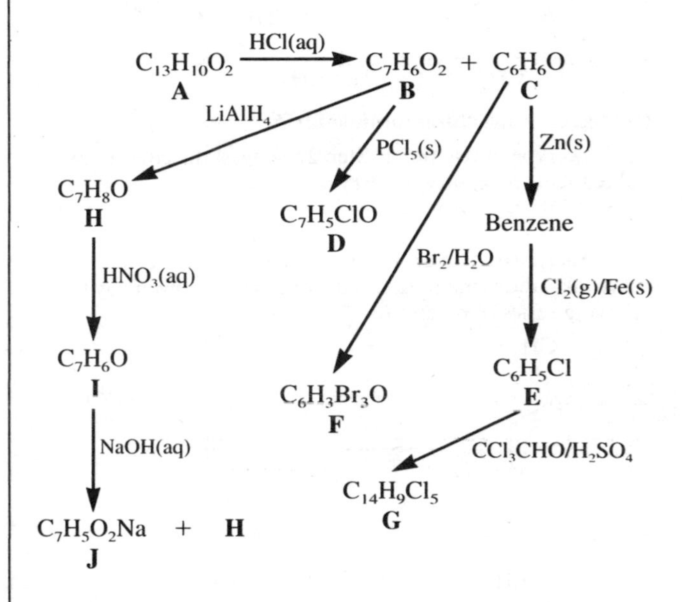

(AEB 92, S)

13. A substance **A**, $C_4H_9NO_2$, is a solid with a high melting point. It dissolves in water to give a nearly neutral solution. It can be resolved into optical isomers.
On treating **A** with aqueous $NaNO_2$/HCl it gives **B**, $C_4H_8O_3$.
On warming **B** with ethanol and concentrated sulphuric acid, a liquid ester **C**, $C_6H_{10}O_2$, is formed which, on exposure to light, slowly gives a glassy solid **D**. **B** reacts with PCl_5 to give **E**, which rapidly reacts with water to give **F**, $C_4H_7ClO_2$.
Treatment of **F** with alcoholic KCN, followed by hydrolysis, gives **G**.

(*a*) Deduce a possible identity of each substance **A** to **G** and comment on their properties given above.

(*b*) State, with a reason, whether you would expect **G** to be readily capable of dehydration to the compound $C_5H_6O_3$.

(C 90, S)

APPENDIX 3

A SELECTION OF QUESTIONS ON ORGANIC CHEMISTRY

1. (*a*) From your knowledge of organic chemistry, choose *one* reaction to illustrate each of the following types of mechanism:

(i) nucleophilic addition

(ii) electrophilic substitution

For each reaction,

(iii) state the reagents and conditions

(iv) draw the structure of the organic starting material and of the organic product.

(*b*) Compound **A** undergoes the following reactions:

$$\mathbf{A} \xrightarrow[\text{trichloroethane}]{\text{bromine in; Step 1}} CH_3CHBrCHBrCH_3 \xrightarrow{\text{Step 2}}$$

$$CH_2{=}CH{-}CH{=}CH_2$$

(i) Suggest a structural formula for **A**.

(ii) What type of reaction is Step 2? Suggest the conditions and reagents necessary to carry it out.

(C 91)

2. (*a*) *Saccharin* is an artificial sweetening agent used in some soft drinks and is manufactured from methylbenzene by the following series of reactions:

Methylbenzene ($C_6H_5CH_3$) $\xrightarrow[\text{Step 1}]{ClSO_3H}$ 2-methylbenzenesulphonyl chloride (CH_3, SO_2Cl)
$\xrightarrow{\text{Step 2}}$ 2-methylbenzenesulphonamide (CH_3, SO_2NH_2) $\xrightarrow{\text{Step 3}}$ (CO_2H, SO_2NH_2) $\xrightarrow{-H_2O}$ Saccharin (O=C—N(H)—SO_2 ring fused to benzene)

Saccharin

(i) What *types of reaction* do Steps 1, 2 and 3 illustrate?

(ii) Suggest reagents for Steps 2 and 3.

(*b*) When methylbenzene is nitrated by a mixture of concentrated nitric acid and concentrated sulphuric acid, the product consists largely of two isomers of formula $C_7H_7NO_2$.

(i) Draw the structural formulae of the two isomers.

(ii) Briefly outline the mechanism of this reaction.

(C 90)

3. (*a*) Give a concise account of oxidation and reduction in organic chemistry, considering in your answer the reactions which may be carried out, the ease with which they occur, and the reagents and conditions used.

(*b*) (i) Describe the methods available (excluding polymerisation) for increasing the length of the carbon chain in an organic compound.

(ii) Suggest a possible route for converting ethanoic acid into methanoic acid, stating the reagents and essential conditions required.

(*c*) With the help of the data given below discuss the general effects of (i) increasing carbon chain length, and (ii) the degree of chain branching on the physical properties of organic compounds.

State, giving your reasons, whether such changes, (i) and (ii), in the carbon chain will have a significant effect on the chemical behaviour of any functional group in the molecule.

Compound	*Melting point/°C*	*Boiling point/°C*
Hexane	−95	69
2-Methylpentane	−154	60
2,2-Dimethylbutane	−99	50
Heptane	−91	98
Octane	−57	126
2-Methylheptane	−109	118
2,2,3,3-Tetramethylbutane	+101	107
Nonane	−54	151
Decane	−32	173

(WJEC 90, S)

4. Answer one of the following ((*a*)–(*d*)):

(*a*) Give an account of the factors which influence the rates of chemical reactions and explain their effects at the molecular level.

Discuss how rate studies on chemical reactions enable reaction mechanisms to be proposed, illustrating your answer with a specific example. (Experimental details are *not* expected.)

(*b*) By drawing on a wide range of examples from both organic and inorganic chemistry, illustrate the statement that 'the physical properties of elements and compounds depend on the nature of their inter- and intramolecular bonding'.

(*c*) Discuss the application of the theories of chemical equilibrium to any *two* industrial processes of your choice. In addition, you should consider the importance of economic and environmental factors in the successful commercial operation of the two processes.

(*d*) Discuss and explain the effects of the structure of organic compounds on the reactions which they undergo, selecting appropriate illustrative examples from among the following pairs of substances: ethane and ethene; propene and benzene; hexane and benzene; propanal and propanone; chloroethane and ethanoyl chloride; 1-bromopropane and 2-bromo-2-methylpropane; phenylamine and ethylamine.

Details of mechanisms should be given where appropriate.

(L 92)

5. (*a*) Give and name the mechanism of each of the following reactions:

(i) bromoethane with sodium cyanide in a suitable solvent

(ii) propanone with sodium cyanide in the presence of a little concentrated sulphuric acid

(iii) ethene with bromine

(iv) benzene with bromine in the presence of iron(III) bromide

(*b*) When one mole of chloroethanoyl chloride ($ClCH_2COCl$) is treated with an excess of aqueous ammonia, one mole of chloride ions is quickly liberated. Over a period of a few days, a second mole of chloride ions is liberated. Indicate the products which are formed at each stage, giving your reasoning.

(O & C 91)

6. Outline suitable reaction schemes by which each of the following conversions may be brought about, giving any necessary reagents and conditions. (Full practical details are not required. You must use the starting substance in each conversion as the only source of any organic reagent required. More than one step may be needed.)

(*a*) $C_2H_5OH \rightarrow CH_3CO_2C_2H_5$

(*b*) $CH_3CH{=}CH_2 \rightarrow CH_3COCH_3$

(*c*) $C_6H_5CO_2H \rightarrow (C_6H_5CO)_2O$

(O & C 91)

7. A series of synthetic reactions beginning at ethanol is outlined below, with the reagents and conditions indicated by numbers:

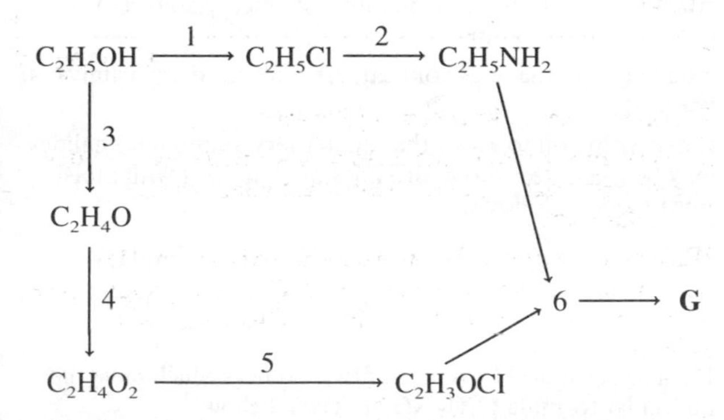

(*a*) For each of the Steps 1–3 indicated above, suggest the appropriate reagents and/or conditions for the maximum yield of product.

(*b*) For each of the Steps 4 and 5 indicated above, suggest the type of reaction taking place.

(*c*) Step 6 involves the two organic chemicals simply coming into contact with each other. The product is indicated as **G**.

Draw the structural formula of **G**.

(*d*) Which *two* of the chemicals shown could be reacted together to produce an ester?

(*e*) (i) Give one use of ethanol other than as a drink.

(ii) Why might a chemical company use ethanol as a starting point for synthetic reactions such as those outlined above, rather than sell it for drinks?

(O & C 92, AS)

8. The following questions relate to organic reaction mechanisms.

(*a*) Comment on each of the following:

(i) When ethene is bubbled into bromine water the main product obtained is 2-bromoethanol. Ethene also reacts with a mixture of bromine and sodium chloride to give a mixture of 1,2-dibromoethane and 1-bromo-2-chloroethane, but no 1,2-dichloroethane.

(ii) The optically active bromoalkane $CH_3{-}CH(C_6H_5){-}Br$ undergoes nucleophilic substitution with aqueous hydroxide ions by an S_N2 mechanism to give an optically active alcohol. A similar reaction with the compound $CH_3{-}CBr(C_6H_5){-}C_2H_5$, which proceeds by an S_N1 mechanism, leads to a mixture of two alcohols which is optically inactive.

(*b*) In an attempt to prepare 1-phenylpropane, $C_6H_5{-}CH_2{-}CH_2{-}CH_3$, benzene was reacted with 1-chloropropane in the presence of an aluminium chloride catalyst under anhydrous conditions. A vigorous reaction occurred and hydrogen chloride was evolved, but the organic product isolated was 2-phenylpropane, $CH_3{-}CH(C_6H_5){-}CH_3$ instead of the expected product. Under these conditions, the electrophile is, initially, $CH_3{-}CH_2{-}CH_2^+$.

Suggest a mechanism for the electrophilic substitution reaction and explain why 2-phenylpropane was formed, rather than the expected product.

(L 91, S)

9. (*a*) (i) Define K_a and pK_a for an acid. Explain why methanoic (formic) acid is described as a *weak acid.*

(ii) Using the data given below, discuss the variation of acid strength within the compounds shown.

		pK_a
Methanoic (formic) acid	HCO_2H	3.76
Propanoic acid	$CH_3CH_2CO_2H$	4.87
2-Chloropropanoic acid	$CH_3CHClCO_2H$	2.83
3-Chloropropanoic acid	$CH_2ClCH_2CO_2H$	4.10
Benzenecarboxylic (benzoic) acid	$C_6H_5CO_2H$	4.20

(*b*) Explain the following order of increasing base strength:

$C_6H_5NH_2$	$<$	NH_3	$<$	CH_3NH_2
Phenylamine (aniline)		Ammonia		Methylamine

(*c*) Given that (I) below is a much weaker base than phenylamine, predict with a reason whether you would expect (II) below to be a stronger or weaker acid than benzenecarboxylic acid.

$O_2N{-}C_6H_4{-}NH_2$ (I) $\qquad$ $O_2N{-}C_6H_4{-}CO_2H$ (II)

(*d*) A sample of 2-aminopropanoic acid (alanine) was added to an excess of hydrochloric acid to form a solution containing the organic species **A**. When sodium hydroxide solution was slowly added to the solution, **A** was converted first into species **B**, and then, with excess sodium hydroxide, into species **C**. Explain the changes described and draw the structures of **A**, **B** and **C**.

(*e*) Outline a method for separating a mixture of phenylamine and methylbenzene other than by fractional distillation.

(NEAB 92)

10. The cyanide ion is a versatile species in organic reactions. This can be demonstrated by comparing the reactions of sodium cyanide under suitable conditions with 1-bromopropane and with propanone.

(*a*) State and explain the role of the cyanide ion in these two reactions and point out any differences in the type of reaction.

(*b*) Outline a mechanism for *each* of these reactions and state any reagents that may be necessary.

(*c*) Give the reagents, conditions and equations to show how the following can be synthesised from appropriate nitriles:

(i) propylamine, starting from propanenitrile

(ii) 2-hydroxypropanoic (lactic) acid, $CH_3CH(OH)CO_2H$, starting with 2-hydroxypropanenitrile

(*d*) When 3.22 g of the organic compound below was treated with an excess of cyanide ions and the chloride ions liberated collected as silver chloride, 2.87 g of silver chloride resulted. Explain this observation and give the structural formula of the organic product of the reaction.

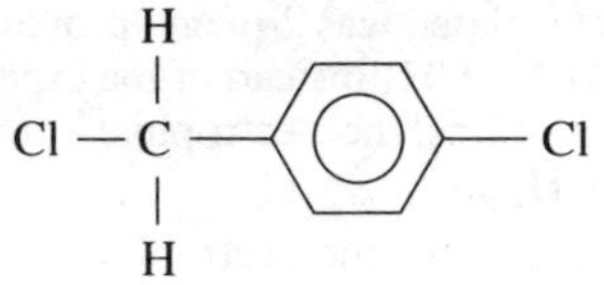

H
Cl — C — (benzene ring) — Cl
H

(NEAB 92)

11. Ethanoyl chloride reacts with benzene in the presence of an aluminium chloride catalyst to give phenylethanone according to the equation:

$$C_6H_6 + CH_3COCl \rightarrow C_6H_5COCH_3 + HCl$$

The reaction is an electrophilic substitution.

(*a*) (i) What is the formula of the electrophile in the reaction?

(ii) Give an equation to show the formation of the electrophile from ethanoyl chloride and the overall reaction mechanism.

(iii) Draw diagrams showing the electronic structures of this electrophile and $CH_3CH_2^+$ and discuss their relative stabilities in the light of their structures.

(iv) Explain why, in a similar experiment to prepare 1-phenylpropane, the yield of the desired product was negligible, though a good yield of 2-phenylpropane resulted.

(*b*) When 2,4-dinitrophenylhydrazine reacts with phenylethanone two 2,4-dinitrophenylhydrazone derivatives result which, when pure, melt at 238 °C and 250 °C.

(i) Explain why two 2,4-dinitrophenylhydrazone derivatives result.

(ii) Which derivative would you expect to be formed in greater yield? Give a reason for your choice.

(iii) Describe an experiment to separate a mixture of the two derivatives in the laboratory, giving the principles on which it is based.

(*c*) (i) Explain why the reaction of phenylethanone with potassium cyanide occurs only in alkaline solution

(ii) Comment on the optical activity of the product of this reaction.

(iii) Suggest an explanation for the failure of alkenes to react with potassium cyanide solution under alkaline conditions.

(L 92, S)

12. (*a*) (i) Benzene undergoes *electrophilic substitution* reactions. Explain what is meant by the terms in italics.

(ii) Draw the structures of *three* substances which could be prepared from benzene using this type of reaction. In *each* case, give the necessary reagents. For *one* of the preparations, outline the mechanism of the reaction.

(*b*) Propene, $CH_3{-}CH{=}CH_2$, reacts with benzene in the presence of phosphoric acid to yield 2-phenylpropane, $CH_3CH(C_6H_5)CH_3$. Comment on the mechanism and product of this reaction.

(*c*) State the conditions required for chlorine to react with the methyl group in methylbenzene and write the structures of *two* of the possible products. Give the mechanism of the reaction.

(AEB 91)

13. Choline, an important naturally occurring compound, has the structure:

$$[(CH_3)_3NCH_2CH_2OH]^+ OH^-$$

(*a*) Account for the fact that choline is water soluble and very strongly basic (similar in base strength to sodium hydroxide; much stronger than ammonia or simple amines).

(*b*) When boiled with barium hydroxide (catalyst), choline gives neurine (which can be isolated from brain tissue):

$$[(CH_3)_3NCH_2CH_2OH]^+ OH^- \rightarrow [C_5H_{12}N]^+OH^- + H_2O$$

Suggest a structure for neurine, and state what type of reaction has occurred here.

(*c*) Suggest a possible synthesis for choline, selecting your starting materials from the following substances:

ethane, ethene (ethylene), ammonia, trimethylamine, iodomethane, bromine

You may also make use of heat, air, water and any catalyst(s) you wish.

It may help you to know that quaternary ammonium halides may be converted into hydroxides by treatment with moist silver oxide as follows:

$$2R_4N^+I^- + Ag_2O + H_2O \rightarrow 2R_4N^+OH^- + 2AgI\,(s)$$

(O & C 91, S)

14. The structural formulae of two isomers which have the molecular formula C_7H_7NO are given below

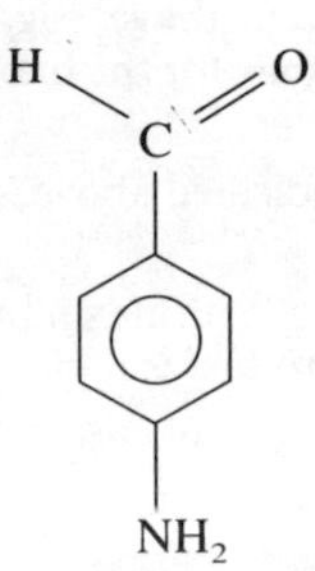

H, C=O, (benzene ring), NH_2 — **I**

O=C, NH_2, (benzene ring) — **II**

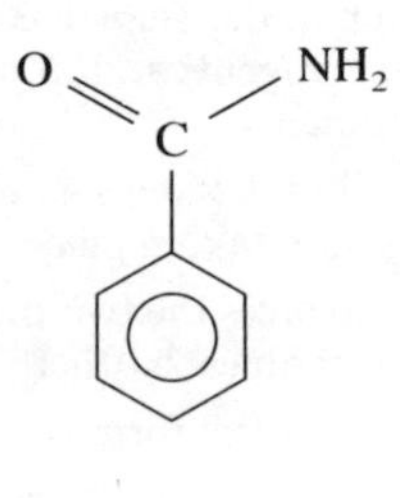

Name *each* of the isomers and describe their reactions (if any) with

(*a*) hydrochloric acid

(*b*) 2,4-dinitrophenylhydrazine
(*c*) nitrous acid
(*d*) sodium hydroxide.

In *each* case you should
(i) give a balanced equation for the reaction
(ii) state the reaction conditions
(iii) describe what you would expect to observe.

(AEB 90)

Relative atomic mass — 1

H
Hydrogen

Proton (atomic) number — 1

TRANSITION ELEMENTS

1	2											3	4	5	6	7	0
																	4 **He** Helium 2
7 **Li** Lithium 3	9 **Be** Beryllium 4											11 **B** Boron 5	12 **C** Carbon 6	14 **N** Nitrogen 7	16 **O** Oxygen 8	19 **F** Fluorine 9	20 **Ne** Neon 10
23 **Na** Sodium 11	24 **Mg** Magnesium 12											27 **Al** Aluminium 13	28 **Si** Silicon 14	31 **P** Phosphorus 15	32 **S** Sulphur 16	35.5 **Cl** Chlorine 17	40 **Ar** Argon 18
39 **K** Potassium 19	40 **Ca** Calcium 20	45 **Sc** Scandium 21	48 **Ti** Titanium 22	51 **V** Vanadium 23	52 **Cr** Chromium 24	55 **Mn** Manganese 25	56 **Fe** Iron 26	59 **Co** Cobalt 27	59 **Ni** Nickel 28	63.5 **Cu** Copper 29	65 **Zn** Zinc 30	70 **Ga** Gallium 31	73 **Ge** Germanium 32	75 **As** Arsenic 33	79 **Se** Selenium 34	80 **Br** Bromine 35	84 **Kr** Krypton 36
85 **Rb** Rubidium 37	88 **Sr** Strontium 38	89 **Y** Yttrium 39	91 **Zr** Zirconium 40	93 **Nb** Niobium 41	96 **Mo** Molybdenum 42	99 **Tc** Technetium 43	101 **Ru** Ruthenium 44	103 **Rh** Rhodium 45	106 **Pd** Palladium 46	108 **Ag** Silver 47	112 **Cd** Cadmium 48	115 **In** Indium 49	119 **Sn** Tin 50	122 **Sb** Antimony 51	128 **Te** Tellurium 52	127 **I** Iodine 53	131 **Xe** Xenon 54
133 **Cs** Caesium 55	137 **Ba** Barium 56	139 **La** Lanthanum 57 *	178 **Hf** Hafnium 72	181 **Ta** Tantalum 73	184 **W** Tungsten 74	186 **Re** Rhenium 75	190 **Os** Osmium 76	192 **Ir** Iridium 77	195 **Pt** Platinum 78	197 **Au** Gold 79	201 **Hg** Mercury 80	204 **Tl** Thallium 81	207 **Pb** Lead 82	209 **Bi** Bismuth 83	210 **Po** Polonium 84	210 **At** Astatine 85	222 **Rn** Radon 86
223 **Fr** Francium 87	226 **Ra** Radium 88	227 **Ac** Actinium 89 †	261 **Unq** Unnil-quadium 104	262 **Unp** Unnil-pentium 105	263 **Unh** Unnil-hexium 106												

140 **Ce** Cerium 58	141 **Pr** Praseo-dymium 59	144 **Nd** Neodymium 60	147 **Pm** Promethium 61	150 **Sm** Samarium 62	152 **Eu** Europium 63	157 **Gd** Gadolinium 64	159 **Tb** Terbium 65	162 **Dy** Dysprosium 66	165 **Ho** Holmium 67	167 **Er** Erbium 68	169 **Tm** Thulium 69	173 **Yb** Ytterbium 70	175 **Lu** Lutetium 71
232 **Th** Thorium 90	231 **Pa** Protactinium 91	238 **U** Uranium 92	237 **Np** Neptunium 93	242 **Pu** Plutonium 94	243 **Am** Americium 95	247 **Cm** Curium 96	245 **Bk** Berkelium 97	251 **Cf** Californium 98	254 **Es** Einsteinium 99	253 **Fm** Fermium 100	256 **Md** Mendelevium 101	254 **No** Nobelium 102	257 **Lr** Lawrencium 103

*58–71 Lanthanum series

†90–103 Actinium series

BASIC SI UNITS

Physical Quantity	*Name of unit*	*Symbol*
Length	metre	m
Mass	kilogram	kg
Time	second	s
Electric current	ampere	A
Temperature	kelvin	K
Amount of substance	mole	mol
Light intensity	candela	cd

DERIVED SI UNITS

Physical Quantity	*Name of unit*	*Symbol*	*Definition*
Energy	joule	J	$kg\,m^2\,s^{-2}$
Force	newton	N	$J\,m^{-1}$
Electric charge	coulomb	C	$A\,s$
Electric potential difference	volt	V	$J\,A^{-1}\,s^{-1}$
Electric resistance	ohm	Ω	$V\,A^{-1}$
Area	square metre		m^2
Volume	cubic metre		m^3
Density	kilogram per cubic metre		$kg\,m^{-3}$
Pressure	newton per square metre or pascal		$N\,m^{-2}$ or Pa
Molar mass	kilogram per mole		$kg\,mol^{-1}$

With all these units, the following prefixes (and others) may be used.

Prefix	*Symbol*	*Meaning*
deci	d	10^{-1}
centi	c	10^{-2}
milli	m	10^{-3}
micro	μ	10^{-6}
nano	n	10^{-9}
kilo	k	10^{3}
mega	M	10^{6}
giga	G	10^{9}
tera	T	10^{12}

ANSWERS TO NUMERICAL PROBLEMS AND SELECTED QUESTIONS

The University of Cambridge Local Examinations Syndicate bears no responsibility for the example answers to questions taken from its past question papers which are contained in this publication.

The University of London Examinations and Assessment Council accepts no responsibility whatsoever for the accuracy or method of working in the answers given.

PART 1: THE FOUNDATION

CHAPTER 1: THE ATOM

Checkpoint 1B **[p. 10]**

1. (*a*) 19p, 19e, 20n (*b*) 13p, 13e, 14n (*c*) 56p, 56e, 81n (*d*) 88p, 88e, 138n
2. (*a*) 6p, 6e, 6n (*b*) 6p, 6e, 8n (*c*) 1p, 1e (*d*) 1p, 1e, 1n (*e*) 1p, 1e, 2n (*f*) 38p, 38e, 49n (*g*) 38p, 38e, 52n (*h*) 92p, 92e, 143n (*i*) 92p, 92e, 146n

Checkpoint 1C: Mass Spectrometry **[p. 13]**

2. 35.45
3. $CH_2{}^{35}Cl_2$, $CH_2{}^{35}Cl^{37}Cl$, $CH_2{}^{37}Cl_2$, ^{35}Cl is 3 × as abundant as ^{37}Cl.
4. 6 5. 91.3 u and 207.2 u

Checkpoint 1D: Nuclear Reactions I **[p. 21]**

3. (*a*) $^{14}_{7}N$ (*b*) $^{19}_{10}Ne$ $^{19}_{9}F$ (*c*) $^{226}_{88}Ra$ $^{222}_{86}Rn$ (*d*) $^{73}_{33}As$ $^{73}_{32}Ge$ (*e*) $^{27}_{14}Si$ (*f*) $^{1}_{0}n$

Questions on Chapter 1 **[p. 29]**

2. 1 = β particle, 2 = α particle, Pb in Gp 4, **X** in Gp 5, **Y** in Gp 3, **Z** in Gp 4
3. 1.88 and 1.94; $^{32}_{15}P \rightarrow {}^{4}_{2}He + {}^{28}_{13}Al$; 1.87
6. 64 and 73; $^{13}C_2{}^{1}H_5{}^{35}Cl$, $^{12}C_2{}^{1}H_5{}^{37}Cl$, $^{12}C_2{}^{1}H_3{}^{2}H_2{}^{35}Cl$, $^{12}C^{13}C^{1}H_4{}^{2}H^{35}Cl$
8. (*a*) $a = 35, b = 16$, **X** = S; (*b*) $c = 4, d = 2$, **Y** = He
9. (*c*) C_4H_8O. Peaks: 15 = CH_3, 29 = C_2H_5, CHO, 43 = C_3H_7, 57 = C_2H_5CO, CH_2CH_2CHO, 72 = C_4H_8O. Molecular formula is $CH_3CH_2CH_2CHO$.
11. $^{228}_{90}$**Z**
12. (*a*) $3Cl_2 + 2NH_4Cl \rightarrow 2NCl_3 + 8HCl$
    $NCl_3 + 3NaOH \rightarrow NH_3 + 3NaClO$
    (*b*)

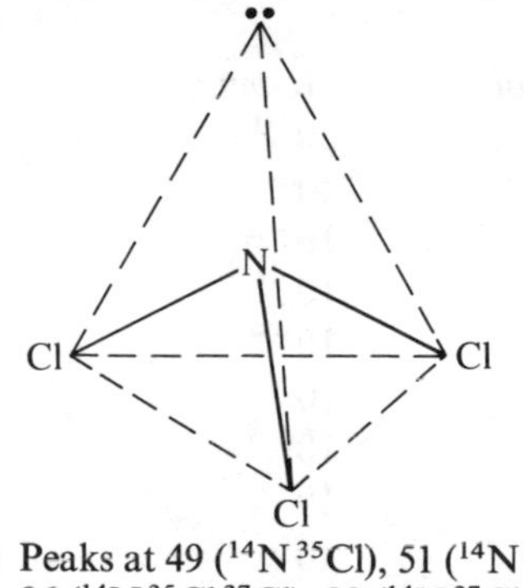

N
Cl Cl
Cl

   (*c*) Peaks at 49 ($^{14}N^{35}Cl$), 51 ($^{14}N^{37}Cl$), 84 ($^{14}N^{35}Cl_2$), 86 ($^{14}N^{35}Cl^{37}Cl$), 88 ($^{14}N^{37}Cl_2$), 119 ($^{14}N^{35}Cl_3$), 121 ($^{14}N^{35}Cl_2{}^{37}Cl$), 123 ($^{14}N^{35}Cl^{37}Cl_2$), 125 ($^{14}N^{37}Cl_3$).
13. (*b*) $^{232}_{90}Th \rightarrow {}^{4}_{2}He + {}^{228}_{88}\mathbf{X} \rightarrow {}^{0}_{-1}e + {}^{228}_{89}\mathbf{Y} \rightarrow {}^{0}_{-1}e + {}^{228}_{90}\mathbf{Z}$
    Mass numbers: **X** = 228, **Y** = 228, **Z** = 228
    Atomic numbers: **X** = 88, **Y** = 89, **Z** = 90
    (*c*) $^{12}C^{16}O_2$, $^{12}C^{16}O^{18}O$, $^{12}C^{18}O_2$
    M_r = 44, 46, 48
    Relative intensities = 16 : 4 : 1
14. (*b*) $^{79}_{35}Br$ $^{81}_{35}Br$
    (*c*) 80.0
    (*d*) Peaks are observed at m/e = 158 ($^{79}Br^{79}Br$), 160 ($^{79}Br^{81}Br$), 162 ($^{81}Br^{81}Br$).

CHAPTER 2: THE ATOM: THE ARRANGEMENT OF ELECTRONS

Questions on Chapter 2 **[p. 55]**

3. BCl_3
4. (*a*) (i) Frequency increases from left to right.
   (ii) Energy increases from left to right.
   (*c*) principal quantum number
   (*d*) All involve transitions from higher levels down to the same energy level.
   (*e*) (i) B
   (ii) None; transitions to level 2 are Balmer series.
   (iii) None; an emission spectrum is caused by a transition from an orbit of higher n to an orbit of lower n.
6. (*d*) 25 mg of $^{24}_{11}Na$ + 175 mg of $^{24}_{12}Mg$

CHAPTER 3: EQUATIONS AND EQUILIBRIA

Checkpoint 3C: Relative Molecular Mass [p. 61]

1. 40, 74.5, 40, 74, 63, 123.5, 80, 159.5, 249.5, 146

Checkpoint 3D: The Avogadro Constant [p. 62]

1. £150 million million
2. (*a*) 39 g (*b*) 1300 g 3. 2.7×10^{-7} p

Checkpoint 3E: The Mole [p. 63]

1. (*a*) 72 g (*b*) 8 g (*c*) 16 g (*d*) 8 g (*e*) 64 g
2. (*a*) 0.33 mol (*b*) 0.25 mol (*c*) 2.0 mol (*d*) 0.010 mol (*e*) 0.33 mol
3. (*a*) 88 g (*b*) 980 g (*c*) 117 g (*d*) 37 g
4. (*a*) 482 g mol^{-1} (*b*) 342 g mol^{-1} (*c*) 368 g mol^{-1}
5. (*a*) 2.50×10^{-3} mol (*b*) 5.00×10^{-2} mol (*c*) 5.40×10^{-2} mol
6. (*a*) 3.0×10^{19} (*b*) 7.5×10^{16} (*c*) 3.0×10^{11}

Checkpoint 3F: Formulae and Percentage Composition [p. 65]

1. (*a*) 72% (*b*) 39% (*c*) 80%
2. (*a*) CO_2 (*b*) C_3O_2 (*c*) $Na_2S_2O_3$ (*d*) Na_2SO_4
3. (*a*) MgO (*b*) $CaCl_2$ (*c*) $FeCl_3$
4. $a = 7, b = 2, c = \frac{1}{2}$

Checkpoint 3G: Masses of Reacting Solids [p. 66]

1. 94 tonnes
2. 25.4%
3. 21.0 g, 93%
4. 2.12 g, 96%

Checkpoint 3H: Reacting Volumes of Gases [p. 67]

1. 2.80 dm^3 at stp
2. 26.8 g
3. 26.1 g NaCl, 43.8 g H_2SO_4

Checkpoint 3I: Concentration [p. 68]

1. (*a*) 0.20 M (*b*) 0.020 M (*c*) 0.20 M (*d*) 8 M
2. (*a*) 10 g (*b*) 0.365 g (*c*) 4.9 g (*d*) 0.14 g

Checkpoint 3J: Solutions [p. 70]

1. (*a*) burette (*b*) measuring cylinder
3. (*a*) 6 dm^3 (*b*) 1 dm^3 (*c*) measuring cylinder
   (*d*) Add the measured volume of acid to distilled water with stirring, make up to 6 dm^3. Approximate volumes will suffice for approximately 2 M bench acid.
4. 140 cm^3
5. (*b*) 0.42 g
   (*c*) Weigh out 0.42 g, dissolve in distilled water in a 500 cm^3 volumetric flask, make up to the mark.
6. (*a*) Measure 62.5 cm^3 of 4.00 M acid in a burette. Dissolve in distilled water and make up to 1.00 dm^3 in a volumetric flask.
   (*b*) Measure 235 cm^3 of glacial ethanoic acid in a measuring cylinder. Pour gradually into distilled water. Make up to 2 dm^3 in a graduated beaker.

Checkpoint 3K: Titration [p. 73]

1. 0.15 M 2. (*a*) 0.25 M (*b*) 10.0 cm^3
3. (*a*) 2 (*b*) 0.24 M
4. (*a*) Stopit (*b*) speed of action, taste
5. 1.0×10^{-5} M

Checkpoint 3L: Titrations [p. 74]

1. 98.9% 2. $n = 10$ 3. 15.0%
4. $4CuSO_4 + 6NaOH \rightarrow Cu_4(OH)_6SO_4 + 3Na_2SO_4$

Checkpoint 3N: Oxidation Number [p. 78]

1. 0, +1, 0, +2, +1, 0, 0, 0, −3, −1, 0, +1, 0, −1
2. +2, +1, −2, +4, +6, +2, +4, +3, +6, +2, +4, +5
3. (*a*) +2, +4, +4, +1, +3, +5, +5
   (*b*) +2, +3, +4, +7, +6
   (*c*) +3, +3, +5, +3
   (*d*) +6, +6, +6
   (*e*) −1, +1, +5, 0, +3, +1

Checkpoint 3O: Equations and Oxidation Numbers [p. 80]

2. $ICl_3 + 3KI \rightarrow 2I_2 + 3KCl$; +3
3. (*a*) −2, −1, −2, 0
   (*b*) 0, +5, +2, +2
   (*c*) +2, +6, +3, +3
   (*d*) +2, 0, +2.5, −1
   (*e*) +5, −1, −1, −1, +1, −1
   (*f*) +6, +6

Checkpoint 3P: Redox Titrations [p. 83]

1. 96% 2. 5.50×10^{-2} mol dm^{-3}, 30.8 cm^3 O_2
3. 95.0% 4. 60.0 cm^3
5. $6Hg + 2KMnO_4 + H_2O \rightarrow 2MnO_2 + 3Hg_2O + 2KOH$

Checkpoint 3Q: Equilibrium [p. 86]

1. Left to right, as OH^-(aq) ions remove H^+(aq) ions to form H_2O. The brown colour fades as more Br_2 molecules react with water. Add a little acid to reverse the change.
2. (*a*) To take up water, the equilibrium moves from left to right, and BiOCl is precipitated.
   (*b*) Adding a little concentrated hydrochloric acid (on the right-hand side of the equation) drives the equilibrium from right to left.

Questions on Chapter 3 [p. 88]

3. 30.0 cm^3 5. 20.0% 6. 20.1%
7. (*b*) (i) **A, B, E, H**
   (ii) **A** H disproportionates: $-1 + 2(+1) \rightarrow +1 + 2(0)$
   B Ca(0) → Ca(+2); H(+1) → H(0)
   E Fe(+2) → Fe(+3); Mn(+7) → Mn(+2)
   H Cl(−1) → Cl(0); Pb(+4) → Pb(+2)
   (iii) e.g. $Fe^{3+} + e^- \rightarrow Fe^{2+}$
   $MnO_4^- + 8H^+ + 5e^- \rightarrow Mn^{2+} + 4H_2O$
8. (*b*) (ii) +6 (iii) H_2SO_4
   $Cr_2O_7^{2-} + 3SO_2 + 8H^+ \rightarrow 2Cr^{3+} + H_2O + 3H_2SO_4$
   (*d*) (i) 1.16×10^{-4} mol (ii) 2.9×10^{-3} mol
   (iii) 2.1×10^{-3} mol (iv) 2.1×10^{-3} mol
   (*e*) (i) 3.20×10^{8} mol (ii) 7.7×10^{9} dm^3
9. (*c*) (i) 3.0×10^{-3} mol
   (ii) 1.5×10^{-3} mol
   (Amount of copper(II) = 3.0×10^{-2} mol);
   % of copper = 93%
10. (*c*) 11.1%
11. (*a*) (i) 2.5 mol (ii) 6.0×10^{-4} mol
    (iii) 1.5×10^{-3} mol (iv) 189 ppm
12. (*b*) 4.0×10^{-3} mol (*c*) 4.0×10^{-2} mol
    (*d*) 1.0×10^{-3} mol (*e*) 2.00×10^{-2} mol
    (*f*) 2.00×10^{-2} mol (*g*) 5.88 g
    (*h*) 2.16 g, 0.120 mol; $MgSO_4 \cdot K_2SO_4 \cdot 6H_2O$
13. (*a*) $x = 2$; volume = 600 cm^3
    (*b*) Li_2O is formed; with water this forms LiOH. Na_2O_2 is formed; it is a peroxide.
    (*c*) **A** = KBr, **B** = HBr,
    C = AgBr, **D** = Br_2, **E** =
    F = $[Ag(NH_3)_2]^+$

OH
Br Br
Br

14. $2SO_4^{2-} \rightarrow S_2O_8^{2-} + 2e^-$

$$\left[O\overset{O}{\underset{O}{=}}S-O-O-S\overset{O}{\underset{O}{=}}O \right]^{-}$$

Either S(+7) and O(−2) or S(+6) and O($-1\frac{3}{4}$)

Amount of $S_2O_8^{2-}$ = 1.25×10^{-3} mol
Amount of KI = 1.00×10^{-2} mol
Amount of thio = 2.5×10^{-3} mol
Amount of I_2 = 1.25×10^{-3} mol
$S_2O_8^{2-} + 2I^- \rightarrow I_2 + 2SO_4^{2-}$
No, the $E^\ominus$ for I_2/I^- has a far lower positive value than $E^\ominus$ for I_2/IO_3^-

CHAPTER 5: THE SHAPES OF MOLECULES

Questions on Chapter 5 [p. 143]

5. (*b*) F_2O similar to H_2O, Figure 5.6; H_3O^+ similar to NH_3, Figure 5.6; ClF_4^- similar to ICl_4^-, Figure 5.13
   (*c*) 12 electrons, making an octahedral distribution of electron pairs, $n = 2$, Ox. No. = +3
8. (*a*) (i) See Figure 5.13 (ii) see Figure 5.12
   (iii) see Figure 5.6 (PF_3 is similar to NH_3)
   (*b*) F — N(=O) (with lone pair on N) planar; N ≡ S — F linear
   (*c*) $[O=N=O]^+$ linear
9. (*d*) NH_3: tetrahedral distribution of electron pairs, BF_3: trigonal planar; in the compound the distributions of bonds about N and B are both tetrahedral.

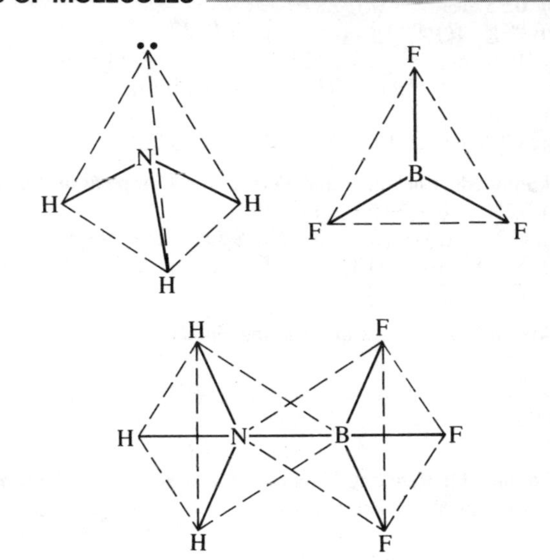

PART 2: PHYSICAL CHEMISTRY

CHAPTER 7: GASES

Checkpoint 7A: Correcting Gas Volumes [p. 166]

1. (*a*) 224 cm^3 (*b*) 70.2 cm^3 (*c*) 54.3 dm^3
   (*d*) 3.21 dm^3
2. (*a*) 13.7 dm^3 (*b*) 439 cm^3

Checkpoint 7B: Diffusion and Effusion [p. 167]

1. 8.28 cm^3 2. 1520 s (25.3 min)
3. 44 g mol^{-1}, C_3H_8

Checkpoint 7C: Gas Molar Volume [p. 169]

1. 64 g mol^{-1} 2. 71 g mol^{-1}
3. 4.65×10^{-2} mol 4. 27.6 dm^3

Checkpoint 7D: Partial Pressures [p. 170]

1. 2.5×10^4 N m^{-2}
2. (*a*) 3.03×10^4 N m^{-2} CO, 5.05×10^4 N m^{-2} O_2, 2.02×10^4 N m^{-2} CO_2
   (*b*) 3.03×10^4 N m^{-2} CO, 5.05×10^4 N m^{-2} O_2
3. 5.33×10^5 N m^{-2}
4. (*a*) $p(NH_3) = 1.96 \times 10^4$ N m^{-2}; $p(H_2) = 5.39 \times 10^4$ N m^{-2}; $p(N_2) = 2.45 \times 10^4$ N m^{-2}
   (*b*) No change
5. 32% oxygen, 68% nitrogen

Checkpoint 7E: The Kinetic Theory and the Ideal Gas Equation [p. 172]

3. $c_{rms}(H_2)/c_{rms}(O_2) = 4.00$
4. $c_{rms}(^{235}UF_6)/c_{rms}(^{238}UF_6) = 1.004$
5. 1.34×10^{19} molecules

Questions on Chapter 7 [p. 174]

1. No, *PV* is not constant.
3. 3.40 kJ mol^{-1}
4. 65.3 g mol^{-1}; NO_2 molecules are dimerised; degree of dimerisation = 0.59
5. 1.22×10^{17}
6. 126 g mol^{-1}; PF_5
7. (*a*) 24.9 dm^3
   (*b*) 46
   (*c*) (i) C_2H_6O (ii) C_2H_5OH, CH_3OCH_3
   (iii) $C_2H_6O(l) + 3O_2(g) \rightarrow 2CO_2(g) + 3H_2O(l)$
   (*d*) (i) C_2H_5OH
   (ii) $2C_2H_5OH(l) + 2Na(s) \rightarrow H_2(g) + 2C_2H_5ONa(s)$
8. (*a*) (i) The graph shows a sudden decrease in *PV* at 10 atm as ammonia liquefies.
   (ii) a horizontal straight line at *PV* = 24.5 atm dm^3
   (*b*) See Figure 8.4.
9. (*c*) (v) 30.5 mol

CHAPTER 8: LIQUIDS

Checkpoint 8A: Vapour Pressure [p. 180]
7. 0.013 g

Checkpoint 8B: Molar Mass of Volatile Liquids [p. 181]
1. 86 $g mol^{-1}$; 84 $g mol^{-1}$ 2. 90 $g mol^{-1}$; 99 $g mol^{-1}$
3. 53 $g mol^{-1}$ (experimental); 56 $g mol^{-1}$ (correct)
4. 153 $g mol^{-1}$

Checkpoint 8C: Vapour Pressures of Solutions of Two Liquids [p. 186]
1. (*a*) 42 kPa (*b*) 31 kPa 2. 36 $kN m^{-2}$
3. (*a*) 21 $kN m^{-2}$ (*b*) 0.29

Checkpoint 8D: Steam Distillation [p. 189]
1. 156 $g mol^{-1}$ 2. 72%

Checkpoint 8E: Partition [p. 191]
1. B 2. 1.88 g 3. (*a*) 3.57 g (*b*) 4.01 g

Questions on Chapter 8 [p. 195]
7. 78% CH_3OH, 22% C_2H_5OH 8. (*a*) 98.5 °C (*b*) 23% 10. 3.56 g
11. (*a*) (i) 87 (ii) 5.5 (iii) 5
(*b*) $4C_5H_{11}O + 29O_2 \rightarrow 20CO_2 + 22H_2O$
12. (*c*) p(hexane) = 27.0 kPa, p(heptane) = 11.0 kPa
(*d*) (ii) Volume of gas dissolved at pressure p_2 and measured at p_1 is V_2p_2/p_1
Therefore $m_1/V_1 = m_2p_1/V_2p_2$
or $m_1V_2/m_2V_1 = p_1/p_2$
(iii) 2.96 atm

CHAPTER 9: SOLUTIONS

Checkpoint 9B: Osmotic Pressure [p. 201]
1. 1.42×10^5 $N m^{-2}$
2. 1.62×10^4 $g mol^{-1}$
3. 1.60×10^4 $g mol^{-1}$
4. 2.51×10^4 $g mol^{-1}$

Questions on Chapter 9 [p. 203]
4. (*c*) $M \doteq 51\,600$

CHAPTER 10: THERMOCHEMISTRY

Checkpoint 10A: Combustion [p. 210]
1. 3.36 GJ, 18.9 GJ 2. 983 kJ; $\Delta H_c^{\ominus}$ at 37 °C rather than at 25 °C 3. Propane 4. 12.0 g

Checkpoint 10B: Enthalpy Changes [p. 213]
1. (*a*) − (*b*) − (*c*) − (*d*) + (*e*)
3. $(d) < (e) < (c) < (f) < (b) < (a)$

Checkpoint 10C: Standard Enthalpy of Reaction and Average Standard Bond Enthalpies [p. 218]
1. (*a*) −85 $kJ mol^{-1}$ (*b*) +33 $kJ mol^{-1}$
(*c*) −484 $kJ mol^{-1}$ (*d*) −246 $kJ mol^{-1}$
(*e*) −202 $kJ mol^{-1}$
2. −474, −246, −484, −286 and −30 $kJ mol^{-1}$
3. −604 $kJ mol^{-1}$
4. (*a*) −1560 $kJ mol^{-1}$ (*b*) −1370 $kJ mol^{-1}$
(*c*) −286 $kJ mol^{-1}$ (*d*) −5520 $kJ mol^{-1}$
5. −372 $kJ mol^{-1}$

Checkpoint 10D: *Entropy [p. 221]
2. (*a*) − (*b*) + (*c*) − (*d*) − (*e*) +
3. $b < c < e < a < d < f$
7. (*a*) + (*b*) + (*c*) + (*d*) − (*e*) −

Checkpoint 10E: *Free Energy [p. 223]
3. (*a*) $\Delta G^{\ominus}$ for the reaction
$2Mg(s) + CO_2(g) \rightarrow 2MgO(s) + C(s)$ is negative.
(*b*) It melts. (*c*) 1900 K upwards

Questions on Chapter 10 [p. 224]
3. −436 $kJ mol^{-1}$ 4. (*a*) 331 $kJ mol^{-1}$
(*b*) 3990 $kJ mol^{-1}$ 5. 2346 $kJ mol^{-1}$
6. (*b*) −140 $kJ mol^{-1}$
7. (*a*) C_6H_6 64 $kJ mol^{-1}$; C_6H_{12} −154 $kJ mol^{-1}$
8. 87 $kJ mol^{-1}$ 9. (*b*) both −240 $kJ mol^{-1}$
10. −57 $kJ mol^{-1}$ 11. 103 $kJ mol^{-1}$
14. (*b*) −896.4 $kJ mol^{-1}$; strongly exothermic
15. (*b*) −2250 $kJ mol^{-1}$
(*c*) Ionisation energies are larger for the smaller Mg atom. The lattice enthalpy of MgO is greater than that of CaO because Mg^{2+} and O^{2-} approach more closely than Ca^{2+} and O^{2-}.
(*d*) (i) The third ionisation energy of calcium is very high; therefore $CaCl_3$ is not formed.
(ii) To form Ca^{2+} requires an input of the second ionisation energy and is more endothermic than the formation of Ca^+. However, the formation of $CaCl_2$ includes the term (2 × electron affinity of Cl) which is a highly exothermic term. Also the exothermic term, lattice enthalpy of $CaCl_2$ is much greater than that of CaCl. The sum of the terms makes the formation of $CaCl_2$ more exothermic and therefore preferred.
16. (*a*) Na^+(2.8) and F^-(2.8) have the same number of electrons, but the positive charge in Na^+ holds the electrons closer to the nucleus. Cs^+(2.8.8.18.18) has more shells of electrons than Na^+ and a larger radius.
(*b*) Na^+ can approach more closely to F^- than can Cs^+, and the lattice enthalpy of NaF is therefore greater than that of CsF. The smaller ion, Na^+, has a higher enthalpy of hydration.
(*c*) ΔH_{soln}(CsF) = 42 $kJ mol^{-1}$;
ΔH_{soln}(NaF) = 71 $kJ mol^{-1}$
(*d*) ClO_4^- is larger than F^-; therefore the lattice enthalpy of $NaClO_4$ is less than that of NaF, and the solubility of $NaClO_4$ is greater than that of NaF. The hydration enthalpy of ClO_4^- is less than that of F^-; therefore the solubility of $CsClO_4$ is less than that of CsF.
(*e*) $NaClO_4$ would be more stable to heat because the smaller ion, Na^+, approaches more closely to ClO_4^- and the lattice enthalpy is higher.
17. (*a*) (i) −380 $kJ mol^{-1}$ (ii) −311 $kJ mol^{-1}$.
The smaller ion has the more exothermic enthalpy of hydration.
(*b*) NaBr salty, KI bitter
18. (*a*) (i) coal −394 $kJ mol^{-1}$, gas −891 $kJ mol^{-1}$
(ii) (*1*) natural gas (*2*) natural gas
(iii) Natural gas produces more energy per mole of CO_2.

(*b*) (i) electron pair repulsion
(*c*) (i) covalent, polarised covalent, double covalent
(ii) The bonds in H_2O are much more polar than those in CH_4.
(*d*) hydrogen bonding

19. (*c*) The first ionisation energy of Xe is only 13 kJ mol^{-1} greater than that for O_2. The first ionisation energy of Rn is less than that for Xe or O_2, and Rn might form compounds. For Kr, the first ionisation energy is 213 kJ mol^{-1} higher than that for O_2 and Kr is less likely to form compounds.

20. (*a*) (i) 127.5 kJ mol^{-1}
(*1*) Increase in temperature increases yield because the reaction is endothermic.
(*2*) Increase in pressure decreases yield because reaction involves an increase in the number of moles of gas.
(ii) Reaction 2 is more endothermic: the cost of fuel is higher. The plant must be able to withstand attack by the HF produced.
(*b*) (i) (C—F) = 426.8 kJ mol^{-1}, (C—Cl) = 329.5 kJ mol^{-1}. The C—F bond is the least reactive of the three. The C—Cl bond is the easiest to break.
(ii) The C—Cl bond is likely to break to give Cl· radicals.

21. (*c*) (i) Step 1: $\Delta H = +210$ kJ mol^{-1}; Step 2: $\Delta H = -102$ kJ mol^{-1}
(ii) Step 1: Increase in temperature increases the yield, and increase in pressure decreases the yield. Step 2: Increase in temperature decreases the yield, and increase in pressure increases the yield.
(iii) Heat evolved in Step 2 can be used to heat the reactants in Step 2. The products from Step 1 can be fed into Step 2.

22. (*b*) −74.9 kJ mol^{-1}
(*c*) (i) 55.4 kJ mol^{-1}
(ii) no heat loss to the surroundings
(iii) The same reaction occurs: $H^+(aq) + OH^-(aq) \rightarrow H_2O(l)$

CHAPTER 11: CHEMICAL EQUILIBRIUM

Checkpoint 11B: Equilibrium Constants [p. 236]
3. 2.5 4. 0.510 $dm^6\,mol^{-2}$
5. (*a*) 126 $atm^{-\frac{1}{2}}$ (*b*) $1.60 \times 10^4\,atm^{-1}$ 6. 3.8

Checkpoint 11C: Dissociation [p. 239]
1. 0.43 2. $\alpha = 0.405$, $K_p = 1.98 \times 10^4\,N\,m^{-2}$
3. $\left(1 + \frac{\alpha}{2}\right)$ mol, $\alpha = 0.34$, fraction = 0.15
4. $9.99 \times 10^3\,N\,m^{-2}$

Questions on Chapter 11 [p. 240]
1. (*c*) $\alpha = 0.30$, $K_p = 4.0 \times 10^4\,N\,m^{-2}$
2. (*b*) 3.73 atm^{-1} 3. (*b*) $K_p = 4.0$
4. 2.89 $mol^{-1}\,dm^3$ 5. 8.91 mol dm^{-3}
6. (*a*) $Concentration^{-1}$ (*b*) No, ratio = 0.33
(*c*) 18.0 dm^3
7. (*a*) No, ratio = 19.6 (*b*) L → R
9. −196 kJ mol^{-1}
10. (*a*) (i) $K_c = [SO_3]^2/[SO_2]^2\,[O_2]$
(ii) movement left to right
(iii) decrease
(*b*) (i) SO_2 would liquefy.
(ii) Attainment of equilibrium would be slow.
(iii) Equilibrium position will move towards left-hand side.
11. (*b*) (i) $K_p = p_{NO_2}{}^2/p_{N_2O_4}$
(iii) 3.212 g dm^{-3}
(iv) 55.9 kJ mol^{-1}
12. (*c*) (i) $Ni(s) + 4CO(g) \rightleftharpoons Ni(CO)_4(g)$
(ii) $K_c = [Ni(CO)_4(g)]/[CO(g)]^4$
(iii) increases 16-fold.
(iv) High pressure drives the position of equilibrium from left to right. Since [CO] is raised to the power 4, an *n*-fold increase in [CO] has an n^4-fold effect on the position of equilibrium.
13. (*a*) (i) The reaction is exothermic.
(ii) Attainment of equilibrium would be slow.
(iii) The forward reaction involves a halving of volume.
(iv) Plant that will withstand high pressure is costly.
(*b*) (i) cost of N_2 and H_2
(ii) cost of fuel, cost of plant
(*d*) (i) $K_p = p_{NH_3}{}^2/p_{N_2}\,p_{H_2}{}^3 = 1.07 \times 10^{-3}\,MPa^{-2}$
14. (*a*) $K_p = p_{NO_2(g)}{}^2/p_{N_2O_4(g)}$
(*b*) 77.4
(*c*) (i) increases as pressure increases
(ii) decreases as temperature increases
(*d*) $NO_2 + SO_2 + H_2O \rightarrow H_2SO_4 + NO$
$2NO + O_2 \rightarrow 2NO_2$
a catalyst
15. (*c*) $K_c = [CH_3CO_2C_2H_5][H_2O]/[CH_3CO_2H][CH_3OH] = 3.98$
16. (*b*) (i) $p_{Cl_2} = p_{PCl_3} = 0.342 \times 10^6\,N\,m^{-2}$, $p_{PCl_5} = 0.326 \times 10^6\,N\,m^{-2}$, Total $p = 1.53 \times 10^6\,N\,m^{-2}$
(ii) Yes. The position of equilibrium does not depend on whether it is reached from the left-hand or right-hand side.
17. (*a*) (iii) (*1*) $K_p = p_{H_2O(g)}/p_{H_2(g)}\,p_{O_2(g)}{}^{1/2}$; $atm^{-1/2}$
(*2*) $K_p = p_{CO_2(g)}$; atm
(*3*) $K_c = \dfrac{[Fe^{2+}(aq)]^2\,[Sn^{4+}(aq)]}{[Fe^{3+}(aq)]^2\,[Sn^{2+}(aq)]}$; dimensionless
(*b*) $K = 40^2/(30 \times 90^3) = 7.3 \times 10^{-5}\,atm^{-2}$
This is the value of K_p; therefore equilibrium has been reached.

CHAPTER 12: ELECTROCHEMISTRY

Checkpoint 12A: Electrolysis [p. 252]
1. 1.8 g
2. (*a*) 2.40 g (*b*) 7.10 g (*c*) 6.35 g (*d*) 20.7 g
(*e*) 0.200 g
3. (*a*) Double (*b*) No change (*c*) No change
(*d*) Double
4. 1.23×10^{-3} mol metal, 2.46×10^{-3} mol electrons, +2

Checkpoint 12C: Acids and Bases [p. 259]
5. (*a*) $pK_w = 15$ (*b*) pH = 13

Checkpoint 12D: pH and Dissociation Constants [p. 263]
1. Less 2. CH_3CO_2H is incompletely dissociated
3. HA is the stronger acid.
5. (*a*) 3.0 (*b*) 1.6 (*c*) 4.4 (*d*) 12.4 (*e*) 11.8
6. (*a*) 3.38 (*b*) 2.38
7. (*a*) 3.97×10^{-10} mol dm^{-3} (*b*) 3.02×10^{-11} mol dm^{-3}
8. (*a*) 4.6 (*b*) 3.0

Checkpoint 12E: Titration [p. 270]
4. (*a*) 3.00 (*b*) 3.22 (*c*) 3.70 (*d*) 10.3
(*e*) 11.0 5. 37% Na_2CO_3

Checkpoint 12F: Buffers [p. 272]
4. 1.79×10^{-5} mol dm^{-3} 5. (*a*) 4.76 (*b*) 5.06
6. pH = 4.76

Checkpoint 12G: Salt Hydrolysis [p. 273]
2. (*a*) 8.7 (*b*) 0.167 mol dm^{-3}

Checkpoint 12H: *Solubility Products [p. 240]
2. a^2 mol^2 dm^{-6}, $4b^3$ mol^3 dm^{-9}, $27c^4$ mol^4 dm^{-12}
4. (*a*) 4.5×10^{-3} mol dm^{-3} (*b*) 2.0×10^{-4} mol dm^{-3}
(*c*) 1.0×10^{-4} mol dm^{-3}

Questions on Chapter 12 [p. 277]
5. 287 min 9. 56 cm^3
10. (*a*) 0.11 A (*b*) 6.35×10^{-2} g (*c*) 11.2 cm^3
(*d*) 0.0347 g
11. 292 cm^3
15. (*a*) (ii) 3.2×10^{-4} mol dm^{-3}
(*b*) 0.11 mol dm^{-3}
(*c*) (i) The acid is a weak acid. (ii) K_a
(iii) 9.1×10^{-7} mol dm^{-3}
(*d*) See Figure 12.8(b).
16. (*a*) (i) <7 (ii) 7 (iii) >7
(*b*) (ii) 1.4×10^{-5} mol dm^{-3} (iii) decrease
(*c*) See Chapter 6.
(*d*) See Chapter 23.
17. (*a*) (ii) (*1*) acids: $[Al(H_2O)_6]^{3+}$, NH_4^+, C_6H_5OH;
(*2*) bases: $CH_3CO_2^-$, CH_3NH_2, CO_3^{2-}
(*b*) (*1*) $CH_3CO_2H + HF \rightleftharpoons CH_3CO_2H_2^+ + F^-$
(*2*) $CH_3CO_2H + HF \rightleftharpoons CH_3CO_2^- + H_2F^+$
(*1*) would have the larger equilibrium constant.
(*c*) 11
(*d*) Solubility = $[Pb^{2+}] = K_{sp}/[OH^-]^2 = K_{sp}[H^+]^2/K_w^2$
18. (*c*) (ii) [ethanoate ion] = 0.100 mol dm^{-3},
[ethanoic acid] = 0.050 mol dm^{-3}
(iii) $K_a = 1.74 \times 10^{-5}$ mol dm^{-3}
(*e*) K_a should (i) increase with temperature and
(ii) remain constant with concentration
19. (*a*) (iv) pH = 2.3
(*b*) CO_2 has dispersed; therefore $\Delta S^{\ominus}_{total}$ is positive.
(*c*) See Chapter 14.
20. (*a*) 1.0×10^{-5} mol dm^{-3}
(*c*) 20.3 g
(*d*) (i) 4.98 (ii) pH = 2.00

CHAPTER 13: OXIDATION–REDUCTION EQUILIBRIA

Checkpoint 13A: Electrode Potentials [p. 285]
3. (*a*) −0.46 V 5. (*b*), (*e*) and (*h*)

Questions on Chapter 13 [p. 290]
5. −0.27 V 6. (*a*) +0.40 V (*b*) +0.26 V
(*c*) −0.29 V (*d*) +0.94 V (*e*) −0.46 V
(*f*) +0.78 V (*g*) +0.63 V 7. (*c*) +0.36 V
8. (*b*) (ii) Al > Ti > Cr > Fe (iii) difficult to operate
(iv) Colour changes as $FeCl_3 \rightarrow FeCl_2 \rightarrow Fe$.
(*c*) a weak acid, a low temperature, a protective layer of oxide on the surface of Cr
9. (*b*) −58 kJ mol^{-1}
(*c*) (i) −60.0 kJ mol^{-1}
(ii) $\Delta G^{\ominus} = -nFE^{\ominus}$ therefore $E^{\ominus} = +0.31$ V
(iii) Since $E^{\ominus}$ for $Cu \rightarrow Cu^{2+}$ is −0.31 V, this reaction does not occur spontaneously.
10. (*b*) (i) Reduction potential of H_2O_2 is below (more positive than) that of HCO_2^-; therefore H_2O_2 will oxidise $(C_4H_4O_6)^{2-}$.
(ii) $3H_2O_2 + (C_4H_4O_6)^{2-} \rightarrow 2HCO_2^- + 2CO_2 + 4H_2O$
(iv) H_2O_2 as oxidant:
$H_2O_2 + 2H^+ + 2e^- \rightleftharpoons 2H_2O$; $E^{\ominus} = -0.68$ V
H_2O_2 as reductant:
$H_2O_2 \rightleftharpoons O_2 + 2H^+ + 2e^-$; $E^{\ominus} = +1.77$ V
Combining half-equations, $2H_2O_2 \rightarrow O_2 + 2H_2O$
$\Delta G^{\ominus} = -nF(1.77 - 0.68)$
Since $\Delta G^{\ominus}$ is negative, the reaction proceeds spontaneously from left to right.
(*c*) See §24.14.4.
11. (*a*) (ii) Cu is positive; electrons flow from Zn to Cu; oxidation occurs at Zn.
(iii) $E^{\ominus} = 0.34 + 0.76 = 1.10$ V
(iv) (*1*) Ag(s) deposited; Cu dissolves as Cu^{2+}(aq)
(*2*) no reaction
(*3*) Br_2 and Fe^{2+} react to form $Br^-(aq) + Fe^{3+}(aq)$
(*b*) See §8.4.
(*c*) (ii) 20
12. (*c*) (ii) 5.6×10^{-12} mol dm^{-3}
(iii) $Cu^{2+}(aq) + H_2(g) \rightarrow Cu(s) + 2H^+(aq)$
(iv) $K_c = 1.8 \times 10^{11}$ mol dm^{-3}

CHAPTER 14: REACTION KINETICS

Checkpoint 14B: Initial Rates [p. 304]
3. (*a*) Rate = $k[A]^2$ (*b*) $k = 5.0$ dm^3 mol^{-1} s^{-1}
(*c*) 1.8 mol dm^{-3} s^{-1}
4. (*a*) 1 wrt **A**; 2 wrt **B**; 3
(*b*) $k = 1.5 \times 10^{-3}$ mol^{-2} dm^6 min^{-1}
(*c*) 8.7×10^{-6} mol dm^{-3} min^{-1}

Checkpoint 14C: First-Order Reactions [p. 307]
1. 8047 s (134 min) 2. 6.25% 3. (*a*) 1
(*b*) 6.3×10^{-4} mol dm^{-3} s^{-1} (*c*) 7.9×10^{-4} s^{-1}
4. 25.5 min 5. (*b*) 14.6 cm^3 (*c*) 10.3 min (618 s)
(*d*) $t_{1/2}$ is the same for different concentrations
(*e*) 0.0673 min^{-1} (1.12×10^{-3} s^{-1})
(*f*) 1.06×10^{-6} mol dm^{-3} s^{-1} (*g*) 9.7×10^{-4} s^{-1}

Checkpoint 14D: Order of Reaction [p. 309]

1. (*a*) $1.7 \times 10^{-4}\,mol\,dm^{-3}\,s^{-1}$
   (*b*) $5.1 \times 10^{-4}\,mol\,dm^{-3}\,s^{-1}$
   (*c*) $15.3 \times 10^{-4}\,mol\,dm^{-3}\,s^{-1}$
2. 1st order; $k = 1.75 \times 10^{-4}\,s^{-1}$
3. 1st order; $k = 2.44 \times 10^{-4}\,s^{-1}$

Checkpoint 14E: Reaction Kinetics [p. 317]

6. Ratio = 1.02   9. (*a*) $52.9\,kJ\,mol^{-1}$
   (*b*) $83.8\,kJ\,mol^{-1}$ 10. $166\,kJ\,mol^{-1}$

Questions on Chapter 14 [p. 321]

5. 50 h   6. 3700 years
7. $\Delta H^{\ominus}_{Reaction} = -19\,kJ\,mol^{-1}$
8. (*a*) 1 (*b*) 1 (*c*) $6.0 \times 10^{-3}\,dm^3\,mol^{-1}\,s^{-1}$; $1.08 \times 10^{-5}\,mol\,dm^{-3}\,s^{-1}$
10. Reaction 1: 1st order, $t_{1/2} = 6.7$ min,
    Reaction 2: zero order, $t_{1/2} = 5.0$ min,
    Reaction 3: 2nd order, $t_{1/2} = 10.0$ min,
18. $E = 107\,kJ\,mol^{-1}$, $A = 9.1 \times 10^{13}\,s^{-1}$
20. (*d*) Order = 1, $k = 6.0 \times 10^{-4}\,s^{-1}$
21. (*a*) (i) $1.49\,mol\,dm^{-3}$
    (ii) first-order since the graph of $\ln c$ against t is linear
    (iii) gradient $= -k = 1.82 \times 10^{-2}\,mol\,dm^{-3}\,min^{-1}$
    (*b*) (i) [water] remains almost constant since [water] is in large excess.
    (ii) The reaction is first-order with respect to [acid].
22. (*a*) (ii) The molecularity of each step is 2.
    (*b*) A plot of time against $[H^+]_{initial}$ shows that when the concentration halves the time is reduced to $\frac{1}{4}$ of its value: therefore the order of reaction with respect to $[H^+]$ is 2.
    (ii) Rate $= k[Br^-][BrO_3^-][H^+]^2$
    (iii) k (iv) $dm^9\,mol^{-3}\,s^{-1}$
23. (*c*) $1226\,kJ\,mol^{-1}$
    (*d*) The energy of activation must be supplied.
    (*e*) $\Delta H_F(H_2O)$ has a more negative value than $\Delta H_F(H_2O_2)$; therefore water is more likely to be formed.
24. (*b*) (i) Plot concentration vs. time and draw tangents at $[C_6H_5NH_2] = 0.0200\,mol\,dm^{-3}$ and $0.0100\,mol\,dm^{-3}$. Since rate at 0.0200 M/rate at 0.0100 M = 4, reaction is second-order.
    (ii) Rate $= k[C_6H_5NH_2]^2$
    Rate $= k[C_6H_5NH_2]^2[C_6H_5COCl]$
    Rate $= k[C_6H_5NH_2]^2[C_6H_5COCl]^2$
    (iii) $0.068\,dm^3\,mol^{-1}\,s^{-1}$
    (iv) increase
25. (*b*) (i) $10.4 \times 10^{-6}\,mol\,dm^{-3}\,s^{-1}$
    (ii) $n = 1$ (Rate $\propto$ Concentration)
    (iii) $5.2 \times 10^{-4}\,s^{-1}$
    (*c*) (ii) 12 minutes
    (*d*) See Figure 14.3.
26. (*b*) (ii) Order with respect to H_2O_2 = 1 (from Expts. 1, 2)
    Order with respect to I^- = 1 (from Expts. 1, 3)
    Order with respect to H^+ = 0 (from Expts. 1, 4)
    Overall order = 2
    (iii) $1.12 \times 10^{-12}\,dm^3\,mol^{-1}\,s^{-1}$
    (iv) $1.34 \times 10^{-2}\,mol\,dm^{-3}$
27. (*b*) (i) The rate-determining step is the reaction of Br and H_2; therefore
    $d[HBr]/dt = 2k_2[H_2][Br] = 2k_2K^{1/2}[H_2][Br_2]^{1/2}$
    since $[Br]^2/[Br_2] = K$
    (ii) Data set 1 is consistent with the law:
    first-order with respect to H_2 (Runs 1, 2 of Set 1),
    1/2 order with respect to Br_2 (Runs 2, 3 of Set 1 and 1, 4 of Set 1)
    (*c*) (1) $d[HI]/dt = 2k_1[H_2][I_2]$
    (2) $d[HI]/dt = 2k_2[H_2][I]^2 = 2k_2K[H_2][I_2]$
    since $[I]^2/[I_2] = K$
    The two routes cannot be distinguished by studies of the effect of concentration on the rate. UV should speed up reaction by Route (2).
28. (*a*) (i) $a = 1, b = 1$
    (ii) $17.5\,mol^{-1}\,dm^3\,s^{-1}$
    (*b*) Collision between $S_2O_8^{2-}$ and I^- is rate-determining.
    (*c*) (i) Rate $\propto [Fe^{2+}]$
    (ii) Fe^{2+} is oxidised by $S_2O_8^{2-}$ to Fe^{3+}; then Fe^{3+} oxidises I^- to I_2.
    (iii) $E^{\ominus}$ values show that $S_2O_8^{2-}$ is a more powerful oxidising agent than Fe^{3+} and Fe^{3+} is a more powerful oxidising agent than I_2. Both reaction steps in (ii) are between oppositely charged ions, whereas the reaction between $S_2O_8^{2-}$ and I^- is between ions of the same charge.
    (iv) No, Sn^{4+}/Sn^{2+} has $E^{\ominus} = +0.15$ V, which is lower than $E^{\ominus}$ for $I_2/I^- = +0.54$ V
29. (*b*) Time ($0\,cm^3$ to $40\,cm^3\,N_2$) = 12 min
    Time ($40\,cm^3$ to $60\,cm^3\,N_2$) = 13 min
    Time ($60\,cm^3$ to $70\,cm^3\,N_2$) = 12 min
    The half-life is independent of concentration, and the reaction is first-order with respect to benzenediazonium chloride.
    (*c*) Repeat at a different temperature. Obtain rate constants at different temperatures; plot $\ln k$ vs. $1/T$; see Figure 14.15.

Appendix 1: A Selection of Questions on Physical Chemistry [p. 327]

1. (*a*) Reaction 1: $\Delta H = +131\,kJ\,mol^{-1}$,
   Reaction 2: $\Delta H = -1220\,kJ\,mol^{-1}$
   (*b*) (i) low pressure (amount of gas increases) and high temperature (endothermic reaction). Carry off products as formed.
   (ii) high pressure (decrease in amount of gas); temperature only moderate (since reaction is exothermic), and a catalyst. Use heat evolved in Reaction 2 to heat the reactants in Reaction 1.
2. (*a*) (ii) $-848\,kJ\,mol^{-1}$
   (*c*) (i) $9.9 \times 10^{-11}\,mol^2\,dm^{-6}$
   (ii) $[Ba^{2+}] = 4.0 \times 10^{-10}\,mol\,dm^{-3}$
3. (*e*) 4.4
4. (*a*) (iv)

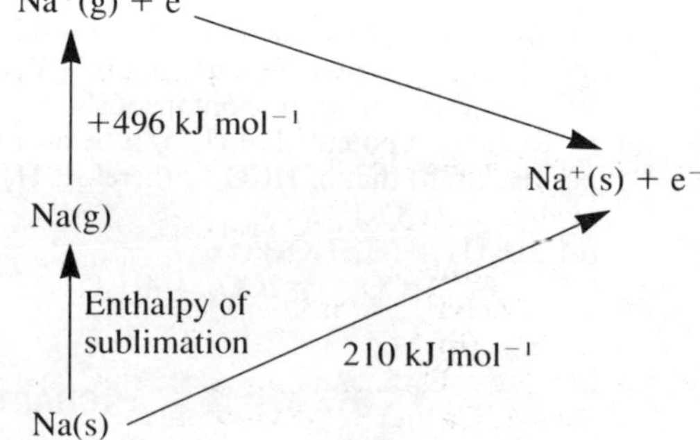

5. (*b*) (i) Both use hydrogen.
   (ii) In each case, high pressure and low temperature favour the right-hand side.
   (iii) The rate of attainment of equilibrium depends on the temperature and the presence of catalysts.
   (iv) 2.28 atm

(*c*) (i) Saving on fuel costs and on the cost of building a plant that can withstand high pressure. Possibly a better catalyst.
(ii) A plot of $\ln k$ vs. $1/T$ gives $E = 20\,\text{kJ mol}^{-1}$
(iii) $120\,\text{kJ mol}^{-1}$
(*d*) Investigate biotechnology.
6. (*c*) (ii) (*1*) **C** (*2*) **B** (*3*) **B, C, D** (iv) 3.5
7. (*a*) (i) Oxidation: $C_6H_{12}O_6 + 6O_2 \rightarrow 6CO_2 + 6H_2O$
Anaerobic decomposition: $C_6H_{12}O_6 \rightarrow 3CH_4 + 3CO_2$
(ii) $60\,m^3$ (iii) 185 MJ
(*b*) $5.55 \times 10^{-2}\,\text{mol dm}^{-3}$
8. (*a*) $M(CH_3CO_2H) = 120$; the molecules are dimers, and association is complete.
$M(PCl_5) = 150$; PCl_5 is partially dissociated into PCl_3 and Cl_2, with $K_p = 45\,\text{kPa}$
(*b*) $H_2(g), H^+(aq, 1.00\,M)$ ¦¦ $Ni^{2+}(aq, 1.00\,M), Ni(s)$
$[Co^{2+}] = 3.6\,\text{mol dm}^{-3}$
(*c*) $Ba(OH)_2$ titre gives total amount of H^+ in $HCl + CH_3CO_2H = 0.0872\,\text{mol}$; $AgNO_3$ titre gives amount of $HCl = 0.0408\,\text{mol}$;
difference $= 0.0464\,\text{mol}$ = amount of CH_3CO_2H = amount of C_2H_5OH. Initial $CH_3CO_2C_2H_5 = 0.1158\,\text{mol}$; residual amount $= 0.0694\,\text{mol}$.
$K_c = 0.230 \times 0.0694/(0.0464)^2 = 7.4$
9. (*a*) 5.6
(*b*) 13.7
(*c*) Phenol is a much weaker acid than carbonic acid. The equilibrium:
$$2C_6H_5OH(aq) + CO_3^{2-}(aq) \rightleftharpoons 2C_6H_5O^-(aq) + H_2O(l) + CO_2(aq)$$
lies very much over to the left-hand side. The equilibria in the solution are:
$C_6H_5OH \rightleftharpoons C_6H_5O^- + H^+$;
$K_a = 1.0 \times 10^{-10}\,\text{mol dm}^{-3}$
$CO_2 + H_2O \rightleftharpoons H^+ + HCO_3^-$;
$K_a = 4.5 \times 10^{-7}\,\text{mol dm}^{-3}$
$HCO_3^- \rightleftharpoons H^+ + CO_3^{2-}$;
$K_a = 2 \times 10^{-4}\,\text{mol dm}^{-3}$
For any small amount of H^+ in the solution, combination with $C_6H_5O^-$ is the most favoured equilibrium.

PART 3: INORGANIC CHEMISTRY

CHAPTER 15: PATTERNS OF CHANGE IN THE PERIODIC TABLE

Questions on Chapter 15 **[p. 341]**
5. (i) F (ii) Na (iii) Li (iv) Mg (v) Cl (vi) F (vii) Al (viii) Al (ix) Al (x) C (xi) B (xii) C
12. (*b*) (i) $-847\,\text{kJ mol}^{-1}$

CHAPTER 17: HYDROGEN

Questions on Chapter 17 **[p. 358]**
6. (*a*) H_2 −143 MJ, CH_4 −55.7 MJ, C_8H_{18} −48.0 MJ
(*b*) (i) more exothermic (ii) greenhouse effect
(iii) danger of explosion (iv) electrolysis
7. (*e*) React with sodium to give deuterium; add this to ethene.
$2D_2O + 2Na \rightarrow D_2 + 2NaOD$
$D_2 + CH_2{=}CH_2 \rightarrow CH_2DCH_2D$
8. (*a*) (i) 1 electron, 1 proton + 2 neutrons in the nucleus
(ii) loss of an electron from the nucleus
(*b*) 49.4 years
(*d*) (i) $3T_2O + PCl_3 \rightarrow 3TCl + T_3PO_3$
(ii) $6T_2O + Mg_3N_2 \rightarrow 2NT_3 + 3Mg(OT)_2$
(iii) $Ca + 2T_2O \rightarrow Ca(OT)_2 + T_2$
9. (*a*) 48 MJ

CHAPTER 18: THE s BLOCK METALS: GROUPS 1 AND 2

Questions on Chapter 18 **[p. 378]**
7. (*a*) (i) Ba
(ii) $H_2(g) + Mg(s) \rightarrow MgH_2(s)$
(iii) electrolysis
(iv) size $Ba^{2+} > Sr^{2+} > Ca^{2+} > Mg^{2+}$
(*b*) (i) Ba (ii) Mg
8. (*a*) (iii) $Ba + O_2$ forms a peroxide; MgO is only slightly soluble. Ca and Sr are possibilities.
(iv) Points to Mg, Ca or Sr. $Ba(OH)_2$ is more soluble.
(v) Solubility should increase down the group (as lattice enthalpy decreases).
(*b*) Points to Ca or Sr as Mg reacts slowly with water.
(*c*) (i) $A_r(X) = 40.1$
(ii) (*1*) 4.99×10^{-5} mol (*2*) $y = 0.100$ g
(*d*) (i) Ca (ii) brick red
9. (*b*) $BaSO_4$ is insoluble.
(*c*) The small Be^{2+} ion has a highly negative enthalpy of hydration.
(*d*) diagonal relationship
10. (*c*) (i) **W** = $^{228}_{89}Ac$, **X** = $^{228}_{90}Th$, **Y** = $^{224}_{88}Ra$, **Z** = $^{220}_{86}Rn$
(iii) 12.5 mg
11. (*c*) $BaMg(CO_3)_2$
12. (*a*) (i) B has 3 valence electrons; N has 5 valence electrons, forming 3 bonds and a lone pair; tetrahedral
(ii) $[Ag(NH_3)_2]^+\ Cl^-$ is soluble.
(*b*) (i) X could be Mg.
(ii) $Mg(OH)_2$ dissociates on heating to give $MgO + H_2O$; $MgCO_3$ is stable up to 800 °C (like limestone).

CHAPTER 19: GROUP 3

Checkpoint 19C: The Aluminium Problem **[p. 390]**
4. 10.7 MC

Questions on Chapter 19 **[p. 391]**
6. (*a*) (iii) 77.4%
(*b*) (i) 2.6×10^{-3} mol (ii) 0.0923 g
(iii) 1.026×10^{-3} mol (iv) Al_2Cl_5
(v) Some Cl^- is not titrated because it is part of a complex $[AlCl_4]^-$ ion.
(*c*) (i)

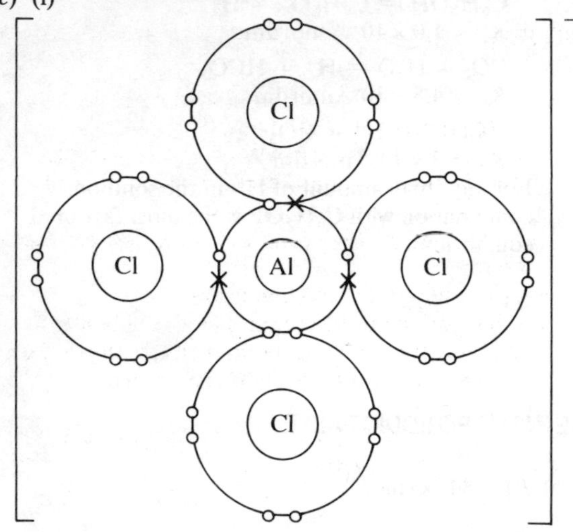

(ii) tetrahedral

7. (*a*) (i) to lower the melting temperature of Al_2O_3
(ii) electrodes
(*b*) (i) Ireland has more land than Anglesey for disposing of the red mud waste. Ireland is keener than Canada to benefit from the employment that the plant creates.
(ii) Any fluoride emission would be in the less densely populated area. It is nearer to the port, but transport to the mainland is needed.
(*c*) (i) The temperature is lower. The consumption of electricity is lower. There is no fluoride emission. Cryolite need not be imported.
(ii) Investment would be needed.
(*d*) (i) hydrolysis (ii) dimer Al_2Cl_6

CHAPTER 20: GROUP 7: THE HALOGENS

Checkpoint 20G: Pesticides **[p. 410]**
4. (*b*) (ii) 430.5 g (3 mol) (iii) silver mirror
(iv) oxidation–reduction

Questions on Chapter 20 **[p. 411]**
7. (*b*) (i) (*1*) +5 (*2*) +1
(ii) $Cl_2(aq) + 2OH^-(aq) \rightarrow Cl^-(aq) + ClO^-(aq) + H_2O(l)$
(iii) $4KClO_3 \rightarrow KCl + 3KClO_4$
8. (*a*) (i) Thermal stability will decrease down the group.
(ii) Acid strength increases down the group.
(*b*) (i) disproportionation
$I_2(aq) + 2NaOH(aq) \rightarrow NaI(aq) + NaIO(aq) + H_2O(l)$
(ii) and (iii) see § 31.6.7.
(*c*) (i) disproportionation
$3I_2(s) + 6NaOH(aq) \rightarrow 5NaI(aq) + NaIO_3(aq) + 3H_2O(l)$
(ii) $5I^-(aq) + IO_3^-(aq) + 6H^+(aq) \rightarrow 3I_2(s) + 3H_2O(l)$
The original amount of iodine is liberated.
9. (*b*) AgCl(s) dissolves as $[Ag(NH_3)_2]^+Cl^-$. AgBr is precipitated because it has a very low solubility product. It dissolves as $Na^+[Ag(CN)_2]^-$.
10. (*a*) (i) $I_2(aq) + 2S_2O_3^{2-}(aq) \rightarrow 2I^-(aq) + S_4O_6^{2-}(aq)$
(ii) Read the burette exactly as e.g. 7.0 cm³. Record the amount of KI added. Add starch indicator near the end-point. Repeat the titration.
(iii) 2.33×10^{-5} mol dm^{-3}
(iv) starch
(*b*) (i) HCl is formed from $Cl_2 + H_2O$
(ii) poisonous (iii) $Ca(ClO)_2$
(iv) $4ClO_2 + 2H_2O \rightarrow HClO + 3HClO_3$
11. (*b*) (i) Na_3AlF_6 (ii) octahedron
(iii) in electrolysis of molten bauxite
(*c*) (i) HF molecules are associated by hydrogen bonding.
(ii) HF forms stronger hydrogen bonds than H_2O. Protein molecules can bond to HF more strongly than to water.
13. (*c*) oxidation–reduction
$2BrO_3^-(aq) + 12H^+(aq) + 5S^{2-}(aq) \rightarrow Br_2(aq) + 5S(s) + 6H_2O(l)$

CHAPTER 21: GROUP 6

Checkpoint 21C: Bonding **[p. 422]**
2. (*a*) +950 kJ mol^{-1} (*b*) +555 kJ mol^{-1}
3. (*a*) +388 kJ mol^{-1} (*b*) +856 kJ mol^{-1}

Questions on Chapter 21 **[p. 441]**
8. (*c*) 1.60 tonne
(*d*) (i) $SO_2 + C \rightarrow S + CO_2$
(ii) 187 g per kg (iv) 9400 litres

11. (*a*) For 1(F—F) bond broken and 2(S—F) bonds made; $\Delta H = -434\,kJ\,mol^{-1}$. The formation of SF_6, $S(s) + 3F_2(g) \rightarrow SF_6(g)$, requires 3(F—F) bonds broken and 6(S—F) bonds made; $\Delta H = 3(-434) = -1302\,kJ\,mol^{-1}$; therefore $\Delta H_F\,(SF_6) = -1302\,kJ\,mol^{-1}$, and the average value for the S—F bond energy is $1302/6 = 217\,kJ\,mol^{-1}$.

CHAPTER 22: GROUP 5B

Questions on Chapter 22 [p. 460]

14. Urea
16. (*a*) (iv) 9%
17. (*c*) **A** = $Pb(NO_3)_2$
18. (*b*) **A** = PF_3Cl_2, **B** = $P_2F_6Cl_4$ = $[PCl_4]^+\,[PF_6]^-$, **C** = $PFCl_4$, **D** = $Na^+PF_6^-$

CHAPTER 23: GROUP 4B

Checkpoint 23B: Carbon [p. 488]

2. (a) $-216\,kJ\,mol^{-1}$ (*b*) $+46\,kJ\,mol^{-1}$
10. (*c*) (i) The formation of 4(Si—O) bonds is more exothermic than the formation of 2(Si=O) bonds; therefore $(SiO_2)_n$ is preferred to SiO_2. The formation of 2(C=O) bonds is more exothermic than the formation of 4(C—O) bonds; therefore the formation of CO_2 is preferred to a macromolecule.
(ii) Silicon does not catenate. The Si—Si bond is weaker than the Si—O bond. In $Si_2H_6 \rightarrow 2SiO_2$,
bonds broken: $176 + 6(318) = 2084\,kJ\,mol^{-1}$
bonds made: $2 \times 638 = 7276\,kJ\,mol^{-1}$;
difference = $-5192\,kJ\,mol^{-1}$
Contrast with $C_2H_6 \rightarrow 2CO_2$:
bonds broken: $348 + 6(412) = 2820\,kJ\,mol^{-1}$,
bonds made: $2 \times 743 = 1486\,kJ\,mol^{-1}$;
difference = $+1334\,kJ\,mol^{-1}$
12. (*e*) $H_2 + SiCl_4 \rightarrow Si$; $LiAlH_4 + SiCl_4 \rightarrow SiH_4$
(*f*) CH_4: no reaction; SiH_4: forms SiO_2; L-shell restriction in carbon
13. (*a*) Change is endothermic and involves a decrease in volume.
(*b*) (i) Energy of activation is lower.
(*c*) Silicon cannot form multiple Si—Si bonds.
(*d*) (i) Polarity of **E**—F bond increases down the group; SnF_4 has some ionic character and has a macromolecular structure.
(ii) It forms PbF_2 which is ionic and much more stable than PbF_4.

CHAPTER 24: THE TRANSITION METALS

Questions on Chapter 24 [p. 530]

14. $0.036\,mol\,dm^{-3}$ 15. Cr^{2+}
25. (*b*) **A** = Zn^{2+}, Cu^{2+}, Fe^{2+}
B = $Zn(OH)_2(s)$, $Cu(OH)_2(s)$, $Fe(OH)_2(s)$
C = $Na_2ZnO_2(aq)$
D = $Zn(OH)_2(s)$
E = $ZnSO_4(aq)$
F = $Cu(OH)_2(s)$, $Fe(OH)_2(s)$
G = $Fe(OH)_2(s)$
H = Fe_2O_3
I = $[Cu(NH_3)_4]^{2+}(aq)$
27. (*a*) **A** = $Cr(OH)_3(s)$, **B** = $[Cr(NH_3)_6]^{3+}$,
C = $CrO_3^{2-}(aq)$, **D** = $CrO_4^{2-}(aq)$
E = $Cr_2O_7^{2-}(aq)$
(*b*) **F** = $[Co(H_2O)_6]^{2+}(aq)$, **G** = $Co(OH)Cl(s)$,
H = $[Co(NH_3)_6]^{2+}(aq)$, **I** = $[Co(NH_3)_6]^{3+}(aq)$
(*c*) **J** = $[Cu(H_2O)_6]^{2+}(aq)$, **K** = $[CuCl_4]^{2-}$,
L = $[CuCl_2]^-$, **M** = $CuCl(s)$
(*d*) $Cu_2O(s) + H_2SO_4(aq) \rightarrow Cu(s) + CuSO_4(aq) + H_2O(l)$
28. (*b*) **A** = $[Co(NH_3)_5(H_2O)]^{3+}\ 3Cl^-$
B = $[Co(NH_3)_5Cl]^{2+}\ 2Cl^-$
C = $[Co(NH_3)_4Cl_2]^+\ Cl^-$
D = $Co(NH_3)_3(NO_2)_3$
(*c*) $[Co(H_2NCH_2CH_2NH_2)_3]^{3+}\ 3Cl^-$; see Figure 24.14.
29. (*c*) (i) Fe(Ar) $4s^23d^6$ might show an oxidation state of +8.
(ii) In practice the maximum oxidation state is +6 (four unpaired d electrons + two s electrons). Fe^{VI} is a highly oxidising state. The formation of Fe^{3+} is preferred because it contains a half-full $3d^5$ shell, which is a stable configuration.
(*d*) (i) ClO^-
(ii) $Fe_2O_3 + 2OH^- + Cl_2 \rightarrow 2FeO_4^{2-} + ClO^- + Cl^- + H_2O$
(iii) $FeO_4^{2-}(aq) + 4H^+(aq) \rightarrow O_2(g) + Fe^{3+}(aq) + 2H_2O(l)$
(iv) At high pH, $[H^+]$ is low and the left-hand side of the equilibrium is favoured.
(*e*) 96%
30. (*c*) (i) $E^\ominus = +0.84\,V$
(ii) $2Fe(s) + O_2(aq) + 2H_2O(l) \rightarrow 2Fe^{2+}(aq) + 4OH^-(aq)$
(iv) OH^- drives the equilibrium to the left-hand side.
31. (*a*) +2
(*b*) octahedron
(*c*) coordination from each $—CO_2^-$ group and from each N atom
(*d*) $—CO_2^-$ groups would form $—CO_2H$ groups.
(*f*) to drive the equilibrium over to formation of the complex
(*h*) 202
33. (*a*) (i) hydrolysis
(ii) The complex ion $[Fe(CNS)(H_2O)_5]^{2+}$ is red. CNS^- is displaced by F^- to form a colourless ion $[FeF(H_2O)_5]^{2+}$.
(*b*) (i) −1 (ii) 2 : 1 (iii) +1; N_2O, dinitrogen oxide
(iv) $2NH_2OH(aq) + 4Fe^{3+}(aq) \rightarrow 4Fe^{2+}(aq) + N_2O(g) + H_2O(l) + 4H^+(aq)$

Appendix 2: Topics which Span Groups of the Periodic Table
A2.1: Some Detective Work [p. 534]

1. (*a*) **A** = CO (*b*) **B** = O_3 (*c*) **C** = PH_3

2. (*a*) **A** = Cr (*b*) **B** = Pb_3O_4 (*c*) **C** = P
   (*d*) **D** = $FeCl_2$
3. **E** = HF, **F** = H_2O, **G** = H_2O_2
4. **J** = K_2CrO_4, **K** = MnO_2, **L** = KI, **M** = $SnCl_2$
5. **P** = $(NH_4)_2Cr_2O_7$, **Q** = $NaNO_2$, **R** = K_2SO_3,
   S = $CrCl_2 \cdot nH_2O$
6. **A** = MnO_2, **B** = K_2MnO_4, **C** = $KMnO_4$,
   D = Mn^{2+}(aq), **E** = MnS
7. **P** = $FeCl_4^-$(aq), **Q** = $Fe(OH)_3$, **R** = $Fe(NO_3)_2$,
   S = $Fe(OH)_2$, **T** = $FeCO_3$

A2.3: Patterns in the Periodic Table

20. (*c*) (i) CO_2(g) and $Al(OH)_3$(s)
21. (*a*) $K^+Fe^{3+}[Fe^{II}(CN)_6]^{4-}$ is formed in both cases.
    (*b*) The yellow product is Cu^{2+} $[CuCl_4]^{2-}$. On dilution, the yellow $[CuCl_4]^{2-}$(aq) ions are converted into blue $[Cu(H_2O)_6]^{2+}$ ions and Cl^- ions. When the solution contains a mixture of yellow and blue ions, the colour is green.
    (*c*) SnI_4 is formed.
    (*d*) BaO_2 + acid gives H_2O_2; this reduces manganate(VII) and also oxidises KI to I_2.
22. **A** = CrO_3, **B** = Cr_2O_3, **C** = $(NH_4)_2Cr_2O_7$,
    D = CrO_2Cl_2
    $4CrO_3 \rightarrow 2Cr_2O_3 + 3O_2$
    $CrO_3 + H_2O \rightarrow H_2CrO_4$
    $(NH_4)_2Cr_2O_7 \rightarrow N_2 + Cr_2O_3 + 4H_2O$
    $CrO_3 + 2HCl \rightarrow CrO_2Cl_2 + H_2O$
23. **W** = $CuSO_4$, **X** = K_2CrO_4, **Y** = $MnCl_2$,
    Z = $(CH_3CO_2)_2Zn$
25. (*a*) **A** = H_2, **B** = Cl_2, **C** = HCl(g), **D** = HCl(aq),
    E = $AlCl_3$, **F** = NaCl, **G** = NaClO
    (*b*) Yellow **E** is Al_2Cl_6 (anhydrous).
    (*c*) White **E** is $AlCl_3 \cdot 6H_2O$; hydrolysed in solution.

PART 4: ORGANIC CHEMISTRY

CHAPTER 25: ORGANIC CHEMISTRY

Questions on Chapter 25 [p. 555]

1. (*a*) (ii) enantiomers

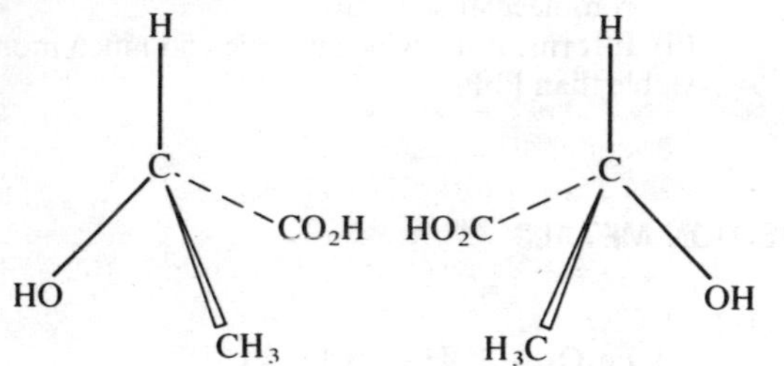

(*b*) **A** *cis*- $H_3C(H)C{=}C(H)CH_3$ and *trans*- $H(H_3C)C{=}C(H)CH_3$

B $H_2C{=}CH{-}CH_2{-}CH_3$
C enantiomers of $C_2H_5CHBrCH_3$. (Arrangement of bonds is similar to that in part (*a*).)

CHAPTER 26: THE ALKANES

Questions on Chapter 26 [p. 569]

10. 6.4 kg
12. (*c*) **B** could be $CH_3CH_2CH{=}C(CH_3)CH_2CH_3$, *cis*- and *trans*-isomers.
    C would be $CH_3CH_2CH_2CH(CH_3)CH_2CH_3$, enantiomers (for bond distribution see answer to Chapter 25, Question 1).
13. (*b*) (i) $CH_3CH(CH_3)CH_2C(CH_3)_3$
    (*d*) C_6H_{12}

CHAPTER 27: ALKENES AND ALKYNES

Questions on Chapter 27 [p. 589]

6. (*b*) (i)

```
   H3C   Br
     \  /
      C
     / \
  H2C   CHBr
   |     |
  H2C   CH2
     \  /
      CH
      |
H3C — C — CH2Br
      |
      Br
```

   (ii) higher M_r
   (*c*) (iii) 4
7. (*d*) (i) C—C (π) = 253 kJ mol^{-1}
    C—C (σ) = 348 kJ mol^{-1}
    (ii) total ($\pi + \sigma$) = 601 kJ mol^{-1}
8. (*a*) 65%
10. (*c*) **X** + $2Br_2$
    (*d*) (i) 2
    (*e*) $CH_2{=}CH{-}CH{=}CH_2$ and $CH_2{=}C{=}CH{-}CH_3$

CHAPTER 28: BENZENE AND OTHER ARENES

Questions on Chapter 28 [p. 613]

7. (*a*) (i) **F** is $C_6H_5-CH_2-CH_3$,

G is cyclohexyl with $CH-CH_2-CH_3$: $H_2C(CH_2-CH_2)_2CH-CH_2-CH_3$

(ii) Bromine adds to ethene to form 1,2-dibromoethane. Bromine with $AlBr_3$ substitutes in benzene to form bromobenzene + HBr.

(iii) Through the intermediate, $D^+AlBr_4^-$, DBr substitutes to form C_6H_5D + HBr.

(*b*) A generator of free radicals

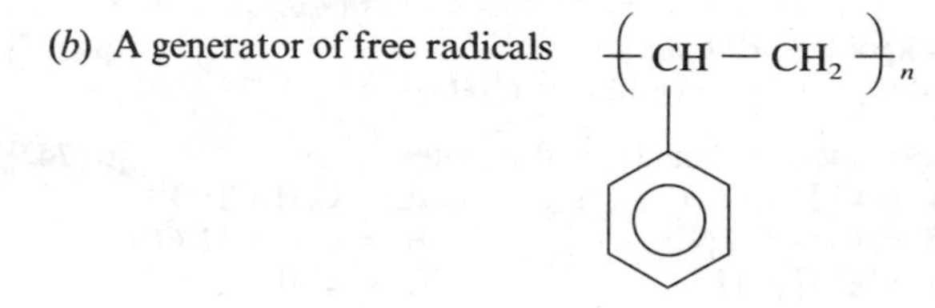

CHAPTER 29: HALOGENOALKANES AND HALOGENOARENES

Questions on Chapter 29 [p. 633]

5. (*d*) (i) Hydrolyse to $CH_3CH_2CH_2OH$; oxidise to $CH_3CH_2CO_2H$
(ii) KCN, followed by hydrolysis of $CH_3CH_2CH_2CN$ to $CH_3CH_2CH_2CO_2H$

6. (*b*) (ii) 7.5 cm^3 of 1.0 M H_2SO_4 because only the $—CH_2Br$ is hydrolysed

CHAPTER 30: ALCOHOLS, PHENOLS AND ETHERS

Questions on Chapter 30 [p. 665]

17. (*b*) **P** is $CH_3CH_2CHOHCH_3$
Q is *cis*- and *trans*-$CH_3CH{=}CHCH_3$
S is $C_2H_5-{}^*C(H)(Br)-CH_3$ with a chiral *C

18. (*a*) For distinguishing between pairs of compounds see Checkpoint 35F.
(*b*) For ethanol see Chapter 30.

19. (*c*) PR: $CH_3CH_2CH_2CH_2OH$ and $CH_3CH(CH_3)CH_2OH$
SEC: $CH_3CH_2{}^*CHOHCH_3$; (tertiary): $(CH_3)_3COH$

CHAPTER 31: ALDEHYDES AND KETONES

Questions on Chapter 31 [p. 690]

7. $C_6H_5COCH_3$, $C_6H_5CH_2CHO$, $CH_3C_6H_4CHO$
8. **P** = cyclohexene, **Q** = hexane-1,6-dial
9. **A** = C_6H_5CHO
11. **A** = $(CH_3)_2CHCH(OH)CH_3$
12. **P** = $CH_3COCH_2CH_2CHO$
13. **Q** = $CH_3CH(OH)CH_2CH_2CH_2OH$
17. **X** = $(CH_3)_3CCHO$ **Y** = $(CH_3)_3CCO_2H$
Z = $(CH_3)_3CCH_2OH$
19. **B** = RCO_2H, **C** = RCH_2OH, **D** = RCO_2CH_2R
20. (*e*) **A** = $CH_3CO_2CH{=}CH_2$, **B** = CH_3CO_2Na,
C = $CH_2{=}CHOH$, **D** = CH_3CHO, **E** = CH_3CO_2H
21. (*b*) (ii) **P** = $CH_3CHOHCH_2NH_2$,
Q = $CH_3CHOHCO_2H$
22. **D** = $CH_3COCH_2CH_3$, **E** and **F** are $CH_3CHOHCH_2CH_3$ (for distribution of bonds in enantiomers see the answer to Chapter 25, Question 1).
23. (*b*) (iii) There are *cis-trans* isomers of $CH_3CH{=}NOH$ but not of $(CH_3)_2C{=}NOH$.

CHAPTER 32: AMINES

Questions on Chapter 32 [p. 713]

8. **A** = $(CH_3)_2CHCN$, **B** = $(CH_3)_2CHCH_2NH_2$,
C = $(CH_3)_2CHCH_2NHCOCH_3$,
D = $(CH_3)_2CHCH_2NHCH_3$,
E = $(CH_3)_2CHCH_2\overset{+}{N}(CH_3)_3I^-$, **F** = $(CH_3)_2CHCO_2H$,
G = $(CH_3)_2CHCH_2OH$, **H** = $(CH_3)_2C{=}CH_2$

11. (*a*) **A** has C 77.4%, H 7.5%, N 15.0%
B has C 78.5%, H 8.4%, N 13.1%
C has values in between those of **A** and **B**.
(*b*) (ii) Conditions: warm with acid chloride or anhydride
(*d*) **B** is a weak base; the salt formed has pH <7, therefore use bromophenol blue.

12. (*b*) (i) $C_2H_5CH(CH_3)NH_2$ (for distribution of bonds see the answer to Chapter 25, Question 1)
(ii) $CH_2{=}CH—CH_2Br$ $CH_2{=}CBr—CH_3$
$CH_2{=}C(CH_3)Br$
$CH_3—CH{=}CHBr$ (*cis*- and *trans*-)
(*c*) (iii) $CH_3CH_2CH_2NH_2$

16. Reduce 5.5 g nitrobenzene to phenylamine; diazotise to give benzenediazonium chloride. Cool the solution and add to a cold solution of 4.0 g of 1-aminonaphthalen-2-ol in sodium hydroxide.

CHAPTER 33: ORGANIC ACIDS AND THEIR DERIVATIVES

Checkpoint 33A: Acids [p. 723]
8. **A** = C_6H_5CHO, **B** = $C_6H_5CH_2OH$, **C** = $C_6H_5CO_2H$

Checkpoint 33D: Acid Chlorides and Anhydrides [p. 732]
4. 0.196 g

Checkpoint 33E: Esters [p. 737]
4. 10.6 g   5. 88 g mol^{-1}, $C_4H_8O_2$

Checkpoint 33F: Amides and Nitriles [p. 742]
5. **A** = C_2H_5COCl
B = $C_2H_5CONH_2$
C = C_2H_5OH
D = $CH_3CO_2C_2H_5$
E = $C_2H_5CO_2CH_3$
x = $Br_2 + CHCl_3 + KOH$
y = HNO_2
z = CH_3OH

6. **A** = C_6H_5CHO
B = $C_6H_5CH_2OH$
C = C_6H_5COCl
D = $C_6H_5COCH_2C_6H_5$
E = $C_6H_5CONH_2$
F = $C_6H_5NH_2$
G = $C_6H_5NHCOC_6H_5$

Checkpoint 33G: Amino Acids and Proteins [p. 746]
5. 166 g mol^{-1} 6. 0.69 g

Questions on Chapter 33 [p. 747]
2. **A** = CH_3CH_2X, **B** = CH_3CH_2OH,
C = $CH_3CH(OH)CN$, **D** = CH_3CH_2CN,
E = $ClCH_2CO_2H$
5. **C** = C_6H_5COCl
9. **A** (benzene ring fused with CO–O–CO, i.e. phthalic anhydride)
B = 1,2-$HO_2CC_6H_4CONH_2$
C = 1,2-$HO_2CC_6H_4NH_2$
D = 1,2-$HO_2CC_6H_4\overset{+}{N}_2Cl^-$
E = 1,2-$HO_2CC_6H_4CN$
F = 1,2-$HO_2CC_6H_4CO_2H$
13. **A** = $CH_3CO_2CH(CH_3)C_2H_5$,
B = $C_2H_5CH(CH_3)CO_2CH_3$
15. $C_2H_5CH(CH_3)CO_2CH(CH_3)_2$
20. (*b*) (i) E.g. the H—O—ring bond angle is 120°. The H—C—H bond angle in CH_3 and the C—N—H bond angle are 109°.
(*i*) £26.23
(*j*) the cost of packaging, distribution, advertising and investment in research
21. (*b*) (i) 180 (ii) 90%
22. (*a*) (ii) 7 volumes of gas → 8 volumes
(iii) Increase in pressure favours the left-hand side of the equilibrium and decreases the yield.
(iv) Use the heat to produce superheated steam to drive a turbine
(*b*) (i) **A** = acidic $K_2Cr_2O_7$, **B** = PCl_3,
C = NH_3, **D** = Br_2 + NaOH(aq)
(ii) fume cupboard
(*c*) $CH_2{=}CHCO_2H$ (with H^+(aq)), $BrCH_2CHBrCN$ (with Br_2), CH_3CH_2CN (with H_2), $CH_2{=}CHCH_2NH_2$ (with $LiAlH_4$)
(*d*) $\text{-(}CH_2{-}CHCN{-}CH_2{-}CHCN{-}CH_2{-}CHCN\text{)-}_n$
23. (*a*) trigonal planar, 120°
(*b*) $COCl_2 + CH_3NH_2 \rightarrow CH_3NCO + 2HCl$
(*d*) The oxygen atom has two unshared pairs of electrons.
(*e*) $CH_3{-}N{=}\overset{\delta+}{C}{=}\overset{\delta-}{O}$ C will be attacked by $:\ddot{O}H_2$
(*f*) $CH_3NCO(l) + H_2O(l) \rightarrow CH_3NH_2(g) + CO_2(g)$
(*g*) 26.7×10^3 dm^3
(*h*) The reaction is exothermic, and it results in an increase in the amount of gas
24. (*b*) (i) **A** $CH_3{-}CH{=}CH{-}CH_2{-}CO_2H$
B $CH_3{-}CH_2{-}CH{=}CH{-}CO_2H$
C $CH_2{=}CH{-}CH_2{-}CH_2CO_2H$
D $CH_2{=}CH{-}{}^*CH(CH_3){-}CO_2H$
E $CH_2{=}C(CH_3){-}CH_2{-}CO_2H$
F $CH_3{-}C(CH_3){=}CH{-}CO_2H$
G $CH_3{-}CH{=}C(CH_3){-}CO_2H$
H $CH_2{=}C(CH_2CH_3){-}CO_2H$
(ii) **D**
(iii) **A, B, G**
(iv) **A** $CH_3CH_2CH_2CH_2CO_2H$
B $CH_3CH_2CH_2CH_2CO_2H$
C $CH_3CH_2CH_2CH_2CO_2H$
D $CH_3CH_2{}^*CH(CH_3)CO_2H$
E $CH_3CH(CH_3)CH_2CO_2H$
F $CH_3CH(CH_3)CH_2CO_2H$
G $CH_3CH_2{}^*CH(CH_3)CO_2H$
H $CH_3{}^*CH(C_2H_5)CO_2H$
(*c*) **G**
25. (*a*) $C_4H_6O_5$
B $CO_2H{-}CHOH{-}CH_2{-}CO_2H$
C $CO_2C_2H_5{-}CHOH{-}CH_2{-}CO_2C_2H_5$
D $CO_2C_2H_5{-}C{=}O{-}CH_2{-}CO_2C_2H_5$
26. (*c*) (i) 0.056%
27. (*a*) **C** = $CH_3CO_2C_3H_7$, **D** = C_2H_5OH, **E** = C_3H_7OH, **F** = CHI_3
(*b*) **G** = $CH_3CH_2CH_3$, **H** = $CH_3CH_2CH_2Cl$, **J** = $(CH_3)_2CHCl$, **K** = $CH_3CH_2CO_2H$, **L** = $(CH_3)_2C{=}O$
28. **A** = $H_2NCH_2CONHCH_2CO_2H$, **B** = $H_2NCH_2CO_2H$, **C** = $H_2NCH_2CH_2OH$, **D** = $H_2NCH_2CH_2Br$, **E** = $H_2NCH_2CH_2NH_2$
E will (i) act as a bidentate ligand to form a complex ion, (ii) form a condensation polymer.

CHAPTER 34: THE IDENTIFICATION OF ORGANIC COMPOUNDS

Checkpoint 34A: Analysis [p. 756]
1. (*a*) CH_2 (*b*) 84 g mol^{-1} (*c*) C_6H_{12}
2. $C_6H_5SO_2Cl$ 3. C_3H_7NO 4. C_4H_{10}

Questions on Chapter 34 [p. 763]
1. (*a*) CHO (*b*) $C_4H_4O_4$, $HO_2CCH{=}CHCO_2H$ (*cis* and *trans* isomers)   3. 56 g mol^{-1}, C_4H_8

8. (*a*) (i) **A** C=C double bond, **B** —C—Br, **C** $—CH_2OH$, **D** —CHO
(iii) 9.16×10^{-3} mol
(iv) 109, **A** = C_2H_4

9. (*b*) (i) $M = 74\,g\,mol^{-1}$
(ii) Empirical formula = molecular formula = $C_4H_{10}O$
(iii) Primary and secondary alcohols with the same carbon chain; dehydrated to different alkenes, hydrogenated to the same alkane
(iv) **A**: 31 = CH_2OH, 43 = C_3H_7
B: 29 = C_2H_5, 45 = CH_3CHOH, 59 = CH_3CH_2CHOH
A is $CH_3CH_2CH_2CH_2OH$; **B** is $CH_3CH_2CHOHCH_3$

10. The spectrum P has a peak at 2800 cm^{-1}, indicating a C—H bond and a peak at 1100 cm^{-1}, indicating a C—O bond in an alcohol, ether or ester. The spectrum Q has a peak at 2800 cm^{-1}(C—H), a peak at 1100 cm^{-1} (a C—O bond) and a peak at 3300 cm^{-1} indicating the —OH group of an alcohol. P corresponds to an ether, e.g. $C_2H_5OC_2H_5$, $CH_3OCH(CH_3)_2$ or $CH_3OCH_2CH_2CH_3$. Q corresponds to an alcohol, one of the isomers of C_4H_9OH. Mass spectrometry or NMR would identify the two compounds by showing which alkyl groups were present.

CHAPTER 35: SOME GENERAL TOPICS

Checkpoint 35G: Questions on Functional Groups and Reaction Mechanisms [p. 785]

10. (*a*) See Checkpoint 35F.
(*b*) (i) Add $NaHSO_3$, react with NaCN; hydrolyse
(ii) With PCl_5 or $SOCl_2$ gives CH_3COCl; allow this to react with phenylamine.
(iii) Boil, pass chlorine through in sunlight to form $C_6H_5CH_2Cl$; then reflux with KOH(aq).
(iv) Diazotise; warm with KI(aq).

12. (*a*) (i) $CH_3CH_2CH_2OH \xrightarrow{Al_2O_3,\ heat} CH_3CH{=}CH_2 \xrightarrow{conc.\ H_2SO_4} CH_3CH(OSO_2OH)CH_3$; boil with water $\rightarrow CH_3CHOHCH_3$
(ii) Nitrate, reduce to $C_6H_5NH_2$, diazotise, warm with KCN(aq) + CuCN(aq).
(iii) $C_6H_5COCH_3 \xrightarrow[acid]{KMnO_4} C_6H_5CO_2H \xrightarrow[H_2SO_4]{CH_3OH} C_6H_5CO_2CH_3$
(*b*) Dissolve in NaOH(aq).
Pass $CO_2 \rightarrow C_6H_5OH + C_6H_5CO_2Na$
Dissolve phenol in ethoxyethane; dry; distil. Filter off sodium benzoate; add HCl(aq) to liberate benzoic acid; recrystallise from water.

Checkpoint 35H: Draw Your Own Conclusions [p. 786]

1. **A** = $CH_3CH_2CH_2CO_2CH_2CH_2CH_3$, **B** = $CH_3CH(OH)CH_3$, **C** = $CH_3CH_2CH_2CONH_2$, **D** = $CH_3CH_2CH_2NH_2$, **E** = $CH_3CH_2CH_2\overset{+}{N}H_3Cl^-$, **F** = $CH_3CH_2CH_2OH$, **G** = CH_3CH_2CHO, **H** = CH_3COCH_3
2. **A** = $C_6H_5CO_2C_2H_5$, **B** = C_2H_5OH, **C** = $C_6H_5CO_2Na$, **D** = C_2H_5Br, **E** = $C_6H_5CO_2H$, **F** = C_6H_6
3. **A** and **B** are *cis* and *trans* $C_6H_5CH{=}CHCO_2H$, **C** = $C_6H_5CH_2CH_2CO_2H$, **D** = $C_6H_5CH_2CH_2COCl$
4. **P** = $H_2NCH_2CO_2H$, **Q** = CH_3NH_2, **R** = $HOCH_2CO_2H$
5. **A** = CH_3CO_2Na, **B** = $(CH_3CO)_2O$, **C** = $C_6H_5NH_2$, **D** = $C_6H_5NH_3{}^+Cl^-$, **E** = 2,4,6-$Br_3C_6H_2NH_2$, **F** = C_6H_5OH, **G** = $C_6H_5OCOC_6H_5$
6. **W** = $(CH_3)_2CHBr$, **X** = NaBr, **Y** = $(CH_3)_2CHOH$, **Z** = $(CH_3)_2CO$
7. **A** = $CH_3C_6H_4NO_2$, **B** = $CH_3C_6H_4NH_2$, **C** = $CH_3C_6H_4OH$, **D** = $CH_3C_6H_4OCH_3$, **E** = $HO_2CC_6H_4OCH_3$, **F** = $HO_2CC_6H_4OH$, **G** = C_6H_5OH
8. **A** = $C_2H_5\overset{+}{N}H_3Cl^-$, **B** = $C_2H_5NH_2$, **C** = C_2H_5NC
10. **X** = CH_3CHO, **Y** = $CH_3CH(OH)SO_3Na$, **Z** = $(CH_3CHO)_3$
11. **H** = CH_2OH—CO_2H **I** = CHO—CO_2H **J** = CO_2H—CO_2H
12. **A** = $CO_2C_6H_5$ (on benzene ring) **B** = CO_2H (on benzene ring) **C** = OH (on benzene ring)
D = COCl (on benzene ring) **E** = Cl (on benzene ring) **F** = OH, Br, Br, Br (on benzene ring)
G = Cl—(benzene ring)—C(H)(CCl_3)—(benzene ring)—Cl
H = CH_2OH (on benzene ring) **I** = CHO (on benzene ring)
J = CO_2Na (on benzene ring) + CH_2OH (on benzene ring)

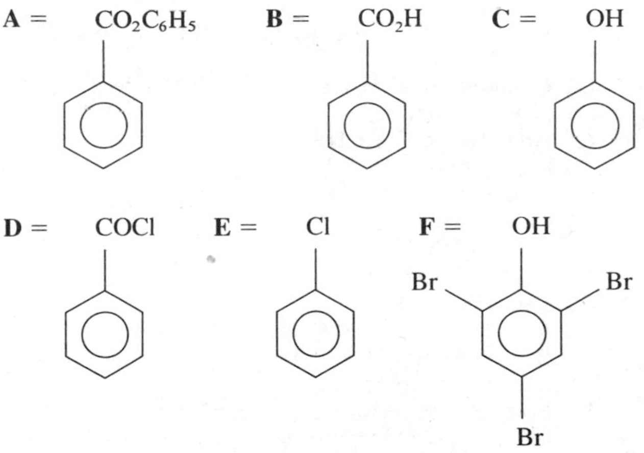

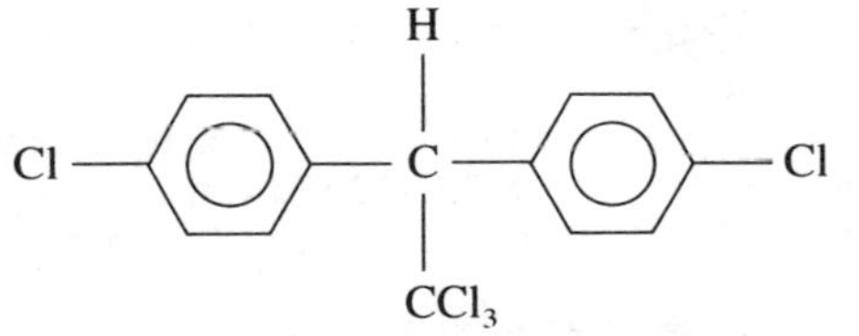

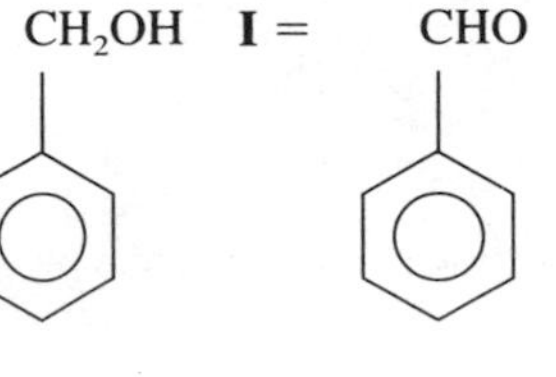

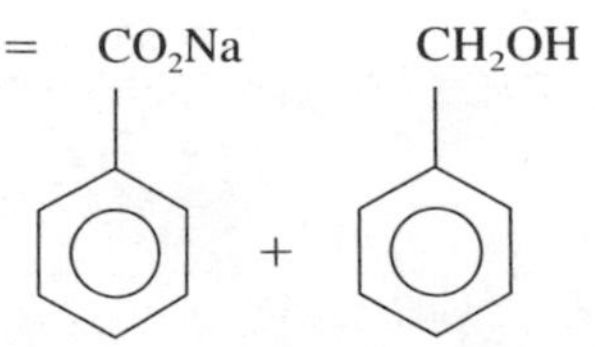

13. **A** = $CH_3CH(NH_2)CH_2CO_2H$
B = $CH_3CHOHCH_2CO_2H$
C = $CH_3CH{=}CHCO_2C_2H_5$
D = $-(CH(CH_3)-CH(CO_2C_2H_5))-_n$
E = $CH_3CHClCH_2COCl$
F = $CH_3CHClCH_2CO_2H$
G = $CH_3CH(CO_2H)CH_2CO_2H$ → (cyclic anhydride: H_3C-CH and H_2C ring with two C=O groups linked by O)

Appendix 3: A Selection of Questions on Organic Chemistry

[p. 788]

1. (*a*) (i) e.g. addition reactions of carbonyl compounds [see § 31.6.4]
(ii) e.g. nitration of benzene [see § 28.7.2] and Friedel–Crafts reactions [see § 28.7.7]
(*b*) (i) **A** = $CH_3CH{=}CHCH_3$
(ii) elimination; heat with KOH in ethanol
2. (*b*) (i) $O_2NC_6H_4CH_3$ (1,2- and 1,4-isomers)
3. (*b*) (ii) $CH_3CO_2H \rightarrow CH_3COCl \rightarrow CH_3CONH_2 \rightarrow CH_3NH_2 \rightarrow CH_3OH \rightarrow HCO_2H$
5. (*b*) First the —Cl in —COCl reacts; the —Cl in $ClCH_2$— reacts more slowly.
6. (*a*) Oxidise some C_2H_5OH to CH_3CO_2H; esterify to give the product.
(*b*) Add HBr → $CH_3CHBrCH_3$;
hydrolyse → $CH_3CH(OH)CH_3$;
oxidise → CH_3COCH_3
(*c*) Convert into C_6H_5COCl and $C_6H_5CO_2Na$. Distil these together.
7. **G** = $C_2H_5NHCOCH_3$
9. (*c*) a stronger acid
(*d*) **A** = $H_3\overset{+}{N}CH(CH_3)CO_2H$
B = $H_2NCH(CH_3)CO_2H$
C = $H_3\overset{+}{N}CH(CH_3)CO_2^-$
(*e*) Form a solid derivative of phenylamine.
10. (*c*) $CH_3CH_2CN \xrightarrow[\text{or } LiAlH_4]{Na + \text{ethanol}} CH_3CH_2CH_2NH_2$
$CH_3CHOHCN \xrightarrow{\text{warm with HCl(aq)}} CH_3CHOHCO_2H$
(*d*) 1 mol CN^- substitutes for 1 mol of Cl in —CH_2Cl and does not attack the —Cl attached to the ring
11. (*a*) (i) CH_3CO^+ [see § 28.7.4 and § 28.7.5]
(iii) In CH_3CO^+ the charge can be delocalised.
(iv) $(CH_3)_2CH^+$ is more stable than $CH_3CH_2CH_2^+$ because the positive charge is more spread out.
(*b*) (i) *cis*- and *trans*-isomers

$(H_3C)(C_6H_5)C{=}N-NHC_6H_3(NO_2)_2$ (two isomers: $NHC_6H_3(NO_2)_2$ on the same side as H_3C, and on the same side as C_6H_5)

(ii) The *trans*-isomer is less sterically hindered. (The large groups are further apart.)
(iii) fractional crystallisation
(*c*) (i) CN^- is the attacking nucleophile. In acid solution the equilibrium $H^+(aq) + CN^-(aq) \rightleftharpoons HCN(aq)$ reduces the concentration of CN^- ions.
(ii) The chiral carbon atom makes the compound optically active.

$C_6H_5-{}^*C(CH_3)(OH)(CN)$

(iii) CN^- attacks the $\delta+$ centre in $-\overset{\delta+}{C}{=}O$. In alkenes the π electrons of the double bond are attacked by electrophiles.

13. (*a*) The equilibrium
$[(CH_3)_3NCH_2CH_2OH]^+\ OH^- \rightleftharpoons (CH_3)_3N^+CH_2CH_2O^- + H_2O$
lies well over to the left-hand side.
(*c*) 1. Ethene + Bromine water → $BrCH_2CH_2OH$; separate the product from $BrCH_2CH_2Br$.
2. $BrCH_2CH_2OH + (CH_3)_3N \rightarrow [(CH_3)_3NCH_2CH_2OH]^+\ Br^-$
3. With $Ag_2O + H_2O$, gives $[(CH_3)_3NCH_2CH_2OH]^+\ OH^-$.

Index

INDEX OF SYMBOLS AND ABBREVIATIONS

A = area
A = mass (nucleon) number
A* = activated/excited A 98,
[A] = concentration/mol dm^{-3} of A
$[A]_o$ = initial concentration of A
Ar = aryl group
A_r = relative atomic mass
AR = anionic radius
c = velocity of light
c = concentration/mol dm^{-3}
C = charge
C = coulomb
C = heat capacity
c = specific heat capacity
cpm = counts per minute
CR = cationic radius
d = distance
e = elementary charge
E = electromotive force (emf)
E = energy
$E^{\ominus}$ = standard electrode potential
E_a = activation energy
EA = electron affinity
f = force
F = Faraday constant
G = free energy
$G^{\ominus}$ = standard free energy
$\Delta G^{\ominus}$ = change in standard free energy
GMV = gas molar volume
h = Planck constant
H = enthalpy
$\Delta H^{\ominus}$ = change in standard enthalpy
I = electric current
IE = ionisation energy
IR = ionic radius
IR = infrared
k = rate/velocity constant
K = equilibrium constant 65,
K_a = acid dissociation constant
K_b = base dissociation constant
K_c = equilibrium constant in concentration terms
K_p = equilibrium constant in partial pressure terms
K_{sp} = solubility product
K_w = ionic product for water
l = length
l = second quantum number
L = Avogadro constant
m = mass
M = molar mass
M_r = relative molar/molecular mass
m_l = third quantum number
m_s = fourth (spin) quantum number
u = atomic mass unit
n = principal quantum number
n_A = amount/mol of A
N = number of molecules
Ox. No. = oxidation number
P = pressure
p_B = partial pressure of B
p_B^0 = vapour pressure of pure B
ppm = parts per million
ΔQ = heat absorbed
R = electric resistance
R = universal gas constant
R = alkyl group
R_F value
r = rate
R.D.S. = rate-determining step
rms = root mean square
rtp = room temperature and pressure
S = entropy
$\Delta S^{\ominus}$ = change in standard entropy
stp = standard temperature and pressure
svp = saturated vapour pressure

t = time
t = temperature/°C
$t_{1/2}$ = half-life
T = temperature/K
T_b = boiling temperature
T_f = freezing temperature
T_m = melting temperature
U = internal energy
ΔU = change in internal energy
UV = ultraviolet
v = velocity/rate
v_o = initial velocity/rate
V = volume
V = potential difference
V = volts
V_m = molar volume
w = work
W = mass
x_A = mole fraction of A
X_2 = halogen
X^- = halide ion
Z = proton (atomic) number

Greek letters:
α = degree of dissociation
α = degree of ionisation
α = position of group
κ = electrolytic conductivity
λ = wavelength
Λ = molar conductivity
Λ_o = molar conductivity at infinite dilution
ν = frequency
π = osmotic pressure
ρ = density
ρ = resistivity